An Exploration of Dynamical Systems and

John Argyris · Gunter Faust
Maria Haase · Rudolf Friedrich

An Exploration of Dynamical Systems and Chaos

Completely Revised and Enlarged
Second Edition

Springer

John Argyris (Deceased)

Gunter Faust
Inst. Statik und Dynamik der
Luft-und Raumfahrtkonstruktionen
Universität Stuttgart
Stuttgart
Germany

Maria Haase
Institut für Höchstleistungsrechnen
Universität Stuttgart
Stuttgart
Germany

Rudolf Friedrich (Deceased)

ISBN 978-3-662-46041-2 ISBN 978-3-662-46042-9 (eBook)
DOI 10.1007/978-3-662-46042-9

Library of Congress Control Number: 2014958980

Springer Heidelberg New York Dordrecht London

Cover Illustration: Renato Vitolo, Paul Glendinning and Jason A. C. Gallas (2011), Global structure of periodicity hubs in Lyapunov phase diagrams of dissipative flows, Phys.Rev. E 84, 016216.

Cover Design: Scientific Publishing Services Pvt. Ltd., Chennai, India.

Printed on acid-free paper

Springer-Verlag GmbH Berlin Heidelberg is part of Springer Science+Business Media
(www.springer.com)

Preface to the Second Edition

In 1994, when the first edition has been published in English and in German, we did not anticipate that this introduction to the field of non-linear dynamics and chaos would meet with such lively interest. Although attention to the subject has lessened in the media and in popular scientific publications over the last 20 years, the fundamental ideas, the theoretical insights and the tools developed for the analysis of non-linear dynamical systems have continued to spread into the most diverse areas of science and technology as well as into the respective literature and are nowadays part of the classical curriculum in many study programmes. Since the English edition sold out many years ago, we decided to follow up the expanded and revised second German edition with another extended edition in English.

After John Argyris passed away in 2004 at the advanced age of 91, we considered it our obligation to continue work on this book in the original sense – first in a second German edition – attempting to explain the complex topic of non-linear dynamics to a wide audience as descriptively and vividly as possible.

We took the opportunity to emphasise and further clarify the differences between purely temporal and spatio-temporal dynamics, in particular between chaos and turbulence. The initial hope – occasionally accompanied by euphoria – that insight into the special features of chaotic systems meant that scientists were about to solve the centuries-old problem of turbulence was not fulfilled. Without doubt, the elucidation of characteristics such as unpredictability and mixing in chaotic systems made a substantial contribution to understanding turbulent flows, yet the fundamental questions concerning fully developed turbulence are unsolved to this day.

By enlarging the circle of authors to include Rudolf Friedrich, we were able to add with Chapter 9 a completely new and more ambitious chapter on *turbulence*, which comprises and critically scrutinises the fundamentals of turbulent flows and a series of classic turbulence models. While there is a multitude of excellent monographs on turbulence from the viewpoint of the engineering sciences, the new chapter mainly discusses the fundamental questions from the physical point of view, highlighting the common ground with concepts existing in chaos theory. To describe and investigate fully developed turbulence, it is indispensable to include probabilistic and stochastic methods. So that the reader can understand the chapter on turbulence better, we thought it appropriate to include two new sections in the mathematical introduction in Chapter 3: *basic concepts of probability theory* and *invariant measure and ergodic orbits*. Our aim in these sections was to present the basic ideas and concepts; we did not, however, strive for a complete description, rather referring the reader to the excellent literature on probability and stochastic processes.

In section 3.8.5, we added a brief introduction to wavelet transformation and outlined its relevance for time-frequency analyses of time series. The use of wavelets is especially favourable for analysing multifractal structures as described in section 8.5.2. In section 3.9.12, we also included a brief description of Markov analysis, a recently developed and far-reaching method which allows us to separate the deterministic part of the dynamics from the dynamical and even measurement noise. In section 9.6.8, this method is applied to velocity increments measured in turbulent flows. Section 8.7 contains an updated and extended section on routes out of chaos. Shilnikov bifurcations and spiral chaos received much attention in various applications, from chemical reactions to the propagation of nerve impulses and epileptic seizures. To understand such systems better, we also included a short overview in section 10.6. In addition, the paragraph on the kinetics of chemical reactions on surfaces has been largely revised in the hope of rendering this widely used field of application more comprehensible to a larger audience of students, see section 10.9. On the other hand, we dispensed with the section on celestial mechanics; this would have gone beyond the scope of this monograph.

Suddenly and unexpectedly, on 16th August 2012, Rudolf Friedrich passed away, taken from the midst of his life and work. His broad knowledge and interdisciplinary expertise made him extraordinarily inventive, not only in theoretical physics, his specialist field, but also in many other areas of science, including experimental and technical applications. We hope we have been able to complete the work on this book as he would have wished.

Without the vigorous support of numerous friends and colleagues from many scientific disciplines, we would not have been able to finish this extended and updated second edition. To implement the original KETEXversion in LaTeX and to generate and reconstruct the figures with the programme system AnT 4.669, we received professional help. We acknowledge the valuable hints for additions and improvements which we obtained in the course of many discussions and during proof-reading. We wish to thank all those who constantly and sedulously helped us and are especially indebted to Viktor Avrutin, Inna Avrutina, Rolf Bader, Anton Daitche, Martin Dziobek, Markus Eiswirth, Jan Friedrich, Jason Gallas, Svetlana Gurevich, Andreas Haase, Marion Hackenberg, Marcus Hauser, Oliver Kamps, David Kleinhans, Yuri A. Kuznetsov, Bernd Lehle, Pedro Lind, Johannes Lülff, Joachim Peinke, Peter Plath, Günter Radons, Michael Schanz, Susanne Schmidt, Daniel Stellbrink, Robert Stresing, Daniel H. Sugondo, Christian Uhl, Hans van den Berg, Judith Vogelsang, Georg Wackenhut and Michael Wilczek. Our special thanks go to Prudence Lawday, the translator of the first English edition, for her excellent help with this extended manuscript. We are also indebted to Michael Resch, the director of the High Performance Computing Center Stuttgart (HLRS) of the University of Stuttgart, for his generous assistance and to the Argyris Foundation for its financial support.

Last but not least, we wish to thank the members of staff at Springer-Verlag for their professional co-operation, and in particular Thomas Ditzinger for his patience and continuous encouragement and support. Our sincere thanks go to everyone who supported us.

Stuttgart, October 2014 M. Haase, G. Faust

Preface to the First Edition

Chaos often breeds life,
when order breeds habit

Henry Brooks Adams, 1838-1918
Education of Henry Adams, 1907

This volume is intended as an introductory textbook on the theory of chaos and is addressed to physicists and engineers who wish to be acquainted with this new and exciting science associated with non-linear deterministic systems. Mathematics are, of course, a pre-requisite tool in such a study, and we did not shirk the task of discussing complex mathematical issues while preferring, in general, through inclination and training to focus the attention of the reader on a physical understanding of phenomena.

Of course, we have to admit that a number of distinguished textbooks incorporating chaos as a primary or secondary subject have appeared in the last few years. In particular, we draw the reader's attention to the treatises of Moon, Thompson & Stewart, Kreuzer, Bergé, Pomeau & Vidal, Schuster and Nicolis & Prigogine, to mention only six recent texts that are not primarily mathematical. The excellent book of F. C. Moon – which has been followed by an expanded exposition – is mainly concerned with experimental techniques and offers interesting insights into the chaotic response of mechanical systems. The book of J. M. T. Thompson and H. B. Stewart is, as to be expected, brilliantly written with a broad outlook on the subject mainly directed at mechanical systems and structures but also including asides into subjects like the Rayleigh-Bénard convection and the Lorenz system of equations. E. Kreuzer's compact book is addressed to non-linear oscillations and mechanical systems; it is based on a sound mathematical foundation and displays a deep knowledge of the dynamic response of non-linear systems. The book by P. Bergé, Y. Pomeau and C. Vidal contains *inter alia* an extensive investigation of the transitions to chaos and their experimental verification on the example of the Rayleigh-Bénard convection. H. G. Schuster's monograph is an outstanding exposition of chaotic manifestations in non-linear physical systems written with precision and economy for theoretical physicists of advanced training. The last of the aforementioned books, that of G. Nicolis and L Prigogine, offers a profound and comprehensible study of the dynamics of non-linear systems far from thermodynamic equilibrium. In addition, we should like to mention the monographs of Hermann Haken on synergetics which – on a mathematically demanding basis – deal with the systematic investigation of structure formation in open dissipative systems.

In this array of textbooks, we now submit our text in the hope that aspiring physicists and engineers will find it of value in their efforts to understand and apply the complex theory of chaos. We have endeavoured not only to expound the general theory as far as possible, but also to include a broad range of physical subjects like fluid mechanics, Rayleigh-Bénard convection, biomechanics, astronomy, physical chemistry and other mechanical and electrical systems represented by the Duffing and van der Pol equations; see our descriptive account in Chapter 1.

Once chaotic manifestations were first perceived consciously in meteorology, the world of classical science as it existed 30 years ago seemed to fade away. Physicists and mechanicians, guided by the epoch-making writings of Kepler and Newton, were unconsciously influenced over centuries to such an extent that they were aware only of regular motions – whether linear or non-linear – and thus were not capable of perceiving irregular phenomena. As early as the turn of the century, Henri Poincaré had indeed drawn attention to the possibility of irregular behaviour in deterministic systems; in the absence of computers, however, this could not be registered directly. While recognising the great contributions of an Osborne Reynolds, we were thus not in a position to penetrate the mysteries of turbulence in a flow or in the atmosphere and in the oceans and hence to comprehend irregularities in natural phenomena. For three centuries, research had been unconsciously directed at regularity.

However, in the early 1970s, an elite of researchers in Europe and the United States initiated a concentrated effort to clear a path of understanding through all these disorders. This group of adventurous and unconventional scientists included physicists, biologists, chemists and mathematicians, all attempting to seek links between different kinds of irregularities in animate and inanimate nature. These irregularities are to be found in the dynamics of our heartbeat and in explosive variations in certain wildlife populations as well as in the turbulence of a flow and the erratic motion of a meteor. Economists were prompted to investigate the theory of economic cycles. All these phenomena and a multitude of others, such as forks of lightning, were observed with curiosity and analysed.

At the same time, mathematicians such as Vladimir Igorevich Arnold made new fundamental contributions to our knowledge of local and global bifurcations in non-linear dynamics. Unavoidably, scientists re-discovered the pioneering work of Henri Poincaré. All these efforts could not have been conceived and realised without the revolution in science and engineering generated by the explosively growing availability and capacity of electronic computers which began a few years after World War II.

A decade later, in the mid-Seventies, the group of scientists working on chaos had established itself as an exponentially growing co-fraternity which was re-shaping the concepts of modern science *nolens volens*. We have now reached the stage at which, in nearly every major university of scientific standing, researchers apply themselves to the manifestations of chaos, irrespective of their formal specialisation. Indeed, at Los Alamos, a special centre for the study of non-linear systems was created and coordinates work on chaos and related manifestations.

Inevitably, the study of chaos has generated new advanced techniques of applying computers and the refined graphic facilities of modern hardware. In this way, we can view displays of delicate, highly imaginative textures which illustrate complexity in a formerly unexpected way. This new science involves the study of fractals, bifurcations, periodicities and intermittency. All these manifestations inspire us to a new understanding of the concept of motion. In all our observations of the world, we now continuously discover manifestations of chaos as, for example, in the rising and quivering column of cigarette smoke which suddenly breaks out into a wild disorder. Similar phenomena may be seen if we look at the complex oscillatory response of a flag fluttering or snapping back and forth in the wind. Observing a dripping tap, we note a transformation from a steady pattern to a random one.

Chaos, in fact, is noticed today – thanks to the revolutionary findings of Edward Lorenz in 1963 who was the first of all modern scientists to comprehend chaotic evolutions and developed a simplified model of chaotic manifestation in the weather. Indeed, in this weather model, he noted the extreme sensitivity of the response arising from small changes in the initial conditions and mentioned the so-called butterfly effect first. But chaos is also contained latently in the response of an airplane in flight and arises *inter alia* through turbulent boundary layers and separation effects as well as other chaotic manifestations. If a dense stream of cars chokes a motorway, this is also an example of a chaotic response. Irrespective of the medium in which these chaotic outbursts take place, the behaviour obeys certain common general rules.

Our incidental remarks are intended to demonstrate that chaos poses problems in most realms of present-day research that do not fit into the traditional patterns of scientific thinking. In contrast, the imaginative study of chaos allows us to discover the universal characteristics of the response of complex non-linear systems. The first chaos researchers who initiated this discipline shared certain preoccupations. They were, for example, fascinated by patterns, especially those that emerge and repeat themselves at the same time on different scales. To the initiated researcher, odd questions like how long the jagged coast of Great Britain is became part of a fundamental enquiry. These early scientists had a talent for exploring complexity and studying jagged edges and sudden leaps. Inevitably, these apostles in chaos speculated about determinism and free will and about the nature of conscious intelligence.

The most profound thinkers of the new science asserted, and still assert today, that 20th century science will be remembered for three great scientific philosophical concepts: relativity, quantum mechanics and chaos. We go one step further and think that the exploration of chaos will determine the mainstream of scientific discovery in the 21st century and shape the evolution of physics, mechanics and also chemistry; naturally, this will also affect engineering. In this way, the preachers of the new science also believe that chaos erodes principles of Newton's physics. As one eminent physicist puts it: relativity eliminated the Newtonian illusion of absolute space and time; the quantum theory eliminated the Newtonian dream of a controllable procedure of measurement and, ultimately, chaos eliminated the Laplacian phantasy (or pipe dream) of deterministic predictability. Of these three scientific revolutions, the revolution in chaos applies to the whole universe as we comprehend and observe it and affects us through manifestations on a human scale.

For all the brilliant achievements of great physicists, we have to ask ourselves today how this edifice of physics could have evolved over so many years into a phantastic mission without providing means of answering some of the most fundamental questions about nature. How does life begin and what is the mystery of turbulence? In a universe ruled by entropy and drawing inexorably towards greater and greater disorder, how does order establish itself? Moreover, objects of everyday life like fluids and non-linear mechanical systems came to be considered as so ordinary that physicists could assume that they were well understood, at least by engineers. Until thirty years or so ago, the reason for our ignorance of such systems was caused by the absence of computer and graphic equipment. In the meantime, experimental and numerical computation has proved that our ignorance was profound.

As the revolution in chaos evolves in a cascade of new surprises, leading physicists find it quite natural to return without embarrassment to problems immediately related to our human nature. They are just as happy studying puffed clouds as galaxies. Leading journals publish articles on the strange dynamics of a ball bouncing on a table as well as on esoteric quantum physics. The simplest non-linear systems – and practically all systems in the real world are non-linear – seem to generate extremely difficult problems of predictability. At the same time, order may suddenly succeed chaos and vice versa. In most systems, we can observe the cohabitation of chaos and order. We are looking at a new kind of science with which we expect to bridge the gulf between knowledge of what one single item can do – e.g. a water molecule or one cell of a heart tissue – and what an assembly of millions of them can create in co-operation. This is one of the main topics to which Hermann Haken dedicated his work.

In the past, physicists traditionally deduced from complex results a hypothesis on complex causes. When they observed a random relationship between what goes into a system and what emerges from it, they assumed that they would have to introduce randomness into a realistic theory by adding the effects of noise or error. In contrast, the modern study of chaos was generated the realisation in the early 1960s that quite simple, but still non-linear deterministic mathematical equations could generate responses every bit as surprising as the phantasmagoric turbulence of a waterfall. Small differences in initial conditions in the input are quickly magnified and produce large differences in the output. Thanks to Lorenz, we now assign this phenomenon to an extremely sensitive dependence on initial conditions. To take the classic example of weather, such effects may be expressed with a mocking smile as part of the butterfly effect. Thus, a butterfly stirring the air in Beijing might generate stormy weather the following month in New York.

We conclude our prologue by offering five very brief illustrations of the revolution that shaped the new theory of chaos.

i. Edward Lorenz was the first inspired scientist who, through his numerical computational experiments, perceived the essence of chaos. He worked as a meteorologist in 1961 at the MIT on an ungainly primitive computer known by the impressive name of Royal McBee. By selecting a simple model of the Rayleigh-Bénard convection in a layer of air activated by a difference of temperature, he proposed to represent its complex response by an innocuous system of three non-linear ordinary differential equations. With this model, he hoped to study the problem of weather forecasting. He used a primitive printing device to produce a graph of the direction and speed of wind. In this way, he made the epoch-making discovery that the smallest difference in the initial conditions may produce diverging courses and patterns which drift apart until they bear no resemblance. This was the clue to the butterfly effect and the ultimate realisation of the unpredictability of weather over longer periods of time. Lorenz's findings were revolutionary and initiated the search for the nature of chaos. Lorenz also discovered a partial view of the corresponding strange attractor in the phase space, but did not name it in this way. Edward Lorenz's work and its influence on the sciences can be illustrated by a quotation from the New Testament:

> *Ἰδοὺ ἡλίκον πῦρ ἡλίκην ὕλην ἀνάπτειν*
> *Καινὴ διαθήκη, Ἐπιστολὴ Ἰακώβου iii, 5*
>
> Behold, how great a matter a little fire kindleth.
> General Epistle of James, *iii*, 5

ii. Mitchell Feigenbaum joined the Los Alamos National Laboratory in 1974. Feigenbaum brought to Los Alamos a conviction that the understanding of non-linear problems was practically non-existent. One of his first investigations concerned a most elementary logistic map with a quadratic form depending on a single parameter. He discovered that this primitive mathematical system produced not only the expected steady answers but also, through a cascade of bifurcations, periodic and also doubly and higher periodic responses leading to chaotic manifestations beyond certain values of the parameter. But this chaos was again interrupted by windows of regular response. Feigenbaum went a step further and proved the universality of his findings which apply with a surprising similarity to different and more complex mathematical expressions.

iii. Let us next consider some of the finer points of the work of the highly gifted Benoît Mandelbrot with particular reference to the Mandelbrot set. This set is possibly the most complex object existing in mathematics. Benoît Mandelbrot initiated his search in 1979 on a generalisation of a certain class of shapes known as Julia sets. These were originally invented by two distinguished French mathematicians, Gaston Julia and Pierre Fatou, during World War I in France. Had these two researchers possessed computers and graphics, they would no doubt have been co-discoverers of chaos. The French mathematician Adrien Douady described the Julia sets as follows: "You obtain an incredible variety of Julia sets: some are a fatty cloud, others are a skinny bush of brambles, some look like meandering sparks which float in the air after a firework has gone off. One has the shape of a rabbit but lots of them have sea-horse tails".

In 1979, Mandelbrot ingeniously created one image in a complex plane that could serve as a dictionary or catalogue of all Julia sets. Scientists like Julia, Fatou, Hubbard, Barnsley and Mandelbrot invented novel rules on how to construct extravagant geometrical shapes that are now called fractals and are ruled by the principle of self-similarity. They produce strange ethereal pictures of great beauty which are seemingly dotted with separate, solitary islands, all giving the impression of a cosmos different from our own. The rules of its construction instruct us how to deduce from the image on one scale the corresponding pictogram on the next level of magnification as it appears to us through a microscope. The two fine mathematicians Douady and Hubbard applied an imaginative chain of subtle mathematics to prove that the aforementioned solitary molecules are, in fact, connected with the mainland by a delicate filigree. It may interest the reader to be informed that Peitgen and Richter, one a mathematician and the other a physicist, have devoted years to the propagation of Mandelbrot sets and their presentation as a new philosophy of art.

iv. Another important research topic associated with the evolution of the chaos theory is that of attractors to which trajectories in the phase space are impelled. For example, a point or a limit is such an important characteristic entity in phase space. This kind of problem – this time in connection with the mystery of turbulent flow – attracted the attention of two distinguished Belgian mathematicians, David Ruelle and Floris Takens. Astonishingly enough, they were unaware of the revolutionary findings of Edward Lorenz in 1963 and his partial presentation of the then unnamed strange attractor. Ruelle and Takens wanted to check the assertion of Landau that turbulence is generated by an infinite sequence of Hopf bifurcations. Using a very demanding mathematical argumentation in association with Poincaré sections, they proved that the assertion of Landau must be erroneous since his scheme does not produce stretching and folding of the trajectories in the phase space and does not reflect a high sensitivity to initial conditions, both of which are expected characteristic features of a turbulent transition. Moreover, the authors construed the first complete strange attractor and named it so appropriately.

v. Let us ultimately remark on another exciting topic, that of the dimension of fractal shapes. In a sense, the degree of irregularity of such shapes corresponds to the ability of such a structure to take up space. Thus, a one-dimensional straight line does not fill any space at all. But a Koch curve – a kind of idealised snowflake – ruled by a fractal construction principle and possessing an infinite length but enclosing a finite area, has no integer dimension. In this way, its dimension exceeds the dimension one of a line but is less than the dimension two of an area. Mandelbrot determined this fractal dimension as 1.2618.

The concept of fractal dimension as expressed today by a number of alternative definitions has now taken hold of physics and non-linear system theory. Thus, we now know that the strange attractor of Lorenz has the dimension 2.06. Fractal non-integer dimensions now appertain to the theory of chaos. They have also taken hold of geo-physicists who have to describe the infinite complexity of the surface of our earth. This and many other developments brought the acceptance of this non-Euclidian and fractal geometry as a tool for solving problems.

Before closing the prologue, let us address our attention and veneration to one of the greatest scientists of this century, Henri Poincaré, who applied his intellectual power with phenomenal success to two topics: topology and dynamical systems.

At the turn of the century, Poincaré was probably the last great mathematician to apply geometric imagination to the laws of motion of our physical world. He was the first to presage the mathematical concept of chaos. His writings and in particular his monumental work "Les méthodes nouvelles de la mécanique céleste" contain more than a profound hint at some kind of unpredictability, nearly as severe as that which Lorenz discovered. He was one of the most brilliant scientists of this century. Following his death, topology flourished, but the modern theory of dynamical systems languished. With this reverence to an inspiring inventive spirit, we conclude the prologue.

Stuttgart, January 1994 *John Argyris*

Contents

1 Descriptive Synopsis of the Text

But thought's the slave of life, and life time's fool;
And time, that takes survey of all the world,
Must have a stop.
William Shakespeare, King Henry IV, Part I, IV, *2*, 81

This book is conceived as an elementary introduction to the modern theory of non-linear dynamical systems with particular emphasis on the exploration of chaotic phenomena. One might ask why yet another book should be published when the literature on chaos and non-linear oscillations already fills shelf after shelf following the stormy developments in this branch of science since the 1970s. The reasons which prompted us have been detailed in the preface.

Our aim was to write an elementary, yet detailed text-book, without assuming an all-too profound mathematical knowledge on the part of the reader. It is aimed not at mathematicians and theoretical physicists, but rather at budding scientists such as chemists, biologists and engineers in all fields, particularly in the aeronautical and electrical engineering, but also in musical acoustics, as well as economic scientists and interested graduate students. Should they be required for a comprehension of the subject, mathematical procedures and tools are introduced in small sub-sections; in all cases, we have attempted to support the text illustratively by graphics, colour plates and additional computer demonstrations. A fairly extensive list of references is intended to help the reader in his quest for further reading.

Chapters 2 and 3 are an elementary introduction to the intellectual and conceptual edifice of non-linear dynamical systems. In *Chapter 2*, we present an outline of the complex of problems investigated in this book. We consider non-linear dynamical processes, restricting ourselves to deterministic and, with few exceptions, finite-dimensional systems which can be described by systems of ordinary differential equations. In contrast to classic treatises on the theory of oscillations which generally devote a great deal of space to the solution of linear problems and mainly consider special solutions of selected problems, the theory of dynamical systems founded by Henri Poincaré allows qualitative statements on the overall set of the solutions of specific as well as broad classes of differential equations by a combination of analytical and geometrical, differential topological methods. Central questions concern the structure of the phase space, the stability behaviour of the solutions, their long-term response and the effects of bifurcations, i.e. the qualitative changes in the flow as a result of variations in the parameters. The fact that even the simplest non-linear systems can generate chaotic behaviour leads to the discussion in section 2.1 of the concept of causality and the connection between determinism and predictability. Subsequently, using the example of single-degree-of- freedom oscillators with and

J. Argyris et al., *An Exploration of Dynamical Systems and Chaos*,
DOI: 10.1007/978-3-662-46042-9_1

without friction, with and without excitation force, we describe possible states of motion of dynamical systems – such as stationary, periodic and chaotic behaviour – and introduce some elementary basic concepts such as phase space, trajectories, manifold, attractors etc.

A central theme in the analysis of non-linear systems is the discussion of the stability characteristics of states of equilibrium, periodic motions etc. *Chapter 3* thus begins with a discussion of the singularities of linear differential equations and a classification of the fixed points on the basis of eigenvalues and eigenvectors for the two-dimensional case as well as the possibilities of an extension of the results to non-linear systems. The introduction of a series of stability concepts is followed by the definition of the invariant manifolds which enable a structuring of the phase space. We then discuss Poincaré sections and discrete maps, followed by a first elementary meeting with the logistic map where we can observe a transition from stationary to periodic, multiple periodic behaviour to chaos through variations of a control parameter. General periodic motions can be easily described in Fourier space. Therefore, we continue with a brief introduction to harmonic analysis, including a short survey of the most important properties of the Fourier transform and a reference to wavelet analysis, which allows us to perform time-frequency analyses of non-stationary processes.

The subsequent section 3.9 presents a summary of the basic principles of the theory of probability. Although this book is devoted to the analysis of deterministic non-linear systems, one often needs a set of fundamental terms and definitions stemming from probability theory to characterise the dynamics of chaotic systems in phase space and to describe turbulence. Among these are concepts such as random variables, probability density, conditional and joint probability and moments of distribution functions. We briefly outline the importance of the central limit theorem and present a short description of the most commonly used density distributions. Finally, we give a short overview of the recently developed Markov analysis, which allows for sequentially measured experimental data to disentangle the underlying deterministic dynamics from the often unavoidable superimposed fluctuating forces and to extract the basic stochastic equations. On the basis of probability theory, we introduce in the final section 3.10 the concepts of invariant measure and ergodicity.

Dynamical systems can be assigned to two basic families, namely conservative systems not exposed to loss of energy and dissipative systems subject to loss of energy. In *Chapter 4*, we consider the conservative systems and attempt to span the gap between integrability and chaos. Within the concept of classical mechanics, the Hamilton equations are derived, taking d'Alembert's principle as a starting point. The theorem of Liouville allows us to investigate the volume conservation of the phase flow of conservative systems and thus to distinguish between conservative and dissipative systems. An analytical solution of non-linear Hamilton differential equations is only possible in exceptional cases, namely if the system can be transformed to action and angle variables and thus described by quasi-periodic motions on a torus in phase space. Such an integration is demonstrated on the example of Kepler's motion of a planet around the sun. In contrast, an analytical solution of the three-body problem is no longer possible. Towards the end of the 19th century,

Poincaré succeeded in proving that small perturbations of the two-body problem throw doubt on the stability of the solutions. He showed that perturbation methods do not necessarily lead to correct results since the series used in the course of the calculation might diverge, thus possibly leading to false conclusions. It was not until 60 years later that, within the framework of the KAM theory, Andrei N. Kolmogorov, Vladimir I. Arnold and Jürgen Moser were able – under certain conditions – to answer the question of the stability of perturbed Hamilton systems. In section 4.4, the KAM theory is outlined and the disintegration of the tori in phase space demonstrated on Poincaré sections of two-dimensional perturbed Hamilton systems. It turns out that the emergence of heteroclinic points generates the irregular, chaotic behaviour in the neighbourhood of hyperbolic points. A detailed discussion of the conservative Hénon map closes this chapter.

Chapter 5 deals with dissipative systems. In the classical mechanics of Galilei and Newton, loss of energy due to friction was considered a disturbing factor and was thus neglected. However, this idealisation can be applied in a good approximation to the orbits of the planets, but only rarely to motions on earth. The theory of dynamical systems shows that the introduction of dissipative terms does not make the problem more complex, but rather simplifies the long-range solution in many cases. The long-term dynamics of high-dimensional dissipative systems is often conditioned by some few fundamental modes; a contraction of the phase space volume takes place and attractors emerge which cannot be generated in conservative systems. Dissipative systems not subject to energy supply come to rest, they approach a state of equilibrium or point attractor. If the system is fed with energy, then, in the simplest case, periodic or quasi-periodic behaviour can occur or, on the other hand, completely irregular behaviour which is described in phase space by a *strange attractor*, an extremely complex structure with zero volume. The typical characteristics of such strange attractors are summarised using the example of the Lorenz attractor, the structure of which is characterised fundamentally by the sensitive dependence of the trajectories on the smallest perturbations of the initial conditions. In the subsequent sections of Chapter 5, we review a series of mathematical tools – namely the power spectrum, autocorrelation, the Lyapunov exponents and the dimensions in phase space – which allow a quantitative characterisation of and differentiation between the individual attractor types.

We begin with two classic criteria, the power spectrum, which supplies information on which basic frequencies dominate, and autocorrelation, which reflects the memory, i.e. the temporal interrelations of a signal. Both criteria are based on Fourier transformation, the most important characteristics of which are summarised previously in section 3.8. While demonstrating a differentiation between regular and irregular behaviour, they are not sufficient to characterise chaotic motion. Further mathematical tools are necessary, such as the Lyapunov exponents, for example, which provide a criterion for the stability behaviour of neighbouring trajectories. The classic Floquet theory, which allows a linear stability analysis of periodic motions, is extended to irregular courses of motion and leads to the spectrum of the Lyapunov exponents. Section 5.4 closes with the description of algorithms which allow a numerical calculation of all Lyapunov exponents.

The concept of dimension represents a further important instrument for the classification of the various attractor types. Damping generally ensures that the number of independent variables determining the dimension of the phase space are reduced considerably during the transient phase. The dimension of an attractor is thus a suitable source of information on the number of fundamental modes participating in the long-term response: a point attractor of dimension 0 corresponds to a state of equilibrium, a limit cycle of dimension 1 to a periodic motion etc. A strange attractor, which, for example, is a sort of hybrid between a surface-like and a spatial structure in a three-dimensional phase space, corresponds to a chaotic motion. Classic dimension concepts are not adequate to describe its intricate, mille-feuille-like structure; it has to be ascribed a non-integral dimension between 2 and 3. The French mathematician Benoît Mandelbrot coined the word "fractals" for such sets. What is responsible for the complexity of fractal sets is their self-similarity, i.e. their invariance with respect to alterations in scale. There is a whole family of suitable dimension concepts – such as the capacity dimension, the information and correlation dimension etc. – which we shall discuss in detail in section 5.5.

Dimensional concepts are often used to characterise experimental time series. If a system is based on deterministic laws, the temporal evolution of a single variable contains all the information on the total dynamics of the system. We discuss a reconstruction technique which offers the opportunity of reconstructing attractors from one-dimensional time series and also applying in this context concepts such as dimension (section 5.5.4) and the Lyapunov exponents (see also section 10.9) to characterise the associated orbits. We conclude Chapter 5 with the discussion of the Kolmogorov-Sinai entropy, based on Shannon's information theory and quantifying the degree of disorder of a system. With the example of a simple map, the Bernoulli shift, it is demonstrated that the KS entropy measures the information transport from a microscopic to a macroscopic scale. Moreover, it allows for the specification of a bound for the time-span within which reliable prognoses on the course of the trajectories are still possible.

Typically, one or more control parameters appear in the model equations describing physical systems. If these differential equations have non-linear character, a variation of the parameters can result in a qualitative change of the topological structure of the solutions. Such changes are referred to as bifurcations. *Chapter 6* contains a detailed discussion of the methods of the local bifurcation theory; we concentrate on local bifurcations of states of equilibrium and periodic motions arising from the variation of one control parameter. As an introduction, we first present some typical bifurcations of fixed points and elucidate the qualitative change of the solutions with corresponding phase portraits.

In the neighbourhood of a hyperbolic fixed point, it suffices to investigate the linearised system in order to establish the stability and solution behaviour. At bifurcation points, however, a fixed point loses its hyperbolic character. In this case, non-linear terms must be included in our considerations in order to clarify the stability and the solution behaviour. To this end, the method of the centre manifold is applied. This approximation technique, presented in section 6.2, leads to a drastic reduction of the number of equations necessary to describe the long-term behaviour in the neighbourhood of non-hyperbolic fixed points and hence the clarification of the stability.

It is the purpose of the local bifurcation theory to classify the local bifurcations of fixed points, limit cycles etc. and to set up their normal forms. In section 6.3, we describe the method of normal forms, a technique with which non-linear differential equations in the neighbourhood of fixed points and limit cycles can be reduced to the simplest form by a series of non-linear transformations; the non-linear terms are eliminated successively whenever possible. A complete linearisation is only possible if no resonances occur. Normal forms of bifurcations, however, necessarily contain resonances.

In section 6.4, we discuss the bifurcations of one-parametric flows, deduce the corresponding normal forms and list the conditions which allow a classification of the individual bifurcation patterns. We then consider the question of the robustness of the individual bifurcations, i.e. we investigate whether small alterations in the model equations generate qualitative changes of the bifurcation patterns. Finally, we discuss the normal forms of bifurcations of fixed points of one-parametric maps. Since periodic orbits appear as fixed points in the Poincaré map, the discussion also includes the bifurcations of limit cycles.

In section 6.7, we demonstrate with the example of the logistic map that a continuous increase of the control parameter can generate a series of local bifurcations, in this case a cascade of period doublings, ultimately leading to chaotic behaviour; this latter is interrupted repeatedly by windows of regular behaviour, however. Mitchell Feigenbaum was the first to prove that this route to chaos is not restricted to the special form of the logistic map, but rather is a universal scenario which can be observed in many systems. In this section, we explore the background to the mechanism of continued period doublings and explain how the self-similarity characteristics of the bifurcation diagram can be decoded with the aid of renormalisation techniques and how the Feigenbaum constants can be determined. We then turn to the link with second-order phase transitions, known to us, for example, from the transition from a liquid to a gaseous state or from the ferromagnetic transition.

We conclude Chapter 6 with an introductory outline of synergetics, a branch of science founded at the end of the 1960s by the physicist Hermann Haken which investigates the dynamics of pattern formation processes. The central question of synergetics is whether and to what extent the formation of structures in the most varied realms of physics, biology, sociology, medicine etc. in animate and inanimate nature is based on generally valid principles and how this self-organisation can be reproduced in a mathematical formalism. On the one hand, we refer to the features in common with the centre manifold theory; on the other, we draw attention to the more extensive framework of synergetics, a result of the inclusion of spatial inhomogeneities and fluctuating forces.

Chapter 7 is devoted to the structure formation in convection flows and the deduction and discussion of the Lorenz system, which stimulated the development of the chaos theory fundamentally. The chapter begins with a qualitative description of the physical laws leading to the formation of the characteristic convection rolls in the Rayleigh-Bénard experiment. There follows a detailed deduction of the basic hydrodynamic equations for convection problems. The continuity equations, the Navier-Stokes equations and the heat equation are deduced from the conservation theorems of mass, impulse and energy. In many cases, these basic equations can be simplified by means of the Boussinesq-Oberbeck approximation, in which the variation of the density is only accounted for in the buoyancy term.

In the early 1960s, the American meteorologist Edward N. Lorenz applied this system of basic equations to study the accuracy of weather forecasts. While his highly simplified model for the Bénard problem cannot stand up to an experimental verification describing the onset of turbulence, it nevertheless led to a new paradigm in the natural sciences: *On the one hand, there are basic limits to the predictability of a dynamical system in the latent presence of chaotic behaviour; on the other hand, however, time series which have been classified as purely stochastic in the past have turned out to describe in certain cases chaotic motions, the essential definition of which can often be construed with a restricted number of non-linear equations.*

In section 7.3, we follow the original publications of Saltzman and Lorenz and deduce the three ordinary differential equations of the Lorenz system. Subsequently, the evolution of the solutions of this system is described for variations of the free parameter which controls the temperature difference between the upper and the lower plate in the Rayleigh-Bénard problem; the emergence of local and global bifurcations is observed. A series of colour plates demonstrates on the one hand the spatial and temporal evolution of the convection rolls in the Lorenz model and on the other hand the significant changes of the phase space structure due to an increase of the control parameter; these are illustrated by consideration of the evolution of the stable manifolds.

In *Chapter 8*, we present a series of mathematical models describing the transition from regular to chaotic motions. In the early years of chaos research, the hope was that these scenarios could describe the transition to turbulent behaviour, thus making a substantial contribution to the solution of the turbulence problem.

The first model describing the onset of turbulence goes back to Lev D. Landau (1944), who assumed that the transition from laminar to turbulent behaviour is generated by an endless series of (Hopf) bifurcations. It was not until a quarter of a century later that the deficiencies of this model were recognised: Landau's model cannot explain the sensitive dependence of the flow on small perturbations nor generate a mixing of the trajectories in the phase space. In 1971, David Ruelle and Floris Takens proved that generally quasi-periodic motions on higher-dimensional tori are not stable, as Landau had presumed, but rather that the quasi-periodic motion becomes chaotic abruptly after the emergence of the third incommensurable frequency, i.e. the motion on a T^3-torus collapses into a *strange attractor*, a designation coined by Ruelle and Takens. A great step forward had been taken and a window opened towards the description of the characteristics of turbulence – such as unpredictability of the motion and mixing – on the basis of deterministic equations. With the aid of refined measuring techniques, Gollub and Swinney succeeded in achieving an experimental verification for the Bénard problem based on the analysis of the power spectra and supported by the determination of the dimension of reconstructed attractors.

Sections 8.3 to 8.5 contain a discussion of the transition from quasi-periodic to chaotic motions using the circle map as a model. Two different scenarios are conceivable. Either chaos occurs following a synchronisation of the frequencies and a subsequent series of period doublings – this is the Feigenbaum route already discussed in section 6.7 for the logistic map – or there is a direct transition from quasi-periodicity to chaos. The second route requires the tuning of two control parameters and can also be interpreted by means of the circle map. Both transitions to chaos are of universal character and the scaling characteristics can be decoded in both cases with the aid of the renormalisation theory. A selection of figures aims at facilitating access to this complex material. There follows a description of the work of Albert Libchaber who succeeded in confirming the theoretical prognoses quantitatively by experiments which were highly precise at that time. Section 8.6 contains a discussion of intermittent transitions which, following Yves Pomeau and Paul Manneville (1980), can be classified in three groups.

Having presented a series of scenarios describing transitions from regular to chaotic behaviour, we present in section 8.7 an overview of various strategies allowing the reverse, a conversion of chaotic motions to regular regions. Following the seminal paper of Edward Ott, Celso Grebogi and James A. Yorke from 1990, this issue became a field of non-linear dynamics which experienced a rapid development. This is because it was recognised in many areas like medicine, chemistry, laser technology and telecommunication, as well as in other fields of engineering science, that chaotic behaviour contains a high potential with regard to a possible control and regulation of chaos. To this end, one takes advantage of a characteristic feature of chaos, the sensitive dependence of the dynamics on small perturbations, by imposing minimal control impulses which stabilise the unstable orbits inherently contained in strange attractors. In section 8.7 we give a brief survey of the most effective strategies and provide information on a multitude of applications with the potential to lead to innovative developments.

In the beginning, it was hoped that chaos research would help uncover the basic mechanisms leading to fully developed turbulence. These high expectations have not yet been fulfilled. The reason is that chaos theory researchers initially confined themselves to a purely temporal description of the dynamics, which is not sufficient for an understanding of turbulent flow. It is rather the additional spatial dependence of all quantities which leads to complex interrelations and to an extremely high dimensionality of the corresponding phase space.

Chapter 9 presents an overview of the state of turbulence research from the present point of view, almost 40 years after the paradigm shift in the understanding of non-linear processes. On the one hand, we identify the issues in which chaos theory has made substantial contributions to an understanding of turbulent flow. On the other hand, it becomes evident that the spatio-temporal complexity and the extraordinarily high number of excited modes in fully developed turbulence makes the use of statistical methods unavoidable.

In the chapter on turbulence, the main focus is on various vortex models and the discussion of their influence on the dynamics of fluid flow. A detailed description of the characteristic scales, of the energy cascade and of Heisenberg's turbulence model follows. Particular emphasis is put on the derivation of the Kármán-Howarth equation and an exposition of the underlying assumptions. Subsequently, we describe the classical turbulence models of Kolmogorov from 1941 and 1962 as well as more recent multifractal models presented by Frisch and She-Leveque and discuss the strengths and weaknesses of the respective models. We also report on an application of the Markov analysis presented in section 3.9 to the velocity increments of a turbulent jet. The fact that the turbulent data can be described by a Markovian process in scale allows us to express the full n-point statistics by the 2-point transition probability densities and to reproduce the intermittent behaviour accurately. Finally, we add an overview of some open questions and possible issues relevant for future research.

In the final *Chapter 10*, we present some fields in which the application of the theory of dynamical systems is very promising. With the aid of computer experiments we discuss a variety of classical models displaying features which are characteristic of a multitude of dynamical systems.

For non-linear systems, analytical solutions can only be given in exceptional cases; in the case of chaotic behaviour, we generally have to rely completely on numerical investigations. This is the explanation for the fact that, as a result of efficient computers, the theory of dynamical systems – once the domain of the mathematical and physical faculties – is now receiving increasing notice in the most varied disciplines such as biology, chemistry, meteorology, electrical engineering and hydrodynamics, as well as in economic sciences, as demonstrated by the flood of publications from these fields.

The discovery of universal characteristics, i.e. of fundamental principles common to all non-linear systems of a whole class of problems, was without doubt a milestone in the more recent history of the theory of dynamical systems and justified the intensive study of seemingly over-simplified systems. The investigations were accelerated greatly by the swift development of efficient computers – indeed, they were only made possible through this explosive growth of computer power.

The first section of Chapter 10 offers an insight into bone remodelling processes. In order to achieve a durable anchoring of endoprostheses, models are being developed with the aim of describing bone retrogression or bone formation on the basis of dynamical systems. Should this succeed, computer simulations could speed up the development of more reliable types of endoprosthesis. In a first study, two very simple dynamical systems for the modelling of bone adaptation processes are discussed.

Specially constructed recursion rules are particularly suitable for elucidating general characteristics of dynamical systems due to the low computational effort. In particular, the dissipative Hénon map was conceived as a model for a Poincaré map of a three-dimensional system. With the aid of this map, we demonstrate in section 10.2 that a mixing of the trajectories in the phase space results from a continuous repetition of stretching and folding processes. This mechanism is the key to the understanding of the exponential drifting apart of neighbouring trajectories in the phase space; it is illustrated with the example of the Lorenz attractor in a series of colour plates.

In section 10.3, we return to the Lorenz equations and discuss some further characteristics of the system. A number of figures demonstrate the emergence of cascades of period doublings within regular regions and the coexistence of attractors. The bifurcation diagram of the Poincaré sections is highly reminiscent of that of the logistic map, thus offering evidence of universal laws. Conclusions about the system behaviour can be drawn by calculating the Lyapunov exponents. The section ends with the determination of the capacity dimension of the Lorenz attractor according to Hunt and Sullivan's procedure.

In section 10.4, we turn to the van der Pol equation, originally set up to describe electrical circuits. The system possesses a non-linear damping term with the characteristic that the system is supplied with energy for small amplitudes, which manifests itself in a negative damping term, whereas energy is dissipated for large amplitudes. Typically, self-excited oscillations occur in such systems, i.e. although no external periodic forces are applied, the system tends towards a limit cycle on which energy supply and dissipation are in balance in the temporal mean. We first consider the van der Pol equation without external excitation. In the case of a small damping parameter, i.e. weak non-linearity, we observe, as expected, almost sinusoidal oscillations; in the case of large damping values, on the other hand, we observe self-excited relaxation oscillations. Computer investigations suggest that exactly one limit cycle exists in the phase space; the strict mathematical proof of this observation with the aid of the Poincaré-Bendixson theorem is laborious. If small periodic excitation forces are applied in the case of low damping, the averaging method can be used; here, by taking a time average, the initial system is replaced by an autonomous system involving the neglect of higher harmonics. Large damping parameters respectively strong non-linearities together with periodic external excitation are discussed for a modification of the van der Pol equation. On the basis of Poincaré sections and phase portraits, the character of the chaotic solutions on the Birkhoff-Shaw attractor is illustrated.

The Duffing equation considered in the subsequent section represents one of the simplest dynamical systems in which chaotic behaviour can be observed. Differential equations of its type are obtained in the formulation of non-linear pendulum oscillations or for electrical circuits and other problems, for example. It approximates the behaviour of an externally excited non-linear oscillator with one degree of freedom. We deduce the Duffing equation for the example of an externally excited buckled beam. In this case, the equation of motion is a non-linear partial differential equation, the solution of which is approximated with the aid of the Galerkin procedure taking the first fundamental mode into account. As a result, we obtain the Duffing equation with negative stiffness. According to the selection of the control parameters in the Duffing equation, we observe a great variety of system response. The Duffing equation is integrated numerically for a series of parameter values and the results presented in the form of phase portraits and Poincaré sections.

A surprising phenomenon can emerge in dissipative systems when several attractors coexist in the phase space. Even for regular behaviour, the boundaries between the basins of attraction can assume fractal character so that predictions on the long-term behaviour are no longer possible in this domain. With the aid of the Melnikov method, bounds for the control parameters for which fractal boundaries emerge can be calculated explicitly. For the example of the Duffing equation, the calculation of the so-called Holmes-Melnikov boundary is carried out explicitly.

In section 10.6, we briefly discuss some bifurcations of flow involving homoclinic orbits in three-dimensional vector fields. The most complex case, the so-called Shilnikov bifurcation, leads to a highly complex topological structure of the flow and can be observed, for example, in the propagation of nerve impulses along axons or during epileptical seizures.

By now, computer experiments play an important role in non-linear dynamics. In general, theorems cannot be proven with the aid of computers; nevertheless, numerical investigations in association with modern computer graphics reveal unexpected characteristics and links in non-linear systems which had previously not been taken into account. Seen in this way, numerical experiments are valuable aids nowadays. The two sections 10.7 and 10.8 illustrate the fact that graphic representations can be extraordinarily helpful for the comprehension of some phenomena occurring in non-linear recursion rules.

Shortly after the First World War, the two French mathematicians Gaston Julia and Pierre Fatou studied the iteration of rational functions. They discovered that, depending on the initial value, the iteration sequences converged to different solutions; the boundaries between the basins of attraction of these attractors exhibit highly peculiar characteristics and are of unsuspected multifariousness. Due to the unavailability of computers, what was disclosed to the inner eye of Julia or Fatou remained hidden to the average scientist until, around 1980, Mandelbrot again turned to one of the simplest non-linear iteration rules in the complex plane, exploiting the opportunities offered by high-performance computers. Section 10.7 offers an insight into the manifold structures of Julia sets and the fractal boundaries of the Mandelbrot set. Calculation of the basin boundary of the point attractor $z = \infty$ for constant values of the control parameter of the mapping rule yield, depending on the parameter value, completely differing limit sets, the so-called Julia sets. What is fascinating about Mandelbrot's recursion formula is that, in spite of its simplicity, such structural diversity emerges. A series of computer graphics shows connected fractal sets, dendritic forms as well as burst sets. The Mandelbrot set, on the other hand, encompasses all values of the complex control parameter for which the Julia sets are connected. A series of colour plates presenting various selective enlargements depicts the richness of structures in the fractal boundaries of the Mandelbrot set.

In section 10.8, we again turn to the circle map of Chapter 8 and, by means of the Lyapunov exponents, gain an overview of the inner structure of the Arnold tongues. The results agree very well with the theoretical results. On the basis of the Lyapunov exponents, the different transitions to chaos can be discussed very simply. Their self-similar structure is reflected in the corresponding bifurcation diagrams.

In the last section of Chapter 10, section 10.9, we decided to present an overview of some interesting dynamic manifestations from the field of physical chemistry which display chaotic outbursts. We concentrate on the dynamic phenomena arising in the oxidation of CO on the surface of platinum crystals. Significantly enough, detailed experiments have been performed under isothermal low-pressure conditions. Extensive theoretical and experimental research work of high originality and distinction has been performed on this subject at the Fritz Haber Institute of the Max-Planck-Gesellschaft in Berlin under the direction of Gerhard Ertl.

In these extensive and varied numerical investigations and parallel experiments, a broad spectrum of complex temporal behaviour was observed, such as mixed-mode oscillations, a transition via period doublings to chaos and even hyperchaos, characterised by at least two positive Lyapunov exponents.

All these phenomena represent purely temporal evolutions as they would arise, for example, in a well-stirred medium; note, however, that in a heterogeneous reaction on a surface, stirring is not feasible. As a result, in heterogeneous reactions, more complex phenomena involving spatial dependencies will arise and complicate the model and the numerical analysis of such reactions considerably.

In a survey on CO oxidation on platinum crystals, Georg Ertl also discussed extensive experimental studies on spatio-temporal phenomena observed by him and his collaborators at the aforementioned Fritz Haber Institute. The formation of stationary, propagating as well as soliton waves and even spiral chemical waves was recorded. Irregular behaviour, such as that of *chemical turbulence*, also emerged.

All the numerical investigations of the Fritz Haber Institute are based on kinetic models evolved for the description of the temporal dynamics on a platinum (110) single crystal surface. The rate constants and parameters arising in the equations have been studied extensively using a variety of surface science techniques in conjunction with complex detection and recording devices. We mention only parenthetically the application of low-energy electronic diffraction (LEED) studies of oscillations on platinum crystals, based on the motion of slow electrons; here, the researchers also observed periodic changes on the surface structure which oscillated between the (1×1) and the respective reconstructed (1×2) phase. Such structural surface changes in turn influence the evolution of other elementary processes and demand additional refinements of the model.

In view of the increasing complexity, the researchers at the Fritz Haber Institute refined the analytical model of oxidation stepwise, starting with the simplest simulation and then refining it step by step to account for additional effects and variables in order to describe the dynamical behaviour of increasing complexity of the CO oxidation on platinum (110) crystals. We elaborate on the importance of the individual steps since they clarify in an examplary manner the stepwise development of models by illustrating the extent of validity and the limits of the respective models.

We close this section with a short look at recent applications in biochemistry. Biochemical reactions involving enzymes which play the role of catalysts are, of course, much more multifarious and complex than the previously mentioned catalytical procedures. However, since biological systems are open systems exchanging energy and mass with their environment and since the occurring processes have a non-linear character, it is once again indispensable to apply methods from non-linear dynamics and synergetics in order to take at least a small step towards understanding the complex interrelations of the system.

2 Preliminaries

> If therefore, those cultivators of the physical science from whom the intelligent public deduce their conception of the physicist ... are led in pursuit of the arcanas of science to the study of the singularities and instabilities, rather than the continuities and stabilities of things, the promotion of natural knowledge may tend to remove that prejudice in favour of determinism which seems to arise from assuming that the physical science of the future is a mere magnified image of that of the past.
>
> *James Clerk Maxwell, 1873*

This book is an attempt to convey concepts of methods evolved in the field of non-linear dynamics to budding physicists and engineers and to illustrate them using simple examples. The basis for these new ideas on dynamics is the topological or geometrical view of temporal processes which leads to a representation in phase space.

In the following, we shall restrict ourselves mainly to those non-linear processes of evolution, the dynamics of which can be described by ordinary differential equations or difference equations. We consider systems with the following characteristics:

i. they are *finite-dimensional*, i.e. each state is determined by a point in a finite-dimensional phase space;

ii. they are *differentiable*, i.e. the temporal change of state is modelled by differentiable functions;

iii. they are *deterministic*, i.e. the equations contain no stochastic terms. Moreover, the initial state determines the solution uniquely at least locally. This means in particular that two trajectories can never intersect.

The differential-topological method offers the opportunity of using the structure of these differential equations to investigate them geometrically in phase space. What is novel in our present-day understanding is that we do not concentrate on a single trajectory but consider the solutions as a whole and analyse their stability globally. In the case of dissipative systems, we observe the ensemble of all the trajectories and classify the long-term behaviour via the stability behaviour of individual groups. This type of analysis is called qualitative dynamics or, in mathematical terminology, the modern theory of dynamical systems. The logical consequence is the investigation of the stability of states of equilibrium or periodic solutions, for example, and, moreover, the study of bifurcations which occur for critical control parameters where the dynamics abruptly undergoes structural changes. The decisive factor here is the phenomenon of instability. This plays the central role in the theory of dynamical systems. Instabilities are responsible for the creation of new temporal patterns as a result of bifurcations and furthermore for the occurrence of

J. Argyris et al., *An Exploration of Dynamical Systems and Chaos*,
DOI: 10.1007/978-3-662-46042-9_2

irregular motions. The consequence is that slight changes in the initial conditions or the system parameters can lead to unpredictable magnification effects. This revolutionary and at the same time startling discovery is fundamental and undermines some supposedly irrefutable principles of science (Cvitanović et al., 2008); this is the subject of the following section.

Prior to this, we would like to point out explicitly which systems are excluded in the following considerations as a consequence of the above-mentioned three properties. These are, in particular, partial differential equations, systems with time delay as well as differential equations with an additional stochastic term which occur for example in synergetics, which we address briefly in section 6.8. In addition, it would go far beyond the scope of this book to discuss dynamical systems which are only piece-wise continuous or piece-wise smooth, i.e. systems which contain jump discontinuities in the function or in the derivative. This includes, for example, mechanical systems with dry friction or systems where shocks or pulses occur, systems with electronic switches, biological systems with thresholds or economic systems. In practice, such systems occur often and can cause complex dynamical behaviour as well as intricate bifurcation scenarios. Recently, these systems have become the subject of intensive research, see (Zhusubaliyev and Mosekilde, 2003; di Bernardo et al., 2008).

2.1 Causality – Determinism

Classical physics was based on the assumption that the future is determined by the present and that, in this way, an accurate knowledge of the present can reveal the future. This belief in a fundamental limitless predictability was expressed in perhaps the most elegant and comprehensible way by Laplace in 1814 through his fictional demon:

> "Une intelligence qui pour un instant donné, connaîtrait toutes les forces dont la nature est animée, et la situation respective des êtres qui la composent, si d'ailleurs elle était assez vaste pour soumettre ces données à l'analyse, embrasserait dans la même formule, les mouvements des plus grands corps de l'univers et ceux du plus léger atome: rien ne serait incertain pour elle, et l'avenir comme le passé, serait présent à ses yeux." (Laplace, 1812)

Anyone expressing the certitude of natural laws so pointedly, who, indeed, interprets causality so narrowly as to mean determinism, is asking to be contradicted, particularly at a time when the consequences of the quantum theory and the development of atomic physics affect not only philosophy but our thought processes.

This confrontation makes it clear how generation-dependent the *principle of causality* or the law of cause and effect is and how it changes in the course of developments in thinking and language. We must conclude that the concept of causality and our understanding of it is linked with the concepts and the meanings associated with them at anyone time! Formulations such as causality, causal law, causal explanation or causal principle are always confusing as long as their explanations are not transparent. In this connection, it appears helpful to begin with the historical development of these concepts.

The fact that causality sets up a link between cause and effect is, historically speaking, relatively recent. For Aristotle, the originator of many scientific disciplines, the word *causa* had a much more general meaning than nowadays. He expressed his thoughts in the two pithy concepts "*τὸ αἴτιον καὶ τὸ αἰτιατὸν*". With direct reference to Aristotle, scholasticism, the philosophy of the Middle Ages, taught that there are are four types of cause: 1. *causa materialis* (= material cause); 2. *causa formalis* (= formal cause); 3. *causa finalis* (= final cause); 4. *causa efficiens* (= efficient cause). In addition, Aristotle differentiated between internal and external causes, the first two mentioned above being internal, the second two external. The one we roughly associate with the word "cause" nowadays is *causa efficiens*, the initiating or efficient cause.

With the emergence of the Renaissance and the birth of modern scientific thinking – usually associated with names such as Nicolaus Copernicus, Galileo Galilei and Johannes Kepler – a change in the concept of *causa* also came about. As man turned away from metaphysics and towards physics, i.e. towards quantification and metrology, the word causa was linked with the material incident which preceded and somehow caused the phenomenon to be explained. At first, it was held that if the cause had been discovered, the natural phenomenon could be explained. In this connection, cause was synonymous with explanation. Gradually, however, the understanding took hold that it was the natural laws, the interconnection of cause and effect, which clearly determine and thus explain natural processes. The culmination of this development was the epoch-making work of Isaac Newton. Immanuel Kant, who, in his "Critique of Pure Reason", equated science as such with the science of Newton, found a formulation for causality which was still valid in the 19th century: "Wenn wir erfahren, daß etwas geschieht, so setzen wir dabei jederzeit voraus, daß etwas vorhergehe, woraus es nach einer Regel folgt."

Einstein praised Newton's outstanding intellectual achievement as "perhaps the greatest progress in thinking that a single individual ever had the privilege to accomplish". His most eminent achievement was his comprehensive mathematical theory of mechanics which remained the basis of scientific thinking until well into the 20th century. It is hardly surprising that his general laws of motion – obeyed by all objects in the solar system from the apple falling from the tree to the planets – or their universal applicability fed the expectation that processes in nature were in principle clearly determined, assuming they are known as a whole or in part after careful study. Newton's universe was a grand mechanical system which functioned in accordance with exact deterministic laws so that the motion of a system in the future could be calculated in advance from the state of this system at any given time. If one is convinced that nature fundamentally behaves in this way, the next logical step is the one Laplace formulated for his fictional supernatural intelligence mentioned above: all the future as well as the past of our universe is calculable from the precise knowledge of the position and velocity of all atoms at any one instant.

This concept of pre-calculability in all eternity is not consistent with 20th century physics and particularly not with quantum mechanics. It is not that physics is in fundamental opposition to man's longing for predictablity but that the two, quantum mechanics and determinism, are incompatible with one another in their philosophical approaches. The atomistic ideas of Democritus and Leucippus handed down from classical antiquity assume that regular, ordered processes on a global level derive their morphology from the irregular, random behaviour on a local level. Such considerations are indeed plausible, as is substantiated by numerous examples from everyday life. A fisherman, for example, fighting against the wind and the waves, only has to ascertain the rhythm of the waves in order to be able to react to them; he does not have to know the motion of each individual drop of water. If we explain the processes we can perceive with our senses by the interaction of many individual processes on the local level, is it not an inevitable consequence to regard the overall regularities of nature as statistical? Statistical laws may provide reliable statements on the whole set of possible events such as, for example, on mean values and possible deviations, yet there will never be a totally reliable prediction for a particular event; this will always remain indeterminate.

In spite of such conceptual difficulties, we are continually setting up statistical laws in day-to-day life as a basis for our practical actions. Engineers, for example – whether they are acting in the design of aircraft, buildings or machines – cannot base their work on precise loads or material data. They have to rely as a matter of course on mean, that is to say statistical characteristic values. Yet nevertheless, when our attention is drawn to such "semi-exact" regularities, we feel uneasy and consider them less trustworthy. We would prefer either precisely definable processes in nature or, on the other hand, chaotic, totally irregular ones. Is there an explanation for such an attitude? One generally uses statistical laws when the physical system in question is only partially known. The simplest example of this is the game of heads or tails. Since no one side of the coin has an advantage over the other, we have to come to terms with the fact that – when playing a large number of games – we can only predict one of the two results with 50% certainty. Games of dice are similar, except that the number of possible results has increased to six. The likelihood of predicting a chosen number is thus reduced to 1/6. If the die is thrown often enough, then the number of throws resulting in a 1 is approximately one-sixth of all the throws. This is, of course, only true if we assume a perfect die and an identical throwing technique; only then can all six possible results be considered equally probable. What we assume is that these conditions are satisfied approximately.

In modern times, it was Robert Boyle who took up the idea from classical antiquity by not only describing the material behaviour on the macroscopic level of observation qualitatively as the result of the statistical behaviour of the molecules in a gas but by also quantifying the well-known relationship between pressure and volume. His idea was that the pressure, a quantity measurable on the macroscopic level, was built up by the numerous impacts of molecules on the wall of the vessel; this interpretation was an miraculous brainwave at the time. In a similar way, the laws of thermodynamics could be conceived when it proved possible to state in a mathematically precise form the fact that atoms move more violently in hot bodies than in cold ones.

While Laplace raised the hopes that one day, the whole world could, in principle, be calculable, in the second half of the last century, the idea spread that Newton's mechanics might be unrestrictedly valid, but that systems within the kinetic gas theory could never be completely determinable due to the immense number of gas molecules. It was mainly Josiah Willard Gibbs and Ludwig Boltzmann who put the incomplete knowledge of these systems into mathematical language by using statistical laws. Gibbs went one step further by introducing temperature as a physical concept for the first time; this only makes sense when the knowledge of the system is incomplete. If indeed the velocity and the position of all the molecules in a gas were known, it would be completely superfluous or meaningless to talk of the temperature of the gas. Hence, the concept of temperature can only be used meaningfully when the system on the microscopic level of observation is incompletely known and one nevertheless does not want to forgo a "qualitative" statement on the macroscopic level. Using such a concept, one does not describe the behaviour of a system by taking an increasing number of degrees of freedom into account – in the ideal case, infinitely many – but by proceeding to new, essential and thus considerably fewer degrees of freedom on a more general level of observation. The motto is not "more precise, more detailed and infinitely many" but "more global, fewer and nevertheless informative". In this highly simplified statement, we share the view of Max Born (1959) who says that absolute accuracy is not a physically meaningful concept and can only be found in the conceptual world of mathematicians. Felix Klein called for the application of "approximation mathematics" side by side with the usual "precision mathematics". Since his suggestion remained without response at the time, the physicists at the turn of the preceding century solved their problems in their own conceptual framework, using methods of probability and statistical laws. What Max Born's statement basically implied was that non-linear laws and deterministic equations may supply unpredictable answers. At the same time, he in effect pointed out that non-linear equations often react with unexpected sensitivity to the slightest changes in the initial conditions and thus suddenly supply unexpectedly differing answers. This was a revolutionary perception, first expressed by Poincaré at the turn of the preceding century but unnoted for a long time:

> "Une cause très petite, qui nous échappe, détermine un effet considérable que nous ne pouvons pas ne pas voir, et alors nous disons que cet effet est dû au hasard. Si nous connaissions exactement les lois de la nature et la situation de l'univers à l'instant initial, nous pourrions prédire exactement la situation de ce même univers à un instant ultérieur. Mais, lors même que les lois naturelles n'auraient plus de secret pour nous, nous ne pourrons connaître la situation initiale qu'*approximativement.* Si cela nous permet de prévoir la situation ultérieure *avec la même approximation*, c'est tout ce qu'il nous faut, nous disons que le phénomène a été prévu, qu'il est régi par des lois; mais il n'en est pas toujours ainsi, il peut arriver que de petites différences dans les conditions initiales en engendrent de très grandes dans les phénomènes finaux; une petite erreur sur les premières produirait une erreur énorme sur les derniers. La prédiction devient impossible et nous avons le phénomène fortuit." (Poincaré, 1908)

In the first two decades of the 20th century, the attempt was initally made to explain the motion of the atoms and molecules according to the basic precepts of classical mechanics, in the spirit of Newton. The result, however, was an entanglement of inextricable contradictions. For example, in accordance with classical physics, a charged electron was supposed to orbit around the atomic nucleus and constantly emit radiation until it collapsed into the nucleus due to loss of energy. However, the electron paths devised in the model were unstable. It was obvious to Niels Bohr that, on the basis of Planck's theory, the paths of the electrons are stationary, that the electron, as long as it sticks to its path, does not emit any radiation, but that a change of path is accompanied by a loss of energy. Bohr solved this contradiction not by changing his concept of the model, as could have been expected, but rather by maintaining that classical physics was not applicable to the description of the dynamical behaviour of atomic relations. The fact that loss of energy which takes place in irregular bursts and in stages when an electron changes its path should lead to the assumption that the radiation emittance of atoms is a statistical phenomenon was acceptable. It is also possible to tolerate, at least temporarily, a new type of mechanics, quantum mechanics, albeit at the expense of determinacy, when the stability of the atoms can thus be mathematically ensured. But the bold assertion that it is fundamentally impossible to know all the necessary defining elements in order to achieve a complete determination of the processes – even Albert Einstein could not and would not accept this as true, even for atomic phenomena. For him, the quantum theory was only an instrument, temporarily necessary; it had emerged due to our lack of knowledge of all the canonical variables of the atomic process but could be suspended as soon as all these unknowns had been clarified. His opinion of quantum mechanics culminated in the statement, "God does not play dice", to which Niels Bohr replied, "It would be presumptuous of us human beings to prescribe to the Almighty how he is to take his decisions."

What is it about the quantum theory that is so challengingly unfamiliar? It is the fact that the concept of the trajectory has been banned from a mechanical theory of the atom shell and replaced by irreducible probabilistic elements. This was imperative in order to ensure discrete energy levels, whose existence had been revealed by spectroscopy, in a mathematical equation for systems such as the atom. It was Werner Heisenberg who took this radical step in 1925 in his decisive work with the title "Über quantentheoretische Umdeutungen kinematischer und mechanischer Beziehungen" (Heisenberg, 1925). Here, the canonical variables position and impulse become non-commutative quantities; Max Born and Pascual Jordan recognised their matrix character. Although the fundamental features of this mathematical theory had already been set up, Heisenberg described the problem of atomic physics in a letter to Pauli in 1926 as completely unsolved. What caused him to make this negative statement? What was still missing was the link to the physical experiment (Heisenberg, 1969).

The solution was provided, again by Heisenberg, by the indeterminacy relation in 1927 (Heisenberg, 1927). Position and velocity or momentum, the classical quantities for the determination of a particle trajectory, can be stated *individually* with arbitrary accuracy; *simultaneously*, however, this is impossible in the quantum philosophy, at least in the microworld of atomic physics. This is not an assumption in quantum mechanics, but the consequence of the laws of the quantum theory. It was thus clear that Newton's mechanics, which assumes the exact knowledge of both position and momentum in order to calculate a mechanical course of motion, is not applicable to the atomic world. Although more than seventy years have passed since Bohr, Heisenberg, Born and others came to the conclusion that the quantum theory forces us to formulate the laws on the quantum level as statistical laws and to bid farewell to determinism, it is difficult to incorporate this into our general philosophy. For the atomic field, this call to abandon pure determinism may be valid. But to speak of absolute randomness and to assert that the classical idea of predictability is invalid in principle, although it was so successful in its search for order, regularity and natural laws, merely due to spontaneous nuclear disintegration for which there is neither cause nor explanation – this is something we still cannot and do not want to bring ourselves to accept.

In contrast to this, it should be remembered that classical physics tacitly complements the principle of "identical" causality with the principle of "similar" causality. Laplace's principle, "identical causes have identical effects", was extended by the principle "similar causes have similar effects". The reason is that it is a justifiable expectation to demand exactly identical causes so as to guarantee the reproducibility of physical processes, but that this is not feasible in practice. Any experimenter knows that, although he is at pains to obtain the same results for repeated measurings under the same test conditions, exactly identical repetitions of the test conditions are basically impossible. The accuracy of measurements is limited, even though it may be extremely high with today's technical sophistication. When the errors in the measurement results are of the same order as the inaccuracies of the experimental set-up, they are not exactly the same, but similar. In spite of statistical laws which we take as the basis of our experimenting, we still speak of the reproducibility of physical behaviour. Due to this discrepancy between abstract mathematical precision and unavoidable physical approximation, Max Born (1959) demands the re-formulation of the question of determinability in mechanics and, in its place, the differentiation between stable and unstable motions.

Heisenberg and Born's renunciation of the dogma of predictability may be valid for the field of atomics, but appears strange when viewed on the macroscopic scale which is directly accessible to our senses. We are, of course, aware that, in weather forecasting, we have to rely on probability estimates although the motion of the earth's atmosphere follows exactly the same physical laws as the motion of the planets. Nevertheless, the weather retains a considerable amount of randomness. The reason for this is the exceptionally high number of unknowns necessary for a detailed description of the dynamics involving correlations and feedback reactions far beyond our knowledge. Until not so very long ago, there was little reason to doubt that, in this case, at least in principle, accurate predictions would ultimately be possible. In the spirit of Laplace's demon, it was assumed that it was only necessary to gather sufficient information about the system and to process it with the necessary effort.

This mechanical conception of the world which was first shaken by quantum mechanics received a second blow as a result of an astonishing discovery, namely that even simple non-linear systems may generate irregular behaviour. We have to come to terms with the idea that such unpredictable behaviour is an intrinsic reality as it does not disappear, even after more information has been gathered. Such seemingly random behaviour which is generated in non-linear deterministic dynamics is called "deterministic chaos". The fact that deterministic laws without stochastic elements cause chaos sounds as paradoxical as the concept of deterministic chaos itself. At this point, we would like to stress that there are various manifestations of chaotic behaviour between which we must carefully distinguish. In a vessel filled with gas, the atoms fly around wildly and collide and only in this case, as Boltzmann already established, does microscopic chaos prevail. In contrast to this, macroscopic or deterministic chaos dominates when purely random oscillations occur although the laws which describe the dynamics are deterministic.

We have repeatedly mentioned the diametrically opposing conceptions of scientific knowledge: on the one hand, the idea of the atomists who stress random collision and, on the other hand, the mechanistic view of the world which is based on timeless dynamic laws. Both conceptions fail when it is a question of explaining both spatial and temporal structures, i.e. as occurring in undamped oscillations. Equilibrium thermodynamics and conventional statistical physics do not provide the methods to deal with such an oscillatory behaviour. A way out of this dilemma begins to emerge.

What is this hope based on? It is the study of the *physics of non-equilibrium states* and on the other hand of the *theory of dynamical systems*. The physics of non-equilibrium states deals with systems far from thermodynamic equilibrium where chaos prevails on the microscopic level but is completely concealed on the macroscopic level by well-organised patterns. In the second discipline, the theory of dynamical systems, it is the instabilities that play the central role. At the point of instability, the system must "choose" between various possibilities. A small, non-predictable fluctuation decides which course the dynamical process finally takes. This occurrence of unpredictability forces us to re-think our understanding of deterministic systems and the "long-term" prognoses we have won.

Processes which are sustained by a continuous flow of energy and possibly matter from outside and which are characterised by their ordered, self-organised, collective behaviour are called evolutionary. A significant aspect of evolutionary processes is their causal coherence, although they may be interspersed with random outbursts. Anyone would reject the idea of regarding the random occurrences of a lottery as an evolutionary development. Intuitively, we demand that the current state of a system is fundamentally moulded, if not determined, by the preceding causes. This does not mean, however, that this demand for causality totally excludes random occurrences; evolutionary processes without mutation, symmetry breaking etc. are unthinkable. Random occurrences reduce the chance of exact predictions, they cannot be specified by laws, they weaken the deterministic net of cause and effect; yet it would be wrong to draw the conclusion that the pattern of reality were chaotic.

Thus, it is not only quantum physics, but also the *chaos theory* which throws considerable doubt on Laplace's demon of absolute determinism. "Für manchen mag dies eine Enttäuschung sein. Aber vielleicht ist ja eine Welt sogar menschlicher, in der nicht alles determiniert und nicht alles berechenbar ist, eine Welt, in der es – dank der Quantenereignisse – Zufall, und damit auch Glück, gibt, in der – weil nicht alle Probleme algorithmisch lösbar sind – Phantasie und Einfallsreichtum, Raten und Probieren, Kreativität und Originalität noch gefragt sind und in der man, wie die Chaos-Theorie zeigt, auch bei chaotischem Verhalten immer noch sinnvoll nach einfachen Grundgesetzen suchen kann." (Vollmer, 1988, p. 350)

It is certainly a result of the historical development of the sciences that the unpredictable (and thus also the chaotic state) was first explicitly suspected and then formulated mathematically in the microworld of atomic physics. This does not mean that these phenomena do not appear in our daily macroscopic life. Chaotic responses of non-linear dynamical systems are almost a matter of course to us today. We draw attention to the appearance of such phenomena in a pendulum with a large amplitude and particularly so in the case of spatial oscillations. The well-known Duffing equation is another example. A further historical example is the well-known anorganic chemical reaction of Belousov-Zhabotinsky which offers a memorable display of colour. Yet the most impressive and still the most mysterious is the onset and development of turbulence in a fluid. Research into this problem – which is highly important in technology and of decisive significance in meteorology and combustion processes – has inspired many famous physicists since Osborne Reynolds to look for a physical-mathematical solution. In spite of admirable contributions of many scientists, even Heisenberg (Heisenberg, 1948) did not succeed in solving the problem.

For fully developed turbulence, however, today's chaos theory is still not broadly enough developed. In Chapter 9, we attempt to give an overview of the current status of turbulence research and to show that further extensive physical considerations and mathematical tools are necessary to model turbulent phenomena. Apart from the difficulties involved in an adequate comprehension of a complex process, there is a further obstacle that everyone breaking new ground must surmount, even – or especially – in science. No one but Werner Heisenberg could have expressed this so memorably: "Wenn wirkliches Neuland betreten wird, kann es aber vorkommen, daß nicht nur neue Inhalte aufzunehmen sind, sondern daß sich die Struktur des Denkens ändern muß, wenn man das Neue verstehen will. Dazu sind offenbar viele nicht bereit oder nicht in der Lage", see (Heisenberg, 1969), p. 102.

2.2 Dynamical Systems – Examples

Before we move on to a detailed description of the chaotic behaviour of systems possessing deterministic equations of motion, we should like to lead up to this subject by beginning with four typical examples.

Our first example is a mechanical system with a single degree of freedom and no loss of energy due to friction. Let a small mass be hung on a spring and allowed to oscillate vertically (fig. 2.2.1). For small displacements, the motion of this undamped spring pendulum can be expressed by the linear differential equation in the two variables displacement x and its second derivative with respect to time t

$$\ddot{x} + \omega_0^2 x = 0 \tag{2.2.1}$$

where ω_0 is the frequency of the oscillation.

Equation (2.2.1) has the general solution

$$x = A \sin \omega_0 t + B \cos \omega_0 t \tag{2.2.2}$$

where the integration constants A and B are determined by the initial conditions. Denoting the initial values of x and $\dot{x}$ at the instant $t = 0$ by x_0 and $\dot{x}_0$, we obtain for the displacement x

$$x = \frac{\dot{x}_0}{\omega_0} \sin \omega_0 t + x_0 \cos \omega_0 t \tag{2.2.3}$$

and for the velocity $\dot{x}$

$$\dot{x} = \dot{x}_0 \cos \omega_0 t - x_0 \omega_0 \sin \omega_0 t \tag{2.2.4}$$

Eliminating the time t from eqs. (2.2.3) and (2.2.4), we find

$$x^2 + \frac{\dot{x}^2}{\omega_0^2} = x_0^2 + \frac{\dot{x}_0^2}{\omega_0^2} \tag{2.2.5}$$

The family of all ellipses in the $x, \dot{x}$-plane designated by eq. (2.2.5) is called a phase portrait. On the right of fig. 2.2.1, the displacement x as a function of time for a concrete initial condition is shown and on the left, the trajectories in the two-dimensional phase space for three different initial conditions. The state variables or coordinates x and $\dot{x}$ which span the phase space characterise the single-mass system uniquely.

In this idealised case where the energy is preserved, the mass returns periodically to its initial position. Also, the states of maximum deflection and zero velocity are repeated with a period of $T = 2\pi/\omega_0$.

In natura, it is practically impossible to realise the assumption "no loss of energy due to friction". Loss of energy, for example due to air resistance, continuously decreases the deflection of the pendulum until at some stage, the mass remains in a state of equilibrium. During the transient phase, the aforementioned elliptical trajectories in the phase space then become spirals which all end in a point, the state of equilibrium (fig. 2.2.2). This point which captures all spirals is called an attractor, in this particular case a point attractor.

When the loss of energy of the pendulum due to friction is of a viscous nature, it can be reproduced by a damping term which depends to a first approximation linearly on the velocity. The linear differential equation (2.2.1) for a conservative system then takes the following standard form, again a linear system,

$$\ddot{x} + 2\zeta\omega_0 \dot{x} + \omega_0^2 x = 0 \tag{2.2.6}$$

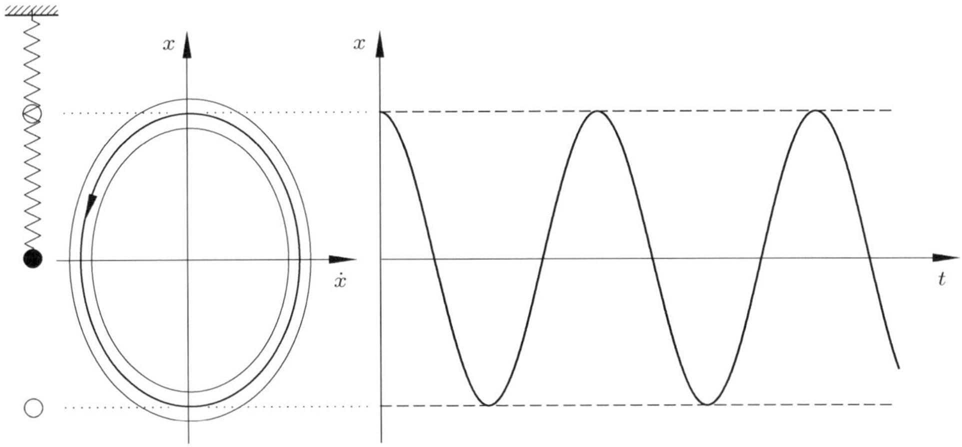

Fig. 2.2.1: Single-degree-of-freedom oscillator without friction

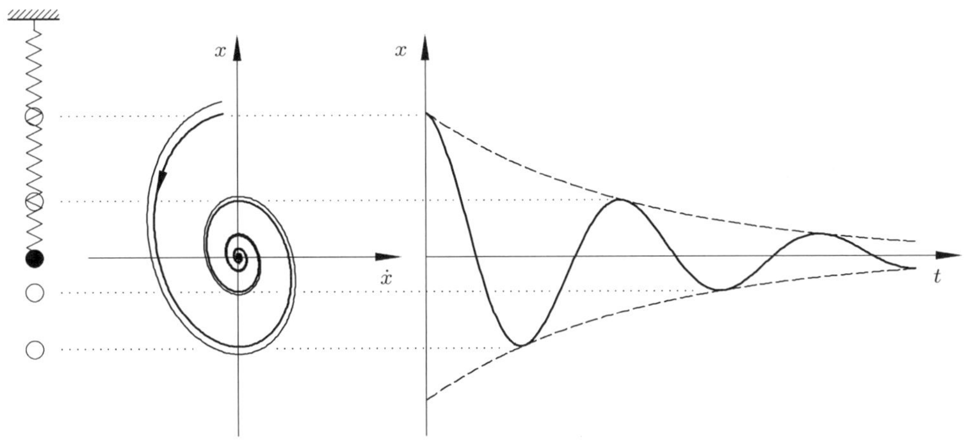

Fig. 2.2.2: Single-degree-of-freedom oscillator with friction

where the damping factor $\zeta > 0$ controls the fading of the periodic transient response. In the case of damping $\zeta > 1$, the system tends aperiodically towards the state of equilibrium $x = 0$. For $0 < \zeta < 1$, the case of sub-critical damping, the amplitudes also decrease; but the motion retains qualitatively the appearance of an oscillatory process.

The loss of energy in a system can be compensated by means of the continuous supply of energy, e.g. a periodic external excitation (fig. 2.2.3). After a certain transient phase, the trajectories in the two-dimensional phase space approach asymptotically a closed curve, the so-called limit cycle, which is run through periodically. Beside the point attractor, the limit cycle is the only possible attractor in the two-dimensional phase space.

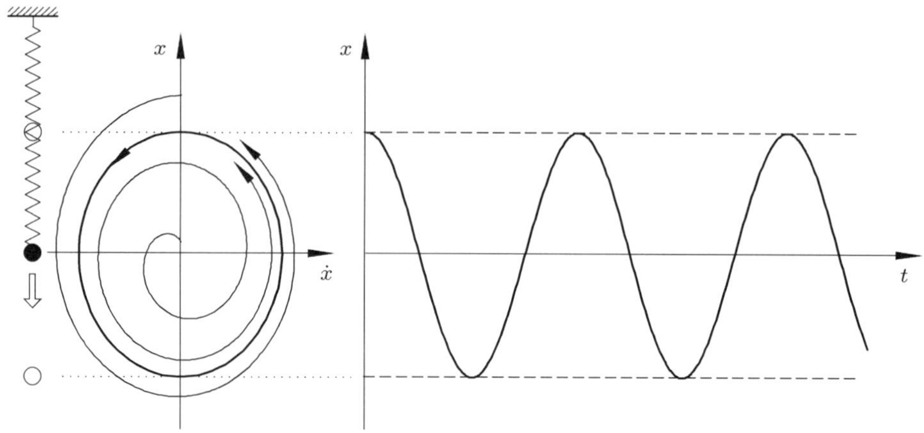

Fig. 2.2.3: Single-degree-of-freedom oscillator with friction, periodically excited; limit cycle

For a harmonically excited, viscously damped oscillator, the equation of motion in linearised form becomes

$$\ddot{x} + 2\zeta\omega_0\dot{x} + \omega_0^2 x = \omega_0^2 x_0 \sin\omega_E t \tag{2.2.7}$$

Here, the periodic external excitation is reproduced by a sinus term on the right-hand side. Linear equations of this type are integrable and lead to a family of solutions, the individual curves of which are determined by the initial conditions. Since the dependence of a specific solution is relatively insensitive with regard to the initial conditions, small changes in the latter cause only small changes in the solution. Thus, our physical system is characterised by a "similar" connection between cause and effect.

The situation in the case of non-linear equations of motion is completely different. A simple dynamical system which leads to chaotic patterns of motion is the Duffing equation (see section 10.5). In its modified form, the restoring force is approximated by a polynomial of the third order, thus allowing, for example, the reproduction of larger deflections in an externally excited beam under shear and normal force. In this special case, the differential equation takes the form

$$\ddot{x} + 2\zeta\omega_0\dot{x} - \omega_0^2 x + \delta\omega_0^2 x^3 = \omega_0^2 x_0 \sin\omega_E t \tag{2.2.8}$$

where the cubic term x^3 possesses a stiffening effect. Due to its simple morphology, the Duffing equation and its modifications led to numerous investigations on the behaviour of dynamical systems both in mechanics and electrical engineering (Ueda, 1980a; Ueda, 1980b; Moon and Holmes, 1979; Seydel, 1980). We stress that the non-linearity – in x, but not in time t – is a necessary condition for chaotic behaviour, though not sufficient in itself.

If the substitution $x_1 = x$, $x_2 = \dot{x}$, $x_3 = t$ is carried out, the non-autonomous differential eq. (2.2.8) with the second derivative with respect to time, the right-hand side of which depends explicitly on the time t, can be re-written as a system of non-linear, so-called autonomous first-order differential equations

$$\begin{aligned}\dot{x}_1 &= x_2\\ \dot{x}_2 &= -2\zeta\omega_0 x_2 + \omega_0^2 x_1 - \delta\omega_0^2 x_1^3 + \omega_0^2 x_0 \sin(\omega_E x_3)\\ \dot{x}_3 &= 1\end{aligned} \tag{2.2.9}$$

This autonomous set of equations is equivalent to the non-autonomous eq. (2.2.8). As a generalisation, we can state that an autonomous differential equation may be put in the form

$$\dot{\boldsymbol{x}} = \boldsymbol{F}(\boldsymbol{x}) \tag{2.2.10}$$

where the non-linear vector function $\boldsymbol{F}(\boldsymbol{x})$ does not depend explicitly on the time.

Thus, the introduction of additional variables transforms the non-autonomous eq. (2.2.8) with second-order derivatives with respect to time into a system of three first-order differential equations. Such a set of equations which is expressed directly in terms of displacement and velocity considerably simplifies a qualitative and quantitative discussion of the trajectories in phase space (the concept of the phase space is elucidated in section 2.3).

The phase portrait of eq. (2.2.9) for the variables $x_1, \dot{x}_1$ shows intersecting trajectories corresponding to specific choices of control parameters and initial conditions (fig. 2.2.4). Trajectories of the autonomous type of eq. (2.2.9) which do not intersect in the extended phase space with the coordinates $x_1, \dot{x}_1$ and t do so in the projection onto the $(x, \dot{x})$-plane, the customary phase space. It is easy to imagine that for a

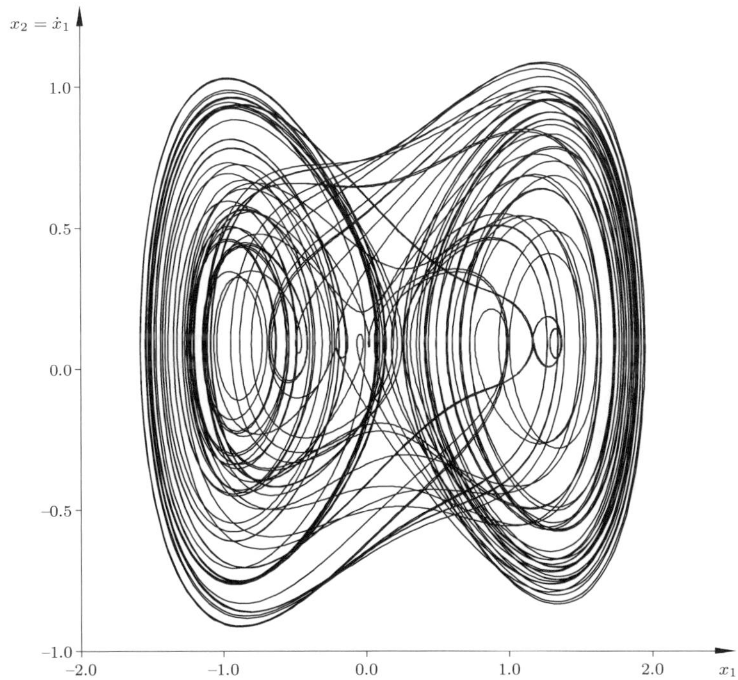

Fig. 2.2.4: Phase portrait of the Duffing equation (chaotic)

non-periodic solution (i.e. an irregular or chaotic motion) for $t \to \infty$, the phase portrait will be blackened to unrecognisability in a relatively short time. For this reason, this representation must be ruled out as an indicator of chaos since it would be futile to try to identify the corresponding trajectories which remain adjacent over a long time but diverge exponentially from one another at some stage.

The most widespread form of a display of the equations of motion consists of plotting the deflection, for example, over time (fig. 2.2.5). However, to deduce a chaotic behaviour by means of this method is not realistic and is doomed to failure because the observed period of time is necessarily finite. Although the bizarre, non-periodic course of the curve in fig. 2.2.5 strongly suggests an erratic course of motion, it remains uncertain whether a periodic motion might not be established after all in case one observed a longer time interval.

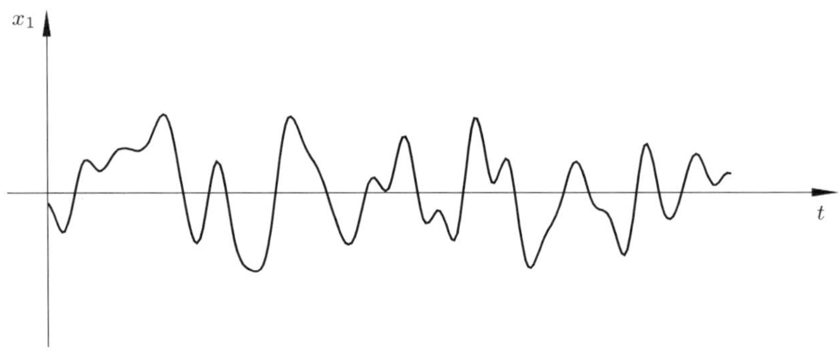

Fig. 2.2.5: Temporal evolution of the motion

It may require skill and effort to distinguish between regular and irregular behaviour on the basis of the transient response of the variables of state of non-linear dynamical models; it is, however, even more difficult to differentiate in an experimentally established transient response between background noise and deterministic chaos. In order to assess dynamical behaviour, it is also possible to apply, apart from the representation in phase space and the temporal change of individual variables of state, the power spectrum method, well known to the engineer (see section 3.8). The recorded time series, which are evaluated in this method by means of a Fourier analysis, yield clearly defined peaks in the power spectrum diagram for periodic or quasi-periodic motions; in contrast, continuous curves or curves with a high noise level indicate chaotic or stochastic behaviour. This means that in the erratic case, the measured variables can no longer be represented as a discrete superposition of oscillations: from a critical value of this state variable in the Fourier-transformed signal onwards, a very high noise component in the spectrum emerges. Since these power spectra can easily be determined experimentally, however, they offer themselves as a means of characterising the irregular behaviour of real systems. We can illustrate this with an example taken from fluid dynamics (fig. 2.2.6). As a consequence, simple spectral analysis does not suffice as a single tool for describing

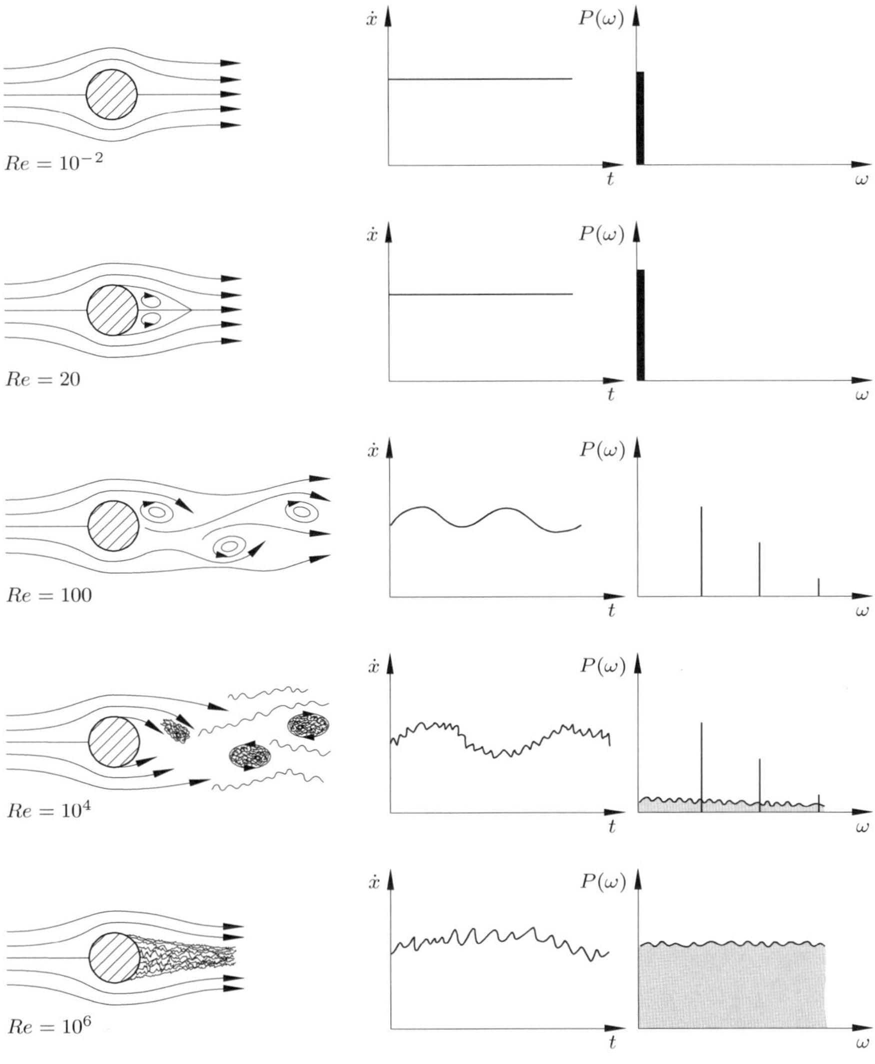

Fig. 2.2.6: Stream lines, temporal evolution and power spectrum in dependence on the Reynolds number Re; cf. (Feynman et al., 1987)

erratic behaviour in chaotic physical systems. Other methods of measuring must be found in order to describe the wide spectrum of regular and irregular motions qualitatively.

In recent years, a number of procedures have been proposed for noise reduction in measured data sets (see (Kantz and Schreiber, 1997) and references cited therein). Of particular interest is the method developed recently for analysing stochastic data which is based on the theory of Markov processes and which allows the determination of the model equations underlying the experimentally measured data (Friedrich and Peinke, 1997; Friedrich et al., 2000). In contrast to other methods, this is a parameter-free approach for reconstructing the model equations, i.e. a priori, because no assumption needs to be made about the type of these equations. Moreover, in many cases different types of noise can be separated from the deterministic part (Siefert and Peinke, 2004; Böttcher et al., 2006; Lehle, 2011; Friedrich et al., 2011). As a result, one obtains stochastic differential equations which contain both a deterministic and a stochastic part. However, in the following, we restrict our considerations to dynamical systems which can be described solely by deterministic equations.

In the literature, a multitude of possible methods for characterising the occurrence of chaotic motions is presented. Apart from the power spectrum and autocorrelation, the two classic tools, we propose to concentrate on the following criteria: Lyapunov exponents, dimensions and Kolmogorov-Sinai entropy. An extensive discussion of these concepts can be found in the context of dissipative systems and attractors in Chapter 5.

The examples of mechanical systems mentioned here make it clear that the concept of the attractor plays a central role in the description of the behaviour of damped systems subject to deterministic equations of motion. We basically differentiate between two types: regular attractors and strange attractors. On the one hand, there are three classic types of motion: equilibrium, periodic motion and quasi-periodic motion. All three states are associated with regular attractors since, in the case of damping, the system tends towards one of these three states after the transient phase. The point attractor corresponds to the state of equilibrium, the limit cycle to periodic motion and the torus to quasi-periodic motion. On the other hand, there is a class of deterministic, but erratic, i.e. chaotic, motions which are not predictable if the initial conditions are subject to small fluctuations. Such long-term behaviour ($t \to \infty$) is associated with the concept of the strange attractor.

2.3 Phase Space

In the case of evolutionary processes which can be analysed quantitatively, we can record the modification of particular quantities as a function of time. These variations must stem from specific causes. If the state is defined completely at a given instant by n variables $x_1, \ldots, x_n$, the evolution of the process may be described by a system of n ordinary differential equations

$$\begin{aligned} \frac{\mathrm{d}x_1}{\mathrm{d}t} &= F_1(x_1, x_2, \ldots, x_\mathrm{n}) \\ \frac{\mathrm{d}x_2}{\mathrm{d}t} &= F_2(x_1, x_2, \ldots, x_\mathrm{n}) \\ &\vdots \\ \frac{\mathrm{d}x_\mathrm{n}}{\mathrm{d}t} &= F_\mathrm{n}(x_1, x_2, \ldots, x_\mathrm{n}) \end{aligned} \tag{2.3.1}$$

The n time-dependent variables represent physical quantities such as location, velocity, temperature, pressure etc. Not only in physics, but also in other branches of science such as biology, chemistry, economic sciences and other fields, many real processes in time can be described by systems of ordinary differential equations. If we formally introduce the column vectors

$$\boldsymbol{x} = \{x_1 \quad x_2 \quad \ldots\ldots \quad x_{\mathrm{n}}\}$$
$$\boldsymbol{F} = \{F_1 \quad F_2 \quad \ldots\ldots \quad F_{\mathrm{n}}\} \tag{2.3.2}$$

we can write the system eq. (2.3.1) in a simplified form (see also eq. (2.2.10))

$$\frac{\mathrm{d}\boldsymbol{x}}{\mathrm{d}t} = \boldsymbol{F}(\boldsymbol{x}) \tag{2.3.3}$$

where n is the number of equations defining the whole system. The fact that the system consists exclusively of first-order differential equations is not a restriction since any system of ordinary differential equations of higher order can be transformed into a system of first order by the introduction of additional variables.

We wish to stress once again that the system eq. (2.3.1) or (2.3.3) is *autonomous* since the right-hand side does not depend explicitly on the independent variable t. This is not a restriction either, as each non-autonomous system of equations can be transformed into an autonomous one by introducing an additional variable $x_{\mathrm{n}+1} = t$ and the trivial relationship $\dot{x}_{\mathrm{n}+1} = 1$.

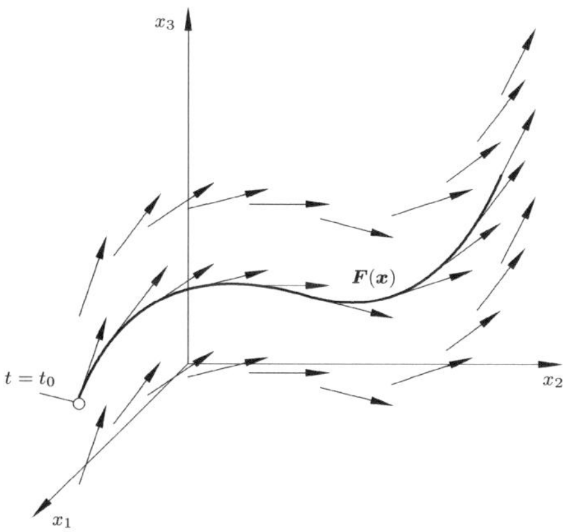

Fig. 2.3.1

Trajectory and velocity field in the three-dimensional phase space

It is very useful to represent the temporal evolution of a system in an abstract space, the phase space. It is n-dimensional and is spanned by the state variables or coordinates $x_1, x_2, \ldots, x_{\mathrm{n}}$. In the phase space, the state of the system at a given time is represented by a point. This point moves with time and its velocity is specified by the vector $\boldsymbol{F}$. Since the velocity $\boldsymbol{F}$ is known from eq. (2.3.3), the velocity field can be represented immediately in the phase space (see fig. 2.3.1). The whole set of directions specifies a direction field, reminiscent of the stream lines in a liquid. A point chosen arbitrarily at the instant $t = t_0$ describes a curve in phase space which at each point is tangential to the vector field of velocities. The graph of the motion in the phase space is called trajectory, phase line or orbit and the total set of possible motions is denoted phase flow ϕ_t. The autonomous set of first-order equations (2.3.1) allows us to display the velocity field without integration directly in the phase space. We thus already obtain a first impression of the form of the solution.

Through each point of the phase space, exactly one trajectory runs. Physically speaking, this means that if a state is known at a particular instant, both the future and the past are determined by integration. This also means that trajectories which represent a unique solution can never intersect. In the special case of the mechanics of a particle where the dynamical state of a mass point is specified in three-dimensional space by its position (three space coordinates) and by its velocity (three velocity components), the phase space is six-dimensional. For a single-degree-of-freedom oscillator, the phase space degenerates to the phase plane with the components displacement and velocity (fig. 2.2.1).

2.4 First Integrals and Manifolds

In the distant past, when mathematical tradition demanded the solution of differential equations by analytical integration, the somewhat strange concept of "first integrals" was coined. What was then understood by an integral is nowadays called solution.

The definition of the first integral used here has been adopted from Arnold (Arnold, 1980). Let $\boldsymbol{F}$ be a vector field, and the individual components $F_1, F_2 \dots F_\mathrm{n}$ differentiable functions.

Definition: a function $I(\boldsymbol{x})$ is called the *first integral* of the differential equation

$$\dot{\boldsymbol{x}} = \boldsymbol{F}(\boldsymbol{x}) \tag{2.4.1}$$

if its so-called Lie derivative $L_{\boldsymbol{F}}$ (Olver, 1986) along the vector field $\boldsymbol{F}$ vanishes

$$L_{\boldsymbol{F}} \equiv F_1 \frac{\partial I}{\partial x_1} + \dots + F_\mathrm{n} \frac{\partial I}{\partial x_\mathrm{n}} = \boldsymbol{F}^\mathrm{t} \frac{\partial I}{\partial \boldsymbol{x}^\mathrm{t}} = \frac{\partial I}{\partial \boldsymbol{x}} \boldsymbol{F} = 0 \tag{2.4.2}$$

where the notation adopts the following conventions:

$$\frac{\partial I}{\partial \boldsymbol{x}} = \left[\frac{\partial I}{\partial x_1} \quad \frac{\partial I}{\partial x_2} \cdots\cdots \frac{\partial I}{\partial x_\mathrm{n}} \right] \tag{2.4.3}$$

denotes a row vector and, correspondingly,

$$\left(\frac{\partial I}{\partial \boldsymbol{x}} \right)^\mathrm{t} = \frac{\partial I}{\partial \boldsymbol{x}^\mathrm{t}} = \left\{ \frac{\partial I}{\partial x_1} \quad \frac{\partial I}{\partial x_2} \cdots\cdots \frac{\partial I}{\partial x_\mathrm{n}} \right\} \tag{2.4.4}$$

a column vector. Equation (2.4.2) implies the following characteristics of the first integral: on the one hand, the function $I(x_1, x_2 \dots x_\mathrm{n})$ remains constant along each given trajectory; on the other hand, each trajectory lies on a hypersurface in the phase space. The hypersurface is defined by $I(\boldsymbol{x}) = C$, where C is a constant (see

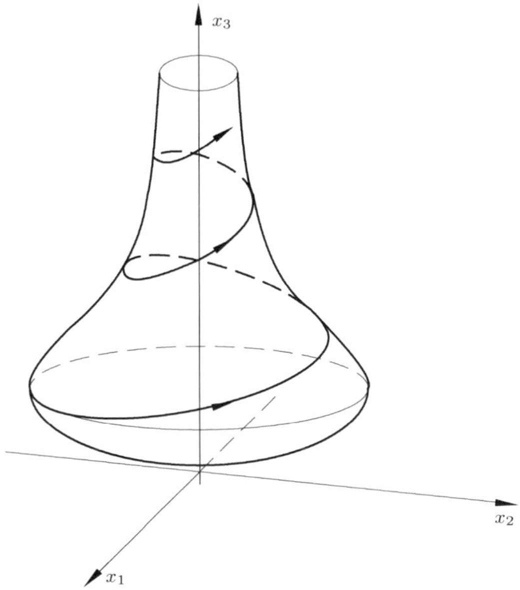

Fig. 2.4.1

Trajectory on the hypersurface $I(x_1, x_2, x_3) = C$ in the three-dimensional phase space

fig. 2.4.1 for a three-dimensional phase space). Each trajectory defined by the initial condition thus lies entirely on a smooth hypersurface or generates it.

The hypersurface defined by $I(\boldsymbol{x}) = C$ designates an (n − 1)-dimensional manifold (see also the discussion at the end of this section). If all possible values are assigned to the parameter C, this leads to a one-parameter family of manifolds which fills the phase space completely. If a first integral I is known, the equation $I(\boldsymbol{x}) = C$ for a given C can be solved with respect to x_i. If this x_i is substituted into the remaining (n − 1) equations of eq. (2.3.1), the whole set of equations is reduced to (n − 1) equations. The more first integrals are known, the lower the number of equations and the lower the dimension of the manifold generated by the trajectory.

In mechanics, it is the conservation theorems which often supply first integrals; unfortunately, however, no systematic rule is known which yields an easier derivation of first integrals. The Hamilton function which represents a first integral for conservative Hamilton systems is a stroke of luck. In addition, it can be shown that in the case of conservative forces, the Hamilton function – if it is not time-dependent – corresponds to the total energy, i.e. the sum of kinetic and potential energy (see also section 4.1).

It is necessary to associate many phenomena with geometrical models which cannot be described in a simple form. In the case of dynamical systems, the geometrical models form manifolds. Manifolds play an important role in the determination of the global behaviour of trajectories. For example, non-chaotic attractors which specify the asymptotic state of many dynamical systems lie on manifolds or mould them.

An n-dimensional manifold $\mathbb{M}$ is a topological space with the property that each point and its neighboorhood can be mapped by a one-to-one continuous bijective mapping that has a continuous inverse function, a so-called homeomorphism, onto

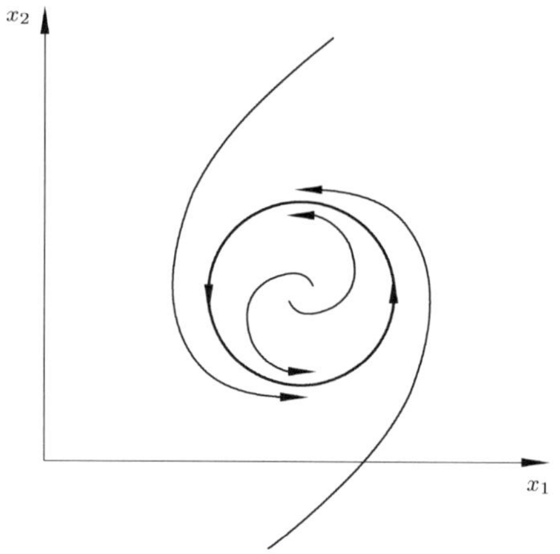

Fig. 2.4.2

Limit cycle or two-dimensional periodic attractor

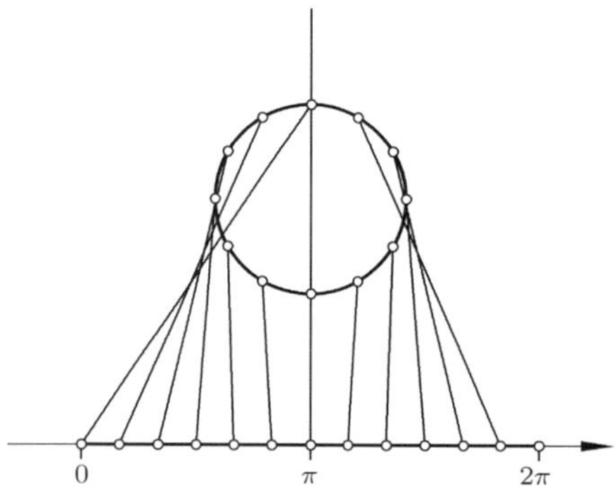

Fig. 2.4.3

Pointwise one-to-one mapping of the limit cycle onto a line

an open subset of the n-dimensional Euclidean space $\mathbb{E}$. Therefore, it is possible to use the coordinates of $\mathbb{E}$ as local coordinates on $\mathbb{M}$, see (Abraham et al., 1993; Bröcker and Jänich, 1990) and fig. 2.4.3. The underlying idea is to map globally complex, but still sufficiently smooth structures onto elementary structures in order to apply, for example, methods from differential geometry. It is well known that the earth's surface cannot be treated as a Euclidean space; measurements on it must be performed in accordance with the laws of spherical trigonometry. Yet parts of the earth can be mapped on charts of an atlas and, when approaching a chart's boundary, one has to switch to the next one.

Geometries which may not admit an overall analytical representation are considered locally in their tangential plane, i.e. the neighborhood of each point can be described by local coordinates. By joining overlapping neighborhoods, so-called charts, the entire manifold and the corresponding mapping can be described analytically. We illustrate this approach using the limit cycle in fig. 2.4.2 as a simple example of a manifold. The limit cycle of fig. 2.4.2 can be subdivided into an assembly of overlapping subsets which can be mapped onto subintervals between 0 and 2π by a continuous one-to-one map. Thus, the φ-coordinate can be used directly as a local coordinate on the manifold, independent of the phase space coordinates which describe the limit cycle.

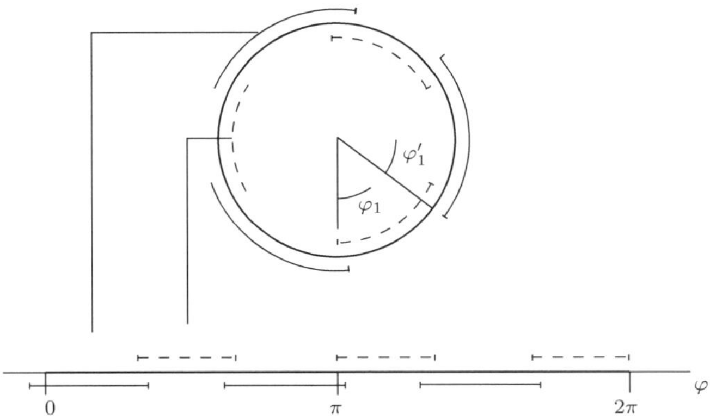

Fig. 2.4.4: Overlapping limit cycle in local coordinates and its mapping onto a line

A further example of a manifold in three dimensions is the two-dimensional torus (fig. 2.4.5). For such a torus, each surface element can be mapped one-to-one onto an element in the plane. The assembly of all the overlapping surface elements once again forms a torus. In a corresponding manner, the surface elements on the associated plane can be joined together. Their assembly yields the square in fig. 2.4.6.

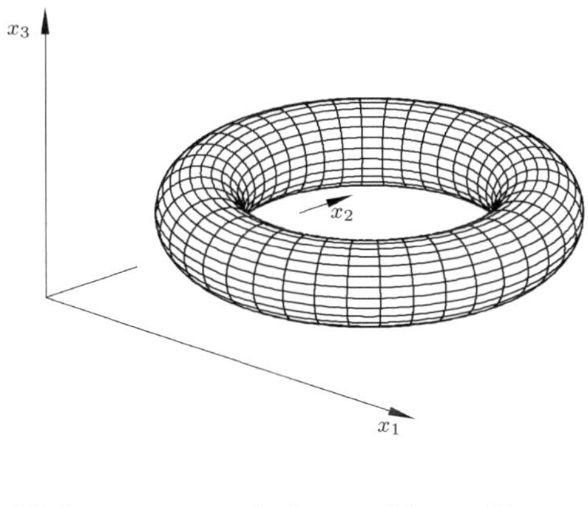

Fig. 2.4.5

Two-dimensional torus in the three-dimensional phase space

With scissors and glue and by rolling and bending, this square can be turned back into the torus; in this way, it becomes clear that each single point of the torus is defined by the coordinates φ_1 and φ_2. A "helical line" on the torus corresponds to a general straight line on the mapping plane. Figure 2.4.6 demonstrates both coordinate lines $\varphi_1 = \text{const}$ and $\varphi_2 = \text{const}$. Since both the limit cycle and the torus are hypersurfaces in the phase space and a velocity vector $\boldsymbol{F}$ exists at each point, they form differentiable manifolds.

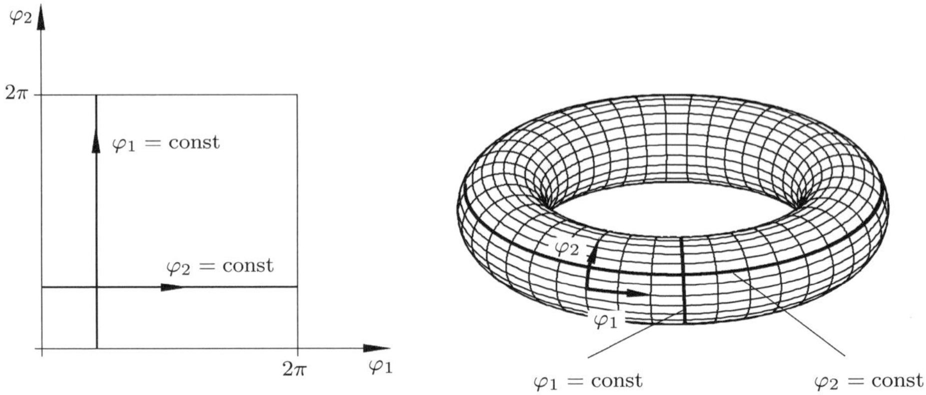

Fig. 2.4.6: The two-dimensional manifold torus in the local coordinates φ_1 and φ_2 bijectively mapped onto a square

2.5 Qualitative and Quantitative Approach

There are basically two contrary approaches to the study and understanding of dynamical systems. In the first case, we concretise a particular problem as a dynamical system and gather as much information as possible about its behaviour. The logical consequence is a complicated set of equations, particularly since the equations must be formulated as realistically as possible in order to incorporate all the involved effects. In the second case, we are interested in the characteristics of dynamical systems in general and not in entering into detail. Here, too, we must differentiate between two cases:

i. A mathematical approximation in the classical sense which evolves and elucidates this new qualitative analysis of differential equations and develops it on the basis of assumptions and strict argumentation. Qualitative approaches are based on geometrical or topological methods which the great mathematician Henri Poincaré (1854 – 1912) first applied to the investigation of the stability of our solar system. Topology, the study of qualitative geometry, has become an indispensible mathematical tool for describing the behaviour of dynamical systems in all their complexity.

ii. An approximation or simulation in the sense of experimental mathematics, an approach made possible by the computer. In this case, the aim is to arrive at generally valid statements on dynamical systems on the basis of simple nonlinear dynamical systems, the behaviour of which is studied numerically. The choice of representative examples in dynamical systems has to be guided by intuition and experience in order to discover characteristic patterns of response and thus obtain informative results for a broader class of dynamical systems.

3 Mathematical Introduction to Dynamical Systems

No roote, no fruite

Jeremiah Dyke, A worthy communicant,

(1640)

In this chapter, we present in the simplest possible manner a survey of some of the fundamental mathematical concepts and tools which are required for the qualitative analysis of the long-term behaviour of dynamical systems. A knowledge of the theory of linear differential equations is a pre-requisite for the comprehension of non-linear dynamics. The reader can find more detailed discussions in Chapters 5 and 6 of this book.

3.1 Linear Autonomous Systems

As we already mentioned in Chapter 2, we consider dynamical processes of a finite dimension n which are differentiable and deterministic.

In the simplest case, the system of differential equations (2.3.3) is linear with constant coefficients

$$\dot{\boldsymbol{x}} = \boldsymbol{L}\boldsymbol{x} \tag{3.1.1}$$

where $\boldsymbol{L}$ denotes the constant, non-singular (n×n)-matrix of the coefficients. Using this example, we now demonstrate which solutions can occur and which patterns the relevant trajectories form in the corresponding phase space.

We introduce the ansatz

$$\boldsymbol{x} = \mathrm{e}^{\lambda t}\boldsymbol{y}, \qquad \boldsymbol{y} = \{y_1 \quad y_2 \quad \ldots\ldots \quad y_{\mathrm{n}}\} \tag{3.1.2}$$

into eq. (3.1.1), thus obtaining a homogeneous linear system of equations

$$[\boldsymbol{L} - \lambda \boldsymbol{I}]\boldsymbol{y} = \boldsymbol{o} \tag{3.1.3}$$

which only possesses non-trivial solutions if the system determinant vanishes

$$P(\lambda) = \det[\boldsymbol{L} - \lambda \boldsymbol{I}] = 0 \tag{3.1.4}$$

or

$$P(\lambda) = A_{\mathrm{n}}\lambda^{\mathrm{n}} + A_{\mathrm{n}-1}\lambda^{\mathrm{n}-1} + \ldots + A_0 = 0 \tag{3.1.5}$$

$P(\lambda)$ is a polynomial of the nth degree in λ and is called characteristic or secular equation. The zeroes of $P(\lambda)$ are the eigenvalues of $\boldsymbol{L}$. A non-vanishing vector $\boldsymbol{y}$

J. Argyris et al., *An Exploration of Dynamical Systems and Chaos*,
DOI: 10.1007/978-3-662-46042-9_3

which satisfies eq. (3.1.3) is called the eigenvector of $\boldsymbol{L}$ appertaining to the eigenvalue λ. If λ and $\boldsymbol{y}$ satisfy eq. (3.1.3), eq. (3.1.2) is a solution of eq. (3.1.1). For each pair $\lambda_i, \boldsymbol{y}_i$, we obtain in accordance with eq. (3.1.2) a solution of the form

$$\boldsymbol{x}_i = e^{\lambda_i t} \boldsymbol{y}_i \tag{3.1.6}$$

If all n eigenvalues λ_i are different, n linearly independent eigenvectors $\boldsymbol{y}_i$ exist and the general solution of eq. (3.1.1) can be expressed by the linear combination

$$\boldsymbol{x}(t) = \sum_{i=1}^{n} C_i e^{\lambda_i t} \boldsymbol{y}_i \tag{3.1.7}$$

with n integration constants C_i which are determined by the initial conditions $\boldsymbol{x}(t_0)$ (Arnold, 1980; Braun, 1979).

From a physical point of view, finding all the states of equilibrium $\boldsymbol{x}_s$ is of interest. These are stationary states of the system in which the dynamical process does not undergo any change; they are thus characterised by $\dot{\boldsymbol{x}}_s = \boldsymbol{o}$. If we observe the corresponding point in the phase space, the vector $\boldsymbol{F}(\boldsymbol{x}_s)$ which defines the change of

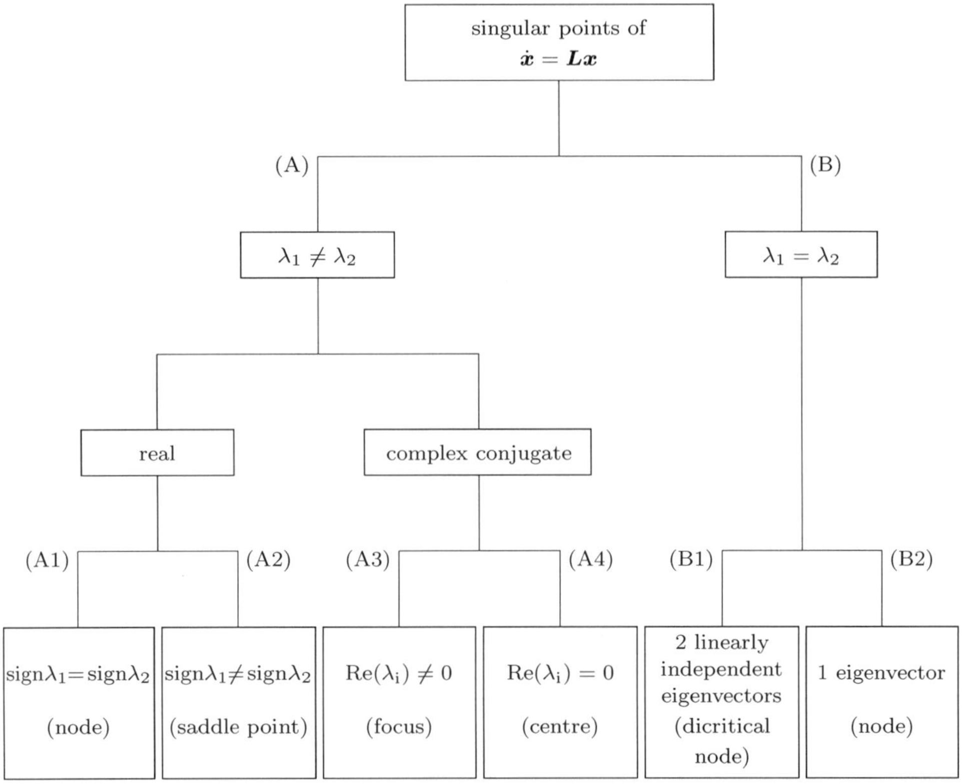

Fig. 3.1.1: Survey of the eigenvalues of singular points for n=2

the trajectory vanishes here. For this reason, $\boldsymbol{x}_s$ is thus also denoted singular point. The decisive factor in characterising the stationary state is the behaviour of the trajectories in the neighbourhood of the singularity. We speak of stable, unstable or neutral equilibrium. If all the trajectories within certain neighbourhoods of $\boldsymbol{x}_s$ are captured, the singular point is asymptotically stable (sink); if, one the other hand, all the trajectories that come close enough to $\boldsymbol{x}_s$ are repelled, $\boldsymbol{x}_s$ is asymptotically unstable (source). On the basis of a system of two differential equations, we now demonstrate the classification of singularities

$$\dot{\boldsymbol{x}} = \boldsymbol{L}\boldsymbol{x}\,, \quad \boldsymbol{L} = \begin{bmatrix} L_{11} & L_{12} \\ L_{21} & L_{22} \end{bmatrix}, \quad \boldsymbol{x} = \{x_1 \; x_2\} \tag{3.1.8}$$

with the origin $\boldsymbol{x}_s = \boldsymbol{o}$ as the singular point. The eigenvalues of the corresponding eq. (3.1.3) serve as a basis to make the following six distinctions, illustrated in the diagram fig. 3.1.1; here, we ignore the degenerate case of a zero eigenvalue.

(A) $\lambda_1 \neq \lambda_2$

Initially, we assume that $\boldsymbol{L}$ possesses different eigenvalues $\lambda_1 \neq \lambda_2$ and denote the corresponding eigenvectors as $\boldsymbol{y}_1$, $\boldsymbol{y}_2$ which we can presume to be normalised without loss of generality.

We now introduce a new system of coordinates

$$\overline{\boldsymbol{x}} = \{\bar{x}_1 \quad \bar{x}_2\} \tag{3.1.9}$$

which is defined by the transformation

$$\boldsymbol{x} = \boldsymbol{T}\bar{\boldsymbol{x}} \qquad \text{with} \qquad \boldsymbol{T} = [\boldsymbol{y}_1 \quad \boldsymbol{y}_2] \tag{3.1.10}$$

Thus, the eigenvectors $\boldsymbol{y}_1$, $\boldsymbol{y}_2$ are the basis vectors of the new coordinate system. Substitution of the transformation eq. (3.1.10) into the given system eq. (3.1.8) yields

$$\dot{\bar{\boldsymbol{x}}} = \boldsymbol{D}\bar{\boldsymbol{x}} \qquad \text{with} \qquad \boldsymbol{D} = \boldsymbol{T}^{-1}\boldsymbol{L}\boldsymbol{T} \tag{3.1.11}$$

Due to

$$\lambda_i \boldsymbol{y}_i = \boldsymbol{L}\boldsymbol{y}_i$$

$\boldsymbol{D}$ can be simplified as follows

$$\begin{aligned} \boldsymbol{D} &= [\boldsymbol{y}_1 \quad \boldsymbol{y}_2]^{-1}\boldsymbol{L}[\boldsymbol{y}_1 \quad \boldsymbol{y}_2] = [\boldsymbol{y}_1 \quad \boldsymbol{y}_2]^{-1}[\boldsymbol{y}_1 \quad \boldsymbol{y}_2]\lceil\lambda_1 \quad \lambda_2\rfloor \\ &= \lceil\lambda_1 \quad \lambda_2\rfloor \end{aligned} \tag{3.1.12}$$

and is thus reduced to diagonal form. The transformed system equivalent to eq. (3.1.8) now becomes

$$\begin{aligned} \dot{\bar{x}}_1 &= \lambda_1 \bar{x}_1 \\ \dot{\bar{x}}_2 &= \lambda_2 \bar{x}_2 \end{aligned} \tag{3.1.13}$$

Here, we can distinguish for $\lambda_1 \neq \lambda_2$ a total of four possibilities (see fig. 3.1.1). If the eigenvalues are real, the solution of eq. (3.1.13) according to eq. (3.1.7) is

$$\bar{x}_1 = C_1 \mathrm{e}^{\lambda_1 t}, \qquad \bar{x}_2 = C_2 \mathrm{e}^{\lambda_2 t} \tag{3.1.14}$$

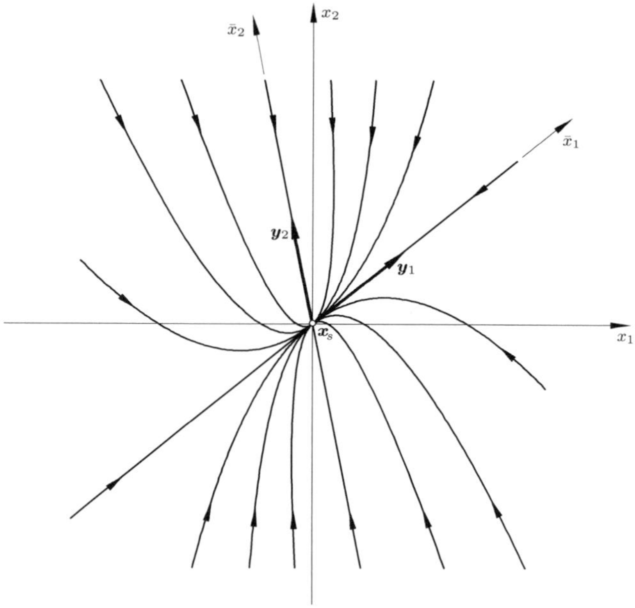

Fig. 3.1.2: Case **A1**: stable node (λ_1, λ_2 real; $\lambda_2 < \lambda_1 < 0$)

or, after eliminating the time parameter t,

$$\bar{x}_2 = C\bar{x}_1^{\lambda_2/\lambda_1} \tag{3.1.15}$$

We next differentiate between two cases having different signs (see fig. 3.1.1);

(**A1**) $\quad \text{sign}\lambda_1 = \text{sign}\lambda_2$

If the eigenvalues have the same sign, eq. (3.1.15) describes a family of parabolae of the order λ_2/λ_1 which have a common tangent at their origin (see fig. 3.1.2). We then speak of a node. If the eigenvalues are negative, the node is stable, as can be seen in the parameter representation of eq. (3.1.14) since all trajectories tend towards the singular point for $t \to \infty$.

(**A2**) $\quad \text{sign}\lambda_1 \neq \text{sign}\lambda_2$

If λ_1 and λ_2 possess different signs, the solution is

$$\bar{x}_2\bar{x}_1^{|\lambda_2/\lambda_1|} = C \tag{3.1.16}$$

i.e. the trajectories are hyperbolae. The corresponding singular point is then called a saddle point or hyperbolic point (see fig. 3.1.3).

If λ_1, λ_2 are complex conjugate, $\boldsymbol{L}$ can always be reduced by applying a (real) linear coordinate transformation $\hat{\boldsymbol{T}}$ to the simple form

$$\dot{\boldsymbol{x}} = \hat{\boldsymbol{L}}\boldsymbol{x} \quad \text{with} \quad \hat{\boldsymbol{L}} = \begin{bmatrix} \alpha & -\omega \\ \omega & \alpha \end{bmatrix}$$

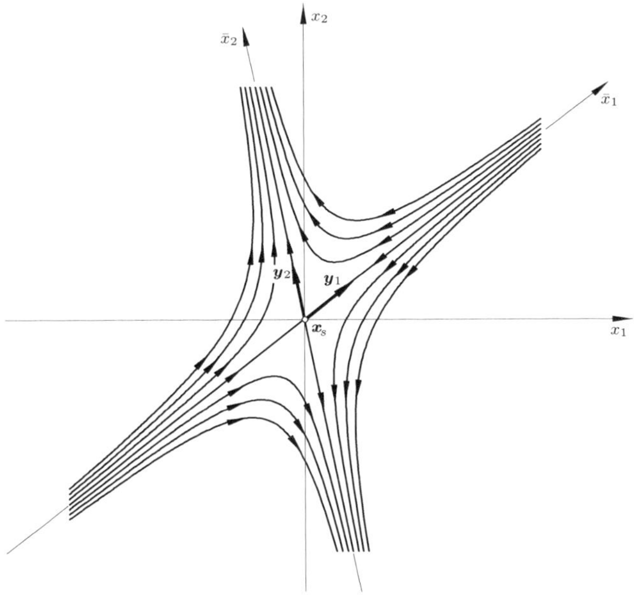

Fig. 3.1.3: Case **A2**: saddle point or hyperbolic point (λ_1, λ_2 real; $\lambda_1 < 0 < \lambda_2$)

The eigenvectors $\boldsymbol{y}_1$ und $\boldsymbol{y}_2$ appertaining to the eigenvalues $\lambda_{1,2} = \alpha \pm i\omega$ are also complex conjugate

$$\boldsymbol{y}_1 = \{1 \quad -i\} \qquad \text{and} \qquad \boldsymbol{y}_2 = \{1 \quad i\} = \boldsymbol{y}_1^*$$

and reduce in accordance with eq. (3.1.10) the transformation of $\hat{\boldsymbol{L}}$ to a diagonal form; in accordance with standard usage, $\boldsymbol{y}_1^*$ denotes the complex conjugate vector of $\boldsymbol{y}_1$. Thus, points $\hat{\boldsymbol{x}}$ of the real phase space are transformed into points $\bar{\boldsymbol{x}}$ of a two-dimensional complex phase space $\mathbb{C}^2$. The solution of eq. (3.1.13) can be expressed immediately as

$$\bar{x}_{1,2} = C_{1,2}\mathrm{e}^{(\alpha \pm i\omega)t} = C_{1,2}\mathrm{e}^{\alpha t}(\cos\omega t \pm i \sin\omega t) \tag{3.1.17}$$

where the complex constants C_1 and C_2 coincide with the initial values $\bar{x}_1(0)$ and $\bar{x}_2(0)$. It is now easy to see that C_1 and C_2 are complex conjugate. From the inverse transformation of eq. (3.1.10) for an initial vector $\hat{\boldsymbol{x}}(0)$

$$\bar{\boldsymbol{x}}(0) = \boldsymbol{T}^{-1}\hat{\boldsymbol{x}}(0) = [\boldsymbol{y}_1 \quad \boldsymbol{y}_1^*]^{-1}\hat{\boldsymbol{x}}(0) = \frac{1}{2}\begin{bmatrix} \boldsymbol{y}_1^{*t} \\ \boldsymbol{y}_1^t \end{bmatrix}\hat{\boldsymbol{x}}(0)$$

we obtain the relation $\bar{x}_2(0) = \bar{x}_1^*(0)$, i.e. $C_2 = C_1^*$. Thus, the solutions according to eq. (3.1.17) are also complex conjugate, i.e. $\bar{x}_2(t) = \bar{x}_1^*(t)$ applies. The inverse transformation of the initial system hence yields a real solution

$$\hat{\boldsymbol{x}}(t) = \mathrm{e}^{\alpha t}\begin{bmatrix} \cos\omega t & -\sin\omega t \\ \sin\omega t & \cos\omega t \end{bmatrix}\hat{\boldsymbol{x}}(0) \tag{3.1.18}$$

We observe that the initial vector $\hat{\boldsymbol{x}}(0)$ is rotated about an angle ωt and stretched with the factor $\mathrm{e}^{\alpha t}$.

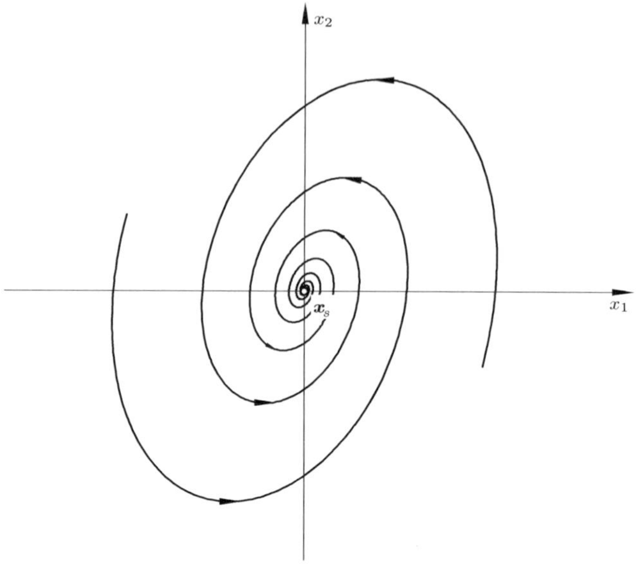

Fig. 3.1.4: Case **A3**: stable focus or spiral point (λ_1, λ_2 complex conjugate; $\mathrm{Re}(\lambda_i) < 0$)

We once again distinguish between two cases (see fig. 3.1.1):

(**A3**) $\alpha = \mathrm{Re}(\lambda_i) \neq 0$

Equation (3.1.18) suggests a representation in polar coordinates for which we set

$$r = r_0 e^{\alpha t}, \qquad \varphi = \omega t + \varphi_0 \tag{3.1.19}$$

with $r_0 = |\hat{\boldsymbol{x}}(0)|$ and $\varphi_0 = \tan^{-1}(\hat{x}_2(0)/\hat{x}_1(0))$. This formulation describes a family of logarithmic spirals. In accordance with an Anglo-Saxon convention, we use the inverse function sign for arcus. If the real part of λ_i is negative, i.e. $\mathrm{Re}(\lambda_i) = \alpha < 0$, we have a stable focus or spiral point; on the other hand, for $\alpha > 0$, the singular point is unstable. Figure 3.1.4 shows the spirals in the original x_1, x_2 coordinate system. They are affinely distorted by the inverse transformation.

(**A4**) $\alpha = \mathrm{Re}(\lambda_i) = 0$

In the special case of $\alpha = 0$, the phase curves in the $\hat{x}_1, \hat{x}_2$ coordinate system become concentric circles in accordance with eq. (3.1.18), the centre of which is the singular point. We then speak of a centre. Figure 3.1.5 shows the trajectories in the original x_1, x_2 coordinate system in which $\boldsymbol{L}$ possesses a general form. Here, we observe concentric ellipses.

(**B**) $\lambda_1 = \lambda_2 = \lambda$

We now move on to a discussion of the case of identical eigenvalues, $\lambda_1 = \lambda_2 = \lambda$. Here, we distinguish between two cases (fig. 3.1.1):

(**B1**) There are *two* linearly independent eigenvectors $\boldsymbol{y}_1, \boldsymbol{y}_2$ for λ.

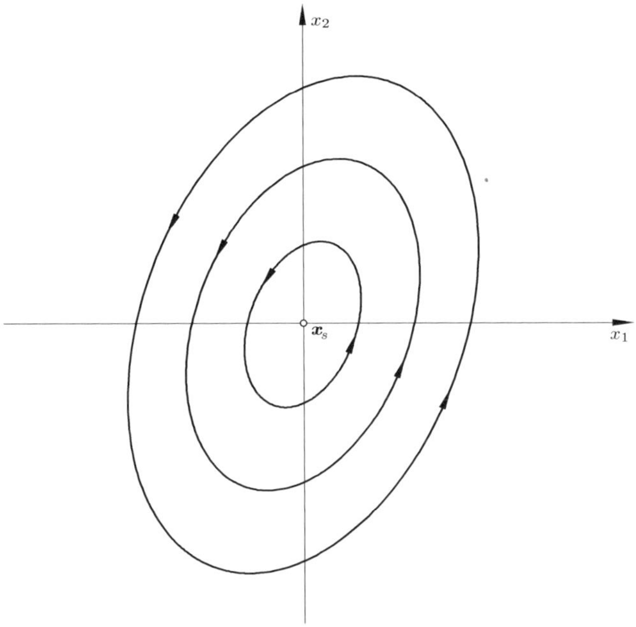

Fig. 3.1.5: Case **A4**: Centre (λ_1, λ_2 complex conjugate; $\mathrm{Re}(\lambda_\mathrm{i}) = 0$)

In analogy to eqs. (3.1.11), (3.1.12), the system $\dot{\boldsymbol{x}} = \boldsymbol{L}\boldsymbol{x}$ can be transformed into a diagonal form, $\dot{\bar{x}}_\mathrm{i} = \lambda \bar{x}_\mathrm{i}$ (i = 1, 2), with the solutions

$$\bar{x}_\mathrm{i} = C_\mathrm{i} \mathrm{e}^{\lambda t} \qquad (\mathrm{i} = 1, 2) \tag{3.1.20}$$

or

$$\bar{x}_2 = C \bar{x}_1 \tag{3.1.21}$$

We thus obtain a bundle of straight lines and speak of a dicritical node (see fig. 3.1.6).

In this case, a sufficient number of eigenvectors does not exist to span the whole phase space; thus, $\boldsymbol{L}$ can no longer be transformed to diagonal form. However, in matrix theory, it is shown that any non-diagonalisable (n × n) matrix can always be transformed uniquely to the so-called Jordan normal form where each Jordan block is associated with an eigenvector (Zurmühl, 1964).

For n = 2, there is thus a specific $\bar{x}_1, \bar{x}_2$ coordinate system in which the system of differential equations (3.1.8) can be transformed to the Jordan normal form

$$\dot{\bar{\boldsymbol{x}}} = \bar{\boldsymbol{L}} \bar{\boldsymbol{x}} \qquad \text{with} \qquad \bar{\boldsymbol{L}} = \begin{bmatrix} \lambda & 1 \\ 0 & \lambda \end{bmatrix} \tag{3.1.22}$$

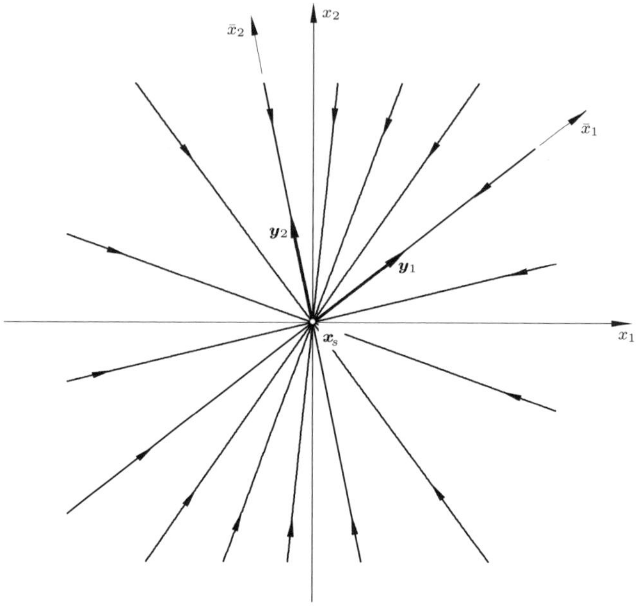

Fig. 3.1.6: Case **B1**: stable dicritical node ($\lambda_1 = \lambda_2 = \lambda < 0$)

The corresponding coordinate transformation is

(B2) There is only *one* linearly independent eigenvector $\boldsymbol{y}$ for λ.

$$\boldsymbol{x} = \boldsymbol{T}\bar{\boldsymbol{x}} \qquad \text{with} \qquad \boldsymbol{T} = [\boldsymbol{y} \quad \boldsymbol{z}] \tag{3.1.23}$$

where a second vector $\boldsymbol{z}$ linearly independent of $\boldsymbol{y}$ must be established. This vector $\boldsymbol{z}$ is denoted generalised eigenvector. Pre-multiplying

$$\bar{\boldsymbol{L}} = \boldsymbol{T}^{-1}\boldsymbol{L}\boldsymbol{T}$$

with $\boldsymbol{T}$ yields

$$\boldsymbol{T}\bar{\boldsymbol{L}} = \boldsymbol{L}\boldsymbol{T}$$

Evaluating the matrix products, we find

$$\boldsymbol{T}\bar{\boldsymbol{L}} = [\boldsymbol{y} \quad \boldsymbol{z}]\begin{bmatrix} \lambda & 1 \\ 0 & \lambda \end{bmatrix} = [\lambda\boldsymbol{y} \quad (\boldsymbol{y} + \lambda\boldsymbol{z})]$$

and

$$\boldsymbol{L}\boldsymbol{T} = \boldsymbol{L}[\boldsymbol{y} \quad \boldsymbol{z}] = [\lambda\boldsymbol{y} \quad \boldsymbol{L}\boldsymbol{z}]$$

Comparing the second columns and resolving for $\boldsymbol{y}$ yields

$$[\boldsymbol{L} - \lambda\boldsymbol{I}]\boldsymbol{z} = \boldsymbol{y} \tag{3.1.24}$$

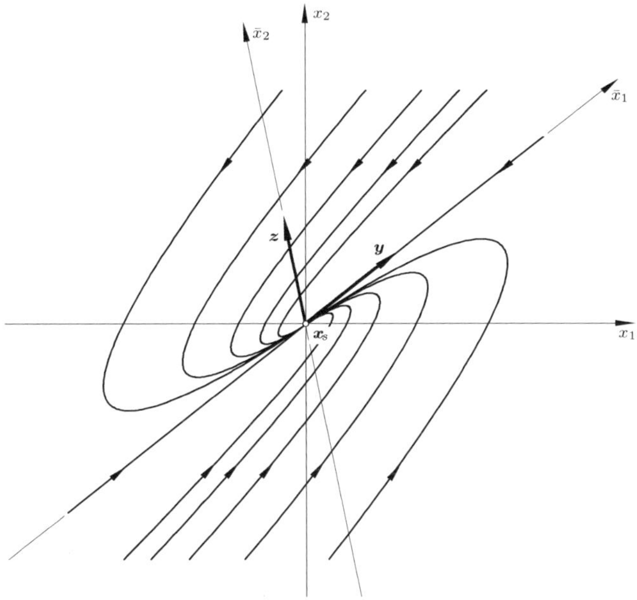

Fig. 3.1.7: Case **B2**: stable improper node ($\lambda_1 = \lambda_2 = \lambda < 0$). Construction of the generalised eigenvector $\boldsymbol{z}$

which determines the unknown vector $\boldsymbol{z}$. The solution of the system eq. (3.1.22) can be written in the form (Braun, 1979)

$$\bar{x}_1 = \mathrm{e}^{\lambda t}(C_1 + C_2 t)$$

$$\bar{x}_2 = C_2 \mathrm{e}^{\lambda t} \tag{3.1.25}$$

or, after eliminating t,

$$\bar{x}_1 = \frac{1}{\lambda}\bar{x}_2(\ln|\bar{x}_2| + C) \tag{3.1.26}$$

The trajectories in fig. 3.1.7 coalesce in the singular point $\boldsymbol{x}_s$, which is called a stable improper node. In the point $\boldsymbol{x}_s$, the trajectories possess a common tangent which coincides with the direction of the eigenvector $\boldsymbol{y}$.

Equation (3.1.25) demonstrates that the general solution of a linear system of differential equations cannot always be expressed as in eq. (3.1.7). In fact, this formulation only applies when all the linearly independent eigenvectors do indeed span the whole phase space; this is always the case when all n eigenvalues are different. If, however, λ_k is a r-fold root, then the coefficients $C_\mathrm{i}(t)$ may take the form of polynomials of order $\mathrm{m} \leqslant \mathrm{r} - 1$.

This completes our discussion of the behaviour of the trajectories in the neighbourhood of singular points in the two-dimensional case. Figure 3.1.8 surveys once again the various types of singularities.

In the two-dimensional case, we clearly recognised that the signs of the real parts of the eigenvalues of $\boldsymbol{L}$ determine whether the observed state of equilibrium is

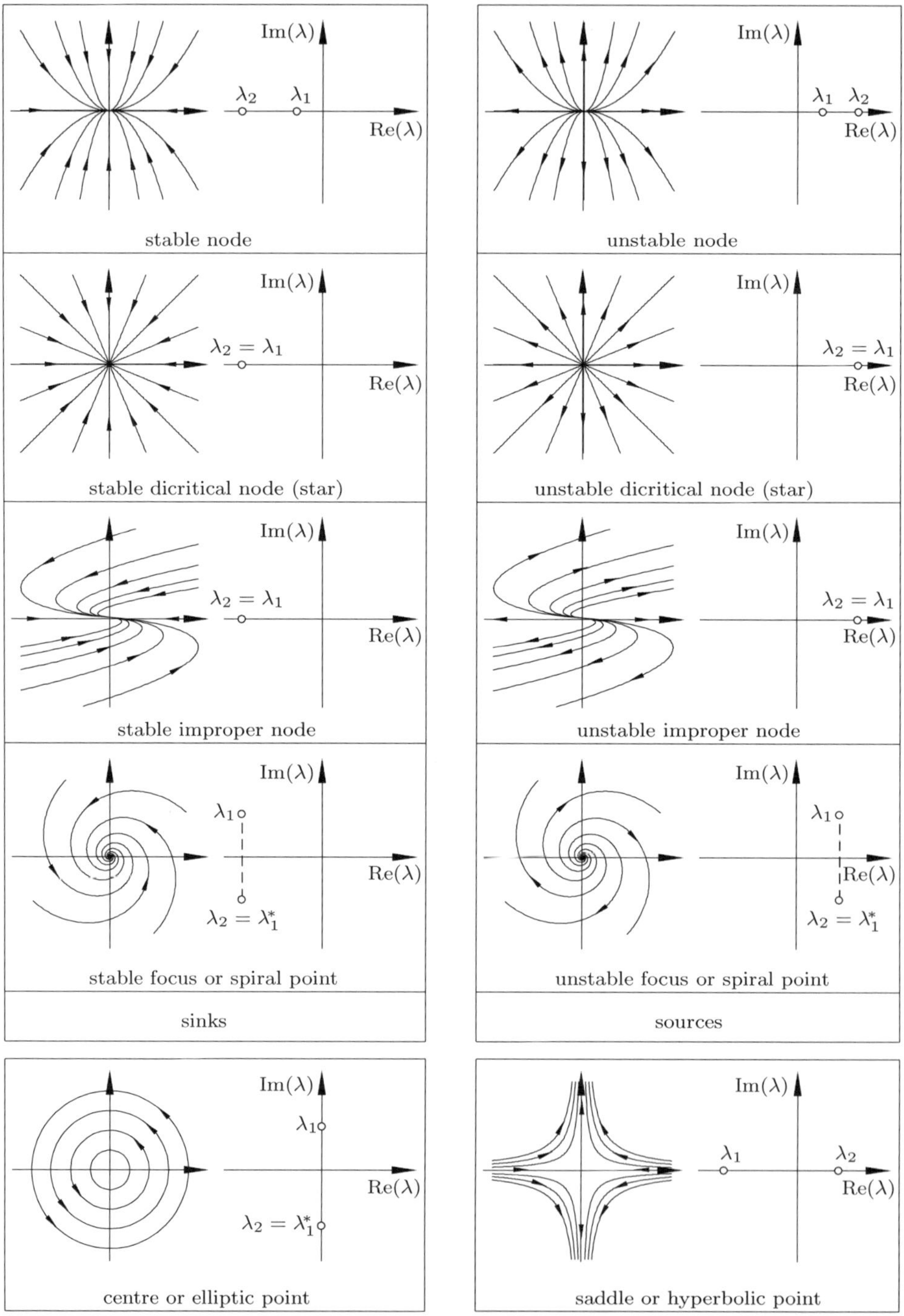

Fig. 3.1.8: Survey of the singularities in a two-dimensional phase space, phase portraits and eigenvalues for n=2

stable, unstable or neutral. The phase portrait alters fundamentally if the eigenvalues change their sign or become purely imaginary.

Let us now return to the general n-dimensional case. If the real part of a single eigenvalue is positive, we can always find one trajectory which diverges from the singular point so that it is unstable. The phase portraits are then either a saddle point or a source.

On the basis of the characteristic polynomial of eq. (3.1.5), it is now easy to decide whether $\mathrm{Re}(\lambda_\mathrm{i}) < 0$ for all $\mathrm{i} = 1, 2, \ldots, \mathrm{n}$. We can deduce this information from the theorem of Hurwitz (Bronstein et al., 1995) which is often used in the classical study of oscillations and reads as follows:

Let the polynomial

$$P(\lambda) = A_0 + A_1\lambda + \ldots + A_\mathrm{n}\lambda^\mathrm{n}$$

have real coefficients, and $A_0 > 0$. All roots λ_i of the equation $P(\lambda) = 0$ contain negative real parts if and only if all determinants H_i $(\mathrm{i} = 1, 2, \ldots, \mathrm{n})$ specified by the coefficients A_j are positive

$$\begin{aligned}
H_1 &= A_1 \\
H_2 &= \begin{vmatrix} A_1 & A_0 \\ A_3 & A_2 \end{vmatrix} \\
H_3 &= \begin{vmatrix} A_1 & A_0 & 0 \\ A_3 & A_2 & A_1 \\ A_5 & A_4 & A_3 \end{vmatrix} \\
&\vdots \\
H_\mathrm{n} &= \begin{vmatrix} A_1 & A_0 & 0 & \ldots & 0 & 0 \\ A_3 & A_2 & A_1 & \ldots & 0 & 0 \\ \vdots & \vdots & \vdots & & \vdots & \vdots \\ A_{2\mathrm{n}-3} & A_{2\mathrm{n}-4} & A_{2\mathrm{n}-5} & \ldots & A_{\mathrm{n}-1} & A_{\mathrm{n}-2} \\ A_{2\mathrm{n}-1} & A_{2\mathrm{n}-2} & A_{2\mathrm{n}-3} & \ldots & A_{\mathrm{n}+1} & A_\mathrm{n} \end{vmatrix}
\end{aligned} \tag{3.1.27}$$

Here, all coefficients A_k which do not appear in $P(\lambda)$ vanish

$$A_\mathrm{k} = 0 \qquad (\text{for } \mathrm{k} > \mathrm{n})$$

We now determine the character of the singular point using a linearised pendulum with friction as an example.

Unforced Pendulum with Viscous Damping

We write the linearised equation of motion in the form

$$\ddot{x} + 2\zeta\omega_0\dot{x} + \omega_0^2 x = 0 \tag{3.1.28}$$

and transform it by the substitution $x_1 = x$, $x_2 = \dot{x}$ into the linear system

$$\begin{bmatrix} \dot{x}_1 \\ \dot{x}_2 \end{bmatrix} = \boldsymbol{L}\begin{bmatrix} x_1 \\ x_2 \end{bmatrix} \quad \text{with} \quad \boldsymbol{L} = \begin{bmatrix} 0 & 1 \\ -\omega_0^2 & -2\zeta\omega_0 \end{bmatrix} \tag{3.1.29}$$

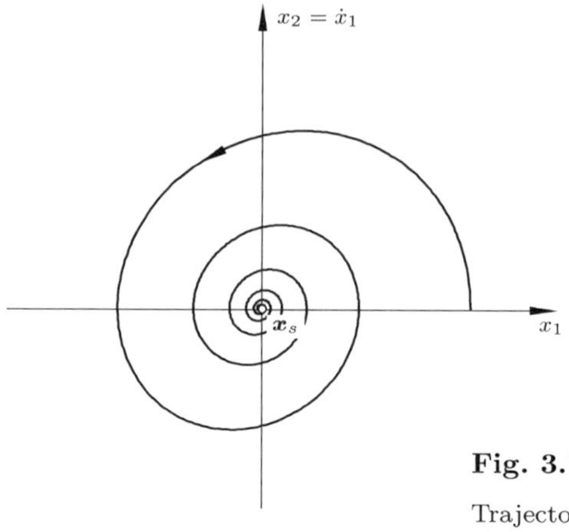

Fig. 3.1.9

Trajectory of an unforced pendulum with viscous damping; $\boldsymbol{x}_s$ is a point attractor

For the sub-critically damped case $0 < \zeta < 1$, we obtain two complex conjugate eigenvalues $\lambda_{1,2} = \omega_0(-\zeta \pm i\sqrt{1-\zeta^2})$. We again write the solution in the transformed system in polar coordinates (case **A3**) to gain a simple interpretation of the trajectories

$$\begin{aligned} r &= r_0 \mathrm{e}^{-\zeta\omega_0 t} \\ \varphi &= \omega_0\sqrt{1-\zeta^2}\, t + \varphi_0 \end{aligned} \tag{3.1.30}$$

i.e. the trajectories are logarithmic spirals which converge towards the singular point which, in the case of $\zeta < 1$, is called focus or point attractor (fig. 3.1.9). Since energy is continuously lost due to damping, the amplitude of the pendulum decreases continuously until it comes to rest in the state of equilibrium $\boldsymbol{x}_s$. In the case of vanishing damping, the spirals degenerate into closed circular orbits about $\boldsymbol{x}_s$.

It is sometimes advantageous to write the solution of the system of linear differential equations (3.1.1) for an initial condition $\boldsymbol{x}(t_0) = \boldsymbol{x}_0$ in the compact form (Arnold, 1980; Braun, 1979)

$$\boldsymbol{x}(t) = \mathrm{e}^{\boldsymbol{L}t}\boldsymbol{x}_0 \tag{3.1.31}$$

where the exponential function of the (n × n)-matrix $\boldsymbol{L}t$ is defined by the series expansion

$$\mathrm{e}^{\boldsymbol{L}t} = \boldsymbol{I} + \boldsymbol{L}t + \frac{(\boldsymbol{L}t)^2}{2!} + \ldots = \sum_{\mathrm{k}=0}^{\infty} \frac{1}{\mathrm{k}!}(\boldsymbol{L}t)^{\mathrm{k}} \tag{3.1.32}$$

Due to the uniform convergence of this series expansion, the sequential order of differentiation and summation can be interchanged and it can be shown immediately that eq. (3.1.31) is the solution of the original eq. (3.1.1). Indeed, the following relation applies

$$\dot{\boldsymbol{x}}(t) = \boldsymbol{L}\mathrm{e}^{\boldsymbol{L}t}\boldsymbol{x}_0 = \boldsymbol{L}\boldsymbol{x}(t) \tag{3.1.33}$$

3.2 Non-linear Systems and Stability

For non-linear systems of ordinary differential equations of the type of eq. (2.3.3), no general methods of establishing explicit solutions exist except in special cases. Interest is often directed less at quantitative statements, such as specific trajectories, than at qualitative ones, e.g. the classification of singular points. This becomes particularly clear if we consider, for example, the infinite-dimensional problem of liquid flowing about a body. Answering the question as to where vortices occur or when the flow becomes turbulent is of greater significance than statements regarding the position and velocity of individual liquid particles. A further example is the Duffing equation, discussed in detail in section 10.5. Here, our interest is aimed for example at the question at which amplitudes of the driving force period doubling occurs and when the motion finally becomes chaotic. Geometrical and topological aspects of the trajectories play an important role here, i.e. global characteristics of the phase portrait, like, for example, the structural stability of a trajectory. As we shall see later in Chapter 6 – well beyond our preceding discussion on the elementary linear case – the behaviour of a dynamical process can change fundamentally if the real part of an eigenvalue undergoes a zero passage due to a modification of a control parameter.

Many authors have dealt in detail with the theory of systems of ordinary differential equations; we refer the interested reader to Arnold (1980), Haken (1983a), Guckenheimer and Holmes (1983) and Wiggins (1990).

In many physical applications, the first aim is directed towards qualitative statements, i.e. for example, finding states of equilibrium characterised by $\dot{\boldsymbol{x}} = \boldsymbol{o}$. Moreover, we are interested in whether the behaviour for $t \to \infty$ – which, for example, can be a state of equilibrium, a periodic solution or an irregular motion – is stable. As we saw in the introduction, the question regarding stability, especially that of our solar system, in the case of $t \to \infty$ has posed a challenge to many mathematicians and physicists.

We first draw some conclusions from the form and characteristic properties of the set of differential equations. Concerning the existence and uniqueness of solutions, the following statement applies to the initial value problem

$$\dot{\boldsymbol{x}} = \boldsymbol{F}(\boldsymbol{x}), \qquad \boldsymbol{x}(t_0) = \boldsymbol{x}_0 \tag{3.2.1}$$

where $\boldsymbol{F}$ is a non-linear vector function. Thus, eq. (3.2.1) has exactly one solution if the partial derivatives of the functions F_{j} with respect to the variables $x_1, x_2, \ldots, x_{\mathrm{n}}$ exist and are continuous. Consequently, exactly one trajectory runs through each (non-singular) point $\boldsymbol{x}$ of the phase space which belongs to an autonomous system of differential equations, i.e. the trajectories cannot intersect.

Since time does not appear explicitly in the autonomous formulation of a system, the system is invariant with respect to a time shift, i.e. with respect to a transformation $\tau = t + c$. For this reason, one can always set $t = 0$ in an initial value problem.

Singular points or stationary solutions $\boldsymbol{x}_s$, characterised by $\dot{\boldsymbol{x}}_s = \boldsymbol{o}$, and the trajectories in the neighbourhood of such states of equilibrium provide a first overview of the solution behaviour of a non-linear system eq. (3.2.1). For this reason, we consider in the phase space a point $\boldsymbol{x}$ adjacent to the singular point $\boldsymbol{x}_s$ following the superimposition of a small perturbation $\tilde{\boldsymbol{x}}$ on the state of equilibrium $\boldsymbol{x}_s$

$$\boldsymbol{x} = \boldsymbol{x}_s + \tilde{\boldsymbol{x}} \qquad \text{with} \qquad |\tilde{\boldsymbol{x}}| \ll 1 \tag{3.2.2}$$

Substituting eq. (3.2.2) into eq. (3.2.1) and expanding $\boldsymbol{F}$ into a Taylor series in the neighbourhood of $\boldsymbol{x}_s$, we obtain

$$\dot{\boldsymbol{x}} = \dot{\boldsymbol{x}}_s + \dot{\tilde{\boldsymbol{x}}} = \boldsymbol{F}(\boldsymbol{x}_s) + \left.\frac{\partial \boldsymbol{F}}{\partial \boldsymbol{x}}\right|_{\boldsymbol{x}_s} \tilde{\boldsymbol{x}} + \mathcal{O}(\tilde{\boldsymbol{x}}^2) \tag{3.2.3}$$

where

$$\left.\frac{\partial \boldsymbol{F}}{\partial \boldsymbol{x}}\right|_{\boldsymbol{x}_s} = \boldsymbol{D}(\boldsymbol{x}_s)$$

is the Jacobian matrix which contains the partial derivatives of the functions F_{j} with respect to the variables x_{k}. In the literature, the symbol $\boldsymbol{J}$ is occasionally also used for the Jacobian matrix, in particular the determinant $\det \boldsymbol{J}$ is used when applying coordinate transformations; see eq. (4.2.37). In $\mathcal{O}(\tilde{\boldsymbol{x}}^2)$, all higher-order terms in $\tilde{\boldsymbol{x}}$ are included. In a first approximation, we may consider the linearised system

$$\dot{\tilde{\boldsymbol{x}}} = \boldsymbol{D}(\boldsymbol{x}_s)\tilde{\boldsymbol{x}} \tag{3.2.4}$$

For singular points $\boldsymbol{x}_s$, $\boldsymbol{D}$ is a constant matrix. We saw in section 3.1 that the sign of the real parts of the eigenvalues of $\boldsymbol{D}$ determines the stability of the solution.

According to a theorem of Hartman and Grobman (Guckenheimer and Holmes, 1983), the classification of the singular points of the non-linear system eq. (3.2.1) can be based on the *linearised* system eq. (3.2.4) as long as *no* eigenvalue of $\boldsymbol{D}(\boldsymbol{x}_s)$ vanishes or no eigenvalue is purely imaginary. We refer the reader to section 5.4.1 for a more detailed discussion.

A singular point $\boldsymbol{x}_s$, for which the eigenvalues of the corresponding Jacobian matrix $\boldsymbol{D}(\boldsymbol{x}_s)$ do not possess vanishing real parts, is called a *hyperbolic* or *non-degenerate fixed point*. If, on the other hand, an index j exists for which $\mathrm{Re}(\lambda_{\mathrm{j}}) = 0$, we speak of a *non-hyperbolic* or *degenerate fixed point*. In this case, the non-linear terms determine the stability behaviour (see section 6.2 for particulars). Figure 3.2.1 shows the phase portraits of the non-degenerate fixed points of non-linear systems for $\mathrm{n} = 2$ which can be classified according to the Hartman-Grobman theorem in accordance with the eigenvalues of the linearised system.

Up to now, we have applied the concept of stability in a local sense in order to characterise the behaviour of the solutions in the neighbourhood of a point of equilibrium. This concept can, however, be understood much more generally.

To this end, we consider a reference solution $\boldsymbol{x}_r = \boldsymbol{x}_r(t)$ of the initial value problem eq. (3.2.1) with the initial condition $\boldsymbol{x}_r(0) = \boldsymbol{x}_{r0}$. Due to imprecision of measurement, the initial conditions cannot be prescribed in physical problems with any desired precision. In the same way, the accuracy of the initial conditions in a numerical calculation performed on a computer is limited in that $\boldsymbol{x}_{r0}$ can only be

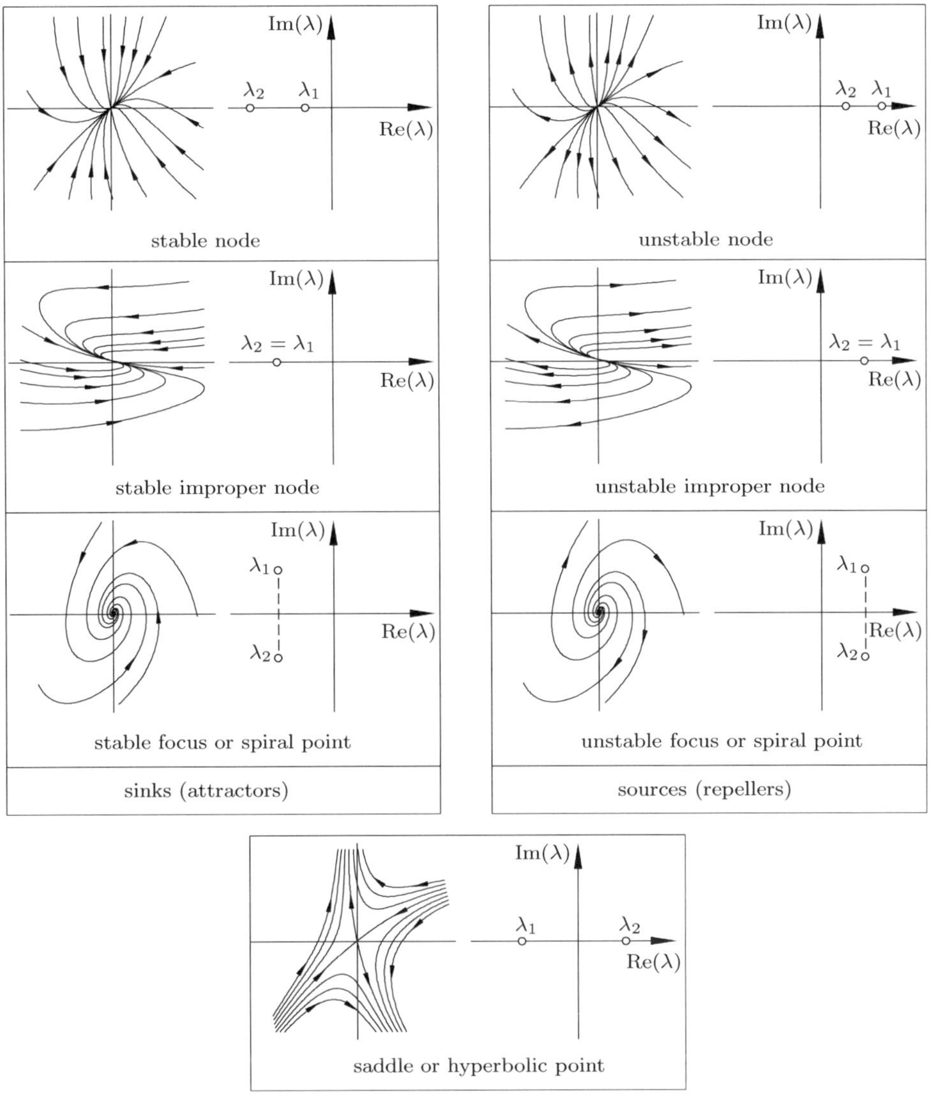

Fig. 3.2.1: Classification of the non-degenerate fixed points for n = 2

specified by a finite number of digits. Thus, it is of primary importance to ascertain to what extent the reference solution reacts sensitively to small deviations in the initial conditions.

One intuitively denotes a trajectory through an initial point $\boldsymbol{x}_{r0}$ as stable if all the curves which start out near $\boldsymbol{x}_{r0}$ stay in the immediate neighbourhood of the initial trajectory forever. More precisely: the solution $\boldsymbol{x}_r(t)$ is *stable in the Lyapunov sense* if, for an arbitrarily small $\varepsilon > 0$, a $\delta > 0$ ($\delta = \delta(\varepsilon)$ exists, i.e. δ is only a function of

ε and not of the time t), so that for each solution $\boldsymbol{x}(t)$ of eq. (3.2.1) with an initial condition satisfying

$$|\boldsymbol{x}(t_0) - \boldsymbol{x}_r(t_0)| < \delta(\varepsilon)$$

the solution $\boldsymbol{x}_r(t)$ satisfies

$$|\boldsymbol{x}(t) - \boldsymbol{x}_r(t)| < \varepsilon \qquad \text{for all } t > t_0 \tag{3.2.5}$$

(see fig. 3.2.2). If one cannot find such a δ-neighbourhood for a prescribed ε-neighbourhood, $\boldsymbol{x}_r(t)$ is called *unstable*.

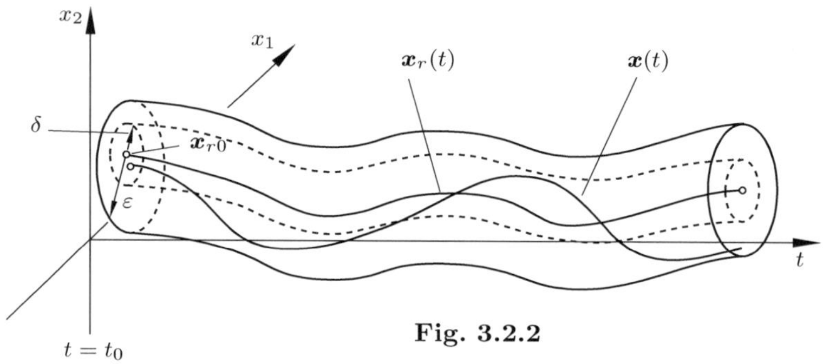

Fig. 3.2.2

On the definition of Lyapunov stability

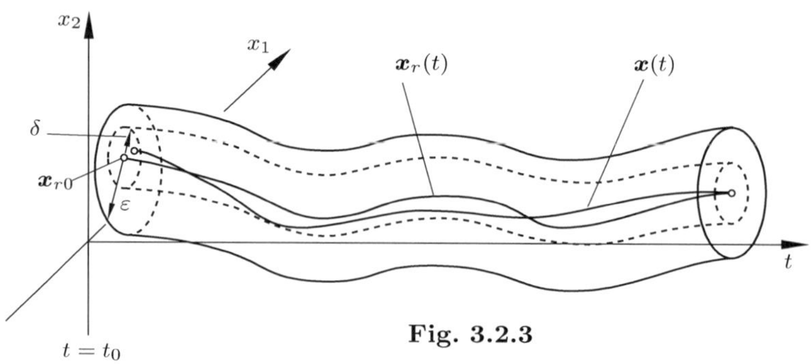

Fig. 3.2.3

On the definition of asymptotic stability

This definition contains no information as to whether adjacent trajectories come arbitrarily close to the reference trajectory for $t \to \infty$. If so, this special case is designated as one of *asymptotic stability* (see fig. 3.2.3). A stable orbit is called asymptotically stable if, in addition to the Lyapunov stability criterion,

$$\lim_{t\to\infty} |\boldsymbol{x}(t) - \boldsymbol{x}_r(t)| = 0 \tag{3.2.6}$$

applies. An example with asymptotic stability for a single-degree-of-freedom oscillator is the fixed point (fig. 2.2.2) or the limit cycle (fig. 2.2.3).

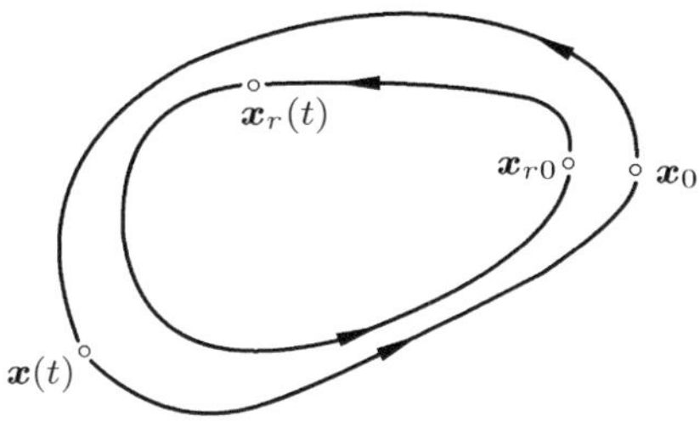

Fig. 3.2.4

Example of orbital stability

It is also possible that adjacent trajectories as a whole remain adjacent, but the orbital periods of the points deviate greatly from one another. We then speak of *orbital stability*, fig. 3.2.4 (Haken, 1990).

We now illustrate the concept of the asymptotic stability of a solution on an example. To this end, we consider a linear oscillator with viscous damping subject to a periodic excitation force; the equation of motion for this is, see eq. (2.2.3),

$$\ddot{x} + 2\zeta\omega_0\dot{x} + \omega_0^2 x = C\sin\omega_E t \qquad (0 < \zeta < 1,\ \omega_0 \neq 0,\ \omega_E \neq \omega_0) \tag{3.2.7}$$

The solution of this second-order differential equation is obtained by superposition of the general solution of the unforced (homogeneous) system and a particular integral of the inhomogeneous differential equation. We first determine the general solution of the homogeneous differential equation. Using as before the substitution $x_1 = x$, $x_2 = \dot{x}$, the corresponding first-order system can be written in the form

$$\dot{\boldsymbol{x}} = \boldsymbol{L}\boldsymbol{x} \qquad \text{with} \qquad \boldsymbol{L} = \begin{bmatrix} 0 & 1 \\ -\omega_0^2 & -2\zeta\omega_0 \end{bmatrix} \tag{3.2.8}$$

see eq. (3.1.33). For $0 < \zeta < 1$, $\boldsymbol{L}$ possesses two complex conjugate eigenvalues

$$\lambda_1 = \omega_0(-\zeta + i\sqrt{1-\zeta^2}\,), \qquad \lambda_2 = \lambda_1^* \tag{3.2.9}$$

The corresponding eigenvectors $\boldsymbol{y}$ derive from the relation

$$[\boldsymbol{L} - \lambda_\mathrm{i}\boldsymbol{I}]\boldsymbol{y}_\mathrm{i} = \boldsymbol{o} \qquad (\mathrm{i} = 1, 2) \tag{3.2.10}$$

Disregarding a proportionality factor, we find

$$\boldsymbol{y}_1 = \begin{bmatrix} 1 \\ \lambda_1 \end{bmatrix} \quad , \quad \boldsymbol{y}_2 = \begin{bmatrix} 1 \\ \lambda_1^* \end{bmatrix} = y_1^* \tag{3.2.11}$$

In accordance with eq. (3.1.7), the general solution of the homogeneous system reads

$$\boldsymbol{x}(t) = \sum_{\mathrm{i}=1}^{2} C_\mathrm{i}\mathrm{e}^{\lambda_\mathrm{i} t}\boldsymbol{y}_\mathrm{i} = C_1\mathrm{e}^{\lambda_1 t}\begin{bmatrix} 1 \\ \lambda_1 \end{bmatrix} + C_1^*\mathrm{e}^{\lambda_1^* t}\begin{bmatrix} 1 \\ \lambda_1^* \end{bmatrix} \tag{3.2.12}$$

Here, we have used the relation $C_2 = C_1^*$ which we referred to in case **A3**, section 3.1. Setting

$$C_1 = \frac{B}{2}\mathrm{e}^{-i\beta} \qquad \text{and} \qquad \omega_D = \omega_0\sqrt{1-\zeta^2} \tag{3.2.13}$$

we deduce from eq. (3.2.12)

$$\boldsymbol{x}(t) = 2\,\mathrm{Re}\big(C_1 \mathrm{e}^{\lambda_1 t}\boldsymbol{y}_1\big) = B\mathrm{e}^{-\zeta\omega_0 t}\,\mathrm{Re}\left(\mathrm{e}^{i(\omega_D t-\beta)}\begin{bmatrix} 1 \\ -\zeta\omega_0 + i\omega_D \end{bmatrix}\right)$$

or, using Euler's formula,

$$\boldsymbol{x}(t) = B\mathrm{e}^{-\zeta\omega_0 t}\begin{bmatrix} \cos(\omega_D t-\beta) \\ -\zeta\omega_0\cos(\omega_D t-\beta) - \omega_D\sin(\omega_D t-\beta) \end{bmatrix} \tag{3.2.14}$$

where the integration constants B and β derive from the initial condition $\boldsymbol{x}_0$ and ω_D is the damped eigenfrequency. In order to determine the general solution of eq. (3.2.7), we still require a particular integral of the inhomogeneous differential equation. This can either be established by variation of the constants B and β, see (Braun, 1979), or by application of an ansatz for x

$$x = A\sin(\omega_E t - \alpha) \tag{3.2.15}$$

where A and α are deduced by substitution of eq. (3.2.15) into eq. (3.2.7). In this way, we obtain the relation

$$A(\omega_0^2 - \omega_E^2)\sin(\omega_E t-\alpha) + 2A\zeta\omega_0\omega_E\cos(\omega_E t-\alpha) = C\sin\omega_E t$$

From this, the constants A and α are found to be

$$\begin{aligned} \alpha &= \tan^{-1}\frac{2\zeta\omega_0\omega_E}{\omega_0^2-\omega_E^2} \\ A &= \frac{C}{2\zeta\omega_0\omega_E}\sin\alpha \end{aligned} \tag{3.2.16}$$

Thus, expressed in x_1, x_2 coordinates, the complete solution becomes

$$\boldsymbol{x}(t) = \begin{bmatrix} B\mathrm{e}^{-\zeta\omega_0 t}\cos(\omega_D t-\beta) + A\sin(\omega_E t-\alpha) \\ -B\mathrm{e}^{-\zeta\omega_0 t}\big(\zeta\omega_0\cos(\omega_D t-\beta) + \omega_D\sin(\omega_D t-\beta)\big) + A\omega_E\cos(\omega_E t-\alpha) \end{bmatrix} \tag{3.2.17}$$

Figure 3.2.5 displays the trajectories in the extended phase space as well as their projection onto the x_1, x_2 phase plane. In the extended phase space, the particular solution is a helix; its projection onto the x_1, x_2 plane is an ellipse. From eq. (3.2.17),

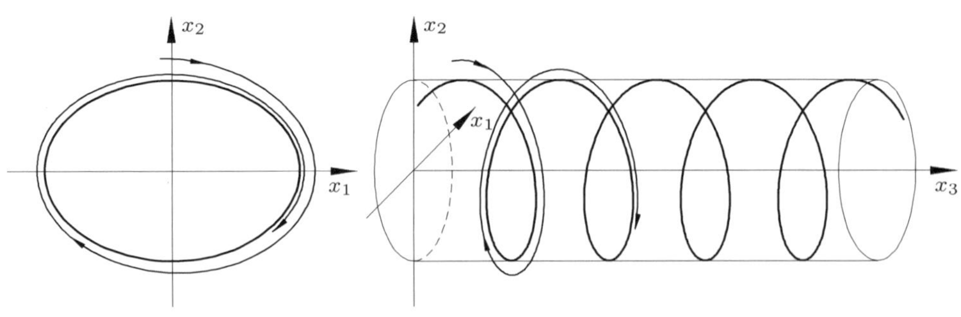

Fig. 3.2.5: Asymptotic stability, limit cycle

we deduce that, for $0 < \zeta < 1$ and $t \to \infty$, all trajectories converge towards this ellipse which thus represents a limit cycle, an asymptotically stable orbit. Here again, it is confirmed that the sign of the real part of the eigenvalues of eq. (3.2.9) determines the stability behaviour.

For the analysis of the asymptotic stability, we exclusively used properties of the linearised system eq. (3.2.4). This means that we could only allow for small perturbations of the initial system. However, in practice, the question often arises whether stability can still be guaranteed for large perturbations where the non-linear terms have to be taken into account. For the investigation of this global stability problem, Alexander Michailowitsch Lyapunov (1857-1918), a Russian mathematician and physicist to whom we owe major contributions to the understanding of the stability of dynamical systems, proposed a direct geometrical method which allows us to analyse the variation of a function $V(\boldsymbol{x})$ along a trajectory $\boldsymbol{x}(t)$.

On the assumption that the fixed point coincides with the origin, the Lyapunov function $V(\boldsymbol{x})$ must have the following characteristics: an open neighbourhood U of the fixed point $\boldsymbol{x} = \boldsymbol{0}$ exists, where $V(\boldsymbol{x})$ is continuous together with its first partial derivatives. In addition, $V(\boldsymbol{0}) = 0$ and $V(\boldsymbol{x}) > 0$ hold for all $\boldsymbol{x} \in U$. Following eq. (2.4.2), the derivative with respect to time of the Lyapunov function along a trajectory of the system $\dot{\boldsymbol{x}} = \boldsymbol{F}(\boldsymbol{x})$ is calculated as the Lie derivative

$$\dot{V}(\boldsymbol{x}(t)) = \operatorname{grad} V \cdot \dot{\boldsymbol{x}} = \operatorname{grad} V \cdot \boldsymbol{F}(\boldsymbol{x}) \tag{3.2.18}$$

According to Lyapunov's stability theorem the sign of $\dot{V}(\boldsymbol{x})$ decides the stability of the fixed point: if for all $\boldsymbol{x} \in U$ with the exception of $\boldsymbol{x} = \boldsymbol{0}$

$$\dot{V}(\boldsymbol{x}(t)) \leqslant 0\,, \quad \text{then } \boldsymbol{x} \text{ is stable}$$

$$\dot{V}(\boldsymbol{x}(t)) < 0\,, \quad \text{then } \boldsymbol{x} \text{ is aymptotically stable}$$

$$\dot{V}(\boldsymbol{x}(t)) > 0\,, \quad \text{then } \boldsymbol{x} \text{ is unstable}$$

see (Haken, 1990; Hale and Koçak, 1991). We emphasise that, as long as such a Lyapunov function is known, it is not necessary to calculate the trajectory explicitly in order to decide about its stability. In special cases, a Lyapunov function can be determined explicitly: for example, if the dynamical system has a potential U, i.e. if $\boldsymbol{F}(\boldsymbol{x}) = -\operatorname{grad} U$. In this case, the potential represents a natural Lyapunov function for conservative systems.

3.3 Invariant Manifolds

Invariant manifolds play a special role in the analysis of systems of linear and non-linear differential equations. In geometrical terms, they are objects in the phase space which, taken as a whole, remain invariant with respect to the dynamics of the system. Trajectories offer a trivial example for such objects; fixed point, limit cycle and torus are also invariant manifolds.

In order to describe invariant manifolds, we call upon some definitions with which we wish to illustrate our argument in a two- and three-dimensional space using two examples. Let us first consider the linear approximation of the system of differential equations in the neighbourhood of a saddle point (see fig. 3.1.3). If we choose as an inital condition an arbitrary point on one of the axes $\bar{x}_1, \bar{x}_2$ which are given by the eigenvectors $\boldsymbol{y}_1, \boldsymbol{y}_2$, then the solutions remain per definition on these straight lines for all time. We denote the linear subspace appertaining to the eigenvector $\boldsymbol{y}_1$ $(\lambda_1 < 0)$ as the stable and that appertaining to $\boldsymbol{y}_2$ $(\lambda_2 > 0)$ as the unstable *invariant manifold.* We can define corresponding subspaces for the saddle point of a non-linear system (fig. 3.3.1).

A *stable* invariant manifold $\mathbb{W}^s$ of a singular point $\boldsymbol{x}_s$ is defined as the set of all points $\boldsymbol{x}$ which determine initial values of trajectories converging towards $\boldsymbol{x}_s$ for $t \to \infty$. The eigenvalues appertaining to the corresponding eigenvectors possess negative real parts. An *unstable* invariant manifold $\mathbb{W}^u$ of a singular point $\boldsymbol{x}_s$ is defined as the set of all points $\boldsymbol{x}$ which form the initial values of trajectories converging towards $\boldsymbol{x}_s$ for $t \to -\infty$. The eigenvectors corresponding to an unstable manifold appertain to eigenvalues with positive real parts. All initial points of the trajectories which are neither attracted nor repelled by $\boldsymbol{x}_s$ lie on a so-called invariant *centre manifold* $\mathbb{W}^c$ (see section 6.2 for a detailed discussion). In the example in fig. 3.3.1, the centre manifold is degenerate and is identical with the saddle point $\boldsymbol{x}_s$.

Let us now consider two closely adjacent initial points $P(0)$ and $Q(0)$ which lie on different sides of the invariant manifold $\mathbb{W}^s$ (see fig. 3.3.1). For $t \to \infty$, the associated trajectories will proceed to completely different parts of the phase space. The stable invariant manifolds of a saddle point are hence occasionally also referred to as *separatrices.*

In analogy, we can define invariant manifolds associated with limit cycles in a three-dimensional phase space (see fig. 3.3.2) and with three-dimensional tori in a four-dimensional phase space.

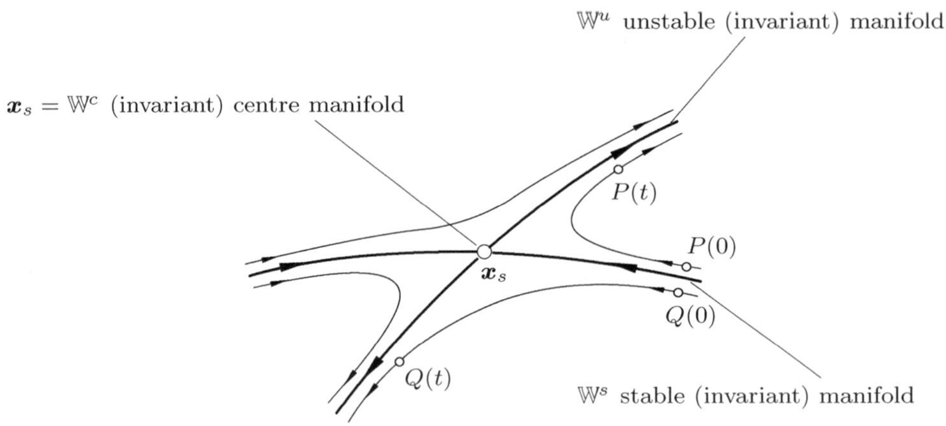

Fig. 3.3.1: Invariant manifolds of a saddle point

Up to now, our discussion has shown that one can achieve a satisfactory global overview of the structure of the phase space if one determines the attractors and the corresponding invariant stable and unstable manifolds for a system of differential equations eq. (2.3.3).

Finally, we note that, due to the uniqueness of the solutions of a system eq. (2.3.3), a stable manifold $\mathbb{W}^s$ cannot cross itself, nor can two different $\mathbb{W}^s$ intersect. However, intersections of stable $\mathbb{W}^s$ and unstable $\mathbb{W}^u$ are possible. Intersections of $\mathbb{W}^s$ and $\mathbb{W}^u$ that belong to the same fixed point are denoted *homoclinic* points. If they belong to different fixed points, they are denoted *heteroclinic* points. In the neighbourhood of such intersections, these points can cause a highly complex behaviour of the trajectories. We shall return to this issue in sections 4.5 and 10.5.

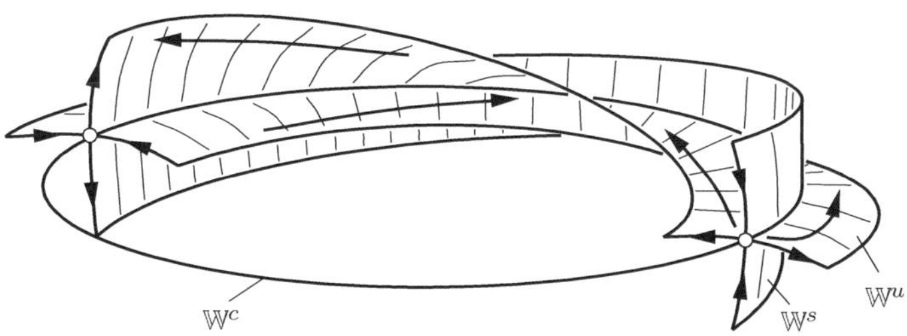

Fig. 3.3.2: Invariant manifolds of a limit cycle

3.4 Discretisation in Time

Up to now, we have described dynamical processes as continuous evolutions in time. An alternative approach is to observe the position vectors of the trajectories only at discrete time intervals τ. This technique is applied in the process of filming a sequence, for example, where a continuous motion is simulated approximately by means of snapshots at constant time intervals. In this way, a continuous function $\boldsymbol{x}(t)$ which contains an "infinite" amount of information is replaced by a discrete sequence of points

$$\boldsymbol{x}(t_0),\, \boldsymbol{x}(t_1), \ldots \boldsymbol{x}(t_\mathrm{k}), \ldots$$

where (3.4.1)

$$t_\mathrm{k} = t_0 + \mathrm{k}\tau \qquad (\mathrm{k} = 1, 2, \ldots)$$

In order to ensure that no essential information on the evolution of the trajectory is lost, it is important to choose a suitable discretisation parameter τ.

Let us first consider a periodically excited oscillator. Selecting the period T_E of the excitation force as the time interval τ and plotting the sequence of points of eq. (3.4.1) in the $x, \dot{x}$-plane, we initially obtain a sequence of scattered points in the transient phase. If a stationary solution has been attained, successive recordings

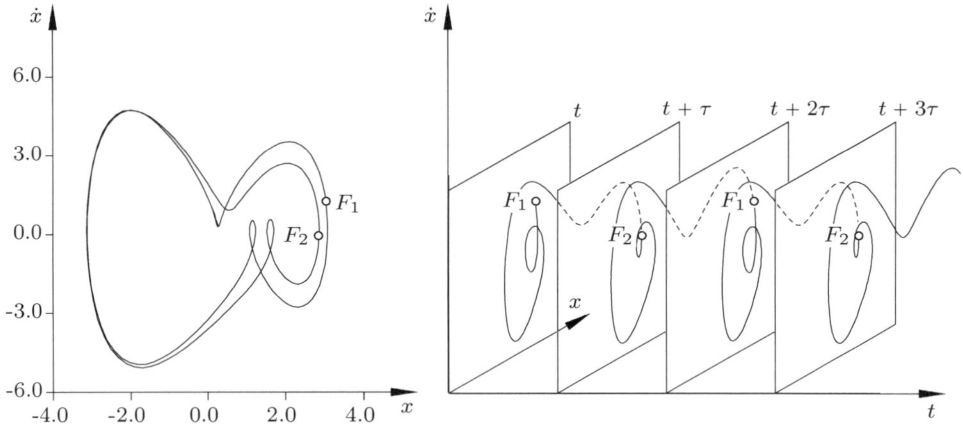

Fig. 3.4.1: Stroboscopic method (second-order subharmonic oscillation)

above a certain index k will coincide. In the case of a subharmonic oscillation of order m with the period $\mathrm{m}T_E$, m fixed points emerge. This method of representation is often called the stroboscopic method. Its advantage is obvious: the stationary solutions are represented by fixed points. In fig. 3.4.1, the stroboscopic method is demonstrated using the Duffing equation as an example. After the transient phase, the stationary solution is represented in the present case by a second-order subharmonic oscillation.

Another interpretation may be construed by considering the sequence of points of eq. (3.4.1) as an array of entries in which each $\boldsymbol{x}(t_{\mathrm{k}+1})$ is an image of its pre-image $\boldsymbol{x}(t_{\mathrm{k}})$. In this way, a generally non-linear recursive mapping rule is defined

$$\boldsymbol{x}(t_{\mathrm{k}+1}) = \boldsymbol{f}\big(\boldsymbol{x}(t_{\mathrm{k}})\big)\,, \qquad \mathrm{k} = 1, 2, \ldots \tag{3.4.2}$$

which, due to the determinacy of the original differential equation, is one-to-one. Equation (3.4.2) can also be written

$$\boldsymbol{x}(t+\tau) = \boldsymbol{f}\big(\boldsymbol{x}(t)\big) \tag{3.4.3}$$

This relation describes a non-linear feedback so that a repeated application leads to

$$\boldsymbol{x}(t+2\tau) = \boldsymbol{f}\Big(\boldsymbol{f}\big(\boldsymbol{x}(t)\big)\Big) = \boldsymbol{f}^2\big(\boldsymbol{x}(t)\big) \tag{3.4.4}$$

It should be noted, however, that it will only be possible to specify the mapping function explicitly in special cases. In general, it is necessary to construct the image point by means of numerical integration.

On the right-hand side of fig. 3.4.2, we display the amplitude of a periodically excited damped oscillator as a function of time (see also fig. 2.2.2). The left-hand side shows the sequence of points resulting from the mapping rule (3.4.3) for a fixed interval τ. It can be clearly seen that the solution spirals towards a stable limit cycle after an initial transient phase.

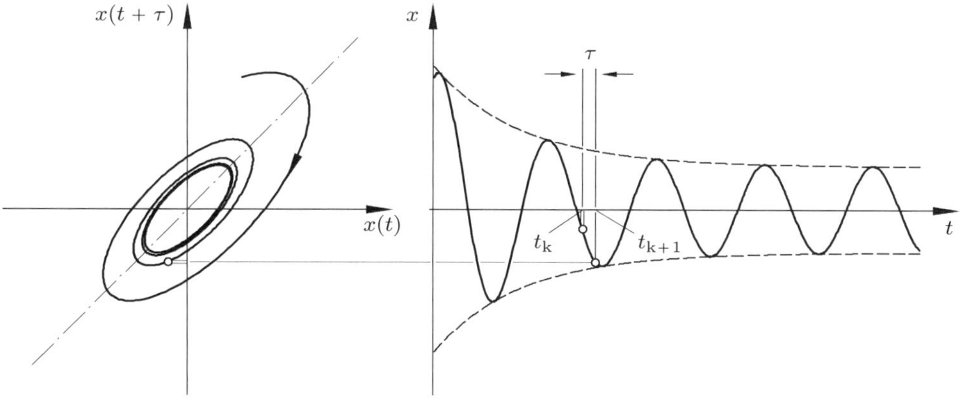

Fig. 3.4.2: Illustration of the discrete map using the example of a periodically excited damped oscillator

3.5 Poincaré Map

In the preceding section 3.4, we approximated a continuous trajectory $\boldsymbol{x} = \boldsymbol{x}(t)$ as a sequence of discrete points sampled at constant time intervals. With this technique, the system of differential equations is reduced to a system of difference equations of the same order. The significance of the following discretisation method, based on an idea of Henri Poincaré, lies in the fact that the dimension of the resulting system is reduced by one.

We consider a recurrent trajectory c in the n-dimensional phase space. We now choose a hypersurface Σ of the dimension $\mathrm{n}-1$

$$g(x_1, x_2, \ldots, x_\mathrm{n}) = 0 \tag{3.5.1}$$

so that it is sequentially intersected transversally by the trajectory. For a system of differential equations (2.3.3), the right-hand side $\boldsymbol{F}(\boldsymbol{x})$ defines the direction of the tangent to the trajectory c at point $\boldsymbol{x}$. The surface normal $\boldsymbol{n}$ of Σ is given by

$$\boldsymbol{n} = |\mathrm{grad}\ g|^{-1}\mathrm{grad}\ g$$

with

$$\mathrm{grad}\ g = \left\{ \frac{\partial g}{\partial x_1} \quad \frac{\partial g}{\partial x_2} \quad \cdots\cdots \quad \frac{\partial g}{\partial x_\mathrm{n}} \right\} \tag{3.5.2}$$

If we require that Σ intersects the trajectory c transversally, then the following condition must hold at the points of intersection

$$\boldsymbol{F}^\mathrm{t}\boldsymbol{n} \neq 0 \tag{3.5.3}$$

If possible, we select a hyperplane as the intersecting surface. The points of c where the trajectory intersects Σ in a well-defined direction, i.e. which are generated after a full orbit, are denoted $P^{(0)}, P^{(1)}, P^{(2)}$ etc. Since the preceding course of the trajectory is determined by the current state, one can also ascertain the former points $P^{(-1)}, P^{(-2)}, P^{(-3)}$ etc. Figure 3.5.1 shows such a so-called *Poincaré section* for a three-dimensional phase space. Thus, in this method, the discretisation is not implemented over a fixed time interval but contains only those points of the trajectory which lie on a given hypersurface Σ. If we interpret each point $P^{(i+1)}$ as an image of its pre-image $P^{(i)}$

$$P^{(i+1)} = f(P^{(i)}) \tag{3.5.4}$$

we obtain a generally non-linear iterative mapping of the hypersurface onto itself which, in honour of its originator, is called a Poincaré map or, alternatively, *first return map*. Since the map is one-to-one,

$$P^{(i)} = f^{-1}(P^{(i+1)}) \tag{3.5.5}$$

applies. If we introduce a coordinate system ξ_k ($k = 1, 2, \ldots, n-1$) on the hypersurface and assign the position vector

$$\boldsymbol{\xi}^{(i)} = \{\xi_1^{(i)} \quad \xi_2^{(i)} \quad \ldots\ldots \quad \xi_{n-1}^{(i)}\} \tag{3.5.6}$$

to the point $P^{(i)}$, the n-dimensional system of differential equations (2.3.3) can be reduced to an $(n-1)$-dimensional system of difference equations

$$\boldsymbol{\xi}^{(i+1)} = \boldsymbol{f}(\boldsymbol{\xi}^{(i)}) \tag{3.5.7}$$

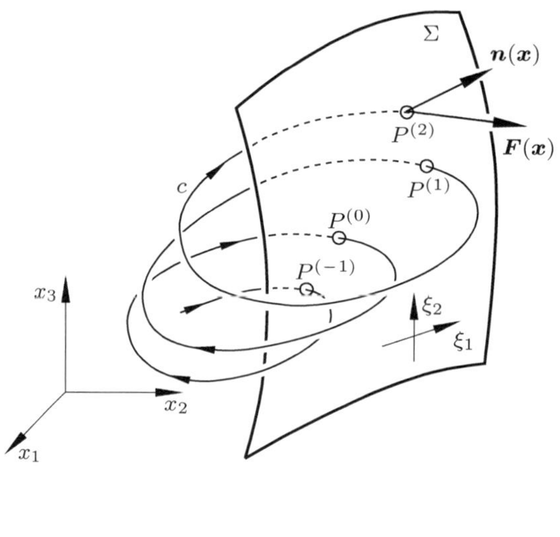

Fig. 3.5.1

Poincaré section

It should be noted, however, that the mapping function $\boldsymbol{f}$ cannot usually be established explicitly. We thus have to continue to rely on numerical integration and cannot expect a significant saving in computing time. The advantage of the Poincaré map derives mainly from the fact that the dimension of the system is reduced by one without, as experience has shown, any significant loss of information concerning the long-term behaviour of the trajectories.

However, there is a loss of topological information on how often the trajectory has rotated about its own tangent during one revolution. This local torsion rate per cycle is a topological invariant of the dynamics and can be analysed by means of the knot theory. A second topological invariant is the degree of braid-like interlacings of neighboring orbits (Tufillaro et al., 1992; Arnold and Khesin, 1998). This topological characterisation will be illustrated in section 10.5 using the example of the Duffing equation, see Colour Plate XXIV.

The theory of dynamical systems received a further decisive impetus by another abstraction step. If we are not interested in the detailed response of a particular dynamic problem, but rather in general qualitative characteristics of whole classes of dynamical systems, we can put aside the study of the differential equations or their Poincaré maps and investigate the general characteristics of explicit recursion formulæ directly. Since the mapping function is known explicitly, numerical integrations are no longer necessary; hence, the computing time between two Poincaré sections is reduced to the insignificant time required for one iteration step. In this way, it is possible to obtain an overview of the long-term behaviour of a system very quickly whilst also gaining increased numerical accuracy since errors due to numerical integration do not occur. Particularly if one chooses a one-to-one mapping, the procedure can be interpreted as a Poincaré map of a dynamical system. In section 4.6, we shall deal with such a recursion rule in more detail, using the example of a conservative system with two degrees of freedom, the so-called Hénon map.

3.6 Fixed Points and Cycles of Discrete Systems

The fixed points P_s of a Poincaré map, eq. (3.5.4), result from the relationship

$$P_s = f(P_s) \tag{3.6.1}$$

The corresponding trajectory c in the phase space returns periodically to its initial position with a period T and thus forms a closed curve (see fig. 3.6.1a). If we apply the Poincaré map to a point $P^{(1)}$ m times and only the result $P^{(\mathrm{m}+1)}$ coalesces again with the initial point $P^{(1)}$, i.e.

$$P^{(2)} = f(P^{(1)}),\ P^{(3)} = f(P^{(2)}),\ \ldots,\ P^{(\mathrm{m}+1)} = f(P^{(\mathrm{m})}) = P^{(1)} \tag{3.6.2}$$

we speak of a cycle of the period m or an m-cycle. The corresponding trajectory in the phase space is a subharmonic solution of the period mT which returns to the original point after m orbits and continues on the same path as before (see fig. 3.6.1b). If we denote the mapping function iterated m times as f^{m}, then, for an m-cycle,

$$P^{(1)} = f^{\mathrm{m}}(P^{(1)}) \tag{3.6.3}$$

applies, i.e. $P^{(1)}$ is a fixed point of the map f^{m}. Consequently, no fundamental difference arises between fixed points and cycles and we can restrict ourselves to fixed points when investigating the stability. We emphasise once more that it is very easy to recognise periodic respectively subharmonic solutions with the help of Poincaré maps. It is only necessary to locate the fixed points of the mapping function f or of the iterated function f^{m} ($\mathrm{m} = 2, 3, \ldots$).

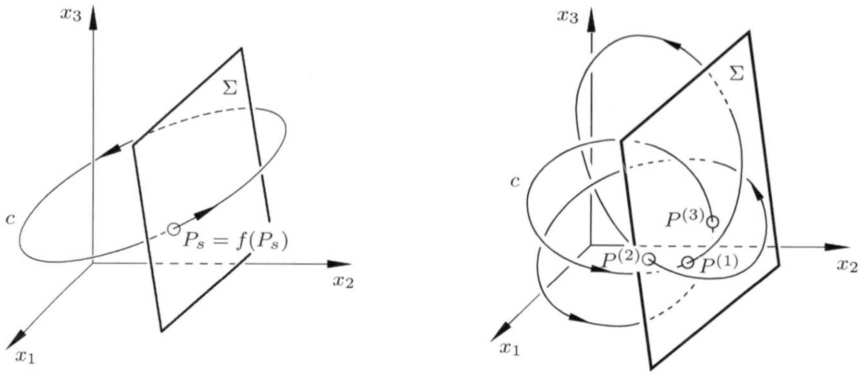

a) Fixed point b) Cycle of period 3

Fig. 3.6.1: Definition of fixed point and cycle of a Poincaré map

We next consider the behaviour of the trajectories in the neighbourhood of a fixed point P_s in order to determine the stability characteristics of P_s. To this end, we observe a point P adjacent to P_s. Let P have the coordinate vector

$$\boldsymbol{\xi} = \boldsymbol{\xi}_s + \boldsymbol{\eta}, \quad |\boldsymbol{\eta}| \ll 1 \tag{3.6.4}$$

where $\boldsymbol{\xi}_s$ is the vector appertaining to P_s. The image of the i-th iteration is

$$\boldsymbol{\xi}^{(\mathrm{i+1})} = \boldsymbol{f}(\boldsymbol{\xi}^{(\mathrm{i})}) \tag{3.6.5}$$

Expanding the mapping function $\boldsymbol{f}$ into a Taylor series in the neighbourhood of $\boldsymbol{\xi}_s$ yields

$$\boldsymbol{\xi}_s + \boldsymbol{\eta}^{(\mathrm{i+1})} = \boldsymbol{f}(\boldsymbol{\xi}_s + \boldsymbol{\eta}^{(\mathrm{i})}) = \boldsymbol{f}(\boldsymbol{\xi}_s) + \left.\frac{\partial \boldsymbol{f}}{\partial \boldsymbol{\xi}}\right|_{\boldsymbol{\xi}_s} \boldsymbol{\eta}^{(\mathrm{i})} + \mathcal{O}(\boldsymbol{\eta}^{(\mathrm{i})^2}) \tag{3.6.6}$$

Since P_s is a fixed point, we obtain, ignoring higher-order terms, the linear mapping rule in $\boldsymbol{\eta}^{(\mathrm{i})}$

$$\boldsymbol{\eta}^{(\mathrm{i+1})} = \boldsymbol{D}(\boldsymbol{\xi}_s)\boldsymbol{\eta}^{(\mathrm{i})} \tag{3.6.7}$$

where

$$\boldsymbol{D}(\boldsymbol{\xi}_s) = \left.\frac{\partial \boldsymbol{f}}{\partial \boldsymbol{\xi}}\right|_{\boldsymbol{\xi}_s} \tag{3.6.8}$$

is a constant (n − 1) × (n − 1)-matrix, the Jacobian matrix of the map. Following section 3.1, we investigate the stability behaviour on the basis of the eigenvalues of $\boldsymbol{D}(\boldsymbol{\xi}_s)$ which determine which sequence the points $P^{(\mathrm{i})}$ run through when successive iterations of the mapping $\boldsymbol{f}$ are applied and which stability characteristics P_s thus possesses.

We do not wish to enter here into a detailed classification of fixed points, but rather prefer to discuss two specific but illustrative cases. For this purpose, we assume that the eigenvalues $\boldsymbol{D}(\boldsymbol{\xi}_s)$ are all different.

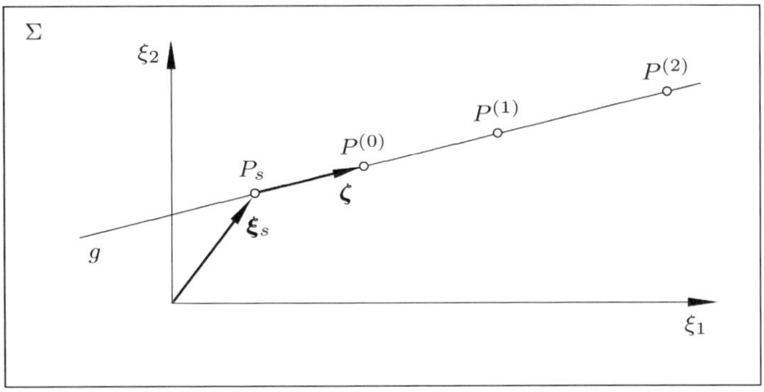

λ		
$\lambda < -1$	$P^{(3)}$ $P^{(1)}$ P_s $P^{(0)}$ $P^{(2)}$	P_s unstable (source) alternating point sequence
$\lambda = -1$	$P^{(1)} = P^{(3)} = \ldots P^{(2k+1)}$ $P^{(0)} = P^{(2)} = \ldots P^{(2k)}$ P_s	each point has period 2 (P_s only has period 1)
$-1 < \lambda < 0$	$P^{(1)}$ $P^{(3)}$ P_s $P^{(2)}$ $P^{(0)}$	P_s asymptotically stable (sink) alternating point sequence
$\lambda = 0$	$P^{(0)}$ $P_s = P^{(1)} = \ldots P^{(k)}$	all points are mapped on P_s
$0 < \lambda < 1$	P_s $P^{(3)}$ $P^{(2)}$ $P^{(1)}$ $P^{(0)}$	P_s asymptotically stable (sink)
$\lambda = 1$	P_s $P^{(0)} = P^{(1)} = \ldots P^{(k)}$	each point of the straight line is a fixed point
$\lambda > 1$	P_s $P^{(0)}$ $P^{(1)}$ $P^{(2)}$	P_s unstable (source)

Fig. 3.6.2: Stability behaviour of a fixed point (real eigenvalue λ)

If λ is a real eigenvalue appertaining to the eigenvector $\boldsymbol{\zeta}$ of $\boldsymbol{D}$, we observe the point $P^{(0)}$ with the position vector

$$\boldsymbol{\xi}^{(0)} = \boldsymbol{\xi}_s + \boldsymbol{\zeta} \tag{3.6.9}$$

Equations (3.6.6) and (3.6.8) yield for the image $P^{(1)}$

$$\boldsymbol{\xi}^{(1)} \approx \boldsymbol{\xi}_s + \boldsymbol{D}(\boldsymbol{\xi}_s)\boldsymbol{\zeta} = \boldsymbol{\xi}_s + \lambda\boldsymbol{\zeta}$$

Following the next iteration, we find

$$\boldsymbol{\xi}^{(2)} \approx \boldsymbol{\xi}_s + \lambda \boldsymbol{D}(\boldsymbol{\xi}_s)\boldsymbol{\zeta} = \boldsymbol{\xi}_s + \lambda^2 \boldsymbol{\zeta}$$

or, after i iterations (see fig. 3.6.2),

$$\boldsymbol{\xi}^{(\mathrm{i})} \approx \boldsymbol{\xi}_s + \lambda^{\mathrm{i}} \boldsymbol{\zeta} \tag{3.6.10}$$

The points P_s as well as $P^{(0)}, P^{(1)}, P^{(2)}$ all lie on a straight line and λ determines whether the sequence of points is attracted by P_s or repelled by it. Figure 3.6.2 summarises the possibilities for a real eigenvalue.

In the case of negative eigenvalues, the vector $\lambda^{\mathrm{i}}\boldsymbol{\zeta}$ in eq. (3.6.10) changes its sign with each iteration, i.e. for negative λ, alternating sequences of points result. If $|\lambda| > 1$, the points $P^{(\mathrm{i})}$ with increasing i tend away from P_s, i.e. P_s is unstable; for $|\lambda| < 1$, P_s is stable. In the marginal case $\lambda = 1$, each point of the straight line is mapped onto itself, while for $\lambda = -1$, each point has the period 2 and is reflected at the point P_s at each iteration step.

We next consider the case of two complex conjugate eigenvalues

$$\lambda = |\lambda| \mathrm{e}^{i\varphi} \quad \text{and} \quad \lambda^* = |\lambda| \mathrm{e}^{-i\varphi} \tag{3.6.11}$$

The corresponding eigenvectors are then also complex conjugate

$$\boldsymbol{\zeta} = \boldsymbol{\zeta}_1 + i\boldsymbol{\zeta}_2 \quad \text{and} \quad \boldsymbol{\zeta}^* = \boldsymbol{\zeta}_1 - i\boldsymbol{\zeta}_2 \tag{3.6.12}$$

The two real vectors

$$\boldsymbol{\zeta}_1 = \frac{1}{2}[\boldsymbol{\zeta}^* + \boldsymbol{\zeta}] \quad \text{and} \quad \boldsymbol{\zeta}_2 = \frac{i}{2}[\boldsymbol{\zeta}^* - \boldsymbol{\zeta}] \tag{3.6.13}$$

span a plane Φ. The image of all points P on the plane Φ also lies on Φ. In order to simplify the notation, we move the origin of the coordinate system to the fixed point P_s. Denoting the position vector of an arbitrary point $P^{(0)}$ on Φ by

$$\boldsymbol{\xi}^{(0)} = \alpha\boldsymbol{\zeta}_1 + \beta\boldsymbol{\zeta}_2 \tag{3.6.14}$$

and applying the mapping function $\boldsymbol{f}$ k times, we obtain the relationships

$$\boldsymbol{\xi}^{(0)} = \tfrac{1}{2}\alpha[\boldsymbol{\zeta}^* + \boldsymbol{\zeta}] + \tfrac{1}{2}i\beta[\boldsymbol{\zeta}^* - \boldsymbol{\zeta}]$$

$$\begin{aligned}\boldsymbol{f}^{\mathrm{k}}(\boldsymbol{\xi}^{(0)}) &\approx \tfrac{1}{2}\alpha\left[(\lambda^*)^{\mathrm{k}}\boldsymbol{\zeta}^* + \lambda^{\mathrm{k}}\boldsymbol{\zeta}\right] + \tfrac{1}{2}i\beta\left[(\lambda^*)^{\mathrm{k}}\boldsymbol{\zeta}^* - \lambda^{\mathrm{k}}\boldsymbol{\zeta}\right] \\ &= \tfrac{1}{2}|\lambda|^{\mathrm{k}}\left[\mathrm{e}^{-i\mathrm{k}\varphi}(\alpha + i\beta)\boldsymbol{\zeta}^* + \mathrm{e}^{i\mathrm{k}\varphi}(\alpha - i\beta)\boldsymbol{\zeta}\right]\end{aligned}$$

Using Euler's formula

$$\mathrm{e}^{i\mathrm{k}\varphi} = \cos(\mathrm{k}\varphi) + i\sin(\mathrm{k}\varphi) \tag{3.6.15}$$

we finally find

$$\boldsymbol{f}^{\mathrm{k}}(\boldsymbol{\xi}^{(0)}) \approx |\lambda|^{\mathrm{k}}\left[(\alpha\boldsymbol{\zeta}_1 + \beta\boldsymbol{\zeta}_2)\cos(\mathrm{k}\varphi) + (\beta\boldsymbol{\zeta}_1 - \alpha\boldsymbol{\zeta}_2)\sin(\mathrm{k}\varphi)\right] \tag{3.6.16}$$

In the case of $|\lambda| = 1$, $P^{(0)}$ and its image points $P^{(\mathrm{k})}$ move on an ellipse about the fixed point P_s; in the case of $|\lambda| \neq 1$, the points $P^{(\mathrm{k})}$ are located on a spiral and, in the case of $|\lambda| < 1$, move towards the fixed point while, in the case of $|\lambda| > 1$, tending away from P_s (see fig. 3.6.3).

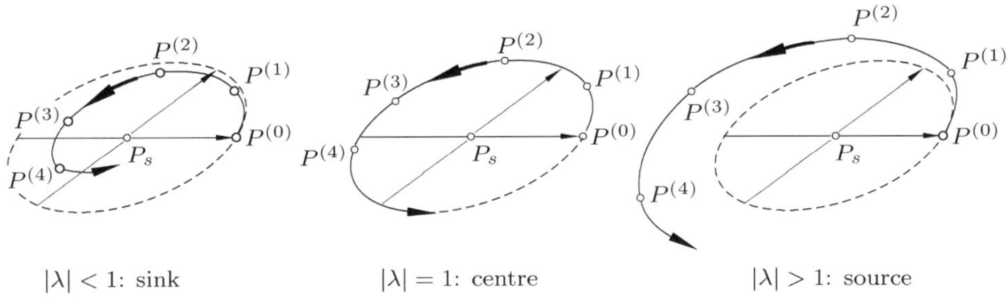

Fig. 3.6.3: Stability behaviour of a fixed point (two complex conjugate eigenvalues)

Hartman and Grobman extended their theorem on the stability behaviour of trajectories in the phase space to incorporate general discrete maps (Guckenheimer and Holmes, 1983). The extended theorem includes the case of multiple eigenvalues and enables us to apply the stability characteristics of the linearised system to the original non-linear system if $|\lambda_{\mathrm{i}}| \neq 1$ is true for all eigenvalues. P_s is then called a hyperbolic fixed point. If eigenvalues exist for which $|\lambda_{\mathrm{i}}| = 1$ holds, further investigations are necessary in order to ascertain the stability of the fixed points; see also section 6.6 as well as (Guckenheimer and Holmes, 1983). A more detailed discussion of iterated mapping rules can be found in sections 6.7 and 8.3 to 8.6.

3.7 An Example of Discrete Dynamics – The Logistic Map

In this section, we first propose to investigate empirically a special discrete non-linear map which, in spite of its simple form, exhibits a multitude of phenomena typical for non-linear behaviour. At the beginning of the seventies of the last century, the biologist Robert M. May researched into the dynamics of animal populations. In order to describe the simplest case of non-overlapping generations, he applied a first-order difference equation

$$x_{\mathrm{n}+1} = f(x_{\mathrm{n}}) \tag{3.7.1}$$

which determines the population $x_{\mathrm{n}+1}$ in the year n+1 from the population of the previous year (May, 1976). Evidently one has to introduce a non-linear ansatz for the function $f(x)$, since linear difference equations can only reproduce the extinction or the unrestricted growth of a population. In order to simulate a saturation effect, May thus introduced a quadratic term and investigated the dynamical behaviour of the simplest non-trivial recursion rule, the so-called logistic map,

$$x_{\mathrm{n}+1} = \alpha x_{\mathrm{n}}(1 - x_{\mathrm{n}}) \tag{3.7.2}$$

Surprisingly, a whole range of dynamical behaviour patterns can be simulated with this simple discrete model, depending on the choice of the parameter α which takes the influence of climatic conditions, food supply or similar aspects into account.

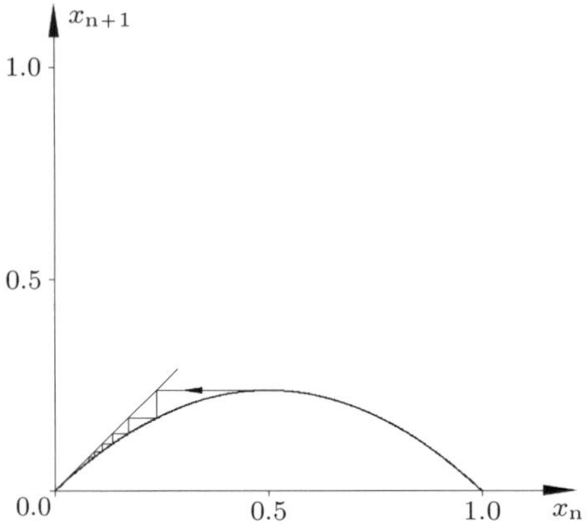

a) Graph of the logistic map $f(x_n)$

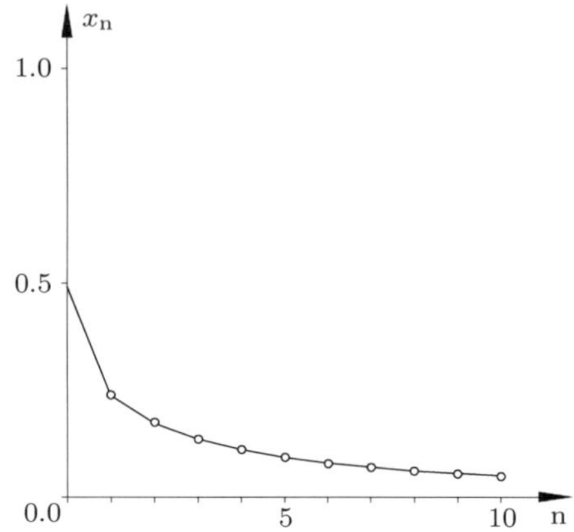

c) Sequence of points x_n

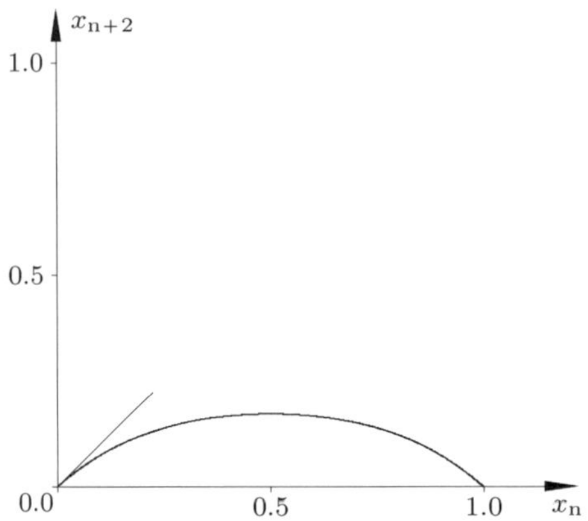

b) Graph of the map $f^2(x_n)$

Fig. 3.7.1

Logistic map for $\alpha = 0.95$, fixed point $x_s = 0$

In the following, we wish to investigate – with the help of a pocket calculator, as it were – the phenomena which arise for the logistic map

$$x \longmapsto f(x) = \alpha x(1 - x) \tag{3.7.3}$$

in the interval $0 \leqslant x \leqslant 1$. The function $f(x)$ is a parabola with its vertex at $x = 1/2$ and open at the bottom; it crosses the x-axis at $x = 0$ and $x = 1$. The constant α is the system parameter which lies between 0 and 4 in order to retain the value of x within the interval [0,1] (fig. 3.7.1a).

Using this parabola, we now propose to study the topological changes that are generated by bifurcations or period doublings when varying the system parameter α. To this end, we first determine possible fixed points or point attractors. If the motion tends towards a fixed point x_s, the following condition must apply

$$x_s = f(x_s) \tag{3.7.4}$$

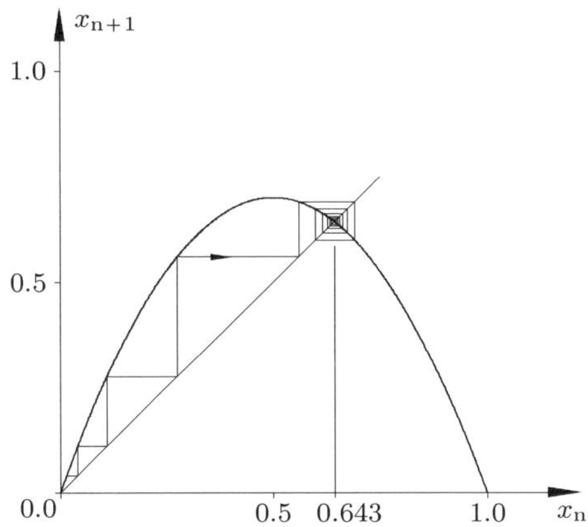

a) Graph of the logistic map $f(x_n)$

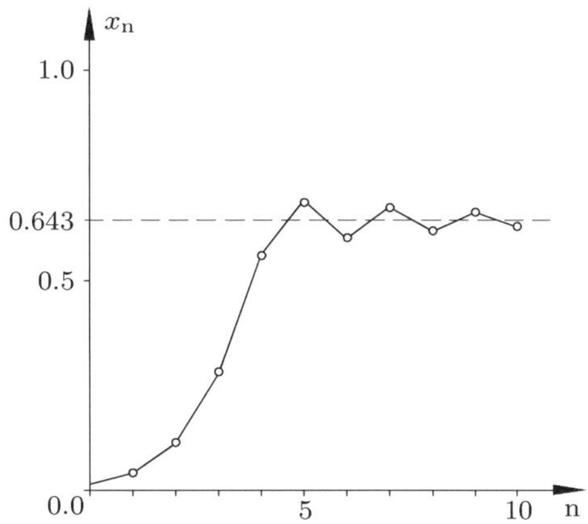

c) Sequence of points x_n

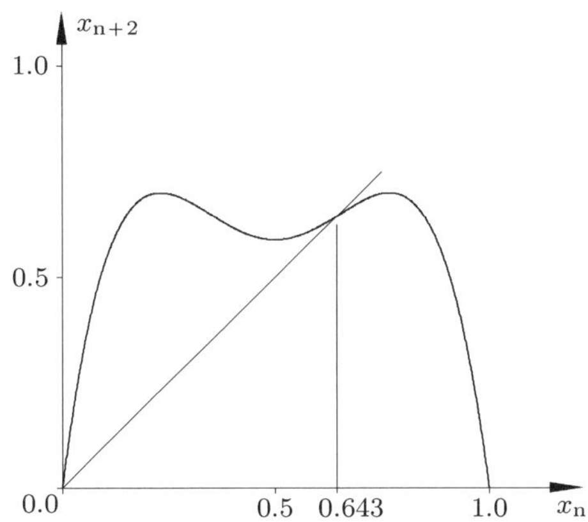

b) Graph of the map $f^2(x_n)$

Fig. 3.7.2

Logistic map for $\alpha = 2.8$, fixed point $x_s = 0.642857$

For the parabola, two fixed points are obtained

$$x_s = 0 \qquad \text{and} \qquad x_s = 1 - 1/\alpha$$

In order to determine the basin of attraction of both fixed points, we start from a point x_n which deviates from the fixed point x_s by ε, i.e. from $x_n = x_s + \varepsilon$. Substituting this starting point into eq. (3.7.1), we obtain for its subsequent image point

$$\begin{aligned} x_{n+1} &= f(x_n) = f(x_s + \varepsilon) \\ &\approx f(x_s) + \varepsilon f'(x_s) = x_s + \varepsilon f'(x_s) \end{aligned} \tag{3.7.5}$$

This means that the perturbation fades away if $|f'(x_s)| < 1$. In this case, the fixed point is stable and we then speak of a point attractor. For $|f'(x_s)| > 1$, however, the fixed point is unstable.

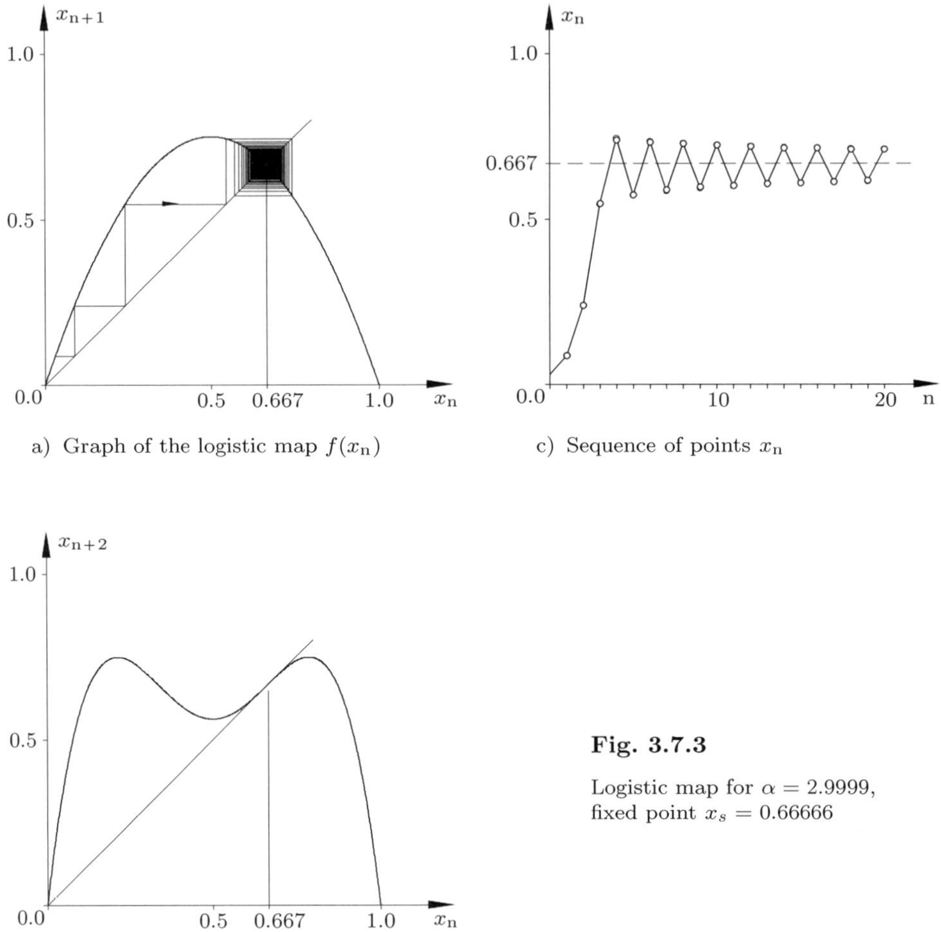

a) Graph of the logistic map $f(x_n)$

c) Sequence of points x_n

b) Graph of the map $f^2(x_n)$

Fig. 3.7.3

Logistic map for $\alpha = 2.9999$, fixed point $x_s = 0.66666$

We can thus summarise the conditions for a point attractor as follows

$$x_s = f(x_s) \,, \qquad |f'(x_s)| < 1 \tag{3.7.6}$$

For our parabola, we obtain for example

$$f'(x_n) = \alpha(1 - 2x_n)$$

and for the two fixed points of the logistic map

$$f'(0) = \alpha \qquad \text{and} \qquad f'(1 - 1/\alpha) = 2 - \alpha \tag{3.7.7}$$

This means that, for $0 < \alpha < 1$, each motion ends in $x_s = 0$ (fig. 3.7.1) and, for $1 < \alpha < 3$, the point attractor $x_s = 1 - 1/\alpha$ takes over (figs. 3.7.2 and 3.7.3). For $\alpha > 3$, neither of the two fixed points acts as an attractor and a new type of behaviour can be discerned. In this case, the series of points oscillates between the

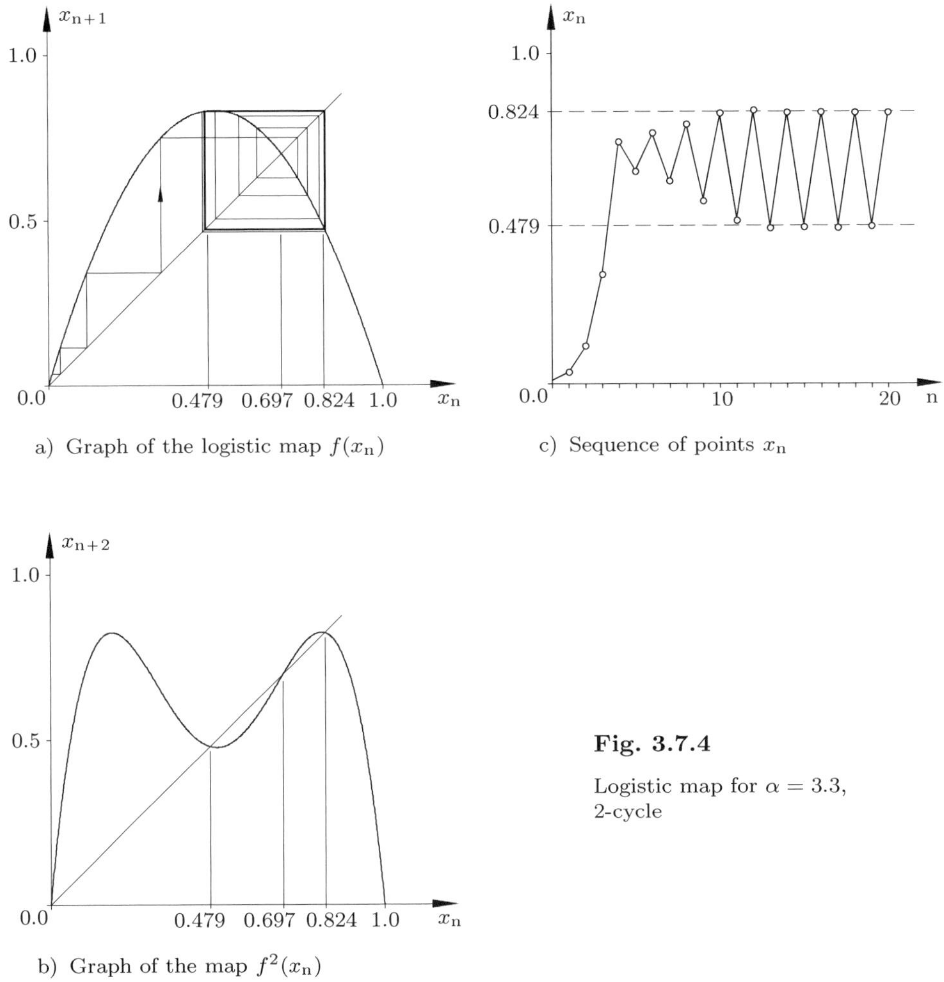

Fig. 3.7.4

Logistic map for $\alpha = 3.3$, 2-cycle

two values $x_{s1} = f(x_{s2})$ and $x_{s2} = f(x_{s1})$ in which process each point in itself does not act as an attractor, but the two as a whole form a 2-cycle (fig. 3.7.4). Following each iteration course for x_{s1} und x_{s2}, we obtain

$$x_{s1} = f\big(f(x_{s1})\big)\,, \qquad x_{s2} = f\big(f(x_{s2})\big)$$

Accordingly, neither point is a fixed point of the function $f(x_n)$ but they are fixed points of the function $f^2 = f(f(x_n))$. The mapping function f^2 describes a motion if two steps are carried out subsequently. For our parabola, for example, $f^2(x_n)$ becomes

$$\begin{aligned} f^2(x_n) &= \alpha f(x_n)[1 - f(x_n)] \\ &= \alpha^2 x_n(1 - x_n)[1 - \alpha x_n(1 - x_n)] \end{aligned} \tag{3.7.8}$$

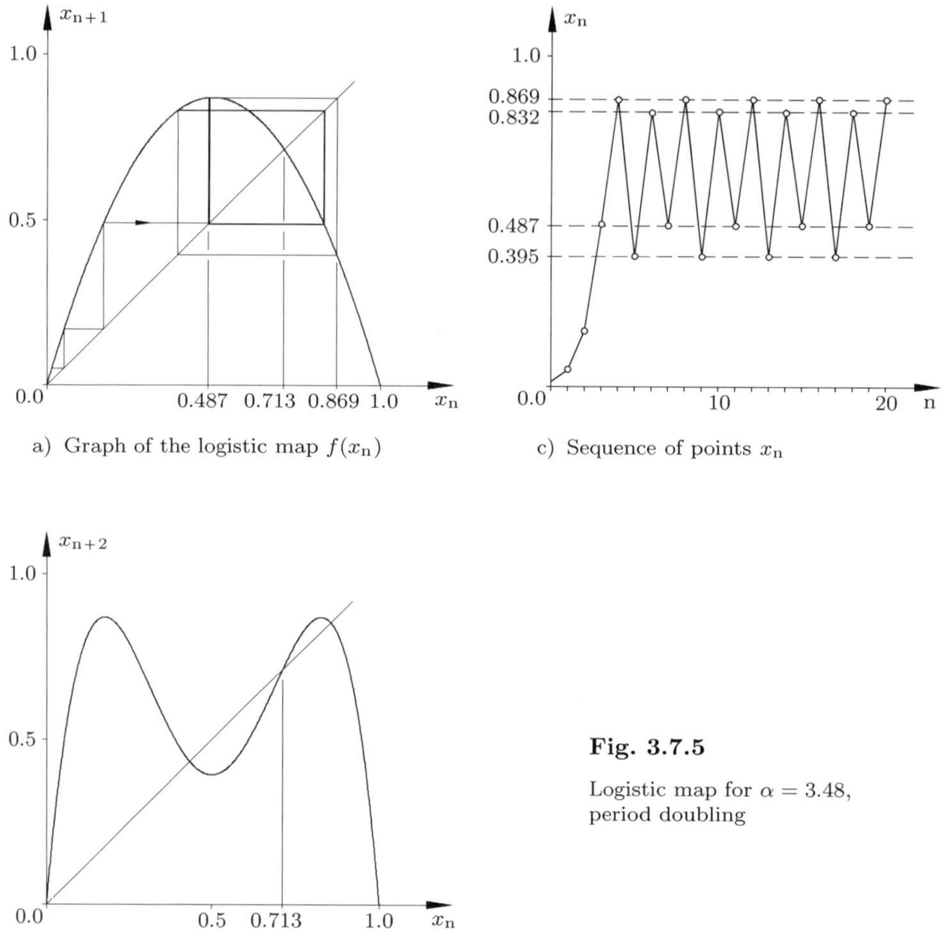

a) Graph of the logistic map $f(x_n)$

c) Sequence of points x_n

b) Graph of the map $f^2(x_n)$

Fig. 3.7.5

Logistic map for $\alpha = 3.48$, period doubling

For the f^2-function, the attractor condition

$$x_s = f^2(x_s)\,, \qquad |f^{2'}(x_s)| < 1$$

naturally also applies.

Figure 3.7.2 presents the curves of $f(x_n)$ and $f^2(x_n)$ for $\alpha < 3$. Both curves intersect the bisector of the first quadrant in the point attractor $x_s = 1 - 1/\alpha$. With increasing α, the curve at the point of intersection with f^2 becomes steeper and steeper so that for $\alpha = 3$, the point of intersection becomes the inflection point. Following a further increase in α, three points of intersection emerge, namely the two new point attractors x_{s1} and x_{s2} and the unstable fixed point x_s (fig. 3.7.4). If α increases up to the value $\alpha = 1 + \sqrt{6} \approx 3.449$, then the same happens for the outer points of intersection of $f^2(x_n)$ as for the point of intersection of $f(x_n)$

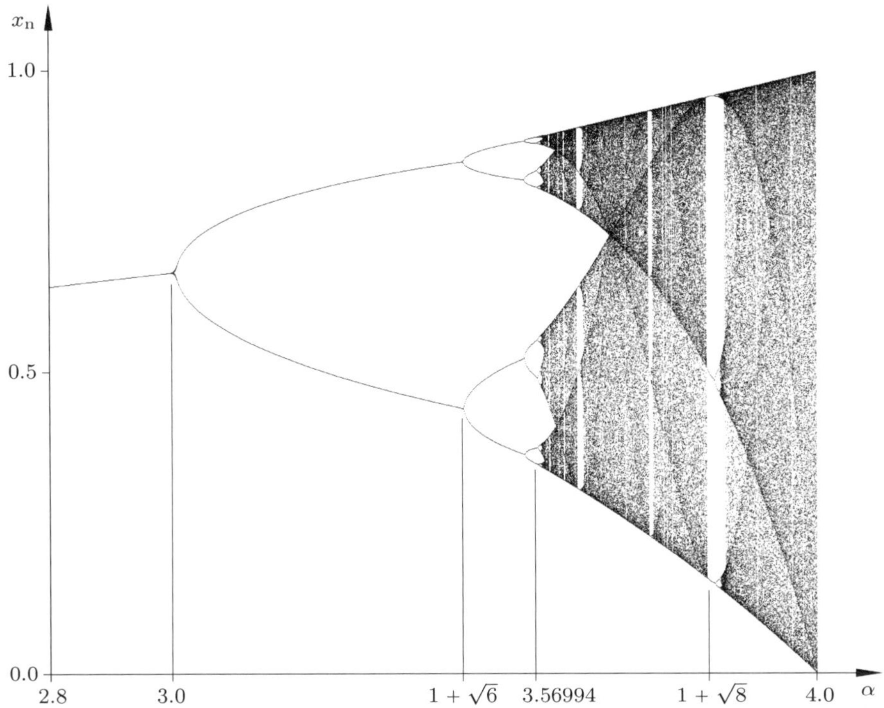

Fig. 3.7.6: Bifurcation cascade of the logistic map

at $\alpha = 3$. We observe a period doubling, i.e. a cycle of the period 4; the points of intersection of $f^2(x_n)$ are now fixed points of the function (fig. 3.7.5)

$$f^4(x_n) = f^2(f^2(x_n)) \tag{3.7.9}$$

For increasing values of the system parameter α, further instabilities occur one after the other, leading to 8-cycles, 16-cycles and generally to 2^n-cycles. The intervals between successive α_k-values at which period doublings occur decrease rapidly and the series $\{\alpha_k\}$ tends towards an accumulation point α_∞ (fig. 3.7.6). For $\alpha > \alpha_\infty$, we observe irregular, chaotic behaviour; this is, however, repeatedly interrupted by ranges of regular behaviour, so-called periodic windows.

The value α_∞ which characterises the transition to chaos via a cascade of period doublings was first calculated for the logistic map by Großmann and Thomae (1977)

$$\alpha_\infty = 3.5699456\ldots \tag{3.7.10}$$

Considering the sequence of critical α_k-values, we confirm that the distances $\alpha_{k+1} - \alpha_k$ decrease like a geometric progression where

$$\lim_{k\to\infty} \frac{\alpha_{k+1} - \alpha_k}{\alpha_{k+2} - \alpha_{k+1}} = \delta \tag{3.7.11}$$

It was Mitchell Feigenbaum (1978) who, applying renormalisation techniques used for second-order phase transitions, succeeded in proving that these laws are of universal character and are thus valid for a whole class of discrete mapping functions with a single quadratic maximum; only shortly afterwards, Coullet and Tresser (1978) were able to confirm this. For this reason, δ is often designated as the Feigenbaum constant.

In section 6.7, we shall discuss in detail the mechanism which leads to a cascade of period doublings and elucidate the background to the universality of this transition. Moreover, in section 8.4.1, we shall again deal with Feigenbaum's route to chaos and show how the scaling laws on which it is based are reflected in the power spectra.

3.8 Fourier Series and Fourier Transform

3.8.1 Fourier Series

Until now, we have studied dynamical systems primarily on the basis of their temporal evolution, i.e. we were interested in the long-time behaviour of displacements, velocities and accelerations as well as in their transient phases. Often, however, another form of information is needed. If one is interested in the oscillation components of a periodical motion, for example, or asks whether there are dominant components, the system has to be split up into its sine and cosine components. This decomposition is denoted harmonic analysis or Fourier analysis. The theoretical basis of the Fourier analysis and the Fourier transform to be discussed in the next section was developed by the French mathematician Joseph de Fourier (1768 - 1830), who described the problem of heat transfer by means of trigonometric series and presented his work in 1822 in the monograph *"Théorie analytique de la chaleur"*. Since that time, Fourier techniques play an exceptionally important role in many areas of science. Applications range from mathematics (approximation theory, probability theory) and physics (classical mechanics, acoustics, electromagnetism) to electrical engineering and signal processing. Many technical inventions are hardly conceivable without Fourier techniques. The reason is the particular properties of trigonometric functions such as orthogonality and their role as eigenfunctions of linear differential operators.

Fourier analysis allows us to expand a *periodic function* to a trigonometric series of the form

$$f(\tau) = \frac{a_0}{2} + \sum_{\mathrm{k}=1}^{\infty} (a_{\mathrm{k}} \cos \mathrm{k}\tau + b_{\mathrm{k}} \sin \mathrm{k}\tau) \tag{3.8.1}$$

Expansions of $f(\tau)$ into infinite series of sine and cosine oscillations are known as *Fourier series.* In 1807, Fourier stated that each 2π-periodic function $f(\tau)$, whether continuous or not, could be represented as an infinite sum of trigonometric functions, i.e. as a superposition of fundamental frequencies and harmonics (see fig. 3.8.1). Although this statement is mathematically not quite correct, his conclusions were of far-reaching importance at a time when there were not even clear definitions of concepts such as function or integral.

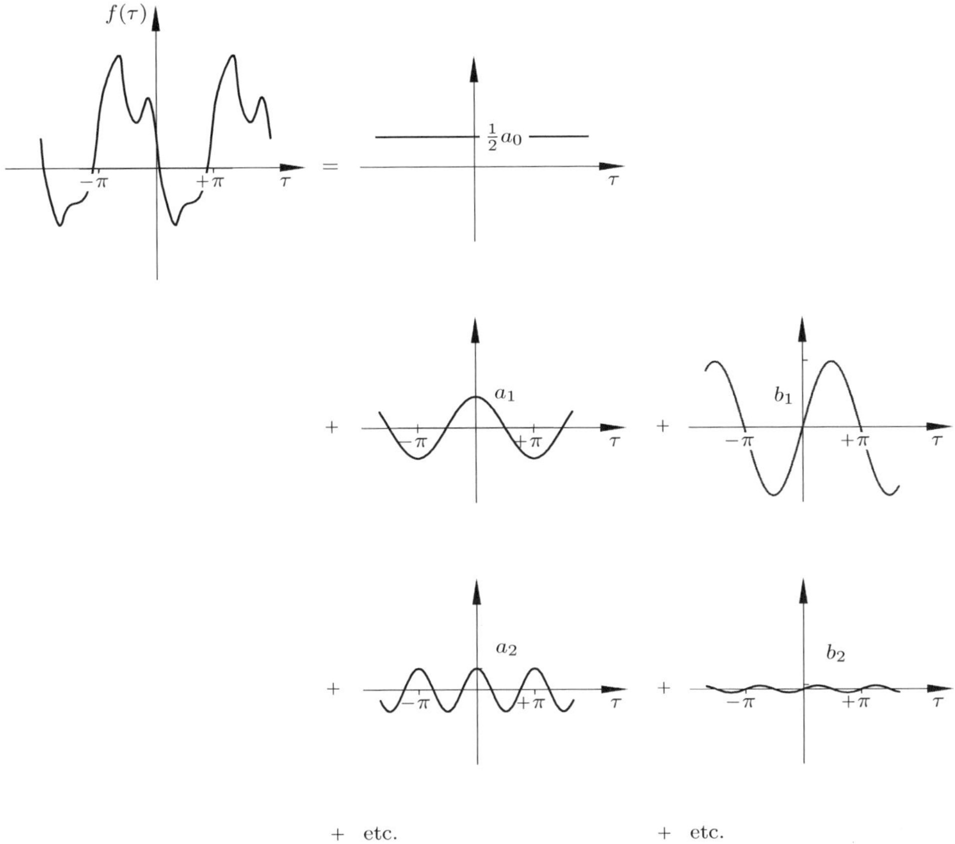

Fig. 3.8.1: Fourier analysis

If we denote the n-th partial sum

$$s_{\mathrm{n}}(\tau) = \frac{a_0}{2} + \sum_{\mathrm{k}=1}^{\mathrm{n}} (a_{\mathrm{k}} \cos \mathrm{k}\tau + b_{\mathrm{k}} \sin \mathrm{k}\tau) \tag{3.8.2}$$

then

$$f(\tau) = \lim_{\mathrm{n}\to\infty} s_{\mathrm{n}}(\tau) \tag{3.8.3}$$

If $f(\tau)$ is given, it is the task of the Fourier analysis to determine the so-called Fourier coefficients a_0, a_{k}, b_{k} ($\mathrm{k} = 1, 2, \ldots, \mathrm{n}$) in eq. (3.8.2). These are obtained from the condition that, for each fixed n, the mean squared error

$$E^2 = \int_{-\pi}^{+\pi} [f(\tau) - s_{\mathrm{n}}(\tau)]^2 \mathrm{d}\tau \tag{3.8.4}$$

is reduced to a minimum. Since the trigonometric functions form an orthogonal set, the integrals

$$
\begin{aligned}
a_0 &= \frac{1}{\pi} \int\limits_{-\pi}^{+\pi} f(\tau)\,\mathrm{d}\tau \\
a_\mathrm{k} &= \frac{1}{\pi} \int\limits_{-\pi}^{+\pi} f(\tau) \cos \mathrm{k}\tau\,\mathrm{d}\tau \\
b_\mathrm{k} &= \frac{1}{\pi} \int\limits_{-\pi}^{+\pi} f(\tau) \sin \mathrm{k}\tau\,\mathrm{d}\tau \qquad (\mathrm{k} = 1, 2, \dots, \mathrm{n})
\end{aligned} \tag{3.8.5}
$$

determine the real Fourier coefficients. If $f(\tau)$ is a piecewise continuous and differentiable function which is at first only defined in the interval [-π, $+\pi$] it can be extended periodically beyond the initial interval with the aid of the relation $f(\tau + 2\pi) = f(\tau)$ (see fig. 3.8.2). In addition, at each jump, we assign to the function the mean value of the function values on either side of the jump.

One important issue in the theory of Fourier series is the question of which functions can be expanded into Fourier series and whether pointwise or uniform convergence can be guaranteed. Here, we cite only some basic results (Courant and Hilbert, 1993). Each function which is piecewise continuous in the interval [-π, $+\pi$] and periodic with period 2π can be expanded into a Fourier series. In each closed interval in which $f(\tau)$ is continuous, the series converges uniformly. If discontinuities arise only in the first derivatives, then absolute and uniform convergence of the Fourier series is assured. This leads to a considerable simplification of computational rules because limit processes can then be performed term by term. For example, in this case, a limit process and integration or summation processes can be interchanged. In addition, infinite series which are uniformly convergent in the interval [-π, $+\pi$] can be differentiated or integrated term by term. By contrast, $f(\tau)$ is no longer uniformly convergent in the direct neighbourhood of jump discontinuities. Approximations overshoot the function value, leading to the so-called *Gibbs phenomenon.* Here, spurious oscillations arise in the partial sums which do not die out even if more Fourier terms are taken into account (Papoulis, 1962; Beerends et al., 2003).

A clearer and more compact representation of the Fourier analysis is obtained by switching to complex numbers with the help of Euler's formula

$$
\mathrm{e}^{i\mathrm{k}\tau} = \cos \mathrm{k}\tau + i \sin \mathrm{k}\tau \tag{3.8.6}
$$

Introducing the relations

$$
\cos \mathrm{k}\tau = \frac{1}{2}\left(\mathrm{e}^{i\mathrm{k}\tau} + \mathrm{e}^{-i\mathrm{k}\tau}\right) \tag{3.8.7}
$$

as well as

$$
\sin \mathrm{k}\tau = \frac{1}{2i}\left(\mathrm{e}^{i\mathrm{k}\tau} - \mathrm{e}^{-i\mathrm{k}\tau}\right) \tag{3.8.8}
$$

into eq. (3.8.2) and defining the quantities

$$
\alpha_0 = \tfrac{1}{2} a_0, \quad \alpha_\mathrm{k} = \tfrac{1}{2}(a_\mathrm{k} - i b_\mathrm{k}), \quad \alpha_{-\mathrm{k}} = \tfrac{1}{2}(a_\mathrm{k} + i b_\mathrm{k}) \tag{3.8.9}
$$

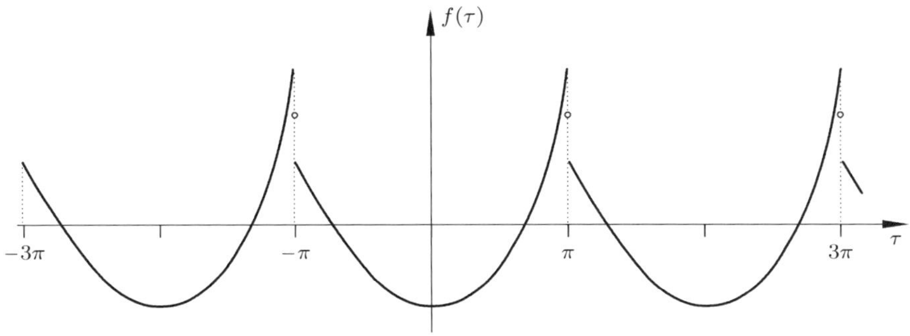

Fig. 3.8.2: Periodic continuation of a function

as *complex* Fourier coefficients, the Fourier series can be written in the form

$$f(\tau) = \sum_{\mathrm{k}=-\infty}^{+\infty} \alpha_{\mathrm{k}}\, \mathrm{e}^{i\mathrm{k}\tau} \tag{3.8.10}$$

with

$$\alpha_{\mathrm{k}} = \frac{1}{2\pi} \int_{-\pi}^{+\pi} f(\tau)\, \mathrm{e}^{-i\mathrm{k}\tau} \mathrm{d}\tau \qquad (\mathrm{k} = 0,\, \pm 1,\, \pm 2,\, \dots) \tag{3.8.11}$$

If the function f has the period T, we may replace τ by ωt in the above relations, where ω is the so-called circular or angular frequency

$$\omega = \frac{2\pi}{T} \tag{3.8.12}$$

The Fourier series and the coefficients are then

$$f(t) = \sum_{\mathrm{k}=-\infty}^{+\infty} \alpha_{\mathrm{k}} \mathrm{e}^{i\mathrm{k}\omega t} \tag{3.8.13}$$

with

$$\alpha_{\mathrm{k}} = \frac{1}{T} \int_{-T/2}^{+T/2} f(t)\, \mathrm{e}^{-i\mathrm{k}\omega t} \mathrm{d}t \qquad (\mathrm{k} = 0,\, \pm 1,\, \pm 2,\, \dots) \tag{3.8.14}$$

3.8.2 Fourier Integral and Fourier Transform

An obvious question is whether arbitrary functions $f(t)$, although non-periodic in the time domain $-\infty < t < +\infty$, can also either be decomposed into sine and cosine oscillations or be represented as linear combinations of exponential functions. The first idea is to establish such a decomposition from a Fourier series, eqs. (3.8.13, 3.8.14), by approaching the limit $T \to \infty$. Thus, if $f(t)$ is a non-periodic function, one can define a function $f_T(t)$ which is identical with $f(t)$ within the interval

$(-\frac{T}{2}, \frac{T}{2})$ and is expanded periodically outside it. In accordance with eq. (3.8.13), $f_T(t)$ can be expanded to a trigonometric series

$$f(t) = \sum_{\mathrm{k}=-\infty}^{+\infty} F(\mathrm{k}\omega_T)\,\mathrm{e}^{i\mathrm{k}\omega_T t}\,\frac{1}{T} \tag{3.8.15}$$

where $\omega_T = \frac{2\pi}{T}$ and

$$F(\mathrm{k}\omega_T) = \int_{-T/2}^{+T/2} f(t)\,\mathrm{e}^{-i\mathrm{k}\omega_T t}\mathrm{d}t \tag{3.8.16}$$

We pass to the limit $T \to \infty$ and substitute $\mathrm{k}\omega_T = \frac{2\pi\mathrm{k}}{T}$ by ω and the difference of two consecutive ω-values by $\mathrm{d}\omega$, i.e. $\mathrm{d}\omega = \frac{2\pi}{T}$. Then, if the summation is replaced by an integral symbol, we obtain by purely formal reasoning from eq. (3.8.15)

$$f(t) = \frac{1}{2\pi}\int_{-\infty}^{+\infty} F(\omega)\,\mathrm{e}^{i\omega t}\mathrm{d}\omega \tag{3.8.17}$$

with

$$F(\omega) = \int_{-\infty}^{+\infty} f(t)\,\mathrm{e}^{-i\omega t}\mathrm{d}t \tag{3.8.18}$$

$F(\omega)$ is denoted the Fourier integral or Fourier transform of $f(t)$. Thus, for non-periodic functions, the frequency spectrum is no longer discrete but continuous. Introducing "negative" frequencies by means of Euler's formula allows a compact representation of the transform.

Two points in the above approach are not satisfactory. Firstly, basic functions such as $f(t) = c$ or $f(t) = \sin t$ do not possess a Fourier transform in the ordinary sense. In fact, one has to use a generalised concept for functions which will be introduced in section 3.8.4. In addition, such a formal passage to the limit leaves open which conditions have to be imposed on the function $f(t)$ in order to guarantee the existence of its Fourier transform. Therefore, in the literature, a direct introduction of the Fourier transform is preferred (Papoulis, 1962; Lighthill, 1966), where it is presupposed that $f(t)$ is absolutely integrable, i.e.

$$\int_{-\infty}^{+\infty} |f(t)|\mathrm{d}t < \infty$$

The classical Fourier transform can therefore only be applied to functions which tend to zero for $t \to \infty$. In the literature, other definitions can sometimes be found which, for example, use a factor $1/\sqrt{2\pi}$ for both the transform and its inverse, or which change the sign in the exponent. However, all these definitions are equivalent.

It should be mentioned that absolute integrability is only a sufficient condition, but is not necessary for the existence of $F(\omega)$. For example, periodic functions are not absolutely integrable, yet a Fourier transform can be assigned to them, see section 3.8.4. If generalised functions are admitted for $f(t)$ and $F(\omega)$ like the Dirac δ-function for example, the condition of absolute integrability can be withdrawn.

The Fourier transform belongs to the large family of integral transforms which also include, for instance, the Laplace and the Hilbert transform, see for example (Bracewell, 1986). All such transforms are linear and possess a unique inverse transform. This means for example for the Fourier transform that both representations, in the time as well as in the frequency domain, are completely equivalent although they highlight quite different aspects of the dynamics. Hence, $f(t)$ and $F(\omega)$ can be considered alternative but equivalent representations of the same dynamical process; the symbol

$$f(t) \Longleftrightarrow F(\omega) \tag{3.8.19}$$

is used for pairs of Fourier transforms. Each of the two formulations described above spotlights particular characteristics of the process in the time and the frequency domain.

From eq. (3.8.18), it follows that the Fourier transform of a real function $f(t)$ is in general complex, i.e.

$$F(\omega) = F_{\mathrm{R}}(\omega) + iF_{\mathrm{I}}(\omega) = A(\omega)\mathrm{e}^{i\phi(\omega)} \tag{3.8.20}$$

The amplitude $A(\omega)$ is designated as the Fourier spectrum, the square

$$A^2(\omega) = |F(\omega)|^2 = F(\omega)F^*(\omega) \tag{3.8.21}$$

energy spectrum and $\phi(\omega)$ as the phase angle. Complex conjugate quantities are denoted by a superior asterisk.

3.8.3 Characteristics of the Fourier Transform

The Fourier transformation is a linear operation. This leads to a number of fundamental relations, listed in table 3.8.1. They derive from the definitions (3.8.17) and (3.8.18). Detailed proofs can be found in (Papoulis, 1962), for example. It is assumed that the following transformation pairs correspond

$$f_1(t) \Longleftrightarrow F_1(\omega) \quad \text{and} \quad f_2(t) \Longleftrightarrow F_2(\omega)$$

$$\text{resp.} \qquad f(t) \Longleftrightarrow F(\omega)$$

One important operation is the so-called *convolution* of two functions f_1, f_2, marked by an intermediate asterisk and defined in the time domain as well as in the frequency domain by

$$f_1(t) * f_2(t) = \int_{-\infty}^{+\infty} f_1(\tau)f_2(t-\tau)\mathrm{d}\tau \tag{3.8.22}$$

$$F_1(\omega) * F_2(\omega) = \int_{-\infty}^{+\infty} F_1(y)F_2(\omega-y)\mathrm{d}y \tag{3.8.23}$$

$a_1 f_1(t) + a_2 f_2(t)$	$\Longleftrightarrow$	$a_1 F_1(\omega) + a_2 F_2(\omega)$	(superposition)
$f(at)$	$\Longleftrightarrow$	$\frac{1}{\lvert a\rvert} F(\frac{\omega}{a})$	(alteration of time scale)
$f(t - t_0)$	$\Longleftrightarrow$	$F(\omega)\,\mathrm{e}^{-\mathrm{i}\omega t_0}$	(time shift)
$f(t)\,\mathrm{e}^{i\omega_0 t}$	$\Longleftrightarrow$	$F(\omega - \omega_0)$	(frequency shift)
$f_1(t) * f_2(t)$	$\Longleftrightarrow$	$F_1(\omega)\,F_2(\omega)$	(convolution in time domain)
$f_1(t)\,f_2(t)$	$\Longleftrightarrow$	$\frac{1}{2\pi} F_1(\omega) * F_2(\omega)$	(convolution in frequency domain)
$\mathrm{corr}\,(f_1, f_2)$	$\Longleftrightarrow$	$F_1(\omega)\,F_2^*(\omega)$	(correlation)
$\mathrm{corr}\,(f, f)$	$\Longleftrightarrow$	$\lvert F(\omega)\rvert^2 = A^2(\omega)$	$(\int_{-\infty}^{+\infty} \lvert f(t)\rvert^2 \mathrm{d}t < \infty)$ (autocorrelation)
$\frac{\mathrm{d}^{\mathrm{n}} f}{\mathrm{d}t^{\mathrm{n}}}$	$\Longleftrightarrow$	$(i\omega)^{\mathrm{n}}\,F(\omega)$	(derivative in time domain)
$t^{\mathrm{n}} f(t)$	$\Longleftrightarrow$	$i^{\mathrm{n}} \frac{\mathrm{d}^{\mathrm{n}} F(\omega)}{\mathrm{d}\omega^{\mathrm{n}}}$	(derivative in frequency domain)

Table 3.8.1: Fundamental characteristics of the Fourier transform

i.e. a convolution is established as the integral over the product of f_1 with the reflected and shifted representation of f_2.

It can now be shown that the product of two functions in the time domain corresponds to the convolution of the corresponding Fourier transforms in the frequency domain and vice versa (see table 3.8.1). With the aid of these relations, we can prove the so-called Parseval's identity (Papoulis, 1962)

$$\text{total energy} \equiv \int_{-\infty}^{+\infty} \lvert f(t)\rvert^2 \mathrm{d}t = \frac{1}{2\pi} \int_{-\infty}^{+\infty} A^2(\omega)\mathrm{d}\omega \tag{3.8.24}$$

This states that, but for a factor $1/2\pi$, the calculation of the total energy of a signal yields the same value whether evaluated in the time domain or in the frequency domain. In accordance with eq. (3.8.21), the expression $A^2(\omega)$ is designated as the energy spectrum.

A further tool in signal processing and its interpretation is the correlation of two functions f_1 and f_2

$$\mathrm{corr}(f_1, f_2) = \int_{-\infty}^{+\infty} f_1(\tau) f_2(t + \tau)\mathrm{d}\tau \tag{3.8.25}$$

or the autocorrelation, i.e. the correlation of a function f with itself

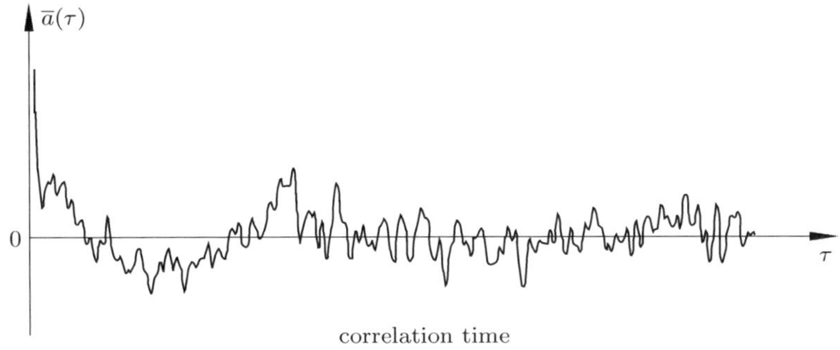

Fig. 3.8.3: Autocorrelation of a chaotic motion

$$\mathrm{corr}(f,f) = a(\tau) = \int_{-\infty}^{+\infty} f(\tau) f(t+\tau)\mathrm{d}\tau \tag{3.8.26}$$

This explores the similarity or agreement of one and the same signal at two distinct instants of time, t and $t+\tau$, and is a function of the time shift τ. In contrast to the convolution, eq. (3.8.22), none of the functions is reflected in this operation.

There are functions $f(t)$ defined over the whole interval $-\infty < t < +\infty$, which, while exhibiting a total energy, eq. (3.8.24), which is not finite, have a finite temporal mean value over an interval of length $2T$. The total power is then defined as

$$\text{total power} \equiv \lim_{T\to\infty} \frac{1}{2T} \int_{-T}^{+T} |f(t)|^2 \mathrm{d}t = \overline{f^2} \tag{3.8.27}$$

This applies, for example, for all periodic functions, the Fourier coefficients of which satisfy the condition $\sum_{\mathrm{k}=-\infty}^{+\infty} |\alpha_{\mathrm{k}}|^2 < \infty$. For such signals with finite power, the autocorrelation is defined as the temporal mean value

$$\bar{a}(\tau) = \lim_{T\to\infty} \frac{1}{2T} \int_{-T}^{+T} f(t+\tau) f(t)\mathrm{d}t \tag{3.8.28}$$

It can be shown, see (Papoulis, 1962), that the autocorrelation $\mathrm{corr}(f,f) = a(\tau)$ and the energy spectrum $A^2(\omega)$, eq. (3.8.24), represent a Fourier transform pair (see table 3.8.1). For functions $f(t)$, the total energy of which is not finite but the total power is $\overline{f^2} < \infty$ in accordance with eq. (3.8.27), we must turn to mean values. By analogy with $a(\tau) \Longleftrightarrow A^2(\omega)$, the power spectrum $P(\omega)$ is defined as the Fourier transform of the autocorrelation as mean value $\bar{a}(\tau)$, eq. (3.8.28),

$$\bar{a}(\tau) \Longleftrightarrow P(\omega) \tag{3.8.29}$$

There is also another way of deriving $P(\omega)$ directly from $f(t)$. We have

$$P(\omega) = \lim_{T\to\infty} \frac{1}{2T} \left| \int_{-T}^{+T} f(t)\mathrm{e}^{-i\omega t}\mathrm{d}t \right|^2 \tag{3.8.30}$$

For a large enough interval $-T \leqslant t \leqslant +T$, we calculate the Fourier transform $F_T(\omega)$ of the "truncated" function

$$f_T(t) = \begin{cases} f(t) & \text{for } |t| \leqslant T \\ 0 & \text{for } |t| > T \end{cases} \tag{3.8.31}$$

and determine the mean power

$$P_T(\omega) = \frac{1}{2T}|F_T(\omega)|^2 \tag{3.8.32}$$

The power spectrum then derives as the limit value

$$P(\omega) = \lim_{T\to\infty} P_T(\omega) \tag{3.8.33}$$

In principle, the autocorrelation $a(\tau)$ and the energy spectrum $A^2(\omega)$ or the averaged autocorrelation $\bar{a}(\tau)$ and the power spectrum $P(\omega)$ respectively are equivalent representations of one and the same evolution process. However, they each reveal quite different characteristics of the initial signal $f(t)$: the autocorrelation reflects the memory and internal relations of the system whereas the power spectrum informs us about dominant basic frequencies. This is elucidated in the following section 3.8.4.

The rules as to how the superposition of functions, scaling, shifting and convolution affect the Fourier transform already show a number of important properties of the Fourier transform. However, the great success and the enormous range of applications of the Fourier transform becomes completely evident only when the impact of derivatives in the time and in the frequency domain are taken into account.

If we calculate the derivative of the exponential function

$$\frac{\mathrm{d}}{\mathrm{d}t}\mathrm{e}^{i\omega t} = i\omega\mathrm{e}^{i\omega t} \tag{3.8.34}$$

we observe that $\lambda = i\omega$ is the eigenvalue of the differential operator $\frac{\mathrm{d}}{\mathrm{d}t}$ and $\mathrm{e}^{i\omega t}$ the associated eigenfunction. From this, the rules for differentiation in the time and frequency domain can be deduced. From eq. (3.8.17), we derive

$$\frac{\mathrm{d}f(t)}{\mathrm{d}t} = \frac{1}{2\pi} \int_{-\infty}^{+\infty} F(\omega)\, i\omega\, \mathrm{e}^{i\omega t}\mathrm{d}\omega \tag{3.8.35}$$

and hence obtain the Fourier transform pairs

$$\frac{\mathrm{d}f}{\mathrm{d}t} \Longleftrightarrow i\omega\, F(\omega) \quad \text{resp.} \quad \frac{\mathrm{d}^{\mathrm{n}}f}{\mathrm{d}t^{\mathrm{n}}} \Longleftrightarrow (i\omega)^{\mathrm{n}}\, F(\omega)$$

Knowing the Fourier transform $F(\omega)$ of a function $f(t)$, we can obtain the Fourier transform of the nth derivative by multiplication with a factor $(i\omega)^{\mathrm{n}}$. In the same way, the rule for differentiation in the frequency domain is deduced from eq. (3.8.18)

$$\frac{\mathrm{d}F(\omega)}{\mathrm{d}\omega} = \int\limits_{-\infty}^{+\infty} f(t)\,(-it)\mathrm{e}^{-i\omega t}\mathrm{d}t \tag{3.8.36}$$

In summary, we obtain the following Fourier transform pairs

$$t\,f(t) \Longleftrightarrow i\frac{\mathrm{d}F(\omega)}{\mathrm{d}\omega} \quad \text{resp.} \quad t^{\mathrm{n}}\,f(t) \Longleftrightarrow i^{\mathrm{n}}\,\frac{\mathrm{d}^{\mathrm{n}}F(\omega)}{\mathrm{d}\omega^{\mathrm{n}}}$$

With the help of the Fourier transform, ordinary differential equations can be converted to algebraic equations and partial differential equations to ordinary ones. The solution of linear differential equations in particular can thus be significantly simplified; see for example (Bracewell, 1986; Courant and Hilbert, 1993). A further field of application for the Fourier transform is the characteristic functions of probability distributions, to be discussed in section 3.9.8.

3.8.4 Elementary Fourier Transform, Line Spectra, Dirac δ-Function

In the following, we present a small catalogue of Fourier transforms of simple functions. Our interest is directed particularly at the Fourier transform of periodic and quasi-periodic functions which enable us to distinguish between regular periodic and quasi-periodic motions and aperiodic, possibly chaotic ones. If we restrict ourselves to the classical concept of functions, it is not possible to establish the Fourier transform of periodic functions because they do not fulfil the precondition of absolute integrability. A way out is offered by the concept of generalised functions introduced by the theory of distributions. In this connection, a particularly important role is played by the Dirac δ function (Dirac, 1935; Lighthill, 1966).

Distributions are generalised functions not defined via their function values but via certain characteristics of an integral nature which do not emerge until they are applied to test functions. If we consider, for example, an arbitrary function $f(t)$ which must be continuous at the point $t = 0$, the δ-function can be regarded as an instruction which assigns the value $f(0)$ to the test function $f(t)$

$$\int\limits_{-\infty}^{+\infty} \delta(t)f(t)\mathrm{d}t = f(0) \tag{3.8.37}$$

However, the δ-function can also – as is customary – be defined as the generalised limit value of a series of impulses of area 1

$$\delta(t) = \lim_{\varepsilon\to 0} f_\varepsilon(t) \tag{3.8.38}$$

where the impulses or spike functions $f_\varepsilon(t)$, for example, can have the form of rectangles or Gaussian distributions (see fig. 3.8.4).

By definition, the area is given as $\int_{-\infty}^{+\infty}\delta(t)\,\mathrm{d}t = 1$. The meaning of the generalised integral

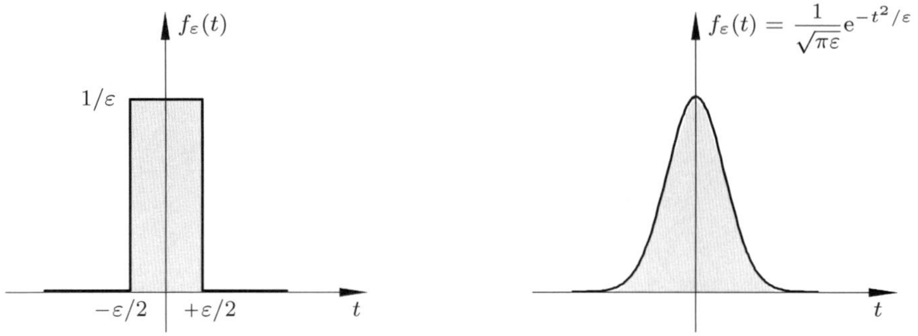

Fig. 3.8.4: On the definition of the Dirac δ-function

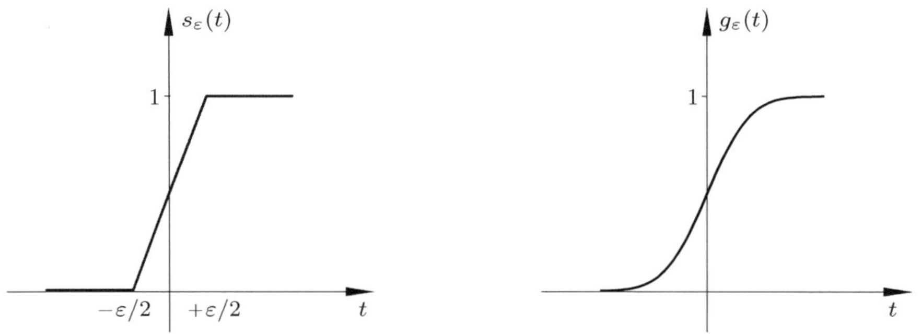

Fig. 3.8.5: On the approximation of the Heaviside function $H(t)$

$$H(t) = \int_{-\infty}^{t} \delta(\tau)\,\mathrm{d}\tau \tag{3.8.39}$$

can be considered, which is again given by the limit $H(t) = \lim_{\varepsilon\to 0} \int_{-\infty}^{t} f_\varepsilon(\tau)\,\mathrm{d}\tau$. Inserting the rectangular function of fig. 3.8.4 for f_ε, the integral reads

$$s_\varepsilon(t) = \int_{-\infty}^{t} f_\varepsilon(\tau)\,\mathrm{d}\tau = \begin{cases} 0 & \text{for} \quad t < -\varepsilon/2 \\ 1/2 + t/\varepsilon & \text{for} \quad -\varepsilon/2 \leqslant t < \varepsilon/2 \\ 1 & \text{for} \quad t \geqslant \varepsilon/2 \end{cases} \tag{3.8.40}$$

Inserting the Gauss function instead, we obtain with the help of the Gauss error function (Bronstein et al., 1995) $\mathrm{erf}(t) = \frac{2}{\sqrt{\pi}} \int_0^t \mathrm{e}^{-\tau^2} \mathrm{d}\tau$

$$g_\varepsilon(t) = \lim_{\varepsilon\to 0} \frac{1}{\sqrt{\pi\varepsilon}} \int_{-\infty}^{t} \mathrm{e}^{-\tau^2/\varepsilon}\,\mathrm{d}\tau = \lim_{\varepsilon\to 0} \frac{1}{2}\left(1 + \mathrm{erf}(\frac{t}{\sqrt{\varepsilon}})\right) \tag{3.8.41}$$

Fig. 3.8.5 presents the functions $s_\varepsilon(t)$ and $g_\varepsilon(t)$. In the limit $\varepsilon \to 0$, both curves converge towards the unit step function, also denoted Heaviside function

$$H(t) = \begin{cases} 0 & \text{for} \quad t < 0 \\ 1/2 & \text{for} \quad t = 0 \\ 1 & \text{for} \quad t > 0 \end{cases} \tag{3.8.42}$$

We have thus calculated the generalised integral $\int_{-\infty}^{t} \delta(\tau)\,\mathrm{d}\tau$ from eq. (3.8.39). Conversely, in the sense of distributions, we can also formulate the generalised derivative

$$H'(t) = \delta(t) \tag{3.8.43}$$

which is expedient for the description of many technical and physical applications.

It is certainly no coincidence that the foundation stone for the theory of distributions was laid at a time when physicists were exploring the structure of the atom and the nature of quantum mechanics. Many physical quantities cannot be measured directly; one can only deduce their existence by observing the totality of their effects and by using their interactions to draw conclusions regarding the quantities themselves. In a way, the distributions thus represent an appropriate mathematical tool for describing such physical quantities.

An easily comprehensible introduction to the theory of generalised functions can be found, for example, in (Papoulis, 1962) or (Lighthill, 1966). The concept of distributions enables us to operate with generalised integrals, derivatives and limit values which are, in the classical sense, meaningless. In the following, we apply this theory in order to establish the Fourier transform pairs most important for us.

If we consider the Fourier transform of the δ-function, then, in accordance with the definition of eq. (3.8.18), we obtain

$$F(\omega) = \int_{-\infty}^{+\infty} \delta(t)\,\mathrm{e}^{-i\omega t}\mathrm{d}t \tag{3.8.44}$$

In accordance with eq. (3.8.37), however, this relation assigns the test function $f(t) = \mathrm{e}^{-i\omega t}$ the value of $f(0) = 1$ at $t = 0$, so that the constant $F(\omega) = 1$ results as the Fourier transform of the δ-function

$$\delta(t) \Longleftrightarrow 1 \tag{3.8.45}$$

In accordance with table 3.8.1, a shift in the time domain then yields

$$\delta(t - t_0) \Longleftrightarrow \mathrm{e}^{-i\omega t_0} \tag{3.8.46}$$

Conversely, the Fourier transform of the constant $f(t) = 1$ can be easily obtained. After introduction of $F(\omega) = 2\pi\delta(\omega)$ into eq. (3.8.17), the integral reads, in accordance with eq. (3.8.37),

$$f(t) = \frac{1}{2\pi} \int_{-\infty}^{+\infty} 2\pi\delta(\omega)\,\mathrm{e}^{i\omega t}\mathrm{d}\omega = \mathrm{e}^{i\omega t}\Big|_{\omega=0} = 1 \tag{3.8.47}$$

or

$$1 \Longleftrightarrow 2\pi\delta(\omega) \tag{3.8.48}$$

pairs of Fourier transforms	time domain	frequency domain
$\delta(t) \Longleftrightarrow 1$	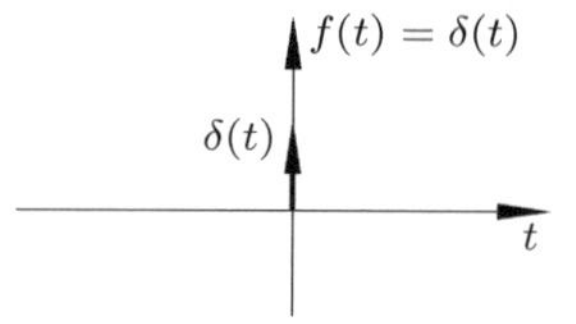	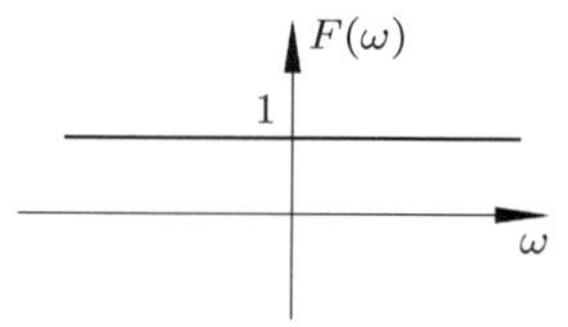
$1 \Longleftrightarrow 2\pi\delta(\omega)$	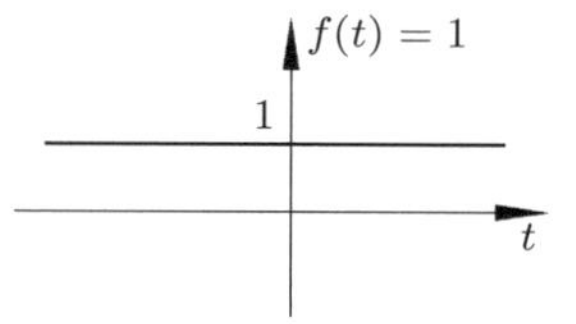	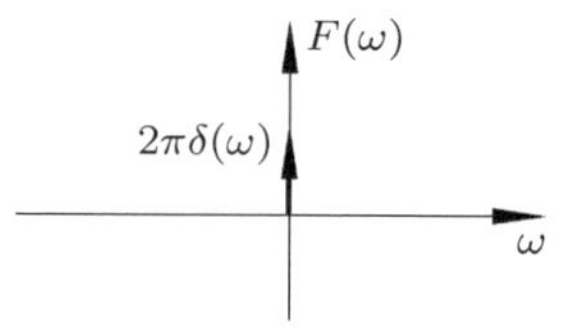
$\sin \omega_0 t$ $\Updownarrow$ $i\pi\left[\delta(\omega + \omega_0) - \delta(\omega - \omega_0)\right]$	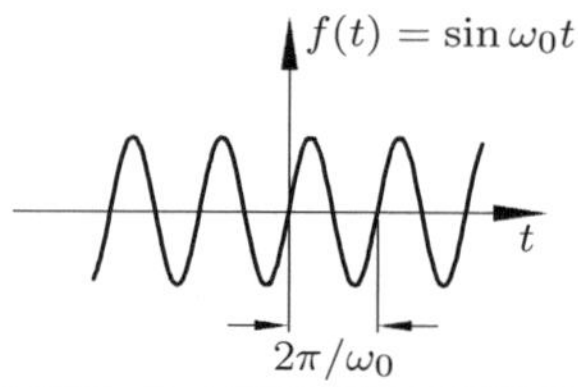	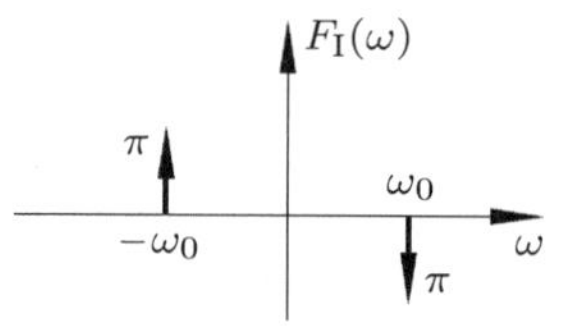
$\cos \omega_0 t$ $\Updownarrow$ $\pi\left[\delta(\omega + \omega_0) + \delta(\omega - \omega_0)\right]$	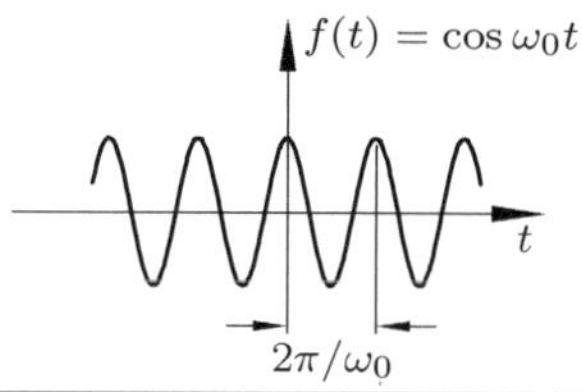	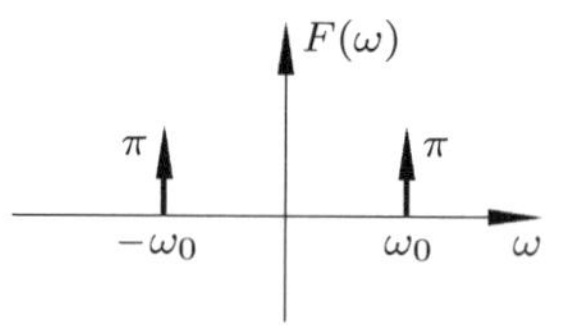

Table 3.8.2: The simplest Fourier transforms

If we perform a shift in the frequency domain, we immediately obtain

$$e^{i\omega_0 t} \Longleftrightarrow 2\pi\delta(\omega - \omega_o) \tag{3.8.49}$$

in accordance with table 3.8.1. Thus, on the basis of the concept of generalised functions, we are also able to apply the Fourier transform to the continuous periodic functions $\sin \omega_0 t$ and $\cos \omega_0 t$, although they do not fulfil the condition of absolute integrability. With the help of eqs. (3.8.7) and (3.8.8), we obtain

$$\sin \omega_0 t \Longleftrightarrow i\pi[\delta(\omega + \omega_0) - \delta(\omega - \omega_0)] \tag{3.8.50}$$

and

$$\cos \omega_0 t \Longleftrightarrow \pi[\delta(\omega + \omega_0) + \delta(\omega - \omega_0)] \tag{3.8.51}$$

i.e. in the frequency domain, a simple line spectrum consisting of two spikes for $\omega = \pm\omega_0$ represents in the time domain a sine or cosine oscillation.

In section 5.3, we make use of the Fourier transform to distinguish between regular and chaotic or random motions. In that section, we will also briefly outline the numerical calculation of the Fourier transform in practical applications.

3.8.5 Wavelet Transform

In practice, Fourier techniques play an important role. Applications range from accoustics, optics and signal processing to probability theory, which will be discussed in section 3.9, and to the solution of differential equations. Often, the Fourier transform is compared with an optical prism which breaks light up into its constituent spectral colours. Although the function $f(t)$ and its Fourier transform $F(\omega)$ are corresponding representations in the time and frequency domain, they nevertheless provide different information. While $f(t)$ displays the temporal evolution without providing any information on the frequency content, $F(\omega)$ contains information on how strongly single frequencies act in the signal while information on their temporal occurrence is hidden in their phases.

Fourier techniques are most successful when applied to linear problems with solutions of *(quasi-)stationary* character. Predicting non-linear behaviour, if possible at all, is much more difficult and generally leads to *unstationary* solutions displaying spikes and irregularities or solutions in which uniform smooth phases are suddenly interrupted by sections containing high frequencies. In such cases, one is interested in *simultaneous* information on the time and frequency content of the signal. Since irregularities, abrupt changes, peaks and jumps contain the most interesting information, one is interested in the frequency score, i.e. one is keen to know the precise instant at which a clash of drums occurs or one would like to discover irregularities in a ECG or edges in an image. In short, one would like to provide information – as with musical notes – on which frequency is played at which instant.

However, Fourier techniques are not qualified to provide such information. Theoretically, it would be possible to gain information on the time instants from the phases of the complex Fourier transform; in practice, this implies inaccuracies which would be too high (Ciarlet and Lions, 1997). The reason is that each *local* change, each jump and peak has a *global* impact on *all* frequencies, since the trigonometric functions used for the Fourier transform oscillate over the whole domain of integration, i.e. they have an infinite support.

For a long time, people tried to perform local time-frequency analyses with the help of the so-called *windowed* or *short-term Fourier transform*, defined as follows

$$Sf(\omega, t_0) = \int_{-\infty}^{+\infty} f(t)\, g(t - t_0)\, \mathrm{e}^{-i\omega t} \mathrm{d}t \tag{3.8.52}$$

Here, a function $f(t)$ is multiplied with a window function $g(t - t_0)$ of *constant* width centered at $t = t_0$, which is only non-zero in a finite interval. In this window, a Fourier transform is applied locally, leading to a function of the two variables frequency ω and time t_0. A disadvantage of this method is that choosing too large a window width means that jumps still cannot be localised accurately enough while, in small windows, low-frequency perturbations cannot be detected (Press et al., 1986).

A way out of this dilemma is offered by the wavelet transform, in which functions $\psi(t)$ which are localised in time and have a width which can be adapted to the signal are used for the decomposition. The wavelet transform, developed in the 1980s, is a natural far-reaching extension of the Fourier transform and is again a linear integral transform. The signal $f(t)$ is multiplied by a complex conjugate function $\psi^*(t)$ which is shifted about a real value b and spread out or squeezed by a factor $a \neq 0$. The continuous wavelet transform is then defined as

$$Wf(a,b) = \int_{-\infty}^{+\infty} f(t)\,\psi^*\left(\frac{t-b}{a}\right)\,\mathrm{d}t \tag{3.8.53}$$

In accordance with eq. (3.8.23), the wavelet transform can be interpreted as a convolution. In comparison with the windowed Fourier transform, the essentially new feature is the scaling parameter a which allows us to vary the width of the wavelet and to perform an adaptive decomposition of the signal into wavelet components.

The wavelet functions can be chosen almost freely. The only condition is that a unique reconstruction of the original function $f(t)$ from $Wf(a,b)$ must be guaranteed. This condition is fulfilled if

$$\int_{-\infty}^{+\infty} \psi(t)\,\mathrm{d}t = 0 \tag{3.8.54}$$

i.e. $\psi(t)$ must take positive and negative values and hence be oscillating. This is also the reason for the designation *wavelet*, the small wave. Usually, basis wavelets $\psi(t)$ are chosen to be different from zero only in a finite t-interval. Depending on the choice of a, one can zoom into the signal at an arbitrary position b as if under a microscope and thus exactly identify local jumps and peaks. For real values $a \neq 0$ and b, the wavelet transform $Wf(a,b)$ is redundant, i.e. there are correlations between its values, which can easily be seen from the fact that $f(t)$ is a function of one variable t while $Wf(a,b)$ depends upon the two variables a, b.

As in the case of the Fourier transform, there is also a discrete formulation beside the continuous wavelet transform. Discrete parameter values

$$\begin{aligned} a_\mathrm{m} &= a_0^\mathrm{m} && \text{with} \quad a_0 > 1, \quad \mathrm{m} \in \mathbb{Z} \\ b_\mathrm{n} &= \mathrm{n} b_0 a_0^\mathrm{m} && \text{with} \quad b_0 > 0, \quad \mathrm{m,n} \in \mathbb{Z} \end{aligned}$$

are chosen such that the original signal $f(t)$ can be uniquely reconstructed from the discrete wavelet transform

$$Wf(\mathrm{m,n}) = \int_{-\infty}^{+\infty} f(t)\,\psi^*_{\mathrm{m,n}}\,\mathrm{d}t \quad \text{with} \quad \psi_{\mathrm{m,n}}(t) = \psi\left(\frac{t-b_\mathrm{n}}{a_\mathrm{m}}\right) \tag{3.8.55}$$

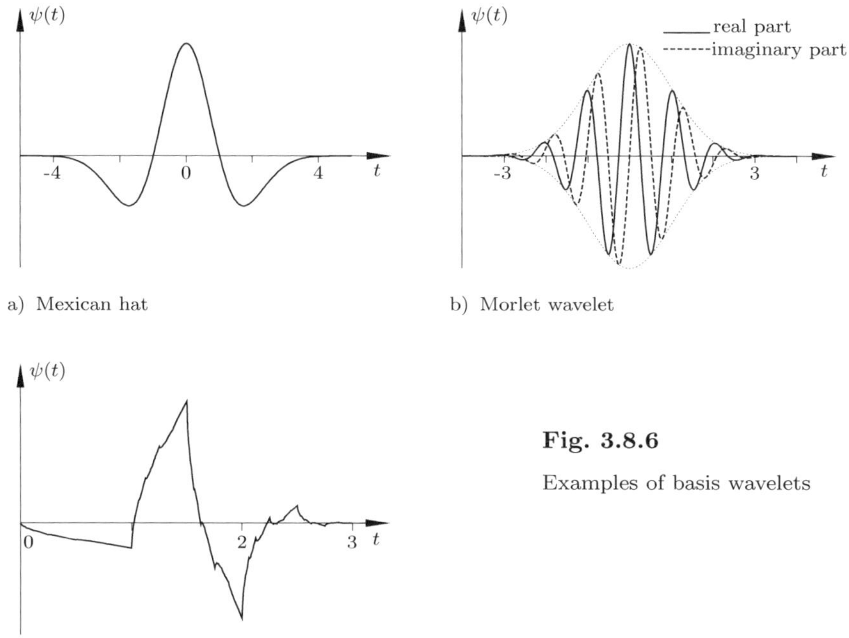

a) Mexican hat

b) Morlet wavelet

c) Daubechies D4 wavelet

Fig. 3.8.6

Examples of basis wavelets

In general, systems of discrete wavelets are also redundant and are denoted as *frames*. However, for special basis wavelets and special values a_0, b_0, it is possible to construct non-redundant orthonormal wavelet bases $\psi_{\mathrm{m,n}}$ with the help of what is called a multiresolution analysis (Daubechies, 1992; Mallat, 1998). As in the case of Fourier series, the original function can then be expanded into an infinite series of basis functions $\psi_{\mathrm{m,n}}$

$$f(t) = \sum_{\mathrm{m,n}}^{\infty} Wf(\mathrm{m,n})\, \psi_{\mathrm{m,n}}(t) \tag{3.8.56}$$

The choice of a specific wavelet is neither trivial nor unique. It depends primarily on the respective problem under consideration. Examples for wavelets are the real-valued *Mexican hat* (fig. 3.8.6a) and the complex-valued *Morlet wavelet* (fig. 3.8.6b). The Mexican hat wavelet is particularly suitable for localising and characterising singularities, spikes and irregularities in signals (Mallat, 1998), while the Morlet wavelet is generally preferred for analysing non-linear oscillations with time-dependent frequencies (Staszewski, 1998; Haase and Widjajakusuma, 2003). In fig. 3.8.6c, a *Daubechies* wavelet is shown which belongs to a family of discrete orthonormal wavelet basis functions (Daubechies, 1992). It can only be defined iteratively and has a fractal character. Wavelets of this type are used preferably in data compression. For example, the JPEG2000 standard for compressing image data makes use of this type of discrete wavelet. It is particularly interesting that, with orthonormal wavelets, the wavelet transform is faster than the fast Fourier transform. For a wavelet transform, the number of operations for a signal of length

N scales only like $\mathcal{O}(\mathrm{N})$, while $\mathrm{N}\log_2 \mathrm{N}$ operations are necessary for a fast Fourier transform, see section 5.3. Wavelet frames, i.e. redundant discrete wavelets, are very useful for pattern recognition. For example, two-dimensional Gabor wavelets with directional dependence, an extension of Morlet wavelets, are the basis for iris recognition in biometric identification (Daugman, 2004).

In the meantime, there is a huge list of books and articles on wavelets, which are extremely successfully applied in many areas, for example in data compression, in the analysis of medical data or in the evaluation of transient oscillations. We would like to mention only a small selection (Daubechies, 1992; Mallat, 1998; Addison, 2002; van den Berg, 1999). In the area of non-linear dynamics, wavelets are most interesting for the analysis of multifractal structures, discussed in section 8.5.1.

3.9 Basic Concepts of Probability Theory

> Random events designate procedures which obey the laws of probability. These laws specify possibilities and just because they are possibilities it remains undetermined, which of them will occur.
>
> *C.F. von Weizsäcker, Der Garten des Menschlichen*

In section 3.7, we discussed the logistic map. Beyond the accumulation point α_∞, the cascade changes from period doublings to chaotic, irregular behaviour (see fig. 3.7.6). A more rigorous study of the chaotic regime requires the application of statistical, probabilistic or stochastic methods. It is precisely for this reason that this section presents an overview of these methods in connection with the invariant measure, section 3.10, the methods characterising dissipative systems, Chapter 5, and turbulence, Chapter 9. The theory of probability is a discipline that has grown over centuries; its beginnings date back to the 17th century. A full list of references would be correspondingly long and it would be a hopeless task to attempt to present it here. Hence, in this section, the following four books (Fisz, 1976; Haken, 1990; Kreyszig, 1965; Papoulis, 1965) have been chosen to serve as a basis for the subsequent text; references to these books are only given explicitly in special cases.

The origin of the *theory of probability* is considered to be the empirical question the gambler Chevalier de Méré (1607–1687) posed to Blaise Pascal (1623–1662) in 1654. The ensuing correspondence between Pascal and Pierre de Fermat (1607/1608–1665), see (Tannery and Henry, 1894), pp. 288–314, documents a solution of the problem (see section 3.9.3). De Méré's problem was that, despite his calculations, he consistently lost a lot of money in a certain game of dice. According to his mathematical reasoning, it should have been equally likely to get either a "six" on four rolls of one die or a "double six" on 24 rolls of a pair of dice. However, in the latter case, he often lost the game. His argumentation was the following: the probability of getting a "six" when he threw one die is $\frac{1}{6}$, for 4 rolls it is $4 \cdot (\frac{1}{6}) = \frac{2}{3}$. The probability of getting a "double six" when he threw two dice once is $\frac{1}{36}$; for 24 rolls, it should be $24 \cdot (\frac{1}{36}) = \frac{2}{3}$.

The work of Pascal and Fermat or the arithmetic probability calculations in gambling can be assigned to the era preceding the modern theory of probability; this actually started with the formulation of the limit theorems by Jakob Bernoulli in *Ars conjectandi* (1713) and by Abraham de Moivre in *Miscellanea analytica* (1730). The significance of de Moivre's results for the probability theory was not realised until Pierre Simon de Laplace published his *Théorie Analytique des Probabilités* in 1812. The next three limit theorems were formulated by Poisson, one of them generalised Bernoulli's limit theorem, a second the theorem of de Moivre-Laplace and the third led to Poisson's distribution law. Further milestones in the development of modern probability theory were a method for the determination of the moments developed by Chebychev (1887) and the method of characteristic functions published in 1900 by Lyapunov. In the 1920s, Paul Pierre Lévy contributed publications on asymptotically stable distributions. Around the same time, Richard von Mises presented his frequentist definition of the concept of probability in "Grundlagen der Wahrscheinlichkeitsrechnung" (1919). The axiomatic definition of the probability theory by A. N. Kolmogorov, published in 1933 in "Foundations of the theory of probability" (English translation 1950) put a temporary end to these developments.

In this section, our intention is to give just an introduction to the rather complex material of modern probability theory. From the multitude of publications on real data analysis of experimentally generated or observed random variables and the check of statistical hypotheses, we mention only the monographs (Honerkamp, 1990; Kantz and Schreiber, 1997). Here, we only aim to focus on the recently developed Markov analysis (Friedrich and Peinke, 1997; Friedrich et al., 2009). In section 3.9.12, we briefly review the basic concepts of this method which allows the stochastic model equations to be extracted from the measured or computed time series, thereby disentangling the underlying deterministic dynamics and the fluctuating forces.

Beck-Bornholdt and Dubben succeeded in presenting probability calculations in a clear manner, using examples of general interest from case law, medical research etc. by means of errors and misjudgments arising in everyday life (Beck-Bornholdt and Dubben, 2003).

3.9.1 Experiment with Random Outcome

A typical example of an experiment with random outcome is tossing one or more dice. The number of spots thrown with an ideal die is random and it is impossible to predict the outcome unambigously. Not only in gambling but also in natural and engineering sciences, experiments are performed which in fact produce random results.

Definition: A well-defined experiment which can be repeated arbitrarily often and has an uncertain, random *outcome* ω is called *experiment with random outcome.*

- **Sample space, event, elementary event**

In the game of dice, one of six possible outcomes is expected; usually, the die will roll to a standstill so that one of the possible numbers $1, 2, 3, 4, 5, 6$ appears. It is common to introduce the notation $\omega_1, \omega_2, \ldots$ for the *possible* outcomes. The total set of n possible outcomes is assembled in the set $\Omega = \{\omega_1, \omega_2, \ldots, \omega_n\}$, the so-called *sample space.* When a coin is tossed, one of the simplest examples of a random experiment, the sample space is given by $\Omega = \{0, 1\}$, where zero is used for "heads" and one for 'tails".

It is only in rare cases that one is interested in a single outcome; rather, it is the combination of outcomes which is of interest, such as, for example, all even numbers when a die is rolled. Such a subset $\{\omega_2, \omega_4, \omega_6\}$ of the sample space Ω is called a random *event* A if it is clear whether the event occurs or does not occur. The occurrence or non-occurrence of an event A means that the complement $A^c = \{\omega_i \in \Omega : \omega_i \notin A\}$ must also be an event. If an event $A = \{\omega_i\}$ contains only one single element ω_i, it is designated an *elementary event.*

A special situation occurs if the total sample space Ω is an event; for instance, when a die is rolled, any number from 1 to 6 is a possible outcome, i.e. $A = \Omega$. In this case, the sample space Ω is called a *certain event.* The complement of the certain event is the *empty set* $\emptyset$, which is also an event which, however, will never occur. This situation is called an *impossible event.*

- **Event relations**

By definition, events are sets or subsets. This is the reason why notations and laws from set theory can be transferred to events. As in the case of sets, relations between events may exist. Let us consider two events A and B where A is assumed to be contained in B:

$$A \subseteq B \tag{3.9.1}$$

Therefore, each elementary event $\omega \in A$ is automatically an elementary event of B ($\omega \in B$). Hence, in an experiment, the occurrence of B follows from the occurrence of A. If both events contain the same elementary events, then

$$A = B \tag{3.9.2}$$

- **Event operations**

Following the theory of sets, we can define operations for events; the most important ones are listed in the following.

1. *Sum*: The sum $A + B$ of two events is the event which contains all elementary events which are elements of A *or* B. In the context of set theory, the sum operation corresponds to the *union set* $A \cup B$, see fig. 3.9.1a

$$A + B = A \cup B = \{\omega \in \Omega \ ; \ \omega \in A \text{ or } \omega \in B\} \tag{3.9.3}$$

2. *Product*: The product AB of two events is the event consisting of all elementary events which belong both to event A *and* to event B. According to set theory, the product AB is the *intersection* of A and B, see fig. 3.9.1b

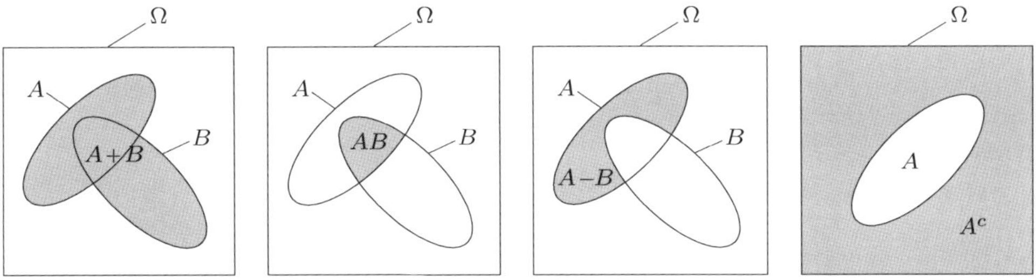

a) b) c) d)

Fig. 3.9.1: Event operations ($A,B \subseteq \Omega$): a) sum $A+B = A \cup B$; b) product $AB = A \cap B$; c) difference $A - B = A \backslash B$; d) complement A^c

$$AB = A \cap B = \{\omega \in \Omega \ ; \ \omega \in A \text{ and } \omega \in B\} \tag{3.9.4}$$

3. *Difference*: The difference $A - B$ designates an event consisting of all elementary events of A which are *not* contained in B, see fig. 3.9.1c

$$A - B = A \setminus B = \{\omega \in \Omega \ ; \ \omega \in A \ ; \ \omega \notin B\} \tag{3.9.5}$$

4. *Complement*: The complement A^c of an event A is defined as the event consisting of all elements of the sample space Ω that are *not* contained in A, see (fig. 3.9.1d)

$$A^c = \{\omega \in \Omega \ ; \ \omega \notin A\} \tag{3.9.6}$$

5. *Disjoint events*: If two events A and B have no elementary events ω in common, i.e. if their intersection is empty, then A and B are called *disjoint* (or mutually exclusive)

$$AB = A \cap B = \emptyset \tag{3.9.7}$$

Example: In dice, the two events $A = \{1, 3, 5\}$ and $B = \{4\}$ have no elements in commons, i.e. the events A and B are disjoint.

• Event space/probability space

The sample space Ω contains all possible experimental outcomes as elementary events. As mentioned previously, various combinations of the elementary events ω can be assembled in a subset A respectively in the event A. For example, in dice, the subset $A = \{\omega_4\}$ (ω_4 corresponds to rolling a "4") or the subset $A = \{\omega_2, \omega_4, \omega_6\}$ corresponds to tossing an "even" number. What is of interest now is the set of all subsets, i.e. the set of all events which can occur under specified experimental conditions. This set is denoted *probability space* $\mathcal{E}$. Playing dice with the sample space $\Omega = \{\omega_1, \omega_2, \omega_3, \omega_4, \omega_5, \omega_6\}$, the probability space $\mathcal{E} = \{\emptyset, \{\omega_\infty\}, ..., \{\omega\}, \{\omega_\infty, \omega_\in\}, ..., \otimes\}$ contains $2^6 = 64$ events respectively subsets. For a finite number n of elementary events, the number of possible events is generally 2^{n}.

3.9.2 Random Variables

In order to establish a quantitative theory of probability, the properties of the outcomes ω of the sample space Ω have to be described quantitatively. As an example, consider a classification of humans ω with respect to their weight X. In this case, it is possible to assign a numerical function X (the weight) to the sample point ω. Thus, for each ω, a real value $X(\omega)$ is known which can be written in mathematical terms as $\omega \to X(\omega)$. This function $X(\omega)$ is designated a *random variable*, simply because the sample point ω is picked at random.

A more rigorous formulation is the following: the real random variable is defined as a function $X : \Omega \to \mathbb{R}$, which assigns a real number to every event $\omega \in \Omega$ fulfilling the following condition of measurability: if a real number $x \in \mathbb{R}$ is chosen, the set of all events for which $X(\omega) \leqslant x$ holds must also be an event of the probability space $\mathcal{E}$, i.e.

$$\forall\, x \in \mathbb{R}: \quad \{\omega \,|\, X(\omega) \leqslant x\} \in \mathcal{E} \tag{3.9.8}$$

As soon as a sample point ω is chosen, $X(\omega)$ is determined. If we count the number of cars on a highway during one year ω and measure their velocity $X(\omega)$, the definition eq. (3.9.8) allows us to specify the probability of a particular driving speed $X(\omega) \leqslant x$ occurring.

The random variable can be discrete, as in the example of tossing a coin, where zero is assigned to the discrete outcome "heads" or one to "tails". On the other hand, the range of values of $X(\omega)$ can also be continuous, as in the case of a javelin throw, where the distance $X(\omega)$ is assigned to each throw ω.

Up to now, we have assumed that one random variable $X(\omega)$ is assigned to each ω. It is, of course, also possible to assign several values $X_1(\omega), X_2(\omega), ...$ to the sample point ω. In the case of throwing a javelin, for example, the three random variables distance of throw, launch speed and launch angle can be assigned to each javelin throw ω. So far, we have thought of experiments which are repeated n times. However, one can also consider n experiments to be one single experiment, where all n experiments are performed at the same time, but only once, as, for example, in the case of tossing n dice. The set of the n measured random variables $X_1, X_2, ..., X_{\mathrm{n}}$ is denoted *ensemble*.

It should be noted that the linear combination $\alpha X + \beta Y$ of the random variables X and Y is also a random variable. The same applies to the product $X \cdot Y$ and the quotient X/Y, $(Y \neq 0)$ or to the general function $\varphi(X(\omega))$.

3.9.3 Probability

- **Classic definition** (P. S. de Laplace)

P. S. de Laplace (1749–1827) presented a summary of contemporary knowledge on probability theory in his work (Laplace, 1812). His considerations intended to make the probability of favorable cases predictable for gamblers. Laplace's definition of probability – also denoted *Laplace experiment* – is based on the two assumptions that the outcomes ω are *equiprobable* and that the number of all possible outcomes n_Ω is finite, i.e. the sample space $\Omega = \{\omega_1, \omega_2, ...\omega_{n_\Omega}\}$ is finite. The probability of one elementary event ω_i occurring is then

$$p_{\mathrm{i}} = \frac{1}{\mathrm{n}_\Omega} \tag{3.9.9}$$

In the case of rolling dice, for example, it is $p_{\mathrm{i}} = \frac{1}{6}$. The question now is how to determine the probability $P(A)$ of an event A with $A \subseteq \Omega$. To do this, one undertakes the thought experiment of counting the number n_A of all "favorable events" or of the possible realisations of the event A and relates this to the number n_Ω of all possible events; this corresponds to the sum of all elementary probablities p_{i} of A according to eq. (3.9.9)

$$P(A) = \frac{|A|}{|\Omega|} = \frac{\text{number } \mathrm{n}_A \text{ of outcomes in } A}{\text{number } \mathrm{n}_\Omega \text{ of all outcomes in } \Omega} = \frac{\mathrm{n}_A}{\mathrm{n}_\Omega} = \sum_{\omega_{\mathrm{i}} \in A} p_{\mathrm{i}} \tag{3.9.10}$$

Example 1: In the case of a perfect die, the probability of rolling an "even number of spots" is to be calculated. For the event $A = \{2, 4, 6\}$ and the set of all possible outcomes $\Omega = \{1, 2, 3, 4, 5, 6\}$, we obtain according to eq. (3.9.10)

$$P(A) = \sum_{\omega_{\mathrm{i}} \in A} p_{\mathrm{i}} = 3 \cdot \frac{1}{6} = \frac{1}{2} \tag{3.9.11}$$

Example 2: Consider the random experiment of tossing a coin three times. How great is the probability that "tails" will come up exactly twice? For three tosses, there are exactly $\mathrm{n}_\Omega = 2^3 = 8$ possible outcomes. If we denote the realisation "tails" by T and "heads" by H, then the following eight outcomes ω are possible: $\Omega = \{TTT, TTH, THH, HHH, HHT, HTT, THT, HTH\}$. The coin is assumed to be perfect, so each outcome is equally probable

$$P(\omega) = \frac{1}{\mathrm{n}_\Omega} = \frac{1}{2^3} \tag{3.9.12}$$

The event "exactly two T" occurs three times in Ω: $A = \{TTH, HTT, THT\}$. According to eq. (3.9.10), the probability of A appearing is $P(A) = \frac{\mathrm{n}_A}{\mathrm{n}_\Omega} = \frac{3}{8}$.

We generalise the example of tossing a coin to n throws and ask about the probability of T occurring x times. For n tosses, the number of possible outcomes is $\mathrm{n}_\Omega = 2^{\mathrm{n}}$. The number n_A of elements in subset A corresponds to the combination of x elements out of n elements without repetitions: $\mathrm{n}_A = \binom{\mathrm{n}}{\mathrm{x}} = \frac{\mathrm{n}!}{\mathrm{x}!(\mathrm{n}-\mathrm{x})!}$ (binomial coefficient). For the Laplacian probability, we obtain

$$P(A) = \frac{\mathrm{n}_A}{\mathrm{n}_\Omega} = \binom{\mathrm{n}}{\mathrm{x}} \frac{1}{2^{\mathrm{n}}} \tag{3.9.13}$$

Strictly speaking, the Laplacian concept is inappropriate for a definition of probability because it is assumed that the outcomes are *equiprobable*. However, since this notion is itself part of the definition, such a definition is circular.

The probability rule according to Laplace provides a purely theoretical (and not experimental) definition of probability. Despite the shortcomings with regard to circularity mentioned above, the Laplacian calculation rule is nevertheless justified in cases where, due to symmetry conditions (e.g. rolling a fair die), equiprobability is satisfied.

- **Frequentist definition (R. von Mises)**

In 1919, von Mises (1883–1953) presented a completely different approach to the definition of probability (von Mises, 1919). In his view, probability is a measurable quantity that can be determined empirically. If a random experiment is carried out n times and if n_A is the number of times or the frequency the event A occurs in n random experiments, then the quotient

$$h_A(\mathrm{n}) = \frac{\mathrm{n}_A}{\mathrm{n}} \tag{3.9.14}$$

is denoted the *relative frequency* of A. Because of $0 \leqslant \mathrm{n}_A \leqslant \mathrm{n}$, we have

$$0 \leqslant h_A(\mathrm{n}) \leqslant 1 \tag{3.9.15}$$

The frequentist notion of probability is now introduced as the limit

$$P(A) = \lim_{\mathrm{n}\to\infty} \frac{\mathrm{n}_A}{\mathrm{n}} \tag{3.9.16}$$

where the number n of experiments carried out tends $\to \infty$.

As early as 1713, Jakob Bernoulli formulated the *"law of large numbers"* in his work *"Ars conjectandi"*, published eight years after his death, and stated that the relative frequency of a random event gets closer to the probability value, the more frequently an experiment is performed. The law of large numbers is applied in various disciplines: in engineering, for example, to determine service life, fatigue and reliability, in the insurance business to estimate possible damages and thus to set premiums, in economics, in game theory, for instance, as well as in sociological studies and other assessments of large populations.

The advantage of the frequentist definition of probability is the practicality of the formulation. It avoids Laplace's circular definition and takes the actual number of experiments performed into account. However, this definition is based on the hypothesis that the limit really exists. The limit is "almost certain" – but only almost. Such a formulation of relative frequencies is certainly not suitable as a basis for a mathematical theory formulating the central limit theorem, to be presented in section 3.9.10.

- **Axiomatic definition** (A.N. Kolmogorov)

We have discussed the advantages and disadvantages of the probability definitions of Laplace and von Mises. The Russian mathematician A. N. Kolmogorov (1903–1987) took a third path, striving to include the basic notions of probability theory – hitherto "strange" for mathematicians – in the conceptional world of modern mathematics (Kolmogorov, 1933). He formulated the basic concepts axiomatically on the basis of set and measure theory. In his opinion, it would have been "pretty hopeless" to cope with this task without the previous development of the Lebesgue measure and the integration theory. The task he set himself was to formulate the theory of probability as a mathematical discipline on the basis of axioms in the same way as geometry or algebra. The following three Kolmogorov axioms should be understood in this context. They form the basis for deductive conclusions and provable theorems of a mathematical theory of probability, which implies a generalisation of Laplace's definition of probability.

Axioms:

Each event A, i.e. a certain subset of Ω, is assigned a probability measure P. P is a function defined on Ω, which satisfies the following three axioms:

(I) $P(A) \geqslant 0$

(II) $P(\Omega) = 1$ (certain event, normalisation)

(III) for a finite or a countably infinite number of pairwise disjoint sets A_{i}, the probability is given as

$$P(\bigcup_{\mathrm{i}}^{\infty} A_{\mathrm{i}}) = \sum_{\mathrm{i}}^{\infty} P(A_{\mathrm{i}}) \tag{3.9.17}$$

(Pairwise disjoint means incompatible: if one event occurs, the other one cannot occur.)

Firstly, the Kolmogorov axioms have to include the two previous definitions; secondly, they should extend Laplace's restriction of countably infinitely many events to uncountably many, and thirdly, they have to ensure that the unprovable and therefore hypothetical frequentist definition of probability according to von Mises is still valid as an empirical limit.

In agreement with the definitions of Laplace and von Mises, solely non-negative real numbers are assigned to the probabilities by means of axiom (I).

Axiom (III), also denoted the *axiom of countable additivity*, extends Laplace's definition of a finite number n_Ω to a countably infinite number of elementary probabilities $P(A_{\mathrm{i}})$. In addition, the number n_Ω of all possible elementary probabilities $P(A_{\mathrm{i}})$ need not necessarily be known in order to determine of the probability of events according to axiom (III); this is illustrated by the example presented below. What is important is that the elementary probabilities are known. If the elementary probabilities $P(A_{\mathrm{i}})$ cannot be determined either by combinatorics or by deduction from the certain event ($\sum_{\mathrm{i}} P(A_{\mathrm{i}}) = 1$) in accordance with axiom (II), the experimental concept of von Mises, see eq. (3.9.16), is still an option.

Example 1 (countably infinitely many events): Let us suppose that two people P_1 and P_2 alternately throw a perfect coin and P_1 starts. Question: what is the probability of P_1 tossing "tails" (T) first? What can happen? Either P_1 tosses T with the first throw or – if P_2 does not toss T with the second throw – with the third or with the fifth (if all four previous throws were "heads") and so on. For the probability of each event, $\frac{1}{2}, \frac{1}{2^3}, \frac{1}{2^5}, \ldots$ applies. Following axiom (III), the probability of P_1 throwing T first is the sum of the infinite geometric series: $\frac{1}{2} + \frac{1}{2^3} + \frac{1}{2^5} + \ldots = \frac{2}{3}$.

Example 2 (Problem of de Méré): We are now able to solve the problem of de Méré, presented at the beginning of section 3.9. In the first case, four fair dice are thrown and the players bet that the "six" (event A) will not appear. Let us suppose that only one die is thrown and no "six" appears: the probability of this elementary event is obviously $P(\omega) = (1 - \frac{1}{6})$ (certain event minus probability that a "six" appears). When four dice are thrown, the probability is then $P(A) = (1 - \frac{1}{6})^4 = (\frac{5}{6})^4 = \frac{625}{1296} \approx 0.482 < \frac{1}{2}$. In the second case, two (fair) dice are thrown 24 times and the bet is that no "double six" will occur. The probability of throwing a "double six" with two dice on one roll is $P(\omega) = \frac{1}{36}$. Not throwing a "double six" means: $P(\omega) = (1 - \frac{1}{36})$. In 24 throws with two dice, the probability is therefore $P(A) = (1 - \frac{1}{36})^{24} = (\frac{35}{36})^{24} \approx 0.508 > \frac{1}{2}$. This result – that the chances of winning are higher in the second case – confirms de Méré's practical experience and disproves his theoretical calculations that there are equal chances of winning in the two cases.

So far, the number of possible events was either finite or countably infinite, i.e. the probability could be determined by counting. Now we want to determine the probability for an uncountably infinite number of possible events, a task which cannot be performed by simple counting. A new approach is needed; in this case, it is provided by measure theory (Kolmogorov, 1933). Classic examples of measures are the area in a plane or the volume in a three-dimensional space (cf. integral calculus due to Leibniz and Newton). Measure theory generalises elementary geometric terms and definitions such as the length of an interval (distance), area or volume, thus allowing us to assign a measure even to more complicated sets like fractals. According to Komogorov's axioms, the probabilities are normalised measures (Bronstein et al., 1995). Measure theory thus forms the basis for modern probability theory. We illustrate the transition from discrete to continuous, from countable to uncountable or the extension of the Laplacian definition with the aid of Kolmogorov's axioms with an example.

Example 3 (uncountably many events): The task is to hit a target of radius R. Each point of the target is assigned its distance r from the centre as a random variable $X(\omega)$. The set of elementary events is $\Omega = \{\omega \mid r(\omega) \leqslant R\}$. We assume that the probability of hitting each point is equal. We further assume that the target is divided into four parts, to which we assign the following four events: $A_1 = \{\omega \mid \omega \leqslant \frac{R}{4}\}$ and $A_\mathrm{k} = \{\omega \mid (\mathrm{k}-1)\frac{R}{4} < r \leqslant \mathrm{k}\frac{R}{4}\}$ for $(\mathrm{k} = 2, 3, 4)$. Since the subsets do not overlap and $\cup_{\mathrm{k}=1}^{4} A_\mathrm{k} = \Omega$ holds, the probability $p_\mathrm{k} = P(A_\mathrm{k})$ of hitting the ring A_k can be expressed as the ratio of the areas $F(A_\mathrm{k}) \ / \ F(\Omega)$. The vector of the elementary probabilities is then

$$[p_1, p_2, p_3, p_4] = [\tfrac{1}{16}, \tfrac{3}{16}, \tfrac{5}{16}, \tfrac{7}{16}]$$

where the sum $\sum_1^4 p_\mathrm{i} = 1$ means the certain event.

Further probability measures and *calculation rules* can be derived from Kolmogorov's axioms (I)–(III).

- *Complementary probability* (cf. fig. 3.9.1d): The event A and the complementary event A^c are disjoint sets. For the certain event, we then have: $1 = P(\Omega) = P(A \text{ or } A^c) = P(A \cup A^c) = P(A) + P(A^c)$ or

$$P(A^c) = 1 - P(A) \tag{3.9.18}$$

A special case of complementary probability is that the impossible event $\emptyset$, the empty set, as the complement of the certain event, has the probability zero

$$P(\emptyset) = 0 \tag{3.9.19}$$

- *Addition theorem for arbitrary events* (formula of Sylvester): The summation rule is valid for disjoint events; the addition theorem is an extension to arbitrary, not necessarily disjoint events (fig. 3.9.1).

$$P(A \cup B) = P(A) + P(B) - P(A \cap B) \tag{3.9.20}$$

So far, we assumed that a probability can be assigned to all events. However, there may be cases where it is not possible to assign a probability unambiguously to all subsets Ω. In order to combine the possible subsets, the concept of σ-field (also denoted σ-algebra or Borel algebra) is introduced. The concept of σ-algebra far exceeds the introductory character of this section; we refer the interested reader to literature such as (Elstrodt, 2005; Bronstein et al., 1995).

3.9.4 Conditional Probability and Joint Probability

Conditional probability is of particular importance because, in practice, experiments are carried out which have to fulfil certain conditions. In a game of dice, for example, the question could be: what is the probability of rolling $A = 6$ under the condition that the outcome is an "even number".

The conditional probability $P(A|B)$ is the probability that an event A will occur "under the condition" that B has happened or that the outcomes A are conditioned by B. The conditional probability is defined as follows: given two events A and B with $P(B) > 0$, then

$$P(A \mid B) = \frac{P(A \cap B)}{P(B)} = \frac{P(AB)}{P(B)} \tag{3.9.21}$$

Here, $P(A \cap B)$ is the *joint probability* expressing the joint occurrence of the events A and B. For the second part of the definition, $A \cap B = AB$ is used according to eq. (3.9.4).

In the previous example of rolling fair dice, $B = \{2, 4, 6\}$ is the set of "even number" events and the required outcome is $A = \{6\} \subset B$. Therefore, we have $P(A \cap B) = P(A) = 1/6$ and $P(B) = 1/2$. According to eq. (3.9.21), the conditional probability is then $P(A \mid B) = \frac{P(A \cap B)}{P(B)} = 1/3$.

Rearranging eq. (3.9.21), we obtain the *multiplication theorem*

$$P(AB) = P(A \cap B) = P(B \cap A) = P(B \mid A)P(A) = P(A \mid B)P(B) \quad (3.9.22)$$

from which *Bayes' theorem* immediately follows (Thomas Bayes (1763), English mathematician). Given two events A and B with $P(A) > 0$ and $P(B) > 0$, then the following relation between the conditional probabilities $P(A \mid B)$ and $P(B \mid A)$ holds

$$P(A \mid B) = \frac{P(B \mid A)P(A)}{P(B)} \quad (3.9.23)$$

The Bayesian relationship is in some sense an inversion of inferences. It is useful if one conditional probability, $P(B \mid A)$, can be easily determined while the "inverted' one, $P(A \mid B)$, is being looked for.

Joint probability refers to at least two random variables. In the case of a javelin throw, this could be the length of throw, the initial throwing speed and the launch angle. If several random variables appear at the same time, i.e. in combination, the probability that both a given throw length and a prescribed launch angle occur could be of interest. In the following, we restrict our considerations to two random variables (X, Y) and demonstrate the formalism by means of two distinguishable coins which we designate red and black. In this case, four events $\{(H^r, H^b), (H^r, T^b), (T^r, H^b), (T^r, T^b)\}$ are possible, to which the following pairs of values (x^r, y^b) can be assigned: $(0^r, 0^b)$, $(0^r, 1^b)$, $(1^r, 0^b)$, $(1^r, 1^b)$. For a pair of values $(x_{\mathrm{i}}^r, y_{\mathrm{j}}^b)$, the joint probability p_{ij} is given as

$$P(X = x_{\mathrm{i}}^r, Y = y_{\mathrm{j}}^b) = p_{\mathrm{ij}}$$

For equiprobable events, we have $p_{\mathrm{ij}} = 1/4$. As an example, we are now interested in the probability that "heads" appears in the first throw of the red coin, i.e. $P(X = x_{\mathrm{i}}^r = 0^r)$. According to the axioms (I)–(III), see eq. (3.9.17), the probability can be determined as a joint probability

$$P(X = x_{\mathrm{i}}^r) = \sum_{\mathrm{i}=1}^{2} P(X = x_{\mathrm{i}}^r, Y = y_{\mathrm{j}}^s) = \frac{1}{4} + \frac{1}{4} = \frac{1}{2} \quad (3.9.24)$$

Therefore, according to eq. (3.9.17), joint probabilities of disjoint events can be reduced to elementary probabilities by summation (or, as we shall see later, by integration).

Let us add a further example to illustrate Bayes' theorem. Consider two identical containers A and B, each of which contains ten balls. Assume that A contains seven red and three white balls and B one red and nine white balls and that one ball is taken randomly from one of the two containers. We further assume that the first ball selected is red. The task is now to find the probability that this ball is taken from A. The following events are defined by the task: event A: "the ball is taken from A", event B: "the ball is taken from B", and event R: "the ball is red". We are then looking for the conditional probability $P(A \mid R)$. The following probabilities are known: $P(A) = P(B) = 1/2$ (both containers have the same probability), $P(R \mid A) = 7/10$ (there are ten balls in A, 7 of them are red), $P(R \mid B) = 1/10$ (there are ten balls in B, one of them is red). With eq. (3.9.21), the probability of picking a red ball is $P(R) = P(R \mid A) \cdot P(A) + P(R \mid B) \cdot P(B) = \frac{7}{10} \cdot \frac{1}{2} + \frac{1}{10} \cdot \frac{1}{2} = \frac{2}{5}$ and the conditional probability $P(A \mid R)$ is hence given as

$$P(A \mid R) = \frac{P(R \mid A)P(A)}{P(R)} = \frac{\frac{7}{10} \cdot \frac{1}{2}}{\frac{2}{5}} = \frac{7}{8} = 87.5\%$$

i.e. the red ball is picked from container A with a 87.5% (conditional) probability.

3.9.5 Distribution Function and Probability Density Function

• Distribution function

In section 3.9.2, we introduced the concept of random variables which assigns a real number to the outcome ω: $\omega \to X(\omega)$. In this section, we specify an interval as a random variable for the throwing distance and ask how probable it is that the throwing distance will lie in this interval (see fig. 3.9.2). Generally, the problem can be specified as follows: let us assume that the random variable X lies in an interval with the constant limits a and b, where $a < b$, and we are looking for sample points ω for which the random variable $X(\omega)$ lies in the interval $[a, b]$. In short, we are looking for sample points ω which fulfil

$$\{a \leqslant X \leqslant b\} = \{\omega \,|\, a \leqslant X(\omega) \leqslant b\} \tag{3.9.25}$$

According to axiom (III), any subset of the sample space Ω as defined in eq. (3.9.25) can be assigned a probability

$$P(a \leqslant X(\omega) \leqslant b) \tag{3.9.26}$$

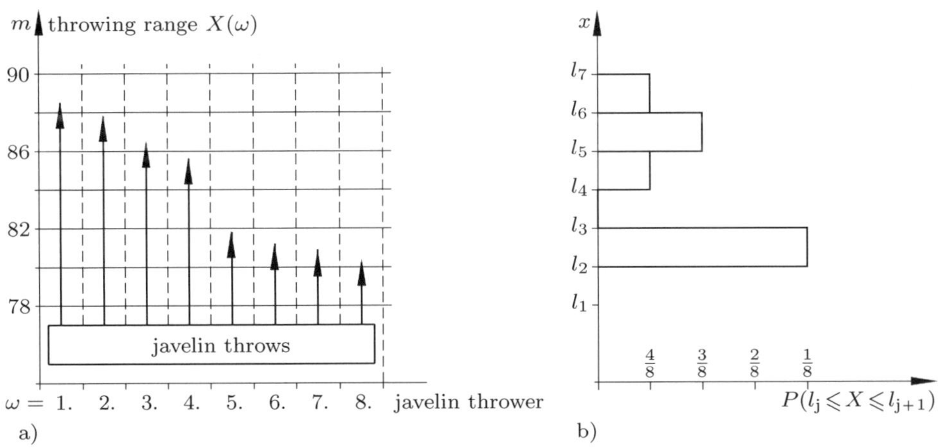

Fig. 3.9.2: a) Distances of javelin throws, b) probability of hitting the interval $[l_j, l_{j+1}]$

Figure 3.9.2b shows the probabilities of a javelin thrower hitting the respective intervals. According to figs. 3.9.3a,b, the probability of hitting the interval [80 m, 82 m] is $P(80 \leqslant X(\omega) \leqslant 82) = P(X = m_2) = p_1 = \frac{4}{8}$, where $m_2 = 81$ m is the centre of the interval. The probability of hitting the larger range [80 m, 88 m] can be determined by means of axiom (III) as the sum of the elementary probabilities p_n

$$P(a \leqslant X(\omega) \leqslant b) = \sum_{a \leqslant m_\mathrm{n} \leqslant b} p_\mathrm{n} \tag{3.9.27}$$

In general, the probability for the subset A is

$$P(X \in A) = \sum_{m_\mathrm{n} \in A} p_\mathrm{n} \tag{3.9.28}$$

In the case of the javelin throw, the throwing distance $X(\omega)$ is a real number of the interval $[-\infty, x]$, i.e. for every real number x, a probability $P(X \leqslant x)$ is defined. In fact, the probability is a function of x and therefore the associated random variable X is defined as the so-called *distribution function* $F_X(x)$ or simply $F(x)$ as follows (cf. fig. 3.9.3c)

$$F(x) = P(X \leqslant x) = \sum_{m_\mathrm{n} \leqslant x} p_\mathrm{n} \tag{3.9.29}$$

According to axioms (I) and (II), the elementary probabilities fulfil $p_\mathrm{n} \geqslant 0$ and $\sum_\mathrm{n} p_\mathrm{n} = 1$.

- **Probability density function**

Random variables can be discrete, as in games such as tossing coins or rolling dice. On the other hand, there are many practical applications where the random variables are continuous, for example, the concentration of pollutants in rivers. In this case, the sample space Ω is a subrange of the space of real numbers. Instead of discrete outcomes like the elementary events $\{\omega_\mathrm{i}\}$, we now consider subintervals $[a, b] \in \mathbb{R}$ together with their unions and intersections. The transition from discrete to continuous is achieved by replacing the sum in the general formula, eq. (3.9.27), by an integral. To this end, we introduce a function $f(x)$ defined on the real axis and with the following properties:

$$f(x) \geqslant 0 \qquad \text{and} \qquad \int_{-\infty}^{+\infty} f(x) = 1 \qquad \text{(normalisation)} \tag{3.9.30}$$

If we describe the probability of the event $\{x \leqslant x_\mathrm{i}\}$ as

$$P(\{x \leqslant x_i\}) = \int_{-\infty}^{x_\mathrm{i}} f(x)\mathrm{d}x \tag{3.9.31}$$

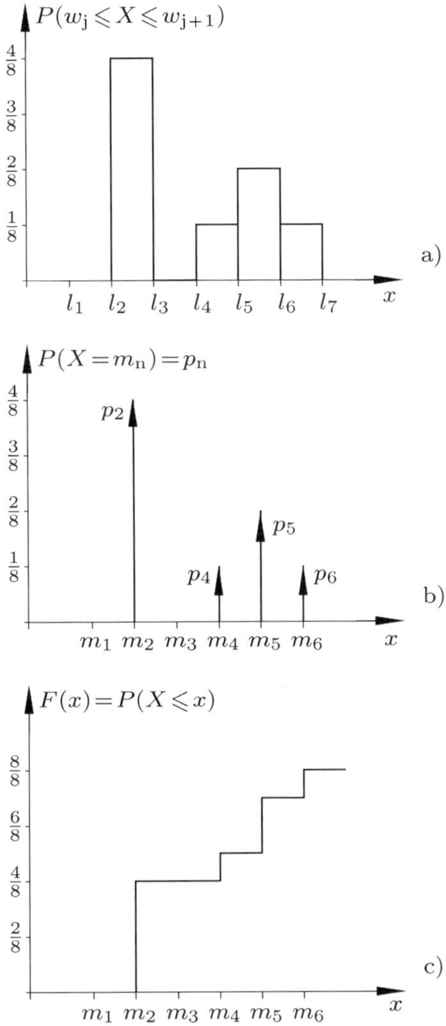

Fig. 3.9.3

a) Histogram of fig. 3.9.2,
b) probability measure p_n of a),
c) distribution function $F(x)$ of b)

the probability is defined for all events. The function $f(x)$ is referred to as *density function*, or as *probability density function*, often abbreviated to "pdf".

In analogy to eq. (3.9.27), the distribution function $F(x)$ can be written for the interval $[a, b]$ (see fig. 3.9.4a) as

$$F(x) = P(\{a \leqslant x \leqslant b\}) = \int_a^b f(x)\mathrm{d}x \tag{3.9.32}$$

For sufficiently small $\mathrm{d}x$, $p(x)\mathrm{d}x$ is the probability that an event assumes a value $\bar{x} \in [x, x + \mathrm{d}x]$.

If the probability density function $f(x)$ is continuous, we find for the derivative of the distribution function

$$f(x) = \frac{\mathrm{d}F(x)}{\mathrm{d}x} \tag{3.9.33}$$

The Dirac δ function, or δ-function, is a generalised function, a distribution, which is defined by its integral properties, see eq. (3.8.37). By definition, the shifted function $\delta(x - x_0)$ has the following two properties

$$\delta(x - x_0) = 0 \qquad \text{for} \qquad x \neq x_0 \tag{3.9.34}$$

and

$$\int\limits_{x_0-\varepsilon}^{x_0+\varepsilon} \delta(x - x_0)\mathrm{d}x = 1 \qquad \text{for} \qquad \varepsilon > 0 \tag{3.9.35}$$

Previously, we only roughly described the transition from a discrete distribution function $F(x)$, eq. (3.9.29), to a continuous one, eq. (3.9.31). We now use the function $\delta(x - x_0)$ to formulate this transition in mathematical terms. To this end, we write the probability density function $f(x)$ in the following form

$$f(x) = \sum_{\mathrm{i}=1}^{\mathrm{n}} p_{\mathrm{i}}\, \delta(x - x_{\mathrm{i}}) \qquad \text{with} \qquad \sum_{\mathrm{i}=1}^{\mathrm{n}} p_{\mathrm{i}} = 1 \tag{3.9.36}$$

which means that $f(x)\mathrm{d}x$ has the value p_{i} at $x = x_{\mathrm{i}}$ and is zero everywhere else. For instance, following eq. (3.9.36), the probability density function $f(x)$ of p_{n} shown in fig. 3.9.3b is

$$f(x) = \frac{4}{8}\delta(x - m_2) + \frac{1}{8}\delta(x - m_4) + \frac{2}{8}\delta(x - m_5) + \frac{1}{8}\delta(x - m_6) \tag{3.9.37}$$

According to eq. (3.9.33), $F(x)$ is given as

$$F(x) = \int\limits_{-\infty}^{x} f(x')\, \mathrm{d}x' \tag{3.9.38}$$

With eq. (3.9.36), it follows

$$F(x) = \int\limits_{-\infty}^{x} \sum_{\mathrm{i}=1}^{\mathrm{n}} p_{\mathrm{i}}\, \delta(x' - x_{\mathrm{i}})\, \mathrm{d}x' \tag{3.9.39}$$

or, using the property of eq. (3.9.35), we obtain

$$F(x) = \sum_{\mathrm{i}=1}^{\mathrm{n}} p_{\mathrm{i}} \underbrace{\int\limits_{-\infty}^{x} \delta(x' - x_{\mathrm{i}})\, \mathrm{d}x'}_{=1\ \forall x_{\mathrm{i}} \leqslant x} \tag{3.9.40}$$

A comparison of eq. (3.9.40) and eq. (3.9.38) makes it clear that the sum is replaced by an integral for the transition from the discrete to the continuous $F(x)$.

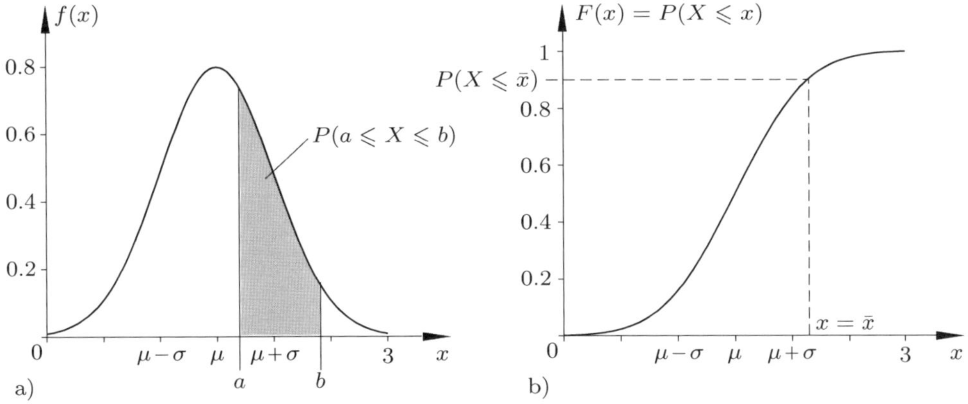

Fig. 3.9.4: a) Probability density function $f(x)$ and
b) distribution function $F(x)$ (normal distribution with $\mu = 1.5$, $\sigma = 0.5$)

3.9.6 Characteristic Values of a Distribution

A complete overview over all features of a random variable $X(\omega)$ is obtained from the probability density function $f(x)$, also denoted *distribution* for short. It can be described by a number of characteristic parameters or measures which can be deduced from the probability density function. The most important ones are the mean μ and the variance σ^2 or the standard deviation σ. Further characteristics are the k-th and the central moments, skewness and kurtosis, which will be presented here. In the following, we assume that all integrals and sums exist and are finite. Exceptions are discussed in section 3.9.11.

- **Mean** μ, **variance** σ^2 and **standard deviation** σ

The *mean* of a distribution, or the mean of the corresponding random variable X, is commonly denoted μ. The definition is best illustrated by an example. In the game of dice, the random variable $X(\omega) = x_{\mathrm{n}}$ is defined on the discrete sample space $\Omega = \{1, 2, 3, 4, 5, 6\}$. The mean is simply the sum of x_{n} divided by the total number of elements in Ω: $\mu = \frac{1}{6}(1 + 2 + 3 + 4 + 5 + 6) = 3.5$. With the elementary probabilities p_{n} and eq. (3.9.17), the mean μ of a discrete random variable $X = x_{\mathrm{n}}$ is defined as

$$\mu = \sum_{\mathrm{n}} x_{\mathrm{n}}\, p_{\mathrm{n}} = \sum_{\mathrm{n}} x_{\mathrm{n}} f(x_{\mathrm{n}}) \qquad (3.9.41)$$

With a game of dice, for example, the mean can be used to calculate the average sum of spots resulting from 1000 throws as $1000 \cdot 3.5 = 3500$.

If the random variable is continuous and has the probability density $f(x)$, we obtain for μ

$$\mu = \int_{-\infty}^{+\infty} x\, f(x)\, \mathrm{d}x \qquad (3.9.42)$$

In analogy to mechanics, the mean corresponds to the centre of gravity of a distribution of mass.

The *variance* σ^2 is the mean square deviation from the mean μ. Using eq. (3.9.41), we have

$$\sigma^2 = \sum_{\mathrm{n}} (x_{\mathrm{n}} - \mu)^2 \, f(x_{\mathrm{n}}) \tag{3.9.43}$$

For a continuous random variable X, the variance σ^2 is defined as

$$\sigma^2 = \int_{-\infty}^{+\infty} (x - \mu)^2 \, f(x) \, \mathrm{d}x \tag{3.9.44}$$

If the value of σ^2 is small, the distribution is highly concentrated around the mean μ. The *standard deviation* σ is defined as the positive square root of the variance; it is a measure of the "statistical dispersion" or "spread".

If the probability density function $f(x)$ of the measured data is unknown, the following approximations can be used

$$\bar{\mu} = \frac{1}{\mathrm{N}} \sum_{\mathrm{n}=1}^{\mathrm{N}} x_{\mathrm{n}} \tag{3.9.45}$$

$$\bar{\sigma}^2 = \frac{1}{\mathrm{N}-1} \Big[\sum_{\mathrm{n}=1}^{\mathrm{N}} x_{\mathrm{n}}^2 - \mathrm{N} \, \bar{\mu}^2 \Big] \tag{3.9.46}$$

where N is the number of measurements.

- **Mathematical expectation, k-th moments, skewness and kurtosis**

In eq. (3.9.41), we introduced the mean μ for the discrete random variable $X = x_{\mathrm{n}}$. Analogously, for a function $g(X)$ of the random variable X, the mean or the (mathematical) expectation $\langle g(X) \rangle$ can be specified as follows

$$\langle g(X) \rangle = \sum_{\mathrm{n}} g(x_{\mathrm{n}}) \, p_{\mathrm{n}} = \sum_{\mathrm{n}} g(x_{\mathrm{n}}) f(x_{\mathrm{n}}) \tag{3.9.47}$$

Correspondingly, for continuous random variables, we have according to eq. (3.9.42)

$$\langle g(X) \rangle = \int_{-\infty}^{+\infty} g(x) \, f(x) \, \mathrm{d}x \tag{3.9.48}$$

where $f(x)$ is the probability density function associated with X.

Mean $\mu = \langle X \rangle$ and variance σ^2 belong to a group of parameters referred to as *moments*, i.e. μ and σ^2 are obtained from the first and second moment of $f(x)$. Moments of order k, written as $m_{\mathrm{k}} = \langle X^{\mathrm{k}} \rangle$, provide a further characterisation of $f(x)$. Thus, for discrete samples $X = x_{\mathrm{n}}$, we have

$$m_{\mathrm{k}} = \langle X^{\mathrm{k}} \rangle = \sum_{\mathrm{n}} x_{\mathrm{n}}^{\mathrm{k}} f(x_{\mathrm{n}}) \tag{3.9.49}$$

If the random variable X is continuous with the probability density $f(x)$, the k-th moment, according to eq. (3.9.48), is given as

$$m_{\mathrm{k}} = \langle X^{\mathrm{k}} \rangle = \int_{-\infty}^{+\infty} x^{\mathrm{k}} \, f(x) \, \mathrm{d}x \tag{3.9.50}$$

Furthermore, we define moments with respect to the mean $\langle X \rangle = m_1 = \mu$, called *central moments of order* k and denoted ζ_k

$$\zeta_{\mathrm{k}} = \langle (X - m_1)^{\mathrm{k}} \rangle \tag{3.9.51}$$

Since the central moments play an important role, we quote the first three of them, expressed in terms of the ordinary moments

$$\begin{aligned} \zeta_1 &= \langle X - m_1 \rangle = 0 \\ \zeta_2 &= \langle (X - m_1)^2 \rangle = \langle X^2 \rangle - 2m_1 \langle X \rangle + m_1^2 = m_2 - m_1^2 \\ \zeta_3 &= \langle (X - m_1)^3 \rangle = \langle X^3 \rangle - 3m_1 \langle X^2 \rangle + 3m_1^2 \langle X \rangle - m_1^3 \\ &= m_3 - 3m_1 m_2 + 3m_1^3 - m_1^3 = m_3 - 3m_1 m_2 + 2m_1^3 \end{aligned} \tag{3.9.52}$$

The central moment of second order, ζ_2, of eq. (3.9.52) corresponds to the definition of the variance given in eq. (3.9.43), i.e. the variance can be calculated by means of moments

$$\sigma^2 = m_2 - m_1^2 = m_2 - \mu^2 \tag{3.9.53}$$

A further characteristic of a distribution is the *skewness* $S(X)$. It characterises the asymmetry of the distribution and is obtained from the third central moment. $S(X)$ is zero if the distribution is symmetrical. The skewness of the random variable X with mean μ and variance σ^2 is defined as

$$S(X) = \frac{1}{\sigma^3} \zeta_3 \tag{3.9.54}$$

The last parameter we introduce here is the *kurtosis* $K(X)$. The kurtosis is a measure for the curvature respectively peakedness, i.e it measures how shallow or peaked the distribution is in the vicinity of the centre μ and is defined as follows

$$K(X) = \frac{1}{\sigma^4} \zeta_4 \tag{3.9.55}$$

A comparison with eq. (3.9.51) shows that the curvature correponds to a central moment of fourth order. For a normally distributed random variable, the kurtosis has the value $K(X) = 3$.

If all the moments $\langle X^{\mathrm{k}} \rangle, \mathrm{k} = 1, 2, \ldots$ exist and are known, then they allow a complete description of the probability density function $f(x)$. This is illustrated in section 3.9.8.

So far, we have specified the expectation values $\langle g(X)\rangle$ for functions of one random variable. Extended to functions of two random variables $\langle g(X)\rangle$, the expectation value $\langle Z\rangle = \langle g(X,Y)\rangle$ can be determined according to the procedure given in sections 3.9.4, 3.9.5 directly from the joint probability $f(x,y)$ of X and Y, see eq. (3.9.48)

$$\langle g(X,Y)\rangle = \int_{-\infty}^{+\infty}\int_{-\infty}^{+\infty} g(x,y)\, f(x,y)\, \mathrm{d}x\, \mathrm{d}y \tag{3.9.56}$$

For the discrete case, we have

$$\langle g(X,Y)\rangle = \sum_{\mathrm{i,j}} g(x_{\mathrm{i}}, y_{\mathrm{j}})\, p_{\mathrm{ij}} \tag{3.9.57}$$

The joint moments m_{ij} of two random variables X and Y are defined as

$$m_{\mathrm{ij}} = \langle X^{\mathrm{i}} Y^{\mathrm{j}}\rangle \tag{3.9.58}$$

and, in analogy to eq. (3.9.50), the expectation value is given as

$$\langle X^{\mathrm{i}} Y^{\mathrm{j}}\rangle = \int_{-\infty}^{+\infty}\int_{-\infty}^{+\infty} x^{\mathrm{i}}\, y^{\mathrm{j}}\, f(x,y)\, \mathrm{d}x\, \mathrm{d}y \tag{3.9.59}$$

3.9.7 Independent and Dependent Events

A central concept of the theory of probability is the independence of two events. Intuitively, we designate the two events A and B as independent if the occurrence of A has no influence on the occurrence of B. In such a case, the conditional probability $P(A \mid B)$ is independent of B, i.e. $P(A \mid B) = \frac{P(A\cap B)}{P(B)} = P(A)$. The multiplication theorem, eq. (3.9.22), leads to the definition that two events are independent if and only if the following holds

$$P(A\cap B) = P(A)\cdot P(B) \tag{3.9.60}$$

The rule can be generalised, i.e. the random variables $X_1, X_2, ..., X_{\mathrm{n}}$ are independent if and only if the product theorem is valid for arbitrary real numbers $x_1, x_2, ..., x_{\mathrm{n}}$

$$P(X_1 = x_1, ..., X_{\mathrm{n}} = x_{\mathrm{n}}) = P(X_1 = x_1)P(X_2 = x_2)\, ...\, P(X_{\mathrm{n}} = x_{\mathrm{n}}) \tag{3.9.61}$$

Following eq. (3.9.61), we have for the distribution function

$$F_{xy}(x,y) = F_x(x)F_y(y) \tag{3.9.62}$$

and X, Y are *independent* if and only if

$$f_{xy}(x,y) = f_x(x)f_y(y) \tag{3.9.63}$$

The variables X, Y are called *uncorrelated* if factorisation is possible

$$\langle XY \rangle = \langle X \rangle \langle Y \rangle \tag{3.9.64}$$

The variables are orthogonal if

$$\langle XY \rangle = 0 \tag{3.9.65}$$

The question now arises whether independent random variables are also uncorrelated. With eqs. (3.9.59) and (3.9.64), we obtain for the expectation value

$$\langle XY \rangle = \int_{-\infty}^{+\infty} \int_{-\infty}^{+\infty} x\, y\, f(x,y)\, \mathrm{d}x\, \mathrm{d}y = \int_{-\infty}^{+\infty} x\, f(x)\, \mathrm{d}x \int_{-\infty}^{+\infty} y\, f(y)\, \mathrm{d}y = \langle X \rangle \langle Y \rangle \tag{3.9.66}$$

i.e. uncorrelatedness follows from independence. In the opposite case where X and Y are not independent, but dependent respectively correlated, a measure for their dependence respectively correlation is required. The degree of dependence between two random variables is called *degree of correlation*. If, in contrast to eq. (3.9.64), we have $\langle X, Y \rangle - \langle X \rangle \langle Y \rangle \neq 0$, then the difference between $\langle X, Y \rangle$ and $\langle X \rangle \langle Y \rangle$ measures the correlation between X and Y. To this end, we determine from eq. (3.9.52) the central moment of second order for the two random variables X and Y

$$\begin{aligned} \Big\langle \big(X - \langle X \rangle\big)\big(Y - \langle Y \rangle\big) \Big\rangle &= \Big\langle XY - X\langle Y \rangle - Y\langle X \rangle + \langle X \rangle \langle Y \rangle \Big\rangle \\ &= \langle XY \rangle - \langle X \rangle \langle Y \rangle - \langle Y \rangle \langle X \rangle + \langle X \rangle \langle Y \rangle \\ &= \langle XY \rangle - \langle X \rangle \langle Y \rangle = \mathrm{Cov}(X,Y) \end{aligned} \tag{3.9.67}$$

and obtain as a result the *covariance* of X and Y; for identical random variables, we obviously have $\mathrm{Cov}(X,X) = \sigma^2(X)$.

The covariance respectively the central moment of second order for the two random variables X and Y is then determined from the joint probability $f(x,y)$ according to eqs. (3.9.52) and (3.9.59) as

$$\mathrm{Cov}(X,Y) = \int_{-\infty}^{+\infty} \int_{-\infty}^{+\infty} (x - \mu_x)\,(y - \mu_y)\, f(x,y)\, \mathrm{d}x\, \mathrm{d}y \tag{3.9.68}$$

Dividing eqs. (3.9.67) respectively (3.9.68) by the product $\sigma(X)\sigma(Y)$ of the standard deviations, we obtain the so-called *correlation coefficient*

$$\rho(X,Y) = \frac{\langle XY \rangle - \langle X \rangle \langle Y \rangle}{\sigma(X)\sigma(Y)} = \frac{\mathrm{Cov}(X,Y)}{\sigma(X)\sigma(Y)} \tag{3.9.69}$$

If X and Y are independent, then $\mathrm{Cov}(X,Y) = \rho(X,Y) = 0$. This means, as mentioned above, that independent random variables are always uncorrelated. However, the opposite is not true, see (Papoulis, 1965) for details.

3.9.8 Moment-Generating and Characteristic Functions

Section 3.9.5 shows how distribution functions and probability densities can be determined from experimental data. In section 3.9.6, we illustrate that the probability density functions can be characterised by characteristic quantities or moments. If all moments exist, the associated probability density function $f(x)$ can be fully described by the so-called *generating function*, an ingenious idea ascribed to the Swiss mathematician Leonard Euler (1707–1783).

Let us consider a random variable X which takes only non-negative integer values and has the following probability density function, see fig. 3.9.3b (probability measures p_n for discrete variables)

$$P(X = \mathrm{n}) = \mathrm{p_n} \qquad \text{with} \quad \mathrm{n} = 0,1,2,... \tag{3.9.70}$$

The idea is to describe eq. (3.9.70) with a single function. To this end, an auxilary variable z is introduced and the generating function is written as a power series

$$g(z) = p_0 + p_1 z + p_2 z^2 + \cdots = \sum_{\mathrm{n}=0}^{\infty} p_\mathrm{n} z^\mathrm{n} \tag{3.9.71}$$

The coefficients of this power series can be specified by means of the Taylor series

$$g(z) = g(0) + \left.\frac{g'}{1!}\right|_{z=0} z + \left.\frac{g''}{2!}\right|_{z=0} z^2 + \cdots \tag{3.9.72}$$

and we obtain the p_n values as

$$p_\mathrm{n} = \left.\frac{1}{\mathrm{n}!}\frac{\mathrm{d}^\mathrm{n} g}{\mathrm{d}z^\mathrm{n}}\right|_{z=0} \tag{3.9.73}$$

i.e. the coefficients p_n can be completely determined from the derivatives of the generating function $g(z)$. The derivatives

$$\begin{aligned} g'(z) &= p_1 + 2p_2 z + 3p_3 z^2 + \cdots = \sum_{\mathrm{n}=1}^{\infty} \mathrm{n} p_\mathrm{n} z^{\mathrm{n}-1} \\ g''(z) &= 2p_2 + 6p_3 z + \cdots = \sum_{\mathrm{n}=2}^{\infty} \mathrm{n}(\mathrm{n}-1) p_\mathrm{n} z^{\mathrm{n}-2} \\ g^{(\mathrm{j})}(z) &= \sum_{\mathrm{n}=\mathrm{j}}^{\infty} \mathrm{n}(\mathrm{n}-1)\cdots(\mathrm{n}-\mathrm{j}+1) p_\mathrm{n} z^{\mathrm{n}-\mathrm{j}} = \sum_{\mathrm{n}=\mathrm{j}}^{\infty} \binom{\mathrm{n}}{\mathrm{j}} \mathrm{j}!\, p_\mathrm{n} z^{\mathrm{n}-\mathrm{j}} \end{aligned} \tag{3.9.74}$$

are used to calculate the values $g'(z)$ and $g''(z)$ for $z = 1$ which leads to the following moments respectively expectation values according to eq. (3.9.49)

$$\begin{aligned} g'(1) &= \sum_{\mathrm{n}=0}^{\infty} \mathrm{n} p_\mathrm{n} = m_1 = \langle X \rangle \\ g''(1) &= \sum_{\mathrm{n}=0}^{\infty} \mathrm{n}^2 p_\mathrm{n} - \sum_{\mathrm{n}=0}^{\infty} \mathrm{n} p_\mathrm{n} = m_2 - m_1 = \langle X^2 \rangle - \langle X \rangle \end{aligned} \tag{3.9.75}$$

Another moment-generating function $g(z)$ of the associated distribution is the expectation value of e^{zX}. Following the two definitions in eqs. (3.9.47) and (3.9.48) for discrete and continuous random variables X, we can write

$$g(z) = \langle \mathrm{e}^{zX} \rangle = \sum_{\mathrm{n}} \mathrm{e}^{z\,x_\mathrm{n}} f(x_\mathrm{n}) \tag{3.9.76}$$

and

$$g(z) = \langle \mathrm{e}^{zX} \rangle = \int\limits_{-\infty}^{+\infty} \mathrm{e}^{z\,x}\, f(x)\,\mathrm{d}x \tag{3.9.77}$$

Differentiating the two eqs. (3.9.76) and (3.9.77) with respect to z yields

$$g'(z) = \sum_{\mathrm{n}} x_\mathrm{n} \mathrm{e}^{z\,x_\mathrm{n}} f(x_\mathrm{n}) \qquad \text{resp.} \qquad g'(z) = \int\limits_{-\infty}^{+\infty} x\mathrm{e}^{z\,x}\, f(x)\,\mathrm{d}x \tag{3.9.78}$$

and k-fold differentiation with respect to z yields

$$\frac{\mathrm{d}^\mathrm{k} g}{\mathrm{d}z^\mathrm{k}} = \sum_{\mathrm{n}} x_\mathrm{n}^\mathrm{k} \mathrm{e}^{z\,x_\mathrm{n}} f(x_\mathrm{n}) \qquad \text{resp.} \qquad \frac{\mathrm{d}^\mathrm{k} g}{\mathrm{d}z^\mathrm{k}} = \int\limits_{-\infty}^{+\infty} x^\mathrm{k}\mathrm{e}^{z\,x}\, f(x)\,\mathrm{d}x \tag{3.9.79}$$

If we substitute $z = 0$ into eq. (3.9.79), the right-hand sides show exactly the k-th moment

$$\langle X^\mathrm{k} \rangle = g^{(\mathrm{k})}(0) \qquad \text{with} \qquad g^{(\mathrm{k})}(0) = \left.\frac{\mathrm{d}^\mathrm{k} g}{\mathrm{d}z^\mathrm{k}}\right|_{z=0} \tag{3.9.80}$$

For $\mathrm{k} = 1$, we obtain the first moment respectively the mean μ

$$\langle X \rangle = g'(0) = \mu \tag{3.9.81}$$

Convergence problems that may occur in eq. (3.9.76) can be avoided by replacing e^{zX} by its complex function. Substituting $z = i\omega$, we obtain the expectation value $\langle \mathrm{e}^{i\omega X} \rangle$, denoted the *characteristic function* of the distribution.

Laplace was the first to use the characteristic function as a tool for treating problems in probability theory. The method was generalised by Lyapunov; nowadays, it is an important tool in the modern theory of probability. Many tasks of probability theory, in particular those which are based on the summation of independent random variables, can be solved much more simply and elegantly by using characteristic functions. Moreover, the characteristic functions can be applied in more general cases than the moment-generating functions.

According to eq. (3.9.48), the characteristic function $\varphi(\omega)$ for the random variable X with the probability density function $f(x)$ is

$$\varphi(\omega) = \langle \mathrm{e}^{i\omega X} \rangle = \int\limits_{-\infty}^{+\infty} \mathrm{e}^{i\omega x}\, f(x)\, \mathrm{d}x \tag{3.9.82}$$

According to eq. (3.8.18), the characteristic function $\varphi(\omega)$ – apart from the sign in the exponent of e – corresponds to the Fourier transform of the probability density function $f(x)$. Hence, the inverse respectively the probability density $f(x)$ is the Fourier integral of the characteristic function $\varphi(\omega)$

$$f(x) = \frac{1}{2\pi} \int\limits_{-\infty}^{+\infty} \mathrm{e}^{-i\omega x}\, \varphi(\omega)\, \mathrm{d}\omega \tag{3.9.83}$$

In analogy to eq. (3.9.79) and in accordance with table 3.8.1, we obtain by k-fold differentiation with respect to ω

$$\varphi^{(\mathrm{k})}(\omega) = \int\limits_{-\infty}^{+\infty} (ix)^{\mathrm{k}}\, \mathrm{e}^{i\omega x}\, f(x)\, \mathrm{d}x = i^{\mathrm{k}} \langle X^{\mathrm{k}} \mathrm{e}^{i\omega X} \rangle \tag{3.9.84}$$

and, for $\omega = 0$, we have

$$\varphi^{(\mathrm{k})}(0) = i^{\mathrm{k}} \int\limits_{-\infty}^{+\infty} x^{\mathrm{k}}\, f(x)\, \mathrm{d}x = i^{\mathrm{k}} \langle X^{\mathrm{k}} \rangle \tag{3.9.85}$$

or, for the moment of order k,

$$m_{\mathrm{k}} = \langle X^{\mathrm{k}} \rangle = \frac{1}{i^{\mathrm{k}}} \varphi^{(\mathrm{k})}(0) \tag{3.9.86}$$

Expanding the characteristic function $\varphi(\omega)$ into a Taylor series at $\omega = 0$

$$\varphi(\omega) = \varphi(0) + \left.\frac{\varphi'}{1!}\right|_{\omega=0} \omega + \left.\frac{\varphi''}{2!}\right|_{\omega=0} \omega^2 + \cdots \tag{3.9.87}$$

we obtain with eq. (3.9.86) the following series in the moments

$$\varphi(\omega) = 1 + i\omega\langle X \rangle + \frac{(i\omega)^2}{2!}\langle X^2 \rangle + \cdots = \sum_{\mathrm{k}=0}^{\infty} \frac{(i\omega)^{\mathrm{k}}}{\mathrm{k}!} \langle X^{\mathrm{k}} \rangle \tag{3.9.88}$$

As soon as the moments of the probability density function $f(x)$ are known, we also know the characteristic function and $f(x)$ can be determined from eq. (3.9.83) as the Fourier integral. Conversely, according to eq. (3.9.86), the k-th-order moment can be directly calculated from the k-th-order derivatives of the characteristic function $\varphi(\omega)$ at $\omega = 0$

$$m_{\mathrm{k}} = \langle X^{\mathrm{k}} \rangle = \left.\frac{1}{i^{\mathrm{k}}} \frac{\mathrm{d}^{\mathrm{k}} \varphi(\omega)}{\mathrm{d}\omega^{\mathrm{k}}}\right|_{\omega=0} \tag{3.9.89}$$

Comment 1: For the validity of eq. (3.9.86), it is assumed that all derivatives and the corresponding integrals exist, an assumption which is not evident. An exception, for example, is the Cauchy distribution $f(x) = \frac{1}{\pi(1+x^2)}$, for which neither mean nor moments of higher order exist, although its characteristic function $\varphi(\omega) = e^{-|\omega|}$ is well defined (see section 3.9.11).

Comment 2: As already mentioned, the characteristic functions are an essential tool of the modern theory of probability. Due to the multiplicative property of the exponential function $e^{i\omega(x_1+x_2)} = e^{i\omega x_1} e^{i\omega x_2}$, it can be shown that the density distribution of the sum of independent random variables X_j with the probability densities $f_\mathrm{j}(x_\mathrm{j})$, $\mathrm{j} = 1, \cdots, \mathrm{n}$ can be written as a convolution

$$f(x_1 + x_2 + \ldots + x_\mathrm{n}) = f_1(x_1) * f_2(x_2) * \ldots * f_\mathrm{n}(x_\mathrm{n}) \tag{3.9.90}$$

This feature simplifies, for example, the proof of limit theorems considerably.

3.9.9 Special Probability Distributions

• Binomial distribution

The binomial distribution is a special discrete distribution which is used for repetitive experiments with two possible outcomes $\{\omega_1, \omega_2\}$. Called after Jakob Bernoulli, such experiments are referred to as Bernoulli experiments. The most common example is the toss of a coin. Let us assume that the probability of throwing "tails" is p and "heads" q, where $p + q = 1$ holds according to axiom (II). We ask for the probability of throwing tails exactly x times in n tosses. In section 3.9.3, example 2, we already calculated this probability using eq. (3.9.13) for perfect coins as $p = q = 1/2$. A second way of calculating the probability is by using the elementary probabilities of success (tails) and failure (heads). In the case of success, we

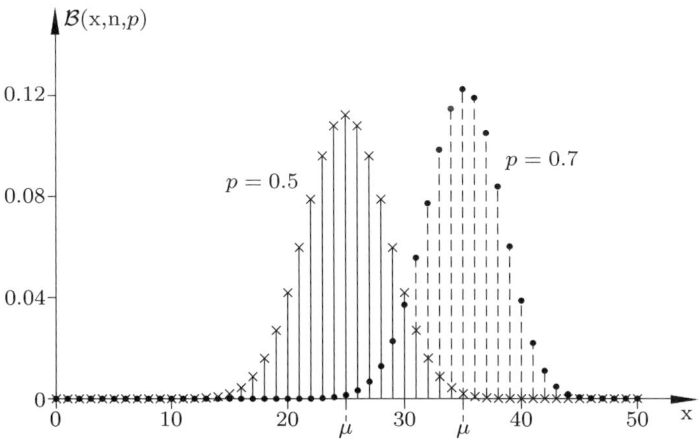

Fig. 3.9.5: Binomial distribution $\mathcal{B}(\mathrm{x}, \mathrm{n}, p)$ for the parameter values n=50 and $p = 0.5$ ($\mu = 25$), $p = 0.7$ ($\mu = 35$)

assume that the random variable X has the value $k = 1$ and in the opposite case $k = 0$. The corresponding elementary probabilities are then $P(X = \mathrm{k}) = p^{\mathrm{k}}$ and $P(X = \mathrm{k}) = q^{1-\mathrm{k}}$. Both results are independent, so that the elementary probabilities of eq. (3.9.60) can be multiplied; hence, for a single toss we obtain $P = p^{\mathrm{k}} q^{1-\mathrm{k}}$. Consequently, for n independent tosses, the probability of throwing a particular sequence, namely x times tails and (n − x)-times heads, is $P = p^{\mathrm{x}} q^{\mathrm{n}-\mathrm{x}}$. Since the sequential order of the single tosses is not of interest and since the probability of success p is realised x times in $\binom{\mathrm{n}}{\mathrm{x}}$ sequences, see eq. (3.9.13), the total probability distribution is

$$\mathcal{B}(\mathrm{x}, \mathrm{n}, p) = \binom{\mathrm{n}}{\mathrm{x}} p^{\mathrm{x}} q^{\mathrm{n}-\mathrm{x}} \tag{3.9.91}$$

The discrete probability distribution $\mathcal{B}(\mathrm{x}, \mathrm{n}, p)$ of eq. (3.9.91) is called *binomial distribution*.

The mean $\mu = \mathrm{n}p$ of the binomial distribution $\mathcal{B}(\mathrm{x}, \mathrm{n}, p)$ can be determined directly from the definition given in eq. (3.9.41)

$$\begin{aligned}
\mu &= \sum_{\mathrm{x}=0}^{\mathrm{n}} \mathrm{x} \binom{\mathrm{n}}{\mathrm{x}} p^{\mathrm{x}} q^{\mathrm{n}-\mathrm{x}} = \mathrm{n}p \sum_{\mathrm{x}=1}^{\mathrm{n}} \mathrm{x} \frac{(\mathrm{n}-1)!}{\mathrm{x}!(\mathrm{n}-\mathrm{x})!} p^{\mathrm{x}-1} q^{(\mathrm{n}-1)-(\mathrm{x}-1)} \\
&= \mathrm{n}p \sum_{\mathrm{x}=1}^{\mathrm{n}} \frac{(\mathrm{n}-1)!}{(\mathrm{x}-1)!(\mathrm{n}-\mathrm{x})!} p^{\mathrm{x}-1} q^{(\mathrm{n}-1)-(\mathrm{x}-1)} \\
&= \mathrm{n}p \sum_{\mathrm{x}=1}^{\mathrm{n}} \binom{\mathrm{n}-1}{\mathrm{x}-1} p^{\mathrm{x}-1} q^{(\mathrm{n}-1)-(\mathrm{x}-1)} \\
&= \mathrm{n}p \sum_{\mathrm{k}=0}^{\mathrm{n}-1} \binom{\mathrm{n}-1}{\mathrm{k}} p^{\mathrm{k}} q^{(\mathrm{n}-1)-\mathrm{k}} = \mathrm{n}p \sum_{\mathrm{k}=0}^{\mathrm{m}} \binom{\mathrm{m}}{\mathrm{k}} p^{\mathrm{k}} q^{\mathrm{m}-k} = \mathrm{n}p \underbrace{(p+q)^{\mathrm{m}}}_{1} \\
&= \mathrm{n}p
\end{aligned} \tag{3.9.92}$$

The variance σ^2 can be immediately calculated using eqs. (3.9.43) and (3.9.53)

$$\sigma^2 = \sum_{\mathrm{x}=0}^{\mathrm{n}} \mathrm{x}^2 \binom{\mathrm{n}}{\mathrm{x}} p^{\mathrm{x}} q^{\mathrm{n}-\mathrm{x}} - \mathrm{n}^2 p^2 = \mathrm{n}p(1-p) \tag{3.9.93}$$

The determination of the mean m_1 and variance $\sigma^2 = m_2 - m_1^2$ of the binomial distribution using the moment-generating function $g(z) = (p\mathrm{e}^z + q)^{\mathrm{n}}$ according to eq. (3.9.76) or the characteristic function $\varphi(\omega) = (p\mathrm{e}^{i\omega} + q)^{\mathrm{n}}$ defined in eq. (3.9.83) is left to the reader. In fig. 3.9.5, two binomial distributions are presented.

• Poisson distribution

In a number of applications of the Bernoulli experiment, the probability of success p is quite small in the individual realisations, i.e. success is rather a rare event, whereas the number n of experiments is very large. In principle, this is the classical case of a binomial distribution, though with the restriction that the calculation of the binomial coefficients is rather cumbersome for very large n. In this case, it is advantageous to approximate the binomial distribution by a distribution which is obtained when n tends to infinity and p to zero while the mean $\mu = \mathrm{n}p$, eq. (3.9.92), remains finite. If the Bernoulli experiment fulfils the conditions $\mathrm{n} \to \infty$, $\mu = \mathrm{const}$ and $p \to 0$, the binomial distribution, eq. (3.9.91), can be replaced by the Poisson distribution. In 1837, S. D. Poisson introduced this distribution which was later named after him.

In order to take the limit formally, we rewrite eq. (3.9.91) as follows

$$\begin{aligned}\mathcal{B}(\mathrm{x},\mathrm{n},p) &= \frac{\mathrm{n}!}{\mathrm{x}!(\mathrm{n}-\mathrm{x})!}\cdot p^{\mathrm{x}}(1-p)^{\mathrm{n}-\mathrm{x}} = \frac{\mathrm{n}(\mathrm{n}-1)\cdots(\mathrm{n}-\mathrm{x}+1)}{\mathrm{x}!}\cdot p^{\mathrm{x}}(1-p)^{\mathrm{n}-\mathrm{x}} \\ &= \frac{\mathrm{n}^{\mathrm{x}}\cdot p^{\mathrm{x}}(1-p)^{\mathrm{n}}}{\mathrm{x}!}\cdot\frac{(1-\frac{1}{\mathrm{n}})(1-\frac{2}{\mathrm{n}})\cdots(1-\frac{\mathrm{x}-1}{\mathrm{n}})}{(1-p)^{\mathrm{x}}}\end{aligned} \tag{3.9.94}$$

For constant $\mu^{\mathrm{x}} = \mathrm{n}^{\mathrm{x}}p^{\mathrm{x}}$ and fixed x, we obtain for $\mathrm{n} \to \infty$ the following expressions for the limit

$$\begin{aligned}\lim_{\mathrm{n}\to\infty}\mathcal{B}(\mathrm{x},\mathrm{n},p) &= \frac{\mu^{\mathrm{x}}}{\mathrm{x}!}\cdot\underbrace{\lim_{\mathrm{n}\to\infty}\left(1-\frac{\mu}{\mathrm{n}}\right)^{\mathrm{n}}}_{e^{-\mu}} \\ &\cdot\underbrace{\lim_{\mathrm{n}\to\infty}\left(1-\frac{1}{\mathrm{n}}\right)\left(1-\frac{2}{\mathrm{n}}\right)\cdots\left(1-\frac{\mathrm{x}-1}{\mathrm{n}}\right)\left\{1-\frac{\mu}{\mathrm{n}}\right\}^{-\mathrm{x}}}_{1}\end{aligned} \tag{3.9.95}$$

The Poisson distribution respectively the Poisson probability density is thus obtained as a function of x and μ

$$\mathcal{P}(\mathrm{x},\mu) \equiv \lim_{\mathrm{n}\to\infty}\mathcal{B}(\mathrm{x},\mathrm{n},p) = \frac{\mu^{\mathrm{x}}}{\mathrm{x}!}\mathrm{e}^{-\mu} \tag{3.9.96}$$

The moment-generating function of the Poisson distribution is

$$g(z) = \mathrm{e}^{-\mu}\mathrm{e}^{\mu\mathrm{e}^{z}} \tag{3.9.97}$$

respectively its characteristic function

$$\varphi(\omega) = \mathrm{e}^{-\mu}\mathrm{e}^{\mu\mathrm{e}^{i\omega}} \tag{3.9.98}$$

Using the definitions of eqs. (3.9.43, 3.9.47) to determine the characteristic parameters of the Poisson distribution makes tedious transformations necessary. The calculation is much more elegant if we use, as mentioned in section 3.9.8, the derivatives of the moment-generating function $g(z)$, eq. (3.9.97).

We first determine the mean value and, to do this, we calculate the first derivative $g'(z) = g(z)\mu\mathrm{e}^{z}$. For $z = 0$, we obtain together with $g(0) = 1$

$$\langle X\rangle = g'(0) = \mu \tag{3.9.99}$$

as expected following eq. (3.9.81). To calculate the variance, we determine the second derivative $g''(z) = g'(z)(1+\mu\mathrm{e}^{z})$, i.e. $g''(0) = \mu^2 + \mu$. Using eq. (3.9.53) immediately yields

$$\sigma^2 = \mu \tag{3.9.100}$$

i.e. mean and variance of the Poisson distribution are identical. The Poisson distribution is the only distribution with this property. The following statement thus holds: a random variable X_{n}, for which the variance is equal to the mean ($\sigma^2 = \mu$) is Poisson-distributed.

To determine the skewness $S(X)$, eq. (3.9.54), we first calculate the third derivative of the moment-generating function for $\langle X^3 \rangle = g'''(0)$. From equation $g'''(z) = g'(z)\left[(1+\mu e^z)^2 + \mu e^z\right]$, $\langle X^3 \rangle = \mu\left[(\mu+1)^2 + \mu\right]$ follows and, using eq. (3.9.52), we obtain

$$S(X) = \frac{1}{\sqrt{\mu}} \tag{3.9.101}$$

For increasing μ, the skewness $S(X)$ tends to zero, i.e. for large μ, the Poisson distribution is approximately symmetrical. Figure 3.9.6 shows two Poisson distributions.

The kurtosis $K(X)$, eq. (3.9.55), can be calculated in analogy to the skewness and we obtain

$$K(X) = \frac{1}{\sigma^4}\langle (X-\mu)^4 \rangle = \frac{1}{\mu} \tag{3.9.102}$$

The Poisson distribution is applied in physics, for example. For a Boltzmann gas, the classical ideal gas, the probability of large fluctuations can be described by means of the Poisson distribution, whereas the Gaussian probability density applies for small fluctuations. In order to derive the Poisson distribution, we assume V_0 to be the total volume and N_0 the number of particles contained in this volume. In comparison to the total volume V_0, we further assume that V is a small sub-volume containing N particles. Provided that the gas is homogeneously distributed, the probability of finding a given particle in V is equal to $\frac{V}{V_0}$. Hence, the probability of simultaneously finding N particles in V is $(\frac{V}{V_0})^{\mathrm{N}}$. Conversely, the probability that a particle is not in volume V is given by $\frac{V_0 - V}{V_0}$ respectively $(\frac{V_0-V}{V_0})^{\mathrm{N}_0 - \mathrm{N}}$ if we assume that $\mathrm{N}_0 - \mathrm{N}$ particles are not in V. The number of possibilities that N particles out of N_0 different realisations are in the sub-volume can be expressed by the binomial coefficients; therefore, according to eq. (3.9.91), we can specify the probability P_{N} of encountering N particles in sub-volume V as

$$P_{\mathrm{N}} = \frac{\mathrm{N}_0!}{\mathrm{N}!(\mathrm{N}_0 - \mathrm{N})!}\left(\frac{V}{V_0}\right)^{\mathrm{N}}\left(1 - \frac{V}{V_0}\right)^{\mathrm{N}_0 - \mathrm{N}} \tag{3.9.103}$$

where $p = V/V_0$, $\mathrm{n} = \mathrm{N}_0$ and $\mathrm{x} = \mathrm{N}$.

We are now interested in the case with the following assumptions: $V \ll V_0$ and $\mathrm{N} \ll \mathrm{N}_0$, though N may differ significantly from the mean $\bar{\mathrm{N}}$. According to eq. (3.9.92), the mean is given as

$$\mu = p\,\mathrm{n} = \frac{V\mathrm{N}_0}{V_0} = \bar{\mathrm{N}} = \mathrm{const} \tag{3.9.104}$$

Taking the limit of eq. (3.9.95), we obtain in analogy to eq. (3.9.96) the Poisson distribution

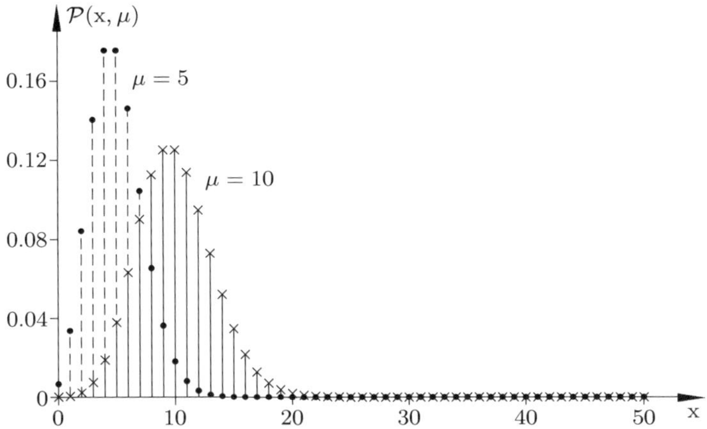

Fig. 3.9.6: Poisson distribution $\mathcal{P}(\mathrm{x},\mu)$ for $\mu = 5$ and $\mu = 10$

$$P_{\mathrm{N}} = \frac{\bar{\mathrm{N}}^{\mathrm{N}}}{\mathrm{N}!}\mathrm{e}^{-\bar{\mathrm{N}}} \tag{3.9.105}$$

Mean and variance are equal. As mentioned above, this property is typical of a Poisson distribution, see (Nicolis and Prigogine, 1987).

Remark: With eq. (3.9.34), the δ-function $\delta(x - x_0)$ can be interpreted as *one-point distribution* respectively one-point probability density $f(x)$

$$f(x) = \delta(x - x_0) \tag{3.9.106}$$

with the mean $m_1 = x_0$, variance $\sigma^2 = 0$, and characteristic function $\varphi(\omega) = \mathrm{e}^{i\omega x_0}$ in accordance with section 3.8.4.

With eq. (3.9.36), we can write the probability density for the n-point distribution as

$$f(x) = \sum_{\mathrm{i}=1}^{\mathrm{n}} p_{\mathrm{i}}\delta(x - x_{\mathrm{i}}) \tag{3.9.107}$$

• Normal distribution or Gaussian distribution

For many reasons, the normal distribution or Gaussian distribution is of particular importance. In natural and engineering science, but also in the field of economics, many processes can be described by means of the normal distribution either exactly or at least in a good approximation. As an example, we mention series of experiments where random measurement and observation errors have to be taken into account. With an appropriate centering and scaling of the binomial distribution, eq. (3.9.91), the normal distribution is obtained as the limit for $\mathrm{n} \to \infty$. In contrast to the Poisson distribution, eq. (3.9.96), p and q remain finite here. The normal distribution gains more importance with regard to the central limit theorem (see section 3.9.10).

Let us start with the Gaussian distribution (after Carl Friedrich Gauß, 1777–1855) respectively with the Gaussian probability density $f(x)$ in the following form

$$f(x) = \frac{1}{\sigma\sqrt{2\pi}} \mathrm{e}^{-\frac{(x-\mu)^2}{2\sigma^2}} = \mathcal{G}(x, \mu, \sigma^2) \qquad -\infty < x < \infty \tag{3.9.108}$$

where $X \sim \mathcal{G}(x, \mu, \sigma^2)$, indicating that the probability density of X corresponds to the normal distribution with mean μ as a shift parameter and variance σ^2 as the scale parameter. In the literature, one often finds the notation $\mathcal{N}(\mu, \sigma^2)$ for the normal distribution. In fig. 3.9.7a, the graphs of eq. (3.9.108) are presented for for $\mu = 0$ and three different σ-values. The smaller the standard deviation σ, the greater the maximum of $f(x)$ and the steeper the decay of the curve. The graph of $f(x)$, which is symmetric with respect to the mean μ, is referred to as a bell-shaped curve (cf. fig. 3.9.7a).

Following eq. (3.9.108), the distribution function of the random variable X with $f(x)$ is given by

$$P(X < x) = F(x) = \frac{1}{\sigma\sqrt{2\pi}} \int\limits_{-\infty}^{x} \mathrm{e}^{-\frac{(y-\mu)^2}{2\sigma^2}} \mathrm{d}y = \frac{1}{2}\left[1 + \mathrm{erf}\left(\frac{x-\mu}{\sqrt{2}\sigma}\right)\right] \tag{3.9.109}$$

where the error integral $\mathrm{erf}(x)$ is defined in section 3.8.4. Inserting $\sigma = 1$ and $\mu = 0$ into eq. (3.9.108), we obtain the normalised form of the Gaussian or normal distribution

$$f(x) = \frac{1}{\sqrt{2\pi}} \mathrm{e}^{-\frac{x^2}{2}} = \mathcal{N}(0, 1) \qquad -\infty < x < \infty \tag{3.9.110}$$

The characteristic function of the scaled Gaussian distribution is obtained from eq. (3.9.82) by completing the square of the exponent

$$\begin{aligned}\varphi(\omega) &= \frac{1}{\sqrt{2\pi}} \int\limits_{-\infty}^{+\infty} \mathrm{e}^{i\omega x} \mathrm{e}^{-\frac{x^2}{2}} \mathrm{d}x = \frac{1}{\sqrt{2\pi}} \int\limits_{-\infty}^{+\infty} \mathrm{e}^{-\frac{1}{2}(x-i\omega)^2} \mathrm{e}^{-\frac{\omega^2}{2}} \mathrm{d}x = \frac{1}{\sqrt{2\pi}} \mathrm{e}^{-\frac{\omega^2}{2}} \sqrt{2\pi} \\ &= \mathrm{e}^{-\frac{\omega^2}{2}}\end{aligned} \tag{3.9.111}$$

Applying the properties of the Fourier transform of table 3.8.1 to the transformation $x \to \frac{x-\mu}{\sigma}$ and replacing $\omega \to -\omega$ in the characteristic function, we obtain

$$\varphi(\omega) = \mathrm{e}^{i\mu\omega} \mathrm{e}^{-\frac{\sigma^2\omega^2}{2}} \qquad \text{with} \qquad \varphi(0) = 1 \tag{3.9.112}$$

With the help of the characteristic function, we now show that $\langle X \rangle = \mu$ holds for the mean. According to eq. (3.9.89), we have

$$\langle X \rangle = \frac{1}{i} \frac{\mathrm{d}\varphi(\omega)}{\mathrm{d}\omega}\bigg|_{\omega=0} = \frac{1}{i}(i\mu - \sigma^2\omega)\varphi(\omega)\bigg|_{\omega=0} = \mu \tag{3.9.113}$$

In the same way, we determine the variance σ^2 from $\varphi(\omega)$. The second moment is then

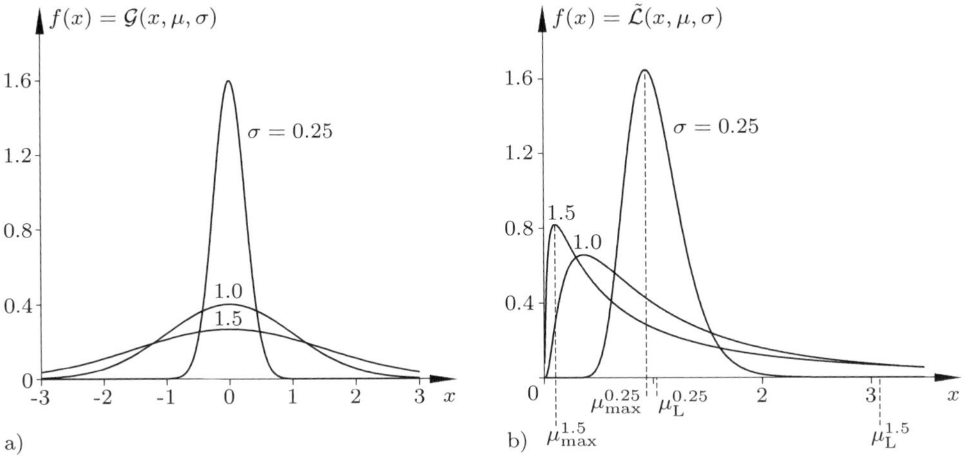

Fig. 3.9.7: a) Gaussian or normal distribution, b) log-normal distribution for $\mu = 0$, $\sigma \in \{0.25, 1.0, 1.5\}$

$$m_2 = -1\left[-\sigma^2\varphi(\omega) + (i\mu - \sigma^2\omega)^2\varphi(\omega)\right]_{\omega=0} = \sigma^2 + \mu^2 \tag{3.9.114}$$

and, according to eq. (3.9.53), we finally obtain the variance σ^2, as expected.

• Log-normal distribution

Further continuous distributions are the *Student's t-distribution*, the *exponential distribution*, the *Weibull distribution*, the *log-normal distribution* and others. With regard to Kolmogorov's refined model of the energy cascade and in view of the modelling of intermittent turbulent flows presented in chapter 9, we present as a final example of distributions the logarithmic normal distribution or log-normal distribution for short.

The log-normal distribution describes the distribution of the random variable X when the random variable $Y = \ln X$ is normally distributed. The log-normal distributed probability density function $f(x)$ can now be constructed by means of the normalisation condition, eq. (3.9.30), together with the Gaussian probability density, eq. (3.9.108). We obtain

$$\int_{-\infty}^{+\infty} f(x)\mathrm{d}x = \frac{1}{\sigma\sqrt{2\pi}}\int_{-\infty}^{+\infty} \mathrm{e}^{-\frac{(y-\mu)^2}{2\sigma^2}}\mathrm{d}y = 1 \tag{3.9.115}$$

Replacing the variable y by $\ln x$ and $\mathrm{d}y$ by $\frac{\mathrm{d}y}{\mathrm{d}x}\mathrm{d}x = \frac{1}{x}\mathrm{d}x$, we can write the log-normally distributed probability density $f_{\mathrm{L}}(x)$ in the form

$$f_{\mathrm{L}}(x) = \frac{1}{x\,\sigma\sqrt{2\pi}}\mathrm{e}^{-\frac{(\ln x-\mu)^2}{2\sigma^2}} = \tilde{\mathcal{L}}(x, \mu, \sigma) \tag{3.9.116}$$

Figure 3.9.7b shows the log-normal distribution for the normally distributed standard deviations $\sigma \in \{1.5, 1.0, 0.5\}$. The figure illustrates that the log-normal distribution $\tilde{\mathcal{L}}(x, \mu, \sigma)$ is skewed with a tail on the right-hand side which becomes heavier for increasing σ. Next, we determine the variance $\sigma_{\rm L}^2$, the maximum frequency value $\mu_{\max}$ and the skewness $S_{\rm L}(X)$ of the log-normal distribution. Since all moments $\langle X^{\rm k}\rangle_{\rm L}$ exist which can be written as

$$\langle X^{\rm k}\rangle_{\rm L} = {\rm e}^{{\rm k}\mu + \frac{{\rm k}^2\sigma^2}{2}} \tag{3.9.117}$$

the expectation value $\langle X\rangle_{\rm L}$ respectively the mean $\mu_{\rm L}$ is according to eq. (3.9.50)

$$\langle X\rangle_{\rm L} = {\rm e}^{\mu + \frac{\sigma^2}{2}} = \frac{1}{\sigma\sqrt{2\pi}} \int_0^\infty {\rm e}^{-\frac{(\ln x - \mu)^2}{2\sigma^2}}\, {\rm d}x = \mu_{\rm L} \tag{3.9.118}$$

Following eq. (3.9.53), the variance $\sigma_{\rm L}^2$ is finally given as

$$\sigma_{\rm L}^2 = {\rm e}^{2\mu+\sigma^2}({\rm e}^{\sigma^2} - 1) = \frac{1}{\sigma\sqrt{2\pi}} \int_0^\infty \frac{1}{x}\left(x - {\rm e}^{\mu+\frac{\sigma^2}{2}}\right)^2 {\rm e}^{-\frac{(\ln x - \mu)^2}{2\sigma^2}}\, {\rm d}x \tag{3.9.119}$$

The mean represents the abscissa of the centre of gravity of the area enclosed by $f_{\rm L}(x)$. Figure 3.9.7b shows that the difference between mean $\mu_{\rm L}$ and $\mu_{\max}$ increases for increasing σ-values, where the probability density $f_{\rm L}(x)$ has, by definition, its maximum at the most probable $\mu_{\max}$-value. The values of $x = \mu_{\max}$ and $f_{\rm L}(\mu_{\max})$ are calculated as

$$\mu_{\max} = {\rm e}^{\mu - \sigma^2} \qquad \text{with} \qquad f_{\rm L}(\mu_{\max}) = \frac{1}{\sigma\sqrt{2\pi}} {\rm e}^{-\mu + \frac{1}{2}\sigma^2} \tag{3.9.120}$$

The skewness $S_{\rm L}(X)$ of eq. (3.9.54) is given as

$$S_{\rm L}(X) = ({\rm e}^{\sigma^2} + 2)\sqrt{{\rm e}^{\sigma^2} - 1} > 0 \tag{3.9.121}$$

i.e. the log-normal distribution is said to be right-skewed.

The characteristic function $\langle {\rm e}^{i\omega X}\rangle_{\rm L}$ of the log-normal distribution cannot be specified explicitly because the moment-generating function does not converge. An approximation can be found in (Beaulieu and Xie, 2004).

For $\sigma \ll 1$, both the log-normal distribution of the random variable X and the normal distribution of $Y = \ln X$, eq. (3.9.113), are very narrow and resemble a Gaussian distribution. For an increasing standard deviation σ, however, the log-normally distributed values $\mu_{\rm L}$ and $x = \mu_{\max}$ drift further and further apart. This in turn means that the standard deviation $\sigma_{\rm L}$ increases considerably and, at the same time, the right tail $f_{\rm L}(x)$ gains in influence with regard to probability statements. The tails describe rare events, heavier tails provide information on the frequency of supposedly rare events.

3.9.10 Central Limit Theorem

The central limit theorem makes an important statement on the sum of N random variables in the limit of large N. First, we illustrate the central limit theorem using

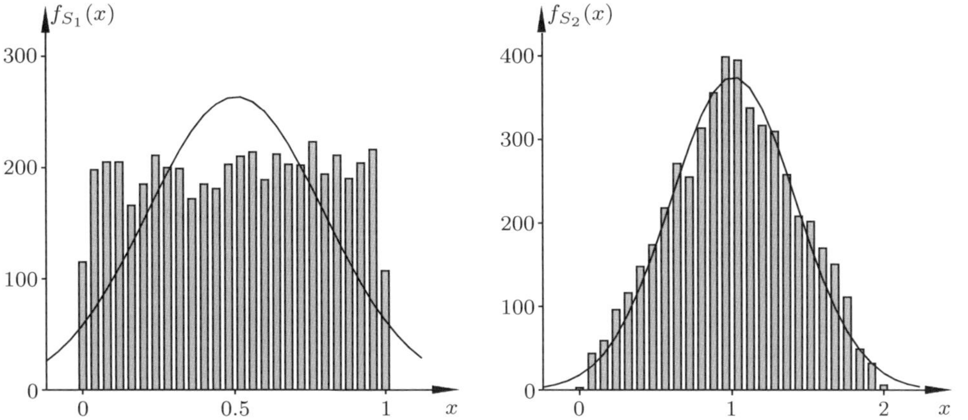

a) $S_1 = X_1$

b) $S_2 = X_1 + X_2$

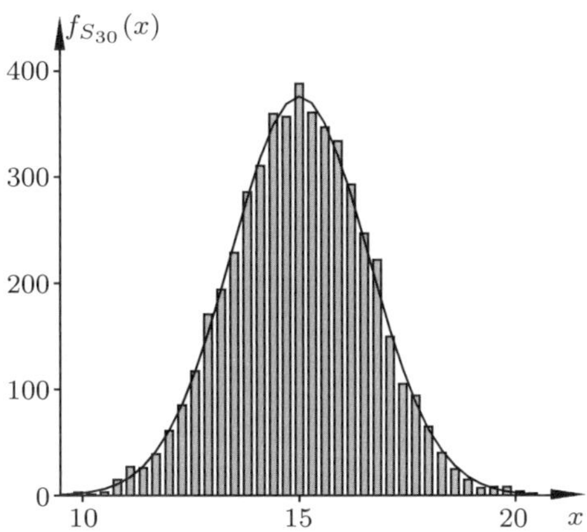

c) $S_{30} = X_1 + ... + X_{30}$

Fig. 3.9.8

Histogram and normal distribution for the sum S_{N} of independent identically distributed random variables

a simple example. In the interval [0,1], we produce 5000 independent, identically distributed random variables each for $X_{\mathrm{j}}, \mathrm{j} = 1, \ldots, 30$. Then we consider the random variables X_1, the sum $S_2 = X_1 + X_2$ and so on up to $S_{30} = X_1 + \ldots + X_{30}$. Figure 3.9.8a shows the approximately rectangular histogram of 5000 random numbers. Figure 3.9.8b shows the distribution of the sum S_2. It is surprising that a unimodal distribution with a distinct maximum is already formed at the sum $S_2 = X_1 + X_2$ of two rectangular distributions. In fig. 3.9.8c, the number of summands of random variables increases to 30: a well-defined approximation to a normal or Gaussian distribution can be clearly seen.

The central limit theorem reads as follows (Gnedenko and Kolmogorov, 1960):

Suppose $\{X_1, X_2, \ldots X_{\mathrm{N}}\}$ *is a sequence of N independent random variables with an identical distribution function* $f(x)$ *which has a finite mean value* $\mu = \langle X_j \rangle$ *and variance* $\sigma^2 < \infty$ *and denote the sample average* $S_{\mathrm{N}} = (X_1 + X_2 + \ldots + X_{\mathrm{N}})/\mathrm{N}$. *Then, as* $\mathrm{N} \to \infty$, *the distribution of* $\sqrt{(\mathrm{N})}(S_{\mathrm{N}} - \mu)$ *approaches the normal distribution* $\mathcal{N}(0, \sigma^2)$.

Let us now take a brief look at a derivation of the central limit theorem. We start from the joint probability distribution of the random variables X_1, X_2, $f_{X_1,X_2}(x_1, x_2)$. We invoke the assumption of statistical independence of the random variables X_1, X_2 which implies that the joint probability density factorises

$$f_{X_1,X_2}(x_1, x_2) = f_{X_1}(x_1) f_{X_2}(x_2) \tag{3.9.122}$$

Furthermore, we assume that the random variables obey the same statistics

$$F_{X_1}(x) = F_{X_2}(x) = F_X(x) \tag{3.9.123}$$

We now consider the sum $S = X_1 + X_2$ of the random variables and perform a coordinate transformation

$$X_2 = S - X_1 \tag{3.9.124}$$

The joint probability distribution of the random variables S, X_1 is then given by

$$F_{S,X_1}(s, x_1) = f_{X_1,X_2}(x_1, s - x_1) = f_X(x_1) f_X(s - x_1) \tag{3.9.125}$$

We are interested in the distribution $f_S(s)$ of the random variable S. This quantity is obtained by summing the joint distribution of S and X_1 over x_1

$$f_S(s) = \int F_{S,X_1}(s, x_1)\, \mathrm{d}x_1 = \int f_X(x_1) f_X(s - x_1)\, \mathrm{d}x_1 = (f_X * f_X)(s) \tag{3.9.126}$$

The probability distribution of the random variable S is obtained as a convolution of the probability distributions $f_X(x)$ of the random variable X.

We can now formulate this result for the characteristic functions $\varphi_S(\omega)$ and $\varphi_X(\omega)$, which are defined according to eq. (3.9.82) by the Fourier transforms of the distributions $f_S(s)$ and $f_X(s)$. The Fourier transform of a convolution, however, is the product of the Fourier transforms

$$\varphi_S(\omega) = \varphi_X(\omega) \varphi_X(\omega) \tag{3.9.127}$$

Taking the logarithm of this relation and defining the functions $W_S = \ln \varphi_S$, $W_X = \ln \varphi$, we end up with the relation

$$W_S(\omega) = 2 W_X(\omega) \tag{3.9.128}$$

Analogously, we can consider the sum of three identically distributed and statistically independent solutions

$$S = X_1 + X_2 + X_3 \tag{3.9.129}$$

Similar reasoning leads us to the representation of the distribution of the random variable S in terms of convolutions of the distribution $f_X(x)$

$$\begin{aligned} f_S(s) &= \int f_{X_1,X_2,X_3}(x_1, x_2, s - x_1 - x_2)\, \mathrm{d}x_1 \mathrm{d}x_2 \\ &= \int\int f_X(x_1) f_X(x_2) f_X(s - x_1 - x_2)\, \mathrm{d}x_2 \mathrm{d}x_1 \\ &= (f_X * (f_X * f_X))(s) \end{aligned} \tag{3.9.130}$$

Fourier transform yields

$$\varphi_S(\omega) = 3\varphi_X(\omega) \tag{3.9.131}$$

The extension to N identically distributed random variables is straightforward. The characteristic function of the random variable $S = \sum_{\mathrm{i}=1}^{\mathrm{N}} X_{\mathrm{i}}$ is given by a product of the characteristic functions

$$\varphi_S(\omega) = \mathrm{N}\varphi_X(\omega) \tag{3.9.132}$$

or, in terms of the logarithms of the characteristic functions, the so-called cumulant generating functions

$$W_S(\omega) = \sum_{\mathrm{i}=1}^{\mathrm{N}} W_X(\omega) = \mathrm{N}W_X(\omega) \tag{3.9.133}$$

The last step is to introduce a scaled random variable $S = C(\mathrm{N})S'$, where the constant $C(\mathrm{N})$ is a function of the number N. The meaning of the scaling of the random variable is evident

$$S' = \frac{1}{C(\mathrm{N})}[X_1 + X_2 + \ldots + X_{\mathrm{N}}] \tag{3.9.134}$$

Using eq. (3.9.82), we set up the relation between the characteristic functions $\varphi_S(\omega)$ and $\varphi_{S'}(\omega)$

$$\varphi_{S'}(\omega) = \langle \mathrm{e}^{i\omega S'} \rangle = \langle \mathrm{e}^{i\omega \frac{S}{C(N)}} \rangle = \varphi_S(\frac{\omega}{C(N)}) \tag{3.9.135}$$

which leads us to the equation

$$W_{S'}(\omega) = \mathrm{N}W_X\left(\frac{\omega}{C(\mathrm{N})}\right) \tag{3.9.136}$$

We are now in a position to discuss the central limit theorem. To this end we assume that the first moment of the random variable X vanishes, whereas the second moment is finite. Then, we are able to expand the cumulant generating function $W_X(\omega)$ into a Taylor series:

$$W_X(\omega) = -\frac{1}{2}\langle X^2 \rangle \omega^2 + \frac{1}{6} W_X''' \omega^3 + \ldots \tag{3.9.137}$$

As a result, the cumulant generating function of the random variable S' is given according to relation (3.9.136)

$$W_{S'}(\omega) = -\frac{1}{2}\frac{\mathrm{N}}{C(\mathrm{N})^2}\langle X^2\rangle\omega^2 + \frac{1}{6}W_X'''\frac{\mathrm{N}}{C(\mathrm{N})^3}\omega^3 + \ldots \tag{3.9.138}$$

If we now take

$$C(\mathrm{N}) = c\sqrt{\mathrm{N}} \tag{3.9.139}$$

all higher-order terms vanish in the limit $\mathrm{N} \to \infty$, since the n-th-order term in the Taylor expansion is proportional to $\mathrm{N}/C(\mathrm{N})^{\mathrm{n}} \propto 1/N^{n/2-1}$ and hence tends to zero in the large-N limit. In this limit, the characteristic function of $\varphi_{S'}(\omega)$ tends towards the Gaussian

$$\varphi_{S'}(\omega) = \mathrm{e}^{-\frac{c^2\langle X^2\rangle}{2}\omega^2} \tag{3.9.140}$$

As a consequence, the probability distribution is Gaussian

$$f_{S'}(s) = \frac{1}{\sqrt{2\pi c^2\langle X^2\rangle}}\mathrm{e}^{-\frac{s^2}{2c^2\langle X^2\rangle}} \tag{3.9.141}$$

The central limit theorem is one of the most powerful theorems in probability theory. The fact that the mean of a large number of independent, identically distributed random variables approaches the normal distribution irrespective of the specific density distribution explains the importance and ubiquitousness of Gaussian distribution. The central limit theorem also plays a prominent role in synergetics, which will be outlined briefly in section 6.8. In synergetics, non-linear open dissipative systems operating far from the thermodynamic equilibrium are studied. The evolution of such systems is described by stochastic differential equations where the microscopic interaction with the environment is modelled by fluctuating forces resulting from a very large number of independent random kicks with identical variance. In a short time period, these fluctuations add up to a normally distributed sum according to the central limit theorem.

3.9.11 Cauchy Distribution and α-Stable Lévy Distribution

This introductory section 3.9 is structured as in many other textbooks on probability theory; this might give the impression that the central limit theorem is a rather theoretical amendment and therefore of only little practical relevance. However, as a matter of fact, the intrinsic importance of probability theory only becomes evident on the basis of the central limit theorems. The three different definitions of the concept of probability presented in section 3.9.3 reveal that the essence of the probability becomes accessible only by means of the central limit theorem. It is decisive, especially for practical implementations, that large data sets of purely random events do not follow random laws, but are in fact even predictable. This in itself would be a purely theoretical finding if the theory of probability were not corroborated by the frequency of particular findings resulting from an experiment repeated x times under the same experimental conditions.

• Cauchy Distribution

Before we turn to the α-stable Lévy distribution, it is appropriate to present the essential properties of the Cauchy distribution. The Cauchy distribution is due to Augustin Louis Cauchy (1789 - 1857), a student of Laplace, and is also known among physicists as Lorentz or Breit-Wigner distribution. It is a continuous probability distribution with the following probability density function $f(x)$ (see figure 3.9.9a)

$$f(x) = \frac{1}{\pi} \cdot \frac{\bar{\sigma}}{\bar{\sigma}^2 + (x - \bar{\mu})^2} = \mathcal{C}(\bar{\mu}, \bar{\sigma}) \tag{3.9.142}$$

with the shift or location parameter $-\infty < \bar{\mu} < \infty$ and the scale parameter $\bar{\sigma} > 0$. The location parameter $\bar{\mu}$ specifies the location $x_{\max} = \bar{\mu}$ where the Cauchy distribution has its maximum $f_{\max} = 1/(\bar{\sigma}\pi)$. The scale parameter $\bar{\sigma}$ specifies the half-width of the distribution at half-maximum. What is new here compared with the distributions presented previously is that in the case of the Cauchy distribution, the characteristic parameters such as mean μ, variance σ^2 (though $m_2 = \langle X^2 \rangle$ is equal to zero), skewness $S(X)$ and kurtosis do not exist since the associated integrals are divergent.

However, the Cauchy distribution possesses a characteristic function

$$\varphi(\omega) = \exp\{i\bar{\mu}\omega - \bar{\sigma}\,|\,\omega\,|\} \tag{3.9.143}$$

and is invariant under convolutions, i.e. this type of probability distribution is stable because all convolutions belonging to this type again yield the same distribution after convolution.

The graphs in figure 3.9.9a show significant differences in the extreme events between the Gaussian, Cauchy and Lévy distribution; this latter is presented in the following. The normalised and centered Gaussian distribution has the value of $\mathcal{N}(0, 1) = 1.338 \cdot 10^{-4}$ at the point $x = 4$, whereas the Cauchy distribution has a value at the same point which is 140 times higher, namely $f(4) = 1.872 \cdot 10^{-2}$.

• α-stable Lévy distribution

We now turn to Lévy's contribution to modern probability theory. As we know, a preconditon of the central limit theorem is that mean and variance exist and are finite, which is not the case for the Cauchy distribution. In the early 1920s, the French mathematician Paul Pierre Lévy (1886–1971) asked himself to what extent the condition of finite variance can be dropped or whether classes of distributions exist like the Cauchy distribution with stretched heavy tails for which a generalised central limit theorem can be formulated.

We know from the central limit theorem that the sum of n random variables with rectangular distribution does not remain rectangular for large n, but leads to a Gaussian distribution, see fig. 3.9.8. Such behaviour – where the (rectangular) form of the distribution is not maintained – is called unstable. In contrast, the Gaussian distribution is stable under summation, i.e. the sum of two independent Gaussian distributed random variables is again Gaussian distributed. This phenomenon of the sum of two independent random variables with appropriate scaling and shifting having the same form of distribution is by no means self-evident, but rather an exception.

Following Lévy (Lévy, 1954), a probability distribution $f(x)$ is called *stable* if, for identically distributed independent random variables X, X_1, X_2, arbitrary parameters $a_1 > 0,\ a_2 > 0$ and constants $a > 0,\ b$ exist so that (Metzler and Klafter, 2000; Nolan, 1997)

$$a_1 X_1 + a_2 X_2 \overset{\mathrm{d}}{=} aX + b \tag{3.9.144}$$

The equality sign $\overset{\mathrm{d}}{=}$ states that the random variables of both sides possess the same probability distribution, i.e. if we consider independent random variables X_i having the same probability distribution $f(x)$, then $f(x)$ is stable if any linear combination of X_i possesses the same probability distribution $f(x)$ with the exception of shifts and scalings.

Let us consider the notion of a stable distribution in more detail. For the sake of simplicity, we consider identically distributed random variables X_i, $i = 1, \mathrm{N}$ and define the random variable, see eq. (3.9.134)

$$S' = \frac{1}{C(\mathrm{N})} \sum_{i=1}^{\mathrm{N}} X_i \tag{3.9.145}$$

As we showed above in our discussion of the central limit theorem, the cumulant generating functions are related by (3.9.136). Stable distributions (with zero mean) are defined by the condition that $W_{S'}(\omega) = W_X(\omega) = W(\omega)$. Hence, they have to be solutions of the functional equation

$$W(\omega) = \mathrm{N} W\left(\frac{\omega}{C(\mathrm{N})}\right) \tag{3.9.146}$$

This equation can be solved by the ansatz

$$W(\omega) = -w_0 |\omega|^{\alpha} \tag{3.9.147}$$

provided we take the constant $C(\mathrm{N})$ as

$$C(\mathrm{N}) = (\mathrm{N})^{1/\alpha} \tag{3.9.148}$$

Up to now, the coefficient α has not yet been specified. However, the condition that the cumulant generating function leads to a positive, normalisable probability distribution

$$f(x) = \int d\omega e^{-i\omega x - w_0 |\omega|^{\alpha}} \tag{3.9.149}$$

leads to the restriction

$$0 < \alpha \leq 2 \tag{3.9.150}$$

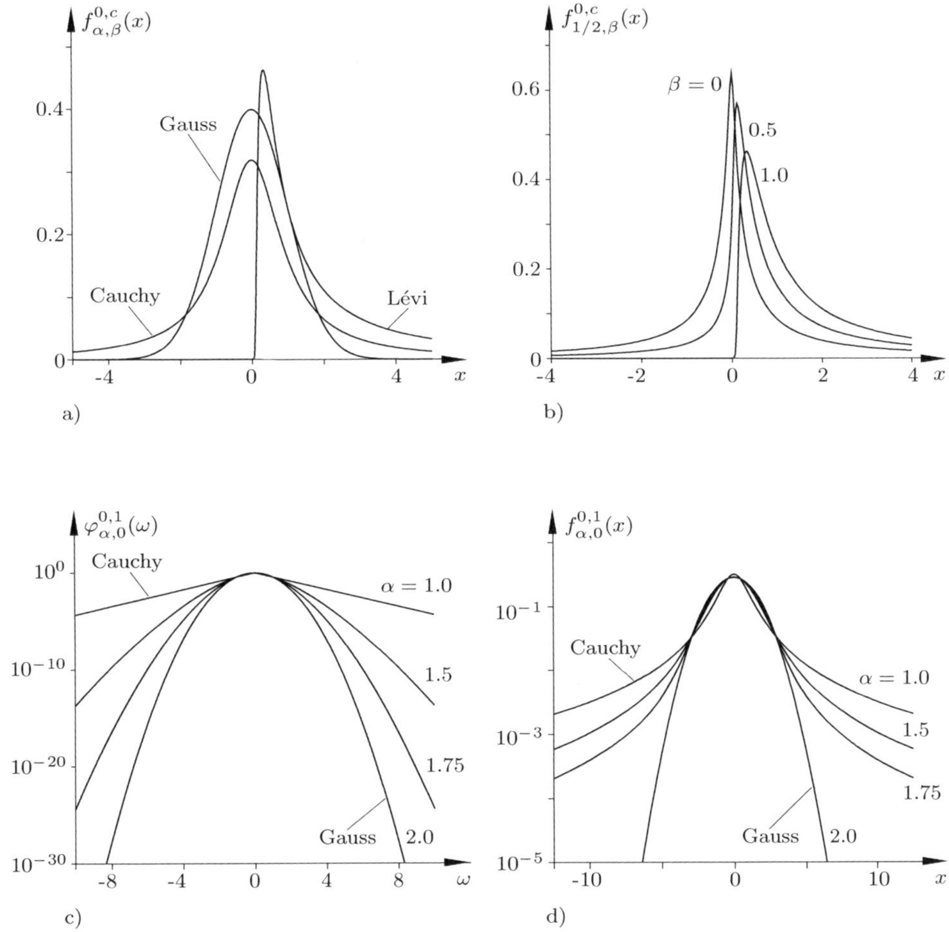

Fig. 3.9.9: α-stable Lévy distributions:
a) the three explicitly representable distributions: Gaussian distribution $\alpha = 2$, $\beta = 0$, Cauchy distribution $\alpha = 1$, $\beta = 0$, Lévy distribution $\alpha = 0.5$, $\beta = 1$,
b) Lévy distributions $f^{0,c}_{1/2,\beta}(x)$,
c) characteristic functions $\varphi^{0,1}_{\alpha,0}(\omega)$,
d) the α-stable Lévy disributions corresponding to c) $f^{0,1}_{\alpha,0}(x)$.

Lévy now succeeded in constructing a four-parametric class of skewed α-stable Lévy distrubutions $f^{\gamma,c}_{\alpha,\beta}(x)$, the distribution function of which remains stable under summation. Although these distributions cannot be specified explicitly with the exception of three special cases, their characteristic functions $\varphi^{\gamma,c}_{\alpha,\beta}(\omega)$ can be given as (cf. section 3.9.8)

$$\log \varphi^{\gamma,c}_{\alpha,\beta}(\omega) = i\gamma\omega - c|\omega|^{\alpha}\left\{1 - i\beta\frac{\omega}{|\omega|}\nu(\omega,\alpha)\right\} \tag{3.9.151}$$

$$\text{with} \quad \nu(\omega,\alpha) = \begin{cases} \tan(\frac{\pi\alpha}{2}) & \text{for} \quad \alpha \neq 1 \\ \frac{2}{\pi}\log|\omega| & \text{for} \quad \alpha = 1 \end{cases}$$

The inverse Fourier transforms

$$f_{\alpha,\beta}^{\gamma,c}(x) = \frac{1}{2\pi} \int_{-\infty}^{+\infty} \varphi_{\alpha,\beta}^{\gamma,c}(\omega) \mathrm{e}^{-i\omega x} \mathrm{d}\omega \tag{3.9.152}$$

are denoted α-stable Lévy distributions. For them, a generalised central limit theorem can be formulated, see (Gnedenko and Kolmogorov, 1960).

For the four parameters α, β, γ, c, it can be stated:

(1) the Lévy index or characteristic exponent $\alpha \in (0, 2]$ controls the decay of the *heavy tails* (fat tails);
(2) $\beta \in [-1, 1]$ is called the skewness parameter, see eq. (3.9.54); for $\beta = 0$, the distribution $f_{\alpha,0}^{\gamma,c}(x)$ is symmetrical;
(3) $\gamma \in \mathbb{R}$ controls the shift, and
(4) $c > 0$ the scaling of the distribution.

It should be noted at this point that there are various parametrisations in the literature for the Lévy distributions, see, for instance, (Nolan, 1997; Weron, 2001). Here, apart from the sign of β, we essentially use the notation of (Gnedenko and Kolmogorov, 1960; Metzler and Klafter, 2000). In order to use the free software in the package Mathematica of Rimmer and Nolan (Rimmer and Nolan, 2005) for the calculation of distributions, one has only to modify the parameter c (Kleinhans, 2008).

Only two of the four parameters, namely α and β, are important because they influence the distribution function qualitatively. For simplicity's sake, we focus in the following on the work of (Metzler and Klafter, 2000; Kleinhans, 2008) and set the shift parameter $\gamma = 0$; moreover, we restrict ourselves to $\alpha \neq 1$ and eliminate c. Equation (3.9.151) then simplifies to

$$\log \varphi_{\alpha,\beta}^{0,c}(\omega) = -c|\omega|^{\alpha} \left\{ 1 - i\beta \frac{\omega}{|\omega|} \tan \frac{\pi\alpha}{2} \right\} \tag{3.9.153}$$

For the elimination of c, we introduce the complex exponential function

$$\log \varphi_{\alpha,\beta}^{0,c}(\omega) = -c\sqrt{1+\beta^2 \tan^2 \frac{\alpha}{2}}\; |\omega|^{\alpha} \exp\left[-i\frac{\omega}{|\omega|} \arctan\left(\beta \tan \frac{\pi\alpha}{2} \right) \right] \tag{3.9.154}$$

For c, we choose

$$c = 1 \Big/ \sqrt{1+\beta^2 \tan^2 \frac{\alpha}{2}} \tag{3.9.155}$$

and obtain the Lévy distribution for the parameters α and β

$$\log \varphi_{\alpha,\beta}(\omega) = -|\omega|^{\alpha} \exp\left[-i\frac{\omega}{|\omega|} \arctan\left(\beta \tan \frac{\pi\alpha}{2} \right) \right] \tag{3.9.156}$$

Because of $\varphi_{\alpha,\beta}(-\omega) = \varphi_{\alpha,\beta}^{*}(\omega)$, the inverse Fourier transform $f_{\alpha,\beta}(x)$ is real and is given as

$$f_{\alpha,\beta}(x) = \frac{1}{\pi}\mathrm{Re}\left\{\int\limits_0^\infty \exp\left[-i\omega x - |\omega|^\alpha \exp\left(-i\arctan\left(\beta\tan\frac{\pi\alpha}{2}\right)\right)\right] \mathrm{d}\omega\right\} \tag{3.9.157}$$

The parametrisation of eq. (3.9.157) corresponds to the version implemented in Mathematica by (Rimmer and Nolan, 2005) if $\delta = 0$ and $\gamma = (1 + \beta^2 \tan^2 \frac{\alpha}{2})^{-\frac{1}{2\alpha}}$ are inserted for the parameters.

As an illustration, some α-stable Lévy distributions are shown in fig. 3.9.9. As mentioned above, apart from three exceptions presented in fig. 3.9.9a, the stable distributions cannot be written in an explicit form. The exceptions are:

(1) for $\alpha = 2$ and $\beta = 0$, we obtain the normal distribution $\mathcal{N}(0, 1)$, eq. (3.9.110),
(2) for $\alpha = 1$ and $\beta = 0$, we obtain the Cauchy distribution $\mathcal{C}(0, 1)$, eq. (3.9.142)
(3) and for $\alpha = 0.5$ and $\beta = 1.0$ the Lévy distribution

$$f^{0,1}_{\alpha,\beta}(x) = \sqrt{\frac{1}{2\pi}} x^{-3/2} \mathrm{e}^{-1/2x} \tag{3.9.158}$$

In fig. 3.9.9b, the Lévy distributions for $\alpha = 1/2$ and different β-values of $[0, 1]$ are displayed. The tendency of the skewness to lean to the right for increasing $\beta > 0$ and the symmetry properties for $\beta = 0$ become clear.

Figure 3.9.9c shows the graphs of the characteristic functions for $\beta = 0$ and $\alpha \in [1.0, 2.0]$ and fig. 3.9.9d the corresponding α-stable Lévy distributions. For $\alpha = 2$, we obtain the normal distribution and, for $\alpha = 1$, the Cauchy distribution.

According to eq. (3.9.85), the k-th moment $\langle X^{\mathrm{k}} \rangle$ can be determined by differentiating the characteristic function $\varphi_{\alpha,\beta}(\omega)$, eq. (3.9.156): $\varphi^{(\mathrm{k})}_{\alpha,\beta}(0) = i^{(\mathrm{k})} \langle x^{\mathrm{k}} \rangle$. From this, it follows immediately that the variance is infinite for $\alpha < 1$ and the variance and mean are infinite for $\alpha < 2$.

As already mentioned, the characteristic exponent α controls the asymptotic behaviour of $f_{\alpha,\beta}(x)$ for $|x| \gg 1$. For $\alpha < 2$, the decay is described by the following power law

$$f_{\alpha,\beta}(x) \sim \frac{1}{|x|^{1+\alpha}} \tag{3.9.159}$$

As early as 1963, Mandelbrot published an article on the development of wool prices on the stock market (Mandelbrot, 1963). He compared the histograms of wool price differences $Z(t+T) - Z(t)$ with Gaussian distributions. The agreement was not very convincing; it was the strongly fluctuating prices and the large number of "outliers" which led to wide heavy tails in the relevant distributions. Mandelbrot was able to show that this behaviour can be modelled quite well with α-stable Lévy distributions. The essential point in the Lévy statistics is that the probability density for $\alpha < 2$ does not decay exponentially as for Gaussian distributions, but rather follows the power law $\sim \frac{1}{|x|^{1+\alpha}}$ of eq. (3.9.159), cf. (Gnedenko and Kolmogorov, 1960; Nolan, 1997). There are numerous examples from the fields of economics and finance, as well as biology and physiology, where α-stable Lévy distributions can be observed (West, 2006).

3.9.12 Analysis of Stochastic Processes

The central aim of this volume is to characterise non-linear deterministic systems which can be described by ordinary differential equations. The basic assumption for determining the characteristic quantitites of the system dynamics described in the following chapters, such as dimensions, Lyapunov exponents and entropy, is that the system is deterministic and can be described by differential equations which do not contain any stochastic part. However, if we analyse real-world data such as measurement data from experiments, it is unavoidable that we are confronted by noisy time series. Typically, non-linearities and superimposed noise caused by the fluctuating forces acting on the system due to coupling with the environment (possibly added by measurement noise) contribute to the complexity of the dynamics, see, for example, the order parameter equations of synergetics presented in section 6.8.

Given a fluctuating set of experimental data, it is a challenge to determine the underlying unspoiled dynamics and to assess the characteristics of the superimposed fluctuations. In recent years, methods have been developed which allow for a wide class of stochastic processes to disentangle the basic dynamics and the noisy part. In particular, methods for parameter-free estimations of the underlying stochastic equations have been proposed. In this section, we give a very brief overview of the basic concepts and numerical methods used. For further reading, we recommend the review articles (Friedrich et al., 2009; Friedrich et al., 2011), in which an extended list of literature can also be found.

The temporal evolution of a random variable $\boldsymbol{X}(t)$ is denoted a stochastic process. In the following, we consider a class of systems for which the evolution of a vector $\boldsymbol{X}(t)$ in a d-dimensional state space can be written as a non-linear stochastic differential equation of first order

$$\frac{\mathrm{d}\boldsymbol{X}(t)}{\mathrm{d}t} = \boldsymbol{N}(\boldsymbol{X}(t), t) + \boldsymbol{F}(\boldsymbol{X}(t), t) \tag{3.9.160}$$

where $\boldsymbol{N}$ is a non-linear function depending on the stochastic variable $\boldsymbol{X}(t)$ and time t, while the second part $\boldsymbol{F}$ fluctuates on a fast time scale. We assume that this term can be written as

$$F_{\mathrm{i}}(\boldsymbol{X}(t), t) = \sum_{\mathrm{j}=1}^{d} g_{\mathrm{i,j}}(\boldsymbol{X}(t), t)\Gamma_{\mathrm{j}}(t) \tag{3.9.161}$$

where $\Gamma_{\mathrm{j}}(t)$ are random functions. Obviously, for such systems, the future is no longer uniquely determined by the initial conditions. As a special case, the dynamics of purely deterministic systems is also included in eq. (3.9.160) for $\boldsymbol{F} = \boldsymbol{0}$.

In 1908, Paul Langevin modelled Brownian motion - a random movement of a "large" particle immersed in a fluid resulting from collisions with liquid molecules due to their thermal motion - by an overdamped stochastic differential equation. The general form of a non-linear Langevin equation for a random variable $\boldsymbol{X}(t)$ is given as

$$\frac{\mathrm{d}\boldsymbol{X}(t)}{\mathrm{d}t} = \boldsymbol{N}(\boldsymbol{X}(t), t) + \boldsymbol{g}(\boldsymbol{X}(t), t)\boldsymbol{\Gamma}(t) \tag{3.9.162}$$

where $\boldsymbol{N}$ is the drift vector and $\boldsymbol{\Gamma}(t)$ describes the rapidly fluctuating Langevin forces, assumed in the following to be Gaussian distributed and δ-correlated, fulfilling

$$\langle \boldsymbol{\Gamma}(t) \rangle = \mathbf{0} \quad \text{and} \quad \langle \Gamma_{\rm i}(t)\Gamma_{\rm j}(t') \rangle = \delta_{\rm i,j}\delta(t - t') \tag{3.9.163}$$

$\boldsymbol{\Gamma}(t)$ is denoted white noise because it has constant spectral density with an infinite variance.

Some explanations are added to help us understand the stochastic differential equation (3.9.162) and Gaussian white noise, eq. (3.9.163). We write the Langevin equation in an incremental form using Itô's definition, see (Risken, 1996; Gardiner, 2004)

$$X_{\rm i}(t + \tau) - X_{\rm i}(t) \approx N_{\rm i}(\boldsymbol{X}(t), t)\tau + \sum_{\rm j} g_{\rm i,j}(\boldsymbol{X}(t), t)\Delta W_{\rm j} \tag{3.9.164}$$

Here, the first term describes a shift, while the second term is the random part; $\Delta W_{\rm j}$ are increments of independent Wiener processes which satisfy the relation

$$W_{\rm j}(t + \tau) - W_{\rm j}(t) = \int_t^{t+\tau} \Gamma_{\rm j}(t')\mathrm{d}t' \tag{3.9.165}$$

By means of the central limit theorem, cf. section 3.9.10, it can be shown that the stochastic variable $\Delta W_{\rm j}$ possesses a Gaussian probability distribution with zero mean $\langle \Delta W_{\rm j} \rangle = 0$ and variance $\langle \Delta W_{\rm j}^2 \rangle = \tau$. Equation (3.9.164) has to be interpreted in the Itô sense, i.e. the value $\boldsymbol{X}$ at time t_0 has drifted to the position $\boldsymbol{X}(t_1) = \boldsymbol{X}(t_0) + \tau \boldsymbol{N}(\boldsymbol{X}(t_0), t_0)$ superimposed by an increment of a random variable with centre $\boldsymbol{X}(t_0)$ and standard deviation $\sqrt{\tau}$ times $\boldsymbol{g}(\boldsymbol{X}(t_0), t_0)$. From the factor $\sqrt{\tau}$ it can be seen that this is a normal diffusion process. Exact definitions and proofs can be found for example in (Risken, 1996; Gardiner, 2004; Kloeden and Platen, 1999; Hänggi and Thomas, 1982).

Equation (3.9.164) can therefore be written as

$$\boldsymbol{X}(t + \tau) - \boldsymbol{X}(t) = \tau \boldsymbol{N}(\boldsymbol{X}(t), t) + \sqrt{\tau}\, \boldsymbol{g}(\boldsymbol{X}(t), t)\, \boldsymbol{\Gamma}(t) \tag{3.9.166}$$

Einstein was the first to explain Brownian motion in his pioneering work (Einstein, 1905). In contrast to the later work of Langevin, he described the motion by statistical methods on a macroscopic level. About 10 years later, Adriaan Fokker, a Dutch physicist and musician, and Max Planck, the originator of quantum physics, developed an evolution equation for the probability density function. This view leads to the Fokker-Planck equation, an equation of motion for the probability distribution of the fluctuating variables and a description equivalent to the Langevin equation (Risken, 1996).

In the Langevin equation (3.9.162), the fluctuating force has been modelled by δ-correlated white noise which has independent increments on subsequent time intervals. This has consequences for the selected class of underlying stochastic processes described in the following. Let us consider the evolution of the state vector $\boldsymbol{X}(t)$ at discrete time steps t_0, $t_1 = t_0 + \tau$, $t_2 = t_0 + 2\tau$, $\ldots$, where the increment $\tau \neq 0$ should be large compared with the time scale of the fluctuations. A complete knowledge of the stochastic process would be ensured if the infinite hierarchy of all n-point joint probability densities

$$f(\boldsymbol{x}_{\mathrm{n}}, t_{\mathrm{n}}; \ldots; \boldsymbol{x}_1, t_1; \boldsymbol{x}_0, t_0) = \left\langle \delta(\boldsymbol{x}_{\mathrm{n}} - \boldsymbol{X}(t_{\mathrm{n}})) \ldots \delta(\boldsymbol{x}_1 - \boldsymbol{X}(t_1)) \delta(\boldsymbol{x}_0 - \boldsymbol{X}(t_0)) \right\rangle \tag{3.9.167}$$

of the process were known. The probability of $\boldsymbol{X}$ having the value x_{i} at time t_{i} on condition that $\boldsymbol{X}$ take the values $x_{\mathrm{i}-1}, \ldots, x_0$ at times $t_{\mathrm{i}-1}, \ldots, t_0$ is then given by the conditional probability $p(x_{\mathrm{i}}, t_{\mathrm{i}} | x_{\mathrm{i}-1}, t_{\mathrm{i}-1}, \ldots, x_0, t_0)$. As described in section 3.9.4, the conditional probability for a general stochastic process is then given as

$$p(x_{\mathrm{i}}, t_{\mathrm{i}} | x_{\mathrm{i}-1}, t_{\mathrm{i}-1}, \ldots x_0, t_0) = \frac{f_{\mathrm{i}}(x_{\mathrm{i}}, t_{\mathrm{i}}, x_{\mathrm{i}-1}, t_{\mathrm{i}-1}, \ldots, x_0, t_0)}{f_{\mathrm{i}-1}(x_{\mathrm{i}-1}, t_{\mathrm{i}-1}, \ldots, x_0, t_0)} \tag{3.9.168}$$

Here, we concentrate on Markov processes, the simplest processes which are continuous in time. Their characteristic property is that there is no memory in the process in the sense that predictions for the future of the process are solely dependent on its current state. Therefore, the Markov property is equivalent to the following expression for the multiple conditioned probabilities

$$p(\boldsymbol{x}_{\mathrm{i}}, t_{\mathrm{i}} | \boldsymbol{x}_{\mathrm{i}-1}, t_{\mathrm{i}-1}; \ldots; \boldsymbol{x}_0, t_0) = p(\boldsymbol{x}_{\mathrm{i}}, t_{\mathrm{i}} | \boldsymbol{x}_{\mathrm{i}-1}, t_{\mathrm{i}-1}) \tag{3.9.169}$$

which is the probability of the system being in volume ΔV_{i} at time t_{i} given that it was in $\Delta V_{\mathrm{i}-1}$ at time $t_{\mathrm{i}-1}$. Thus, for Markov processes, the n-point pdf factorises as

$$f(\boldsymbol{x}_{\mathrm{n}}, t_{\mathrm{n}}; \ldots; \boldsymbol{x}_1, t_1; \boldsymbol{x}_0, t_0) = p(\boldsymbol{x}_{\mathrm{n}}, t_{\mathrm{n}} | \boldsymbol{x}_{\mathrm{n}-1}, t_{\mathrm{n}-1}) \ldots p(\boldsymbol{x}_1, t_1 | \boldsymbol{x}_0, t_0) f(\boldsymbol{x}_0, t_0) \tag{3.9.170}$$

and it suffices to know the transition probability and the initial distribution $f(\boldsymbol{x}_0, t_0)$ to define the n-point probability distribution. A characteristic feature of a Markov process is the so-called Chapman-Kolmogorov equation

$$p(\boldsymbol{x}_{\mathrm{j}}, t_{\mathrm{j}} | \boldsymbol{x}_{\mathrm{i}}, t_{\mathrm{i}}) = \int p(\boldsymbol{x}_{\mathrm{j}}, t_{\mathrm{j}} | \boldsymbol{x}_{\mathrm{k}}, t_{\mathrm{k}}) p(\boldsymbol{x}_{\mathrm{k}}, t_{\mathrm{k}} | \boldsymbol{x}_{\mathrm{i}}, t_{\mathrm{i}}) \mathrm{d}\boldsymbol{x}_{\mathrm{k}}. \tag{3.9.171}$$

for $t_{\mathrm{i}} < t_{\mathrm{k}} < t_{\mathrm{j}}$, which follows directly from

$$f(\boldsymbol{x}_{\mathrm{j}}, t_{\mathrm{j}}; \boldsymbol{x}_{\mathrm{i}}, t_{\mathrm{i}}) = \int f(\boldsymbol{x}_{\mathrm{j}}, t_{\mathrm{j}}; \boldsymbol{x}_{\mathrm{k}}, t_{\mathrm{k}}; \boldsymbol{x}_{\mathrm{i}}, t_{\mathrm{i}}) \mathrm{d}\boldsymbol{x}_{\mathrm{k}} \tag{3.9.172}$$

since the left-hand side is equal to $p(\boldsymbol{x}_{\mathrm{j}}, t_{\mathrm{j}} | \boldsymbol{x}_{\mathrm{i}}, t_{\mathrm{i}}) f(\boldsymbol{x}_{\mathrm{i}}, t_{\mathrm{i}})$; using eq. (3.9.169), we can write the integrand on the right-hand side as

$$p(\boldsymbol{x}_{\mathrm{j}}, t_{\mathrm{j}} | \boldsymbol{x}_{\mathrm{k}}, t_{\mathrm{k}}; \boldsymbol{x}_{\mathrm{i}}, t_{\mathrm{i}}) f(\boldsymbol{x}_{\mathrm{k}}, t_{\mathrm{k}}; \boldsymbol{x}_{\mathrm{i}}, t_{\mathrm{i}}) = p(\boldsymbol{x}_{\mathrm{j}}, t_{\mathrm{j}} | \boldsymbol{x}_{\mathrm{k}}, t_{\mathrm{k}}) p(\boldsymbol{x}_{\mathrm{k}}, t_{\mathrm{k}} | \boldsymbol{x}_{\mathrm{i}}, t_{\mathrm{i}}) f(\boldsymbol{x}_{\mathrm{i}}, t_{\mathrm{i}})$$

Finally, the factor $f(\boldsymbol{x}_{\mathrm{i}}, t_{\mathrm{i}})$ can be omitted because the initial distribution can be chosen arbitrarily.

We will now see that Gaussian fluctuations are decisive for the correspondence between the Langevin and the Fokker-Planck equation. The starting point is the definition

$$f(\boldsymbol{x}, t + \tau) = \int p(\boldsymbol{x}, t + \tau | \boldsymbol{x}', t) f(\boldsymbol{x}', t) \mathrm{d}\boldsymbol{x}' \tag{3.9.173}$$

Assuming, that the n-th order tensors of the conditional moments

$$\boldsymbol{M}^{(\mathrm{n})}(\boldsymbol{x}, t, \tau) = \int (\boldsymbol{x}' - \boldsymbol{x})^{\mathrm{n}} p(\boldsymbol{x}, t + \tau | \boldsymbol{x}', t) \mathrm{d}\boldsymbol{x}' \tag{3.9.174}$$

are known for $\mathrm{n} \geq 1$, the characteristic function of $p(\boldsymbol{x}, t+\tau|\boldsymbol{x}', t)$ can be written as a moment expansion according to eq. (3.9.82) as follows

$$\begin{aligned} \varphi(\boldsymbol{\omega}, \boldsymbol{x}, t, \tau) &= \int_{-\infty}^{\infty} p(\boldsymbol{x}, t+\tau|\boldsymbol{x}', t) \mathrm{e}^{i\boldsymbol{\omega}\cdot(\boldsymbol{x}'-\boldsymbol{x})} \mathrm{d}\boldsymbol{x}' \\ &= \sum_{\mathrm{n}=0}^{\infty} \frac{1}{\mathrm{n}!} (i\boldsymbol{\omega})^{\mathrm{n}} \boldsymbol{M}^{(\mathrm{n})}(\boldsymbol{x}, t, \tau) \end{aligned} \tag{3.9.175}$$

where the power n has to be interpreted as a tensor product. By means of the inverse Fourier transform, the conditional pdf in eq. (3.9.173) is then expressed as a series in terms of the moments. Using $\frac{\partial f(\boldsymbol{x},t)}{\partial t} = \lim_{\tau\to\infty}\big(f(\boldsymbol{x}, t+\tau) - f(\boldsymbol{x}, t)\big)$, the so-called Kramers-Moyal expansion is obtained

$$\frac{\partial f(\boldsymbol{x}, t)}{\partial t} = \sum_{\mathrm{n}=1}^{\infty} \frac{1}{\mathrm{n}!} (-\boldsymbol{\nabla})^{\mathrm{n}} \boldsymbol{D}^{(\mathrm{n})}(\boldsymbol{x}, t) f(\boldsymbol{x}, t) \tag{3.9.176}$$

where the Kramers-Moyal coefficients are defined as

$$\boldsymbol{D}^{(\mathrm{n})}(\boldsymbol{x}, t) = \lim_{\tau\to\infty} \frac{1}{\tau} \boldsymbol{M}^{(\mathrm{n})}(\boldsymbol{x}, t, \tau) \tag{3.9.177}$$

For stochastic processes with Gaussian fluctuations, the Pawula theorem states that the series must stop after the second term (Risken, 1996). Further, the coefficents $\boldsymbol{D}^{(1)}$ and $\boldsymbol{D}^{(2)}$, denoted drift vector and diffusion matrix, are related to the coefficients $\boldsymbol{N}$ and $\boldsymbol{g}$ of the Langevin equation (3.9.162) by

$$D_{\mathrm{i}}^{(1)}(\boldsymbol{x}, t) = N_{\mathrm{i}}(\boldsymbol{x}, t) \tag{3.9.178}$$

$$D_{\mathrm{i,j}}^{(2)}(\boldsymbol{x}, t) = \sum_{\mathrm{k}} g_{\mathrm{i,k}}(\boldsymbol{x}, t) g_{\mathrm{k,j}}(\boldsymbol{x}, t) \tag{3.9.179}$$

Thus, for Gaussian fluctuations which are completely defined by mean and variance, the Kramers-Moyal expansion reduces to the Fokker-Planck equation

$$\frac{\partial f(\boldsymbol{x}, t)}{\partial t} = \boldsymbol{L}^{(\mathrm{KM})} f(\boldsymbol{x}, t) \tag{3.9.180}$$

with the linear Kramers-Moyal differential operator

$$\boldsymbol{L}^{(\mathrm{KM})} = -\sum_{\mathrm{i}} \frac{\partial}{\partial x_{\mathrm{i}}} D_{\mathrm{i}}^{(1)}(\boldsymbol{x}, t) + \frac{1}{2} \sum_{\mathrm{i,j}} \frac{\partial^2}{\partial x_{\mathrm{i}} \partial x_{\mathrm{j}}} D_{\mathrm{i,j}}^{(2)}(\boldsymbol{x}, t) \tag{3.9.181}$$

The conditional pdf also obeys the so-called second Fokker-Planck equation

$$\frac{\partial p(\boldsymbol{x}, t|\boldsymbol{x}', t')}{\partial t} = \boldsymbol{L}^{(\mathrm{KM})} p(\boldsymbol{x}, t|\boldsymbol{x}', t') \tag{3.9.182}$$

The Fokker-Planck equation is a deterministic partial differential equation which describes the evolution of the (conditional) pdf over time. Numerical solutions can be obtained by Monte Carlo simulation or path-integral algorithms. The Monte Carlo method is based on a large number of simulations of eq. (3.9.166), which are used to generate the probability density function, see (Landau and Binder, 2005; Ermak and Buckholz, 1980). On the other hand, path integral solutions are obtained by repeated application of the Chapman-Kolmogorov equation (3.9.171), see (Risken, 1996). Figure 3.9.10 illustrates the evolution of the conditional probability density modelled in the Fokker-Planck equation (3.9.182) and its relation to sample paths generated by the Langevin equation (3.9.162).

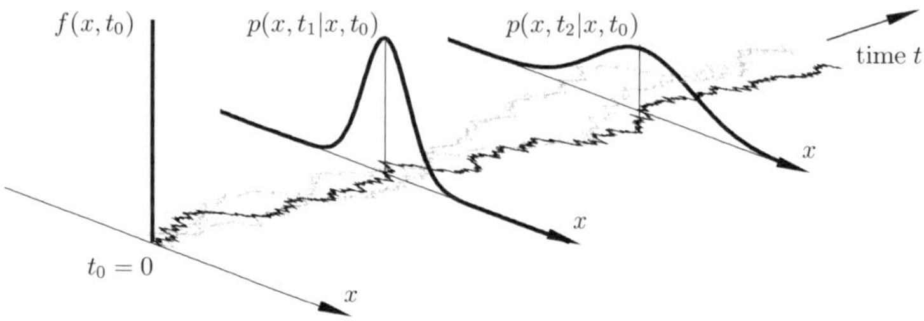

Fig. 3.9.10: Illustration of the evolution of the conditional probability density together with three sample paths generated by the Langevin equation. The initial probability density function $f(x, t_0) = \delta(x)$ has a sharp peak at $x = 0$ which widens and decays over time due to the random fluctuating forces and is shifted away from $x = 0$ due to the deterministic drift.

The ultimate aim of the analysis of time series is to uncover the underlying deterministic and stochastic dynamics – where one also has to distinguish between internal dynamic and external measurement noise – and to characterise the underlying dynamics by means of non-linear time series analysis described in the following chapters. The review papers (Friedrich et al., 2009; Friedrich et al., 2011) contain the details of the operational approach of the so-called Markov analysis which are summarised here in a few keywords: identify and separate dynamic noise from measurement noise, evaluate the Markov properties, partition the phase space into finite bins, estimate drift vector $\boldsymbol{D}^{(1)}$ and diffusion matrix $\boldsymbol{D}^{(2)}$ in these bins according to the above equations and finally verify the estimated Fokker-Planck equations.

A multitude of phenomena have been investigated by means of the Markov method; these include electrical circuits, wind-energy convertors, traffic flow, heart rhythms, epileptic brain dynamics and earthquake prediction, to name just a few examples. A far-reaching finding was the realisation that one can also formulate a Fokker-Planck equation in scale for scale-dependent processes which display a high complexity in space. Applications range from turbulent flows, financial data and surface roughness to earthquakes and cosmic background radiation. In Chapter 9, we present an application of the Markov analysis to velocity increments measured in the turbulent region of a round free jet.

3.10 Invariant Measure and Ergodic Orbits

3.10.1 Natural Invariant Density of the Logistic Map

In section 3.7, using the example of the logistic map, eq. (3.7.2), we demonstrated that a change of the system parameters in non-linear systems may trigger a series of bifurcations and hence lead to highly variant long-time behaviour resulting even in chaos. Fig. 3.10.1 shows the chaotic evolution of a sequence of points $\{x_0, x_1, x_2, \ldots\}$ generated by the logistic map for $\alpha = 4.0$ for the inital value $x_0 = 0.1$. Other initial values lead to different point sequences which are unpredictable as well. Nevertheless, there are statistical properties of the chaotic behaviour of such sequences which can indeed be reproduced. For their definiton, we make use of some concepts from probability theory introduced in the previous section 3.9.

If we consider the cascade of bifurcations displayed in fig. 3.7.6 for $2.8 \leqslant \alpha \leqslant 4.0$ and if we disregard the transient behaviour at the beginning of the iteration, we obtain the same diagram for almost all[1] initial values x_0 with $0 \leqslant x_0 \leqslant 1$. The diagram shows period doubling, periodic windows and chaotic domains. Different densities of the point distributions indicate how often certain subsets of the interval [0, 1] are visited. In fig. 3.10.2, a normalised histogram of the logistic map is shown for $\alpha = 4.0$. For this purpose, the interval [0, 1] is sub-divided into M = 200 sub-intervals of the same size. Following eq. (3.9.14), the relative frequency of occurrences $p_i = N_i/N$

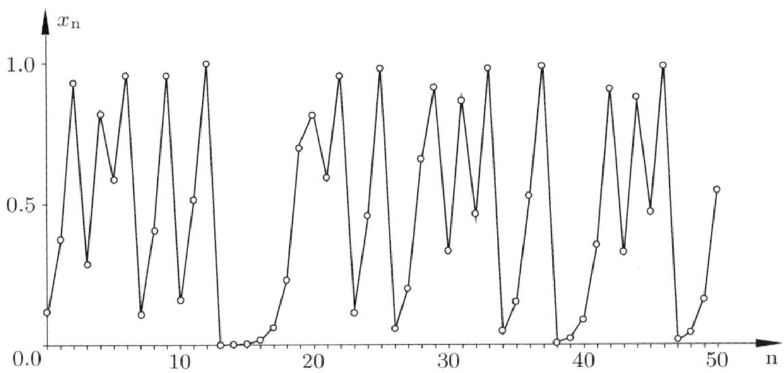

Fig. 3.10.1: Sequence of points of the logistic map for $x_0 = 0.1$

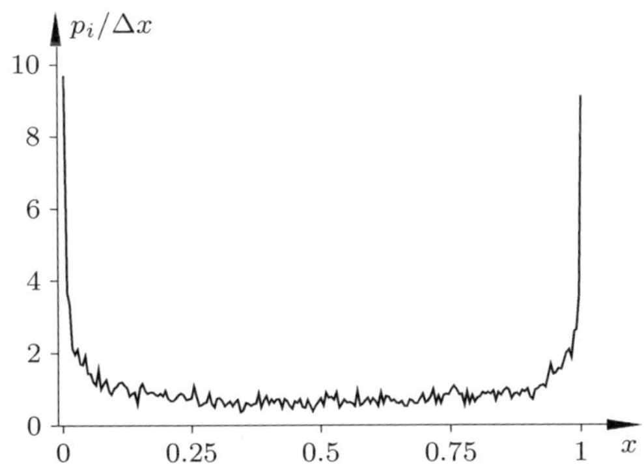

Fig. 3.10.2

Normalised histogram of the logistic map for $\alpha = 4.0$: N = 10^4 time steps, M = 200 subintervals

[1] In mathematics, the phrase *almost all* has a special meaning. It expresses that there are exceptions, which are, however, relatively rare or, more precisely, have Lebesgue measure zero. Fixed points, periodic points and their pre-images are such exceptions, like, for example, $x_0 = 0, \frac{1}{2}, \frac{3}{4}, 1$.

in the i-th sub-interval is displayed for an arbitrarily chosen sequence of $\mathrm{N} = 10^4$ points. Here, $\mathrm{N_i}$ denotes the number of points in the i-th interval. The longer the sequence of points and the higher the resolution, the smoother the distribution becomes. In the limit $\mathrm{N} \to \infty$, $\mathrm{M} \to \infty$, we obtain for almost all initial conditions x_0 a probability density distribution $\rho(x)$, which is invariant under the logistic map, cf. section 3.9.5. Formally, according to eq. (3.9.36), the density distribution of a discrete map $x_\mathrm{n} = f(x_{\mathrm{n}-1}) = f^\mathrm{n}(x_0)$ defined for the unit interval can be written as follows

$$\rho(x) = \lim_{\mathrm{N}\to\infty} \frac{1}{\mathrm{N}} \sum_{\mathrm{n}=0}^{\mathrm{N}} \delta[x - f^\mathrm{n}(x_0)] \tag{3.10.1}$$

where δ denotes the Dirac δ function. For an infinitely high resolution and an orbit of infinite length, all hits $x = x_\mathrm{n}$ are counted and the time average is taken. Usually, the density distribution is normalised, i.e. $\int_0^1 \rho(x)\,\mathrm{dx} = 1$ holds, and denoted *natural invariant density*. The integral over the density distribution

$$\mu(B) = \int_a^b \rho(x)\,\mathrm{d}x \tag{3.10.2}$$

is a probability measure for the sub-interval $B = [a, b] \subseteq [0, 1]$ which specifies the average fraction of time the orbit visits the interval B. While $\rho(x)$ is a function of the spatial coordinate x, the natural invariant density $\mu(B)$ is the probability assigned to a set of points. If $\mu(B)$ is independent of the arbitrarily chosen initial conditions, it is denoted *natural invariant measure*. There are only a few special systems for which the density distribution $\rho(x)$ can be determined analytically. In the particular case of the logistic map for $\alpha = 4$, $\rho(x)$ can be derived successfully, as will be shown next.

The idea is to transform the logistic map

$$x_{\mathrm{n}+1} = f(x_\mathrm{n}) \quad \text{with} \quad f(x) = 4x_\mathrm{n}(1 - x_\mathrm{n}) \tag{3.10.3}$$

by a transformation of variables $x \to y$ to a particularly simple piecewise linear map, the so-called *tent map* (see fig. 3.10.3) which, like the logistic map, has only one single maximum. At the very beginning of research in chaos theory, the properties of such maps were already studied intensively (Shaw, 1981b). In what follows, we will see that a characteristic feature of the tent map is that the density distribution can be calculated immediately independently of x.

We first show that the following transformation of variables (Ott, 1993)

$$x = \sin^2(\frac{\pi y}{2}) = \frac{1}{2}(1 - \cos \pi y) \quad \text{with} \quad 0 \leqslant y \leqslant 1 \tag{3.10.4}$$

converts the logistic map eq. (3.10.3) to the tent map. The transformation which maps the x-values of the unit interval [0, 1] to y-values of [0, 1] is one-to-one and, together with its inverse map, continuous. Such a transformation is denoted a homeomorphism. It guarantees that both maps have topologically equivalent properties. Introducing eq. (3.10.4) into both sides of eq. (3.10.3) yields

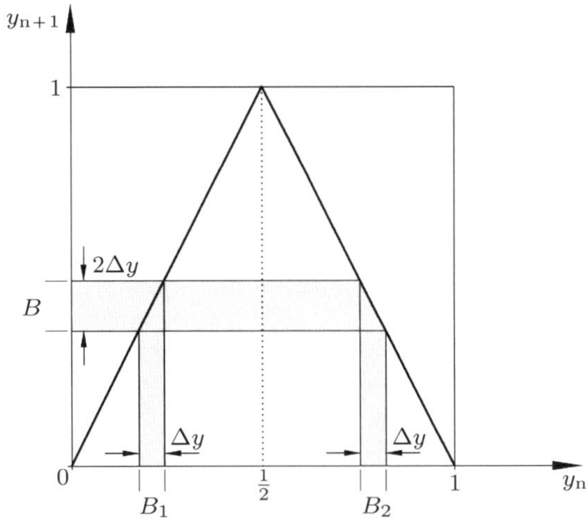

Fig. 3.10.3

Doubling of an interval of length Δy for the tent map

$$
\begin{aligned}
\sin^2\left(\frac{\pi y_{n+1}}{2}\right) &= 4\,\frac{1}{2}[\,1-\cos\pi y_n\,][\,1-\frac{1}{2}(1-\cos\pi y_n)\,] \\
&= 1-\cos^2\pi y_n \;=\; \sin^2\pi y_n
\end{aligned}
\tag{3.10.5}
$$

Hence, the argument has the form $\dfrac{\pi y_{n+1}}{2} = \pm\pi y_n + k\pi$, where k is an integer. Since y is taken from $[0,\,1]$, different signs for k have to be chosen in the subintervals $[0,\,\frac{1}{2}]$ and $[\frac{1}{2},\,1]$. As a result, one obtains the tent map described by a piecewise constant linear function

$$
y_{n+1} = g(y_n) \quad \text{with} \quad g(y) = \begin{cases} 2y & \text{for} \quad 0 \leqslant y \leqslant \frac{1}{2} \\ 2(1-y) & \text{for} \quad \frac{1}{2} \leqslant y \leqslant 1 \end{cases}
\tag{3.10.6}
$$

The characteristic properties of this function can be analysed much more easily than those of the topologically equivalent logistic map. By means of the tent map, all mechanisms leading to chaos in the case of one-dimensional maps can be studied.

Our next steps are as follows: first, we show that the orbits of the tent map also depend sensitively on the initial conditions, a property well-known from the logistic equation. We then determine the natural invariant density $\rho(y)$ of the tent map. Since eq. (3.10.4) is a homeomorphic transform, the inverse mapping finally leads to the desired density distribution $\rho(x)$ of the logistic map for $\alpha = 4$.

Let us turn first to the sensitivity of the initial conditions. The emergence of a constant factor 2 in eq. (3.10.6) suggests using a binary decomposition of an arbitrary initial condition y_0 between 0 and 1

$$
y_0 = \sum_{i=1}^{\infty} a_i\,\frac{1}{2^i} \;\;\hat{=}\;\; (\,.\,a_1\,a_2\,a_3\,a_4\ldots)
\tag{3.10.7}
$$

where the digits a_i can only take the values 0 or 1. Points lying in the left semi-interval characterised by $a_1 = 0$ are mapped to $y_1 = \sum_{i=1}^{\infty} a_{i+1}\,2^{-i}$, while points in the right semi-interval identified by $a_1 = 1$ are mapped to $y_1 = \sum_{i=1}^{\infty} T(a_{i+1})\,2^{-i}$, where the operator $T(a)$ simply exchanges zeroes and ones. As in the example of the Bernoulli shift, to be discussed in detail in section 5.6.1, the first binary digit is lost in each iteration, i.e. small deviations Δy are doubled, see fig. 3.10.3. Evidently, the increase of their magnitude depends on the modulus of the slope of the underlying map $|\frac{\mathrm{d}f}{\mathrm{d}y}|$ which has the constant value 2 in the case of the tent map. On the assumption that the initial value y_0 in eq. (3.10.7) has an accuracy $\varepsilon = 2^{-\mathrm{N}}$, the total information about its initial condition is obviously lost after N iterations. This immediately shows the sensitive dependency of the tent map on initial conditions, a typical characteristic of chaotic behaviour. The underlying mechanism leading to this sensitivity is a universal one, arising from the *stretching and folding* of intervals. In each iteration of the map, intervals are either stretched to double their length or folded back if they contain $y = 1/2$.

In the next step, we determine the density distribution $\rho(y)$ of the tent map. Examining the impact of one iteration step, we easily see that

$$\rho(y) = 1 \quad \text{for} \quad y \in [0, 1] \tag{3.10.8}$$

must hold. Indeed, considering an arbitrary set of initial values y_0 uniformly distributed over the unit interval, we can express the probability that $y_0 \in B = [a, b]$ lies in a subinterval of [0, 1] as

$$P(y_0 \in [a, b]) = \mu(B) = \int_a^b \rho(y)\,\mathrm{d}y = b - a \tag{3.10.9}$$

i.e. because of $\rho(y) = 1$, the natural measure just coincides with the length of the interval B. In particular, $\mu([0, 1]) \equiv 1$ holds for the unit interval in accordance with the certain event of axiom 2, section 3.9.3. In order to determine the effect of the tent map $g(y)$ on the density distribution, we make use of the relation

$$P\big(g(y_0) \in B\big) = P\left(y_0 \in g^{-1}(B)\right) \tag{3.10.10}$$

which states that the probability is conserved because, for typical orbits, the intervals of pre-images and images are visited with the same frequency. The pre-image of B is composed of two disjoint intervals B_1 and B_2, cf. fig. 3.10.3

$$g^{-1}(B) = B_1 \cup B_2 \tag{3.10.11}$$

each half the length of the original interval B, leading to

$$P\left(y_0 \in g^{-1}(B)\right) = \mu(B_1) + \mu(B_2) = b - a \tag{3.10.12}$$

It follows immediately that the probability density $\rho(y)$ is conserved under iterations. Although the impact of one iteration step on the density distribution cannot in general be specified explicitly, it is possible to formulate an evolution equation for the density distribution, the so-called Frobenius-Perron equation, which will be introduced in the next section 3.10.2.

Finally, we apply the inverse map $y \to x$ according to eq. (3.10.4). Since $\rho(x)\,\mathrm{d}x = \rho(y)\,\mathrm{d}y$ is conserved, inserting the transformation of variables given in eq. (3.10.4) and using the invariant density of the logistic map for $\alpha = 4$ leads to

$$\rho(x) = \frac{1}{\frac{\pi}{2}\sin\pi y} = \frac{2}{\pi\sqrt{1-(1-2x)^2}}$$

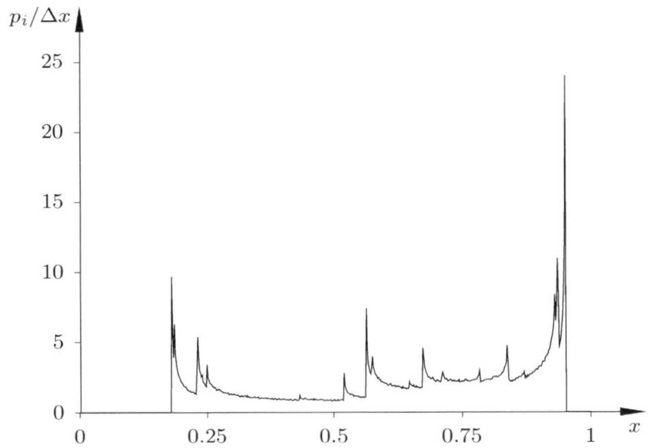

Fig. 3.10.4: Normalised histogram of the logistic map for $\alpha = 3.8$: $\mathrm{N} = 10^6$ time steps, $\mathrm{M} = 2000$ subintervals

or

$$\rho(x) = \frac{1}{\pi\sqrt{x(1-x)}} \tag{3.10.13}$$

which is normalised as well. Apart from a constant factor, $\rho(x)$ coincides with the histogram in accordance with eq. (3.10.13) if one approaches the distribution in the limit $\mathrm{N} \to \infty, \quad \mathrm{M} \to \infty$.

Equation (3.10.13) shows that $\rho(x)$ has singularities for $x = 0$ and $x = 1$, i.e. the probability that points visit the neighborhood of the unstable fixed points $x = 0$ or $x = 1$ in the course of iterations is much higher than hits in the middle of the interval.

In general, for other α-values, it is not possible to find an explicit variable transformation which transforms the logistic map to the tent map. Hence, the density distribution can no longer be specified explicity. The numerical computation of, for example, $\rho(x)$ for $\alpha = 3.8$ as shown in fig. 3.10.4 demonstrates that, with increasing resolution, more and more singularities emerge. Detailed studies of the density distribution of the logistic map for general parameter values α can be found in (Collet and Eckmann, 1980; Ott, 1993), see also (Tél and Gruiz, 2006).

3.10.2 Frobenius-Perron Equation and Ergodicity

In the previous sections 3.1 – 3.7, we used *geometrical* or *topological* concepts to describe the long-term behaviour of dynamical systems. This view, introduced by Poincaré, has been extraordinarily successful and has led among other things to the development of the theory of dynamical systems. Graphical displays of strange attractors in low-dimensional phase spaces provide very clear information on the long-term behaviour of chaotic systems and allow a classification of the systems.

However, for motions in higher-dimensional spaces, such graphical illustrations are no longer possible. In this case, concepts based on measure theory and probability theory combined in the so-called *ergodic theory* are much better suited. The intended purpose of this theory is to use *statistical* features to characterise the orbits. This enables us to distinguish attractors and motions by means of their characteristic quantities and to assign different complexity levels to them (Schuster and Just, 2006).

In chapter 5, we discuss such characteristic quantities like, for example, Lyapunov exponents and KS entropy, which describe the long-time behaviour of dynamical systems and are both based on statistical concepts. Lyapunov exponents quantify the average divergence of orbits starting from adjacent initial conditions while the KS entropy measures the average information produced by the system (more details on the concept of *information* are given in section 5.5.3).

When striving to make statistical statements, one first has to define how mean values are to be calculated, assuming that transients have already died out. Either time averages are taken over an individual sample trajectory or spatial respectively ensemble averages taken over the whole distribution of orbits at a particular time. For the quantities mentioned previously, time averages are taken since, for numerical calculations, it is much easier to confine oneself to one single orbit than to an whole ensemble of orbits. Then, however, the question arises whether the result depends upon the choice of the specific initial condition or whether it is characteristic of the whole dynamical system. This exactly defines the problem ergodic theory deals with.

In the previous section, we derived the invariant density distribution explicitly for the logistic map in the case of $\alpha = 4$. Generally, an infinite set of initial conditions with an arbitrary distribution $\rho_0(x)$ can be selected for the determination of the invariant density $\rho(x)$. The map $f(x)$ can then be applied to each initial condition, leading to a new distribution $\rho_1(x)$

$$\rho_1(x) = \int \rho_0(y)\, \delta[x - f(y)]\, \mathrm{d}y \tag{3.10.14}$$

Repeating this procedure n times yields an evolution equation for the density

$$\rho_{\mathrm{n}+1}(x) = \int \rho_{\mathrm{n}}(y)\, \delta[x - f(y)]\, \mathrm{d}y \tag{3.10.15}$$

which is denoted the *Frobenius-Perron equation* (Farmer, 1982c; Ott, 1993). In the limit $\mathrm{n} \to \infty$, one obtains a stationary distribution as the fixed-point solution of eq. (3.10.15), i.e. the invariant density $\rho(x)$ which was already introduced in the previous section 3.10.1 for the logistic equation.

In general, the invariant density is *not uniquely* determined. Fortunately, however, not all stationary solutions of eq. (3.10.15) are relevant in practice. For example, an unstable fixed point x^* of a dynamical system has the invariant density $\rho(x) = \delta(x - x^*)$. Since it is generally not possible to specify a fixed point x^* with infinite precision and small deviations are amplified over time, the corresponding invariant density cannot be observed in reality, for example in computer calculations or measurements, and is therefore physically irrelevant (Eckmann and Ruelle, 1985). The Frobenius-Perron equation is frequently applied for the numerical calculation of $\rho(x)$. Starting from a constant initial distribution $\rho_0(x)$, one obtains a good approximation of the physically relevant invariant density $\rho(x)$ after only a few iterations.

The probability measure $\mu(B)$ of the set B which was defined in eq. (3.10.2) is a much more general concept than that of the density distribution $\rho(x)$. If the point set has, for example, a fractal structure like the Cantor set to be discussed in section 5.5.1, it is no longer possible to specify a density distribution, whereas a singular probability measure can be determined (Farmer et al., 1983; Halsey et al., 1986). If, moreover, the invariant measure exhibits locally strong fluctuations, it can be characterised by means of a *multifractal spectrum*, which will be discussed in detail in section 8.5.1.

The question now is whether the invariant measure which can be observed also possibly depends on the chosen *initial condition*. Obviously, this is always the case when the dynamical system possesses more than one attractor. We only mention here the system $\dot{x_1} = x_1 - {x_1}^3$, $\dot{x_2} = -x_2$ (Eckmann and Ruelle, 1985). It has two stable fixed points $x_1 = \pm 1, x_2 = 0$ together with an unstable fixed point at the origin. Consequently, for initial conditions chosen from different basins of attraction, one obtains different probability measures. In this, case the invariant measure is denoted *decomposable measure*.

If, however, the invariant measure is not decomposable, i.e. if the time average does not depend on the choice of initial conditions, it is denoted *ergodic* or *natural invariant measure*. This case is the very one which is of practical importance for the calculation of characteristic quantities like the Lyapunov exponents and the KS entropy.

The *ergodic theorem of Birkhoff-Khinchin* formulates the interchangeability of time averages and ensemble averages (Billingsley, 1965; Eckmann and Ruelle, 1985) as follows: Let μ be an ergodic measure. Then, for every continuous function $\varphi(x)$ and for almost all initial conditions, we have

$$\lim_{\mathrm{N}\to\infty} \frac{1}{\mathrm{N}} \sum_{\mathrm{n}=0}^{\mathrm{N}} \varphi(x_{\mathrm{n}}) = \int_0^1 \varphi(x)\,\mathrm{d}\mu(x) \tag{3.10.16}$$

The theorem is analogous to the thermodynamic mean value in classical statistical mechanics. In 1871, Boltzmann formulated the ergodic hypothesis, which states that, for systems in thermodynamic equilibrium, averages in time can be replaced by ensemble averages (Boltzmann, 1871), cf. also (Schuster and Just, 2006; Tél and Gruiz, 2006). In ergodic systems, a phase trajectory approaches every arbitrarily small neighboorhood of any point in the phase space within finite time.

At the end of this chapter, we return to the logistic map for $\alpha = 4$ and demonstrate that the system is ergodic and that the stationary solution of the Frobenius-Perron equation is identical to the invariant density distribution found in eq. (3.10.13).

It is sufficient to verify that the tent map is ergodic. We have to show that the iterated point sequence $\{y_0, y_1, y_2, \ldots\}$ approaches every randomly chosen point y in the unit interval arbitrarily closely. To put it more precisely: if we choose a real initial value y_0 for the iteration and arbitrarily select a point $y = (.a_1, a_2, \ldots, a_{\mathrm{N}}) \in (0,1)$ with a precision $\varepsilon = 2^{-\mathrm{N}}$ in phase space we will make clear that an infinite number of iterations of y_0 lie in a ε-neighborhood of y. Periodic points and their pre-images are exceptions because their orbits contain only a finite number of points. One can show that the tent map contains a countably infinite set of periodic orbits. However, since there is an uncountable number of real initial conditions, choosing an inital value randomly yields a non-periodic orbit with probability 1.

The proof of ergodicity is provided by number theory (Hardy and Wright, 1979). It can be shown that almost all irrational numbers (all but a set of Lebesgue measure zero) contain in their binary expansion any arbitrarily chosen finite sequence of zeroes and ones infinitely often, i.e. in particular they contain the sequence $\{a_0, a_1, \ldots, a_N\}$ for the binary representation of y. In each iteration of y_0, these sequences are shifted one digit to the left until $g^{\mathrm{n}}(y_0)$ coincides with y up to the order $\varepsilon = 2^{-\mathrm{N}}$.

Finally, we want to show that the invariant density distribution of eq. (3.10.13) for the logistic map $f(x) = 4x(1-x)$ satisfies the Frobenius-Perron equation. As an abbreviation, we use $y = f(x)$. According to eq. (3.10.13), the probability of finding points of the map in a small interval $I(y)$ of length Δy is given by

$$P[I(y)] = \rho(y)\Delta y = \frac{\Delta y}{\pi\sqrt{y(1-y)}} \tag{3.10.17}$$

Pre-images of points in the interval $I(y)$ are contained either in $I(x)$ or in $I(1-x)$, both of which have the length Δx given as

$$\Delta y = |f'(x)|\Delta x = |f'(1-x)|\Delta x \tag{3.10.18}$$

The Frobenius-Perron equation (3.10.15) can thus be written in the form

$$\rho(y)\Delta y = \rho(x)\Delta x + \rho(1-x)\Delta x$$

or, together with eq. (3.10.18),

$$\rho(y) = \frac{\rho(x)}{|f'(x)|} + \frac{\rho(1-x)}{|f'(1-x)|} \tag{3.10.19}$$

Inserting eq. (3.10.13) for $\rho(x)$ and $|f'(x)| = |f'(1-x)| = 4|(1-2x)|$ for the derivatives, one obtains for the right-hand side *rhs* of eq. (3.10.19)

$$rhs = \frac{2}{4|1-2x|}\,\frac{1}{\pi\sqrt{x(1-x)}} \tag{3.10.20}$$

On account of $1 - y = (1-2x)^2$, the left-hand side of eq. (3.10.19) yields

$$\rho(y) = \frac{1}{\pi\sqrt{y(1-y)}} = \frac{1}{\pi|1-2x|\sqrt{x(1-x)}} \tag{3.10.21}$$

i.e. eqs. (3.10.20) and (3.10.21) are in agreement. Equation (3.10.13) thus satisfies the Frobenius-Perron equation.

4 Dynamical Systems without Dissipation

Scientia et potentia humana in idem coincidunt,
quia ignoratio causae destituit effectum
Francis Bacon, Novum Organum (1620)

In the following chapter, our intention is to present an introductory discussion of conservative or Hamilton systems. By this, we mean those systems in which the total mechanical energy is preserved, i.e. no loss of energy due to friction or damping occurs. We direct our attention in particular to so-called multi-body problems. This category includes most of the standard problems of celestial mechanics since loss of energy, caused, for example, by tidal friction, can generally be ignored here. It should be noted, however, that the assumed motion of the planets in our solar system over aeons, apparently unshakeable in their temporal invariance and predictability, does not necessarily reproduce a generally valid behaviour of the stars and other celestial bodies. To illustrate such subtle considerations, we draw the reader's attention to one of Saturn's outer moons, Hyperion, which experiences a tumbling, irregular motion (probably chaotic) over long periods. Similar complex dynamic aspects generate the appearance of meteors. It may also be expedient to consider the motion of sub-microscopic particles as occurring, for example, in fusion reactors, using the Hamiltonian methodology. In the course of this and the subsequent chapter, Chapter 5, we shall observe that the characteristics of conservative systems differ fundamentally from those of dissipative systems.

4.1 Hamilton Equations for Conservative Systems

Hamilton's brilliant methodology has repeatedly given the formal development of classic dynamics decisive impulses. Seen historically, the realisation that the determination of a single function, namely the Hamilton function H, is sufficient to generate the equations of motion and thus to describe a deterministic motion was in accordance with the laws of reason; hence, one often spoke of rational mechanics. Once the Hamilton function can be transformed into a normal form with so-called action and angle variables, the integration of the system becomes trivial. Throughout the 19th century, mathematicians and physicists attempted to reduce the Hamilton function to this normal form and thus to establish the integrability of dynamical systems. However, the great Henri Poincaré proved in a celebrated treatise (Poincaré, 1899) that even the limited three-body problem was not integrable; this put an end to these hopes. Hamilton's formalism flourished anew with the development of Bohr's atomic model. With the help of the action

J. Argyris et al., *An Exploration of Dynamical Systems and Chaos*,
DOI: 10.1007/978-3-662-46042-9_4

and angle variables, it proved possible within the framework of classical mechanics to quantify the motion directly and thus gain a deeper understanding of the wave and corpuscle character of light. Such successes did not, however, exhaust the study of classical dynamics. In the middle of the last century, in particular, another stormy development took place, influenced decisively by the distinguished mathematicians Kolmogorov, Siegel, Arnold and Moser. The greatest advantage of Hamilton's formulation lies in the fact that it yields a deeper insight into the fundamental framework of mechanics and into its structure; as an elegant calculation tool for quantitative considerations, however, it plays a minor role. On the other hand, we should remember that Hamilton's approach is of fundamental importance for qualitative investigations of deterministic motion including chaos.

The authors apologise to the readers that this single volume cannot give a complete overview of the methodology of Hamilton's equations. From the wide range of textbooks on mechanics which discuss equations of motion and particularly their Hamiltonian formulation, we refer to two classics (Sommerfeld, 1964; Goldstein, 1978).

In the present textbook, we take d'Alembert's principle as a starting point and develop a concise overview of Hamilton's equations of motion. We consider a system of N mass points containing the masses m_k ($\mathrm{k} = 1, 2, \ldots, \mathrm{N}$) having the position vectors $\boldsymbol{r}_\mathrm{k}$. We assume that the coordinates of the position vectors are subject to b constraints of the form

$$g_\mathrm{i}(\boldsymbol{r}_1, \boldsymbol{r}_2, \ldots, \boldsymbol{r}_\mathrm{N}) = 0 \qquad \mathrm{i} = 1, 2, \ldots \mathrm{b} \tag{4.1.1}$$

These are expressed exclusively in components of the position vectors, not their derivatives, and time t does not appear explicitly. According to a classification of Hertz respectively Boltzmann such constraints are called holonomic-scleronomic, see (Päsler, 1968). Each of the equations (4.1.1) forces the mass points to move on a specific rigid hypersurface. Let us consider, for example, a mass point on an inclined plane which, in this case, defines a primitive hypersurface. We denote the number of degrees of freedom of this system by f; thus,

$$\mathrm{f} = 3\mathrm{N} - \mathrm{b} \tag{4.1.2}$$

Newton's equations of motion for a system of N mass points subject to external forces $\boldsymbol{R}_\mathrm{k}$

$$\boldsymbol{R}_\mathrm{k} - m_\mathrm{k}\ddot{\boldsymbol{r}}_\mathrm{k} = \boldsymbol{o}, \qquad \mathrm{k} = 1, 2, \ldots \mathrm{N} \tag{4.1.3}$$

only apply if no constraints are imposed. Introducing the auxiliary forces

$$\overline{\boldsymbol{R}}_\mathrm{k} = \boldsymbol{R}_\mathrm{k} - m_\mathrm{k}\ddot{\boldsymbol{r}}_\mathrm{k}, \qquad \mathrm{k} = 1, 2, \ldots \mathrm{N} \tag{4.1.4}$$

we can regard Newton's equations of motion formally as N equilibrium conditions

$$\overline{\boldsymbol{R}}_\mathrm{k} = \boldsymbol{o} \tag{4.1.5}$$

i.e. the introduction of d'Alembert's inertial forces $-m_\mathrm{k}\ddot{\boldsymbol{r}}_\mathrm{k}$, formally reduces the dynamical problem to a static equilibrium one. Lagrange had the brilliant idea (Lagrange, 1788) of applying the principle of virtual work to the forces $\overline{\boldsymbol{R}}_\mathrm{k}$ and so established d'Alembert's principle in the Lagrange formulation

$$\delta W = \sum_{k=1}^{N} \bar{\boldsymbol{R}}_k^t \delta \boldsymbol{r}_k = \sum_{k=1}^{N} [\boldsymbol{R}_k - m_k \ddot{\boldsymbol{r}}_k]^t \delta \boldsymbol{r}_k = 0$$

or, in vector formulation,

$$\delta W = \bar{\boldsymbol{R}}^t \delta \boldsymbol{r} = [\boldsymbol{R} - \boldsymbol{m}\ddot{\boldsymbol{r}}]^t \delta \boldsymbol{r} = 0 \tag{4.1.6}$$

where

$$\begin{aligned} \boldsymbol{r} &= \{\ \boldsymbol{r}_1 \quad \boldsymbol{r}_2 \quad \dots \quad \boldsymbol{r}_k \quad \boldsymbol{r}_N\ \} \\ \boldsymbol{R} &= \{\ \boldsymbol{R}_1 \quad \boldsymbol{R}_2 \quad \dots \quad \boldsymbol{R}_k \quad \boldsymbol{R}_N\ \} \\ \boldsymbol{m} &= \lceil\ \boldsymbol{m}_1 \quad \boldsymbol{m}_2 \quad \dots \quad \boldsymbol{m}_k \quad \boldsymbol{m}_N\ \rfloor \end{aligned} \tag{4.1.7}$$

and

$$\boldsymbol{m}_k = m_k \boldsymbol{I}_3 \tag{4.1.8}$$

In contrast to Newton's equations of motion, eq. (4.1.6) is always valid, irrespective of whether constraints exist or not. In fact, if one wants to deduce the equations of motion for a constrained system of mass points, then one should start from d'Alembert's principle and establish in this way the Lagrange equations of the first kind. D'Alembert's principle is usually expressed as a variational principle in which an arbitrarily chosen state of the system $\boldsymbol{r}(t)$ at a given instant t is compared with neighbouring trajectories $\boldsymbol{r}(t) + \delta\boldsymbol{r}(t)$ which evolve from $\boldsymbol{r}(t)$ through virtual displacements, time not being varied ($\delta t = 0$). Due to the vector formulation of eq. (4.1.6), one is not restricted to a particular system of coordinates.

In contrast to our previous differential approach, variational principles in integral form prove to be particularly suitable for many applications. One such principle is Hamilton's principle in which the state of the overall system over a finite period of time from t_0 to t_1 is compared with virtual neighbouring states.

Hamilton's principle is based on the assumption that the trajectory which is actually adopted and its neighbouring virtual trajectories are traversed synchronously from $t = t_0$ to $t = t_1$, recollecting that the initial and final points of the trajectory are not varied (Sommerfeld, 1964), i.e.

$$\begin{aligned} \delta \boldsymbol{r}_k &= \boldsymbol{o} \quad \text{for} \quad t = t_0,\ t = t_1 \qquad \mathrm{k} = 1, 2, \dots \mathrm{N} \\ \delta t &= 0 \end{aligned} \tag{4.1.9}$$

In order to deduce Hamilton's principle, we start from d'Alembert's principle, eq. (4.1.6), which, for this purpose, is best expressed in Cartesian coordinates. Transforming in conjunction with eq. (4.1.9) and integrating over the fixed time interval $[t_0, t_1]$, we obtain the statement (Sommerfeld, 1964)

$$\int_{t_0}^{t_1} (\delta T + \delta W) \mathrm{d}t = 0 \tag{4.1.10}$$

where T is the kinetic energy of the system and W the work of the external forces on the virtual displacements. Since we are only considering conservative systems in this chapter, the external forces are derivable from a potential U; thus,

$$\int_{t_0}^{t_1} \delta W \mathrm{d}t = -\int_{t_0}^{t_1} \delta U \mathrm{d}t \tag{4.1.11}$$

Noting that time may not be varied, we can then interchange the integration and variation operations and deduce Hamilton's principle for conservative systems

$$\delta \int_{t_0}^{t_1} (T - U)\mathrm{d}t = 0 \tag{4.1.12}$$

This contains only physical quantities and does not depend on a particular system of coordinates. The integrand is the so-called Lagrange function

$$L = T - U \tag{4.1.13}$$

occasionally also called "free energy". Hamilton's principle, eq. (4.1.12), states that the motion takes place in such a way that the variation of the time integral over the Lagrange function vanishes for fixed values t_0 and t_1, i.e. the time integral of the Lagrange function forms an extremum. Since the quantity which has to be varied has the dimension of an *action*, i.e. energy × time, Hamilton's principle used to be referred to as the "principle of least action". Equation (4.1.12) can also be written in the form

$$\delta\left(\frac{1}{t_1 - t_0}\int_{t_0}^{t_1} L\mathrm{d}t\right) = 0 \tag{4.1.14}$$

This formulation means that, of all conceivable motions in nature, those favoured satisfy the condition of a stationary value for the temporal mean value of the free energy. All equations of motion for conservative systems of mass points can be deduced from both d'Alembert's and Hamilton's principle, i.e. eqs. (4.1.6) and (4.1.12) may be considered the fundamental expressions for all rigid-body mechanics.

If b constraints are imposed on a system with N mass points, then one can introduce, instead of 3N dependent coordinates, f = 3N − b new *independent* coordinates q_k which are assembled in the vector

$$\boldsymbol{q} = \{q_1 \quad q_2 \quad \cdots\cdots \quad q_\mathrm{f}\} \tag{4.1.15}$$

and which are called generalised coordinates. The constraint conditions are then complied with automatically. If the generalised velocities

$$\dot{\boldsymbol{q}} = \{\dot{q}_1 \quad \dot{q}_2 \quad \cdots\cdots \quad \dot{q}_\mathrm{f}\} \tag{4.1.16}$$

are also known, the state of the system is established completely. The Lagrange equations of the second kind follow immediately from Hamilton's principle if one applies generalised coordinates $\boldsymbol{q}$ and $\dot{\boldsymbol{q}}$ and the calculus of variations (Päsler, 1968). The Euler equations

$$\frac{\mathrm{d}}{\mathrm{d}t}\left(\frac{\partial L}{\partial \dot{\boldsymbol{q}}^\mathrm{t}}\right) - \frac{\partial L}{\partial \boldsymbol{q}^\mathrm{t}} = \boldsymbol{o} \tag{4.1.17}$$

appertaining to the variational problem

$$\delta \int_{t_0}^{t_1} L(\boldsymbol{q},\dot{\boldsymbol{q}})\mathrm{d}t = 0 \tag{4.1.18}$$

are known as the Lagrange equations of the second kind for conservative systems, which lead to a stationary value of the variational problem, eq. (4.1.18). The knowledge of the Lagrange function $L(\boldsymbol{q},\dot{\boldsymbol{q}})$ thus suffices to define the motion of the mechanical system completely. The Lagrange equations of motion of the second kind form a set of f ordinary second-order differential equations in the generalised coordinates. The solution of this set requires 2f initial conditions

$$\boldsymbol{q}(t_0) = \boldsymbol{q}_0 \quad \text{and} \quad \dot{\boldsymbol{q}}(t_0) = \dot{\boldsymbol{q}}_0 \tag{4.1.19}$$

Although constraints are included *ab initio* in the Lagrange equations, conservation laws (energy, impulse) and symmetry properties of the system do not lead to a reduction of the problem size; this is, on the other hand, possible when applying Hamilton's formalism, to which we now turn our attention. In order to establish Hamilton's equations of motion, we first reduce the set of equations (4.1.17) to a set of 2f first-order differential equations. If we were, as is usual, to include the velocity $\dot{\boldsymbol{q}}$ as a new variable, the resulting system would not acquire a particularly concise form. If, however, new variables p_i are introduced via the relationship

$$\boldsymbol{p} = \frac{\partial L}{\partial \dot{\boldsymbol{q}}^\mathrm{t}}, \qquad \boldsymbol{p} = \{p_1 \quad p_2 \quad \ldots\ldots \quad p_\mathrm{f}\} \tag{4.1.20}$$

we obtain mnemonically convenient symmetrical first-order differential equations. Since the quantities p_i are momenta when orthogonal coordinates are used, they are called generalised coordinates.

While the Lagrange equations of motion of the second kind are deduced from the Lagrange function $L = T - U$, the so-called Hamilton function H plays a central role in Hamilton's equations; it is defined by

$$H(\boldsymbol{p},\boldsymbol{q}) = \boldsymbol{p}^\mathrm{t}\dot{\boldsymbol{q}} - L(\boldsymbol{q},\dot{\boldsymbol{q}}) \tag{4.1.21}$$

We now form on the one hand the total derivative of the left-hand side

$$\mathrm{d}H = \frac{\partial H}{\partial \boldsymbol{p}}\mathrm{d}\boldsymbol{p} + \frac{\partial H}{\partial \boldsymbol{q}}\mathrm{d}\boldsymbol{q} \tag{4.1.22}$$

and on the other hand that of the right-hand side of eq. (4.1.21)

$$\mathrm{d}H = \dot{\boldsymbol{q}}^\mathrm{t}\mathrm{d}\boldsymbol{p} + \boldsymbol{p}^\mathrm{t}\mathrm{d}\dot{\boldsymbol{q}} - \frac{\partial L}{\partial \boldsymbol{q}}\mathrm{d}\boldsymbol{q} - \frac{\partial L}{\partial \dot{\boldsymbol{q}}}\mathrm{d}\dot{\boldsymbol{q}} \tag{4.1.23}$$

If we substitute the generalised momenta $\boldsymbol{p}$ of eq. (4.1.20) into eq. (4.1.23), the second and fourth summands cancel one another out. Moreover, the Lagrange equations of the second kind, eq. (4.1.17), yield in conjunction with eq. (4.1.20)

$$\dot{\boldsymbol{p}} = \frac{\partial L}{\partial \boldsymbol{q}^\mathrm{t}} \tag{4.1.24}$$

so that finally, a comparison of both expressions for the total differential of H, eqs. (4.1.22) and (4.1.23), yields a completely symmetrical, so-called canonical form of the equations of motion

$$\dot{\boldsymbol{p}} = -\frac{\partial H}{\partial \boldsymbol{q}^{\mathrm{t}}}$$
$$\dot{\boldsymbol{q}} = \frac{\partial H}{\partial \boldsymbol{p}^{\mathrm{t}}} \tag{4.1.25}$$

These are called Hamilton's equations of motion. The variables p_{i} and q_{i} are denoted canonically conjugate. If we form the time derivative of the Hamilton function $H(\boldsymbol{p}, \boldsymbol{q})$

$$\dot{H} = \frac{\mathrm{d}H}{\mathrm{d}t} = \frac{\partial H}{\partial \boldsymbol{p}}\dot{\boldsymbol{p}} + \frac{\partial H}{\partial \boldsymbol{q}}\dot{\boldsymbol{q}} = \dot{\boldsymbol{q}}^{\mathrm{t}}\dot{\boldsymbol{p}} - \dot{\boldsymbol{p}}^{\mathrm{t}}\dot{\boldsymbol{q}} = 0 \tag{4.1.26}$$

then, due to eq. (4.1.25), we recognise that H is a constant and can show that the Hamilton function coincides with the total energy E of the system (Päsler, 1968).

The characteristic quantity in Hamilton's equations is thus the total energy

$$H = T + U = E \tag{4.1.27}$$

if, as we always presume here, $H(\boldsymbol{p}, \boldsymbol{q})$ does not depend explicitly on time.

In the case of Hamilton systems with f degrees of freedom, the phase space is spanned by the conjugate variables q_{i} and p_{i} and has the dimension 2f. The motion of a conservative system differs fundamentally from those of dissipative systems. The main cause of this lies in the fact that the phase flow of the Hamilton equations (4.1.25) preserves the volume of an element in the phase space. This has far-reaching consequences. For example, Hamilton systems can never be asymptotically stable; in particular, no point attractors exist, no limit cycles and no strange attractors. Nevertheless, "irregular'" or "chaotic" regions may occur which are, however, closely interwoven with regular regions.

In the case of a system with one degree of freedom, the property of area preservation can be directly illustrated geometrically. As before, we presume that Hamilton's function $H(\boldsymbol{p}, \boldsymbol{q})$ does not depend explicitly on time. Since $H = E$ is a constant of motion, the trajectories coincide with the contour lines $H = \text{const}$ (see fig. 4.1.1).

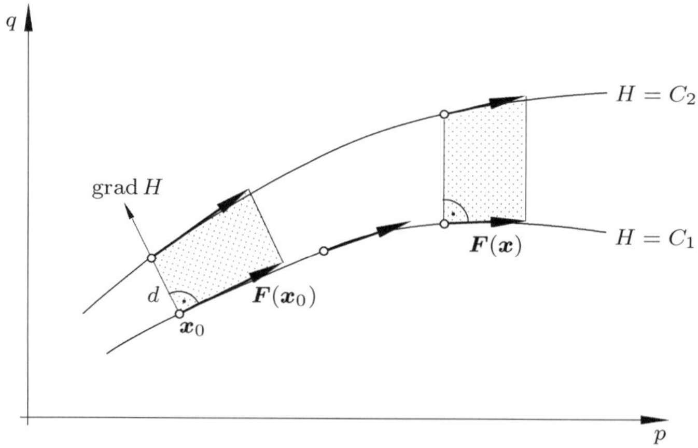

Fig. 4.1.1

On the area preservation of conservative systems

We consider a small rectangular surface element in the phase space (p, q) between two neighbouring contour lines. The length of one side is equal to the modulus of the phase velocity $\boldsymbol{F}$. From Hamilton's equations, eq. (4.1.25), we then obtain

$$|\boldsymbol{F}| = \left|\left\{-\frac{\partial H}{\partial q} \quad \frac{\partial H}{\partial p}\right\}\right| = \left[\left(\frac{\partial H}{\partial q}\right)^2 + \left(\frac{\partial H}{\partial p}\right)^2\right]^{1/2} \tag{4.1.28}$$

The distance d between two contour lines is, on the other hand, inversely proportional to the modulus of the gradient of Hamilton's function, i.e.

$$d = \left|\left\{\frac{\partial H}{\partial p} \quad \frac{\partial H}{\partial q}\right\}\right|^{-1} = \left[\left(\frac{\partial H}{\partial p}\right)^2 + \left(\frac{\partial H}{\partial q}\right)^2\right]^{-1/2} \tag{4.1.29}$$

Thus, at every instant t of the motion, the area of the rectangular element remains constant. In fluid mechanics, this corresponds to an incompressible flow.

In the general case of an n-dimensional phase space, the volume preservation of Hamilton systems is expressed by the theorem of Liouville which we shall discuss in more detail in section 5.1. If we denote the volume of a region $B(t)$ by $V(t)$, the theorem of Liouville can be written in the form, see, for example, (Arnold, 1978)

$$\left.\frac{\mathrm{d}V(t)}{\mathrm{d}t}\right|_{t=t_0} = \int\limits_{B(t_0)} \operatorname{div} \boldsymbol{F}\, \mathrm{d}p_1 \ldots \mathrm{d}p_{\mathrm{f}} \mathrm{d}q_1 \ldots \mathrm{d}q_{\mathrm{f}} \tag{4.1.30}$$

If we calculate the divergence (i.e. the "vector field source") of the phase velocity for Hamilton's equations (4.1.25)

$$\operatorname{div} \boldsymbol{F} = \frac{\partial}{\partial \boldsymbol{p}}\left(-\frac{\partial H}{\partial \boldsymbol{q}^{\mathrm{t}}}\right) + \frac{\partial}{\partial \boldsymbol{q}}\left(\frac{\partial H}{\partial \boldsymbol{p}^{\mathrm{t}}}\right) \equiv 0 \tag{4.1.31}$$

we deduce the volume preservation directly from the particular symmetrical form of the canonical equations of Hamilton.

4.2 Canonical Transformations, Integrability

In physics, and particularly in analytical mechanics, Hamilton's equations are of great significance as a theoretical tool. There are a number of applications, e.g. the motion of a body in the gravitational field generated by two masses at rest, which can be solved elegantly and easily with the help of the Hamiltonian formulation. For the majority of applications, however, this solution technique proves rather clumsy. The main advantages of the method are principally to be found in the fact that it allows us to gain deeper insight into the domain of mechanics. Since the generalised momenta $\boldsymbol{p}$ and coordinates $\boldsymbol{q}$ are of equal importance and appear independent of one another in eq. (4.1.25), we obtain a greater flexibility in the selection of the physical quantities and in their interpretation. In this way, more abstract formulations are possible; these serve as a starting point in statistical mechanics and quantum mechanics.

Hamilton's equations are particularly suitable for our account since, with their help, we can establish qualitative characteristics of the courses of motion. An example of this is the theorem of Liouville. From the structure of Hamilton's equations, we deduce div $\boldsymbol{F} = 0$ and thus immediately the volume preservation of the phase flow. Moreover, we shall see in this section that the integrability of a system is reflected very succinctly in the form of the Hamilton function. For this reason, this formulation was used as the basis of the stability investigations which led to the KAM theorem (see section 4.4).

Using an elementary example, Kepler's laws of motion for a planet around the sun, we now demonstrate how the Hamilton equations of motion can be set up and integrated.

We consider the plane motion of a planet with mass m in the gravitational field of the sun with mass M where the mass of the planet and the sun are assumed to be concentrated in two points, see fig. 4.2.1.

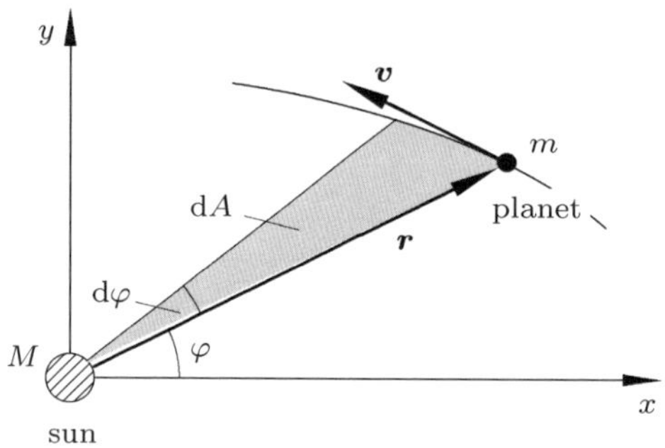

Fig. 4.2.1

On the interpretation of Kepler's Law of Areas: conservation of area

In order to describe the motion in the x, y-plane, we introduce the polar coordinates

$$x = r\cos\varphi, \quad y = r\sin\varphi \tag{4.2.1}$$

and can thus determine the velocity v of the planet. From

$$\begin{bmatrix} \dot{x} \\ \dot{y} \end{bmatrix} = \begin{bmatrix} \cos\varphi & -\sin\varphi \\ \sin\varphi & \cos\varphi \end{bmatrix} \begin{bmatrix} \dot{r} \\ r\dot{\varphi} \end{bmatrix} \tag{4.2.2}$$

we obtain

$$\begin{aligned} v^2 &= \dot{x}^2 + \dot{y}^2 \\ &= \dot{r}^2 + r^2\dot{\varphi}^2 \end{aligned} \tag{4.2.3}$$

as well as the kinetic energy for the planet

$$T = \tfrac{1}{2}m(\dot{r}^2 + r^2\dot{\varphi}^2) \tag{4.2.4}$$

and its potential energy

$$U = -G\frac{mM}{r} \tag{4.2.5}$$

where G denotes the gravitational constant. Since the motion takes place in a plane, the system has the two degrees of freedom r and φ which we introduce as generalised coordinates

$$q_1 = r, \quad q_2 = \varphi \tag{4.2.6}$$

In accordance with eq. (4.1.13), the Lagrange function expressed in the generalised coordinates $\boldsymbol{q}$ and $\dot{\boldsymbol{q}}$ is

$$\begin{aligned} L &= L(\boldsymbol{q}, \dot{\boldsymbol{q}}) = T - U \\ &= \tfrac{1}{2} m \left(\dot{q}_1^2 + q_1^2 \, \dot{q}_2^2 \right) + G \frac{mM}{q_1} \end{aligned} \tag{4.2.7}$$

In order to deduce Hamilton's equations, we require the variables $\boldsymbol{p}$, conjugate to $\boldsymbol{q}$, which we establish with the help of eq. (4.1.20)

$$\begin{aligned} p_1 &= \frac{\partial L}{\partial \dot{q}_1} = m\dot{q}_1 \\ p_2 &= \frac{\partial L}{\partial \dot{q}_2} = m q_1^2 \dot{q}_2 \end{aligned} \tag{4.2.8}$$

Equation (4.1.21) now enables us to specify the Hamilton function $H(\boldsymbol{p}, \boldsymbol{q})$ immediately and we obtain

$$H = \frac{1}{2m} \left(p_1^2 + \frac{p_2^2}{q_1^2} \right) - G \frac{mM}{q_1} = E \tag{4.2.9}$$

It can be seen from this relationship that H does not depend on q_2. Thus, one of Hamilton's equations, namely

$$\dot{p}_2 = -\frac{\partial H}{\partial q_2} = 0 \tag{4.2.10}$$

can be integrated immediately; applying eq. (4.2.8), we obtain

$$p_2 = m q_1^2 \dot{q}_2 = mC \tag{4.2.11}$$

Reverting to polar coordinates and using eq. (4.2.6), we confirm Kepler's Second Law (Law of Areas)

$$r^2 \dot{\varphi} = C \tag{4.2.12}$$

which yields a simple geometrical interpretation. If we observe the change in the area which the radius vector $\boldsymbol{r}$ sweeps over for an incremental angle $\mathrm{d}\varphi$, we deduce immediately from fig. 4.2.1

$$\mathrm{d}A = \frac{1}{2} r^2 \mathrm{d}\varphi + \frac{1}{2} r \mathrm{d}\varphi \mathrm{d}r$$

Thus, in conjunction with eq. (4.2.12), we obtain in a first approximation

$$\frac{\mathrm{d}A}{\mathrm{d}t} \approx \frac{1}{2} r^2 \frac{\mathrm{d}\varphi}{\mathrm{d}t} = \text{const}$$

Hamilton's equations thus directly confirm Kepler's Second Law which states that the radius vector from the sun to the planet sweeps over equal areas during equal intervals of time.

In accordance with eq. (4.1.25), the remaining Hamilton equations are now

$$\begin{aligned}\dot{p}_1 &= -\frac{\partial H}{\partial q_1} = \frac{p_2^2}{mq_1^3} - G\frac{mM}{q_1^2}\\ \dot{q}_1 &= \frac{\partial H}{\partial p_1} = \frac{p_1}{m}\\ \dot{q}_2 &= \frac{\partial H}{\partial p_2} = \frac{p_2}{mq_1^2}\end{aligned} \tag{4.2.13}$$

We are mainly interested in the trajectories $r = r(\varphi)$ respectively $q_1 = q_1(q_2)$, eq. (4.2.6), which we deduce by integration from the second and third expressions of eq. (4.2.13)

$$\frac{\mathrm{d}q_2}{\mathrm{d}q_1} = \frac{\dot{q}_2}{\dot{q}_1} = \frac{p_2}{p_1 q_1^2} \tag{4.2.14}$$

Here, we substitute for p_1 and p_2 expressions gained from eq. (4.2.9) and eq. (4.2.11) respectively and take eq. (4.2.12) into account. We obtain

$$q_2 = \int \frac{mC}{q_1[2mq_1(Eq_1 + GmM) - m^2C^2]^{1/2}}\mathrm{d}q_1 \tag{4.2.15}$$

This integral can be evaluated directly. If we adopt polar coordinates once more, the solution is

$$\varphi = \cos^{-1}\frac{m(C^2 - GMr)}{r\left[m(2EC^2 + G^2mM^2)\right]^{1/2}} \tag{4.2.16}$$

where we have set the integration constant to zero without loss of generality. Introducing the abbreviations

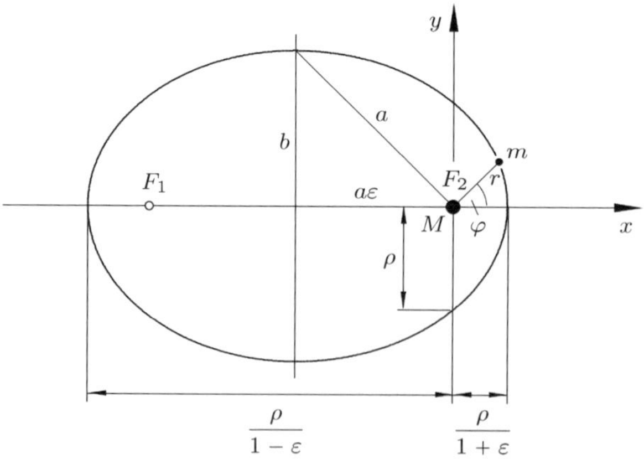

Fig. 4.2.2

Kepler's ellipse (Kepler's First Law)

$$\rho = \frac{C^2}{GM} \quad \text{and} \quad \varepsilon = \left(1 + \frac{2EC^2}{G^2mM^2}\right)^{1/2}, \quad (E < 0) \tag{4.2.17}$$

we find that the trajectory of the planet, interpreted as a function $r = r(\varphi)$, becomes

$$r = \frac{\rho}{1 + \varepsilon \cos \varphi} \tag{4.2.18}$$

This is the polar equation of an ellipse (see fig. 4.2.2) expressed in dependence of the eccentricity ε and the parameter ρ. The sun resides in one of the focal points of the ellipse and the planet describes an elliptical orbit in the ecliptic.

In Kepler's problem, we saw that the Hamilton function H, eq. (4.2.9), does not explicitly contain the variable q_2. Therefore, one of the equations of motion, eq. (4.2.10), can be integrated immediately. Thus, besides the total energy E, eq. (4.1.27), we obtain a further quantity of conservation from which, as stated before, Kepler's Second Law may be deduced. According to Helmholtz, a variable q_i, which does not appear explicitly in the Hamilton function, is denoted a cyclic variable. Since, in this case,

$$\dot{p}_i = -\frac{\partial H}{\partial q_i} = 0 \tag{4.2.19}$$

there follows

$$p_i(t) = p_i(0) = \alpha \tag{4.2.20}$$

If we substitute the constant parameter α in place of p_i in the Hamilton function, then H depends on only $(2f - 2)$ variables and the number of degrees of freedom is reduced by one, the number of equations of motion by two. In the case of a conservative system with two degrees of freedom, the existence of one cyclic variable is thus sufficient to render the set of Hamilton equations integrable since, together with the total energy E of the system, two constants of motion respectively two first integrals have been found.

Should the special case occur that all variables q_i are cyclic, i.e. if H depends only on the f variables p_i

$$H = H(\boldsymbol{p}) \tag{4.2.21}$$

the set of Hamilton equations is reduced to the *normal form* and can be integrated immediately. As a result, we obtain

$$p_i(t) = p_i(0) = \alpha_i, \qquad i = 1, 2, \ldots, f \tag{4.2.22}$$

and, from eqs. (4.2.21) and (4.1.25), it follows

$$\dot{q}_i(t) = \frac{\partial H}{\partial p_i} = \omega_i(\alpha_1, \alpha_2, \ldots, \alpha_f)$$

or

$$q_i(t) = \omega_i t + \beta_i, \qquad i = 1, 2, \ldots, f \tag{4.2.23}$$

In the literature, the p_i in this case are called *action variables*, the q_i *angle variables* and the ω_i *frequencies*. In order to stress the particular significance of the canonical variables which lead to an immediate direct integration of the Hamilton equations, we introduce the notation

$$\boldsymbol{I}_H = \{I_1 \quad I_2 \quad \ldots\ldots \quad I_{\mathrm{f}}\} \tag{4.2.24}$$

for the action variables $\boldsymbol{p}$ and

$$\boldsymbol{\theta} = \{\theta_1 \quad \theta_2 \quad \ldots\ldots \quad \theta_{\mathrm{f}}\} \tag{4.2.25}$$

for the angle variables $\boldsymbol{q}$. We deduce from eq. (4.2.22) that each action variable is a constant of motion and, according to eq. (4.2.23), each angle variable is linearly dependent on the time. One particular advantage of applying action and angle variables is that they allow immediate evaluation of the frequencies of periodic motions without a prior determination of the complete solution. When the total energy respectively the Hamilton function H has been expressed in action variables I_{i}, the frequencies follow immediately by differentiation

$$\omega_{\mathrm{i}} = \frac{\partial H}{\partial I_{\mathrm{i}}} \tag{4.2.26}$$

Elementary physical variables with which a given problem can be formulated with suggestive clarity as an assembly of equations are not necessarily the most suitable for establishing a solution in practice. It is in general preferable to attempt to transform the equations of motion, eq. (4.1.25), into the simplest possible form in order to ease the integration. Our task is thus:

(a) to look for transformations

$$\begin{aligned} \boldsymbol{p} &= \boldsymbol{p}(\boldsymbol{P}, \boldsymbol{Q}) \\ \boldsymbol{q} &= \boldsymbol{q}(\boldsymbol{P}, \boldsymbol{Q}) \end{aligned} \tag{4.2.27}$$

which maintain the form of the Hamilton equations and

(b) to find special transformations which reduce the Hamilton function to a simpler form, possibly even to the normal form.

In this introductory account, we restrict ourselves to time-independent transformations. The value of the Hamilton function is thus preserved

$$H(\boldsymbol{p}, \boldsymbol{q}) = H\big(\boldsymbol{p}(\boldsymbol{P}, \boldsymbol{Q}), \boldsymbol{q}(\boldsymbol{P}, \boldsymbol{Q})\big) = H^*(\boldsymbol{P}, \boldsymbol{Q}) = E \tag{4.2.28}$$

On the other hand, eq. (4.2.28) does not apply for time-dependent transformations. In the case of a transformation of an inertial system into a co-moving (accelerated) coordinate system, for example, additional forces occur which modify the value of H. Time-dependent Hamilton functions are used in section 9.2.2 to describe the dynamics of point vortices in two-dimensional ideal flows.

If the form of the Hamilton equations is not destroyed in the transition $\boldsymbol{p}, \boldsymbol{q} \rightarrow \boldsymbol{P}, \boldsymbol{Q}$

$$\dot{\boldsymbol{P}} = -\frac{\partial H^*}{\partial \boldsymbol{Q}^{\mathrm{t}}} \tag{4.2.29}$$

$$\dot{\boldsymbol{Q}} = \frac{\partial H^*}{\partial \boldsymbol{P}^{\mathrm{t}}} \tag{4.2.30}$$

the transformation is called *canonical.* To establish the transformation equations, we presume that the old and the new equations of motion can be deduced from

Hamilton's principle, eq. (4.1.18) (Päsler, 1968). In the case of a non-identical coordinate transformation, the two Lagrange functions $L(\boldsymbol{q},\dot{\boldsymbol{q}})$ and $L^*(\boldsymbol{Q},\dot{\boldsymbol{Q}})$ are seen to agree but for an addition of a total time derivative of an arbitrary function

$$F = F(\boldsymbol{p},\boldsymbol{q},\boldsymbol{P},\boldsymbol{Q})$$

the so-called generating function.With eq. (4.1.21), we thus obtain

$$\boldsymbol{p}^{\mathrm{t}}\dot{\boldsymbol{q}} - H = \boldsymbol{P}^{\mathrm{t}}\dot{\boldsymbol{Q}} - H^* + \frac{\mathrm{d}F}{\mathrm{d}t} \tag{4.2.31}$$

However, the transformation equations (4.2.27) imply that only 2f of the 4f variables of the function F are mutually independent. Figure 4.2.3 shows the possible combinations. Let us assume, for example, that F has the form

$$F = F(\boldsymbol{q},\boldsymbol{Q}) \tag{4.2.32}$$

Introducing the total derivative with respect to time

independent variables	dependent variables			
	$\boldsymbol{p}$	$\boldsymbol{q}$	$\boldsymbol{P}$	$\boldsymbol{Q}$
$\boldsymbol{q},\boldsymbol{P}$	$\frac{\partial F}{\partial \boldsymbol{q}^{\mathrm{t}}}$			$\frac{\partial F}{\partial \boldsymbol{P}^{\mathrm{t}}}$
$\boldsymbol{q},\boldsymbol{Q}$	$\frac{\partial F}{\partial \boldsymbol{q}^{\mathrm{t}}}$		$-\frac{\partial F}{\partial \boldsymbol{Q}^{\mathrm{t}}}$	
$\boldsymbol{p},\boldsymbol{P}$		$-\frac{\partial F}{\partial \boldsymbol{p}^{\mathrm{t}}}$		$\frac{\partial F}{\partial \boldsymbol{P}^{\mathrm{t}}}$
$\boldsymbol{p},\boldsymbol{Q}$		$-\frac{\partial F}{\partial \boldsymbol{p}^{\mathrm{t}}}$	$-\frac{\partial F}{\partial \boldsymbol{Q}^{\mathrm{t}}}$	

Fig. 4.2.3

Possible combinations of variables for the generating function F

$$\frac{\mathrm{d}F}{\mathrm{d}t} = \frac{\partial F}{\partial \boldsymbol{q}}\dot{\boldsymbol{q}} + \frac{\partial F}{\partial \boldsymbol{Q}}\dot{\boldsymbol{Q}} \tag{4.2.33}$$

into eq. (4.2.31), we obtain the transformation equations as a consequence of the independence of the variables $\boldsymbol{q}$ and $\boldsymbol{Q}$

$$\boldsymbol{p} = \frac{\partial F}{\partial \boldsymbol{q}^{\mathrm{t}}}$$

$$\boldsymbol{P} = -\frac{\partial F}{\partial \boldsymbol{Q}^{\mathrm{t}}} \tag{4.2.34}$$

The remaining transformations given in fig. 4.2.3 are deduced by means of appropriate Legendre transformations; see section 8.5.1 as well as (Goldstein, 1978).

If the first of the eqs. (4.2.34) is inserted into the Hamilton function, we deduce the so-called Hamilton-Jacobi differential equation

$$H(\boldsymbol{p},\boldsymbol{q}) = H\left(\frac{\partial F}{\partial \boldsymbol{q}^{\mathrm{t}}},\boldsymbol{q}\right) = E \tag{4.2.35}$$

This is an in general non-linear partial differential equation of first order for the function F. Finding the general solution of this equation is normally also difficult; by separating the variables, however, we may occasionally determine a complete integral.

An alternative way of establishing a transformation to normal form is offered in some cases by the theorem of Liouville, eq. (4.1.30), which proves the volume preservation of Hamilton systems. Obviously, canonical transformations must also always preserve the volume. In the two-dimensional phase space, this means that the determinant of the Jacobian matrix $\boldsymbol{J}$ has the value 1, i.e.

$$\det \boldsymbol{J} = \det \frac{\partial(p,q)}{\partial(P,Q)} = \begin{vmatrix} \dfrac{\partial p}{\partial P} & \dfrac{\partial p}{\partial Q} \\ \dfrac{\partial q}{\partial P} & \dfrac{\partial q}{\partial Q} \end{vmatrix} = 1 \qquad (4.2.36)$$

The transformation of polar coordinates to Cartesian coordinates

$$p = P\cos Q, \quad q = P\sin Q$$

is, for example, *not* a canonical transformation since, in this case,

$$\det \boldsymbol{J} = \begin{vmatrix} \cos Q & -P\sin Q \\ \sin Q & P\cos Q \end{vmatrix} = P \neq 1$$

applies. On the other hand, the linear transformation

$$p = aP + bQ, \quad q = cP + dQ$$

with

$$ad - bc = 1$$

is a canonical area-preserving coordinate transformation.

Using the example of the undamped spring pendulum, we now set up the Hamilton function $H(\boldsymbol{p},\boldsymbol{q})$, eq. (4.1.21), and subsequently transform it into the normal form $H^*(P)$, eq. (4.2.21), applying the theorem of Liouville. Let the pendulum have the mass m and spring stiffness k (fig. 4.2.4a).

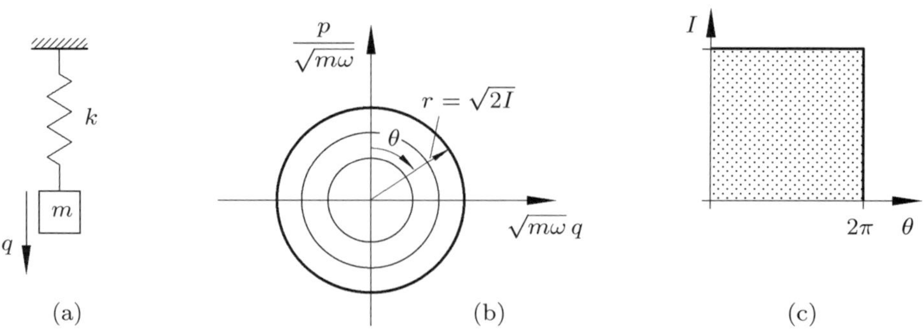

Fig. 4.2.4: Action-angle variables for an undamped spring pendulum

From the kinetic energy

$$T = \tfrac{1}{2} m \dot{q}^2$$

and the potential energy

$$U = \tfrac{1}{2} k q^2$$

the free energy results as

$$L = T - U = \tfrac{1}{2}(m\dot{q}^2 - kq^2)$$

Following eq. (4.1.20), we introduce the generalised momentum

$$p = \frac{\partial L}{\partial \dot{q}} = m\dot{q}$$

and obtain the Hamilton function $H(p,q)$

$$H = T + U = \tfrac{1}{2}\Big(\frac{p^2}{m} + kq^2\Big) = E \tag{4.2.37}$$

In order to find the transformation to the normal form $H^*(I)$ which depends only on the action variable $P = I$, we make use of the volume preservation of Hamilton systems. We calculate the surface S which is enclosed by a closed curve c in the phase space (see fig. 4.2.5). According to Stoke's theorem, we obtain

$$S = \iint \mathrm{d}p\,\mathrm{d}q = \oint p(q)\mathrm{d}q \tag{4.2.38}$$

Substitution of $p(q)$ from eq. (4.2.37) yields

$$S = \oint \sqrt{m(2E - kq^2)}\mathrm{d}q \tag{4.2.39}$$

Calculating on the other hand the area in the I, θ-plane, we have

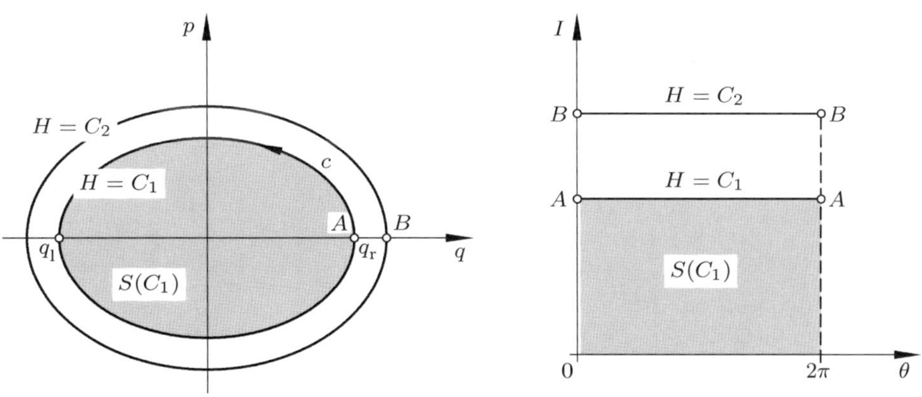

Fig. 4.2.5: Transformation $(p,q) \to (I,\theta)$

$$S = \int_0^{2\pi} I \mathrm{d}\theta = 2\pi I \tag{4.2.40}$$

Considering the symmetry, we deduce from the last two equations

$$I = \frac{1}{\pi} \int_{q_\mathrm{l}}^{q_\mathrm{r}} \sqrt{m(2E - kq^2)} \mathrm{d}q = \frac{\sqrt{mk}}{2\pi} \left[q\sqrt{\frac{2E}{k} - q^2} + \frac{2E}{k} \sin^{-1} q\sqrt{\frac{k}{2E}} \right]_{q_\mathrm{l}}^{q_\mathrm{r}}$$

We can now obtain for $p = 0$ the limits of integration $q_\mathrm{r}, q_\mathrm{l}$ from eq. (4.2.37) (see fig. 4.2.5)

$$q_{\mathrm{r,l}} = \pm\sqrt{\frac{2E}{k}}$$

and thus find the following expression for the action variable I

$$I = \sqrt{\frac{m}{k}} E = \frac{1}{\omega} H \tag{4.2.41}$$

where we have introduced the frequency

$$\omega = \sqrt{k/m} \tag{4.2.42}$$

The normal form of the Hamilton function of the undamped spring pendulum is hence in accordance with eq. (4.2.28) simply

$$H^*(I) = \omega I \tag{4.2.43}$$

According to eqs. (4.2.29) and (4.2.30), the equations of motion in the action-angle variables thus take the form

$$\begin{aligned} \dot{I} &= -\frac{\partial H^*}{\partial \theta} = 0\,, && \text{i.e. } I = \mathrm{const} \\ \dot{\theta} &= \frac{\partial H^*}{\partial I} = \omega\,, && \text{i.e. } \theta = \omega t + \beta \end{aligned} \tag{4.2.44}$$

We next set up the link between the conjugate variables (p, q) and the action-angle variables (I, θ). Replacing the variable p in eq. (4.2.37) by $\partial F/\partial q$, in accordance with the table in fig. 4.2.3, and taking eq. (4.2.28) into account, we obtain for the generating function $F = F(q, I)$ the Hamilton-Jacobi differential equation

$$\frac{1}{2}\left[\frac{1}{m}\left(\frac{\partial F}{\partial q}\right)^2 + kq^2\right] = H^* = \omega I \tag{4.2.45}$$

Since we do not need the generating function F itself, however, but rather the coordinate transformation $(p, q) \to (P, Q) \equiv (I, \theta)$, we calculate following the table in fig. 4.2.3 the derivative

$$\theta = \frac{\partial F}{\partial I} = \int \frac{m\omega}{\sqrt{m(2\omega I - kq^2)}} \mathrm{d}q$$

and obtain by integration

$$\theta = \sin^{-1} \sqrt{\frac{k}{2\omega I}} q$$

or

$$q = \sqrt{\frac{2I}{m\omega}} \sin\theta \tag{4.2.46}$$

Equation (4.2.45) finally yields

$$p = \frac{\partial F}{\partial q} = \sqrt{2m\omega I} \cos\theta \tag{4.2.47}$$

Figure 4.2.4b shows that the trajectories in the p, q-phase plane are closed curves; by an appropriate normalisation, these can be reduced to circles.

If we extend the system to f uncoupled oscillators with masses m_{i} and spring stiffnesses k_{i}, then, in analogy to eq. (4.2.44), the equations of motion can also be expressed in action-angle variables

$$\dot{I}_{\mathrm{i}} = -\frac{\partial H}{\partial \theta_{\mathrm{i}}} = 0$$

$$\dot{\theta}_{\mathrm{i}} = \frac{\partial H}{\partial I_{\mathrm{i}}} = \omega_{\mathrm{i}} \ , \qquad \mathrm{i} = 1, \ldots \mathrm{f} \tag{4.2.48}$$

with $\omega_{\mathrm{i}} = \sqrt{k_{\mathrm{i}}/m_{\mathrm{i}}}$.

This system can be integrated directly, cf. eqs. (4.2.22), (4.2.23). Conversely, each conservative system in which the equations of motion can be transformed to normal form can be interpreted as a system of f uncoupled oscillators. In contrast to our simple example, in which all the frequencies are constant, the frequencies generally depend on the action variables: $\omega_{\mathrm{i}} = \omega_{\mathrm{i}}(I_1, \ldots I_{\mathrm{f}})$, i.e. they differ from trajectory to trajectory.

Summing up, we note that the Hamilton systems are integrable if $\mathrm{f} = \mathrm{n}/2$ integrals are known. The reason for this lies in the particular structure of the Hamilton equations. In the case of general, dissipative systems, however, n integrals are necessary.

4.3 f-Dimensional Tori and Trajectories

The search of mathematicians and physicists in the 19th century for the normal form of a given Hamilton system proved successful only in some exceptional cases. Poincaré showed that, even in the case of the apparently straightforward three-body problem, too few first integrals exist. If even this "simple" problem is not integrable, it will certainly only prove possible to construct analytical solutions of higher-dimensional Hamilton systems in exceptional cases. The study of integrable systems is nevertheless of interest in order to understand the general case of non-integrable systems for which chaotic and regular domains are closely interwoven.

We first deal with the characteristics of recurrent trajectories of integrable systems. In this case, we can assume that the Hamilton function H – expressed in action-angle variables – is given in its normal form, eq. (4.2.44). The f action variables

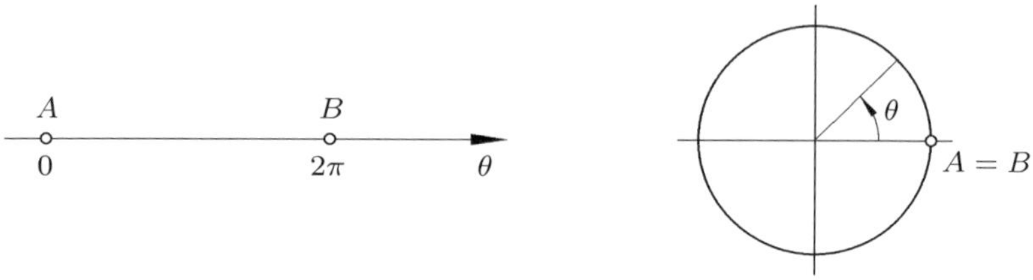

Fig. 4.3.1: Trajectories for a single-degree-of-freedom system

then coincide with the f first integrals $I_k = C_k$ (k = 1, 2, ..., f). The trajectories in the n=2f-dimensional phase space thus lie on an f-dimensional manifold.

First taking f = 1, we again consider the undamped spring pendulum. The trajectories in the (p, q)-phase plane are closed curves. By means of a transformation into the action-angle variables (fig. 4.2.4c), they are mapped onto line segments in the (θ, I)-plane. The parameter pairs (θ, I) and $(\theta + 2\pi, I)$ of a periodic motion describe the same state of the system, initial and final points of the line segment being identical so that the trajectories in the (θ, I)-plane can be considered closed orbits. If initial and final points are joined together, we observe that the trajectories are topologically equivalent to circles (fig. 4.3.1).

For an integrable system with f = 2 degrees of freedom, the angle variables are, in accordance with eq. (4.2.23), linear functions of time

$$\begin{aligned} \theta_1 &= \omega_1 t + \beta_1 \\ \theta_2 &= \omega_2 t + \beta_2 \end{aligned} \tag{4.3.1}$$

The trajectories lie in a two-dimensional subspace $(\theta_1, \theta_2, I_1 = C_1, I_2 = C_2)$ of the four-dimensional phase space $(\theta_1, \theta_2, I_1, I_2)$. They can be reproduced in a (θ_1, θ_2)-plane as straight lines (see fig. 4.3.2). Since in this case each pair of canonical variables (p_i, q_i) also depends periodically on time – though possibly with differing frequencies ω_i – the trajectories are, assuming a suitable normalisation, limited to a square of a side length 2π, subject to the understanding that opposite edges define identical states: thus, if the trajectory reaches any edge $\theta_i = 2\pi$, it jumps back to $\theta_i = 0$. In order to circumvent discontinuities in the representation in the (θ_1, θ_2)-plane, opposite sides can be joined together; bending then results in a two-dimensional torus. Mathematically, this means that the square is mapped onto a torus (see section 2.4). Both "radii" of the torus are given by the integrals $I_1 = C_1$ and $I_2 = C_2$ (fig. 4.3.2). A modification of the initial conditions leads to changes in these radii, i.e. each trajectory then runs on its own torus, the frequencies also being altered in general since $\omega_i = \omega_i(I_1, I_2)$ (i = 1, 2).

In the two-dimensional case, the gradient of the straight lines in the (θ_1, θ_2)-plane

$$\nu = \omega_2 / \omega_1 \tag{4.3.2}$$

– also known as the rotation number – determines whether the trajectory is closed or not. Figures 4.3.3 and 4.3.4 show the course of the trajectories and irrational ratio of the frequencies respectively. In the case of rational rotation numbers, we

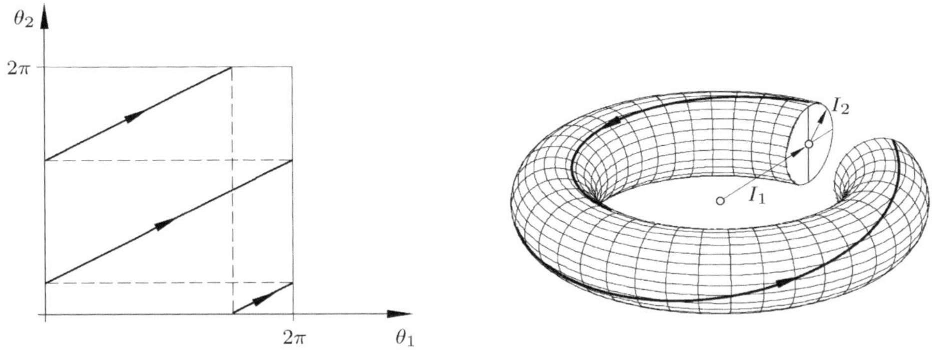

Fig. 4.3.2: Trajectories for an integrable two-degree-of-freedom system, $\omega_2 : \omega_1 = 1 : 2$

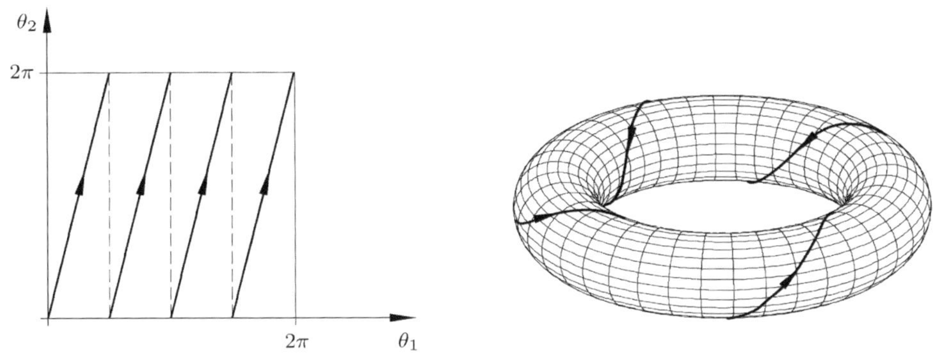

Fig. 4.3.3: Closed trajectories for a rational frequency ratio, $\omega_2 : \omega_1 = 4 : 1$

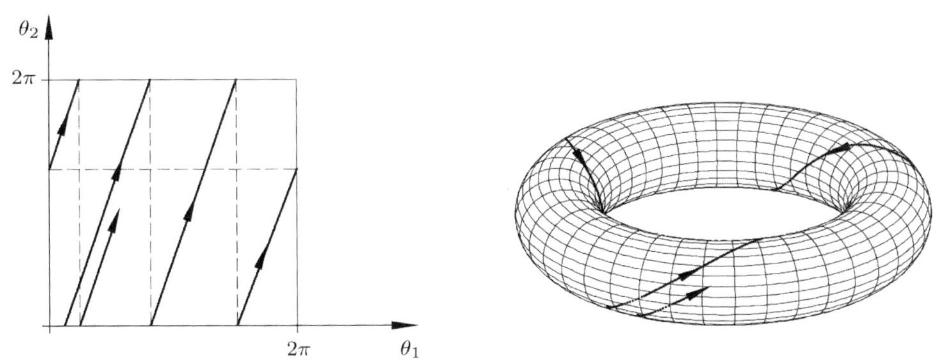

Fig. 4.3.4: Non-closed trajectory for an irrational frequency ratio

observe trajectories which – possibly only after a number of orbits – are closed, i.e. the motion is periodic. In the case of irrational rotation numbers, on the other hand, the trajectory never closes and in the course of time completely covers the square in the (θ_1, θ_2)-plane respectively torus. We then speak of a quasi-periodic motion.

The argument is also applicable to Hamilton systems with f degrees of freedom. Closed trajectories result if the frequencies ω_i are commensurable, i.e. if they satisfy a (non-trivial) linear relationship

$$\mathbf{n}^\mathrm{t}\boldsymbol{\omega} = \mathrm{n}_1\omega_1 + \mathrm{n}_2\omega_2 + \ldots + \mathrm{n}_\mathrm{f}\omega_\mathrm{f} = 0 \tag{4.3.3}$$

with integral coefficients n_i. If such a rational dependence does not exist, the trajectories fill the f-dimensional tori everywhere densely.

Summing up, we can state that the trajectories of integrable Hamilton systems lie on tori which are encapsulated into each other; each trajectory stays on its own torus for ever. This also means that trajectories which start on adjacent tori always stay on these tori.

4.4 An Outline of the KAM Theory

In the preceding section, we described the trajectories of integrable Hamilton systems. The question as to the stability of these trajectories has to be considered next. In this connection, the perturbation theory examines the course of the trajectories if a small perturbation is applied to the originally regular motion. The perturbation theory originated in astronomy and evolved from the necessity of predicting the positions of the planets as precisely as possible for navigational purposes. It was also used by mathematicians – such as Lagrange, Laplace and Poisson – to answer questions regarding the stability of the orbits of the planets. The approximations they used to prove the stability only enabled them to make prognoses for a limited period of several hundred years. Poincaré, who introduced a new qualitative approach to differential equations, believed that a proof of stability for the trajectories on a perturbed torus was not possible for $t \to \infty$. He based this conjecture on the fact that the three-body problem is not integrable as he was able to demonstrate that a sufficient number of first integrals does not exist for the solution of this problem. It was not until about 60 years later that the mathematicians Kolmogorov (Kolmogorov, 1954), Arnold (Arnold, 1963) and Moser (Moser, 1967) succeeded in setting up the KAM theory named after them, in which they made statements on the stability of Hamilton systems. In the following, we sketch the fundamentals of their proof for a system with *two* degrees of freedom (Arnold, 1978; Arnold et al., 1988; Tabor, 1989; Lichtenberg and Lieberman, 1983).

The reasoning can be summarised as follows. In order to investigate the stability of a trajectory c_0, a perturbation is applied to the initial integrable system. If the equations of motion of the perturbed system can be transformed to normal form, a torus which has been distorted by the perturbation exists on which the perturbed trajectory runs, and its stability is guaranteed. According to section 4.2, the transformation to normal form is equivalent to the solution of the Hamilton-Jacobi differential equation. Thus, the question regarding the stability of the trajectory c_0 is reduced to the question of the solvability of the Hamilton-Jacobi equation.

We presume an integrable system, the Hamilton function of which, using action and angle variables $(\boldsymbol{I}, \boldsymbol{\theta})$, assumes the normal form $H_0(\boldsymbol{I})$. A small perturbation $\varepsilon H_1(\boldsymbol{I}, \boldsymbol{\theta})$ is applied so that

$$H(\boldsymbol{I}, \boldsymbol{\theta}) = H_0(\boldsymbol{I}) + \varepsilon H_1(\boldsymbol{I}, \boldsymbol{\theta}) \tag{4.4.1}$$

is the Hamilton function of the disturbed system whereby, however,

$$\boldsymbol{I} = \{I_1 \quad I_2\}, \qquad \boldsymbol{\theta} = \{\theta_1 \quad \theta_2\}$$

now no longer represent the action-angle variables of the perturbed system. For the perturbed system, the Hamilton equations in the general canonically conjugate variables $\boldsymbol{I}$ and $\boldsymbol{\theta}$ are

$$\begin{aligned}\dot{\boldsymbol{I}} &= -\frac{\partial}{\partial \boldsymbol{\theta}^{\mathrm{t}}} H(\boldsymbol{I}, \boldsymbol{\theta}) \\ \dot{\boldsymbol{\theta}} &= \frac{\partial}{\partial \boldsymbol{I}^{\mathrm{t}}} H(\boldsymbol{I}, \boldsymbol{\theta})\end{aligned}$$

where, in accordance with eq. (4.2.44),

$$\begin{aligned}\dot{\boldsymbol{I}} &= -\frac{\partial}{\partial \boldsymbol{\theta}^{\mathrm{t}}} H_0(\boldsymbol{I}) = \boldsymbol{o} \\ \dot{\boldsymbol{\theta}} &= \frac{\partial}{\partial \boldsymbol{I}^{\mathrm{t}}} H_0(\boldsymbol{I}) = \boldsymbol{\omega}_0(\boldsymbol{I})\end{aligned} \tag{4.4.2}$$

applies for the perturbed system. If we now consider a trajectory c_0 of the unperturbed system, the initial conditions $I_1 = C_1$ and $I_2 = C_2$ define the "radii" of a torus M_0 on which c_0 lies while, according to eq. (4.2.44), the frequencies result as functions of the initial conditions

$$\boldsymbol{\omega}_0 = \boldsymbol{\omega}_0(\boldsymbol{I})$$

The question regarding the stability of the trajectory c_0 can now be reduced to the question as to the existence of a canonical transformation of the perturbed system into new action-angle variables $(\boldsymbol{I}^*, \boldsymbol{\theta}^*)$ so that the transformed Hamilton function H^* now only depends on $\boldsymbol{I}^*$. Hence, what we are looking for is a canonical transformation $(\boldsymbol{I}, \boldsymbol{\theta}) \to (\boldsymbol{I}^*, \boldsymbol{\theta}^*)$ with a generating function $F(\boldsymbol{\theta}, \boldsymbol{I}^*)$, and the question is whether the corresponding Hamilton-Jacobi differential equation (4.2.35)

$$H\left(\frac{\partial F}{\partial \boldsymbol{\theta}^{\mathrm{t}}}, \boldsymbol{\theta}\right) = H^*(\boldsymbol{I}^*) \tag{4.4.3}$$

possesses a solution. According to the table in fig. 4.2.3, the transformation equations are

$$\begin{aligned}\boldsymbol{I} &= \frac{\partial}{\partial \boldsymbol{\theta}^{\mathrm{t}}} F(\boldsymbol{\theta}, \boldsymbol{I}^*) \\ \boldsymbol{\theta}^* &= \frac{\partial}{\partial \boldsymbol{I}^{*\mathrm{t}}} F(\boldsymbol{\theta}, \boldsymbol{I}^*)\end{aligned} \tag{4.4.4}$$

The generating function F is now expanded into a power series in terms of a perturbation variable ε. Since the canonical transformation for the unperturbed motion ($\varepsilon = 0$) should yield the identity $\boldsymbol{I} = \boldsymbol{I}^*$, $\boldsymbol{\theta} = \boldsymbol{\theta}^*$, the generating function must, according to eq. (4.4.4), be expressed in the form

$$F(\boldsymbol{\theta}, \boldsymbol{I}^*) = \boldsymbol{\theta}^{\mathrm{t}} \boldsymbol{I}^* + \varepsilon F_1(\boldsymbol{\theta}, \boldsymbol{I}^*) + \mathcal{O}(\varepsilon^2)$$

Introducing eq. (4.4.1) into the Hamilton-Jacobi differential equation, we obtain

$$H_0\Big(\frac{\partial F}{\partial \boldsymbol{\theta}^{\mathrm{t}}}\Big) + \varepsilon H_1\Big(\frac{\partial F}{\partial \boldsymbol{\theta}^{\mathrm{t}}}, \boldsymbol{\theta}\Big) = H^*(\boldsymbol{I}^*) \tag{4.4.5}$$

Inserting the above series expansion of $F(\boldsymbol{\theta}, \boldsymbol{I}^*)$, the individual terms in eq. (4.4.5) can be expanded into Taylor series in ε and we find

$$\begin{aligned} H_0\Big(\frac{\partial F}{\partial \boldsymbol{\theta}^{\mathrm{t}}}\Big) &= H_0\Big(\boldsymbol{I}^* + \varepsilon \frac{\partial}{\partial \boldsymbol{\theta}^{\mathrm{t}}} F_1(\boldsymbol{\theta}, \boldsymbol{I}^*) + \mathcal{O}(\varepsilon^2)\Big) \\ &= H_0(\boldsymbol{I}^*) + \varepsilon \frac{\partial H_0}{\partial \boldsymbol{I}} \frac{\partial F_1}{\partial \boldsymbol{\theta}^{\mathrm{t}}} + \mathcal{O}(\varepsilon^2) \end{aligned}$$

$$\varepsilon H_1\Big(\frac{\partial F}{\partial \boldsymbol{\theta}^{\mathrm{t}}}, \boldsymbol{\theta}\Big) = \varepsilon H_1(\boldsymbol{I}^*, \boldsymbol{\theta}) + \mathcal{O}(\varepsilon^2)$$

If we substitute these expressions into eq. (4.4.5), we obtain

$$H_0(\boldsymbol{I}^*) + \varepsilon \frac{\partial H_0}{\partial \boldsymbol{I}} \frac{\partial F_1}{\partial \boldsymbol{\theta}^{\mathrm{t}}} + \varepsilon H_1(\boldsymbol{I}^*, \boldsymbol{\theta}) + \mathcal{O}(\varepsilon^2) = H^*(\boldsymbol{I}^*) \tag{4.4.6}$$

Since the left-hand side of this equation should be independent of $\boldsymbol{\theta}$, cf. (Schuster and Just, 2006), and, according to eq. (4.4.2),

$$\frac{\partial H_0}{\partial \boldsymbol{I}} = \boldsymbol{\omega}_0^{\mathrm{t}} = [\omega_1 \quad \omega_2]$$

are the frequencies of the undisturbed system, we obtain the condition

$$\boldsymbol{\omega}_0^{\mathrm{t}} \frac{\partial F_1}{\partial \boldsymbol{\theta}^{\mathrm{t}}} = -H_1(\boldsymbol{I}^*, \boldsymbol{\theta}) \tag{4.4.7}$$

The perturbations H_1 and F_1 can now be set up as periodic functions in the angle variables θ_1, θ_2 using a generalised Fourier series

$$H_1(\boldsymbol{I}^*, \boldsymbol{\theta}) = \sum_{\substack{\mathrm{n}_1 \quad \mathrm{n}_2 \\ \text{not both}=0}}\sum H_{1\mathrm{n}_1\mathrm{n}_2}(\boldsymbol{I}^*)\, \mathrm{e}^{i(\mathrm{n}_1\theta_1 + \mathrm{n}_2\theta_2)}$$

$$F_1(\boldsymbol{I}^*, \boldsymbol{\theta}) = \sum_{\substack{\mathrm{n}_1 \quad \mathrm{n}_2 \\ \text{not both}=0}}\sum F_{1\mathrm{n}_1\mathrm{n}_2}(\boldsymbol{I}^*)\, \mathrm{e}^{i(\mathrm{n}_1\theta_1 + \mathrm{n}_2\theta_2)}$$

Introducing these series into eq. (4.4.7) and comparing corresponding Fourier coefficients, we finally obtain the following series expansion for the generating function F

$$F(\boldsymbol{\theta}, \boldsymbol{I}^*) = \boldsymbol{\theta}^{\mathrm{t}} \boldsymbol{I}^* + i\varepsilon \sum_{\substack{\mathrm{n}_1 \quad \mathrm{n}_2 \\ \text{not both}=0}}\sum \frac{H_{1\mathrm{n}_1\mathrm{n}_2}}{\mathrm{n}_1\omega_1 + \mathrm{n}_2\omega_2}\, \mathrm{e}^{i(\mathrm{n}_1\theta_1 + \mathrm{n}_2\theta_2)} \tag{4.4.8}$$

A glimpse at the Fourier coefficients of this series expansion shows that this will certainly lead to a "dangerous" response (in fact, a resonance) if the frequencies of the unperturbed system are commensurable, i.e. if a relationship

$$\mathrm{n}_1\omega_1 + \mathrm{n}_2\omega_2 = 0 \tag{4.4.9}$$

exists with integers $\mathrm{n}_1, \mathrm{n}_2$, which can also be negative, and the corresponding coefficients $H_{1\mathrm{n}_1\mathrm{n}_2}$ do not vanish. In this case, the Hamilton-Jacobi differential equation cannot be solved and hence, no statement can be made concerning the stability of the initial trajectory c_0. We are thus forced to exclude from the outset the initial conditions which lead to trajectories with rational frequency ratios. Yet even if we exclude commensurable frequencies, integers $\mathrm{n}_1, \mathrm{n}_2$ will exist for which the denominator of the coefficients in the Fourier series, eq. (4.4.8), is reduced to arbitrarily small values. The reason is that any irrational number can be approximated arbitrarily exactly by rational numbers since both rational and irrational numbers are dense in the set of all points on the real axis. This means that in any neighbourhood, however small it may be, of a rational number, there is always an irrational number and *vice versa*. The question thus arises whether the series of eq. (4.4.8) converge at all.

Series expansions of the form of eq. (4.4.8) were already familiar to Weierstrass, see (Moser, 1973); he could not, however, provide the proof of their convergence in the absence of commensurable frequencies. He nevertheless believed that a proof of convergence was possible. He based this on a statement of Dirichlet who told his pupil Kronecker shortly before his death that he had found the proof of stability. Since Dirichlet was known for his extraordinarily strict argumentation, but had left no written evidence, Weierstrass tried to find the lost proof by formulating the famous prize question set by the Swedish King Oscar II in 1885 regarding the stability of the solar system. Poincaré, who was finally awarded the prize for his great work, was able to show that not even the three-body problem was integrable and came to the conclusion that no quasi-periodic solutions exist and that the series expansions, eq. (4.4.8), do not converge. This contradicted Weierstrass' expectations, however. The latter (rightly) held such a deduction to be unconvincing. The work of Kolmogorov, Arnold and Moser, summarised in the so-called KAM theory, brought a final answer to this "problem of small denominators".

Presuming a non-zero condition for the determinant

$$\det \frac{\partial \boldsymbol{\omega}}{\partial \boldsymbol{I}} = \det \begin{bmatrix} \dfrac{\partial\omega_1}{\partial I_1} & \dfrac{\partial\omega_1}{\partial I_2} \\ \dfrac{\partial\omega_2}{\partial I_1} & \dfrac{\partial\omega_2}{\partial I_2} \end{bmatrix} \neq 0$$

the series, eq. (4.4.8), converges if the rotation number $\nu = \omega_2/\omega_1$ satisfies the KAM condition (Moser, 1973)

$$\left|\nu - \frac{\mathrm{i}}{\mathrm{j}}\right| > k(\varepsilon)\,\mathrm{j}^{-5/2} \tag{4.4.10}$$

where i and j are arbitrary integers which posses no common factors and k is a constant for a given perturbation $\varepsilon \ll 1$, with $k(\varepsilon) \to 0$ for $\varepsilon \to 0$. The expression

$\delta(\mathrm{j}) = 2k(\varepsilon)\,\mathrm{j}^{-5/2}$ represents the length of the interval about each rational number i/j for which *no* statement of convergence can be made; the exponent $^{-5/2}$ only applies for the dimension two. One might intuitively think that all the irrational numbers ν were thus excluded. This is not correct, however. Figure 4.4.1 illustrates how fast – for a sufficiently small perturbation – the length of the excluded intervals is reduced in dependence on j. Since we can restrict ourselves to values in the range of $0 < \nu < 1$, only values $\mathrm{i} = 0, 1, \ldots, (\mathrm{j}-1)$ are admissible for any chosen j. We can now calculate the total length of these excluded intervals and obtain

$$l(\mathrm{j}) = \mathrm{j}\,\delta(\mathrm{j}) = 2k(\varepsilon)\,\mathrm{j}^{-3/2} \tag{4.4.11}$$

It is now only necessary to sum over all j's. In doing so, however, we will overestimate the *total length of the excluded intervals* since many intervals are counted several times. Surprisingly, we obtain

$$L < \sum_{\mathrm{j}=1}^{\infty} l(\mathrm{j}) = 2k(\varepsilon)\sum_{\mathrm{j}=1}^{\infty}\mathrm{j}^{-3/2} \approx 5.21k(\varepsilon) \tag{4.4.12}$$

This means that, in the case of small perturbations, most values of ν fulfil the convergence condition of eq. (4.4.10), i.e. most initial conditions will lead to stable trajectories. Due to the countability of the rational numbers, the total length of the "holes" about each rational number could hence be estimated.

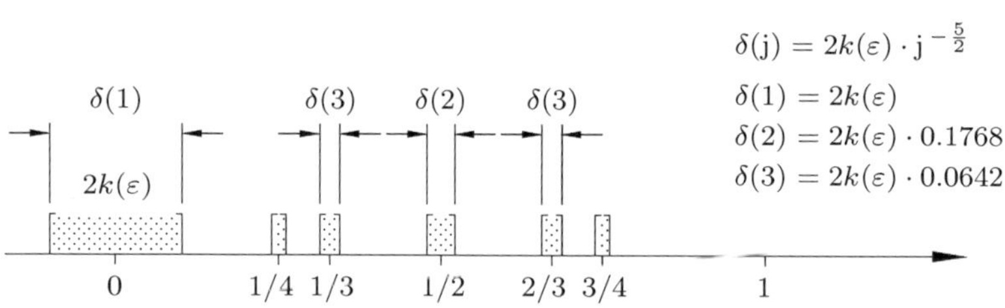

Fig. 4.4.1: Illustration of the KAM condition, eq. (4.4.10): stable tori in the gaps

From a physical point of view, it is not possible to differentiate between rational and irrational numbers due to the limited precision of any measurement. For the physicist, the statement of eq. (4.4.12) means that "most" tori are preserved in the case of small perturbations. The KAM theory thus introduces a weakened concept of stability. The tori which are deformed due to the perturbation but are still stable are called KAM surfaces.

It should be noted that the KAM theory only provides *sufficient* conditions for the existence of invariant tori. Nothing is said about the behaviour of the excluded trajectories. Our aforementioned considerations, however, lead us to assume that the tori in the neighbourhood of rational frequency ratios disintegrate. This subject will be discussed in section 4.5.

The inequality relation (4.4.10) supplies a condition for ν being sufficiently far removed from any rational number i/j. For small perturbations, eq. (4.4.10) can easily be satisfied. However, for increasing ε, the lengths of the excluded intervals, i.e. the dotted areas in fig. 4.4.1, will increase, i.e. all the tori will gradually become unstable. One might be tempted to guess that the last torus to disintegrate appertains to a quasi-periodic motion with the golden mean as the rotation number

$$\nu = \tfrac{1}{2}(\sqrt{5} - 1)$$

Of all the irrational numbers, this one is "furthest" from the rational numbers i/j. On the basis of numerical experiments, J. M. Greene set up criteria (Greene, 1979) which indicate when the KAM surfaces disintegrate and "stochastic" motion sets in. He was able to prove the special role of the golden mean numerically. For arbitrary perturbation formulations, however, the most stable torus is no longer associated with the golden mean (Lichtenberg and Lieberman, 1983). In section 4.6, using a numerical example, we demonstrate that it depends greatly on the initial conditions whether chaotic or regular behaviour occurs. Moreover, in sections 8.3 and 8.5, we discuss quasi-periodic motions and their transition to chaotic states.

One might consider the search for statements concerning the stability of our solar system over infinitely long time intervals to be an idle mathematical game. In the past millennia, no planet has escaped the solar system or collided with another planet. There is nevertheless a significant practical application of the results of the KAM theory, namely the fusion reactor of CERN in Geneva. Here, electrons and protons are accelerated in a circular ring until they have almost reached the speed of light. In the so-called storage rings, 10^{10} to 10^{11} revolutions are necessary to achieve maximum velocity. The question regarding the stability of the trajectories is thus of fundamental importance for the construction of such particle accelerators. Should chaotic motion occur, the particles would hit the wall of the ring, thus uncontrollably losing a great part of their energy. Since not even the largest computers could pursue the trajectories over 10^{10} revolutions, it is necessary to rely on theoretical assertions. Thanks to the KAM theory, the stability of most of the trajectories has been guaranteed. If one equates one revolution of the protons in the storage ring with one year in astronomical problems, the number of 10^{10} to 10^{11} revolutions would represent a time interval which far surpasses the age of our planets; this justifies the apparently unrealistic question regarding stability for infinite ranges of time.

4.5 Unstable Tori, Chaotic Regions

In this section, we direct our attention to those tori which the KAM theory does not include in its statements on stability, i.e. we investigate perturbed trajectories, the initial conditions of which have been chosen so as to fall into the gaps between the invariant tori. In order to elucidate this more clearly, we restrict ourselves in the following to Hamilton systems with two degrees of freedom.

Since the Hamilton function is itself a first integral, the trajectories in the four-dimensional phase space lie on a family of three-dimensional hypersurfaces with H = const. Introducing a Poincaré section, we obtain a two-dimensional Poincaré map which is particularly suitable for investigating the motions on perturbed and unperturbed tori. If we again denote the angle variables on the unperturbed torus as θ_1, θ_2, the planes $\theta_1 = \text{const}$ respectively $\theta_2 = \text{const}$ offer themselves as Poincaré sections (cf. fig. 2.4.6). Here, we select a meridian section $\theta_1 = \text{const}$.

The advantages of investigating Hamilton systems on a Poincaré section are evident; we thus first wish to enumerate some of the general characteristics of such sections and include a classification of possible fixed points. It can be shown that the Poincaré map of Hamilton systems is symplectic (Arnold, 1978). A symplectic structure in a vector space is characterised by a skew-symmetric bilinear form, a Euclidean one, on the other hand, by a symmetrical form. Although the geometry of a symplectic space is different from that of a Euclidean one, there are nevertheless many similarities. Simplecticity implies the following:

(a) Each Poincaré map of Hamilton systems is area-preserving (cf. theorem of Liouville, section 4.1). Among other things, we can then immediately conclude that no attractors can occur and asymptotically stable states of equilibrium are precluded.

(b) Periodic orbits correspond to fixed points of the Poincaré map. The stability behaviour of the fixed points can be investigated on the basis of the eigenvalues of the associated Jacobian matrix (cf. section 3.6). In the case of symplectic maps which are always of an even-numbered dimension 2(f − 1), the eigenvalues can always be arranged in pairs, the product of which is 1. Since the characteristic equation for the determination of the eigenvalues possesses real coefficients, the following three possibilities result for the distribution of the eigenvalues (see fig. 4.5.1):

i. There are two complex conjugate eigenvalues

$$\lambda_{1,2} = |\lambda| e^{\pm i\varphi} \qquad \text{with} \qquad |\lambda| \overset{!}{\neq} 1$$

Hence,

$$\lambda_{3,4} = \left|\frac{1}{\lambda}\right| e^{\pm i\varphi}$$

are also eigenvalues of this map. They are deduced from $\lambda_{1,2}$ by mirroring at the unit circle (see fig. 4.5.1). This case can obviously only occur for Hamilton systems having at least three degrees of freedom and will not be discussed here.

ii. There are two complex conjugate eigenvalues

$$\lambda_{1,2} = |\lambda| e^{\pm i\varphi} \qquad \text{with} \qquad |\lambda| \overset{!}{=} 1$$

which lie on the unit circle in the plane of complex numbers. We have already dealt with this case in section 3.6. The linearised system possesses a degenerate stable fixed point P_s, also called elliptic point since, in a linear approximation, the image points of a neighbouring trajectory lie on an ellipse about P_s (see figs. 3.6.3 and 4.5.2). If the neighbouring trajectory describes a periodic motion, there arise a finite number of fixed points on the ellipse, whereas, in the case of a quasi-periodic motion, the complete ellipse is swept over in the course of

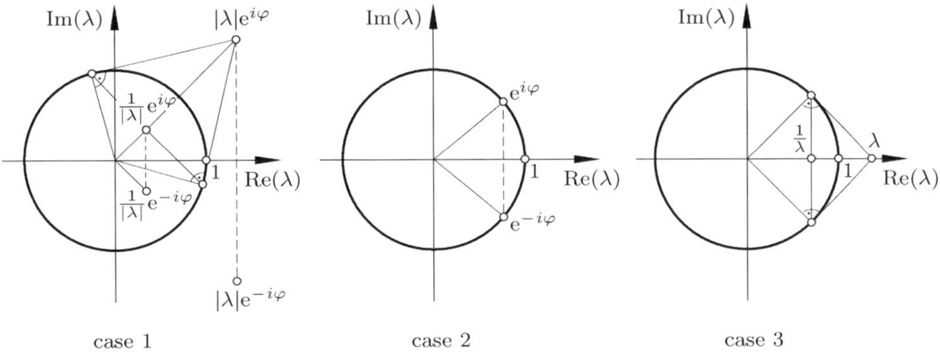

Fig. 4.5.1: Possible eigenvalues of Poincaré maps of Hamilton systems (f = 2)

time. On the basis of a linear stability analysis, we cannot at first establish the stability of P_s with respect to the actual non-linear map. This only becomes possible within the framework of the KAM theory since the latter asserts that, apart from a few special cases (Hénon, 1983), we can always find a neighbourhood of an elliptic point in which closed invariant curves exist.

iii. There are two real eigenvalues λ and $1/\lambda$ with $|\lambda| \neq 1$.

In this case, the image points of neighbouring trajectories lie on hyperbolic branches where the eigenvectors give the directions of the asymptotes (see fig. 4.5.2). Such fixed points are unstable (cf. section 5.4.1) and are called hyperbolic. In order to assess the stability of a fixed point P_s of a conservative map, the existence of a single eigenvalue λ of the Jacobian matrix with $|\lambda| \neq 1$ is thus sufficient to establish the instability of P_s.

The special case of coincident real eigenvalues (parabolic fixed points) is more complicated and is not dealt with here.

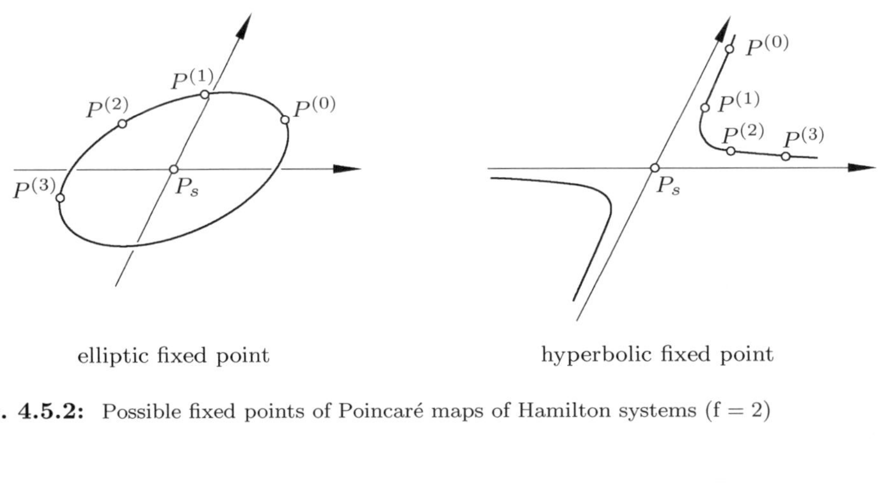

Fig. 4.5.2: Possible fixed points of Poincaré maps of Hamilton systems (f = 2)

We now turn our attention to our actual subject matter concerning unstable tori respectively chaotic regions, and discuss the behaviour of a periodic orbit (rational frequency ratio) to which a perturbation is applied. To this end, we first study the Poincaré map of an integrable system. If we introduce polar coordinates r, θ in the meridian section of the torus, the map T of the unperturbed system, known as the standard map or *Moser's twist map* (Moser, 1973), is

$$\begin{aligned} r_{\mathrm{i}+1} &= r_{\mathrm{i}} \\ \theta_{\mathrm{i}+1} &= \theta_{\mathrm{i}} + 2\pi\nu(r_{\mathrm{i}}) \end{aligned} \tag{4.5.1}$$

The invariant curves are thus circles and the rotation number $\nu = \omega_2/\omega_1$ depends on their radius. For quasi-periodic motions, ν is irrational and the image points $P^{(\mathrm{i})}$ fill in the circle everywhere densely for $\mathrm{i} \to \infty$. In the case of periodic motions with $\nu = \mathrm{k/m}$ (k, m have no common divisor), the trajectory returns to its initial point after precisely m orbits. We denote the m-fold iterated map as T^{m}; it has the form

$$\begin{aligned} r_{\mathrm{i}+1} &= r_{\mathrm{i}} \\ \theta_{\mathrm{i}+1} &= \theta_{\mathrm{i}} + 2\pi\nu(r_{\mathrm{i}})\mathrm{m} \end{aligned} \tag{4.5.2}$$

Introducing $\nu = \mathrm{k/m}$, we recognise that each point of the circle c with radius r is a fixed point of the map T^{m}.

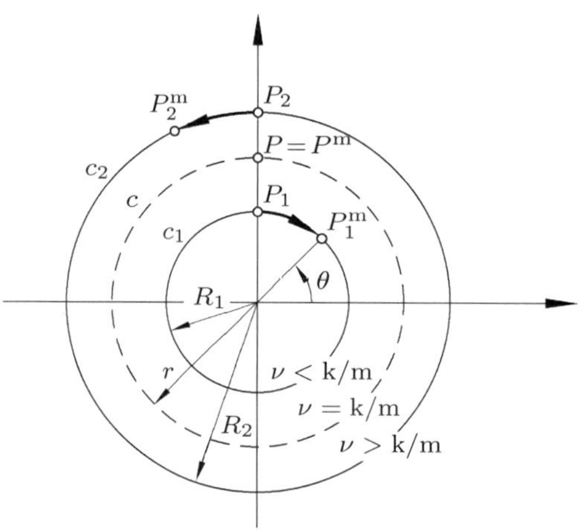

Fig. 4.5.3

Direction of rotation of the unperturbed map T^{m}

We now consider a circle c_1 respectively c_2 inside respectively outside the circle c (see fig. 4.5.3) having the radii $R_1 < r < R_2$. Moreover, we presume that the function $\nu(r)$ in the neighbourhood of r increases monotonically so that

$$\nu(R_1) < \frac{\mathrm{k}}{\mathrm{m}} < \nu(R_2)$$

holds. Applying the m-fold iterated map T^{m} to all three circles, we recognise that points on the smaller circle c_1 turn clockwise, points on the larger circle c_2 anti-clockwise, while all the points on c are fixed points.

We now apply a small perturbation on the radius and the angle of the integrable system, eq. (4.5.1), i.e. we perturb the Hamilton function H_0 of the initial system by a function εH_1 and designate the perturbed map T_ε

$$r_{\mathrm{i}+1} = r_{\mathrm{i}} + \varepsilon f(r_{\mathrm{i}}, \theta_{\mathrm{i}})$$
$$\theta_{\mathrm{i}+1} = \theta_{\mathrm{i}} + 2\pi\nu(r_{\mathrm{i}}) + \varepsilon g(r_{\mathrm{i}}, \theta_{\mathrm{i}}) \tag{4.5.3}$$

According to the theorem of Liouville, eq. (4.1.30), the map T_ε is also area-preserving. We now apply T_ε m times to the three circles, assuming that c_1 and c_2 are KAM curves, i.e. their rotation numbers fulfil the condition of convergence, eq. (4.4.10). According to the KAM theory, both the inner and the outer circles retain their topology and deviate only little from their original circular form (see fig. 4.5.4). In the case of sufficiently small perturbations and an m-fold iteration of the map T_ε, the points on the inner KAM curve will again rotate clockwise, points in the outer KAM curve anti-clockwise. Consequently, between the curves $T_\varepsilon^{\mathrm{m}}(c_1)$ and $T_\varepsilon^{\mathrm{m}}(c_2)$, a curve exists ($c_\varepsilon$ in fig. 4.5.4) on which the points do not rotate, but are only shifted in radial direction. If the perturbed map T_ε is applied m times to the points on c_ε, a curve $T_\varepsilon^{\mathrm{m}}(c_\varepsilon)$ arises, illustrated in fig. 4.5.4 by a broken line. Due to the area-preservation of the mapping, c_ε and $T_\varepsilon^{\mathrm{m}}(c_\varepsilon)$ must enclose the same area. This is only possible, however, if both curves intersect in an even number of points which are then fixed points of the map $T_\varepsilon^{\mathrm{m}}$. This means that the original torus with a rational rotation number is unstable but does not disintegrate completely, while the Poincaré map preserves an even number of fixed points.

This theorem was formulated by Poincaré shortly before his death for systems with two degrees of freedom and proven by G. D. Birkhoff; it is known as the Poincaré-Birkhoff theorem, see (Birkhoff, 1927).

If we observe the fixed points in fig. 4.5.4 more closely, we recognise two different types. At the beginning of this section, we saw that a two-dimensional symplectic map having no coinciding eigenvalues (fig. 4.5.2) only possesses elliptic or hyperbolic points as fixed points. In the case of the elliptic points E_s, the radial displacements and the displacements along the KAM curves combine to form circles, whereas this is not true for hyperbolic points H_s. In this latter case, neighbouring points are carried away from the immediate neighbourhood of the fixed point if the mapping is applied repeatedly.

Let us first consider the *elliptic* points. Each elliptic fixed point E_s is again surrounded by closed curves which are themselves Poincaré sections of invariant tori. The KAM theory can now be applied to these tori, i.e. there are stable, but also unstable tori in the neighbourhood of rational frequency ratios; these in turn disintegrate while, according to the Poincaré-Birkhoff theorem, an even number of points is preserved in each case, namely as many elliptic as hyperbolic points. These considerations can now be repeated infinitely often. Figure 4.5.5 is a schematic representation of the structure of the tori for different magnification scales. Regular trajectories (KAM curves) and irregular regions are nested in one another on any scale, resulting in an immensely complex image with a self-similar structure: each section, however small it may be, resembles the whole.

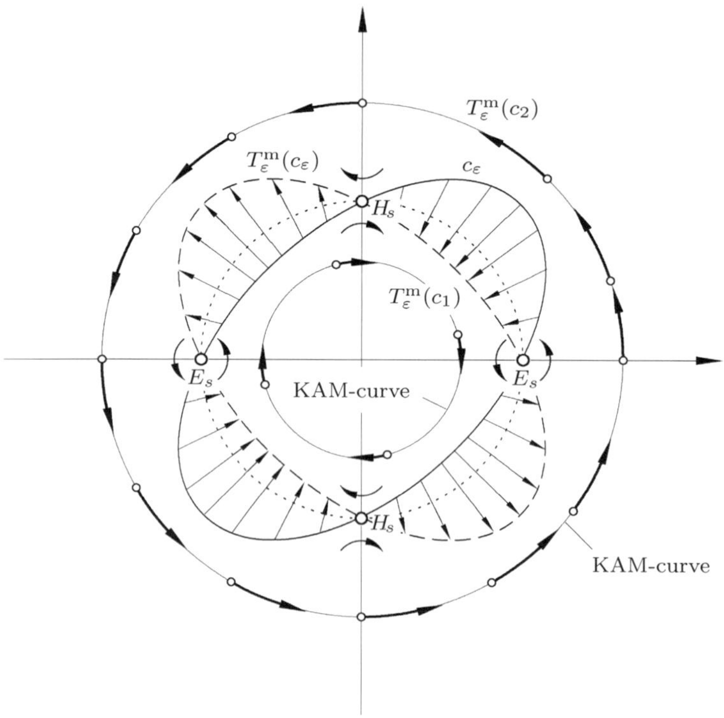

Fig. 4.5.4: Effect of the perturbed map $T_\varepsilon^{\mathrm{m}}$ (Poincaré-Birkhoff theorem)

We now investigate the behaviour of the Poincaré map in the neighbourhood of the *hyperbolic* fixed points H_s. In contrast to the stable behaviour of the elliptic points, the motion in the neighbourhood of hyperbolic points becomes unstable and is determined by the invariant manifolds (see section 3.3 and Chapter 6). Since elliptic points are related to periodic and quasi-periodic motions, it is to be expected that hyperbolic points are associated with chaotic behaviour.

In order to render this more easily comprehensible, we again return to the integrable case and consider a classic non-linear pendulum (see fig. 4.5.6). Let the length of the pendulum be l, its mass m and the acceleration due to gravity g. If we denote the angle which the pendulum forms with the vertical by q, the Hamilton function is

$$H = \frac{1}{2ml^2}p^2 - mgl\cos q \tag{4.5.4}$$

For simplicity's sake, we assume $ml^2 = 1$ and $\alpha^2 = mgl$ and obtain as the total energy

$$H = \tfrac{1}{2}p^2 - \alpha^2 \cos q = E \tag{4.5.5}$$

Fig. 4.5.5: Self-similar structure of the tori in the neighbourhood of elliptic points

The equations of motion

$$\begin{aligned}\dot{q} &= p \\ \dot{p} &= -\alpha^2 \sin q\end{aligned} \tag{4.5.6}$$

can be integrated with the aid of elliptic functions. Figure 4.5.6 shows the flow in the (p, q)-phase plane. The trajectories are the lines of constant energy and are given by eq. (4.5.5). We obtain singular points for $q = 0, \pm 2\pi, \dots$ (elliptic points E_s) and for $q = \pm\pi, \pm 3\pi, \dots$ (hyperbolic points H_s). The state of rest of the pendulum corresponds to the elliptic point ($E = 0$). For energy values $0 < E < \alpha^2$, the pendulum swings to and fro and the point in the phase space follows a closed curve about E_s. For $E > \alpha^2$, the pendulum possesses so much energy that it turns right over and rotates. The dashed lines (separatrices) correspond to the limiting case $E = \alpha^2$; they separate the regions of oscillatory and rotational behaviour from one another. The points of intersection of the separatrices are hyperbolic points H_s, where in each case, two stable manifolds $\mathbb{W}^s$ and two unstable manifolds $\mathbb{W}^u$ meet. If we follow the separatrix from one hyperbolic point to the next, we first move along an unstable manifold $\mathbb{W}^u$ which then merges continously into one of the two stable manifolds $\mathbb{W}^s$ of the next hyperbolic point H_s.

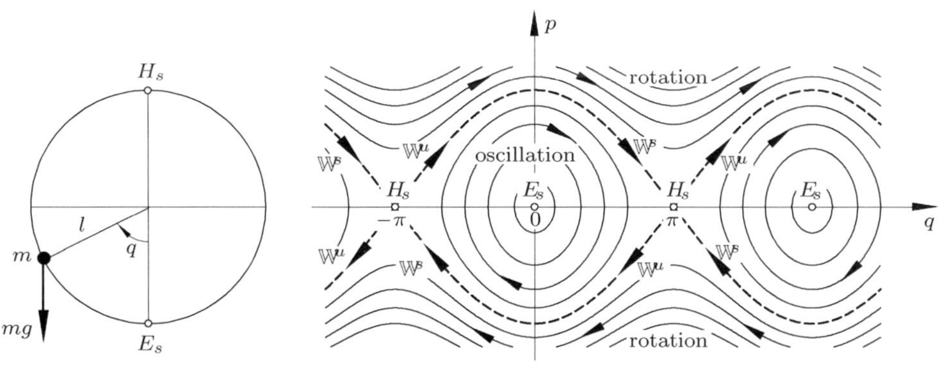

Fig. 4.5.6: Phase portrait of a non-linear pendulum without damping

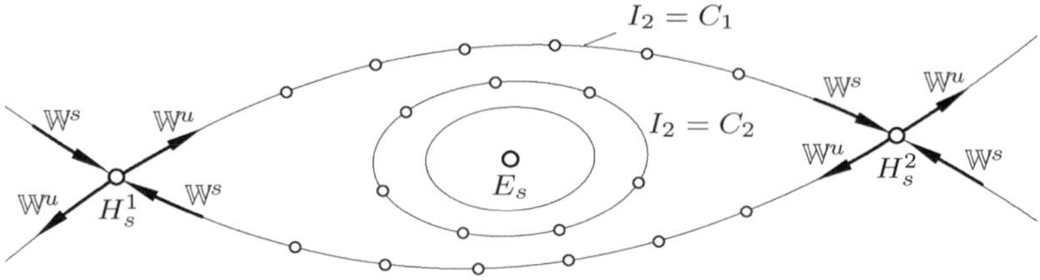

Fig. 4.5.7: Poincaré section of a two-dimensional integrable Hamilton system

Let us now return to the Poincaré map. For an integrable Hamilton system with two degrees of freedom, an image corresponding to the phase portrait of fig. 4.5.6 would result, though a sequence of points would replace the continuous curve. In the integrable case, besides the Hamilton function

$$H(p_1, p_2, q_1, q_2) = E \tag{4.5.7}$$

there is a further integral

$$I_2(p_1, p_2, q_1, q_2) = C \tag{4.5.8}$$

If we introduce a Poincaré section, the surface of intersection is defined by

$$g(p_1, p_2, q_1, q_2) = 0 \tag{4.5.9}$$

Selecting p_1 and q_1, for example, as coordinates in the Poincaré section, we can deduce p_2 and q_2 from eqs. (4.5.7) and (4.5.9) for fixed energy values and, after substitution into eq. (4.5.8), obtain

$$I_2(p_1, q_1) = C \tag{4.5.10}$$

An invariant curve then corresponds to each value of the constant C; on this curve lie all points of intersection of the trajectory with the Poincaré section (small circles in fig. 4.5.7) appertaining to an initial condition determined by C and E. If we imagine a three-dimensional space with the axes p_1, q_1 and C, these invariant curves can be interpreted as "contour lines". As can be seen from fig. 4.5.7, in the integrable case, it is possible to move along a contour line from one hyperbolic point to the next.

We next consider a Poincaré map of a non-integrable Hamilton system (see fig. 4.5.4) for which a relationship eq. (4.5.10) no longer exists. Starting from a hyperbolic fixed point H_s^1 on an unstable manifold $\mathbb{W}^u$, we can presuppose that the image points generally do not adhere to the level $C = \text{const}$. If we follow the stable and unstable manifolds (fig. 4.5.8), we observe that they will not merge continuously into each other but will intersect at a point P_h^0 at a certain angle. Such points of intersection of $\mathbb{W}^u$ and $\mathbb{W}^s$ which are not fixed points are called heteroclinic points. It is easy to conjecture that the existence of an *infinite* number of such points of intersection of a stable and an unstable manifold derives from the existence of *one single* heteroclinic point P_h^0. Indeed, if we apply the perturbed map $T_\varepsilon^{\mathrm{m}}$ repeatedly to P_h^0 (which, as we presume, is not a fixed point) as well as to its image points P_h^1 etc., the result, since each image point must lie both on a stable and an unstable

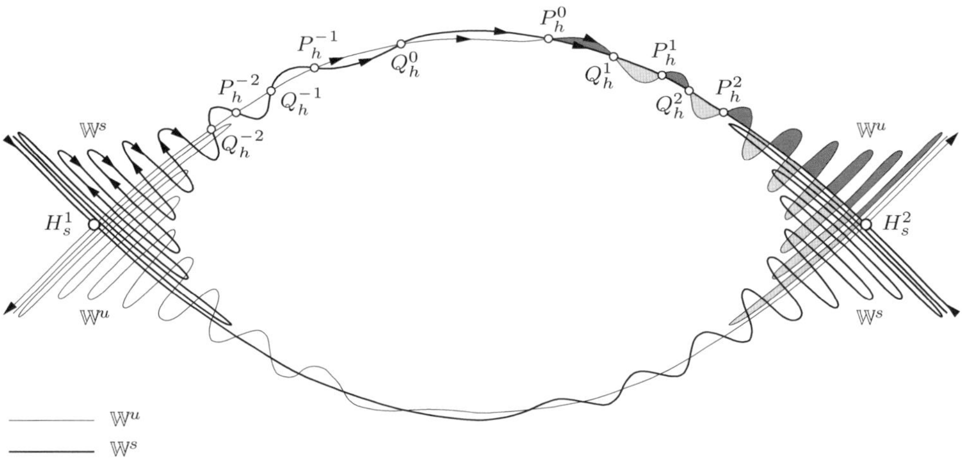

Fig. 4.5.8: Heteroclinic points and irregular behaviour in the neighbourhood of hyperbolic points

manifold, is seen to be an infinite sequence of heteroclinic points $\{P_h^{\mathrm{k}}\}$ which converge towards the hyperbolic point for $k \to \infty$. $\mathbb{W}^u$ and $\mathbb{W}^s$ can only intersect in an infinite sequence if $\mathbb{W}^u$ oscillates about $\mathbb{W}^s$, as shown schematically in fig. 4.5.8 (Moser, 1973). We shall return to this phenomenon in section 10.5 in connection with the disintegration of homoclinic orbits in the case of the Duffing equation.

We stress that the existence of heteroclinic points is *not* contradictory to the uniqueness of the solutions of continuous dynamical systems since the invariant manifolds $\mathbb{W}^u$ and $\mathbb{W}^s$ of the Poincaré map are not trajectories in the phase space.

We now remind the reader that $T_\varepsilon^{\mathrm{m}}$ is area- and orientation-preserving. Hence, besides the sequence of points $\{P_h^{\mathrm{k}}\}$, there must be a second sequence $\{Q_h^{\mathrm{k}}\}$ consisting of heteroclinic points which alternate with P_h^{k}. Due to the area-preserving character of $T_\varepsilon^{\mathrm{m}}$, successive loops of $\mathbb{W}^u$ which lie on one side of the invariant manifold $\mathbb{W}^s$ must enclose the same areas since these emerge successively from the preceding area by the mapping $T_\varepsilon^{\mathrm{m}}$ (fig. 4.5.8). The image points P_h^{k} now approach each other at an increasing rate when approaching H_s^2, the loops of $\mathbb{W}^u$ thus become longer and more slender and cling more closely to the unstable manifold $\mathbb{W}^u$ which originates in the fixed point H_s^2 but without ever intersecting it. As was demonstrated in section 3.3, invariant manifolds of the same type can never intersect since they derive from a deterministic set of differential equations.

We can follow exactly the same argument for an infinite sequence of heteroclinic points when retracing the invariant manifolds for $t \to -\infty$. Again, two infinite sequences of heteroclinic points arise, $\{P_h^{-\mathrm{k}}\}$ alternating with $\{Q_h^{-\mathrm{k}}\}$. Proceeding in the same way for all invariant manifolds yields new points of intersection of the loops of $\mathbb{W}^u$ and $\mathbb{W}^s$ which are also heteroclinic points. As the length of the loops increases, we observe oscillations of higher order, so that the whole area between the KAM curves is gradually filled with heteroclinic points (Hénon, 1983).

We then speak of stochastic layers between the invariant KAM curves. The result is an extraordinarily complex image of intersecting invariant manifolds; the area-preserving characteristic of the map $T_\varepsilon^{\mathrm{m}}$ must always be maintained and two stable or two unstable manifolds may never cross. Dragt and Finn (1976) reconstructed the oscillations of the invariant manifolds numerically and published diagrams of the heteroclinic points in the neighbourhood of a hyperbolic fixed point. If, in addition, we take into account the alternating occurrence of new elliptic and hyperbolic points as illustrated in fig. 4.5.5, we are scarcely able to reconstruct the course of the trajectories between two invariant tori. Regular and irregular behaviour of the trajectories is closely interwoven.

Up to now, we have considered Hamilton systems with two degrees of freedom. Each invariant two-dimensional torus divides the three-dimensional manifold $H = \text{const}$ into two domains. This means that irregular trajectories are trapped between two invariant tori and cannot escape into any other regions of the energy hypersurface (fig. 4.5.5).

In the case of systems with more than two degrees of freedom, this changes radically. For f $= 3$, the energy hypersurface $H = \text{const}$ has the dimension $2\mathrm{f} - 1 = 5$ whereas invariant tori have the dimension f $= 3$. In this case, the gaps which correspond to the disintegrated tori can communicate with one another (this is more easily understood if one imagines a three-dimensional space in which curves represent the invariant tori). The stable rings cannot prevent the irregular trajectories from meandering into the whole five-dimensional space of the energy hypersurface. This phenomenon is known as Arnold diffusion (Arnold, 1964).

Put another way, this means that, in the case of a two-degree-of-freedom system subject to a small enough perturbation, the action variables along a trajectory change only slightly and will deviate even for $t \to \infty$ only a little from their initial values. This changes radically in the case of systems with more than two degrees of freedom. Although the KAM theory still applies under certain conditions and most initial conditions thus lead to stable trajectories, we observe that, for some initial values, trajectories exist whose angle variables along the curve depart steadily from the original values. The chaotic regions where resonance occurs wander aimlessly between the invariant tori through the space.

It should be added that many questions are still unanswered for systems with f $\geqslant 3$. The Poincaré-Birkhoff theorem cannot simply be extended to higher dimensions, for example, and it is still not known whether in this case, one can presume the existence of fixed points in chaotic regions (Arnold, 1978).

4.6 A Numerical Example: The Hénon Map

In a qualitative approach, it is mainly the long-term behaviour of the trajectory for $t \to \infty$ that is of interest. In this case, it is sufficient to take "random samples" of the current position of the trajectories from time to time with the help of a Poincaré section Σ; all the fundamental characteristics of the system are reflected in the Poincaré map, see section 3.5. The discussion of the dynamics is then made considerably easier by the reduction of the dimension to 2(f − 1). However, it is not generally possible to specify the mapping function explicitly so that we still have to rely on a highly time-consuming numerical integration of the Hamilton equations of motion in order to establish the points of intersection of the trajectory with Σ.

If we are not interested in a specific dynamic problem but rather in the general characteristics of Hamilton systems, it is possible to study non-linear, area-preserving iterated maps directly. In this way, we can dispense with time integrations, the calculation time is reduced drastically (normally by a factor of 1 000!) and we can thus extend the observation period of the system behaviour considerably. Moreover, the numerical accuracy is increased so that only rounding-off errors occur.

In the literature, several two-dimensional recursive mapping rules of this sort have been studied intensively. We mention the work of J. M. Greene (Greene, 1979) who, on the basis of a standard map which results directly from Moser's perturbed twist map, eq. (4.5.3), carried out a great many numerical explorations in order to study when and where KAM curves occur and according to which laws they vanish as the perturbation increases, thus making room for stochastic regions.

In the following, we deal in more detail with a different map, the study of which was originally proposed by C. L. Siegel (Siegel, 1956) but which was studied extensively by Michel Hénon in 1969 (Hénon, 1969). The consequences resulting from the KAM theory and the Poincaré-Birkhoff theorem can be demonstrated geometrically particularly well on the basis of this map.

As the prototype of a Poincaré map of a Hamilton system with two degrees of freedom, let us construct an area-preserving map T of the x, y-plane onto itself

$$\begin{aligned} x' &= g(x, y) \\ y' &= h(x, y) \end{aligned} \tag{4.6.1}$$

for which the origin should be an elliptic point. For numerical investigations, it is most appropriate to select polynomials for g and h.

If g and h are linear functions and the origin a fixed point, due to the postulated area-preservation, we obtain a rotation R as a map

$$\begin{bmatrix} x' \\ y' \end{bmatrix} = \boldsymbol{R} \begin{bmatrix} x \\ y \end{bmatrix} \qquad \text{with} \qquad \boldsymbol{R} = \begin{bmatrix} \cos\alpha & -\sin\alpha \\ \sin\alpha & \cos\alpha \end{bmatrix} \tag{4.6.2}$$

where α denotes the angle of rotation.

In order to create the pre-requisite for chaotic behaviour, however, g and h must be non-linear. The most general quadratic map T with the origin as a fixed point can be expressed as follows

$$\begin{aligned} x' &= x\cos\alpha - y\sin\alpha + ax^2 + bxy + cy^2 \\ y' &= x\sin\alpha + y\cos\alpha + dx^2 + exy + fy^2 \end{aligned} \tag{4.6.3}$$

Since T is area-preserving for a Hamilton system, the determinant of the Jacobian matrix, eq. (4.2.36), must possess the value of 1, i.e.

$$\det\left[\frac{\partial(x', y')}{\partial(x, y)}\right] = \begin{vmatrix} (\cos\alpha + 2ax + by) & (-\sin\alpha + bx + 2cy) \\ (\sin\alpha + 2dx + ey) & (\cos\alpha + ex + 2fy) \end{vmatrix} = 1 \tag{4.6.4}$$

This identity provides five relationships for the seven coefficients. If we also introduce a linear coordinate transformation (Hénon, 1969), we finally obtain the following normal form of the quadratic map

$$\begin{aligned} x' &= x\cos\alpha - (y - x^2)\sin\alpha \\ y' &= x\sin\alpha + (y - x^2)\cos\alpha \end{aligned} \tag{4.6.5}$$

which contains only one free system parameter α.

In the following, we investigate the behaviour of the map T in dependence on α. The origin is a fixed point and its stability behaviour is determined in accordance with section 4.5 by means of the eigenvalues of the appertaining Jacobian matrix. If we exclude the case $\alpha = 0$, we obtain the complex conjugate eigenvalues

$$\lambda_{1,2} = \cos\alpha \pm i\sin\alpha = \mathrm{e}^{\pm i\alpha}$$

with modulus 1, i.e. the origin is indeed an elliptic fixed point. This is immediately understood if we observe points which are close enough to the origin O and for which we can thus ignore the perturbation due to the quadratic expressions. These points are rotated about O through the angle α so that the origin is surrounded in its immediate neighbourhood by concentric circles.

If the map T is to be a meaningful model for the Poincaré map of a general continuous Hamilton system with two degrees of freedom, then, due to the determinacy of the differential equations from which it is derived, the inverse map T^{-1} must also exist. It can easily be ascertained that the inverse map T^{-1} is also of second order and can be written

$$\begin{aligned} x &= x'\cos\alpha + y'\sin\alpha \\ y &= -x'\sin\alpha + y'\cos\alpha + (x'\cos\alpha + y'\sin\alpha)^2 \end{aligned} \tag{4.6.6}$$

We first investigate some geometrical characteristics of the map T, subsequently determine further fixed points of T – which exist in addition to the origin – and their iterated maps T^{m} and finally present and interpret a series of numerical results for various system parameters α.

Let us begin with the geometrical characteristics of T. From eq. (4.6.5), we can see that T can be interpreted as a compound mapping; first the transformation S

$$\begin{aligned} \overline{x} &= x \\ \overline{y} &= y - x^2 \end{aligned} \tag{4.6.7}$$

is applied to the point (x, y) and subsequently a rotation R in accordance with eq. (4.6.2). Here, S imposes an area-preserving shearing parallel to the y-axis. Successive application of S and R yields the transformation

$$T = RS \quad \text{or} \quad T^{-1} = S^{-1}R^{-1} \tag{4.6.8}$$

from which we can immediately deduce the inverse map, eq. (4.6.6). By this decomposition of the mapping T, we can now conclude that T is in a sense "symmetrical" with respect to a straight line w through the origin, this line enclosing the angle $\alpha/2$ with the x-axis. To confirm this, we write the inverse map T^{-1} of eq. (4.6.8) in the form

$$T^{-1} = R^{-1}(RS^{-1}R^{-1}) \tag{4.6.9}$$

and first set up a relation between the shear S and the partial map $RS^{-1}R^{-1}$.

In fig. 4.6.1, we have first subjected a point P to the mapping T and then to T^{-1} where we have used the expressions

$$P^{(1)} = S(P) \quad , \qquad P^{(2)} = T(P) = R(P^{(1)})$$

$$P^{(3)} = RS^{-1}R^{-1}(P^{(2)}) \, , \qquad P^{(4)} = R^{-1}(P^{(3)}) = T^{-1}T(P) = P$$

to represent the image points of the individual partial maps. We can see from this diagram that the triangle $OP^{(2)}P^{(3)}$ evolves from the triangle $OP^{(1)}P$ by means of a rotation through the angle α. If we interpret the vector $\overrightarrow{PP^{(1)}}$ as a beam of light and observe its reflection on the straight line w, we obtain a vector parallel to $P^{(2)}P^{(3)}$ which represents the map $RS^{-1}R^{-1}$ and may again be interpreted as a shear.

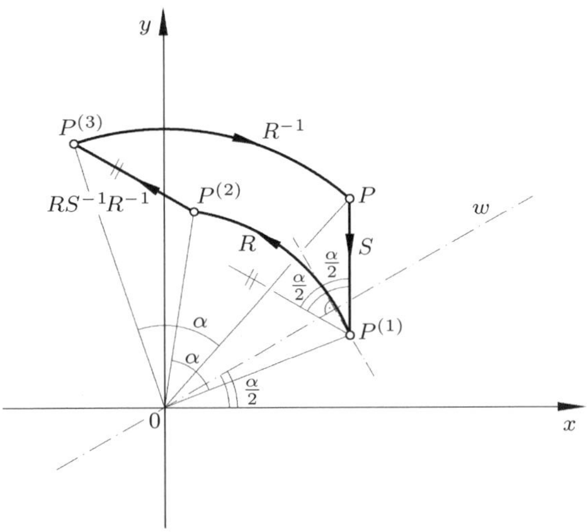

Fig. 4.6.1

Geometrical characteristics of T and T^{-1}

If we mark the points which are generated through a reflection on the w-line with an asterisk, we obtain from the geometrical considerations above

$$T^{-1}(P^*) = [T(P)]^* \tag{4.6.10}$$

i. e. T^{-1} is a map symmetrical to T with respect to the straight line w. This is again made clear in fig. 4.6.2. If we consider the sequence of image points which results from the repeated application of T

$$P^0, P^1 = T(P^0), \; P^2 = T(P^1), \ldots$$

there is a sequence of points placed symmetrically to the axis w which is generated by successive application of the inverse map T^{-1}. Thus,

$$P^{0*}, \; P^{1*} = T^{-1}(P^{0*}), \; P^{2*} = T^{-1}(P^{1*}), \ldots$$

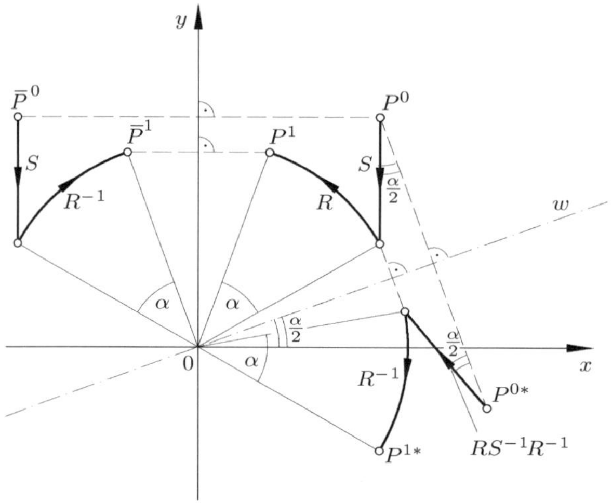

Fig. 4.6.2

Geometrical interpretation of the relationship $T^{-1}(P^*) = [T(P)]^*$

This elucidates the symmetry with respect to the axis w in figs 4.6.4 to 4.6.18. It can now easily be seen that it suffices to restrict the system parameter α to the interval

$$0 < \alpha < \pi \tag{4.6.11}$$

If we consider a point $\bar{P}^0$ symmetrical to P^0 with respect to the y-axis and subject it to the mapping $T_{-\alpha}$ (see fig. 4.6.2) – where the lower index indicates that α in eq. (4.6.5) is replaced by $-\alpha$ (respectively $2\pi - \alpha$) – then the image point $\bar{P}^1 = T_{-\alpha}(\bar{P}^0)$ is obviously symmetrical to the point $P^1 = T_\alpha(P^0)$. Thus, if we replace α by $-\alpha$ (respectively $2\pi - \alpha$), the map T is replaced by a map symmetrical to it with respect to the y-axis so that a restriction to $0 \leqslant \alpha \leqslant \pi$ is justified. Since T for $\alpha = 0$ coincides with the shear S and T^2 for $\alpha = \pi$ reproduces the identity, a restriction to the interval $0 < \alpha < \pi$ proves sufficient.

Our next step is to determine the fixed points of T and of their iterations T^{m} and to investigate their stability behaviour. We designate the fixed points respectively invariant points of the m-fold iterated map T^{m} as I_{mj} (see fig. 4.6.3). I_{11} and I_{12} result from eq. (4.6.5) as fixed points of T with the coordinates

$$\begin{aligned} I_{11} &: x_{11} = 0, && y_{11} = 0 \\ I_{12} &: x_{12} = 2\tan(\alpha/2), && y_{12} = 2\tan^2(\alpha/2) \end{aligned} \tag{4.6.12}$$

Apart from the origin I_{11}, a stable elliptic point, a further fixed point I_{12} is generated which lies on the axis of symmetry w and whose stability behaviour we investigate in accordance with section 3.6. To this end, we determine the eigenvalues of the Jacobian matrix $\boldsymbol{D}$ for the point I_{12}, i.e. the eigenvalues λ_1, λ_2 of the matrix

$$\boldsymbol{D} = \begin{bmatrix} \cos\alpha + 4\sin\alpha\tan(\alpha/2) & -\sin\alpha \\ \sin\alpha - 4\cos\alpha\tan(\alpha/2) & \cos\alpha \end{bmatrix} \tag{4.6.13}$$

Noting the condition of eq. (4.6.11),

$$\operatorname{tr}\boldsymbol{D} = \lambda_1 + \lambda_2 = 2\cos\alpha + 4\sin\alpha\tan(\alpha/2) = 2 + 4\sin^2(\alpha/2) > 2 \tag{4.6.14}$$

applies for the eigenvalues and, because of the area-preserving character of the map, $\lambda_1\lambda_2 = 1$ must also be fulfilled (see section 4.5). Were λ_1, λ_2 complex conjugate, they would lie on the unit circle of the complex plane, so that $\lambda_1 + \lambda_2 < 2$ (cf. fig. 4.5.1). It then follows that λ_1 and λ_2 are real and that the invariant point I_{12} is an unstable hyperbolic fixed point.

In order to determine the fixed points of the iterated map T^{m}, the following set of 2m non-linear equations has to be solved

$$\begin{aligned} x_{\mathrm{k}+1} &= x_{\mathrm{k}}\cos\alpha - (y_{\mathrm{k}} - x_{\mathrm{k}}^2)\sin\alpha \\ y_{\mathrm{k}+1} &= y_{\mathrm{k}}\sin\alpha + (y_{\mathrm{k}} - x_{\mathrm{k}}^2)\cos\alpha \end{aligned} \qquad (\mathrm{k} = 1, 2, \ldots, \mathrm{m}) \tag{4.6.15}$$

where $x_{\mathrm{m}+1} = x_1$ and $y_{\mathrm{m}+1} = y_1$ must apply. From the first of the mapping equations for the inverse map T^{-1}, eq. (4.6.6), we obtain

$$y_{\mathrm{k}} = \frac{1}{\sin\alpha}(x_{\mathrm{k}-1} - x_{\mathrm{k}}\cos\alpha) \tag{4.6.16}$$

Substitution into eq. (4.6.15) leads to m quadratic equations for the x-coordinates of the fixed points

$$x_{\mathrm{k}}^2\sin\alpha + 2x_{\mathrm{k}}\cos\alpha - x_{\mathrm{k}+1} - x_{\mathrm{k}-1} = 0 \qquad \begin{aligned} &(\mathrm{k} = 1, 2, \ldots, \mathrm{m}) \\ &(x_{\mathrm{m}+1} = x_1,\ x_{\mathrm{m}} = x_0) \end{aligned} \tag{4.6.17}$$

Equations (4.6.16) and (4.6.17) yield a maximum of $\mathrm{n} = 2^{\mathrm{m}}$ fixed points for the map T^{m}.

For $\mathrm{m} = 2$, the x-coordinates of the invariant points $I_{2\mathrm{n}}$ ($\mathrm{n} = 1, \ldots, 4$) are determined from the system

$$\begin{aligned} x_1^2\sin\alpha + 2x_1\cos\alpha - 2x_2 &= 0 \\ x_2^2\sin\alpha + 2x_2\cos\alpha - 2x_1 &= 0 \end{aligned} \tag{4.6.18}$$

where, of course, both invariant points of T ($I_{21} = I_{11}$ and $I_{22} = I_{12}$) are again fixed points. In order to calculate the remaining two fixed points, we add respectively subtract the two equations (4.6.18) and obtain

$$\begin{aligned} x_1 + x_2 &= -2\cot(\alpha/2) \\ x_1^2 + x_2^2 &= -4 \end{aligned} \tag{4.6.19}$$

It follows from the last equation that the invariant points I_{23} and I_{24} have complex coordinates and are not of interest for our investigation.

Figure 4.6.3 shows the invariant points for T, T^2, T^3 and T^4 according to (Hénon, 1969); in each case, the ranges of α are given, these being the intervals for which the fixed points possess real coordinates and stable respectively unstable behaviour occurs. We dispense with the calculation of the coordinates and the stability investigations – which may be carried out according to the pattern above – and refer the reader to (Hénon, 1969).

iterated Hénon map	number of fixed points	fixed points			
		stable	unstable	real for	always complex
T	2	I_{11}	I_{12}		
T^2	4	$I_{21} = I_{11}$	$I_{22} = I_{12}$		I_{23}, I_{24}
T^3	8	$I_{31} = I_{11}$	$I_{32} = I_{12}$		
			$I'_{31}, I'_{32}, I'_{33}$		
		$I''_{31}, I''_{32}, I''_{33}$ for $-1/2 < \cos\alpha < 1-\sqrt{2}$	$I''_{31}, I''_{32}, I''_{33}$ for $-1 < \cos\alpha \leqslant -1/2$	$\cos\alpha \leqslant 1-\sqrt{2}$	
T^4	16	$I_{41} = I_{11}$	$I_{42} = I_{12}$		$I_{43} = I_{23}$ $I_{44} = I_{24}$
			$I'_{41}, I'_{42}, I'_{43}, I'_{44}$		
		$I''_{41}, I''_{42}, I''_{43}, I''_{44}$ for $-0.10336 < \cos\alpha < 0$	$I''_{41}, I''_{42}, I''_{43}, I''_{44}$ for $-1 < \cos\alpha \leqslant -0.10336$	$\cos\alpha \leqslant 0$	
					I'''_{41}, I'''_{42} I'''_{43}, I'''_{44}

Fig. 4.6.3: Invariant points of the Hénon map

We now reproduce a series of numerical results. Figures 4.6.4 to 4.6.18 show the development and modifications of the map in dependence on the system parameter α. In each case, the dashed line represents the axis of symmetry w which encloses with the x-axis the angle $\alpha/2$.

We first look at fig. 4.6.6 for $\cos\alpha = 0.24$ more closely since it contains all the fundamental characteristics of Poincaré maps of Hamilton systems with two degrees of freedom. In the neighbourhood of the origin, the linear terms predominate so that the sequence of points which appertain to a fixed initial condition lie on closed, almost circular curves about O. If the initial point is at a sufficiently large distance from the origin I_{11}, however, the quadratic terms become dominant. Equation (4.6.5) then yields the approximate expressions

$$x_{\mathrm{i}+1} \approx x_{\mathrm{i}}^2 \sin\alpha$$
$$y_{\mathrm{i}+1} \approx -x_{\mathrm{i}}^2 \cos\alpha$$

Multiplying the first relationship with $\sin\alpha$

$$x_{\mathrm{i}+1} \sin\alpha \approx (x_{\mathrm{i}} \sin\alpha)^2 \tag{4.6.20}$$

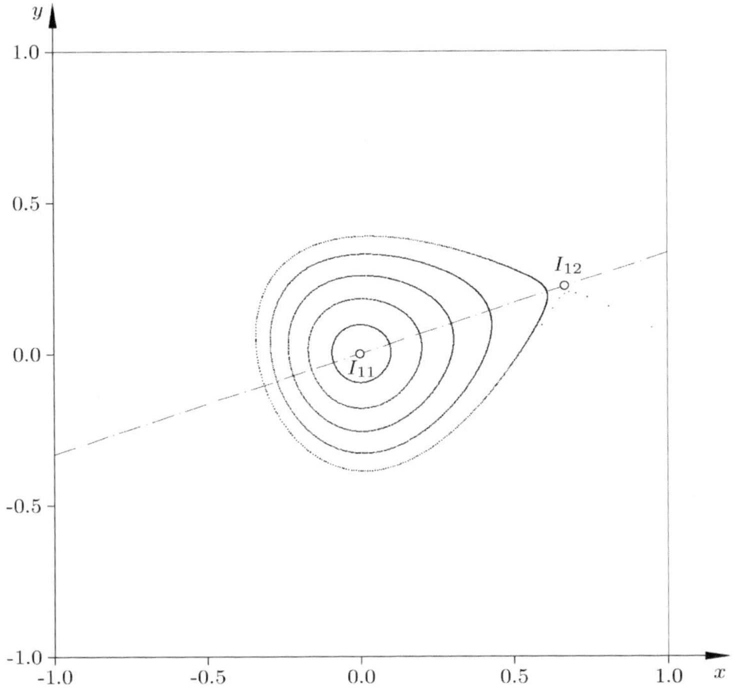

Fig. 4.6.4

Hénon map for $\cos\alpha = 0.8$

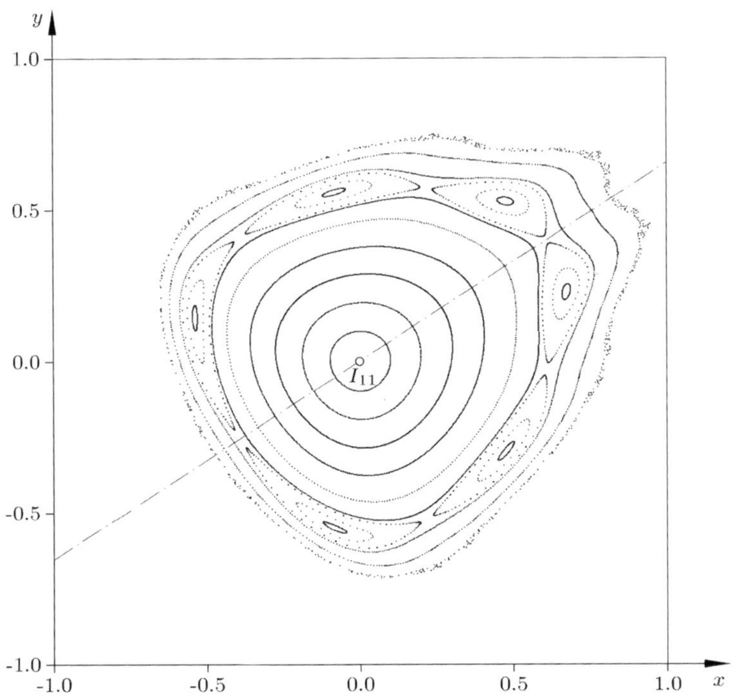

Fig. 4.6.5

Hénon map for $\cos\alpha = 0.4$

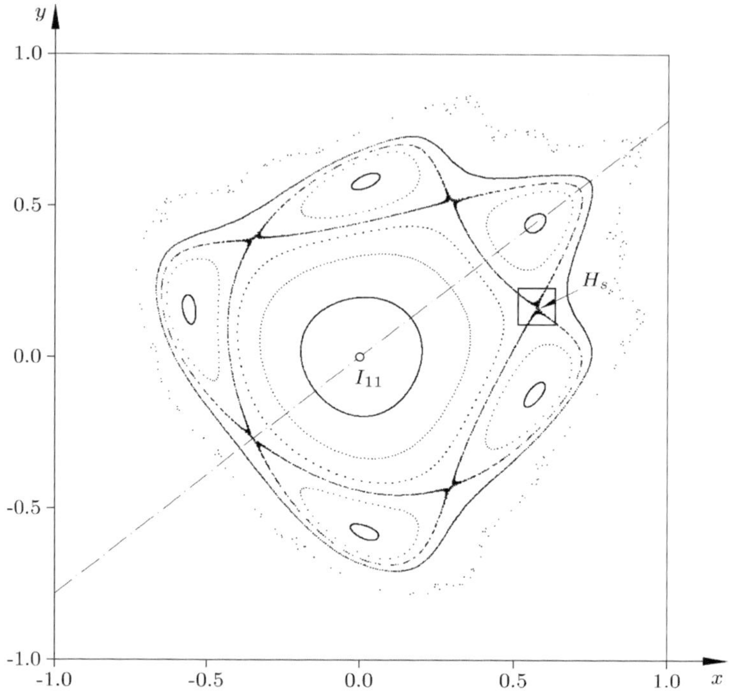

Fig. 4.6.6

Hénon map for $\cos\alpha = 0.24$

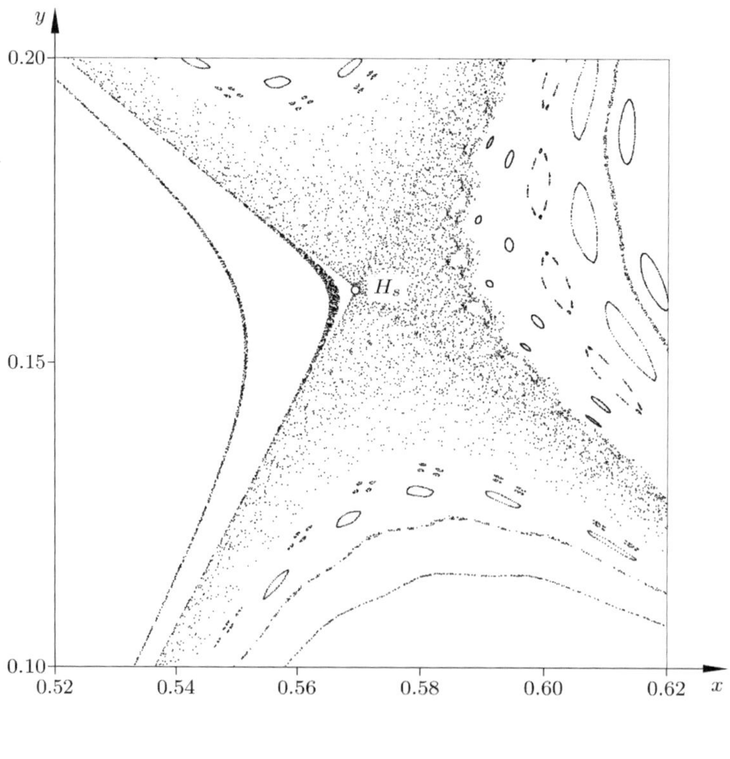

Fig. 4.6.7

Enlarged section of the region around a hyperbolic fixed point we recognise that the quantity $x_i \sin\alpha$ is approximately squared at each iteration so that for

$$|x_i| > \frac{1}{\sin\alpha} \tag{4.6.21}$$

the sequence of image points tends towards infinity very fast. On each closed curve, the topological order of successive image points is characterised by the corresponding rotation number ν, eq. (4.3.2), which varies continuously from one curve to another. If ν is irrational, the image points gradually fill the whole curve. For rational values of ν, the theorem of Poincaré-Birkhoff, section 4.5, takes effect. The curve disintegrates with the exception of an even number of 2k fixed points, i. e. a chain of k islands emerges in the map. In the centre of each island lies an elliptic fixed point, again surrounded by closed irrational KAM curves. Between each two elliptic fixed points, one observes a hyperbolic fixed point (cf. fig. 4.5.4).

In order to elucidate the behaviour of the map T in the neighbourhood of a hyperbolic fixed point, fig. 4.6.7 presents an enlargement of the region around such a point H_s shown in fig. 4.6.6. This figure indicates the extremely complex behaviour of the map on any scale considered. The image points are scattered irregularly in the neighbourhood of the hyperbolic point H_s; chaotic behaviour prevails. A clear limitation of the chaotic area to the right cannot be detected. Figure 4.6.6 shows, however, that such points cannot escape from the domain between two closed curves. Points which seem to lie on closed curves in fig. 4.6.6 turn out in the enlargement to be points lying scattered within narrow strips. In fig. 4.6.7, we can recognise further island chains of second and third order; the extent of the islets tends very fast towards zero. These islets are formed when a curve with a rational rotation number disintegrates. Each elliptic point is surrounded by closed curves which in turn partially disintegrate, thus resulting in further elliptic and hyperbolic points (see section 4.5, particularly the schematic representation in fig. 4.5.5). The self-similarity of the map can be observed clearly in these island chains.

Let us return to fig. 4.6.4 which shows the Hénon map for $\cos\alpha = 0.8$. We have displayed the elliptic point I_{11} which is surrounded by concentric, almost circular KAM curves. One sequence of points in the neighbourhood of the unstable fixed point I_{12} tends towards infinity on a hyperbola branch.

In fig. 4.6.5 ($\cos\alpha = 0.4$), we can recognise a chain of 6 islands. Successive image points carry out a compound motion. At each application of T, they jump anticlockwise onto the next island. If we observe the sixfold iterated map T^6, however, the sequence of points moves on concentric curves about the corresponding elliptic fixed point. The island chain clearly appertains to a rotation number $\nu = 1/6$. In the same way, the island chain in fig. 4.6.6 for $\cos\alpha = 0.24$ appertains to a rotation number $\nu = 1/5$.

For $\cos\alpha = 0.02$, fig. 4.6.8 shows two island chains. In the case of the outer chain of 13 islands, two islands are skipped over at each iteration of T so that this chain appertains to $\nu = 3/13$. Similarly, we can determine the rotation number $\nu = 4/17$ by calculating the image point of a point on an island on the inner chain.

Figure 4.6.9 ($\cos\alpha = 0.01$) clearly illustrates the hierarchy of the island chains. About the origin, we observe closed curves which disintegrate into chains of 21 respectively 17 islets; the outer chain appears to be surrounded by yet smaller island chains. It is easy to imagine that the "discovery" of island chains of higher order by the adoption of suitable starting points becomes increasingly difficult, the smaller the extent of the islets is.

According to the table in fig. 4.6.3, $\alpha = \pi/2$ represents a borderline case. For $\cos\alpha \leqslant 0$, four unstable fixed points $I'_{41}, I'_{42}, I'_{43}, I'_{44}$ appear additionally which, however, for $\alpha = \pi/2$, all coincide with the origin, i. e. a stable elliptic point; thus, the origin is a degenerate point (Moser, 1958). The consequence is that, in the case of this α-value, the closed curves about O are no longer circular but are formed by 4 hyperbolic branches joined to one another (fig. 4.6.10).

For slightly larger α-values, 4 distinct stable and 4 unstable fixed points emerge (see figs. 4.6.11 to 4.6.14). At first, four large islands are formed which are separated from the outer region by closed curves ($\cos\alpha = -0.01$). If the system parameter α increases further, these outer curves disintegrate until, finally, all 4 islands lie outside the KAM curves about the origin ($\cos\alpha = -0.05$).

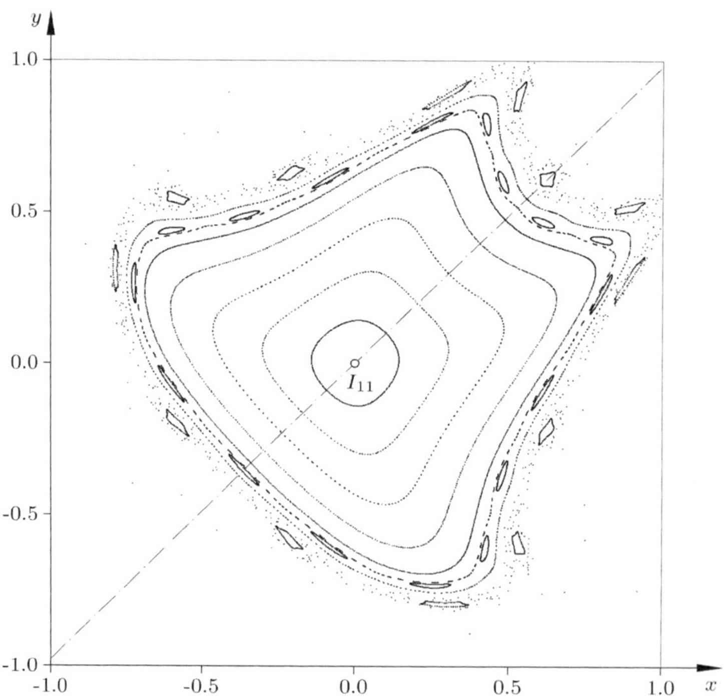

Fig. 4.6.8

Hénon map for $\cos\alpha = 0.02$

For $\cos\alpha \leq 1 - \sqrt{2} \approx -0.414$, two new additional groups with three fixed points each arise (fig. 4.6.3). The first group is always unstable, whereas the second group is stable for $-1/2 < \cos\alpha < 1 - \sqrt{2}$ and unstable for $-1 < \cos\alpha \leqslant -1/2$. The is-

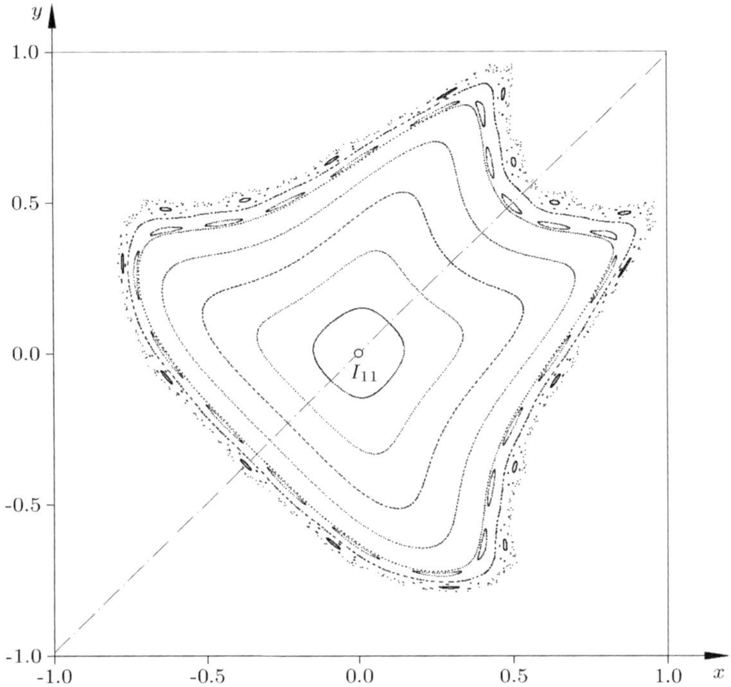

Fig. 4.6.9

Hénon map for $\cos\alpha = 0.01$

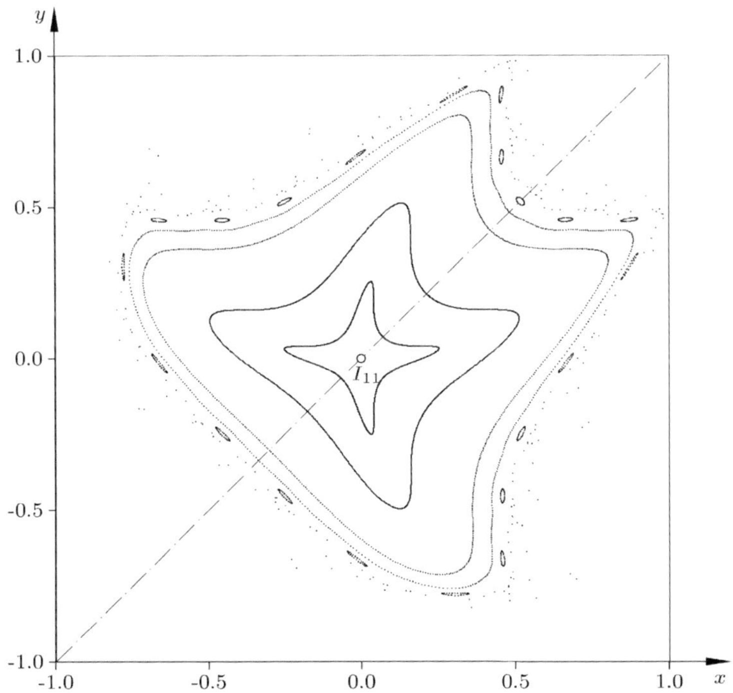

Fig. 4.6.10

Hénon map for $\cos\alpha = 0$

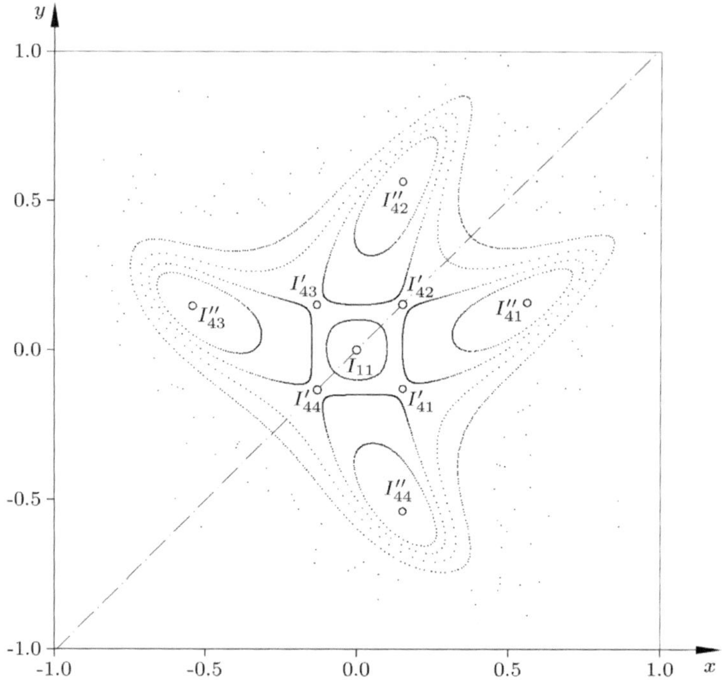

Fig. 4.6.11

Hénon map for $\cos\alpha = -0.01$

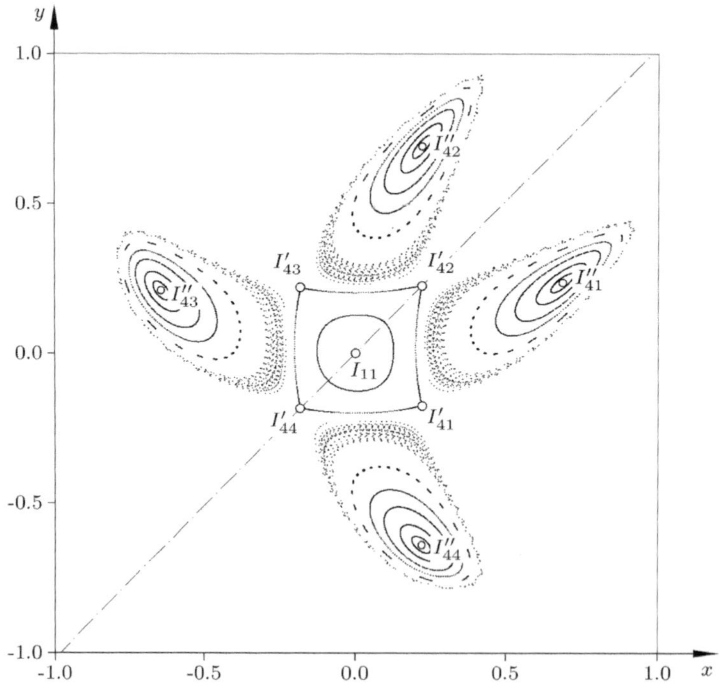

Fig. 4.6.12

Hénon map for $\cos\alpha = -0.02$

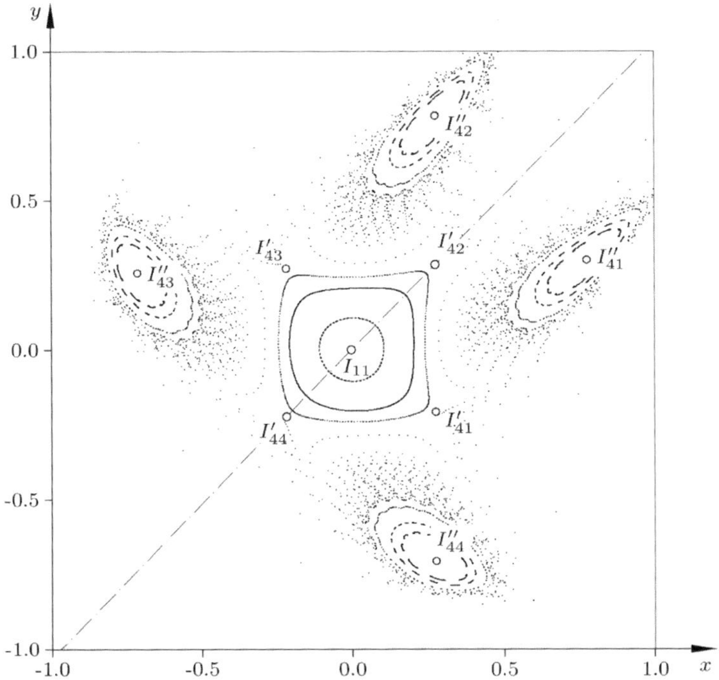

Fig. 4.6.13

Hénon map for $\cos\alpha = -0.03$

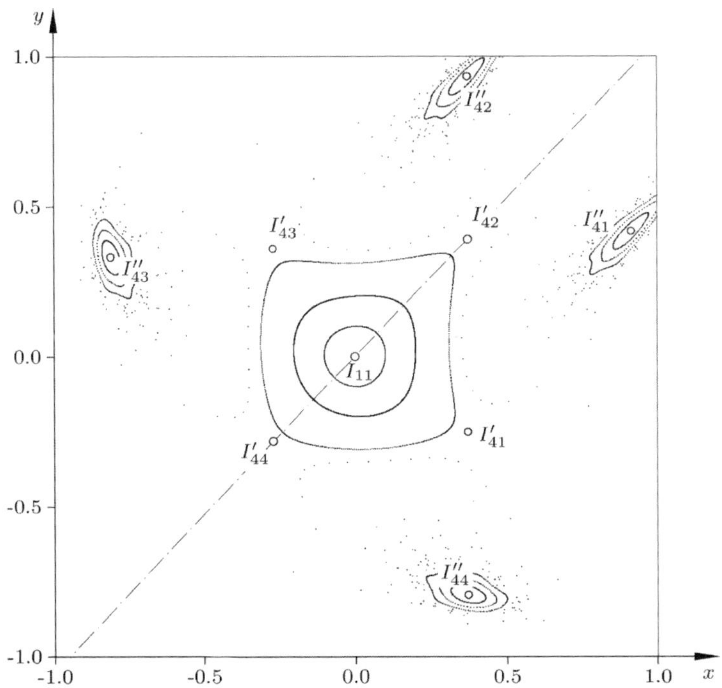

Fig. 4.6.14

Hénon map for $\cos\alpha = -0.05$

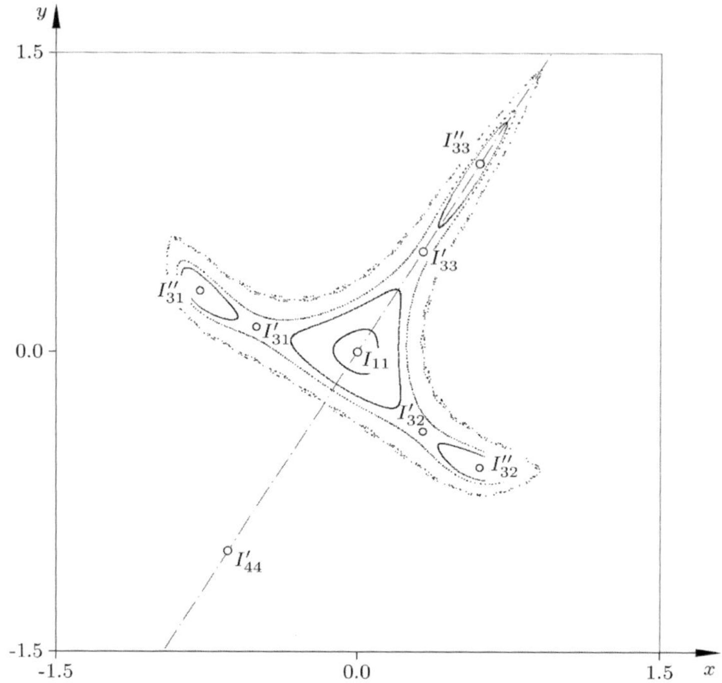

Fig. 4.6.15

Hénon map
for $\cos\alpha = -0.42$

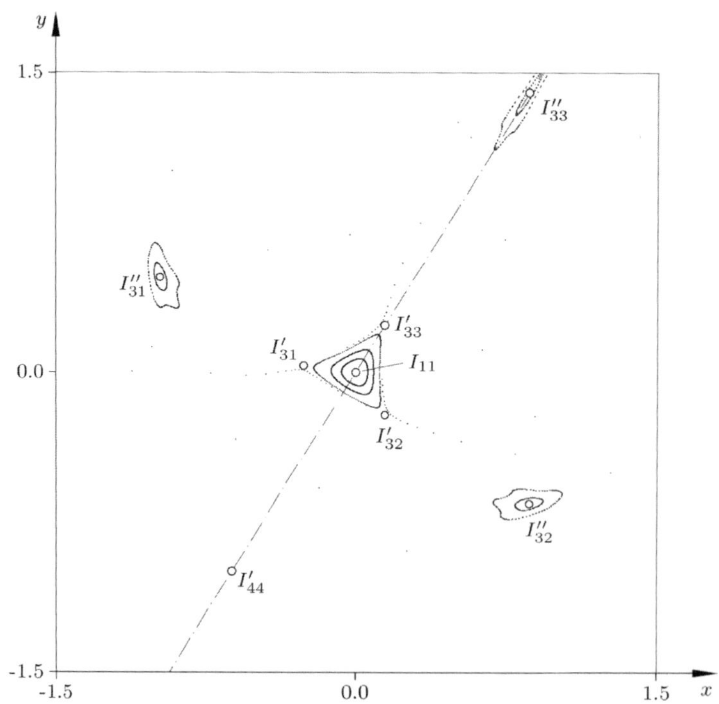

Fig. 4.6.16

Hénon map
for $\cos\alpha = -0.45$

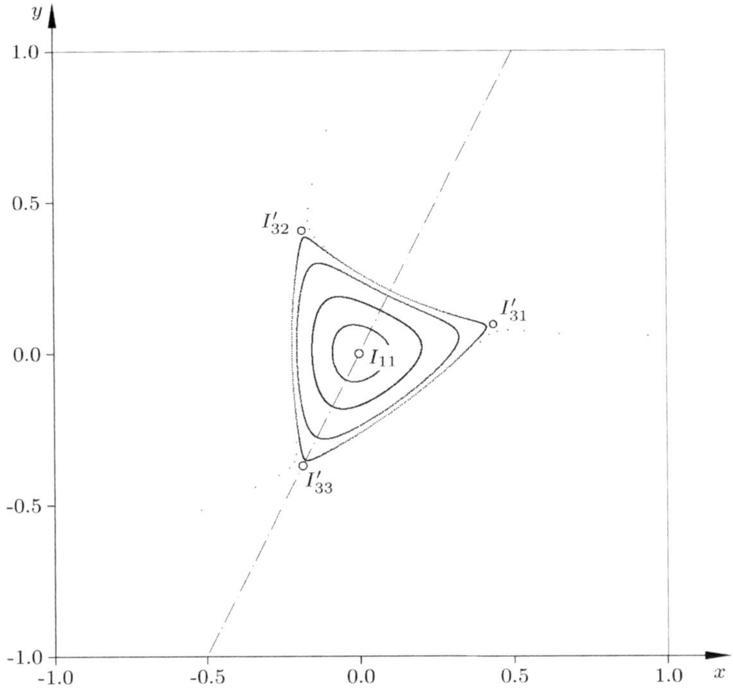

Fig. 4.6.17

Hénon map for $\cos\alpha = -0.6$

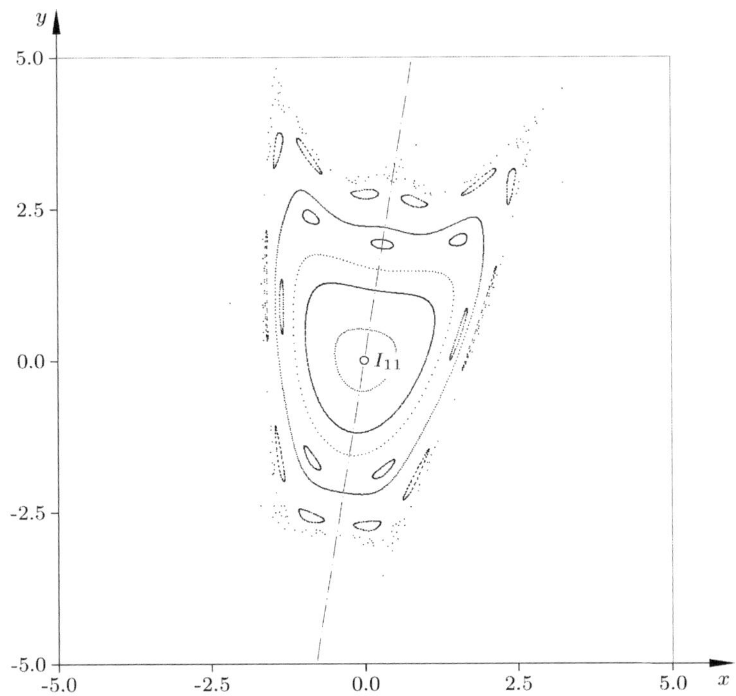

Fig. 4.6.18

Hénon map for $\cos\alpha = -0.95$

lands about the elliptic fixed points $I''_{31}, I''_{32}, I''_{33}$ lie at first for $\cos\alpha = -0.42$ inside a closed curve about the origin (fig. 4.6.15); in the case of larger angles α, however, they lie outside (fig. 4.6.16 for $\cos\alpha = -0.45$). It should be noted that a larger scale had to be adopted for these two maps.

Moser showed (Moser, 1958) that $\alpha = 2\pi/3$ ($\cos\alpha = -0.5$) is a further special case where no invariant curves exist at all. If we observe the behaviour of the unstable fixed points in the neighbourhood of this critical α-value, we recognise that the hyperbolic points approach the origin (see fig. 4.6.15 for $\cos\alpha = -0.42$ and fig. 4.6.16 for $\cos\alpha = -0.45$) while the area containing the invariant curves about the origin clearly has to shrink. For $\cos\alpha = -0.5$, the unstable fixed points coincide with the origin and move away from it again in the case of $\cos\alpha < -0.5$. In this way, the triangular region of closed curves contracts to a point and appears again in a mirrored form for $\alpha > 2\pi/3$ (fig. 4.6.17). For $\alpha \to \pi$, the area in which the closed curves exist expands so that, for $\cos\alpha = -0.95$ (fig. 4.6.18), we show a considerably enlarged section.

On the basis of the numerical results of the Hénon map, we were able to show the fundamental characteristics of Poincaré maps of conservative dynamical systems with two degrees of freedom, these being the existence of closed curves about elliptic points, the disintegration of curves with rational frequency ratios into island chains, where as many stable as unstable points are preserved, the chaotic regions in the neighbourhood of hyperbolic points and, finally, the self-similarity of the whole structure.

A final remark: all the calculations were carried out on a VAX 785 in single precision (ca. 6 decimal points). This explains certain deviations of our figures from those in (Hénon, 1969) where the precision of 16 decimal points was employed.

5 Dynamical Systems with Dissipation

Mit dem Unerwarteten
ist schwer zu rechnen
Paul Klee

In the preceding chapter, we discussed conservative, i.e. dissipation-free, dynamical systems. This category of physical or mechanical systems is characterised by the fact that a volume of an element in the phase space remains invariant, (see eq. (4.1.30), Liouville theorem). In this chapter, we expand our theoretical considerations and deal with non-linear dissipative systems, i.e. dynamical systems experiencing a loss of energy. If we again observe a volume element in the phase space, then, for $t \to \infty$, it will contract to a subset of an attractor, the dimensionality of which is lower than that of the phase space.

The latent tendency of non-linear dissipative systems to irregular motions remained unobserved for a long time. The study of such systems was of necessity restricted to the range of regular behaviour characterised by the regular attractors (such as fixed points, limit cycles or tori). In the study of dissipative systems, a whole new world was revealed when Edward N. Lorenz (Lorenz, 1963) developed a highly simplified hydrodynamic model for meteorology which derives from an extreme simplification of the Navier-Stokes equation. These equations lead to an "attractive" structure in the phase space which has the characteristics of neither a fixed point, a limit cycle nor a torus but discloses an irregular, seemingly erratic behaviour of the trajectories. However, this phenomenon remained unnoted until, independent of Lorenz, Ruelle and Takens (1971) set up a model to describe turbulence and discovered, in the phase space, a limited domain which attracts the trajectories globally and causes them to diverge locally from one another exponentially. They denoted this odd and mysterious part of the phase space a *strange attractor*.

In the following, we first describe the basic characteristics of dissipative systems, cf. also (Abarbanel et al., 1993). This includes the volume contraction in the phase space and the concept of simple attractors which characterise regular motion. We then deal with the phenomenon of chaotic motion using the example of the Lorenz attractor and go on to discuss both the classic tools – such as the power spectrum and autocorrelation – as well as more recent geometrical and numerical methods – such as Lyapunov exponents, fractal dimensions and Kolmogorov entropy – which are used for the characterisation of the strange attractor. As mentioned previously in section 3.4, we restrict our considerations in the following to purely deterministic systems without any stochastic terms.

J. Argyris et al., *An Exploration of Dynamical Systems and Chaos*,
DOI: 10.1007/978-3-662-46042-9_5

5.1 Volume Contraction – A Basic Characteristic of Dissipative Systems

In order to describe the processes of motion, we consider both a set of n non-linear autonomous first-order differential equations, see eq. (2.3.3),

$$\dot{\boldsymbol{x}} = \boldsymbol{F}(\boldsymbol{x}) \tag{5.1.1}$$

where $\boldsymbol{x}$ and $\boldsymbol{F}$ are n-dimensional vectors and $\boldsymbol{F}$ does not depend explicitly on the independent variable t, as well as a discrete system of iterative equations

$$\boldsymbol{x}_{\mathrm{k}+1} = \boldsymbol{f}(\boldsymbol{x}_{\mathrm{k}}) \tag{5.1.2}$$

which can be deduced, for example, by a discretisation in time or by Poincaré sections from eq. (5.1.1) (cf. sections 3.4 and 3.5).

We first consider the volume contraction for a dissipative flow, eq. (5.1.1). Our discussion follows that of Lichtenberg and Lieberman (1983). For an n-dimensional non-linear flow, we determine the temporal change of a small volume element $\Delta V(\boldsymbol{x})$ in the phase space by following ΔV along the trajectory which originates in $\boldsymbol{x}_0$ for $t = 0$. ΔV can be expressed in Cartesian coordinates in the form

$$\Delta V(\boldsymbol{x}) = \prod_{\mathrm{i}} \Delta x_{\mathrm{i}}(\boldsymbol{x}) \tag{5.1.3}$$

where we apply the classical product definition. For the temporal change of ΔV, there applies

$$\begin{aligned}\frac{\mathrm{d}(\Delta V(\boldsymbol{x}))}{\mathrm{d}t} &= \frac{\mathrm{d}(\Delta x_1(\boldsymbol{x}))}{\mathrm{d}t}\prod_{\mathrm{i}\neq 1}\Delta x_{\mathrm{i}}(\boldsymbol{x}) + \frac{\mathrm{d}(\Delta x_2(\boldsymbol{x}))}{\mathrm{d}t}\prod_{\mathrm{i}\neq 2}\Delta x_{\mathrm{i}}(\boldsymbol{x}) + \dots \\ &= \left(\prod_{\mathrm{i}}\Delta x_{\mathrm{i}}(\boldsymbol{x})\right)\frac{\mathrm{d}(\Delta x_1(\boldsymbol{x}))}{\mathrm{d}t}\Delta x_1^{-1}(\boldsymbol{x}) + \left(\prod_{\mathrm{i}}\Delta x_{\mathrm{i}}(\boldsymbol{x})\right)\frac{\mathrm{d}(\Delta x_2(\boldsymbol{x}))}{\mathrm{d}t}\Delta x_2^{-1}(\boldsymbol{x}) + \dots\end{aligned} \tag{5.1.4}$$

From eqs. (5.1.4) and (5.1.3), we obtain

$$\frac{\mathrm{d}(\Delta V(\boldsymbol{x}))}{\mathrm{d}t} = \Delta V(\boldsymbol{x})\sum_{\mathrm{i}}\Delta x_{\mathrm{i}}^{-1}(\boldsymbol{x})\frac{\mathrm{d}(\Delta x_{\mathrm{i}}(\boldsymbol{x}))}{\mathrm{d}t} \tag{5.1.5}$$

The length $\Delta x_{\mathrm{i}}(\boldsymbol{x})$ of the i-th edge of the volume element can be approximated at $\boldsymbol{x}_0$ by the linear term of the Taylor series

$$\Delta x_{\mathrm{i}}(\boldsymbol{x}) = \Delta x_{\mathrm{i}}(\boldsymbol{x}_0, t) = \Delta x_{\mathrm{i}}(\boldsymbol{x}_0)\frac{\partial x_{\mathrm{i}}(\boldsymbol{x})}{\partial x_{\mathrm{i}0}} = \Delta x_{\mathrm{i}0}\frac{\partial x_{\mathrm{i}}(\boldsymbol{x})}{\partial x_{\mathrm{i}0}} \tag{5.1.6}$$

Adopting this expression, the temporal change of $\Delta x_{\mathrm{i}}(\boldsymbol{x})$ can now be approximated as

$$\frac{\mathrm{d}(\Delta x_{\mathrm{i}}(\boldsymbol{x}))}{\mathrm{d}t} = \Delta x_{\mathrm{i}0}\frac{\mathrm{d}}{\mathrm{d}t}\frac{\partial x_{\mathrm{i}}(\boldsymbol{x})}{\partial x_{\mathrm{i}0}} \tag{5.1.7}$$

Taking eq. (5.1.7) into account, we deduce from eq. (5.1.5) for the volume change rate per unit volume, Λ, following an interchange of the spatial and temporal derivatives, the following relation

$$\Lambda(\boldsymbol{x}) = \frac{1}{\Delta V(\boldsymbol{x})}\frac{\mathrm{d}(\Delta V(\boldsymbol{x}))}{\mathrm{d}t} = \sum_{\mathrm{i}} \frac{\Delta x_{\mathrm{i}0}}{\Delta x_{\mathrm{i}}(\boldsymbol{x})}\frac{\partial}{\partial x_{\mathrm{i}0}}\frac{\mathrm{d}x_{\mathrm{i}}(\boldsymbol{x})}{\mathrm{d}t} \tag{5.1.8}$$

Consideration of eq. (5.1.1) for the phase flow in expression eq. (5.1.8) yields

$$\Lambda(\boldsymbol{x}) = \sum_{\mathrm{i}} \frac{\Delta x_{\mathrm{i}0}}{\Delta x_{\mathrm{i}}(\boldsymbol{x})}\frac{\partial F_{\mathrm{i}}(\boldsymbol{x})}{\partial x_{\mathrm{i}0}} \tag{5.1.9}$$

For $t \to t_0$, $\Delta x_{\mathrm{i}}(\boldsymbol{x})$ can be replaced by $\Delta x_{\mathrm{i}0}$; thus, eq. (5.1.9) for an arbitrary phase point $\boldsymbol{x}$ is reduced to

$$\Lambda(\boldsymbol{x}) = \sum_{\mathrm{i}} \frac{\partial F_{\mathrm{i}}}{\partial x_{\mathrm{i}}} \tag{5.1.10}$$

Equation (5.1.10) thus yields the condition for the volume contraction of dynamical systems in the form

$$\Lambda(\boldsymbol{x}) = \sum_{\mathrm{i}} \frac{\partial F_{\mathrm{i}}}{\partial x_{\mathrm{i}}} = \operatorname{div} \boldsymbol{F} = \operatorname{tr} \boldsymbol{D} < 0 \tag{5.1.11}$$

where $\boldsymbol{D}$ is the Jacobian matrix, see eq. (3.2.3).

Thus, the Liouville formula, eq. (4.1.30), is confirmed as a function of the phase velocity $\boldsymbol{F}$. If $\Lambda(\boldsymbol{x}) > 0$ in eq. (5.1.11), the dynamical process is expanding and for $\Lambda(\boldsymbol{x}) < 0$ contracting. The statement of eq. (5.1.11) on the initial volume in the phase space is of local nature; if we wish to make statements on contraction respectively expansion for dissipative systems as a whole, we must specify a mean rate $\bar{\Lambda}$ taken along each trajectory which, in the case of dissipation, must also be negative. In section 5.4.4, we shall show that the n-dimensional Lyapunov exponent $\sigma^{(\mathrm{n})}$ agrees exactly with the mean exponential change rate $\bar{\Lambda}$ of the volume elements.

Using a simple two-dimensional example, we now demonstrate the characteristics of volume contraction. To this end, we select the linearised equation of a pendulum with a viscous damping ζ. This is written following eq. (2.2.6) as

$$\ddot{x} + 2\zeta\omega_0\dot{x} + \omega_0^2 x = 0 \tag{5.1.12}$$

Applying the substitution $x_1 = x$, $x_2 = \dot{x}$, we can transform this equation into an autonomous first-order system

$$\begin{bmatrix} \dot{x}_1 \\ \dot{x}_2 \end{bmatrix} = \begin{bmatrix} 0 & 1 \\ -\omega_0^2 & -2\zeta\omega_0 \end{bmatrix} \begin{bmatrix} x_1 \\ x_2 \end{bmatrix}$$

In compact matrix notation, see also eq. (3.1.1), we may write

$$\dot{\boldsymbol{x}} = \boldsymbol{L}\boldsymbol{x}$$

Applying eq. (5.1.11) for the rate of volume change Λ to the pendulum equation (5.1.12), we obtain

$$\Lambda = \text{div}\, \boldsymbol{F} = \frac{\partial F_1}{\partial x_1} + \frac{\partial F_2}{\partial x_2} = \text{tr}\, \boldsymbol{L} = -2\zeta\omega_0$$

If $\zeta > 0$, the system is dissipative due to $\omega_0 > 0$; consequently $\Lambda < 0$ and an area contraction occurs. If, on the other hand, $\zeta < 0$, the area expands. This assumption of a "negative" damping may be represented physically by an external energy input. These associated phenomena are an important characteristic of the van der Pol equation to be discussed in section 10.4. The case of negative damping corresponds to the phenomenon of flutter in aeronautics which involves an energy input into the dynamical system and causes the expansion of the volume.

Many processes of evolution can be expressed directly as discrete-time feedback equations. The best-known and at the same time simplest example which does not lead to trivial results is the logistic map on the unit interval [0,1] defined by

$$x_{k+1} = \alpha x_k (1 - x_k) \tag{5.1.13}$$

see also sections 3.7 and 6.7.

In the case of n-dimensional discrete systems, the updated volume element ΔV_{k+1} is determined by multiplication of the current local volume element ΔV_k by the modulus of the Jacobian determinant of the mapping rule taken at the k-th iteration (Lichtenberg and Lieberman, 1983)

$$\Delta V_{k+1} = |\det \boldsymbol{D}_k| \Delta V_k \tag{5.1.14}$$

where $\boldsymbol{D}_k$ designates the Jacobian matrix at $\boldsymbol{x}_k$, i.e. after the k-th iteration. Thus, for the volume contraction rate as arising for the k-th volume element, we obtain

$$\begin{aligned} \Lambda(\boldsymbol{x}_k) &= \frac{1}{\Delta V_k} \frac{d(\Delta V_k)}{dk} = \frac{|\det \boldsymbol{D}_k| \Delta V_k - \Delta V_k}{\Delta V_k} \\ &= |\det \boldsymbol{D}_k| - 1 \end{aligned} \tag{5.1.15}$$

Here again, as we shall see, the mean exponential change rate $\bar{\Lambda}$ coincides with the n-dimensional Lyapunov exponent $\sigma^{(n)}$, eq. (5.4.98), and is negative in the case of dissipative systems.

It should be noted that volume contraction does not automatically infer a length contraction in all directions. It is in fact possible that the distance between two at first arbitrarily close points may increase although the volume contracts to zero.

5.2 Strange Attractor: Lorenz Attractor

In this section, we discuss the manifestation of the *strange attractor* respectively chaotic determinism, using the now famous Lorenz attractor as an example. The concept "strange" attractor was coined by Ruelle and Takens (1971) in their paper "On the Nature of Turbulence" in which they expounded the reasons why the "Landau scenario" is inappropriate as an interpretation of the onset of turbulence. It was their intention to describe turbulence not as the superposition of an infinite number of modes but as a "strange" attractor. The word "strange" is intended to make it clear that this attractor is neither a fixed point, a limit cycle nor a torus and that it does not form a manifold. What does define it? We shall attempt to explain this using the example of the Lorenz attractor.

Relying on prior work of Saltzman (Saltzman, 1962), Lorenz, a meteorologist working on weather models, reduced the Navier-Stokes equations in the Boussinesq approximation for a layer of liquid heated from below (for a detailed description, see sections 7.3, 7.4) to a set of three ordinary non-linear differential equations representing the temporal change of three essential modes, one velocity and two temperature distributions (Lorenz, 1963). The equations are

$$\begin{aligned} \dot{X} &= -\sigma X + \sigma Y \\ \dot{Y} &= \quad rX - Y - XZ \\ \dot{Z} &= -bZ + XY \qquad\qquad \text{for } \sigma, r, b > 0 \end{aligned} \tag{5.2.1}$$

They contain three parameters: the Prandtl number σ, the relative Rayleigh number r proportional to the applied temperature difference ΔT and the geometrical measure b, derived from the geometry of the convection cells.

We first determine the divergence of the system in order to gain a global impression of the flow respectively of the motion in the phase space. We obtain the volume contraction rate Λ in accordance with eq. (5.1.11) as the trace of the Jacobian matrix

$$\Lambda = \frac{\partial}{\partial X}(-\sigma X + \sigma Y) + \frac{\partial}{\partial Y}(rX - Y - XZ) + \frac{\partial}{\partial Z}(-bZ + XY) = -(\sigma + 1 + b) \tag{5.2.2}$$

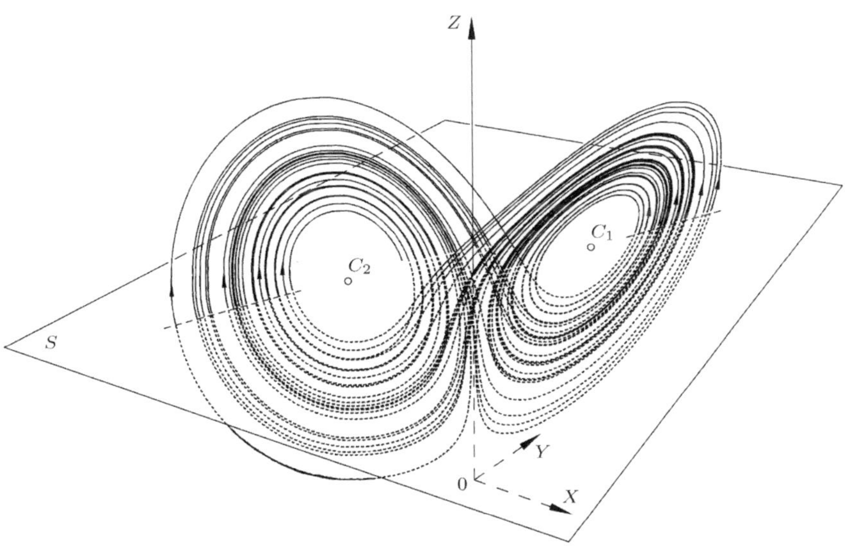

Fig. 5.2.1: Lorenz attractor for $r = 28, \sigma = 10$ and $b = 8/3$. The trajectory range behind the plane $Z = r - 1 = 27$ is dotted (Lanford, 1977)

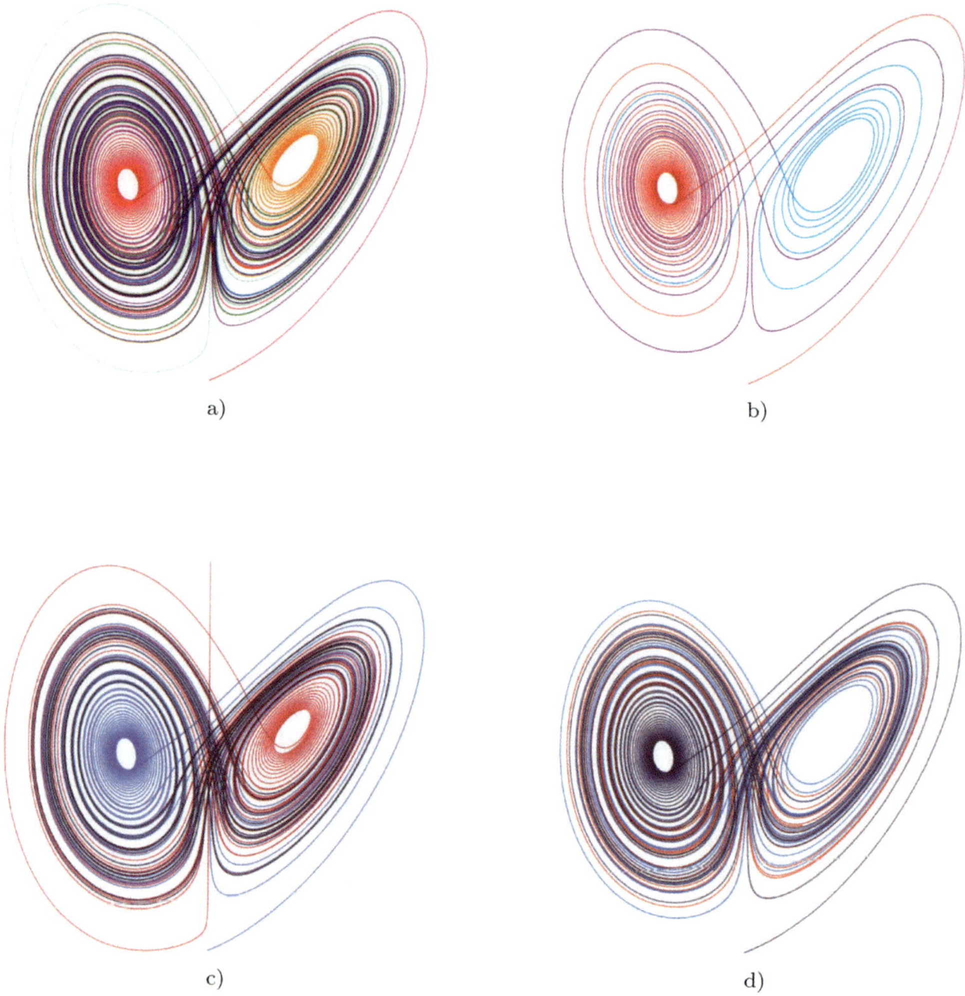

Plate I Lorenz attractor ($r = 28$, $\sigma = 10$, $b = 8/3$):

a) initial condition $X_0 = Y_0 = Z_0 = 0.001$

b) erratic switching

c) basin of attraction for two initial conditions, $X_0 = Y_0 = Z_0 = 0.001$ and $\overline{X}_0 = \overline{Y}_0 = 0.001$, $\overline{Z}_0 = 60$.

d) sensitiviy to initial conditions, $X_0 = Y_0 = Z_0 = 0.001$ and $\overline{X}_0 = 0.0011$, $\overline{Y}_0 = \overline{Z}_0 = 0.001$

The volume contraction rate Λ is negative, i.e. a volume element $V(0)$ contracts exponentially in time in accordance with eq. (3.1.31)

$$V(t) = V(0)\,e^{\Lambda t} = V(0)\,e^{-(\sigma+1+b)t} \tag{5.2.3}$$

Lorenz selected for the Prandtl number $\sigma = 10$ and for $b = 8/3$. For these values, the exponential factor of the volume reduction is $e^{-41/3} \sim 10^{-6}$ per unit time. The Lorenz system is hence strongly dissipative.

Lorenz attributed the generation of an instability of the stationary convection flow to a sub-critical Hopf bifurcation (see section 7.4) which arises at $r_{cr} = 470/19 \approx 24.7368$. For the numerical integration of the set of equations (5.2.1), he selected a somewhat higher value, namely $r = 28$. If we follow a trajectory starting almost at the origin with the initial condition ($X = 0.001$, $Y = 0.001$, $Z = 0.001$), the strange geometrical figure which emerges for $t \to \infty$ in the phase space spanned by the coordinates X, Y, Z is the so-called Lorenz attractor (see fig. 5.2.1 as well as Colour Plates Ia and Ib, p. 194). At first, it sounds astonishing to assert that a configuration which is so structured and bounded in the phase space should reflect chaotic, non-predictable motions.

The two points C_1 and C_2, fig. 5.2.1, with the coordinates

$$\left(\pm\sqrt{b(r-1)},\ \pm\sqrt{b(r-1)},\ r-1\right)$$

are unstable fixed points respectively saddle foci, each with one negative eigenvalue and a pair of complex conjugate eigenvalues with a positive real part (cf. section 10.3). This eigenvalue constellation determines the paths of the trajectories in the neighbourhood of C_1 and C_2. The stable manifold of C_1 induces an inflow for C_1, the unstable manifold of C_1 (a surface) a spiral-shaped outflow of C_1. The flow around C_2 is analogous.

If we choose another form of representation, e.g. the temporal evolution of the independent variables $X(t)$, the chaotic character of the Lorenz system is more readily recognisable. Figure 5.2.2 shows the transient response of X over a finite period of observation for $\sigma = 10$, $r = 28$ and $b = 8/3$. Within this limited period up to $t = 200$, we observe no periodicity and see that the change from C_1 to C_2 is of a random, stochastic nature. If we differentiate between spatial-chaotic and temporal-chaotic patterns, the Lorenz system demonstrates chaotic behaviour in time while the spatial pattern in the phase space is globally ordered. The colour plate Ib makes it clear that the trajectory starting close to the origin immediately jumps to the hemisphere of the unstable fixed point respectively saddle focus C_2, remains there for a while, circling round C_2, and then suddenly and unpredictably returns to the sphere of influence of C_1. The attraction of C_1, however, proves insufficient to trap the trajectory for ever; in fact, it stays there only for a limited period and then escapes once more into the domain of attraction of C_2. This bouncing to and fro from left to right and from right to left is repeated irregularly. Colour Plate I illustrates this; to clarify matters, the colour of the trajectory is changed following each jump.

A fundamental characteristic of the Lorenz attractor is its property of attraction in phase space. In order to demonstrate this, two trajectories with clearly differing initial conditions (0.001, 0.001, 0.001) and (0.001, 0.001, 60) are shown in the Colour

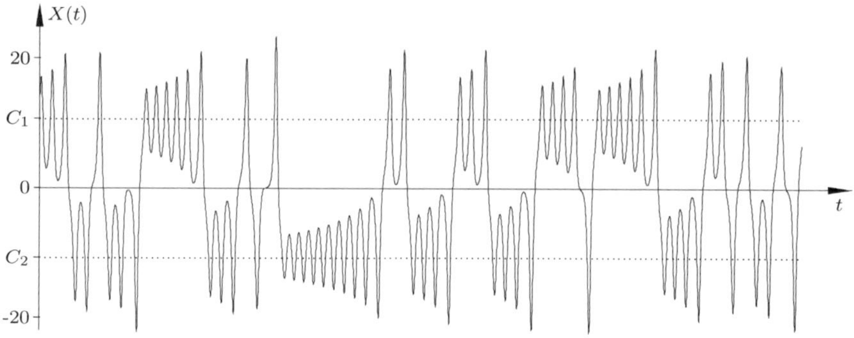

Fig. 5.2.2: Transient time response $X(t)$ for $r = 28$, $\sigma = 10$ and $b = 8/3$

plate Ic. Both trajectories tend very fast towards the attractor, are sucked up by it and remain captured by it for all time. Each individual trajectory possesses the same global structure whereas the local structure is different.

Colour Plate Id exemplifies the sensitive behaviour of a strange attractor with respect to the initial conditions. A minimal deviation of $1/10\,000$ is seen to increase explosively. Both trajectories which – due to an inevitable imprecision in our graphics – appear to follow an identical course branch after a finite number of orbits into two clearly recognisable trajectories and form, in spite of their locally differing courses, the globally characteristic structure of the Lorenz attractor.

Summarising, we can say of the Lorenz system that for $r = 28$

1) for $t \to \infty$, a trajectory will run through a bounded domain in the phase space;
2) the motion is erratic, i.e. the trajectories behave chaotically and a jump from the C_1-hemisphere to the C_2-hemisphere cannot be predicted;
3) the trajectories react extremely sensitively to the initial conditions and
4) the characteristics of the attractor are evident for differing initial conditions.

In order to describe the qualitative characteristics of the Lorenz attractor, the Poincaré section and the Poincaré map are very helpful. Of the multitude of sections, the plane section $Z = r - 1$ is particularly illustrative since it contains the two unstable fixed points C_1 and C_2. The sequence of points shown in fig. 5.2.3 confirms that the trajectories jump stroboscopically from one fixed point to the other. What is astonishing is the branching into two line segments which do not lie exactly on a straight line but which form an almost straight line. In the phase space, a surface of dimension two generates an exact line in the Poincaré section. A dimension of almost two for such a complex structure respectively for a strange attractor is what one would have expected least. And in fact, numerically determined values of the capacity dimension D_c (see section 5.5 where the various dimension concepts are defined) show for the Lorenz attractor that it is very close to, but nevertheless higher than two. The quantity D_c (see section 10.3) was determined for the parameters $r = 28$, $\sigma = 10$ and $b = 8/3$ as (cf. Colour Plate III, p. 198),

$$D_c = 2.06$$

$\gamma\lambda\alpha\tilde{v}\kappa$' '$A\vartheta\acute{\eta}\nu\alpha\zeta\epsilon$ $\kappa o\mu\acute{\iota}\zeta\epsilon\iota\nu$;

Plate II Lorenz attractor ($r = 28$)

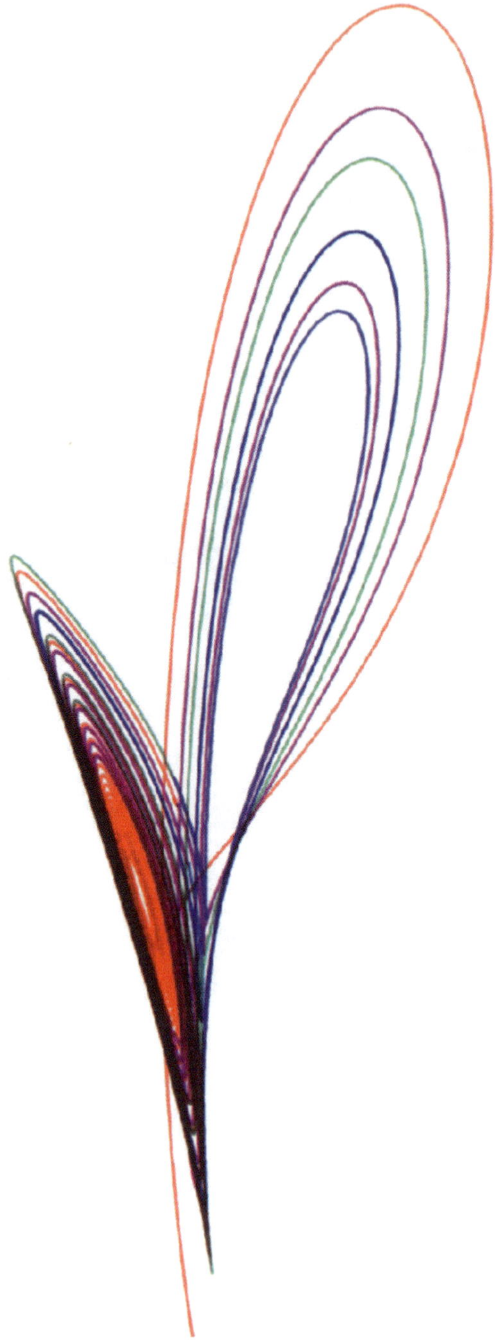

Plate III Lorenz attractor: as an illustration of the capacity dimension $D_c = 2.06$

If we consider the geometry of the Lorenz attractor, it can be shown (Sparrow, 1982; Guckenheimer and Holmes, 1983), that it consists of an infinite number of closely packed layers having a mille-feuille structure; this filigree construction possesses, however, zero volume, but does not form a surface of the dimension $D_c = 2$ in the phase space (Colour Plate II, p. 197).

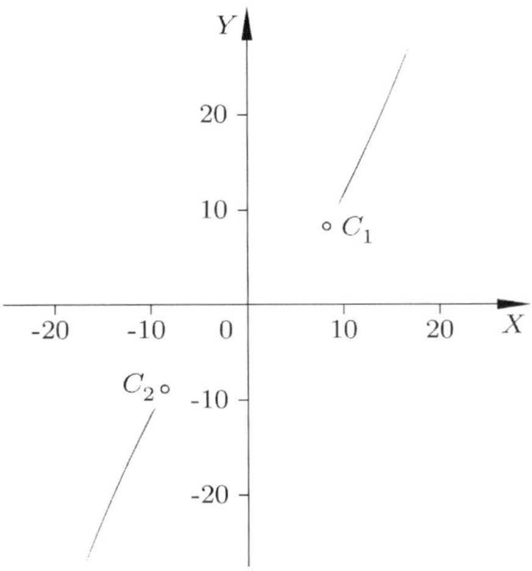

Fig. 5.2.3

Poincaré section $Z = r - 1$ of the Lorenz attractor for $r = 28$, $\sigma = 10$ and $b = 8/3$; the intersection points in the positive Z-direction are shown

Lorenz already used the linear character of the intersection points in the Poincaré section, fig. 5.2.3, to reduce the description of the complex geometry of the Lorenz attractor to a single variable. He obtained the map of a variable by representing in a graphic form the relative maximum $Z_{\mathrm{k}+1}$-values as a function of the preceding maximum Z_{k}-values. As a condition for the maximum Z-values, we have

$$\frac{\mathrm{d}Z}{\mathrm{d}X} = \frac{\mathrm{d}Z}{\mathrm{d}t} : \frac{\mathrm{d}X}{\mathrm{d}t} = \frac{XY - bZ}{\sigma(Y - X)} = 0 \tag{5.2.4}$$

Figure 5.2.4a shows the discrete sequence of points for the parameters $\sigma = 10$, $b = 8/3$, $r = 28$ that could be determined by Lorenz with the computer resources available in 1963 (Lorenz, 1963). Nowadays, the return map of the maximum Z-values can be evaluated for much longer orbits. One obtains a continuous map shown in fig. 5.2.4b, possessing a cusp where the modulus of the gradient of each point on the curve exceeds 1. This means that for each initial value Z_0 in the respective interval, the map cannot possess a stable fixed point and, consequently, no stable periodic orbit exists for the original system (c.f. section 3.7).

A dissipative system is characterised by a volume contraction, i.e. as a result of the phase flow, its volume contracts for $t \to \infty$ to an attractor of a dimension lower than that of the phase space. This again underlines the significance of attractors for dissipative systems. The following definition of attractors goes back to Lanford (1981).

An attractor A of a phase flow ϕ_t is a compact (closed and bounded) set with the following characteristics (Guckenheimer and Holmes, 1983):

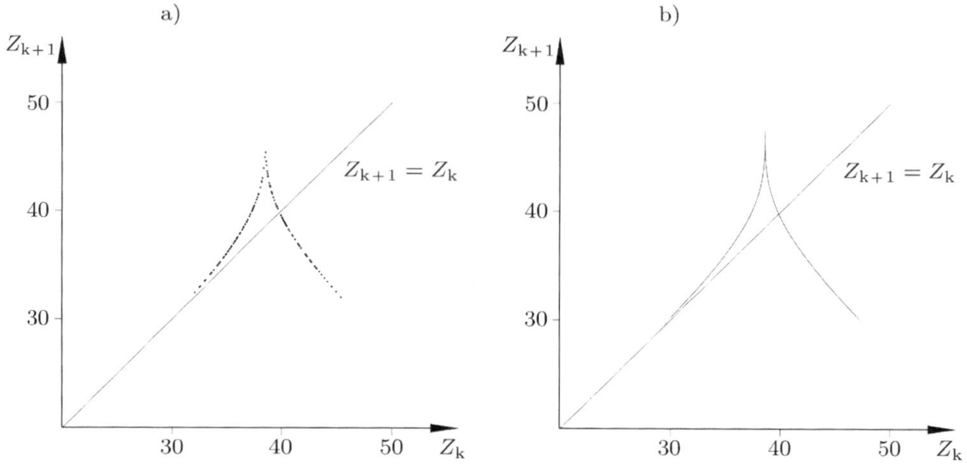

Fig. 5.2.4: Discrete map of the relative maximal Z-values: a) for a shorter orbit, c.f. (Lorenz, 1963) and b) for $t \to \infty$.

i. for all t-values, the attractor A is invariant under the influence of the phase flow ϕ_t ($\phi_t A = A$);
ii. the attractor A has an open neighbourhood U (basin of attraction) which contracts under the influence of the phase flow ϕ_t on A;
iii. the flow ϕ_t on the attractor is non-wandering and it is not possible to decompose the attractor A into two closed, non-overlapping invariant sets.

The attractors already known to us – such as stable fixed points, limit cycles and tori – are manifolds; a strange attractor, on the other hand, is not.

iv. *Strange attractors* are characterised by their unpredictable, chaotic behaviour; nevertheless, they occupy a subspace of lower dimension in the phase space. If we observe neighbouring trajectories on the attractor, we note their characteristic exponential divergence. Due to different dilatations in individual directions, the strange attractor reacts extremely *sensitively to small changes in the initial conditions.* Thus, predictability is excluded since the motion is not periodic, nor are states which are separated from one another over long periods of time correlated. Although long-term predictions are impossible for chaotic motions, the strange attractor retains its topological structure and is invariant with respect to the phase flow ϕ_t.

5.3 Power Spectrum and Autocorrelation

In the previous section, we considered a strange attractor, the Lorenz attractor, a structure which has been intensively researched in the literature. The question now arises which criteria allow us to distinguish between regular and strange attractors in practice. In this and the three subsequent sections, we present various methods and mathematical tools which can be used to characterise regular or chaotic motion. We first apply the knowledge about Fourier series and Fourier transform acquired in section 3.8 to distinguish between regular and chaotic attractors.

A limit cycle of period T can be described by a periodic function $f(t) = f(t+T)$ and expanded into a Fourier series according to eq. (3.8.13) as follows

$$f(t) = \sum_{\mathrm{k}=-\infty}^{+\infty} \alpha_{\mathrm{k}} e^{ik\omega_0 t}, \qquad \omega_0 = \frac{2\pi}{T} \tag{5.3.1}$$

Due to eq. (3.8.49), the Fourier transform $F(\omega)$ can be written as a superposition of equidistant impulses

$$F(\omega) = 2\pi \sum_{\mathrm{k}=-\infty}^{+\infty} \alpha_{\mathrm{k}} \delta(\omega - \mathrm{k}\omega_0) \tag{5.3.2}$$

where the Fourier coefficients are given by eq. (3.8.14). We thus establish a means of recognising limit cycles, i.e. regular periodic attractors. Indeed, a transformation of the corresponding periodic motion $f(t)$ into the frequency domain should yield a spectrum of equidistant lines (see table 5.3.1).

How do we recognise a motion on a torus? Quasi-periodic motions with two incommensurable frequencies ω_1 and ω_2 can be expanded into twofold Fourier series

$$f(t) = \sum_{\mathrm{k}_1=-\infty}^{+\infty} \sum_{\mathrm{k}_2=-\infty}^{+\infty} \alpha_{\mathrm{k}_1 \mathrm{k}_2} e^{i(\mathrm{k}_1\omega_1 + \mathrm{k}_2\omega_2)t} \tag{5.3.3}$$

Applying eq. (3.8.49), we thus immediately deduce the Fourier transform

$$F(\omega) = 2\pi \sum_{\mathrm{k}_1=-\infty}^{+\infty} \sum_{\mathrm{k}_2=-\infty}^{+\infty} \alpha_{\mathrm{k}_1 \mathrm{k}_2} \delta(\omega - \mathrm{k}_1\omega_1 - \mathrm{k}_2\omega_2) \tag{5.3.4}$$

A line spectrum with peaks at the basic frequencies ω_1 and ω_2 as well as at all possible integer linear combinations hence corresponds to a motion on a torus (c.f. table 5.3.1).

Applying the knowledge we have gained up to now about Fourier transforms by means of the Dirac δ-function, it is now possible to distinguish between regular and strange attractors; moreover, we can contrast them with purely random, featureless behaviour like "white noise". Since, according to eq. (3.8.18), the Fourier transform $F(\omega)$ of a real signal $f(t)$ is generally a complex-valued function, e.g. eq. (3.8.50), it is more expedient to specify not $F(\omega)$ directly but rather the power spectrum $P(\omega)$, eq. (3.8.33), and its Fourier transform, the time average of the autocorrelation $\bar{a}(\tau)$, eq. (3.8.28). Table 5.3.1 presents a survey of the characteristic graphs of $P(\omega)$ and $\bar{a}(\tau)$ for the various attractor types. For regular attractors such as point attractor, limit cycle and torus, characterised by motions which are constant in time, periodic and quasi-periodic, the power spectrum and autocorrelation can be given explicitly, see (Papoulis, 1962). In the following, we give a short overview of the corresponding signal types

<table>
<tr><th>attractor</th><th>power spectrum</th><th>autocorrelation</th></tr>
<tr><td>point attractor
</td><td>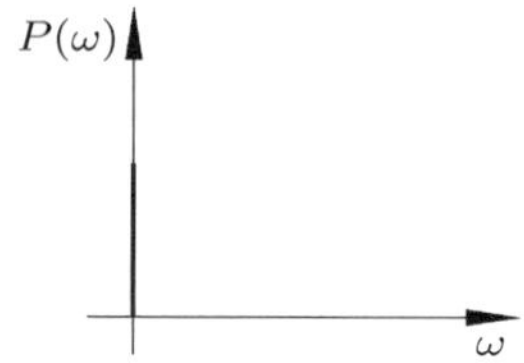</td><td>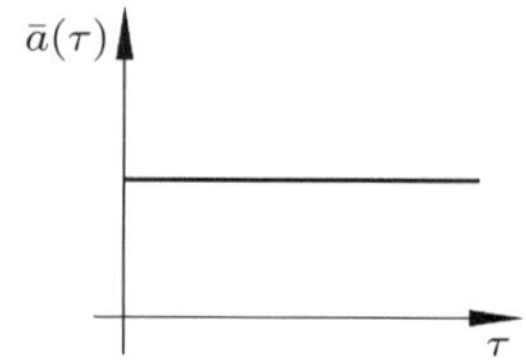</td></tr>
<tr><td>limit cycle
</td><td>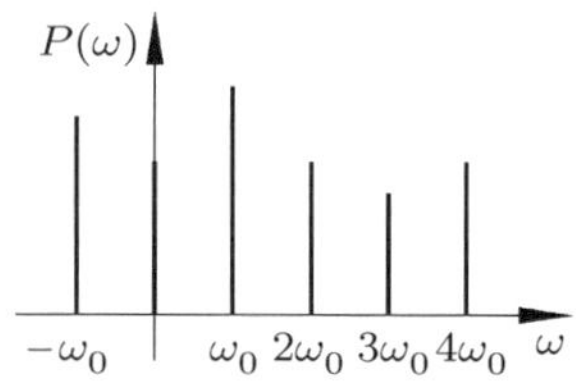</td><td>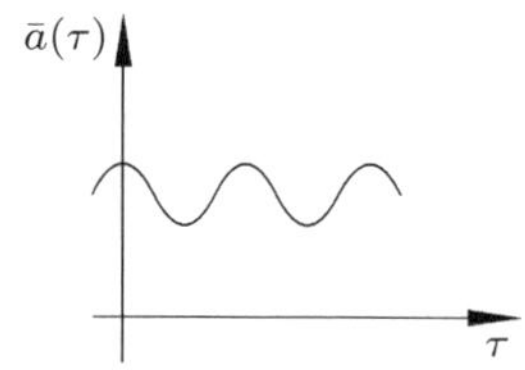</td></tr>
<tr><td>torus
</td><td>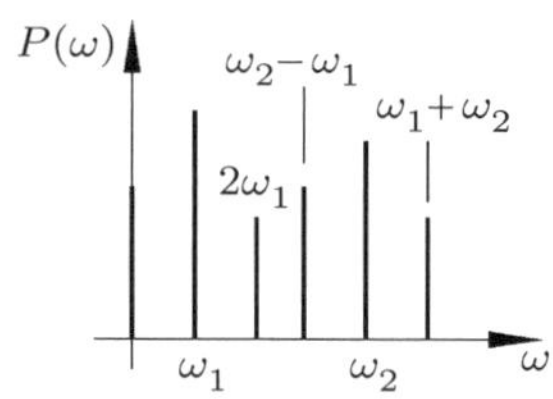</td><td>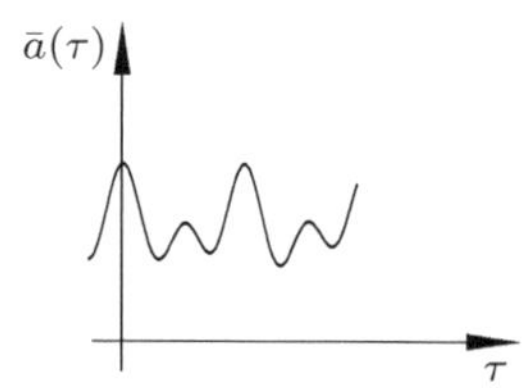</td></tr>
<tr><td>strange attractor
</td><td>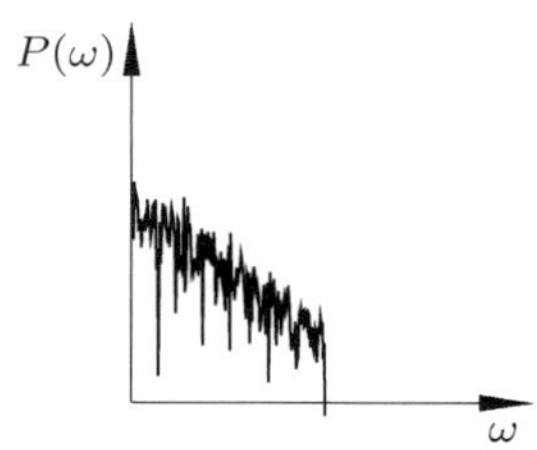</td><td>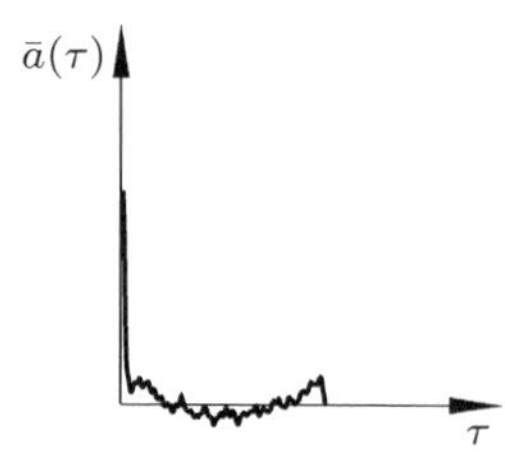</td></tr>
<tr><td>“white noise”</td><td>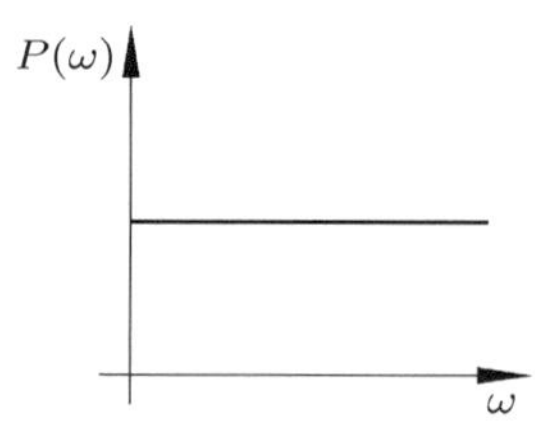</td><td>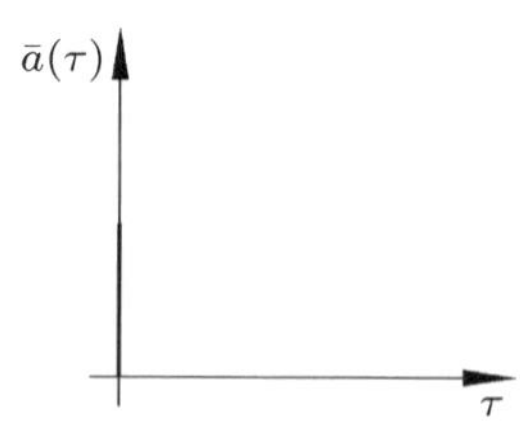</td></tr>
</table>

Table 5.3.1: Characterisation of attractors

i) **Point attractor**: (5.3.5)

Time signal: $f(t) \;= c$
(constant)

Power spectrum: $P(\omega) = 2\pi c^2 \delta(\omega)$
(single impulse)

Autocorrelation: $\overline{a}(\tau) \;= c^2$
(constant)

ii) **Limit cycle**: (5.3.6)

Time signal: $f(t) \;= \sum\limits_{\mathrm{k}=-\infty}^{+\infty} \alpha_{\mathrm{k}} \mathrm{e}^{i\mathrm{k}\omega_0 t}$
(periodic)

Power spectrum: $P(\omega) = 2\pi \sum\limits_{\mathrm{k}=-\infty}^{+\infty} |\alpha_{\mathrm{k}}|^2 \delta(\omega - \mathrm{k}\omega_0)$
(equidistant impulses)

Autocorrelation: $\overline{a}(\tau) \;= \sum\limits_{\mathrm{k}=-\infty}^{+\infty} |\alpha_{\mathrm{k}}|^2 \cos \mathrm{k}\omega_0 \tau$
(periodic)

iii) **Torus**: (5.3.7)

Time signal: $f(t) \;= \sum\limits_{\mathrm{k}_1=-\infty}^{+\infty} \sum\limits_{\mathrm{k}_2=-\infty}^{+\infty} \alpha_{\mathrm{k}_1\mathrm{k}_2} \mathrm{e}^{i(\mathrm{k}_1\omega_1 + \mathrm{k}_2\omega_2)t}$
(quasi-periodic)

Power spectrum: $P(\omega) = 2\pi \sum\limits_{\mathrm{k}_1=-\infty}^{+\infty} \sum\limits_{\mathrm{k}_2=-\infty}^{+\infty} |\alpha_{\mathrm{k}_1\mathrm{k}_2}|^2 \delta(\omega - \mathrm{k}_1\omega_1 - \mathrm{k}_2\omega_2)$
(integer linear combinations of two basic impulses)

Autocorrelation: $\overline{a}(\tau) \;= \sum\limits_{\mathrm{k}_1=-\infty}^{+\infty} \sum\limits_{\mathrm{k}_2=-\infty}^{+\infty} |\alpha_{\mathrm{k}_1\mathrm{k}_2}|^2 \cos(\mathrm{k}_1\omega_1 + \mathrm{k}_2\omega_2)\tau$
(quasi-periodic)

If we wish to analyse and classify a dynamical process which depends on one control parameter, the power spectrum is particularly appropriate for recognising possible period doublings or the emergence of new incommensurable basic frequencies. In Chapter 8, using the Rayleigh-Bénard experiment as an example, we demonstrate various routes to "turbulent" behaviour and establish, evaluating the power spectra, a connection with the corresponding mathematical models. Specifically, in section 8.4, we apply the Fourier transform to a discrete system, the logistic equation, and elucidate the cascade of period doublings with the aid of the power spectrum.

How can we distinguish deterministic chaos from regular periodic processes on the one hand and white noise on the other? In the case of completely random behaviour, there is per definition no relation between the signals at various points in time, i.e. the autocorrelation consists of a single impulse for $\tau = 0$. In accordance with eq. (3.8.45), the Fourier transform of a δ-function, and thus also the power spectrum in the case of white noise, is a constant (see table 5.3.1).

On the other hand, chaotic behaviour is characterised by the fact that the power spectrum consists of a wide continuous band of frequencies, possibly accompanied by single characteristic peaks (Farmer et al., 1980). At the same time, the autocorrelation makes it clear that the memory of the initial state is only gradually lost, i.e. the correlation between the signal at time t and the previous time $t - \tau$ tends towards zero for large time differences τ. Therefore, the autocorrelation of a chaotic motion reflects the memory of the system of former states.

In practice, $f(t)$ is in general not given in analytical form, but rather measurements or computer simulations supply only discrete values of the signal $f_i = f(t_i)$, $i = 0, 1, \ldots, \mathrm{N}-1$ in a finite time interval $t_0 \leqslant t \leqslant t_{\mathrm{N}-1}$. Therefore, when calculating the power spectrum, we generally have to rely on numerical procedures for the Fourier transform; however, a straightforward application of the discretised form of eq. (3.8.18) proves very time-consuming. For a time series with N equidistant sampling points, the computational effort normally increases by $\mathcal{O}(\mathrm{N}^2)$. A standardised application only became possible in 1965 after the introduction of the effective *fast Fourier transform* (FFT) by Cooley and Tukey (Cooley and Tukey, 1965; Press et al., 1986), which had already been discovered by Danielson and Lanczos in 1942 (Danielson and Lanczos, 1942). With the FFT, the number of necessary operations increases by only $\mathcal{O}(\mathrm{N}\log_2 \mathrm{N})$ where the number of points must be an integer power of 2: $\mathrm{N} = 2^{\mathrm{n}}$. For $\mathrm{N} = 2^{10} = 1024$ sampling points, the calculation effort can already be reduced to approximately 1% of the former effort! In the mid-1960s, this was enormous progress and enabled the breakthrough of the Fourier transform in many areas, such as signal processing, physics and electrical engineering. Selecting $\mathrm{N} = 2^{\mathrm{n}}$ points $f_0, f_1, \ldots, f_{\mathrm{N}-1}$ with fixed time intervals Δt and discretising eq. (3.8.18) – where the integral is approximated by a sum – we obtain

$$F_{\mathrm{k}} = \Delta t \sum_{\mathrm{j}=0}^{\mathrm{k}-1} \mathrm{e}^{-2\pi i \mathrm{jk}/\mathrm{N}} f_{\mathrm{j}} \tag{5.3.8}$$

as discrete values of the Fourier transform. This calculation rule can also be applied directly to discrete systems such as the logistic map (cf. section 8.4).

Finally, it should be pointed out that certain difficulties can arise in the course of the calculation and evaluation of the power spectrum. Applying a numerical procedure, one will always calculate $P(\omega)$ according to equ (3.8.30) for a fixed but finite time interval of the length $2T$. However, since the possible period length of the signal is not known at the outset, there is always a danger of selecting too short an interval of length T so that a periodicity is not recognisable, especially when a longer transient phase precedes the periodicity. A further problem will certainly arise when the motion is characterised by three or more basic frequencies. In such a case, it is difficult to discern the basic frequencies from the complicated line spectrum, particularly since the integer linear combinations are also present.

5.4 Lyapunov Exponents

As we have seen, there are attractors of the most varying types, such as the stable focus (point attractor), the limit cycle, the torus and the strange attractor. Following our discussion on autocorrelation and the power spectrum, we now present a further criterion for distinguishing between the individual attractors: the Lyapunov exponents. In the case of autocorrelation and the power spectrum, the basic information is supplied by time series; the method for determining Lyapunov exponents makes use of the exponentially divergent or convergent behaviour of neighbouring trajectories in the phase space. In the following, we elucidate this method for both continuous and discrete systems.

The concept of Lyapunov exponents allows us to make statements regarding the stability of given trajectories. The regular attractors – such as the stable focus, the limit cycle and the torus – possess such a strong attractive force that all neighbouring trajectories within a certain neighbourhood converge asymptotically to the attractor and neighbouring orbits remain so. Strange attractors also attract all the trajectories in a particular basin of attraction; on the attractor, however, trajectories which were originally close to one another diverge exponentially. It is thus clear that it is possible to distinguish between the various attractor types on the basis of the stability behaviour of a dynamical system.

In 1892, the Russian mathematician Aleksandr M. Lyapunov published an extensive work in which he discussed the problem of the stability of motions and developed the fundamental characteristics of so-called characteristic exponents, thus leading to the concept of Lyapunov exponents (Lyapunov, 1892). This had been preceded by works by G. Floquet (Floquet, 1883) on the stability of periodic motions and by Henri Poincaré who had investigated the stability of special systems of second-order differential equations and whose methods were adopted and generalised by Lyapunov. It was his aim to make statements on the stability of a motion even in cases in which no closed-form solutions could be given. Lyapunov compiled the conditions and stated a procedure with which it is nevertheless possible to make a stability statement. We follow Lyapunov's procedure and, in section 5.4.1, answer the question regarding the stability of a state of equilibrium. In section 5.4.2, we go on to discuss the stability of a periodic, stationary state, finally arriving at the criterion of the Lyapunov exponents for the case of an arbitrary but bounded motion (sections 5.4.3 to 5.4.6).

5.4.1 Linear Stability Analysis of Non-linear Systems: State of Equilibrium

The central problem of this section is the asymptotic stability of non-linear systems. We shall see that here, the statements on the stability of linear systems play a fundamental role. It thus seems appropriate to reiterate in a condensed form our comments on the asymptotic stability of linear systems in section 3.1 and to formulate them more precisely for a concrete application.

In accordance with eq. (3.1.1), the linear system is

$$\dot{\boldsymbol{x}} = \boldsymbol{L}\boldsymbol{x} \tag{5.4.1}$$

Since the set of linear equations (5.4.1) can be solved exactly, it is not surprising that, in this case, the stability problem can also be solved explicitly. In the mathematical textbooks, cf. (Arnold, 1980; Braun, 1979), the following theorem on the stability of linear systems is found:

i. each solution $\boldsymbol{x}(t)$ of eq. (5.4.1) is stable if all eigenvalues of $\boldsymbol{L}$ possess negative real parts;
ii. each solution $\boldsymbol{x}(t)$ of eq. (5.4.1) is unstable if at least one eigenvalue of $\boldsymbol{L}$ possesses one positive real part;
iii. if the real part of one or more eigenvalues of $\boldsymbol{L}$ is equal to zero and if there are exactly $\mathrm{k_j}$ linearly independent eigenvectors for each such eigenvalue $\lambda_\mathrm{j} = i\sigma_\mathrm{j}$ having a multiplicity $\mathrm{k_j}$ and if all other real parts are negative, so-called marginal stability exists.

In the special case of linear systems, eq. (5.4.1), the trivial solution $\boldsymbol{x}(t) \equiv \boldsymbol{o}$ is always the equilibrium solution and correspondingly, the statement on stability respectively asymptotic stability is immediately applicable to a corresponding arbitrary solution $\boldsymbol{x}(t)$ of eq. (5.4.1). It can now be shown (Braun, 1979) that each solution $\boldsymbol{x}(t)$ of the linear system of eq. (5.4.1) tends towards zero for $t \to \infty$ if all eigenvalues of $\boldsymbol{L}$ possess negative real parts. This in turn means that in this case the equilibrium solution $\boldsymbol{x}(t) \equiv \boldsymbol{o}$ is not only stable but asymptotically stable.

The following examples serve to illustrate the three distinctions *i.* to *iii.*

Example 1

We consider the linearised equation for the pendulum with friction (without external forcing), which we write here in the form

$$\ddot{x} + \delta\dot{x} + x = 0$$

Transformed into an autonomous first-order system, this yields the set of two equations

$$\begin{aligned} \dot{x}_1 &= x_2 \\ \dot{x}_2 &= -x_1 - \delta x_2 \end{aligned}$$

If we select $\delta = 3$, the system matrix $\boldsymbol{L}$ is

$$\boldsymbol{L} = \begin{bmatrix} 0 & 1 \\ -1 & -3 \end{bmatrix}$$

and its characteristic equation is quadratic and reads

$$\lambda^2 + 3\lambda + 1 = 0$$

Both roots, $\lambda_1 = (-3+\sqrt{5})/2$, $\lambda_2 = (-3-\sqrt{5})/2$ are negative, i.e. the equilibrium solution $\boldsymbol{x}(t) \equiv \boldsymbol{o}$ respectively each solution $\boldsymbol{x}(t)$ is asymptotically stable.

Example 2

We consider a pendulum with a driving restoring force (inverted pendulum)

$$\ddot{x} + \delta\dot{x} - x = 0 \qquad (\delta > 0)$$

Written as an autonomous system, we have

$$\dot{x}_1 = x_2$$
$$\dot{x}_2 = x_1 - \delta x_2$$

The system matrix $\boldsymbol{L}$

$$\boldsymbol{L} = \begin{bmatrix} 0 & 1 \\ 1 & -\delta \end{bmatrix}$$

leads to the characteristic equation

$$\lambda^2 + \delta\lambda - 1 = 0$$

For positive δ, one eigenvalue is always positive, i.e. each solution $\boldsymbol{x}(t)$ is unstable.

Example 3

If we consider a single-degree-of-freedom oscillator without damping (see section 2.2), the following equation of motion holds

$$\ddot{x} + x = 0$$

Written as an autonomous system, we have

$$\dot{x}_1 = x_2$$
$$\dot{x}_2 = -x_1$$

The system matrix $\boldsymbol{L}$

$$\boldsymbol{L} = \begin{bmatrix} 0 & 1 \\ -1 & 0 \end{bmatrix}$$

leads to the characteristic equation

$$\lambda^2 + 1 = 0$$

and yields the eigenvalues $\lambda_{1,2} = \pm i$. According to item *iii.* of the stability theorem, any solution of the single-degree-of-freedom oscillator is then stable respectively marginally stable, but not asymptotically stable. This becomes clear if we determine the general solution. The eigenvectors appertaining to the two eigenvalues $\lambda_{1,2}$ are

$$\boldsymbol{y}_1 = \begin{bmatrix} 1 \\ i \end{bmatrix} \quad \text{and} \quad \boldsymbol{y}_2 = \begin{bmatrix} 1 \\ -i \end{bmatrix}$$

and, since they are linearly independent, the general solution is

$$\boldsymbol{x}(t) = C_1 \mathrm{e}^{it} \begin{bmatrix} 1 \\ i \end{bmatrix} + C_2 \mathrm{e}^{-it} \begin{bmatrix} 1 \\ -i \end{bmatrix}$$

or, in accordance with Euler's formula for complex numbers,

$$\boldsymbol{x}(t) = C_1(\cos t + i\sin t)\begin{bmatrix} 1 \\ i \end{bmatrix} + C_2(\cos t - i\sin t)\begin{bmatrix} 1 \\ -i \end{bmatrix}$$

With $C_1 = a_1 + ib_1$ and $C_2 = a_1 - ib_1$, we obtain as the general solution

$$\boldsymbol{x}(t) = 2a_1\begin{bmatrix} \cos t \\ -\sin t \end{bmatrix} - 2b_1\begin{bmatrix} \sin t \\ \cos t \end{bmatrix}$$

which represents concentric circles about the origin in the phase space (if t is eliminated) (cf. 3.1). The radii of the circles are determined by the initial conditions for $t = 0$. For example, the solution curve for the initial point $\{1\ 0\}$ is given in the form $\boldsymbol{x}(t) = \{\cos t \quad -\sin t\}$ and is presented in fig. 5.4.1. The equation of a circle $x_1^2 + x_2^2 = 1$ characterises the trajectory in the phase space x_1, x_2 (fig. 5.4.2). The solution is thus periodic and has the period 2π, i.e. no trajectory tends towards zero for $t \to \infty$ with the exception of $\boldsymbol{x}(t) \equiv \boldsymbol{o}$. The system behaviour is, as already determined on the basis of the eigenvalue analysis, stable respectively marginally stable. (The equation of a circle in the phase space can be confirmed with considerably greater simplicity directly from the initial autonomous equations: $x_1\dot{x}_1 + x_2\dot{x}_2 = x_2x_1 - x_1x_2 = 0$ and consequently $x_1^2 + x_2^2 = \text{const.}$)

Summing up, we can state that the long-term response of a linear system is characterised by one of the following four response patterns:

i. the solution $\boldsymbol{x}(t)$ is constant in time;
ii. the solution $\boldsymbol{x}(t)$ is periodic in time;
iii. the solution $\boldsymbol{x}(t)$ tends to a state of equilibrium for $t \to \infty$, or
iv. the solution $\boldsymbol{x}(t)$ tends to infinity for $t \to \infty$.

Thus, in the case of linear *autonomous* systems, it is impossible for a trajectory to spiral towards a simple closed curve, i.e. a limit cycle. Such periodic long-term behaviour is restricted exclusively to non-linear systems, as is also quasi-periodic or even erratic respectively chaotic long-term behaviour.

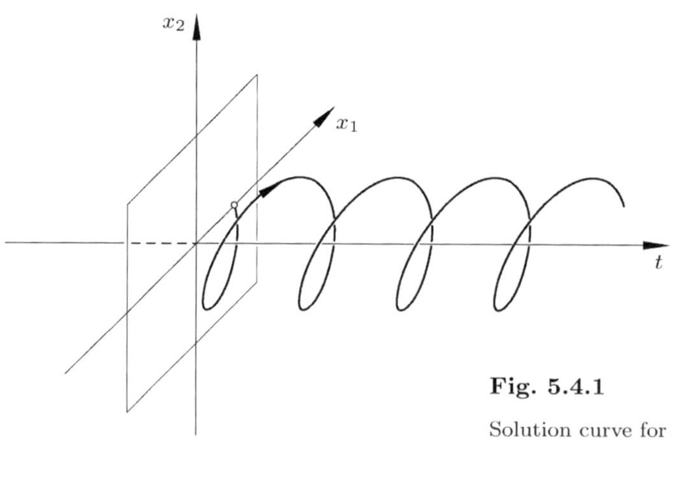

Fig. 5.4.1

Solution curve for $\boldsymbol{x}(t) = \{\cos t \quad -\sin t\}$

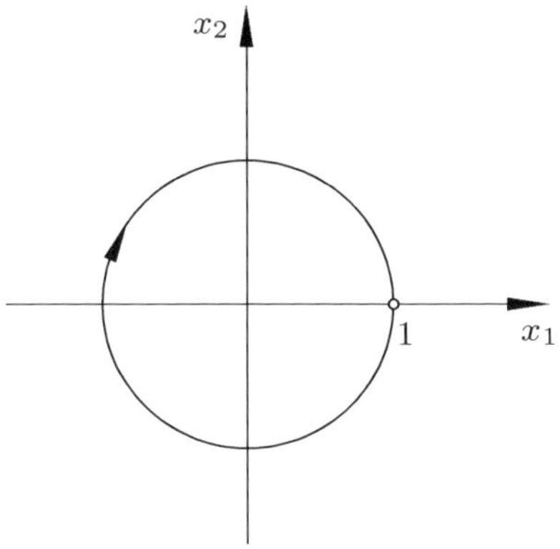

Fig. 5.4.2

Trajectory of the solution $\boldsymbol{x}(t) = \{\cos t \ \ -\sin t\}$ through the point $\{1 \ 0\}$

We now sketch the procedure for calculating the asymptotic stability of non-linear systems and consider the set of autonomous equations

$$\dot{\boldsymbol{x}} = \boldsymbol{F}(\boldsymbol{x}) \tag{5.4.2}$$

The temporal evolution is determined by the vector field $\boldsymbol{F}$, a non-linear function of $\boldsymbol{x}$.

What can the linear stability analysis contribute to the description of the long-term behaviour of non-linear systems? If we consider a reference state $\boldsymbol{x}_r$, which may at first be time-dependent and is a particular solution of eq. (5.4.2), we have

$$\dot{\boldsymbol{x}}_r = \boldsymbol{F}(\boldsymbol{x}_r) \tag{5.4.3}$$

Perturbing the state $\boldsymbol{x}_r$, we observe a neighbouring state $\boldsymbol{x}$ of the form

$$\boldsymbol{x} = \boldsymbol{x}_r + \widetilde{\boldsymbol{x}} \tag{5.4.4}$$

where $\widetilde{\boldsymbol{x}}$ represents the perturbation. Inserting eq. (5.4.4) in eq. (5.4.2) and rearranging according to the perturbation $\widetilde{\boldsymbol{x}}$, we find

$$\frac{\mathrm{d}\widetilde{\boldsymbol{x}}}{\mathrm{d}t} = \boldsymbol{F}(\boldsymbol{x}_r + \widetilde{\boldsymbol{x}}) - \boldsymbol{F}(\boldsymbol{x}_r) \tag{5.4.5}$$

Assuming that the perturbation is small

$$|\widetilde{\boldsymbol{x}}|/|\boldsymbol{x}_r| \ll 1 \tag{5.4.6}$$

then each component F_j of $\boldsymbol{F}(\boldsymbol{x}_r + \widetilde{\boldsymbol{x}})$ in the near neighbourhood of $\boldsymbol{x}_r$ can be expanded into a Taylor series

$$F_\mathrm{j}(\boldsymbol{x}_r + \widetilde{\boldsymbol{x}}) = F_\mathrm{j}(\boldsymbol{x}_r) + \left.\frac{\partial F_\mathrm{j}}{\partial \boldsymbol{x}}\right|_{\boldsymbol{x}_r} \widetilde{\boldsymbol{x}} + \frac{1}{2}\widetilde{\boldsymbol{x}}^\mathrm{t} \left.\frac{\partial^2 F_\mathrm{j}}{\partial \boldsymbol{x}\partial \boldsymbol{x}^\mathrm{t}}\right|_{\boldsymbol{x}_r} \widetilde{\boldsymbol{x}} + \ldots \tag{5.4.7}$$

After the introduction of abbreviations for the Jacobian matrix

$$\boldsymbol{D}(\boldsymbol{x}_r) = \left.\frac{\partial \boldsymbol{F}}{\partial \boldsymbol{x}}\right|_{\boldsymbol{x}_r} \tag{5.4.8}$$

and for the higher-order terms

$$N_{\mathrm{j}}(\widetilde{\boldsymbol{x}}) = \frac{1}{2}\widetilde{\boldsymbol{x}}^{\mathrm{t}} \left. \frac{\partial^2 F_{\mathrm{j}}}{\partial \boldsymbol{x} \partial \boldsymbol{x}^{\mathrm{t}}} \right|_{\boldsymbol{x}_r} \widetilde{\boldsymbol{x}} + \ldots \tag{5.4.9}$$

the set of equations eq. (5.4.5) can be expressed succinctly as

$$\frac{\mathrm{d}\widetilde{\boldsymbol{x}}}{\mathrm{d}t} = \boldsymbol{D}(\boldsymbol{x}_r)\widetilde{\boldsymbol{x}} + \boldsymbol{N}(\widetilde{\boldsymbol{x}}) \tag{5.4.10}$$

respectively in its linearised form

$$\frac{\mathrm{d}\widetilde{\boldsymbol{x}}}{\mathrm{d}t} = \boldsymbol{D}(\boldsymbol{x}_r)\widetilde{\boldsymbol{x}} \tag{5.4.11}$$

It should be pointed out that eqs. (5.4.10) and (5.4.11) satisfy the trivial solution $\widetilde{\boldsymbol{x}}(t) \equiv \boldsymbol{o}$ and that both equations are homogeneous.

For the originally non-linear system, the following stability statement can be deduced from the behaviour of the linearised system:

i. if the trivial solution $\widetilde{\boldsymbol{x}}(t) \equiv \boldsymbol{o}$ of the *linearised* system eq. (5.4.11) is asymptotically stable, then $\widetilde{\boldsymbol{x}}(t) \equiv \boldsymbol{o}$ is also an asymptotically stable solution of the *non-linear* system eq. (5.4.10);
ii. if the trivial solution $\widetilde{\boldsymbol{x}}(t) \equiv \boldsymbol{o}$ of the *linearised* problem eq. (5.4.11) is unstable, then $\widetilde{\boldsymbol{x}}(t) \equiv \boldsymbol{o}$ is also an unstable solution of the *non-linear* problem, eq. (5.4.10).

Note that this statement is still generally valid since no particular assumptions on the reference state have as yet been made.

Asymptotic stability respectively instability can be recognised particularly easily in two special cases, namely if the reference state

(a) $\boldsymbol{x}_r = \boldsymbol{x}_s$, $\dot{\boldsymbol{x}}_s = \boldsymbol{o}$ is a state of equilibrium or
(b) $\boldsymbol{x}_r(t+T) = \boldsymbol{x}_r(t)$ represents a periodic solution of the system eq. (5.4.2).

In this section, we discuss case (a) and formulate conditions for the stability of the state of equilibrium $\boldsymbol{x}_s$ of a non-linear system. In the following section 5.4.2, we investigate the stability of periodic solutions.

The trivial solution $\widetilde{\boldsymbol{x}} = \boldsymbol{o}$ of eq. (5.4.11) is equivalent to $\boldsymbol{x} = \boldsymbol{x}_r$. If we investigate the stability of a state of equilibrium $\boldsymbol{x}_s$, the coefficients of the matrix $\boldsymbol{D}(\boldsymbol{x}_s)$ in eq. (5.4.11) are constant and we can deduce from the eigenvalues of the matrix $\boldsymbol{D}$ a stability criterion. In accordance with the Hartman-Grobman theorem, the following holds:

i. the state of equilibrium $\widetilde{\boldsymbol{x}}(t) \equiv \boldsymbol{o}$ of eq. (5.4.11) respectively $\widetilde{\boldsymbol{x}}(t) \equiv \boldsymbol{x}_s$ of eq. (5.4.1) is asymptotically stable if all eigenvalues of $\boldsymbol{D}$ possess *negative real parts* ($\mathrm{Re}\,\lambda < 0$);
ii. the equilibrium solution $\widetilde{\boldsymbol{x}}(t) \equiv \boldsymbol{o}$ of eq. (5.4.11) respectively $\widetilde{\boldsymbol{x}}(t) \equiv \boldsymbol{x}_s$ of eq. (5.4.1) is unstable if at least one eigenvalue of $\boldsymbol{D}$ possesses one *positive* real part;
iii. the stability of the state of equilibrium $\boldsymbol{x}_s$ for the *non-linear* system *cannot* be deduced from the stability of the equilibrium solution of the linearised system

$\dot{\boldsymbol{x}} = \boldsymbol{D}\boldsymbol{x}$ if at least one eigenvalue with a vanishing real part exists, even if the real parts of all the other eigenvalues are negative.

In case *iii.*, the state of temporal marginal stability, the linearised system is too rough an approximation of the non-linear one. If we still wish to enquire into the stability behaviour of the initial system, the dominant non-linear terms must also be taken into account. The theory of the centre manifold, section 6.2, and the technique of the normal form, section 6.3, are methods which help us in such an enquiry.

In order to enhance the clarity of our remarks on linear stability analysis, we end this section with a simple example. We select the set of non-linear equations of the Lorenz model (see also sections 5.2, 7.3, 10.3). The equations for the temperature and velocity distributions are

$$\begin{aligned} \dot{X} &= -\sigma X + \sigma Y \\ \dot{Y} &= rX - Y - XZ \\ \dot{Z} &= -bZ + XY \qquad\qquad \text{for} \quad \sigma, r, b > 0 \end{aligned} \tag{5.4.12}$$

We illustrate our preceding arguments with a discussion of the stability behaviour at the fixed point $X = Y = Z = 0$. To this end, we calculate the Jacobian matrix $\boldsymbol{D}$ of eq. (5.4.11) of the linearised system at the origin

$$\boldsymbol{D} = \left.\frac{\partial \boldsymbol{F}}{\partial \boldsymbol{X}}\right|_{\boldsymbol{X}=\boldsymbol{0}} = \begin{bmatrix} -\sigma & \sigma & 0 \\ r-Z & -1 & -X \\ Y & X & -b \end{bmatrix}_{\boldsymbol{X}=\boldsymbol{0}} = \begin{bmatrix} -\sigma & \sigma & 0 \\ r & -1 & 0 \\ 0 & 0 & -b \end{bmatrix} \tag{5.4.13}$$

The eigenvalues of the $\boldsymbol{D}$-matrix, eq. (5.4.13), are

$$\lambda_{1,2} = -\frac{\sigma+1}{2} \pm \frac{1}{2}\sqrt{(\sigma+1)^2 + 4(r-1)\sigma}\ , \qquad \lambda_3 = -b \tag{5.4.14}$$

For $r < 1$, all eigenvalues are negative and all trajectories in the neighbourhood of the origin thus tend towards zero for $t \to \infty$. The trivial solution $\boldsymbol{X}(t) \equiv \boldsymbol{o}$ is asymptotically stable, i.e. the origin is a point attractor.

5.4.2 Stability of Periodic Solutions: Floquet Theory

Having investigated states of equilibrium with regard to their stability in the previous section, we now discuss the stability of periodic processes. The Floquet theory, which we present in the following, represents a linear stability analysis of periodic processes and goes back to Floquet (Floquet, 1883).
Periodic motions of the period length T, $\boldsymbol{x}_r(t) = \boldsymbol{x}_r(t+T)$, can represent the solutions of a variety of classes of differential equations. In the main, we distinguish between two cases:

i. $\boldsymbol{x}_r(t) = \boldsymbol{x}_r(t+T)$ is the solution of a periodically driven system, defined by an inhomogeneous set of non-autonomous differential equations

$$\dot{\boldsymbol{x}} = \boldsymbol{F}(\boldsymbol{x}, t) = \boldsymbol{F}(\boldsymbol{x}, t+T), \quad \text{where} \qquad \boldsymbol{F}(\boldsymbol{o}, t) \neq \boldsymbol{o} \tag{5.4.15}$$

As an example, we mention the Duffing equation in its non-autonomous form

$$\begin{aligned} \dot{x}_1 &= \quad x_2 \\ \dot{x}_2 &= -cx_2 + \beta x_1 - \alpha x_1^3 + f \cos \omega t \end{aligned} \tag{5.4.16}$$

which we shall discuss in more detail in section 10.5. It is shown that the system has a stable limit cycle for the parameter values $c = 0.55$, $\beta = -8$, $\alpha = 2$, $f = 24$, $\omega = 1$ (fig. 10.5.4).

ii. $\boldsymbol{x}_r(t) = \boldsymbol{x}_r(t+T)$ is the solution of an autonomous system

$$\dot{\boldsymbol{x}} = \boldsymbol{F}(\boldsymbol{x}) \tag{5.4.17}$$

As an example of the second case, we refer to the Lorenz system, eq. (5.4.12), which, for example, possesses a limit cycle for $r = 151.32$, $\sigma = 10$, $b = 8/3$ (see fig. 8.4.4). Cases *i.* and *ii.* possess different eigenvalues of the characteristic matrix, the so-called monodromy matrix. We shall return to this later.

In order to investigate the stability of the periodic solution $\boldsymbol{x}_r(t)$, we again consider the influence of a small perturbation $\widetilde{\boldsymbol{x}}$

$$\boldsymbol{x}(t) = \boldsymbol{x}_r(t) + \widetilde{\boldsymbol{x}}(t) \tag{5.4.18}$$

In case *i.* of a periodically driven system, eq. (5.4.15) yields

$$\dot{\boldsymbol{x}}_r(t) + \dot{\widetilde{\boldsymbol{x}}}(t) = \boldsymbol{F}(\boldsymbol{x}_r + \widetilde{\boldsymbol{x}}, t) = \boldsymbol{F}(\boldsymbol{x}_r, t) + \dot{\widetilde{\boldsymbol{x}}}(t)$$

or

$$\dot{\widetilde{\boldsymbol{x}}}(t) = \boldsymbol{F}(\boldsymbol{x}_r + \widetilde{\boldsymbol{x}}, t) - \boldsymbol{F}(\boldsymbol{x}_r, t) = \widetilde{\boldsymbol{F}}(\widetilde{\boldsymbol{x}}, t) = \widetilde{\boldsymbol{F}}(\widetilde{\boldsymbol{x}}, t+T) \tag{5.4.19}$$

where, for the phase flow of the perturbation $\widetilde{\boldsymbol{x}}$, we have

$$\widetilde{\boldsymbol{F}}(\boldsymbol{o}, t) = \boldsymbol{o}$$

i.e. the perturbation $\widetilde{\boldsymbol{x}}$ is the solution of a *homogeneous* non-autonomous system. For small perturbations, $\widetilde{\boldsymbol{F}}$ can be expanded into a Taylor series

$$\dot{\widetilde{\boldsymbol{x}}}(t) = \left. \frac{\partial \widetilde{\boldsymbol{F}}(\widetilde{\boldsymbol{x}}, t)}{\partial \widetilde{\boldsymbol{x}}} \right|_{\widetilde{\boldsymbol{x}}=\boldsymbol{o}} \widetilde{\boldsymbol{x}}(t) + \text{ higher-order terms} \tag{5.4.20}$$

where we neglect the higher-order terms. Thus, to analyse the perturbation, we have to consider a set of linear differential equations with a time-dependent coefficient matrix $\boldsymbol{D}(t)$

$$\dot{\widetilde{\boldsymbol{x}}}(t) = \boldsymbol{D}(t)\widetilde{\boldsymbol{x}}(t) \tag{5.4.21}$$

In accordance with eqs. (5.4.19) and (5.4.20), $\boldsymbol{D}(t)$ is determined as

$$\boldsymbol{D}(t) = \left. \frac{\partial \widetilde{\boldsymbol{F}}}{\partial \widetilde{\boldsymbol{x}}} \right|_{\widetilde{\boldsymbol{x}}=\boldsymbol{o}} = \left. \frac{\partial \boldsymbol{F}(\boldsymbol{x}, t)}{\partial \boldsymbol{x}} \right|_{\boldsymbol{x}_r(t)} \tag{5.4.22}$$

Since $\boldsymbol{x}_r(t)$ is a periodic function, $\boldsymbol{D}(t)$ is also periodic with the same period T

$$\boldsymbol{D}(t) = \boldsymbol{D}(t+T) \tag{5.4.23}$$

In the second case *ii.* of the periodic solution of an autonomous system, we obtain a system corresponding to eq. (5.4.21) for the perturbation where we have now

$$\boldsymbol{D}(t) = \left.\frac{\partial \boldsymbol{F}(\boldsymbol{x})}{\partial \boldsymbol{x}}\right|_{\boldsymbol{x}_r(t)} \tag{5.4.24}$$

Due to the periodicity of $\boldsymbol{x}_r(t)$, $\boldsymbol{D}$ is again a matrix with periodic coefficients. The solution $\boldsymbol{x}_r(t) = \boldsymbol{x}_r(t+T)$ of the initial system eq. (5.4.15) respectively eq. (5.4.17) can only be expressed analytically in special cases. Hence, the period length T is usually not known explicitly at the outset, but must generally be determined numerically. In each case, we must investigate whether the solutions of the set of linear differential equations eq. (5.4.21) with a periodic coefficient matrix fade away, display marginal behaviour or are amplified.

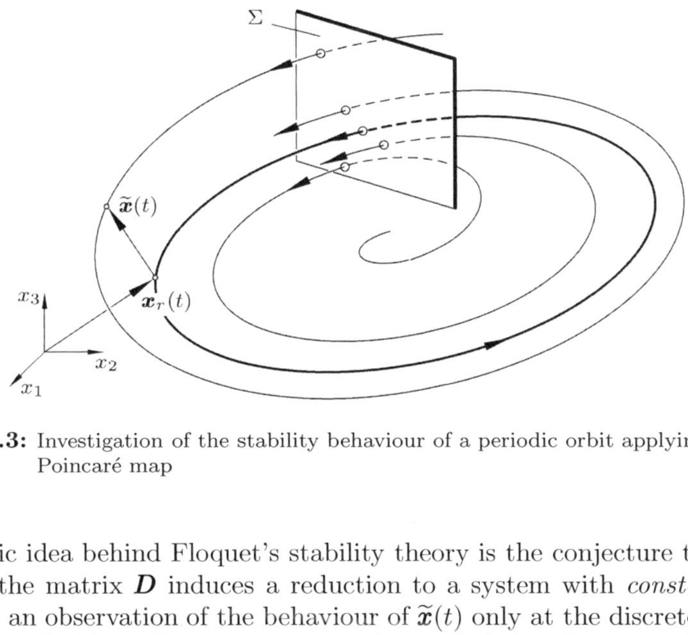

Fig. 5.4.3: Investigation of the stability behaviour of a periodic orbit applying the Poincaré map

The basic idea behind Floquet's stability theory is the conjecture that the periodicity of the matrix $\boldsymbol{D}$ induces a reduction to a system with *constant* coefficients, allowing an observation of the behaviour of $\widetilde{\boldsymbol{x}}(t)$ only at the discrete points in time $t = 0, T, 2T, 3T, \ldots$. From a geometrical point of view, however, this corresponds exactly to a Poincaré section (see fig. 5.4.3). For this reason, we are encouraged to draw conclusions from the characteristics of the appertaining Poincaré map as to the behaviour of the continuous system.

We first consider the n-dimensional system of linear differential equations eq. (5.4.21) for the perturbation $\widetilde{\boldsymbol{x}}(t)$ in case *i.* for an externally driven system. This system possesses n linearly independent solutions $\widetilde{\boldsymbol{x}}_1(t), \widetilde{\boldsymbol{x}}_2(t), \cdots \widetilde{\boldsymbol{x}}_\mathrm{n}(t)$, the so-called fundamental solutions, and any solution $\widetilde{\boldsymbol{x}}(t)$ can be represented as a superposition of these functions

$$\widetilde{\boldsymbol{x}}(t) = c_1\widetilde{\boldsymbol{x}}_1(t) + c_2\widetilde{\boldsymbol{x}}_2(t) + \cdots + c_\mathrm{n}\widetilde{\boldsymbol{x}}_\mathrm{n}(t) \tag{5.4.25}$$

Assembling the fundamental system of solutions in a (n × n)-matrix $\widetilde{\boldsymbol{X}}(t)$

$$\widetilde{\boldsymbol{X}}(t) = [\widetilde{\boldsymbol{x}}_1(t) \quad \widetilde{\boldsymbol{x}}_2(t) \ \dots \ \widetilde{\boldsymbol{x}}_\mathrm{n}(t)] \tag{5.4.26}$$

and introducing the constant column vector $\boldsymbol{c}$

$$\boldsymbol{c} = \{c_1 \quad c_2 \quad \dots \quad c_\mathrm{n}\} \tag{5.4.27}$$

we can write eq. (5.4.25) as

$$\widetilde{\boldsymbol{x}}(t) = \widetilde{\boldsymbol{X}}(t)\boldsymbol{c} \tag{5.4.28}$$

If $\widetilde{\boldsymbol{x}}(t)$ is a solution of the system of eq. (5.4.21), then, trivially, $\widetilde{\boldsymbol{x}}(t+T)$ is also a solution since $\boldsymbol{D}(t) = \boldsymbol{D}(t+T)$. It should be noted that generally, the periodicity of $\widetilde{\boldsymbol{x}}(t)$ *cannot* be deduced from this statement, but only the fact that $\widetilde{\boldsymbol{x}}(t+T)$ can be represented as a linear combination of the fundamental solutions as in eq. (5.4.25).

We now select the system of fundamental solutions so that the initial conditions lie on a hypersphere of radius 1 centered at the origin. For the matrix of the corresponding fundamental solutions, we introduce the notation

$$\boldsymbol{\Phi}(t) = [\boldsymbol{\varphi}_1(t) \quad \boldsymbol{\varphi}_2(t) \ \dots \ \boldsymbol{\varphi}_\mathrm{n}(t)] \tag{5.4.29}$$

with

$$\boldsymbol{\Phi}(0) = \boldsymbol{I} \tag{5.4.30}$$

If we apply the above argument that for each $\widetilde{\boldsymbol{x}}(t)$, $\widetilde{\boldsymbol{x}}(t+T)$ is also a solution to the system of fundamental solutions itself, we establish a constant (n $\times$ n)-matrix $\boldsymbol{C}$ so that

$$\boldsymbol{\Phi}(t+T) = \boldsymbol{\Phi}(t)\boldsymbol{C} \tag{5.4.31}$$

holds, i.e. $\boldsymbol{C}$ maps $\boldsymbol{\Phi}(t)$ onto $\boldsymbol{\Phi}(t+T)$, that is to say $\boldsymbol{C}$ exactly describes a Poincaré map. Due to the special initial conditions of eq. (5.4.30), we deduce from eq. (5.4.31) in particular

$$\boldsymbol{\Phi}(T) = \boldsymbol{\Phi}(0)\boldsymbol{C} = \boldsymbol{C} \tag{5.4.32}$$

which determines $\boldsymbol{C}$, the so-called *monodromy matrix*. In general, $\boldsymbol{C}$ cannot be defined analytically either, but must be set up numerically. If the period T is known, we can integrate the system eq. (5.4.21) over a period T, applying the initial conditions of eq. (5.4.30). This yields the monodromy matrix $\boldsymbol{C}$ directly.

Figure 5.4.4 shows a geometrical interpretation of eq. (5.4.32) for the case n = 3. The matrix $\boldsymbol{C}$ defines a linear Poincaré map and, in the following, we assume that $\boldsymbol{C}$ possesses three linearly independent eigenvectors. If we apply this transformation to the coordinates of a vector $\boldsymbol{x}$, the image vector $\boldsymbol{x}'$ is given by

$$\boldsymbol{x}' = \boldsymbol{C}\boldsymbol{x}$$

or conversely

$$\boldsymbol{x} = \boldsymbol{C}^{-1}\boldsymbol{x}' \tag{5.4.33}$$

The points of the hypersphere S of radius 1 on which the initial conditions $\boldsymbol{\varphi}_\mathrm{j}(0)$ (j = 1, 2, ... n) at $t = 0$ of the fundamental solutions lie are thus mapped onto an ellipsoid

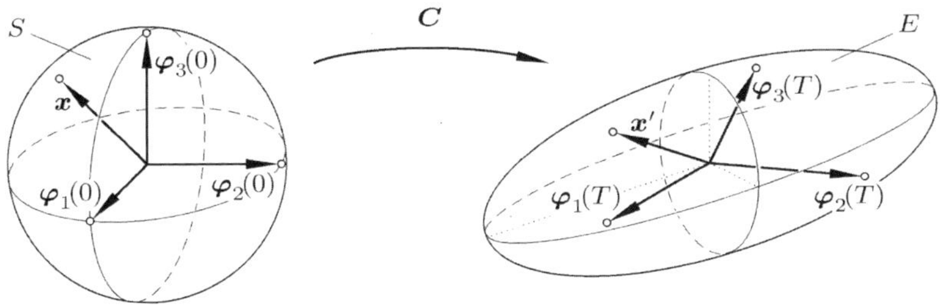

Fig. 5.4.4: Geometrical interpretation of the monodromy matrix $\boldsymbol{C}$ in the Euclidean space

E which contains the fundamental solutions $\boldsymbol{\varphi}_{\mathrm{j}}(T)$ after one period. The equation of the hypersphere is

$$S: \qquad \boldsymbol{x}^{\mathrm{t}}\boldsymbol{x} = 1 \tag{5.4.34}$$

Applying eq. (5.4.33), we obtain the equation of the ellipsoid

$$E: \qquad \boldsymbol{x}'^{\mathrm{t}}\boldsymbol{A}\boldsymbol{x}' = 1 \qquad \text{with} \quad \boldsymbol{A} = \boldsymbol{C}^{-\mathrm{t}}\boldsymbol{C}^{-1} \tag{5.4.35}$$

If we now transform this quadratic form to principal axes, the eigenvalues of $\boldsymbol{A}$ are seen to coincide with the reciprocal values of the square of the semi-axes of the ellipsoid. Consequently, we deduce from eq. (5.4.35) that the semi-axes of the ellipsoid agree with the eigenvalues of $\boldsymbol{C}$.

It is thus immediately clear that the eigenvalues λ_{j} of the monodromy matrix, the so-called *Floquet multipliers*, provide the necessary information on the stability of the limit cycle $\boldsymbol{x}_r(t)$. If $|\lambda_{\mathrm{j}}| < 1$ holds for all $\mathrm{j} = 1, 2, \ldots \mathrm{n}$, the ellipsoid E lies inside the sphere S, i.e. after one cycle, the perturbations $\widetilde{\boldsymbol{x}}(t)$ have faded away in every direction. Since the same transformation matrix $\boldsymbol{C}$ operates at every period, the perturbations tend asymptotically towards zero. The limit cycle is hence asymptotically stable if all Floquet multipliers are $|\lambda_{\mathrm{j}}| < 1$. Moreover, from eq. (5.4.31) together with eq. (5.4.32), there follows

$$\boldsymbol{\Phi}(2T) = \boldsymbol{\Phi}^2(T) = \boldsymbol{C}^2$$

or generally

$$\boldsymbol{\Phi}(\mathrm{k}T) = \boldsymbol{\Phi}^{\mathrm{k}}(T) = \boldsymbol{C}^{\mathrm{k}} \tag{5.4.36}$$

We can now deduce from this functional equation that the eigenvalues of the monodromy matrix possess a special structure. This is seen by the following argument. We designate with $\boldsymbol{\xi}$ the eigenvector of $\boldsymbol{\Phi}(T)$ corresponding to the eigenvalue $\lambda(T)$,

$$\boldsymbol{\Phi}(T)\boldsymbol{\xi} = \lambda(T)\boldsymbol{\xi}$$

Multiplying this equation from the left with $\boldsymbol{\Phi}(T)$, we recognise that $\boldsymbol{\xi}$ is also the eigenvector of $\boldsymbol{\Phi}^2(T) = \boldsymbol{\Phi}(2T)$ corresponding to the eigenvalue $\lambda(2T)$

$$\boldsymbol{\Phi}^2(T)\boldsymbol{\xi} = \lambda(T)\boldsymbol{\Phi}(T)\boldsymbol{\xi} = \lambda^2(T)\boldsymbol{\xi}$$

or

$$\boldsymbol{\Phi}(2T)\boldsymbol{\xi} = \lambda(2T)\boldsymbol{\xi}$$

Generalising, we note the relation

$$\boldsymbol{\Phi}^{\mathrm{k}}(T)\boldsymbol{\xi} = \lambda^{\mathrm{k}}(T)\boldsymbol{\xi} = \lambda(\mathrm{k}T)\boldsymbol{\xi} \tag{5.4.37}$$

We have thus found a scalar function equation for the Floquet multipliers

$$\lambda^{\mathrm{k}}(T) = \lambda(\mathrm{k}T) \tag{5.4.38}$$

Solutions of this functional equation are exponential functions; thus, the eigenvalues of the monodromy matrix can be expressed in the form

$$\lambda(T) = \mathrm{e}^{\sigma T} \tag{5.4.39}$$

where $\sigma = \sigma_1 + i\sigma_2$ represents the so-called (complex) *Floquet exponent* which can be deduced, but for multiples of the ratio $2\pi i/T$, from the Floquet multipliers, eq. (5.4.39)

$$\sigma = \frac{1}{T}\ln\lambda(T) + \frac{2\pi\mathrm{k}}{T}i \qquad (\mathrm{k} = 0, \pm 1, \pm 2, \ldots) \tag{5.4.40}$$

Note that negative λ-values also allow a representation in the exponential form of eq. (5.4.39) since complex exponents $\sigma = \sigma_1 + i\sigma_2$ are admitted. If we specify $\sigma_2 T = \pi$, there applies

$$\begin{aligned} \mathrm{e}^{(\sigma_1 + i\sigma_2)T} = \mathrm{e}^{\sigma_1 T}\mathrm{e}^{i\sigma_2 T} &= \mathrm{e}^{\sigma_1 T}(\cos\sigma_2 T + i\sin\sigma_2 T) \\ &= -\mathrm{e}^{\sigma_1 T} \end{aligned}$$

The solution eq. (5.4.39) of the scalar functional equation (5.4.38) for the Floquet multipliers can also be extended to the functional equation (5.4.36) for the monodromy matrix. The ansatz

$$\boldsymbol{C} = \mathrm{e}^{\boldsymbol{S}T} \tag{5.4.41}$$

satisfies the equation (5.4.36) since, for example for $\mathrm{k} = 2$,

$$\boldsymbol{\Phi}(T)\boldsymbol{\Phi}(T) = \boldsymbol{C}^2 = \mathrm{e}^{\boldsymbol{S}T}\mathrm{e}^{\boldsymbol{S}T} = \mathrm{e}^{2\boldsymbol{S}T} = \boldsymbol{\Phi}(2T)$$

Moreover, it can be shown, see e.g. (Hartman, 1964; Haken, 1987), that the eigenvalues of $\boldsymbol{S}$ coincide with the Floquet exponents σ. Using the ansatz for the matrix of fundamental solutions,

$$\boldsymbol{\Phi}(t) = \boldsymbol{U}(t)\,\mathrm{e}^{\boldsymbol{S}t} \tag{5.4.42}$$

then, following eqs. (5.4.41) and (5.4.31), we find

$$\boldsymbol{\Phi}(t+T) = \boldsymbol{U}(t+T)\,\mathrm{e}^{\boldsymbol{S}(t+T)} = \boldsymbol{U}(t+T)\,\mathrm{e}^{\boldsymbol{S}t}\boldsymbol{C} = \boldsymbol{\Phi}(t)\boldsymbol{C}$$

Noting eq. (5.4.42), we thus infer the periodicity of the solution matrix $\boldsymbol{U}(t)$

$$\boldsymbol{U}(t+T) = \boldsymbol{U}(t) \tag{5.4.43}$$

Let us summarise the results for case *i.* for periodically externally driven systems. Each solution $\widetilde{\boldsymbol{x}}(t)$ of the linearised system eq. (5.4.21) can be written as a linear combination of fundamental solutions which in turn are composed of a periodic term $\boldsymbol{U}(t)$ together with an exponential function $\mathrm{e}^{\boldsymbol{S}t}$. The perturbation $\widetilde{\boldsymbol{x}}(t)$ tends

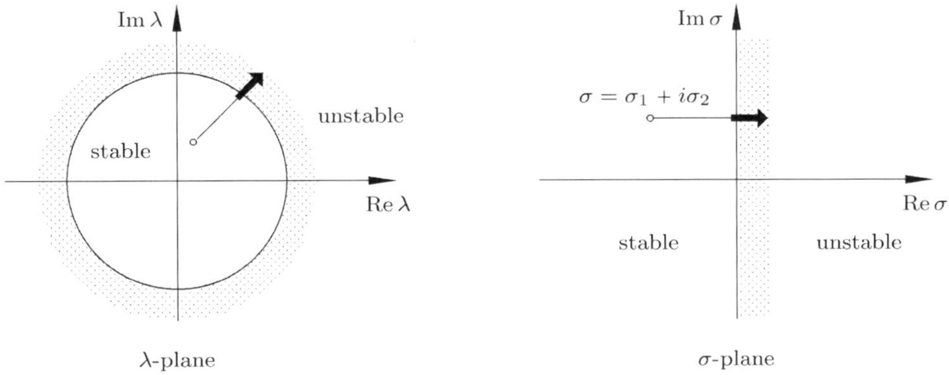

Fig. 5.4.5: Stability regions for Floquet multipliers λ and Floquet exponents σ

towards zero for $t \to \infty$ if $|\lambda_{\mathrm{j}}| = |\mathrm{e}^{\sigma_{\mathrm{j}} T}| < 1$ holds for all Floquet multipliers respectively if the real parts of all Floquet exponents are $\mathrm{Re}(\sigma_{\mathrm{j}}) < 0$ (see fig. 5.4.5). In this case, the limit cycle is asymptotically stable.

If $|\lambda_{\mathrm{j}}| > 1$ respectively $\mathrm{Re}(\sigma_{\mathrm{j}}) > 0$ hold for an index j, the perturbations $\widetilde{\boldsymbol{x}}(t)$ increase exponentially and the corresponding orbit is unstable. According to the general theorem on the stability of linearised problems stated in section 5.4.1, the Floquet multipliers provide information on the stability of the non-linear system as long as none of the Floquet multipliers has the absolute value 1. Indeed, should there be a Floquet multiplier for which $|\lambda| = 1$ applies, we can no longer – in analogy with the equilibrium states – apply the stability statement for the linearised system to the non-linear one so that further considerations are necessary.

A periodic orbit can lose its stability in three different ways (see fig. 5.4.6):

(a) $\lambda = 1\,,\quad \mathrm{Im}(\lambda) = 0$
(b) $\lambda = -1\,,\quad \mathrm{Im}(\lambda) = 0$
(c) $|\lambda| = 1$ and $\mathrm{Im}(\lambda) \neq 0$ (a pair of complex conjugate eigenvalues).

In section 6.4, we shall discuss the corresponding bifurcations in detail.

In contrast to periodically externally driven systems, in which the appearance of a Floquet multiplier $\lambda = 1$ always indicates a bifurcation, the monodromy matrix of eq. (5.4.17) always possesses an eigenvalue $\lambda = 1$; in these circumstances, the limit cycle $\boldsymbol{x}_r(t)$ is the solution of an *autonomous* system. In this case, only the set of remaining eigenvalues can provide information on the stability behaviour of the periodic motion. In order to prove that the monodromy matrix in case *ii.* possesses the eigenvalue 1, we differentiate eq. (5.4.17) with respect to time and obtain

$$\ddot{\boldsymbol{x}} = \frac{\partial \boldsymbol{F}}{\partial \boldsymbol{x}} \dot{\boldsymbol{x}}$$

In conjunction with eq. (5.4.24), we obtain in particular for the solution $\boldsymbol{x}_r(t)$

$$\ddot{\boldsymbol{x}}_r = \boldsymbol{D}(t) \dot{\boldsymbol{x}}_r \tag{5.4.44}$$

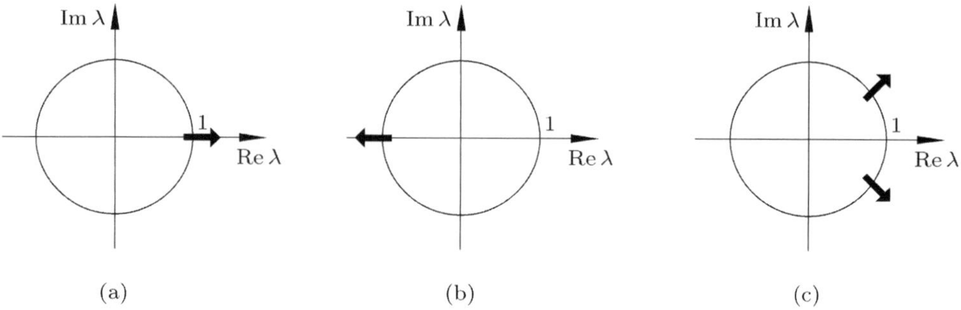

Fig. 5.4.6: Three possibilities for the loss of stability of a periodic orbit when crossing the unit circle

i.e. the perturbation $\widetilde{\boldsymbol{x}}(t) \equiv \dot{\boldsymbol{x}}_r(t)$ which is tangent to the original orbit $\boldsymbol{x}_r(t)$ is indeed the solution of the linearised problem. The function $\dot{\boldsymbol{x}}_r(t)$ can be written as a linear combination of the fundamental solutions

$$\dot{\boldsymbol{x}}_r(t) = \boldsymbol{\Phi}(t)\dot{\boldsymbol{x}}_r(0)$$

Since $\dot{\boldsymbol{x}}_r(t)$ is a periodic function, there applies especially for $t = T$

$$\dot{\boldsymbol{x}}_r(T) = \boldsymbol{\Phi}(T)\dot{\boldsymbol{x}}_r(0) = \dot{\boldsymbol{x}}_r(0) \tag{5.4.45}$$

i.e. $\lambda = 1$ is indeed the eigenvalue of the monodromy matrix $\boldsymbol{\Phi}(T) = \boldsymbol{C}$.

In conclusion, we demonstrate the stability of periodic motions for a simple example. To this end, we consider the following set of differential equations

$$\begin{aligned} \dot{x} &= x[\mu - (x^2 + y^2)] - y \\ \dot{y} &= y[\mu - (x^2 + y^2)] + x \end{aligned} \tag{5.4.46}$$

which possesses the periodic solution

$$\boldsymbol{x}_r(t) = \sqrt{\mu}\{\cos t \quad \sin t\} \tag{5.4.47}$$

with period $T = 2\pi$. In the case of $\mu > 0$, eq. (5.4.47) reproduces a limit cycle, a circle of radius $\sqrt{\mu}$. In order to investigate its stability, we observe a neighbouring trajectory

$$\boldsymbol{x}(t) = \boldsymbol{x}_r(t) + \widetilde{\boldsymbol{x}}(t)$$

Following eqs. (5.4.21) and (5.4.24), we now set up the associated linear system for the perturbation $\widetilde{\boldsymbol{x}}(t)$

$$\dot{\widetilde{\boldsymbol{x}}} = \boldsymbol{D}(\boldsymbol{x}_r)\widetilde{\boldsymbol{x}}$$

with

$$\boldsymbol{D}(\boldsymbol{x}_r) = \begin{bmatrix} (\mu - 3x^2 - y^2) & -(1 + 2xy) \\ (1 - 2xy) & (\mu - x^2 - 3y^2) \end{bmatrix}_{\boldsymbol{x}_r(t)}$$

In the last two relations, we have used the abbreviation $\boldsymbol{x}_r = \boldsymbol{x}_r(t)$ for typographical simplicity. Introduction of eq. (5.4.47) yields

$$\dot{\tilde{\boldsymbol{x}}}(t) = -\begin{bmatrix} 2\mu\cos^2 t & (\mu\sin 2t + 1) \\ (\mu\sin 2t - 1) & 2\mu\sin^2 t \end{bmatrix} \tilde{\boldsymbol{x}}(t) \tag{5.4.48}$$

By substitution, we immediately verify that

$$\boldsymbol{\varphi}_1 = \mathrm{e}^{-2\mu t}\begin{bmatrix} \cos t \\ \sin t \end{bmatrix} \quad \text{and} \quad \boldsymbol{\varphi}_2 = \begin{bmatrix} -\sin t \\ \cos t \end{bmatrix} \tag{5.4.49}$$

form a system of fundamental solutions $\boldsymbol{\Phi}(t) = [\boldsymbol{\varphi}_1 \quad \boldsymbol{\varphi}_2]$ which already satisfies the initial conditions $\boldsymbol{\Phi}(0) = \boldsymbol{I}$.

In accordance with eq. (5.4.32), the monodromy matrix $\boldsymbol{C}$, which defines a linear Poincaré map, becomes

$$\boldsymbol{C} = \boldsymbol{\Phi}(T) = \boldsymbol{\Phi}(2\pi) = \begin{bmatrix} \mathrm{e}^{-4\pi\mu} & 0 \\ 0 & 1 \end{bmatrix} \tag{5.4.50}$$

with the eigenvalues (Floquet multipliers)

$$\lambda_1 = \mathrm{e}^{-4\pi\mu}, \quad \lambda_2 = 1 \tag{5.4.51}$$

Since the set of equations (5.4.46) is autonomous, $\lambda_2 = 1$ emerges as an eigenvalue and appertains to a perturbation $\tilde{\boldsymbol{x}} = \boldsymbol{\varphi}_2$ in tangential direction. The corresponding Floquet exponent σ_2 is zero. The Floquet exponent appertaining to λ_1 is $\sigma_1 = -2\mu < 0$ (for $\mu > 0$), i.e. we have a stable limit cycle. In section 6.1, we investigate the set of equations (5.4.46) in dependence of the control parameter μ and show that it describes a so-called Hopf bifurcation for $\mu = 0$.

If we introduce the polar coordinates

$$x = r\cos\theta, \quad y = r\sin\theta$$

eqs. (5.4.46) go over into

$$\begin{aligned} \dot{r}\cos\theta - (r\sin\theta)\dot{\theta} &= r(\mu - r^2)\cos\theta - r\sin\theta \\ \dot{r}\sin\theta + (r\cos\theta)\dot{\theta} &= r(\mu - r^2)\sin\theta + r\cos\theta \end{aligned}$$

or

$$\begin{aligned} \dot{r} &= (\mu - r^2)r \\ \dot{\theta} &= 1 \end{aligned} \tag{5.4.52}$$

These two uncoupled differential equations can be integrated easily

$$\int\limits_{r_0}^{r} \frac{\mathrm{d}r}{r(\mu - r^2)} = \int\limits_{0}^{t} \mathrm{d}t \quad , \qquad \theta = t + \theta_0$$

$$\frac{1}{2\mu}\ln\frac{r^2}{\mu - r^2}\bigg|_{r_0}^{r} = t \quad , \qquad \theta = t + \theta_0$$

Following a number of algebraic operations, we finally obtain

$$r = \sqrt{\mu}\left[\left(\frac{\mu}{r_0^2} - 1\right)\mathrm{e}^{-2\mu t} + 1\right]^{-1/2} \tag{5.4.53}$$

If we now record the trajectory at fixed time intervals $\Delta t = 2\pi$, i.e. after each cycle, we fix a Poincaré section (e.g. $\theta = 0$) and can easily specify the corresponding Poincaré map (see fig. 5.4.7) in terms of the r-coordinate r_1 of the trajectory and in dependence of the initial value $r = r_0$ after one cycle. Applying eq. (5.4.53), we obtain

$$r_1 = f(r_0) = \sqrt{\mu}\left[\left(\frac{\mu}{r_0^2} - 1\right)\mathrm{e}^{-4\pi\mu} + 1\right]^{-1/2} \tag{5.4.54}$$

If we set $\mu = 1$, $r_0 = 10$, for example, this yields $r_1 = 1.00000173$ as r-coordinate of the image point. If we start in the interior of the limit cycle with $r_0 = 1/2$, we obtain $r_1 = 0.99999477$, i.e. the neighbouring trajectories converge to the periodic orbit, a circle with radius $r = \sqrt{\mu} = 1$, very fast.

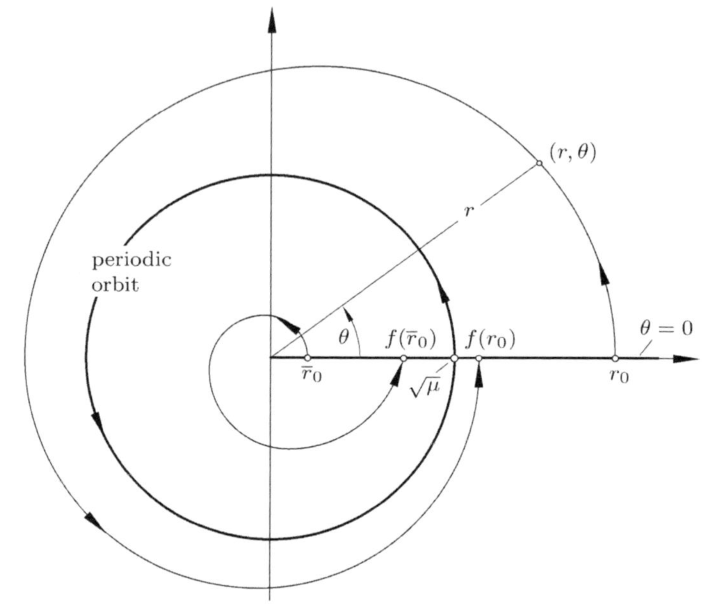

Fig. 5.4.7: Poincaré section $\theta = 0$ of the neighbouring trajectories of a periodic orbit

We recognise from eq. (5.4.53) that $r_0 = \sqrt{\mu}$ is a fixed point of the Poincaré map. In order to investigate the stability of this fixed point, we form the derivative in accordance with section 3.6

$$\lambda = \left.\frac{\mathrm{d}f}{\mathrm{d}r_0}\right|_{r_0=\sqrt{\mu}} = -\frac{\sqrt{\mu}}{2}\left[\left(\frac{\mu}{r_0^2} - 1\right)\mathrm{e}^{-4\pi\mu} + 1\right]^{-3/2}\left.\left(\frac{-2\mu}{r_0^3}\mathrm{e}^{-4\pi\mu}\right)\right|_{r_0=\sqrt{\mu}} = \mathrm{e}^{-4\pi\mu}$$

This value coincides with the first eigenvalue of the monodromy matrix, see eq. (5.4.51). Due to $\mathrm{e}^{-4\pi\mu} < 1$ for $\mu > 0$, $r_0 = \sqrt{\mu}$ is, as expected, a stable fixed point of the Poincaré map.

5.4.3 Lyapunov Exponent of One-Dimensional Maps

In section 3.7, we discussed the logistic equation $x_{n+1} = \alpha x_n(1 - x_n)$. For $\alpha = 2.8$, the sequence of points $\{x_n\}$ converges for all initial values $x_0 \neq 0$ to the fixed point $x_s = 1 - 1/2.8 = 0.64286\ldots$, see fig. 3.7.2. If we restrict ourselves to the interval $0.5 \leqslant x \leqslant 0.76726$, then the map in this range is uniquely invertible and can be interpreted as a Poincaré map of a dynamical system having a limit cycle for $\alpha = 2.8$. Its stability is determined by the Floquet multiplier

$$\lambda = \left.\frac{\mathrm{d}f}{\mathrm{d}x_n}\right|_{x_s} = \alpha(1 - 2x_s) = -0.8 \tag{5.4.55}$$

Due to $-1 < \lambda < 0$, x_s is hence a stable fixed point which the sequence of points $\{x_n\}$ approaches alternatingly. The corresponding Floquet exponent σ for a normalised period length $T = 1$ is then

$$\sigma = \frac{1}{T}\ln\lambda = -0.223 + \pi i \quad , \quad \text{i.e.} \quad \mathrm{Re}\,\sigma < 0 \tag{5.4.56}$$

In the previous section, it was shown that the Floquet exponents determine the behaviour of neighbouring trajectories of a limit cycle in the Poincaré section. Lyapunov extended this stability consideration to any recurrent trajectory which does not necessarily have to be closed. In order to clarify the basic idea, we consider a one-dimensional map

$$\begin{aligned} &x \longmapsto f(x) \\ \text{resp.}\quad & \\ &x_{n+1} = f(x_n) \end{aligned} \tag{5.4.57}$$

where we assume that the values x_n lie in a finite interval $[a, b]$ for $n \to \infty$. We consider a sequence of points $\{\bar{x}_n\}$ for a perturbed initial condition $x_0 + \delta x_0$ this sequence being adjacent to a sequence of points $\{x_n\}$ with the initial condition x_0. After k iterations, we obtain

$$\begin{aligned} \bar{x}_k &= x_k + \delta x_k = f(x_{k-1} + \delta x_{k-1}) \\ &= f(x_{k-1}) + f'(x_{k-1})\delta x_{k-1} \quad + \text{ terms of higher order} \end{aligned} \tag{5.4.58}$$

By linearisation, this yields for the perturbation

$$\delta x_k = f'(x_{k-1})\delta x_{k-1} \tag{5.4.59}$$

where, in contrast to eq. (5.4.55), $f'(x_{k-1})$ generally depends on k and is not constant. If we repeatedly apply the recursion rule, eq. (5.4.59), we obtain

$$\begin{aligned} \delta x_k &= f'(x_{k-1})f'(x_{k-2})\ldots f'(x_0)\delta x_0 \\ &= \prod_{k=0}^{k-1} f'(x_k)\delta x_0 \end{aligned} \tag{5.4.60}$$

The Lyapunov exponent σ is now defined as the mean exponential divergence or convergence of neighbouring sequences of points

$$|\delta x_k| \approx \mathrm{e}^{k\sigma}\,|\delta x_0| \tag{5.4.61}$$

for $\mathrm{k} \to \infty$ and $|\delta x_0| \to 0$. Thus, we determine the Lyapunov exponent for one-dimensional maps in conjunction with eq. (5.4.60) as

$$\sigma = \lim_{\mathrm{k}\to\infty} \lim_{|\delta x_0|\to 0} \frac{1}{\mathrm{k}} \ln \frac{|\delta x_\mathrm{k}|}{|\delta x_0|} \tag{5.4.62}$$

or

$$\sigma = \lim_{\mathrm{k}\to\infty} \frac{1}{\mathrm{k}} \sum_{\mathrm{i}=0}^{\mathrm{k}-1} \ln |f'(x_\mathrm{i})| \tag{5.4.63}$$

and it can be regarded in this formulation as the generalisation of the Floquet exponent in eq. (5.4.40) for periodic motions, although the Lyapunov exponent per definition is real. If this limit value does not exist, it has to be replaced by the *limes superior*, i.e. the largest accumulation point of the sequence $\{a_\mathrm{k}\}$ where

$$a_\mathrm{k} = \frac{1}{\mathrm{k}} \sum_{\mathrm{i}=0}^{\mathrm{k}-1} \ln |f'(x_\mathrm{i})|$$

see also (Haken, 1987), and hence

$$\sigma = \limsup_{\mathrm{k}\to\infty} \frac{1}{\mathrm{k}} \sum_{\mathrm{i}=0}^{\mathrm{k}-1} \ln |f'(x_\mathrm{i})| \tag{5.4.64}$$

It follows from this definition of the Lyapunov exponent as a limit for $\mathrm{k} \to \infty$ that the influence of the transient phase is eliminated. Hence, the Lyapunov exponent for a sequence of points tending towards a fixed point coincides with the real part of the Floquet exponent and is negative.

As an illustration, we have calculated the Lyapunov exponent as specified in eq. (5.4.63) for the logistic map $x_{\mathrm{n}+1} = \alpha x_\mathrm{n}(1 - x_\mathrm{n})$ numerically. The upper part of fig. 5.4.8 shows the bifurcation diagram, i.e. the long-term behaviour of x_n as a function of the control parameter α, cf. fig. 3.7.6. In the lower part, we present the corresponding evolution of the Lyapunov exponent σ in dependence of α. In agreement with eq. (5.4.56), we obtain for $\alpha = 2.8$, for example, the value $\sigma = -0.223$. The range of regular behaviour $0 < \alpha < \alpha_\infty$ is characterised by Lyapunov exponents $\sigma \leqslant 0$, where σ assumes a zero value only for the critical values α_k. At $\alpha_1 = 3$, the first bifurcation takes place, a periodic motion sets in, the originally stable fixed point becomes unstable. In the iterated map $f^2(x_\mathrm{n}) = f(f(x_\mathrm{n}))$, two new stable fixed points emerge which coincide for $\alpha_1 = 3$. Calculating these fixed points, we obtain from $x_\mathrm{n} = f(f(x_\mathrm{n}))$ the value $x_s = 2/3$ with $f'(2/3) = -1$. For α-values lying slightly below the critical value α_1, i.e. for example $\alpha = \alpha_1 - \varepsilon = 3 - \varepsilon$, the fixed point $x_s = 1 - 1/(3 - \varepsilon)$ is stable and, following eq. (5.4.55), we obtain

$$\lambda(\varepsilon) = \left.\frac{\mathrm{d}f}{\mathrm{d}x_\mathrm{n}}\right|_{x_s} = (3-\varepsilon)\left[1 - 2\left(1 - \frac{1}{3-\varepsilon}\right)\right] = -1 + \varepsilon \tag{5.4.65}$$

Taking the logarithm, we obtain the corresponding Lyapunov exponent

$$\sigma(\varepsilon) = \ln|1 - \varepsilon| \tag{5.4.66}$$

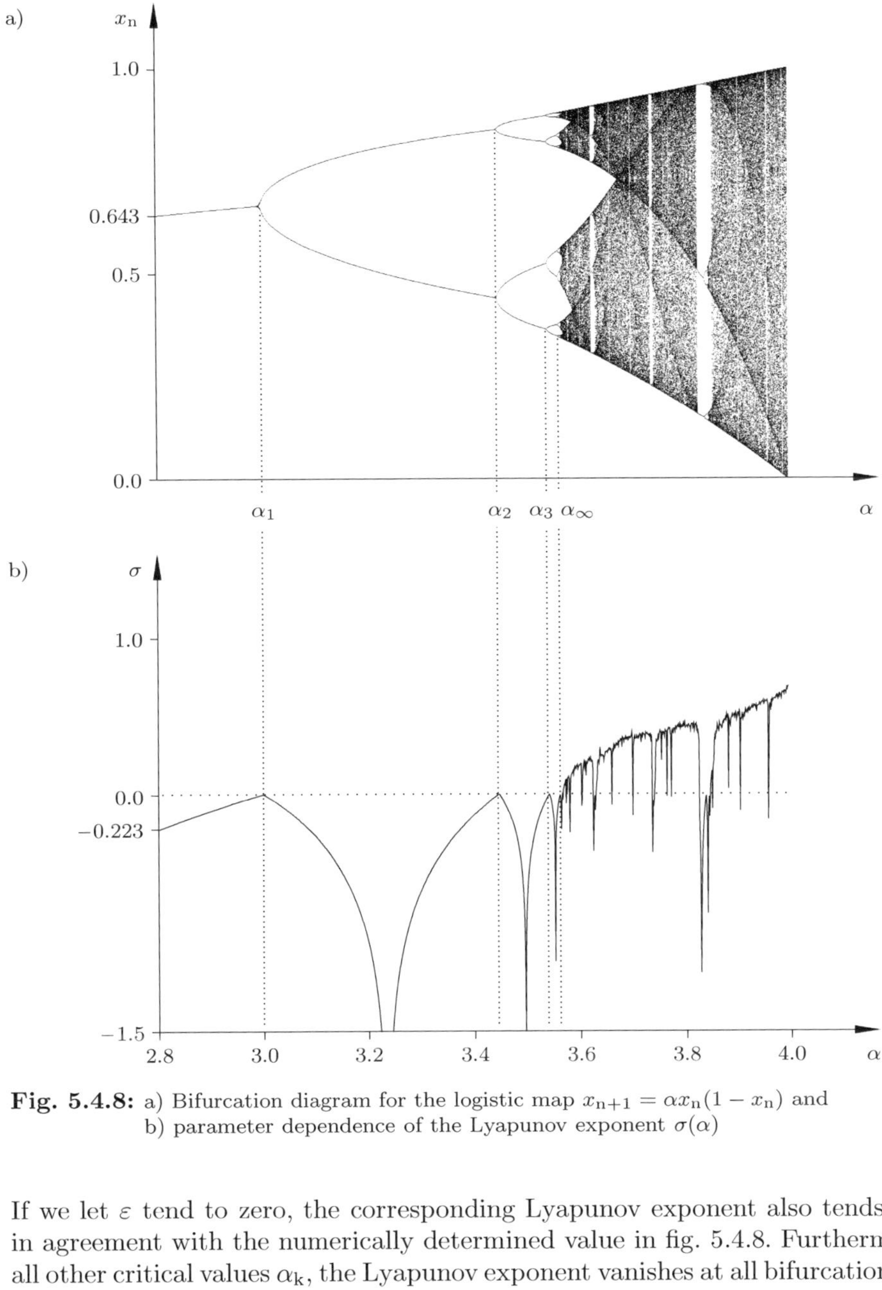

Fig. 5.4.8: a) Bifurcation diagram for the logistic map $x_{n+1} = \alpha x_n(1 - x_n)$ and b) parameter dependence of the Lyapunov exponent $\sigma(\alpha)$

If we let ε tend to zero, the corresponding Lyapunov exponent also tends to zero in agreement with the numerically determined value in fig. 5.4.8. Furthermore, for all other critical values α_k, the Lyapunov exponent vanishes at all bifurcation points.

For $\alpha > \alpha_\infty$, chaotic behaviour sets in, characterised by positive σ-values. It can be shown that the stable periodic cycles in the whole parameter range $0 \leqslant \alpha \leqslant 4$ are everywhere dense, i.e. in any neighbourhood of an α-value, such a cycle exists (Lichtenberg and Lieberman, 1983). The representation of the Lyapunov exponent in fig. 5.4.8b is hence highly complex. The chaotic region is repeatedly interrupted by periodic windows, clearly recognisable by the negative σ-values in the lower diagram. It is particularly interesting that the self-similarity of the bifurcation cascades – which we shall discuss in more detail in section 6.7 – is reflected in a self-similarity of the Lyapunov exponent. Were we to single out and enlarge the largest periodic window at $\alpha = 1 + \sqrt{8}$, for example, we would discover a similar σ-distribution to that for $\alpha < \alpha_\infty$.

5.4.4 Lyapunov Exponents of n-Dimensional Continuous Systems

The concept of the Lyapunov exponents can now be applied to n-dimensional continuous systems, eq. (5.4.2). The theoretical basis for this was developed by Oseledec (Oseledec, 1968). Once more, we investigate the stability behaviour of a reference trajectory $\boldsymbol{x}_r(t)$ and, with the aid of the Lyapunov exponents, can assess the sensitivity of the system to small perturbations in the initial conditions.

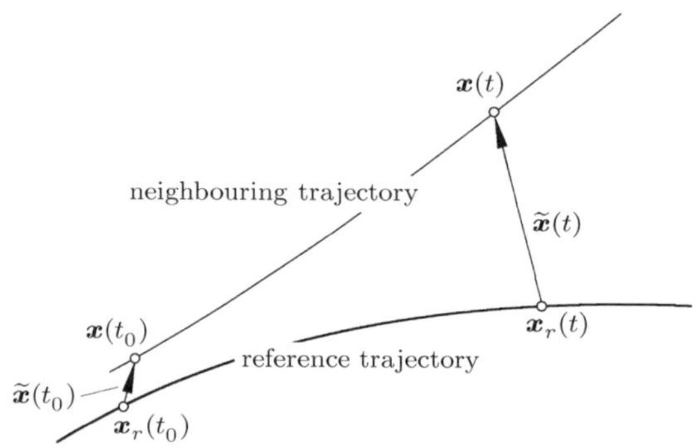

Fig. 5.4.9

On the definition of the Lyapunov exponent

Taking a reference trajectory $\boldsymbol{x}_r(t)$ for which $\dot{\boldsymbol{x}}_r = \boldsymbol{F}(\boldsymbol{x}_r)$ is satisfied, we observe a neighbouring trajectory at an infinitesimal distance which appertains to a small initial perturbation $\widetilde{\boldsymbol{x}}(t_0)$ (fig. 5.4.9). Our interest is directed at the asymptotic behaviour ($t \to \infty$) of the perturbation $\widetilde{\boldsymbol{x}}(t)$; as usual, we expand $\boldsymbol{F}(\boldsymbol{x})$ into a Taylor series in the neighbourhood of $\boldsymbol{x}_r(t)$

$$\begin{aligned}\dot{\boldsymbol{x}}(t) = \dot{\boldsymbol{x}}_r(t) + \dot{\widetilde{\boldsymbol{x}}}(t) &= \boldsymbol{F}(\boldsymbol{x}_r(t) + \widetilde{\boldsymbol{x}}(t)) \\ &= \boldsymbol{F}(\boldsymbol{x}_r(t)) + \left.\frac{\partial \boldsymbol{F}}{\partial \boldsymbol{x}}\right|_{\boldsymbol{x}_r(t)} \widetilde{\boldsymbol{x}}(t) \qquad + \cdots\end{aligned}$$

Ignoring higher-order terms in $\widetilde{\boldsymbol{x}}(t)$, we obtain an n-dimensional linear system of homogeneous differential equations having a time-dependent coefficient matrix

$$\dot{\widetilde{\boldsymbol{x}}}(t) = \left.\frac{\partial \boldsymbol{F}}{\partial \boldsymbol{x}}\right|_{\boldsymbol{x}_r(t)} \widetilde{\boldsymbol{x}}(t) \tag{5.4.67}$$

It can be shown (Arnold, 1980) that a solution $\widetilde{\boldsymbol{x}}(t)$ exists for every initial condition $(t_0, \widetilde{\boldsymbol{x}}(t_0))$ as long as $\boldsymbol{F}(\boldsymbol{x})$ is assumed continuously differentiable. This system possesses a fundamental system of n solutions and a fundamental matrix $\mathbf{\Phi}(t, t_0)$ so that the solution of eq.(5.4.67) can be expressed formally as

$$\widetilde{\boldsymbol{x}}(t) = \mathbf{\Phi}(t, t_0)\widetilde{\boldsymbol{x}}(t_0) \tag{5.4.68}$$

This relation can also be interpreted as a linear map of two vector spaces. By means of the matrix $\mathbf{\Phi}(t, t_0)$, the vector $\widetilde{\boldsymbol{x}}(t_0)$ appertaining to the tangent space $\mathbb{E}_0$ of the flow at the point $\boldsymbol{x}_r(t_0)$ is mapped onto the image vector $\widetilde{\boldsymbol{x}}(t)$ associated with the corresponding tangent space $\mathbb{E}_t$ at the point $\boldsymbol{x}_r(t)$, where

$$\mathbf{\Phi}(t_2, t_0) = \mathbf{\Phi}(t_2, t_1)\mathbf{\Phi}(t_1, t_0) \tag{5.4.69}$$

evidently applies. In accordance with eq. (5.4.62), the Lyapunov exponent of the vector $\widetilde{\boldsymbol{x}}(t)$ with respect to the reference trajectory $\boldsymbol{x}_r(t)$ is now given by

$$\sigma_{\boldsymbol{x}_r}(\widetilde{\boldsymbol{x}}) = \limsup_{t\to\infty} \frac{1}{t} \ln \frac{|\widetilde{\boldsymbol{x}}(t)|}{|\widetilde{\boldsymbol{x}}(t_0)|} = \limsup_{t\to\infty} \frac{1}{t} \ln |\widetilde{\boldsymbol{x}}(t)| \tag{5.4.70}$$

In order to guarantee that $\sigma_{\boldsymbol{x}_r}$ remains finite, we assume that

$$\limsup_{t\to\infty} \frac{1}{t} \ln \|\mathbf{\Phi}(t, t_0)\| < \infty$$

Lyapunov exponents can, in principle, be defined for conservative as well as for dissipative systems; we first concentrate on dissipative systems. Here, too, the basic idea is that the influence of the transient phase becomes irrelevant once we take the mean of the infinitesimal perturbations with respect to time for $t \to \infty$ and note that only the long-term behaviour remains significant.

Two important characteristics follow from the definition in eq. (5.4.70):

i. if we select any real number $c \neq 0$, there applies

$$\sigma_{\boldsymbol{x}_r}(c\widetilde{\boldsymbol{x}}) = \sigma_{\boldsymbol{x}_r}(\widetilde{\boldsymbol{x}}) \tag{5.4.71}$$

ii.

$$\sigma_{\boldsymbol{x}_r}(\widetilde{\boldsymbol{x}}_1 + \widetilde{\boldsymbol{x}}_2) \leqslant \max\{\sigma_{\boldsymbol{x}_r}(\widetilde{\boldsymbol{x}}_1), \sigma_{\boldsymbol{x}_r}(\widetilde{\boldsymbol{x}}_2)\} \tag{5.4.72}$$

Equation (5.4.71) follows immediately from the definition, eq. (5.4.70),

$$\sigma_{\boldsymbol{x}_r}(c\widetilde{\boldsymbol{x}}) = \limsup_{t\to\infty} \frac{1}{t} \ln |c\widetilde{\boldsymbol{x}}| = \limsup_{t\to\infty} \frac{1}{t}(\ln |c| + \ln |\widetilde{\boldsymbol{x}}|) = \sigma_{\boldsymbol{x}_r}(\widetilde{\boldsymbol{x}})$$

The relation eq. (5.4.72) follows from the characteristics of linear vector spaces. Assume for example that

$$\max\{\sigma_{\boldsymbol{x}_r}(\widetilde{\boldsymbol{x}}_1), \sigma_{\boldsymbol{x}_r}(\widetilde{\boldsymbol{x}}_2)\} = \sigma_{\boldsymbol{x}_r}(\widetilde{\boldsymbol{x}}_1)$$

applies. There then exists an instant of time $\bar{t}$ so that, for all later instants $t > \bar{t}$, the following estimate can be established using the triangle inequality

$$\frac{1}{t} \ln |\widetilde{\boldsymbol{x}}_1 + \widetilde{\boldsymbol{x}}_2| \leqslant \frac{1}{t} \ln(|\widetilde{\boldsymbol{x}}_1| + |\widetilde{\boldsymbol{x}}_2|) = \frac{1}{t} \ln |\widetilde{\boldsymbol{x}}_1| + \frac{1}{t} \ln\left(1 + \frac{|\widetilde{\boldsymbol{x}}_2|}{|\widetilde{\boldsymbol{x}}_1|}\right)$$

where $|\widetilde{\boldsymbol{x}}_1| \geqslant |\widetilde{\boldsymbol{x}}_2|$ holds. Proceeding to the limit $t \to \infty$, there immediately follows the characteristic *ii.*

The definition eq. (5.4.70) is still unsatisfactory in two respects. Firstly, it would he preferable to replace the *limes superior*, which can still experience oscillations (Haken, 1987), with the standard definition of *limes*. Secondly, due to the relations eqs. (5.4.67) and (5.4.68), σ does not only depend on the direction of the initial perturbation $\widetilde{\boldsymbol{x}}(t_0)$ but also on the selected reference trajectory $\boldsymbol{x}_r(t)$. An answer to the question regarding the conditions under which the limiting value exists and – with only few exceptions – is independent of the selected initial trajectory is provided by the ergodic theory, an important part of statistical mechanics, and particularly by the multiplicative ergodic theory of Oseledec (Oseledec, 1968); see also (Benettin et al., 1980; Shimada and Nagashima, 1979).

There are basically two approaches when defining the long-term behaviour of dynamical systems. Up to now, we have used the *global geometrical* characteristics of the solutions of differential equations and the theory of linear stability, i.e. we have investigated the geometrical structure of attractors respectively the appertaining trajectories. If we wish to study very complex, higher-dimensional systems, we soon reach the limits of our imagination. In order to extract meaningful information from the long-term evolution of dynamical systems in these cases as well and in order to neglect the influences of non-typical trajectories, Oseledec and Ruelle *et al.* (Oseledec, 1968; Eckmann and Ruelle, 1985; Ruelle, 1989) applied the ergodic theory and Shaw (Shaw, 1981b) methods of the information theory. In both cases, our fundamental consideration is addressed to the characteristics of the invariant natural measure or probability measure μ, which is a measure for the density of the points respectively trajectories in the phase space, cf. section 3.10. We shall return to this concept in section 5.5.3 in connection with entropy respectively the information dimension. However, it would go beyond the scope of this book were we to deal in detail with the mathematical background of the ergodic theory. We refer the interested reader to (Sinai, 1977; Cornfeld et al., 1982), for example.

If the motion in one part of the phase space is ergodic, then, according to the ergodic theorem, section 3.10.2, the temporal mean value over an individual trajectory can be replaced by the mean value over the spatial distribution where the invariant measure μ acts like a weight function. This means that the total information of the dynamical system is already contained in (almost) every arbitrarily selected trajectory. For a large category of dynamical systems, Oseledec was able to prove (Oseledec, 1968) that the limit in the definition eq. (5.4.70) exists for σ, is finite and assumes the same value for almost all reference trajectories $\boldsymbol{x}_r(t)$. (Exceptional trajectories, such as, for example, those which end in an unstable fixed point for $t \to -\infty$, possess other Lyapunov exponents, but are not generally typical for the attractor and are of measure zero.) This is evident since the influence of the transient phase is eliminated by the definition of eq. (5.4.70). If we restrict our attention to the basin of attraction of an attractor, then (almost) all trajectories end in this attractor for $t \to +\infty$ so that the choice of the reference trajectory is arbitrary. Hence, the Lyapunov exponents do not only characterise the stability behaviour of single orbits, but that of the whole dynamical system respectively the corresponding attractor. Thus, in the following, we can dispense with the index $\boldsymbol{x}_r$ for σ.

Let us return to the two characteristics of eqs. (5.4.71) and (5.4.72) for the Lyapunov exponent σ, cf. (Benettin et al., 1980). It follows directly from these relations that all vectors $\widetilde{\boldsymbol{x}}$ for which the Lyapunov exponent is smaller than or equal to a given real number r, i.e. $\sigma(\widetilde{\boldsymbol{x}}) \leqslant r$, form a subspace of the vector space $\mathbb{E}_0$. From this, it can be further deduced that the Lyapunov exponent in dependence on $\widetilde{\boldsymbol{x}}$ can assume at most n different values where n= $\dim \mathbb{E}_0$ coincides with the dimension of the phase space. The n values are denoted as the spectrum of the Lyapunov exponents and can be ordered according to their magnitude

$$\sigma_1 \geqslant \sigma_2 \geqslant \ldots \geqslant \sigma_{\mathrm{n}} \tag{5.4.73}$$

We assume that this sequence contains precisely s different values which we denote with ν_{k}

$$\nu_1 > \nu_2 > \ldots > \nu_{\mathrm{s}} \qquad (1 \leqslant \mathrm{s} \leqslant \mathrm{n}) \tag{5.4.74}$$

Following the above considerations, there appertains to each ν_{k} a linear vector space

$$\mathbb{L}_{\mathrm{k}} = \{\widetilde{\boldsymbol{x}} \in \mathbb{E}_0 \quad \text{with} \quad \sigma(\widetilde{\boldsymbol{x}}) \leqslant \nu_{\mathrm{k}}\} \tag{5.4.75}$$

These s linear subspaces are nested, and we have

$$\mathbb{E}_0 = \mathbb{L}_1 \supset \mathbb{L}_2 \supset \ldots \supset \mathbb{L}_{\mathrm{s}}$$

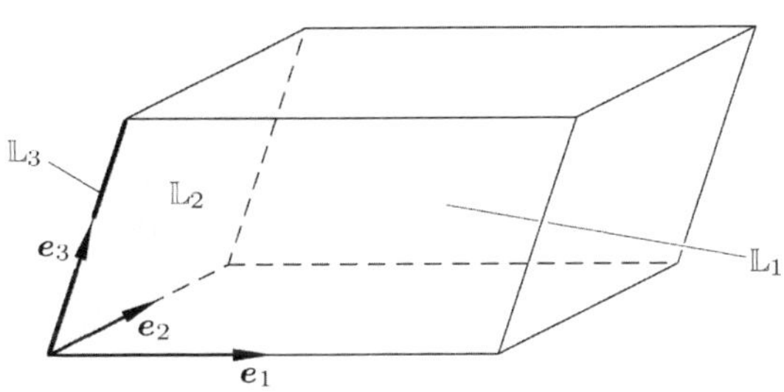

Fig. 5.4.10

Linear subspaces and basis vectors for the calculation of the Lyapunov exponents

We can then determine a basis $\{\boldsymbol{e}_1, \boldsymbol{e}_2, \ldots, \boldsymbol{e}_{\mathrm{n}}\}$ in $\mathbb{E}_0$ in such a way that the basis vectors span the subspaces (see fig. 5.4.10). Obviously, this basis is not uniquely defined; it only has to satisfy the condition that for perturbations in the direction of the basis vectors,

$$\sigma(\boldsymbol{e}_{\mathrm{i}}) = \sigma_{\mathrm{i}} \qquad (\mathrm{i} = 1, \ldots, \mathrm{n}) \tag{5.4.76}$$

holds. Starting from the subspace $\mathbb{L}_{\mathrm{s}}$ with the lowest dimension, we can select the basis vectors in particular in such a way that they form an orthonormal system. It is now easy to deduce that, in general, for any perturbation, the largest Lyapunov exponent σ_1 emerges. To this end, we express the perturbation in the form

$$\widetilde{\boldsymbol{x}} = c_1 \boldsymbol{e}_1 + c_2 \boldsymbol{e}_2 + \cdots + c_{\mathrm{n}} \boldsymbol{e}_{\mathrm{n}} \tag{5.4.77}$$

and, due to eqs. (5.4.72) and (5.4.76), we obtain $\sigma(\widetilde{\boldsymbol{x}}) = \sigma_1$. We can recognise from fig. 5.4.10 that $\sigma_2 < \sigma_1$, for example, would only emerge if $\widetilde{\boldsymbol{x}} \in \mathbb{L}_2$ holds respectively $\sigma_3 < \sigma_2$ only if $\widetilde{\boldsymbol{x}} \in \mathbb{L}_3$ holds. However, this can never be attained exactly in numerical calculations on which we must rely when determining the Lyapunov exponents; this will be described in more detail in section 5.4.6. Inevitable numerical errors will always generate a small component in $\boldsymbol{e}_1$-direction which is then amplified exponentially.

The question thus arises how it is possible to determine the other Lyapunov exponents $\sigma_\mathrm{i} < \sigma_1$ ($\mathrm{i} = 2, 3, \ldots, \mathrm{n}$) at all. To this end, we introduce the concept of the *Lyapunov exponents of p-th order*, proposed by Oseledec (Oseledec, 1968). The quantity σ defined in eq. (5.4.70) is designated as the first-order Lyapunov exponent. In this context, we observed the mean expansion or contraction in one direction, i.e. along a vector. Now, in the case of the generalised concept of Lyapunov exponents of p-th order, we study the mean exponential growth rate of the volume V_p of a p-dimensional parallelepiped in the tangent space $\mathbb{E}_0$ and define

$$\sigma_{\boldsymbol{x}_r}^{(\mathrm{p})}(V_\mathrm{p}) = \limsup_{t\to\infty} \frac{1}{t} \ln \frac{V_\mathrm{p}(t)}{V_\mathrm{p}(0)} \tag{5.4.78}$$

We again assume that the limit exists, is finite and assumes the same value for almost all trajectories appertaining to the basin of attraction of an attractor so that we can again dispense with the index $\boldsymbol{x}_r$. If we consider a parallelpiped spanned by the basis vectors $\boldsymbol{e}_1$, $\boldsymbol{e}_2$, ... $\boldsymbol{e}_\mathrm{p}$, there applies

$$\sigma^{(\mathrm{p})}(V_\mathrm{p}) = \sigma(\boldsymbol{e}_1) + \sigma(\boldsymbol{e}_2) + \cdots + \sigma(\boldsymbol{e}_\mathrm{p}) \tag{5.4.79}$$

i.e. the Lyapunov exponent of p-th order is equal to the sum of the corresponding σ-values of first order. This relation forms the basis for the calculation of the Lyapunov exponents of p-th order (Benettin et al., 1980). Depending on the choice of the p basis vectors, there exist for $\mathrm{p} < \mathrm{n}$ alternative possibilities for spanning p-dimensional parallelepipeds. For $\mathrm{p} = 2$ and $\mathrm{n} = 3$, for example, three possibilities arise (see fig. 5.4.10), namely the three lateral faces of the parallelepiped; thus

$$\sigma^{(2)}(V_2) = \text{one of the values from the set } \{\sigma_1 + \sigma_2,\ \sigma_2 + \sigma_3,\ \sigma_3 + \sigma_1\}$$

Here again, we observe that, in the course of a numerical calculation, the largest possible value emerges so that $\sigma^{(\mathrm{p})}(V_\mathrm{p})$ is seen to depend only on the dimension p, not on the arbitrarily selected parallelepiped. We can thus express eq. (5.4.79) in the preciser form

$$\sigma^{(\mathrm{p})} = \sigma_1 + \sigma_2 + \cdots + \sigma_\mathrm{p} \tag{5.4.80}$$

It follows that if we know the Lyapunov exponents of order for $\mathrm{p} = 1, \ldots, \mathrm{n}$, we can determine all the Lyapunov exponents of first order successively from the following set of equations

$$\begin{aligned}
\sigma^{(1)} &= \sigma_1 \\
\sigma^{(2)} &= \sigma_1 + \sigma_2 \quad \Rightarrow \quad \sigma_2 = \sigma^{(2)} - \sigma^{(1)} \\
&\vdots \\
\sigma^{(\mathrm{n})} &= \sigma_1 + \sigma_2 + \cdots + \sigma_\mathrm{n} \quad \Rightarrow \quad \sigma_\mathrm{n} = \sigma^{(\mathrm{n})} - \sigma^{(\mathrm{n}-1)}
\end{aligned} \tag{5.4.81}$$

We can now draw several important conclusions from the characteristics of the Lyapunov exponents:

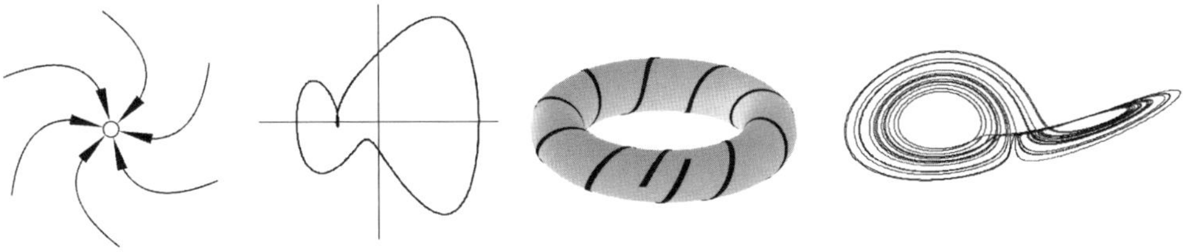

Fig. 5.4.11: Characterisation of the attractor types in the three-dimensional phase space by means of the Lyapunov exponents

i. We can show that for all trajectories not ending in a fixed point, at least one Lyapunov exponent is zero (Benettin et al., 1980; Haken, 1983). To prove this, we observe a perturbation along a tangent to the trajectory $\widetilde{\boldsymbol{x}}(t) = \dot{\boldsymbol{x}}(t)$.

ii. The Lyapunov exponents enable us to distinguish between the various attractors. Figure 5.4.11 presents the possible attractor types for a three-dimensional phase space as well as the corresponding Lyapunov exponents $(\sigma_1, \sigma_2, \sigma_3)$. A fixed point possesses three negative Lyapunov exponents; a limit cycle is characterised by $\sigma_1 = 0$, $\sigma_3 < \sigma_2 < 0$ whereas, for a torus, we obtain $\sigma_1 = \sigma_2 = 0$, $\sigma_3 < 0$. A chaotic, irregular motion on a strange attractor is characterised by one positive Lyapunov exponent $\sigma_1 > 0$ as well as by $\sigma_2 = 0, \sigma_3 < 0$.

A positive Lyapunov exponent $\sigma_1 > 0$ characterises the exponential divergence of the trajectories. Since a strange attractor occupies only a restricted part of the phase space, however, exponential growth is not possible in all directions; contractions and folding must of necessity occur. This behaviour is reflected in a negative Lyapunov exponent $\sigma_3 < 0$. As trajectories which were originally neighbouring diverge exponentially, a chaotic system proves very sensitive to small deviations in the initial conditions.

In multi-dimensional phase spaces, cases also occur in which two or more Lyapunov coefficients are positive. We then occasionally speak of hyperchaos; cf. (Rössler, 1979b; Brun, 1989).

iii. The spectrum of the Lyapunov exponents can be used to reveal instabilities and bifurcations of the system in dependence of a control parameter. If a stable torus evolves into a strange attractor, for example, there is a critical point at which the spectrum of Lyapunov exponents switches from $\{0, 0, -\}$ to $\{+, 0, -\}$.

iv. In section 5.1, we discussed the volume contraction of dissipative systems which, in accordance with eqs. (5.1.8, 5.1.11), is defined by div $\boldsymbol{F}$ (cf. also equ. (4.1.30), Liouville theorem)

$$\frac{\mathrm{d}V(t)}{\mathrm{d}t} = \int \mathrm{div}\, \boldsymbol{F}(\boldsymbol{x}) \mathrm{d}V$$

If, for example, div $\boldsymbol{F} = C = \text{const} < 0$, there follows

$$\frac{\mathrm{d}V(t)}{\mathrm{d}t} = C \cdot V(t) \qquad \text{or} \qquad V(t) = V(0)\,\mathrm{e}^{Ct}$$

From this, the Lyapunov exponent of n-th order can be deduced

$$\sigma^{(\mathrm{n})} = \lim_{t\to\infty} \frac{1}{t} \ln |V(t)| = C = \operatorname{div} \boldsymbol{F} \tag{5.4.82}$$

For an arbitrary dissipative system, however, div $\boldsymbol{F} < 0$ is in general not constant. The exponent $\sigma^{(\mathrm{n})}$ can then be interpreted as the temporal mean value

$$\sigma^{(\mathrm{n})} = \lim_{t\to\infty} \frac{1}{t} \int_0^t \operatorname{div} \boldsymbol{F}(\boldsymbol{x}(\tau))\mathrm{d}\tau = \sigma_1 + \sigma_2 + \cdots + \sigma_\mathrm{n} \tag{5.4.83}$$

Dissipative systems are hence characterised by the fact that the sum of their Lyapunov exponents is negative

$$\sigma^{(\mathrm{n})} = \sigma_1 + \sigma_2 + \cdots + \sigma_\mathrm{n} < 0 \tag{5.4.84}$$

Let us consider, for example, the Lorenz system eq. (5.2.1). For the divergence, we obtain a constant negative value

$$\operatorname{div} \boldsymbol{F} = \frac{\partial F_1}{\partial X} + \frac{\partial F_2}{\partial Y} + \frac{\partial F_3}{\partial Z} = -\sigma - 1 - b = \sigma^{(3)} \tag{5.4.85}$$

(here, σ is the Prandtl number!). Since $\sigma_2 = 0$, it suffices in the present case to calculate the largest Lyapunov exponent σ_1. From eqs. (5.4.80) and (5.4.85) then follows

$$\sigma_3 = -(\sigma + 1 + b) - \sigma_1 \tag{5.4.86}$$

v. In section 4.1, we demonstrated the volume preservation of conservative systems, using for this purpose the Liouville theorem, eq. (4.1.30). From eq. (5.4.83), due to div $\boldsymbol{F} = 0$, there thus immediately follows

$$\sigma^{(\mathrm{n})} = \sigma_1 + \sigma_2 + \cdots + \sigma_\mathrm{n} = 0 \tag{5.4.87}$$

If the equations of motion exist in the canonical form eq. (4.1.25)

$$\dot{\boldsymbol{p}} = -\frac{\partial H}{\partial \boldsymbol{q}^\mathrm{t}} \quad , \qquad \dot{\boldsymbol{q}} = \frac{\partial H}{\partial \boldsymbol{p}^\mathrm{t}}$$

we can prove, cf. (Benettin et al., 1980), the appearance of pairs of Lyapunov exponents with opposite signs

$$\{\sigma_1, \sigma_2, \ldots, \sigma_{f-1}, \sigma_f, -\sigma_f, -\sigma_{f-1}, \ldots, -\sigma_2, -\sigma_1\} \tag{5.4.88}$$

For periodic motions, this is immediately clear and is a direct result of the symplectic structure of the Poincaré map of Hamilton systems, see section 4.5, conclusion (b). In this case, all the eigenvalues of the corresponding Jacobian matrix can be ordered in pairs, the product of which is 1.

The symmetry in the Hamilton equations is thus directly reflected in the Lyapunov exponents. In accordance with *i.*, two Lyapunov exponents must have the value zero in this case. The corresponding perturbation directions are firstly the direction of the tangent to the trajectory and secondly a further direction on the hypersurface of constant energy on which all trajectories run.

5.4.5 Lyapunov Exponents of n-Dimensional Discrete Systems

By analogy with the definition of the Lyapunov exponents for continuous systems, we now briefly introduce the Lyapunov exponents for n-dimensional discrete systems. In section 5.4.3, we defined the Lyapunov exponent for one-dimensional maps. We now consider an n-dimensional discrete system

$$\boldsymbol{x}_{\mathrm{k}+1} = \boldsymbol{f}(\boldsymbol{x}_{\mathrm{k}}) \tag{5.4.89}$$

and investigate the stability behaviour of a sequence of points $\{\boldsymbol{x}_{r\mathrm{k}}\}$ by again observing the evolution of a small perturbation $\widetilde{\boldsymbol{x}}_{\mathrm{k}}$ for $\mathrm{k} \to \infty$

$$\boldsymbol{x}_{\mathrm{k}} = \boldsymbol{x}_{r\mathrm{k}} + \widetilde{\boldsymbol{x}}_{\mathrm{k}} \tag{5.4.90}$$

In analogy to eqs. (5.4.11, 5.4.59), we are led for a linear stability analysis to the following set of linear homogeneous equations

$$\widetilde{\boldsymbol{x}}_{\mathrm{k}+1} = \boldsymbol{D}_{\mathrm{k}}\widetilde{\boldsymbol{x}}_{\mathrm{k}} \tag{5.4.91}$$

where the Jacobian matrix

$$\boldsymbol{D}_{\mathrm{k}} = \left.\frac{\partial \boldsymbol{f}}{\partial \boldsymbol{x}}\right|_{\boldsymbol{x}_{r\mathrm{k}}} \tag{5.4.92}$$

depends on the k-th iteration of the reference point $\boldsymbol{x}_{r\mathrm{k}}$. Successive application of eq. (5.4.91) yields the product expression

$$\widetilde{\boldsymbol{x}}_{\mathrm{k}+1} = \boldsymbol{D}_{\mathrm{k}}\,\boldsymbol{D}_{\mathrm{k}-1}\ldots\boldsymbol{D}_1\widetilde{\boldsymbol{x}}_1 \tag{5.4.93}$$

Corresponding to eq. (5.4.70), we can define the Lyapunov exponent of the perturbation vector $\widetilde{\boldsymbol{x}}_{\mathrm{k}}$ appertaining to the sequence of reference point $\{\boldsymbol{x}_{r\mathrm{k}}\}$ as

$$\sigma_{\boldsymbol{x}_r}(\widetilde{\boldsymbol{x}}_1) = \limsup_{\mathrm{k}\to\infty} \frac{1}{\mathrm{k}} \ln \frac{|\widetilde{\boldsymbol{x}}_{\mathrm{k}+1}|}{|\widetilde{\boldsymbol{x}}_1|} \tag{5.4.94}$$

As in the previous section, a limit exists for a class of maps and is in general independent of the selected sequence of points $\{\boldsymbol{x}_{r\mathrm{k}}\}$. In dependence on the initial perturbations $\widetilde{\boldsymbol{x}}_1$, there again arises a maximum of n different Lyapunov exponents which can be deduced from the n eigenvalues $\lambda_{\mathrm{i}}(\mathrm{k})$ $(\mathrm{i} = 1, \ldots, \mathrm{n})$ of the matrix

$$\boldsymbol{J}_{\mathrm{k}} = \boldsymbol{D}_{\mathrm{k}}\,\boldsymbol{D}_{\mathrm{k}-1}\ldots\boldsymbol{D}_1 \tag{5.4.95}$$

where both the eigenvalues as well as the directions of the corresponding eigenvectors vary from iteration to iteration. The Lyapunov exponents are given by

$$\sigma_{\mathrm{i}} = \lim_{\mathrm{k}\to\infty} \frac{1}{\mathrm{k}} \ln|\lambda_{\mathrm{i}}(\mathrm{k})| \quad (\mathrm{i} = 1, \ldots, \mathrm{n}) \tag{5.4.96}$$

If we interpret the n-dimensional map (5.4.89) as a Poincaré map of a (n+1)-dimensional continuous system, the perturbation tangent to the flow drops out for the discrete system and thus the corresponding zero Lyapunov exponent. If we denote the Lyapunov exponents of the (n+ 1)-dimensional continuous system by $\bar{\sigma}_{\mathrm{i}}$ and those of the corresponding Poincaré map by σ_{i}, we can establish the relation

$$\sigma_{\mathrm{i}} = \tau\bar{\sigma}_{\mathrm{i}} \tag{5.4.97}$$

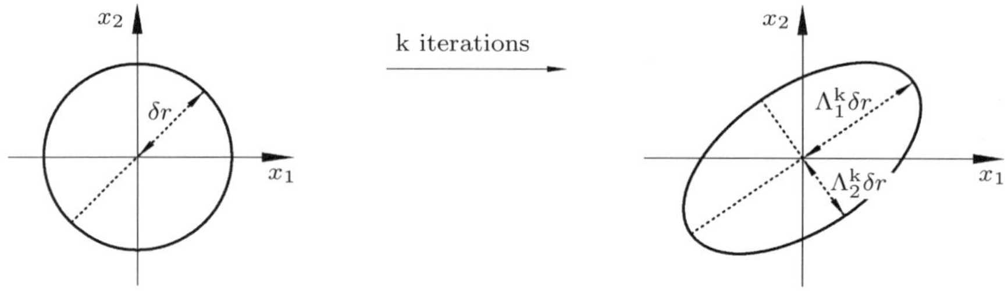

Fig. 5.4.12: Geometrical interpretation of the Lyapunov numbers Λ_1, Λ_2 of a 2d map

where τ is a mean time between two successive Poincaré sections (Eckmann and Ruelle, 1985).

Since in all cases the largest Lyapunov exponent σ_1 emerges irrespective of the adopted initial perturbation $\widetilde{\boldsymbol{x}}_1$, we again require the Lyapunov exponents of p-th order to determine the whole spectrum of the exponents. Here again, the exponents are defined by the mean exponential expansion or contraction of p-dimensional parallelepipeds of volume V_p, cf. eq. (5.4.78)

$$\sigma^{(\mathrm{p})} = \lim_{\mathrm{k}\to\infty} \frac{1}{\mathrm{k}} \ln \frac{V_\mathrm{p}(\mathrm{k})}{V_\mathrm{p}(1)} \tag{5.4.98}$$

In the literature, e.g. (Farmer et al., 1983), the concept of Lyapunov numbers $\Lambda_1, \ldots \Lambda_\mathrm{n}$ is often adopted in place of Lyapunov exponents. The Lyapunov exponents are simply the natural logarithms of the Lyapunov numbers. We thus immediately deduce from eq. (5.4.96)

$$\sigma_\mathrm{i} = \ln \Lambda_\mathrm{i} \qquad (\mathrm{i} = 1, \ldots, \mathrm{n}) \tag{5.4.99}$$

or

$$\Lambda_\mathrm{i} = \lim_{\mathrm{k}\to\infty} |\lambda_\mathrm{i}(\mathrm{k})|^{1/\mathrm{k}} \tag{5.4.100}$$

The Lyapunov numbers admit – as do the Floquet multipliers in section 5.4.2, fig. 5.4.4 – a simple geometrical interpretation. For a two-dimensional map, for example, Λ_1 and Λ_2 are the mean principal elongation ratios which, after k iterations, map a small circle of radius δr approximately on an ellipse with the semi-major axis $\Lambda_1^\mathrm{k} \delta r$ respectively semi-minor axis $\Lambda_2^\mathrm{k} \delta r$ (see fig. 5.4.12).

5.4.6 Numerical Calculation of Lyapunov Exponents

For a numerical determination of the Lyapunov exponents, we have to integrate the linear system eq. (5.4.67)

$$\dot{\widetilde{\boldsymbol{x}}}(t) = \left.\frac{\partial \boldsymbol{F}}{\partial \boldsymbol{x}}\right|_{\boldsymbol{x}_r(t)} \widetilde{\boldsymbol{x}}(t)$$

where the coefficient matrix depends on the reference trajectory $\boldsymbol{x}_r(t)$. In order to calculate $\boldsymbol{x}_r(t)$, we must thus simultaneously also integrate the initial non-linear system eq. (5.4.2)

$$\dot{\boldsymbol{x}}_r(t) = \boldsymbol{F}(\boldsymbol{x}_r(t))$$

An analytical solution of these differential equations is only possible in exceptional cases so that we generally have to rely on a numerical integration.

(a) *Determination of the largest Lyapunov exponent*

For given initial conditions $\boldsymbol{x}_r(t_0) = \boldsymbol{x}_{r0}$ and $\widetilde{\boldsymbol{x}}(t_0) = \widetilde{\boldsymbol{x}}_0$, we calculate the evolution of the perturbation $\widetilde{\boldsymbol{x}}(t)$ along the reference trajectory and determine its magnitude

$$d(t) = |\widetilde{\boldsymbol{x}}(t)| \tag{5.4.101}$$

The difficulty is that in the case of chaotic behaviour – which is of particular interest to us – $d(t)$ increases exponentially with time so that, in addition to numerical inaccuracies, an "overflow" can quickly occur in the computer. This problem can be avoided by using for the construction of the fundamental matrix $\boldsymbol{\Phi}(t, t_0)$ the characteristic transition property, eq. (5.4.69).

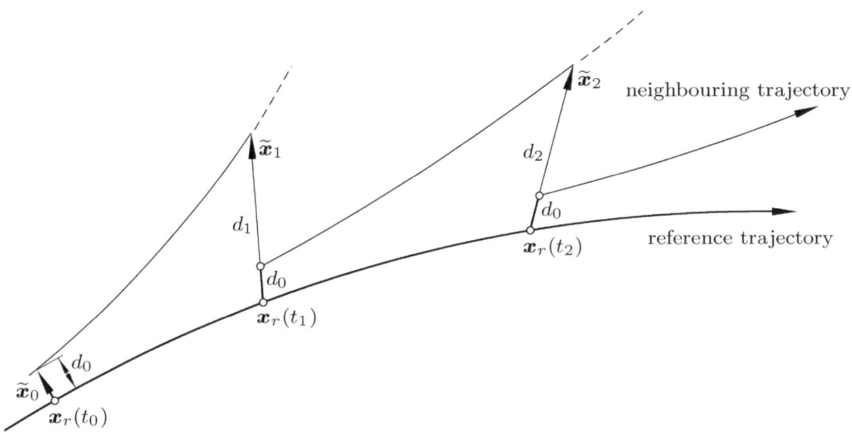

Fig. 5.4.13: Sequential normalisation procedure for the numerical calculation of σ_1

Selecting a fixed time increment Δt and designating the corresponding discrete moments in time as

$$t_{\mathrm{m}} = t_0 + \mathrm{m}\Delta t$$

we can reformulate the relation eq. (5.4.68) in the form of a k-fold product

$$\widetilde{\boldsymbol{x}}(t_{\mathrm{k}}) = \boldsymbol{\Phi}(t_{\mathrm{k}}, t_{\mathrm{k}-1})\boldsymbol{\Phi}(t_{\mathrm{k}-1}, t_{\mathrm{k}-2}) \ldots \boldsymbol{\Phi}(t_1, t_0)\widetilde{\boldsymbol{x}}(t_0) \tag{5.4.102}$$

Following each time increment Δt, we now restore the length of the vector $\widetilde{\boldsymbol{x}}$ to its original length $d_0 = |\widetilde{\boldsymbol{x}}_0|$ (see fig. 5.4.13) where, for convenience, we select $d_0 = 1$. We thus successively calculate the vectors

$$\begin{aligned}
\text{k}=1: \quad & \boldsymbol{\Phi}(t_1,t_0)\widetilde{\boldsymbol{x}}(t_0)=\widetilde{\boldsymbol{x}}_1 \\
& |\widetilde{\boldsymbol{x}}_1|=d_1,\ \bar{\widetilde{\boldsymbol{x}}}_1=\frac{\widetilde{\boldsymbol{x}}_1}{d_1} \\
\text{k}=2: \quad & \boldsymbol{\Phi}(t_2,t_1)\bar{\widetilde{\boldsymbol{x}}}_1=\widetilde{\boldsymbol{x}}_2 \\
& |\widetilde{\boldsymbol{x}}_2|=d_2,\ \bar{\widetilde{\boldsymbol{x}}}_2=\frac{\widetilde{\boldsymbol{x}}_2}{d_2}
\end{aligned} \tag{5.4.103}$$

etc. The length $|\widetilde{\boldsymbol{x}}(t_\text{k})|$ can then be deduced from eq. (5.4.102) as follows

$$\begin{aligned}
|\widetilde{\boldsymbol{x}}(t_\text{k})| &= |\boldsymbol{\Phi}(t_\text{k},t_{\text{k}-1})\boldsymbol{\Phi}(t_{\text{k}-1},t_{\text{k}-2})\ldots\underbrace{\boldsymbol{\Phi}(t_1,t_0)\widetilde{\boldsymbol{x}}(t_0)}_{=d_1\bar{\widetilde{\boldsymbol{x}}}_1}| \\
&= |\boldsymbol{\Phi}(t_\text{k},t_{\text{k}-1})\boldsymbol{\Phi}(t_{\text{k}-1},t_{\text{k}-2})\ldots\underbrace{\boldsymbol{\Phi}(t_2,t_1)\bar{\widetilde{\boldsymbol{x}}}_1}_{=d_2\bar{\widetilde{\boldsymbol{x}}}_2}|d_1 \\
&\ \ \vdots \\
&= |\underbrace{\boldsymbol{\Phi}(t_\text{k},t_{\text{k}-1})\bar{\widetilde{\boldsymbol{x}}}_{\text{k}-1}}_{=d_\text{k}\bar{\widetilde{\boldsymbol{x}}}_\text{k}}|\,d_{\text{k}-1}d_{\text{k}-2}\ldots d_2d_1
\end{aligned}$$

Altogether, observing the relation $|\bar{\widetilde{\boldsymbol{x}}}_\text{k}|=1$, we thus obtain

$$|\widetilde{\boldsymbol{x}}(t_\text{k})|=\prod_{\text{i}=1}^{\text{k}}d_\text{i} \tag{5.4.104}$$

If a sufficiently small time increment Δt is selected, we can avoid an "overflow" on the computer and derive in accordance with eq. (5.4.70) the largest Lyapunov exponent

$$\sigma_1=\lim_{t\to\infty}\frac{1}{t}\ln|\widetilde{\boldsymbol{x}}(t)|=\lim_{\text{k}\to\infty}\frac{1}{\text{k}\Delta t}\ln\prod_{\text{i}=1}^{\text{k}}d_\text{i}$$

This yields the simple formula

$$\sigma_1=\lim_{\text{k}\to\infty}\frac{1}{\text{k}\Delta t}\sum_{\text{i}=1}^{\text{k}}\ln d_\text{i} \tag{5.4.105}$$

For practical calculations, it is expedient to integrate the 2n equations

$$\begin{aligned}
&\dot{\boldsymbol{x}}_r=\boldsymbol{F}\boldsymbol{x}_r && \text{with} && \boldsymbol{x}_r(t_0)=\boldsymbol{x}_{r0} \\
&\dot{\widetilde{\boldsymbol{x}}}=\left.\frac{\partial \boldsymbol{F}}{\partial \boldsymbol{x}}\right|_{\boldsymbol{x}_r(t)}\widetilde{\boldsymbol{x}} && \text{with} && \widetilde{\boldsymbol{x}}(t_0)=\widetilde{\boldsymbol{x}}_0
\end{aligned} \tag{5.4.106}$$

simultaneously, where the initial instant t_0 should be chosen so that transient modes have already faded away. For a fixed time increment, we deduce the lengths d_k (k $=1,2,\ldots$) from eq. (5.4.103) and obtain successive approximations for the largest Lyapunov exponent

$$\sigma_{1,\text{k}}=\frac{1}{\text{k}\Delta t}\sum_{\text{i}=1}^{\text{k}}\ln d_\text{i} \tag{5.4.107}$$

The (k + 1)-th approximation $\sigma_{1,k+1}$ is established recursively from the preceding ones as follows

$$\sigma_{1,k+1} = \frac{1}{(k+1)\Delta t}\left(\sum_{i=1}^{k} \ln d_i + \ln d_{k+1}\right)$$

or

$$\sigma_{1,k+1} = \frac{k}{k+1}\sigma_{1,k} + \frac{1}{(k+1)\Delta t}\ln d_{k+1} \tag{5.4.108}$$

An account of specific computer experiments is given in section 10.3, fig. 10.3.3, using the Lorenz system as an example.

(b) *Determination of all Lyapunov exponents*

In section 5.4.4, we already pointed out that, for the numerical calculation of all first-order Lyapunov exponents $\sigma_1, \ldots, \sigma_n$, we first have to determine all Lyapunov exponents of p-th order $\sigma^{(p)}$ ($p = 1, \ldots, n$). Here, too, the problem of "overflow" may arise in the case of an exponential increase of the perturbations. We also meet with an additional difficulty. In accordance with eq. (5.4.77), each vector can be represented as a linear combination of the vectors $\boldsymbol{e}_i$ of an orthonormal basis, i.e.

$$\tilde{\boldsymbol{x}}(0) = c_1\boldsymbol{e}_1 + c_2\boldsymbol{e}_2 + \ldots + c_n\boldsymbol{e}_n \tag{5.4.109}$$

where we presume $c_1 \neq 0$. For extended intervals of time t, the perturbation vector $\tilde{\boldsymbol{x}}(t)$ can be represented with the aid of the approximate values $\tilde{\sigma}_i \simeq \sigma_i$ for the Lyapunov exponents in the form

$$\begin{aligned}\tilde{\boldsymbol{x}}(t) &\approx c_1 e^{\tilde{\sigma}_1 t}\boldsymbol{e}_1 + c_2 e^{\tilde{\sigma}_2 t}\boldsymbol{e}_2 + \ldots + c_n e^{\tilde{\sigma}_n t_1}\boldsymbol{e}_n \\ &= c_1 e^{\tilde{\sigma}_1 t}\Big(\boldsymbol{e}_1 + \underbrace{\frac{c_2}{c_1}e^{(\tilde{\sigma}_2-\tilde{\sigma}_1)t}\boldsymbol{e}_2 + \ldots + \frac{c_n}{c_1}e^{(\tilde{\sigma}_n-\tilde{\sigma}_1)t}\boldsymbol{e}_n}_{\to 0 \quad \text{for} \quad t\to\infty}\Big)\end{aligned} \tag{5.4.110}$$

Were $c_1 = 0$ for $t = 0$, numerical inaccuracies would soon generate a small component in $\boldsymbol{e}_1$-direction which increases exponentially and becomes dominating after a short interval. In agreement with the characteristic rule of eq. (5.4.72), this means that each vector will become parallel to $\boldsymbol{e}_1$ in the course of time. Thus, if two vectors are subjected to the dynamics of the system, the angle enclosed by the two vectors will tend towards zero for increasing t. This means that the parallelepipeds observed for the calculation of the $\sigma^{(p)}$ collapse like a house of cards.

Yet there is a way out of this difficulty, as described in detail in (Benettin et al., 1980), part II, and (Shimada and Nagashima, 1979). In addition to the aforementioned renormalisation, after each fixed period of time Δt, the vectors spanning the parallelepiped are replaced by a system of orthonormalised vectors again spanning the same subspace. A classic construction procedure is the orthonormalisation procedure according to Gram-Schmidt (Golub and van Loan, 1983) which can be illustrated graphically for the three-dimensional case (see fig. 5.4.14).

We start from three linearly independent vectors $\{\boldsymbol{f}_1 \ \boldsymbol{f}_2 \ \boldsymbol{f}_3\}$. Step by step, we then construct a new orthonormalised basis $\{\boldsymbol{e}_1 \ \boldsymbol{e}_2 \ \boldsymbol{e}_3\}$ which spans the same space

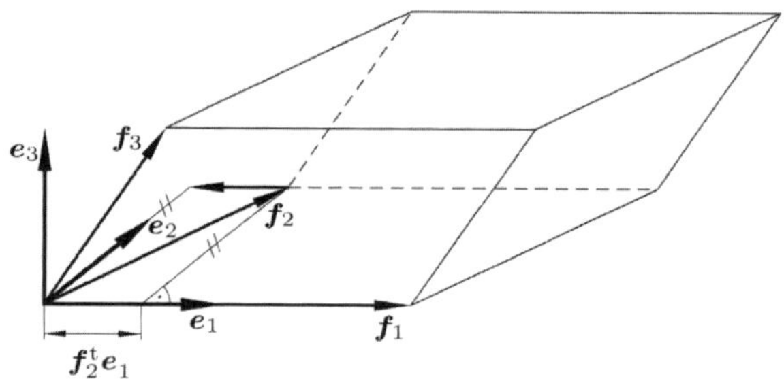

Fig. 5.4.14: Gram-Schmidt orthonormalisation procedure for n=3

$$e_1 = \frac{f_1}{|f_1|} \tag{5.4.111}$$

$$\bar{e}_2 = f_2 + c_{21} e_1 \quad , \qquad e_2 = \frac{\bar{e}_2}{|\bar{e}_2|} \tag{5.4.112}$$

where c_{21} is selected so that $\bar{e}_2 \perp e_1$, i.e.

$$c_{21} = -f_2^t e_1 \tag{5.4.113}$$

Correspondingly, we adopt an ansatz for $\bar{e}_3$ with one non-vanishing component in the f_3-direction and two components in the e_1-respectively e_2-direction

$$\bar{e}_3 = f_3 + c_{31} e_1 + c_{32} e_2 \quad , \qquad e_3 = \frac{\bar{e}_3}{|\bar{e}_3|} \tag{5.4.114}$$

The constants c_{31} and c_{32} are selected so that the conditions $\bar{e}_3 \perp e_1$ and $\bar{e}_3 \perp e_2$ are ensured. Due to the orthonormality of $\{e_1, e_2\}$, we obtain simply

$$c_{31} = -f_3^t e_1 \quad , \qquad c_{32} = -f_3^t e_2 \tag{5.4.115}$$

This construction procedure can be easily progressively continued. In the case of a p-dimensional basis $\{f_1, \ldots, f_p\}$, which is to be replaced by an orthonormalised basis $\{e_1, \ldots, e_p\}$, the k-th step is

$$\begin{aligned} \bar{e}_k &= f_k + \sum_{i=1}^{k-1} c_{ki} e_i \quad \text{with} \quad c_{ki} = -f_k^t e_i \\ e_k &= \frac{\bar{e}_k}{|\bar{e}_k|} \end{aligned} \tag{5.4.116}$$

The calculation of the Lyapunov exponents of p-th order ($p = 1, \ldots, n$) can now be divided into the following steps:

i. select an orthonormal basis $\{e_1, \ldots, e_n\}$. In each case, $p \leqslant n$ vectors $e_1, \ldots, e_p$ span a p-dimensional hypercube $\mathbb{H}_p$;
ii. elect a fixed time increment Δt;
iii. subject the basis set $\{e_i\}$ to the flow of the dynamical system and determine by numerical integration after a time increment Δt the distorted set of vectors at the instant $t_k = t_0 + k\Delta t$

$$\boldsymbol{f}_{\mathrm{i}} = \boldsymbol{\Phi}(t_{\mathrm{k}}, t_{\mathrm{k}-1})\boldsymbol{e}_{\mathrm{i}}$$

iv. calculate the volume $V_{\mathrm{k}}^{(\mathrm{p})}$ of the parallelepiped spanned by $\boldsymbol{f}_1, \ldots, \boldsymbol{f}_{\mathrm{p}}$ for $\mathrm{p} = 1, \ldots, \mathrm{n}$;

v. according to the Gram-Schmidt orthonormalisation procedure, determine new basis vectors $\{\boldsymbol{e}_1, \ldots, \boldsymbol{e}_{\mathrm{n}}\}$ so that for $\mathrm{p} = 1, \ldots, \mathrm{n}$ each subset $\{\boldsymbol{e}_1, \ldots, \boldsymbol{e}_{\mathrm{p}}\}$ spans the same space as $\{\boldsymbol{f}_1, \ldots, \boldsymbol{f}_{\mathrm{p}}\}$, i.e.

$$\mathrm{span}\{\boldsymbol{e}_1, \ldots, \boldsymbol{e}_{\mathrm{p}}\} = \mathrm{span}\{\boldsymbol{f}_1, \ldots, \boldsymbol{f}_{\mathrm{p}}\}$$

Back to *iii.*, loop over k.

As in the case of the determination of the Lyapunov exponents of first order, we can, in accordance with eq. (5.4.104), represent the volume $V_{\mathrm{p}}(t_{\mathrm{k}})$ as the product of the volumes $V_{\mathrm{i}}^{(\mathrm{p})}$ calculated in step *iv.*

$$V_{\mathrm{p}}(t_{\mathrm{k}}) = \prod_{\mathrm{i}=1}^{\mathrm{k}} V_{\mathrm{i}}^{(\mathrm{p})} \tag{5.4.117}$$

The Lyapunov exponent of p-th order can now be calculated in accordance with eq. (5.4.78) applying the following scheme

$$\sigma^{(\mathrm{p})} = \lim_{\mathrm{k}\to\infty} \sigma_{\mathrm{k}}^{(\mathrm{p})} \qquad \text{with} \quad \sigma_{\mathrm{k}}^{(\mathrm{p})} = \frac{1}{\mathrm{k}\Delta t} \sum_{\mathrm{i}=1}^{\mathrm{k}} \ln V_{\mathrm{i}}^{(\mathrm{p})}$$

If we know the k-th approximation $\sigma_{\mathrm{k}}^{(\mathrm{p})}$, we can set up a recursion rule in analogy to eq. (5.4.108)

$$\sigma_{\mathrm{k}+1}^{(\mathrm{p})} = \frac{\mathrm{k}}{\mathrm{k}+1}\, \sigma_{\mathrm{k}}^{(\mathrm{p})} + \frac{1}{(\mathrm{k}+1)\Delta t} \ln V_{\mathrm{k}+1}^{(\mathrm{p})} \tag{5.4.118}$$

The calculation of the Lyapunov exponents for discrete systems is simplified considerably by the fact that numerical integrations are not necessary, but that only iterations have to be carried out. Instead of eqs. (5.4.2) and (5.4.67), the equations

$$\widetilde{\boldsymbol{x}}_{\mathrm{k}+1} = \left.\frac{\partial \boldsymbol{f}}{\partial \boldsymbol{x}}\right|_{\boldsymbol{x}_{\mathrm{k}}} \widetilde{\boldsymbol{x}}_{\mathrm{k}} \tag{5.4.119}$$

$$\boldsymbol{x}_{\mathrm{k}+1} = \boldsymbol{f}(\boldsymbol{x}_{\mathrm{k}})$$

are now iterated. If a renormalisation respectively orthonormalisation is performed after each iteration step, we obtain equations corresponding to eqs. (5.4.105) and (5.4.118) once the orbital period Δt is specified and k denotes the current iteration.

With the aid of the Lyapunov exponents, a number of important statements can be made on dynamical systems. In section 5.5.6, we shall, on the one hand, discuss the connection between the Lyapunov exponents and the so-called Lyapunov dimension of the attractor; on the other hand, in section 5.6.3, we shall estimate the Kolmogorov-Sinai entropy with the help of the positive Lyapunov exponents.

At this point, we should like to point out a third application. The knowledge of the largest Lyapunov exponent σ_1 of a chaotic motion allows us to make an estimation

– albeit very rough – of the so-called relaxation time t^*. This is the time after which there is no longer any correlation between the initial condition $\boldsymbol{x}(0)$ and the current state $\boldsymbol{x}(t)$, i.e. t^* is the time after which all information about the position of the initial point in the phase space has been lost. In practice, the initial value $\boldsymbol{x}_r(0)$ of a reference trajectory $\boldsymbol{x}_r(t)$ can only be specified to a finite number of digits, i.e. $\boldsymbol{x}_r(0)$ contains an error ε. Let us now consider a neighbouring trajectory $\boldsymbol{x}(t)$ with the initial condition $\boldsymbol{x}(0) = \boldsymbol{x}_r(0) + \delta\boldsymbol{x}(0)$ where we presume $|\delta\boldsymbol{x}(0)| < \varepsilon$. We have already demonstrated the sensitive dependence on small errors in the inital conditions using the Lorenz attractor as an example (see Colour Plate Id, p. 194). Two trajectories, originally close to one another, remain close for some time. The local instability of the system, reflected in a positive Lyapunov exponent σ_1, leads, however, to an exponential divergence of the two trajectories which finally reach domains of the attractor far apart from one another. If we designate a characteristic length of the attractor in the phase space by L, the relaxation time t^* can be roughly estimated. If the initial conditions contain an error $|\delta\boldsymbol{x}(0)|$, this error increases exponentially

$$|\delta\boldsymbol{x}(t)| \sim |\delta\boldsymbol{x}(0)|\mathrm{e}^{\sigma_1 t}$$

If we disregard folding, all the information about the initial conditions is lost as soon as

$$|\delta\boldsymbol{x}(0)|\mathrm{e}^{\sigma_1 t^*} \sim L$$

applies. For the relaxation time, i.e. the period for which short-term predictions are possible, we thus obtain

$$t^* \sim \frac{1}{\sigma_1} \ln \frac{L}{|\delta\boldsymbol{x}(0)|} \tag{5.4.120}$$

This estimate shows that the smaller $|\delta\boldsymbol{x}(0)|$ is, i.e. the more accurately the initial value can be localised, the larger t^* becomes. On the other hand, the larger σ_1 is, i.e. the stronger the local divergence characteristics of the system are, the smaller t^* will be.

5.5 Dimensions

The long-term behaviour of dissipative systems is characterised by attractors of the most varied types if the trajectories are recurrent and do not tend towards infinity. After a transient phase, in which some modes of motion fade away and finally vanish due to damping, the state of the system approaches an attractor where the number of independent variables which determine the dimension of the phase space is generally reduced considerably. This means that for the description of the long-term behaviour, few, but essential variables are sufficient, certainly fewer than are required to describe the initial state. In section 5.1, we discovered the reason for this. There, we showed that the contraction of a volume element in the phase space is characteristic for dissipative systems and that hence, each attractor in its respective phase space must have zero volume.

Following our characterisation of various attractors with respect to their stability behaviour in section 5.4, we now ask ourselves how we can achieve a further classification of the different attractor types using the concept of dimension. Due to the non-intersection constraint of the trajectories, only regular attractors, namely fixed point and limit cycle, can occur in a two-dimensional phase space. In the three-dimensional phase space, motions with two independent frequencies lead to quasi-periodic long-term behaviour described by two-dimensional tori. Since the non-intersection feature in three-dimensional space does not impose such a severe restriction as on a plane, a completely new attractor type can emerge, such as the Lorenz attractor, for example, a complex, tangled, filigree-like structure which no longer forms a surface or manifold. These strange attractors which reflect chaotic, irregular behaviour can, however, never fill the whole phase space due to the contraction of the volume elements. The only way out of this apparently paradox situation is that the strange attractor in a three-dimensional phase space is a hybrid form between a surface and a spatial object, thus of necessity possessing a dimensionality between 2 and 3.

In order to characterise it, we therefore require a generalised concept of dimension which, in the case of the regular attractors fixed point, limit cycle and torus, coincides with the conventional dimension originating in Euclidean geometry – i.e. 0, 1 or 2 – and yet provides a non-integer quantity in the case of strange attractors. Classical geometry, established didactically more than 2000 years ago by Euclid, deals with regular structures (e.g. circles, triangles and spheres), i.e. objects of integer dimension, radical abstractions not to be observed in nature. Is it possible to imagine a complex structure of non-integer dimensionality?

Benoît Mandelbrot investigated this problem in connection with the question of the length and nature of the coast of Britain. The answer to this apparently so trivial question is by no means elementary. It emerges that measuring the length of the coast depends on the scale used. If we begin with a rough map (i.e. one with a large scale), we obtain our first approximation of the length. With more accurate cartographic material, in which more bays and headlands emerge, a greater total length will be recorded. The same is true if we walk around the island with a surveyor's staff in the hope of finally being able to determine the precise length. The length measured again depends on the size of the scale (see fig. 5.5.1). Although the coast surrounds a finite area of land, measurements of the length of the encircling line, carried out with increasing precision, do not approach a limit value but increase towards infinity. The reason is that for every scale, however small it may be, indentations and crumplings continuously arise, similar to those we already recorded on the larger scale. Expressed mathematically, at no point of the coast is its differentiability guaranteed.

Again, we have a paradox situation, this time taken directly from nature, from reality. Mandelbrot found the solution. The coast must be something between a line and a surface, a "monster", as he called it, of non-integer dimensionality. Indeed, a generalised dimensionality of about 1.2 can be assigned to the coast of Britain (Barnsley, 1988). For such complex structures of non-integer dimension which, moreover, can be described in different ways depending on the distance

Fig. 5.5.1: How long is the coast of Britain?

of the observer from the object, Mandelbrot coined the concept of the "fractal" (Mandelbrot, 1982). As an introduction, (Tél, 1988) can be recommended.

5.5.1 Cantor Set

Towards the end of the nineteenth century, the mathematician Georg Cantor (1845 -1918), the founder of the set theory, worked intensively on the concept of infinity. For example, questions such as the cardinality of sets occupied him. Although each interval – however small it may be – contains an infinite number of rational and irrational numbers, the rational numbers are countable whereas the irrational are not. In this connection, he constructed a set which fits into the framework of our topic. This set, called Cantor set in his honour, is a hybrid between an assembly of discrete points and a continuous line: it is constructed in the following way (see fig. 5.5.2).

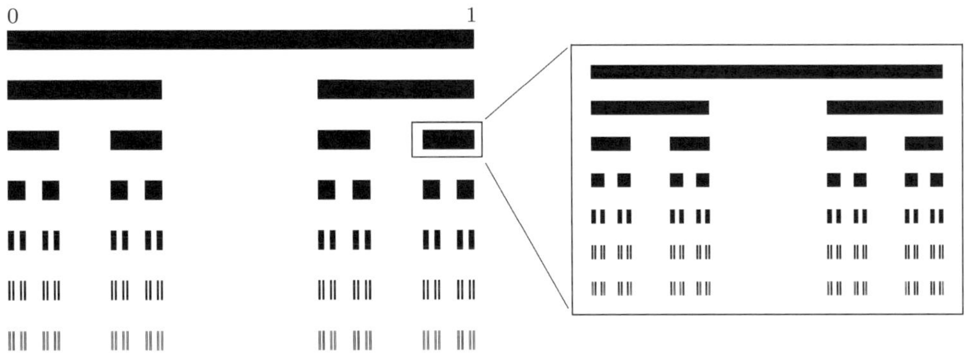

Fig. 5.5.2: Cantor set

We consider the closed interval [0,1] and remove the middle third (1/3, 2/3) without, however, removing the end points 1/3 and 2/3. We proceed in the same way with the two remaining intervals [0,1/3] and [2/3,1]: we remove the open middle third of each and obtain four closed sub-intervals. If we continue this procedure ad infinitum, the limit set results as an infinite, uncountable set of limit points which are not connected.

In order to obtain an idea of the dimensionality of the Cantor set, we calculate the length L of the intervals removed. From fig. 5.5.2, we deduce

$$L = 1 \cdot \frac{1}{3} + 2 \cdot \frac{1}{9} + 4 \cdot \frac{1}{27} + \cdots + 2^{\mathrm{n}-1} \cdot \frac{1}{3^{\mathrm{n}}} + \cdots = \frac{1}{3} \sum_{\mathrm{n}=0}^{\infty} \left(\frac{2}{3}\right)^{\mathrm{n}} \tag{5.5.1}$$

This, however, is a geometrical series of the form $\sum_{\mathrm{n}=0}^{\infty} x^{\mathrm{n}} = 1/(1-x)$ with $x = 2/3$ so that, as the length, $L = 1$ results, i.e. exactly the length of the initial interval. Hence, the Cantor set itself must have the length zero. Nevertheless, we can assign to it a generalised dimension which lies between that of a point and that of a line.

In the same way, we can, of course, construct sets with a dimension lying between that of a curve and that of a surface. An example is the Koch snowflake curve, an idealised model for demonstrating the problem of the length of the coast of Britain.

The construction of the Koch curve proceeds via an iteration rule as in the case of the Cantor set. Starting with an equilateral triangle, we divide each side into three equal sections, remove the middle third of each and put in its place a correspondingly smaller equilateral triangle (see fig. 5.5.3). If we repeat

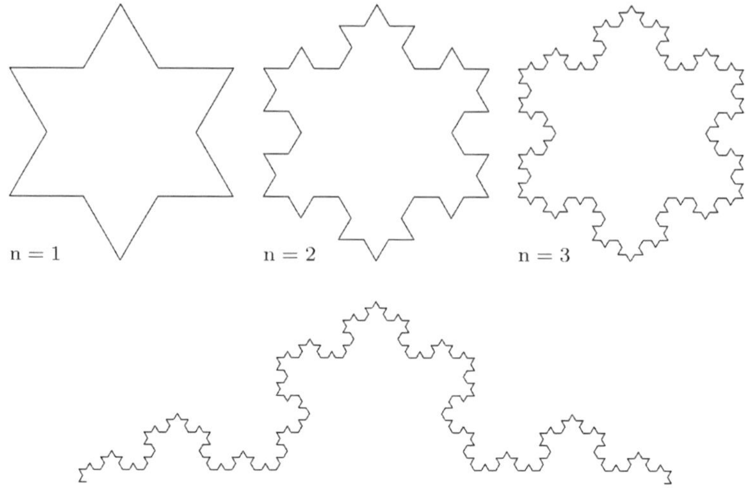

Fig. 5.5.3: On the construction of the Koch snowflake curve

this an infinite number of times, there emerges in the limit an infinitely jagged continuous curve which is not differentiable anywhere and possesses infinite length while enclosing a finite area. In the following, we shall see that we can assign a non-integer dimension between 1 and 2 to this fractal structure.

Before we introduce a more generalised dimension concept, we first wish to point out an important fundamental characteristic of the Cantor set and the Koch curve which sets up a link to the problem of the length of the coast of Britain. In the same way as the coast appears "crumpled" on any scale, however small it may be, both the Cantor set and the snowflake curve are characterised by the fact that, as a result of the principle governing their construction, each cutaway portion maintains the form of the whole (see figs. 5.5.2 and 5.5.3), i.e. both sets are self-similar. Whereas in the nineteenth and at the beginning of the twentieth century, such sets were considered pathological phenomena serving to explore the concept of infinity, for example, they are today viewed with new interest as sets which can be used to describe realistic phenomena. And indeed, Mandelbrot came across the Cantor set when, commissioned by IBM, he was investigating the occurrence of *flicker noise* in telephone lines used for data transmission between computers. It emerged that the flaws always appeared in bursts: trouble-free intervals were interrupted by periods with noise which in turn could be divided up into smaller intervals without disturbance and shorter periods with. Mandelbrot succeeded in showing that the bursts of noise possess the self-similar structure of a Cantor set (Mandelbrot, 1977).

A further example is Saturn's ring (fig. 5.5.4), discovered in 1610 by Galileo Galilei and described in more detail in 1655 by Christian Huygens. At first, it was believed that there was only a single ring. In 1675, however, an improved telescope revealed to Cassini a subdivision into a so-called A ring and a B ring, known as the Cassini division. The pictures transmitted back to Earth over 1500 million kilometres by the spacecraft Pioneer 11 at the end of August 1979 show, however,

Fig. 5.5.4

The rings of Saturn. Reproduced with kind permission of NASA (NASA, 1996)

that there is not only one, but an extraordinary number of gaps in Saturn's ring, the structure of which, according to Mandelbrot's conjecture, again resembles a Cantor set. In September 2006, the Saturn probe Cassini discovered a further, hitherto unknown outer ring, probably consisting of tiny ice particles. The Cantor set appears to reflect a basic principle of order in nature. Indeed, in the case of dissipative systems, it plays a fundamental role in the description of attractors.

Let us now turn to the definition of a generalised dimension. There are various possiblities for assigning a non-integer dimension to attractors or fractal sets. These quantities reflect various attempts to quantify the intuitive idea of how "densely" a set fills the space in which it lies. The abundance of different dimension definitions can basically be divided into two groups, see (Farmer, 1982b; Farmer et al., 1983), each of which emphasises different mathematical and physical characteristics of the set, as well as a further definition which is based on the dynamic characteristics of a deterministic process. In the following sub-sections, we present various definitions, in some cases in a somewhat simplified form. We recommend (Mandelbrot, 1982; Farmer et al., 1983; Hentschel and Procaccia, 1983), for example, for further reading.

5.5.2 Fractal Dimensions: Capacity Dimension and Hausdorff-Besicovitch Dimension

The definition of this category of dimensions is based only on metric concepts, i.e. our considerations rely on a concept of space in which the distance between two points is defined, as, for example, in the Euclidean space. We hence also speak of metric dimensions.

Figure 5.5.5 shows a line segment, a square and a cube, the length, surface respectively volume of which we can determine by dividing the respective set into n-dimensional "boxes" of edge length ε where n is the dimension of the underlying Euclidean space.

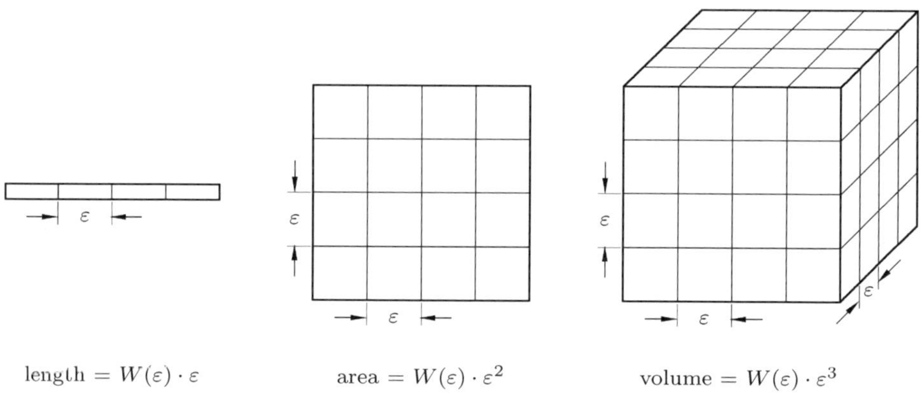

Fig. 5.5.5: On the calculation of the n-dimensional volume

If we denote the number of boxes required by $W(\varepsilon)$, the respective "n-dimensional volume" is

$$V = W(\varepsilon)\varepsilon^D \tag{5.5.2}$$

where the exponent D coincides with the dimension of the measured set.

We can proceed in the same way when calculating an arbitrary area A (fig. 5.5.6). We cover the area with squares of edge length ε and determine the number $W(\varepsilon)$ of boxes in which there are points of the set. If ε is small enough, the number of populated cells will increase in inverse proportion to ε^2, i.e.

$$W(\varepsilon) \sim \frac{1}{\varepsilon^2} \tag{5.5.3}$$

and, in the limit for $\varepsilon \to 0$, we obtain the area

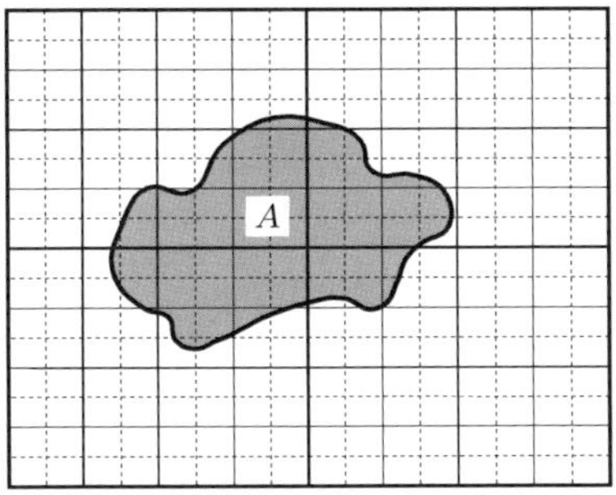

Fig. 5.5.6

Calculation of area

$$A = \lim_{\varepsilon \to 0} W(\varepsilon)\varepsilon^2 \tag{5.5.4}$$

i.e. the exponent of ε reproduces once more the Euclidean dimension of the set.

This procedure can also be adopted for general fractal sets in an n-dimensional space. We cover the set with n-dimensional hypercubes of edge length ε and determine the minimum number $W(\varepsilon)$ of hypercubes required to cover the set completely. If ε is small enough, $W(\varepsilon)$ will be, in analogy with eq. (5.5.3), inversely proportional to a power of ε

$$W(\varepsilon) \propto \frac{1}{\varepsilon^{D_c}} \tag{5.5.5}$$

The symbol $\propto$ indicates that $W(\varepsilon)$ is proportional to $1/\varepsilon^{D_c}$ *in the limit* $(\varepsilon \to 0)$. For small ε, the still unknown proportionality factor $V(\varepsilon)$ plays the role of a generalised volume

$$W(\varepsilon) = V(\varepsilon)\frac{1}{\varepsilon^{D_c}} \tag{5.5.6}$$

Solving this relation with respect to the unknown exponent D_c, we obtain

$$D_c = \frac{\ln W(\varepsilon) - \ln V(\varepsilon)}{\ln(1/\varepsilon)} \tag{5.5.7}$$

Proceeding to the limit $\varepsilon \to 0$, $\ln V(\varepsilon)/\ln(1/\varepsilon)$ tends towards zero since $V(\varepsilon)$ is bounded and we find

$$D_c = \lim_{\varepsilon \to 0} \frac{\ln W(\varepsilon)}{\ln(1/\varepsilon)} \tag{5.5.8}$$

D_c is called the *capacity dimension*, a definition which was introduced by Kolmogorov as early as 1958 (Kolmogorov, 1958). In an n-dimensional space, $W(\varepsilon)$ denotes the minimum number of n-dimensional hypercubes of edge length ε required to cover the set. Occasionally, as a result of counting the populated cells, D_c is also denoted *box dimension*, see (Peitgen et al., 1992).

As an example, we consider the Cantor set (fig. 5.5.2) and calculate its capacity dimension according to the abovementioned rule. If we divide the initial interval in the first construction step into three equal sections of length $\varepsilon_1 = 1/3$, we require $W_1 = 2$ such intervals to cover the set. If we again divide the resulting intervals into thirds at each subsequent construction step, in the k-th step, $W_\mathrm{k} = 2^\mathrm{k}$ intervals of length $\varepsilon_{\mathrm{k}} = 1/3^{\mathrm{k}}$ are necessary to cover the set. In this way, eq. (5.5.8) yields the capacity dimension

$$D_c = \lim_{\mathrm{k}\to\infty} \frac{\ln W_{\mathrm{k}}}{\ln(1/\varepsilon_{\mathrm{k}})} = \lim_{\mathrm{k}\to\infty} \frac{\ln 2^{\mathrm{k}}}{\ln 3^{\mathrm{k}}}$$

i.e. for the fractal structure of the "Cantor dust", we obtain the non-integer value

$$D_c = \frac{\ln 2}{\ln 3} = 0.6309\ldots \tag{5.5.9}$$

The capacity dimension of the Koch snowflake curve can also be determined very easily in this way. In the first construction step, we require $W_1 = 4$ intervals of length $\varepsilon_1 = 1/3$ to cover one side, and in the k-th step, $W_{\mathrm{k}} = 4^{\mathrm{k}}$ of length $\varepsilon_{\mathrm{k}} = 1/3^{\mathrm{k}}$. This yields the capacity dimension

$$D_c = \lim_{\mathrm{k}\to\infty} \frac{\ln 4^{\mathrm{k}}}{\ln 3^{\mathrm{k}}} = \frac{\ln 4}{\ln 3} = 1.2618\ldots$$

The capacity dimension is a special case of the so-called *Hausdorff-Besicovitch dimension* D_H, a more general, also not necessarily integer dimension concept introduced by Felix Hausdorff in 1918 (Hausdorff, 1918) and written in its final form by A.S. Besicovitch, see (Mandelbrot, 1982). In an n-dimensional space, we consider covers of the set with a countable number of n-dimensional hypercubes of *variable* edge length ε_{i} which may not exceed a certain length ε; we look for the cover which minimises a generalised volume. It is easy to imagine how difficult, if not impossible, it is to determine D_H in the case of real examples and numerical experiments, since of all the possible ways of covering the set, we have to find a particular optimum one (Umberger et al., 1986b). Only recently has a sophisticated method been introduced, based on the critical behaviour of a spectrum of measure functions (HMSF: Hausdorff measure spectrum functions) which allows a reliable numerical determination of the Hausdorff-Besicovitch dimension (Li et al., 2004).

It is easy to convince oneself, however, that the capacity dimension D_c can never be smaller than the Hausdorff-Besicovitch dimension

$$D_H \leqslant D_c \tag{5.5.10}$$

Due to the minimum condition in the definition of D_H, a special cover consisting of hypercubes of equal edge length, as is required for D_c, can never lead to values smaller than D_H.

Only metric characteristics were used to define the two dimensions D_c and D_H. We shall, as is customary, call dimensions of this type *fractal dimensions* (Mandelbrot, 1982).

5.5.3 Information Dimension

The capacity dimension depends only on the metric characteristics of the attractor since, for its calculation, all the hypercubes required to cover the attractor are evenly weighted, irrespective of how many points lie in them respectively how often they are crossed by a trajectory. However, fractal sets such as strange attractors,

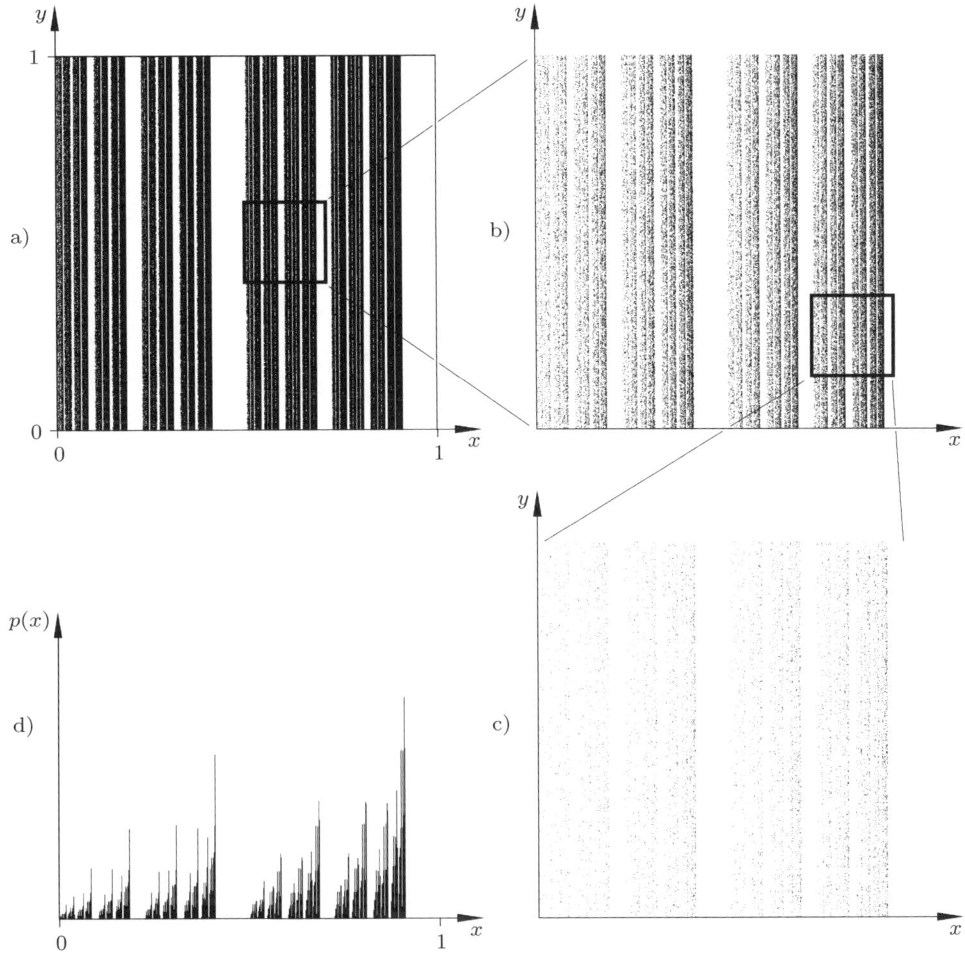

Fig. 5.5.7: Density distribution of the image points for the generalised baker's map, eq. (5.5.11); $\lambda_a = \lambda_b = 0.45,\ \alpha = 0.35$

for example, often have a complex, inhomogeneous structure: different parts of the phase space are occupied with varying density by trajectories.

Such density distributions can be made visible with the aid of Poincaré sections. As a two-dimensional model, we consider the so-called generalised baker's map

$$x_{\mathrm{n}+1} = \begin{cases} \lambda_a x_{\mathrm{n}} & \text{for } y_{\mathrm{n}} < \alpha \\ \frac{1}{2} + \lambda_b x_{\mathrm{n}} & \text{for } y_{\mathrm{n}} > \alpha \end{cases}$$

and

$$y_{\mathrm{n}+1} = \begin{cases} y_{\mathrm{n}}/\alpha & \text{for } y_{\mathrm{n}} < \alpha \\ (y_{\mathrm{n}} - \alpha)/(1-\alpha) & \text{for } y_{\mathrm{n}} > \alpha \end{cases} \tag{5.5.11}$$

for the unit square $0 \leqslant x_{\mathrm{n}} \leqslant 1$, $0 \leqslant y_{\mathrm{n}} \leqslant 1$. Figure 5.5.7a shows the distribution of the image points after ca. 4 million mapping steps for the parameter values $\lambda_a = \lambda_b = 0.45$, $\alpha = 0.35$. If we observe in succession different magnifications of the set of points in an image section (fig. 5.5.7b,c), we recognise the self-similar structure as well as an irregular fractal density distribution of the image points which is apparent on every scale. Figure 5.5.7d shows the relative density distribution $p(x)$. To demonstrate this, the unit square $0 \leqslant x \leqslant 1$, $0 \leqslant y \leqslant 1$ was decomposed into narrow strips of width ε parallel to the y-axis and the relative proportion of the image points in each such strip was counted.

The baker's map can be interpreted as a model of a physical system on which, due to its sensitive dependence on the initial conditions, we have only very incomplete information on a local level since it is not possible to make predictions about the exact position of the subsequent image points. Methods taken from statistics and the information theory are particularly suitable for dealing with systems where such detailed knowledge is missing. Thus, should we want to take the frequency with which a trajectory visits various parts of the attractor into account when creating a dimensional concept, then, in addition to the metric characteristics, we must also consider the density or probability distributions for its definition. In this way, a new family of dimension concepts emerges, called the "dimensions of the natural measure".

Theoretically, the variables describing the phase space are continuous quantities. In practice, however, for example in the case of a measurement, due to the finite resolution of the measuring instrument, a partitioning of the phase space takes place since each reading is inevitably a rational number with a finite number of digits. The same naturally applies to each value determined numerically on a computer since in this case as well, only a finite number of digits is at our command. If ε is the finest resolution, this leads to a uniform sub-division of the phase space into hypercubes C_{i} of edge length ε (see fig. 5.5.8). Thus, we cannot locate the position of a measuring point respectively a numerically calculated point in the phase space with arbitrary accuracy, but can only pinpoint the i-th hypercube in which the relevant point lies.

If we now carry out a series of N measurements over a certain period of time for a dissipative system, the number $\mathrm{N_i}$ of the measuring points in the hypercube C_{i} can be counted. We can also enquire about the probability of finding a measuring point in the cube C_{i}. To this end, we remind ourselves of the frequentist definition of probability presented in section 3.9.3. We can assign to each hypercube a relative frequency $p_{\mathrm{i}} = \mathrm{N_i}/\mathrm{N}$ which naturally also depends on the reading accuracy ε. If we allow $\varepsilon \to 0$ and simultaneously let the number of the measured points tend to infinity, we can assign to each attractor a probability density distribution which is denoted a natural measure, see section 3.10 and (Farmer et al., 1983; Eckmann and Ruelle, 1985; Ruelle, 1989).

The next question is how much information can be won about the system on average per individual measurement. In order to answer this, we first have to turn our attention to the basic concepts of the information theory.

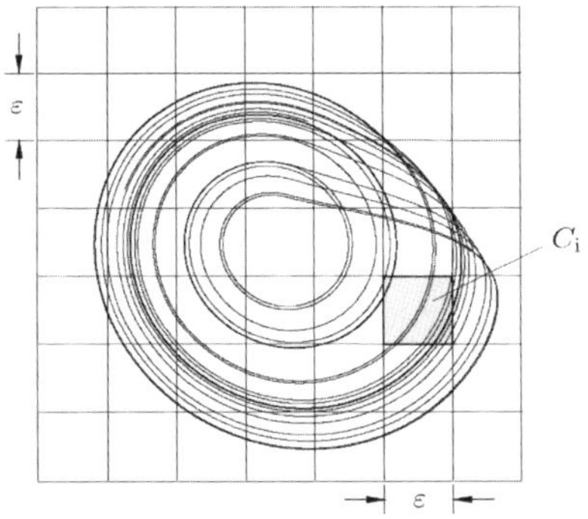

Fig. 5.5.8

Uniform sub-division of the phase space

Without communication, without the exchange of information, human coexistence – indeed, any interaction in animate and inanimate nature – is unthinkable. In our daily life, pieces of information are constantly being passed on in conversations, newspapers, news and television broadcasts, letters and measurements. This throws up some fundamental questions, e.g. how much information is contained in a measurement or in a signal sequence, how precisely data can be transmitted, how messages can be coded particularly effectively or how ancient characters can be deciphered.

At the end of the 1940s, the Nobel Prize winner Claude E. Shannon, a research mathematician at the Bell Telephone Laboratories, worked intensively on such problems of communication theory (Shannon, 1948). He is rightly considered the founder of the information theory (Shannon and Weaver, 1949) which spread extraordinarily quickly and was applied in diverse scientific branches. Today, for example, it plays a fundamental role in all areas of data communication: in the capacity of telephone lines, for instance, or the optimum electronic storage and reproduction of music or images. But it is also one of the basic tools in information technology, theoretical physics, biology as well as in the field of non-linear dynamics (Haken, 1988).

When dealing with a new theory, the basic concepts must first be defined precisely. Terms are often introduced which have a completely different meaning in everyday usage. The concept "information" possesses a specific significance in the communication theory and may not be confused in particular with meaning, value or content (Shannon and Weaver, 1949; Brillouin, 1962). The information theory is *not* interested in the subjectively loaded content of a message, i.e. in semantics. As far as the theory is concerned, it is irrelevant whether a telegram contains an important message for the recipient while being nonsense for anybody else. The concept information refers only to what one *could say*, not what one *does say*, i.e. information is an objective measure of the number of realisations or alternatives available for the choice of a message. The concept of information never refers to a special signal but always to all the possible signals in the given situation.

We now introduce Shannon's measure for the information (Shannon and Weaver, 1949). To this end, we first consider a system in which R realisations occur with equal probability. In the initial situation, we do not yet have any information about the system, i.e. $I = 0$. If the number R of possible realisations is only one, no selection can be made, i.e. no information gain can be expected. If, however, $R > 1$, then, after the occurrence or arrival of an event or a message, a certain amount of information about the system is gained; the greater the uncertainty, i.e. the number R of possible realisations at the outset, the greater the gain. It is thus expedient to select for I a function which is positive for $R > 1$ and increases monotonically with R. Moreover, the information I should be additive. If, for example, we toss an ideal coin, there is an equal probability of getting heads or tails or, expressed in a binary system, 0 or 1. By tossing a coin, we can thus make a decision "0" or "1". If we toss two ideal coins, $2^2 = 4$ results are possible; tossing two coins means that two decisions are made, i.e. the information gained is equal to the sum of the two individual pieces of information resulting from two separate tosses.

If we prescribe a general adherence to these characteristics of additivity and monotonicity, the information must follow a logarithmic law

$$I = K \ln R \tag{5.5.12}$$

where K is at first an arbitrary constant. Since, however, there is a close connection between the information content and the storage capacity of computers, it has become customary to specify the information I in "yes-no"-decisions or "bits" (binary digits). If, in a binary system, we form a symbol of the length n, there are $R = 2^{\mathrm{n}}$ possible representations. For example, for $\mathrm{n} = 3$, there are the following $2^3 = 8$ realisations: 000, 001, 010, 100, 011, 101, 110, 111 where each number is determined by three decisions, i.e. the information is $I = 3$. From the requirement $I = \mathrm{n}$, we can determine the constant K

$$I = K \ln R = K \ln 2^{\mathrm{n}} = \mathrm{n} \tag{5.5.13}$$

This leads to

$$K = \frac{1}{\ln 2} = \log_2 \mathrm{e} \tag{5.5.14}$$

Thus, the information can also be expressed as the logarithm to the base 2 of the number of possibilities

$$I = \log_2 \mathrm{e} \cdot \ln R = \log_2 R \tag{5.5.15}$$

Up to now, we have assumed that all occurrences take place with equal probability. In reality, however, deviations and accumulations can be observed: e.g. the coin might be slightly bent, for instance. If we now toss this flawed coin very often, the result might be heads in 40% of the cases and tails in 60%. We now calculate the information content for this case. We assume that for N tosses, heads appears N_1 times and tails N_2 times. Two games with N tosses each will certainly differ in the order of the appearance of heads and tails. How many possible orders of outcome are there for a fixed N_1 and N_2? We know from combinatorics that the total number of possible combinations is

$$R = \frac{\mathrm{N}!}{\mathrm{N}_1!\mathrm{N}_2!} \tag{5.5.16}$$

from which, following eq. (5.5.16) and in accordance with eq. (5.5.13), we deduce the information

$$I = K(\ln \mathrm{N}! - \ln \mathrm{N}_1! - \ln \mathrm{N}_2!) \tag{5.5.17}$$

In order to make statements on the probability, the coin must be tossed very often per game, i.e. N ought to be very large. Since, however, the information increases with the number N of the realisations and thus with the number of tosses, only the mean value is of interest, i.e. the information per toss

$$\bar{I} = \frac{I}{\mathrm{N}} \tag{5.5.18}$$

In order to calculate $\bar{I}$, we use a simplification of Stirling's formula

$$\ln \mathrm{N}! \approx \mathrm{N}(\ln \mathrm{N} - 1) \tag{5.5.19}$$

which represents a good approximation for values $\mathrm{N} > 100$, see (Sommerfeld, 1977). We obtain this approximation formula very easily by approximating the integral of the logarithm function by the area under a step function

$$\begin{aligned} \ln \mathrm{N}! &= \ln 1 + \ln 2 + \ldots + \ln \mathrm{N} \approx \int\limits_1^{\mathrm{N}} \ln x \mathrm{d}x \\ &= x(\ln x - 1)\Big|_{x=1}^{x=\mathrm{N}} = \mathrm{N}(\ln \mathrm{N} - 1) + 1 \end{aligned}$$

This yields the average information per toss

$$\bar{I} = \frac{K}{\mathrm{N}} \left[\mathrm{N}(\ln \mathrm{N} - 1) - \mathrm{N}_1(\ln \mathrm{N}_1 - 1) - \mathrm{N}_2(\ln \mathrm{N}_2 - 1)\right]$$

or, since $\mathrm{N}_1 + \mathrm{N}_2 = \mathrm{N}$,

$$\bar{I} = -K\left(\frac{\mathrm{N}_1}{\mathrm{N}} \ln \frac{\mathrm{N}_1}{\mathrm{N}} + \frac{\mathrm{N}_2}{\mathrm{N}} \ln \frac{\mathrm{N}_2}{\mathrm{N}}\right) \tag{5.5.20}$$

Introducing the relative frequencies or probabilities with which N_1 respectively N_2 are realised in the form

$$p_\mathrm{i} = \frac{\mathrm{N}_\mathrm{i}}{\mathrm{N}} \quad , \quad \mathrm{i} = 1, 2 \tag{5.5.21}$$

we ultimately obtain the information per toss as

$$\bar{I} = -K(p_1 \ln p_1 + p_2 \ln p_2) \tag{5.5.22}$$

where

$$p_1 + p_2 = 1$$

holds.

Up to now, we have assumed that N_1 and N_2 are known fixed numbers and that the relative frequencies p_1 and p_2 and the mean information $\bar{I}$ are thus also fixed. We can also consider $\bar{I}$ a function of p_i, however and ask, for example, what the probability density must be in order to make $\bar{I}$ a maximum.

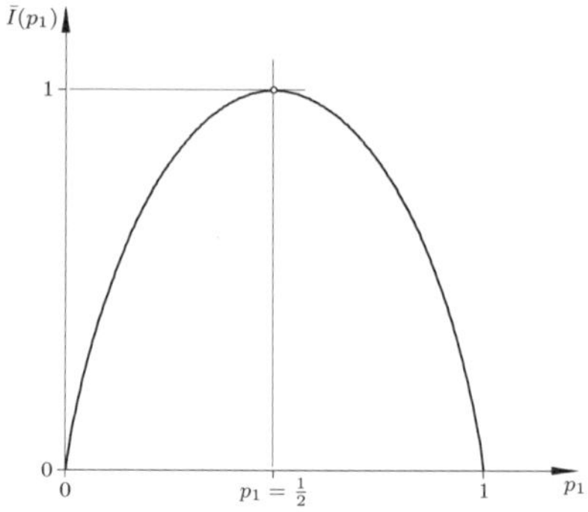

Fig. 5.5.9

Mean information when tossing a coin

Figure 5.5.9 shows the function $\bar{I}(p_1)$. Shannon designated as the mean information $\bar{I}(p_1)$, eq. (5.5.22), as entropy $H(p_1)$ in analogy to the concept in statistical mechanics, see Boltzmann's H-theorem, (Sommerfeld, 1977); see also the subsequent eq. (5.5.38) for the binary entropy function $H(p)$. For $p_1 = 1/2$, i.e. the case of the perfect coin, the information, and thus the number of possibilities, is the greatest. Each imperfection means a restriction and thus lower average information. In the limiting case $p_1 = 0$, the result is clear at the outset, i.e. the information is zero. This is certainly correct. In a conversation, someone who answers every question with "yes" is definitely not a source of information.

How can we use this theory to form a dimension concept? A coupled system of deterministic equations describing a dynamical process establishes certain correlations between the variables. Such links can limit the possibilities of a system for entering certain domains of the phase space considerably. The average information will thus certainly be a measure of how "densely" the attractor fills the phase space and is thus suitable for formulating a definition for a dimension.

In order to determine the dimension of the attractor, we must assume that the transient part of the process has already faded away. We consider a uniform subdivision of the phase space into hypercubes of edge length ε. Let $W(\varepsilon)$ be the number of boxes which contain measuring points and assume that, of N measurements, $\mathrm{N_k}$ measuring points lie in the k-th hypercube. The relative frequency of finding a point in the k-th box depends on the measuring accuracy ε, i.e. on the partitioning of the phase space, and is expressed by

$$p_{\mathrm{k}} = \frac{\mathrm{N_k}}{\mathrm{N}} \tag{5.5.23}$$

The number of possibilities for distributing the measurement outcomes in accordance with their respective probabilities among the W boxes is

$$R = \frac{\mathrm{N}!}{\mathrm{N_1!N_2!\ldots N_{\mathit{W}}!}} \tag{5.5.24}$$

with $\sum_{\mathrm{k}=1}^{W(\varepsilon)} \mathrm{N_k} = \mathrm{N}$ (see fig. 5.5.10). Generalising eq. (5.5.22), we obtain

$$\bar{I}(\varepsilon) = \frac{I(\varepsilon)}{\mathrm{N}} = -K \sum_{\mathrm{k}=1}^{W(\varepsilon)} p_{\mathrm{k}} \ln p_{\mathrm{k}} \tag{5.5.25}$$

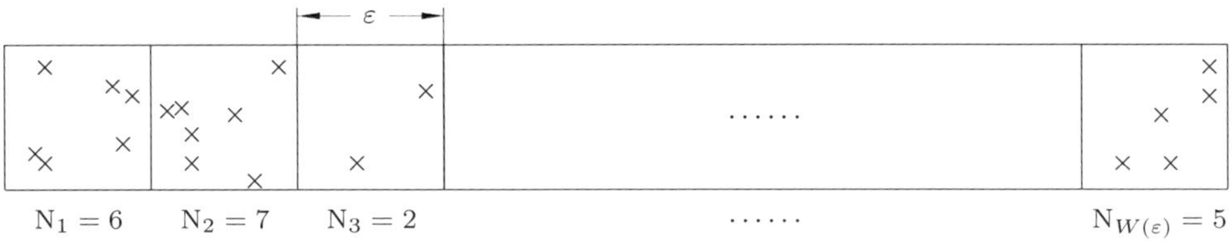

Fig. 5.5.10: On the definition of the average information of a series of measurements

as the average information per single measurement. Since for all relative frequencies $p_k < 1$ holds, the overall result is a positive expression. The *information dimension* D_I, originally introduced by Balatoni and Rényi, see (Balatoni and Rényi, 1957; Rényi, 1959), is now defined as follows

$$D_I = \lim_{\varepsilon \to 0} \frac{\bar{I}(\varepsilon)}{\log_2(1/\varepsilon)} \tag{5.5.26}$$

D_I is a measure for how fast the information which is required to define a point on an attractor increases as $\varepsilon \to 0$. If all hypercubes possess the same probability $p_k = 1/W(\varepsilon)$, $\bar{I}(\varepsilon)$ from eq. (5.5.25) assumes its maximum value $\bar{I}(\varepsilon) = K \ln W(\varepsilon) = \log_2 W(\varepsilon)$. In this case, the capacity dimension D_c, eq. (5.5.8), coincides with D_I. Since the information gain is smaller for different probability distributions, i.e. $\bar{I}(\varepsilon) < \log_2 W(\varepsilon)$, there generally applies

$$D_I \leqslant D_c \tag{5.5.27}$$

Instead of $1/\ln 2$, we can select the Boltzmann constant $k_B = 1.38 \cdot 10^{-23}\,\mathrm{J/K}$ for the constant K. In this case, the information concept coincides exactly with the entropy $S = k_B \ln W$ as it is defined in thermodynamics, where W denotes the number of different microstates, see (Brillouin, 1962; Haken, 1988). If a system is well-ordered and structured, the number of free possibilities is low, i.e. the information respectively the entropy is small. It is thus understandable that the concept of information, taken as a measure for the number of possible realisations or as a measure for the internal structure, is particularly suitable for introducing a dimension concept.

For the experimenter, application of the information concept certainly often involves an amount of re-thinking. Normally, he tries to verify a theoretical model with the aid of an experiment. With a series of measurements, he will, however, never realise precisely the same value but will always be faced with deviations and fluctuations, i.e. instead of one point, he will observe a whole cloud of points where the state corresponding to the theoretical model will lie somewhere in that cloud. When determining the information of a system, we follow exactly the reverse route. The originally infinite possibilities of the phase space will be limited considerably by the measurement results, i.e. the information decreases significantly. In the case of one point, the system does not even contain any information at all.

In fig. 5.5.11, we have calculated the information dimension for the simplest attractors, namely a fixed point and a limit cycle, described by a straight line, the

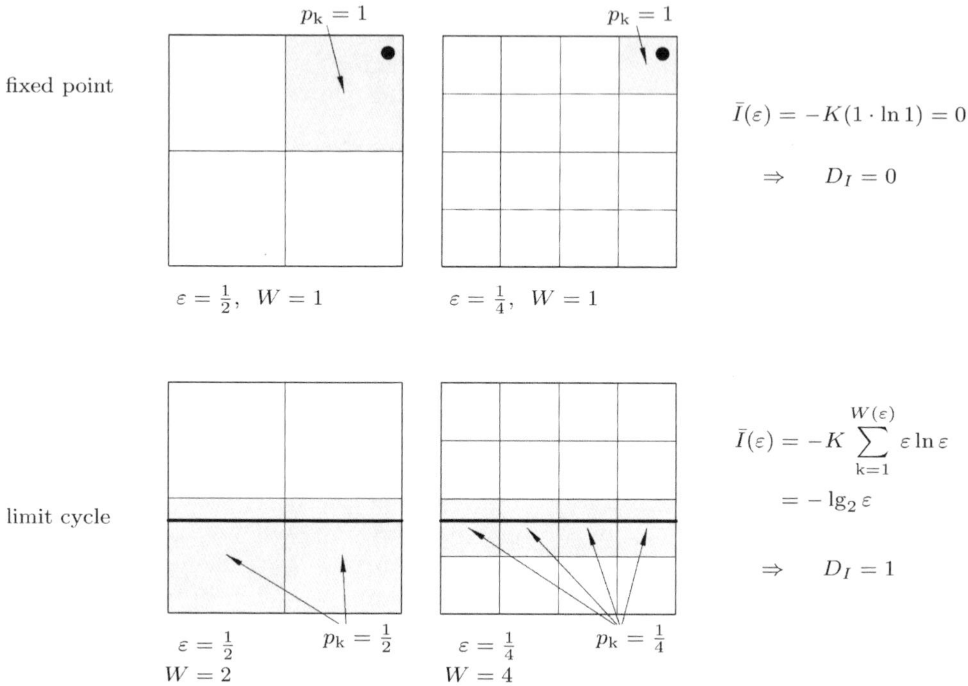

Fig. 5.5.11: Information dimension D_I for regular attractors

end points of which should coincide, and have shown the agreement of D_I with the usual dimension concept.

We next determine the capacity dimension D_c and the information dimension D_I using an asymmetrical Cantor set as an example, see (Farmer, 1982c). An asymmetrical Cantor set is generated, for example, when, starting with a unit interval, we omit each third quarter at every construction step. Figure 5.5.12 reproduces the result of the individual steps.

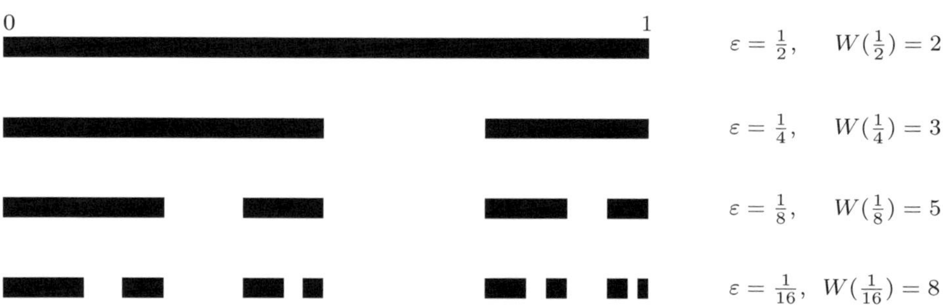

Fig. 5.5.12: Asymmetrical Cantor set for the unit interval $[0, 1]$

We first calculate the capacity dimension. We cover the initial unit interval with $W_0 = 2$ intervals of length $\varepsilon_0 = 1/2$. In the first construction step, we require $W_1 = 3$ intervals of length $\varepsilon_1 = 1/4$ in order to cover the set; in the second step, we need $W_2 = 5$ intervals of length $\varepsilon_2 = 1/8$, in the third step, $W_3 = 8$ intervals of length $\varepsilon_3 = 1/16$ etc. The number of intervals which are needed to cover the set follows a sequence of Fibonacci numbers $\{2, 3, 5, 8, \ldots\}$, where the relation

$$W_{n+2} = W_{n+1} + W_n \tag{5.5.28}$$

holds with the initial conditions $W_0 = 2, W_1 = 3$. In the n-th step, W_n intervals of length $\varepsilon_n = (1/2)^{n+1}$ are needed to cover the set; in accordance with eq. (5.5.8), this yields

$$D_c = \lim_{n\to\infty} \frac{\ln W_n}{\ln 1/\varepsilon_{n+1}} = \lim_{n\to\infty} \frac{\ln(W_{n-1} + W_{n-2})}{\ln 2^{n+1}} \approx 0.6945 \tag{5.5.29}$$

When determining the information dimension, we select a different sequence of sub-divisions of length ε to establish the probability distribution (see fig. 5.5.13). We designate the probability of finding a point in the i-th interval after the n-th construction step by $p_i^{(n)}$.

If we divide the initial interval in two sub-sections in the first step, the probability of the point being in the left half is twice as large as its being in the right half, i.e. since $p_1^{(1)} + p_2^{(1)} = 1$, we obtain $p_1^{(1)} = 2/3$ and $p_2^{(1)} = 1/3$. In the second step, the length of the interval is again halved. The probability $p_1^{(2)}$ of finding a point in the interval [0, 1/4] is twice as large as the probability $p_2^{(2)}$ for the interval [1/4, 1/2] where obviously $p_1^{(2)} + p_2^{(2)} = p_1^{(1)}$ must hold. In the interval [1/2, 3/4], there are no points of the set; it is thus not observed any longer. All points to the right of the centre are thus concentrated in the interval [3/4, 1] so that $p_3^{(2)} = p_2^{(1)}$ holds. Hence, the (non-zero) probabilities in the second step result as

$$\{p_i^{(2)}\} = \tfrac{1}{9}\{\,4\,,\,2\,;\,3\,\} \qquad (i = 1, 2, 3) \tag{5.5.30}$$

Continuing this procedure, we obtain in the third step $k_3 = 5$ intervals of length $\varepsilon_3 = 1/8$ with the following non-zero probabilities

$$\{p_i^{(3)}\} = \tfrac{1}{27}\{\,8\,,\,4\,,\,6\,;\,6\,,\,3\,\} \qquad (i = 1, \ldots, k_3) \tag{5.5.31}$$

and in the fourth step $k_4 = 8$ non-zero probabilities

$$\{p_i^{(4)}\} = \tfrac{1}{81}\{\,16\,,\,8\,,\,12\,,\,12\,,\,6\,;\,12\,,\,6\,,\,9\,\} \qquad (i = 1, \ldots, k_4) \tag{5.5.32}$$

From these relations and from fig. 5.5.13, we can recognise a general rule. In the n-th step, the probability distribution in the left half [0, 1/2] is deduced from the probabilities in the (n − 1)-th step by multiplication with the factor 2/3. For the right half [3/4, 1], the probability distribution of the (n − 2)-th step is multiplied by the factor 1/3, i.e. we have

$$p_i^{(n)} = \tfrac{2}{3}\, p_i^{(n-1)} \qquad \text{for} \qquad i = 1, \ldots, k_{n-1} \tag{5.5.33}$$

$$p_{k_{n-1}+i} = \tfrac{1}{3}\, p_i^{(n-2)} \qquad \text{for} \qquad i = 1, \ldots, k_{n-2} \tag{5.5.34}$$

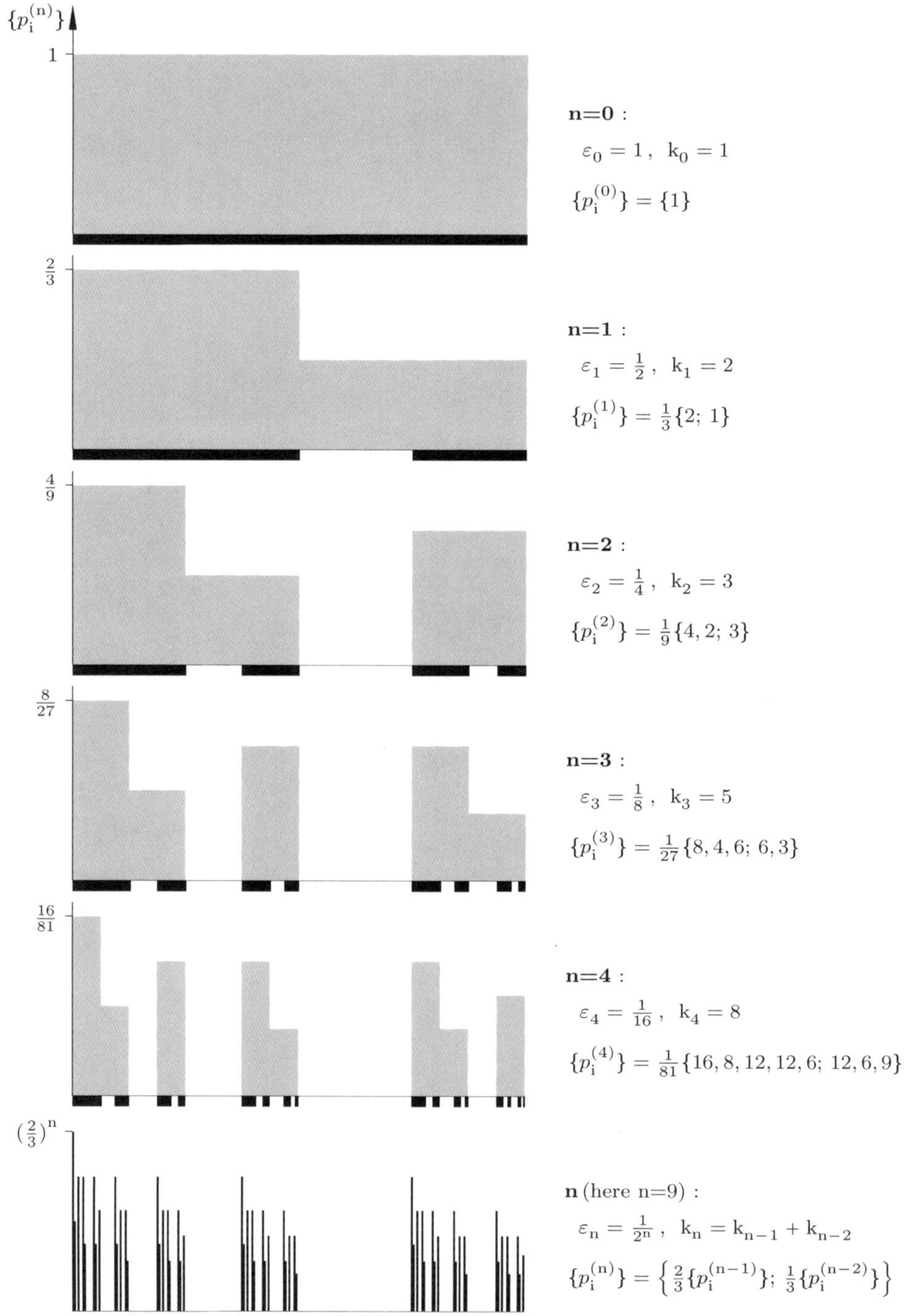

Fig. 5.5.13: Approximation of the probability distribution for the asymmetrical Cantor set

This construction rule implies that the number $\mathrm{k_n}$ of those intervals possessing non-zero probabilities coincides with the Fibonacci numbers

$$\{\mathrm{k_n}\} = \{1, 2, 3, 5, 8, 13, \ldots\} \tag{5.5.35}$$

i.e. it obeys the following rule

$$\mathrm{k_n} = \mathrm{k_{n-1}} + \mathrm{k_{n-2}} \tag{5.5.36}$$

In accordance with eqs. (5.5.15, 5.5.25), we can now calculate the average information in the n-th step

$$\begin{aligned}\bar{I}(\mathrm{n}) &= -\sum_{\mathrm{i=1}}^{\mathrm{k_n}} p_{\mathrm{i}}^{(\mathrm{n})} \log_2 p_{\mathrm{i}}^{(\mathrm{n})} \\ &= -\sum_{\mathrm{i=1}}^{\mathrm{k_{n-1}}} p_{\mathrm{i}}^{(\mathrm{n})} \log_2 p_{\mathrm{i}}^{(\mathrm{n})} - \sum_{\mathrm{i=k_{n-1}+1}}^{\mathrm{k_n}} p_{\mathrm{i}}^{(\mathrm{n})} \log_2 p_{\mathrm{i}}^{(\mathrm{n})}\end{aligned}$$

Introducing eqs. (5.5.33) and (5.5.34) and applying eq. (5.5.36), we obtain

$$\begin{aligned}\bar{I}(\mathrm{n}) &= -\sum_{\mathrm{i=1}}^{\mathrm{k_{n-1}}} \tfrac{2}{3} p_{\mathrm{i}}^{(\mathrm{n-1})} \log_2\left(\tfrac{2}{3} p_{\mathrm{i}}^{(\mathrm{n-1})}\right) - \sum_{\mathrm{i=1}}^{\mathrm{k_{n-2}}} \tfrac{1}{3} p_{\mathrm{i}}^{(\mathrm{n-2})} \log_2\left(\tfrac{1}{3} p_{\mathrm{i}}^{(\mathrm{n-2})}\right) \\ &= -\tfrac{2}{3} \sum_{\mathrm{i=1}}^{\mathrm{k_{n-1}}} p_{\mathrm{i}}^{(\mathrm{n-1})} \left(\log_2 \tfrac{2}{3} + \log_2 p_{\mathrm{i}}^{(\mathrm{n-1})}\right) \\ &\qquad - \tfrac{1}{3} \sum_{\mathrm{i=1}}^{\mathrm{k_{n-2}}} p_{\mathrm{i}}^{(\mathrm{n-2})} \left(\log_2 \tfrac{1}{3} + \log_2 p_{\mathrm{i}}^{(\mathrm{n-2})}\right)\end{aligned}$$

i.e.

$$\bar{I}(\mathrm{n}) = \tfrac{2}{3}\bar{I}(\mathrm{n}-1) + \tfrac{1}{3}\bar{I}(\mathrm{n}-2) - \tfrac{2}{3}\log_2 \tfrac{2}{3} \sum_{\mathrm{i=1}}^{\mathrm{k_{n-1}}} p_{\mathrm{i}}^{(\mathrm{n-1})} - \tfrac{1}{3}\log_2 \tfrac{1}{3} \sum_{\mathrm{i=1}}^{\mathrm{k_{n-2}}} p_{\mathrm{i}}^{(\mathrm{n-2})}$$

However, since the sum of the probabilities in each step must be equal to one, we may easily establish the following simple recursion formula

$$\bar{I}(\mathrm{n}) = \tfrac{2}{3}\bar{I}(\mathrm{n}-1) + \tfrac{1}{3}\bar{I}(\mathrm{n}-2) + H(\tfrac{1}{3}) \tag{5.5.37}$$

where $H(1/3)$ results from the so-called binary entropy function

$$H(p) = -p \log_2 p - (1-p)\log_2(1-p) \tag{5.5.38}$$

for $p = 1/3$; see also eq. (5.5.22) and fig. 5.5.9. With the initial values $\bar{I}(0) = 0$ and $\bar{I}(1) = H(1/3)$, we can thus deduce each $\bar{I}(\mathrm{n})$ stepwise from eq. (5.5.37).

Our interest is directed at the behaviour of $\bar{I}(\mathrm{n})$ for large values of n. In order to determine the asymptotic behaviour of $\bar{I}(\mathrm{n})$, we interpret the relation eq. (5.5.37) as a difference equation, the general solution of which must depend on two constants as in the case of a second-order differential equation. This general solution can either be guessed or determined with the aid of the calculus of finite differences (Boole, 1867). We obtain

$$\bar{I}(\mathrm{n}) = C_1 + (-\tfrac{1}{3})^{\mathrm{n}} C_2 + \tfrac{3}{4}\,\mathrm{n}\,H(\tfrac{1}{3}) \tag{5.5.39}$$

Thus, for large n-values, the average information will be proportional to $3/4\,\mathrm{n}\,H(1/3)$ independent of the constants C_1, C_2, i.e. independent of any initial condition. Hence, applying eq. (5.5.26), the information dimension is obtained

$$D_I = \lim_{\mathrm{n}\to\infty} \frac{\bar{I}(\mathrm{n})}{\log_2(1/\varepsilon_{\mathrm{n}})} = \lim_{\mathrm{n}\to\infty} \frac{1}{\log_2 2^{\mathrm{n}}} \Big(C_1 + (-\tfrac{1}{3})^{\mathrm{n}} C_2 + \tfrac{3}{4}\,\mathrm{n}\,H(\tfrac{1}{3})\Big)$$

or

$$D_I = \tfrac{3}{4}\,H(\tfrac{1}{3}) = \tfrac{3}{4}\,\frac{1}{\ln 2}\big(\tfrac{2}{3}\ln\tfrac{3}{2} + \tfrac{1}{3}\ln 3\big) \approx 0.6887 \tag{5.5.40}$$

Comparing with the capacity dimension of eq. (5.5.29), $D_c \approx 0.6945$, we confirm

$$D_I < D_c$$

i.e. the capacity dimension of the asymmetrical Cantor set is larger than its information dimension. This is always the case when the density distribution of the points is non-homogeneous but itself possesses a fractal character, i.e. if the probabilities $p_{\mathrm{i}}^{(\mathrm{n})}$ are not constant for all sub-intervals i; see figs. 5.5.7d and 5.5.13 as well as (Farmer, 1982c).

A wide range of possible dimension concepts based on the natural measure respectively probability distributions still exists; we turn to them in the following two sub-sections.

5.5.4 Correlation Dimension, Pointwise Dimension and Reconstruction of Attractors

῎Εξ ὄνυχος τὸν λέοντα γράφειν
Alkaios, ca. 620 v. Chr.

We next consider the so-called correlation dimension D_K. It is particularly suitable for practical calculations respectively the evaluation of experimentally measured time series and was proposed by Grassberger and Procaccia (Grassberger and Procaccia, 1983a; Grassberger and Procaccia, 1984). According to a result gained by Takens, (Takens, 1981), the dimension of an attractor can be determined from the temporal evolution of a *single* component x_{i} of the phase space vector $\boldsymbol{x}(t) = \{x_1(t) \ldots x_{\mathrm{i}}(t) \ldots x_{\mathrm{n}}(t)\}$.

In a first step, however, let us initially consider the complete phase space. After the fading of the transient phase, we determine for a fixed time increment Δt a set of N points $\boldsymbol{x}_1$, $\boldsymbol{x}_2$, ..., $\boldsymbol{x}_{\mathrm{N}}$ of the attractor according to the rule $\boldsymbol{x}_{\mathrm{k}} = \boldsymbol{x}(t + \mathrm{k}\Delta t)$ $(\mathrm{k} = 1, \ldots, \mathrm{N})$. If we calculate the autocorrelation function eq. (3.8.28) for the trajectory of a strange attractor, we observe that there is no temporal correlation between most of the pairs of points $(\boldsymbol{x}_{\mathrm{i}}, \boldsymbol{x}_{\mathrm{j}})$, i.e. they are *dynamically uncorrelated*. Nevertheless, all points reside on an attractor which displays a distinct spatial structure in the phase space. Consequently, in order to determine the dimension of the attractor, we thus direct our attention to the *spatial correlations* of pairs of points $(\boldsymbol{x}_{\mathrm{i}}, \boldsymbol{x}_{\mathrm{j}})$.

To this end, we decompose that part of the phase space which contains the set of points $\{\boldsymbol{x}_1, \boldsymbol{x}_2, \ldots, \boldsymbol{x}_{\mathrm{N}}\}$ uniformly into $W(r)$ hypercubes of edge length r. The number of points in the i-th hypercube is denoted $\mathrm{N_i}$, then, for $\mathrm{N} \to \infty$, $p_{\mathrm{i}} = \mathrm{N_i/N}$ describes the probability to find an arbitrarily selected point $\boldsymbol{x}_{\mathrm{k}}$ in the i-th cell. Let us recall the game with two ideal dice. The probability of throwing a six twice, for example, is $p_6{}^2 = (\frac{1}{6})^2$. This product rule holds since the two rolls are independent of one other. If this is applied to the set of points, we infer that the probability of finding two points in one and the same cell i is $p_{\mathrm{i}}{}^2 = (\mathrm{N_i/N})^2$. Considering all W cells, we find that $\sum\limits_{\mathrm{i}=1}^{W} p_{\mathrm{i}}{}^2$ is the probability that two arbitrarily chosen points of the attractor both fall into one single hypercube. However, apart from a factor which is independent of r, this is also the probability for two points $\boldsymbol{x}_{\mathrm{k}}$ and $\boldsymbol{x}_{\mathrm{m}}$ having a distance $\leqslant r$. Designating this probability $C(r)$, we obtain

$$C(r) = \frac{1}{\mathrm{N}^2} \times \{\text{ number of pairs } (\boldsymbol{x}_{\mathrm{k}}, \boldsymbol{x}_{\mathrm{m}}) \quad \text{with} \quad |\boldsymbol{x}_{\mathrm{k}} - \boldsymbol{x}_{\mathrm{m}}| \leqslant r \quad (\mathrm{k} \neq \mathrm{m})\} \tag{5.5.41}$$

Formally, this relation can be reformulated using the so-called Heaviside function H, eq. (3.8.42), as follows

$$C(r) = \frac{1}{\mathrm{N}^2} \sum_{(\mathrm{k}\neq\mathrm{m})} H(r - |\boldsymbol{x}_{\mathrm{k}} - \boldsymbol{x}_{\mathrm{m}}|) \tag{5.5.42}$$

If the number of points, i.e. N, is large enough, we expect that for small r-values $C(r)$ will behave like $C(r) \propto r^{D_K}$ where the exponent indicates the dimensionality and derives from

$$D_K = \lim_{r\to 0} \frac{\ln C(r)}{\ln r} \tag{5.5.43}$$

The function $C(r)$ thus expresses spatial coherences or correlations, i.e. statistically reproducible relations, and is called the *correlation integral* of the attractor (Grassberger and Procaccia, 1983b). The exponent D_K is denoted the *correlation dimension.*

Grassberger and Procaccia (1983a) have shown that the correlation dimension represents a lower bound for both the information dimension and the capacity dimension, i.e. with eq. (5.5.27), we obtain the relation

$$D_K \leqslant D_I \leqslant D_c \tag{5.5.44}$$

which we shall discuss again in sub-section 5.5.5.

We add three comments regarding the practical determination of the dimension. Firstly, for the calculation of the distances in eq. (5.5.41) respectively eq. (5.5.42), the Euclidean concept of distance need not necessarily be adopted. One can also simply determine the sum of the absolute values of the differences of the individual terms and take this as the measure for the distance. Secondly, the radii of the spheres should be selected so that r is on the one hand considerably smaller than the radius of a sphere enclosing the whole attractor and on the other hand large enough so that at least one pair of points is contained in the corresponding sphere. What is essential, however, is a reduction of the high computational effort which is needed when directly implementing eq. (5.5.42). If N denotes the length of the data set $\boldsymbol{x}_1, \boldsymbol{x}_2, \ldots, \boldsymbol{x}_\mathrm{N}$, the computational time for calculating the distances between all pairs of points increases for a fixed radius r like $\mathcal{O}(\mathrm{N}^2)$. Several methods have been proposed in order to reduce the computational time. All algorithms take advantage of the fact that only pairs of points with a sufficiently small distance need to be considered to estimate the correlation dimension. An implementation can be found in (Kantz and Schreiber, 1997) and in particular in (Theiler, 1987; Grassberger, 1990). With the help of such algorithms, it is possible to reduce the computational time drastically; in the best-case scenario, the effort scales like $\mathcal{O}(\mathrm{N}) + \mathcal{O}(\mathrm{N}_1)$, where N_1 denotes the number of pairs of points with a distance $< r$.

At this point, we introduce the alternative measure of a *pointwise dimension* D_p which is based on a similar calculation method and is also used in practice to characterise series of experimental measurements.

We again consider a set of N points in the phase space (or in a Poincaré section) $\boldsymbol{x}_1, \boldsymbol{x}_2, \ldots \boldsymbol{x}_\mathrm{N}$ and ask what the probability $p_\mathrm{i}(r)$ is of finding an arbitrary point of the set in a hypersphere of radius r about the point $\boldsymbol{x}_\mathrm{i}$. To this end, we simply determine the number $\mathrm{N}_\mathrm{i}(r)$ of all points of the set lying within the hypersphere and obtain

$$p_\mathrm{i}(r) = \frac{\mathrm{N}_\mathrm{i}(r)}{\mathrm{N}} \tag{5.5.45}$$

It is easy to understand (see fig. 5.5.14) that $p_\mathrm{i}(r)$ will be proportional to r if the points lie uniformly distributed on a line respectively that $p_\mathrm{i}(r)$ will be proportional to r^2 if the points lie uniformly distributed on a plane. In the general case of a fractal set, for small radii r, $p_\mathrm{i}(r)$ will scale like r^{D_p} where the exponent represents a measure for the dimensionality and is obtained as the limit

$$D_p(\boldsymbol{x}_\mathrm{i}) = \lim_{r\to 0} \frac{\ln p_\mathrm{i}(r)}{\ln r} \tag{5.5.46}$$

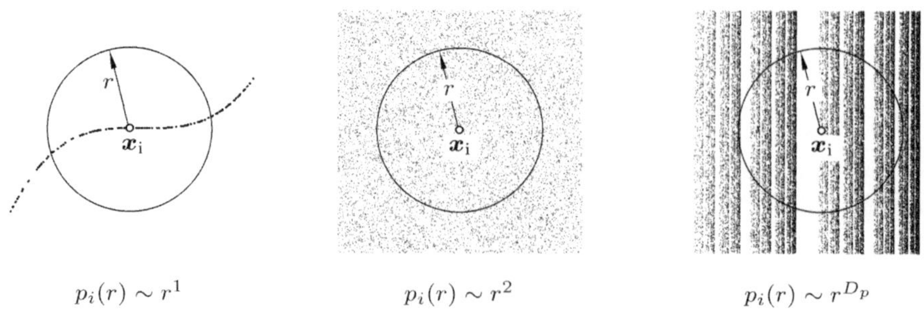

Fig. 5.5.14: On the definition of the pointwise dimension

The quantity $D_p(\boldsymbol{x}_{\mathrm{i}})$ is denoted *pointwise dimension.* $D_p(\boldsymbol{x}_{\mathrm{i}})$ is a measure for the relative percentage density with which points of the set lie in a small neighbourhood of the point $\boldsymbol{x}_{\mathrm{i}}$; it will generally vary from point to point. If the system is ergodic, the same value D_p is obtained for almost all N points of the set (Ott, 1993). Otherwise, one can, for example, take the mean value over M randomly selected points of the set and designate the quantity

$$\bar{D}_p = \frac{1}{\mathrm{M}} \sum_{\mathrm{i}=1}^{\mathrm{M}} D_p(\boldsymbol{x}_{\mathrm{i}}) \tag{5.5.47}$$

as the average pointwise dimension. The difference between the mean pointwise dimension $\bar{D}_p$ and the correlation dimension D_K lies in the fact that taking the mean and evaluating the limit are executed in an opposite sequential order.

We now come to an important application of the correlation dimension respectively the mean pointwise dimension. Up to now, we have assumed that the time evolution of *all* the variables involved in the dynamics is known. In a multitude of experiments, however, only the time evolution of some few – and occasionally of only one single – state variables can be measured. Moreover, in the case of continuous systems – in hydrodynamics or continuum mechanics, for example – it is generally not known whether a finite number of degrees of freedom is sufficient to describe the long-term behaviour of the system.

In the following, we consider dissipative systems based on deterministic laws and ask ourselves how much information is contained in the time evolution of a single state variable. Is it possible, for example, to determine an attractor and its dimension or its Lyapunov exponents with the aid of *one* specific measurement series? At first, it seems surprising that it should be possible to obtain information about additional variables involved in the dynamics from the temporal evolution of a single state variable and to determine characteristics of the attractor, i.e. invariants of the flow. Yet the couplings in the underlying deterministic equations cause each individual component to contain essential information about the total dynamics.

The following example illustrates how a strange attactor – in this case the Lorenz attractor – can be reconstructed from a single component without losing the topological, qualitative characteristics of the attractor. The underlying system of differential equations is (see also sections 5.2 and 10.3)

$$\begin{aligned} \dot{X} &= -\sigma X + \sigma Y \\ \dot{Y} &= \quad rX - Y - XZ \\ \dot{Z} &= -bZ + XY \end{aligned} \tag{5.5.48}$$

Differentiation of the first equation in conjunction with the second yields the following system

$$\begin{aligned} X &= \quad X \\ \dot{X} &= -\sigma X + \sigma Y \\ \ddot{X} &= -\sigma \dot{X} + \sigma(rX - Y - XZ) \end{aligned} \tag{5.5.49}$$

or, resolving with respect to X, Y, Z,

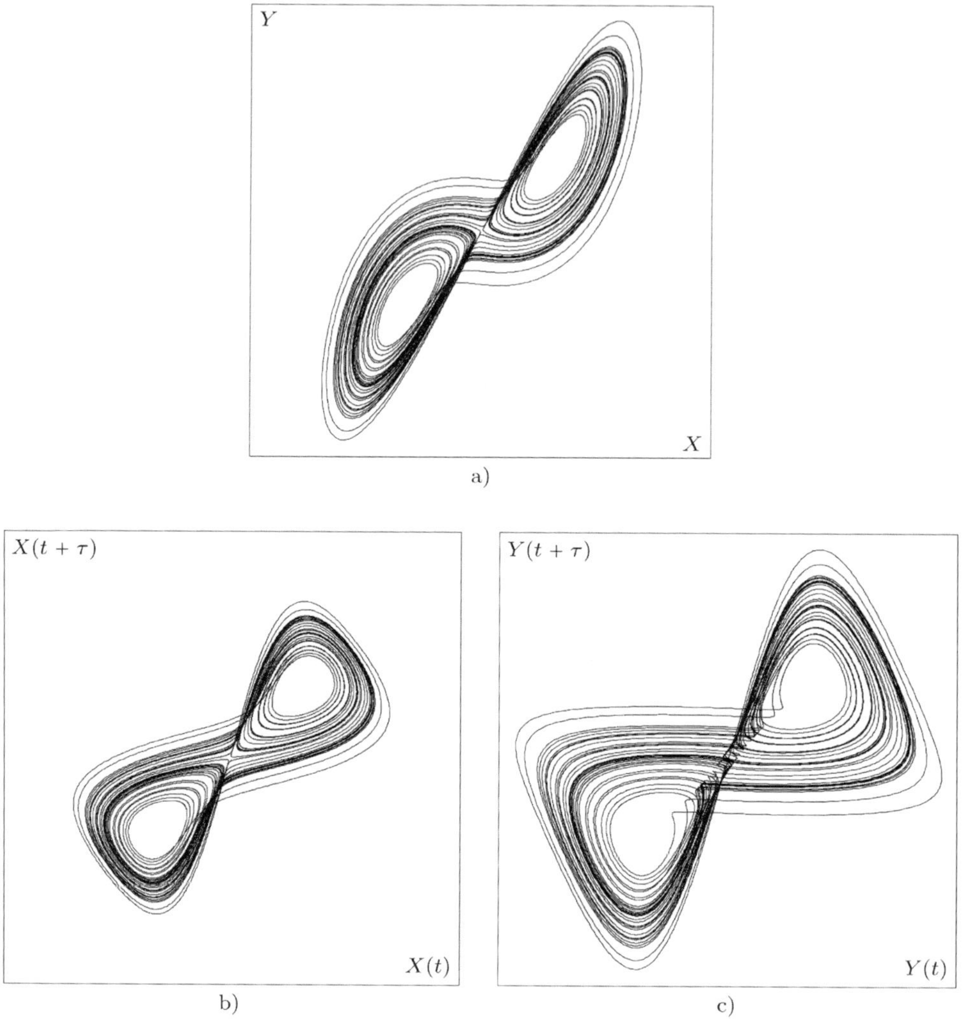

Fig. 5.5.15: Lorenz attractor and (a) reconstruction from X-component (b) resp. Y-component (c); $\sigma = 10,\ b = 8/3,\ r = 28,\ \tau = 0.13$

$$\begin{aligned} X &= X \\ Y &= X + (1/\sigma)\dot{X} \\ Z &= r - 1 - [(\sigma+1)\dot{X} + \ddot{X}]/(\sigma X) \end{aligned} \tag{5.5.50}$$

i.e. the independent variables $X(t), Y(t), Z(t)$ can be replaced by a set of new independent variables, in this case $X(t), \dot{X}(t), \ddot{X}(t)$. This transition can be interpreted as a transformation to curvilinear coordinates in the phase space which, with certain assumptions of continuity, does not alter the topology of the attractor. The disadvantage of this particular transformation is obvious. Normally, with the reconstruction method, one assumes one-dimensional time series $x(t_0), x(t_1), \ldots, x(t_k), \ldots$. The numerical determination of derivatives from the discrete series $\{x(t_k)\}$ is, however, generally very problematic and imprecise. According to a suggestion of Ruelle, see (Packard et al., 1980), footnote 8, resp. (Ruelle, 1989), it is more expedient to select a fixed time delay τ and $x(t), x(t+\tau), x(t+2\tau), \ldots$ as the new independent variables.

As an illustration, we have integrated the Lorenz equations numerically and reconstructed the attractor from the X- respectively Y-component alone. Figure 5.5.15 presents the results: in (a) the projection of the original attractor on the XY-plane in the XYZ-phase space is plotted; (b) shows a projection of the attractor on the $X(t), X(t+\tau)$-plane reconstructed from the quantities $X(t)$, $X(t+\tau)$, $X(t+2\tau)$ and (c) the corresponding reconstruction from the time-history of the Y-component. Comparing the three presentations, we recognise that the topological characteristics and, in the main, also the geometrical form of the attractor have been preserved.

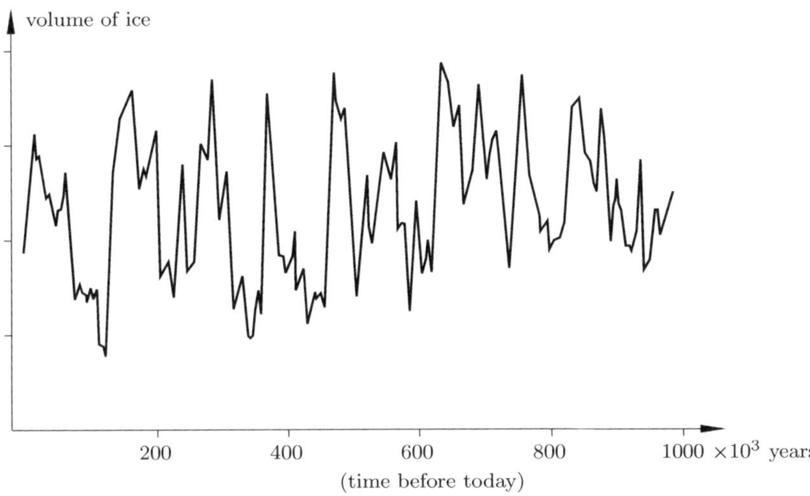

Fig. 5.5.16: Variation of the global ice volume (Nicolis and Prigogine, 1987)

After this heuristic introduction, we now explain a reconstruction method, the so-called *time delay method*, using an example in which the dimension of the phase space is not known at the outset, unlike in the case of the Lorenz system. We start from discrete measurements of a state variable $x(t)$, such as, for example, the variation of the global volume of ice on the Earth during the past million years, see fig. 5.5.16, after (Nicolis and Nicolis, 1984; Nicolis and Prigogine, 1987). In this case, we do not know at the outset how many independent variables are required for the reconstruction. We assume that the x-values of the original series of measurements were determined at fixed time intervals Δt

$$x_0 = x(t_0),\ x_1 = x(t_0+\Delta t),\ x_2 = x(t_0+2\Delta t),\ \ldots\ , x_k = x(t_0+k\Delta t),\ \ldots \tag{5.5.51}$$

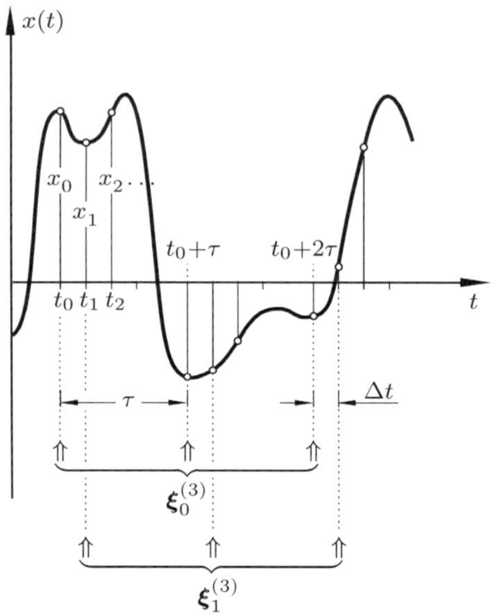

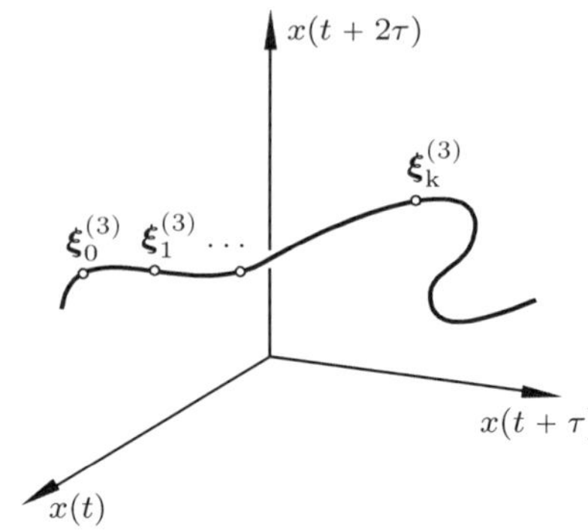

Fig. 5.5.17

Reconstruction of a trajectory in the three-dimensional embedding space from a time series

We next select a fixed time delay or time shift τ (generally a multiple of Δt) and construct from the series eq. (5.5.51) a sequence of vectors of fixed length m (m $= 2, 3, \ldots$)

$$\begin{aligned}
\boldsymbol{\xi}_0^{(\mathrm{m})} &= \{x(t_0) \quad x(t_0+\tau) \ \ldots \ x(t_0+(\mathrm{m}-1)\tau)\} \\
\boldsymbol{\xi}_1^{(\mathrm{m})} &= \{x(t_1) \quad x(t_1+\tau) \ \ldots \ x(t_1+(\mathrm{m}-1)\tau)\} \\
&\ \ \vdots \\
\boldsymbol{\xi}_{\mathrm{k}}^{(\mathrm{m})} &= \{x(t_{\mathrm{k}}) \quad x(t_{\mathrm{k}}+\tau) \ \ldots \ x(t_{\mathrm{k}}+(\mathrm{m}-1)\tau)\} \\
&\ \ \vdots
\end{aligned} \tag{5.5.52}$$

where $t_{\mathrm{k}} = t_0 + \mathrm{k}\Delta t$. These vectors can be plotted in an m-dimensional embedding space. The procedure for m = 3 is illustrated in fig. 5.5.17.

Figure 5.5.18 shows as an example a three-dimensional embedding of the climate attractor for a delay time τ of 2000 years. In this representation, the structure of the attractor is not yet recognisable so that we can assume a dimension 3 of the embedding space is not yet sufficient.

How large a dimension for the embedding space must we select in order to gain the desired information from the reconstruction? The dimension must certainly be high enough to guarantee that the trajectories do not intersect. Depending on the dimension m and the time shift τ, we obtain different coordinate transformations and thus different structures in the corresponding embedding spaces. Let us assume that the time series $\{x_{\mathrm{k}}\}$ is based on a dissipative dynamical system described by a set of differential equations $\dot{\boldsymbol{x}} = \boldsymbol{F}(\boldsymbol{x})$ and let n be the dimension of the

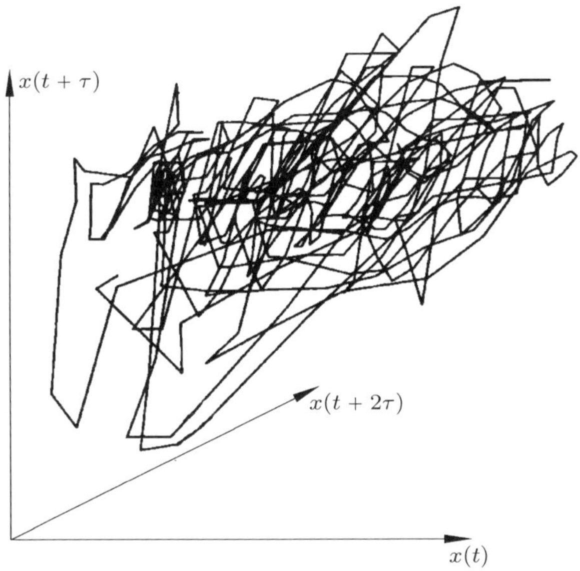

Fig. 5.5.18

Three-dimensional embedding of the climate attractor. Reconstruction from the time series of fig. 5.5.16 (Nicolis and Nicolis, 1984)

subspace respectively of the manifold $\mathbb{M}$ containing the attractor; then, in the case of a quasi-periodic motion on a two-dimensional torus, the dimension would be $\mathrm{n} = 2$. If we now consider all diffeomorphic, i.e. uniquely reversible, continuously differentiable maps of $\mathbb{M}$, the inverse maps of which are also continuously differentiable, then the question as to the dimension of the required embedding space for which an intersection of the trajectories is precluded is by no means trivial. If we take diffeomorphic maps of a circle, for example, a closed helical line on a torus would require a three-dimensional embedding space. Equally, there is a classic example for the fact that not all two-dimensional surfaces can be represented in a three-dimensional space without self-intersection. If we take a piece of a cylinder and attach the ends to one another with opposite orientation, the result is a so-called Klein bottle requiring a four-dimensional embedding space. According to a theorem of Whitney, (Whitney, 1936), an embedding dimension of $2\mathrm{n} + 1$ is certainly sufficient; for more details, see e.g. (Guillemin and Pollack, 1974) or (Bröcker and Jänich, 1990).

Based on this, Takens created the mathematical background for the reconstruction method by proving the following theorem (Takens, 1981):

> If the deterministic system $\dot{\boldsymbol{x}} = \boldsymbol{F}(\boldsymbol{x})$ generates a flow on an n-dimensional manifold $\mathbb{M}$, then
>
> $$\boldsymbol{\xi}(t) = \{x(t)\ x(t+\tau)\ \ldots\ x(t+2\mathrm{n}\tau)\} \tag{5.5.53}$$
>
> represents a continuously differentiable embedding, where x can be any component of the vector $\boldsymbol{x}$.

This means that, in the reconstruction, the geometrical invariants of the dynamics – such as the dimension of the attractor and the positive Lyapunov exponents, for example – are certainly preserved if

$$m \geqslant 2n + 1 \tag{5.5.54}$$

is selected as the embedding dimension. According to the theorem of Whitney, m = 2n + 1 is a sufficient condition for the embedding dimension. Depending on the geometrical complexity of the attractor A respectively the n-dimensional manifold $\mathbb{M}$, which contains A, a smaller embedding dimension suffices in many cases to guarantee the conservation of the geometrical invariants.

In practice, we determine the embedding dimension from a given time series by calculating for increasing dimensions m of the embedding space in accordance with eq. (5.5.42) the correlation integral $C(r)$ of the set of points defined by eq. (5.5.52). According to eq. (5.5.43), the dimensionality d of an attractor is then approximated by the gradient of the curves

$$\ln C(r) = d|\ln r| \tag{5.5.55}$$

in a particular r-range. If we increase the embedding dimension m and the gradients of the curves reach a saturation for relatively small m-values, then an attractor exists for the given time series.

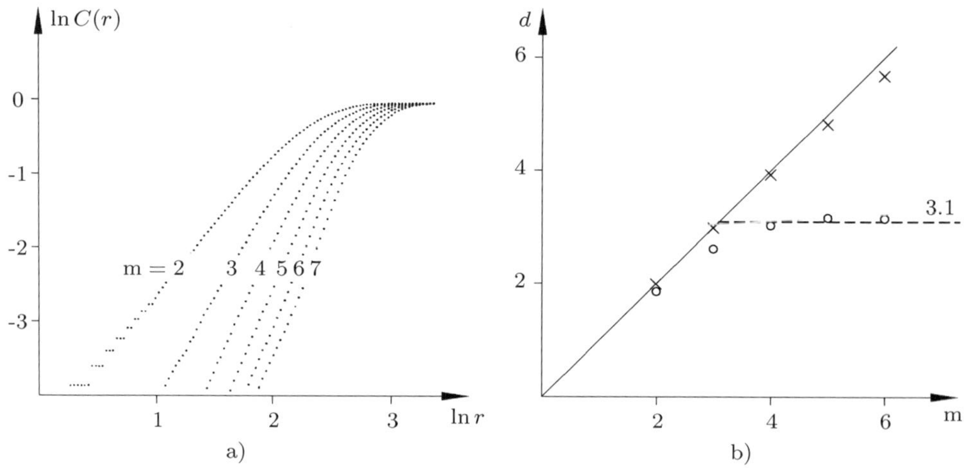

Fig. 5.5.19: Determination of the correlation function $C(r)$ and the dimension d of the climate attractor for increasing embedding dimension m (Nicolis and Prigogine, 1987)

To clarify matters, we briefly reproduce in fig. 5.5.19 the results from the analysis of the climate data of fig. 5.5.16. Figure 5.5.19a displays the function $C(r)$ for varying values of the embedding dimension. The family of curves indicates that a linear interrelation exists with the gradient d over a wide range. If we plot the slopes d as a function of the dimension m of the embedding space, we note a saturation for m $\approx$ 5 to 6 and obtain the correlation dimension $D_K \approx 3.1$. On the other hand, were we to evaluate a random signal corresponding to white noise, there would be no saturation and d would increase proportionally to m (×-signs in fig. 5.5.19b). This result suggests that the evolution of the ice volume takes place on a low-dimensional strange attractor. On the one hand, this confirms the unpredictability of climate evolution; on the other hand, the result $D_K \approx 3.1$ means that the underlying dynamical process can be described by a deterministic system of ordinary differential equations involving max. 9 fundamental variables; for further discussions, see (Grassberger, 1986; Nicolis and Nicolis, 1987); cf. also (Fraedrich, 1986).

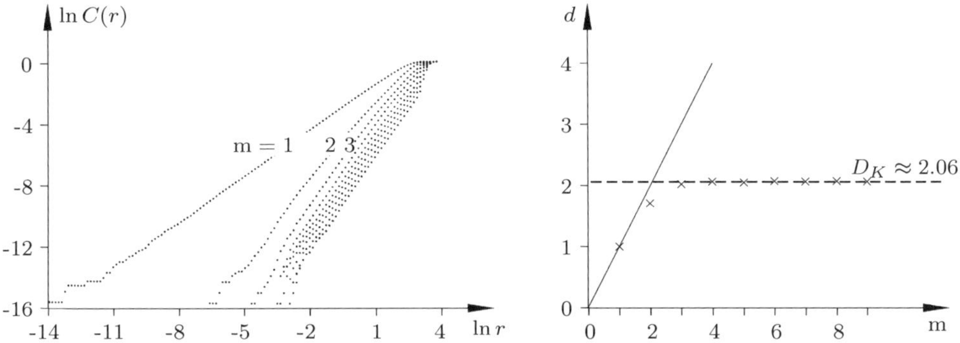

Fig. 5.5.20: Correlation dimension D_K of the Lorenz attractor calculated from the X-component ($D_K \approx 2.06$)

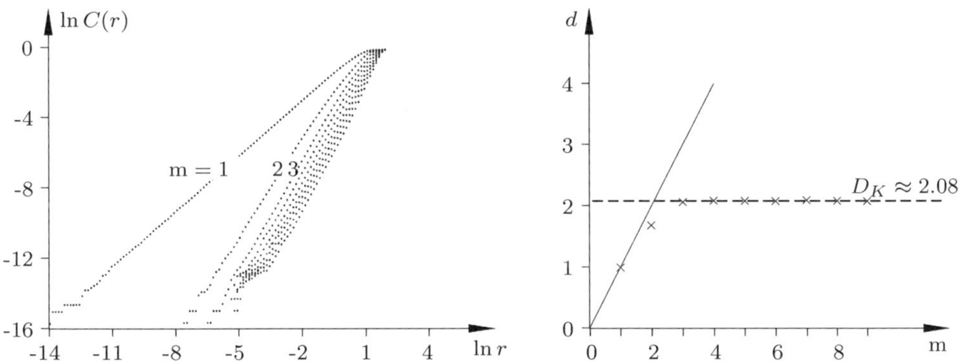

Fig. 5.5.21: Correlation dimension D_K of the Lorenz attractor calculated from the Z-component ($D_K \approx 2.08$)

Using the Lorenz equations (for $\sigma = 10,\ b = 8/3,\ r = 28$) as an example, we have, in order to determine the dimension, calculated the correlation function $C(r)$ from the time-history of the X- respectively Z-component alone (fig. 5.5.20 respectively fig. 5.5.21); here, the parameter r in $C(r)$ is the aforementioned distance of a pair of points. In the evaluation of the curves $\ln C(r)$ as a function of $\ln r$, it should be noted that only a central linear range provides meaningful statements. In the case of r-values which are too small, the curves are distorted by the noise that arises due to the measuring respectively rounding errors in the computer. If the r-values are of the magnitude of the global dimensions of the attractor, the corresponding $C(r)$-values are not meaningful either and lead to a flattening of the curves due to the finite number of the measuring points.

From the diagrams, we observe that $D_K \approx 2.06$ when operating exclusively with the X-component and $D_K \approx 2.08$ when using the Z-component, i.e. almost the same values which agree well with the capacity dimension $D_c \approx 2.06$ for the attractor in the full XYZ-phase space (cf. section 10.3).

Although the correlation dimension deduced from the time-history of the Z-component appears reasonable, caution is called for. The reason is that the Lorenz system possesses two (unstable) fixed points, the Z-coordinates of which coincide. In this case, a reconstruction from the Z-component does not supply an image which is topologically equivalent to the Lorenz attractor, see fig. 5.5.22 and cf. (Takens, 1981).

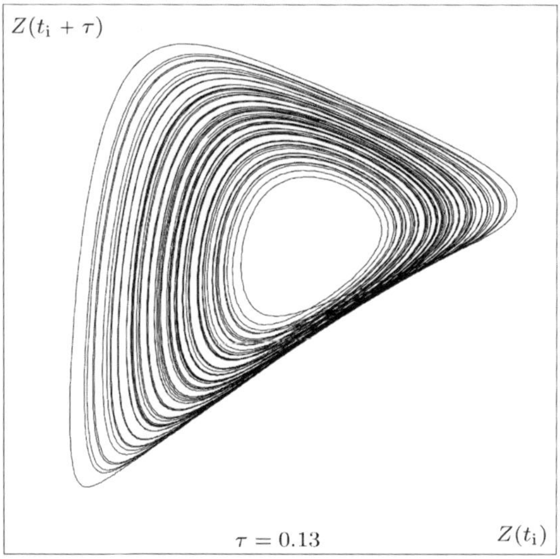

Fig. 5.5.22

Reconstruction of the Lorenz attractor from the Z-component: no topological equivalence

The reconstruction method has been applied to determine the dimension and the Lyapunov exponents in many experimental investigations, cf. section 10.9.3 and, for example, (Peinke et al., 1992). An analysis of the dynamics at the onset of turbulence in the Rayleigh-Bénard experiment is of particular interest. For small Rayleigh numbers, a strange attractor of the correlation dimension $D_K \approx 2.8$ was deduced (Malraison et al., 1983); cf. also section 8.2.

Concepts such as the correlation dimension and Lyapunov exponents have also been used to characterise musical sounds and to explain basic musical perception. It is the non-linear driving mechanisms and self-organisation which are the basic principles of the tone production of musical instruments and the human voice.

Together, they are crucial ingredients in producing a tight overtone spectrum in the first place, but sudden phase changes or the ability to play different tones on an instrument at the same time can also be explained by these concepts. For a deeper insight into the vast field of musical acoustics, the basic sound-production mechanisms of musical instruments and music psychology such as timbre, rhythm, pitch and tonality, we refer the reader to the monograph (Bader, 2013).

Apart from basic methods like Fourier spectra, autocorrelation and wavelet analysis, embedding representations in a reconstructed phase space, correlation dimensions and Lyapunov exponents turn out to be most suitable tools for understanding and measuring the complexity of sounds. In fig. 5.5.23, the time series and delay embeddings of four musical sounds are displayed. The saxophone produces a harmonic spectrum with only a few overtones (subharmonics) leading to a limit cycle; the bass drum displays an inharmonic spectrum with strong dissipation quickly tending to a fixed point; a cello tone with hard attack shows a rich overtone spectrum characteristic of its typical timbre and, depending on the bow pressure, changes its dissipation. The most complex dynamics can be seen in the crash cymbal of a drum kit. As in fig. 5.5.19, the embedding dimension is not sufficient; this is due to the presence of inharmonic components. In such cases, the correlation dimension and the Lyapunov exponents provide further information.

The integer part of the correlation dimension measures the number of all inharmonic components including the respective harmonic spectra of a musical sound. For example, a violin tone with up to 80 overtones has a correlation dimension $D_K = 1$. Harmonic overtone spectra are fused by the auditory system into one pitch sensation which we call a musical note. Playing the chord C-E-G on a piano thus results in $D_K = 3$. However, more complex behaviour of musical tones can also be investigated. When saxophones or clarinets are played at the threshold lung pressure for tone production, they vibrate in a bifurcation regime producing more than one tone at the same time, a sound known in music as multiphonic. In (Gibiat and Castellengo, 2000), these multiphonics are measured using the correlation dimension. As the correlation dimension is a measure of the complexity, it also serves as a tool for analysing music without a score, like contemporary classical music or free improvised pieces (Borgo, 2005).

(Touzé and Chaigne, 2000) determine the Lyapunov exponents from the time series of a cymbal used in the drum sets of jazz and rock music. Cymbals are highly non-linear percussion instruments. Drummers often hit cymbals with strong strokes using a wooden stick, causing oscillations with an amplitude larger than the thickness of the instrument, leading to the typical cymbal sound known as "crash". When the cymbals are played in this way with increasing amplitude, there comes a point beyond which the sound spectrum shows sudden phase transitions: additional spectral peaks occur which do not correspond to the eigenmodes of the cymbal. Moreover, mode coupling can be observed with a transference of energy from the lower to the higher modes, as with the Chinese Tam-Tam. In (Touzé and Chaigne, 2000), a positive Lyapunov exponent is calculated in the transition scenario for the cymbal, providing an additional characterisation of highly complex "chaotic" behaviour.

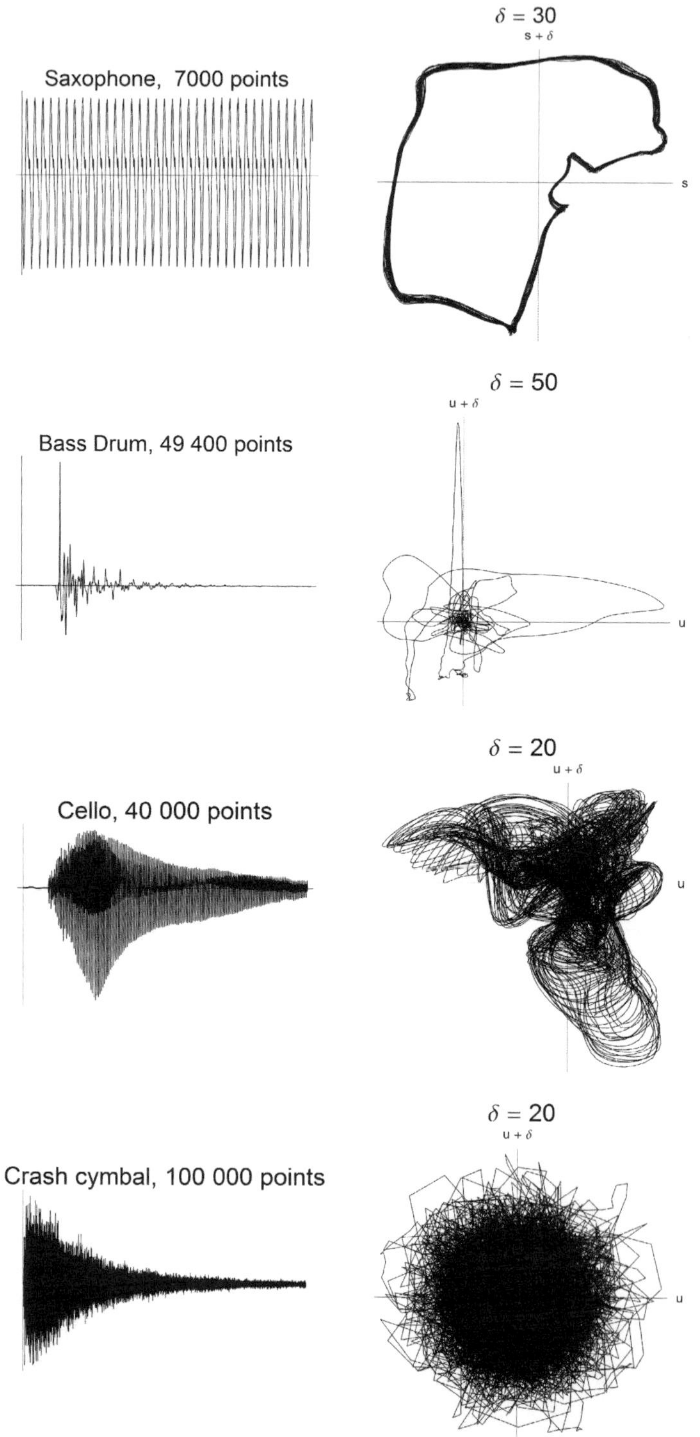

Fig. 5.5.23: Four examples of musical sounds. Time series (left side) and reconstruction of the phase space with different delays δ (right side). From: (Bader, 2013).

We finally wish to add some comments on the numerical calculation of the correlation dimension from one-dimensional time series. In the case of an infinite set of noise-free data, on account of the underlying theorems, there is no restriction with regard to the selection of the time delay τ for the reconstruction of an attractor according to the time delay method. In practice, however, we have only records of finite length which are, moreover, noise-infested due to measurement imprecisions respectively rounding errors. Even if we succeeded in separating measurement noise from the intrinsic dynamical process, see (Böttcher et al., 2006), suitable parameters must be chosen for the reconstruction method and the dimension calculation in order to gain reliable results (Kantz and Schreiber, 1997; Schreiber, 1999).

An investigation of Eckmann and Ruelle (1992) shows which minimum magnitude for the length N of the record must be selected in order to establish the dimension of a set of points. They obtained the rough estimate

$$D_K < 2\log_{10}\mathrm{N}$$

For example, at least 10^5 measurement points are necessary in order to calculate dimensions of the order $D_K \leqslant 10$; cf. discussion on the dimension of the climate attractor: (Grassberger, 1986; Nicolis and Nicolis, 1987).

The second comment concerns the choice of the time delay τ which is responsible for the "quality" of the reconstruction and thus also for the accuracy of the invariants extracted from it such as dimension and Lyapunov exponents. If the values selected for τ are too small, the random errors in the data respectively the limited number of digits cause consecutive values $x(t)$ and $x(t+\tau)$ to be almost identical so that consecutive vectors $\boldsymbol{\xi}_\mathrm{k}$ and $\boldsymbol{\xi}_{\mathrm{k}+1}$ in eq. (5.5.52) are almost linearly dependent. On the other hand, the value of τ must not be too large, otherwise there is no longer a causal link between consecutive sampling points, a confusion of lines results and the structure of the attractor is lost. Figure 5.5.24 illustrates this situation for the Y-component of the Lorenz attractor (Kugiumtzis, 1996).

In the literature, there is a whole series of suggestions for the choice of τ. According to (Packard et al., 1980) $\tau \ll \varepsilon/\Lambda$ should be chosen where ε denotes the accuracy with which a state of the system can be specified and $\Lambda = \sum^+ \sigma_\mathrm{i}$ is the sum over the positive Lyapunov exponents. Another possibility for avoiding linear dependencies of the vectors $\boldsymbol{\xi}_\mathrm{k}$ and $\boldsymbol{\xi}_{\mathrm{k}+1}$ is choosing the decay time of the autocorrelation function (Schuster and Just, 2006) for τ respectively the first zero of this function. One particularly promising suggestion is to select the first zero of the so-called transinformation or mutual information, see (Jumarie, 1990), and to minimise for higher-dimensional reconstructions a generalised mutual information, the so-called redundancy (Fraser, 1989b; Fraser, 1989a). The mutual information states how much information can be gained on average about a state at the time $t+\tau$ knowing the state at the time t; this information offers the possibility of revealing *non-linear* interrelations.

Finally, we point out a further method which can be used to determine the essential modes describing the long-term behaviour of a dynamical system and to reconstruct attractors from one-dimensional time series; this method is based on the singular value decomposition (Broomhead and King, 1986; King et al., 1987). From the

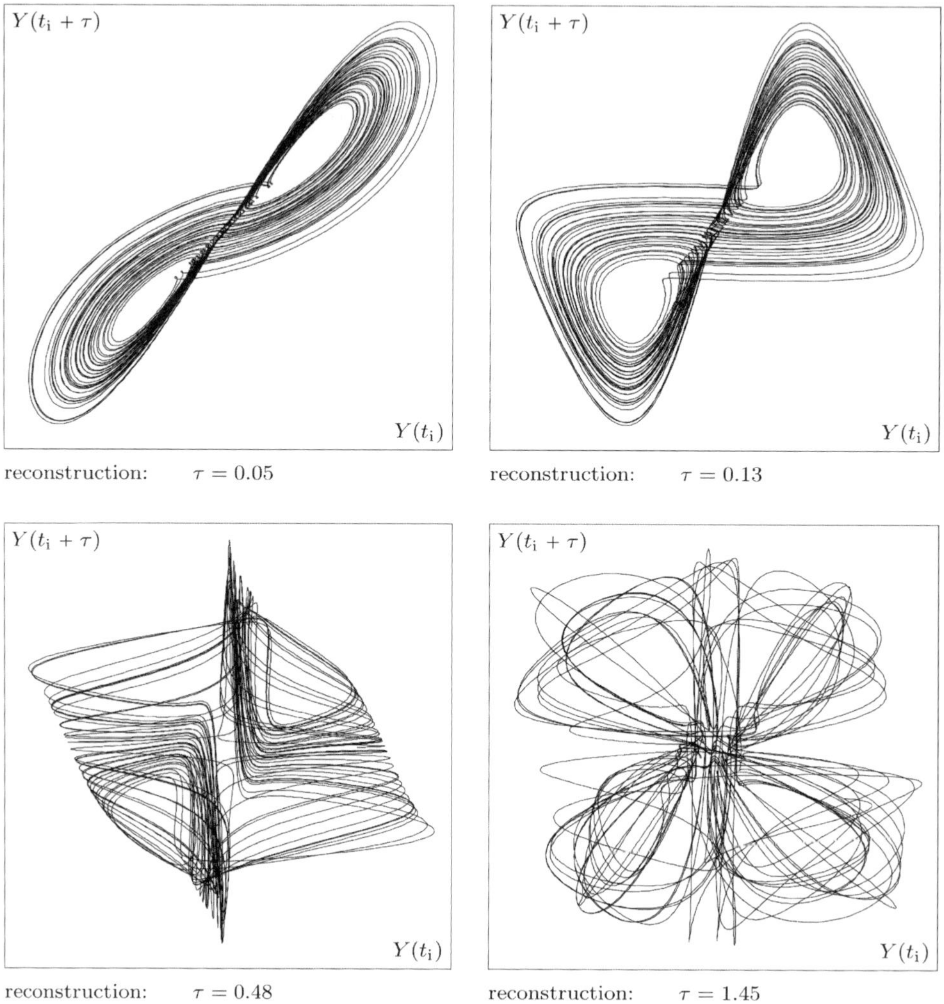

Fig. 5.5.24: Reconstruction of the Lorenz attractor (from the Y-component) in dependence of the time delay τ

one-dimensional data record $x_0(t_0)$, $x_1(t_0 + \Delta t)$, $x_2(t_0 + 2\Delta t), \ldots$, eq. (5.5.51), we again form in accordance with eq. (5.5.52) vectors $\boldsymbol{\xi}_0$, $\boldsymbol{\xi}_1$, $\ldots\boldsymbol{\xi}_\mathrm{N}$ of length m, but select here for the shift τ the fixed time increment Δt of the initially measured data set. From this set of vectors, we form the $(\mathrm{N} \times \mathrm{m})$-dimensional trajectory matrix

$$\boldsymbol{X} = \begin{bmatrix} \boldsymbol{\xi}_0^\mathrm{t} \\ \boldsymbol{\xi}_1^\mathrm{t} \\ \vdots \\ \boldsymbol{\xi}_\mathrm{N}^\mathrm{t} \end{bmatrix}$$

Broomhead and King's basic idea (1986) is that the smallest sufficient embedding dimension coincides with the number k of the linearly independent vectors of the trajectory matrix $\boldsymbol{X}$ for sufficiently large m. In order to determine the rank k of the matrix $\boldsymbol{X}$, we use the singular value decomposition (Golub and van Loan, 1983), i.e. we form the (m×m)-dimensional covariance matrix

$$\boldsymbol{C} = \frac{1}{\mathrm{N}}\boldsymbol{X}^{\mathrm{t}}\boldsymbol{X}$$

which also has the rank k, and determine its eigenvalues

$$\lambda_1 \geqslant \lambda_2 \geqslant \ldots \geqslant \lambda_{\mathrm{k}} > \lambda_{\mathrm{k}+1} = \ldots \lambda_{\mathrm{m}} = 0$$

and the appertaining eigenvectors. In practice, however, noise and the limited accuracy of the data recordings cause a shift of the eigenvalues

$$\lambda_1 \geqslant \lambda_2 \geqslant \ldots \geqslant \lambda_{\mathrm{k}} \gg \lambda_{\mathrm{k}+1} \geqslant \ldots \geqslant \lambda_{\mathrm{m}} > 0$$

The number k of the significant eigenvalues above the noise level yields a sufficiently large embedding dimension and a projection of the time series on the space spanned by the appertaining eigenvectors can be considered a noise-free reconstruction of the attractor.

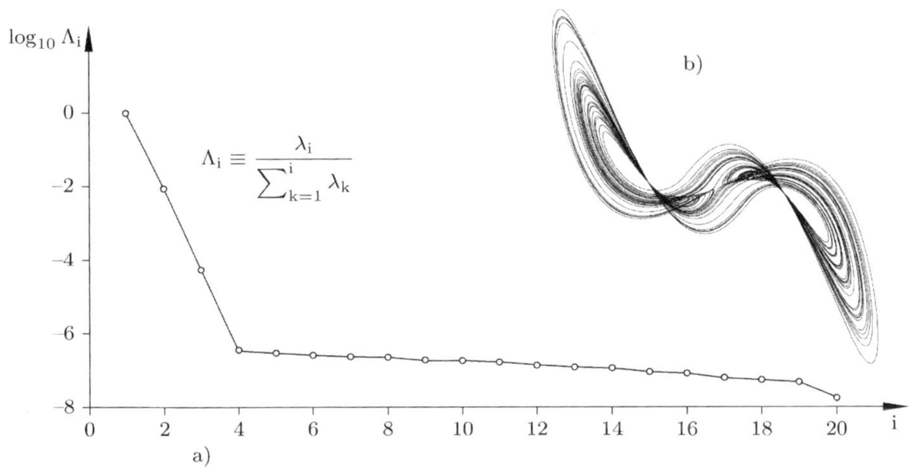

Fig. 5.5.25: Singular value decomposition, example from (Berndes, 1992):
a) spectrum of the eigenvalues;
b) reconstruction of the Lorenz attractor from the X-component

Figure 5.5.25 shows the spectrum of the eigenvalues and a reconstruction of the Lorenz attractor from the X-component according to the singular value decomposition of Broomhead and King.

The advantage of this procedure compared with the time delay method with subsequent determination of the correlation dimension lies in the fact that we avoid the selection of a suitable time delay τ and can automatically establish a sufficient embedding dimension by means of the number of significant eigenvalues; moreover, a filter for reducing the noise interference is built into the procedure. However, here too, the number k of the significant eigenvalues depends on the window width $\tau_\omega = \mathrm{m}\Delta t$ (King et al., 1987; Kugiumtzis, 1996). A further disadvantage of the singular value decomposition is that only an optimal *linear* embedding can be found since the covariance matrix only reflects linear, but not general dependencies (Fraser, 1989b).

What is the significance of the various reconstruction methods and the determination of the dimension from a single time series? Due to its complicated temporal evolution, a time series, as shown in fig. 5.5.16, for example, at first seems to be completely irregular. It cannot be guessed from this representation whether the dynamics is based on any deterministic laws and how many independent degrees of freedom are causing the erratic behaviour. It is known – and we shall discuss this in detail in section 6.8 and Chapter 8 – that many dissipative systems, such as, for example, the flow of fluids which are subject to an energy flow and which are described by partial differential equations with an infinite number of degrees of freedom, display a long-term behaviour on a macroscopic scale which is determined by a limited number of modes and the dynamics of which can be described by low-dimensional attractors.

The reconstruction method provides us with a mathematical procedure for ascertaining whether a given time series, either measured experimentally or produced in the computer, is based on some few determinsitic laws or whether the dynamics is generated by a high number of degrees of freedom. *Long-term* predictions are certainly not possible in the case of chaotic behaviour. Nevertheless, chaos theory and fractal geometry in the case of low-dimensional dynamics enable us to make short-term predictions since the dynamics rely in this case on only few deterministic laws (Farmer and Sidorowich, 1988).

5.5.5 Generalised Dimension D_q

According to a conjecture of (Farmer et al., 1983), basically only two independent dimension concepts exist, namely metric dimensions (such as the capacity dimension D_c) and probabilistic dimensions (such as the information dimension D_I). All other dimensions should hence be equivalent to either D_c or D_I.

In the same year, however, a publication by Hentschel and Procaccia appeared which brought order to the different dimension concepts (Hentschel and Procaccia, 1983). The authors showed that an infinite number of generalised dimensions D_q is required in order to describe the fractal characteristics of a strange attractor with an arbitrary inhomogeneous density distribution.

As indicated in fig. 5.5.8 we again consider a strange attractor in an n-dimensional phase space and divide the phase space evenly into n-dimensional hypercubes of edge length ε. If $W(\varepsilon)$ denotes the number of hypercubes containing attractor points, N the total number of measurement points on the attractor and $\mathrm{N_i}$ the number of measurement points in the i-th cell, then $p_\mathrm{i} = \mathrm{N_i/N}$ determines the probability of a measurement point lying in the i-th cube. According to Balatoni and Renyi, eq. (5.5.25) can be generalised and we can define a mean information of q-th order, the so-called Rényi information, (Balatoni and Rényi, 1957)

$$\bar{I}_{\mathrm{q}}(\varepsilon) = \frac{1}{1-\mathrm{q}} \ln \sum_{\mathrm{i}=1}^{W(\varepsilon)} (p_{\mathrm{i}})^{\mathrm{q}} \tag{5.5.56}$$

and, in analogy to this, as in eq. (5.5.26), we can introduce the generalised dimension of q-th order

$$D_{\mathrm{q}} = \lim_{\varepsilon \to 0} \frac{\bar{I}_{\mathrm{q}}(\varepsilon)}{\ln 1/\varepsilon} \tag{5.5.57}$$

We first show that this generalised dimension concept encompasses the aforementioned capacity, information and correlation dimensions as special cases.

a) $\mathrm{q} = 0$ (*capacity dimension*)

From eqs. (5.5.56) and (5.5.57), there results

$$D_0 = \lim_{\varepsilon \to 0} \frac{\ln \sum_{\mathrm{i}=1}^{W(\varepsilon)} 1}{\ln 1/\varepsilon} = \lim_{\varepsilon \to 0} \frac{\ln W(\varepsilon)}{\ln 1/\varepsilon} \tag{5.5.58}$$

i.e. D_0 coincides with the capacity dimension in accordance with eq. (5.5.8)

$$D_0 = D_c \tag{5.5.59}$$

b) $\mathrm{q} = 1$ (*information dimension*)

Due to $\sum_{\mathrm{i}=1}^{W(\varepsilon)} p_{\mathrm{i}} = 1$, numerator and denominator in eq. (5.5.56) are equal to zero. Applying the l'Hospital rule, however, the limit value for $\mathrm{q} \to 1$ can be derived by differentiation of numerator and denominator

$$\lim_{\mathrm{q} \to 1} \bar{I}_{\mathrm{q}}(\varepsilon) = \lim_{\mathrm{q} \to 1} \frac{\frac{\mathrm{d}}{\mathrm{dq}} \left(\ln \sum_{\mathrm{i}=1}^{W(\varepsilon)} p_{\mathrm{i}}^{\mathrm{q}} \right)}{-1} = \lim_{\mathrm{q} \to 1} \frac{-\sum_{\mathrm{i}=1}^{W(\varepsilon)} p_{\mathrm{i}}^{\mathrm{q}} \ln p_{\mathrm{i}}}{\sum_{\mathrm{i}=1}^{W(\varepsilon)} p_{\mathrm{i}}^{\mathrm{q}}} \tag{5.5.60}$$

Noting eqs. (5.5.25), (5.5.26), we thus obtain

$$D_1 = -\lim_{\varepsilon \to 0} \frac{\sum_{\mathrm{i}=1}^{W(\varepsilon)} p_{\mathrm{i}} \ln p_{\mathrm{i}}}{\ln 1/\varepsilon} = D_I \tag{5.5.61}$$

i.e. for $\mathrm{q} = 1$, we obtain exactly the information dimension.

c) $q = 2$ (*correlation dimension*)

In this case, $\sum_{\mathrm{i}=1}^{W(\varepsilon)} p_{\mathrm{i}}^2$ expresses the probability that two points of the attractor lie in a cube of edge length ε; then the sum coincides with the correlation integral $C(r)$ in accordance with eq. (5.5.41). For $\mathrm{q} = 2$, this yields the correlation dimension

$$D_2 = D_K \tag{5.5.62}$$

If q is an arbitrary natural number, D_{q} is evaluated as the mean value of the q-th powers (moments) of p_{i} applied to all those cells of the partitioning containing attractor points. Thus, $\sum_{\mathrm{i}=1}^{W(\varepsilon)} p_{\mathrm{i}}^{\mathrm{q}}$ expresses the probability of q points having a distance $\leqslant \varepsilon$. The larger the exponent q, the greater the weighting effect of the highly frequented areas of the attractor.

We can, however, generalise the significance of q and also define it for arbitrary positive and negative numbers. It may then be shown that D_{q} decreases monotonically with increasing order q (Hentschel and Procaccia, 1983)

$$D_{\mathrm{q}'} \leqslant D_{\mathrm{q}} \quad \text{for } \mathrm{q}' > \mathrm{q} \tag{5.5.63}$$

This relation obviously incorporates eq. (5.5.44), $D_K \leqslant D_I \leqslant D_0$, as a special case. If the distribution of points on the attractor is homogeneous, we have

$$D_{\mathrm{q}'} = D_{\mathrm{q}} \quad \text{for } \mathrm{q}' \lesseqgtr \mathrm{q} \tag{5.5.64}$$

In order to establish a complete two-point statistic of the fractal characteristics of inhomogeneous sets of points, we have to consider an infinite hierarchy of dimensions D_{q}. The limit values for $\mathrm{q} \to +\infty$ and $\mathrm{q} \to -\infty$ are of particular interest here. $D_{+\infty}$ describes the dimension of the most frequently visited region of the attractor while $D_{-\infty}$ is the dimension of the least visited region. In section 8.5.1, we shall discuss such a *multifractal* set of points in detail, using the example of the critical attractor of the circle map.

5.5.6 Lyapunov Dimension and Kaplan-Yorke Conjecture

The Lyapunov exponents which we discussed in section 5.4 characterise the dynamics of the attractor. They describe the stability behaviour of the system on the basis of the *temporal* mean value of the exponential convergence respectively divergence of neighbouring trajectories. The dimension concept, on the other hand, deals with the static characteristics of the attractor and derives from *spatial* mean values in the phase space. The link between these two approaches is provided by the ergodic theory. The existence of an invariant natural measure μ (cf. section 3.10) enables us to replace temporal mean values by spatial ones. For this reason, as early as 1979, attempts were made to develop a link between the Lyapunov exponents and the dimensionality of an attractor; these were mainly based on works of J.L. Kaplan and J.A. Yorke, (Kaplan and Yorke, 1979a; Farmer, 1982a; Frederickson et al., 1983).

In the following, we give a heuristic reasoning for the relationship between the Lyapunov exponents and the so-called Lyapunov dimension for the case of two-dimensional maps; cf. (Farmer et al., 1983). To this end, we first consider the Hénon map (Hénon, 1976), cf. also section 10.2,

$$\begin{aligned} x_{\mathrm{k}+1} &= y_{\mathrm{k}} + 1 - a x_{\mathrm{k}}^2 \\ y_{\mathrm{k}+1} &= b x_{\mathrm{k}} \end{aligned} \tag{5.5.65}$$

which possesses a strange attractor for the parameter values $a = 1.4$, $b = 0.3$. The determinant of the Jacobian matrix eq. (5.4.92) is constant and negative here

$$\det(\boldsymbol{D}_{\mathrm{k}}) = -b = -0.3 \tag{5.5.66}$$

i.e. the phase space shrinks uniformly and, in accordance with eqs. (5.4.81) and (5.4.98), we have

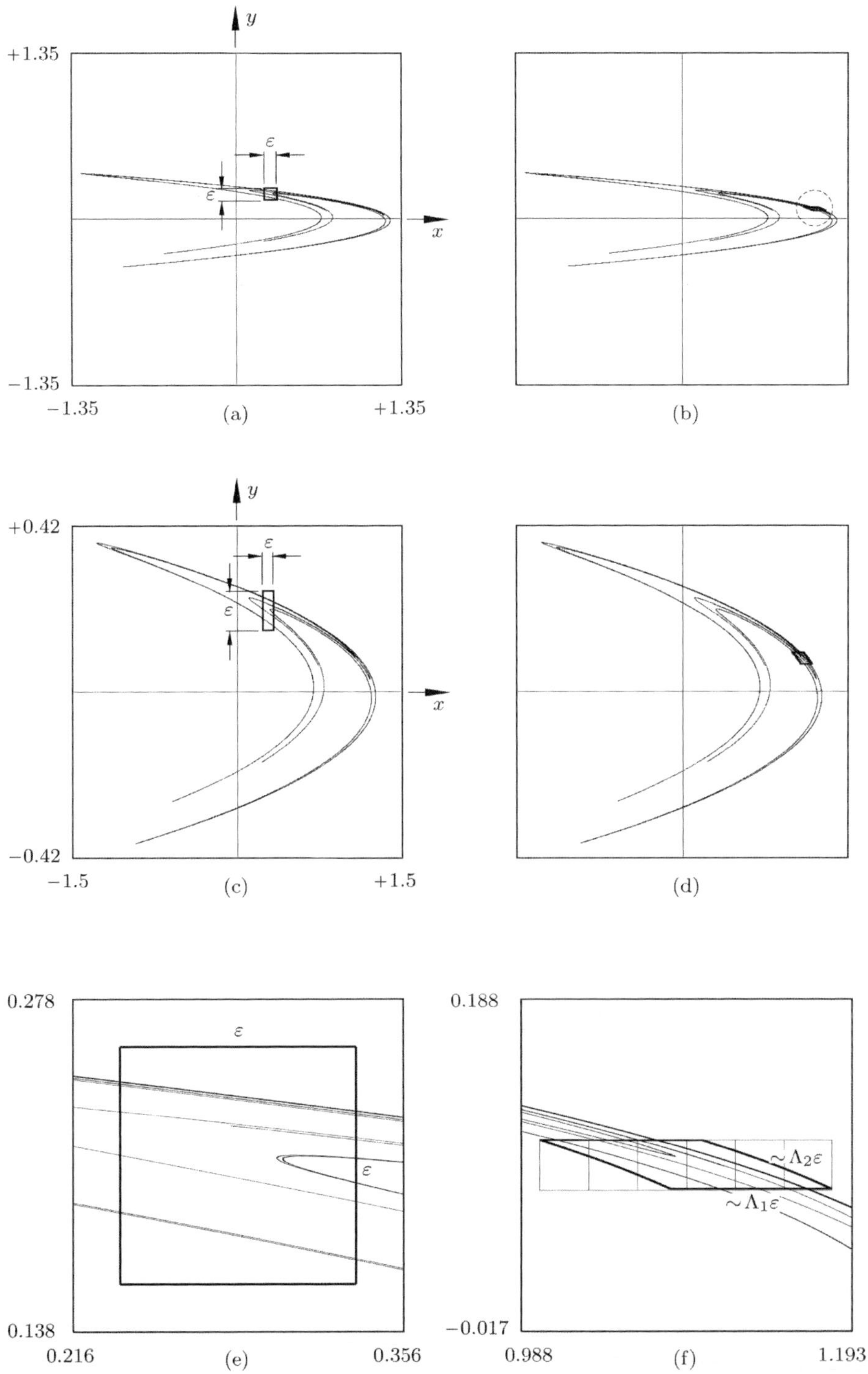

Fig. 5.5.26: Hénon attractor: on the interpretation of the Lyapunov dimension

$$\sigma_1 + \sigma_2 = \ln|\det(\boldsymbol{D}_\mathrm{k})| = \ln 0.3 \approx -1.204 \tag{5.5.67}$$

In this case, it thus suffices to establish the largest Lyapunov exponent; this yields $\sigma_1 \approx 0.42$ and, due to eq. (5.5.67), $\sigma_2 \approx -1.62$. Figure 5.5.26a presents this strange attractor, the Hénon attractor. In order to determine its dimension, we cover it with a uniform mesh with a spacing ε. If we subject the whole configuration to the map of eq. (5.5.65), the attractor basically retains its form while the mesh is deformed step by step (fig. 5.5.26b). Since the original square of edge length ε is already considerably distorted after one iteration step (fig. 5.5.26b), we have replaced the mappings figs. 5.5.26(a) and (b) with (c) and (d) where, in each step, we have adopted a new scale on the y-axis. In analogy to fig. 5.4.12, a small square of edge length ε has been deformed after one iteration approximately to an elongated parallelogram of side length $\Lambda_1\varepsilon$ and height $\Lambda_2\varepsilon$, where $\Lambda_\mathrm{i} = \mathrm{e}^{\sigma_\mathrm{i}}$ are the Lyapunov numbers. Figure 5.5.26(e) shows an enlarged detail of the attractor together with the cell shown in (a). After one iteration, this cell is mapped approximately onto a parallelogram (fig. 5.5.25f) which is elongated horizontally by the factor $\Lambda_1 = \mathrm{e}^{\sigma_1} \approx 1.52$ and flattened vertically by $\Lambda_2 = \mathrm{e}^{\sigma_2} \approx 0.20$.

Following this example, we now turn to a general two-dimensional map, the long-term behaviour of which is governed by a strange attractor so that for the Lyapunov numbers, we can assume

$$\Lambda_1 > 1 > \Lambda_2 \tag{5.5.68}$$

We now cover the attractor with a grid of squares of side length ε and determine the number $W(\varepsilon)$ of the boxes containing points of the attractor. If we iterate the map q times, the attractor as a whole remains invariant (see fig. 5.5.26) while the grid is distorted by the dynamics. If ε is small enough, after q mapping steps, a quadratic box is mapped approximately onto an elongated parallelogram (fig. 5.5.27), where the average side length is $\Lambda_1^\mathrm{q}\varepsilon$ and the average height $\Lambda_2^\mathrm{q}\varepsilon$. Thus, by subjecting the grid used to cover the attractor to the dynamics of the mapping, we obtain via the Lyapunov numbers further information about the number of boxes required. Were we to use a finer mesh with boxes of side length $\Lambda_2^\mathrm{q}\varepsilon$ to cover the attractor (fig. 5.5.27), about $(\Lambda_1/\Lambda_2)^\mathrm{q}$ smaller boxes would be necessary to cover each previous parallelogram. Assuming that *all* the boxes of the attractor are distorted uniformly in the manner described above, we obtain the relation

$$W(\Lambda_2^\mathrm{q}\varepsilon) \approx (\Lambda_1/\Lambda_2)^\mathrm{q} W(\varepsilon) \tag{5.5.69}$$

Let us recall the definition of the capacity dimension eq. (5.5.6). In that case, the number of cubes required to cover the attractor

$$W(\varepsilon) \approx V/\varepsilon^{D_c}$$

Substitution of this relation into both sides of eq. (5.5.69) yields

$$V\left(\frac{1}{\Lambda_2^\mathrm{q}\varepsilon}\right)^{D_c} = \left(\frac{\Lambda_1}{\Lambda_2}\right)^\mathrm{q} V\frac{1}{\varepsilon^{D_c}} \tag{5.5.70}$$

After cancellation, we take the logarithm and obtain

$$\mathrm{q}D_c \ln\frac{1}{\Lambda_2} = \mathrm{q}\big(\ln\Lambda_1 + \ln\frac{1}{\Lambda_2}\big)$$

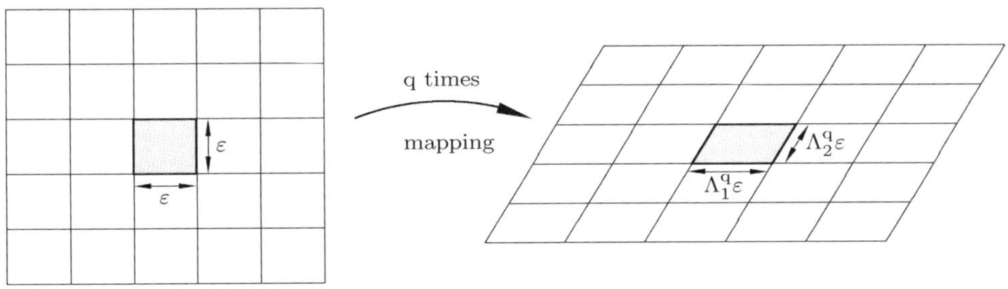

Fig. 5.5.27: Distortion of the mesh after q iterations of the map

or

$$D_c = 1 + \frac{\ln \Lambda_1}{(-\ln \Lambda_2)} \tag{5.5.71}$$

If we introduce the Lyapunov exponents σ_1, σ_2 as specified by eq. (5.4.99) and note that the relation eq. (5.5.68) leads to

$$\sigma_1 > 0 > \sigma_2 \tag{5.5.72}$$

we obtain

$$D_c = 1 + \frac{\sigma_1}{|\sigma_2|} \tag{5.5.73}$$

For cases in which the heuristic reasoning is not generally valid, Kaplan and Yorke (Kaplan and Yorke, 1979a) introduced a dimension concept based on the relation eq. (5.5.73), the so-called Lyapunov dimension D_L. In the two-dimensional case, this is given by

$$D_L = 1 + \frac{\sigma_1}{|\sigma_2|} \tag{5.5.74}$$

and generally, in the n-dimensional space,

$$D_L = \mathrm{m} + \frac{\sum_{\mathrm{i}=1}^{\mathrm{m}} \sigma_{\mathrm{i}}}{|\sigma_{\mathrm{m}+1}|} = \mathrm{m} + \frac{\sigma^{(\mathrm{m})}}{|\sigma_{\mathrm{m}+1}|} \tag{5.5.75}$$

where $\mathrm{m} \leqslant \mathrm{n}$ is the largest index for which the Lyapunov exponent of m-th order is $\sigma^{(\mathrm{m})} = \sum_{\mathrm{i}=1}^{\mathrm{m}} \sigma_{\mathrm{i}} \geqslant 0$ where we presume once more $\sigma_1 \geqslant \sigma_2 \geqslant \ldots \geqslant \sigma_{\mathrm{n}}$. If $\sigma_1 < 0$, we define $D_L = 0$.

It is easy to see that at least the integer term m in eq. (5.5.75) is reasonable. The condition $\sigma^{(\mathrm{m})} \geqslant 0$ means that the volume of the m-dimensional subspaces increases or at least remains constant in the mean whereas the volume of the (m + 1)-dimensional subspaces tends towards zero due to $\sigma^{(\mathrm{m}+1)} < 0$ (cf. section 5.4.4). Thus, each dimension D of the attractor must at least satisfy the relation $\mathrm{m} \leqslant D < \mathrm{m} + 1$. The term $\sigma^{(\mathrm{m})}/|\sigma_{\mathrm{m}+1}|$ represents a correction term in the presence of a strange attractor for which at least one Lyapunov exponent is positive.

In the original publication (Kaplan and Yorke, 1979a), the authors at first express the conjecture that the Lyapunov dimension coincides with the capacity dimension: $D_c = D_L$. This hypothesis was, however, refuted by means of counterexamples. Since the Lyapunov exponents are temporal mean values of the trajectories, the corresponding spatial averages must, in accordance with the ergodic theory, be weighted with the invariant natural measure; consequently, the density of the orbits on the attractor respectively the frequency with which parts of the phase space are visited will play a role. A more recent conjecture is indeed that the Lyapunov dimension coincides with the information dimension for a typical attractor (Frederickson et al., 1983)

$$D_L = D_I \tag{5.5.76}$$

This conjecture proved to be correct in a number of concrete examples (Farmer et al., 1983) and can also be confirmed for two-dimensional bijective mapping rules. As far as higher-dimensional systems are concerned, it has up to now only been possible to prove that the Lyapunov dimension forms an upper limit for the Hausdorff-Besicovitch dimension

$$D_H \leqslant D_L \tag{5.5.77}$$

A more detailed discussion of the Lyapunov dimension as well as further literature can be found in (Grassberger and Procaccia, 1984) or (Temam, 1988).

We close this section with a discussion of some aspects of the *numerical calculation of dimensions*. In order to determine the capacity dimension, the part of the phase space in which the attractor lies is covered uniformly with hypercubes of edge length ε and the number $W(\varepsilon)$ of cubes containing attractor points is counted. Here, the mesh size ε must be small enough to satisfy the condition $W(\varepsilon) \sim \varepsilon^{-D_c}$ i.e. the number of cubes required to cover the subspace increases exponentially with the dimension (Greenside et al., 1982; Peitgen et al., 1992). If we proceed directly according to this definition, the calculation effort is unjustifiably high for dimensions greater than three: 1 million cubes are needed for $\varepsilon = 0.01$ and $D_c = 3$! In section 10.3, we present a method of Hunt and Sullivan for determining D_c which considerably reduces the computational effort for the calculation of dimensions between 2 and 3 by means of Monte-Carlo simulations together with an effective programming code (Hunt and Sullivan, 1986; Hunt and Sullivan, 1988).

If the distribution density of the points on the attractor is highly inhomogeneous, a further problem occurs. In order to determine the capacity dimension, *all* the cubes containing attractor points must be counted. Some of them – their number does, however, increase rapidly with ε – are only rarely visited, i.e. in order to register them, very many attractor points are required. For this reason, the determination of the information dimension is much more reliable (Farmer et al., 1983) since the less probable cubes are irrelevant.

Due to the lower computational effort, either the mean pointwise dimension $\overline{D}_p$, the correlation dimension D_K (particularly in the case of reconstructions from an experimental time series) or the Lyapunov dimension D_L are usually preferred to all others. In the case of the Lyapunov dimension, the computational effort increases only approximately with the square of the number of dimensions. In the case of higher dimensions, this means a drastic reduction of computing time and storage requirements.

What is the *practical significance of the dimension concept*? Firstly, the dimension is a criterion for characterising the different attractor types: integer dimension values mean regular motions, non-integer values, on the other hand, dynamical processes which are irregular and unpredictable in the long term. Such considerations do not call upon a high degree of accuracy which is certainly not important in practice. If, however, the value of the dimension lies near an integer, as in the case of the Lorenz attractor, for example, where $D_c \approx 2.06$, the dimension concept does not suffice to characterise the motion unambiguously. In such cases, further criteria, such as the power spectrum or the Lyapunov exponents, must be called upon.

A further and perhaps the fundamental consideration of the significance of the dimension concept is that it provides the means of assessing the number of effective independent variables determining the long-term behaviour of a motion. The intensive research carried out in the field of non-linear processes in the last 30 years has aroused great interest in coherent structures or patterns in space and time, particularly in connection with the problem of turbulence. With the aid of the dimension concept, it could be shown (Malraison et al., 1983) that the onset of turbulence in the Rayleigh-Bénard experiment occurs due to the interaction of a few non-linear modes, although the underlying Navier-Stokes equations possess strictly an infinite number of degrees of freedom. The dimensionality of a system allows us to decide whether its complex behaviour requires a complex, high-dimensional description or, at least in principle, a few modes suffice to describe its long-term behaviour approximately.

5.6 Kolmogorov-Sinai Entropy

In Chapter 5, our attention is directed mainly at dissipative systems, their special characteristics and, above all, the quantitative characterisation of their attractors. The final criterion we wish to elucidate in this connection is the concept of entropy, as introduced by Andrei Nikolaevič Kolmogorov (Kolmogorov, 1958) and Yakov Grigor'evič Sinai (Sinai, 1959; Sinai, 1977). Originally, the metrical entropy was introduced by Shannon in the framework of his communication theory (Shannon and Weaver, 1949). Kolmogorov and Sinai applied this concept to dynamical systems and were able to prove that the KS entropy named after them is a topological invariant. It quantifies the degree of disorder, the degree of our ignorance of the system, and thus enables us to state the fundamental bounds of the predictability of a dynamical system. As in the case of the Lyapunov exponents, this criterion is, however, by no means applicable only to dissipative systems but can also be used to characterise conservative systems.

It is the aim and the strength of the natural sciences to set up a link between cause and effect, i.e. to find natural laws which allow us to make objective predictions about processes in nature. On the basis of Newton's laws, it is possible to make very precise predictions about the planetary orbits or the return of Halley's comet, for example, over long periods of time. Ultimately, the great successes in astronautics are based on the predictability of trajectories. There are also, moreover, other natural phenomena such as the motion of the atmosphere or the flow of a river which display such irregularities. The complex behaviour of these non-linear systems has its origin in the high number of coupled sub-systems, for the dynamics of which we know neither the initial conditions nor the details of their microscopic interactions with sufficient precision. Moreover, by means of the theory of dynamical systems, additional, thus far unknown mechanisms have been revealed which may lead on principle to bounds of predictability, even in systems with a low number of degrees of freedom, and even if the system obeys classical deterministic laws.

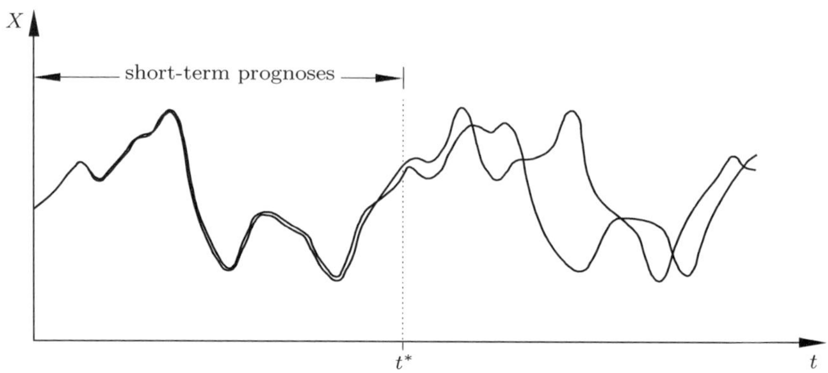

Fig. 5.6.1: Evolution of two trajectories of the Lorenz equations with almost identical initial conditions

The American meteorologist Edward Lorenz was the first to recognise that determinism does not necessarily imply predictability. In 1963, in the course of an attempt to improve weather forecasting, he developed a model encompassing three deterministic laws, i.e. containing no stochastic quantities, which nevertheless causes erratic behaviour (Lorenz, 1963), see also sections 5.2, 7.3 and 10.3. Even the smallest deviations in the initial conditions increase exponentially to such a degree that, after a short time, the curves bear no resemblance to one another (fig. 5.6.1).

The predictability of motions obeying deterministic laws would only be guaranteed for infinitely long periods of time if the initial conditions could be prescribed with an infinitely high degree of accuracy. In practice, however, any physical quantity can only be measured with a certain finite precision. In the same way, any number in the computer can only be represented by a limited number of digits. In the case of stable motions, inaccuracies in the initial values generally have no dramatic effect on the course of the motion. In chaotic states, however, these perturbations increase within a short time to such an extent that predictions can only be made for limited periods of time.

5.6.1 The Bernoulli Shift

One of the simplest deterministic systems which causes irregular unpredictable behaviour is a one-dimensional map of the unit interval onto itself, the so-called Bernoulli shift (fig. 5.6.2)

$$x_{n+1} = 2x_n \quad (\text{mod}1) \tag{5.6.1}$$

or, written in full,

$$x_{n+1} = \begin{cases} 2x_n & \text{for } 0 \leqslant x_n < \frac{1}{2} \\ 2x_n - 1 & \text{for } \frac{1}{2} \leqslant x_n < 1 \end{cases} \tag{5.6.2}$$

Like the tent map introduced in section 3.10.1, this map generates a sequence of numbers $x_0, x_1, x_2, \ldots$, the characteristics of which are particularly easy to grasp if we use a binary representation.

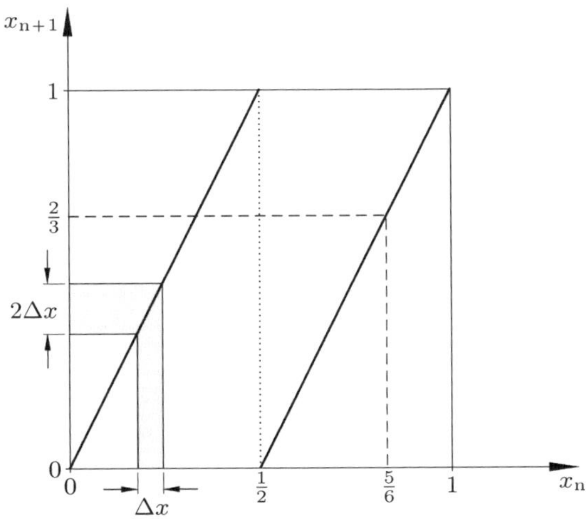

Fig. 5.6.2

The Bernoulli shift

To simplify understanding, we use the notation

$$x_0 = \sum_{i=1}^{\infty} a_i \frac{1}{2^i} \quad \hat{=} \quad (\,.\, a_1\, a_2\, a_3\, a_4 \ldots) \tag{5.6.3}$$

where the individual digits a_i can only assume the values 0 or 1. Points in the left-hand side of the unit interval with $0 \leqslant x_0 < \frac{1}{2}$ are characterised in this representation by $a_1 = 0$, points in the right-hand side with $\frac{1}{2} \leqslant x_0 < 1$ by $a_1 = 1$. Applying the mapping rule eq. (5.6.2), we obtain for $0 \leqslant x_0 < \frac{1}{2}$ (since $a_1 = 0$)

$$x_1 = 2\sum_{i=2}^{\infty} a_i \frac{1}{2^i} = \sum_{k=1}^{\infty} a_{k+1} \frac{1}{2^k} \quad \hat{=} \quad (\,.\, a_2\, a_3\, a_4 \ldots)$$

and for $\frac{1}{2} \leqslant x_0 < 1$ (since $a_1 = 1$)

$$x_1 = 2\left[\frac{1}{2} + \sum_{i=2}^{\infty} a_i \frac{1}{2^i}\right] - 1 = \sum_{k=1}^{\infty} a_{k+1} \frac{1}{2^k} \mathrel{\hat{=}} (\,.\,a_2\,a_3\,a_4\ldots)$$

In both cases, we thus obtain the same value, the whole sequence of digits being shifted one position to the left and the first digit dropped. An example illustrates this

$$x_0 = (.\ 1\ 0\ 0\ 1\ 1\ 0\ 1\ 1\ 1\ \ldots)$$

$$\swarrow\swarrow\swarrow\swarrow$$

$$x_1 = (.\ 0\ 0\ 1\ 1\ 0\ 1\ 1\ 1\ ?\ \ldots)$$

$$\swarrow\swarrow\swarrow$$

$$x_2 = (.\ 0\ 1\ 1\ 0\ 1\ 1\ 1\ ?\ ?\ \ldots)$$

ect.

If the initial value is a rational number, the sequence of numbers either tends towards a fixed point or becomes periodic as the following examples demonstrate. We assume that the state of the system can be expressed to four binary digits precisely.

(i)

$$\begin{aligned} x_0 &= \tfrac{1}{4} \mathrel{\hat{=}} (.0100) \\ x_1 &= \tfrac{1}{2} \mathrel{\hat{=}} (.1000) \\ x_2 &= 0 \mathrel{\hat{=}} (.0000) \\ x_3 &= 0 \mathrel{\hat{=}} (.0000) \\ &\vdots \end{aligned}$$

(ii)

$$\begin{aligned} x_0 &= \tfrac{1}{3} \mathrel{\hat{=}} (.0101) \\ x_1 &= \tfrac{2}{3} \mathrel{\hat{=}} (.1010) \\ x_2 &= \tfrac{1}{3} \mathrel{\hat{=}} (.0101) \\ x_3 &= \tfrac{2}{3} \mathrel{\hat{=}} (.1010) \\ &\vdots \end{aligned}$$

These two examples show that an observer receives no new information after a number of iterations respectively after the end of the first period. No surprises are in store for him; he could easily leave the room without missing anything.

If, on the other hand, the initial value is an irrational number, the map produces an infinite sequence of digits 0 or 1, i.e. per iteration, the system yields 1 bit of additional information about the position of the initial value. Figure 5.6.3 illustrates this zooming process, again assuming that we can read off the numbers precisely to four digits. The shifting of a zero into the four-digit domain indicates that the number lies in the left half of the interval currently considered, whereas a 1 characterises the right half.

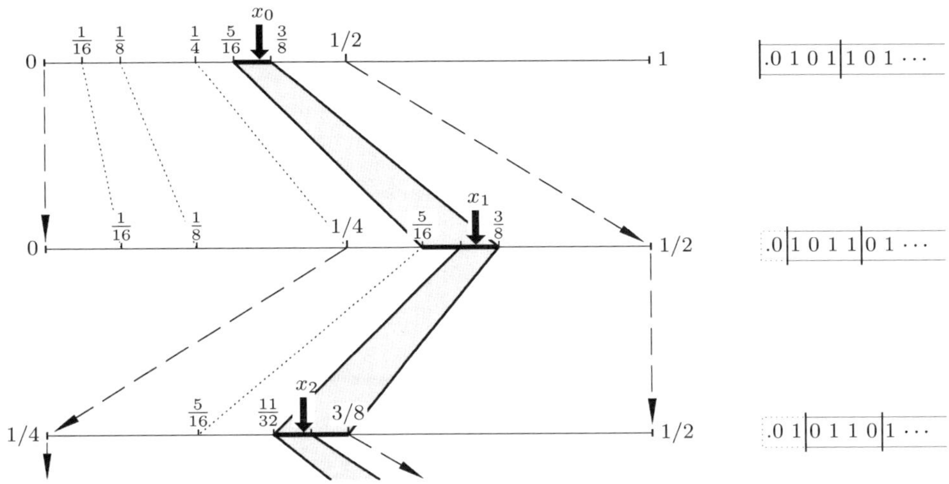

Fig. 5.6.3: Information gain in the Bernoulli shift

The information we gain per time unit – or here, per iteration step – is exactly the KS entropy h. In the case of the Bernoulli shift, it is

$$h = 1 \text{ bit / iteration}$$

There is also another interpretation of the KS entropy. We recall that, in thermodynamics, the concept of entropy S was introduced in order to characterise the disorder of a system.

Let us imagine a vessel containing gas molecules. By means of a piston, we ensure that, at the outset, all the gas molecules are, for example, in the right-hand side of vessel. The position of each individual molecule is uniquely determined by a yes-no statement (no $\hat{=}$ left, yes $\hat{=}$ right). If we remove the piston, then the molecules will be distributed in the whole vessel after a short time; the process is irreversible, i.e. the entropy of the system has increased, and increased entropy means more disorder. Simultaneously, however, our ignorance of the state of the system, i.e. of the position – left or right – of each molecule has increased. This caused Kolmogorov to transfer the concept of entropy to the ignorance of the microscopic details of chaotic systems.

As illustrated in fig. 5.6.2, any measurement x is only possible with a certain accuracy Δx . If we can specify the initial condition x_0 for the Bernoulli shift precisely to n binary digits, for example, this corresponds to a partitioning of the continuous unit interval into sub-intervals of length $2^{-\mathrm{n}}$. For each point, we can only state in which sub-interval it lies. As the inaccuracy is doubled with each iteration, the point can theoretically be found in double the number of sub-intervals after each step, cf. also fig. 3.10.3. After n iterations, the uncertainty, our ignorance, has spread over the whole interval.

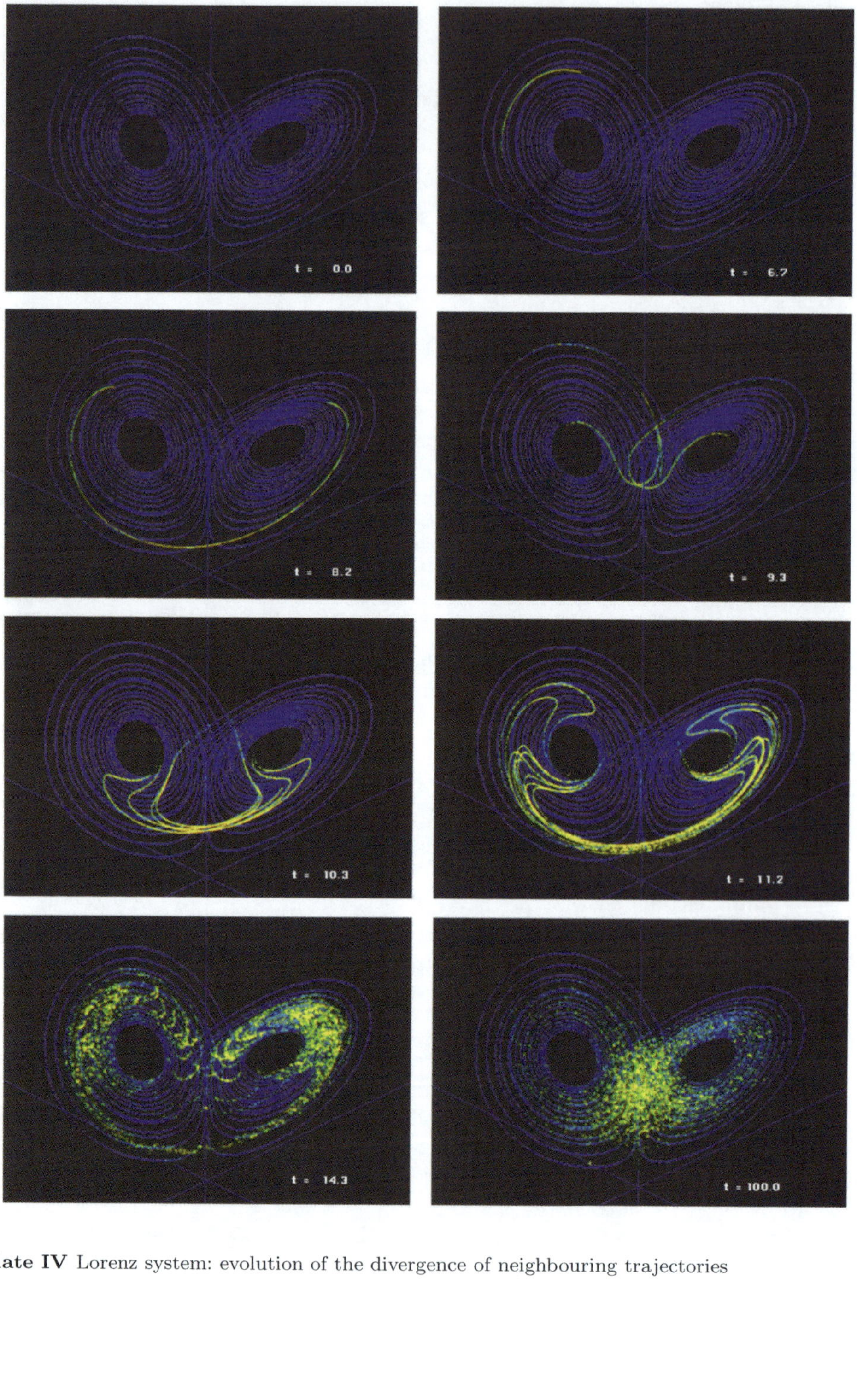

Plate IV Lorenz system: evolution of the divergence of neighbouring trajectories

The KS entropy hence measures the transport of information from a microscopic to a macroscopic scale, the latter being accessible for our measuring devices. Here, a constant flow of information from a microscopic to a macroscopic level takes place so that a series of *ongoing* measurements provides information about the dynamics of the system.

Robert Shaw (1981a) considers the Bernoulli shift a model of a physical process which is connected on a microscopic scale with a "heat bath", as physicists call it. After a finite number of steps, the experimenter should only observe fluctuations which occur on a considerably smaller time scale than the motion on the macroscopic observation level. Each chaotic system can thus be considered a vehicle for magnifying microscopic fluctuations exponentially in a short time and transferring them to a macroscopic scale accessible to our senses. This mechanism can be demonstrated very impressively using the Lorenz attractor (see Colour Plate IV, p. 286, "Lorenz dust"). In this example, we have selected 15000 initial conditions on a very small line segment, calculated the trajectories and taken snapshots at various instants of time. At first, the segment line of points stretches out to a long line; this is folded and stretched again and again and is finally distributed over the whole attractor like dust. Clearly, it is now impossible to make predictions over long periods, the final state can be anywhere on the attractor. With light-hearted exaggeration, meteorologists have denoted this the "butterfly effect": a single butterfly, moving the air in Peking by beating its wings, would he able to cause a storm in New York.

5.6.2 Definition of KS Entropy

The definition of the information dimension D_I in section 5.5.3 was based on the information gain which can be won by an observer with a single measurement. In contrast to this, KS entropy measures the information per time unit for a series of consecutive measurements. The information dimension is hence a static quantity which emerges from a "snapshot" of the system, whereas KS entropy takes the dynamics into account by observing a "film" of the system (Farmer, 1982c). If the dynamics is based on deterministic laws, then – at least for a certain period – correlations exist between the individual measurements, i.e. the probability that a particular reading will occur depends on the previous readings.

To discuss this aspect, we now present two examples. If we throw an (ideal) die a number of times, each throw is completely independent of the previous one. If, on the other hand, we take a pack of skat cards and take one without looking, the probability of drawing a heart, for example, is 1/4. If we were to replace the card and shuffle the pack well, the probability of again getting a heart remains unchanged. If we do not replace the card, however, our chances of drawing a heart are reduced to 7/31 since, of the 31 remaining cards, only seven are hearts. On the other hand, it has become more likely that we will now draw a spade (8/31).

The theory of conditional probabilities – in which correlations between single events are taken into account – was developed by the Russian mathematician Andrej Markov and is known as the theory of Markov chains. This theory forms the basis for many investigations in meteorology, physics and linguistics and is particularly important for the definition of KS entropy.

In the following, we can only give a rough overview of the basic ideas required for the definition of KS entropy. For details, we refer the interested reader to (Haken, 1988; Shaw, 1981b; Farmer, 1982c; Walters, 1982; Leven et al., 1989; Kantz and Schreiber, 1997), for example.

We first deal with the gain respectively loss of information in connection with measurements or calculations of dynamical processes. In eq. (5.5.25), we defined the information gain per single measurement as

$$\bar{I}(\varepsilon) = -K \sum_{i=1}^{W(\varepsilon)} p_i(\varepsilon) \ln p_i(\varepsilon) \,, \quad K = \frac{1}{\ln 2} \tag{5.6.4}$$

Here, $W(\varepsilon)$ was the number of hypercubes of edge length ε required to cover the attractor. Since each measuring device only provides a finite resolution, we automatically obtain a partitioning of the phase space in cells or blocks of edge length ε which we denote as macrostates of the system. Obviously, $\bar{I}(\varepsilon)$ increases with a refinement of the resolution, i.e. a reduction of ε.

Using three examples, we now wish to demonstrate the effects contractions respectively expansions of the phase space volume have on the individual cells and thus on the information gain or loss in the case of measurements taken at two different instants.

Example 1:

We consider a damped single-degree-of-freedom oscillator having the equation of motion $\ddot{x} - 2\zeta\dot{x} + x = 0$ or

$$\begin{aligned} \dot{x}_1 &= -x_2 \\ \dot{x}_2 &= x_1 - 2\zeta x_2 \qquad\qquad (\zeta = 0.2) \end{aligned} \tag{5.6.5}$$

Let a partitioning of the phase space in discrete macrostates be given by the measurement precision. At the point in time $t_0 = 0$, let the system be in cell Z_0, at a later instant t_1 in cell Z_1 (see fig. 5.6.4). Since the system is dissipative, the phase space volume shrinks and the flow Φ_t maps the original cell Z_0 on a smaller domain $\Phi_{t_1}(Z_0) \subset Z_1$. The observer at time t_1 has thus lost information about the initial conditions since conversely, $\Phi_{-t_1}(Z_1) \supset Z_0$ holds, i.e. there are still trajectories which start outside Z_0 at time $t_0 = 0$ and are in Z_1 at time t_1. The contraction of the phase space volume hence causes a loss of information.

Example 2:

The following example demonstrates the reverse case (see fig. 5.6.5). We consider the van der Pol equation $\ddot{x} + \mu(1 - x^2)\dot{x} + x = 0$ (see also section 10.4)

$$\begin{aligned} \dot{x}_1 &= +x_2 \\ \dot{x}_2 &= -x_1 - \mu(1 - x_1^2)x_2 \qquad\qquad (\mu = -0.4) \end{aligned} \tag{5.6.6}$$

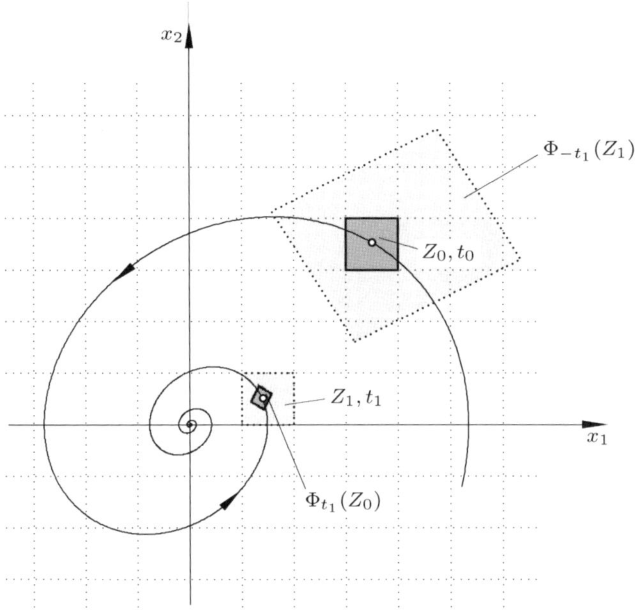

Fig. 5.6.4: Information loss for a damped single-degree-of-freedom oscillator

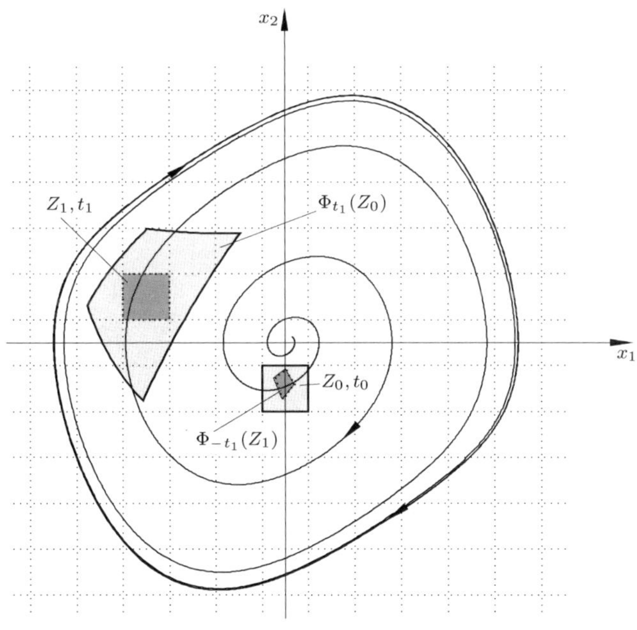

Fig. 5.6.5: Information gain with the van der Pol equation

and select an initial condition for $t_0 = 0$ inside the limit cycle in a cell Z_0 with $|x| \ll 1$, $|\dot{x}| \ll 1$. At the time t_1, the trajectory is in cell Z_1. If we subject the initial cell Z_0 to the flow Φ_t, then it will be enlarged due to the expansion of the phase space volume and encompass Z_1 at the time t_1: $\Phi_{t_1}(Z_0) \supset Z_1$. In this case, $\Phi_{-t_1}(Z_1) \subset Z_0$, i.e. the second measurement at the time t_1 enables us to localise the initial condition more accurately in Z_0. In this case, we have thus gained additional information; the expansion of the phase space volume acts as a source of information.

It should be noted, however, that in both cases, the information gain respectively loss can only be observed during the transient phase. Both the regular attractors – fixed point and limit cycle – are static objects in terms of the information theory. As soon as we have attained them – within our measuring precision – the system produces no new information, there are no more surprises. In our two examples, the cells displayed the same behaviour in *all* directions: they were either stretched or shortened. We know, however, that in the case of dissipative chaotic (or mixing) systems, volume elements expand in some directions while shrinking in others. The following example illustrates this behaviour.

Example 3:

We consider a simplified version of the dissipative baker's map, eq. Gl. (5.5.11),

$$x_{n+1} = 2x_n \pmod 1$$

$$y_{n+1} = \begin{cases} a y_n & \text{for } 0 \leqslant x_n < \frac{1}{2} \\ \frac{1}{2} + a y_n & \text{for } \frac{1}{2} \leqslant x_n < 1, \quad a < \frac{1}{2} \end{cases} \tag{5.6.7}$$

The mapping rule for the x-component corresponds exactly to the Bernoulli shift, eq. (5.6.1), i.e. a perturbation in x-direction will be doubled at each iteration step. Thus, the Lyapunov exponent in x-direction results as $\sigma_1 = \ln 2 > 0$ which leads to a sensitive dependence on the initial conditions. In contrast, the segments in y-direction are shortened and we obtain $\sigma_2 = \ln a < 0$ for the Lyapunov exponent in y-direction. The baker's map thus represents a chaotic mapping. The long-term behaviour is determined by a strange attractor consisting of two mille feuille layers between $0 \leqslant y < \frac{1}{4}$ and $\frac{1}{2} \leqslant y < \frac{3}{4}$ (see fig. 5.6.6b).

If we select an initial condition (x_0, y_0) in the cell Z_0 and carry out one iteration step, the point (x_1, y_1) is found in the cell Z_1 (see fig. 5.6.6a) where the image of the cell Z_0 extends over subsets of Z_1 and a neighbouring cell Z_1', i.e. $\Phi_1(Z_0) \subset (Z_1 \cup Z_1')$. After each iteration step, the number of cells in which the point might reside (in Z_1 or Z_1') has thus been doubled. On the other hand, there are additional initial conditions outside Z_0 which are mapped to the cell Z_1 after one step. The specific calculation result for $n = 1$ can thus be interpreted as an information gain. If we apply the inverse transformation to Z_1, the initial point must lie in the intersection $\Phi_{-1}(Z_1) \cap Z_0$, i.e. in a subset of Z_0, and can thus be localised more accurately. The baker's map eq. (5.6.7) is thus an inexhaustible source of information which does not dry up even in the case of $n \to \infty$. If

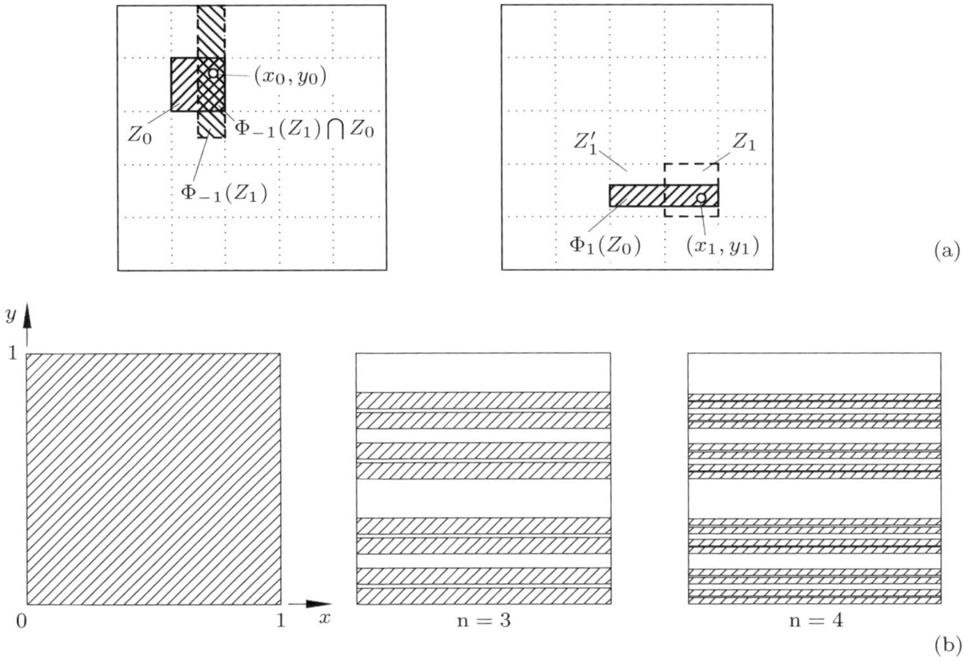

Fig. 5.6.6: a) Information gain in the case of the baker's map, $a = 0.4$, b) emergence of the mille feuille layers

we consider that the long-term behaviour of the baker's map is irregular, this demonstrates that a chaotic attractor continuously produces new information (Ott, 1993; Farmer, 1982c).

If we have a dynamical system in which – as in our example – the number of cells in which the image point may reside increases exponentially with the number of iterations respectively with time, long-term prognoses become impossible (Shaw, 1981b). Each tiny inaccuracy spreads exponentially over the whole attractor so that, after a finite period of time, no causal relation can be made between the initial and final point.

Following these examples, it is now possible to discuss the introduction of the KS entropy to deterministic systems in order to answer the question as to how much information can be gained about the initial state by a series of n consecutive measurements or, conversely, how much uncertainty as to the future state remains if n previous states are known. Expressed more precisely, KS entropy provides the upper limit of the mean information production per time unit if both measurement precision and recording frequency of the data are varied. In the case of regular, predictable behaviour, further measurements provide no new information from a particular instant onwards, i.e. the information gain is zero. If, on the other hand, the dynamics of the system is chaotic, the trajectories diverge exponentially in the mean and each subsequent measurement provides a further contribution to the information.

Our measuring respectively reading precision ε prescribes a partitioning of the phase space into a finite number of macrostates, namely $\mathrm{M}(\varepsilon)$ cells which we number consecutively $Z_1, Z_2, \ldots, Z_\mathrm{M}$ where, in contrast to the three previous examples, the lower index merely identifies the cells. If we start with an initial condition $\boldsymbol{x}_0$ in cell $Z_{\mathrm{i}(0)}$ and measure at fixed time intervals Δt each position $\boldsymbol{x}_1, \boldsymbol{x}_2, \boldsymbol{x}_3, \ldots$ of the image points of $\boldsymbol{x}_0$, the measuring device registers a sequence of cells $Z_{\mathrm{i}(1)}, Z_{\mathrm{i}(2)}, Z_{\mathrm{i}(3)}, \ldots$ in which the image points can be localised. The lower indices $\mathrm{i(k)} \in \{1, 2, \ldots, \mathrm{M}\}$ identify the cell $Z_{\mathrm{i(k)}}$ in which the image point $\boldsymbol{x}_\mathrm{k} = \boldsymbol{x}(t_0 + \mathrm{k}\Delta t)$ of $\boldsymbol{x}_0$ lies after k mapping steps. Since the system evolves in accordance with deterministic laws, we can conversely calculate the pre-image $\Phi_{-\Delta t}(Z_{\mathrm{i}(1)})$ of the cell $Z_{\mathrm{i}(1)}$. The second "compute measurement" has thus provided further information about the position of the original point $\boldsymbol{x}_0$: it must be in the intersection set $Z_{\mathrm{i}(0)} \cap \Phi_{-\Delta t}(Z_{\mathrm{i}(1)})$ (see fig. 5.6.7). The next measurement yields a point $\boldsymbol{x}_2$ in cell $Z_{\mathrm{i}(2)}$, thus providing information enabling us to localise the initial point more precisely, namely $\boldsymbol{x}_0 \in Z_{\mathrm{i}(0)} \cap \Phi_{-\Delta t} Z_{\mathrm{i}(1)} \cap \Phi_{-2\Delta t} Z_{\mathrm{i}(2)}$. If we continue this procedure, $\boldsymbol{x}_0$ must lie somewhere within n such intersection sets after n measurements.

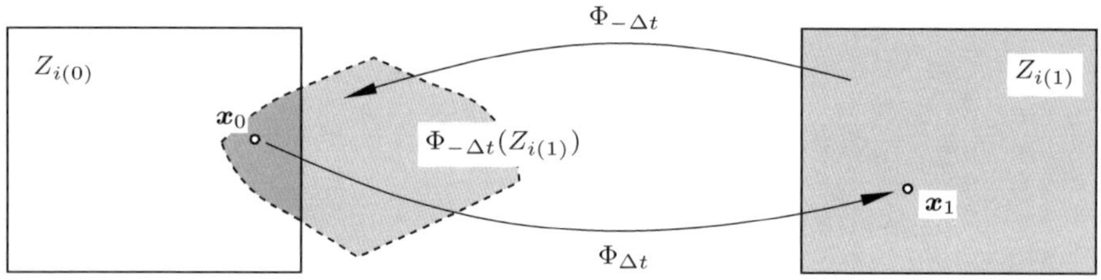

Fig. 5.6.7: On the definition of the KS entropy

Moreover, in order to obtain a mean rate for the information gain for a fixed ε, we average over all initial conditions. To this end, we consider simultaneously the intersection of all cells of a given partitioning with their own pre-images. Designating a given partitioning

$$Z = \{Z_\mathrm{i};\ \mathrm{i} = 1, \ldots, \mathrm{M}(\varepsilon)\} \tag{5.6.8}$$

and the corresponding pre-images

$$Y = \Phi_{-\Delta t} Z = \{Y_\mathrm{i};\ Y_\mathrm{i} = \Phi_{-\Delta t} Z_\mathrm{i},\ \mathrm{i} = 1, 2, \ldots \mathrm{M}(\varepsilon)\} \tag{5.6.9}$$

we introduce the following notation for all possible intersections

$$X^{(1)} := Z \wedge Y := \{X^{(1)}_{\mathrm{jk}};\ X^{(1)}_{\mathrm{jk}} = Z_\mathrm{j} \cap Y_\mathrm{k},\ \text{with } \mathrm{j,k} = 1, \ldots, \mathrm{M}(\varepsilon)\} \tag{5.6.10}$$

where $X^{(1)}$ comprises $\mathrm{M}^2(\varepsilon)$ subsets; we then call this set the first refinement of the partitioning (see fig. 5.6.8). In accordance with eq. (5.5.25), we can determine the corresponding information gain

$$\bar{I}(X^{(1)}) = -\sum_{\mathrm{j,k}=1}^{\mathrm{M}} p(X^{(1)}_{\mathrm{jk}}) \ln p(X^{(1)}_{\mathrm{jk}}) \tag{5.6.11}$$

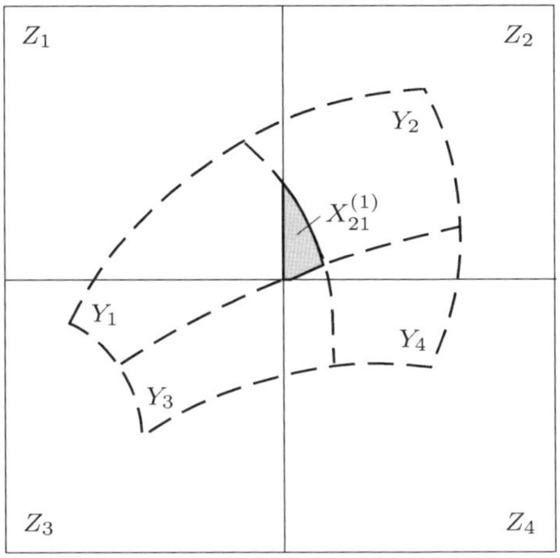

Fig. 5.6.8

Given partitioning $\{Z_i\}$, pre-images $\{Y_i\} = \{\Phi_{-\Delta t} Z_i\}$ and first refinement $\{X_{jk}^{(1)}\} = \{Z_j \cap Y_k\}$

where $p(X_{jk}^{(1)})$ is the probability of locating a point in cell $X_{jk}^{(1)}$ of the first refinement of the partitioning. We denote the n-th refinement as

$$X^{(n)} = Z \wedge \Phi_{-\Delta t} Z \wedge \Phi_{-2\Delta t} Z \wedge \ldots \wedge \Phi_{-n\Delta t} Z \tag{5.6.12}$$

In accordance with eq. (5.6.11), we can now calculate the information $\bar{I}(X^{(n)})$ appertaining to $X^{(n)}$. The KS entropy h is then defined as the upper limit of the average information gain per unit time produced by the dynamical system

$$h(\mu) = \sup_{Z,\Delta t} \left(\lim_{n\to\infty} \frac{\bar{I}(X^{(n)})}{n\Delta t} \right) \tag{5.6.13}$$

The supremum formation refers here to all possible partitionings Z and all possible time increments Δt. The KS entropy h depends on the invariant natural measure μ (cf. section 3.10) since, in accordance with eq. (5.6.11), we require the probability density in order to calculate the entropy.

As we have seen in the first two examples, for regular attractors – such as fixed point, limit cycle or torus – $h(\mu) = 0$. On the other hand, strange attractors are characterised by a positive finite value of KS entropy: $h(\mu) > 0$. If the dynamics of the system is completely random or "noisy", i.e. if there are no correlations in the n-dimensional phase space, then $h(\mu) \to \infty$ for $n \to \infty$. The KS entropy can thus also be applied for the qualitative characterisation of the motion. However, a calculation according to the abovementioned formula would scarcely be practicable, particularly due to the required determination of the supremum. Fortunately, Ya.B. Pesin (Pesin, 1977) has succeeded in setting up a link between h and the Lyapunov exponents which yields a practical calculation tool for h.

5.6.3 Link between KS Entropy and Lyapunov Exponents

Typical of chaotic systems is their sensitive dependence on the initial conditions caused by the exponential divergence of neighbouring trajectories. Two initial conditions which are at first minimally apart and cannot be distinguished within a certain accuracy of measurement diverge and can be distinguished from one another after a finite period of time. Microscopic deviations become visible on a macroscopic scale after a short time and the dynamical system thus acts as a source of information. It is hence possible to imagine intuitively that there must be a link between the KS entropy $h(\mu)$ and the *positive* Lyapunov exponents which cause the drifting apart.

Let us again first consider a one-dimensional map $x_{k+1} = f(x_k)$ and assume that we carry out a single measurement with a certain accuracy $\varepsilon \ll 1$. In accordance with eq. (5.5.26) and in a good approximation, we obtain for the mean information content in the initial condition

$$\bar{I}(\varepsilon) \approx D_I \log_2 \frac{1}{\varepsilon} \tag{5.6.14}$$

where D_I is the information dimension. After one iteration step, the interval of length ε is stretched on average by the factor $\Lambda = \mathrm{e}^{\sigma} > 1$ (see fig. 5.4.12) where Λ is the Lyapunov number respectively σ the positive Lyapunov exponent.

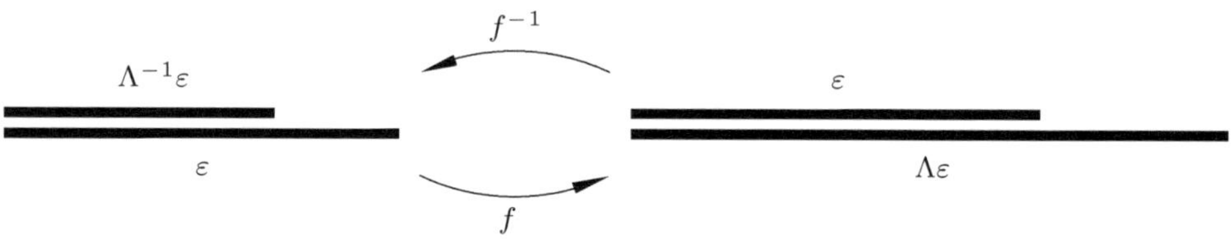

Fig. 5.6.9: On the change of the information content per iteration for one-dimensional maps

A second measurement with the same measuring device (i.e. the same resolution ε) yields the same amount of information as would have been obtained from a shorter interval ε/Λ (see fig. 5.6.9), i.e.

$$\bar{I}\Big(\frac{1}{\Lambda}\,\varepsilon\Big) \approx D_I \log_2 \frac{\Lambda}{\varepsilon} \tag{5.6.15}$$

After one iteration step, we thus obtain an average information gain of

$$\bar{I}\Big(\frac{1}{\Lambda}\,\varepsilon\Big) - \bar{I}(\varepsilon) \approx D_I \Big(\log_2 \frac{\Lambda}{\varepsilon} - \log_2 \frac{1}{\varepsilon}\Big)$$

This mean rate of increase, however, determines approximately the KS entropy and yields

$$h(\mu) \approx D_I \log_2 \Lambda = D_I \frac{\ln \Lambda}{\ln 2} \approx \frac{\sigma D_I}{\ln 2} \tag{5.6.16}$$

The factor $1/\ln 2$ stems from the fact that $h(\mu)$ is measured in bits per unit time. It could be avoided if this measurement unit were also used for the Lyapunov exponents. In the case of the Bernoulli shift eq. (5.6.1), we obtain $D_I = 1$ and $\sigma = \ln 2$, i.e. we have confirmed our earlier result of $h = 1$ bit/iteration. In general, however, we will have $0 \leqslant D_I \leqslant 1$ for one-dimensional maps so that the relation

$$h(\mu) \leqslant \frac{\sigma}{\ln 2} \tag{5.6.17}$$

respectively

$$h(\mu) \leqslant \sigma \tag{5.6.18}$$

holds when we measure σ in bit/unit time. For n-dimensional dynamical systems, Pesin (1977) was able to show that under certain conditions – for more details, see for example (Leven et al., 1989; Ruelle, 1989) – the following link exists between the KS entropy of a domain V of the phase space and the Lyapunov exponents

$$h(\mu) = \int\limits_V \sum{}^{+} \sigma_{\mathrm{i}}(\boldsymbol{x})\mu(\boldsymbol{x})\mathrm{d}\boldsymbol{x} \tag{5.6.19}$$

where the upper index $^+$ indicates that we only sum over all the positive Lyapunov exponents and $\mu(\boldsymbol{x})$ is the invariant natural measure, i.e. the probability density. This relation is simplified if we select the basin of attraction of a strange attractor A, for example, as the domain of integration. Under these conditions, the Lyapunov exponents are the same for almost all initial conditions, i.e. independent of $\boldsymbol{x}$, and due to

$$\int\limits_A \mu(\boldsymbol{x})\mathrm{d}\boldsymbol{x} = 1 \tag{5.6.20}$$

eq. (5.6.19) can be simplified to

$$h(\mu) = \sum_{\mathrm{i}}{}^{+} \sigma_{\mathrm{i}} \tag{5.6.21}$$

It should be noted that every Lyapunov exponent must be included in the sum in accordance with its multiplicity. If the conditions for eq. (5.6.19) are not satisfied, then, according to Ruelle (1989), at least the inequality

$$h(\mu) \leqslant \sum_{\mathrm{i}}{}^{+} \sigma_{\mathrm{i}} \tag{5.6.22}$$

holds as long as the natural measure μ is continuous in all directions with positive σ_{i}-values. This condition appears to be satisfied in most of the cases that interest us.

We can also give a lower limit, the so-called correlation entropy $h_2(\mu)$, for the KS entropy $h_2(\mu)$

$$h_2(\mu) \leqslant h(\mu) \tag{5.6.23}$$

see e.g. (Schuster and Just, 2006). This relation is of interest because the calculation of $h_2(\mu)$ is based on the correlation integral, eq. (5.5.41), and can thus be determined experimentally with relative ease, particularly when reconstructing attractors. It would exceed the scope of this book, however, to reproduce the full theory. We refer the interested reader to publications such as (Grassberger and Procaccia, 1983b; Grassberger and Procaccia, 1983c; Grassberger and Procaccia, 1984; Ruelle, 1989).

5.6.4 Time Span for Reliable Prognoses

Figure 5.6.1 presents two time histories for the Lorenz equation, originating from two initial conditions which cannot be distinguished within our reading precision. Within a certain time span, up to the point in time t^*, the curves stay together; beyond a threshold t^*, however, the two curves drift apart and no resemblance can be discerned, i.e. predictions can no longer be made.

Even for deterministic systems, prognoses are only conditionally reliable. In section 5.4.6, we gave an estimation of the relaxation time, i.e. we quantified this time, with the help of the largest Lyapunov exponent, after which the total information regarding the initial conditions is lost. In connection with KS entropy, we now again ask whether it is possible to give or even to quantify limits for the predictability. Can we determine a time span or a maximum number of iterations after which all efforts to improve the predictability within a certain measurement precision are fruitless?

Obviously, the more "ordered" the motion, the easier it is to set up long-term prognoses. We can thus suppose that it is possible to make a connection between the KS entropy $h(\mu)$ – which, after all, is a yardstick for how "chaotic" a motion is – and the time threshold t^*. The greater the disorder, the shorter t^* is. On the other hand, the reading precision also plays a role. The more accurately we can localise an initial condition, the larger t^* will be. It is without doubt one of the most important practical applications of KS entropy $h(\mu)$ that it offers a possibility of estimating t^*.

With the aid of the information dimension, the information content of an initial condition can be estimated and, with the help of KS entropy, the speed with which this information decreases (Farmer, 1982c).

Let us return once more to the Bernoulli shift eq. (5.6.1). If we divide the unit interval uniformly into 2^n sub-intervals, we can reproduce the initial conditions in a binary system accurately to n places. In section 5.6.1, we observed that, at each iteration step, all digits move one place to the left while the first digit is dropped. This means that inaccuracies in the representation of the numbers, i.e. microscopic fluctuations, overwrite all the information contained in the initial conditions after n steps.

The estimate of t^* is based on this idea. We first require the amount of information $\bar{I}(\varepsilon)$ contained in the initial condition. We assume that the measurement accuracy determines a partitioning of the phase space in cells of edge length ε. In accordance with eq. (5.6.14), we then have

$$\bar{I}(\varepsilon) \approx D_I \log_2 \frac{1}{\varepsilon}$$

If the system produces information, $h(\mu) > 0$, part of the original information $\bar{I}(\varepsilon)$ will be overwritten at every time-step respectively iteration step. Since $h(\mu)$ is the mean rate of information production measured in bit/unit time, we can thus estimate when, in the mean, all the information in the initial condition will have been erased. Thus, there must hold

$$\bar{I}(\varepsilon) - t^* h(\mu) = 0 \tag{5.6.24}$$

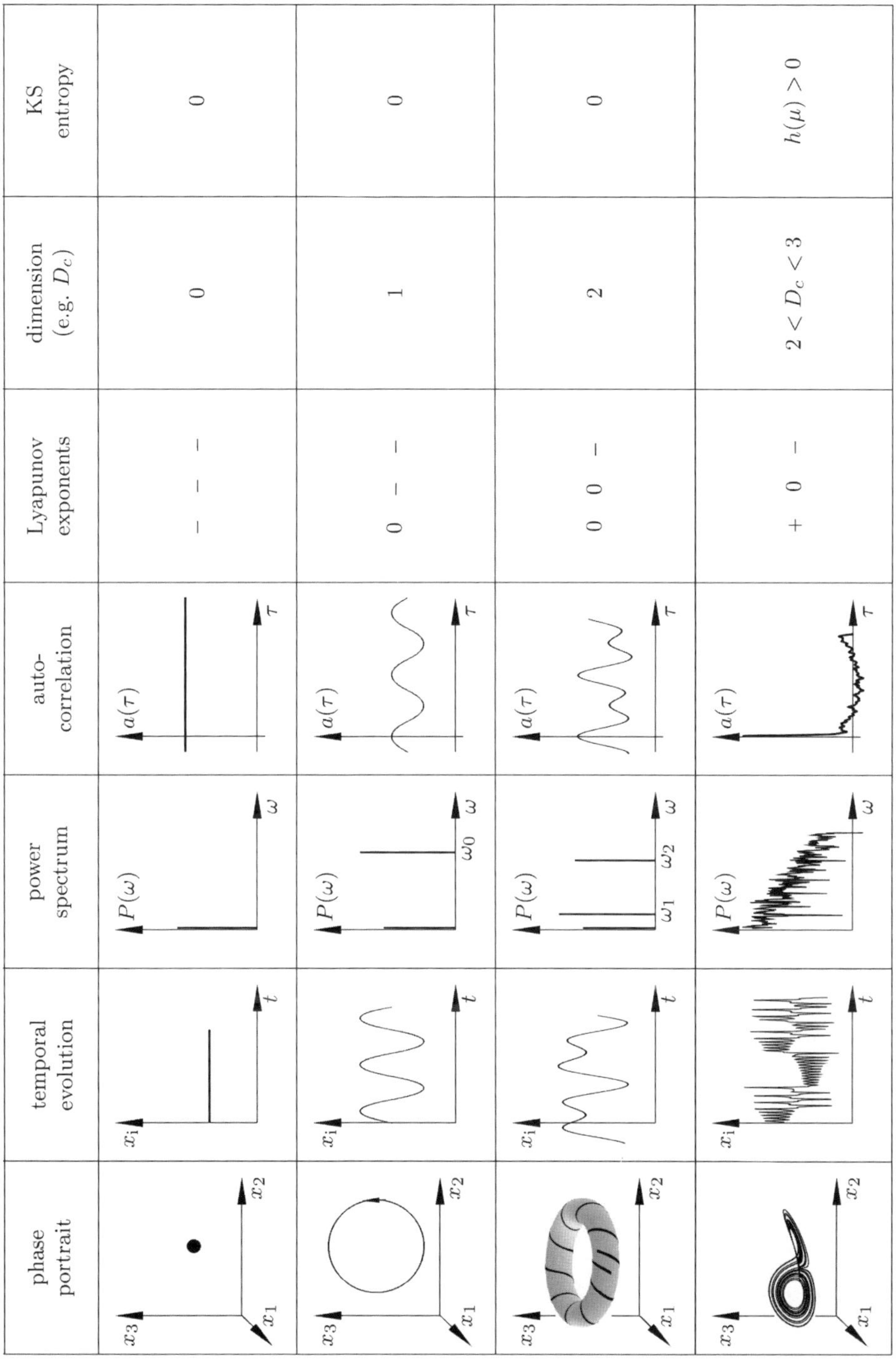

Fig. 5.6.10: Characterisation of various attractor types in a three-dimensional phase space

Consequently, using relation eq. (5.6.14),

$$t^* = \frac{D_I \log_2 (1/\varepsilon)}{h(\mu)} \tag{5.6.25}$$

If we apply this relation to the Bernoulli shift and reproduce the initial condition precisely to n places, we obtain with $\varepsilon = 2^{-\mathrm{n}}$, $D_I = 1$, $h(\mu) = 1$ bit/iteration as expected

$$\mathrm{n}^* = \frac{1 \cdot \log_2 2^{\mathrm{n}}}{1} = \mathrm{n}$$

i.e. after n steps, the system becomes oblivious of the starting value.

In Chapter 5, we have presented a number of criteria – such as power spectrum, autocorrelation, Lyapunov exponent, dimensions and KS entropy – which allow us to characterise regular and strange attractors. For numerical reasons, several criteria will generally have to be used in order to provide clear assertions. Figure 5.6.10 presents an overview of the individual characterisation possibilities.

In conclusion, we wish to refer to a number of freely available software packages which are very useful for the analysis of non-linear systems. For the fractal analysis of signals and digital images, the software package *FracLab*, developed by a team headed by J. Lévy Véhel, offers a number of tools running under MATLAB (FracLab, 2012). Among other things, this package incorporates recently developed methods for the characterisation of multifractal structures (see section 8.5) and the pointwise regularity of functions, where wavelet methods (see section 3.8.5) are also used. In addition to a series of tools based on Fourier methods and filter techniques, the software package *TISEAN*, developed by R. Hegger, H. Kantz and T. Schreiber (TISEAN, 2007), offers a large number of programmes for reconstructing attractors and calculating Lyapunov exponents and dimensions; it is very useful as an accompaniment to their book „Nonlinear Time Series Analysis "(Kantz and Schreiber, 1997). The software package *AnT 4.669* (AnT 4.669, 2009) is a particularly comprehensive and elaborate programme system developed by M. Schanz, V. Avrutin and collaborators. It not only provides tools for simulating non-linear dynamical systems, both discrete and continuous, with and without time delay, but also primarily for the analysis of their dynamics. To this end, a large number of investigation methods are made available, such as, for example, programmes for spectral and period analysis, calculation of Lyapunov exponents and fractal dimensions, determination of unstable orbits and manifolds, analysis of multiband chaotic attractors, generalised Poincaré sections as well as analyses based on symbolic dynamics. Furthermore, one of the strengths of the programme package is that it provides the possibilty of distributing time-consuming parameter studies among several computers and to enable real-time visualisations in two and three dimensions, which might promote teaching and research activities. A number of figures in this volume have been created with the help of AnT 4.669.

6 Local Bifurcation Theory

Yes, I will be thy priest, and build a fane
In some untrodden region of my mind,
Where branched thoughts, new grown with pleasant pain,
Instead of pines shall murmur in the wind.
John Keats (1795-1821), Ode to Psyche

Variations in the control parameters of dynamical systems generate completely new long-term patterns of motion. The Duffing equation (2.2.8) previously mentioned (see also Colour Plates XXVII, XXV, pp. 761, 751 and section 10.5) illustrates how small changes in the frequency or amplitude of the driving force as well as in the damping can cause qualitative changes in the physical behaviour. The investigation of bifurcations in the field of natural and engineering sciences is of great importance because each bifurcation is accompanied by a split into new states of equilibrium and drastic qualitative changes in the motion. Bifurcations can have destructive effects, if, for example, they trigger the dangerous aeroelastic fluttering of wings in aircraft. However, they may also be desirable, as in the case of Rayleigh-Bénard convection (see Chapter 7), where they contribute to the improvement of the heat and mass transport.

Qualitative changes of the topological structure of the trajectories in the phase space as a result of the change of one or more control parameters are called bifurcations and the appertaining critical parameters bifurcation values. For example, an originally stable state of equilibrium can become unstable for a critical value μ_{cr} of the control parameter μ and give rise to two new stable states of equilibrium.

It is the purpose of bifurcation analysis to determine the bifurcation values μ_{cr} and to construct the new bifurcating solutions for $\mu > \mu_{cr}$. A systematic analysis of the bifurcating solutions as is possible for one-dimensional systems with one control parameter is much more complicated for higher-dimensional ones with several parameters.

In this chapter, we therefore focus on the simplest *local bifurcations* of states of equilibrium and periodic motions for systems depending on *one* control parameter only. Applying a linear stability analysis, we can determine the bifurcation value μ_{cr} for which the fixed point or periodic orbit takes on a non-hyperbolic character. In this case, we speak of local bifurcations because we study the vector field of the dynamical system in the neighbourhood of the degenerate fixed point or degenerate closed orbit and determine the new solutions for $\mu > \mu_{cr}$ in the direct vicinity of these limit sets. Non-linear coordinate transformations subsequently allow a transformation to a so-called normal form and thus an assignment of the bifurcations to a few typical basic patterns.

J. Argyris et al., *An Exploration of Dynamical Systems and Chaos*,
DOI: 10.1007/978-3-662-46042-9_6

Apart from the local ones, *global bifurcations* also arise, in the case of which, as a result of the variation of control parameters, global qualitative changes occur in the dynamic characteristics which cannot be deduced from local information. We shall not undertake a systematic analysis of these bifurcations in this book, but restrict ourselves to a discussion of some typical special examples, see sections 10.5 and 10.6.

For the undamped Duffing equation not subject to an excitation force, see eq. (2.2.8), for example, a global qualitative change comes about when a small damping term is added. As a result, the topology of the invariant manifolds at the saddle point in the origin (see fig. 10.5.9 as well as Colour Plate XXVI, p. 753) changes and thus also the basins of attraction of the two stable states of equilibrium. If, in addition to the damping, we introduce a periodic excitation force, homoclinic points emerge for a critical excitation amplitude; in their immediate neighbourhood, the dynamics of the process becomes exceedingly complex and causes the boundary between the two basins of attraction to develop a fractal character. In section 10.5, we shall discuss this so-called global homoclinic bifurcation in detail (see also section 10.1).

In section 10.6, we present a brief overview of a homoclinic bifurcation in three-dimensional phase space, the so-called Shilnikov bifurcation, together with some applications. The bifurcation is either accompanied by the appearance of stable periodic orbits or by spiralling chaotic behaviour. The homoclinic solutions are of considerable interest because they represent solitary pulses and wave trains.

A further example of a global bifurcation occurs in the Lorenz system for $r_{cr} \approx 13.92$ (cf. section 7.4). At this critical parameter value, two homoclinic orbits are generated at the origin, i. e. the two branches of the unstable manifold of the origin return to the origin as stable manifolds for $t \to \infty$. In this case, all the attractors are preserved; neither their type nor their stability behaviour changes. What does change fundamentally, however, is the basins of attraction of the two stable fixed points.

In this connection, so-called *border-collision bifurcations* are worth mentioning; they occur in dynamical systems with piecewise continuous or piecewise smooth functions $F(x)$. Although we excluded such systems from our considerations at the beginning of Chapter 2, they play an important role, for example in many technical systems with switching processes. Border-collision bifurcations occur when fixed points or periodic orbits collide with a surface in phase space, on which the function value of the derivative of $F(x)$ is discontinuous. An overview of the multitude of atypical bifurcations and scenarios that may arise can be found in (di Bernardo et al., 2008), see also (Nusse et al., 1994; Avrutin and Schanz, 2004).

Topological changes – such as the intersection, tangency or avoidance of the invariant manifolds of fixed points and limit cycles or the spontaneous emergence, disintegration or qualitative change of attractors – are typical features of global qualitative alterations of the dynamics and are thus typical of *global* bifurcations. Further examples can be found in (Thompson and Stewart, 2001); for a study of the fundamental theory, we recommend (Guckenheimer and Holmes, 1983; Wiggins, 1988; Kuznetsov, 1998).

6.1 Motivation

Before discussing the local bifurcation theory in detail, we first present some simple mathematical models which demonstrate bifurcations of fixed points as a result of the variation of *one* control parameter leading to a qualitative change of the dynamics.

a) *Saddle-node bifurcation*

We first consider the evolution equation of a single variable x. Let the dissipative system be

$$\dot{x} = F(x, \mu) = \mu - x^2 \tag{6.1.1}$$

where the control parameter is denoted μ. We first determine the states of equilibrium respectively the fixed points x_s

$$x_s = \pm\sqrt{\mu} \tag{6.1.2}$$

and plot their coordinates in dependence on μ, see fig. 6.1.1. The figure is a typical example of a bifurcation diagram. For $\mu < 0$, no real solutions respectively equilibria exist. A stability analysis

$$\left.\frac{\partial F}{\partial x}\right|_{x_s} = -2x|_{x_s}$$

shows that the equilibrium branch $x_s = \sqrt{\mu}$ is stable for $\mu > 0$ whereas the branch $x_s = -\sqrt{\mu}$ is unstable.

For $\mu \to +0$, the stable and the unstable solution branch approach one another until they merge at $\mu = 0$ and neutralise one another. This is the reason why we also speak of a fold at $\mu = 0$.

In order to elucidate the phase flow of this bifurcation, we expand the one-dimensional system eq. (6.1.1) into a two-dimensional one without inducing a

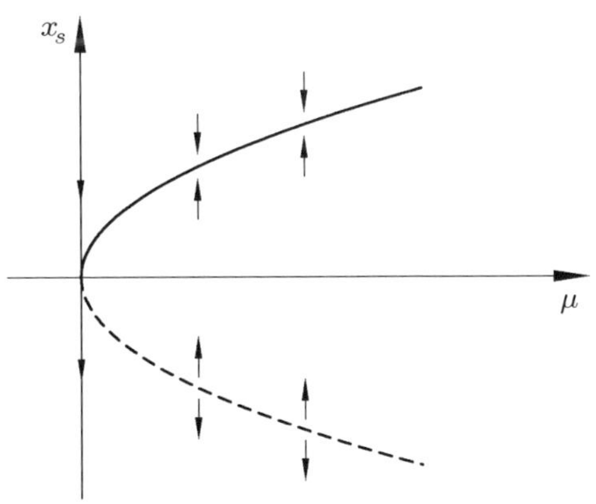

Fig. 6.1.1

Bifurcation diagram of a saddle-node bifurcation: $\dot{x} = \mu - x^2$

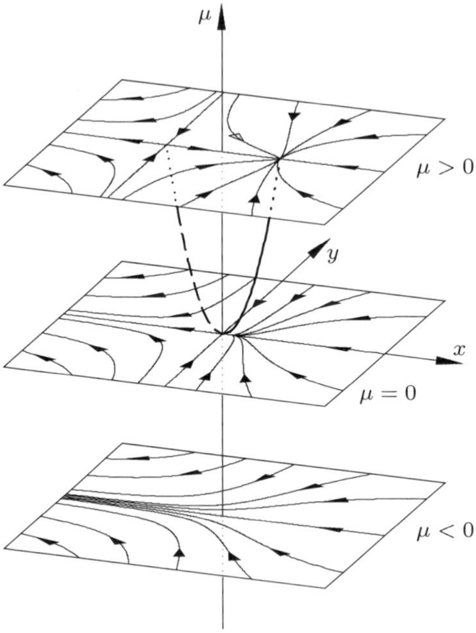

Fig. 6.1.2

Two-dimensional saddle-node bifurcation: $\dot{x} = \mu - x^2, \;\; \dot{y} = -y$

change in the bifurcation character. To illustrate the argument, we consider the uncoupled system

$$\begin{aligned} \dot{x} &= \mu - x^2 \\ \dot{y} &= -y \end{aligned} \tag{6.1.3}$$

Along the x-axis, the two-dimensional problem is reduced to the aforementioned one-dimensional one, eq. (6.1.1). For varying values of $\mu > 0$, we obtain two sets of fixed points $\boldsymbol{x}_{s1} = \{\sqrt{\mu} \quad 0\}$ and $\boldsymbol{x}_{s2} = \{-\sqrt{\mu} \quad 0\}$. The Jacobian matrix

$$\left.\frac{\partial \boldsymbol{F}}{\partial \boldsymbol{x}}\right|_{\boldsymbol{x}_s} = \begin{bmatrix} \frac{\partial F_1}{\partial x} & \frac{\partial F_1}{\partial y} \\ \frac{\partial F_2}{\partial x} & \frac{\partial F_2}{\partial y} \end{bmatrix}_{\boldsymbol{x}_s} = \begin{bmatrix} -2x & 0 \\ 0 & -1 \end{bmatrix}_{\boldsymbol{x}_s} \tag{6.1.4}$$

has the eigenvalues $\lambda_1 = -2x_s$ and $\lambda_2 = -1$. Since both eigenvalues are negative for the fixed points $\boldsymbol{x}_{s1}$ $(\lambda_1 = -2\sqrt{\mu},\ \lambda_2 = -1)$, we observe stable nodes (fig. 3.1.8). For $\boldsymbol{x}_{s2}$ we obtain eigenvalues with opposite signs $\boldsymbol{x}_{s2}$ $(\lambda_1 = 2\sqrt{\mu},\ \lambda_2 = -1)$ which indicates saddle points (fig. 3.1.8). Figure 6.1.2 presents the numerically determined phase flow and the bifurcation in dependence on the control parameter μ. For $\mu < 0$, a state of equilibrium does not exist. If μ is positive, stable nodes form the stable solution branch and saddle points the unstable one. Both types of fixed points merge for $\mu = 0$ and half of the phase portrait is characterised by the phase portrait of the *saddle* point, the other half by that of the *node*. This is the reason why we speak of a saddle-node bifurcation here.

b) *Pitchfork bifurcation*

A second basic bifurcation pattern is the pitchfork bifurcation. In order to illustrate this, we consider the following equation

$$\dot{x} = F(x, \mu) = \mu x - x^3 \tag{6.1.5}$$

Again, we first determine the equilibrium solutions from eq. (6.1.5)

$$\mu x_s - x_s^3 = 0 \tag{6.1.6}$$

Independent of μ, there exists the trivial solution

$$x_{s0} = 0 \tag{6.1.7}$$

For $\mu > 0$, we obtain an additional pair of equilibrium solutions

$$x_{s1,2} = \pm\sqrt{\mu} \tag{6.1.8}$$

All three solution sets are shown in the bifurcation diagram of fig. 6.1.3. In accordance with eq. (6.1.7), $x_{s0} = 0$ represents an equilibrium solution for all μ-values (μ-axis). At $\mu = 0$ and for positive μ-values, the two solutions $x_{s1,2}$ branch off the x_{s0}-solution; we thus speak of a pitchfork bifurcation. The stability analysis expressed through $F' = -3x^2 + \mu$ demonstrates that the solution $x_{s0} = 0$ is asymptotically stable for $\mu < 0$ and unstable for $\mu > 0$. The solution pair $x_{s1,2} = \pm\sqrt{\mu}$ remains asymptotically stable for all positive μ-values. This bifurcation splitting at the critical value $\mu_{cr} = 0$ into two stable branches is also called a supercritical bifurcation (fig. 6.1.3).

How does the phase flow evolve for the supercritical bifurcation in a two-dimensional phase space? In analogy to the saddle-node bifurcation, we again expand eq. (6.1.5) by $\dot{y} = -y$ and obtain the following uncoupled set of equations

$$\begin{aligned} \dot{x} &= \mu x - x^3 \\ \dot{y} &= -y \end{aligned} \tag{6.1.9}$$

The equilibrium condition $\boldsymbol{F}(\boldsymbol{x}_s) = \boldsymbol{F}(x_s, y_s) = \boldsymbol{o}$ yields the fixed points

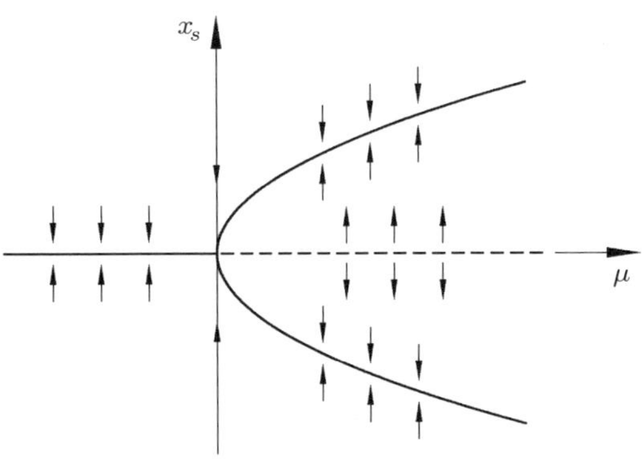

Fig. 6.1.3

Supercritical pitchfork bifurcation: $\dot{x} = \mu x - x^3$

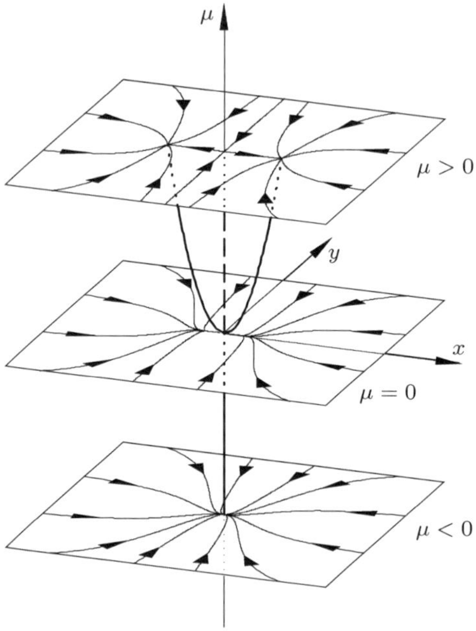

Fig. 6.1.4

Two-dimensional pitchfork bifurcation: $\dot{x} = \mu x - x^3, \;\; \dot{y} = -y$

$$\boldsymbol{x}_{s0} = \{0 \quad 0\} \qquad \text{and} \qquad \boldsymbol{x}_{s1,2} = \{\pm\sqrt{\mu} \quad 0\} \qquad \text{for } \mu \geqslant 0$$

From the Jacobian matrix

$$\left.\frac{\partial \boldsymbol{F}}{\partial \boldsymbol{x}}\right|_{\boldsymbol{x}_s} = \begin{bmatrix} \frac{\partial F_1}{\partial x} & \frac{\partial F_1}{\partial y} \\ \frac{\partial F_2}{\partial x} & \frac{\partial F_2}{\partial y} \end{bmatrix}_{\boldsymbol{x}_s} = \begin{bmatrix} -3x^2 + \mu & 0 \\ 0 & -1 \end{bmatrix}_{\boldsymbol{x}_s} \tag{6.1.10}$$

we obtain the eigenvalues $\lambda_1 = -3x_s^2 + \mu$ and $\lambda_2 = -1$. For fixed points on the μ-axis, $\boldsymbol{x}_{s0} = \{0 \quad 0\}$, both eigenvalues are negative for $\mu < 0$ and we have a stable node (fig. 6.1.4). If μ changes its sign, the first eigenvalue λ_1 becomes positive so that the state of equilibrium loses its asymptotic stability and becomes unstable in the x-direction, i. e. for $\mu > 0$, $\boldsymbol{x}_{s0}$ is a saddle point (fig. 6.1.4). The eigenvalues of the fixed points $\boldsymbol{x}_{s1,2} = \{\pm\sqrt{\mu} \quad 0\}$ are $\lambda_1 = -2\mu$ and $\lambda_2 = -1$. Since $\mu > 0$, they remain negative and thus, both fixed points are asymptotically stable (fig. 6.1.4).

The numerically determined phase portraits in fig. 6.1.4 illustrate this at a glance. At the critical value $\mu_{cr} = 0$, we observe a qualitative change in the course of the trajectory. For negative μ-values, the only equilibrium solution $\boldsymbol{x}_{s0}$ is a stable node. After the bifurcation, a saddle point flanked by two stable nodes $\boldsymbol{x}_{s1,2} = \{\pm\sqrt{\mu} \quad 0\}$ takes the place of the stable node.

c) *Transcritical bifurcation*

We now turn to a further basic pattern, the so-called transcritical bifurcation. We consider the equation

$$\dot{x} = F(x, \mu) = \mu x - x^2 \tag{6.1.11}$$

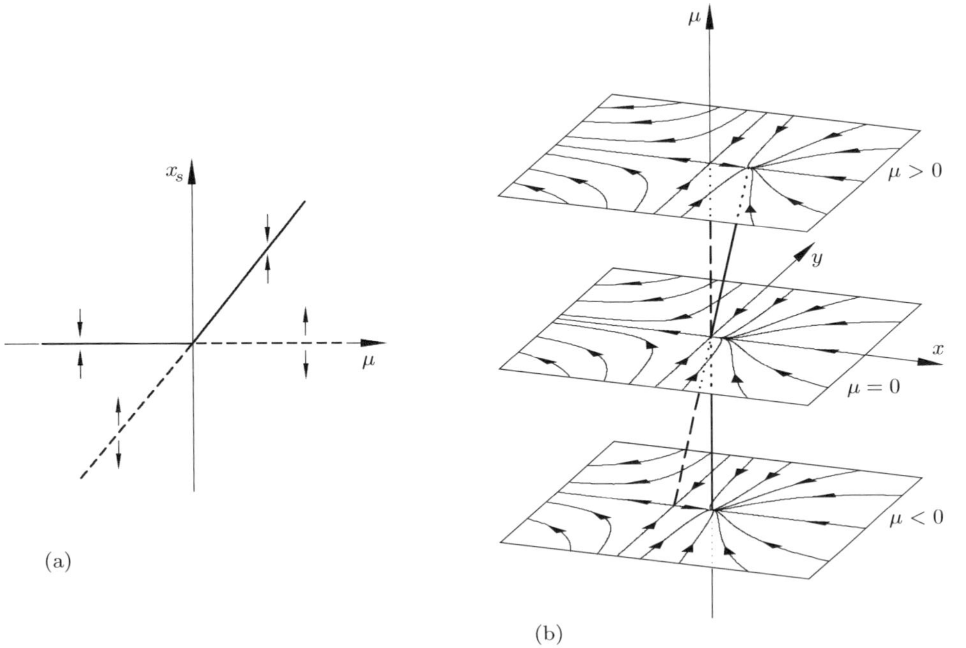

Fig. 6.1.5: Transcritical bifurcation: $\dot{x} = \mu x - x^2,\ \ \dot{y} = -y$
a) one-dimensional, b) two-dimensional

The equilibrium solutions are the two straight lines $x_{s0} = 0$ and $x_{s1} = \mu$. For $\mu < 0$, the first one is asymptotically stable ($\frac{\partial F}{\partial x} = -2x+\mu$), the second unstable. At $\mu = 0$, the equilibrium solutions interchange their stability characteristics. For $\mu > 0$, the first one is now unstable, the second asymptotically stable (see fig. 6.1.5a). In fig. 6.1.5b, the phase trajectories of the extended uncoupled dissipative system

$$\dot{x} = \mu x - x^2, \quad \dot{y} = -y$$

illustrate the switch of stability: the stable node develops into a saddle point and, conversely, the saddle point into a node.

d) *Hopf bifurcation*

All the aforementioned examples have one aspect in common: the eigenvalues of the Jacobian matrix are real at equilibrium. Up to now, we have avoided discussing the case of complex conjugate eigenvalues (fig. 3.1.8) although this is in fact the most interesting. It leads to a classic among the bifurcations, the so-called Hopf bifurcation, named after the mathematician Eberhard Hopf who studied this type of bifurcation for the n-dimensional case (Hopf, 1942). To illustrate the Hopf bifurcation, we consider the two-dimensional set of equations

$$\begin{aligned} \dot{x} &= -y + x\,[\mu - (x^2 + y^2)] \\ \dot{y} &= x + y\,[\mu - (x^2 + y^2)] \end{aligned} \tag{6.1.12}$$

For all μ-values, $x_0 = y_0 = 0$ is the only equilibrium point. A bifurcation into other states of equilibrium in dependence on the single control parameter μ does not exist. The Jacobian matrix

$$\left.\frac{\partial \boldsymbol{F}}{\partial \boldsymbol{x}}\right|_{\boldsymbol{x}_0=\boldsymbol{o}} = \begin{bmatrix} \mu & -1 \\ 1 & \mu \end{bmatrix} \tag{6.1.13}$$

possesses the eigenvalues $\lambda_{1,2} = \mu \pm i$. For $\mu < 0$, the state of equilibrium is a stable focus, $\mathrm{Re}(\lambda_{1,2}) < 0$, and for $\mu > 0$ an unstable one, $\mathrm{Re}(\lambda_{1,2}) > 0$ (fig. 3.1.8), i. e. for $\mu = 0$, the point attractor changes its stability behaviour.

It is still unclear which state the trajectories departing from the unstable focus tend towards for $t \to \infty$. Do they move to infinity or are they attracted by another type of attractor, such as a limit cycle, for example? In order to answer this question, we transform the initial system eq. (6.1.12) into polar coordinates. With the transformation rule

$$x = r\cos\varphi\,,\ y = r\sin\varphi \tag{6.1.14}$$

we obtain the system

$$\begin{aligned} \dot r\cos\varphi - r\dot\varphi\sin\varphi &= -r\sin\varphi + r(\mu - r^2)\cos\varphi \\ \dot r\sin\varphi + r\dot\varphi\cos\varphi &= r\cos\varphi + r(\mu - r^2)\sin\varphi \end{aligned} \tag{6.1.15}$$

Multiplying the first equation by $\cos\varphi$ respectively $(-\sin\varphi)$ and the second with $\sin\varphi$ respectively $\cos\varphi$ and adding the resulting expressions, we obtain the following uncoupled system in polar coordinates

$$\begin{aligned} \dot r &= -r^3 + \mu r \\ \dot\varphi &= 1 \end{aligned} \tag{6.1.16}$$

where the cubic equation in r resembles eq. (6.1.5) in its structure. The states of equilibrium ($\dot r = 0$) are the trivial solution $r_{s0} = 0$ (as above) and, for positive radii r, $r_{s1} = \sqrt{\mu}$ (fig. 6.1.6). The solution r_{s1} leads in the x, y-phase space to a closed curve, a limit cycle, i. e. a periodic orbit, the amplitude of which increases with $\sqrt{\mu}$. The following stability analysis (see also section 5.4.2)

$$\begin{aligned} &r < \sqrt{\mu} \text{ inside the limit cycle} && \Rightarrow \dot r > 0 \\ &r > \sqrt{\mu} \text{ outside the limit cycle} && \Rightarrow \dot r < 0 \end{aligned} \tag{6.1.17}$$

shows that this limit cycle is stable (fig. 6.1.6b).

It can clearly be seen in the phase portrait (fig. 6.1.6b) that the trajectories spiral towards the only fixed point for $\mu < 0$ and away from it for $\mu > 0$, converging to the closed curve $r_s = \sqrt{\mu}$, the limit cycle. At the bifurcation point $\mu = 0$, not only does the fixed point become unstable, a limit cycle also emerges. This bifurcation of a focus into a limit cycle, a state of equilibrium into a periodic state or a static into a dynamic state is a completely new type of bifurcation, known as the Hopf bifurcation.

In the examples we have considered so far, the control parameter determines both the existence of new states of equilibrium as well as their stability behaviour beyond a critical value μ_{cr}. The variation of the control parameter μ causes – when crossing

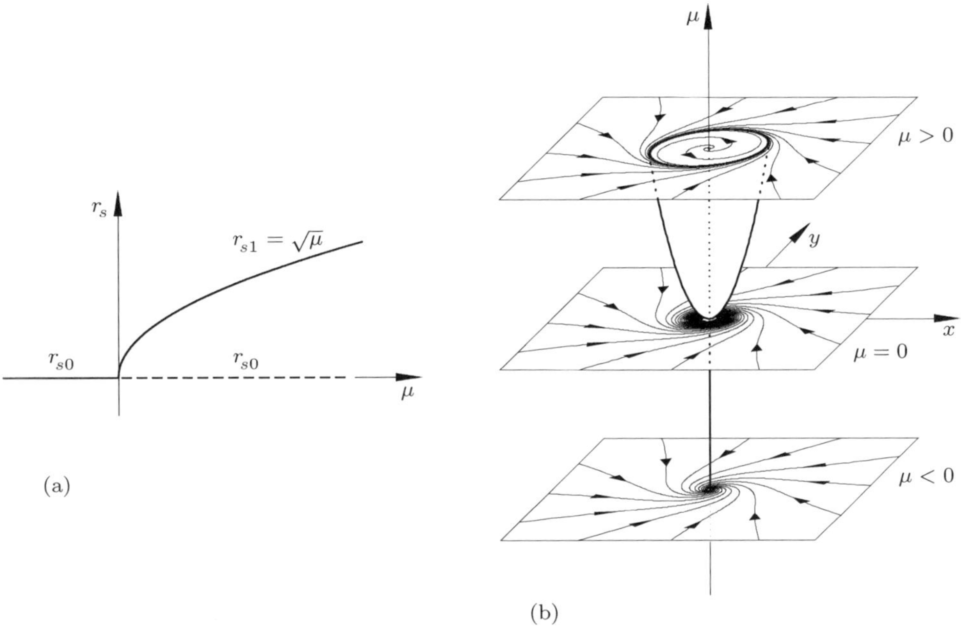

Fig. 6.1.6: Hopf bifurcation: a) bifurcation diagram $r_S(\mu)$, b) in x, y-phase space

the μ_{cr}-value – a change of sign in the real part of one or more eigenvalues of the Jacobian matrix from negative to positive. However, we should not draw the conclusion that this must always be so. Let us consider the following set of two-dimensional non-linear equations (Carr, 1981)

$$\dot{x} = \mu x^3 + x^2 y$$
$$\dot{y} = -y + y^2 + xy - x^3 \tag{6.1.18}$$

The trivial solution $\boldsymbol{x}_{s0} = \{0 \quad 0\}$ is, independent of the μ-value, an equilibrium solution. For the Jacobian matrix at the point $\boldsymbol{x}_{s0}$, we obtain

$$\left.\frac{\partial \boldsymbol{F}}{\partial \boldsymbol{x}}\right|_{\boldsymbol{x}_{s0}} = \begin{bmatrix} 3\mu x^2 + 2xy & x^2 \\ y - 3x^2 & -1 + 2y + x \end{bmatrix}_{\boldsymbol{x}_{s0}} = \begin{bmatrix} 0 & 0 \\ 0 & -1 \end{bmatrix} \tag{6.1.19}$$

having the eigenvalues $\lambda_1 = 0$ and $\lambda_2 = -1$.

An eigenvalue of the linearised matrix $\frac{\partial \boldsymbol{F}}{\partial \boldsymbol{x}}$ is zero for the equilibrium solution $\boldsymbol{x}_{s0}$ not on account of the variations of μ but independently of them. We are confronted with the presence of a *non*-hyperbolic fixed point which possesses at least one eigenvalue of the Jacobian matrix $\frac{\partial \boldsymbol{F}}{\partial \boldsymbol{x}}$ with a vanishing real part. In accordance with the stability statement for linearised systems given in section 5.4.1, the *linear* approximation can*not* in the present case supply a statement on the stability of non-hyperbolic fixed points. In this case, it is the *non*-linear terms of the set of equations which determine the stability of the equilibrium solution.

Figure 6.1.7 presents the numerically determined phase flow for $\mu < 0$, $\mu = 0$ and $\mu > 0$ of eq. (6.1.18). The three phase portraits demonstrate that the stability

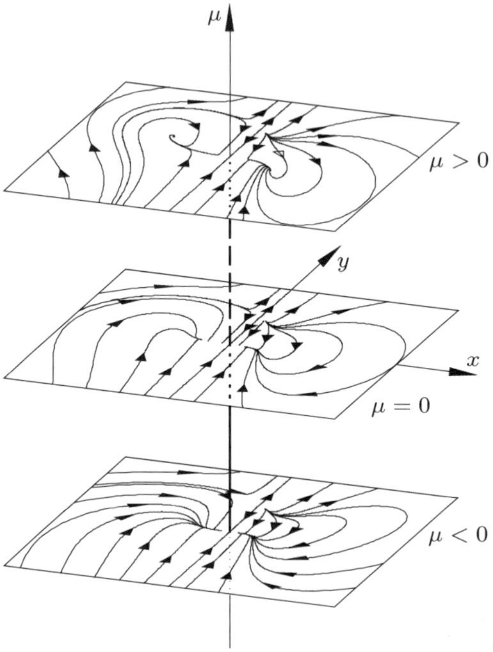

Fig. 6.1.7

Numerically determined phase flow of the non-linear system eq. (6.1.18)

behaviour of the fixed point $\boldsymbol{x}_{s0}$ in the x-direction nonetheless does change radically in dependence on μ at $\mu = 0$. It changes from stable ($\mu < 0$) to unstable ($\mu > 0$). The question is whether there is a possibility of determining the flow of this degenerate state of equilibrium $\boldsymbol{x}_{s0}$ analytically. The answer is yes, and one technique is that of the centre manifold theory which we elucidate in the following section.

6.2 Centre Manifold

Let us recall our remarks on the classification of non-linear systems in sections 3.2 and 5.4.1. Two fundamental statements of the theory of dynamical systems are firstly the Hartman-Grobman theorem on the stability of hyperbolic fixed points and secondly the theorem of invariant manifolds of hyperbolic fixed points. In order to understand the centre manifold and the bifurcation theory better, we now state and elucidate these two theorems. The examples selected for this section are largely based on those dealt with in (Guckenheimer and Holmes, 1983; Carr, 1981), as they are particularly suitable for illustrating the essential characteristics of the problem. However, we describe the steps for approximating the centre manifolds in more detail and elucidate them by presenting many illustrations.

We again consider the non-linear system of ordinary differential equations (see sections 3.2 and 5.4.1)

$$\dot{\boldsymbol{x}} = \boldsymbol{F}(\boldsymbol{x}) \, , \qquad \boldsymbol{x}(t_0) = \boldsymbol{x}_0 \tag{6.2.1}$$

In order to investigate the stability of a fixed point, it is helpful to observe the temporal evolution of the point $\boldsymbol{x} = \boldsymbol{x}_s + \widetilde{\boldsymbol{x}}$ adjacent to $\boldsymbol{x}_s$. Expanding the function $\boldsymbol{F}(\boldsymbol{x}_s + \widetilde{\boldsymbol{x}})$ in the neighbourhood of $\boldsymbol{x}_s$ into a Taylor series, we obtain

$$\boldsymbol{F}(\boldsymbol{x}_s + \widetilde{\boldsymbol{x}}) = \boldsymbol{F}(\boldsymbol{x}_s) + \left.\frac{\partial \boldsymbol{F}}{\partial \boldsymbol{x}}\right|_{\boldsymbol{x}_s} \widetilde{\boldsymbol{x}} + \mathcal{O}(\widetilde{\boldsymbol{x}}^2) \tag{6.2.2}$$

Noting that $\dot{\boldsymbol{x}} = \dot{\boldsymbol{x}}_s + \dot{\widetilde{\boldsymbol{x}}}$ and that $\dot{\boldsymbol{x}}_s = \boldsymbol{F}(\boldsymbol{x}_s) = \boldsymbol{o}$ (fixed point), we obtain for the temporal evolution of the small perturbation $\dot{\widetilde{\boldsymbol{x}}}$,

$$\dot{\widetilde{\boldsymbol{x}}} = \left.\frac{\partial \boldsymbol{F}}{\partial \boldsymbol{x}}\right|_{\boldsymbol{x}_s} \widetilde{\boldsymbol{x}} + \mathcal{O}(\widetilde{\boldsymbol{x}}^2) \tag{6.2.3}$$

If $\mathcal{O}(\widetilde{\boldsymbol{x}}^2)$ is small compared with $\widetilde{\boldsymbol{x}}$, it suffices to consider the linearised system

$$\dot{\widetilde{\boldsymbol{x}}} = \boldsymbol{D}(\boldsymbol{x}_s)\widetilde{\boldsymbol{x}} \qquad \text{with} \qquad \left.\frac{\partial \boldsymbol{F}}{\partial \boldsymbol{x}}\right|_{\boldsymbol{x}_s} = \boldsymbol{D}(\boldsymbol{x}_s) \tag{6.2.4}$$

However, we know from section 6.1 that the linearised matrix $\boldsymbol{D}(\boldsymbol{x}_s)$ does not supply the information for the stability analysis of the example eq. (6.1.18), because one eigenvalue of the Jacobian matrix is zero. For this reason, it is necessary to go into more detail, starting with the Hartman-Grobman theorem (Guckenheimer and Holmes, 1983). It states:

Theorem 1 (Hartman-Grobman):

> If the matrix $\boldsymbol{D}$ of the linearised form eq. (6.2.4) of the non-linear differential equation (6.2.1) at the fixed point $\boldsymbol{x}_s$ has no eigenvalue with a vanishing real part, then there is a bijective (one-to-one and onto) and bi-continuous map h (homeomorphism) defined on some neighbourhood $U(\boldsymbol{x}_s)$ of the fixed point $\boldsymbol{x}_s$ which locally maps the trajectories of the non-linear flow ϕ_t of eq. (6.2.1) onto those of the linear flow of eq. (6.2.4), $\mathrm{e}^{\boldsymbol{D}t}$. The homeomorphism h preserves the sense of the trajectories and, moreover, h can be chosen so as also to preserve the parametrisation by time t.

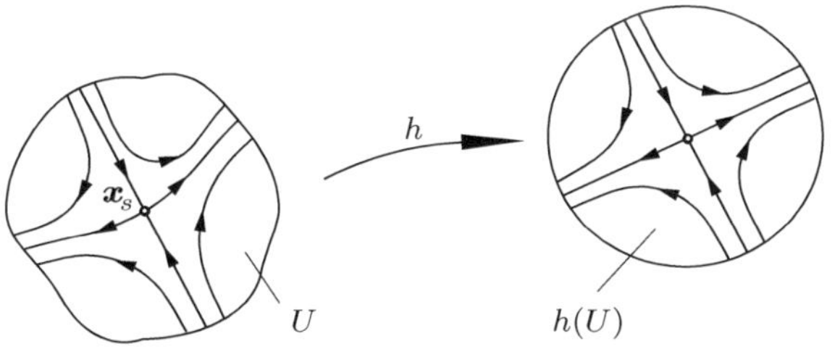

Fig. 6.2.1

Hartman-Grobman theorem on the topological equivalence of flows in the neighbourhood $U(\boldsymbol{x}_s)$

What does this theorem mean? It means that, for hyperbolic fixed points, we can deduce significant information on the stability and the behaviour of the solutions in the neighbourhood of the fixed point from the matrix $\boldsymbol{D}$ of the linearised system. In this case, in a neighbourhood $U(\boldsymbol{x}_s)$, the phase flow of the non-linear system eq. (6.2.1) is topologically equivalent to the flow of the linearised system eq. (6.2.4), i. e. the two flows can be transformed into one another by a non-linear coordinate transformation (cf. fig. 6.2.1).

In particular, assuming that the real parts of all eigenvalues of the matrices $\boldsymbol{D}_1$ and $\boldsymbol{D}_2$ of two linearised systems are non-zero, the two flows $\mathrm{e}^{\boldsymbol{D}_1 t}$ and $\mathrm{e}^{\boldsymbol{D}_2 t}$ are topologically equivalent if the eigenvalues of both matrices possess an equal number of positive and an equal number of negative real parts. This means that a stable node is topologically equivalent to a stable focus, but that neither of these is equivalent to a saddle point (see fig. 3.1.8).

We should remember, however, that our interest is directed at the stability behaviour of a degenerate fixed point (real part of an eigenvalue equals zero). Bearing this in mind, we first supply some additional definitions.

In section 3.1, we discussed the various phase flows of singularities (such as node, focus, saddle point etc.) in the space spanned by the eigenvectors. Certain solutions $\mathrm{e}^{\boldsymbol{D}t}$ of the flow play a special role, namely those lying in the linear subspace spanned by the eigenvectors. These subspaces of $\mathrm{e}^{\boldsymbol{D}t}$ are invariant with respect to the flow, i. e. all trajectories with starting points in these subspaces remain there for all time.

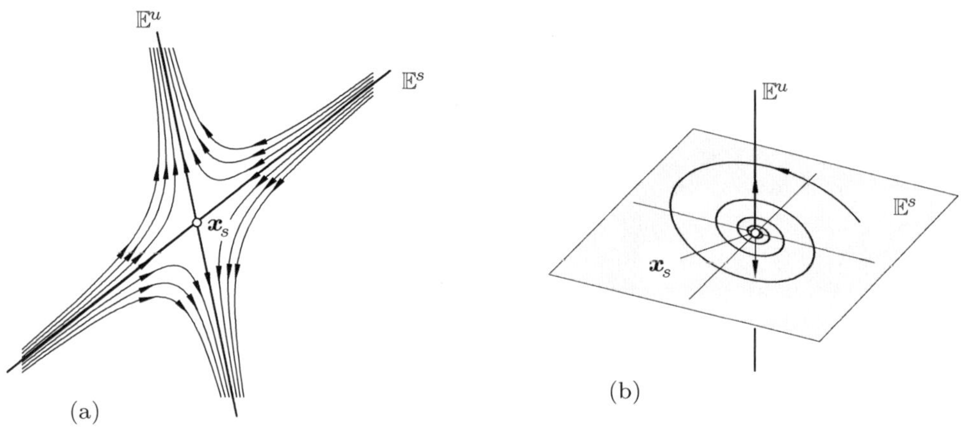

Fig. 6.2.2: Invariant subspaces: a) $\mathbb{E}^u = \text{span}\left\{\begin{bmatrix}0\\1\end{bmatrix}\right\}$, $\mathbb{E}^s = \text{span}\left\{\begin{bmatrix}1\\0\end{bmatrix}\right\}$

b) $\mathbb{E}^u = \text{span}\left\{\begin{bmatrix}0\\0\\1\end{bmatrix}\right\}$, $\mathbb{E}^s = \text{span}\left\{\begin{bmatrix}1\\0\\0\end{bmatrix}, \begin{bmatrix}0\\1\\0\end{bmatrix}\right\}$

In a nutshell, the eigenspaces of the linear matrix $\boldsymbol{D}$ are invariant subspaces of the flow $\phi_t = \mathrm{e}^{\boldsymbol{D}t}$. Since we know by now that there are stable and unstable subspaces, we can classify them correspondingly. If we denote the n_s eigenvectors respectively principal vectors appertaining to eigenvalues of $\boldsymbol{D}$ with a negative real part as $\boldsymbol{u}_1, \ldots, \boldsymbol{u}_{\mathrm{n}_s}$ and denote correspondingly the n_u eigenvectors respectively principal vectors appertaining to eigenvalues of $\boldsymbol{D}$ with a positive real part as $\boldsymbol{v}_1, \ldots, \boldsymbol{v}_{\mathrm{n}_u}$, the space $\mathbb{E}^s$ respectively $\mathbb{E}^u$ spanned by them is called stable respectively unstable invariant subspace (fig. 6.2.2):

$$\begin{aligned}\mathbb{E}^s &= \operatorname{span}\{\boldsymbol{u}_1, \ldots, \boldsymbol{u}_{\mathrm{n}_s}\}\\ \mathbb{E}^u &= \operatorname{span}\{\boldsymbol{v}_1, \ldots, \boldsymbol{v}_{\mathrm{n}_u}\}\end{aligned}$$

In analogy to the linear invariant eigenspaces $\mathbb{E}^s$ and $\mathbb{E}^u$, we introduce non-linear local stable and unstable manifolds at the fixed point $\boldsymbol{x}_s$. We designate these $\mathbb{W}^s_{\mathrm{loc}}(\boldsymbol{x}_s), \mathbb{W}^u_{\mathrm{loc}}(\boldsymbol{x}_s)$ and define

$$\begin{aligned}\mathbb{W}^s_{\mathrm{loc}}(\boldsymbol{x}_s) &:= \{\boldsymbol{x} \in U; \phi_t(\boldsymbol{x}) \to \boldsymbol{x}_s \text{ for } t \to +\infty \text{ and } \phi_t(\boldsymbol{x}) \in U, \forall\, t \geqslant 0\}\\ \mathbb{W}^u_{\mathrm{loc}}(\boldsymbol{x}_s) &:= \{\boldsymbol{x} \in U; \phi_t(\boldsymbol{x}) \to \boldsymbol{x}_s \text{ for } t \to -\infty \text{ and } \phi_t(\boldsymbol{x}) \in U, \forall\, t \leqslant 0\}\end{aligned} \tag{6.2.5}$$

where U is a neighbourhood of $\boldsymbol{x}_s$. It is now important to set up a link between the invariant manifolds $\mathbb{W}^s_{\mathrm{loc}}, \mathbb{W}^u_{\mathrm{loc}}$ and the invariant eigenspaces $\mathbb{E}^s, \mathbb{E}^u$. The following theorem tells us that the subspaces $\mathbb{E}^s, \mathbb{E}^u$ are tangent to the local manifolds $\mathbb{W}^s_{\mathrm{loc}}, \mathbb{W}^u_{\mathrm{loc}}$ in $\boldsymbol{x}_s$.

Theorem 2 (Stable manifold theorem for a fixed point):

> On the assumption that $\dot{\boldsymbol{x}} = \boldsymbol{F}(\boldsymbol{x})$ has a hyperbolic fixed point $\boldsymbol{x}_s$, there exist local stable and unstable manifolds $\mathbb{W}^s_{\mathrm{loc}}(\boldsymbol{x}_s)$, $\mathbb{W}^u_{\mathrm{loc}}(\boldsymbol{x}_s)$ of the same dimensions $\mathrm{n}_s, \mathrm{n}_u$ as those of the eigenspaces $\mathbb{E}^s, \mathbb{E}^u$ of the linearised eq. (6.2.4) and tangent to $\mathbb{E}^s, \mathbb{E}^u$ at $\boldsymbol{x}_s$. The manifolds $\mathbb{W}^s_{\mathrm{loc}}(\boldsymbol{x}_s)$, $\mathbb{W}^u_{\mathrm{loc}}(\boldsymbol{x}_s)$ are as smooth (i. e. from the same differentiability class) as the function $\boldsymbol{F}$.

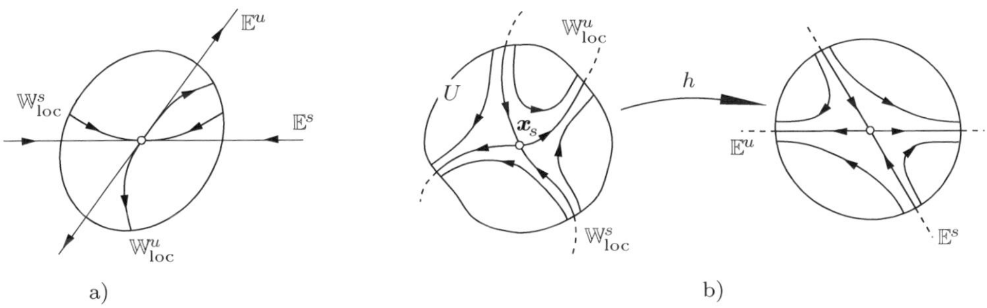

Fig. 6.2.3: Linearisation and subspaces:
a) approximation of $\mathbb{W}^s_{\mathrm{loc}}, \mathbb{W}^u_{\mathrm{loc}}$ by $\mathbb{E}^s, \mathbb{E}^u$
b) Hartman-Grobman theorem on topological equivalence

Summarising, we can say that the eigenspaces $\mathbb{E}^s$ and $\mathbb{E}^u$ approximate the local manifolds $\mathbb{W}^s_{\mathrm{loc}}(\boldsymbol{x}_s)$ and $\mathbb{W}^u_{\mathrm{loc}}(\boldsymbol{x}_s)$ in $\boldsymbol{x}_s$ and that the homeomorphism h (according to Hartman-Grobman) maps the non-linear manifolds $\mathbb{W}^s_{\mathrm{loc}}, \mathbb{W}^u_{\mathrm{loc}}$ onto the linear eigenspaces $\mathbb{E}^s, \mathbb{E}^u$ one-to-one. The result of both theorems is illustrated in fig 6.2.3.

If the local invariant manifolds are known, we enquire into their continuation beyond the neighbourhood U. This leads to the definition of *global* stable and unstable manifolds of $\boldsymbol{x}_s$: for the solution of eq. (6.2.1), we introduce the local manifolds as

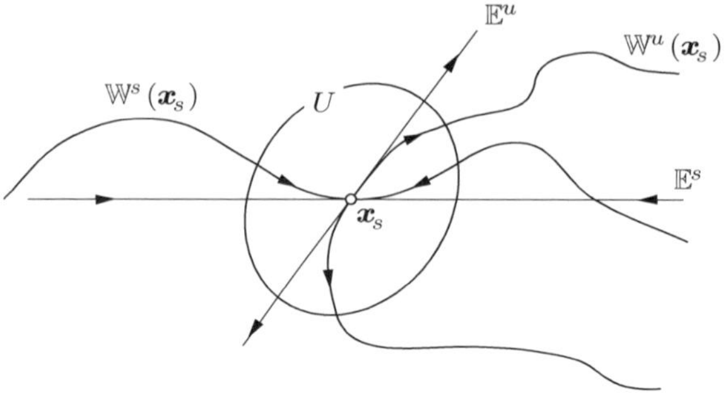

Fig. 6.2.4

Global stable and unstable manifold for $\boldsymbol{x}_s$

initial conditions and select the time direction so that points emanating from $\mathbb{W}^s_{\text{loc}}$ and $\mathbb{W}^u_{\text{loc}}$ move away from the fixed point $\boldsymbol{x}_s$

$$\mathbb{W}^s(\boldsymbol{x}_s) = \bigcup_{t\leqslant 0} \phi_t\big(\mathbb{W}^s_{\text{loc}}(\boldsymbol{x}_s)\big)$$

$$\mathbb{W}^u(\boldsymbol{x}_s) = \bigcup_{t\geqslant 0} \phi_t\big(\mathbb{W}^u_{\text{loc}}(\boldsymbol{x}_s)\big) \tag{6.2.6}$$

Figure 6.2.4 presents the global manifolds $\mathbb{W}^s$ and $\mathbb{W}^u$ and illustrates the difference to the local $\mathbb{W}^s_{\text{loc}}$ and $\mathbb{W}^u_{\text{loc}}$ of fig. 6.2.3a.

The existence of solutions of the system eq. (6.2.1) and their uniqueness ensure that the invariant manifolds of two fixed points $\boldsymbol{x}_{s1}, \boldsymbol{x}_{s2}$ of the same respectively different stability never intersect, let alone intersect themselves (fig. 6.2.5a). However, this does not preclude the smooth merging of invariant manifolds of different stability of different fixed points or of the same fixed point (fig. 6.2.5b). Note that in the case of continuous systems, it follows from the existence of an intersection point P of $\mathbb{W}^u(\boldsymbol{x}_{s\text{i}})$ with $\mathbb{W}^s(\boldsymbol{x}_{s\text{j}})$ and the uniqueness of the solutions that whole parts of the manifolds must be identical since P would otherwise have two different possibilities for future and past. (Transversal intersections of stable and unstable manifolds, i. e. homoclinic respectively heteroclinic points, can only arise in discrete systems, e. g. in Poincaré sections, which are then responsible for an extraordinarily complex behaviour of the dynamical system, cf. figs. 4.5.8 and 10.5.9c).

Let us illustrate the previous statements on the linearised eigenspaces, the local manifolds and the global manifolds by means of a simple example in the two-dimensional phase space

$$\begin{aligned} \dot{x} &= \quad x \\ \dot{y} &= -y + x^2 \end{aligned} \tag{6.2.7}$$

The only fixed point is the origin $\boldsymbol{x}_s = \{0 \;\; 0\}$. For the linearised system

$$\begin{bmatrix} \dot{x} \\ \dot{y} \end{bmatrix} = \begin{bmatrix} 1 & 0 \\ 0 & -1 \end{bmatrix} \begin{bmatrix} x \\ y \end{bmatrix} \tag{6.2.8}$$

we determine the eigenvalues $\lambda_u = 1$ (unstable) and $\lambda_s = -1$ (stable) and the appertaining eigenvectors $\boldsymbol{v} = \{1 \;\; 0\}$ and $\boldsymbol{u} = \{0 \;\; 1\}$ respectively the invariant subspaces (fig. 6.2.6a)

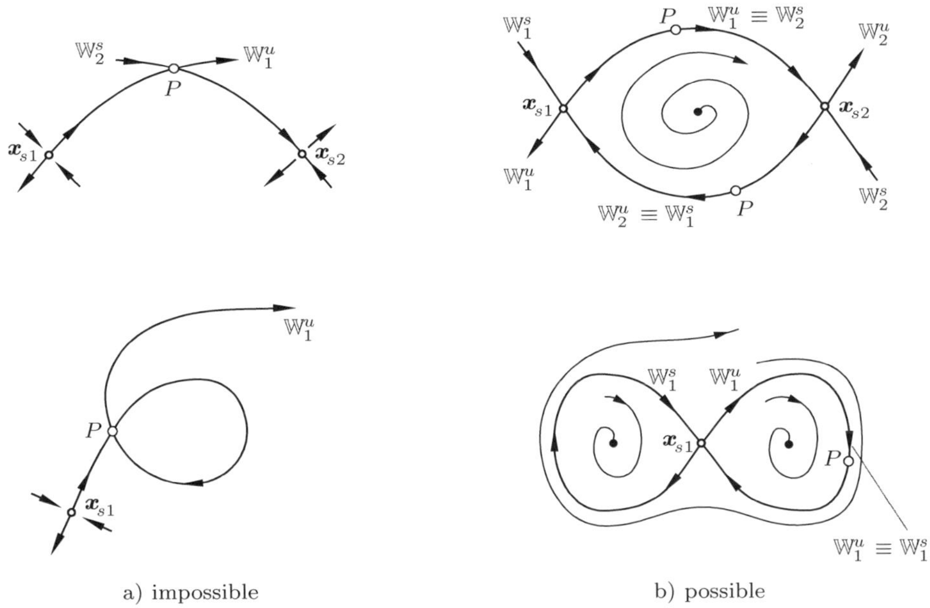

Fig. 6.2.5: Common points P of stable and unstable manifolds of one respectively two fixed points

$$\mathbb{E}^s = \operatorname{span}\left\{\begin{bmatrix} 0 \\ 1 \end{bmatrix}\right\} \quad \text{or} \quad x = 0$$
$$\mathbb{E}^u = \operatorname{span}\left\{\begin{bmatrix} 1 \\ 0 \end{bmatrix}\right\} \quad \text{or} \quad y = 0 \tag{6.2.9}$$

Eliminating time, eq. (6.2.7) may be expressed as

$$\frac{\mathrm{d}y}{\mathrm{d}x} = -\frac{y}{x} + x \qquad \text{or} \qquad x\frac{\mathrm{d}y}{\mathrm{d}x} + y = x^2 \tag{6.2.10}$$

Applying direct integration, we obtain the general solution

$$y(x) = \frac{x^2}{3} + \frac{c}{x} \tag{6.2.11}$$

where c is determined by the initial condition. In accordance with eq. (6.2.9), the x-axis is an unstable subspace $\mathbb{E}^u$ tangent to the unstable local manifold $\mathbb{W}^u_{\text{loc}}(0,0)$ following Theorem 2 (the stable manifold theorem for a fixed point). It follows from eq. (6.2.11) that $\mathbb{W}^u_{\text{loc}}(0,0)$ can be represented as the graph $y = h(x)$ where, according to Theorem 2, the function $h(x)$ at the fixed point $\boldsymbol{x}_s = \{0\ \ 0\}$ is tangent to the x-axis and thus satisfies the condition $h'(0) = h(0) = 0$. Since $y(0) = 0$ holds in accordance with eq. (6.2.11) only for $c = 0$, the unstable global manifold satisfies the equation $y = x^2/3$, i. e.

$$\mathbb{W}^u(0,0) = \left\{(x,y) \in \mathbb{R}^2\,;\; y = \frac{x^2}{3}\right\} \tag{6.2.12}$$

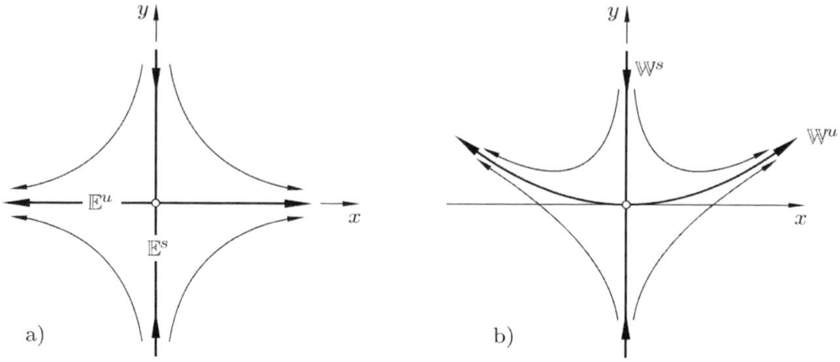

Fig. 6.2.6: Stable and unstable manifolds of the systems eq. (6.2.7); a) linearised, b) non-linearised

For the stable global manifold $\mathbb{W}^s(0,0)$, we consider the initial system eq. (6.2.7). If $x(0) = 0$, then $x(t) \equiv 0$ and the solution $x(t) \equiv 0$, $y(t) = y_0 \mathrm{e}^{-t}$ is stable so that $\mathbb{W}^s(0,0) \equiv \mathbb{E}^s$ holds.

Figure 6.2.6 presents the stable and unstable manifolds of both the linearised and the non-linear system and shows the topological equivalence of the phase flows in the linearised and non-linearised cases as well as the considerably improved description of the phase flow in the neighbourhood of the fixed point achieved by the non-linear manifolds.

In our argumentation and definitions describing the behaviour of the system near the fixed points, we have intentionally ignored the degenerate case "real part equals zero" in the eigenvalues. However, nothing can prevent us from introducing a central invariant subspace $\mathbb{E}^c$ in analogy to the stable respectively unstable linear eigenspaces $\mathbb{E}^s$ respectively $\mathbb{E}^u$. We define as the centre eigenspace

$$\mathbb{E}^c = \mathrm{span}\{\boldsymbol{w}_1, \ldots, \boldsymbol{w}_{\mathrm{n}_c}\}$$

the eigenvectors $\boldsymbol{w}_1, \ldots, \boldsymbol{w}_{\mathrm{n}_c}$ of which appertain to eigenvalues with a zero real part. Since $\mathrm{n} = \mathrm{n}_s + \mathrm{n}_u + \mathrm{n}_c$ must hold (n is the dimension of the system), all solutions which neither decrease exponentially (in a monotonic or oscillating manner) ($\mathbb{E}^s$) nor increase exponentially ($\mathbb{E}^u$) lie in the invariant subspace $\mathbb{E}^c$. In the case of *simple* eigenvalues, the solutions in $\mathbb{E}^c$ remain constant ($\lambda = 0$), fig. 6.2.7, or oscillate with constant amplitude ($\lambda_{1,2} = \pm i\omega$).

Corresponding to the invariant subspace $\mathbb{E}^c$, we introduce the non-linear analogue $\mathbb{W}^c$. When the meaning of $\mathbb{W}^c$ is clear, we refrain from differentiating between local and global and write simply $\mathbb{W}^c$. The definition of $\mathbb{W}^c$ is as follows:

> Let $\boldsymbol{x}_s$ be the fixed point of eq. (6.2.1). The invariant manifold is called the centre manifold $\mathbb{W}^c$ if it is tangent at $\boldsymbol{x}_s$ to the eigenspace $\mathbb{E}^c$ appertaining to the eigenvalues with zero real part.

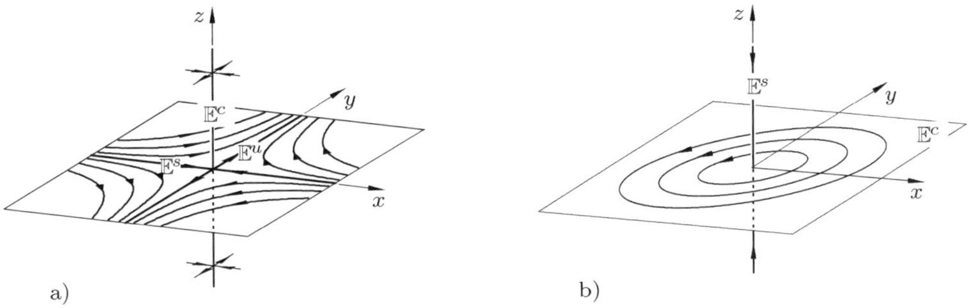

Fig. 6.2.7: Invariant eigenspaces $\mathbb{E}^s, \mathbb{E}^u$ and $\mathbb{E}^c$ of the set of equations $\dot{x} = -x,\ \dot{y} = y,\ \dot{z} = 0$ and $\dot{x} = y,\ \dot{y} = -x,\ \dot{z} = 0$

To clarify matters, we consider the following two-dimensional example

$$\begin{aligned}\dot{x} &= \ \ x^2 \\ \dot{y} &= -y\end{aligned} \tag{6.2.13}$$

the solution of which is

$$x(t) = \frac{x_0}{1 - x_0 t} \qquad \text{and} \qquad y(t) = y_0 \mathrm{e}^{-t}$$

If we eliminate the time t, the function

$$y(x) = (y_0 \mathrm{e}^{-1/x_0})\mathrm{e}^{1/x}$$

describes the course of the trajectories. For $x < 0$, the x-axis at the origin is the common tangent to all solution curves. For $x \geqslant 0$, the x-axis is the only solution curve which approaches the origin for $t \to -\infty$. Since the x-axis is the eigendirection appertaining to the eigenvalue $\lambda_1 = 0$, each trajectory forms a centre manifold $\mathbb{W}^c$ for $x < 0$ while the x-axis alone represents this for positive x. Figure 6.2.8 illustrates the above and makes it clear that the centre manifold is not necessarily unique.

The example just described serves to elucidate the theorem of the centre manifold for flows. We can now formulate the following existence theorem (Guckenheimer and Holmes, 1983):

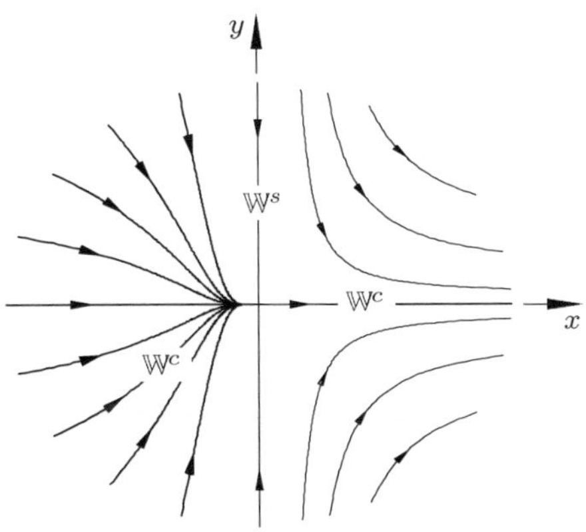

Fig. 6.2.8

Centre manifold of the system $\dot{x} = x^2,\ \dot{y} = -y$

Centre manifold theorem:

> Let $\boldsymbol{x}_s$ be a fixed point of the system $\dot{\boldsymbol{x}} = \boldsymbol{F}(\boldsymbol{x})$ and $\boldsymbol{D} = \frac{\partial \boldsymbol{F}}{\partial \boldsymbol{x}}\Big|_{\boldsymbol{x}_s}$ where we presume that $\boldsymbol{F}(\boldsymbol{x})$ is r-times continuously differentiable; the eigenvalue spectrum of $\boldsymbol{D}$ can then be divided into three sub-sets $\{\lambda_{si}\}$, $\{\lambda_{ci}\}$ and $\{\lambda_{ui}\}$ defined by
>
> $$\mathrm{Re}(\lambda)\begin{cases} < 0, & \text{if } \lambda \in \{\lambda_{si}\} \\ = 0, & \text{if } \lambda \in \{\lambda_{ci}\} \\ > 0, & \text{if } \lambda \in \{\lambda_{ui}\} \end{cases}$$
>
> The generalised eigenspaces assigned to the eigenvalues of $\{\lambda_{si}\}$, $\{\lambda_{ci}\}$ and $\{\lambda_{ui}\}$ are $\mathbb{E}^s, \mathbb{E}^c$ and $\mathbb{E}^u$. There then exist r-times differentiable stable and unstable manifolds $\mathbb{W}^s$ and $\mathbb{W}^u$ tangent to $\mathbb{E}^s$ and $\mathbb{E}^u$ in $\boldsymbol{x}_s$ and a (r − 1)-times differentiable centre manifold $\mathbb{W}^c$ tangent to $\mathbb{E}^c$ in $\boldsymbol{x}_s$. All three manifolds are invariant under the flow of $\boldsymbol{F}$. The stable and unstable manifolds are unique; this does not necessarily apply for $\mathbb{W}^c$.

Figure 6.2.9 illustrates the geometrical interpretation of the theorem of the centre manifold. In order to decide whether the flow on $\mathbb{W}^c$ moves towards the origin or away from it, higher-order terms must be considered in the stability analysis (see also the previous example, fig. 6.2.8).

If we now introduce curvilinear "manifold coordinates" $\bar{\boldsymbol{x}}, \bar{\boldsymbol{y}}$ and $\bar{\boldsymbol{z}}$ locally at the origin where $\bar{\boldsymbol{x}}$, $\bar{\boldsymbol{y}}$, $\bar{\boldsymbol{z}}$ are assigned to the manifolds $\mathbb{W}^c$, $\mathbb{W}^s$, $\mathbb{W}^u$ respectively, then, as a consequence of the centre manifold theorem, the initial system $\dot{\boldsymbol{x}} = \boldsymbol{F}(\boldsymbol{x})$ is in a neighbourhood of the singular point topologically equivalent to the system

$$\begin{aligned} \dot{\bar{\boldsymbol{x}}} &= \bar{\boldsymbol{f}}(\bar{\boldsymbol{x}}) \\ \dot{\bar{\boldsymbol{y}}} &= -\bar{\boldsymbol{y}} \\ \dot{\bar{\boldsymbol{z}}} &= \bar{\boldsymbol{z}} \end{aligned} \quad , \qquad (\bar{\boldsymbol{x}}, \bar{\boldsymbol{y}}, \bar{\boldsymbol{z}}) \in \mathbb{W}^c \times \mathbb{W}^s \times \mathbb{W}^u \tag{6.2.14}$$

i. e. the asymptotic behaviour of the initial system can be reduced in the vicinity of the origin to the behaviour of the vector field $\bar{\boldsymbol{f}}(\bar{\boldsymbol{x}})$. The problem is to find a suitable coordinate transformation which converts the initial system to the system eq. (6.2.14) and allows a specification of the function $\bar{\boldsymbol{f}}$. In order to elucidate the individual calculation steps, we assume for simplicity that no unstable manifold $\mathbb{W}^u$ exists and consider the following system (fig. 6.2.10)

$$\begin{aligned} \dot{\boldsymbol{x}} &= \boldsymbol{A}\boldsymbol{x} + \boldsymbol{f}(\boldsymbol{x}, \boldsymbol{y}) \\ \dot{\boldsymbol{y}} &= \boldsymbol{B}\boldsymbol{y} + \boldsymbol{g}(\boldsymbol{x}, \boldsymbol{y}) \end{aligned} \tag{6.2.15}$$

where $\boldsymbol{x} \in \mathbb{R}^c, \boldsymbol{y} \in \mathbb{R}^s$. The sub-division into the constant matrices $\boldsymbol{A}$ and $\boldsymbol{B}$ derives from the real parts of their eigenvalues. All eigenvalues of $\boldsymbol{A}$ have a real part equal to zero whereas all eigenvalues of $\boldsymbol{B}$ have negative real parts.

The functions $\boldsymbol{f}$ and $\boldsymbol{g}$ are non-linear and vanish together with their first derivatives at the origin, i. e. $\boldsymbol{f}(\boldsymbol{o}, \boldsymbol{o}) = \boldsymbol{o}$, $\boldsymbol{f}'(\boldsymbol{o}, \boldsymbol{o}) = \boldsymbol{o}$ and $\boldsymbol{g}(\boldsymbol{o}, \boldsymbol{o}) = \boldsymbol{o}$, $\boldsymbol{g}'(\boldsymbol{o}, \boldsymbol{o}) = \boldsymbol{o}$ should hold

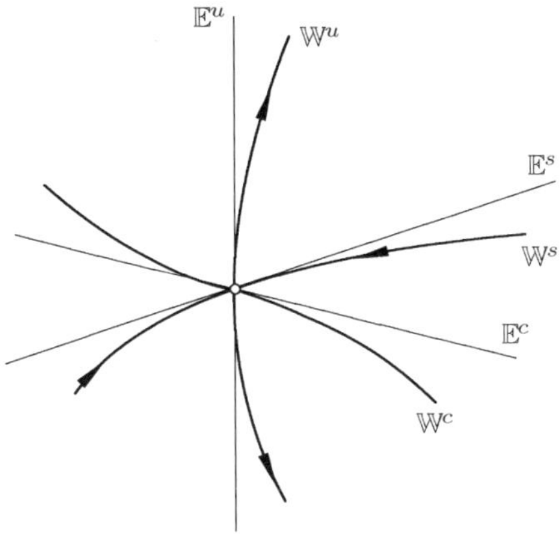

Fig. 6.2.9

The manifolds $\mathbb{W}^s$, $\mathbb{W}^u$ and $\mathbb{W}^c$

where $\boldsymbol{f}'$ and $\boldsymbol{g}'$ are the derivatives of $\boldsymbol{f}$ and $\boldsymbol{g}$ with respect to $\boldsymbol{x}$ and $\boldsymbol{y}$ and are given as follows

$$\boldsymbol{f}' = \begin{bmatrix} \dfrac{\partial \boldsymbol{f}}{\partial \boldsymbol{x}} & \dfrac{\partial \boldsymbol{f}}{\partial \boldsymbol{y}} \end{bmatrix}$$

We now proceed by taking up the idea of reducing the initial system to "fundamentals" – without a loss of quality in the dynamic behaviour – and clarify the argumentation leading to the reduction by observing the transition from the linear to the non-linear system.

In the special linear case – the non-linear components $\boldsymbol{f}$ and $\boldsymbol{g}$ of eq. (6.2.15) are equal to zero – we obtain the two invariant manifolds $\mathbb{W}^s : \boldsymbol{x} = \boldsymbol{o}$ and $\mathbb{W}^c : \boldsymbol{y} = \boldsymbol{o}$. They are identical with the respective eigenspaces $\mathbb{E}^s$ respectively $\mathbb{E}^c$. For the flow on the centre manifold $\mathbb{W}^c$, the linearised initial system eq. (6.2.15) is reduced to

$$\dot{\boldsymbol{x}} = \boldsymbol{A}\boldsymbol{x} \tag{6.2.16}$$

Since the solutions of $\dot{\boldsymbol{y}} = \boldsymbol{B}\boldsymbol{y}$ of the system eq. (6.2.15) decay exponentially fast, the solutions of eq. (6.2.16) determine the system behaviour of the complete

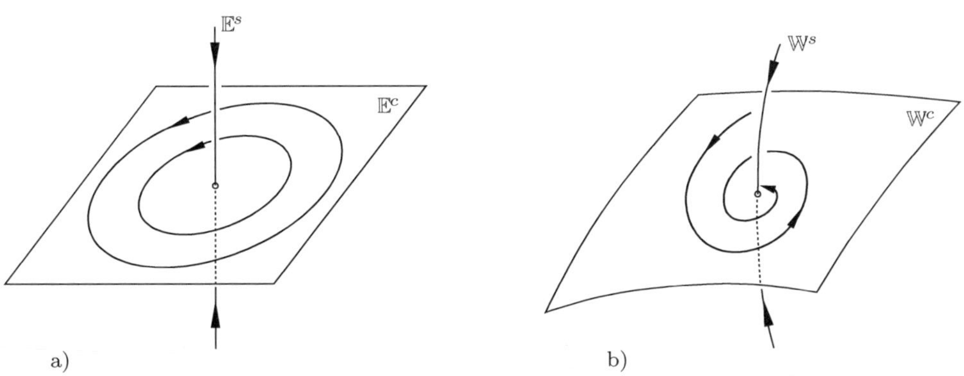

Fig. 6.2.10: Stable and centre manifold of a
a) linear system and a b) non-linear system

linearised system eq. (6.2.15) for $t \to \infty$. This means that eq. (6.2.16) respectively the flow on $\mathbb{W}^c$ determines whether the fixed point $\{\boldsymbol{x}\ \boldsymbol{y}\} = \{\boldsymbol{o}\ \boldsymbol{o}\}$ is stable or unstable. We now ask to what extent the reduction to $\mathbb{W}^c$ is applicable to the non-linear system eq. (6.2.15).

The existence theorem of the centre manifold implies that $\mathbb{W}^c$ can be expressed by a function $\boldsymbol{h}(\boldsymbol{x})$

$$\boldsymbol{y} = \boldsymbol{h}(\boldsymbol{x}) \tag{6.2.17}$$

for which the conditions

$$\boldsymbol{h}(\boldsymbol{o}) = \boldsymbol{o} \qquad \text{and} \qquad \boldsymbol{h}'(\boldsymbol{o}) = \left.\frac{\partial \boldsymbol{h}}{\partial \boldsymbol{x}}\right|_{\boldsymbol{o}} = \boldsymbol{o}$$

must be satisfied so that the centre manifold $\boldsymbol{y} = \boldsymbol{o}$ of the linearised system is tangent to the non-linear $\mathbb{W}^c$. If we substitute eq. (6.2.17) into the initial system eq. (6.2.15), we obtain the projection of the vector field on the centre manifold $\mathbb{W}^c$ onto $\mathbb{E}^c$

$$\dot{\boldsymbol{x}} = \boldsymbol{A}\boldsymbol{x} + \boldsymbol{f}(\boldsymbol{x}, \boldsymbol{h}(\boldsymbol{x})) \tag{6.2.18}$$

Equation (6.2.18) is the non-linear analogue to the linear case, eq. (6.2.16), and corresponds to the first equation of eq. (6.2.14). The fact that the stability behaviour of eq. (6.2.18) instructs us, without loss of information, how the non-linear eq. (6.2.15) behaves at $\{\boldsymbol{x}\ \boldsymbol{y}\} = \{\boldsymbol{o}\ \boldsymbol{o}\}$ is the fundamental result of the centre manifold theory. Carr (1981) and Henry (1981) formulated the *equivalence theorem* as follows:

> If the origin $\boldsymbol{x} = \boldsymbol{o}$ of eq. (6.2.18) is locally asymptotically stable (respectively unstable), then the zero solution of the non-linear initial system of eq. (6.2.15) is also locally asymptotically stable (respectively unstable).

The solutions of eq. (6.2.18) represent in the neighbourhood of the origin a good approximation for the flow of eq. (6.2.14), $\dot{\bar{\boldsymbol{x}}} = \bar{\boldsymbol{f}}(\bar{\boldsymbol{x}})$, on the centre manifold.

In order to be able to determine the function $\boldsymbol{h}(\boldsymbol{x})$, we differentiate $\boldsymbol{y}(t) = \boldsymbol{h}(\boldsymbol{x}(t))$ according to the chain rule with respect to t and take account of the second equation of eq. (6.2.15). We obtain

$$\dot{\boldsymbol{y}} = \frac{\partial \boldsymbol{h}}{\partial \boldsymbol{x}}\dot{\boldsymbol{x}} = \boldsymbol{B}\boldsymbol{h}(\boldsymbol{x}) + \boldsymbol{g}(\boldsymbol{x}, \boldsymbol{h}(\boldsymbol{x})) \tag{6.2.19}$$

Substituting the first expressions of eq. (6.2.15) for $\dot{\boldsymbol{x}}$, we deduce a system of partial differential equations for the centre manifold

$$\frac{\partial \boldsymbol{h}}{\partial \boldsymbol{x}}\left[\boldsymbol{A}\boldsymbol{x} + \boldsymbol{f}(\boldsymbol{x}, \boldsymbol{h}(\boldsymbol{x}))\right] - \boldsymbol{B}\boldsymbol{h}(\boldsymbol{x}) - \boldsymbol{g}(\boldsymbol{x}, \boldsymbol{h}(\boldsymbol{x})) = \boldsymbol{o} \tag{6.2.20}$$

with the boundary conditions $\boldsymbol{h}(\boldsymbol{o}) = \boldsymbol{o}$ and $\left.\frac{\partial \boldsymbol{h}}{\partial \boldsymbol{x}}\right|_{\boldsymbol{o}} = \boldsymbol{o}$. It is, of course, generally impossible to solve this partial differential equation exactly; if this were the case, we would already have established the solution of the initial system eq. (6.2.15), since eq. (6.2.20) is equivalent to it. If, however, $\mathbb{W}^c$ in the origin can be described by an analytic function, the solution of eq. (6.2.20) can be approximated arbitrarily precisely by a Taylor series. Selecting an approximation function $\boldsymbol{\psi}$ for $\boldsymbol{h}$, we obtain for the residuum of eq. (6.2.20)

$$\boldsymbol{R}(\boldsymbol{\psi}(\boldsymbol{x})) = \frac{\partial \boldsymbol{\psi}}{\partial \boldsymbol{x}} \left[\boldsymbol{A}\boldsymbol{x} + \boldsymbol{f}(\boldsymbol{x}, \boldsymbol{\psi}(\boldsymbol{x}))\right] - \boldsymbol{B}\boldsymbol{\psi}(\boldsymbol{x}) - \boldsymbol{g}(\boldsymbol{x}, \boldsymbol{\psi}(\boldsymbol{x})) \tag{6.2.21}$$

The degree of accuracy of the approximation is determined by the following theorem (Henry, 1981; Carr, 1981):

> If a function $\boldsymbol{\psi}(\boldsymbol{x})$ with $\boldsymbol{\psi}(\boldsymbol{o}) = \boldsymbol{o}$ and $\boldsymbol{\psi}'(\boldsymbol{o}) = \boldsymbol{o}$ can be found for which $\boldsymbol{R}(\boldsymbol{\psi}(\boldsymbol{x})) = \mathcal{O}(|\boldsymbol{x}|^{\mathrm{q}})$ with q > 1 is satisfied when $|\boldsymbol{x}| \to \boldsymbol{o}$, then the approximation degree of $\boldsymbol{h}$ is also of q-th order, i. e. we have $\boldsymbol{h}(\boldsymbol{x}) = \boldsymbol{\psi}(\boldsymbol{x}) + \mathcal{O}(|\boldsymbol{x}|^{\mathrm{q}})$ for $|\boldsymbol{x}| \to \boldsymbol{o}$.

We point out two further aspects: firstly, the residuum in eq. (6.2.20) vanishes, i.e. $\boldsymbol{R}(\boldsymbol{h}(\boldsymbol{x})) = \boldsymbol{o}$, and secondly, the reduction and approximation principles naturally also hold for the general case of an additional unstable manifold, eq. (6.2.14).

The method of the centre manifold hence offers a systematic technique for determining the stability behaviour of degenerate fixed points and approximating the dynamics in the neighbourhood of these fixed points.

We now demonstrate the application of this method with three simple examples.

Example 1

We first consider the two-dimensional system of non-linear autonomous differential equations

$$\begin{aligned} \dot{x} &= -xy \\ \dot{y} &= -y + x^2 \end{aligned} \tag{6.2.22}$$

There is a single fixed point at the origin $\{x \;\; y\} = \{0 \;\; 0\}$. To determine its stability, we write eq. (6.2.22) in the standard form of eq. (6.2.15)

$$\begin{bmatrix} \dot{x} \\ \dot{y} \end{bmatrix} = \begin{bmatrix} 0 & 0 \\ 0 & -1 \end{bmatrix} \begin{bmatrix} x \\ y \end{bmatrix} + \begin{bmatrix} -xy \\ x^2 \end{bmatrix} \tag{6.2.23}$$

where $A = 0$, $B = -1$, $f = -xy$ and $g = x^2$.

The eigenvalues $\lambda_1 = 0$ and $\lambda_2 = -1$ of the linearised matrix and their respective eigenvectors yield the one-dimensional eigenspaces $\mathbb{E}^c = \{(x, y) \in \mathbb{R}^2; y = 0\}$ and $\mathbb{E}^s = \{(x, y) \in \mathbb{R}^2; x = 0\}$. According to the existence theorem, eq. (6.2.22) possesses a centre manifold $y = h(x)$. In order to approximate h by a function $\psi(x)$, we determine the residuum in accordance with eq. (6.2.21)

$$R(\psi(x)) = \frac{\mathrm{d}\psi}{\mathrm{d}x}\big(-x\psi(x)\big) + \psi(x) - x^2$$

Choosing $\psi(x) = \mathcal{O}(x^2)$, we obtain for $R(\psi(x))$

$$R(\psi(x)) = \left[\mathcal{O}(x^4)\right] + \psi(x) - x^2$$

Consequently, the approximation theorem $(R(\psi(x)) = \mathcal{O}(x^4))$ leads to

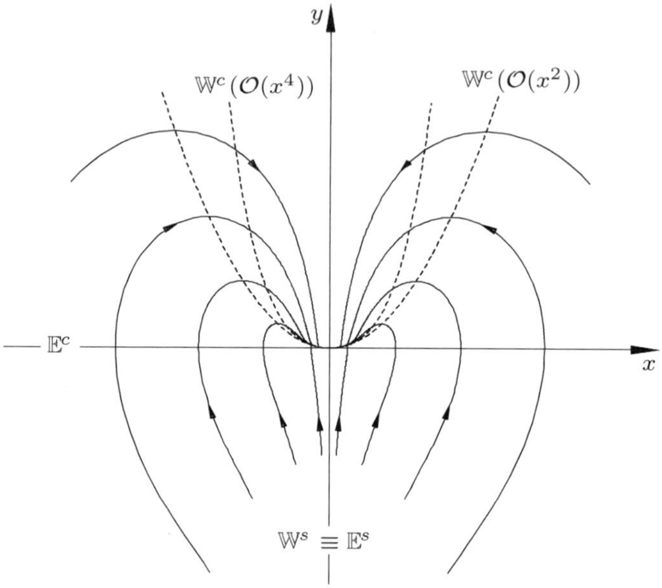

Fig. 6.2.11: Trajectories and two alternative approximations $\mathbb{W}^c(\mathcal{O}(x^2))$ resp. $\mathbb{W}^c(\mathcal{O}(x^4))$ of the centre manifold

$$\psi(x) = x^2 \qquad \text{and} \qquad h(x) = x^2 + \mathcal{O}(x^4)$$

Substituting $h(x)$ into the first equation of eq. (6.2.22), we obtain a relation which, in accordance with the equivalence theorem, determines the stability of the initial system and the dynamical behaviour in the neighbourhood of the origin

$$\dot{x} = -x^3 + \mathcal{O}(x^5) \tag{6.2.24}$$

Equation (6.2.24) demonstrates that the zero solution of eq. (6.2.22) is asymptotically stable. Figure 6.2.11 presents the numerically calculated course of the trajectories and the approximated centre manifold $\psi(x) = x^2$. Both the approximation by $\psi(x)$ in the neighbourhood of the origin as well as the limited range of validity of $\mathbb{W}^c(\mathcal{O}(x^2))$ can be clearly recognised.

We next enquire whether an approximation of higher order for h will significantly extend the range of validity of $\mathbb{W}^c$. To this end, we choose $\psi(x) = x^2 + \theta(x)$ with the ansatz $\theta(x) = ax^3 + bx^4$. The residuum is then

$$R(\psi(x)) = -2x^4 + \mathcal{O}(x^5) + x^2 + \theta(x) - x^2$$

or

$$\theta(x) = 2x^4 \quad \text{resp.} \quad \psi(x) = x^2 + 2x^4$$

and

$$h(x) = x^2 + 2x^4 + \mathcal{O}(x^5)$$

We finally obtain for the dynamics on $\mathbb{W}^c(\mathcal{O}(x^4))$

$$\dot{x} = -x^3 - 2x^5 + \mathcal{O}(x^6)$$

the zero solution of which also possesses asymptotic stability.

The higher-order approximated centre manifold $\mathbb{W}^c(\mathcal{O}(x^4))$ is also shown in fig. 6.2.11. The higher accuracy of the approximation is accompanied by a sharper boundary of the range of validity.

Example 2

Our second example is the set of equations (6.1.18). We first write the system of equations (6.1.18) in the standard form

$$\begin{bmatrix} \dot{x} \\ \dot{y} \end{bmatrix} = \begin{bmatrix} 0 & 0 \\ 0 & -1 \end{bmatrix} \begin{bmatrix} x \\ y \end{bmatrix} + \begin{bmatrix} \mu x^3 + x^2 y \\ y^2 + xy - x^3 \end{bmatrix}$$

Following eq. (6.2.15), $A = 0$, $B = -1$, $f = \mu x^3 + x^2 y$ and $g = y^2 + xy - x^3$.

We are interested exclusively in the stability behaviour of the fixed point $\{x \quad y\} = \{0 \quad 0\}$. From the eigenvalues and eigenvectors, we determine the eigenspaces $\mathbb{E}^s = \{(x, y) \in \mathbb{R}^2; x = 0\}$ and $\mathbb{E}^c = \{(x, y) \in \mathbb{R}^2; y = 0\}$. For the residuum of the approximation function $\psi(x)$ for $h(x)$, we obtain

$$R(\psi(x)) = \psi'(x)\big(\mu x^3 + x^2\psi(x)\big) + \psi(x) - \psi^2(x) - x\psi(x) + x^3$$

Introducing $\psi(x) = \mathcal{O}(x^2)$, we find

$$R(\psi(x)) = \big(\mathcal{O}(x^4) + \mathcal{O}(x^5)\big) + \psi(x) + \mathcal{O}(x^4) + \mathcal{O}(x^3) + \mathcal{O}(x^3)$$

According to the approximation theorem, we have

$$\psi(x) = 0 \qquad \text{resp.} \qquad h(x) = 0 + \mathcal{O}(x^3)$$

The flow $\mathbb{W}^c$ is then approximated by

$$\dot{x} = \mu x^3 + \mathcal{O}(x^5)$$

For $\mu < 0$, the initial system (6.1.18) is asymptotically stable and for $\mu > 0$ unstable (fig. 6.1.7). The question as to the stability for $\mu = 0$ remains open. A polynomial ansatz of higher order for $\psi(x)$ may provide us with further insight.

We choose $\psi(x) = \mathcal{O}(x^3)$. The residuum results as

$$R(\psi(x)) = \big(\mathcal{O}(x^5) + \mathcal{O}(x^7)\big) + \psi(x) + \mathcal{O}(x^6) + \mathcal{O}(x^4) + x^3$$

or

$$\psi(x) = -x^3 \qquad \text{resp.} \qquad h(x) = -x^3 + \mathcal{O}(x^4)$$

The dynamics on $\mathbb{W}^c$ is expressed by the differential equation

$$\dot{x} = \mu x^3 - x^5 + \mathcal{O}(x^6)$$

From this approximation, it becomes clear that the system eq. (6.1.19) is asymptotically stable for $\mu = 0$. The numerically calculated trajectories in fig. 6.1.7 are thus correctly reproduced at the origin.

In the aforementioned examples, the eigenspaces coincided with the Cartesian coordinate system. Also, the linearised matrix $\boldsymbol{D}$ appeared in diagonal form, i. e. uncoupled. Such conditions may not always prevail, as the next example shows.

Example 3

We consider the set of two non-linear differential equations

$$\begin{aligned} \dot{x} &= y \\ \dot{y} &= -y + ax^2 + bxy \end{aligned} \tag{6.2.25}$$

respectively

$$\begin{bmatrix} \dot{x} \\ \dot{y} \end{bmatrix} = \begin{bmatrix} 0 & 1 \\ 0 & -1 \end{bmatrix} \begin{bmatrix} x \\ y \end{bmatrix} + \begin{bmatrix} 0 \\ ax^2 + bxy \end{bmatrix}$$

or

$$\dot{\boldsymbol{x}} = \boldsymbol{D}\boldsymbol{x} + \boldsymbol{N}(\boldsymbol{x})$$

The eigenvalues for the only fixed point $\{x \;\; y\} = \{0 \;\; 0\}$ are $\lambda_1 = 0$ and $\lambda_2 = -1$. The eigenvectors $\boldsymbol{y}_1 = \{1 \quad 0\}$ and $\boldsymbol{y}_2 = \{1 \quad -1\}$ constitute the basis vectors of the transformed system. Applying the transformation, see eq. (3.1.10),

$$\begin{bmatrix} x \\ y \end{bmatrix} = \boldsymbol{T} \begin{bmatrix} \bar{x} \\ \bar{y} \end{bmatrix} , \quad \text{where}$$

$$\boldsymbol{T} = [\boldsymbol{y}_1 \quad \boldsymbol{y}_2] = \begin{bmatrix} 1 & 1 \\ 0 & -1 \end{bmatrix} = \boldsymbol{T}^{-1} \quad \text{and} \quad \boldsymbol{\Lambda} = \boldsymbol{T}^{-1}\boldsymbol{D}\boldsymbol{T} = \begin{bmatrix} \lambda_1 & 0 \\ 0 & \lambda_2 \end{bmatrix}$$

we obtain the standard form of eq. (6.2.25)

$$\begin{bmatrix} \dot{\bar{x}} \\ \dot{\bar{y}} \end{bmatrix} = \boldsymbol{\Lambda} \begin{bmatrix} \bar{x} \\ \bar{y} \end{bmatrix} + \boldsymbol{T} \begin{bmatrix} 0 \\ a(\bar{x} + \bar{y})^2 - b(\bar{x}\bar{y} + \bar{y}^2) \end{bmatrix}$$

Multiplying out the matrices and omitting the bars to simplify notation, we find

$$\begin{aligned} \dot{x} &= a(x+y)^2 - b(xy + y^2) = Ax + f(x,y) \\ \dot{y} &= -y - a(x+y)^2 + b(xy + y^2) = By + g(x,y) \end{aligned} \tag{6.2.26}$$

Noting that $A = 0$ and $B = -1$, the residuum of the system eq. (6.2.26) becomes

$$R(\psi(x)) = \psi'(x)\Big(a\big(x + \psi(x)\big)^2 - b\big(x\psi(x) + \psi^2(x)\big)\Big) + \psi(x) + a\big(x + \psi(x)\big)^2 - b\big(x\psi(x) + \psi^2(x)\big)$$

Choosing for the order of approximation $\psi(x) = \mathcal{O}(x)$, we obtain

$$R(\psi(x)) = \psi(x) + \mathcal{O}(x^2)$$

Since $R(\psi(x)) = \mathcal{O}(x^2)$, we find for $\psi(x)$ respectively $h(x)$

$$\psi(x) = 0 \qquad \text{resp.} \qquad h(x) = 0 + \mathcal{O}(x^2)$$

The dynamics in the origin is represented to a first approximation by the flow on $\mathbb{W}^c \equiv \mathbb{E}^c$; we insert $y = h(x) = 0$ into the first equation of eq. (6.2.26) and obtain

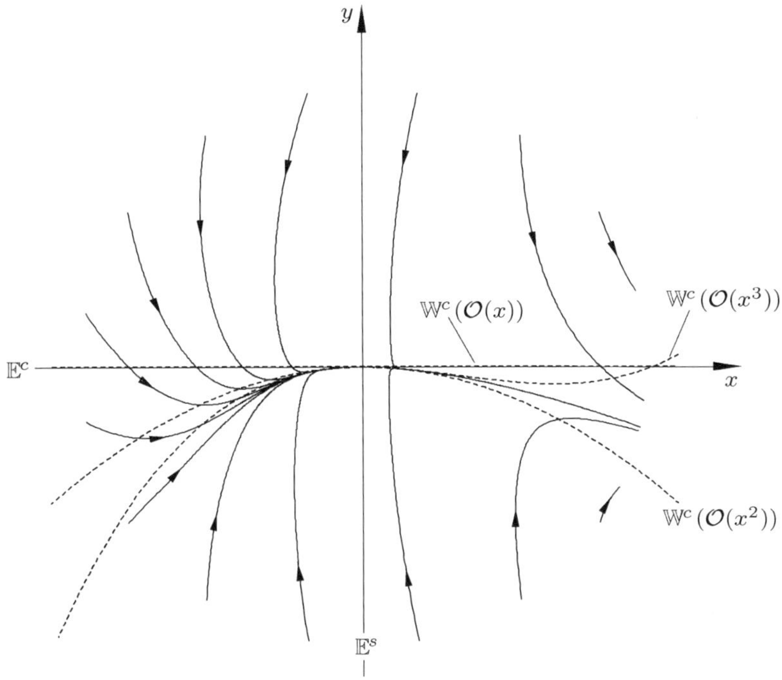

Fig. 6.2.12: Different approximations of the centre manifold of eq. (6.2.26) for $a > 0$

$$\dot{x} = ax^2 + \mathcal{O}(x^3)$$

Direct integration yields $x = \dfrac{1}{x_0^{-1} - at}$ and we therefore have the following stability behaviour for $a > 0$

$$x \to -0, \quad \text{for } t \to +\infty \text{ (asymptotically stable)} \qquad \text{and}$$
$$x \to +0, \quad \text{for } t \to -\infty \text{ (unstable)}$$

i. e. the trajectories approach the origin asymptotically for $x < 0$ and tend away from it for $x > 0$. If we choose $\psi(x) = \mathcal{O}(x^2)$ respectively $\psi(x) = \mathcal{O}(x^3)$, we obtain for $h(x)$

$$\begin{aligned} h(x) &= -ax^2 + \mathcal{O}(x^3) \qquad \text{resp.} \\ h(x) &= -ax^2 + a(4a - b)x^3 + \mathcal{O}(x^4) \end{aligned}$$

and for the flow on $\mathbb{W}^c$

$$\begin{aligned} \dot{x} &= a\big(x^2 + (b - 2a)x^3\big) + \mathcal{O}(x^4) \qquad \text{resp.} \\ \dot{x} &= a\big(x^2 + (b - 2a)x^3 + (9a^2 - 7ab + b^2)x^4\big) + \mathcal{O}(x^5) \end{aligned}$$

In order to analyse the stability behaviour, it suffices to restrict ourselves to the linear approximation of the eigenspace $\psi(x) = 0$. Figure 6.2.12 illustrates the asymptotically stable left and unstable right half-space for the system eq. (6.2.26).

After these detailed comments on the centre manifold and the demonstrations with examples, which provide an opportunity of assessing the stability of degenerate fixed points, we have to focus our attention on the real subject matter of this chapter. We know that the qualitative character of a flow alters at the critical value μ_{cr} respectively that new solution branches may emerge. Our main concern in this section is to show that the centre manifold represents a suitable technique for describing bifurcating solutions. To illustrate this aspect, we consider two more examples.

Example 4

We investigate the set of non-linear differential equations (Carr, 1981)

$$\begin{aligned} \dot{x} &= \mu x - x^3 + xy \\ \dot{y} &= -y + y^2 - x^2 \end{aligned} \tag{6.2.27}$$

in which the control parameter is denoted μ. The phase portraits for three different μ-values are shown in fig. 6.2.13. If we consider the linearised part of the system eq. (6.2.27), the eigenvalues are $\lambda_1 = \mu$ and $\lambda_2 = -1$. We are interested in solutions for small $|\mu|$, but the centre manifold theory is not directly applicable to this case. This problem can be circumvented if we increase the dimension of eq. (6.2.27) by one and interpret the control parameter μ as a system variable. The expanded system is then written in the form

$$\begin{bmatrix} \dot{x} \\ \dot{\mu} \\ \dot{y} \end{bmatrix} = \begin{bmatrix} 0 & 0 & 0 \\ 0 & 0 & 0 \\ 0 & 0 & -1 \end{bmatrix} \begin{bmatrix} x \\ \mu \\ y \end{bmatrix} + \begin{bmatrix} \mu x - x^3 + xy \\ 0 \\ y^2 - x^2 \end{bmatrix} \tag{6.2.28}$$

What have we achieved? The decisive characteristic of eq. (6.2.28) is that the formerly linear term μx is transferred to the non-linear group. The linearised system then has at the origin the eigenvalues $\lambda_1 = 0$, $\lambda_2 = 0$, $\lambda_3 = -1$ and the centre manifold theory becomes applicable. According to the existence theorem, eq. (6.2.28) has a two-dimensional centre manifold of the form

$$y = h(x, \mu)$$

For the approximation of h, we use the sub-division in eq. (6.2.15) together with eq. (6.2.21); the following residuum in terms of $\psi(x, \mu)$ can then be formulated

$$R\big(\psi(x,\mu)\big) = \begin{bmatrix} \dfrac{\partial \psi}{\partial x} & \dfrac{\partial \psi}{\partial \mu} \end{bmatrix} \begin{bmatrix} \mu x - x^3 + x\psi \\ 0 \end{bmatrix} + \psi - \psi^2 + x^2$$

If $\psi(x, \mu) = -x^2$, the residuum is of the order $R\big(\psi(x,\mu)\big) = \mathcal{O}(3)$ where $\mathcal{O}(3)$ contains only cubic terms of the form $x^3, x^2\mu, x\mu^2, \mu^3$. Hence, the approximation of h may be expressed as

$$h(x, \mu) = -x^2 + \mathcal{O}(3)$$

Application of the equivalence theorem and substitution of h in the first equation of eq. (6.2.28) finally yields for the flow on $\mathbb{W}^c$

$$\begin{aligned} \dot{x} &= \mu x - 2x^3 + \mathcal{O}(4) \\ \dot{\mu} &= 0 \end{aligned} \tag{6.2.29}$$

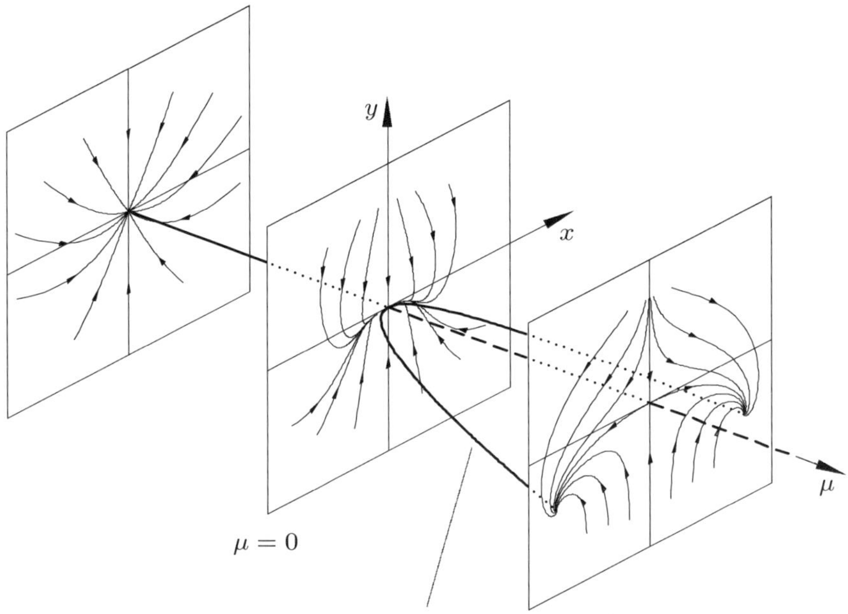

Fig. 6.2.13: Phase portraits of eq. (6.2.27) in dependence on μ

If $\mu \leqslant 0$, the solution $(x, \mu) = (0, 0)$ is asymptotically stable and consequently, as a result of the equivalence theorem, the initial system eq. (6.2.28) is stable as well. For $\mu > 0$, the zero solution becomes unstable. For small positive μ-values, the additional fixed points $\pm\sqrt{\mu/2}$ exist, both of which are also asymptotically stable, i. e. we have a pitchfork bifurcation, see eq. (6.1.5). Figure 6.2.14 illustrates the centre manifold $\mathbb{W}^c$, the flow on $\mathbb{W}^c$ and the pitchfork bifurcation. The deviation of the approximated pitchfork solutions from the exact ones is also revealed.

Example 5

In our next example, we consider a quadratic variant of the Duffing equation (Guckenheimer and Holmes, 1983)

$$\begin{aligned} \dot{x} &= \ y \\ \dot{y} &= \beta x - x^2 - \delta y \end{aligned} \tag{6.2.30}$$

We choose $\delta > 0$ and β as the control parameter near zero. The system

$$\begin{bmatrix} \dot{x} \\ \dot{y} \end{bmatrix} = \begin{bmatrix} 0 & 1 \\ \beta & -\delta \end{bmatrix} \begin{bmatrix} x \\ y \end{bmatrix} + \begin{bmatrix} 0 \\ -x^2 \end{bmatrix}$$

possesses at the origin a degenerate fixed point for $\beta = 0$ and the eigenvalues of the linearised system are $\lambda_1 = 0$, $\lambda_2 = -\delta$. In order to analyse the system in the directions of the eigenvectors, we use the following transformation

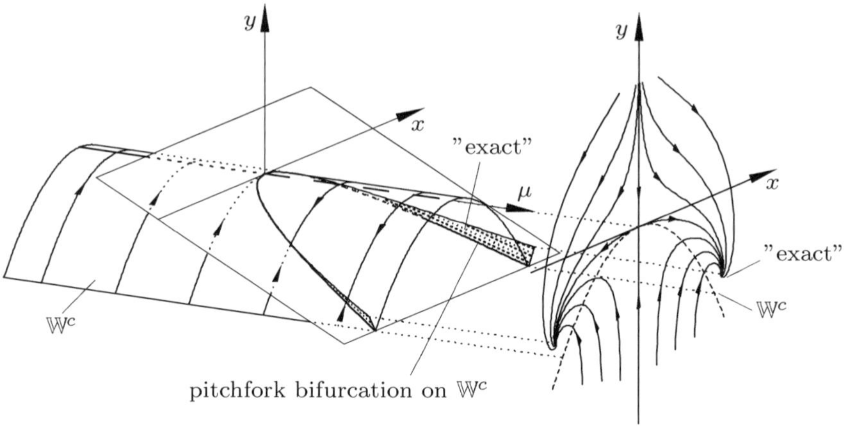

Fig. 6.2.14: Approximation of eq. (6.2.28) by the centre manifold $h(x,\mu) = -x^2 + \mathcal{O}(3)$

$$\begin{bmatrix} x \\ y \end{bmatrix} = \begin{bmatrix} 1 & 1 \\ 0 & -\delta \end{bmatrix} \begin{bmatrix} \bar{x} \\ \bar{y} \end{bmatrix} \quad \text{or} \quad \begin{bmatrix} \bar{x} \\ \bar{y} \end{bmatrix} = \begin{bmatrix} 1 & 1/\delta \\ 0 & -1/\delta \end{bmatrix} \begin{bmatrix} x \\ y \end{bmatrix}$$

and establish the system extended by the control parameter β. For simplicity's sake, we again dispense with the bars for the transformed system

$$\begin{aligned} \dot{x} &= \frac{\beta}{\delta}(x+y) - \frac{1}{\delta}(x+y)^2 \\ \dot{\beta} &= 0 \\ \dot{y} &= -\delta y - \frac{\beta}{\delta}(x+y) + \frac{1}{\delta}(x+y)^2 \end{aligned} \tag{6.2.31}$$

Splitting this system once more into a linear and a non-linear contribution yields

$$\begin{bmatrix} \dot{x} \\ \dot{\beta} \\ \dot{y} \end{bmatrix} = \begin{bmatrix} 0 & 0 & 0 \\ 0 & 0 & 0 \\ 0 & 0 & -\delta \end{bmatrix} \begin{bmatrix} x \\ \beta \\ y \end{bmatrix} + \frac{1}{\delta}(x+y) \begin{bmatrix} \beta - (x+y) \\ 0 \\ -\beta + (x+y) \end{bmatrix}$$

The linearised system indicates that the centre manifold can be expressed in the two variables

$$y = h(x,\beta)$$

Correspondingly, we approximate h by $\psi(x,\beta)$ and obtain as the residuum in accordance with eq. (6.2.21)

$$R\big(\psi(x,\beta)\big) = \begin{bmatrix} \dfrac{\partial\psi}{\partial x} & \dfrac{\partial\psi}{\partial\beta} \end{bmatrix} \begin{bmatrix} \frac{\beta}{\delta}(x+\psi) - \frac{1}{\delta}(x+\psi)^2 \\ 0 \end{bmatrix} + \delta\underline{\psi} + \frac{\beta}{\delta}(\underline{x}+\psi) - \frac{1}{\delta}(\underline{x}+\psi)^2$$

We choose for $\psi(x,\beta)$ in this expression a polynomial ansatz of second order in x and β. All the terms which are not underlined are then of the order $\mathcal{O}(3)$ in x and β. With $\psi(x,\beta) = x(x-\beta)/\delta^2$, the residuum becomes $R(\psi(x,\beta)) = \mathcal{O}(3)$. Thus, the approximate expression for the manifold becomes

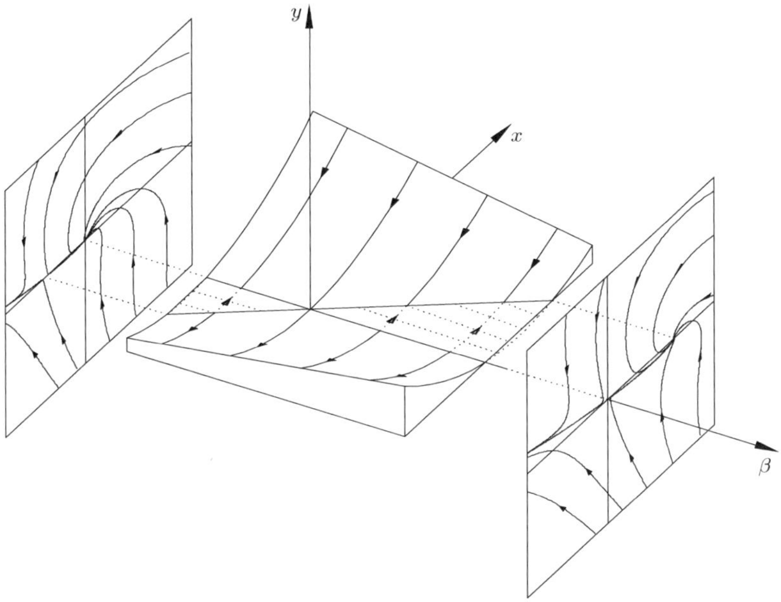

Fig. 6.2.15: Centre manifold in the origin and numerically calculated trajectories for the quadratic Duffing equation (6.2.30)

$$y = h(x,\beta) = \frac{1}{\delta^2}(x^2 - \beta x) + \mathcal{O}(3) \tag{6.2.32}$$

Substituting eq. (6.2.32) in eq. (6.2.31), we obtain for the flow on $\mathbb{W}^c$

$$\begin{aligned} \dot{x} &= \frac{\beta}{\delta}\left[x + \frac{1}{\delta^2}(x^2 - \beta x)\right] - \frac{1}{\delta}\left[x + \frac{1}{\delta^2}(x^2 - \beta x)\right]^2 + \mathcal{O}(4) \\ \dot{\beta} &= 0 \end{aligned} \tag{6.2.33}$$

If we take only the quadratic terms in the variables x, β into consideration, eq. (6.2.33) reduces to

$$\begin{aligned} \dot{x} &= \frac{\beta}{\delta}x - \frac{1}{\delta}x^2 \\ \dot{\beta} &= 0 \end{aligned} \tag{6.2.34}$$

If $\delta \gg \beta$ and β is small enough, we recognise for a fixed δ the transcritical bifurcation type of eq. (6.1.11) already shown in fig. 6.1.5. Naturally, the centre manifold must reflect this bifurcation type in the neighbourhood of the origin; this basic characteristic is preserved in the polynomial approximation. Figure 6.2.15 illustrates this type of bifurcation and its solution and stability behaviour. The centre manifold of eq. (6.2.32), an oblique parabolic cylinder, intersects the x,β-plane at $x = 0$ and $x = \beta$.

6.3 Normal Forms

The method of normal forms offers us a technique for reducing non-linear differential equations in the neighbourhood of fixed points and limit cycles to their "simplest" or essential form by means of a non-linear transformation technique without distorting the main characteristics of the local solution. This reduction technique originates from Poincaré and is a fundamental result of his dissertation. The idea behind it involves the successive elimination of as many non-linear terms of the vector field as possible by a series of non-linear coordinate transformations. Although the power series representing the coordinate tranformation does not always converge, the method of normal forms is a highly efficient tool since often, only few low-order terms suffice to reproduce essential information about the solution behaviour in the neighbourhood of fixed points respectively limit cycles.

In this chapter, our interest is directed mainly at the phenomenon of local bifurcations. With the aid of the technique of normal forms, we can bring order to the multitude of possible bifurcations by assigning all bifurcations possessing the same normal form and thus qualitatively equivalent dynamics to a single family respectively bifurcation type. We shall return to this in more detail in the subsequent sections 6.4 and 6.6.

The mathematical formalism underlying the method of normal forms renders its representation and thus the accessibility to this – in general computationally very laborious – technique difficult. In the following, we elucidate the fundamental ideas, citing illustrative examples. Our presentation relies in the main on four monographs on the subject of non-linear dynamical systems, namely those of

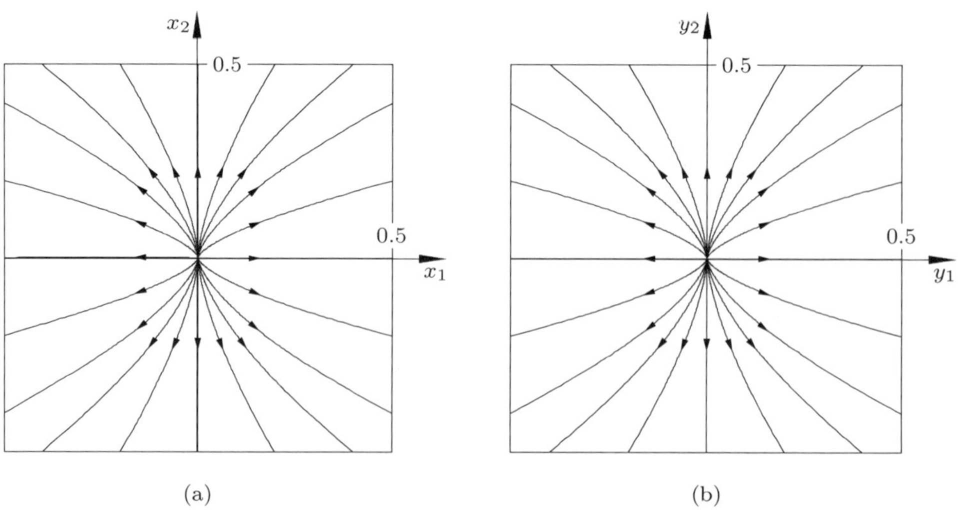

Fig. 6.3.1: Comparison of a system of non-linear equations with its normal form in the non-resonant case (for $a = 0.05,\ \lambda_1 = 0.9,\ \lambda_2 = 0.6$):
(a) $\dot{x}_1 = \lambda_1 x_1,\ \dot{x}_2 = \lambda_2 x_2 + a x_1^3$ and (b) $\dot{y}_1 = \lambda_1 y_1,\ \dot{y}_2 = \lambda_2 y_2$

Arnold (1988), Guckenheimer and Holmes (1983), Verhulst (1990) and Wiggins (1990).

The effects of a non-linear transformation can be demonstrated most vividly by means of a simple example. We consider the set of two-dimensional non-linear equations

$$\begin{aligned}\dot{x}_1 &= \lambda_1 x_1 \\ \dot{x}_2 &= \lambda_2 x_2 + a x_1^3\end{aligned} \tag{6.3.1}$$

The non-linear transformation

$$\begin{aligned}x_1 &= y_1 \\ x_2 &= y_2 + \frac{a}{3\lambda_1 - \lambda_2} y_1^3\end{aligned} \tag{6.3.2}$$

reduces the initial system eq. (6.3.1) to the simplest of all forms, the diagonal form

$$\begin{aligned}\dot{y}_1 &= \lambda_1 y_1 \\ \dot{y}_2 &= \lambda_2 y_2\end{aligned} \tag{6.3.3}$$

on the one condition – as the transformation eq. (6.3.2) shows – that $3\lambda_1 \neq \lambda_2$. The special case, which precludes diagonalisation, is called resonant. The definition of resonance will emerge subsequently according to eq. (6.3.56) when we discuss the general formulation of the technique of normal forms. Figure 6.3.1 contrasts the solution curves in the neighbourhood of the origin of the non-linear system eq. (6.3.1) and the diagonalised system eq. (6.3.3). For small x_i- respectively y_i-values, no differences can be discerned.

The generalisation of the result of our example represents the essence of Poincaré's dissertation. We start with an n-dimensional dynamical system $\dot{\boldsymbol{x}} = \boldsymbol{F}(\boldsymbol{x})$ possessing a fixed point $\boldsymbol{x}_s$ which is displaced to the origin by a translation; let the vector field $\boldsymbol{F}(\boldsymbol{x})$ be r-times continuously differentiable. If we expand the vector field in the neighbourhood of $\boldsymbol{x}_s = \boldsymbol{o}$ into a Taylor series and split it once more into linear and non-linear components

$$\dot{\boldsymbol{x}} = \boldsymbol{D}\boldsymbol{x} + \boldsymbol{N}(\boldsymbol{x}), \qquad \boldsymbol{x} \in \mathbb{R}^{\mathrm{n}} \tag{6.3.4}$$

where $\boldsymbol{D}$ is the constant Jacobian matrix at the origin, the following *theorem of Poincaré* applies:

Theorem of Poincaré:

> If the eigenvalues of the matrix $\boldsymbol{D}$ exclude the case of resonance, eq. (6.3.4) can be reduced by means of a non-linear *near-identity transformation*
>
> $$\boldsymbol{x} = \boldsymbol{y} + \boldsymbol{h}(\boldsymbol{y}) \tag{6.3.5}$$
>
> to the linear system
>
> $$\dot{\boldsymbol{y}} = \boldsymbol{D}\boldsymbol{y} \tag{6.3.6}$$

At this point, we only cite this basic theorem and trust that the statement will become plausible in the course of the subsequent more detailed description of the technique of normal forms.

We now begin by simplifying the non-linear term $\boldsymbol{N}(\boldsymbol{x})$. We first expand $\boldsymbol{N}(\boldsymbol{x})$ into a Taylor series around the origin and obtain for eq. (6.3.4)

$$\dot{\boldsymbol{x}} = \boldsymbol{D}\boldsymbol{x} + \boldsymbol{N}_2(\boldsymbol{x}) + \boldsymbol{N}_3(\boldsymbol{x}) + \cdots + \boldsymbol{N}_{\mathrm{r}-1}(\boldsymbol{x}) + \mathcal{O}(|\boldsymbol{x}|^{\mathrm{r}}) \tag{6.3.7}$$

where the series expansion begins with the second-order term $\boldsymbol{N}_2(\boldsymbol{x})$ and continues with all terms of i-th order being collected in $\boldsymbol{N}_{\mathrm{i}}(\boldsymbol{x})$. Correspondingly, we also expand the coordinate transformation eq. (6.3.5) into a Taylor series

$$\boldsymbol{x} = \boldsymbol{y} + \boldsymbol{h}_2(\boldsymbol{y}) + \boldsymbol{h}_3(\boldsymbol{y}) + \ldots + \boldsymbol{h}_{\mathrm{r}-1}(\boldsymbol{y}) + \mathcal{O}(|\boldsymbol{y}|^{\mathrm{r}}) \tag{6.3.8}$$

and attempt to eliminate respectively simplify the quadratic terms, the cubic terms etc. successively in eq. (6.3.7). We begin with a quadratic ansatz

$$\boldsymbol{x} = \boldsymbol{y} + \boldsymbol{h}_2(\boldsymbol{y}) \tag{6.3.9}$$

Here, $\boldsymbol{h}_2(\boldsymbol{y})$ represents a homogeneous vector-valued monomial of second degree. We substitute eq. (6.3.9) in eq. (6.3.7) and find

$$\begin{aligned}\dot{\boldsymbol{x}} &= \dot{\boldsymbol{y}} + \frac{\partial \boldsymbol{h}_2}{\partial \boldsymbol{y}}\dot{\boldsymbol{y}} = \left[\boldsymbol{I} + \frac{\partial \boldsymbol{h}_2}{\partial \boldsymbol{y}}\right]\dot{\boldsymbol{y}} \\ &= \boldsymbol{D}[\boldsymbol{y} + \boldsymbol{h}_2(\boldsymbol{y})] + \boldsymbol{N}_2\big(\boldsymbol{y} + \boldsymbol{h}_2(\boldsymbol{y})\big) + \boldsymbol{N}_3\big(\boldsymbol{y} + \boldsymbol{h}_2(\boldsymbol{y})\big) + \\ &\qquad \cdots + \boldsymbol{N}_{\mathrm{r}-1}\big(\boldsymbol{y} + \boldsymbol{h}_2(\boldsymbol{y})\big) + \mathcal{O}(|\boldsymbol{y}|^{\mathrm{r}})\end{aligned} \tag{6.3.10}$$

The non-linear terms

$$\boldsymbol{N}_{\mathrm{k}}\big(\boldsymbol{y} + \boldsymbol{h}_2(\boldsymbol{y})\big), \quad 2 \leqslant \mathrm{k} \leqslant \mathrm{r} - 1 \tag{6.3.11}$$

can also be expanded into Taylor series

$$\boldsymbol{N}_{\mathrm{k}}(\boldsymbol{y}) + \mathcal{O}(|\boldsymbol{y}|^{\mathrm{k}+1}) + \cdots + \mathcal{O}(|\boldsymbol{y}|^{2\mathrm{k}}) \tag{6.3.12}$$

so that eq. (6.3.10) takes the form

$$\left[\boldsymbol{I} + \frac{\partial \boldsymbol{h}_2}{\partial \boldsymbol{y}}\right]\dot{\boldsymbol{y}} = \boldsymbol{D}\boldsymbol{y} + \boldsymbol{D}\boldsymbol{h}_2(\boldsymbol{y}) + \boldsymbol{N}_2(\boldsymbol{y}) + \widetilde{\boldsymbol{N}}_3(\boldsymbol{y}) + \cdots + \widetilde{\boldsymbol{N}}_{\mathrm{r}-1}(\boldsymbol{y}) + \mathcal{O}(|\boldsymbol{y}|^{\mathrm{r}}) \tag{6.3.13}$$

where the $\widetilde{\boldsymbol{N}}_{\mathrm{k}}$ are the modified $\boldsymbol{N}_{\mathrm{k}}$-terms deduced from the coordinate transformation. We seek to solve eq. (6.3.13) with respect to $\dot{\boldsymbol{y}}$. For small $\boldsymbol{y}$, there exists the inverse matrix

$$\left[\boldsymbol{I} + \frac{\partial \boldsymbol{h}_2}{\partial \boldsymbol{y}}\right]^{-1}$$

It can be expanded into a series so that

$$\left[\boldsymbol{I} + \frac{\partial \boldsymbol{h}_2}{\partial \boldsymbol{y}}\right]^{-1} = \boldsymbol{I} - \frac{\partial \boldsymbol{h}_2}{\partial \boldsymbol{y}} + \mathcal{O}(|\boldsymbol{y}|^2) \tag{6.3.14}$$

Inserting eq. (6.3.14) into eq. (6.3.13) finally yields

$$\dot{\boldsymbol{y}} = \boldsymbol{D}\boldsymbol{y} + \boldsymbol{D}\boldsymbol{h}_2(\boldsymbol{y}) - \frac{\partial \boldsymbol{h}_2}{\partial \boldsymbol{y}}\boldsymbol{D}\boldsymbol{y} + \boldsymbol{N}_2(\boldsymbol{y}) + \widetilde{\boldsymbol{N}}_3(\boldsymbol{y}) + \cdots + \widetilde{\boldsymbol{N}}_{\mathrm{r}-1}(\boldsymbol{y}) + \mathcal{O}(|\boldsymbol{y}|^{\mathrm{r}}) \tag{6.3.15}$$

Our initial task was to select $\boldsymbol{h}_2(\boldsymbol{y})$ in eq. (6.3.9) so as either to eliminate the quadratic terms in eq. (6.3.7) or, should this prove impossible, at least to reduce them to their simplest form. If we consider eq. (6.3.15) from this point of view and consider only the quadratic terms, we obtain the following equation for $\boldsymbol{h}_2(\boldsymbol{y})$

$$\boldsymbol{D}\boldsymbol{h}_2(\boldsymbol{y}) - \frac{\partial \boldsymbol{h}_2}{\partial \boldsymbol{y}}\boldsymbol{D}\boldsymbol{y} = -\boldsymbol{N}_2(\boldsymbol{y}) \tag{6.3.16}$$

Equation (6.3.16) is called the *homological* equation for the unknown function $\boldsymbol{h}_2$ (Arnold, 1988). In Lie algebra, the expression on the left-hand side is denoted the *Lie bracket* of the two vector fields $\boldsymbol{h}_2(\boldsymbol{y})$ and $\boldsymbol{D}\boldsymbol{y}$

$$[\boldsymbol{h}_2(\boldsymbol{y}), \boldsymbol{D}\boldsymbol{y}] \equiv \boldsymbol{D}\boldsymbol{h}_2(\boldsymbol{y}) - \frac{\partial \boldsymbol{h}_2}{\partial \boldsymbol{y}}\boldsymbol{D}\boldsymbol{y} \tag{6.3.17}$$

This operation can be considered as mapping which assigns a quadratic monomial to a given quadratic monomial $\boldsymbol{h}_2(\boldsymbol{y})$. Since differentiation and multiplication with the constant matrix $\boldsymbol{D}$ are linear operations, the mapping is linear in $\boldsymbol{h}_2$ and its properties depend only on $\boldsymbol{D}$, i. e. on the linear part of the initial system eq. (6.3.4). We hence denote this mapping $\boldsymbol{L_D}$ and obtain

$$\boldsymbol{L_D}(\boldsymbol{h}_2(\boldsymbol{y})) = \boldsymbol{D}\boldsymbol{h}_2(\boldsymbol{y}) - \frac{\partial \boldsymbol{h}_2}{\partial \boldsymbol{y}}\boldsymbol{D}\boldsymbol{y} \tag{6.3.18}$$

We note that the operator $\boldsymbol{L_D}$ leaves the space of the homogeneous vector-valued monomials of any degree invariant. Since $\boldsymbol{N}_2(\boldsymbol{y})$ is a vector, the components of which are homogeneous polynomials of second degree in the coordinates of $\boldsymbol{y} = \{y_1 \quad y_2 \quad \dots \quad y_n\}$, it makes sense to regard eq. (6.3.16) as an equation for $\boldsymbol{h}_2(\boldsymbol{y})$. If we thus introduce a linear vector space, the elements of which are vector-valued monomials of a certain fixed degree, eq. (6.3.16) can be interpreted as a *linear* relation on this vector space.

We denote a reference basis in $\mathbb{R}^n$ by $\{\boldsymbol{e}_1 \quad \boldsymbol{e}_2 \quad \dots \quad \boldsymbol{e}_n\}$ and the coordinates of a vector $\boldsymbol{y}$ with respect to this basis by $\{y_1 \quad y_2 \quad \dots \quad y_n\}$. The basis vectors in the space of the vector-valued monomials of k-th degree are of the form

$$y_1^{m_1} y_2^{m_2} \dots y_n^{m_n} \boldsymbol{e}_i \qquad \text{with} \quad \sum_{j=1}^{n} m_j = k \tag{6.3.19}$$

with non-negative integers $m_j \geqslant 0$. The set of all vector-valued monomials of k-th degree forms a linear vector space which we denote $\mathbb{H}_k$. The vectors spanning the space $\mathbb{H}_k$ are the products of all possible monomials of degree k with the basis vectors $\boldsymbol{e}_i$.

We now elucidate the above by means of an example. For the two-dimensional case, we choose second-degree monomials in x and y and use the standard basis

$$\boldsymbol{e}_1 = \begin{bmatrix} 1 \\ 0 \end{bmatrix}, \quad \boldsymbol{e}_2 = \begin{bmatrix} 0 \\ 1 \end{bmatrix} \tag{6.3.20}$$

Then, as detailed above and in accordance with eq. (6.3.19), the space $\mathbb{H}_2$ is spanned by the vectors

$$\mathbb{H}_2 = \text{span}\left\{\begin{bmatrix} x^2 \\ 0 \end{bmatrix}, \begin{bmatrix} xy \\ 0 \end{bmatrix}, \begin{bmatrix} y^2 \\ 0 \end{bmatrix}, \begin{bmatrix} 0 \\ x^2 \end{bmatrix}, \begin{bmatrix} 0 \\ xy \end{bmatrix}, \begin{bmatrix} 0 \\ y^2 \end{bmatrix}\right\} \tag{6.3.21}$$

We next turn to the solution of eq. (6.3.16). Each vector $\boldsymbol{h}_2(\boldsymbol{y})$ can be represented by a linear combination of the basis vectors of $\mathbb{H}_2$, as can $\boldsymbol{N}_2(\boldsymbol{y})$. The linear Lie operator $\boldsymbol{L_D}$ assigns an image vector of $\mathbb{H}_2$ to $\boldsymbol{h}_2(\boldsymbol{y})$

$$\boldsymbol{h}_2(\boldsymbol{y}) \to \boldsymbol{D}\boldsymbol{h}_2(\boldsymbol{y}) - \frac{\partial \boldsymbol{h}_2}{\partial \boldsymbol{y}}\boldsymbol{D}\boldsymbol{y} \tag{6.3.22}$$

If we substitute for $\boldsymbol{h}_2(\boldsymbol{y})$ all basis vectors of $\mathbb{H}_2$ in turn, the image vectors will generally only span a subspace $\boldsymbol{L_D}(\mathbb{H}_2)$ of $\mathbb{H}_2$, i.e. $\mathbb{H}_2$ can be expressed as the *direct sum* of linearly independent subspaces $\boldsymbol{L_D}(\mathbb{H}_2)$ and $\mathbb{G}_2$

$$\mathbb{H}_2 = \boldsymbol{L_D}(\mathbb{H}_2) \oplus \mathbb{G}_2 \tag{6.3.23}$$

where $\mathbb{G}_2$ is a (not uniquely determined) complementary space to $\boldsymbol{L_D}(\mathbb{H}_2)$. It is thus clear that eq. (6.3.18) can be considered as system of linear equations and solved accordingly. If $\boldsymbol{N}_2(\boldsymbol{y})$ is contained in $\boldsymbol{L_D}(\mathbb{H}_2)$, eq. (6.3.16) can be solved, i.e. all $\mathcal{O}(|\boldsymbol{y}|^2)$-terms in eq. (6.3.15) can be eliminated. If this is not the case, $\boldsymbol{h}_2(\boldsymbol{y})$ can be chosen so that the troublesome terms $\mathcal{O}(|\boldsymbol{y}|^2)$ which cannot be eliminated lie in the complementary subspace $\mathbb{G}_2$ (Guckenheimer and Holmes, 1983), theorem 3.3.1. We identify these terms with the upper index *res* (= resonance)

$$\boldsymbol{N}_2^{res}(\boldsymbol{y}) \in \mathbb{G}_2 \tag{6.3.24}$$

The concept of resonance will be explained later. As an intermediate result, we note that eq. (6.3.15) can be transformed by an appropriate non-linear coordinate transformation to the following form of "simplest quadratic terms"

$$\dot{\boldsymbol{y}} = \boldsymbol{D}\boldsymbol{y} + \boldsymbol{N}_2^{res}(\boldsymbol{y}) + \widetilde{\boldsymbol{N}}_3(\boldsymbol{y}) + \cdots + \widetilde{\boldsymbol{N}}_{\mathrm{r}-1}(\boldsymbol{y}) + \mathcal{O}(|\boldsymbol{y}|^{\mathrm{r}}) \tag{6.3.25}$$

The remaining second-order term, $\boldsymbol{N}_2^{res}(\boldsymbol{y})$, lies in the complementary space of $\boldsymbol{L_D}(\mathbb{H}_2)$. If $\mathbb{G}_2$ is empty, i.e. $\boldsymbol{L_D}(\mathbb{H}_2) = \mathbb{H}_2$, all second-order terms can be eliminated.

In the next step, we simplify the third-order terms. In analogy to eq. (6.3.9), we carry out the coordinate transformation

$$\boldsymbol{y} \to \boldsymbol{y} + \boldsymbol{h}_3(\boldsymbol{y}) \tag{6.3.26}$$

where $\boldsymbol{h}_3(\boldsymbol{y}) = \mathcal{O}(|\boldsymbol{y}|^3)$. In accordance with the transformations eqs. (6.3.10) to (6.3.15), we obtain from eq. (6.3.25)

$$\begin{aligned}\dot{\boldsymbol{y}} = \boldsymbol{D}\boldsymbol{y} + \boldsymbol{N}_2^{res}(\boldsymbol{y}) + \boldsymbol{D}\boldsymbol{h}_3(\boldsymbol{y}) - \frac{\partial \boldsymbol{h}_3}{\partial \boldsymbol{y}}\boldsymbol{D}\boldsymbol{y} + \widetilde{\boldsymbol{N}}_3(\boldsymbol{y}) + \widetilde{\widetilde{\boldsymbol{N}}}_4(\boldsymbol{y}) + \quad \cdots \\ + \widetilde{\widetilde{\boldsymbol{N}}}_{\mathrm{r}-1} + \mathcal{O}(|\boldsymbol{y}|^{\mathrm{r}})\end{aligned} \tag{6.3.27}$$

The crucial point is that all linear and quadratic terms in eq. (6.3.27) are unaffected by the transformation eq. (6.3.26), the cubic terms appear in a form similar to the quadratic terms in eq. (6.3.15) and only terms $\widetilde{\widetilde{\boldsymbol{N}}}_{\mathrm{k}}$ for $\mathrm{k} \geqslant 4$ are modified. At this point, it thus becomes clear that this method can be applied to a successive elimination of higher-order terms.

In order to simplify the third-order terms in eq. (6.3.27), the following equation has to be solved

$$\boldsymbol{D}\boldsymbol{h}_3(\boldsymbol{y}) - \frac{\partial \boldsymbol{h}_3}{\partial \boldsymbol{y}}\boldsymbol{D}\boldsymbol{y} = -\widetilde{\boldsymbol{N}}_3(\boldsymbol{y}) \tag{6.3.28}$$

A comparison with eqs. (6.3.16) and (6.3.18) suggests itself. The map

$$\boldsymbol{h}_3(\boldsymbol{y}) \to \boldsymbol{D}\boldsymbol{h}_3(\boldsymbol{y}) - \frac{\partial \boldsymbol{h}_3}{\partial \boldsymbol{y}}\boldsymbol{D}\boldsymbol{y} = \boldsymbol{L_D}(\boldsymbol{h}_3(\boldsymbol{y})) \tag{6.3.29}$$

is a linear map of the space $\mathbb{H}_3$ of vector-valued third-degree monomials into $\mathbb{H}_3$. For $\mathbb{H}_3$, we again perform a decomposition

$$\mathbb{H}_3 = \boldsymbol{L_D}(\mathbb{H}_3) \oplus \mathbb{G}_3 \tag{6.3.30}$$

where $\mathbb{G}_3$ is a space complementary to $\boldsymbol{L_D}(\mathbb{H}_3)$. If $\boldsymbol{L_D}(\mathbb{H}_3) = \mathbb{H}_3$, all the third-order terms can be eliminated. If $\widetilde{\widetilde{\boldsymbol{N}}}_3(\boldsymbol{y})$ does not lie in the subspace $\boldsymbol{L_D}(\mathbb{H}_3)$, the simplified third-order terms remain

$$\boldsymbol{N}_3^{res}(\boldsymbol{y}) \in \mathbb{G}_3 \tag{6.3.31}$$

For the iteration process, the following *theorem of normal forms*, cf. (Wiggins, 1990) can be formulated:

> The initial system eq. (6.3.4) respectively eq. (6.3.7) can be transformed by a sequence of analytic non-linear transformation steps into the following simpler form:
>
> $$\dot{\boldsymbol{y}} = \boldsymbol{D}\boldsymbol{y} + \boldsymbol{N}_2^{res}(\boldsymbol{y}) + \cdots + \boldsymbol{N}_{\mathrm{r}-1}^{res}(\boldsymbol{y}) + \mathcal{O}(|\boldsymbol{y}|^{\mathrm{r}}) \tag{6.3.32}$$
>
> where $\boldsymbol{N}_{\mathrm{k}}^{res} \in \mathbb{G}_{\mathrm{k}}$, $2 \leqslant \mathrm{k} \leqslant \mathrm{r}-1$ and $\mathbb{G}_{\mathrm{k}}$ denotes a space complementary to $\boldsymbol{L_D}(\mathbb{H}_{\mathrm{k}})$. Equation (6.3.32) is said to be in normal form.

Before demonstrating the above technique on an example, we first add some pertinent comments:

i. The solvability of the homological equations for $\boldsymbol{h}_2(\boldsymbol{y})$, $\boldsymbol{h}_3(\boldsymbol{y})$ etc. and consequently the structure of the remaining non-linear terms in eq. (6.3.32) depends on the linear part of the vector field alone, i. e. on the constant matrix $\boldsymbol{D}$.

ii. The determination of the normal form can be reduced to the solution of a series of linear equation systems. A successive elimination of non-resonant non-linear terms is possible because the k-th degree terms already determined remain unaffected by a coordinate transformation of (k+1)-th degree.

iii. The question regarding the condition under which resonant terms appear in eq. (6.3.32) will be dealt with following the next example.

iv. If we determine the normal form of a dynamical system in the neighbourhood of a hyperbolic fixed point, we have seen that non-linear terms can indeed emerge. How can we reconcile this with the Hartman-Grobman theorem (theorem 1, section 6.2), according to which a coordinate transformation always exists which maps the non-linear flow in the neighbourhood of a hyperbolic fixed point onto the linear flow? According to this theorem, should not a linearisation always be possible as long as no eigenvalue of $\boldsymbol{D}$ has a vanishing real part? There is in fact no contradiction. The decisive point is that the Hartman-Grobman theorem only guarantees the existence of a homeomorphism, i.e. a bijective *continuous* map. The construction of such a map with the aid of the technique of normal forms is generally not possible since in this case, *differentiable* coordinate transformations are constructed by means of a power series ansatz.

Example:

We now demonstrate the technique of the method of normal forms on an example (Wiggins, 1990). We are looking for the normal form of a two-dimensional vector field (in the neighbourhood of a fixed point), the linear part of which is given in the form

$$\boldsymbol{D} = \begin{bmatrix} 0 & 1 \\ 0 & 0 \end{bmatrix} \tag{6.3.33}$$

We limit our account to the determination of the second-order term. In accordance with eq. (6.3.21), we obtain for $\mathbb{H}_2$

$$\mathbb{H}_2 = \text{span}\left\{ \begin{bmatrix} x^2 \\ 0 \end{bmatrix}, \begin{bmatrix} xy \\ 0 \end{bmatrix}, \begin{bmatrix} y^2 \\ 0 \end{bmatrix}, \begin{bmatrix} 0 \\ x^2 \end{bmatrix}, \begin{bmatrix} 0 \\ xy \end{bmatrix}, \begin{bmatrix} 0 \\ y^2 \end{bmatrix} \right\} \tag{6.3.34}$$

We now apply the operator $\boldsymbol{L_D}$, eq. (6.3.18), to all basis vectors spanning $\mathbb{H}_2$ in accordance with eq. (6.3.34)

$$\begin{aligned}
\boldsymbol{L_D}\begin{bmatrix} x^2 \\ 0 \end{bmatrix} &= \begin{bmatrix} 0 & 1 \\ 0 & 0 \end{bmatrix}\begin{bmatrix} x^2 \\ 0 \end{bmatrix} - \begin{bmatrix} 2x & 0 \\ 0 & 0 \end{bmatrix}\begin{bmatrix} y \\ 0 \end{bmatrix} = -2\begin{bmatrix} xy \\ 0 \end{bmatrix} \\
\boldsymbol{L_D}\begin{bmatrix} xy \\ 0 \end{bmatrix} &= \begin{bmatrix} -y^2 \\ 0 \end{bmatrix} \\
\boldsymbol{L_D}\begin{bmatrix} y^2 \\ 0 \end{bmatrix} &= \begin{bmatrix} 0 \\ 0 \end{bmatrix} \\
\boldsymbol{L_D}\begin{bmatrix} 0 \\ x^2 \end{bmatrix} &= \begin{bmatrix} x^2 \\ -2xy \end{bmatrix} = \begin{bmatrix} x^2 \\ 0 \end{bmatrix} - 2\begin{bmatrix} 0 \\ xy \end{bmatrix} \\
\boldsymbol{L_D}\begin{bmatrix} 0 \\ xy \end{bmatrix} &= \begin{bmatrix} xy \\ -y^2 \end{bmatrix} = \begin{bmatrix} xy \\ 0 \end{bmatrix} - \begin{bmatrix} 0 \\ y^2 \end{bmatrix} \\
\boldsymbol{L_D}\begin{bmatrix} 0 \\ y^2 \end{bmatrix} &= \begin{bmatrix} y^2 \\ 0 \end{bmatrix}
\end{aligned} \tag{6.3.35}$$

The space spanned by $\boldsymbol{L_D}(\mathbb{H}_2)$ is the linear combination of all the vectors listed in eq. (6.3.35), of which the following set of four is linearly independent

$$\boldsymbol{L_D}(\mathbb{H}_2) = \text{span}\left\{ \begin{bmatrix} -2xy \\ 0 \end{bmatrix}, \begin{bmatrix} x^2 \\ -2xy \end{bmatrix}, \begin{bmatrix} xy \\ -y^2 \end{bmatrix}, \begin{bmatrix} y^2 \\ 0 \end{bmatrix} \right\} \tag{6.3.36}$$

This means that only those second-order terms can be eliminated which are linear combinations of these four terms. Second-order terms which, according to eq. (6.3.23), lie in the space $\mathbb{G}_2$ complementary to $\boldsymbol{L_D}(\mathbb{H}_2)$ can, on the other hand, not be eliminated. In our case, the $\mathbb{G}_2$-space is two-dimensional. In order to determine $\mathbb{G}_2$, it is expedient to describe the linear map $\boldsymbol{L_D}$ of the space $\mathbb{H}_2$ into itself by a matrix. The six basis vectors are specified in eq. (6.3.34). The operator $\boldsymbol{L_D}$ assigns to each element of $\mathbb{H}_2$ an element of $\mathbb{H}_2$ which can again be represented as a linear combination of the basis vectors, eq. (6.3.34). Using the results of eq. (6.3.35), we obtain for $\boldsymbol{L_D}(\mathbb{H}_2)$ the matrix

$$\begin{array}{cccccccc} & \begin{bmatrix} x^2 \\ \; \end{bmatrix} & \begin{bmatrix} xy \\ \; \end{bmatrix} & \begin{bmatrix} y^2 \\ \; \end{bmatrix} & \begin{bmatrix} \; \\ x^2 \end{bmatrix} & \begin{bmatrix} \; \\ xy \end{bmatrix} & \begin{bmatrix} \; \\ y^2 \end{bmatrix} & \end{array}$$

$$\boldsymbol{L_D} = \begin{bmatrix} 0 & 0 & 0 & 1 & 0 & 0 \\ -2 & 0 & 0 & 0 & 1 & 0 \\ 0 & -1 & 0 & 0 & 0 & 1 \\ 0 & 0 & 0 & 0 & 0 & 0 \\ 0 & 0 & 0 & -2 & 0 & 0 \\ 0 & 0 & 0 & 0 & -1 & 0 \end{bmatrix} \begin{array}{l} \{x^2 \quad \} \\ \{xy \quad \} \\ \{y^2 \quad \} \\ \{ \quad x^2\} \\ \{ \quad xy\} \\ \{ \quad y^2\} \end{array} \tag{6.3.37}$$

Only the eigenvectors respectively principal vectors appertaining to the zero eigenvalues span the $\mathbb{G}_2$-space. Thus, only two linearly independent left eigenvectors appertaining to zero eigenvalues of the matrix eq. (6.3.37) have to be found. The following two left eigenvectors are found to be

$$\{1 \;\; 0 \;\; 0 \;\; 0 \;\; \tfrac{1}{2} \;\; 0\} \quad \text{and} \quad \{0 \;\; 0 \;\; 0 \;\; 1 \;\; 0 \;\; 0\} \tag{6.3.38}$$

Within the set of basis vectors specified by eq. (6.3.34), the two vectors

$$\begin{bmatrix} x^2 \\ \frac{1}{2}xy \end{bmatrix}, \begin{bmatrix} 0 \\ x^2 \end{bmatrix} \tag{6.3.39}$$

span the two-dimensional subspace $\mathbb{G}_2$ of $\mathbb{H}_2$ which, by definition, represents a space complementary to $\boldsymbol{L_D}(\mathbb{H}_2)$. The vectors of the $\mathbb{G}_2$-space are formed by the terms which are resonant, i. e. those which cannot be eliminated by a coordinate transformation, eq. (6.3.9). All vector fields whose linear part is described in the neighbourhood of a fixed point by the matrix eq. (6.3.33) can thus be transformed up to second-order terms to the following normal form

$$\begin{aligned} \dot{x} &= y + 2a_1x^2 + \mathcal{O}(3) \\ \dot{y} &= a_1xy + a_2x^2 + \mathcal{O}(3) \end{aligned} \tag{6.3.40}$$

where a_1 and a_2 are constants. Incorporating all non-linear terms, a complete second-order polynomial is expressed by the following set of equations

$$\begin{aligned} \dot{x} &= y + a_1x^2 + b_1xy + b_2y^2 \\ \dot{y} &= a_2xy + a_3x^2 + b_3y^2 \end{aligned} \tag{6.3.41}$$

The normal form means that the course of the trajectories of eq. (6.3.41) in the neighbourhood of the fixed point is represented qualitatively completely by eq. (6.3.40). In fig. 6.3.2, the trajectories near the fixed point at $\{0 \;\; 0\}$ are displayed for the two systems of eqs. (6.3.40) and (6.3.41); the left-hand side shows

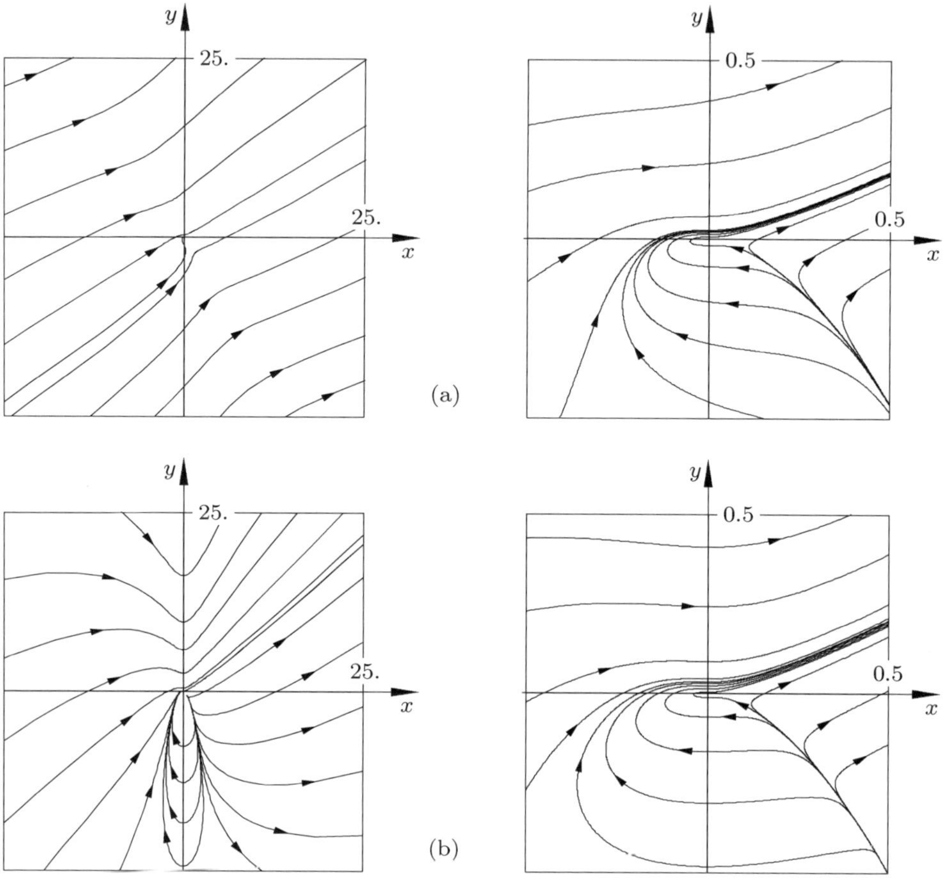

Fig. 6.3.2: Global and local trajectories for the linear part $\boldsymbol{D} = \left[\begin{smallmatrix} 0 & 1 \\ 0 & 0 \end{smallmatrix}\right]$
a) Gl. (6.3.41) with $a_1 = 2$, $a_2 = a_3 = 1$, $b_1 = 1.1$, $b_2 = b_3 = 0.6$ and
b) Gl. (6.3.40) with $a_1 = a_2 = 1$

a large-scale depiction while the right-hand side is a zoom of the neighboorhood of the origin. An astonishing agreement in the course of the trajectories in the vicinity of the origin is clearly recognisable.

We pointed out that the $\mathbb{G}_2$-space is not uniquely determined. From a combination of the second basis vector of eq. (6.3.36) with the first of eq. (6.3.39), a new basis vector for the construction of a $\mathbb{G}_2$-space can be established; we find, for example,

$$\begin{bmatrix} x^2 \\ -2xy \end{bmatrix} + 4\begin{bmatrix} x^2 \\ \frac{1}{2}xy \end{bmatrix} = 5\begin{bmatrix} x^2 \\ 0 \end{bmatrix} \tag{6.3.42}$$

Retaining the second vector of eq. (6.3.39), we can construct a second complementary space $\mathbb{G}_2$ as follows

$$\mathbb{G}_2 = \text{span}\left\{\begin{bmatrix} x^2 \\ 0 \end{bmatrix}, \begin{bmatrix} 0 \\ x^2 \end{bmatrix}\right\} \tag{6.3.43}$$

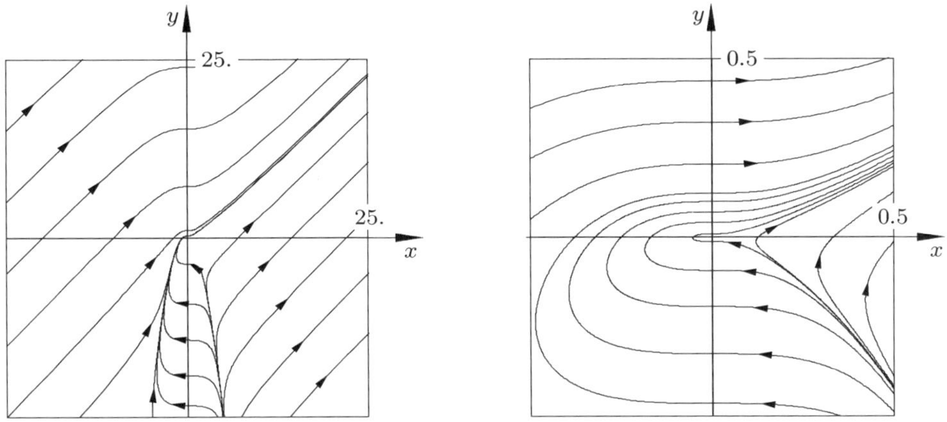

Fig. 6.3.3: Trajectories of the normal form eq. (6.3.44) for the linear part $\boldsymbol{D} = \left[\begin{smallmatrix} 0 & 1 \\ 0 & 0 \end{smallmatrix}\right]$ and for $a_1 = a_2 = 1$, after (Takens, 1974)

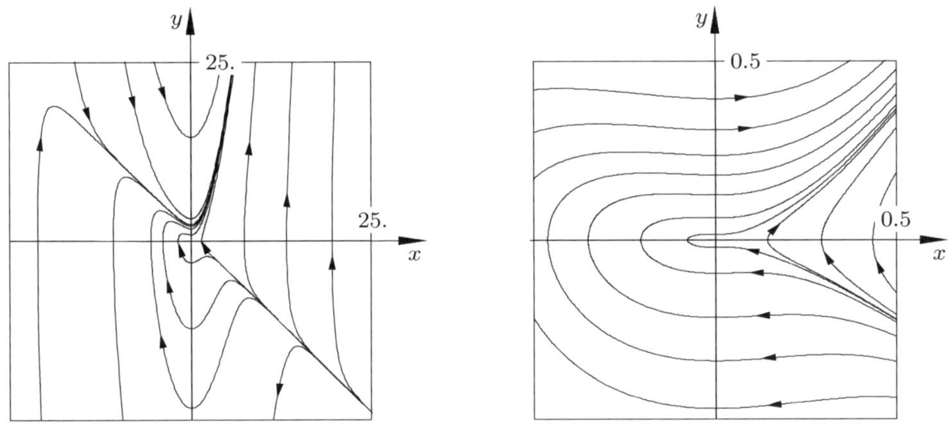

Fig. 6.3.4: Trajectories of the normal form eq. (6.3.47) for the linear component $\boldsymbol{D} = \left[\begin{smallmatrix} 0 & 1 \\ 0 & 0 \end{smallmatrix}\right]$ and for $a_1 = a_2 = 1$, after (Bogdanov, 1975)

The resulting normal form is

$$\begin{aligned} \dot{x} &= y + a_1 x^2 + \mathcal{O}(3) \\ \dot{y} &= a_2 x^2 + \mathcal{O}(3) \end{aligned} \tag{6.3.44}$$

Figure 6.3.3 displays the course of the trajectories in the neighbourhood of the fixed point $\{0 \;\; 0\}$ for this normal form, first studied by Takens (Takens, 1974).

The normal form first investigated by Bogdanov (Bogdanov, 1975) is the third variant we present. With a linear combination of the basis vectors which is different from that in eq. (6.3.42), another linearly independent vector can be formed

$$\begin{bmatrix} x^2 \\ -2xy \end{bmatrix} - \begin{bmatrix} x^2 \\ \frac{1}{2}xy \end{bmatrix} = \begin{bmatrix} 0 \\ -\frac{5}{2}xy \end{bmatrix} = -\frac{5}{2}\begin{bmatrix} 0 \\ xy \end{bmatrix} \tag{6.3.45}$$

As a further possibility for $\mathbb{G}_2$, we thus obtain

$$\mathbb{G}_2 = \operatorname{span}\left\{\begin{bmatrix} 0 \\ x^2 \end{bmatrix}, \begin{bmatrix} 0 \\ xy \end{bmatrix}\right\} \tag{6.3.46}$$

The normal form associated with the $\mathbb{G}_2$ of eq. (6.3.46) is then

$$\begin{aligned} \dot{x} &= y + \mathcal{O}(3) \\ \dot{y} &= a_1 xy + a_2 x^2 + \mathcal{O}(3) \end{aligned} \tag{6.3.47}$$

Figure 6.3.4 illustrates the course of the trajectories.

What is most surprising is firstly the confirmation that the most varying constellations of second-order terms display essentially the same behaviour in the neighbourhood of the fixed point and secondly that such complex behaviour of trajectories can be described by much simpler normal forms. Additional non-linear terms only require more computing effort, but do not yield more information. We should not forget, however, that it is the linear term which is ultimately decisive for the structure of the non-linear terms.

We record that the determination of higher-order terms in the normal form requires a considerable computational effort, particularly when the underlying phase space is of more than two dimensions. The dimension of the sets of linear equations to be solved increases fast so that the basis vectors of $\boldsymbol{L_D}(\mathbb{H}_k)$ and $\mathbb{G}_k$ cannot be determined as easily as in the previous example. In this case, it is advisable to use methods of computer algebra in the setting-up of the sets of linear equations (Rand and Armbruster, 1987) and subsequently to determine the corresponding subspaces with the aid of singular value decomposition (Press et al., 1986). A further very effective method for constructing normal forms is presented in (Elphick et al., 1987).

Finally, as declared previously, we elucidate the concept of "resonance". We recall the initial equation (6.3.7)

$$\dot{\boldsymbol{x}} = \boldsymbol{D}\boldsymbol{x} + \boldsymbol{N}_2(\boldsymbol{x}) + \boldsymbol{N}_3(\boldsymbol{x}) + \cdots + \boldsymbol{N}_{r-1}(\boldsymbol{x}) + \mathcal{O}(|\boldsymbol{x}|^r), \quad \boldsymbol{x} \in \mathbb{R}^n \tag{6.3.48}$$

which is to be transformed into a normal form. A necessary condition for the elimination of the non-linear terms $(\mathcal{O}(|\boldsymbol{x}|^k),\ 2 \leqslant k \leqslant r-1)$ in the above eq. (6.3.48) is the invertibility of the linear operator $\boldsymbol{L_D}(\mathbb{H}_k)$ on $\mathbb{H}_k$. Non-invertibility implies resonance; we now explain briefly when this takes place.

The Lie operator $\boldsymbol{L_D}(\boldsymbol{h}_k)$ is, in accordance with eq. (6.3.18), defined as

$$\boldsymbol{L_D}(\boldsymbol{h}_k(\boldsymbol{x})) \equiv \boldsymbol{D}\boldsymbol{h}_k(\boldsymbol{x}) - \frac{\partial \boldsymbol{h}_k}{\partial \boldsymbol{x}}\boldsymbol{D}\boldsymbol{x} \tag{6.3.49}$$

where $\boldsymbol{h}_k \in \mathbb{H}_k$ and $\mathbb{H}_k$ forms the linear vector space of the vector-valued k-th degree monomials. Simplifying, we first assume that the Jacobian matrix $\boldsymbol{D}$ can be transformed by a linear coordinate transformation to diagonal form and possesses the eigenvalues $\lambda_1, \ldots, \lambda_n$. We denote by $x_1, \ldots, x_n$ the coordinates with respect to the standard basis $\boldsymbol{e}_1, \ldots, \boldsymbol{e}_n$ where $\boldsymbol{e}_i$ represents an n-row vector whose i-th element is 1 and the remaining elements contain only zeroes. We then have

$$\boldsymbol{D}\boldsymbol{e}_i = \lambda_i \boldsymbol{e}_i \tag{6.3.50}$$

In conformity with eq. (6.3.19), we can express the basis vectors of $\mathbb{H}_k$ as follows

$$x_1^{m_1} \cdots x_n^{m_n} \boldsymbol{e}_i, \qquad \sum_{j=1}^{n} m_j = k, \quad m_j \geqslant 0 \qquad\qquad i = 1, \dots n \tag{6.3.51}$$

where each basis vector $\boldsymbol{e}_i$ must be multiplied by all possible k-th degree monomials (see preceding example for $k = 2$). In accordance with eq. (6.3.35), we now apply the operator $\boldsymbol{L_D}(\boldsymbol{h}_k)$ to each basis vector $\boldsymbol{h}_k$ of $\mathbb{H}_k$

$$\boldsymbol{h}_k(\boldsymbol{x}) = x_1^{m_1} \cdots x_n^{m_n} \boldsymbol{e}_i, \qquad \sum_{j=1}^{n} m_j = k, \quad m_j \geqslant 0, \ k \geqslant 2 \tag{6.3.52}$$

We dispense with a further index for $\boldsymbol{h}_k$, but keep in mind that $\boldsymbol{h}_k$ depends on i as well as on the special combination of the exponents $m_1, \dots m_n$.

Substituting $\boldsymbol{h}_k(\boldsymbol{x})$ in eq. (6.3.49) and noting that only its i-th component is non-zero, we observe that only the i-th component does not vanish for the image vector $\boldsymbol{L_D}(\boldsymbol{h}_k)$ as well and contributes the following expression

$$\lambda_i x_1^{m_1} \cdots x_n^{m_n} - \sum_{j=1}^{n} \frac{\partial}{\partial x_j}(x_1^{m_1} \cdots x_j^{m_j} \cdots x_n^{m_n}) \lambda_j x_j = (\lambda_i - \sum_{j=1}^{n} m_j \lambda_j) x_1^{m_1} \cdots x_n^{m_n} \tag{6.3.53}$$

Equation (6.3.49) can thus be expressed in the simplified form

$$\boldsymbol{L_D}(\boldsymbol{h}_k) = \Big[\lambda_i - \sum_{j=1}^{n} m_j \lambda_j\Big] \boldsymbol{h}_k(\boldsymbol{x}) \tag{6.3.54}$$

Equation (6.3.54) shows that each basis vector $\boldsymbol{h}_k$ of $\mathbb{H}_k$ is an eigenvector of the operator $\boldsymbol{L_D}(\boldsymbol{h}_k)$ and the term in brackets

$$\lambda_i - \sum_{j=1}^{n} m_j \lambda_j \tag{6.3.55}$$

the appertaining eigenvalue. Equation (6.3.54) makes it clear that $\boldsymbol{L_D}(\boldsymbol{h}_k)$ is not invertible when the expression eq. (6.3.55) is zero; in this case,

$$\lambda_i = \sum_{j=1}^{n} m_j \lambda_j \tag{6.3.56}$$

Equation (6.3.56) expresses precisely the condition for *resonance*; in this case, the non-linear terms cannot be eliminated and the remaining resonant terms constitute the non-linear part of the normal form. The number

$$k = \sum_{j=1}^{n} m_j \geqslant 2 \tag{6.3.57}$$

is denoted the order of resonance.

Remarks:

a) To simplify the argument, we assumed that the Jacobian matrix $\boldsymbol{D}$ can be transformed to diagonal form. If this is not possible and if $\boldsymbol{D}$ can only be reduced to a Jordan normal form, it can be shown that the operator $L_{\boldsymbol{D}}$ also adopts a Jordan form. For the eigenvalues, however, the same relation is deduced as in the case of diagonalisable matrices (Arnold, 1988).

b) In this chapter, we discuss local bifurcations. It is one of our aims to classify the local bifurcations of fixed points with the aid of normal forms. If we determine the normal form in the neighbourhood of a non-hyperbolic fixed point, then resonant terms will certainly emerge since the Jacobian matrix $\boldsymbol{D}$ in this case possesses at least one eigenvalue with a vanishing real part. We can distinguish two cases:

   i. A real eigenvalue of $\boldsymbol{D}$ vanishes, for example $\lambda_1 = 0$. If we select an arbitrary integer $\mathrm{k} \geqq 2$ and set $\mathrm{m}_1 = \mathrm{k}$, $\mathrm{m}_2 = \ldots = \mathrm{m}_\mathrm{n} = 0$, the resonance condition eq. (6.3.56) is satisfied, i. e. for each order k, resonant terms appear in the normal form.

   ii. Let us assume that $\boldsymbol{D}$ possesses the eigenvalues $\lambda_{1,2} = \pm i$. The condition for resonance is then

$$\mathrm{m}_1 i - \mathrm{m}_2 i = \pm i \quad \text{resp.} \quad \mathrm{m}_1 - \mathrm{m}_2 = \pm 1 \tag{6.3.58}$$

   where $\mathrm{k} = \mathrm{m}_1 + \mathrm{m}_2 \geqslant 2$ must hold. If k is an even number, i. e. $\mathrm{k} = 2\mathrm{k}_1$, then, due to eq. (6.3.58), the following expression

$$2\mathrm{k}_1 = \mathrm{m}_1 + \mathrm{m}_2 = \mathrm{m}_1 + (\mathrm{m}_1 \mp 1)$$

   has no solution, i. e. terms of even order are not resonant. By contrast, all terms of odd order are resonant and cannot be eliminated.

At the end of this section, we refer briefly to the role of the control parameter in the calculation of the normal form. A detailed description of this subject and further literature can be found in the monograph (Wiggins, 1990). We apply the same method that we used to determine the centre manifold of a parametric system. We expand the system $\dot{\boldsymbol{x}} = \boldsymbol{f}(\boldsymbol{x}, \boldsymbol{\mu})$ to

$$\begin{aligned} \dot{\boldsymbol{x}} &= \boldsymbol{f}(\boldsymbol{x}, \boldsymbol{\mu}) \\ \dot{\boldsymbol{\mu}} &= \boldsymbol{o} \end{aligned} \tag{6.3.59}$$

and thus declare the parameters $\boldsymbol{\mu}$ to be independent variables. This system expansion enables us to apply the technique of the normal form without any parameters, as described in this section, directly to systems with parameters; then, the coefficients of the transformation depend on the parameters $\boldsymbol{\mu}$.

6.4 Normal Forms of Bifurcations for One-Parametric Flows

We underline once more that the main topic of this chapter is "local bifurcations"; in the present and following two sections, we turn our attention to the local bifurcations of vector fields and maps. The concept "local" refers to bifurcations in the neighbourhood of fixed points. Fixed points of vector fields mean states of equilibrium respectively stationary solutions. Fixed points of Poincaré maps reflect the periodic orbits of vector flows. Bifurcations of the fixed points of such iterated maps are hence bifurcations in the neighbourhood of periodic orbits. We first consider bifurcations of the fixed points of flows respectively vector fields.

We know that bifurcations of fixed points occur only if the determinant of the Jacobian matrix $\frac{\partial \boldsymbol{F}}{\partial \boldsymbol{x}}(\boldsymbol{x}_s, \mu_{cr})$ is zero or if the fixed point is non-hyperbolic. We restrict ourselves here to the simplest case in which the real part of one eigenvalue becomes zero and the real parts of all other eigenvalues remain non-zero. Such bifurcations are denoted local bifurcations of the codimension 1, because the number of independent conditions needed for the determination of the bifurcation is one. For a general definition of the concept codimension, we refer to (Guckenheimer and Holmes, 1983; Kuznetsov, 1998). We are interested in the change in the dynamical behaviour in the neighbourhood of a fixed point following the alteration of the single parameter μ in the proximity of the critical value μ_{cr} for which the fixed point has non-hyperbolic character. As we know from section 6.2, the centre manifold theory offers an effective method of analysing bifurcations by observing the flow on the centre manifold which is of lower dimension. If the real part of a single eigenvalue vanishes, the vector field on the centre manifold is a one-parametric family of vector fields which can be further simplified by a subsequent transformation to normal form.

Let us first consider one-dimensional vector fields for which the flow on the centre manifold has the form

$$\dot{x} = f(x, \mu) \tag{6.4.1}$$

where we assume that the bifurcation point (x_s, μ_{cr}) coincides with the origin $(0, 0)$. This simplifies the discussion of eq. (6.4.1) considerably and can always be achieved by a translation. For the bifurcation, the following two conditions must necessarily be satisfied

$$f(0,0) = 0, \quad \text{fixed point condition} \tag{6.4.2}$$

$$\frac{\partial f}{\partial x}(0,0) = 0, \quad \text{zero eigenvalue condition} \tag{6.4.3}$$

The examples discussed in section 6.1 satisfy these conditions. Before we specify the concept "bifurcation", we first summarise the bifurcation examples already mentioned and add a further example for which eqs. (6.4.2) and (6.4.3) are satisfied but which nevertheless displays no bifurcation:

1. saddle-node bifurcation (fig. 6.4.1a)

$$\dot{x} = f(x, \mu) = \mu - x^2 \tag{6.4.4}$$

2. transcritical bifurcation (fig. 6.4.1b)

$$\dot{x} = f(x, \mu) = \mu x - x^2 \tag{6.4.5}$$

3. pitchfork bifurcation (fig. 6.4.1c)

$$\dot{x} = f(x, \mu) = \mu x - x^3 \tag{6.4.6}$$

4. no bifurcation (fig. 6.4.1d)

$$\dot{x} = f(x, \mu) = \mu - x^3 \tag{6.4.7}$$

All states of equilibrium respectively fixed points of eq. (6.4.7) are defined by the relation

$$\mu = x^3 \tag{6.4.8}$$

Although the conditions are satisfied in eqs. (6.4.2) and (6.4.3), no qualitative change of the dynamics takes place at the transition $\mu = 0$. For both $\mu < 0$ and $\mu > 0$, a unique fixed point solution persists.

In practice, one-parametric families of one-dimensional vector fields on the centre manifold have a more complex form like $\dot{x} = f_0(\mu) + f_1(\mu)x + f_2(\mu)x^2 + \ldots = f(x, \mu)$. The four examples demonstrate that the conditions $f(0,0) = 0$ and $\frac{\partial f}{\partial x}(0,0) = 0$ represent necessary but not sufficient conditions for a bifurcation. The condition of one eigenvalue being zero for $\mu = 0$ does not suffice to guarantee a fundamental change in the topology of the dynamics. In this section, additional conditions have to be developed which characterise the bifurcation patterns precisely as they are illustrated in fig. 6.4.1a-c. Note that we restrict ourselves in the following to non-degenerate bifurcations, i. e. for each type of bifurcation, we consider vector fields $f(x, \mu)$ which contain only those terms of lowest order which are necessary to describe the respective bifurcation. For example, we do not consider the bifurcation $\dot{x} = f(x, \mu) = \mu - x^4$ (degenerate saddle-node bifurcation) nor $\dot{x} = f(x, \mu) = \mu x - x^4$ (degenerate transcritical bifurcation).

A) Saddle-node bifurcation

We first discuss the saddle-node bifurcation of a one-parametric family of one-dimensional vector fields. Which general conditions characterise this type of bifurcation? In order to answer this question, it is helpful to recall the graph of the fixed points in the μ, x-plane (fig. 6.4.1a). In contrast to the fixed point equation $x = x(\mu)$, the inverse relation $\mu = \mu(x)$ is unique. It thus appears advantageous to study this bifurcation on the basis of the inverse fixed point equation.

The fixed point curve passing through $(\mu, x) = (0,0)$ possesses the two following significant characteristics:

$$\frac{\mathrm{d}\mu}{\mathrm{d}x}(0) = 0 \tag{6.4.9}$$

i. e. the line $\mu = 0$ is a tangent at $x = 0$ and

$$\frac{\mathrm{d}^2\mu}{\mathrm{d}x^2}(0) \neq 0 \tag{6.4.10}$$

i. e. $(\mu, x) = (0,0)$ is a regular point and the fixed point curve lies completely on one side of the tangent $\mu = 0$.

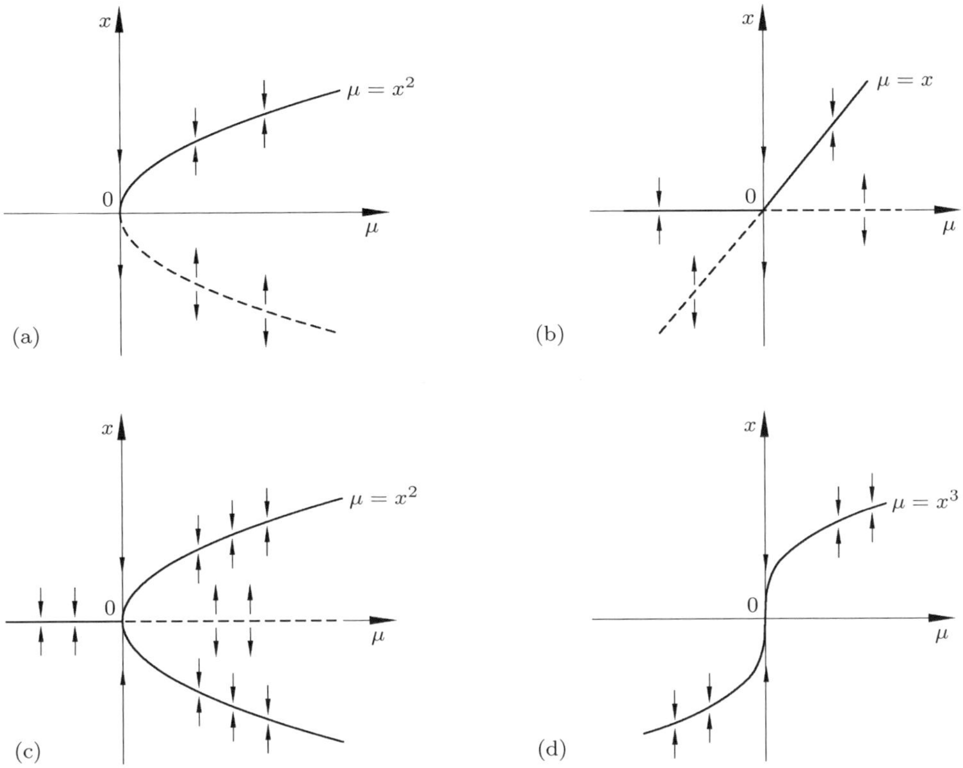

Fig. 6.4.1: Conditions $f(0,0) = 0$ and $\frac{\partial f}{\partial x}(0,0) = 0$ satisfied:
a) saddle-node bifurcation, b) transcritical bifurcation
c) pitchfork bifurcation, d) no bifurcation

We want to deduce conditions under which the saddle-node bifurcation occurs. Some conditions characterising the bifurcations are common to all types, others distinguish them from one another. Despite a possible objection of redundancy, we prefer, for clarity's sake, to reproduce all the conditions for each individual type of bifurcation. The general one-parametric initial equation for one-dimensional flows is

$$\dot{x} = f(x, \mu) \tag{6.4.11}$$

For the fixed point $(x, \mu) = (0, 0)$ of eq. (6.4.11), we have

$$f(0,0) = 0 \tag{6.4.12}$$

We are interested in bifurcations of fixed points. These only take place in the case of non-hyperbolic fixed points such that

$$\frac{\partial f}{\partial x}(0,0) = 0 \tag{6.4.13}$$

The *Theorem on Implicit Functions* which we introduce here is required for the deduction of additional bifurcation conditions under which an implicit function can be resolved locally explicitly with respect to one variable:

> Let $G(y,\alpha)$ be a continuously differentiable function, and let it fulfil at the point (y_0,α_0) the conditions
>
> $$G(y_0,\alpha_0)=0 \qquad \text{and} \qquad \frac{\partial G}{\partial \alpha}(y_0,\alpha_0)\neq 0$$
>
> There then exist open intervals U about y_0 and V about α_0 and a continuously differentiable function $\alpha = A(y)$ so that $\alpha_0 = A(y_0)$ and $G\big(y, A(y)\big) = 0$ for all $y \in U$ (fig. 6.4.2).

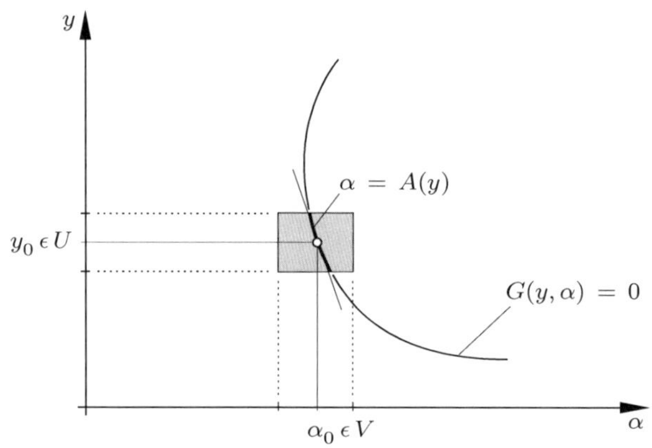

Fig. 6.4.2: Theorem on implicit functions

This theorem implies the following statements for the saddle-node bifurcation (fig. 6.4.1a): under the condition that

$$\frac{\partial f}{\partial \mu}(0,0)\neq 0 \tag{6.4.14}$$

holds, the theorem on implicit functions asserts that, for sufficiently small x, a unique function $\mu = \mu(x)$ with $\mu(0) = 0$ exists and that the fixed point equation then fulfils $f\big(x,\mu(x)\big) = 0$.

Our aim now is to deduce conditions for the derivatives of f in eq. (6.4.11) at the bifurcation point $(\mu, x) = (0,0)$ so that the significant characteristic conditions $\frac{\mathrm{d}\mu}{\mathrm{d}x}(0) = 0$ and $\frac{\mathrm{d}^2\mu}{\mathrm{d}x^2}(0) \neq 0$ of the saddle-node bifurcation are satisfied. Noting eq. (6.4.14), there holds

$$f\big(x,\mu(x)\big) = 0 \tag{6.4.15}$$

Applying implicit differentiation of eq. (6.4.15) with respect to x together with the chain rule – where, for typographical clarity, we omit the argument of eq. (6.4.15) – we obtain

$$\frac{\partial f}{\partial x} + \frac{\partial f}{\partial \mu}\frac{\mathrm{d}\mu}{\mathrm{d}x} = 0 \tag{6.4.16}$$

respectively for $(\mu, x) = (0, 0)$

$$\frac{\mathrm{d}\mu}{\mathrm{d}x}(0) = -\frac{\dfrac{\partial f}{\partial x}(0,0)}{\dfrac{\partial f}{\partial \mu}(0,0)} \tag{6.4.17}$$

Equation (6.4.17) demonstrates that the derivatives specified in eqs. (6.4.13) and (6.4.14) imply the first condition $\frac{\mathrm{d}\mu}{\mathrm{d}x}(0) = 0$, eq. (6.4.9), of a saddle-node bifurcation. A further differentiation of eq. (6.4.16) with respect to x using the chain and product rules yields

$$0 = \frac{\partial^2 f}{\partial x^2} + 2\frac{\partial^2 f}{\partial x \partial \mu}\frac{\mathrm{d}\mu}{\mathrm{d}x} + \frac{\partial f}{\partial \mu}\frac{\mathrm{d}^2\mu}{\mathrm{d}x^2} + \frac{\partial^2 f}{\partial \mu^2}\left(\frac{\mathrm{d}\mu}{\mathrm{d}x}\right)^2 \tag{6.4.18}$$

If we take eq. (6.4.9) into account, eq. (6.4.18) is simplified at the bifurcation point $(x, \mu) = (0, 0)$ to

$$0 = \frac{\partial^2 f}{\partial x^2}(0,0) + \frac{\partial f}{\partial \mu}(0,0)\frac{\mathrm{d}^2\mu}{\mathrm{d}x^2}(0) \tag{6.4.19}$$

respectively

$$\frac{\mathrm{d}^2\mu}{\mathrm{d}x^2}(0) = -\frac{\dfrac{\partial^2 f}{\partial x^2}(0,0)}{\dfrac{\partial f}{\partial \mu}(0,0)} \tag{6.4.20}$$

In order that the second condition for the saddle-node bifurcation $\frac{\mathrm{d}^2\mu}{\mathrm{d}x^2}(0) \neq 0$ be satisfied, $\frac{\partial^2 f}{\partial x^2}(0,0) \neq 0$ must also hold in addition to eq. (6.4.14).

On the basis of eq. (6.4.11), we can now summarise the following four conditions for the saddle-node bifurcation:

$$f(0,0) = 0\,, \quad \text{fixed point}$$

$$\frac{\partial f}{\partial x}(0,0) = 0\,, \quad \text{non-hyperbolic} \tag{6.4.21}$$

$$\frac{\partial f}{\partial \mu}(0,0) \neq 0\,, \quad \text{unique fixed point curve } \mu = \mu(x) \text{ through } (x,\mu) = (0,0) \tag{6.4.22}$$

$$\frac{\partial^2 f}{\partial x^2}(0,0) \neq 0\,, \quad \text{fixed points on one side of } \mu = 0 \tag{6.4.23}$$

We note that the sign of eq. (6.4.20) determines on which side of $\mu = 0$ the fixed point curves lie (fig. 6.4.3). However, this provides no information as to which branch appertains to a stable and which to an unstable fixed point behaviour. In order to clarify this, we rewrite eq. (6.4.16) in the form

$$\frac{\partial f}{\partial x} = -\frac{\partial f}{\partial \mu}\frac{\mathrm{d}\mu}{\mathrm{d}x} \tag{6.4.24}$$

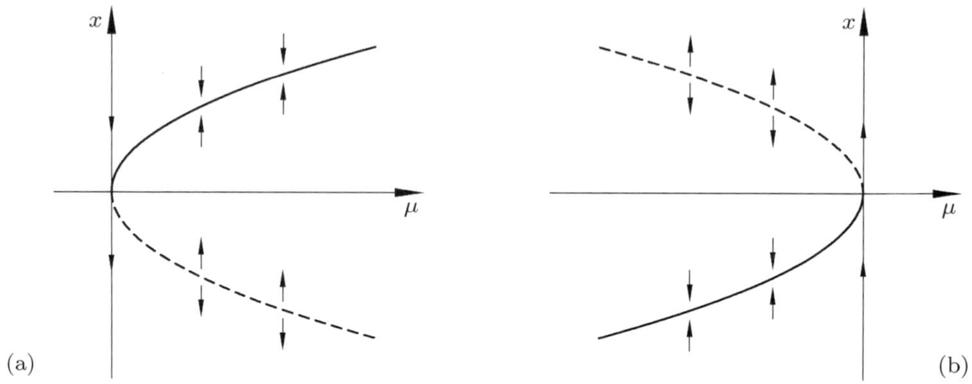

Fig. 6.4.3: Saddle-node bifurcation, eq. (6.4.26); case distinction according to eq. (6.4.20): a) supercritical ($-f_{xx}/f_\mu > 0$), b) subcritical ($-f_{xx}/f_\mu < 0$)

If $\frac{\partial f}{\partial x}(x,\mu) < 0$, the fixed point branch is stable, for $\frac{\partial f}{\partial x}(x,\mu) > 0$ unstable. As we know, $\frac{\mathrm{d}\mu}{\mathrm{d}x}(0)$ alters its sign at the bifurcation point $(x,\mu) = (0,0)$; following eq. (6.4.24), it is thus clear that if one branch is stable, the other must be unstable. Figure 6.4.3 presents two bifurcation types, (a) the supercritical and (b) the subcritical saddle-node bifurcation. Generally, a bifurcation is denoted *supercritical* (*subcritical*) if, for $\mu < 0$ ($\mu > 0$), either no fixed point exists, as in the saddle-node bifurcation, or the fixed point $x = 0$ is stable (unstable).

The derivation of the conditions for the saddle-node bifurcation is influenced by the bifurcation associated with the very simple eq. (6.4.4). We can now enquire which other families of equations also lead to this saddle-node bifurcation. To this end, we consider the general one-parametric family of one-dimensional vector fields at the non-hyperbolic fixed point $(x,\mu) = (0,0)$. Without loss of generality, the Taylor expansion of the vector field at this point may be expressed as

$$f(x,\mu) = a_0\mu + a_1x^2 + a_2\mu x + a_3\mu^2 + \mathcal{O}(3) \tag{6.4.25}$$

Figure 6.4.4 shows that the bifurcation diagram in the neighbourhood of the bifurcation point $(x,\mu) = (0,0)$ is equivalent to that of the vector field

$$\dot{x} = \mu - x^2 \tag{6.4.26}$$

The quadratic terms in eq. (6.4.25) additional to those in eq. (6.4.26) do not lead to a qualitative change of the bifurcation diagram since higher-order terms can be reduced successively to their essential form by non-linear transformations according to the method of normal forms as described in section 6.3. Hence, eq. (6.4.26) can be considered to be the normal form of a saddle-node bifurcation.

B) Transcritical bifurcation

The second basic bifurcation type is the transcritical one. In the case of the saddle-node bifurcation, the theorem on implicit functions enables us to construct conditions for the derivatives of the vector field characterising the bifurcation. We now

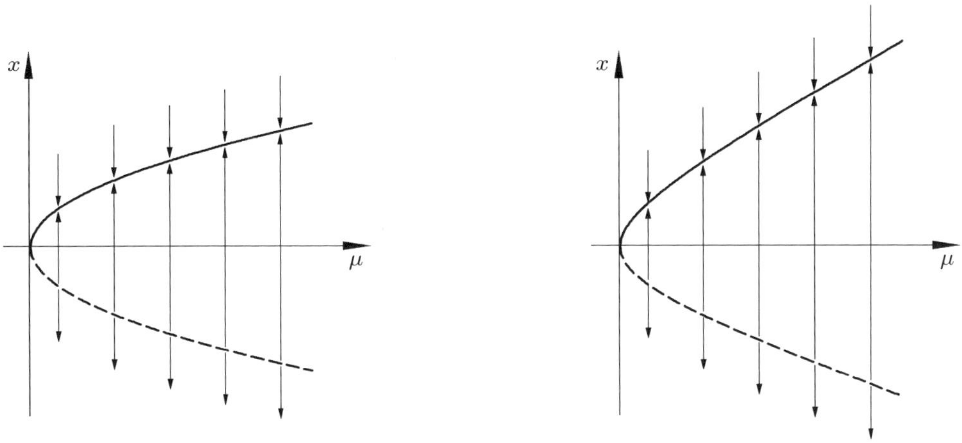

Fig. 6.4.4: Normal form of a saddle-node bifurcation eq. (6.4.25) versus eq. (6.4.26) $(a_0 = 1, a_1 = -1, a_2 = a_3 = 0.2)$

apply this theorem to the transcritical bifurcation. Equation (6.1.11) reproduced in figs. 6.1.5 and 6.4.1b lays out the conditions which have to be satisfied for a transcritical bifurcation. The following features characterise the transcritical bifurcation in the neighbourhood of the fixed point $(x, \mu) = (0, 0)$ (fig. 6.4.1b):

1. to the left and the right of the critical point $(x, \mu) = (0, 0)$, two fixed point curves exist and are given by the equations of straight lines $x = \mu$ and $x = 0$;
2. the stability changes from stable to unstable and vice versa when passing through $\mu = 0$.

We begin the derivation of the conditions for a transcritical bifurcation with a discussion of the one-parametric family of one-dimensional vector fields $\dot{x} = f(x, \mu)$ and assume that $(x, \mu) = (0, 0)$ is a non-hyperbolic fixed point that fulfils the conditions of eqs. (6.4.12, 6.4.13). In contrast to the saddle-node bifurcation, we have two fixed point curves which intersect in $(x, \mu) = (0, 0)$. Hence, the theorem of implicit functions can be applied only in a negated form: as there is no unique fixed point curve, both derivatives $\frac{\partial f}{\partial x}(0, 0)$ and $\frac{\partial f}{\partial \mu}(0, 0)$ have to vanish in this case; this prevents us from proceeding as in the case of saddle-node bifurcations, see eq. (6.4.17). In the following, we show how we can nevertheless use the theorem of implicit functions to derive the conditions for transcritical bifurcations with a simple trick.

We consider the example eq. (6.4.5) and fig. 6.4.1b in which there is a fixed point branch $x = 0$ for all μ which crosses the other branch in the origin. Hence, eq. (6.4.11) can be put in the following form

$$\dot{x} = f(x, \mu) = xF(x, \mu) \tag{6.4.27}$$

with the solution $x = 0$ as a factor. The remaining function $F(x, \mu)$ is defined as

$$\begin{aligned} x \neq 0 : F(x,\mu) &= \frac{f(x,\mu)}{x} \\ x = 0 : F(x,\mu) &= \frac{\partial f}{\partial x}(0,\mu) \end{aligned} \tag{6.4.28}$$

Since the fixed point curve $x = 0$ has been split off from eq. (6.4.27), we concentrate on formulating bifurcation conditions in terms of derivatives of F which can then be transformed into conditions for derivatives of f in accordance with eq. (6.4.28). The derivation now follows the same lines as in the case of saddle-node bifurcations; here, however, we are looking for conditions for F.

Following eq. (6.4.28) respectively eq. (6.4.27), there holds

$$\begin{aligned} F(0,0) &= 0 \\ \frac{\partial F}{\partial x}(0,0) &= \frac{1}{2}\frac{\partial^2 f}{\partial x^2}(0,0) \end{aligned} \tag{6.4.29}$$

and

$$\frac{\partial F}{\partial \mu}(0,0) = \frac{\partial^2 f}{\partial x \partial \mu}(0,0) \tag{6.4.30}$$

If $\partial F/\partial \mu(0,0) \neq 0$, then, for sufficiently small x, according to the theorem on implicit functions, a unique function $\mu = \mu(x)$ exists such that

$$F(x,\mu(x)) = 0 \tag{6.4.31}$$

Which conditions does this $\mu(x)$-fixed point curve of eq. (6.4.27) have to satisfy? Firstly, it must not coincide with the fixed point curve $x = 0$ and secondly, according to fig. 6.4.1b, it must proceed to the right respectively the left of $\mu = 0$, i. e. it must neither touch $x = 0$ nor intersect it orthogonally. These conditions may be expressed as

$$0 < \left| \frac{\mathrm{d}\mu}{\mathrm{d}x}(0) \right| < \infty \tag{6.4.32}$$

After these preliminaries, we can now differentiate eq. (6.4.31) implicitly, in analogy with the strategy for the saddle-node bifurcation. We finally obtain

$$\frac{\mathrm{d}\mu}{\mathrm{d}x}(0) = -\frac{\dfrac{\partial F}{\partial x}(0,0)}{\dfrac{\partial F}{\partial \mu}(0,0)} \tag{6.4.33}$$

Taking eqs. (6.4.29) and (6.4.30) into consideration and substituting the derivative of f in place of that for F, we can write eq. (6.4.33) as

$$\frac{\mathrm{d}\mu}{\mathrm{d}x}(0) = -\frac{1}{2}\frac{\dfrac{\partial^2 f}{\partial x^2}(0,0)}{\dfrac{\partial^2 f}{\partial x \partial \mu}(0,0)} \tag{6.4.34}$$

Equation (6.4.34) satisfies the transversality condition eq. (6.4.32) for the fixed point curve $\mu = \mu(x)$ if both numerator and denominator are non-zero.

Equation (6.4.34) thus represents a condition for the gradient of the inclined fixed point line.

Starting from the general one-parametric family of one-dimensional vector fields in eq. (6.4.11), we can now summarise the following five conditions for the transcritical bifurcation:

$$f(0,0) = 0\ , \quad \text{fixed point}$$
$$\frac{\partial f}{\partial x}(0,0) = 0\ , \quad \text{non-hyperbolic} \tag{6.4.35}$$

$$\frac{\partial f}{\partial \mu}(0,0) = 0\ , \quad \text{intersection of two fixed point curves} \tag{6.4.36}$$

$$\left.\begin{array}{r} \dfrac{\partial^2 f}{\partial x^2}(0,0) \neq 0 \\ \dfrac{\partial^2 f}{\partial x \partial \mu}(0,0) \neq 0 \end{array}\right\} \quad \text{“oblique” transversality of } \mu = \mu(x) \tag{6.4.37}$$

If the conditions of eqs. (6.4.35) to (6.4.37) are satisfied, the sign of $\frac{\mathrm{d}\mu}{\mathrm{d}x}(0)$ in eq. (6.4.34) determines which of the two possible transcritical bifurcation types we are dealing with. The case $\frac{\mathrm{d}\mu}{\mathrm{d}x}(0) > 0$ defines a positive gradient of the inclined fixed point line differing from $x = 0$ and $\frac{\mathrm{d}\mu}{\mathrm{d}x}(0) < 0$ a negative gradient. Both cases are shown in fig. 6.4.5; nothing is said as yet about the stability behaviour of the individual inclined fixed point lines.

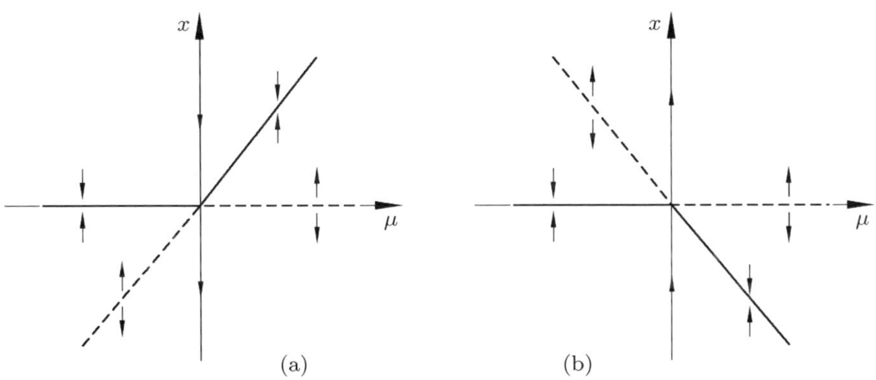

Fig. 6.4.5: Transcritical bifurcation, eq. (6.4.38):
a) supercritical ($-f_{xx}/f_{x\mu} > 0$), b) subcritical ($-f_{xx}/f_{x\mu} < 0$)

We recall eq. (6.4.24) which provides evidence concerning the stability behaviour of a saddle-node bifurcation. We consider the fixed point curve $\mu = \mu(x)$. In the case of the transcritical bifurcation, the $\frac{\partial f}{\partial \mu}$–term changes its sign at the bifurcation point $(x, \mu) = (0, 0)$, whereas $\frac{\mathrm{d}\mu}{\mathrm{d}x}$ keeps its sign; for this reason, we speak of an exchange of stability in this case. Figure 6.4.5 presents two transcritical bifurcations, a supercritical one (a) and a subcritical one (b). The equation

$$\dot{x} = \mu x \mp x^2 \tag{6.4.38}$$

satisfies the condition eqs. (6.4.35) to (6.4.37). Equation (6.4.38) is thus denoted the normal form of the transcritical bifurcation, the minus sign applying for the supercritical and the plus sign for the subcritical case.

C) Pitchfork bifurcation

The derivation of the conditions for a pitchfork bifurcation, known to us from eq. (6.4.6) and fig. 6.4.1c, is very similar to that for the transcritical bifurcation. We begin with a characterisation of the fixed point curves in the neighbourhood of the bifurcation point $(x, \mu) = (0, 0)$:

1. the two curves $x = 0$ and $\mu = x^2$ intersect in the origin $(x, \mu) = (0, 0)$;
2. the fixed point curve $x = 0$ exists for all μ, the curve $\mu = x^2$ remains on one side of $\mu = 0$ and
3. the fixed point solution $x = 0$ changes its stability when crossing $\mu = 0$, whereas that of $\mu = x^2$ remains unaffected for all x-values.

Again, $(x, \mu) = (0, 0)$ is a non-hyperbolic fixed point of $\dot{x} = f(x, \mu)$. As in the case of the transcritical bifurcation, we can split off the solution $x = 0$ and express the initial system eq. (6.4.11) in the form

$$\dot{x} = f(x, \mu) = xF(x, \mu) \tag{6.4.39}$$

where $F(x, \mu)$ satisfies the conditions

$$\begin{aligned} x \neq 0: &\quad F(x, \mu) = \frac{f(x, \mu)}{x} \\ x = 0: &\quad F(x, \mu) = \frac{\partial f}{\partial x}(0, \mu) \end{aligned} \tag{6.4.40}$$

We consider the fixed point branch which passes through $(x, \mu) = (0, 0)$ and is different from $x = 0$. Substituting the conditions for a non-hyperbolic fixed point, eqs.(6.4.12, 6.4.13) into eq. (6.4.40) yields

$$F(0, 0) = 0 \tag{6.4.41}$$

and, since no other solution which differs from $x = 0$ branches off this solution curve at the bifurcation point, we obtain in analogy to the saddle-node bifurcation eq. (6.4.22)

$$\frac{\partial F}{\partial \mu}(0, 0) \neq 0 \tag{6.4.42}$$

The implicit function theorem now states that a unique function $\mu = \mu(x)$ exists in the neighbourhood of $(x, \mu) = (0, 0)$. Hence, it follows

$$F\big(x, \mu(x)\big) = 0 \tag{6.4.43}$$

In analogy to the saddle-node bifurcation, the following conditions apply to $\mu(x)$

$$\frac{\mathrm{d}\mu}{\mathrm{d}x}(0) = 0 \tag{6.4.44}$$

and

$$\frac{\mathrm{d}^2\mu}{\mathrm{d}x^2}(0) \neq 0 \tag{6.4.45}$$

Which derivatives of F, eq. (6.4.43), and f, eq. (6.4.11), now satisfy eqs. (6.4.44) and (6.4.45)? To this purpose, we differentiate eq. (6.4.43) implicitly and finally obtain

$$\frac{\mathrm{d}\mu}{\mathrm{d}x}(0) = -\frac{\dfrac{\partial F}{\partial x}(0,0)}{\dfrac{\partial F}{\partial \mu}(0,0)} = 0 \tag{6.4.46}$$

and

$$\frac{\mathrm{d}^2\mu}{\mathrm{d}x^2}(0) = -\frac{\dfrac{\partial^2 F}{\partial x^2}(0,0)}{\dfrac{\partial F}{\partial \mu}(0,0)} \neq 0 \tag{6.4.47}$$

If we express F by $\frac{\partial f}{\partial x}$ in accordance with eq. (6.4.40), the last two relations become

$$\frac{\mathrm{d}\mu}{\mathrm{d}x}(0) = -\frac{1}{2}\frac{\dfrac{\partial^2 f}{\partial x^2}(0,0)}{\dfrac{\partial^2 f}{\partial x \partial \mu}(0,0)} = 0 \tag{6.4.48}$$

and

$$\frac{\mathrm{d}^2\mu}{\mathrm{d}x^2}(0) = -\frac{1}{3}\frac{\dfrac{\partial^3 f}{\partial x^3}(0,0)}{\dfrac{\partial^2 f}{\partial x \partial \mu}(0,0)} \neq 0 \tag{6.4.49}$$

We now summarise the following result for the pitchfork bifurcation. For the one-dimensional vector field

$$\dot{x} = f(x,\mu)$$

the bifurcation in the origin $(x,\mu) = (0,0)$ is a pitchfork bifurcation if the following six conditions are satisfied:

$$\begin{aligned} f(0,0) &= 0\,, && \text{fixed point} \\ \frac{\partial f}{\partial x}(0,0) &= 0\,, && \text{non-hyperbolic} \end{aligned} \tag{6.4.50}$$

$$\frac{\partial f}{\partial \mu}(0,0) = 0\,, \quad \text{intersection of two fixed point curves} \tag{6.4.51}$$

$$\frac{\partial^2 f}{\partial x^2}(0,0) = 0\,, \quad \text{the branch (different from } x=0\text{) has } \mu = 0 \text{ as the tangent} \tag{6.4.52}$$

$$\left.\begin{array}{l}\dfrac{\partial^2 f}{\partial x \partial \mu}(0,0) \neq 0 \\ \dfrac{\partial^3 f}{\partial x^3}(0,0) \neq 0\end{array}\right\} \quad \begin{array}{l}\text{the branch (different from } x=0\text{)} \\ \text{remains on one side}\end{array} \tag{6.4.53}$$

The sign of $\frac{\mathrm{d}^2\mu}{\mathrm{d}x^2}(0,0)$ in eq. (6.4.49) determines on which side the fixed point equation $F(x,\mu) = 0$ branches off. Both possibilities together with two states of stability are shown in fig. 6.4.6; in particular, a) displays the supercritical and b) the subcritical pitchfork bifurcation.

The bifurcation types associated with the conditions eqs. (6.4.50) to (6.4.53) correspond qualitatively to those of the system

$$\dot{x} = \mu x \mp x^3 \tag{6.4.54}$$

The minus sign applies to the supercritical and the plus sign to the subcritical bifurcation. Equation (6.4.54) is denoted the normal form of the pitchfork bifurcation.

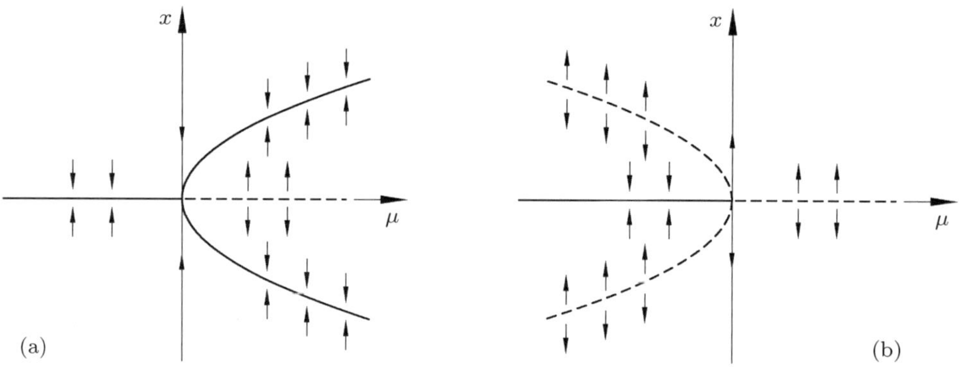

Fig. 6.4.6: Pitchfork bifurcation, see eq. (6.4.54):
a) supercritical ($-f_{xxx}/f_{x\mu} > 0$), b) subcritical ($-f_{xxx}/f_{x\mu} < 0$)

D) Hopf bifurcation

The last local bifurcation of the codimension 1 of a non-hyperbolic fixed point to be described here is the Hopf bifurcation already mentioned in section 6.1. It happens when a simple pair of complex conjugate eigenvalues crosses the imaginary axis at μ_{cr}. The general set of equations

$$\dot{\boldsymbol{x}} = \boldsymbol{F}(\boldsymbol{x},\boldsymbol{\mu}) , \qquad \boldsymbol{x} \in \mathbb{R}^{\mathrm{n}}, \quad \boldsymbol{\mu} \in \mathbb{R}^{\mathrm{p}} \tag{6.4.55}$$

can, in accordance with the centre manifold theory, be reduced to a p-parameter family of a two-dimensional flow. If we keep all parameters fixed with the exception of one, μ, the vector field on the two-dimensional centre manifold can be split into a linear and a non-linear component and can be expressed as

$$\begin{bmatrix} \dot{x} \\ \dot{y} \end{bmatrix} = \begin{bmatrix} \alpha(\mu) & -\omega(\mu) \\ \omega(\mu) & \alpha(\mu) \end{bmatrix} \begin{bmatrix} x \\ y \end{bmatrix} + \begin{bmatrix} g_x(x,y,\mu) \\ g_y(x,y,\mu) \end{bmatrix}, \qquad (x,y,\mu) \in \mathbb{R}^1 \times \mathbb{R}^1 \times \mathbb{R}^1 \tag{6.4.56}$$

where we assume that the fixed point lies in the origin. The values $\lambda(\mu) = \alpha(\mu) \pm i\omega(\mu)$ are the complex conjugate eigenvalues of the linearised system and g_x, g_y are the non-linear terms in the variables x, y. To simplify the argument, we may assume that the bifurcation occurs at $\mu_{cr} = 0$; then $\mathrm{Re}\lambda(0) = 0$, i. e.

$$\begin{aligned} &\alpha(0) = 0 \\ &\omega(0) \neq 0 \end{aligned} \tag{6.4.57}$$

In order to ascertain which higher-order terms do not contribute anything substantially new to the bifurcation, topologically and qualitatively, eq. (6.4.56) is transformed to its normal form. The determination of the normal form by means of successive elimination of higher-order terms was described in detail in section 6.3. The explicit derivation of the normal form for the case of two complex conjugate, purely imaginary eigenvalues is detailed in (Guckenheimer and Holmes, 1983), so that we can reproduce the result here directly

$$\begin{aligned} \dot{x} &= \alpha(\mu)x - \omega(\mu)y + [a(\mu)x - b(\mu)y](x^2 + y^2) + \mathcal{O}(5) \\ \dot{y} &= \omega(\mu)x + \alpha(\mu)y + [b(\mu)x + a(\mu)y](x^2 + y^2) + \mathcal{O}(5) \end{aligned} \tag{6.4.58}$$

In polar coordinates r, θ, eqs. (6.4.58) can be expressed more succinctly as

$$\begin{aligned} \dot{r} &= \alpha(\mu)r + a(\mu)r^3 + \mathcal{O}(r^5) \\ \dot{\theta} &= \omega(\mu) \quad + b(\mu)r^2 + \mathcal{O}(r^4) \end{aligned} \tag{6.4.59}$$

We are interested in the dynamics in dependence on μ in the neighbourhood of $\mu = 0$. To this end, it is appropriate to expand the coefficients of eq. (6.4.59) into Taylor series at $\mu = 0$. Retaining only the leading terms of the Taylor series and considering eq. (6.4.57), we finally approximate eq. (6.4.59) in the form

$$\begin{aligned} \dot{r} &= \frac{\mathrm{d}\alpha}{\mathrm{d}\mu}(0)\,\mu r + a(0)r^3 \\ \dot{\theta} &= \omega(0) + \frac{\mathrm{d}\omega}{\mathrm{d}\mu}(0)\,\mu + b(0)r^2 \end{aligned}$$

respectively, with $d = \frac{\mathrm{d}\alpha}{\mathrm{d}\mu}(0)$ and $c = \frac{\mathrm{d}\omega}{\mathrm{d}\mu}(0)$, we find

$$\dot{r} = d\mu r + ar^3 \quad = f(r) \tag{6.4.60}$$

$$\dot{\theta} = \omega + c\mu + br^2 = g(r) \tag{6.4.61}$$

The expansion of the radius in eq. (6.4.60) is independent of θ. The condition $\dot{r} = 0$ thus determines the long-term behaviour for $t \to \infty$. The trivial solution $r = 0$ is a fixed point solution for all μ and, since $\dot{\theta} \neq 0$, the fixed point can thus only be a focus. The non-trivial solution $r = (-d\mu/a)^{1/2}$ shows that, since $\dot{\theta} \neq 0$, circular periodic orbits with $r > 0$ emerge for $-\infty < d\mu/a < 0$. In the bifurcation diagram, the non-trivial solutions for $a \neq 0$ and $d \neq 0$ lie on the paraboloid $\mu = -ar^2/d$. This means that at $\mu = 0$, depending on the signs of a and d, a limit cycle either disappears or emerges.

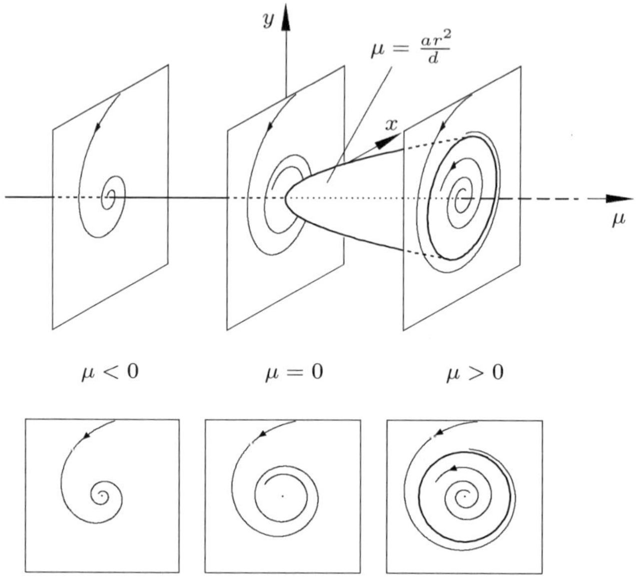

Fig. 6.4.7: Supercritically *stable* periodic bifurcation
= *supercritical* Hopf bifurcation ($d > 0,\ a < 0$)

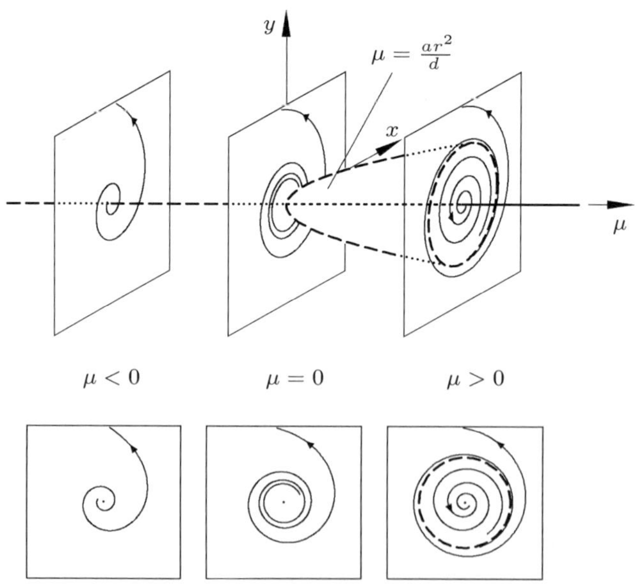

Fig. 6.4.8: Supercritically *unstable* periodic bifurcation ($d < 0,\ a > 0$)

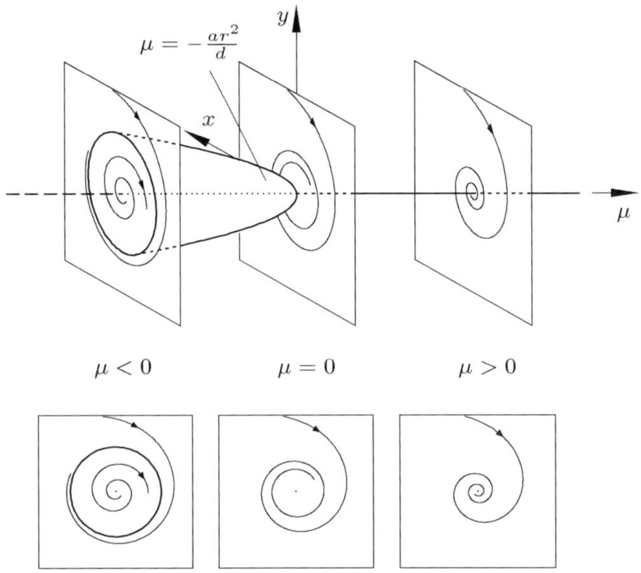

Fig. 6.4.9: Subcritically *stable* periodic bifurcation ($d < 0$, $a < 0$)

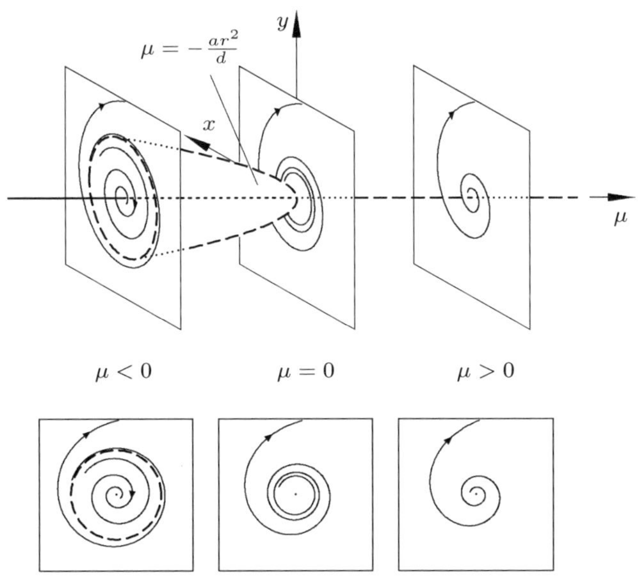

Fig. 6.4.10: Subcritically *unstable* periodic bifurcation
= *subcritical* Hopf bifurcation ($d > 0$, $a > 0$)

If all solutions for $t \to \infty$ have been determined, the next question concerns their stability. The stability of the fixed points and limit cycles is determined by the sign of the derivative of eq. (6.4.60) with respect to r. We obtain

$$\frac{\mathrm{d}f}{\mathrm{d}r} = d\mu + 3ar^2 \tag{6.4.62}$$

For the fixed points $r = 0$, we obtain $\frac{\mathrm{d}f}{\mathrm{d}r} = d\mu$, i. e. at the zero passage of the control parameter μ, a change of sign of the derivative and thus also a stability change of the fixed points will certainly take place. To determine the stability of the periodic orbit, we calculate the derivative $\frac{\mathrm{d}f}{\mathrm{d}r}$ along the paraboloid $\mu = -ar^2/d$ and obtain

$$\frac{\mathrm{d}f}{\mathrm{d}r} = d(-\frac{ar^2}{d}) + 3ar^2 = 2ar^2$$

We note that the limit cycle is asymptotically stable for $a < 0$ and unstable for $a > 0$.

The stability of the fixed point for $\mu = 0$ can be deduced directly from eq. (6.4.62). We again observe that the sign of a determines the stability, namely as follows:

$$\begin{array}{lcr} \mu = 0 & \text{is} & \text{stable for } a < 0 \\ \mu = 0 & \text{is} & \text{unstable for } a > 0 \end{array}$$

Let us summarise and illustrate the results regarding the stable and unstable fixed point solutions and stable and unstable limit cycles. For $-\infty < \mu < +\infty$, four constellations are possible:

1. $d > 0\,,\ a < 0$ (fig. 6.4.7) (supercritical)
2. $d < 0\,,\ a > 0$ (fig. 6.4.8)
3. $d < 0\,,\ a < 0$ (fig. 6.4.9)
4. $d > 0\,,\ a > 0$ (fig. 6.4.10) (subcritical)

ad 1 : $d > 0,\ a < 0$: For $\mu \leqslant 0$, the only existing solution represents an asymptotically stable focus. At $\mu = 0$, a bifurcation occurs and the set of equations generates an additional stable limit cycle and an unstable focus for $\mu > 0$. If we allow μ to evolve from $-\infty$ to $+\infty$, the periodic branch is supercritically stable; this is called the *supercritical Hopf bifurcation* (fig. 6.4.7).

ad 2 : $d < 0,\ a > 0$: The only solution for $\mu \leqslant 0$ is an unstable focus. The limit cycle which emerges at $\mu = 0$ and exists for $\mu > 0$ is also unstable. The focus in the region $\mu > 0$, on the other hand, is stable. There is no special name for this supercritically unstable limit cycle bifurcation of an unstable fixed point solution (fig. 6.4.8).

ad 3 : $d < 0,\ a < 0$: For $\mu < 0$, there are two long-term solutions, a stable limit cycle and an unstable focus. At $\mu = 0$, the limit cycle vanishes; what remains is a stable focus for $\mu \geqslant 0$. We observe a subcritically stable bifurcation (fig. 6.4.9).

ad 4 : $d > 0,\ a > 0$: In this case, there is an unstable limit cycle and a stable focus for $\mu < 0$. The only one solution for $\mu \geqslant 0$ is an unstable focus. The subcritical unstable periodic bifurcation is called the *subcritical* or *inverse* (to the supercritical) *Hopf bifurcation* (fig. 6.4.10).

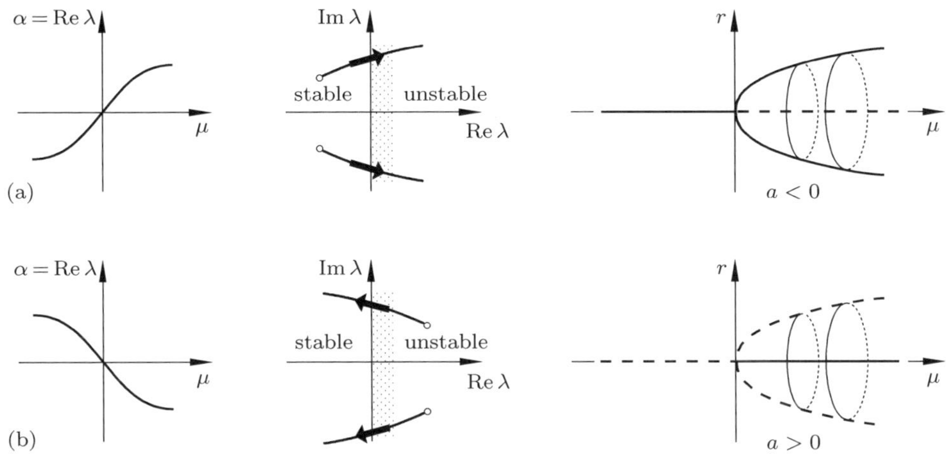

Fig. 6.4.11: Transversality condition $d = \frac{\mathrm{d(Re}\lambda)}{\mathrm{d}\mu}\big|_0 \neq 0$: a) $d > 0$, b) $d < 0$

We have seen that the stability behaviour of the periodic bifurcating solution is controlled by the parameter a. Moreover, we have shown that the stability behaviour of the limit cycle is dependent on the stability behaviour of the enclosed focus. If the limit cycle is stable, the focus is unstable and *vice versa*. We can thus draw conclusions regarding the stability behaviour of the limit cycle if we know the stability behaviour of the focus. If we consider the linear term of eq. (6.4.60) for this purpose, it becomes clear that, for $d > 0$, the focus $r = 0$ is stable for $\mu < 0$ and unstable for $\mu > 0$ (figs. 6.4.7 and 6.4.10). The opposite is true for $d < 0$: for $\mu < 0$, the fixed point is unstable and for $\mu > 0$, asymptotically stable. The switch in stability becomes more evident if we recall the meaning of d, eq. (6.4.60),

$$d = \left.\frac{\mathrm{d}\alpha}{\mathrm{d}\mu}\right|_0 = \left.\frac{\mathrm{d(Re}\lambda)}{\mathrm{d}\mu}\right|_0 \lessgtr 0 \tag{6.4.63}$$

Figure 6.4.11 demonstrates the two possibilities for $d \neq 0$. If the function $\mathrm{Re}\,\lambda(\mu)$ crosses the μ-axis at the origin with a positive slope (fig. 6.4.11a), the eigenvalues change with increasing μ from the negative to the positive half-space and the originally asymptotically stable fixed point of eq. (6.4.60) becomes unstable. If $\mathrm{d(Re}\lambda)/\mathrm{d}\mu|_0 < 0$, the eigenvalues change with increasing μ from the positive to the negative half-space and the previously unstable fixed point of eq. (6.4.60) becomes asymptotically stable.

The same is true for the transcritical bifurcation and also for the pitchfork bifurcation for which $\mathrm{Im}\,\lambda(\mu) = 0$. Should a transversality condition, $\mathrm{d(Re}\lambda)/\mathrm{d}\mu|_0 \neq 0$, apply, the following far-reaching statement can be proven:

> If, in the case of a simple eigenvalue λ, the function $\mathrm{Re}\,\lambda(\mu)$ crosses the μ-axis at $\mu = 0$ with a positive slope $0 < \mathrm{d(Re}\lambda)/\mathrm{d}\mu|_0 < \infty$, a fixed point which is asymptotically stable for $\mathrm{Re}\lambda < 0$ bifurcates supercritically stably

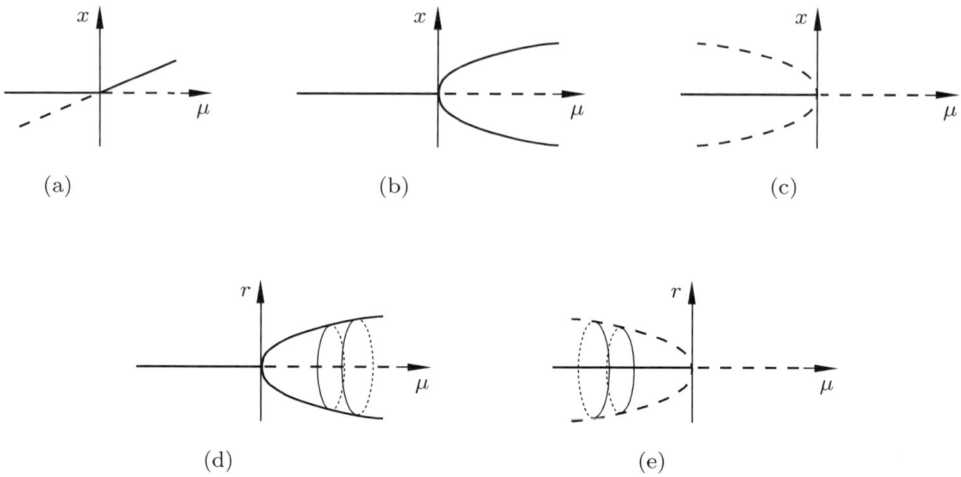

Fig. 6.4.12: Transversality condition $\frac{\mathrm{d}(\mathrm{Re}\lambda)}{\mathrm{d}\mu}\big|_0 \neq 0$ satisfied:
a) transcritical bifurcation: supercritically stable, subcritically unstable
b) supercritically stable pitchfork bifurcation
c) subcritically unstable pitchfork bifurcation
d) supercritically stable Hopf bifurcation
e) subcritically unstable Hopf bifurcation

respectively subcritically unstably; the bifurcating solution is a fixed point (transcritical or pitchfork bifurcation) for $\mathrm{Im}\,\lambda(\mu) = 0$ and a limit cycle (Hopf bifurcation) for $\mathrm{Im}\,\lambda(\mu) \neq 0$ (fig. 6.4.12).

Figure 6.4.13 presents an overview of the transversality conditions in dependence of the control parameter μ, the crossing of the imaginary axis by the real parts of the eigenvalues and the corresponding supercritical bifurcations of the transcritical, the pitchfork and the Hopf bifurcation.

After this brief digression on the transversality condition, we recall that we dispensed with the higher-order terms in the normal form, eq. (6.4.58), leading to the Hopf bifurcation. An obvious question concerns the influence of the higher-order terms on the bifurcation type. A theorem of Hopf from 1942 states that the higher-order terms do not contribute any new qualitative information to the bifurcation; cf. (Guckenheimer and Holmes, 1983), p. 151. The proof of the theorem is extensive and is based on the application of the centre manifold theory and the normal form transformation (Marsden and McCracken, 1976).

In this section, we presented local bifurcations of codimension 1 emerging as a result of the zero-crossing of the real part of the eigenvalue. The analysis is much more complex, if, varying two system parameters, the real parts of two eigenvalues vanish simultaneously. In the simplest case, the two-parametric flow on the centre manifold must have the dimension 2. The Jacobian matrix at the bifurcation point then has a Jordan normal form

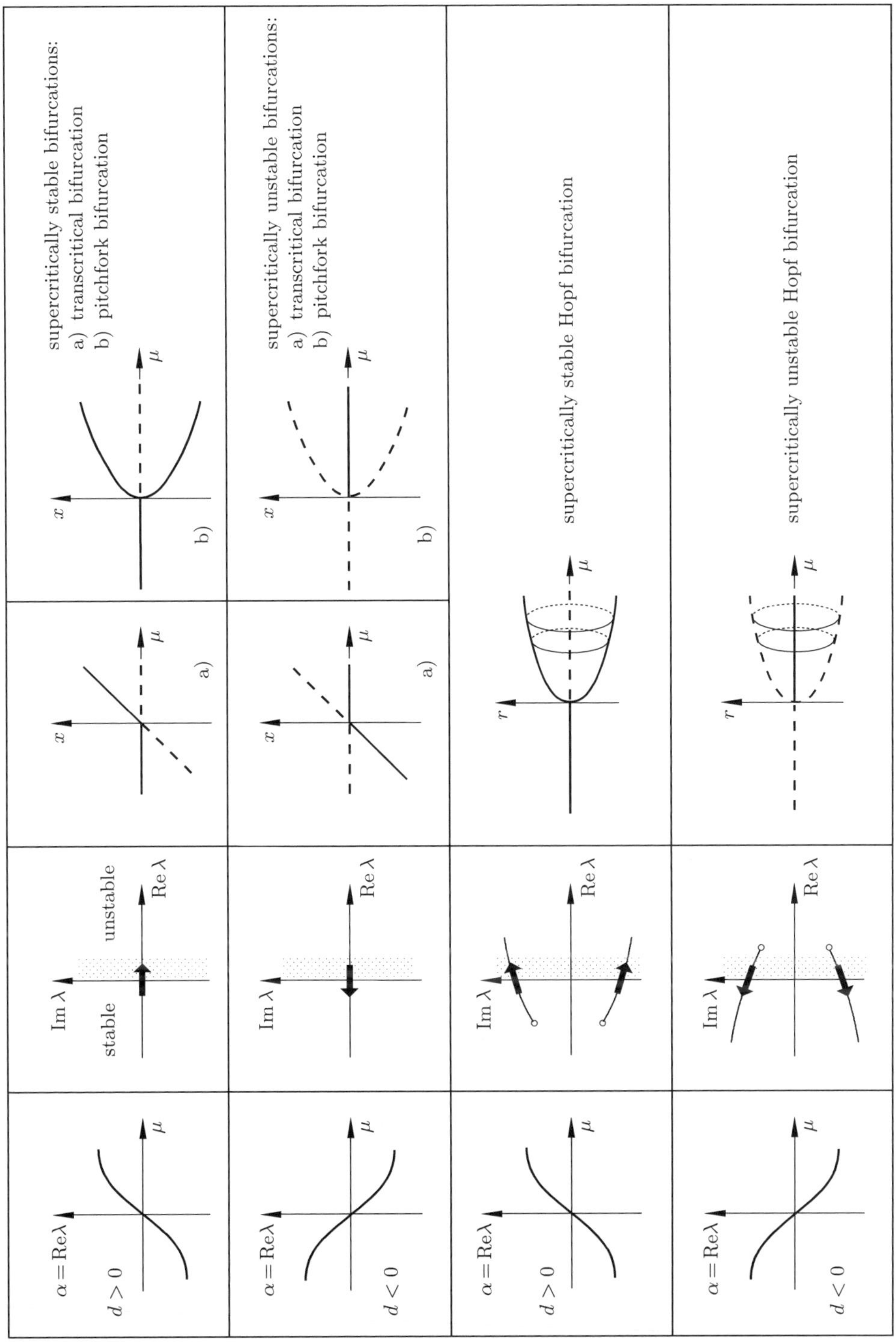

Fig. 6.4.13: Bifurcation types: overview

$$D = \begin{bmatrix} 0 & 1 \\ 0 & 0 \end{bmatrix}$$

This bifurcation is denoted Takens-Bogdanov bifurcation (Dangelmayr and Knobloch, 1987). A discussion of local bifurcations of codimension 2 can be found in (Guckenheimer and Holmes, 1983). We will meet bifurcations of codimension 2 when we discuss the van der Pol equation in section 10.4.

6.5 Stability of Bifurcations Subject to Perturbations

We have seen that in the neighbourhood of a critical value μ_{cr}, small changes of μ can lead to qualitatively new solution patterns. For $\mu > \mu_{cr}$ in the case of the pitchfork bifurcation, for example, a previously stable fixed point becomes unstable and, in addition, two new stable fixed point solutions emerge as if from thin air.

Fortunately, for one-parameter families of one-dimensional flows, the bifurcation patterns are limited to four basic types. The question arises whether these types of bifurcation are stable under perturbations. This is the subject of this section. The question regarding the stability of the bifurcation is as follows: is there a qualitative change in the dynamics of a vector field if the system is subjected to a perturbation and if so, what is the dynamics of the perturbed system? Should perturbations lead to no fundamental change in the dynamics, that is to say the system proves robust with regard to small deviations, we speak of structural stability (Wiggins, 1990). We shall return to this concept in section 8.2. In order to answer our question, we must first define what we mean by "perturbation". A perturbation of the vector field $f(x, \mu)$ can be expressed by

1. a constant term or parameter ε,
2. lower-order terms respectively
3. higher-order terms in the Taylor series of the vector field in the neighbourhood of a non-hyperbolic fixed point.

For the normal forms of the bifurcations already known to us – the saddle-node bifurcation, the transcritical bifurcation, the pitchfork bifurcation and the Hopf bifurcation – we now enquire to what extent a perturbation causes a qualitative change of the dynamics in the neighbourhood of the bifurcation point.

A) Saddle-node bifurcation (structurally stable)

The normal form of a saddle-node bifurcation is, see eq. (6.4.26),

$$\dot{x} = \mu - x^2 \tag{6.5.1}$$

Let us begin with a parameter perturbation. It is obvious that the equation

$$\dot{x} = \mu + \varepsilon - x^2 = \mu' - x^2 \ , \qquad \mu' = \mu + \varepsilon \tag{6.5.2}$$

satisfies the conditions of eqs. (6.4.21) to (6.4.23) for a saddle-node bifurcation at $\mu' = 0$.

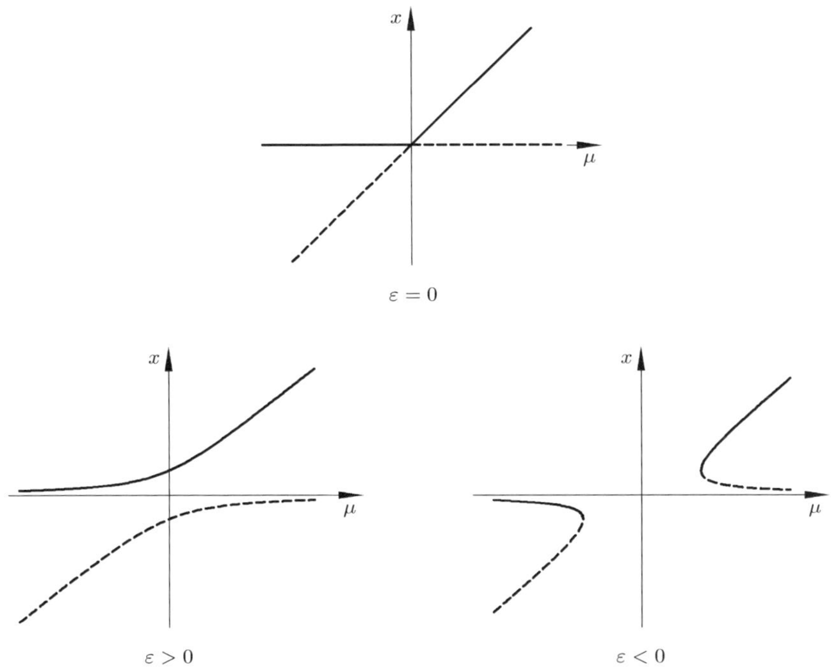

Fig. 6.5.1: Structurally unstable transcritical bifurcation

The vector flow in the neighbourhood of a non-hyperbolic fixed point can be expanded into a Taylor series. The quadratic term in eq. (6.5.2) is the first non-vanishing term for a saddle-node bifurcation. We perturb the vector flow by a term of lower order than $\mathcal{O}(x^2)$ and obtain

$$\dot{x} = \mu + \varepsilon x - x^2 \tag{6.5.3}$$

Equation (6.5.3) represents a two-parameter family of one-dimensional flows with the non-hyperbolic fixed point $(x, \mu, \varepsilon) = (0, 0, 0)$. The bifurcation type is determined completely by the conditions of eqs. (6.4.21) to (6.4.23), i. e. the perturbation term εx does not change the saddle-node bifurcation qualitatively.

How about the perturbation terms $\mathcal{O}(x^3)$ and those of higher order? Since only derivatives up to second order in x appear in the conditions for the saddle-node bifurcation, eqs. (6.4.21) to (6.4.23), we see that the bifurcation type is completely preserved in this case as well. This means that neither perturbations in the parameter μ nor in the function terms alter the nature of the saddle-node bifurcation. The saddle-node bifurcation is hence *structurally stable*.

B) Transcritical bifurcation (structurally unstable)

The normal form of the transcritical bifurcation which satisfies the conditions of eqs. (6.4.35) to (6.4.37) is

$$\dot{x} = \mu x - x^2 \tag{6.5.4}$$

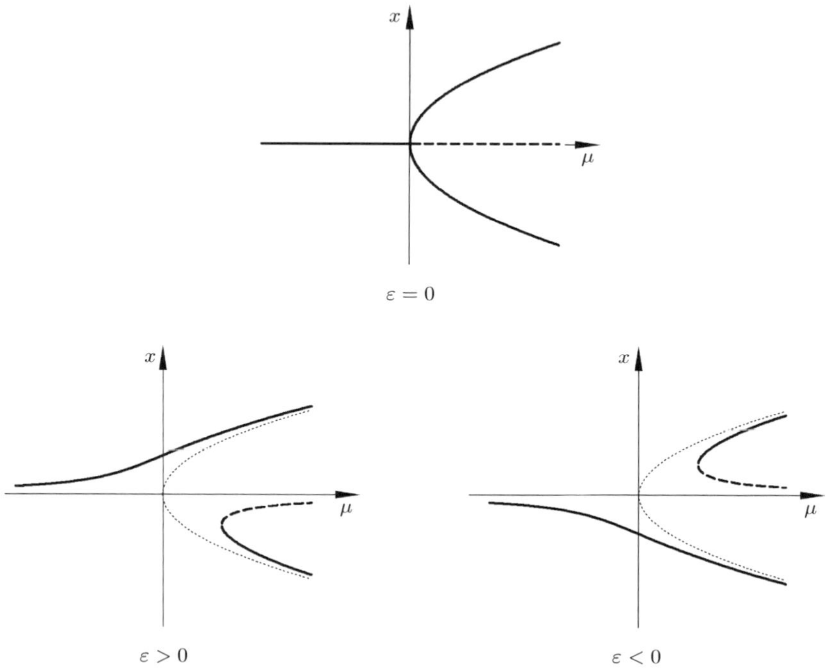

Fig. 6.5.2: Structurally unstable pitchfork bifurcation

In analogy to the argumentation regarding a perturbed saddle-node bifurcation, the perturbation terms here can also be considered to represent terms of a Taylor expansion of the vector field $f(x, \mu)$ in the neighbourhood of $(x, \mu) = (0, 0)$. A constant perturbation term already leads to surprises. The equation

$$\dot{x} = \varepsilon + \mu x - x^2 \tag{6.5.5}$$

displays qualitatively new dynamics. Figure 6.5.1 presents the basic transcritical pattern and the changes for $\varepsilon < 0$ and $\varepsilon > 0$. Depending on the sign of the perturbation term, differing fixed point curves emerge. If $\varepsilon > 0$, the solution curves split into two completely separate branches and, at the crossing of the line $\mu = 0$, no bifurcation takes place. If $\varepsilon < 0$, we observe two saddle-node bifurcations which occur for $\mu \geqslant 2\sqrt{-\varepsilon}$ respectively $\mu \leqslant -2\sqrt{-\varepsilon}$.

We note that this type of bifurcation is *structurally unstable* since a parameter perturbation suffices to cause the disintegration of the basic pattern of the transcritical bifurcation (fig. 6.5.1, $\varepsilon = 0$).

C) Pitchfork bifurcation (structurally unstable)

We know the normal form of the pitchfork bifurcation satisfying eq. (6.4.54). When perturbed only by a single parameter ε, the dynamics is expressed by the equation

$$\dot{x} = \varepsilon + \mu x - x^3 \tag{6.5.6}$$

Figure 6.5.2 illustrates the bifurcation structure of eq. (6.5.6) for $\varepsilon > 0$, $\varepsilon = 0$ and $\varepsilon < 0$.

The symmetrical pitchfork bifurcation already breaks up for small perturbation values $\varepsilon \neq 0$. The result is a non-bifurcating branch of fixed point solutions and, for $\mu_{cr} > 0$, there is an additional saddle-node bifurcation. As in the case of the transcritical bifurcation, the pitchfork bifurcation is *structurally unstable*.

D) Hopf bifurcation (structurally stable)

We are already familiar with the normal form of the Hopf bifurcation in polar coordinates r, θ, cf. section 6.4, eqs. (6.4.60) and (6.4.61)

$$\dot{r} = d\mu r + ar^3 \tag{6.5.7}$$

$$\dot{\theta} = \omega + c\mu + br^2 \tag{6.5.8}$$

Our interest is directed at the influence of a perturbation on the solution behaviour in the neighbourhood of the bifurcation point $(\mu, r) = (0, 0)$. Three aspects are decisive for the argumentation:

1. The Hopf theorem states that, for $a \neq 0$, $d \neq 0$, the higher-order terms do not alter the dynamics topologically in the neighbourhood of $(\mu, r) = (0, 0)$.
2. The bifurcation dynamics is determined by eq. (6.5.7) since eq. (6.5.8) yields $\dot{\theta} \approx \omega = \text{const}$ for small μ and r.
3. We recall point b)*ii.* at the end of section 6.3. There, we showed that if the Jacobian matrix $\boldsymbol{D}$ of a vector field possesses a pair of complex conjugate eigenvalues with a vanishing real part, the resonance condition leads to an elimination restricted to all even-order terms so that exclusively odd-order terms appear in the normal form. It thus becomes clear that perturbation terms of the form ε or εr^2 in eq. (6.5.7) cause no qualitative change of the bifurcation structure.

All in all, this means that the Hopf bifurcation is *structurally stable*.

How can we find out if a bifurcation is structurally stable, i. e. whether it possesses typical, generic characteristics or not? Conditions that have to be exactly zero – for example for the transcritical bifurcation, eq. (6.4.36), or the pitchfork bifurcation, eqs. (6.4.51, 6.4.52) – cannot be satisfied on principle, i. e. small perturbations will always lead to deviations of the zero condition, unless symmetries and conservation laws are decisive (Stewart, 1983a; Stewart, 1983b). Conversely, for each condition that is not equal to zero, such as the transversality condition, i.e. if no tangency exists, there is a small parameter interval where the condition remains non-zero. However, these heuristic considerations can only serve as a guide; exact definitions of the terms *structurally stable, generic, transversal* are detailed in (Wiggins, 1990).

Summarising, we note that, of the four normal forms of bifurcations of one-parametric families, only the saddle-node bifurcation and the Hopf bifurcation maintain structural stability under perturbations. This may explain why these two bifurcations are predominantly observed in nature.

6.6 Bifurcations of the Fixed Points of One-Parametric Maps

In this section, we discuss the simplest bifurcations of fixed points in one-parametric maps. Bearing in mind that periodic orbits appear as fixed points in the Poincaré map, the bifurcations of fixed points of such one-parametric maps represent corresponding bifurcations of periodic orbits. The bifurcations of periodic solutions can thus be studied by observation of the fixed point bifurcations of iterated maps (fig. 6.6.1). The knowledge that iterated maps reflect the dynamic behaviour of periodic orbits was also used in the derivation of the Floquet theory, see section 5.4.2.

In section 6.4, we discussed the normal forms of bifurcations of codimension 1 for continuous systems and the conditions these bifurcations have to satisfy. In this section, we apply those results to iterated maps.

Let us consider a one-parametric mapping of $\mathbb{R}^n$ in $\mathbb{R}^n$

$$\boldsymbol{x}_{k+1} = \boldsymbol{f}(\boldsymbol{x}_k, \mu) \qquad \text{with} \quad \boldsymbol{f}(\boldsymbol{x}, \mu),\, \boldsymbol{x} \in \mathbb{R}^n, \quad \mu \in \mathbb{R}^1 \tag{6.6.1}$$

Our interest is focused on the bifurcations of fixed points. Let the fixed points of eq. (6.6.1) be given by $(\boldsymbol{x}, \mu) = (\boldsymbol{x}_s, \mu_s)$; the fixed point condition is then

$$\boldsymbol{x}_s = \boldsymbol{f}(\boldsymbol{x}_s, \mu_s) \tag{6.6.2}$$

If the fixed points are known, we next enquire into their stability in dependence of the control parameter μ. As we know, a sudden change in the stability behaviour may occur as a result of an alteration of the control parameter μ beyond a critical value μ_{cr}.

In the case of hyperbolic fixed points, we may draw conclusions on the stability by analysing the linearised map as a function of the perturbation $\boldsymbol{\eta}_k$ (see section 3.6). We obtain

$$\boldsymbol{\eta}_{k+1} = \boldsymbol{D}(\boldsymbol{x}_s, \mu_s)\boldsymbol{\eta}_k \ , \qquad \boldsymbol{D} = \left.\frac{\partial \boldsymbol{f}}{\partial \boldsymbol{x}}\right|_{\boldsymbol{x}_s, \mu_s} \tag{6.6.3}$$

where $\boldsymbol{D}(\boldsymbol{x}_s, \mu_s)$ is the Jacobian matrix of the map at the fixed point $(\boldsymbol{x}_s, \mu_s)$. If no eigenvalue of $\boldsymbol{D}(\boldsymbol{x}_s, \mu_s)$ has a modulus equal to 1 ($|\lambda| \neq 1$), the fixed point is hyperbolic and the stability respectively instability of the linearised map implies the stability respectively instability of the non-linear map. It can be shown using the implicit function theorem that, in the direct proximity of $(\boldsymbol{x}_s, \mu_s)$ and for varying μ, the fixed point solution is unique and its stability is preserved. As a consequence, a hyperbolic fixed point cannot experience a bifurcation.

The case which is relevant for this section is hence the non-hyperbolic one. For non-hyperbolic fixed points, however, the linear stability analysis fails. Methods which include the non-linear terms for the determination of the stability – such as the centre manifold theory – have already been discussed in detail for the case of continuous systems. The fundamental characteristic of non-hyperbolic fixed points is their ability to generate new states of equilibrium. But when is a fixed point of a non-linear map non-hyperbolic? We can distinguish three cases (see fig. 5.4.6):

1. $\boldsymbol{D}(\boldsymbol{x}_s, \mu_s)$ has a single eigenvalue equal to $+1$; the remaining eigenvalues have moduli not equal to 1;

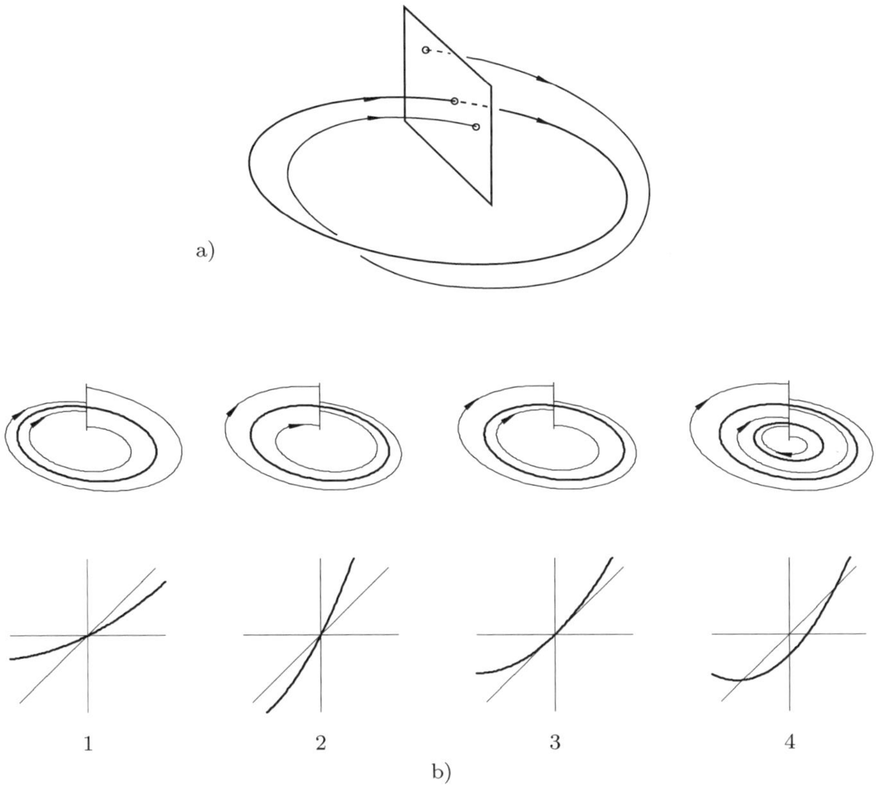

Fig. 6.6.1: a) Poincaré map of continuous recurrent orbits
b) Connection between limit cycle in phase space and fixed point arising in an iterated map:
1. stable limit cycle $\hat{=}$ stable fixed point
2. unstable limit cycle $\hat{=}$ unstable fixed point
3. semi-stable limit cycle $\hat{=}$ saddle node
4. stable and unstable limit cycle $\hat{=}$ stable and unstable fixed point (arising due to a perturbation of 3.)

2. $\boldsymbol{D}(\boldsymbol{x}_s, \mu_s)$ has a single eigenvalue equal to -1; the remaining eigenvalues have moduli not equal to 1;
3. $\boldsymbol{D}(\boldsymbol{x}_s, \mu_s)$ possesses a pair of complex conjugate eigenvalues λ, λ^* with $|\lambda| = 1$; the remaining eigenvalues have moduli not equal to 1.

In all three cases, the centre manifold theory allows a reduction of the n-dimensional iterated map to a one- respectively two-dimensional one. We begin with the one-dimensional map with the eigenvalue $+1$.

Case 1: *Single eigenvalue* $+1$

To analyse the bifurcation, we consider the following one-dimensional map on the centre manifold

$$x_{k+1} = f(x_k, \mu) \tag{6.6.4}$$

A reduction to the centre manifold is accompanied by a transformation of the fixed point (x_s, μ_s) to the origin, so that we have for the non-hyperbolic fixed point

$$f(0,0) = 0 \tag{6.6.5}$$

$$\frac{\partial f}{\partial x}(0,0) = 1 \tag{6.6.6}$$

In analogy to the bifurcations of time-continuous flows discussed in section 6.4, we begin with the saddle-node bifurcation for maps.

A) Saddle-node bifurcation

We first investigate the one-dimensional map of the form

$$x_{k+1} = f(x_k, \mu) \qquad \text{with} \quad f(x,\mu) = x + \mu - x^2 \tag{6.6.7}$$

The conditions in eqs. (6.6.5) and (6.6.6) for a non-hyperbolic fixed point with eigenvalue +1 in the origin $(x, \mu) = (0, 0)$ are satisfied by eq. (6.6.7). Figure 6.6.2 presents the graph of the function eq. (6.6.7); in particular, the iteration process is shown for the three parameter domains ($\mu < 0$, $\mu = 0$, $\mu > 0$) as well as the bifurcation diagram (fig. 6.6.2c), which can be established very easily. Noting the fixed point condition for maps, we obtain for eq. (6.6.7) the expression

$$f(x,\mu) - x = \mu - x^2 = 0 \tag{6.6.8}$$

which represents a parabola

$$\mu = x^2 \tag{6.6.9}$$

in the (x, μ)-space. A stability analysis demonstrates that in the origin, two fixed point solutions of different stability merge and, since no fixed point solutions exist for negative μ-values, we recognise the typical phenomenon of a saddle-node bifurcation. In analogy to the bifurcation of continuous flows, we now set up general conditions governing the evolution of saddle-node bifurcations in maps.

The bifurcation diagram, fig. 6.6.2c, shows that, in the (x, μ)-plane, the map defined by eq. (6.6.7) generates for $\mu \geqslant 0$ a unique fixed point curve $\mu = \mu(x)$ which passes through the origin $(x, \mu) = (0, 0)$. As in the case of time-continuous flows (section 6.4), we want to deduce conditions for the characterisation of the bifurcations for general maps of the form $f(x, \mu) = f_0(\mu) + f_1(\mu)x + f_2(\mu)x^2 + \ldots$. Again, we apply the theorem on implicit functions to iterated maps.

Starting from the fixed point condition for maps

$$f(x,\mu) - x \equiv g(x,\mu) = 0 \tag{6.6.10}$$

we seek conditions which describe a saddle-node bifurcation for the graph, fig. 6.6.2c. Applying the theorem on implicit functions to $g(x, \mu)$ and assuming that

$$g(0,0) = 0 \qquad \text{and} \qquad \frac{\partial g}{\partial \mu}(0,0) = \frac{\partial f}{\partial \mu}(0,0) \neq 0 \tag{6.6.11}$$

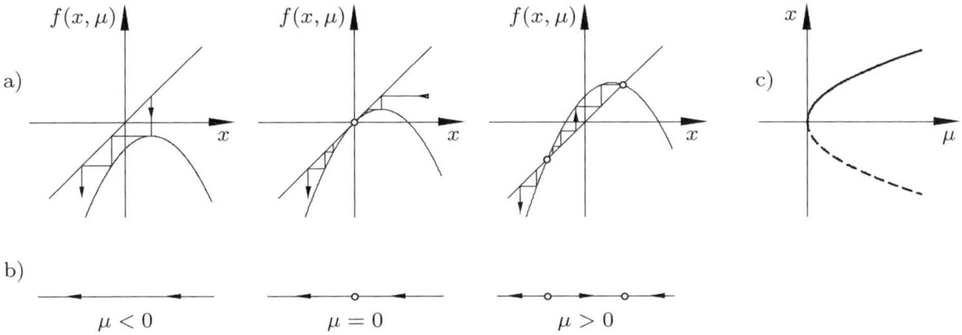

Fig. 6.6.2: Saddle-node bifurcation of a one-dimensional map (eigenvalue 1):
a) iteration process
b) stability behaviour in dependence of μ;
c) bifurcation diagram

then, for small x in the neighbourhood of the fixed point $(x, \mu) = (0, 0)$, a unique function $\mu(x)$ exists which satisfies the equation

$$g\big(x, \mu(x)\big) \equiv f\big(x, \mu(x)\big) - x = 0 \tag{6.6.12}$$

The graph of the bifurcation diagram, fig. 6.6.2c, implies the following two conditions

$$\frac{\mathrm{d}\mu}{\mathrm{d}x}(0) = 0 \qquad \text{and} \qquad \frac{\mathrm{d}^2\mu}{\mathrm{d}x^2}(0) \neq 0 \tag{6.6.13}$$

If this argument is applied to eq. (6.6.12), implicit differentiation with respect to x and the chain rule yields

$$\begin{aligned} \frac{\partial g}{\partial x} \qquad + \frac{\partial g}{\partial \mu}\frac{\mathrm{d}\mu}{\mathrm{d}x} &= 0 \\ \frac{\partial f}{\partial x} - 1 + \frac{\partial f}{\partial \mu}\frac{\mathrm{d}\mu}{\mathrm{d}x} &= 0 \end{aligned} \tag{6.6.14}$$

where we have suppressed the arguments for the sake of typographical clarity. Following a transformation and considering eqs. (6.6.6) and (6.6.11), we obtain at the origin $(x, \mu) = (0, 0)$

$$\frac{\mathrm{d}\mu}{\mathrm{d}x}(0) = -\frac{\dfrac{\partial g}{\partial x}(0,0)}{\dfrac{\partial g}{\partial \mu}(0,0)} = -\frac{\dfrac{\partial f}{\partial x}(0,0) - 1}{\dfrac{\partial f}{\partial \mu}(0,0)} = 0$$

Differentiating eqs. (6.6.14) with respect to x and using eq. (6.6.11) for the second condition in eq. (6.6.13), we obtain the expression

$$\frac{\mathrm{d}^2\mu}{\mathrm{d}x^2}(0) = -\frac{\dfrac{\partial^2 g}{\partial x^2}(0,0)}{\dfrac{\partial g}{\partial \mu}(0,0)} = -\frac{\dfrac{\partial^2 f}{\partial x^2}(0,0)}{\dfrac{\partial f}{\partial \mu}(0,0)} \neq 0 \tag{6.6.15}$$

In order to satisfy eq. (6.6.13), the denominator $\frac{\partial^2 f}{\partial x^2}(0,0)$ must not vanish.

Summarising, we note that a one-parametric family of one-dimensional iterated maps which are two times continuously differentiable

$$x_{\mathrm{k}+1} = f(x_{\mathrm{k}}, \mu) \tag{6.6.16}$$

undergoes a saddle-node bifurcation in the origin $(x, \mu) = (0, 0)$ if the following conditions are satisfied:

$$f(0,0) = 0 \,, \quad \text{fixed point}$$

$$\frac{\partial f}{\partial x}(0,0) = 1 \,, \quad \text{non-hyperbolic} \tag{6.6.17}$$

$$\frac{\partial f}{\partial \mu}(0,0) \neq 0 \,, \quad \text{unique fixed point curve } \mu = \mu(x) \text{ through } (x,\mu) = (0,0) \tag{6.6.18}$$

$$\frac{\partial^2 f}{\partial x^2}(0,0) \neq 0 \,, \quad \text{fixed points on one side of } \mu = 0 \tag{6.6.19}$$

We note in addition that the sign of eq. (6.6.15) determines on which side of the line $\mu = 0$ the bifurcation diagram lies. If the signs of $\frac{\partial^2 f}{\partial x^2}(0,0)$ and $\frac{\partial f}{\partial \mu}(0,0)$ are opposite, i. e. if the curvature and thus $\frac{\mathrm{d}^2 \mu}{\mathrm{d}x^2}(0) > 0$, the bifurcation parabola lies in the positive half-plane $\mu > 0$, otherwise in the negative one. Both possibilities are shown in fig. 6.6.3.

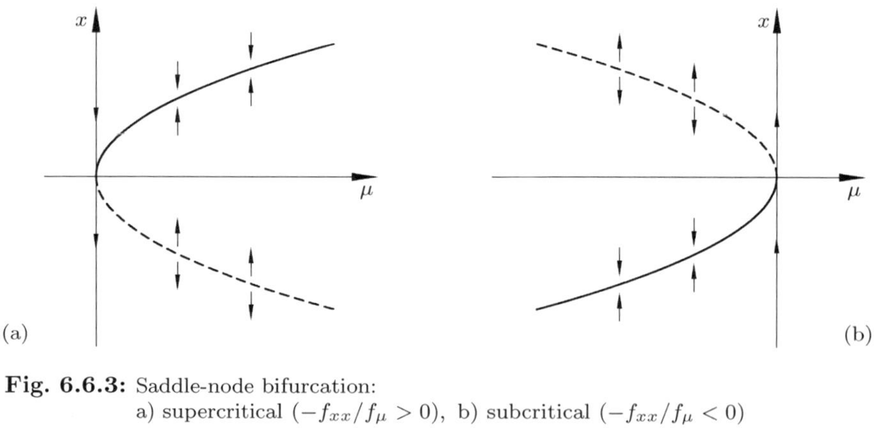

Fig. 6.6.3: Saddle-node bifurcation:
a) supercritical ($-f_{xx}/f_\mu > 0$), b) subcritical ($-f_{xx}/f_\mu < 0$)

The conditions for the saddle-node bifurcation of an iterated map, eqs. (6.6.17) to (6.6.19), agree with the conditions of the time-continuous flows, eqs. (6.4.21) to (6.4.23), with the exception of $\frac{\partial f}{\partial x}(0,0) = 1$. For positive μ, the upper fixed point branch in fig. 6.6.3a is stable due to the condition $\left.\frac{\partial f}{\partial x}\right|_{x_s,\mu_s} < 1$, the lower one is unstable. The normal form eq. (6.6.7) is obviously a supercritical saddle-node bifurcation. On the other hand, the expression

$$f(x, \mu) = x + \mu + x^2$$

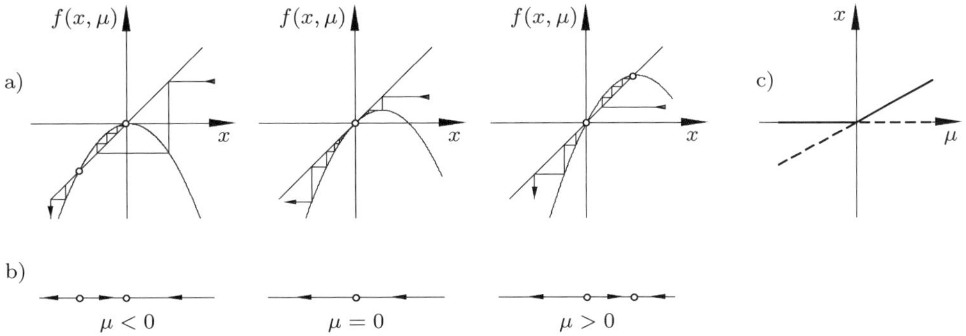

Fig. 6.6.4: Transcritical bifurcation of a one-dimensional map (eigenvalue 1):
a) iteration process and
b) stability behaviour in dependence of μ;
c) bifurcation diagram

is the normal form of a subcritical saddle-node bifurcation and is displayed in fig. 6.6.3b.

B) Transcritical bifurcation

We consider the map

$$x_{k+1} = f(x_k, \mu) \qquad \text{with} \qquad f(x,\mu) = x + \mu x - x^2 \tag{6.6.20}$$

The necessary conditions for a bifurcation are satisfied by eq. (6.6.20). The origin $(x,\mu) = (0,0)$ is a non-hyperbolic fixed point with the eigenvalue $+1$, i.e.

$$f(0,0) = 0 \tag{6.6.21}$$

$$\frac{\partial f}{\partial x}(0,0) = 1 \tag{6.6.22}$$

Let us first concentrate on the iterative mapping. Figure 6.6.4a shows the dependence of the function eq. (6.6.20) on the system parameter μ. Figure 6.6.4b illustrates the stability behaviour of the fixed points for $\mu < 0$, $\mu = 0$ and $\mu > 0$ respectively and is reflected in the iteration process. The bifurcation diagram, fig. 6.6.4c, is determined applying the fixed point condition $x = f(x,\mu)$ to eq. (6.6.20),

$$f(x,\mu) - x = \mu x - x^2 = 0 \tag{6.6.23}$$

The fixed points thus lie on the straight lines

$$x = 0 \qquad \text{and} \qquad \mu = x \tag{6.6.24}$$

which intersect in the origin $(x,\mu) = (0,0)$: a typical transcritical bifurcation. As in the case of the saddle-node bifurcation, we are now interested in the general conditions characterising this special type of bifurcation. The bifurcation diagram, fig. 6.6.4c, illustrates the typical behaviour: two fixed point curves intersect in the origin and exist – albeit with varying stability – on both sides of it.

We now consider the arbitrary one-dimensional iterated map which is two times continuously differentiable

$$x_{\mathrm{k}+1} = f(x_{\mathrm{k}}, \mu) \tag{6.6.25}$$

with the non-hyperbolic fixed point in the origin $(x, \mu) = (0, 0)$. For the fixed points of eq. (6.6.25), we have

$$f(x, \mu) - x = g(x, \mu) = 0 \tag{6.6.26}$$

Applying the condition $\frac{\partial g}{\partial \mu}(0, 0) \neq 0$, the theorem on implicit functions expresses that exactly one curve passes through the bifurcation point $(x, \mu) = (0, 0)$. However, since two curves intersect in the origin, we require that

$$\frac{\partial g}{\partial \mu}(0, 0) = \frac{\partial f}{\partial \mu}(0, 0) = 0 \tag{6.6.27}$$

In analogy to the transcritical bifurcation of time-continuous flows (section 6.4), we note that $x = 0$ is a fixed point solution. By factoring out x, we can express eq. (6.6.26) in the form

$$g(x, \mu) = xG(x, \mu) = x[F(x, \mu) - 1] \tag{6.6.28}$$

where, by definition, we have

$$\begin{aligned} x \neq 0: \qquad & G(x, \mu) = \frac{g(x, \mu)}{x} \\ x = 0: \qquad & G(x, \mu) = \frac{\partial g}{\partial x}(0, \mu) \end{aligned} \tag{6.6.29}$$

respectively

$$\begin{aligned} x \neq 0: \qquad & F(x, \mu) = \frac{f(x, \mu)}{x} \\ x = 0: \qquad & F(x, \mu) = \frac{\partial f}{\partial x}(0, \mu) \end{aligned} \tag{6.6.30}$$

We expect the fixed point branch which differs from $x = 0$ and passes through the origin to be unique and to exist on both sides of $\mu = 0$. Then, the relation

$$\frac{\partial G}{\partial \mu}(0, 0) = \frac{\partial F}{\partial \mu}(0, 0) \neq 0 \tag{6.6.31}$$

must hold. In conjunction with eq. (6.6.30), we finally deduce for eq. (6.6.31) an important condition for the second derivative

$$\frac{\partial^2 f}{\partial x \partial \mu}(0, 0) \neq 0 \tag{6.6.32}$$

The application of the theorem on implicit functions to $G(x, \mu)$ indicates that a unique function $\mu(x)$ exists for small x. The fixed point branch differing from $x = 0$ of eq. (6.6.28) can then be determined as follows

$$G\big(x, \mu(x)\big) = F\big(x, \mu(x)\big) - 1 = 0 \tag{6.6.33}$$

We know from the bifurcation diagram, fig. 6.6.4c, that the fixed point branch which differs from $x = 0$ has to fulfil

$$\frac{\mathrm{d}\mu}{\mathrm{d}x}(0) \neq 0 \tag{6.6.34}$$

Implicit differentiation of eq. (6.6.33) leads together with eq. (6.6.31) to

$$\frac{\mathrm{d}\mu}{\mathrm{d}x}(0) = -\frac{\dfrac{\partial G}{\partial x}(0,0)}{\dfrac{\partial G}{\partial \mu}(0,0)} = -\frac{\dfrac{\partial F}{\partial x}(0,0)}{\dfrac{\partial F}{\partial \mu}(0,0)} \tag{6.6.35}$$

Taking eq. (6.6.30) into account, we can express the condition eq. (6.6.34) as

$$\frac{\mathrm{d}\mu}{\mathrm{d}x}(0) = -\frac{1}{2}\frac{\dfrac{\partial^2 f}{\partial x^2}(0,0)}{\dfrac{\partial^2 f}{\partial x \partial \mu}(0,0)} \neq 0 \tag{6.6.36}$$

The conditions for a transcritical bifurcation can now be summarised on the basis of eqs. (6.6.21), (6.6.22), (6.6.27), (6.6.32) and (6.6.36)

$$\begin{aligned} &f(0,0) = 0\,, \qquad \frac{\partial f}{\partial x}(0,0) = 1\,, \qquad \frac{\partial f}{\partial \mu}(0,0) = 0 \\ &\frac{\partial^2 f}{\partial x^2}(0,0) \neq 0\,, \qquad \frac{\partial^2 f}{\partial x \partial \mu}(0,0) \neq 0 \end{aligned} \tag{6.6.37}$$

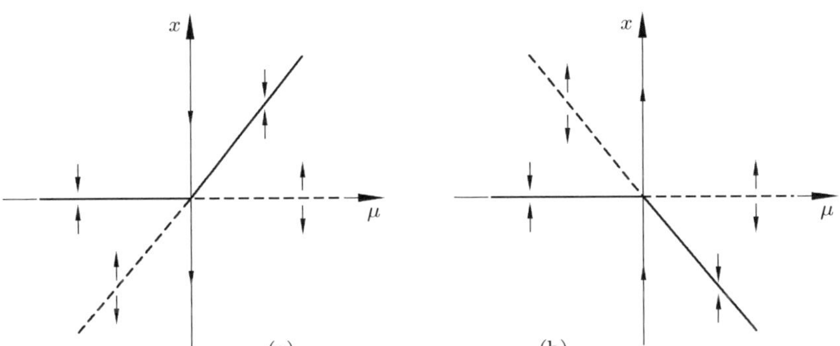

Fig. 6.6.5: Transcritical bifurcation of a one-dimensional map (eigenvalue 1):
a) supercritical ($-f_{xx}/f_{x\mu} > 0$)
b) subcritical ($-f_{xx}/f_{x\mu} < 0$)

The sign of eq. (6.6.36) decides on the gradient of the fixed point line differing from $x = 0$. Figure 6.6.5 presents the two alternatives of a positive and negative slope. For eq. (6.6.20) which satisfies the conditions of eq. (6.6.37) and for which $\frac{\mathrm{d}\mu}{\mathrm{d}x}(0) > 0$, the diagram a) illustrates the change of stability at the origin. Comparing fig. 6.6.5a and fig. 6.6.4c, we observe that eq. (6.6.20) is the normal form of a transcritical bifurcation, often designated supercritical. On the other hand, the normal form

$$f(x,\mu) = x + \mu + x^2$$

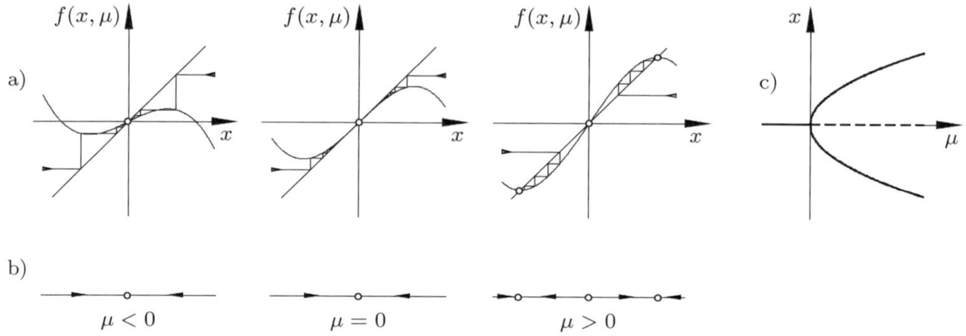

Fig. 6.6.6: Pitchfork bifurcation of a one-dimensional map (eigenvalue 1):
a) iteration process
b) stability behaviour in dependence of μ;
c) bifurcation diagram

leads to a transcritical bifurcation, for which $\frac{d\mu}{dx}(0) < 0$ (diagram b) and which is therefore also called subcritical.

C) Pitchfork bifurcation

We consider the mapping

$$x_{k+1} = f(x_k, \mu) \qquad \text{with} \quad f(x, \mu) = x + \mu x - x^3 \tag{6.6.38}$$

The origin $(x, \mu) = (0, 0)$ is a non-hyperbolic fixed point with eigenvalue $+1$ and thus a “likely candidate” for a bifurcation. The fixed point condition $x = f(x, \mu)$, incorporated in eq. (6.6.38), yields

$$f(x, \mu) - x = \mu x - x^3 = x(\mu - x^2) = 0 \tag{6.6.39}$$

This equation determines two fixed point curves passing through the origin

$$x = 0 \qquad \text{and} \qquad \mu = x^2$$

Figure 6.6.6 reproduces the graph of eq. (6.6.38) for three different μ-domains ($\mu < 0$, $\mu = 0$, $\mu > 0$) and displays the individual iteration steps, the corresponding stability behaviour in dependence of μ in the neighbourhood of the fixed points and the bifurcation diagram.

Next, we deduce the general conditions for one-dimensional maps which characterise a pitchfork bifurcation. The bifurcation diagram, fig. 6.6.6c, shows that the fixed point branch $x = 0$ exists for all μ-values; the fixed point parabola, on the other hand, is only real for positive μ-values.

Let the one-dimensional iterated map

$$x_{k+1} = f(x_k, \mu) \tag{6.6.40}$$

which is three times continuously differentiable, possess a non-hyperbolic fixed point in the origin $(x, \mu) = (0, 0)$, i.e. eqs. (6.6.21), (6.6.22) hold. The fixed points of eq. (6.6.40) are defined by

$$f(x,\mu) - x = g(x,\mu) = 0 \tag{6.6.41}$$

As in the case of the transcritical bifurcation, the fact that two fixed point curves intersect at $(0,0)$ leads to

$$\frac{\partial g}{\partial \mu}(0,0) = \frac{\partial f}{\partial \mu}(0,0) = 0 \tag{6.6.42}$$

As in eq. (6.6.28), we split off the fixed point solution $x = 0$

$$g(x,\mu) = xG(x,\mu) = x[F(x,\mu) - 1] \tag{6.6.43}$$

where eqs. (6.6.29) to (6.6.31) from the transcritical bifurcation can be adopted for $G(x,\mu)$ and $F(x,\mu)$. From eqs. (6.6.30) and (6.6.31), we again obtain the important condition for the derivative of the original function $f(x,\mu)$

$$\frac{\partial^2 f}{\partial x \partial \mu}(0,0) \neq 0 \tag{6.6.44}$$

The bifurcation diagram, fig. 6.6.6c, shows that the following conditions must be satisfied

$$\frac{\mathrm{d}\mu}{\mathrm{d}x}(0) = 0 \qquad \text{and} \qquad \frac{\mathrm{d}^2\mu}{\mathrm{d}x^2}(0) \neq 0 \tag{6.6.45}$$

The condition eq. (6.6.31) suggests the application of the theorem on implicit functions to G. For small x, a single curve $\mu(x)$ thus exists, defined by

$$G\big(x,\mu(x)\big) = F\big(x,\mu(x)\big) - 1 = 0 \tag{6.6.46}$$

After implicit differentiation and transformation, eq. (6.6.46) yields

$$\frac{\mathrm{d}\mu}{\mathrm{d}x}(0) = -\frac{\dfrac{\partial G}{\partial x}(0,0)}{\dfrac{\partial G}{\partial \mu}(0,0)} = -\frac{\dfrac{\partial F}{\partial x}(0,0)}{\dfrac{\partial F}{\partial \mu}(0,0)} \tag{6.6.47}$$

and

$$\frac{\mathrm{d}^2\mu}{\mathrm{d}x^2}(0) = -\frac{\dfrac{\partial^2 G}{\partial x^2}(0,0)}{\dfrac{\partial G}{\partial \mu}(0,0)} = -\frac{\dfrac{\partial^2 F}{\partial x^2}(0,0)}{\dfrac{\partial F}{\partial \mu}(0,0)} \tag{6.6.48}$$

Substituting eqs. (6.6.45), (6.6.30) into eqs. (6.6.47) and (6.6.48), we finally obtain

$$\frac{\mathrm{d}\mu}{\mathrm{d}x}(0) = -\frac{1}{2}\frac{\dfrac{\partial^2 f}{\partial x^2}(0,0)}{\dfrac{\partial^2 f}{\partial x \partial \mu}(0,0)} = 0 \tag{6.6.49}$$

$$\frac{\mathrm{d}^2\mu}{\mathrm{d}x^2}(0) = -\frac{1}{3}\frac{\dfrac{\partial^3 f}{\partial x^3}(0,0)}{\dfrac{\partial^2 f}{\partial x \partial \mu}(0,0)} \neq 0 \tag{6.6.50}$$

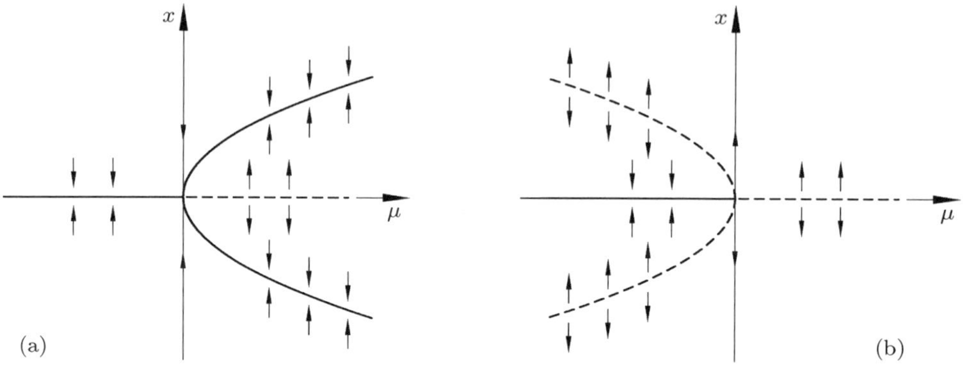

Fig. 6.6.7: Pitchfork bifurcation of a one-dimensional map (eigenvalue 1): a) supercritical ($-f_{xxx}/f_{x\mu} > 0$), b) subcritical ($-f_{xxx}/f_{x\mu} < 0$)

We now summarise the conditions for one-dimensional three times continuously differentiable maps which, if satisfied, induce a pitchfork bifurcation at the origin $(x, \mu) = (0, 0)$

$$\begin{aligned} &f(0,0) = 0\,, && \frac{\partial f}{\partial x}(0,0) = 1\,, && \frac{\partial f}{\partial \mu}(0,0) = 0 \\ &\frac{\partial^2 f}{\partial x^2}(0,0) = 0\,, && \frac{\partial^2 f}{\partial x \partial \mu}(0,0) \neq 0\,, && \frac{\partial^3 f}{\partial x^3}(0,0) \neq 0 \end{aligned} \tag{6.6.51}$$

Figure 6.6.7 presents the two parabolae of the fixed point solutions depending on the sign of eq. (6.6.50). Thus, in the case of a positive curvature, i.e. $\frac{\mathrm{d}^2\mu}{\mathrm{d}x^2}(0) > 0$, the bifurcation parabola exists for $\mu \geqslant 0$, diagram a), otherwise the one of diagram b) for $\mu \leqslant 0$. Equation (6.6.38) fulfils the conditions of eq. (6.6.51) with $\frac{\mathrm{d}^2\mu}{\mathrm{d}x^2}(0) > 0$. An investigation of the stability of the fixed point solutions results in the response shown in diagram a). Equation (6.6.38) is hence the normal form of a supercritical pitchfork bifurcation. The normal form of the subcritical pitchfork bifurcation, shown in diagram b), is

$$f(x, \mu) = x + \mu x + x^3$$

Case 2: *Single eigenvalue* -1: *period doubling or flip bifurcation*

The fact that an eigenvalue of the linearised map possesses the value -1 for a fixed point leads to fundamental consequences. As detailed before, the $+1$-eigenvalues lead to types of bifurcation analogous to those of time-continuous flows. On the other hand, the emergence of bifurcations of non-hyperbolic fixed points with the eigenvalue -1 is a characteristic specific to iterated maps and is not met with in the dynamics of one-dimensional vector fields. An example best illustrates what is novel about this.

We consider the following one-parametric family of one-dimensional maps

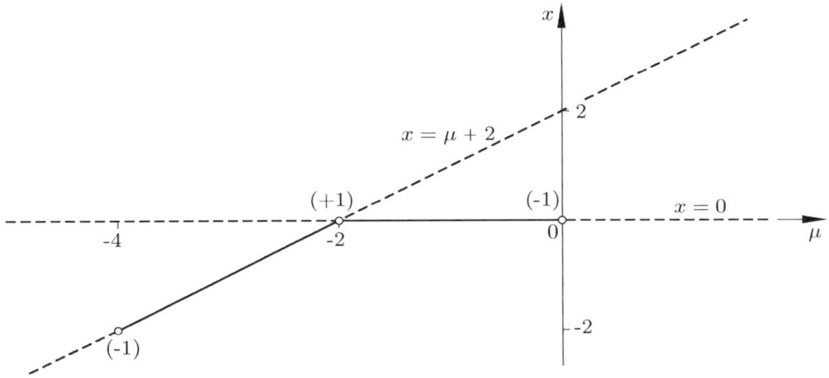

Fig. 6.6.8: Bifurcation diagram of the map $x_{k+1} = -(1+\mu)x_k + x_k^2$; the figures in parentheses apply to the eigenvalues of the respective non-hyperbolic fixed points $f_x(-2,-4) = -1$, $f_x(0,-2) = 1$, $f_x(0,0) = -1$

$$x_{k+1} = f(x_k, \mu) \qquad \text{with} \quad f(x,\mu) = -x - \mu x + x^2 \tag{6.6.52}$$

We immediately recognise that the non-hyperbolic fixed point $(x,\mu) = (0,0)$ of eq. (6.6.52) possesses the eigenvalue -1; we have

$$\begin{aligned} f(0,0) &= 0 \\ \frac{\partial f}{\partial x}(0,0) &= -1 \end{aligned} \tag{6.6.53}$$

We now determine the fixed points for eq. (6.6.52)

$$f(x,\mu) - x = x[x - (2+\mu)] = 0 \tag{6.6.54}$$

Equation (6.6.52) generates two straight lines of fixed points

$$x = 0 \qquad \text{and} \qquad x = 2 + \mu \tag{6.6.55}$$

A stability analysis of the linearised mapping reveals the following behaviour

$$x = 0 \qquad \text{is for} \begin{cases} \mu < -2 & \text{unstable} \\ \mu = -2 & \text{semi-stable (saddle-node)} \\ -2 < \mu \leqslant 0 & \text{stable} \\ \mu > 0 & \text{unstable} \end{cases}$$

and

$$x = 2 + \mu \text{ is for} \begin{cases} \mu < -4 & \text{unstable} \\ -4 \leqslant \mu < -2 & \text{stable} \\ \mu = -2 & \text{semi-stable (saddle-node)} \\ \mu > -2 & \text{unstable} \end{cases}$$

Figure 6.6.8 reproduces the two fixed point lines and their stability behaviour. We can clearly recognise the transcritical bifurcation for $(x,\mu) = (0,-2)$ and, what is

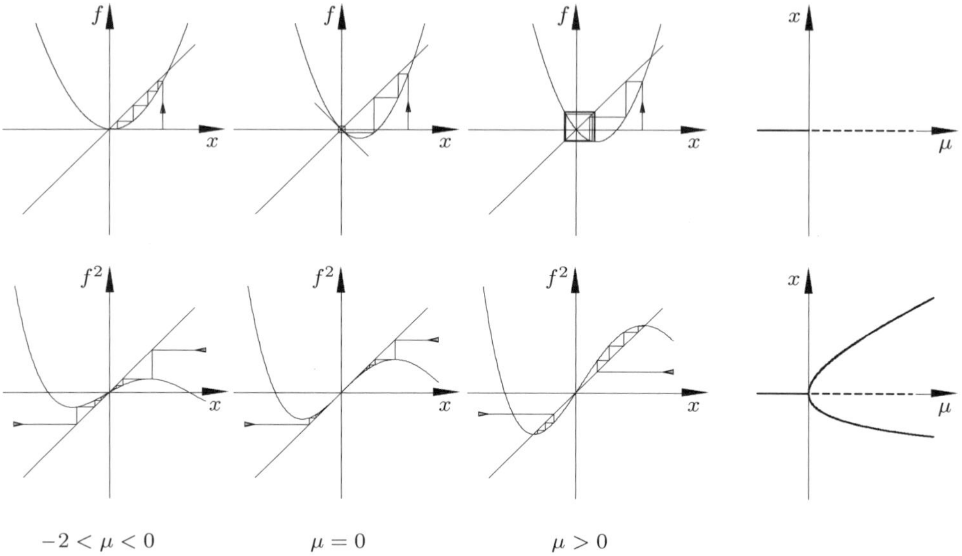

Fig. 6.6.9: Period doubling of a non-hyperbolic fixed point $(x_s, \mu_s) = (0, 0)$ (eigenvalue -1)

most important, all existing fixed points for $\mu > 0$ (in our case two) are unstable, a phenomenon which is not experienced for time-continuous flows.

What is new is the fact that, while the non-hyperbolic fixed point $(x_s, \mu_s) = (0, 0)$ with the eigenvalue -1 experiences a change in its stability behaviour, the otherwise typical phenomenon of a fixed point bifurcation does not occur. If we pursue the iteration process of the first iterate of the map $f(x, \mu)$ for $\mu > 0$, fig. 6.6.9, it becomes clear that the fixed point $x_s = 0$ is indeed unstable, but the functional values of the individual iterations remain bounded and oscillate. If we pick out every second value of the iteration series, it is apparent that this series will converge to a limit x_1^*, the intermediate values on the other hand to $x_2^* = f(x_1^*)$. In this case, $f(x_1^*) \neq x_1^*$, i. e. x_1^* (respectively x_2^*) is a periodic point of period 2. For obvious reasons, this bifurcation type is called period doubling.

The easiest way of checking whether our argumentation is correct is by inspecting our example. To this end, we form the second iteration $f^2(x, \mu)$ of eq. (6.6.52)

$$x_{\mathrm{k}+2} = f^2(x_{\mathrm{k}}, \mu)$$

with

$$f^2(x, \mu) = (1 + \mu)^2 x + \mu(1 + \mu)x^2 - 2(1 + \mu)x^3 + x^4 \qquad (6.6.56)$$

In the case of the second iterate $f^2(x, \mu)$ of the map, the origin $(x, \mu) = (0, 0)$ is a non-hyperbolic fixed point, but now has the eigenvalue $+1$

$$f^2(0, 0) = 0$$

$$\frac{\partial f^2}{\partial x}(0, 0) = 1$$

We expect a pitchfork bifurcation for the second iteration $f^2(x,\mu)$ and check this, applying the conditions deduced in eq. (6.6.51). We find

$$\frac{\partial f^2}{\partial \mu}(0,0) = 0\,, \qquad \frac{\partial^2 f^2}{\partial x^2}(0,0) = 0\,, \qquad \frac{\partial^2 f^2}{\partial x \partial \mu}(0,0) = 2$$

$$\frac{\partial^3 f^2}{\partial x^3}(0,0) = -12 \tag{6.6.57}$$

Analysing the $f^2(x,\mu)$-function with respect to its fixed point behaviour in dependence of μ reveals the typical phenomenon of the supercritical pitchfork bifurcation of the origin $(x,\mu) = (0,0)$. Figure 6.6.9 presents an overview of our findings, showing the f-and f^2-graphs, the iteration process and the corresponding bifurcation diagrams. It becomes apparent that the fixed points of $f^2(x,\mu)$ which emerge for $\mu > 0$ do not represent fixed points of $f(x,\mu)$; therefore, they must be points of period 2 of $f(x,\mu)$. This indicates that the initial map $f(x,\mu)$ bifurcates into a period doubling at $(x,\mu) = (0,0)$.

Summarising, we note the following: a period doubling occurs whenever a non-hyperbolic fixed point with eigenvalue -1 emerges and the second iterated map bifurcates as a pitchfork bifurcation. This occurs if, for a one-parametric family of one-dimensional three times continuously differentiable maps, the following conditions are satisfied

$$\begin{aligned}
&f(0,0) = 0 \\
&\frac{\partial f}{\partial x}(0,0) = -1 \quad \Rightarrow \quad \frac{\partial f^2}{\partial x}(0,0) = +1 \\
&\qquad\qquad \frac{\partial f^2}{\partial \mu}(0,0) = 0\,, \qquad \frac{\partial^2 f^2}{\partial x^2}(0,0) = 0 \\
&\qquad\qquad \frac{\partial^2 f^2}{\partial x \partial \mu}(0,0) \neq 0\,, \qquad \frac{\partial^3 f^2}{\partial x^3}(0,0) \neq 0
\end{aligned} \tag{6.6.58}$$

Figure 6.6.9 illustrates the fact that a stable orbit of period 2 for $\mu > 0$ can only emerge from a fixed point which is stable for $\mu < 0$ if $\frac{\mathrm{d}^2\mu}{\mathrm{d}x^2}\big|_0 > 0$ respectively $\frac{\partial^3 f^2}{\partial x^3}\big|_0 < 0$ holds, i. e. if the pitchfork bifurcation of the iterated map is supercritical.

We end this section on single real eigenvalues $+1$ or -1 with a small example which demonstrates two types of bifurcation for different μ-values. We consider the map

$$x_{\mathrm{k}+1} = f(x_{\mathrm{k}},\mu) \qquad \text{with} \quad f(x,\mu) = (\mu - \tfrac{1}{4}) - x^2 \tag{6.6.59}$$

The fixed points are given by

$$f(x,\mu) - x = (\mu - \tfrac{1}{4}) - x^2 - x = 0$$

respectively

$$x_{1,2} = -\tfrac{1}{2} \pm \sqrt{\mu} \tag{6.6.60}$$

No real fixed point exists for $\mu < 0$, while there are two for $\mu > 0$. This indicates a saddle-node bifurcation of the point $(x,\mu) = (-\frac{1}{2}, 0)$. The first requirement for this case is a non-hyperbolic character of the fixed point $(-\frac{1}{2}, 0)$ with an eigenvalue $+1$; this can be very easily verified for eq. (6.6.59)

$$\frac{\partial f}{\partial x}(-\tfrac{1}{2}, 0) = 1 \tag{6.6.61}$$

Moreover, the conditions of eqs. (6.6.18) and (6.6.19) must be satisfied. We obtain

$$\frac{\partial f}{\partial \mu}(-\tfrac{1}{2}, 0) = 1 \neq 0 \qquad \text{and} \qquad \frac{\partial^2 f}{\partial x^2}(-\tfrac{1}{2}, 0) = -2 \neq 0$$

All the conditions for a saddle-node bifurcation are thus fulfilled. However, the discussion of the bifurcation behaviour is still not complete. At $\mu = 1$, the eigenvalue -1 emerges at the fixed point $x = 1/2$; this means that the fixed point branch which is stable for $0 < \mu < 1$ becomes unstable at $\mu = 1$. This is the first indication of a period doubling, especially since no further fixed point solution appears for $\mu > 1$. The conditions for a period doubling are summarised in eq. (6.6.58). For the example of eq. (6.6.59), the second iteration of the map is

$$f^2(x, \mu) = (\mu - \tfrac{1}{4})(\tfrac{5}{4} - \mu + 2x^2) - x^4$$

We examine $f^2(x, \mu)$ with regard to period doubling on the basis of the conditions in eq. (6.6.58)

$$\begin{aligned}
&f(\tfrac{1}{2}, 1) - \tfrac{1}{2} = 0 \\
&\frac{\partial f}{\partial x}(\tfrac{1}{2}, 1) = -1 \quad \Rightarrow \quad \frac{\partial f^2}{\partial x}(\tfrac{1}{2}, 1) = 1 \\
&\qquad\qquad\qquad\qquad \frac{\partial f^2}{\partial \mu}(\tfrac{1}{2}, 1) = 0\ , \qquad \frac{\partial^2 f^2}{\partial x \partial \mu}(\tfrac{1}{2}, 1) = 2 \neq 0 \\
&\qquad\qquad\qquad\qquad \frac{\partial^2 f^2}{\partial x^2}(\tfrac{1}{2}, 1) = 0\ , \qquad \frac{\partial^3 f^2}{\partial x^3}(\tfrac{1}{2}, 1) = -12 \neq 0
\end{aligned} \tag{6.6.62}$$

The conditions for a period doubling of the fixed point $(x, \mu) = (1/2, 1)$ are thus satisfied. Equation (6.6.59) is an example illustrating that a function can experience completely different bifurcation types for various μ_{cr}-values. Figure 6.6.10 summarises our discussion.

In the following section 6.7, we again discuss period doublings and describe the mechanisms which lead to a cascade of infinitely many period doublings and finally to chaotic motions.

Case 3: *Single pair of complex conjugate eigenvalues with modulus* 1: *Neĭmark-Sacker bifurcation*

We conclude this section on the bifurcation theory of maps with a discussion of the bifurcation due to a single pair of complex conjugate eigenvalues with modulus 1 ($|\lambda| = 1$). This type of bifurcation is often called the "Hopf bifurcation for maps", although the relevant bifurcation theorem was first established by Neĭmark (1959) and Sacker (1965) independently of one another. For this reason, we shall denote it the Neĭmark-Sacker bifurcation.

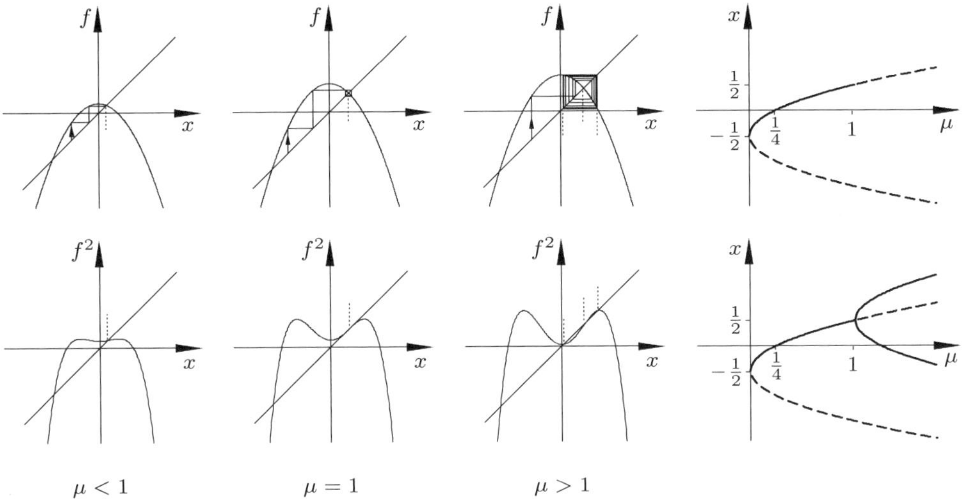

Fig. 6.6.10: Saddle-node bifurcation and period doubling of the mapping $x_{k+1} = (\mu - \frac{1}{4}) - x_k^2$

We assume that the dynamics on the two-dimensional centre manifold in the neighbourhood of the bifurcation point completely reproduces the dynamics of the general eq. (6.6.1). In this case, it suffices to discuss the mapping rule reduced to two dimensions in a one-parametric form

$$\boldsymbol{x}_{k+1} = \boldsymbol{f}(\boldsymbol{x}_k, \mu), \qquad \boldsymbol{x} \in \mathbb{R}^2, \ \mu \in \mathbb{R}^1 \tag{6.6.63}$$

If there is more than one control parameter, the system behaviour can be studied by assigning fixed values to all parameters but one. As in all earlier descriptions of one-dimensional generic bifurcations, we shift the bifurcation point to the origin $(\boldsymbol{x}, \mu) = (\boldsymbol{o}, 0)$ so that the condition

$$\boldsymbol{f}(\boldsymbol{o}, 0) = \boldsymbol{o} \tag{6.6.64}$$

is satisfied for the non-hyperbolic fixed point. Here, the Jacobian matrix $\boldsymbol{D}$

$$\boldsymbol{D}(\boldsymbol{o}, 0) = \frac{\partial \boldsymbol{f}}{\partial \boldsymbol{x}}(\boldsymbol{o}, 0) \tag{6.6.65}$$

possesses two complex conjugate eigenvalues $\lambda(0)$, $\lambda^*(0)$ with modulus

$$|\lambda(0)| = 1 \tag{6.6.66}$$

In addition, for the eigenvalues, we assume the relations

$$\lambda^k(0) \neq 1 \quad , \quad k \leqslant 4 \tag{6.6.67}$$

The special cases which are excluded are denoted *strong resonances*. For $k = 1$ and $k = 2$, two pairs of collapsing real eigenvalues result, $\lambda_{1,2} = 1$ and $\lambda_{1,2} = -1$ respectively. For $k = 3$ and $k = 4$, we obtain two complex conjugate eigenvalues, $\lambda_{1,2} = \frac{1}{2}(-1 \pm i\sqrt{3})$ and $\lambda_{1,2} = \pm i$ respectively. In both these cases, the normal forms of the bifurcations possess a special structure discussed in (Arnold, 1988).

The derivation of the normal form for the case of two complex conjugate eigenvalues λ and λ^* is eased considerably if we change to complex variables $z = x + iy$, $z^* = x - iy$ applying a linear coordinate transformation. The complex representation has the advantage of reducing the dynamics in the case of complex conjugate eigenvalues to *one* equation in the complex variable z, i. e. the map is expressed as

$$z_{\mathrm{n}+1} = f(z_{\mathrm{n}}, \mu) , \qquad z \in \mathbb{C}, \;\; \mu \in \mathbb{R}^1 \tag{6.6.68}$$

The technique of normal forms which we described in detail in section 6.3 for continuous systems can also be extended to maps. To this end, the mapping rule $f(z, \mu)$ of eq. (6.6.68) is expanded into a Taylor series and a successive elimination of higher-order terms is attempted by applying non-linear coordinate transformations. As expected (cf. section 6.3, remark b) *ii.*), all even-order terms can be eliminated and we obtain the following normal form, cf. (Arnold, 1988; Wiggins, 1990)

$$f(z, \mu) = \lambda(\mu) z + c(\mu) z^2 z^* + \mathcal{O}|z|^5 \tag{6.6.69}$$

A transformation of eq. (6.6.69) to polar coordinates facilitates the discussion of this normal form. We have

$$z = r\mathrm{e}^{2\pi i\theta}, \;\; z^2 = r^2\mathrm{e}^{4\pi i\theta}, \;\; z^* = r\mathrm{e}^{-2\pi i\theta} \tag{6.6.70}$$

Noting eq. (6.6.70), we obtain for $f(z, \mu)$, eq. (6.6.69), a new expression in the variables r, θ, μ which is once more denoted as f

$$f(r, \theta, \mu) = r\mathrm{e}^{2\pi i\theta}[\lambda(\mu) + c(\mu) r^2] + \mathcal{O}(r^5) \tag{6.6.71}$$

Let us recall that, apart from z, $\lambda(\mu)$ and $c(\mu)$ are also complex quantities

$$\begin{aligned} \lambda(\mu) &= \lambda_r(\mu) + i\lambda_i(\mu) \\ c(\mu) &= \alpha(\mu) + i\omega(\mu) \end{aligned} \tag{6.6.72}$$

We split eq. (6.6.71) into a mapping rule $r \to f_1(r, \mu)$ for the length r of the radius vector and a mapping rule for $\theta \to f_2(r, \theta, \mu)$ for the phase angle θ. The complex quantities $\lambda(\mu)$ and $c(\mu)$ provide a contribution to the length r, and we obtain for the mapping $f_1(r, \mu)$

$$\begin{aligned} f_1(r, \mu) &= r|\lambda(\mu) + r^2 c(\mu) + \mathcal{O}(r^4)| \\ &= r|\lambda(\mu)|\left|1 + r^2\frac{c(\mu)}{\lambda(\mu)} + \mathcal{O}(r^4)\right| \\ &= r|\lambda(\mu)|\left[1 + 2r^2\mathrm{Re}\frac{c(\mu)}{\lambda(\mu)} + \mathcal{O}(r^4)\right]^{1/2} \end{aligned}$$

Expanding the radicand into a series in powers of r, we obtain for the length of the radius vector

$$f_1(r, \mu) = r|\lambda(\mu)|\left[1 + r^2\mathrm{Re}\frac{c(\mu)}{\lambda(\mu)} + \mathcal{O}(r^4)\right] \tag{6.6.73}$$

In order to determine the mapping expression $f_2(r,\theta,\mu)$ for the phase angle θ, we transform eq. (6.6.71) and obtain

$$f(r,\theta,\mu) = r\mathrm{e}^{2\pi i\theta}\lambda(\mu)\Big[1 + r^2\frac{c(\mu)}{\lambda(\mu)}\Big] + \mathcal{O}(r^5) \tag{6.6.74}$$

respectively

$$f(r,\theta,\mu) = r\mathrm{e}^{2\pi i\theta}|\lambda(\mu)|\,\mathrm{e}^{2\pi i\phi(\mu)}\left|1 + r^2\frac{c(\mu)}{\lambda(\mu)}\right|\mathrm{e}^{2\pi i\varphi(r,\mu)} + \mathcal{O}(r^5) \tag{6.6.75}$$

If we consider only the phase angle in eq. (6.6.75), we find

$$f_2(r,\theta,\mu) = \theta + \phi(\mu) + \varphi(r,\mu) + \mathcal{O}(r^4) \tag{6.6.76}$$

where

$$\phi(\mu) = \frac{1}{2\pi}\tan^{-1}\frac{\lambda_i}{\lambda_r}$$

Due to eq. (6.6.75), the following expression holds for the phase angle $\varphi(r,\mu)$ in eq. (6.6.76)

$$\tan 2\pi\varphi(r,\mu) = \frac{r^2\mathrm{Im}\dfrac{c(\mu)}{\lambda(\mu)}}{1 + r^2\mathrm{Re}\dfrac{c(\mu)}{\lambda(\mu)}} = r^2\mathrm{Im}\frac{c(\mu)}{\lambda(\mu)} + \mathcal{O}(r^4)$$

Expanding $\tan(2\pi\varphi)$ into a power series and solving this equation with respect to $\varphi(r,\mu)$ yields

$$\varphi(r,\mu) = \frac{1}{2\pi}r^2\mathrm{Im}\frac{c(\mu)}{\lambda(\mu)} + \mathcal{O}(r^4)$$

Inserting this expression for $\varphi(r,\mu)$ into eq. (6.6.76), we find

$$f_2(r,\theta,\mu) = \theta + \phi(\mu) + \frac{1}{2\pi}r^2\mathrm{Im}\frac{c(\mu)}{\lambda(\mu)} + \mathcal{O}(r^4) \tag{6.6.77}$$

We now expand the coefficients of eqs. (6.6.73) and (6.6.77) into Taylor series in the neighbourhood of the bifurcation point $\mu = 0$ and obtain, noting that $|\lambda(0)| = 1$,

$$\begin{aligned} f_1(r,\mu) &= r + \frac{\mathrm{d}}{\mathrm{d}\mu}|\lambda(\mu)|\Big|_{\mu=0}\mu r + r^3\mathrm{Re}\frac{c(0)}{\lambda(0)} + \mathcal{O}(\mu^2 r, \mu r^3, r^5) \\ f_2(r,\theta,\mu) &= \theta + \phi(0) + \frac{\mathrm{d}}{\mathrm{d}\mu}\phi(\mu)\Big|_{\mu=0}\mu + \frac{1}{2\pi}r^2\mathrm{Im}\frac{c(0)}{\lambda(0)} + \mathcal{O}(\mu^2, \mu r^2, r^4) \end{aligned} \tag{6.6.78}$$

In order to simplify eq. (6.6.78), we introduce the following notations

$$
\begin{aligned}
d &\equiv \frac{\mathrm{d}}{\mathrm{d}\mu}|\lambda(\mu)|\Big|_{\mu=0} \\
a &\equiv \mathrm{Re}\frac{c(0)}{\lambda(0)} \\
\phi_0 &\equiv \phi(0) \\
\phi_1 &\equiv \frac{\mathrm{d}}{\mathrm{d}\mu}\phi(\mu)\Big|_{\mu=0} \\
b &\equiv \frac{1}{2\pi}\mathrm{Im}\frac{c(0)}{\lambda(0)}
\end{aligned}
$$

In this way, we finally rewrite eqs. (6.6.78) as

$$
\begin{aligned}
f_1(r,\mu) &= (1+d\mu)r + ar^3 + \mathcal{O}(\mu^2 r, \mu r^3, r^5) \\
f_2(r,\theta,\mu) &= \theta + \phi_0 + \phi_1\mu + br^2 + \mathcal{O}(\mu^2, \mu r^2, r^4)
\end{aligned}
\tag{6.6.79}
$$

These mapping equations are very similar to those of the Hopf bifurcation of continuous systems, eq. (6.4.59). In order to study the bifurcation dynamics of eqs. (6.6.79), we follow a strategy analogous to that presented in section 6.4 for the Hopf bifurcation of vector fields. We first consider the dynamics of the normal form without the higher-order terms and then investigate the influence of the higher-order terms not included in the familiar dynamics of the "truncated" normal form.

The normal form without the higher-order terms is

$$
\begin{aligned}
f_1(r,\mu) &= (1+d\mu)r + ar^3 \\
f_2(r,\theta,\mu) &= \theta + \phi_0 + \phi_1\mu + br^2
\end{aligned}
\tag{6.6.80}
$$

If we use the first expression of eq. (6.6.80) for the determination of the fixed points, $f(r,\mu) - r = 0$, we obtain the trivial fixed point solution $r = 0$. The stability of this solution can be verified easily as

asymptotically stable for $d\mu < 0$,
unstable for $d\mu > 0$,
unstable for $\mu = 0$, $a > 0$ and
asymptotically stable for $\mu = 0$, $a < 0$.

The non-trivial fixed point solution leads to the equation of a circle with radius $r = (-d\mu/a)^{1/2}$. The geometrical interpretation corresponds to that of the Hopf bifurcation, whereas the dynamic interpretation differs significantly due to the fact that a circle in the map represents a torus in 3D. The circle is composed of a set of discrete points which is invariant, since initial conditions starting on the circle remain on the circle during the iteration process. The sign constellation of a and d determine for which μ-values a solution $r \neq 0$ exists. If solutions exist, we next enquire into their stability. If we observe that a sequence of mapping points not initiated on the circle converges towards the circle, it seems reasonable to speak of a *stable invariant circle* of the iterated map; mapping points which tend away, on the other hand, indicate an *unstable invariant circle*. This argument, applied to the first expression in eq. (6.6.80), shows that the parameter a controls the stability. We note that the invariant circle of the map is asymptotically stable for $a < 0$ and unstable for $a > 0$.

We now summarise the results regarding the characteristics of the fixed point and invariant circle as well as their stability and present them in bifurcation diagrams. For $-\infty < \mu < +\infty$, there are four cases, specified by the signs of d and a:

1. $d > 0\,,\; a < 0\,,$ fig. 6.6.11
2. $d < 0\,,\; a > 0\,,$ fig. 6.6.12
3. $d < 0\,,\; a < 0\,,$ fig. 6.6.13
4. $d > 0\,,\; a > 0\,,$ fig. 6.6.14

These four cases display the following behaviour:

ad 1 : $d > 0,\, a < 0$: for $\mu < 0$, the trivial solution $r = 0$ is asymptotically stable and unstable for $\mu > 0$. The non-trivial solution $r = (-d\mu/a)^{1/2}$ exists only for $\mu > 0$, and the invariant circle is asymptotically stable (fig. 6.6.11).

ad 2 : $d < 0,\, a > 0$: in this case, the origin is an unstable fixed point for $\mu < 0$. For $\mu > 0$, however, the origin is an asymptotically stable fixed point, enclosed by an unstable invariant circle (fig. 6.6.12).

ad 3 : $d < 0,\, a < 0$: for $\mu < 0$, the origin is an unstable fixed point and the existing invariant circle is asymptotically stable. The trivial solution $r = 0$ is asymptotically stable for $\mu > 0$ (fig. 6.6.13).

ad 4 : $d > 0,\, a > 0$: in this case, the origin is asymptotically stable for $\mu < 0$ and the existing invariant circle is unstable. For $\mu > 0$, the only fixed point solution in the origin is unstable (fig. 6.6.14).

Concluding remarks:

Having detailed and illustrated the significance of the change of signs of a and d in their mutual dependence, we investigate the meaning of each individual parameter. Let us first consider the change of sign of d alone, recalling that d is defined by

$$d = \frac{\mathrm{d}}{\mathrm{d}\mu}|\lambda(\mu)|\bigg|_{\mu=0}$$

The value d is a measure for the velocity with which the eigenvalues pass the unit circle at the zero crossing of μ. For $d > 0$, $|\lambda(\mu)|$ increases if μ changes from negative to positive values, i. e. the complex conjugate eigenvalues cross over the unit circle from the inside to the outside. For $d < 0$, the opposite occurs, namely a crossing of the unit circle from the outside to the inside. The eigenvalues ($|\lambda(\mu)| \lesseqgtr 1$) define the stability; it follows that for $d > 0$ at the zero crossing of μ, the stability of the origin changes from asymptotically stable to unstable. If $d < 0$, the stability change occurs the other way around. The parameter d appertains to the linear r-term and thus controls the stability switch of the fixed point and causes the bifurcation.

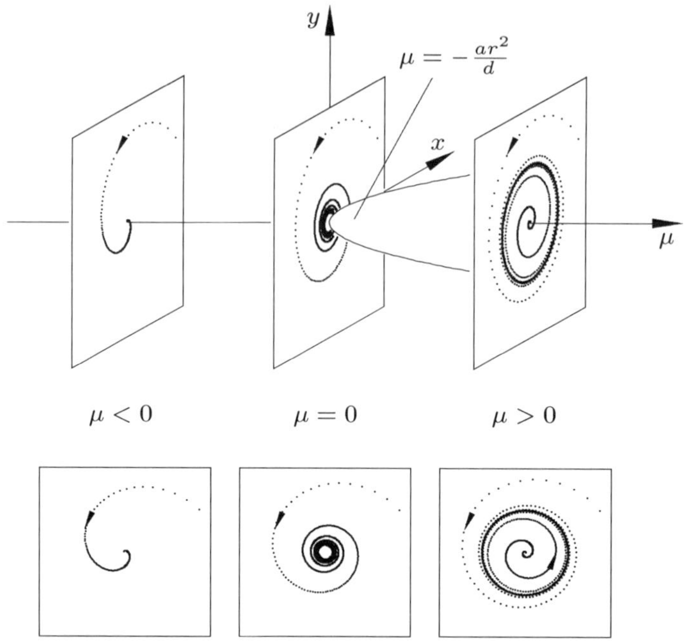

Fig. 6.6.11: Supercritically stable periodic bifurcation = supercritical Neĭmark-Sacker bifurcation ($d > 0$, $a < 0$)

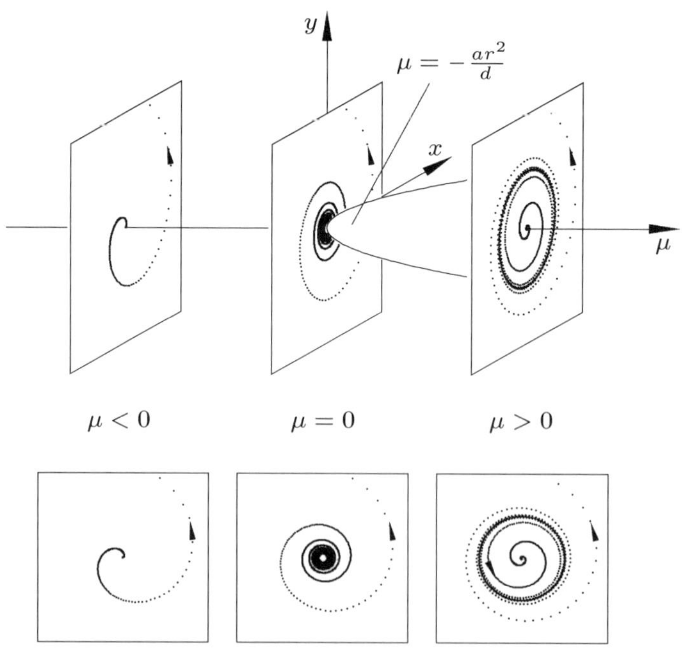

Fig. 6.6.12: Supercritically unstable periodic bifurcation ($d < 0$, $a > 0$)

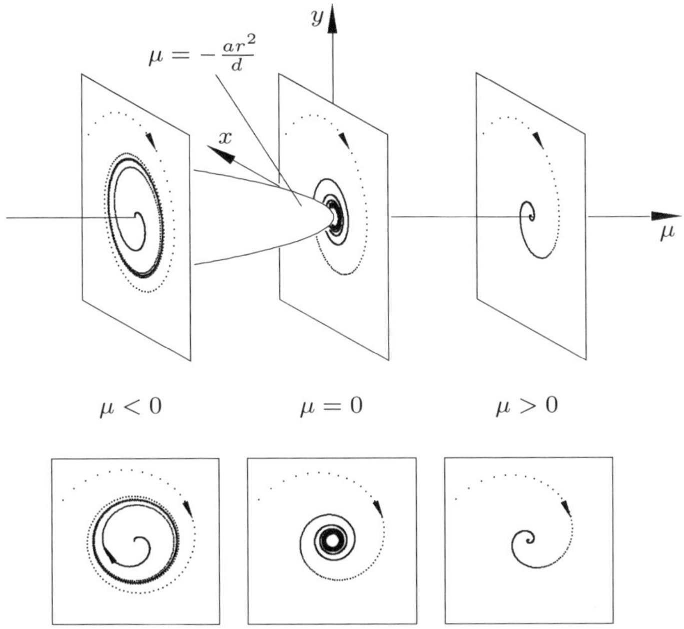

Fig. 6.6.13: Subcritically stable periodic bifurcation ($d < 0$, $a < 0$)

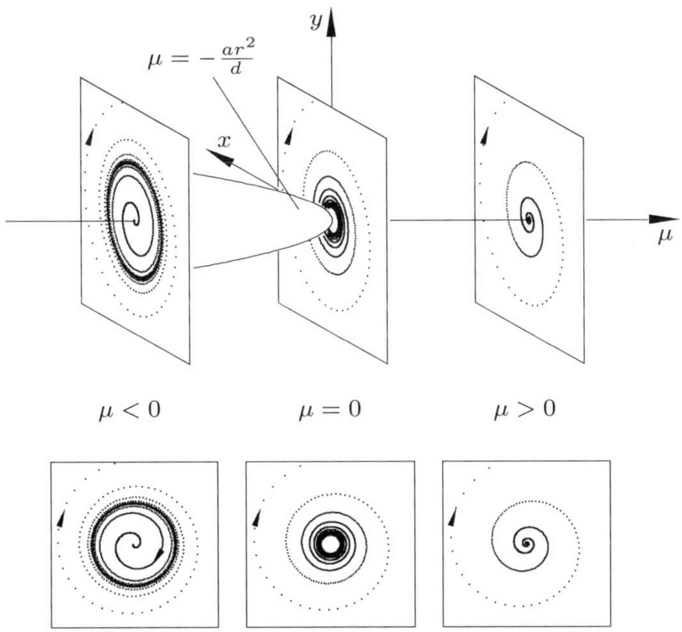

Fig. 6.6.14: Subcritically unstable periodic bifurcation = subcritical Neĭmark-Sacker bifurcation ($d > 0$, $a > 0$)

What influence does the parameter a exert? If $a > 0$, the existing invariant circles are unstable as illustrated in figs. 6.6.12 and 6.6.14. Correspondingly, they are asymptotically stable for $a < 0$ (figs. 6.6.11 and 6.6.13). The parameter a appertains to the cubic term of r and thus controls the stability behaviour of the bifurcating invariant circle.

Up to now, our discussion of the bifurcation due to a single pair of complex eigenvalues with modulus 1 was limited to the normal form with lowest-order terms. The question regarding the influence of the higher-order terms on the type of bifurcation still has to be answered. The theorem of Neĭmark and Sacker provides us with the answer, namely that the higher terms in the normal form eq. (6.6.79) as compared with the truncated normal form eq. (6.6.80) do not cause any significant topological change.

The reader will have noticed the astonishing similarity in the bifurcation diagrams of the Hopf bifurcation of continuous systems, figs. 6.4.7 to 6.4.10, and the Neĭmark-Sacker bifurcation, figs. 6.6.11 to 6.6.14. It would be erroneous, however, to conclude that the dynamics conforms to similarly simple laws in both situations. In the case of the Hopf bifurcation, as with all the other bifurcations of iterated maps we have discussed up to now, attractive sets are generated which form single orbits. In the present case, however, an invariant circle emerges in the Poincaré section; this circle is generated by different orbits. The dynamics of the initial system eq. (6.6.80) is nevertheless reflected in the dynamics of this invariant circle. It thus suffices to study the dynamics on the circle; this can be achieved by restricting the selection to initial conditions lying on the invariant circle. In our case, the radius of the invariant circle is given by

$$r = \sqrt{-\frac{d}{a}\mu}$$

If we take this as the fixed r-value in the expansion for the angle θ, we obtain in accordance with eq. (6.6.80) the associated linear circle map (cf. section 8.3.2)

$$\theta_{\mathrm{n}+1} = f(\theta_{\mathrm{n}}, \mu) \qquad \text{with} \quad f(\theta, \mu) = \theta + \phi_0 + \left(\phi_1 - \frac{bd}{a}\right)\mu \tag{6.6.81}$$

Equation (6.6.81) facilitates the discussion of the dynamics considerably as compared with eq. (6.6.80). The dynamics is shaped here by the expression $\Omega = \phi_0 + (\phi_1 - bd/a)\mu$. If Ω is rational, all the trajectories on the torus are closed curves (fig. 4.3.3), i. e. the motion is periodic. If Ω is irrational, on the other hand, the trajectories on the torus cannot close (fig. 4.3.4). In this case, the torus is completely covered by densely packed orbits. Variations in the parameter μ hence lead to a continual switch of the angle θ from rational to irrational and thus to alternating periodic and quasi-periodic motions.

In conclusion, we note that the motions on the torus described above only hold for the linear circle map eq. (6.6.81) which emanates from the "truncated" normal form, eq. (6.6.80). It is not so easy to answer the question as to which course the orbits on the torus take if higher-order terms are included, eq. (6.6.79). As far as this topic is concerned, we refer to the discussion on the non-linear circle map in section 8.3.2.

6.7 Renormalisation and Self-similarity with the Example of the Logistic Map

6.7.1 The Mechanism of Period Doublings ad Infinitum

In section 3.7, we discussed a particularly simple non-linear recursion relation, the logistic map,

$$\begin{aligned} x_{\mathrm{n}+1} &= f(x_{\mathrm{n}}) \\ f(x) &= \alpha x(1-x) \quad , \qquad 0 \leqslant x \leqslant 1,\ 0 < \alpha \leqslant 4 \end{aligned} \tag{6.7.1}$$

and saw that the character of the long-term behaviour of the sequence $\{x_{\mathrm{n}}\}$ is governed by the value of the control parameter α. For $0 < \alpha \leqslant 1$, the only fixed point $x_s = 0$ in the interval [0,1] is stable, i. e. each sequence tends towards the origin, irrespective of its initial value. For $\alpha > 1$, the origin becomes unstable and, within $0 \leqslant x \leqslant 1$, a new fixed point $x_s = 1 - 1/\alpha$ emerges which, in accordance with eq. (3.7.7), is stable in the region $1 < \alpha < \alpha_1$ ($\alpha_1 = 3$). Following a further increase of the control parameter, this state of equilibrium also loses its stability, a period doubling emerges and the sequence $\{x_{\mathrm{n}}\}$ oscillates periodically between two fixed values x_1^* and x_2^*. If α increases beyond the next critical value a second period doubling takes place. In ever decreasing intervals, further period doublings occur (Metropolis et al., 1973) until, from a limit value $\alpha_\infty = 3.5699\ldots$, the long-term behaviour finally becomes irregular, chaotic; this is, however, repeatedly interrupted by periodic windows (see fig. 3.7.6). Interestingly, as early as 1963, the Finn P.J. Myrberg calculated the accumulation point α_∞ with a precision of 10 digits – incredibly high at the time – on his "electron machine", as he called his computer (Myrberg, 1963). His work was cited in a pioneering article (May, 1976) by R.M. May, see also (Abraham et al., 1997).

Feigenbaum was the first to recognise that this transition from regular to chaotic behaviour is by no means limited to the special form of the logistic map (Feigenbaum, 1978). Indeed, for all recursion relations $x_{\mathrm{n}+1} = f(x_{\mathrm{n}})$, which – after a suitable normalisation – map the unit interval to itself and possess a single maximum in this region, the route to chaos leads via a cascade of period doublings, see also (Ausloos and Dirickx, 2006). In fact, we can observe this transition to irregular behaviour in many non-linearly coupled systems of differential equations and in many experiments (see Colour Plates XXIII, p. 712, and XXV, p. 751, for the Lorenz system and the Duffing equation respectively). If, for example, the Poincaré map of a system is approximately one-dimensional as a result of dissipation and displays only a single maximum, the conditions for this transition to chaos are already satisfied.

The interpretation of a non-monotonic function as a Poincaré map is only approximately permissible, for example in systems which display a strong dissipation. Were the trajectories to lie exactly on a two-dimensional surface, the Poincaré map would be strictly one-dimensional and would have to be one-to-one due to the determinacy of the system implying the non-intersection of the trajectories. One-dimensional one-to-one iteration rules can, however, only describe regular behaviour. This is, of course, in conformity with the fact that no chaotic motions are possible in the two-dimensional phase space.

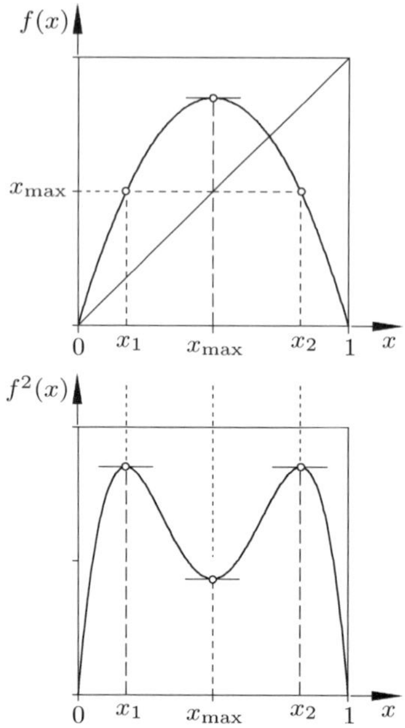

Fig. 6.7.1

Extreme values of $f(x)$ and the iterate $f^2(x)$

The real cause for the emergence of a complex long-term behaviour lies in the double-valuedness of the inverse map. Here, the velocity with which the period doublings follow one another, i. e. the limit value of the relative distances between successive critical values α_i

$$\lim_{\mathrm{k}\to\infty} \frac{\alpha_\mathrm{k} - \alpha_{\mathrm{k}-1}}{\alpha_{\mathrm{k}+1} - \alpha_\mathrm{k}}$$

depends only on the character of the maximum. For maps with a quadratic maximum (such as the logistic map), for example, the limit value is the Feigenbaum constant (Feigenbaum, 1978)

$$\delta = 4.669\,201\ldots$$

In this section, we want to reveal the background to this universal behaviour. A deciphering of the characteristics of self-similarity leads to fascinating scaling laws which enable us to make the universality quantitatively measurable. Ultimately, there are only some few characteristics which occur in the combination or iteration of certain functions – such as the logistic map – and which give rise to an infinite sequence of period doublings and to new cycles, the so-called "Feigenbaum route". We shall investigate these characteristics using the example of the logistic map; at the same time, we shall realise that the results are applicable to a whole class of functions.

In the following, we restrict ourselves to recursion relations which map the unit interval [0,1] to itself and possess their only maximum at the point x_max, i. e. $f'(x_\mathrm{max}) = 0$.

The points x_{cr} for which $f'(x_{cr}) = 0$ holds are also denoted *critical points* of the map $f(x)$. Possible critical points are maxima, minima and saddle points. The orbits of the critical points play an important role in the analysis of the dynamics of a recursion relation, cf. (Devaney, 1987; Abraham et al., 1997). Diffeomorphisms, i. e. bijective maps, which are continuously differentiable together with their inverse mapping, obviously possess no critical points and consequently also no complex long-term behaviour.

In the case of the logistic map, $x_{\max} = \frac{1}{2}$. Applying the chain rule to the iterated map f^2, we obtain

$$\frac{\mathrm{d}}{\mathrm{d}x} f^2(x)\bigg|_{x_{\max}} = f'\big(f(x_{\max})\big) f'(x_{\max}) = 0 \tag{6.7.2}$$

i. e. at the point $x_{\max}$, an extreme value emerges for all iterations. The iterate $f^2(x)$ possesses two further extreme values, exactly at the points $x_{1,2}$ for which $f(x_{1,2}) = x_{\max}$ holds (see fig. 6.7.1). In fact, in analogy to eq. (6.7.2), we have

$$\frac{\mathrm{d}}{\mathrm{d}x} f^2(x)\bigg|_{x_{1,2}} = f'\big(f(x_{1,2})\big) f'(x_{1,2}) = f'(x_{\max}) f'(x_{1,2}) = 0 \tag{6.7.3}$$

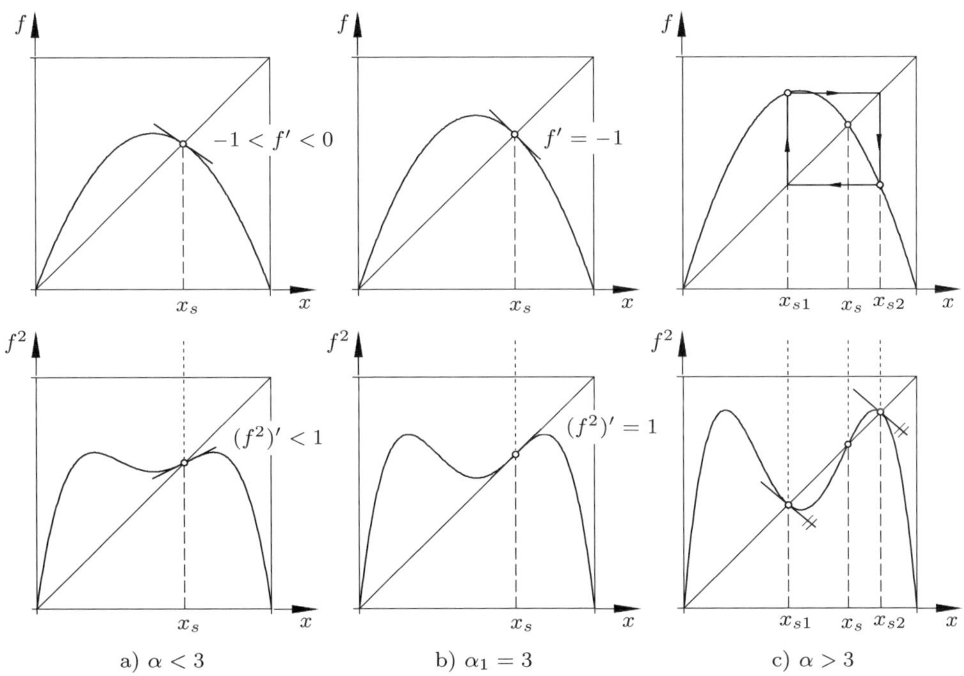

Fig. 6.7.2: Mechanism of period doubling at $\alpha_1 = 3$ for the logistic map
a) stable fixed point x_s of f and f^2
b) period doubling for x_s
c) 2-cycle of f between x_{s1} and x_{s2}

We next return to the stability of the fixed points of $f(x)$ and $f^2(x)$. The value x_s defines a stable fixed point of the map eq. (6.7.1) if

$$\begin{aligned} &x_s = \alpha x_s(1 - x_s) \\ \text{and} \\ &|f'(x_s)| = \alpha|1 - 2x_s| < 1 \end{aligned} \tag{6.7.4}$$

From this, we immediately conclude that, for $0 < \alpha < 1$, $x_s = 0$ is the only point attractor which is replaced by $x_s = 1 - 1/\alpha$ in the parameter range $1 < \alpha < 3$. We now obtain the slope of the function as

$$f'(x_s) = 2 - \alpha \tag{6.7.5}$$

For increasing α, the tangent at the fixed point becomes steeper until it reaches the value $f'(x_s) = -1$ for $\alpha_1 = 3$ (see fig. 6.7.2), i.e. for $\alpha_1 = 3$, the fixed point $x_s = 2/3$ becomes non-hyperbolic.

In the one-dimensional case, this value coincides with the eigenvalue of the linearised map. Using the conditions specified in eq. (6.6.58), we can easily verify that a period doubling occurs at this critical value α_1 for the logistic map: the originally stable fixed point x_s becomes unstable and in the iterated map $f^2(x)$ two new fixed points x_{s1} and x_{s2} emerge, for which the relations hold (see fig. 6.7.2c)

$$x_{s1} = f^2(x_{s1}) \,, \quad x_{s2} = f^2(x_{s2}) \tag{6.7.6}$$

If we again apply the mapping to the first relation, i.e. to $x_{s1} = f\big(f(x_{s1})\big)$, we obtain

$$f(x_{s1}) = f^2\big(f(x_{s1})\big) \tag{6.7.7}$$

i.e. $f(x_{s1})$ is also the fixed point of f^2 and must thus coincide with x_{s2} as there are only two fixed points. Thus, the function f, applied to one of the fixed points x_{s1} and x_{s2} of f^2, generates the other one

$$x_{s1} = f(x_{s2}) \,, \quad x_{s2} = f(x_{s1}) \tag{6.7.8}$$

Since each fixed point of f – irrespective of its stability characteristics – is also a fixed point of the iterated function f^2, $f^2(x)$ possesses four fixed points in all for $\alpha > \alpha_1$, two unstable ones and – in a certain parameter range – two new stable ones (see fig. 6.7.2c). Each iteration sequence $\{x_\mathrm{n}\}$ (with $x_0 \neq 0$, $x_0 \neq x_s$) then tends towards an attractor of period 2 of f, i.e. towards a 2-cycle (x_{s1}, x_{s2}).

We demonstrated the emergence of a period doubling in fig. 6.7.2, exemplifying it on the logistic map. If, in the case of an arbitrary mapping function, a fixed point x_s becomes unstable for a critical value α_{cr} of the control parameter, two alternative types of behaviour emerge for the iterated map $f^2(x)$; these are summarised in fig. 6.7.3. Obviously, only the possibility presented in fig. 6.7.3a generates a period doubling; in case b), x_s indeed becomes unstable, but no new fixed points are generated for increasing α. A necessary condition for the behaviour of $f^2(x)$ as shown in fig. 6.7.3a is that the so-called Schwarzian derivative

$$\mathrm{S}f = \frac{f'''}{f'} - \frac{3}{2}\left(\frac{f''}{f'}\right)^2 \tag{6.7.9}$$

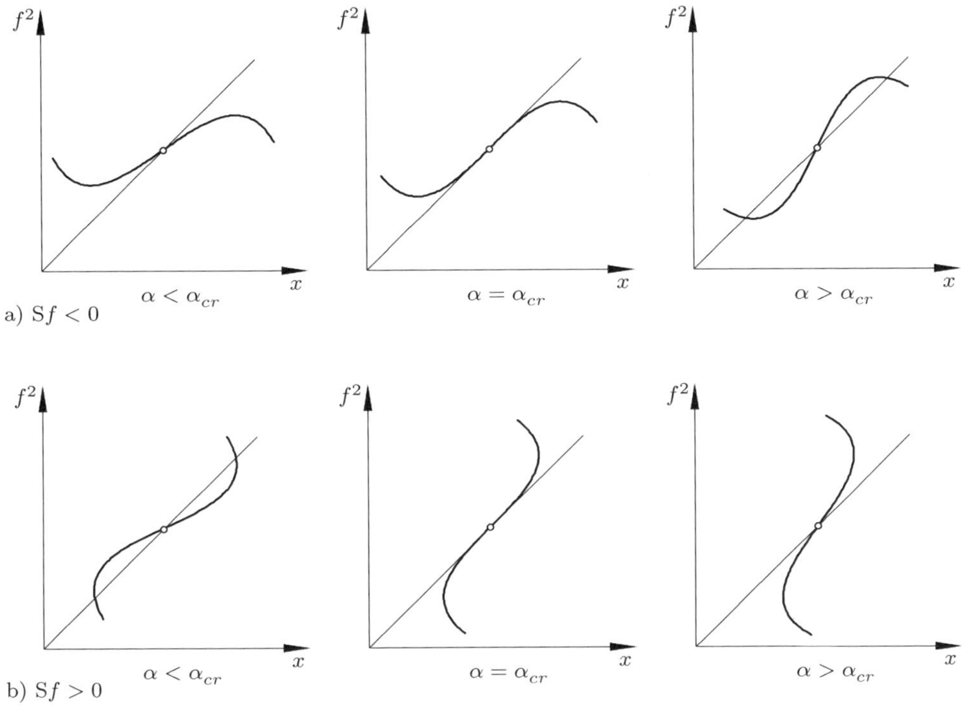

Fig. 6.7.3: Elucidation of the Schwarzian derivative

of the three times continuously differentiable function $f(x)$ in a neighbourhood of the bifurcating fixed point is negative, i. e. a necessary condition for the occurrence of a period doubling is $\mathrm{S}f < 0$ in the vicinity of the bifurcation point. The significance of the Schwarzian derivative for the dynamics of one-dimenisonal maps was first recognised by Singer (1978). By straightforward operations, we can verify that $\mathrm{S}f^{\mathrm{n}} < 0$ follows from $\mathrm{S}f < 0$ for all iterated functions, cf. (Misuriewicz, 1983).

In order to make the condition $\mathrm{S}f < 0$ plausible, we note that $f'(x_s) = -1$, $(f^2)'(x_s) = +1$ holds at the critical α-value and f^2 possesses an inflection point, i. e.}

$$(f^2)''(x_s) = 0 \tag{6.7.10}$$

Since, in fact, $\mathrm{S}f^2 < 0$ follows from $\mathrm{S}f < 0$, the condition $\mathrm{S}f^2 < 0$ is reduced to $(f^2)'''(x_s) < 0$ due to eq. (6.7.10). This exactly corresponds to the case presented in fig. 6.7.3a, however.

One might now suppose that it would suffice to assume that $f'(x)$ and $f'''(x)$ in a neighbourhood of x_s have opposite signs, i. e.

$$\frac{f'''}{f'} < 0$$

This condition would then no longer apply for the iterated function f^2, however. This is the reason why Singer introduced the – at first apparently involved – Schwarzian

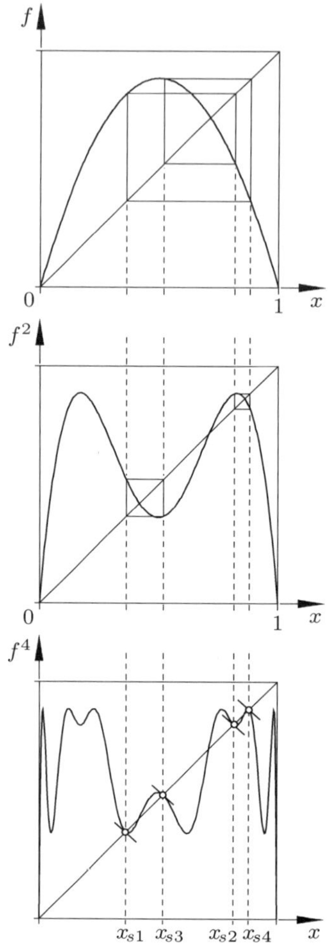

Fig. 6.7.4

Limit cycle of period 4 ($\alpha = 3.54$)

condition which also suffices to guarantee the emergence of new fixed points, which also holds for all iterates, cf. (Collet and Eckmann, 1980; Misuriewicz, 1983).

Mayer-Kress and Haken (Mayer-Kress and Haken, 1982) have shown that, in the case of maps which possess a positive Schwarzian derivative $\mathrm{S}f > 0$ in sub-intervals, attractors of different periods can coexist and that the transition to chaos may be generated in this case through an intermittency of type III (cf. section 8.6.4). In order to guarantee an infinite cascade of period doublings, we presume in the following $\mathrm{S}f(x) < 0$ for the whole interval (with the exception of the critical point $x_{\max}$, for which $f'(x_{\max}) = 0$ holds). This condition is identically satisfied for the logistic map due to $f''' = 0$.

The relation eq. (6.7.6) – which appears so trivial – in fact plays the key role for the emergence of an infinite sequence of period doublings. Application of the chain rule to eq. (6.7.6) and use of eq. (6.7.8) yields

$$\frac{\mathrm{d}}{\mathrm{d}x} f^2(x)\Big|_{x_{s1}} = f'\big(f(x_{s1})\big) f'(x_{s1}) = f'(x_{s2}) f'(x_{s1}) = \frac{\mathrm{d}}{\mathrm{d}x} f^2(x)\Big|_{x_{s2}} \tag{6.7.11}$$

i. e. at the two fixed points x_{s1} and x_{s2}, the iterated map f^2 possesses – independent of the control parameter α – tangents with the same slope (fig. 6.7.2c). A further increase of α means that, for a higher critical value α_2, we have

$$(f^2)'(x_{s1}) = (f^2)'(x_{s2}) = -1 \tag{6.7.12}$$

i. e. the two fixed points x_{s1} and x_{s2} become unstable *simultaneously*. The process starts again: each of the two fixed points x_{s1} and x_{s2} becomes unstable, a period doubling of $f^2(x)$ occurs and, in each case, two new stable fixed points of the iterated function $f^2(f^2) = f^4$ emerge. This leads to four stable point attractors for f^4 which correspond to two cycles of period 2 of f^2, i.e. to a 4-cycle $(x_{s1}, x_{s2}, x_{s3}, x_{s4})$ of f (see fig. 6.7.4) and can be written as

$$\begin{aligned} x_{s\mathrm{i}+1} &= f(x_{s\mathrm{i}})\,, \qquad \mathrm{i} = 1, 2, 3 \\ x_{s1} &= f(x_{s4}) \end{aligned} \tag{6.7.13}$$

Repeated application of the chain rule in conjunction with the relation eq. (6.7.13) yields

$$\begin{aligned} \frac{\mathrm{d}}{\mathrm{d}x} f^4(x)\Big|_{x_{s1}} &= f'\big(f^3(x_{s1})\big) f'\big(f^2(x_{s1})\big) f'\big(f(x_{s1})\big) f'(x_{s1}) \\ &= f'(x_{s4}) f'(x_{s3}) f'(x_{s2}) f'(x_{s1}) \end{aligned} \tag{6.7.14}$$

i. e. the tangents in the fixed points $x_{s1}, x_{s2}, x_{s3}, x_{s4}$ all have the same gradient and are thus coupled; they will hence become unstable again simultaneously at a critical value α_3.

We are now in a position to unravel the mechanism of period doublings *ad infinitum*. A period doubling which occurs at a critical parameter value α_k arises as a result of the birth of a 2^k-cycle of f $(x_{s1}, x_{s2}, \ldots, x_{s2^\mathrm{k}})$, where we have

$$\begin{aligned} x_{s\mathrm{i}+1} &= f(x_{s\mathrm{i}})\,, \qquad \mathrm{i} = 1, 2, \ldots, (2^\mathrm{k} - 1) \\ x_{s1} &= f(x_{s2^\mathrm{k}}) \end{aligned} \tag{6.7.15}$$

This cycle of period 2^k of the map f disintegrates into 2^k fixed points of the iterated function f^{2^k}, the dynamic behaviour of which is coupled as a result of its tangents having the same gradient

$$\frac{\mathrm{d}}{\mathrm{d}x} f^{2^\mathrm{k}}(x)\Big|_{x_{s\mathrm{j}}} = \prod_{\mathrm{i}=1}^{2^\mathrm{k}} f'(x_{s\mathrm{i}})\,, \qquad \mathrm{j} = 1, 2, \ldots, 2^\mathrm{k} \tag{6.7.16}$$

In the following, it thus suffices to consider only the fixed point of f^{2^k} which lies closest to the extremum $x = x_{\max} = 1/2$.

6.7.2 Superstable Cycles

Each cycle of period 2^k of the logistic map $f(x)$ is stable in a certain α-interval. If we first consider a 2-cycle, there is a parameter value $\alpha = A_1$ for which the extreme value at the point $x = 1/2$ is simultaneously the fixed point of the iterated mapping $f^2(x)$ (see fig. 6.7.5a). Thus, we can write $x_{s1} = 1/2$, $x_{s2} = f(x_{s1})$, $x_{s1} = f(x_{s2})$ and $f'(x_{s1}) = 0$. At such a point, due to $\frac{\mathrm{d}}{\mathrm{d}x} f^2(x_{s_i}) = 0$ (i=1,2), the condition for stability, $|\frac{\mathrm{d}}{\mathrm{d}x} f^2(x_{si})| < 1$, is optimally satisfied; we thus speak of superstability and the corresponding 2-cycle is called *superstable*. If we denote the distance $x_{s2} - x_{s1}$ of the two stable fixed points of f^2 as d_1, we have

$$d_1 = f_{A_1}(\tfrac{1}{2}) - \tfrac{1}{2} \tag{6.7.17}$$

We assigned a subscript A_1 to the function f in order to ensure that the parameter value α in f corresponds to the superstable 2-cycle. If the control parameter α is increased further, then, after the period doubling at $\alpha = \alpha_2$, a superstable 2^2-cycle emerges for $\alpha = A_2 > \alpha_2$ (fig. 6.7.5b). The four stable fixed points of $f^4_{A_2}$ on the diagonal then possess horizontal tangents; in the map $f^2_{A_2}$, two superstable periodic cycles have emerged which are coupled to one another. It thus again suffices to consider the cycle containing $x_s = 1/2$. We denote as d_2 the distance of the fixed point of $f^4_{A_2}$ which is nearest to $x_s = 1/2$ and is given as

$$d_2 = f^2_{A_2}(\tfrac{1}{2}) - \tfrac{1}{2} \tag{6.7.18}$$

or for a superstable 2^{k}-cycle generally

$$d_{\mathrm{k}} = f^{2(\mathrm{k}-1)}_{A_{\mathrm{k}}}(\tfrac{1}{2}) - \tfrac{1}{2} \tag{6.7.19}$$

Note that $d_1, d_2, \ldots, d_{\mathrm{k}}$ have alternating signs. The reason is that, due to the iteration of the function f, the original maximum becomes a minimum and flaps down. Each iteration hence involves a scaling as well as a mirroring.

If, in accordance with eq. (5.4.63) for a superstable 2^{k}-cycle, we calculate the Lyapunov exponent σ – which is a measure for its stability – we obtain $\sigma = -\infty$. The upper part of fig. 6.7.6 reproduces the bifurcation diagram of the logistic map, the lower part the corresponding Lyapunov exponents σ in dependence of α. At the critical points α_{k} at which period doublings occur, σ vanishes in accordance with eq. (5.4.63). Between two critical values α_{k} and $\alpha_{\mathrm{k}+1}$, an A_{k} lies at which the 2^{k}-cycle becomes superstable. At these points, the appertaining numerically calculated Lyapunov exponents possess the negative value which is largest in each case. From the bifurcation diagram, we can read off d_{k}, i. e. the distance between $x_s = 1/2$ and the nearest stable fixed point of the map $f^{2^{\mathrm{k}}}(x)$.

Proceeding along the same lines, we have calculated for the recursion relation

$$x_{\mathrm{n}+1} = \frac{\alpha}{4} \sin \pi x_{\mathrm{n}} \,, \quad 0 \leqslant x \leqslant 1 \,,\ 0 < \alpha < 4 \tag{6.7.20}$$

the bifurcation diagram and the Lyapunov exponent shown in fig. 6.7.7. In accordance with eq. (6.7.9), the Schwarzian derivative is

$$\mathrm{S}f = -\pi^2(1 + \tfrac{3}{2}\tan^2 \pi x) < 0 \tag{6.7.21}$$

i. e. an infinite cascade of period doublings occurs here as well. The following considerations show that the overall good agreement of the two maps is based on the fact that both recursion relations possess a quadratic maximum.

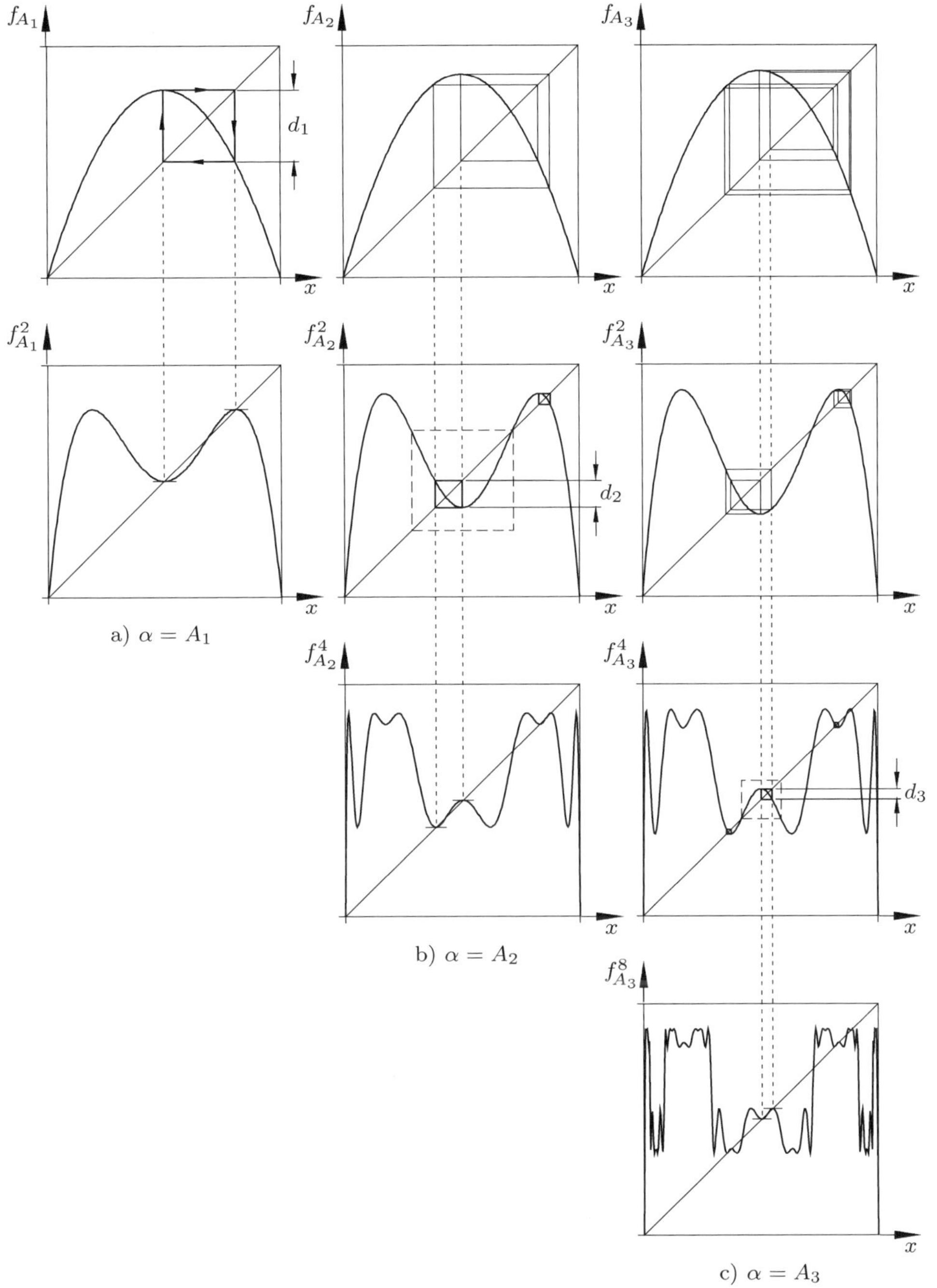

Fig. 6.7.5: Superstable cycles of the logistic map $x_{n+1} = \alpha x_n(1 - x_n)$

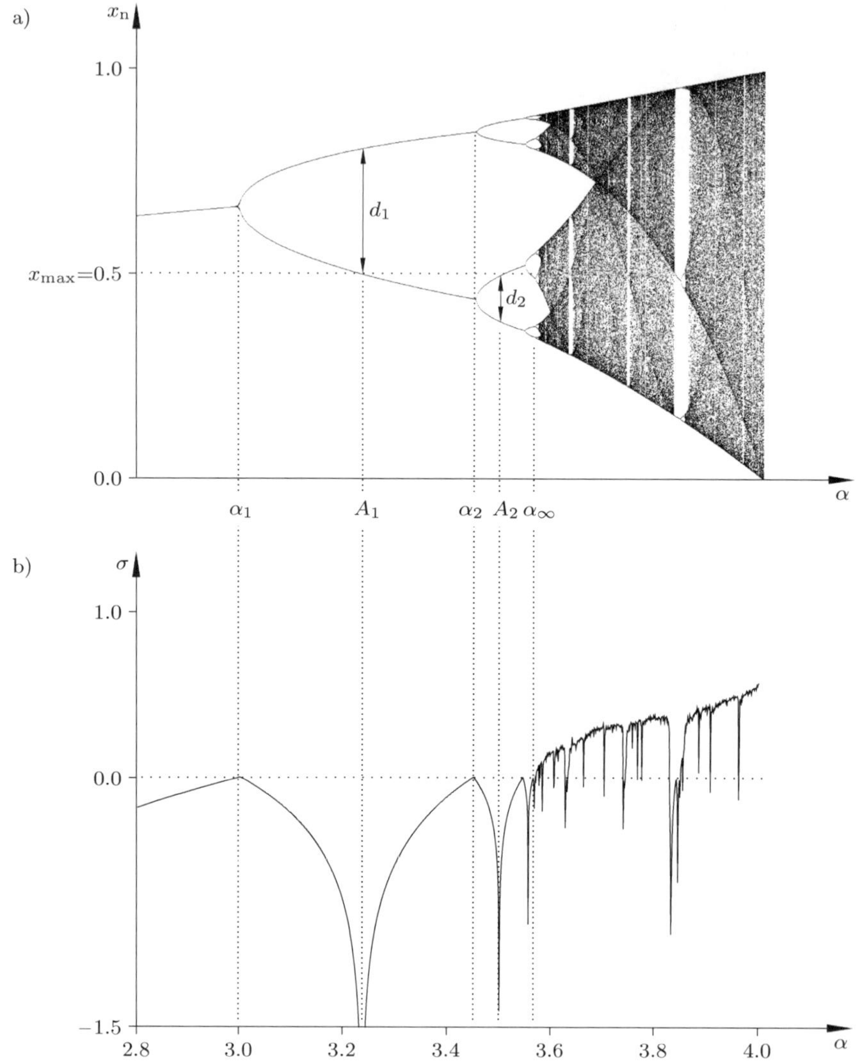

Fig. 6.7.6: Bifurcation diagram and Lyapunov exponent σ for the logistic map $x_{n+1} = \alpha x_n(1 - x_n)$

Both bifurcation diagrams are self-similar with respect to both the variable x and the control parameter α: each new branch resembles the whole picture on a smaller scale. If, on the one hand, we consider the superstable cycles, the distances d_k decrease for a sufficiently large k by a factor of approximately 2.5 at each new bifurcation. On the other hand, we observe that the period doublings at the parameter values α_k occur in decreasing intervals and tend geometrically towards a limit value α_∞.

At this point, we would like to add two comments:

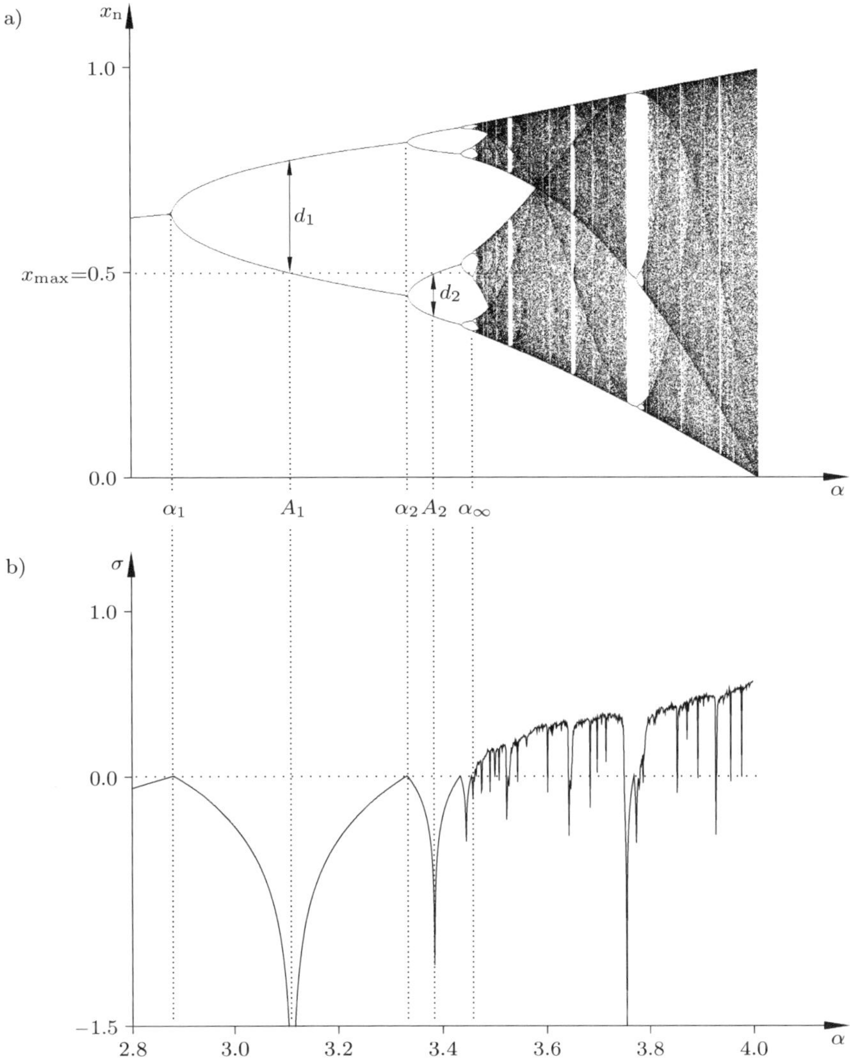

Fig. 6.7.7: Bifurcation diagram and Lyapunov exponent σ for the map $x_{n+1} = \frac{\alpha}{4} \sin \pi x_n$

i. The two bifurcation diagrams in figs. 6.7.6 and 6.7.7 show that the chaotic ranges for values $\alpha > \alpha_\infty$ are repeatedly interrupted by periodic windows of varying width. In section 8.6, we shall elucidate the mechanism responsible for the transition from chaotic to periodic behaviour. Following a far-reaching theorem of Sharkowskii (Sharkowskii, 1964), all positive even integers can be organised as follows

$$3, 5, 7, \ldots,\; 2 \cdot 3, 2 \cdot 5, 2 \cdot 7, \ldots,\; 2^2 \cdot 3, 2^2 \cdot 5, 2^2 \cdot 7, \ldots,\; 2^3, 2^2, 2, 1 \qquad (6.7.22)$$

i. e. firstly, apart from 1, all odd numbers are recorded, then all odd numbers apart from 1 times 2, then all odd numbers apart from 1 times 2^2 etc. At the end of the list are all the powers of 2 in reverse order. The famous theorem of Sharkowskii states the following:

> *If a one-dimensional continuous map of an interval into itself possesses a cycle of period p, then it has cycles of period q for all q-values standing to the right of p in the Sharkowskii order eq. (6.7.22).*

Note that this is only a statement concerning existence; the question regarding the stability of these n-cycles remains open, as well as the question as to how to find these cycles. In fig. 6.7.7, the window of period 3 is particularly striking. We recognise that, following an increase of the control parameter, the cycle of period 3 passes to chaotic behaviour via a cascade of period doublings. According to the theorem of Sharkowskii, the period 3 indeed plays a key role, because for continuous mappings $f(x)$, it follows from the existence of a cycle of period 3 that $f(x)$ must also possess cycles of all other periods n, where n is an arbitrary integer; thus, from the existence of the orbits of period 3, chaos really follows (Li and Yorke, 1975).

ii. A further typical feature of the bifurcation diagrams is the varying density distribution of the points for $\alpha > \alpha_\infty$. The position of the peak values of the probability density distribution alters in dependence of α and forms a network of characteristic lines. We can prove that these exterior border lines are generated from the location of the critical points $x_{cr} = \alpha/4$ and their images $f(x_{cr})$. All further iterations of the critical point $f^{\mathrm{n}}(x_{cr})$ with $\mathrm{n} \geqslant 2$ form the clearly visible network of interior border lines in fig. 6.7.6 (Jensen and Myers, 1985).

We have already stated several times that all the results in this section are only valid for continuously differentiable functions $f(x)$. If we pass on to piecewise smooth functions, completely new scenarios arise in the transition to chaos in which, for example, periodic windows do not occur at all and for which the chaotic attractor is uniquely determined. After (Banerjee et al., 1998; Avrutin and Schanz, 2008), such scenarios are denoted *robust chaos*. With piecewise smooth mappings, it can be guaranteed that a system displays chaotic behaviour even for small alterations of the system parameters and does not jump into periodic cycles. Such systems are of interest in technical systems and are used in the rapid transfer and secure encryption of data streams (Kocarev, 2001).

6.7.3 Self-similarity in the x-Space

Grossmann and Thomae as well as Feigenbaum revealed the laws behind the self-similar structure of the bifurcation diagram of the logistic map respectively a whole class of one-dimensional discrete processes (Großmann and Thomae, 1977; Feigenbaum, 1978; Feigenbaum, 1979b). In the following, we verify the relations for the scaling factors a_{k} and δ_{k}

$$\frac{d_k}{d_{k+1}} = -a_k \,, \qquad \lim_{k\to\infty} a_k = a \tag{6.7.23}$$

$$\frac{\alpha_{k+1} - \alpha_k}{\alpha_{k+2} - \alpha_{k+1}} = \delta_k \,, \qquad \lim_{k\to\infty} a_k = \delta \tag{6.7.24}$$

We also elucidate the universal character of the constants a and δ and evaluate numerically the value of the *Feigenbaum constants*

$$\begin{aligned} a &= 2.502\,907\,875\ldots \\ \delta &= 4.669\,201\,609\ldots \end{aligned} \tag{6.7.25}$$

We first study the self-similarity with regard to the x-coordinate and return to fig. 6.7.5. If we consider the function f^2 within the dashed square for $\alpha = A_2$ (fig. 6.7.5b), we observe that it reflects locally the global evolution of the function f in fig. 6.7.5a for $\alpha = A_1$. In particular, by mirroring at the point $x = 1/2$, $f^2_{A_2} = 1/2$ and magnification by the factor $a_1 = -d_1/d_2$, we can trace and superpose the two superstable 2-cycles on top of the same figure and compare them. In fig. 6.7.8, we have plotted the mirrored and magnified functions f_{A_1}, $f^2_{A_2}$ respectively $f^4_{A_3}$ on one another within the 2-cycles. It is shown that the functions within the squares converge very fast to a limit function $g_1(x)$. Figure 6.7.5 shows that each superstable 2-cycle proceeds on the scale x/a_k. We can thus define an operator T^* which, applied to a function f for $\alpha = A_k$, generates the iterated function f^2 while shifting the control parameter α to the superstable 2-cycle and simultaneously performing a scaling. In this way, the iterated function f^2 can be compared with the original function f. For the operator T^*, we make the following ansatz

$$T^* f_{A_1}(x) = -a_1 f^2_{A_2}(-\frac{x}{a_1})$$

or generally

$$T^* f_{A_k}(x) = -a_k f^2_{A_{k+1}}(-\frac{x}{a_k}) \tag{6.7.26}$$

With the help of T^*, we can thus zoom out and observe increasingly small sections of the sequence of iterated functions $f_{A_1}, f^2_{A_2}, f^4_{A_3}, \ldots f^{2^{(k-1)}}_{A_k}$ as if through a magnifying microscope. We note that the consecutive graphs of these functions cannot be distinguished from one another after only a few iterations, they converge to the limit curve $g_1(x)$ (cf. fig. 6.7.8, left).

In the following, we deduce an analytical expression for the limit function $g_1(x)$. In order to approximate $g_1(x)$ in fig. 6.7.8, we consider the evolution of the sequence of iterated functions within the intervals around $x = 1/2$, the length d_k of which tends towards zero very fast with increasing α. We now carry out a coordinate transformation so that $x = 1/2$ becomes the origin of the new system. In this way, eq. (6.7.19) can be simplified to

$$d_k = f^{2^{(k-1)}}_{A_k}(0) \tag{6.7.27}$$

Considering a scaling expression of the form eq. (6.7.23), we observe the relation

$$d_1 = -a_1 d_2 = a_1 a_2 d_3 = \ldots = (-1)^k \prod_{i=1}^{k} a_i d_{k+1} \qquad \text{with} \quad \lim_{i\to\infty} a_i = +a$$

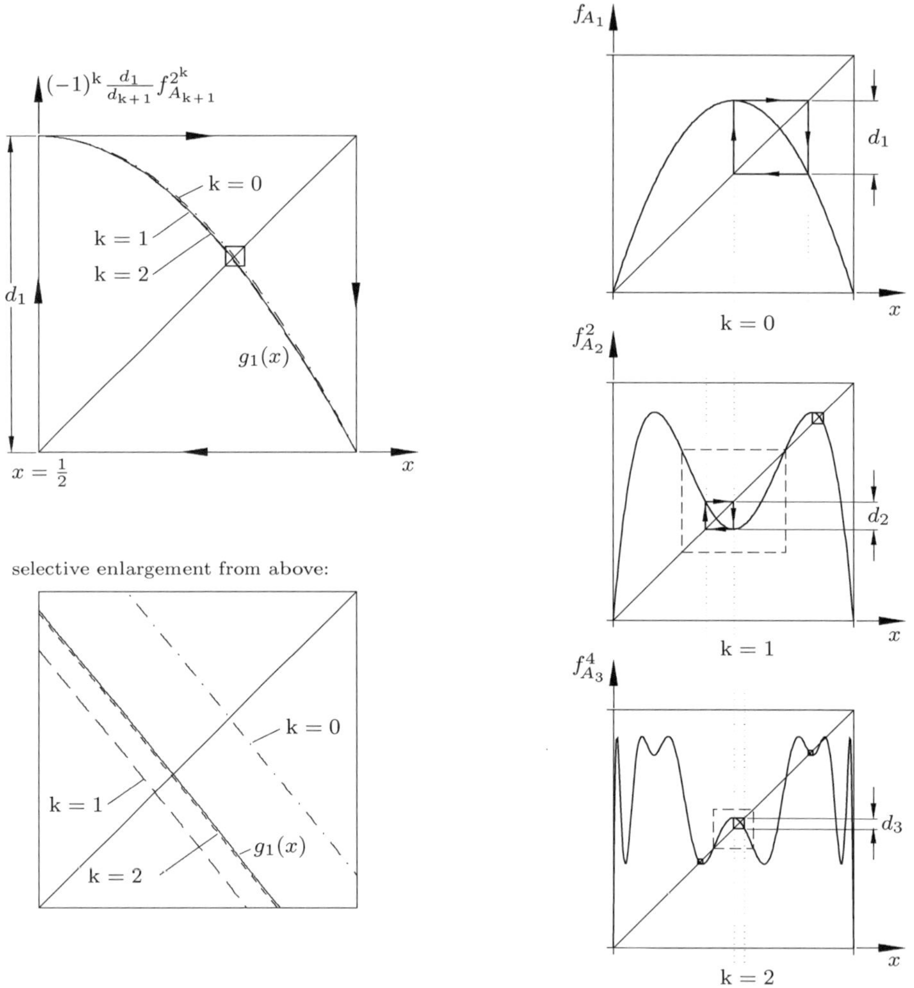

Fig. 6.7.8: Comparison of f_{A_1} and the corresponding magnified and mirrored functions $f^2_{A_2}$ and $f^4_{A_3}$ within the 2-cycles

i.e. we can write $d_1 = \text{const} \cdot (-a)^{\mathrm{k}} d_{\mathrm{k}+1}$. For k $\to \infty$, the renormalised distances $(-a)^{\mathrm{k}} d_{\mathrm{k}+1}$ and thus also, in accordance with eq. (6.7.27), the sequence $(-a)^{\mathrm{k}} f^{2^{\mathrm{k}}}_{A_{\mathrm{k}+1}}(0)$ converge towards a finite limit

$$\lim_{\mathrm{k}\to\infty} (-a)^{\mathrm{k}} f^{2^{\mathrm{k}}}_{A_{\mathrm{k}+1}}(0) = c \tag{6.7.28}$$

The existence of a limit value c thus follows from the existence of the scaling relation eq. (6.7.23); indeed, fig. 6.7.8 reveals still more, namely the existence of a limit function. If we magnify the k-fold iterated function $f^{2^{\mathrm{k}}}_{A_{\mathrm{k}+1}}$ by a factor $(-a)^{\mathrm{k}}$ for sufficiently large values k, the sequence of functions converges towards $g_1(x)$, the value of which, in accordance with eq. (6.7.28), coincides with c at the origin

$$g_1(x) = \lim_{k\to\infty} (-a)^k f^{2^k}_{A_{k+1}}\left(\frac{x}{(-a)^k}\right) \tag{6.7.29}$$

or, using the abbreviation $\xi_k = (-a)^{-k}x$,

$$g_1(x) = \lim_{k\to\infty} (-a)^k f^{2^k}_{A_{k+1}}(\xi_k)$$

The continual zooming out of increasingly small intervals around the point $x = 1/2$ shows that what is decisive for the limit function $g_1(x)$ is only the special nature of the maximum of $f(x)$ and not the shape of the function in the whole interval $0 \leqslant x \leqslant 1$, i. e. for all functions which map the unit interval into itself and display a single maximum of p-th order, we obtain the same limit function; this thus has universal character (Feigenbaum, 1978; Feigenbaum, 1979b).

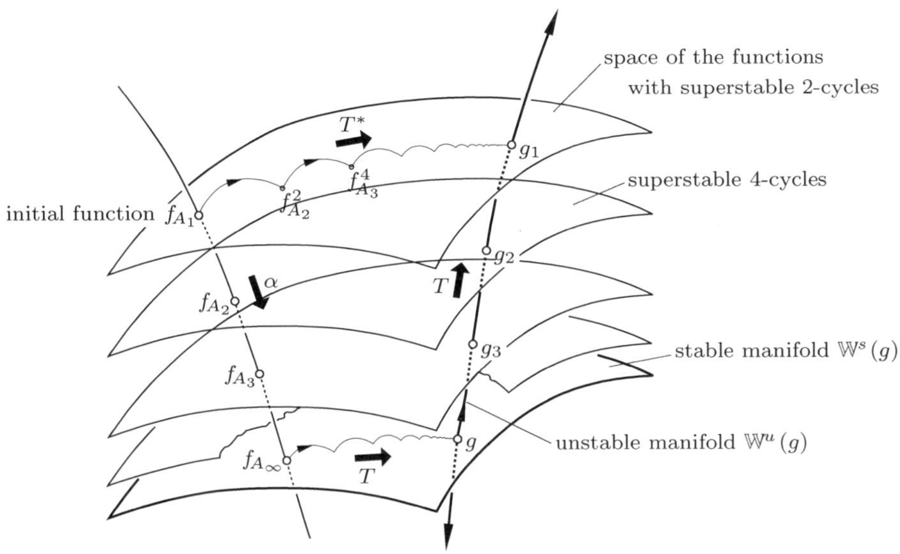

Fig. 6.7.9: Construction of the limit functions g_i (resp. g) through repeated application of the transformation T^* (resp. T)

This universality of the limit function allows us to switch to a higher level of observation: instead of functions and variables, we consider operators and function spaces. This ultimately enables us to determine the Feigenbaum constant a numerically. For example, in the space of all functions which map the unit interval into itself and display a single quadratic maximum and superstable 2-cycle there, the operator T^* possesses a universal limit function $g_1(x)$ which plays the role of a stable point, see fig. 6.7.9. In the long term, a repeated application of T^* causes all the information on the evolution of the function – with the exception of the characteristics of the maximum – to be lost, i. e. the system forgets almost the whole evolution of its initial function.

The limit function $g_1(x)$ is generated through a repeated application of the operator T^* to the function f_{A_1}. Figure 6.7.5 suggests that we could just as well use f_{A_2} as the initial function, that is a function with a superstable 2^2-cycle: again, the functions f_{A_2}, $f^2_{A_3}$ etc. within these cycles resemble one another and tend towards a limit function which displays 4-cycles. Thus, in analogy to eq. (6.7.29), we can define a whole family of limit functions

$$g_\mathrm{i}(x) = \lim_{\mathrm{k}\to\infty} (-a)^\mathrm{k} f^{2^\mathrm{k}}_{A_\mathrm{k+i}}(\xi_\mathrm{k}) \;, \qquad \mathrm{i} = 0, 1, 2, \ldots \tag{6.7.30}$$

which possess 2^i-cycles.

If we apply the operator T^* in particular to the function $f_{A_\infty}(x)$, its effect corresponds to an iteration of f_{A_∞} whereby the control parameter does not vary and where, at the same time, the variable x is scaled by a fixed constant a (cf. figs. 6.7.9 and 6.7.10). It is thus appropriate to introduce a special designation for the transformation in this limiting case. The operator is then denoted the *doubling transformation* T, which has the form

$$\begin{aligned} Tf(x) &= -af\big(f(-\tfrac{x}{a})\big) \quad \text{respectively} \\ Tf(x) &= -af\big(f(\xi)\big) \qquad \text{with} \quad \xi = -\frac{x}{a} \end{aligned} \tag{6.7.31}$$

We can prove that the limit functions follow from one another by application of T

$$g_\mathrm{i}(x) = Tg_\mathrm{i+1}(x) \tag{6.7.32}$$

Using the abbreviation $\xi_\mathrm{k} = (-a)^{-\mathrm{k}} x$, we have in accordance with eq. (6.7.30)

$$\begin{aligned} g_\mathrm{i}(x) &= \lim_{\mathrm{k}\to\infty} (-a)^\mathrm{k} f^{2^\mathrm{k}}_{A_\mathrm{k+i}}(\xi_\mathrm{k}) \\ &= \lim_{\mathrm{k}\to\infty} (-a)(-a)^{\mathrm{k}-1} f^{2^{\mathrm{k}-1+1}}_{A_\mathrm{k+i}}\big(-a^{-1}\xi_{\mathrm{k}-1}\big) \end{aligned}$$

Replacing the index (k − 1) by m, we obtain due to $f^{2^\mathrm{m}}(f^{2^\mathrm{m}}) = f^{2^\mathrm{m+1}}$

$$g_\mathrm{i}(x) = -a \lim_{\mathrm{m}\to\infty} (-a)^\mathrm{m} f^{2^\mathrm{m}}_{A_\mathrm{m+i+1}}\Big((-a)^{-\mathrm{m}}(-a)^\mathrm{m} f^{2^\mathrm{m}}_{A_\mathrm{m+i+1}}\big(-a^{-1}\xi_\mathrm{m}\big)\Big) \tag{6.7.33}$$

On the other hand, in accordance with eq. (6.7.30), replacing i by i + 1 and x by $\xi_1 = -a^{-1}x$ yields

$$g_\mathrm{i+1}(\xi_1) = \lim_{\mathrm{m}\to\infty} (-a)^\mathrm{m} f^{2^\mathrm{m}}_{A_\mathrm{m+i+1}}\big((-a)^{-\mathrm{m}}\xi_1\big)$$

Due to $(-a)^{-\mathrm{m}}\xi_1 = -a^{-1}\xi_\mathrm{m}$, eq. (6.7.33) can be simplified to read

$$g_\mathrm{i}(x) = -a \lim_{\mathrm{m}\to\infty} (-a)^\mathrm{m} f^{2^\mathrm{m}}_{A_\mathrm{m+i+1}}\big((-a)^{-\mathrm{m}} g_\mathrm{i+1}(\xi_1)\big)$$

or finally with eq. (6.7.31)

$$g_\mathrm{i}(x) = -a g_\mathrm{i+1}\big(g_\mathrm{i+1}(\xi_1)\big) = Tg_\mathrm{i+1}(x)$$

thus proving eq. (6.7.32).

Proceeding to the limit i → ∞ in this relation, we obtain a function

$$g(x) = \lim_{\mathrm{i}\to\infty} g_\mathrm{i}(x) = \lim_{\mathrm{k}\to\infty} (-a)^\mathrm{k} f^{2^\mathrm{k}}_{A_\infty}(\xi_\mathrm{k}) \qquad \text{with} \quad \xi_\mathrm{k} = (-a)^{-\mathrm{k}} x \tag{6.7.34}$$

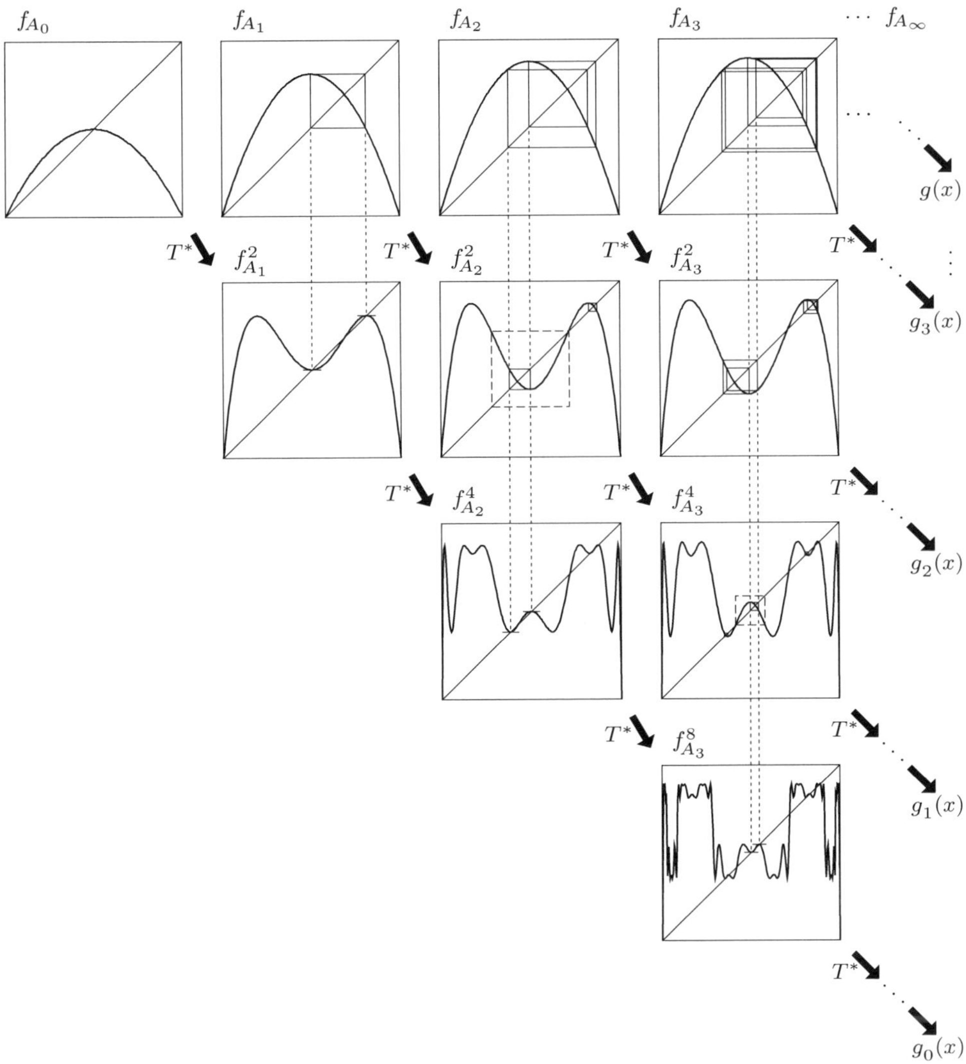

Fig. 6.7.10: Approximation of the limit functions $g_i(x)$

Due to eq. (6.7.32), $g(x)$ is a fixed point with respect to the transformation T, i. e. we have

$$g(x) = Tg(x) = -ag\big(g(-\frac{x}{a})\big) \tag{6.7.35}$$

respectively

$$Tg(x) = -ag\big(g(\xi)\big) \qquad \text{with} \quad \xi = -x/a$$

We obtain $g(x)$ as the limit function of iterated functions f^{2^k} for a *fixed* value A_∞ of the control parameter which, due to $\alpha_k < A_k < \alpha_{k+1}$ and $\lim_{k\to\infty}(\alpha_k) = \alpha_\infty$, coincides with α_∞, i. e. with the α-value for which chaotic behaviour sets in. This is important for the numerical calculation since at first, we do not explicitly know either the bifurcation points α_k or the parameter values A_k appertaining to the superstable cycles.

The existence of a limit value α_∞ follows directly from the relation eq. (6.7.24). Due to its construction, the sequence $\{\alpha_k\}$ is monotonically increasing. Equation (6.7.24) expresses that, for sufficiently large k-values, successive distances $\{\alpha_{k+1}-\alpha_k\}$ deviate arbitrarily little from a geometrical sequence $\{r_k\}$, for which therefore

$$\frac{r_{k+1}}{r_k} = \text{const} \tag{6.7.36}$$

holds. Stated more precisely, for every $\varepsilon > 0$, there is an index $k_0(\varepsilon)$, so that

$$\delta - \varepsilon < \frac{\alpha_{k+1} - \alpha_k}{\alpha_{k+2} - \alpha_{k+1}} < \delta + \varepsilon \tag{6.7.37}$$

holds for all indices $k \geqslant k_0$. Since $\delta > 4$, we can restrict ourselves to values $\varepsilon < \delta - 1$. For k_0, the reciprocal of the relation eq. (6.7.37) is

$$0 < \frac{1}{\delta + \varepsilon} < \frac{\alpha_{k_0+2} - \alpha_{k_0+1}}{\alpha_{k_0+1} - \alpha_{k_0}} < \frac{1}{\delta - \varepsilon} \tag{6.7.38}$$

and thus certainly

$$0 < \frac{1}{\delta + \varepsilon} < \frac{\alpha_{k_0+3} - \alpha_{k_0+2}}{\alpha_{k_0+2} - \alpha_{k_0+1}} < \frac{1}{\delta - \varepsilon} \tag{6.7.39}$$

The product of eq. (6.7.38) and eq. (6.7.39) leads to

$$0 < \frac{1}{(\delta + \varepsilon)^2} < \frac{\alpha_{k_0+3} - \alpha_{k_0+2}}{\alpha_{k_0+1} - \alpha_{k_0}} < \frac{1}{(\delta - \varepsilon)^2}$$

or generally for all $p = 2, 3, \ldots$

$$0 < \frac{1}{(\delta + \varepsilon)^p} < \frac{\alpha_{k_0+p+1} - \alpha_{k_0+p}}{\alpha_{k_0+1} - \alpha_{k_0}} < \frac{1}{(\delta - \varepsilon)^p} \tag{6.7.40}$$

Since $\delta - \varepsilon > 1$, an index p_0 can always be found for any arbitrary $\varepsilon_0 > 0$, so that

$$0 < \alpha_{k_0+p+1} - \alpha_{k_0+p} < \frac{\alpha_{k_0+1} - \alpha_{k_0}}{(\delta - \varepsilon)^p} < \varepsilon_0 \tag{6.7.41}$$

holds for all $p \geqslant p_0$. Thus, the distance of the successive bifurcation points tends towards zero and the sequence $\{\alpha_k\}$ possesses a limit value α_∞.

From eq. (6.7.32), we recognise that $g(x)$ is an unstable fixed point of T. If we select a sufficiently large index i so that $g_i(x)$ deviates arbitrarily little from $g(x)$, then T transports the function $g_i(x)$ away from $g(x)$. To illustrate matters, we present a schematic overview of this argument in fig. 6.7.9, cf. (Collet and Eckmann, 1980; Cvitanović, 1984).

The approximation of the limit functions $g_i(x)$ within the individual function subspaces by repeated application of T^* to f_{A_i} is again demonstrated in fig. 6.7.10. Figure 6.7.11 presents an approximation of the universal function $g_1(x)$.

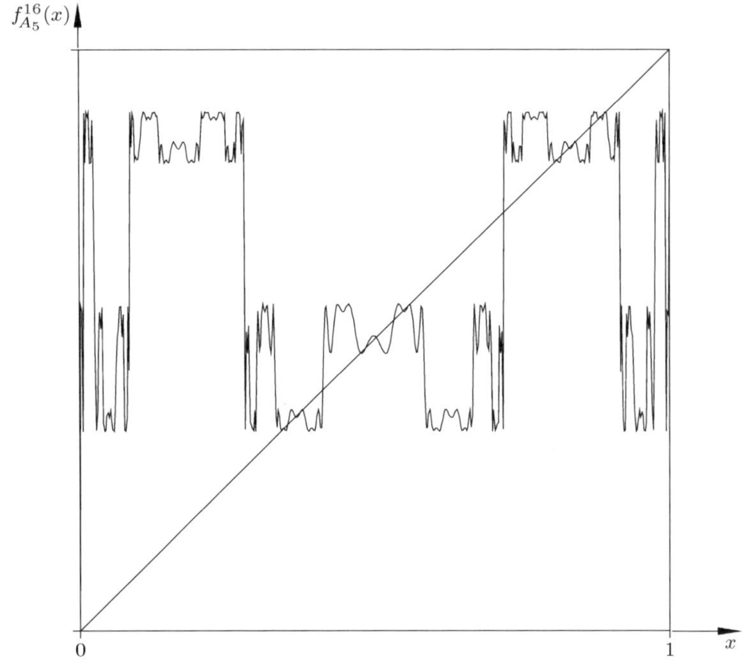

Fig. 6.7.11: Approximation of the limit function $g_1(x)$; the diagram shows $f_{A_5}^{16}(x)$

The fixed point characteristic of $g(x)$ now enables us to demonstrate a very elegant determination of the Feigenbaum constant a which regulates the scaling of the variable x. For $x = 0$, eq. (6.7.35) yields

$$g(0) = -ag\big(g(0)\big) \tag{6.7.42}$$

The following simple calculation shows that our choice of $g(0)$ is still free. Indeed, from the fixed point characteristic eq. (6.7.35) of the function $g(x)$, the fixed point characteristic of all functions of the form $g^*(x) = \mu g(x/\mu)$ follows directly

$$\begin{aligned} -ag^*\big(g^*(-\frac{x}{a})\big) &= -a\mu g\big(\frac{1}{\mu}\mu g(-\frac{x}{\mu a})\big) \\ &= \mu g(\frac{x}{\mu}) = g^*(x) \end{aligned} \tag{6.7.43}$$

If $g(0) = c$ holds at first, the function $g^*(x) = c^{-1}g(cx)$ possesses the value 1 at the point $x = 0$, i. e. without loss of generality, we can dispense with the asterisk * and put

$$g(0) = 1 \tag{6.7.44}$$

Equation (6.7.42) thus yields

$$a = -\frac{1}{g(1)} \tag{6.7.45}$$

We do not know of a general solution of functional equations such as eq. (6.7.35). However, if we specify the characteristics of the maximum and assume that $g(x)$ is r times continuously differentiable, eq. (6.7.35) has a unique solution.

If we assume, for example, that $f(x)$ in the recursion relation eq. (6.7.1) possesses a quadratic maximum, then we can express $g(x)$ as an even analytical function

$$g(x) = 1 + \gamma_1 x^2 + \gamma_2 x^4 + \dots \tag{6.7.46}$$

since, by iterations, the local symmetry of the function $f(x)$ in the neighbourhood of the maximum is carried over to the limit function. If we truncate the power series after the first relevant term and substitute this expression into eq. (6.7.35), we obtain

$$1 + \gamma_1 x^2 = -a\Big[1 + \gamma_1\big(1 + \gamma_1 \frac{x^2}{a^2}\big)^2\Big] + \mathcal{O}(x^4)$$

By comparing the coefficients of the absolute and quadratic terms, we obtain two equations for the first approximations of a and γ_1

$$1 = -a(1 + \gamma_1) \qquad \text{and} \qquad a = -2\gamma_1 \tag{6.7.47}$$

or

$$\begin{aligned} a &= 1 + \sqrt{3} \approx 2.732 \\ \gamma &\approx -1.366 \end{aligned} \tag{6.7.48}$$

We deduce a better approximation of the Feigenbaum constant a and the universal function $g(x)$ if we take the fourth-order terms into account in the power series. The fixed point condition then yields

$$1 + \gamma_1 x^2 + \gamma_2 x^4 = -a\Big[1 + \gamma_1\big(1 + \gamma_1 \frac{x^2}{a^2} + \gamma_2 \frac{x^4}{a^4}\big)^2 + \gamma_2\big(1 + \gamma_1 \frac{x^2}{a^2} + \gamma_2 \frac{x^4}{a^4}\big)^4\Big] + \mathcal{O}(x^6)$$

Comparison of the coefficients now yields three non-linear relations between a, γ_1 and γ_2 which can easily be put into the form

$$\begin{aligned} 1 &= -a(1 + \gamma_1 + \gamma_2) \\ 0 &= \quad a + 2\gamma_1 + 4\gamma_2 \\ 0 &= \quad \gamma_1^2(\gamma_1 + 6\gamma_2) + \gamma_2 a(a^2 - 1) \end{aligned} \tag{6.7.49}$$

This system of equations can be solved, for example, by means of the Newton-Raphson procedure, see (Press et al., 1986). It proves helpful to initiate the process by using the first approximations from eq. (6.7.48) as initial values for a and γ_1 and deduce in this way an initial value for γ_2 from the third relation of eq. (6.7.49). The iteration procedure converges very fast and we obtain a second approximation for the Feigenbaum constant

$$a \approx 2.534 \tag{6.7.50}$$

and for the universal function

$$g(x) \approx 1 - 1.522x^2 + 0.128x^4 + \dots \tag{6.7.51}$$

This procedure can be continued by incorporating higher-order terms of the power series step by step until the desired precision has been achieved. If we calculate the universal function $g(x) = 1 + \Sigma_i \gamma_i x^{2i}$ and the Feigenbaum constant a considering the first seven coefficients $\gamma_1, \dots \gamma_7$, we obtain, using 15 decimal places in the numerical operations, the following constants which we reproduce here with 5 significant digits (cf. table 6.7.1)

$$\begin{aligned} a &= 2.5029 \\ \gamma_1 &= -1.5276 \\ \gamma_2 &= 1.0481 \cdot 10^{-1} \\ \gamma_3 &= 2.6707 \cdot 10^{-2} \\ \gamma_4 &= -3.5284 \cdot 10^{-3} \\ \gamma_5 &= 8.2132 \cdot 10^{-5} \\ \gamma_6 &= 2.5078 \cdot 10^{-5} \\ \gamma_7 &= -2.4989 \cdot 10^{-6} \end{aligned} \tag{6.7.52}$$

in good agreement with the values reported by Feigenbaum (Feigenbaum, 1979b)

$$a = 2.502\,907\,875 \dots \tag{6.7.53}$$

and

$$\begin{aligned} g(x) = 1 &- 1.52763 \cdot 10^{0} x^2 \quad + 1.04815 \cdot 10^{-1} x^4 \\ &+ 2.67057 \cdot 10^{-2} x^6 \quad - 3.52741 \cdot 10^{-3} x^8 \\ &+ 8.15819 \cdot 10^{-5} x^{10} + 2.53684 \cdot 10^{-5} x^{12} \\ &- 2.68777 \cdot 10^{-6} x^{14} + \dots \end{aligned} \tag{6.7.54}$$

We have thus succeeded in deducing the Feigenbaum constant a and the universal function $g(x)$ from the fixed point condition eq. (6.7.35). With this background, we are now in a position to calculate the parameter value α_∞ for which irregular or chaotic behaviour sets in.

It is already recognisable from the two bifurcation diagrams in figs. 6.7.6 and 6.7.7 for the logistic map and the sine function that α_∞ depends on the respective recursion relation

$$\alpha_\infty = \alpha_\infty(f) \tag{6.7.55}$$

a	γ_1	γ_2 $[10^{-1}]$	γ_3 $[10^{-2}]$	γ_4 $[10^{-3}]$	γ_5 $[10^{-5}]$	γ_6 $[10^{-5}]$	γ_7 $[10^{-6}]$
2.5340	−1.5222	1.2761	–	–	–	–	–
2.4789	−1.5218	0.7293	4.5509	–	–	–	–
2.5032	−1.5278	1.0533	2.6309	−3.3438	–	–	–
2.5031	−1.5277	1.0518	2.6417	−3.3882	3.0166	–	–
2.5029	−1.5276	1.0479	2.6733	−3.5451	9.0691	2.1666	–
2.5029	−1.5276	1.0481	2.6707	−3.5284	8.2132	2.5078	−2.4989

Table 6.7.1: Step-wise calculation of a and $g(x) = 1 + \gamma_1 x^2 + \gamma_2 x^4 + \dots$

We can see from the schematic representation in fig. 6.7.9 that $g(x)$ can be approximated starting from $f_{A_\infty} = f_{\alpha_\infty}$ by repeated application of the transformation T. A first rough approximation for $g(x)$ is already the initial function f_{α_∞} itself which still has be transformed and scaled so that – as in the case of $g(x)$ – the maximum occurs at the point $x = 0$ with $f_{\alpha_\infty}(0) = 1$. In the following, we calculate the appertaining α_∞-value numerically for the logistic map (6.7.1) as well as for the sine function eq. (6.7.20).

We first carry out the translation $\bar{x}_{\rm n} = x_{\rm n} - 1/2$ for the logistic map so that the maximum moves to the point $\bar{x} = 0$. Equation (6.7.1) then yields

$$\bar{x}_{\rm n+1} = \tfrac{1}{4}\alpha_\infty(1 - 4\bar{x}_{\rm n}^2) - \tfrac{1}{2} = \bar{f}_{\alpha_\infty}(\bar{x}_{\rm n}) \tag{6.7.56}$$

We now scale with the function value at the point 0

$$\mu = \bar{f}_{\alpha_\infty}(0) = \tfrac{1}{4}\alpha_\infty - \tfrac{1}{2} \tag{6.7.57}$$

where we dispense with the bars

$$\mu x_{\rm n+1} = \tfrac{1}{4}\alpha_\infty(1 - 4\mu^2 x_{\rm n}^2) - \tfrac{1}{2}$$

or

$$x_{\rm n+1} = 1 - \alpha_\infty \mu x_{\rm n}^2 = 1 - \frac{\alpha_\infty}{2}\left(\frac{\alpha_\infty}{2} - 1\right)x_{\rm n}^2 = \hat{f}_{\alpha_\infty}(x_{\rm n}) \tag{6.7.58}$$

$\hat{f}_{\alpha_\infty}$ is now a first approximation for the function $g(x)$, the value of which is $-1/a$ at the point $x = 1$ in accordance with eq. (6.7.45). As we have already calculated the Feigenbaum constant a, we can consider the relation

$$\hat{f}_{\alpha_\infty}(1) = 1 - \frac{\alpha_\infty}{2}\left(\frac{\alpha_\infty}{2} - 1\right) = -\frac{1}{a} \tag{6.7.59}$$

as an equation for the first approximation of α_∞. Solving the quadratic equation and taking $\alpha \in (0, 4]$ into account, we obtain with eq. (6.7.53)

$$\alpha_\infty = 1 + \sqrt{5 + \frac{4}{a}} \approx 3.5687 \tag{6.7.60}$$

This is already a surprisingly good approximation; the value displays a deviation of only 0.04% percent from the exact value

$$\alpha_\infty = 3.569\,945\ldots \tag{6.7.61}$$

Even better approximations are obtained by repeated application of the transformation T using

$$T\hat{f}_{\alpha_\infty}(1) = -\frac{1}{a}\,, \qquad T^2\hat{f}_{\alpha_\infty}(1) = -\frac{1}{a} \quad \text{etc.}$$

In a corresponding manner, we transform the recursion relation eq. (6.7.20). If the maximum is shifted to the origin, we obtain – dispensing *ab initio* with bars –

$$x_{\rm n+1} = \tfrac{1}{4}\alpha_\infty \sin\pi(x_{\rm n} + \tfrac{1}{2}) - \tfrac{1}{2}$$

Scaling with the value of the right-hand side at the point $x_{\rm n} = 0$

$$\mu = \tfrac{1}{4}\alpha_\infty - \tfrac{1}{2}$$

leads to the function

$$\hat{f}_{\alpha_\infty}(x) = \frac{1}{\mu}\left(\tfrac{1}{4}\alpha_\infty \sin\pi(\mu x + \tfrac{1}{2}) - \tfrac{1}{2}\right) \approx g(x) \tag{6.7.62}$$

Again setting $\hat{f}_{\alpha_\infty}(1) = -1/a$, we obtain the transcendental equation

$$\frac{\alpha_\infty}{4}\sin\left(\pi\frac{\alpha_\infty}{4}\right) - \frac{1}{2} = -\frac{1}{2a}\left(\frac{\alpha_\infty}{2} - 1\right) \tag{6.7.63}$$

This equation can be solved using Newton's iteration procedure, yielding

$$\alpha_\infty \approx 3.464 \tag{6.7.64}$$

in good agreement with the α_∞-value obtained from fig. 6.7.7.

6.7.4 Self-similarity in the Parameter Space

The *fixed point property* of $g(x)$ with respect to the operation T enabled us to decode the self-similarity of the bifurcation diagram in the x-space and to determine the Feigenbaum constant a. We next show that the *stability characteristics* of the fixed point yield the second universal constant δ which underlies the self-similar structure of the bifurcation cascade with respect to the control parameter (see fig. 6.7.12). As in rest of this section, we cannot enter into mathematical details here, but only present the basic steps leading to the determination of δ.

As in earlier sections, we again attempt to arrive at statements on the stability of the fixed point $g(x)$ via a linear stability analysis, i. e. via the eigenvalues of the linearised system. What is new is the fact that here, functions adopt the role of points in the

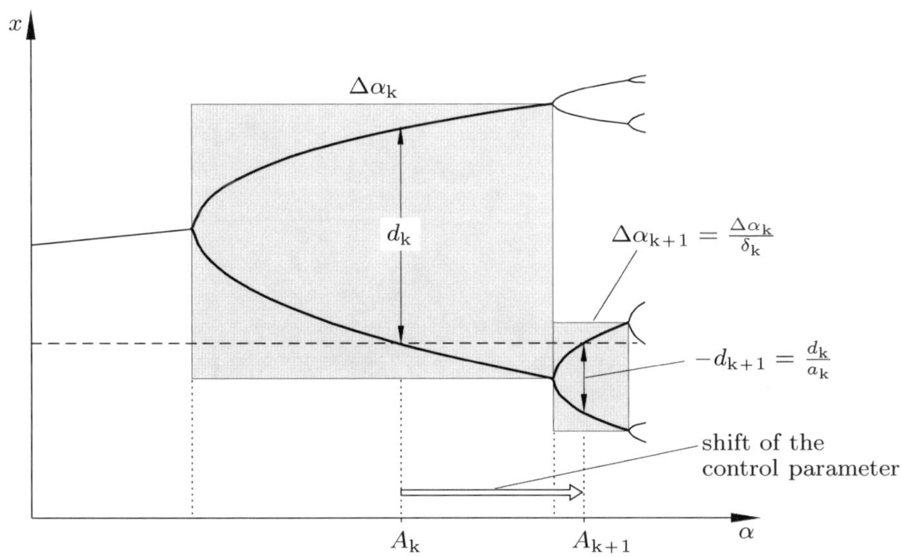

Fig. 6.7.12: Scaling characteristics of the bifurcation diagram

phase space and operators the role of functions. Our aim is to linearise the operator T at the point $g(x)$ and to formulate the corresponding eigenvalue problem.

The instability of the fixed point $g(x)$ which can be recognised from eqs. (6.7.32) and (6.7.34) and is illustrated in fig. 6.7.9 suggests that the largest eigenvalue is positive and characterises the repelling effect of the fixed point. It is thus to be expected that, for sufficiently large k, the largest eigenvalue will provide information on the distance of successive functions g_{k} and $g_{\mathrm{k}+1}$ and thus also on the distance of the parameter values of successive superstable cycles A_{k} and $A_{\mathrm{k}+1}$. Since the sequences $\{\alpha_{\mathrm{k}}\}$ and $\{A_{\mathrm{k}}\}$ are nested within one another

$$\alpha_{\mathrm{k}} < A_{\mathrm{k}} < \alpha_{\mathrm{k}+1} < A_{\mathrm{k}+1} \tag{6.7.65}$$

we expect, due to eq. (6.7.24), to find a connection between δ and the largest eigenvalue of the linearised problem.

We first realise that eq. (6.7.24) yields

$$\lim_{\mathrm{k}\to\infty} (\alpha_\infty - \alpha_{\mathrm{k}})\delta^{\mathrm{k}} = \text{const} \tag{6.7.66}$$

There also follows from eq. (6.7.65)

$$\lim_{\mathrm{k}\to\infty} (A_\infty - A_{\mathrm{k}})\delta^{\mathrm{k}} = \text{const} \tag{6.7.67}$$

In analogy with eq. (6.7.37), there is a k_0 for every $\varepsilon > 0$, so that the inequality

$$0 < (A_{\mathrm{k}+2} - A_{\mathrm{k}+1})(\delta - \varepsilon) < (A_{\mathrm{k}+1} - A_{\mathrm{k}}) < (A_{\mathrm{k}+2} - A_{\mathrm{k}+1})(\delta + \varepsilon) \tag{6.7.68}$$

is valid for all indices $\mathrm{k} \geqslant \mathrm{k}_0$. The following relations are then also valid

$$0 < (A_{\mathrm{k}+3} - A_{\mathrm{k}+2})(\delta - \varepsilon) < (A_{\mathrm{k}+2} - A_{\mathrm{k}+1}) < (A_{\mathrm{k}+3} - A_{\mathrm{k}+2})(\delta + \varepsilon) \tag{6.7.69}$$

etc. If we add the sequence of all following inequalities to eq. (6.7.68), all intermediate terms cancel each other pairwise and only the first and the "last" term are left

$$0 < (A_\infty - A_{\mathrm{k}+1})(\delta - \varepsilon) < (A_\infty - A_{\mathrm{k}}) < (A_\infty - A_{\mathrm{k}+1})(\delta + \varepsilon) \tag{6.7.70}$$

i. e.

$$\lim_{\mathrm{k}\to\infty} \frac{A_\infty - A_{\mathrm{k}}}{A_\infty - A_{\mathrm{k}+1}} = \delta \tag{6.7.71}$$

This means that, for sufficiently large values of k, the sequence $\{A_\infty - A_{\mathrm{k}}\}$ deviates arbitrarily little from a geometrical sequence $\{r_{\mathrm{k}}\}$ for which $r_{\mathrm{k}}/r_{\mathrm{k}+1} = \delta$ respectively $r_{\mathrm{k}} = C/\delta^{\mathrm{k}}$ holds. The verification of the relations eqs. (6.7.66, 6.7.67) follows ultimately from this argument.

Originally, Feigenbaum intuitively assumed the scaling law eq. (6.7.24) and supported his assumption by means of numerical experiments. Later, he succeeded in developing the corresponding theory, in verifying the relation eq. (6.7.24) and thus also eqs. (6.7.66, 6.7.67) and in calculating the constant δ with arbitrary precision. In the following, we present a rough sketch of the necessary operational steps, cf. (Schuster and Just, 2006).

We first linearise the operator T and apply it to a neighbouring function $f(x)$ of $f_0(x)$

$$f(x) = f_0(x) + \tilde{f}(x)$$

Here, $\tilde{f}(x)$ denotes a small deviation. In accordance with eq. (6.7.35), we obtain, applying the abbreviation $\xi = -x/a$,

$$\begin{aligned} Tf(x) = T\big(f_0(x) + \tilde{f}(x)\big) &= -a(f_0 + \tilde{f})\big((f_0 + \tilde{f})(\xi)\big) \\ &= -a\Big[(f_0 + \tilde{f})\big(f_0(\xi)\big) + (f_0 + \tilde{f})'\big(f_0(\xi)\big)\tilde{f}(\xi) + \ldots\Big] \\ &= -a\Big[f_0\big(f_0(\xi)\big) + \tilde{f}\big(f_0(\xi)\big) + f_0'\big(f_0(\xi)\big)\tilde{f}(\xi) + \mathcal{O}(\tilde{f}^2)\Big] \end{aligned}$$

or

$$Tf(x) = Tf_0(x) + L_{f_0}\tilde{f} + \mathcal{O}(\tilde{f}^2)$$

The last two relations yield the linearised operator L_{f_0}

$$L_{f_0}\tilde{f} = -a\big[\tilde{f}\big(f_0(\xi)\big) + f_0'\big(f_0(\xi)\big)\tilde{f}(\xi)\big] \qquad \text{with} \quad \xi = -x/a \tag{6.7.72}$$

Our aim is to linearise the operator T around the function $g(x)$. To this end, we consider a neighbouring function $f_A(x)$ of $f_{A_\infty}(x)$ and expand it into a Taylor series

$$f_A(x) = f_{A_\infty}(x) + (A - A_\infty)\partial_A f(x) + \mathcal{O}(2) \tag{6.7.73}$$

where

$$\partial_A f(x) = \left.\frac{\partial f_A(x)}{\partial A}\right|_{A=A_\infty}$$

We now apply T to $f_A(x)$ and obtain

$$Tf_A(x) = Tf_{A_\infty}(x) + (A - A_\infty)L_{f_{A_\infty}}\partial_A f(x) + \mathcal{O}(2)$$

where, in accordance with eq. (6.7.72), the linearised operator $L_{f_{A_\infty}}$ has the form

$$L_{f_{A_\infty}}\partial_A f(x) = -a\big[\partial_A f\big(f_{A_\infty}(\xi)\big) + f_{A_\infty}'\big(f_{A_\infty}(\xi)\big)\partial_A f(\xi)\big] \tag{6.7.74}$$

again with $\xi = -x/a$. Application of the operator n times yields

$$T^{\mathrm{n}} f_A(x) = T^{\mathrm{n}} f_{A_\infty}(x) + (A - A_\infty)L_{T^{\mathrm{n}-1}f_{A_\infty}} \cdots L_{f_{A_\infty}}\partial_A f(x) + \mathcal{O}(2) \tag{6.7.75}$$

We should like to apologise to the reader at this point for the multitude of indices. For large values n, $T^{\mathrm{n}} f_{A_\infty}(x)$ tends towards $g(x)$ (cf. fig. 6.7.9). Neglecting the higher-order terms in $\partial_A f(x)$, we obtain for $\mathrm{n} \gg 1$ approximately

$$T^{\mathrm{n}} f_A(x) \approx g(x) + (A - A_\infty)L_g^{\mathrm{n}}\, \partial_A f(x) \tag{6.7.76}$$

where L_g follows from T in accordance with eq. (6.7.72) by linearisation around $g(x)$

$$L_g\partial_A f(x) = -a\big[\partial_A f\big(g(\xi)\big) + g'\big(g(\xi)\big)\partial_A f(\xi)\big] \tag{6.7.77}$$

Equation (6.7.76) can be simplified further by expanding $\partial_A f(x)$ with respect to the eigenfunctions of L_g. If we denote the eigenfunctions φ_i and the appertaining eigenvalues λ_i, then we can write by definition

$$L_g\varphi_\mathrm{i} = \lambda_\mathrm{i}\varphi_\mathrm{i} \qquad (\mathrm{i} = 1, 2, 3\ldots) \tag{6.7.78}$$

Since the eigenfunctions form a basis in the associated function space, we can expand $\partial_A f(x)$ with respect to φ_i

$$\partial_A f(x) = \sum_\mathrm{i} a_\mathrm{i}\varphi_\mathrm{i}(x) \tag{6.7.79}$$

and we have

$$L_g\partial_A f(x) = \sum_\mathrm{i} a_\mathrm{i} L_g\varphi_\mathrm{i}(x) = \sum_\mathrm{i} a_\mathrm{i}\lambda_\mathrm{i}\varphi_\mathrm{i}(x) \tag{6.7.80}$$

respectively

$$L_g^\mathrm{n}\partial_A f(x) = \sum_\mathrm{i} a_\mathrm{i}\lambda_\mathrm{i}^\mathrm{n}\varphi_\mathrm{i}(x) \tag{6.7.81}$$

For $\mathrm{n} \gg 1$, we recognise that the influence of all those eigenfunctions for the eigenvalues of which $|\lambda_\mathrm{i}| < 1$ applies decreases rapidly so that, for large n, only the modes appertaining to eigenvalues with $|\lambda_\mathrm{i}| > 1$ are still relevant. We now assume that there is only a single positive eigenvalue λ_1 and denote the corresponding eigenfunction $\varphi_1(x) = h(x)$. (The proof of the existence of a single positive eigenvalue is laborious; we refer the reader to (Collet et al., 1980), for example.) For $\mathrm{n} \gg 1$, eq. (6.7.76) can now be expressed approximately as

$$T^\mathrm{n} f_A(x) \approx g(x) + (A - A_\infty)a_1\lambda_1^\mathrm{n} h(x) \tag{6.7.82}$$

We now evaluate this relation at the point $x = 0$ for $A = A_\mathrm{n}$. The condition for a superstable 2^n-cycle was that the extreme value at $x = 1/2$ is simultaneously the fixed point of the iterated map $f^{2^\mathrm{n}}(x)$, i. e.

$$f_{A_\mathrm{n}}^{2^\mathrm{n}}(\tfrac{1}{2}) = \tfrac{1}{2} \tag{6.7.83}$$

(cf. fig. 6.7.5). Applying the translation which shifts $x = 1/2$ to the origin, we obtain for eq. (6.7.83)

$$f_{A_\mathrm{n}}^{2^\mathrm{n}}(0) = 0 \tag{6.7.84}$$

However, according to eq. (6.7.31), we have

$$T^\mathrm{n} f_{A_\mathrm{n}}(x) = (-a)^\mathrm{n} f_{A_\mathrm{n}}^{2^\mathrm{n}}(\xi) \tag{6.7.85}$$

which, together with eq. (6.7.84), yields

$$T^\mathrm{n} f_{A_\mathrm{n}}(0) = 0 \tag{6.7.86}$$

In accordance with eq. (6.7.44), we can choose $g(0) = 1$. At the point $x = 0$, eq. (6.7.82) thus leads to

$$1 + (A_\mathrm{n} - A_\infty)\lambda_1^\mathrm{n} a_1 h(0) \approx 0$$

or

$$(A_\infty - A_{\rm n})\lambda_1^{\rm n} \approx \frac{1}{a_1 h(0)} = \text{const} \tag{6.7.87}$$

This relation indeed coincides with the scaling law eq. (6.7.67) if we identify δ with the eigenvalue λ_1.

Our only remaining task is the numerical determination of the Feigenbaum constant δ. This can be carried out step-wise in analogy with the calculation of a. If we rewrite the eigenvalue equation

$$L_g h(x) = \delta h(x) \tag{6.7.88}$$

by substitution of eq. (6.7.77) into

$$-a\big[h\big(g(\xi)\big) + g'\big(g(\xi)\big)h(\xi)\big] = \delta h(x) \tag{6.7.89}$$

and insert in this relation the quantity a and the function $g(x)$ already established in eqs. (6.7.53) and (6.7.54), we obtain an expression for δ and $h(x)$. We deduce from eq. (6.7.82) that in the power series ansatz for $h(x)$ as well only even powers in x can appear

$$h(x) = h(0)(1 + \eta_1 x^2 + \eta_2 x^4 + \ldots)$$

Since eq. (6.7.89) is a linear relation in $h(x)$, the value of the function at the point $x = 0$ becomes irrelevant and the factor $h(0)$ drops out.

Step by step, we have calculated approximate values for δ and $h(x)$ by taking only terms up to the second, fourth etc. order into account in the power series expansion of $h(x)$. Table 6.7.2 presents an overview of the results, the calculations being based on fifteen significant digits. Feigenbaum quotes the following results (Feigenbaum, 1978; Feigenbaum, 1979b)

$$\delta = 4.669\,201\,609\ldots \tag{6.7.90}$$

$$\begin{aligned} h(x) = h(0)(1 &-3.25651\cdot 10^{-1}x^2 \quad -5.05539\cdot 10^{-2}x^4 \\ &+1.45598\cdot 10^{-2}x^6 \quad -8.81042\cdot 10^{-4}x^8 \\ &-1.06217\cdot 10^{-4}x^{10} \quad +1.98399\cdot 10^{-5}x^{12}) \end{aligned} \tag{6.7.91}$$

We have now deduced all the essential quantities which characterise the route to chaos via period doublings as arising in the example of the logistic map respectively the recursive equation (6.7.20). In the following, we summarise the basic points:

i. The universal behaviour of one-dimensional dissipative maps relates to the asymptotic range ($\alpha \to \alpha_\infty$). The decisive factor for the type of bifurcation, i. e. the pre-requisite for the occurrence of period doublings, is that $\mathrm{S}f < 0$ holds for all x for the Schwarzian derivative in accordance with eq. (6.7.9).

ii. The self-similarity with respect to the x-space is described by the fixed point equation $g(x) = Tg(x)$ in eq. (6.7.35); this allows us to deduce a and $g(x)$ which are determined by the order of the maximum alone.

δ	η_1 $[10^{-1}]$	η_2 $[10^{-2}]$	η_3 $[10^{-2}]$	η_4 $[10^{-4}]$	η_5 $[10^{-4}]$	η_6 $[10^{-5}]$	η_7 $[10^{-7}]$
4.5409	−3.1136	–	–	–	–	–	–
4.7089	−3.3399	−4.4492	–	–	–	–	–
4.6653	−3.2449	−5.1654	1.5087	–	–	–	–
4.6689	−3.2554	−5.0670	1.4627	−9.0661	–	–	–
4.6693	−3.2568	−5.0508	1.4525	−8.6152	−1.1482	–	–
4.6692	−3.2564	−5.0558	1.4564	−8.8331	−1.0598	2.0278	–
4.6692	−3.2564	−5.0559	1.4564	−8.8360	−1.0583	2.0222	1.5562

Table 6.7.2: Step-wise determination of δ and $h(x) = h(0)(1 + \eta_1 x^2 + \eta_2 x^4 + \ldots)$

iii. The self-similarity with respect to the parameter space is described by the eigenvalue equation $L_g h(x) = \delta h(x)$ in eq. (6.7.88) and allows us to calculate the Feigenbaum constant δ which coincides with the largest (positive) eigenvalue of the linearised operator T at the point $g(x)$ and also to determine the appertaining eigenfunction $h(x)$.

Finally, we demonstrate the universality of the transition to chaos via period doubling and the associated phenomenon of self-similarity on two examples of dynamical systems taken from completely different branches of science, namely the logistic map originally used to describe the dynamics of biological populations and a non-linear, periodically excited electrical circuit. Figure 6.7.13 displays the bifurcation diagrams of both systems, the upper part being that of the logistic map and the lower one the experimentally evaluated bifurcation diagram of (van Buskirk and Jeffries, 1985). In both diagrams, the transition to chaos follows a Feigenbaum cascade with the same scaling laws. The characteristic of self-similarity is clearly illustrated in fig. 6.7.13a: the magnification of a small section within the 3-period window displays a texture which is astoundingly similar to that of the whole bifurcation diagram.

We have seen that the values of the Feigenbaum constants a und δ depend on the order of the maximum of the mapping. Hu and Satija have determined these constants for all maps of the form

$$f(x) = 1 - \alpha|x|^{\mathrm{p}} \qquad (1 < \mathrm{p} \leqslant 12) \tag{6.7.92}$$

(Hu and Satija, 1983), where p is the order of the maximum. We present the results in fig. 6.7.14.

The mappings of eq. (6.7.92) investigated by Hu and Satija are all characterised by a symmetry with respect to the maximum. The question now arises: what influence does an asymmetrical evolution of the function have on the universal function $g(x)$, on its symmetry characteristics and thus on the appertaining constants a and $g(x)$? We consider a typical growth model from the field of population dynamics (cf. (Arnold, 1988; May, 1976))

$$x_{\mathrm{n}+1} = A x_{\mathrm{n}} \mathrm{e}^{-x_{\mathrm{n}}} \tag{6.7.93}$$

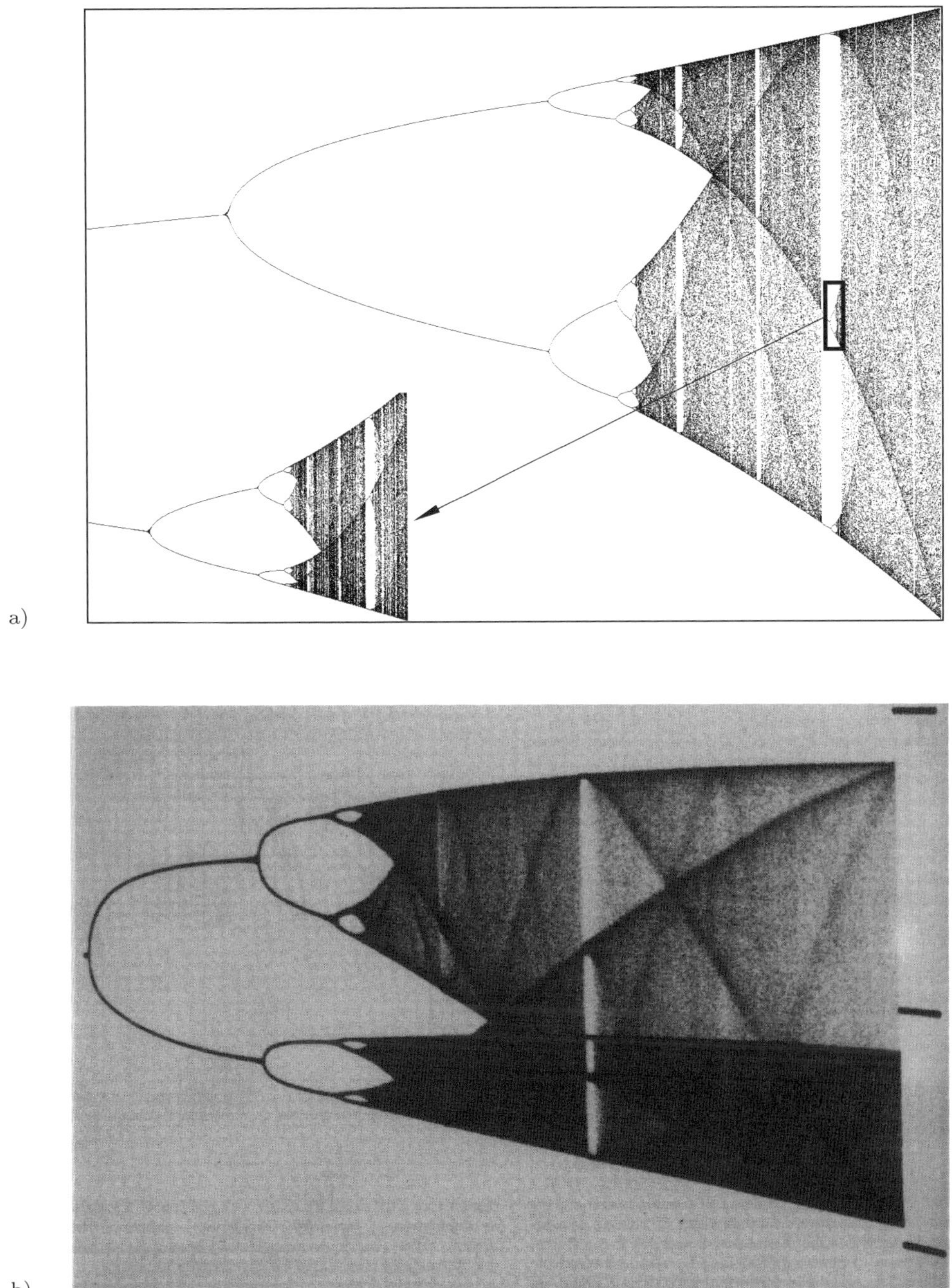

Fig. 6.7.13: Universality and self-similarity – bifurcation diagram
a) of the logistic map and
b) of a non-linear electrical circuit (van Buskirk and Jeffries, 1985)

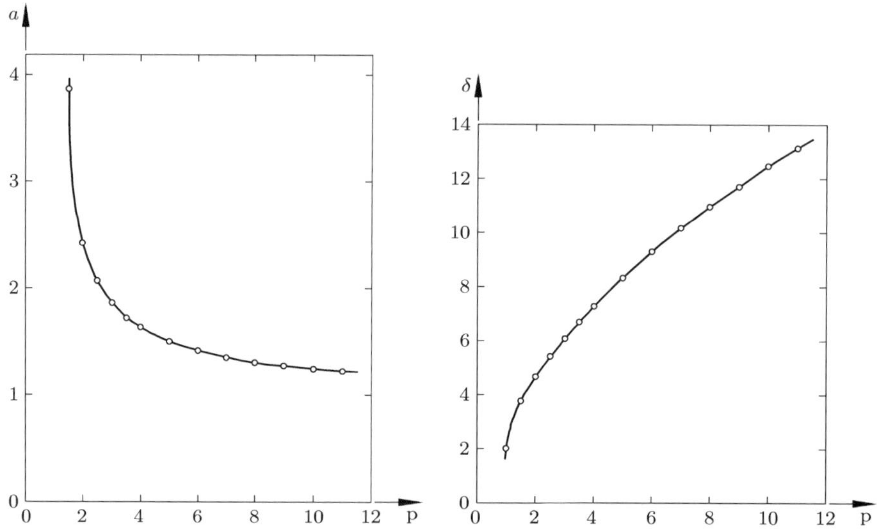

Fig. 6.7.14: The universal constants a and δ in dependence of the order p of the maximum (Hu and Satija, 1983)

Figure 6.7.15 reproduces the graph of the function

$$f(x) = Ax\mathrm{e}^{-x} \qquad (A > 0) \tag{6.7.94}$$

Since

$$f'(x) = A(1 - x)\mathrm{e}^{-x} \tag{6.7.95}$$

the function possesses for all A a single maximum at the point $x = 1$. The fixed points are obtained from $x = Ax\mathrm{e}^{-x}$ as $x = 0$ and $x = \ln A$. From eq. (6.7.95), we find $f'(0) = A$, i. e. $x = 0$ is stable for $A < 1$ and unstable for $A > 1$. Substitution of $x = \ln A$ in eq. (6.7.95) yields

$$f'(\ln A) = 1 - \ln A \tag{6.7.96}$$

i. e. the fixed point $x = \ln A$ is unstable for $A < 1$ and $A > \mathrm{e}^2$ and stable in the intermediate range $1 < A < \mathrm{e}^2$. The corresponding bifurcation diagram is presented

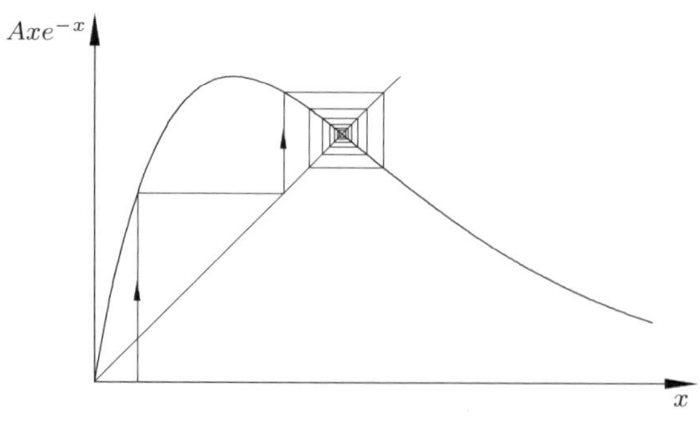

Fig. 6.7.15

The asymmetrical map $f(x) = Ax\mathrm{e}^{-x}$

in fig. 6.7.16a. For $A = 1$, there is a transcritical bifurcation. If we exclude negative x-values, then, for $0 < A < 1$, $x = 0$ is the only relevant fixed point and is stable there. For $A > 1$, $x = 0$ becomes unstable and a new stable fixed point $x = \ln A$ emerges which loses its stability at the point $A = \mathrm{e}^2$ since $f'(2) = -1$ without, however, bifurcating. In order to be able to ascertain if a period doubling occurs at this point, we inspect the condition of eq. (6.6.58) or calculate the Schwarzian derivative. By differentiation, eq. (6.7.95) yields

$$f''(x) = A(x-2)\mathrm{e}^{-x}, \quad f'''(x) = A(-x+3)\mathrm{e}^{-x}$$

In accordance with eq. (6.7.9), we thus obtain the Schwarzian derivative

$$\mathrm{S}f = \frac{f'''}{f'} - \frac{3}{2}\left(\frac{f''}{f'}\right)^2 = \frac{x-3}{x-1} - \frac{3}{2}\left(\frac{x-2}{x-1}\right)^2$$

or

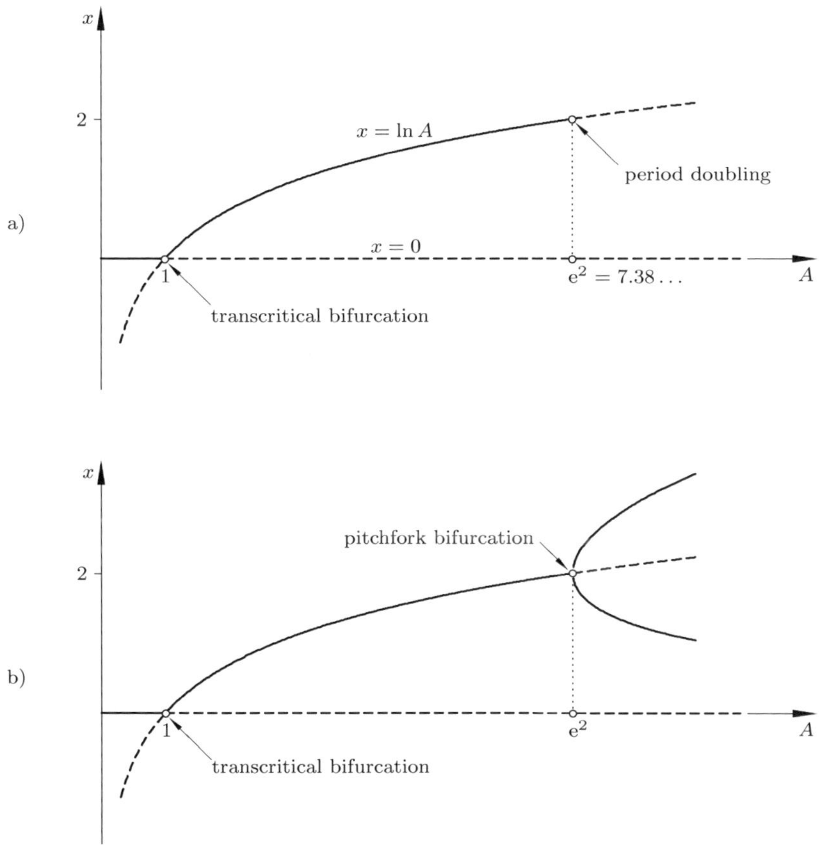

Fig. 6.7.16: Bifurcation diagram
a) of $f(x) = Ax\mathrm{e}^{-x}$ and b) the iterated function $f^2(x)$

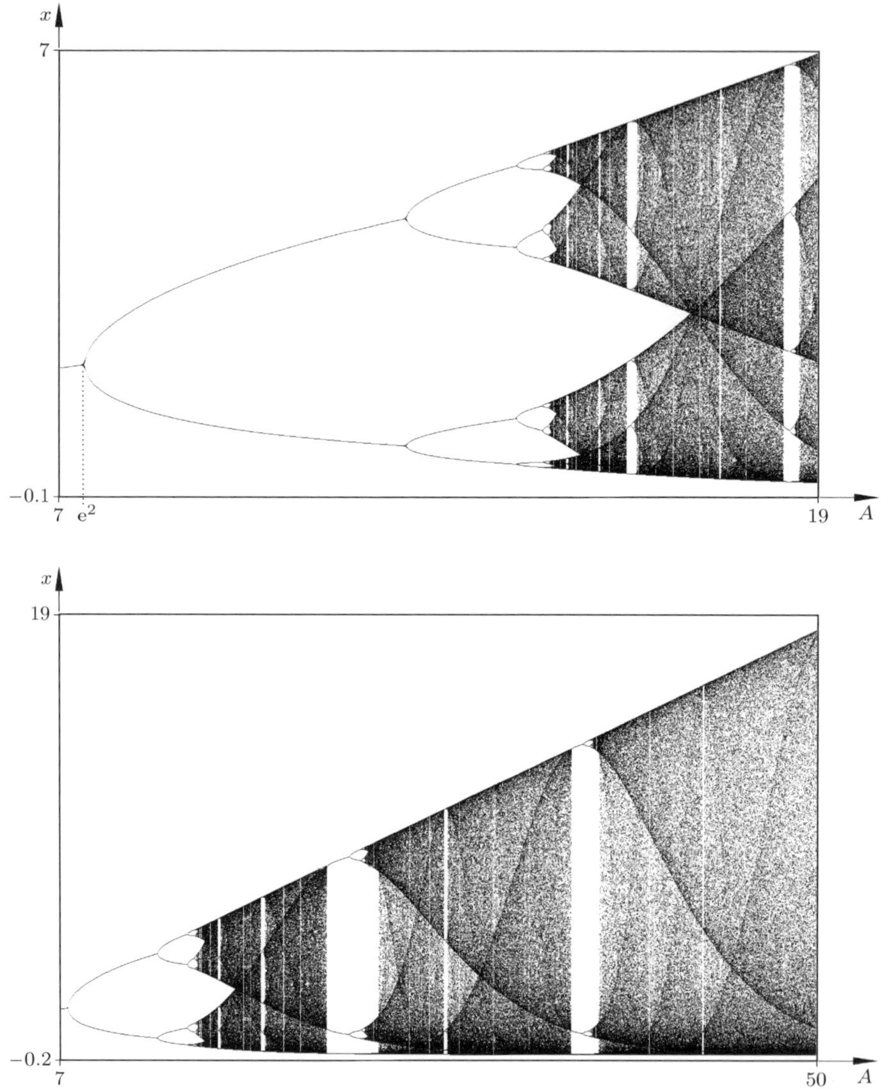

Fig. 6.7.17: Bifurcation diagram of the map $f(x) = Axe^{-x}$

$$\mathrm{S}f = -\frac{1}{2}\frac{x^2 - 4x + 6}{(x-1)^2} \tag{6.7.97}$$

Since $x^2 - 4x + 6 > 0$ holds for all x-values, $\mathrm{S}f$ is consistently negative so that, at the fixed point $x = 2$ which loses its stability at $A = e^2$, a period doubling is generated. For the iterated function $f^2(x)$, a supercritical pitchfork bifurcation takes place at this point (see fig. 6.7.16b). Since $\mathrm{S}f < 0$ applies over the whole x-interval, the route to chaos leads through a sequence of period doublings in this recursive mapping as well.

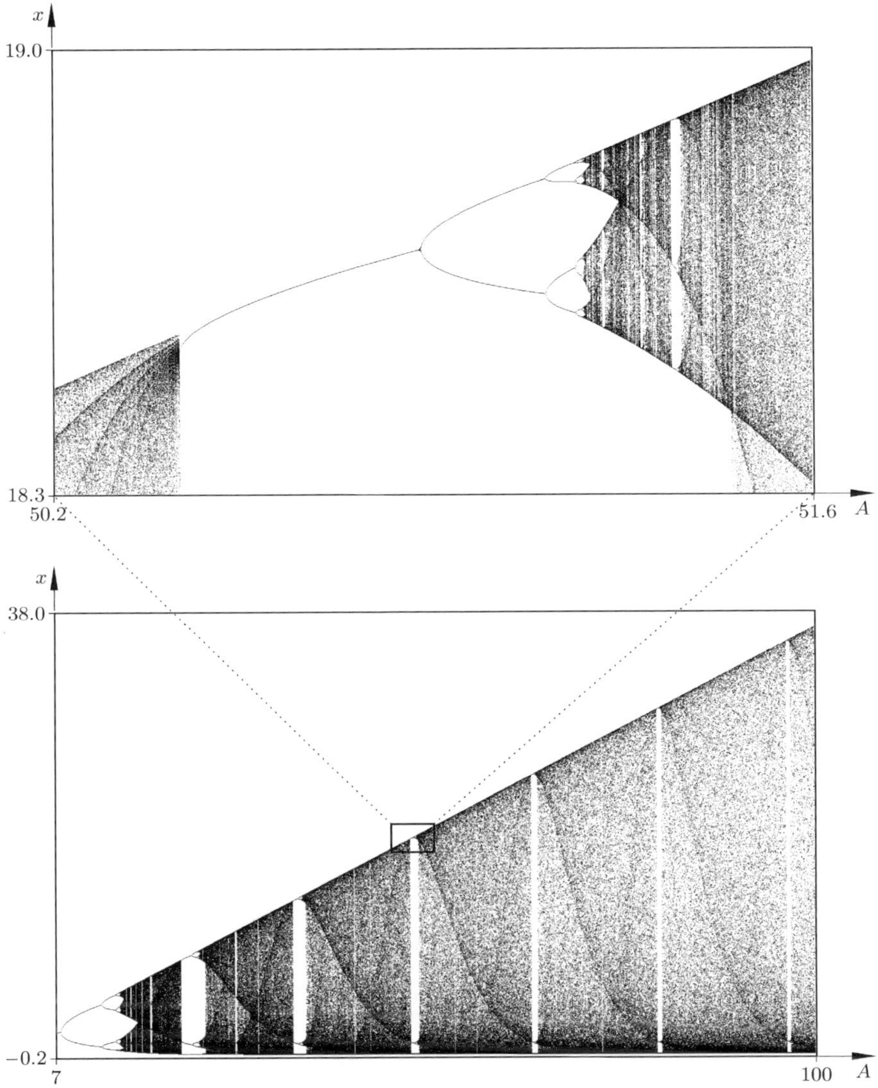

Fig. 6.7.18: Periodic windows: crises and self-similarity

Figure 6.7.17 reproduces the bifurcation diagrams for the parameter ranges $7 \leqslant A \leqslant 19$ and $7 \leqslant A \leqslant 50$. The transition to chaos is very similar to that of the logistic map (fig. 6.7.6) and that of the sine map (fig. 6.7.7). However, since the control parameter A is not bounded from above, arbitrarily large sections of the bifurcation diagram can be plotted. In doing so, the periodic windows can be magnified considerably and we can recognise crises, i. e. abrupt changes in the chaotic attractors, as well as the self-similar structure of repeated period doublings (see fig. 6.7.18).

One can distinguish between three types of crises (Grebogi et al., 1982; Grebogi et al., 1983). If an unstable orbit approaches and then collides with the basin boundary of an chaotic attractor for a critical parameter value, the attractor suddenly disintegrates and one refers to a *boundary crisis* (Grebogi et al., 1982) or *contact bifurcation* (Abraham et al., 1997). If, on the other hand, such a collision takes place within the basin of attraction of a chaotic attractor, then, all of a sudden, the attractor bursts and expands. This case is called an *internal crisis*. If an unstable orbit encounters the basin boundaries of two chaotic attractors, both attractors may eventually merge abruptly (Ott, 1993). All three types can be observed for the logistic map and the previously discussed growth model.

We now turn our attention to the open question: to what extent does the asymmetry in the evolution of the function of Axe^{-x} affect the limit function $g(x)$ and the constants a and δ? Starting from the fixed point condition eq. (6.7.35), we formulate a power series ansatz for $g(x)$ where again, $g(x)$ at the point $x = 0$ should be normalised to 1 and possesses a quadratic maximum there, i. e. we presume

$$g(0) = 1\,, \quad g'(0) = 0\,, \quad g''(0) \neq 0 \tag{6.7.98}$$

However, we cannot assume for reasons of symmetry that only even powers in x appear in the series expansion. Differentiating the relation $g(x) = -ag(g(-x/a))$, see eq. (6.7.35), we obtain in succession

$$g'(x) = \quad g'(g)\cdot g' \tag{6.7.99}$$

$$g''(x) = -\frac{1}{a}\left[g''(g)g'^2 + g'(g)g''\right] \tag{6.7.100}$$

$$g'''(x) = \quad \frac{1}{a^2}\left[g'''(g)g'^3 + 3g''(g)g'g'' + g'(g)g'''\right] \tag{6.7.101}$$

For typographical simplicity, we omitted the arguments of the functions whenever these arc given by $(-x/a)$. In conjunction with eq. (6.7.98), eq. (6.7.100) yields

$$g''(0) = -\frac{1}{a}g'(1)g''(0)$$

or, since $g''(0) \neq 0$,

$$g'(1) = -a \tag{6.7.102}$$

Evaluating eq. (6.7.101) at the point $x = 0$, we deduce, applying the previous expressions,

$$g'''(0) = \frac{1}{a^2}[-ag'''(0)] \tag{6.7.103}$$

We can certainly exclude $a = -1$ since $g(x) \equiv 1$ would otherwise follow from the fixed point expression, eq. (6.7.35). However, this means

$$g'''(0) = 0 \tag{6.7.104}$$

i. e. the cubic term vanishes in the series expansion for $g(x)$. We can now establish the expressions for the higher derivatives of $g(x)$, namely

$$g^{(\mathrm{k})}(x) = (-a)^{1-\mathrm{k}}\sum_{\mathrm{i}=1}^{\mathrm{k}} f_{\mathrm{i}}(\xi)g^{(\mathrm{i})}\big(g(\xi)\big)\,, \qquad \text{with} \quad \xi = -x/a \tag{6.7.105}$$

where

$$g^{(k)}(x) = \frac{d^k g(x)}{dx^k}$$

Here, the factors $f_i(\xi)$ are products, formed from derivatives of $g(\xi)$ and deduced by implicit differentiation of the iterated function $g(g(\xi))$. Moreover, for $i = 1$, $f_1 = g^{(k)}(\xi)$, and at the point $x = 0$, $g'(g(0)) = g'(1) = -a \neq 1$ applies. The decision whether $g^{(k)}(0) \neq 0$ holds thus depends on whether further factors f_i ($i \neq 1$) which do not vanish at the point $x = 0$ appear in the sum of eq. (6.7.105). Let us therefore consider the form of the individual products of f_i for $i \neq 1$

$$c_i (g')^{n_1} (g'')^{n_2} \dots (g^{(k-1)})^{n_{k-1}} \tag{6.7.106}$$

where we have (the $n_j \dots$ are natural numbers)

$$1 \cdot n_1 + 2 \cdot n_2 + \dots + (k-1) \cdot n_{k-1} = k \tag{6.7.107}$$

If a product in eq. (6.7.106) contains the factor g' respectively g''' (i. e. $n_1 \neq 0$ or $n_3 \neq 0$), this term vanishes at the point $x = 0$. Equation (6.7.107) is a diophantine equation for the n_j which implies a series of allowable combinations. If we put $k = 4$, for example, we obtain

$$1 \cdot n_1 + 2 \cdot n_2 + 3 \cdot n_3 = 4$$

where we are only interested in solutions with $n_1 = 0$ and $n_3 = 0$. For $n_2 = 2$ and $n_1 = n_3 = 0$, we obtain a non-vanishing factor in eq. (6.7.107). If we express the 4th derivative explicitly

$$g''''(x) = -\frac{1}{a^3}\left[g''''(g)g'^4 + 6g'''(g)g''g'^2 + 4g''(g)g'''g' + 3g''(g)g''^2 + g'(g)g''''\right] \tag{6.7.108}$$

and evaluate it at the point 0, the penultimate term corresponds exactly to the non-vanishing term for $n_2 = 2$ and we recognise that $g''''(0) \neq 0$ generally holds. On the other hand, for $k = 5$, eq. (6.7.107) yields

$$1 \cdot n_1 + 2 \cdot n_2 + 3 \cdot n_3 + 4 \cdot n_4 = 5$$

Since the sum is odd, this equation can only be satisfied if n_1 and n_3 do not vanish simultaneously. At the point $x = 0$, each factor f_i ($i \neq 1$) will thus either contain $g'(0)$ or $g'''(0)$ so that $g^{V}(0) = 0$ also holds, i. e. fifth-order powers do not appear in the series expansion of $g(x)$ either. Applying mathematical induction, we can ultimately prove that the conditions in eq. (6.7.98) lead to the conclusion that $g(x)$ contains only even powers in x. Hence, $g(x)$ – and thus also the Feigenbaum constants a und δ – depend only on the order of the maximum of the mapping function $f(x)$ but not on the symmetry or asymmetry in $f(x)$. If we consider fig. 6.7.9 or 6.7.10 again, this result implies that the influence of possible asymmetries vanishes as a result of the iteration process. The consequence is that the mapping instruction $f(x) = Axe^{-x}$ possesses the same universal function $g(x)$ and the same Feigenbaum constants a and δ as the logistic map.

6.7.5 Link with Second-Order Phase Transitions and Renormalisation Methods

Feigenbaum's scheme for determining the universal quantities characterising the transition to chaos via period doublings originates in the renormalisation theory developed at the end of the 1960s and beginning of the 1970s to explore second-order phase transitions in which the second-order derivative of the free energy (but not the first order) possesses a discontinuity at the critical point, cf. (Peitgen and Richter, 1986b). We also speak of continuous phase transitions since the entropy is continuous at the critical point (Stanley, 1971). L. P. Kadanoff had already developed the basic ideas on self-similarity as early as 1966, see (Kadanoff, 1966). The explicit formulation of the renormalisation theory goes back to K. G. Wilson, however, who was awarded the Nobel Prize for Physics in 1982 for his work in the field of critical phenomena; for extensive references, see (Wilson, 1983).

In general, each substance can appear in three states: solid, liquid, gaseous. However, numerous other states or phases exist, distinguished from one another in their crystalline structure, their electric or magnetic characteristics. Modifications in the external parameters such as pressure or temperature, for example, can cause the transformation of the varying phases of a material into the others (Bruce and Wallace, 1989; Lipowsky, 1983).

A classic example of a second-order phase transition is the transition of water to vapour. At a pressure of 1 at, the boiling point of water is 100°C. At the transition from the liquid to the vaporous state, a volume of a given mass expands suddenly to about 1600 times its original volume, i. e. the phase transition is accompanied by an abrupt fall in the density. If the external pressure is increased, the boiling point is raised while the difference of the densities $\varrho_F - \varrho_G$ of the liquid and gaseous phase decreases continuously until ultimately, at a critical point, it becomes zero and the boiling point is reached at a critical pressure $p_{cr} = 218$ at and a critical temperature $T_{cr} = 374$°C (cf. fig. 6.7.19).

The expression which interrelates the pressure, density and temperature of a system is called the equation of state: $f(p, \varrho, T) = 0$. In a three-dimensional space with the coordinates p, ϱ and T, the equation of state describes a surface where each point of the surface corresponds to a state of equilibrium of the system. On this surface, we can now mark lines which separate the single phase regions and which are called phase boundaries. Figure 6.7.19a shows the projection onto the p, T-plane of these coexistence curves along which two phases can coexist. For low temperatures and pressures, areas of the solid and vaporous phase are divided by the sublimation curve I which bifurcates at a triple point into the melting pressure curve II and the vapour pressure curve III. In contrast to the melting pressure curve, the phase boundary between the gaseous and liquid phase terminates at a critical point with the coordinates p_{cr}, T_{cr}. For $T < T_{cr}$, the transition from the gaseous to the liquid phase is associated with a jump in the density, $\varrho_G - \varrho_F$. In the critical state, the dry vapour has the same density as the boiling liquid, i. e. the two states cannot be distinguished from one another. For $T > T_{cr}$, the transition from low to high density takes place continuously. For this reason, we speak in this case of a continuous phase transition. Figure 6.7.19b reproduces the projection of the vapour pressure curve

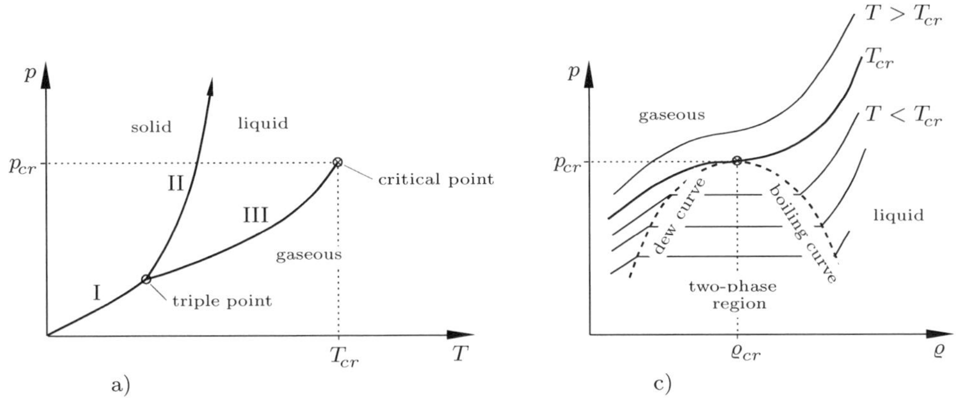

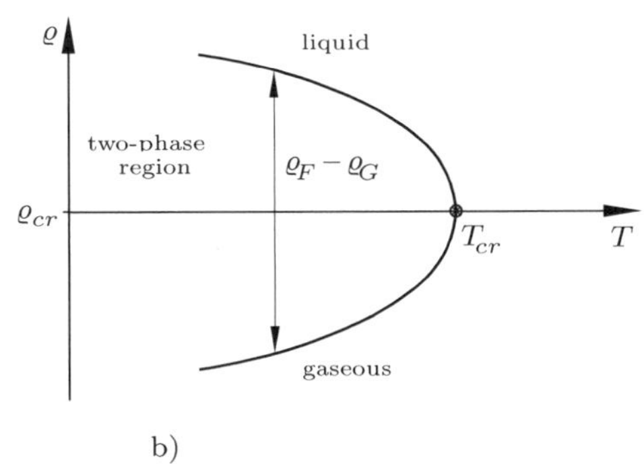

Fig. 6.7.19

Critical point of a liquid
a) coexistence curves: (I) sublimation-curve, (II) melting pressure curve, (III) vapour pressure curve
b) density jump between the liquid and gaseous phase along the coexistence curve
c) isotherms

III, i. e. the coexistence curve between the gaseous and the liquid phase, onto the ϱ, T-plane. It becomes clear that, for $T < T_{cr}$ the transition from one phase to another is accompanied by a jump in the density $\varrho_F - \varrho_G$. In addition, fig. 6.7.19c shows some isotherms in a p, ϱ-diagram. For low temperatures, there is a large difference between the density of the gaseous and that of the liquid phase; this tends, however, towards zero for $T \to T_{cr}$. For $T > T_{cr}$, only a single "fluid" phase exists; it is no longer possible to distinguish between "gaseous" and "liquid". At the critical point, there are vapour bubbles and water droplets mixed together on all scales, from the visible down to the atomic dimension. Near the critical point, the droplets reach a dimension of the order of the wavelength of light so that light is scattered strongly; the mixture of water and vapour becomes milky and we speak of a *critical opalescence* (Stanley, 1971). At the critical point, the substance hence looks alike over a wide spectrum of scales. It then possesses a special symmetry property, the so-called self-similarity, characterised by scale invariance.

Another continuous phase transition takes place in ferromagnets at the so-called Curie point. At room temperature, a ferromagnet – such as the needle of a compass, for example – exhibits permanent magnetisation, i. e. the constituent molecular magnets of which it is composed are basically aligned with one another. Depending on the time history, two ferromagnetic phases may arise, expressed in a positive ($M_\uparrow$) and a negative ($M_\downarrow$) magnetisation M. If no external magnetic field H is applied ($H = 0$), the two phases coexist at temperatures $T < T_{cr}$ (fig. 6.7.20a). An increase in the temperature which is accompanied by an increase of the thermal fluctuations gradually causes more and more constituent molecular magnets to flip. The ordered state is transformed to a disordered one, the magnetisation decreases continuously and – for iron, for example – drops to zero at a Curie temperature of $T_{cr} = 770°\mathrm{C}$ (fig. 6.7.20 a, b). Above T_{cr}, the "communication" of the constituent magnets breaks down, i. e. they can no longer align with one another over large distances and longer periods of time. The magnetisation M is denoted "order parameter" and, when approaching the critical temperature, it obeys a power law

$$M \propto |T - T_{cr}|^{\beta} \tag{6.7.109}$$

or

$$\lim_{T \to T_{cr}} M|T - T_{cr}|^{-\beta} = \mathrm{const}$$

Experimental measurements on three-dimensional magnets yield the value $\beta \approx 1/3$ for the so-called critical exponent. Moreover, it has been observed that some characteristic values of the material, such as the susceptibility or the specific heat, become infinitely large for $T = T_{cr}$.

What is surprising is the fact that completely differing continuous phase transitions display the same characteristics at the critical point: the order parameter tends continuously towards zero, various "material constants" diverge and – this is the astounding aspect – the critical exponents have a universal character, i. e. they coincide for totally different physical systems. This means that, at the critical point, details of the microscopic interactions of different physical systems have no effect on the qualitative behaviour. This also explains the success of particularly simple models such as the Ising model, for example; see e. g. (Bruce and Wallace, 1989). The renormalisation theory has not only provided practical methods – though no instant recipes – for the determination of the critical exponents; with its help, it has become possible to divide a whole series of second-order phase transitions into universality classes. Here, evidently (only) two parameters are essential: the spatial dimension of the physical system under investigation and the number of degrees of freedom, i. e. the number of possible discrete states, of the order parameter (Lipowsky, 1983; Peitgen and Richter, 1986b).

In order to set up a link between the transition to chaos via period doubling bifurcations in one-dimensional maps and the phase transition of a ferromagnet, we return once more to the fundamental ideas of self-similarity which go back to Kadanoff. To this end, we observe the same magnet on various scales.

For $T = 0$, absolute order prevails, all the constituent magnets are aligned in parallel. At very high temperatures, on the other hand, absolute disorder prevails, there are no longer any correlations between neighbouring magnetic dipoles. In both cases, the magnet appears the same on each scale, all the molecular magnets are either aligned in parallel or distributed completely randomly. At low temperatures $T < T_{cr}$, there is a very strong coupling between the spins of neighbouring magnetic

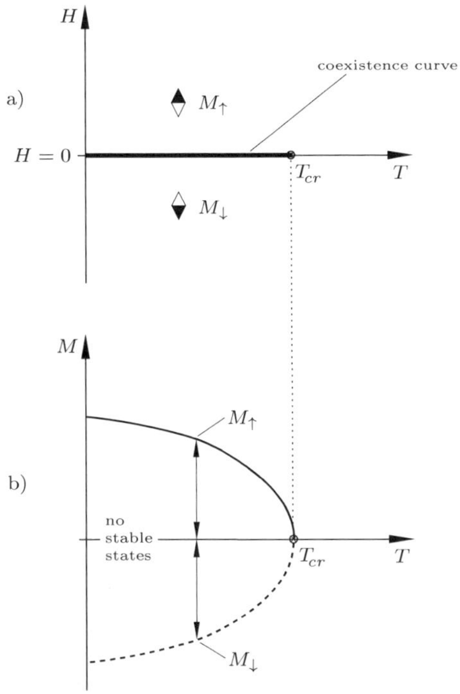

Fig. 6.7.20

Critical point of a ferromagnet
a) coexistence curve of two magnetic phases $M_\uparrow$ and $M_\downarrow$ for $H = 0$
b) spontaneous magnetisation of the two phases (upper branch $M_\uparrow$, lower branch $M_\downarrow$) along the coexistence curve

dipoles. Globally, the magnet appears ordered, i. e. the thermodynamic mean of the spins, its magnetisation, is different from zero. However, if we zoom out smaller sections, we can recognise fluctuations. "Clusters" of correlated spins emerge with a magnetisation deviating from the global orientation. An increase in the temperature is associated with a decrease in the spin coupling and an expansion of the clusters. If we zoom out one section, we obtain the same impression as if we were observing the whole magnet in the state of lower order respectively at a higher temperature. An alteration of the scale thus corresponds to an alteration of the temperature. At the critical point, an enlargement or reduction of the section causes no alteration; there are clusters of correlated spins on every scale, i. e. at the critical point, the magnet is in a scale-invariant state. It is the task of the renormalisation theory to formulate the corresponding transformations and scaling laws. Here, $T = 0$ and $T = \infty$ are stable fixed points, $T = T_{cr}$ an unstable fixed point. The critical exponents can now be deduced from the characteristics of the unstable fixed point. In the case of the logistic map, the task was to set up the transformations T^* respectively T, eqs. (6.7.26) and (6.7.31), and to determine the universal constants. In (Schuster and Just, 2006), the analogy between the phase transition in a ferromagnet and the route to chaos via period doubling is summarised in a table.

The application of renormalisation methods is promising when the physical problem is based on a complex microscopic behaviour influencing the macroscopic behaviour. In problems – such as are often met with in the engineering sciences – the mean of the microscopic fluctuations is zero; a classic description is possible on the basis of continuum equations. If we are interested in the static or dynamic displacement of a beam or in the pressure and velocity profile of a laminar flow, we do not need to consider the fluctuations on a microscopic scale.

6.8 A Descriptive Introduction to Synergetics

In this last section of Chapter 6, we give a short survey of an interdisciplinary branch of science called synergetics. Synergetics has sparked an interesting development in the understanding of self-organising processes in physics, biology, sociology, economics and medicine. The list of publications on this subject is correspondingly extensive. For those wishing a more generally comprehensible overview, we recommend the two books (Haken, 1981) and (Haken and Wunderlin, 1991); the more mathematically inclined reader can turn to the two standard works (Haken, 1990; Haken, 1987).

In two earlier sections on centre manifolds and normal forms (sections 6.2 and 6.3), we presented two methods which enable us to reduce and simplify systems of non-linear ordinary differential equations to essentials in such a way that the qualitative character of the flow in the neighbourhood of a bifurcation is retained. The method of normal forms involves a transformation technique with which the initial non-linear system can be considerably simplified if not even linearised. The centre manifold method, on the other hand, essentially involves an approximation technique for significantly reducing the number of equations required to describe the long-term behaviour. The fundamental principle of synergetics is the reduction of the degrees of freedom at the critical point respectively at the bifurcation point. In the context of synergetics, this is called the *slaving principle* and was formulated in 1975 by Hermann Haken (Haken, 1975b; Haken, 1975a). It states: "In the neighbourhood of critical points where a system can change from one state to another, a small number of collective variables, the *order parameters*, determine the macroscopically self-organised state of an open system by dominating or slaving the other degrees of freedom" (Wunderlin, 1985).

Let us consider non-linear dissipative systems composed of many subsystems far from thermodynamic equilibrium and through which a flow of energy and matter takes place. Changes of the control parameters of such systems can trigger the formation of spatio-temporal structures near instabilities. In general, we observe a clear separation of time scales: the newly occurring unstable modes vary slowly in comparison with the stable modes which, due to dissipation, decay rapidly so that, after only a short time, the unstable modes dominate the dynamics and are therefore denoted order parameters. Nevertheless, the stable modes play an important role: their non-linear coupling with the system prevents the unstable modes from growing limitlessly and thus lead to a stabilisation of the system. One important consequence is generally a drastic, system-immanent reduction in the number of degrees of freedom. In the case of purely temporal deterministic processes, the slaving principle and the method of the centre manifold lead to statements which are qualitatively the same; the claim of synergetics goes further, however. Synergetics deals with equations of motion which, in addition to the temporal evolution, also include spatial inhomogeneities and fluctuating forces.

In the following, we present the basic concepts of synergetics. The word is borrowed from the Greek and suggests "working together". The term synergetics was coined by Hermann Haken for a new area of research describing the co-operation of many sub-systems to a whole.

A popular and often very successful method of understanding processes or structures in nature is the analytical approach of breaking down the whole system into smaller and smaller parts and drawing conclusions about the behaviour of the original system from the behaviour of these component parts. However, nowadays, more than ever, we must recognise that the whole is more than the sum of the parts or, as Goethe put it, *"Dann hat er die Teile in der Hand, fehlt leider! nur das geistige Band"* (Faust I, 1808). What Goethe lamented is the subject of synergetics. It investigates the fundamental principles according to which individual parts co-operate in a self-organising manner in order ultimately to create a structural whole. New structures emerge spontaneously whenever the behaviour alters dramatically; it is precisely in this phase of instability that regulating or self-organising "forces" appear to be at work, causing the emergence of new structures.

We observe the formation of structures respectively organised structures not only in physics, chemistry and biology, but also in sociology, economics and medicine. In spite of the diversity of the component parts, we enquire into uniform fundamental laws which enable us to understand how such new structures come into being. If we restrict ourselves to qualitative alterations on the macroscopic level, synergetics reveals laws based on very few concepts – such as instability, order parameters and slaving principle – which can, moreover, be formulated in mathematical terms.

In summary, it can be said that synergetics is a field of interdisciplinary research which can be applied in the various areas mentioned above (from physics to psychology). It deals with open systems, composed of many sub-systems whose dynamics is fed by a flow of energy and/or matter. If the exchange of energy and matter with the surroundings reaches a critical value μ_{cr}, both spatial and temporal structures respectively patterns can be formed spontaneously on the macroscopic level of observation by the self-organisation of the sub-systems (microscopic level). Synergetics enables us to reconstruct this process of structure formation mathematically, thus rendering it comprehensible. The pre-requisite for structure formation is that the system is far from thermodynamic equilibrium and that the participating sub-systems are non-linearly correlated since only non-linearity allows bifurcations and thus the formation of patterns.

The aim of synergetics is to describe the complex behaviour of a system near an instability (phase transition) by the (spatio-)temporal evolution of some few variables. These quantities are called order parameters. They depend, among other things, on the flows of energy and matter through the system which are characterised by a set of control parameters $\boldsymbol{\mu}$. What is decisive is that the overwhelming majority of degrees of freedom of the system can be expressed uniquely by the order parameters, i. e. they are slaved by the order parameters. The practical consequence of the slaving principle is the reduction of the system to some few essential degrees of freedom since the numerous remaining ones are dominated by the few order parameters determining the total behaviour of the system.

After these very general remarks on synergetics, we now sketch the mathematical framework for the description of structure formation in six steps.

i. Equation of motion

In the first step, our task is to set up the evolution equations respectively the equations of motion on the descriptive level of the sub-systems. We collect the independent variables which determine the state of the spatially extended system in the state vector $\boldsymbol{x}(\boldsymbol{r}, t)$, where $\boldsymbol{r}$ describes the spatial dependency. The equation of motion then takes, for example, the form

$$\dot{\boldsymbol{x}} = \boldsymbol{F}(\boldsymbol{x}, \boldsymbol{r}, \nabla_{\boldsymbol{r}}, \boldsymbol{\mu}, t) + \boldsymbol{f}(t) \tag{6.8.1}$$

Here, $\boldsymbol{F}$ is a non-linear function of the current state $\boldsymbol{x}$ of the whole system. The operator $\nabla_{\boldsymbol{r}}$ expresses the dependence on all spatial derivatives and the fact that we are dealing with an open system is documented by a set of control parameters collected in the vector $\boldsymbol{\mu}$. The influence of stochastic fluctuations is expressed by $\boldsymbol{f}(t)$, which describes the coupling of the system to its environment under the assumption that, on a macroscopic level, the dynamics is independent of microscopic, very rapidly fluctuating states of motion. The fluctuating term can be modelled with delta-correlated white noise (Haken, 1990; Gardiner, 2004).

In order to put our subsequent argumentation in the context of the preceding chapter and to simplify the formalism, we restrict ourselves in the following to purely temporal processes, neglecting the fluctuations but without distorting the essential approach of synergetics. Equation (6.8.1) is then simplified to a deterministic system of ordinary differential equations, which can be written in autonomous form

$$\dot{\boldsymbol{x}} = \boldsymbol{F}(\boldsymbol{x}, \boldsymbol{\mu}) \tag{6.8.2}$$

ii. Stationary basic solution

In the second step, we calculate the stationary basic state of eq. (6.8.2), such as a limit set: fixed point, limit cycle or torus, for example. The simplest case is the state of equilibrium $\boldsymbol{x}_s$ which we obtain by solving the equation

$$\dot{\boldsymbol{x}}_s = \boldsymbol{o} \qquad \text{resp.} \qquad \boldsymbol{F}(\boldsymbol{x}_s, \boldsymbol{\mu}) = \boldsymbol{o} \tag{6.8.3}$$

(cf. sections 3.1 and 3.2).

iii. Linear stability analysis

In the third step, we analyse the dynamic behaviour in the vicinity of the fixed point $\boldsymbol{x}_s$. To this end, we select a small deviation $\widetilde{\boldsymbol{x}}(t)$ from the stationary state $\boldsymbol{x}_s$ and express this as

$$\boldsymbol{x} = \boldsymbol{x}_s + \widetilde{\boldsymbol{x}} \tag{6.8.4}$$

Following the linear stability analysis of section 5.4.1, we establish the following equation of motion in the neighbourhood of a fixed point $\boldsymbol{x}_s$

$$\dot{\widetilde{\boldsymbol{x}}} = \boldsymbol{L}\widetilde{\boldsymbol{x}} + \boldsymbol{N}(\widetilde{\boldsymbol{x}}, \boldsymbol{\mu}) \tag{6.8.5}$$

respectively in the linearised form

$$\dot{\widetilde{\boldsymbol{x}}} = \boldsymbol{L}\widetilde{\boldsymbol{x}} \tag{6.8.6}$$

with

$$\boldsymbol{L} = \left.\frac{\partial \boldsymbol{F}}{\partial \widetilde{\boldsymbol{x}}}\right|_{\boldsymbol{x}_s} = \boldsymbol{L}(\boldsymbol{x}_s, \boldsymbol{\mu}) \tag{6.8.7}$$

Since the linear matrix $\boldsymbol{L}$ does not depend on time, the solution ansatz $\widetilde{\boldsymbol{x}}(t) = \mathrm{e}^{\lambda t}\widetilde{\boldsymbol{x}}_0$, cf. section 3.1, leads to an eigenvalue problem

$$[\boldsymbol{L} - \lambda \boldsymbol{I}]\widetilde{\boldsymbol{x}}_0 = \boldsymbol{o} \tag{6.8.8}$$

which can be solved if the system determinant vanishes, cf. eq. (3.1.4). The eigenvalues λ_i and the appertaining eigenvectors $\boldsymbol{y}_\mathrm{i}$ satisfy the condition

$$\lambda_\mathrm{i}\boldsymbol{y}_\mathrm{i} = \boldsymbol{L}\boldsymbol{y}_\mathrm{i} \tag{6.8.9}$$

(Note, that the summation convention is not used here.) Selecting the eigenvectors $\boldsymbol{y}_\mathrm{i}$ as the basis vectors for the new system of coordinates $\boldsymbol{\xi}$ then, in accordance with eq. (3.1.12) and in the case of linearly independent eigenvectors, we can transform the initial linearised system, eq. (6.8.6), into a decoupled one

$$\dot{\xi}_\mathrm{i} = \lambda_\mathrm{i}\xi_\mathrm{i} \qquad \text{or} \qquad \dot{\boldsymbol{\xi}} = \boldsymbol{\lambda}\boldsymbol{\xi} \tag{6.8.10}$$

where $\boldsymbol{\lambda} = \lceil \lambda_1 \ldots \lambda_\mathrm{n} \rfloor$. In accordance with eq. (3.1.7), eq. (6.8.10) can be solved directly and we obtain

$$\xi_\mathrm{i}(t) = \xi_\mathrm{i}(0)\mathrm{e}^{\lambda_\mathrm{i} t} \tag{6.8.11}$$

The solution eq. (6.8.11) demonstrates that, if the real parts of the eigenvalues λ_i are negative, the perturbation $\widetilde{\boldsymbol{x}}$ decays; consequently, the state of equilibrium $\boldsymbol{x}_s$ is then stable (cf. section 5.4.1). We recall that the linearised system eq. (6.8.7), is dependent on the control parameters $\boldsymbol{\mu}$. The modification of only one control parameter – that of μ_1, for example – can transform a formerly stable state of equilibrium $\boldsymbol{x}_s$ into an unstable one. As we know, the equilibrium solution $\boldsymbol{x}_s$ becomes unstable if the real part of at least one eigenvalue changes from negative to positive (cf. section 6.2). Thus, the point of instability for the first eigenvalue λ_1, for example, is defined by the condition

$$\mathrm{Re}(\lambda_1(\mu_1)) = 0 \tag{6.8.12}$$

Summarising, we note that the linear stability analysis provides the following important findings about the original non-linear system eq. (6.8.2):

a. With its aid, the critical point respectively point of instability is determined. This means that at least one control parameter adopts a critical value and that the state of equilibrium $\boldsymbol{x}_s$ respectively the fixed point is non-hyperbolic for $\boldsymbol{\mu}_{cr}$.
b. It provides the eigenvectors $\boldsymbol{y}_{\mathrm{i}}$ respectively, in the case of spatial systems, if we think of an oscillating string, the eigenforms, also denoted modes.
c. It allows a classification in the equivalence classes stable or unstable, a procedure familiar to us from section 6.2 (centre manifold theory).

The capabilities of linearisation are thus exhausted. A further analysis of the initial system, eq. (6.8.2), requires consideration of the effects of non-linearity.

iv. The complete non-linear equations

The linear stability analysis has provided us with information about the dynamics near the critical point. In the following, our interest is directed at the dynamic behaviour beyond the critical point. In section 6.2, it became clear that this is only possible if the complete non-linear equations, eq. (6.8.5), are considered in the analysis. Furthermore, we know that the space directions defined by the eigenvectors $\boldsymbol{y}_{\mathrm{i}}$ play a primary role in the stability consideration. The transformation of the non-linear equations eq. (6.8.5) into the basis system $\boldsymbol{y}_{\mathrm{i}}$ thus suggests itself. We recall that, due to eq. (3.1.7), the general solution of the linearised system, eq. (6.8.6), can be expressed as

$$\widetilde{\boldsymbol{x}}(t) = \sum_{\mathrm{i}=1}^{\mathrm{n}} \xi_{\mathrm{i}} \mathrm{e}^{\lambda_{\mathrm{i}} t} \boldsymbol{y}_{\mathrm{i}} \tag{6.8.13}$$

In analogy with eq. (6.8.13), we now set up the solution vector $\widetilde{\boldsymbol{x}}(t)$ of the non-linear system, eq. (6.8.5), in the reference system spanned by the eigenvectors $\boldsymbol{y}_{\mathrm{i}}$ and note that here, the time dependence is expressed through the appertaining amplitudes $\xi_{\mathrm{i}}(t)$. We obtain the following ansatz in the basis vectors $\boldsymbol{y}_{\mathrm{i}}$

$$\widetilde{\boldsymbol{x}}(t) = \sum_{\mathrm{i}=1}^{\mathrm{n}} \xi_{\mathrm{i}}(t) \boldsymbol{y}_{\mathrm{i}} \tag{6.8.14}$$

Considering eq. (6.8.9) and the fact that the eigenvalues λ_{i} depend on $\boldsymbol{\mu}$, we substitute the solution ansatz eq. (6.8.14) in the initial equation eq. (6.8.5) and find

$$\sum_{\mathrm{i}=1}^{\mathrm{n}} \dot{\xi}_{\mathrm{i}}(t) \boldsymbol{y}_{\mathrm{i}} = \sum_{\mathrm{i}=1}^{\mathrm{n}} \lambda_{\mathrm{i}}(\boldsymbol{\mu}) \xi_{\mathrm{i}}(t) \boldsymbol{y}_{\mathrm{i}} + \boldsymbol{N}\Big(\sum_{\mathrm{i}=1}^{\mathrm{n}} \xi_{\mathrm{i}} \boldsymbol{y}_{\mathrm{i}}, \boldsymbol{\mu}\Big) \tag{6.8.15}$$

In order to derive an equation of motion for the amplitudes $\xi_{\mathrm{i}}(t)$ of the eigenforms respectively modes, we introduce the adjoint row vectors $\boldsymbol{y}_{\mathrm{i}}^{+}$ which form a biorthogonal system together with the eigenvectors $\boldsymbol{y}_{\mathrm{i}}$

$$\boldsymbol{y}_{\mathrm{i}}^{+} \boldsymbol{y}_{\mathrm{j}} = \delta_{\mathrm{ij}} \tag{6.8.16}$$

where the Kronecker symbol δ_{ij} equals 1 for i = j and zero for i $\neq$ j. Multiplication of eq. (6.8.15) by $\boldsymbol{y}_j^+$ yields the following equations of evolution for the amplitudes $\xi_i(t)$

$$\dot{\boldsymbol{\xi}} = \boldsymbol{\lambda}(\boldsymbol{\mu})\boldsymbol{\xi} + \overline{\boldsymbol{N}}(\boldsymbol{\xi}, \boldsymbol{\mu}) \tag{6.8.17}$$

where the diagonal form in the eigenvalues, for example, holds if all eigenvalues of the eigenvalue problem eq. (6.8.8) are different. The components of $\overline{\boldsymbol{N}}$ are obtained from the scalar product of $\boldsymbol{N}$ and $\boldsymbol{y}_j^+$.

We conclude the fourth step by splitting the components of the $\boldsymbol{\xi}$-vector, eq. (6.8.17), into two groups. The control parameters are selected in such a way that the system is near a state of instability. This means that the real parts of one or more eigenvalues λ_i have just undergone a zero passage. We now assemble all eigenvalues with negative real parts ($\mathrm{Re}\lambda^s < 0$) in $\boldsymbol{\lambda}^s$ and all those with positive real parts ($\mathrm{Re}\lambda^u \gtrsim 0$) in $\boldsymbol{\lambda}^u$. The appertaining eigenvectors are $\boldsymbol{y}^s$ and $\boldsymbol{y}^u$. Correspondingly, we split the components of the amplitude vectors $\boldsymbol{\xi}(t)$ in $\boldsymbol{\xi}^s$ und $\boldsymbol{\xi}^u$. In the following, we dispense with the dependence on $\boldsymbol{\mu}$ since the system, as already mentioned, is near an instability. Due to the splitting, eq. (6.8.17) now becomes, cf. eqs. (6.2.15)

$$\dot{\boldsymbol{\xi}}^u = \boldsymbol{\lambda}^u \boldsymbol{\xi}^u + \overline{\boldsymbol{N}}^u(\boldsymbol{\xi}^u, \boldsymbol{\xi}^s) \tag{6.8.18}$$

$$\dot{\boldsymbol{\xi}}^s = \boldsymbol{\lambda}^s \boldsymbol{\xi}^s + \overline{\boldsymbol{N}}^s(\boldsymbol{\xi}^u, \boldsymbol{\xi}^s) \tag{6.8.19}$$

We stress that the two equations (6.8.18) and (6.8.19) derive directly from the initial non-linear eq. (6.8.5) without any approximation; they are hence exact. The essential step for the solution of the initial equation (6.8.5) respectively eqs. (6.8.18) and (6.8.19) is based on the fundamental concept of synergetics, the slaving principle.

v. Slaving principle

We recall that we were faced with a similar problem when dealing with the centre manifold theory in section 6.2 (there $\lambda^c = 0$, here $\lambda^u \gtrsim 0$): how can the system of non-linear equations (6.2.15) be solved approximately – an exact solution is impossible – without falsifying the character of the solution? There, the idea was (cf. figs. 6.2.11 and 6.2.12) that the centre manifold $\mathbb{W}^c$ can be approximated by a function $\boldsymbol{y} = \boldsymbol{h}(\boldsymbol{x})$ for small $|\boldsymbol{x}|$. For the solution $(\boldsymbol{x}(t), \boldsymbol{y}(t))$ of the system of equations (6.2.15), however, this means that it comes to lie near the critical point on the centre manifold $\boldsymbol{y} = \boldsymbol{h}(\boldsymbol{x})$ and the dynamics may consequently be expressed in the implicit form with respect to time, $\boldsymbol{y}(t) = \boldsymbol{h}(\boldsymbol{x}(t))$. The fact that the centre manifold exists is guaranteed, under the corresponding conditions, by the centre manifold theory.

This somewhat abstract formulation, characterised by the geometrical interpretation of the flow in the phase space, can also be interpreted from the point of view of dynamics. We again consider figs. 6.2.11 and 6.2.12; it becomes clear that the stable trajectories tend to the centre manifold, i. e. the behaviour of the stable trajectories in the neighbourhood of the origin is given by the dynamics on $\mathbb{W}^c$. If we now take into account the classification in amplitudes $\boldsymbol{y}(t)$ of stable eigenforms or modes (real parts of the eigenvalues negative) and amplitudes $\boldsymbol{x}(t)$ of central eigenforms (real parts of the eigenvalues zero), it becomes clear that the amplitudes $\boldsymbol{y}(t)$ of the stable eigenforms can be expressed as a function of the (central) amplitudes $\boldsymbol{y}(t) = \boldsymbol{h}(\boldsymbol{x}(t))$.

After this brief recapitulation of the basic idea of the centre manifold theory, we now come to the central aspect of synergetics, the slaving principle. As mentioned above, additional conditions are necessary to solve the non-linear equations (6.8.18) and (6.8.19). The slaving principle is the very step required to create the necessary condition.

The slaving principle is substantiated by a hierarchy of time scales which the system itself generates near the instability. If we consider eq. (6.8.10), a simple dimensional analysis shows that "1/time" must be the dimension of the eigenvalues λ. Correspondingly, a characteristic time scale can be established in the vicinity of the critical point

$$\tau^u = \frac{1}{|\mathrm{Re}\lambda^u|} \tag{6.8.20}$$

for each amplitude $\boldsymbol{\xi}^u$ of the unstable modes. A corresponding time scale is defined for the amplitudes $\boldsymbol{\xi}^s$ of the stable modes

$$\tau^s = \frac{1}{|\mathrm{Re}\lambda^s|} \tag{6.8.21}$$

Due to the zero passage of the λ^u-eigenvalues ($|\mathrm{Re}\lambda^u| < |\mathrm{Re}\lambda^s|$), the system generates in the critical region the remarkable inequality

$$\tau^u \gg \tau^s \tag{6.8.22}$$

The time scale of the amplitude $\boldsymbol{\xi}^s$ of the stable modes is hence very much smaller than that of the amplitudes $\boldsymbol{\xi}^u$ of the unstable modes. The comparatively fast damping of the amplitudes $\boldsymbol{\xi}^s$ conditions the fact that they follow the amplitudes $\boldsymbol{\xi}^u$ of the unstable modes directly. The temporal dependence of the amplitudes $\boldsymbol{\xi}^s$ is therefore not expressed directly, but formulated implicitly in terms of the time dependence of the amplitudes $\boldsymbol{\xi}^u$ of the unstable modes. This yields the following mathematical formulation of the slaving principle

$$\boldsymbol{\xi}^s(t) = \boldsymbol{\xi}^s(\boldsymbol{\xi}^u(t)) \tag{6.8.23}$$

The slaving principle hence entails the expression of $\boldsymbol{\xi}^s$ as a function of the $\boldsymbol{\xi}^u$. This in turn means deducing $\boldsymbol{\xi}^s$ as a function of $\boldsymbol{\xi}^u$ from eq. (6.8.19). Since this equation is non-linear, it is generally impossible to solve it explicitly. Haken proposes an iterative procedure for the construction of $\boldsymbol{\xi}^s$ (Haken, 1975c; Haken, 1987).

vi. The order parameter equation

We can now use eq. (6.8.23) to reduce the number of the non-linear equations, eqs. (6.8.18) and (6.8.19), drastically. To this end, we substitute eq. (6.8.23) in eq. (6.8.18) and obtain the order parameter equation

$$\dot{\boldsymbol{\xi}}^u = \lambda^u \boldsymbol{\xi}^u + \overline{\boldsymbol{N}}^u(\boldsymbol{\xi}^u, \boldsymbol{\xi}^s(\boldsymbol{\xi}^u)) \tag{6.8.24}$$

which is governed exclusively by the amplitudes $\boldsymbol{\xi}^u$ of the unstable modes. Here also, a comparison with eq. (6.2.18) which describes the dynamics on the centre manifold is admissible. If the solution of eq. (6.8.24) is known, the solution $\boldsymbol{x}(t) = \boldsymbol{x}_s + \widetilde{\boldsymbol{x}}(t)$ of the initial system can be given. We obtain

$$\boldsymbol{x}(t) = \boldsymbol{x}_s + \sum_u \xi^u(t)\boldsymbol{y}^u + \sum_s \xi^s(\xi^u(t))\boldsymbol{y}^s \tag{6.8.25}$$

Summary:

We have offered – as we hope – a systematic introduction to synergetics in the simplest of cases, that of purely temporal processes, where the initial state is presumed to be an equilibrium $\boldsymbol{x}_s$ and where fluctuating forces are neglected, $\boldsymbol{f}(t) = \boldsymbol{o}$. More complex dynamic long-term behaviour, such as limit cycles or tori, may increase the mathematical effort considerably but do not cast doubt on the concept of synergetics.

In this brief overview, we have so far not dealt with spatially extended systems and fluctuations. However, they are necessary to drive the system from equilibrium; otherwise, it remains at rest. Fluctuations always appear in complex systems; examples are hydrodynamic fluctuations in flows or mutations in the evolution of biological systems. They play an essential role in pattern formation and allow the system to explore the phase space and perhaps to visit other nearby states of equilibrium, see for example (Ditzinger and Haken, 1989). The occurrence of fluctuating forces has the consequence that we have to deal with stochastic (partial) differential equations (Haken, 1990; Gardiner, 2004). A prime example for synergetics is the laser; it was here that the general concepts were originally developed (Haken and Graham, 1971).

A further important field of application for synergetic methods is hydrodynamics. The mathematical formulation of the Taylor and Bénard problem, for example, (cf. Chapter 7) is given by partial non-linear differential equations. The emergence of instabilities and the spontaneous, self-organised, spatio-temporal formation of structures connected with this is described in (Friedrich, 1986; Bestehorn, 1988).

Within synergetics, non-physical systems play a significant role. Models of pattern formation in biology, chemistry, sociology and perceptual physiology are investigated. In these cases, the sub-systems are already so complex that the basic equations are normally unknown, unlike in most physical systems. It is therefore necessary to take recourse to a phenomenological model formation. Synergetics also provides systematic access to this (Haken, 1988).

Further research activities in the field of pattern recognition have shown that the principles of synergetics describing structure formation can be successfully applied to pattern recognition (Fuchs and Haken, 1988). The idea that the reverse of pattern formation can be considered pattern recognition could be realised in the meantime by modelling an associative memory – one of the most important concepts of the modern theory of pattern recognition – by means of synergetic systems (Haken, 2004).

7 Convective Flow: Bénard Problem

Though this be madness,
yet there is method in't
William Shakespeare, Hamlet, II, *2*

In this chapter, we analyse in more detail the Lorenz model already discussed in section 5.2 and explore the flows and pattern formations in liquids and gases as they evolve and undergo transformations under the influence of inhomogeneities in the temperature. We then speak of convective flows, thermal convection or simply convection.

Spectacular geophysical examples of such phenomena are the circulation in the atmosphere and the oceans as well as the continental drift, the recordable drifting apart of the continental plates caused by flows in the earth's mantle. Outside our Earth as well – for example in the gaseous atmosphere of Jupiter – we can observe flow patterns which can be explained with the mechanism of the convective mixing of mass and heat.

There are basically two reasons why a theoretical access to such flow processes is impeded. On the one hand, there are the various actions impinging upon the flow behaviour. The determining quantities are the forces which are imposed upon the flow, the heat transport in the flowing medium and its complex material characteristics. The forces are divided into the lift, drag and Coriolis forces, but also cohesive forces at the surfaces. Convection, thermal conduction, thermal diffusion and thermal radiation contribute to the heat transport. On the other hand, the systems under consideration are open systems in a state far from thermodynamic equilibrium. This is a necessary pre-requisite for the formation of patterns in flows (cf. section 6.8), yet simultaneously the reason for increasing difficulties involved in a theoretical treatment. However, this should not prevent us from investigating the problem of "convection". Dealing with even the simple forms of thermal convection offers us a deeper insight into physical laws and enables us to recognise and understand the characteristics of flows and their patterns.

The fundamental manifestation of convection is the rising of warmer liquid in a gravitational field. If a lower layer of liquid or gas is heated – by whatever means – its density is reduced leading to a local rising of lower layers into colder regions whereas, conversely, colder upper layers sink. This interplay of ascending and descending fluid generates an intrinsically unstable motion in a gravitational field or a symmetry breaking leading to pattern formation. If the flow is initiated by forces acting in the medium itself, i.e. only by density differences within the fluid, we speak of free or natural convection. In contrast to this, the convection is denoted "forced" if the

J. Argyris et al., *An Exploration of Dynamical Systems and Chaos*,
DOI: 10.1007/978-3-662-46042-9_7

Fig. 7.0.1

Hexagonal convection cells in a thin layer of liquid with free surface uniformly heated from below: Bénard-Marangoni convection (Koschmieder, 1974)

motion of the liquid is generated and maintained by external influences, for example by a ventilator or a pump.

Descriptions of free convection go back as far as the 18th century; the most important experimental contributions were not made, however, until the turn of the 20th century by the French physicist Henri Bénard (Bénard, 1900). His great achievement was the discovery of a fluid cell structure like a honeycomb (fig. 7.0.1) which is generated when a thin layer of liquid with a free surface is heated from below. Explanations of the structuring in Bénard cells are more recent although their beginnings go back much further to Lord Rayleigh's theory on convection from the year 1916 (Normand et al., 1977; Cross and Hohenberg, 1993). However, the Bénard cells in fig. 7.0.1 cannot be explained by this theory. The reason for the hexagonal Bénard cells is an instability of the fluid surface which occurs due to a dependence of the surface tension on temperature. Convection in fluids with a free surface is denoted Bénard-Marangoni convection.

Nevertheless, a series of amazing phenomena can be demonstrated with a modified Bénard experiment without a free surface. To this end, we imagine the liquid as divided into many thin layers, one representative layer being bounded by two horizontal parallel plates. In order to exclude disturbing boundary effects on the flow behaviour in the centre, we assume the horizontal extension of this layer to be considerably greater than its height. Assuming that all relevant physical quantities, in particular the temperature, are distributed completely homogeneously, it is apparent that no macroscopic (but in contrast microscopic) fluid motions take place as long as the temperature in the layer of liquid and in the top and bottom plate is kept constant. The system is in a state of equilibrium. Moreover, it is easy to imagine that small perturbations – increasing the temperature of the bottom plate locally by 10°C for a short period, for example – are absorbed very fast and the original homogeneous

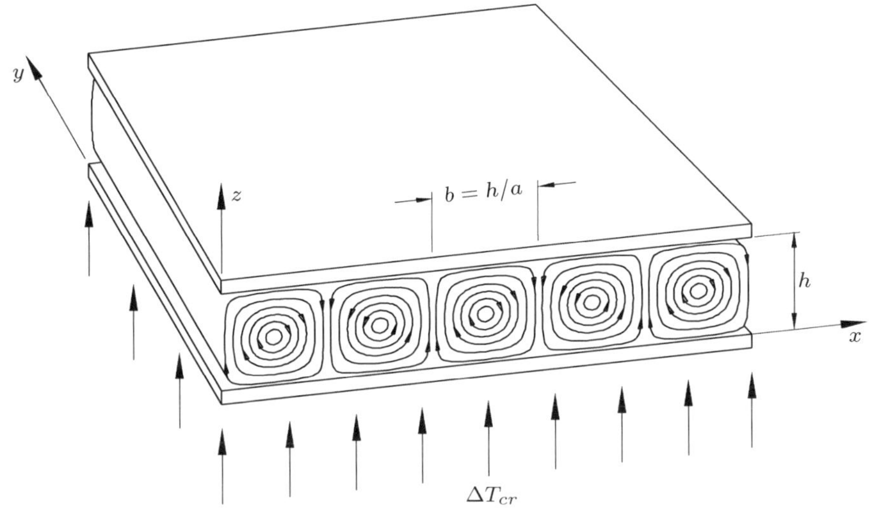

Fig. 7.0.2: Onset of roll-shaped convection cells at ΔT_{cr}

state of equilibrium is restored without leaving any enduring traces in the system. If a system is in a state in which perturbations die out sooner or later, this state is denoted *asymptotically stable.*

We now supply energy to this system externally by heating the bottom plate uniformly. If we keep the energy supply constant, the upper plate will also be heated by the thermal conduction of the liquid. Due to heat loss to the environment, the temperature of the upper plate is lower than that of the bottom plate. The result is a temperature difference between the upper and lower side which is constant and the same everywhere. In contrast to the aforementioned state of equilibrium, the temperature, pressure and density are no longer distributed homogeneously in the medium but vary in a first approximation linearly between top and bottom, between cold and warm. In spite of the external energy supply, this new state is also stable for small temperature differences ΔT, the layer of liquid remains at rest macroscopically; the system behaviour can be described as simply as that of the state of equilibrium. If we increase the heat supply – and thus the temperature difference ΔT between the two plates – suddenly, something surprising occurs. From a particular value ΔT onwards, the liquid begins to flow spontaneously on a macroscopic level and forms itself into small regular rolls. The state of Rayleigh-Bénard convection has now been reached. The flow of the liquid is at first completely regular and laminar (fig. 7.0.2).

In order to give a rough qualitative explanation of this phenomenon, we simplify the problem as follows:

1) the liquid is presumed incompressible;
2) due to the increase in temperature, the density decreases as a result of the expansion of the liquid;
3) the only external force affecting the liquid is gravity.

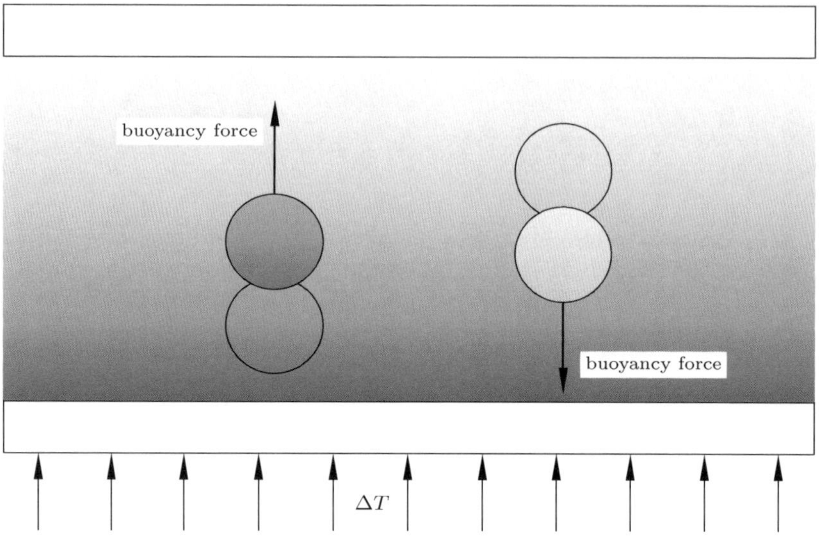

Fig. 7.0.3: Buoyancy force due to the temperature difference $\Delta T = T_{below} - T_{above}$, after (Velarde and Normand, 1980)

If we now enforce a temperature difference between top and bottom, then, due to the thermal expansion of the layer of liquid, a density gradient arises, the liquid elements in the warmer lower zone possessing a lower density than those in the cooler upper area. Since in the gravitational field there are warmer liquid particles with a lower specific density below heavier layers of liquid, the temperature gradient causes an unstable density distribution. If we enquire into the forces maintaining the motion of the liquid, it is expedient to pick out a spherical volume of the liquid and to consider the associated accelerating and retarding physical effects.

Let us first observe a drop of liquid near the bottom of the layer. Here, an increased temperature and thus a lower density in comparison with the layer above prevails. As long as the liquid element remains stationary, the fluid is at rest, since all the forces acting on the volume of the liquid are in equilibrium. On the other hand, however, this liquid element can undergo by chance (due to internal molecular motions) a slight upward displacement. The particle is now in a colder and thus denser environment of liquid, and is consequently subject to an uplift (buoyancy). The buoyancy force acts in the direction of the initial motion and reinforces it. From a random upward motion reinforced by the density gradient, an upward stream sets in. Conversely, if a particle of liquid near the upper plate is forced downwards, the environment with lower density generates in this case a downlift which reinforces the downward motion (fig. 7.0.3). These random upward and downward motions are the triggering "factors" of free or natural convection which finally encompasses the whole layer of liquid.

The rising and descending flows as they evolve in the experiment and form convection cells can, in principle, be explained by the buoyancy force which arises due to the density gradient; what cannot be explained in this way is the fact that the macroscopic

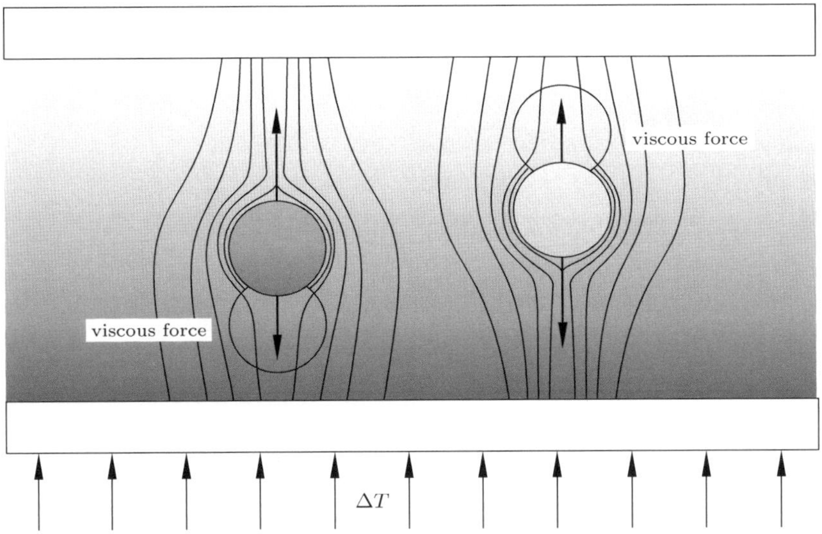

Fig. 7.0.4: Drag force as a result of viscosity resp. internal friction, after (Velarde and Normand, 1980)

flow is not initiated until a certain threshold value of ΔT is attained. Stabilising forces must arise which delay an immediate onset of the macroscopic motion and act against the destabilising effect of the buoyancy force which is generated even at the smallest ΔT. Beside the density gradient, there are indeed at least two further factors influencing the motion of the fluid particles: firstly, the viscosity of the fluid which delays the motion (fig. 7.0.4) and, secondly, the heat transport as a result of thermal conduction respectively thermal diffusion which causes a temperature equalisation between a fluid particle and, for example, the cooler environment (fig. 7.0.5).

If the convection flow triggered by the buoyancy force and reduced by drag (fluid resistance) due to friction is slower than the heat exchange due to diffusion, the convection comes to a halt. The heat supplied to the fluid from below is then dissipated by thermal diffusion alone. This is the explanation for the existence of a critical threshold value ΔT_{cr} known to us from the experiment. Temperatures just above the critical temperature difference ΔT_{cr} generate the formation of motion patterns in the fluid already mentioned. If buoyancy predominates, the fluid rises at any chosen point, moves along the upper plate, cools, sinks downwards, moves along the bottom plate towards its initial state, warms up and then ascends again. The fluid cells which evolve spontaneously in self-organisation rotate, as shown in fig. 7.0.2, clockwise and anti-clockwise alternately.

Summarising, we note that the most important physical processes involved in Bénard convection are the expansion of the fluid when heated, the viscous effects within the fluid and the heat exchange between the various fluid elements.

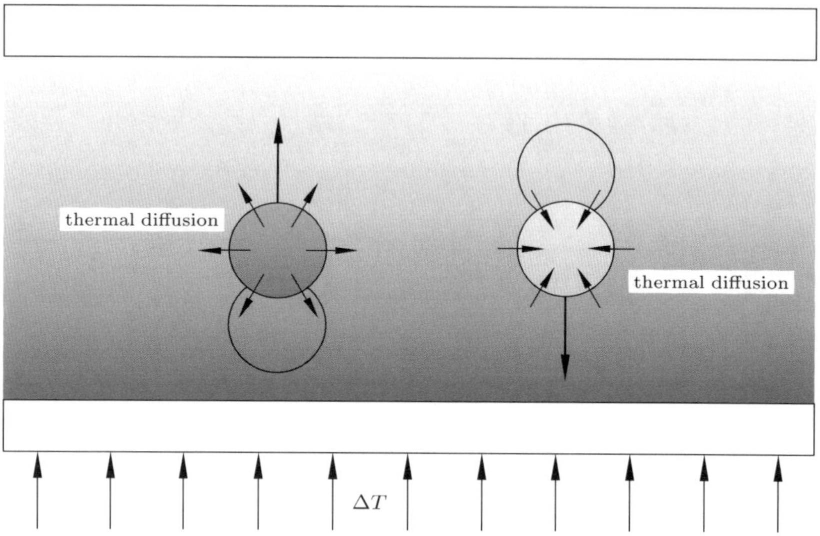

Fig. 7.0.5: Temperature equalisation as a result of thermal diffusion, after (Velarde and Normand, 1980)

The astonishing thing about these Bénard cells is that they evolve – depending on the height of the system – with a characteristic length in the millimetre range (10^{-1}cm), while the mean range of the intermolecular forces is in the order of magnitude of Ångström (10^{-8}cm). If we call to mind that the staggering number of 10^{21} molecules in one Bénard cell demonstrate coherent behaviour in the interplay of thermal motions, it becomes clear why such a complex phenomenon long evaded a classical physical interpretation. The Rayleigh-Bénard experiment confronts us with manifestations which were, in physics, previously the domain of quantum mechanics, namely the duality of chance and determinism. The fact that convection patterns evolve in a completely reproducible manner at the same threshold value ΔT_{cr} is pure determinism. Whether an arbitrarily chosen cell rotates clockwise or anti-clockwise is, however, not predictable and is left exclusively to chance respectively microscopic initial perturbations (symmetry breaking). Rayleigh-Bénard convection is a good example for the interaction of chance and determinism – not only on the atomic level, but also in the macroscopic world accessible to our senses.

We naturally also have to ask ourselves which phenomena occur when the temperature difference between the top and bottom plate is increased still further. In spite of the formation of the Bénard cells, the flow pattern initially remains constant in time. If the temperature difference passes a second critical value, however, the flow begins to pulsate periodically. Subsequently, a secondary periodic flow is superimposed on the first. Although the flow behaviour is now more complex it is nevertheless still regular respectively laminar and thus predictable with the exception of symmetry breaking. Following a further increase in the temperature difference beyond a new critical threshold value, the cellular structure disintegrates, the temporal evolution of the flow velocity possesses no recognisable periodicity, the flow is indeed turbulent, and thus spatially and temporally chaotic (fig. 7.0.6).

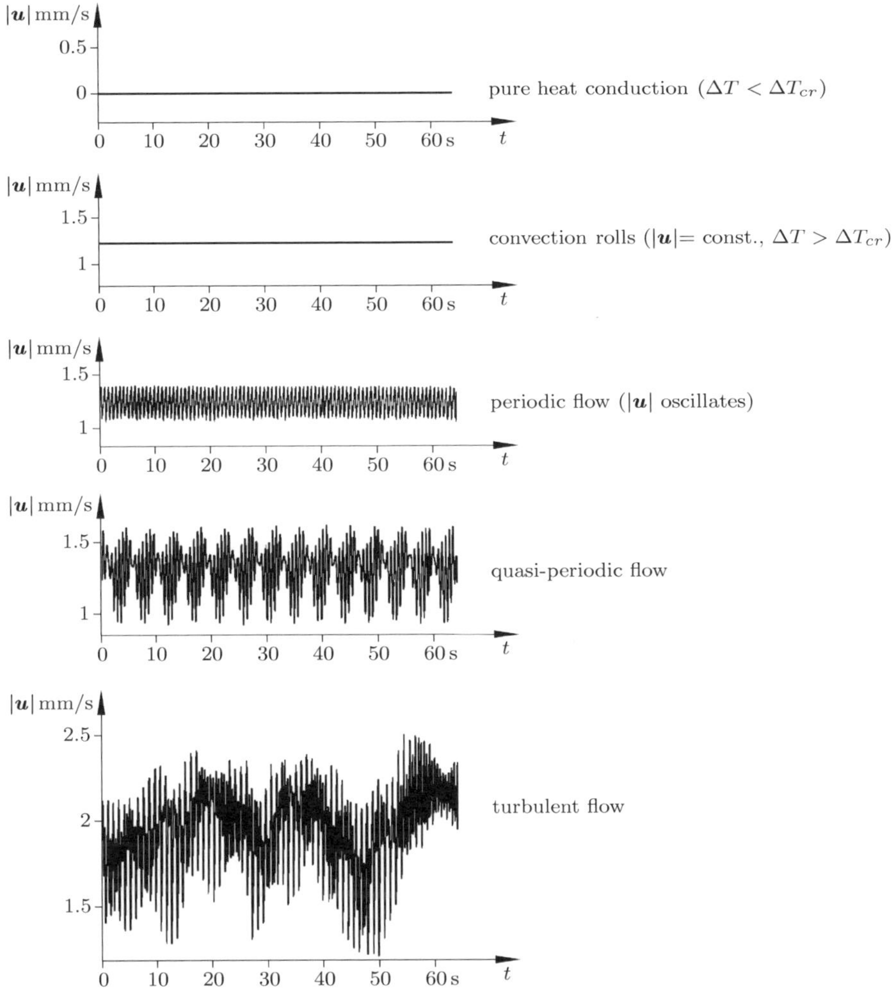

Fig. 7.0.6: Five different flow states of the Rayleigh-Bénard experiment corresponding to increasing temperature difference ΔT, after (Graham, 1982)

After this introductory qualitative description of the Bénard convection problem, we now set up the equations necessary for a mathematical formulation of the model.

7.1 Basic Hydrodynamic Equations

In this section, we establish the basic equations for a one-component fluid. These fundamental equations describe the dynamics of liquid and gaseous systems phenomenologically. The specific structure of the molecules is of no relevance for the characteristics on the macroscopic level. Accordingly, the liquid or gas is considered a continuum and the continuous mass distribution in space and time can be described by the density field $\varrho(\boldsymbol{x}, t)$. In line with thermodynamics, the state is determined by the pressure $p(\boldsymbol{x}, t)$ and the temperature $T(\boldsymbol{x}, t)$. The velocity field $\boldsymbol{u}(\boldsymbol{x}, t)$ describes the state of motion of the fluid at the point $\boldsymbol{x}$ and the instant t. The flow behaviour can now be deduced from the following three conservation theorems: conservation of mass, balance of momentum and conservation of energy (Chandrasekhar, 1961; Landau and Lifschitz, 1991; Chorin and Marsden, 1993).

i. Conservation of mass

The variation of the mass in a given volume $V(\boldsymbol{x})$ *fixed in space* may be expressed by the change of the density field $\varrho(\boldsymbol{x}, t)$ (mass per unit volume)

$$\frac{\mathrm{d}}{\mathrm{d}t}\int\limits_V \varrho(\boldsymbol{x}, t)\mathrm{d}V = \int\limits_V \frac{\partial \varrho}{\partial t}\,\mathrm{d}V \tag{7.1.1}$$

This quantity can only increase in time if more liquid enters the control volume V through the surface Ω than leaves it. The following integral form of the mass balance thus holds

$$\int\limits_V \frac{\partial \varrho}{\partial t}\,\mathrm{d}V = -\int\limits_\Omega \varrho\boldsymbol{u}\cdot \mathrm{d}\boldsymbol{\Omega} \tag{7.1.2}$$

Here, $\mathrm{d}\boldsymbol{\Omega} = \boldsymbol{n}\mathrm{d}\Omega$ defines a vector of length $\mathrm{d}\Omega$ in the direction of the unit vector $\boldsymbol{n}$ normal to the surface Ω pointing out. By definition, the direction pointing from the inside to the outside is positive. Furthermore, $\boldsymbol{u}(\boldsymbol{x}, t)$ is the velocity field in the control volume. Applying the Gauss theorem, the surface integral of eq. (7.1.2) can be transformed into a volume integral and we obtain

$$\int\limits_V \left[\frac{\partial \varrho}{\partial t} + \mathrm{div}\,(\varrho\boldsymbol{u})\right]\mathrm{d}V = 0 \tag{7.1.3}$$

Since the integration volume is chosen arbitrarily, the integral can only become zero if the integrand vanishes. Equation (7.1.3) thus leads to the differential form of the mass balance resp. conservation of mass, the so-called *continuity equation*

$$\frac{\partial \varrho}{\partial t} + \mathrm{div}\,(\varrho\boldsymbol{u}) = 0 \tag{7.1.4}$$

Equation (7.1.4) expresses the fact that the mass is conserved, i.e. that the mass in a volume element can only change if matter flows into or out of the volume element. Splitting div $(\varrho\boldsymbol{u})$, we can write eq. (7.1.4) in the form

$$\frac{\partial \varrho}{\partial t} + \varrho\,\mathrm{div}\boldsymbol{u} + \boldsymbol{u}\cdot\mathrm{grad}\;\varrho = 0 \tag{7.1.5}$$

Next, a transformation into Cartesian notation proves helpful

$$\frac{\partial \varrho}{\partial t} + u_1\frac{\partial \varrho}{\partial x_1} + u_2\frac{\partial \varrho}{\partial x_2} + u_3\frac{\partial \varrho}{\partial x_3} = -\varrho\left(\frac{\partial u_1}{\partial x_1} + \frac{\partial u_2}{\partial x_2} + \frac{\partial u_3}{\partial x_3}\right) \tag{7.1.6}$$

In the case of incompressible flow or flow with constant density, $\varrho = \text{const}$, eq. (7.1.5) is simplified to

$$\text{div}\boldsymbol{u} = 0 \tag{7.1.7}$$

So far in our derivation of the temporal evolution of the density field $\varrho(\boldsymbol{x}, t)$, we have assumed an observer at a fixed point who evidently registers the *local* time derivative $\frac{\partial \varrho(\boldsymbol{x},t)}{\partial t}$. An alternative description is the total one, recorded by an observer moving with a fluid particle. In this second case, we speak of a *substantial* or *material* time derivative, denoted by the symbol $\frac{\mathrm{D}\varrho(\boldsymbol{x},t)}{\mathrm{D}t}$. The connection between the two derivatives is derived as follows: since the spatial coordinates represent explicit functions in time for any scalar field $a(\boldsymbol{x}(t), t)$ moving with the particle, we have according to the chain rule

$$\frac{\mathrm{d}a(\boldsymbol{x}(t),t)}{\mathrm{d}t} = \frac{\partial a}{\partial x_1}\frac{\mathrm{d}x_1}{\mathrm{d}t} + \frac{\partial a}{\partial x_2}\frac{\mathrm{d}x_2}{\mathrm{d}t} + \frac{\partial a}{\partial x_3}\frac{\mathrm{d}x_3}{\mathrm{d}t} + \frac{\partial a}{\partial t} \equiv \frac{\mathrm{D}a(\boldsymbol{x}(t),t)}{\mathrm{D}t} \tag{7.1.8}$$

where

$$\frac{\mathrm{D}}{\mathrm{D}t} = \frac{\partial}{\partial t} + \boldsymbol{u}\cdot\nabla \tag{7.1.9}$$

denotes the substantial derivative. The substantial time derivative of a scalar field a is thus made up of two components: firstly, the local change in time $\frac{\partial a}{\partial t}$ at the point $\boldsymbol{x}$ and secondly, a convective component $\boldsymbol{u}\cdot\text{grad}\,a$ which results from the fact that the material element moves with velocity $\boldsymbol{u}$ along the gradient field of a.

An interesting relation can be deduced if we take the balance equation in the form of eqs. (7.1.4) and (7.1.8) into account for an arbitrary local field $a(\boldsymbol{x}, t)$ (it can be a scalar or a component of a vector or tensor). We find

$$\varrho\frac{\mathrm{D}a}{\mathrm{D}t} = \frac{\partial(a\varrho)}{\partial t} + \text{div}\,(a\varrho\boldsymbol{u}) \tag{7.1.10}$$

ii. Balance of momentum

A fundamental principle of Newton's mechanics of a system of mass points is the law of balance of momentum which states that the whole momentum of the system can only alter under the influence of external forces. In analogy to Newton's mechanics of point masses, Cauchy formulated the conservation law of momentum for a compressible medium (liquid or gas): for a given material element of a liquid or gas, the momentum of this volume can only undergo a substantial change through application of the resultant of all the forces acting on this volume. The forces acting on a compressible medium include the body forces on the one hand and the surface forces on the other. The body forces are external forces such as the gravitational force. They act on every point of the volume and are independent of the state of motion of the material volume. In contrast to them, the surface forces are dependent on the state of motion of the material volume and are responsible for its deformation. Moreover, they are a function of the position, the time and the orientation of the surface element and they possess components normal and tangential to the surface.

We denote the body forces per unit mass by $\boldsymbol{f}$, the body force density by $\varrho\boldsymbol{f}$ and the surface forces per unit area applied to the volume element V_t from the surrounding matter by $\boldsymbol{t}_n(\boldsymbol{x}, t, \boldsymbol{n})$. Here, V_t denotes the transported material volume at time t which has evolved from an arbitrarily chosen initial volume V_0 at time t_0 by the flow. If we consider the rate of change of the momentum at time t for an arbitrary material volume V_t enclosed by a surface Ω_t, we have to take into account that the integrand $\varrho\boldsymbol{u}$ as well as the integration domain V_t respectively Ω_t depend on the time. Hence, one has to make use of Reynolds' transport theorem for the left-hand side of the equation for momentum balance, cf. (Becker and Bürger, 1975; Chorin and Marsden, 1993)

$$\frac{\mathrm{D}}{\mathrm{D}t}\int\limits_{V_t}\varrho\boldsymbol{v}\mathrm{d}V = \int\limits_{V_t}\varrho\frac{\mathrm{D}\boldsymbol{v}}{\mathrm{D}t}\mathrm{d}V = \int\limits_{V_t}\varrho\boldsymbol{f}\mathrm{d}V - \int\limits_{\Omega_t}\boldsymbol{t}_n\mathrm{d}\Omega \tag{7.1.11}$$

If we consider the normal vector $\boldsymbol{n}$ of the surface Ω_t, the surface force $\boldsymbol{t}_n$ can be expressed with the aid of the symmetrical Cauchy stress tensor $\boldsymbol{\sigma}$

$$\int\limits_{\Omega_t}\boldsymbol{t}_n\mathrm{d}\Omega = \int\limits_{\Omega_t}\boldsymbol{\sigma}\boldsymbol{n}\mathrm{d}\Omega \tag{7.1.12}$$

The Gauss theorem can be applied to each component of the surface integral in eq. (7.1.12) and we thus obtain the material derivative of the momentum, eq. (7.1.11), again in vector notation

$$\int\limits_{V_t}[\varrho\frac{\mathrm{D}\boldsymbol{u}}{\mathrm{D}t} - \varrho\boldsymbol{f} + \mathrm{Div}\,\boldsymbol{\sigma}]\mathrm{d}V = \boldsymbol{o} \tag{7.1.13}$$

Here, Div $\boldsymbol{\sigma}$ denotes the vector, the i-th component of which coincides with the divergence of the i-th column of the tensor $\boldsymbol{\sigma}$

$$(\mathrm{Div}\,\boldsymbol{\sigma})_{\mathrm{i}} = \frac{\partial\sigma_{\mathrm{ji}}}{\partial x_{\mathrm{j}}} = \frac{\partial\sigma_{1\mathrm{i}}}{\partial x_1} + \frac{\partial\sigma_{2\mathrm{i}}}{\partial x_2} + \frac{\partial\sigma_{3\mathrm{i}}}{\partial x_3}$$

(note the Einstein summation convention). The volume V_t is again arbitrary; thus, the volume integral in eq. (7.1.13) can only vanish if the integrand is zero. The material derivative of the momentum per unit volume follows directly in differential form for a viscous medium

$$\varrho\frac{\mathrm{D}\boldsymbol{u}}{\mathrm{D}t} = \varrho\boldsymbol{f} - \mathrm{Div}\,\boldsymbol{\sigma} \tag{7.1.14}$$

We next specify the surface forces more precisely. If the medium is at rest, then exclusively normal stresses can arise which can be expressed in the thermodynamic sense as pressure via the equation of state $p = p(\varrho, T)$. The stress tensor $\boldsymbol{\sigma}$ then takes the form

$$\boldsymbol{\sigma} = p\boldsymbol{I} \tag{7.1.15}$$

where $\boldsymbol{I}$ is the unit tensor.

Due to the molecular forces, every real medium (liquid, gas) has a certain viscosity. If the liquid is at rest, the effect of the molecular forces on the stress tensor $\boldsymbol{\sigma}$ is zero.

The situation changes completely if the viscous liquid is in motion and the liquid particles thus move relative to one another. The frictional or viscous forces which then occur are incorporated in the stress tensor $\boldsymbol{\sigma}$ by decomposing the latter into additive components

$$\boldsymbol{\sigma} = p\boldsymbol{I} - \boldsymbol{\sigma}' \tag{7.1.16}$$

Here, $\boldsymbol{\sigma}'$ is called the tensor of the frictional stresses or viscous stress tensor for short. The internal friction due to the molecular forces comes into effect via the term $\boldsymbol{\sigma}'$ in the stress tensor. The minus sign expresses that the direction of action is opposite to the pressure. If we substitute eq. (7.1.16) in eq. (7.1.14), the equation of motion for a viscous medium takes on the form

$$\varrho\frac{\mathrm{D}\boldsymbol{u}}{\mathrm{D}t} = \varrho\boldsymbol{f} - \operatorname{grad} p + \operatorname{Div}\boldsymbol{\sigma}' \tag{7.1.17}$$

The force per unit volume of eq. (7.1.17) is composed of $\varrho\boldsymbol{f}$, the external force per unit volume, $-\operatorname{grad} p$, the pressure gradient per unit volume, and $\operatorname{Div}\boldsymbol{\sigma}'$, the molecular frictional force per unit volume. The viscous stress tensor $\boldsymbol{\sigma}'$, which only exists if the medium is flowing, is expressed for Newtonian fluids as a linear function of the rate of change of the velocity[1]. Applying the tensor $\boldsymbol{\varepsilon}$ of the shear rate[2]

$$\boldsymbol{\varepsilon} = [\varepsilon_{\mathrm{ij}}] \qquad \text{with} \quad \varepsilon_{\mathrm{ij}} = \frac{1}{2}\left(\frac{\partial u_{\mathrm{i}}}{\partial x_{\mathrm{j}}} + \frac{\partial u_{\mathrm{j}}}{\partial x_{\mathrm{i}}}\right) \tag{7.1.18}$$

we use the Navier-Stokes model in the form

$$\sigma'_{\mathrm{ij}} = 2\eta\left(\varepsilon_{\mathrm{ij}} - \frac{1}{3}\delta_{\mathrm{ij}}\frac{\partial u_{\mathrm{k}}}{\partial x_{\mathrm{k}}}\right) + \eta_V\,\delta_{\mathrm{ij}}\frac{\partial u_{\mathrm{k}}}{\partial x_{\mathrm{k}}} \tag{7.1.19}$$

cf. (Landau and Lifschitz, 1991) respectively

$$\boldsymbol{\sigma}' = 2\eta\left[\boldsymbol{\varepsilon} - \frac{1}{3}(\operatorname{div}\boldsymbol{u})\boldsymbol{I}\right] + \eta_V(\operatorname{div}\boldsymbol{u})\boldsymbol{I} \tag{7.1.20}$$

where the coefficient η is denoted shear viscosity or also dynamic viscosity and η_V volume viscosity.

If we apply the relation eq. (7.1.10) to the time derivative of eq. (7.1.18) and also consider eq. (7.1.20) for the friction term $\boldsymbol{\sigma}'$, we deduce the Navier-Stokes equation for viscous flow expressed in the local time derivatives

$$\frac{\partial(\varrho\boldsymbol{u})}{\partial t} + \operatorname{Div}\left(\varrho\boldsymbol{u}\boldsymbol{u}^{\mathrm{t}}\right) = \varrho\boldsymbol{f} - \operatorname{grad} p + \operatorname{Div}\boldsymbol{\sigma}' \tag{7.1.21}$$

where the i-th component of the vector $\operatorname{Div}\boldsymbol{\sigma}'$ is given by

$$(\operatorname{Div}\boldsymbol{\sigma}')_{\mathrm{i}} = \frac{\partial}{\partial x_{\mathrm{j}}}\left[2\eta\left(\varepsilon_{\mathrm{ij}} - \frac{1}{3}\delta_{\mathrm{ij}}\frac{\partial u_{\mathrm{k}}}{\partial x_{\mathrm{k}}}\right) + \eta_V\,\delta_{\mathrm{ij}}\frac{\partial u_{\mathrm{k}}}{\partial x_{\mathrm{k}}}\right]$$

[1] Newtonian fluids, such as water or air, are characterised by a small resistance to the deformation of their elements. Hence, for these fluids the relation between the shear stress and the shear strain rate is linear. There are also non-Newtonian fluids, such as blood, tar and paint, which exhibit a greater resistance to deformation. For these fluids, the deformation depends on the strength, change and duration of the stresses applied (Truckenbrodt, 1996).

[2] In section 9.1.5.2, the shear rate tensor is denoted $\boldsymbol{S} = [S_{\mathrm{ij}}]$, as is usual in physics.

For the case of constant respectively velocity-independent material coefficients η and η_V and with eq. (7.1.18), eq. (7.1.21) can be simplified to

$$\frac{\partial(\varrho\boldsymbol{u})}{\partial t} + \text{Div}\,(\varrho\boldsymbol{u}\boldsymbol{u}^{\text{t}}) = \varrho\boldsymbol{f} - \text{grad}\,p + \eta\nabla^2\boldsymbol{u} + \left(\eta_V + \frac{\eta}{3}\right)\text{grad div}\,\boldsymbol{u} \qquad (7.1.22)$$

where the scalar operator ∇^2, the so-called Laplace operator, possesses the following form in a Cartesian reference system

$$\nabla^2 = \frac{\partial^2}{\partial {x_1}^2} + \frac{\partial^2}{\partial {x_2}^2} + \frac{\partial^2}{\partial {x_3}^2}$$

If the flow is incompressible, div $\boldsymbol{u} = 0$ holds in accordance with eq. (7.1.7). For a divergence-free flow, the equation of motion, eq. (7.1.22), is simplified considerably. After division by the factor ϱ, we obtain the Navier-Stokes equations for incompressible viscous flow

$$\frac{\partial\boldsymbol{u}}{\partial t} + \boldsymbol{u}\cdot\nabla\boldsymbol{u} = \boldsymbol{f} - \frac{1}{\varrho}\nabla p + \nu\nabla^2\boldsymbol{u} \qquad (7.1.23)$$

where $\nu = \eta/\varrho$ denotes the kinematic viscosity and the Nabla operator ∇ is used for concise notation. The viscous stress tensor $\boldsymbol{\sigma}'$, eq. (7.1.20), then also simplifies to

$$\boldsymbol{\sigma}' = 2\eta\boldsymbol{\varepsilon} \qquad (7.1.24)$$

We observe that the viscous behaviour of incompressible media can be expressed by the dynamic viscosity η alone.

iii. Conservation of energy

According to the principle of energy conservation, the total energy in a volume element V fixed in space can only vary if energy flows into or out of the volume V under consideration through its surface Ω, cf. (De Groot and Mazur, 1969)

$$\frac{\text{d}}{\text{d}t}\int_V (\varrho e)\text{d}V = \int_V \frac{\partial(\varrho e)}{\partial t}\text{d}V = -\int_\Omega \boldsymbol{J}^e\cdot\text{d}\boldsymbol{\Omega} \qquad (7.1.25)$$

We denote the energy per unit mass e and the energy flux per unit area and unit time $\boldsymbol{J}^e$. The direction of the vector $\boldsymbol{J}^e$ coincides with the direction of the transfer of energy prevailing at $\boldsymbol{x}$ at the time t. The modulus $|\boldsymbol{J}^e|$ is equal to the quantity of energy transported per unit time through the surface unit normal to the direction. Only the component normal to the surface Ω contributes to the exchange between the volume element and its environment. According to the Gauss theorem (divergence theorem), the surface integral of eq. (7.1.25) can be transformed into a volume integral. In this way, we obtain the energy conservation theorem in differential and local form

$$\frac{\partial(\varrho e)}{\partial t} = -\text{div}\,\boldsymbol{J}^e \qquad (7.1.26)$$

The specific total energy e can be split into the components: kinetic energy e_{kin}, potential energy e_{pot} and internal energy U

$$e = e_{kin} + e_{pot} + U \tag{7.1.27}$$

Three components contribute to the flux of total energy $\boldsymbol{J}^e$: the energy contribution $\varrho e\boldsymbol{u}$ conveyed away by convection, the energy flux $\boldsymbol{\sigma u}$ generated by the stress tensor $\boldsymbol{\sigma}$ at the surface and the heat flux $\boldsymbol{J}^q$

$$\boldsymbol{J}^e = \varrho e\boldsymbol{u} + \boldsymbol{\sigma u} + \boldsymbol{J}^q \tag{7.1.28}$$

What we are now seeking is the material derivative of the internal energy since, in the context of equilibrium thermodynamics, its balance equation represents the first theorem of thermodymanics.

We begin by deducing the local derivative of the specific potential energy e_{pot}. A mass element possesses potential energy if it is exposed to a conservative external force field only dependent on the position. By definition, for the external force $\boldsymbol{f}$ per unit mass, we have

$$\boldsymbol{f} = -\text{grad}\, e_{pot} \tag{7.1.29}$$

Since we assume a time-independent force field, i.e. $\frac{\partial \boldsymbol{f}}{\partial t} = \boldsymbol{o}$, we have

$$\frac{\partial e_{pot}}{\partial t} = 0 \qquad \text{resp.} \qquad \frac{\partial(\varrho\, e_{pot})}{\partial t} = e_{pot}\frac{\partial \varrho}{\partial t} \tag{7.1.30}$$

Applying the continuity condition, eq. (7.1.4), we find

$$\frac{\partial(\varrho\, e_{pot})}{\partial t} = -e_{pot}\,\text{div}\,(\varrho\,\boldsymbol{u}) \tag{7.1.31}$$

In accordance with the calculation rule for the divergence and taking eq. (7.1.29) into account, we obtain the balance of the potential energy expressed as a local time derivative

$$\frac{\partial(\varrho\, e_{pot})}{\partial t} = -\,\text{div}\,(\varrho\, e_{pot}\boldsymbol{u}) - \varrho\,\boldsymbol{u}^{\text{t}}\boldsymbol{f} \tag{7.1.32}$$

We next deduce the balance equation for the kinetic energy. For the material time derivative of the specific kinetic energy $e_{kin} = \frac{1}{2}\boldsymbol{u}^2$, we obtain, using eq. (7.1.14),

$$\varrho\frac{\mathrm{D}(\frac{1}{2}\boldsymbol{u}^2)}{\mathrm{D}t} = \varrho\,\boldsymbol{u}\cdot\frac{\mathrm{D}\boldsymbol{u}}{\mathrm{D}t} = \varrho\,\boldsymbol{u}\cdot\boldsymbol{f} - \boldsymbol{u}\,\text{Div}\,\boldsymbol{\sigma} \tag{7.1.33}$$

If we apply the calculation rule for the divergence of tensors, then

$$\text{div}\,(\boldsymbol{\sigma u}) = \boldsymbol{u}\cdot\text{Div}\,\boldsymbol{\sigma} + \boldsymbol{\sigma} : \text{Grad}\,\boldsymbol{u} \tag{7.1.34}$$

The expression $\text{Grad}\,\boldsymbol{u} = \boldsymbol{u}\cdot\nabla$ is a directional derivative, called vector gradient, and denotes a tensor, the i-th column of which is given by the gradient of the i-th component of $\boldsymbol{u}$

$$(\text{Grad}\,\boldsymbol{u})_{\text{ij}} = \frac{\partial u_{\text{i}}}{\partial x_{\text{j}}}$$

and the scalar term $\boldsymbol{\sigma} : \text{Grad}\,\boldsymbol{u}$ becomes

$$\boldsymbol{\sigma} : \text{Grad}\, \boldsymbol{u} = \sigma_{\text{ij}} \frac{\partial u_{\text{i}}}{\partial x_{\text{j}}}$$

Again, the Einstein summation convention is applied. Taking eq. (7.1.34) into account when considering eq. (7.1.33), we obtain

$$\varrho \frac{\text{D}e_{kin}}{\text{D}t} = \varrho \boldsymbol{u} \cdot \boldsymbol{f} - \text{div}\, (\boldsymbol{\sigma u}) + \boldsymbol{\sigma} : \text{Grad}\, \boldsymbol{u} \tag{7.1.35}$$

In accordance with eq. (7.1.10), the material derivative of the kinetic energy can be written in the local form

$$\frac{\partial(\varrho\, e_{kin})}{\partial t} = -\, \text{div}\, (\varrho\, e_{kin} \boldsymbol{u} + \boldsymbol{\sigma u}) + \boldsymbol{\sigma} : \text{Grad}\, \boldsymbol{u} + \varrho \boldsymbol{u} \cdot \boldsymbol{f} \tag{7.1.36}$$

The sum of potential, eq. (7.1.32), and kinetic energy, eq. (7.1.36), then reads

$$\frac{\partial}{\partial t} \left[\varrho(e_{kin} + e_{pot})\right] = -\text{div} \left\{\varrho(e_{kin} + e_{pot})\boldsymbol{u} + \boldsymbol{\sigma u}\right\} + \boldsymbol{\sigma} : \text{Grad}\, \boldsymbol{u} \tag{7.1.37}$$

We now proceed to deduce the balance equation for the internal energy U, eq. (7.1.27). To this end, we subtract eq. (7.1.37) from eq. (7.1.26), account for eq. (7.1.28) and obtain the balance of the internal energy U in the local form

$$\frac{\partial(\varrho U)}{\partial t} = -\, \text{div}\, (\varrho U \boldsymbol{u} + \boldsymbol{J}^q) - \boldsymbol{\sigma} : \text{Grad}\, \boldsymbol{u} \tag{7.1.38}$$

The relation (7.1.10), applied to eq. (7.1.38), leads to the material balance of the internal energy U

$$\varrho \frac{\text{D}U}{\text{D}t} = -\, \text{div}\, \boldsymbol{J}^q - \boldsymbol{\sigma} : \text{Grad}\, \boldsymbol{u} \tag{7.1.39}$$

Recalling the definition of q, the heat transferred per unit mass,

$$\varrho \frac{\text{D}q}{\text{D}t} = -\, \text{div}\, \boldsymbol{J}^q \tag{7.1.40}$$

we obtain for eq. (7.1.39)

$$\varrho \frac{\text{D}U}{\text{D}t} = \varrho \frac{\text{D}q}{\text{D}t} - \boldsymbol{\sigma} : \text{Grad}\, \boldsymbol{u} \tag{7.1.41}$$

To continue our consideration of eq. (7.1.41), we introduce the splitting of the stress tensor $\boldsymbol{\sigma}$ as in eq. (7.1.16). The term ($\boldsymbol{\sigma}$: Grad $\boldsymbol{u}$) can then be converted to the expression

$$\boldsymbol{\sigma} : \text{Grad}\, \boldsymbol{u} = (p\boldsymbol{I} - \boldsymbol{\sigma}') : \text{Grad}\, \boldsymbol{u} = p\, \text{div}\, \boldsymbol{u} - \boldsymbol{\sigma}' : \text{Grad}\, \boldsymbol{u} \tag{7.1.42}$$

Substituting eq. (7.1.42) into eq. (7.1.41), we find

$$\varrho \frac{\text{D}U}{\text{D}t} = \varrho \frac{\text{D}q}{\text{D}t} - p\, \text{div}\, \boldsymbol{u} + \boldsymbol{\sigma}' : \text{Grad}\, \boldsymbol{u} \tag{7.1.43}$$

Noting the definition of the specific volume $v = \varrho^{-1}$ we derive from eq. (7.1.10) in conjunction with $\varrho v = 1$

$$\varrho \frac{\text{D}v}{\text{D}t} = \text{div}\, \boldsymbol{u} \qquad \text{resp.} \qquad \varrho \frac{\text{D}\varrho^{-1}}{\text{D}t} = \text{div}\, \boldsymbol{u} \tag{7.1.44}$$

With this expression, eq. (7.1.43) can be transformed into a form corresponding to the first law of thermodynamics ($\frac{\mathrm{D}q}{\mathrm{D}t} = \frac{\mathrm{D}U}{\mathrm{D}t} + p\frac{\mathrm{D}\varrho^{-1}}{\mathrm{D}t}$)

$$\varrho \frac{\mathrm{D}U}{\mathrm{D}t} + p\varrho \frac{\mathrm{D}\varrho^{-1}}{\mathrm{D}t} = \varrho \frac{\mathrm{D}q}{\mathrm{D}t} + \boldsymbol{\sigma}' : \mathrm{Grad}\, \boldsymbol{u} \tag{7.1.45}$$

To continue with our investigation of eq. (7.1.45), we consider Fourier's law of heat conduction which states that a heat transfer occurs in the presence of a temperature gradient

$$\boldsymbol{J}^q = -\boldsymbol{\Lambda} \nabla T \tag{7.1.46}$$

where

$$\boldsymbol{\Lambda} = \begin{bmatrix} \lambda_{11} & \lambda_{12} & \lambda_{13} \\ & \lambda_{22} & \lambda_{23} \\ & & \lambda_{33} \end{bmatrix} \tag{7.1.47}$$

is the symmetric matrix of the thermal conductivity coefficients in the anisotropic case. For an isotropic medium, the $\boldsymbol{\Lambda}$-matrix is reduced to

$$\boldsymbol{\Lambda} = \lambda \boldsymbol{I} \qquad \text{resp.} \qquad \Lambda_{\mathrm{ik}} = \lambda \delta_{\mathrm{ik}} \tag{7.1.48}$$

Applying eqs. (7.1.46) and (7.1.48), the amount of the transferred heat, eq. (7.1.40), can be expressed in dependence of the temperature as

$$\varrho \frac{\mathrm{D}q}{\mathrm{D}t} = \mathrm{div}\, (\lambda \nabla T) \tag{7.1.49}$$

The term $\boldsymbol{\sigma}' : \mathrm{Grad}\, \boldsymbol{u}$ in eq. (7.1.45) represents the energy which is transformed into heat as a result of the viscosity. For a subsequent classification into incompressible and quasi-incompressible media, it is expedient to transform this term. For the Navier-Stokes model and constant material coefficients η, η_V, eq. (7.1.19), we have

$$\sigma'_{\mathrm{ij}} \frac{\partial u_{\mathrm{i}}}{\partial x_{\mathrm{j}}} = 2\eta \frac{\partial u_{\mathrm{i}}}{\partial x_{\mathrm{j}}} \left(\varepsilon_{\mathrm{ij}} - \frac{1}{3} \delta_{\mathrm{ij}} \frac{\partial u_{\mathrm{k}}}{\partial x_{\mathrm{k}}} \right) + \eta_V \frac{\partial u_{\mathrm{i}}}{\partial x_{\mathrm{j}}} \delta_{\mathrm{ij}} \frac{\partial u_{\mathrm{k}}}{\partial x_{\mathrm{k}}} \tag{7.1.50}$$

With the expression

$$\bar{\boldsymbol{\varepsilon}} = \boldsymbol{\varepsilon} - \frac{1}{3} (\mathrm{div}\, \boldsymbol{u}) \boldsymbol{I}$$

eq. (7.1.50) is simplified to

$$\boldsymbol{\sigma}' : \mathrm{Grad}\, \boldsymbol{u} = 2\eta \bar{\boldsymbol{\varepsilon}} : \bar{\boldsymbol{\varepsilon}} + \eta_v (\mathrm{div}\, \boldsymbol{u})^2 \tag{7.1.51}$$

Substituting eqs. (7.1.49) and (7.1.51) in eq. (7.1.45), we deduce the heat equation

$$\varrho \frac{\mathrm{D}U}{\mathrm{D}t} + p\varrho \frac{\mathrm{D}\varrho^{-1}}{\mathrm{D}t} = \mathrm{div}\, (\lambda \nabla T) + 2\eta \bar{\boldsymbol{\varepsilon}} : \bar{\boldsymbol{\varepsilon}} + \eta_v (\mathrm{div}\, \boldsymbol{u})^2 \tag{7.1.52}$$

The first law of thermodynamics, cf. (Sommerfeld, 1964; Kubo, 1968; Callen, 1985), instructs us that

$$\varrho T \mathrm{d}s = \varrho\, \mathrm{d}U + \varrho p\, \mathrm{d}\varrho^{-1} \tag{7.1.53}$$

The left-hand side of eq. (7.1.52) can thus be expressed in the entropy density s

$$\varrho T\frac{\mathrm{D}s}{\mathrm{D}t} = \varrho\frac{\mathrm{D}U}{\mathrm{D}t} + \varrho p\frac{\mathrm{D}\varrho^{-1}}{\mathrm{D}t} \tag{7.1.54}$$

In the following, we introduce the heat capacity or thermal capacity of a material. The heat capacity is the measurable physical quantity that characterises the amount of heat required to change the temperature of the material by one temperature unit. The thermal capacity is dependent on the kind of heating. If the heating process takes place under constant pressure, the relevant quantity is the isobaric heat capacity c_p; if the volume is kept constant, the relevant quantity is the isochoric heat capacity c_v. The two thermal capacities c_p and c_v are given by the following relations

$$c_p = T\left(\frac{\partial s}{\partial T}\right)_p \quad , \quad c_v = T\left(\frac{\partial s}{\partial T}\right)_v \tag{7.1.55}$$

The variation of the entropy density $\mathrm{d}s$ is also process-dependent; for p = const respectively v = const, we obtain the following expressions, cf. (Bošnjaković and Knoche, 1988)

$$\mathrm{d}s = \frac{c_p}{T}\mathrm{d}T - \left(\frac{\partial v}{\partial T}\right)_p \mathrm{d}p \tag{7.1.56}$$

and

$$\mathrm{d}s = \frac{c_v}{T}\mathrm{d}T + \left(\frac{\partial p}{\partial T}\right)_v \mathrm{d}v \tag{7.1.57}$$

Using eqs. (7.1.56) and (7.1.57), we can transform the left-hand side of eq. (7.1.52), eq. (7.1.54), to a form from which the special cases incompressible and quasi-incompressible are directly derivable

$$\varrho T\frac{\mathrm{D}s}{\mathrm{D}t} = \varrho c_p\frac{\mathrm{D}T}{\mathrm{D}t} - \varrho\left(\frac{\partial v}{\partial T}\right)_p T\frac{\mathrm{D}p}{\mathrm{D}t} = \varrho c_v\frac{\mathrm{D}T}{\mathrm{D}t} + \varrho\left(\frac{\partial p}{\partial T}\right)_v T\frac{\mathrm{D}v}{\mathrm{D}t} \tag{7.1.58}$$

The following standard approximations of eq. (7.1.52) can be written down immediately using eq. (7.1.54) in conjunction with eqs. (7.1.58), (7.1.8):

i. Incompressible

In the incompressible case, $\frac{\mathrm{D}v}{\mathrm{D}t} = 0$, $c_v = c_p$ and div $\boldsymbol{u} = 0$ applies, i.e. we have

$$\varrho c_v\frac{\mathrm{D}T}{\mathrm{D}t} = \varrho c_v\left(\frac{\partial T}{\partial t} + \boldsymbol{u}\cdot\nabla T\right) = \mathrm{div}\,(\lambda\nabla T) + 2\eta\boldsymbol{\varepsilon} : \boldsymbol{\varepsilon} \tag{7.1.59}$$

ii. Quasi-incompressible

In the quasi-incompressible case, the pressure p is constant by definition, i.e. $\frac{\mathrm{D}p}{\mathrm{D}t} = 0$. In this case we have

$$\varrho c_p\frac{\mathrm{D}T}{\mathrm{D}t} = \varrho c_p\left(\frac{\partial T}{\partial t} + \boldsymbol{u}\cdot\nabla T\right) = \mathrm{div}\,(\lambda\nabla T) + 2\eta\bar{\boldsymbol{\varepsilon}} : \bar{\boldsymbol{\varepsilon}} + \eta_v(\mathrm{div}\,\boldsymbol{u})^2 \tag{7.1.60}$$

where we have substituted

$$\bar{\boldsymbol{\varepsilon}} = \boldsymbol{\varepsilon} - \frac{1}{3}(\mathrm{div}\,\boldsymbol{u})\boldsymbol{I}$$

Summarising, we note that a coupled thermal and fluid flow problem is described completely by the continuity equation, eq. (7.1.4), the equation of motion respectively the Navier-Stokes equations, eq. (7.1.21) respectively (7.1.22), and the energy equation, expressed in the internal energy U, eq. (7.1.52). These five equations are formulated in the seven unknowns $\boldsymbol{u}, T, \varrho, p, U$. It is thus necessary to supplement these five equations by two equations of state, such as

$$\begin{aligned} p &= p(\varrho, T) \\ U &= U(\varrho, T) \end{aligned} \tag{7.1.61}$$

Equations (7.1.4), (7.1.22), (7.1.52) and (7.1.61) describe completely the temporal evolution of the unknown quantities $\boldsymbol{u}, T, \varrho, p, U$ of an isotropic flow for given initial and boundary conditions.

In further considerations, we restrict ourselves to an incompressible viscous medium with constant material coefficients. The relevant equations are:

i. Continuity condition, eq. (7.1.7):

$$\operatorname{div} \boldsymbol{u} = 0 \tag{7.1.62}$$

ii. Navier-Stokes equation, eq. (7.1.23):

$$\varrho \left(\frac{\partial \boldsymbol{u}}{\partial t} + \boldsymbol{u} \cdot \nabla \boldsymbol{u} \right) = \varrho \boldsymbol{f} - \nabla p + \eta \nabla^2 \boldsymbol{u} \tag{7.1.63}$$

iii. Heat equation, eq. (7.1.59):

$$\varrho c_v \left(\frac{\partial T}{\partial t} + \boldsymbol{u} \cdot \nabla T \right) = \lambda \nabla^2 T + 2\eta \boldsymbol{\varepsilon} : \boldsymbol{\varepsilon} \tag{7.1.64}$$

As mentioned in the introductory account of Chapter 7, we speak of free or natural convection if a fluid under gravity balances temperature gradients on its own by means of fluid motion, i.e. without any external influence. If convection sets in, the static equilibrium is perturbed. If the temperature distribution is a function of $\boldsymbol{x}$, $T = T(\boldsymbol{x})$, then no static equilibrium can exist in the medium. But also in the special case $T = T(z)$, in which we assume that the temperature gradient points in the direction of the gravitational acceleration, a state of static equilibrium is impossible above a critical value T_{cr}. The Bénard convection corresponds to just this special case. The fundamental equations (7.1.62) to (7.1.64) describing Rayleigh-Bénard convection can be simplified considerably. These simplifications were first introduced by Oberbeck and Boussinesq at the turn of the last century; this model of free convection is hence called the Boussinesq-Oberbeck approximation.

7.2 Boussinesq-Oberbeck Approximation

It is obvious that, in thermal convection, the buoyancy force initialises respectively maintains the motion of gas or liquid particles. The buoyancy force is generated by density differences in the fluid. The Boussinesq-Oberbeck model, which considerably simplifies the basic equations (7.1.62) to (7.1.64) for the convective motion, is based on the observation that the density variations caused by small temperature differences are also small. In the Boussinesq-Oberbeck model, the density is thus assumed to be constant everywhere except in the term of the volume-specific external force $\varrho \boldsymbol{f}$ since here, the acceleration due to $(\delta\varrho)\boldsymbol{f}$ can be larger than all other terms of the equation. Hence, it is only necessary to formulate the density dependence of the external force term $\varrho \boldsymbol{f}$. In the case of incompressible flow, we can assume that the pressure in the medium varies so little that a density variation due to the pressure variation can be neglected. What does effect the process, however, is the density variation as a result of inhomogeneous temperature distribution. In a first approximation, we can assume that the density variation depends linearly on the temperature difference. We denote the reference temperature T_0 and the isobaric volume expansion coefficient α

$$\alpha \equiv \frac{1}{v}\left(\frac{\partial v}{\partial T}\right)_p = \varrho\left(\frac{\partial \varrho^{-1}}{\partial T}\right)_p = -\frac{1}{\varrho}\left(\frac{\partial \varrho}{\partial T}\right)_p$$

Thus, for small temperature deviations $\Delta T = T - T_0$ from the reference temperature T_0, we obtain for the density $\varrho(T)$

$$\varrho(T) = \varrho_0[1 - \alpha(T - T_0)] \tag{7.2.1}$$

where the constant density value ϱ_0 and the thermal volume expansion coefficient α refers to the reference state T_0. Under the aforementioned conditions that the flow is incompressible and the material data η, c_p, λ and α are independent of the velocity respectively the temperature, we obtain, starting from eqs. (7.1.62) to (7.1.64) and applying eq. (7.2.1), the complete system of equations in the Boussinesq-Oberbeck approximation describing Bénard convection:

i. Continuity equation

$$\operatorname{div} \boldsymbol{u} = 0 \tag{7.2.2}$$

ii. Equation of motion

$$\frac{\partial \boldsymbol{u}}{\partial t} + \boldsymbol{u}\cdot\nabla\boldsymbol{u} = \frac{\varrho}{\varrho_0}\boldsymbol{g} - \frac{1}{\varrho_0}\nabla p + \frac{\eta}{\varrho_0}\nabla^2\boldsymbol{u} \tag{7.2.3}$$

$$= [1 - \alpha(T - T_0)]\boldsymbol{g} - \frac{1}{\varrho_0}\nabla p + \nu\nabla^2\boldsymbol{u} \tag{7.2.4}$$

where the vector of gravitational acceleration is $\boldsymbol{g} = \{0 \quad 0 \; -g\}$ and the kinematic viscosity is given by $\nu = \eta/\varrho_0$.

iii. Heat equation

$$\frac{\partial T}{\partial t} + \boldsymbol{u}\cdot\nabla T = \chi\nabla^2 T \tag{7.2.5}$$

where $\chi = \frac{\lambda}{\varrho_0 c_v}$ denotes the thermal diffusivity.

The term $2\eta\boldsymbol{\varepsilon} : \boldsymbol{\varepsilon}$ which represents the loss of energy due to inner friction in eq. (7.1.64) can be neglected here since, for the Bénard experiment, in the case of the liquids usually used such as water or silicon oil, it is smaller than the thermal conduction term $\chi\nabla^2 T$ by a factor of the order of magnitude 10^{-7}.

The American meteorologist Edward N. Lorenz applied the basic eqs. (7.2.2) to (7.2.5) in the early 1960s in order to examine the precision of weather forecasts (Lorenz, 1963). His simplified model for the Bénard problem, three ordinary differential equations, could not withstand experimental verification with regard to unstable system behaviour; his results nevertheless represent a fundamental contribution to and impetus for the development of the chaos theory. The astonishing and completely unexpected result of his work was that three non-linear deterministic evolution equations can already lead to irregular, chaotic behaviour.

7.3 Lorenz Model

Barry Saltzman reduced the three-dimensional convection problem to a two-dimensional one (Saltzman, 1962) by assuming that the Bénard cells in the x, z-plane evolve independently of the y-coordinate, i.e. that the roll pattern remains constant respectively homogeneous in the y-direction (fig. 7.0.2). Formally, this means that the velocity component in the y-direction and all derivatives $\partial/\partial y$ vanish. With this assumption, the basic hydrodynamic equations, eqs. (7.2.2) to (7.2.5), are simplified considerably. Setting

$$x = x_1, \; z = x_3, \; u = u_1, \; w = u_3$$

we establish for the two-dimensional convection problem the following basic equations in component notation

$$\frac{\partial u}{\partial x} + \frac{\partial w}{\partial z} = 0 \tag{7.3.1}$$

$$\begin{aligned}
\frac{\partial u}{\partial t} + u\frac{\partial u}{\partial x} + w\frac{\partial u}{\partial z} &= -\frac{1}{\varrho_0}\frac{\partial p}{\partial x} + \nu\nabla^2 u \\
\frac{\partial w}{\partial t} + u\frac{\partial w}{\partial x} + w\frac{\partial w}{\partial z} &= -\frac{\varrho}{\varrho_0}g - \frac{1}{\varrho_0}\frac{\partial p}{\partial z} + \nu\nabla^2 w
\end{aligned} \tag{7.3.2}$$

$$\frac{\partial T}{\partial t} + u\frac{\partial T}{\partial x} + w\frac{\partial T}{\partial z} = \chi\nabla^2 T \tag{7.3.3}$$

For this two-dimensional case, the introduction of a stream function $\psi(x, z, t)$ is appropriate and yields the velocities

$$u = -\frac{\partial \psi}{\partial z}\,, \qquad w = \frac{\partial \psi}{\partial x} \tag{7.3.4}$$

The two-dimensional continuity condition, eq. (7.3.1), is thus automatically satisfied. Starting from the boundary conditions for the temperature

$$\begin{aligned}
T(x, z, t)\big|_{z=0} &= T_0 + \Delta T \\
T(x, z, t)\big|_{z=h} &= T_0
\end{aligned} \tag{7.3.5}$$

we postulate a deviation $\theta(x, z, t)$ from a linear temperature profile. This leads to the following expression for the temperature field $T(x, z, t)$

$$T(x, z, t) = T_0 + \Delta T\Big(1 - \frac{z}{h}\Big) + \theta(x, z, t) \tag{7.3.6}$$

In order to eliminate the pressure p in eq. (7.3.2) for the velocity field $\boldsymbol{u}(x,z)$, we establish the difference

$$\frac{\partial}{\partial z}(\text{first equation}) - \frac{\partial}{\partial x}(\text{second equation})$$

which corresponds to taking the curl of eq. (7.3.2)

$$\frac{\partial}{\partial t}\left(\frac{\partial u}{\partial z}\right) + \frac{\partial u}{\partial z}\frac{\partial u}{\partial x} + u\frac{\partial^2 u}{\partial x \partial z} + \frac{\partial w}{\partial z}\frac{\partial u}{\partial z} + w\frac{\partial^2 u}{\partial z^2} + \frac{1}{\varrho_0}\frac{\partial^2 p}{\partial x \partial z} - \nu\frac{\partial}{\partial z}(\nabla^2 u)$$

$$-\frac{\partial}{\partial t}\left(\frac{\partial w}{\partial x}\right) - \frac{\partial u}{\partial x}\frac{\partial w}{\partial x} - u\frac{\partial^2 w}{\partial x^2} - \frac{\partial w}{\partial x}\frac{\partial w}{\partial z} - w\frac{\partial^2 w}{\partial x \partial z} - g\frac{\partial}{\partial x}\left(\frac{\varrho}{\varrho_0}\right) - \frac{1}{\varrho_0}\frac{\partial^2 p}{\partial x \partial z}$$

$$+ \nu\frac{\partial}{\partial x}(\nabla^2 w) = 0 \tag{7.3.7}$$

Applying the incompressibility condition, eq. (7.3.1), together with eq. (7.3.4), eq. (7.3.7) is simplified to

$$-\frac{\partial}{\partial t}(\nabla^2\psi) + \frac{\partial \psi}{\partial z}\frac{\partial}{\partial x}(\nabla^2\psi) - \frac{\partial \psi}{\partial x}\frac{\partial}{\partial z}(\nabla^2\psi) - g\frac{\partial}{\partial x}\left(\frac{\varrho}{\varrho_0}\right) + \nu\nabla^4\psi = 0 \tag{7.3.8}$$

We point out that $\nabla^2\psi = \partial^2\psi/\partial x^2 + \partial^2\psi/\partial z^2 = -(\partial u/\partial z - \partial w/\partial x)$ represents the rotation of the liquid in the x, z-plane. We also note that the following relation holds: $\nabla^4\psi = \nabla^2\nabla^2\psi = \partial^4\psi/\partial x^4 + 2\partial^4\psi/\partial x^2\partial z^2 + \partial^4\psi/\partial z^4$. Inserting eqs. (7.2.1) and (7.3.6) into eqs. (7.3.8) and (7.3.3) yields the basic relations of the convection model

$$\frac{\partial}{\partial t}(\nabla^2\psi) - \frac{\partial \psi}{\partial z}\frac{\partial}{\partial x}(\nabla^2\psi) + \frac{\partial \psi}{\partial x}\frac{\partial}{\partial z}(\nabla^2\psi) - g\alpha\frac{\partial \theta}{\partial x} - \nu\nabla^4\psi = 0 \tag{7.3.9}$$

$$\frac{\partial \theta}{\partial t} - \frac{\partial \psi}{\partial z}\frac{\partial \theta}{\partial x} + \frac{\partial \psi}{\partial x}\left(\frac{\partial \theta}{\partial z} - \frac{\Delta T}{h}\right) - \chi\nabla^2\theta = 0 \tag{7.3.10}$$

Equations (7.3.9) and (7.3.10) can be further simplified if we express the non-linear convective terms using the Jacobi operator

$$\frac{\partial(a,b)}{\partial(x,z)} = \frac{\partial a}{\partial x}\frac{\partial b}{\partial z} - \frac{\partial a}{\partial z}\frac{\partial b}{\partial x}$$

In this way, we obtain the equations which E. N. Lorenz initially used for his numerical experiments and which led him to the revolutionary statement that deterministic equations can indeed exhibit irregular motion behaviour (Saltzman, 1962; Lorenz, 1963). The final expressions are

$$\frac{\partial}{\partial t}\nabla^2\psi = -\frac{\partial(\psi,\nabla^2\psi)}{\partial(x,z)} + \nu\nabla^4\psi + g\alpha\frac{\partial \theta}{\partial x} \tag{7.3.11}$$

$$\frac{\partial}{\partial t}\theta = -\frac{\partial(\psi,\theta)}{\partial(x,z)} + \frac{\Delta T}{h}\frac{\partial \psi}{\partial x} + \chi\nabla^2\theta \tag{7.3.12}$$

Using the following dimensionless quantities marked with an asterisk *

$$\begin{aligned} x &= hx^* \qquad & \nabla^2 &= \left(\frac{1}{h^2}\right)\nabla^{*2} \\ z &= hz^* \qquad & \psi &= \chi\psi^* \\ t &= \frac{h^2}{\chi}t^* \qquad & \theta &= \left(\frac{\chi\nu}{g\alpha h^3}\right)\theta^* \end{aligned} \tag{7.3.13}$$

we obtain from eqs. (7.3.11) and (7.3.12) the following relations for the stream function ψ^* and the temperature deviation θ^*

$$\frac{\partial}{\partial t^*}\nabla^{*2}\psi^* = -\frac{\partial(\psi^*, \nabla^{*2}\psi^*)}{\partial(x^*, z^*)} + \sigma\frac{\partial\theta^*}{\partial x^*} + \sigma\nabla^{*4}\psi^* \tag{7.3.14}$$

$$\frac{\partial}{\partial t^*}\theta^* = -\frac{\partial(\psi^*, \theta^*)}{\partial(x^*, z^*)} + Ra\frac{\partial\psi^*}{\partial x^*} + \nabla^{*2}\theta^* \tag{7.3.15}$$

In the last two equations, two new dimensionless quantities emerge

$$\sigma = \frac{\nu}{\chi}\,, \qquad \text{Prandtl number,} \tag{7.3.16}$$

which plays a subordinate role in this case, and

$$Ra = \frac{g\alpha h^3\Delta T}{\chi\nu}\,, \qquad \text{Rayleigh number} \tag{7.3.17}$$

The Rayleigh number, which depends on the material characteristics α, χ and ν, the geometry h and the prescribed external temperature difference ΔT, plays the role of a control parameter in the Bénard problem.

We next consider the boundary conditions. In our model, it is assumed that the convection layer extends horizontally to infinity and that it is bounded in the z-direction by two planes. The temperature deviation θ resulting from the convection flow vanishes at the boundaries $z = 0$ and $z = h$ since the temperatures T_0 and $T_0 + \Delta T$ are kept constant here.

In the case of the boundary conditions for the velocity vector $\boldsymbol{u}$, we distinguish between two extreme cases, namely that the flow sticks to the boundary respectively that it does not stick. We then speak of a rigid (no-slip) respectively free boundary. In the case of the no-slip boundary plane, $\boldsymbol{u} = \boldsymbol{o}$ applies, while for a free boundary plane, only the normal component of the velocity ($w = 0$) vanishes and, moreover, no shear stresses σ_{zx} can arise.

Lorenz assumed free boundary conditions for the flow at the upper ($z = h$) and lower ($z = 0$) plate (as this is simplest to analyse). This means that the vertical velocity $w|_{z=0,h}$ and the tangential stress $\sigma_{zx}|_{z=0,h}$ vanish. For $z = 0$ and $z = h$, the following boundary conditions can be derived:

$$i. \quad w\big|_{z=0,h} = \frac{\partial\psi}{\partial x}\bigg|_{z=0,h} = 0 \qquad \Longrightarrow \qquad \psi\big|_{z=0,h} = \text{const} = 0 \tag{7.3.18}$$

From eq. (7.1.24), we obtain

$$ii. \quad \sigma'_{zx}|_{z=0,h} = \eta\,\frac{\partial u}{\partial z}\bigg|_{z=0,h} = 0 \qquad \Longrightarrow \qquad \frac{\partial^2\psi}{\partial z^2}\bigg|_{z=0,h} = 0$$

From *i.* and *ii.* there follows

$$\nabla^2\psi\big|_{z=0,h} = \frac{\partial^2\psi}{\partial x^2}\bigg|_{z=0,h} + \frac{\partial^2\psi}{\partial z^2}\bigg|_{z=0,h} = 0 \tag{7.3.19}$$

$$iii. \quad T_0,\, T_0+\Delta T = \text{const} \qquad \Longrightarrow \qquad \theta\big|_{z=0,h} = 0 \tag{7.3.20}$$

For the dimensionless equations in ψ^* and θ^*, the following boundary conditions thus apply

$$\begin{aligned} &\theta^*(x,0,t) = \theta^*(x,h,t) = \psi^*(x,0,t) = \psi^*(x,h,t) = 0 \\ &\nabla^{*2}\psi^*(x,0,t) = \nabla^{*2}\psi^*(x,h,t) = 0 \end{aligned} \tag{7.3.21}$$

Saltzman expanded the stream function ψ^* and the temperature deviation θ^* into a double Fourier series with wavelengths l in x-direction and $2h$ in z-direction (Saltzman, 1962)

$$\begin{aligned} \psi^*(x^*,z^*,t^*) &= \sum_{\mathrm{m}=-\infty}^{\infty}\sum_{\mathrm{n}=-\infty}^{\infty} \psi_{\mathrm{mn}}(\mathrm{m},\mathrm{n},t^*)\exp\Big(2\pi i\left[\mathrm{m}\tfrac{h}{l}x^* + \tfrac{\mathrm{n}}{2}z^*\right]\Big) \\ \theta^*(x^*,z^*,t^*) &= \sum_{\mathrm{m}=-\infty}^{\infty}\sum_{\mathrm{n}=-\infty}^{\infty} \theta_{\mathrm{mn}}(\mathrm{m},\mathrm{n},t^*)\exp\Big(2\pi i\left[\mathrm{m}\tfrac{h}{l}x^* + \tfrac{\mathrm{n}}{2}z^*\right]\Big) \end{aligned} \tag{7.3.22}$$

Saltzman included 52 modes in his numerical analysis. The result of his study was that, in most cases, all modes, with the exception of three, tend to zero and that these three show irregular, non-periodic behaviour for certain ranges of the control parameters. This result inspired Lorenz to express the stream function ψ and the temperature deviation θ in only three amplitudes $X(t)$, $Y(t)$ and $Z(t)$

$$\begin{aligned} \psi(x,z,t) &= \frac{\chi(1+a^2)\sqrt{2}}{a}X(t)\sin\Big(\frac{\pi a}{h}x\Big)\sin\Big(\frac{\pi}{h}z\Big) \\ \theta(x,z,t) &= \frac{\Delta T}{\pi}\frac{Ra_{cr}}{Ra}\left[\sqrt{2}Y(t)\cos\Big(\frac{\pi a}{h}x\Big)\sin\Big(\frac{\pi}{h}z\Big) - Z(t)\sin\Big(\frac{2\pi}{h}z\Big)\right] \end{aligned} \tag{7.3.23}$$

where Ra is the aforementioned dimensionless Rayleigh number

$$Ra = \frac{\alpha g h^3 \Delta T}{\chi\nu}$$

and Ra_{cr} its critical value

$$Ra_{cr} = \frac{\pi^4(1+a^2)^3}{a^2} \tag{7.3.24}$$

Figure 7.3.1 illustrates the significance of the three modes. The X-mode is proportional to the modulus of the convection velocity and thus reproduces the flow pattern. The Y-mode is proportional to the temperature difference between ascending and descending currents and stands for the temperature cells. The Z-mode is proportional to the deviation of the vertical temperature profile from linearity and represents the non-linear temperature stratification.

If we substitute eq. (7.3.23) in the initial relations eqs. (7.3.11) and (7.3.12), then, after algebraic transformations and the introduction of appropriate scaled quantities, we obtain the so-called Lorenz equations

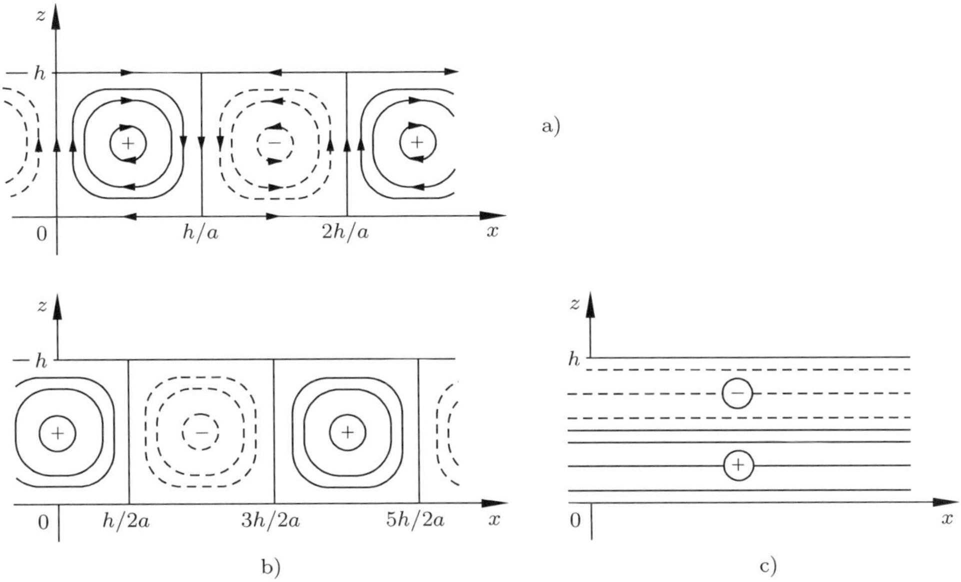

Fig. 7.3.1: The Lorenz modes in eq. (7.3.25):
a) flow pattern (X-mode), b) temperature cells (Y-mode),
c) temperature stratification (Z-mode)

$$\begin{aligned} \dot{X} &= -\sigma X + \sigma Y \\ \dot{Y} &= \quad rX - Y - XZ \\ \dot{Z} &= -bZ + XY \end{aligned} \tag{7.3.25}$$

The dot denotes the derivative with respect to the scaled time $\tau = \pi^2(1+a^2)\chi h^{-2} t$, $\sigma = \nu/\chi$ the aforementioned Prandtl number (where ν is the kinematic viscosity and χ the thermal diffusivity), $r = Ra/Ra_{cr}$ the relative Rayleigh number and $b = 4/(1+a^2)$ a measure for the cell geometry in x-direction. The critical Rayleigh number Ra_{cr} defines the marginal (neutral) stability state derived from a linear stability analysis. In fact, at the threshold $Ra = Ra_{cr}$, the eigenvalue of the critical mode changes its sign. The smallest value of Ra_{cr} possesses the value $27\pi^4/4 \approx 657.5$.

Before we discuss the long-term behaviour of the Lorenz system, eq. (7.3.25), and the corresponding representation in phase space as well as the bifurcations arising due to variations of r, we return to the physical meaning of the Lorenz variables X, Y and Z. Colour Plate V presents the spatial and temporal evolution of the deviation from the linear temperature profile $T^* = \theta$ (figure a), of the magnitude of the velocity $|\boldsymbol{u}|$ (figure b) and of the vertical velocity w (figure c). The colours illustrate the spatial dependency of the quantities. For $T^* = \theta$ in figure a, red was chosen for warmer than the linear temperature profile and blue for colder. The choice of $r = 28$ means that we are in a chaotic regime. Below the colour snapshots, a)

b)

c)

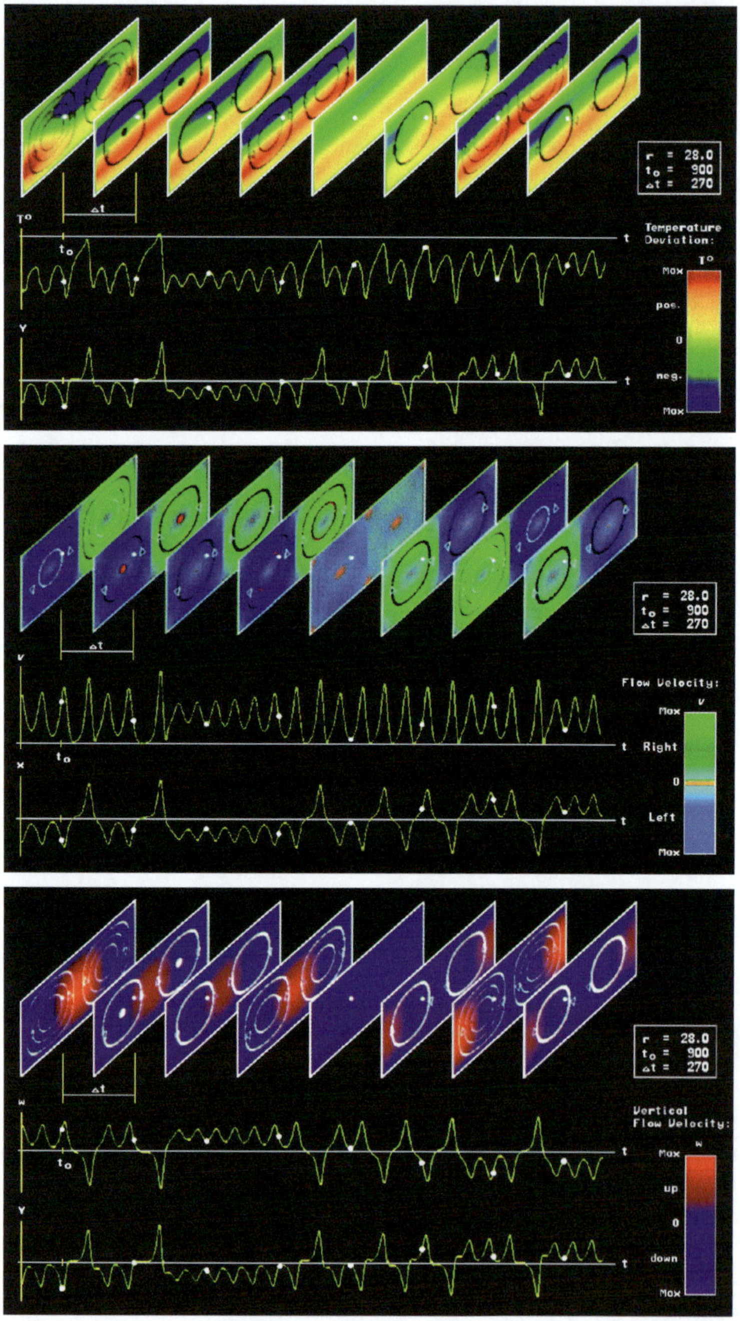

Plate V Evolution of the Lorenz systems ($r = 28$):
a) deviation T^* from the linear temperature profile
b) modulus of the velocity vector $|\boldsymbol{u}|$
c) vertical velocity w

the temporal evolution of $T^* = \theta$ and the Y-mode for a specific white point are plotted. The white point lies in the upper half of the convection cell; consequently, the temporal mean value of $T^* = \theta$ must be negative. The solid black lines within the snapshots are lines of constant modulus of the velocity vector ($|\boldsymbol{u}| = \text{const}$). The arrow indicates the sense of rotation of the convection roll. The temperature evolution of $T^* = \theta$ – cold above, warm below – is clearly recognisable. A spatial pattern which does not repeat itself in the temperature distribution signalises the chaotic regime.

Figures b and c show the quantities characterising the flow pattern. Figure b illustrates the X-mode and the magnitude of the velocity $\boldsymbol{u}$, $|\boldsymbol{u}| = \sqrt{u^2 + w^2}$. The sense of the rotation is characterised by the background colour: blue for anti-clockwise, green for clockwise. The modulus of the velocity is indicated by the colour scale. Deep blue and yellow-green are the maximum modulus of velocity in each case.

Figure c shows the vertical velocity component w primarily responsible for the formation of the convection rolls. The background colour blue represents a downward, red an upward direction. Here, too, the unpredictable change of the roll motion can be clearly recognised.

7.4 Evolution of the Lorenz System

In order to study the Lorenz equations, i.e. to get an impression of the trajectories in the phase space (X, Y, Z), we apply the numerical and mathematical techniques described in the previous sections. Should the computational operations be too extensive, we dispense with a detailed account in favour of a more descriptive interpretation of the Lorenz equations, while offering the reader references for further reading. In order to arrive at a fundamental understanding of the Lorenz system, it is necessary to know the states of equilibrium and, in dependence of the free parameter r which is a measure for the imposed temperature difference ΔT, to determine the response of the system with respect to its local and global bifurcations.

An important piece of preliminary information about a dissipative system is its volume contraction, see eq. (5.1.11). Indeed, the trace of the Jacobian matrix, the divergence of the vector field,

$$\frac{\partial}{\partial X}\big(\sigma(Y - X)\big) + \frac{\partial}{\partial Y}(rX - Y - XZ) + \frac{\partial}{\partial Z}(-bZ + XY) = -(\sigma + 1 + b) \tag{7.4.1}$$

is negative so that, due to the flow, a volume element $V(0)$ contracts exponentially in time to the volume

$$V(t) = V(0)\,\mathrm{e}^{-(\sigma+1+b)t} \tag{7.4.2}$$

cf. eq. (4.1.30). What does this mean for the Lorenz system? Due to eq. (7.4.2), each initial volume element $V(0)$ of the phase space will contract, i.e. the Lorenz system is dissipative. We again note that volume contraction does not necessarily imply that all three manifolds of a fixed point must be stable. This can be demonstrated very

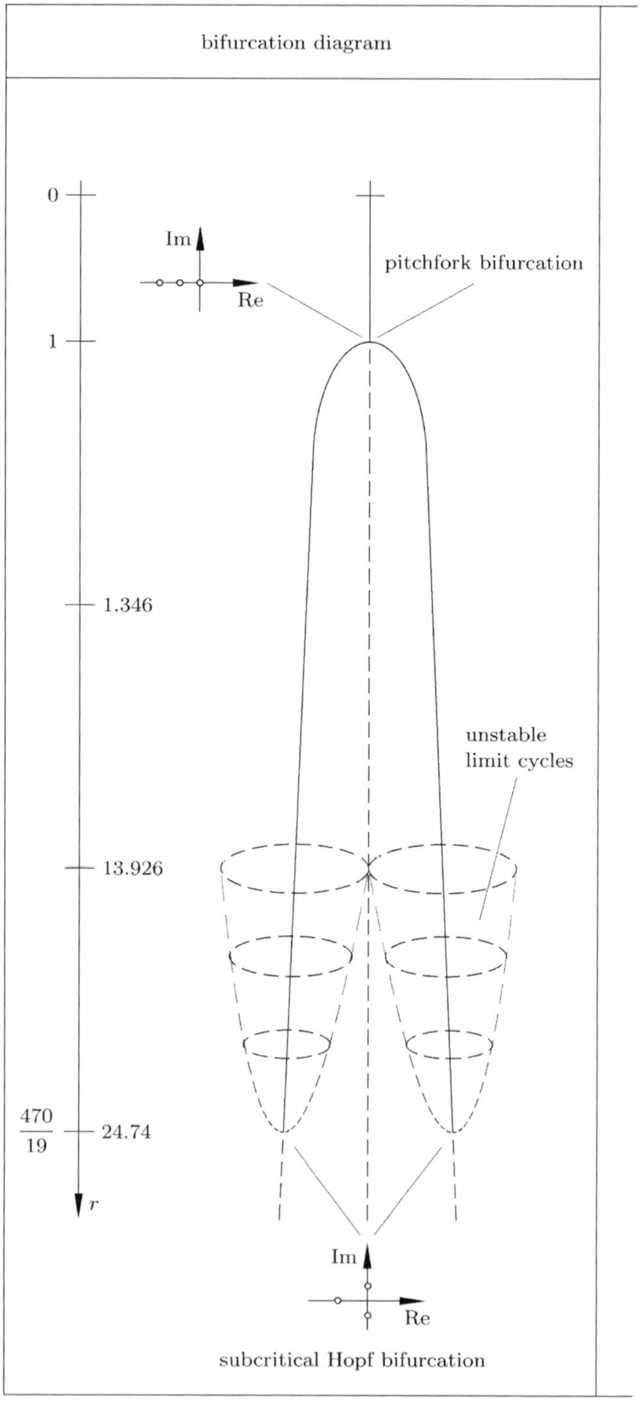

Fig. 7.4.1: Dependence of the Lorenz system on the control parameter r ($\sigma = 10$, $b = 8/3$); *Part 1:* bifurcation diagram

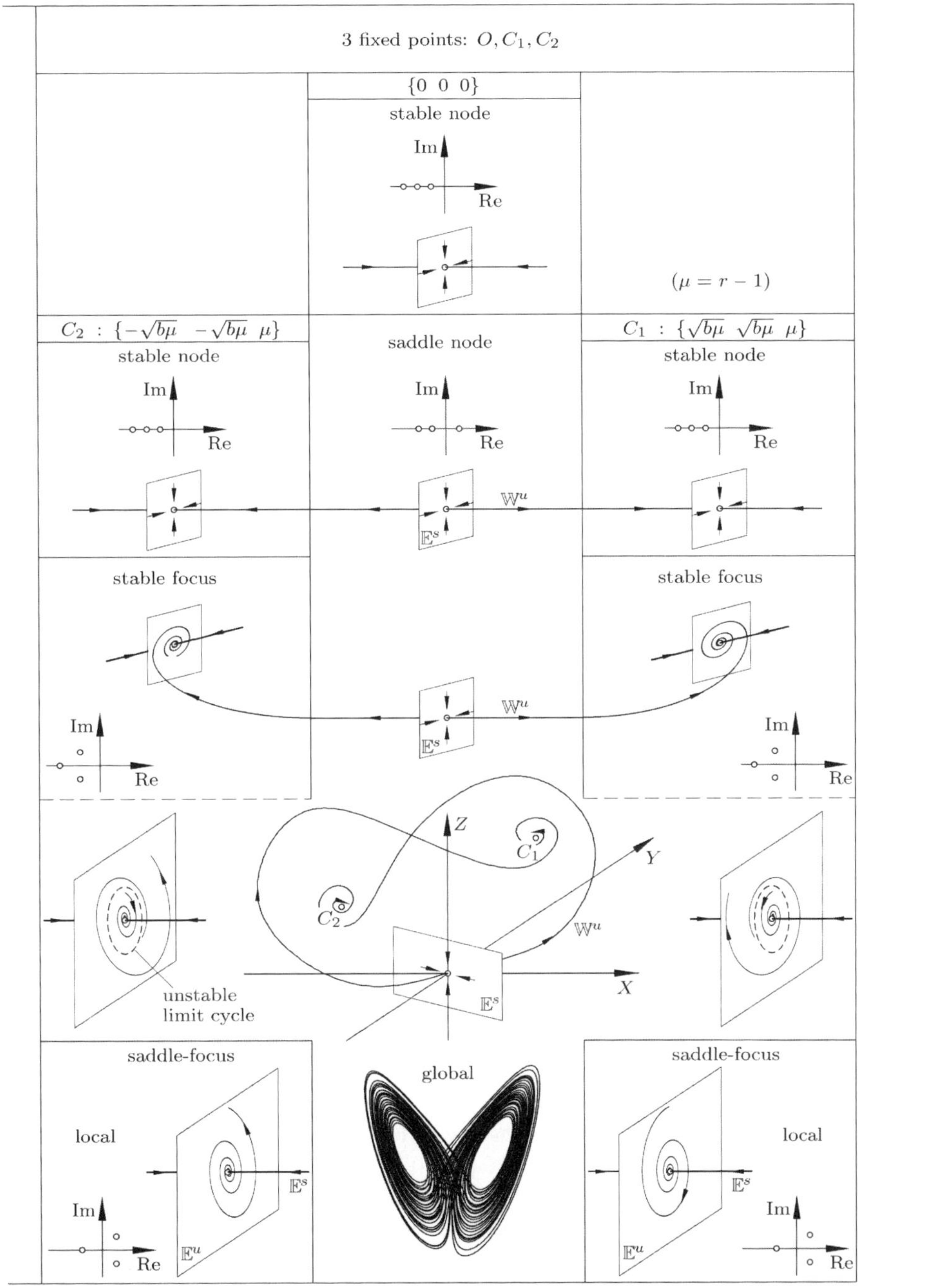

Fig. 7.4.1: Dependence of the Lorenz system on the control parameter r ($\sigma = 10$, $b = 8/3$); *Part 2:* evolution of the fixed points

easily on the two-dimensional example of a saddle-node, for which the sum of the eigenvalues is $\lambda_1 + \lambda_2 < 0$.

We next discuss the dynamics of the Lorenz system in the phase space. In order to determine the states of equilibrium respectively the stationary points, we apply the condition that time derivatives of the independent variables should vanish. The Lorenz eqs. (7.3.25) then become

$$\begin{aligned} 0 &= -\sigma X + \sigma Y \\ 0 &= \quad rX - Y - XZ \\ 0 &= -bZ + XY \end{aligned} \tag{7.4.3}$$

We first consider the trivial solution $X = Y = Z = 0$ and direct our attention to its linear stability characteristics. In accordance with section 5.4.1, we put

$$\begin{aligned} X &= 0 + \tilde{X} \\ Y &= 0 + \tilde{Y} \\ Z &= 0 + \tilde{Z} \end{aligned} \tag{7.4.4}$$

Substituting eq. (7.4.4) in eq. (7.4.3) and ignoring the non-linear terms (in the following, we dispense with the tilde symbol ˜), we obtain

$$\dot{\boldsymbol{X}} = \boldsymbol{L}\boldsymbol{X} \,, \qquad \boldsymbol{X} = \begin{bmatrix} X \\ Y \\ Z \end{bmatrix}, \quad \boldsymbol{L} = \begin{bmatrix} -\sigma & \sigma & 0 \\ r & -1 & 0 \\ 0 & 0 & -b \end{bmatrix} \tag{7.4.5}$$

In accordance with eq. (3.1.2), we use the following ansatz for the set of linear equations in eq. (7.4.5)

$$\boldsymbol{X} = \mathrm{e}^{\lambda t}\boldsymbol{y} \tag{7.4.6}$$

and, in accordance with eq. (3.1.4), calculate the three eigenvalues λ_i of the $\boldsymbol{L}$-matrix

$$\begin{aligned} \lambda_{1,2} &= -\frac{\sigma + 1}{2} \pm \frac{1}{2}\sqrt{(\sigma+1)^2 + 4\sigma(r-1)} \\ \lambda_3 &= -b \end{aligned} \tag{7.4.7}$$

The three eigenvalues are different, i.e. there exist three linearly independent eigenvectors $\boldsymbol{y}_\mathrm{i}$ with $\boldsymbol{X}_\mathrm{i} = \mathrm{e}^{\lambda_\mathrm{i} t}\boldsymbol{y}_\mathrm{i}$ as the solution. The two eigenvalues $\lambda_{1,2}$ depend on the control parameter r, i.e. for $0 < r < 1$, all three eigenvalues are negative and the fixed point (0,0,0) is thus stable. For $r = 1$, the largest eigenvalue becomes $\lambda_1 = 0$, the other two remain negative ($\lambda_2 = -[\sigma + 1]$, $\lambda_3 = -b$). The marginal stability of the origin indicates a qualitative change in the course of the trajectories. For $r > 1$, the largest eigenvalue becomes positive ($\lambda_1 > 0$), the other two remain negative. The fixed point (0,0,0) has changed from a stable node to a saddle point since, for $r > 1$, the flow in the direction of the first two eigenvectors is stable but unstable in the direction of the third eigenvector, see fig. 7.4.1.

We have presented two methods for describing the response of the non-linear dynamics beyond the critical point r_{cr}: on the one hand, the method of the centre manifold, section 6.2, and, on the other hand, the slaving principle within a synergetic approach, section 6.8. For the determination of the basic pattern of the bifurcation at the critical point $r_{cr} = 1$, we restrict ourselves here to the method of the centre manifold.

We first introduce a new control parameter $\mu = r - 1$, following the scheme of eq. (6.2.28), and reformulate the Lorenz system, eq. (7.3.25), as follows

$$\dot{\boldsymbol{X}} = \boldsymbol{L}\boldsymbol{X} + \boldsymbol{N}(\boldsymbol{X}) = \begin{bmatrix} -\sigma & \sigma & 0 \\ 1 & -1 & 0 \\ 0 & 0 & -b \end{bmatrix} \boldsymbol{X} + \begin{bmatrix} 0 \\ \mu X - XZ \\ XY \end{bmatrix}$$

$$\dot{\mu} = 0 \tag{7.4.8}$$

We must now transform these equations into the form given in eq. (6.2.15), i.e. we first transform $\boldsymbol{L}$ to diagonal form. To this end, we determine the set of right eigenvectors

$$\boldsymbol{y}_1 = \begin{bmatrix} 1 \\ 1 \\ 0 \end{bmatrix}, \qquad \boldsymbol{y}_2 = \begin{bmatrix} -\sigma \\ 1 \\ 0 \end{bmatrix}, \qquad \boldsymbol{y}_3 = \begin{bmatrix} 0 \\ 0 \\ 1 \end{bmatrix} \tag{7.4.9}$$

appertaining to the eigenvalues $\lambda_1 = 0$, $\lambda_2 = -(\sigma + 1)$, $\lambda_3 = -b$. These eigenvectors now form the basis vectors of the new $\overline{\boldsymbol{X}}$-coordinate system. In accordance with eqs. (3.1.10) and (3.1.11), we obtain the following coordinate transformation

$$\boldsymbol{X} = \boldsymbol{T}\overline{\boldsymbol{X}} \qquad \text{with} \quad \boldsymbol{T} = [\boldsymbol{y}_1 \quad \boldsymbol{y}_2 \quad \boldsymbol{y}_3] = \begin{bmatrix} 1 & -\sigma & 0 \\ 1 & 1 & 0 \\ 0 & 0 & 1 \end{bmatrix} \tag{7.4.10}$$

We also require the inverse matrix $\boldsymbol{T}^{-1}$

$$\boldsymbol{T}^{-1} = \frac{1}{1+\sigma} \begin{bmatrix} 1 & \sigma & 0 \\ -1 & 1 & 0 \\ 0 & 0 & (1+\sigma) \end{bmatrix} \tag{7.4.11}$$

In the $\overline{\boldsymbol{X}}$-coordinate system, the Lorenz system extended by μ, eq. (7.4.8), is now

$$\begin{aligned} \dot{\overline{\boldsymbol{X}}} &= \boldsymbol{T}^{-1}\boldsymbol{L}\boldsymbol{T}\overline{\boldsymbol{X}} + \boldsymbol{T}^{-1}\boldsymbol{N}(\boldsymbol{T}\overline{\boldsymbol{X}}) = \boldsymbol{D}\overline{\boldsymbol{X}} + \overline{\boldsymbol{N}}(\overline{\boldsymbol{X}}) \\ \dot{\mu} &= 0 \end{aligned} \tag{7.4.12}$$

or, expressed in a rearranged sequential order,

$$\begin{aligned} \begin{bmatrix} \dot{\overline{X}} \\ \dot{\mu} \end{bmatrix} &= \begin{bmatrix} 0 & 0 \\ 0 & 0 \end{bmatrix} \begin{bmatrix} \overline{X} \\ \mu \end{bmatrix} + (\overline{X} - \sigma\overline{Y}) \begin{bmatrix} \frac{\sigma}{\sigma+1}(\mu - \overline{Z}) \\ 0 \end{bmatrix} \\ \begin{bmatrix} \dot{\overline{Y}} \\ \dot{\overline{Z}} \end{bmatrix} &= -\begin{bmatrix} \sigma+1 & 0 \\ 0 & b \end{bmatrix} \begin{bmatrix} \overline{Y} \\ \overline{Z} \end{bmatrix} + (\overline{X} - \sigma\overline{Y}) \begin{bmatrix} \frac{1}{\sigma+1}(\mu - \overline{Z}) \\ (\overline{X} + \overline{Y}) \end{bmatrix} \end{aligned} \tag{7.4.13}$$

A comparison of eq. (7.4.13) and eq. (6.2.15) yields the following linear and non-linear terms with regard to a division into centre and stable manifolds

$$\begin{aligned} \boldsymbol{A} &= \boldsymbol{O}; & \boldsymbol{f}\,(\overline{X}, \mu, \overline{Y}, \overline{Z}) &= (\overline{X} - \sigma\overline{Y}) \begin{bmatrix} \frac{\sigma}{\sigma+1}(\mu - \overline{Z}) \\ 0 \end{bmatrix} \\ \boldsymbol{B} &= \lceil -(\sigma+1) \quad -b \rfloor; & \boldsymbol{g}\,(\overline{X}, \mu, \overline{Y}, \overline{Z}) &= (\overline{X} - \sigma\overline{Y}) \begin{bmatrix} \frac{1}{\sigma+1}(\mu - \overline{Z}) \\ (\overline{X} + \overline{Y}) \end{bmatrix} \end{aligned} \tag{7.4.14}$$

Following the results of section 6.2, a centre manifold $\mathbb{W}^c$ exists which can be described by $\overline{\boldsymbol{X}}_s = \boldsymbol{h}(\overline{\boldsymbol{X}}_c)$ respectively by two equations of the form

$$\overline{\boldsymbol{X}}_s = \begin{bmatrix} \overline{Y} \\ \overline{Z} \end{bmatrix} = \begin{bmatrix} h_1(\overline{X}, \mu) \\ h_2(\overline{X}, \mu) \end{bmatrix} \tag{7.4.15}$$

The centre manifold can be determined from the differential equation (6.2.20) in conjunction with the boundary conditions $\boldsymbol{h}(\boldsymbol{o}) = \boldsymbol{h}'(\boldsymbol{o}) = \boldsymbol{o}$; generally, however, an exact solution of this equation is not possible. However, we can approximate the function $\boldsymbol{h}(\overline{X}, \mu)$ by $\boldsymbol{\psi}(\overline{X}, \mu)$ and, in accordance with eq. (6.2.21), obtain the following relation for the residuum

$$\begin{aligned} \begin{bmatrix} R_1(\psi_1(\overline{X}, \mu)) \\ R_2(\psi_2(\overline{X}, \mu)) \end{bmatrix} &= \begin{bmatrix} \frac{\partial \psi_1}{\partial \overline{X}} & \frac{\partial \psi_1}{\partial \mu} \\ \frac{\partial \psi_2}{\partial \overline{X}} & \frac{\partial \psi_2}{\partial \mu} \end{bmatrix} \left[\begin{bmatrix} 0 & \\ & 0 \end{bmatrix} \begin{bmatrix} \overline{X} \\ \mu \end{bmatrix} + \begin{bmatrix} \frac{\sigma}{\sigma+1}(\overline{X} - \sigma\psi_1)(\mu - \psi_2) \\ 0 \end{bmatrix} \right] \\ &\quad - \begin{bmatrix} -(\sigma+1) & 0 \\ 0 & -b \end{bmatrix} \begin{bmatrix} \psi_1 \\ \psi_2 \end{bmatrix} - \begin{bmatrix} \frac{1}{\sigma+1}(\overline{X} - \sigma\psi_1)(\mu - \psi_2) \\ (\overline{X} - \sigma\psi_1)(\overline{X} + \psi_1) \end{bmatrix} \end{aligned} \tag{7.4.16}$$

We can find the two components R_1 and R_2 of the residuum as

$$\begin{aligned} R_1(\psi_1(\overline{X}, \mu)) &= \frac{\sigma}{\sigma+1} \frac{\partial \psi_1}{\partial \overline{X}} (\overline{X} - \sigma\psi_1)(\mu - \psi_2) + (\sigma+1)\psi_1 \\ &\quad - \frac{1}{\sigma+1}(\overline{X} - \sigma\psi_1)(\mu - \psi_2) \\ R_2(\psi_2(\overline{X}, \mu)) &= \frac{\sigma}{\sigma+1} \frac{\partial \psi_2}{\partial \overline{X}} (\overline{X} - \sigma\psi_1)(\mu - \psi_2) + b\psi_2 - (\overline{X} - \sigma\psi_1)(\overline{X} + \psi_1) \end{aligned} \tag{7.4.17}$$

In a first approximation, we consider only quadratic terms in $\psi_1(\overline{X}, \mu)$ and $\psi_2(\overline{X}, \mu)$, i.e. $\psi_\mathrm{i} = \mathcal{O}(2)$ for i = 1,2 simplifies eq. (7.4.17) to

$$\begin{aligned} R_1 &= (\sigma+1)\psi_1 - \frac{1}{\sigma+1}\mu\overline{X} + \mathcal{O}(3) \\ R_2 &= b\,\psi_2 - \overline{X}^2 + \mathcal{O}(3) \end{aligned} \tag{7.4.18}$$

Following the approximation theorem, section 6.2, we finally deduce from eq. (7.4.18) in a first approximation

$$\begin{aligned} h_1(\overline{X}, \mu) &= \frac{1}{(\sigma+1)^2}\mu\overline{X} + \mathcal{O}(3) \\ h_2(\overline{X}, \mu) &= \frac{1}{b}\overline{X}^2 + \mathcal{O}(3) \end{aligned} \tag{7.4.19}$$

The dynamics on the centre manifold is established if, in the first equation of eq. (7.4.13), we substitute $\overline{Y}$ by h_1 and $\overline{Z}$ by h_2 from eq. (7.4.19), cf. eq. (7.4.15)

$$\begin{aligned} \dot{\overline{X}} &= \frac{\sigma}{\sigma+1}\left[\overline{X} - \frac{\sigma}{(\sigma+1)^2}\mu\overline{X}\right]\left[\mu - \frac{1}{b}\overline{X}^2\right] + \mathcal{O}(4) \\ &= \frac{\sigma}{\sigma+1}\left[\mu\overline{X} - \frac{1}{b}\overline{X}^3 - \frac{\sigma}{(\sigma+1)^2}\mu^2\overline{X}\right] + \mathcal{O}(4) \end{aligned} \tag{7.4.20}$$

On the basis of the conditions eqs. (6.4.50) to (6.4.53), we can easily ascertain that eq. (7.4.20) defines a pitchfork bifurcation. For small μ, we obtain from eq. (7.4.20) the normal form of a pitchfork bifurcation, cf. eq. (6.4.54),

$$\dot{\overline{X}} = \frac{\sigma}{\sigma+1}\left(\mu\overline{X} - \frac{1}{b}\overline{X}^3\right) \ , \qquad \mu = r - 1 \tag{7.4.21}$$

i.e. for $r > 1$, the origin becomes unstable and two new fixed points $\overline{C}_1$ and $\overline{C}_2$ emerge with the coordinates $\overline{X}_{1,2} = \pm\sqrt{b\mu}$ in the $\overline{\boldsymbol{X}}$-system. We also arrive at this result of two new stable fixed points if we investigate the initial system, eq. (7.4.3), with respect to fixed points and their stability. Beside the trivial solution, we obtain two fixed points C_1 and C_2 with the coordinates (see fig. 7.4.1)

$$C_{1,2} : \boldsymbol{X} = \{X \quad Y \quad Z\}_{1,2} = \{\pm\sqrt{b\mu} \quad \pm\sqrt{b\mu} \quad \mu\} \tag{7.4.22}$$

where $\mu = r - 1$.

The coincidence of the $\overline{X}_{1,2}$- and $X_{1,2}$-coordinates of the fixed points $\overline{C}_{1,2}$ and $C_{1,2}$ is purely fortuitous. Nothing has been ascertained as yet concerning the quality of the approximation by the centre manifold.

If we investigate the stability behaviour of the fixed points C_1 und C_2, the Jacobian matrix

$$\left.\frac{\partial \boldsymbol{F}}{\partial \boldsymbol{X}}\right|_{C_1,C_2} = \begin{bmatrix} -\sigma & \sigma & 0 \\ 1 & -1 & \mp\sqrt{b\mu} \\ \pm\sqrt{b\mu} & \pm\sqrt{b\mu} & -b \end{bmatrix} \tag{7.4.23}$$

yields the following third-order characteristic equation for the determination of the eigenvalues

$$P(\lambda) = \lambda^3 + (\sigma + b + 1)\lambda^2 + b(\sigma + r)\lambda + 2\sigma b(r - 1) = 0 \tag{7.4.24}$$

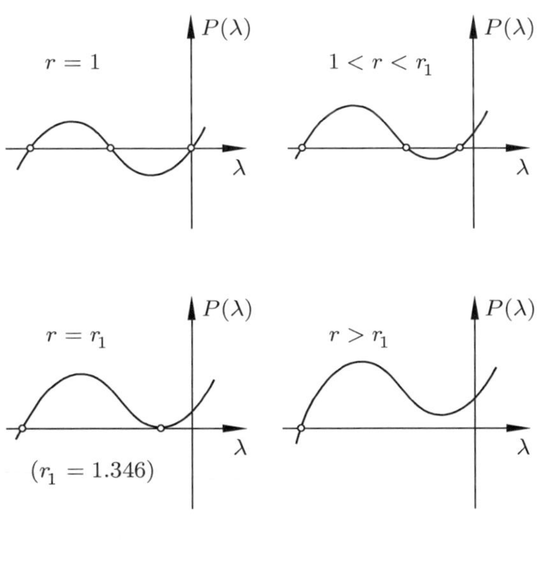

Fig. 7.4.2

Qualitative graph of the polynomial in the characteristic eq. (7.4.24) for C_1 and C_2 in dependence of r

The qualitative representation of the function $P(\lambda)$ in fig. 7.4.2 shows that, for $1 < r < r_1 = 1.346\ldots$, all three eigenvalues are negative. Consequently, the fixed points C_1, C_2 are stable nodes (fig. 7.4.1).

At $r_1 = 1.346\ldots$, two eigenvalues coincide; for $r > r_1$, they occur as pairwise complex conjugate. The stable attractor type of C_1 and C_2 changes from a stable node to a stable node-focus combination (fig 7.4.1). We recall that the classic case of a pitchfork bifurcation occurs for $r = 1$ at the origin: the only stable node becomes unstable in one direction (saddle point) and, simultaneously, the two stable nodes C_1 and C_2 emerge (fig. 7.4.1). Physically, the pitchfork bifurcation ($r = 1$) indicates the onset of the roll patterns, i.e. that the heat resulting from ΔT is carried through the fluid via convection rolls in addition to the influence of the pure thermal conduction.

We return to the fixed point at the origin. Its saddle-point character is maintained for all $r > 1$. Due to the eigenvalue constellation $\mathrm{Re}(\lambda_{1,2}) < 0$ and $\mathrm{Re}(\lambda_3) > 0$, the fixed point possesses a two-dimensional stable manifold and a one-dimensional unstable one, cf. eq. (7.4.21). If we follow the unstable manifold of the origin numerically in positive and negative direction, we recognise that the respective evolutions are captured by the two foci C_1 and C_2 (Colour Plate VI, p. 467, lower figure). The question now arises: in which direction does the two-dimensional stable manifold of the origin evolve respectively where do the trajectories forming the stable manifold of the origin come from? The upper four figures of Colour Plate VI show the extension of the stable manifold in the phase space. For the numerical analysis of the surface formed by stable trajectories, we use the trick of time reversal ($t \rightarrow -t$), thus simulating its withdrawal from the origin. The stable eigenvectors $\boldsymbol{y}_2$ and $\boldsymbol{y}_3$ in eq. (7.4.9) span the initial surface in the immediate neighbourhood of the origin. The numerical reconstruction of the stable and unstable manifolds is summarised in Colour Plate VI. For a more comprehensible reproduction, the front of the surface is depicted red, the back blue.

The four upper figures of Colour Plate VI show that the stable two-dimensional manifold spreads between the two branches of the unstable one-dimensional manifold, twisting about the Z-axis. The unstable branches end in the foci C_1 and C_2. The stable surface embraces both loops about the foci. After twisting round the C_1- and C_2-loops, the surface returns to the origin on both sides of the initial surface, the orginally red front side becoming visible as the blue back and the blue back as the red front. Since trajectories in the phase space cannot intersect for reasons of uniqueness, the blue surface must fold downwards. If we now follow single initial points either near the blue side or near the red side of the stable surface, we obtain information on the basins of attraction of C_1 and C_2: trajectories with initial points in the blue domain always end in the blue focus, initial points in the red domain in the yellow focus. The spiral loops of the unstable manifolds which unwind in very close twists around C_1 and C_2 for r-values somewhat larger than 1 expand continuously with increasing r. At the parameter value $r = 13.926\ldots$, we observe that the unstable manifold tending towards the fixed points C_1 and C_2 in a wide arc returns to the origin (fig. 7.4.3).

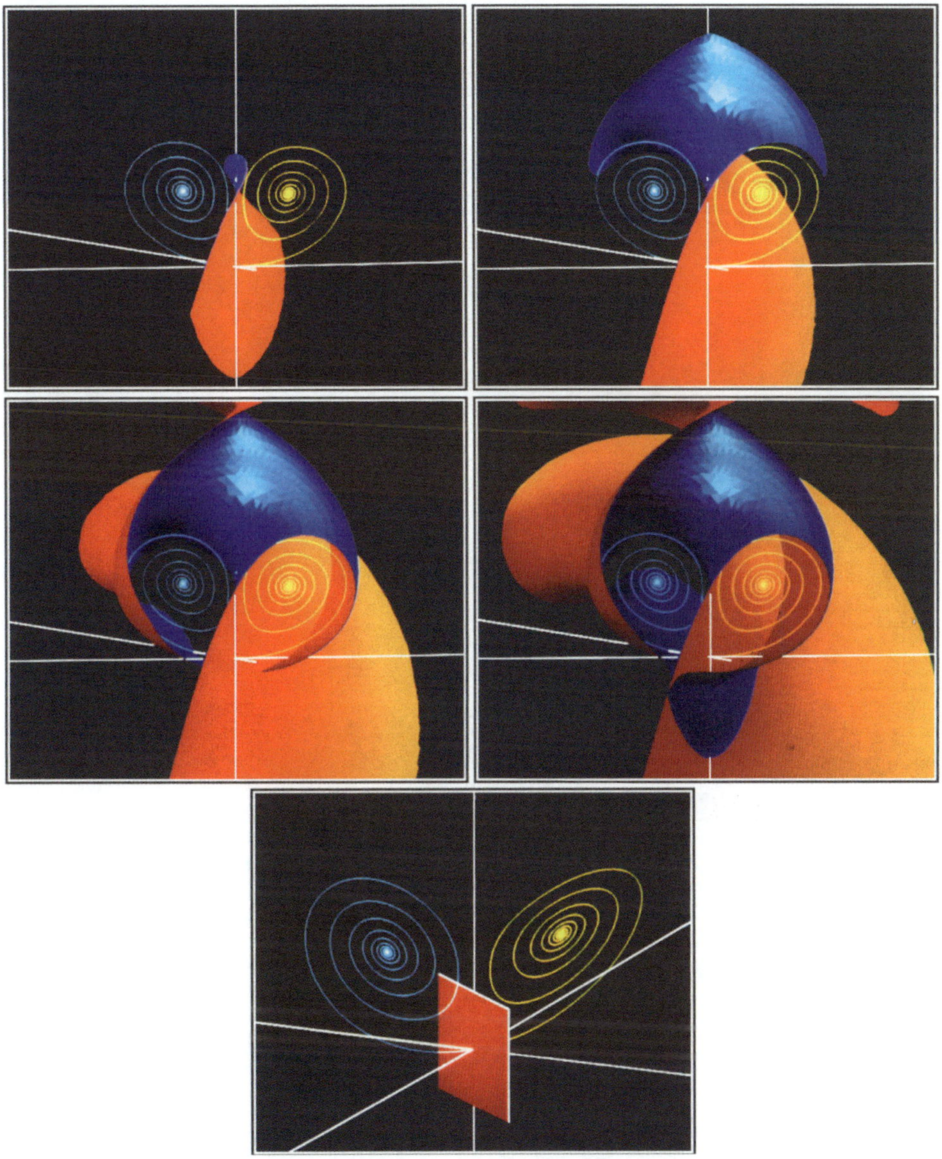

Plate VI Lorenz system ($r = 12$): evolution of the local two-dimensional stable ($\mathbb{W}^s$) and the one-dimensional unstable manifold ($\mathbb{W}^u$) for the saddle point at the origin

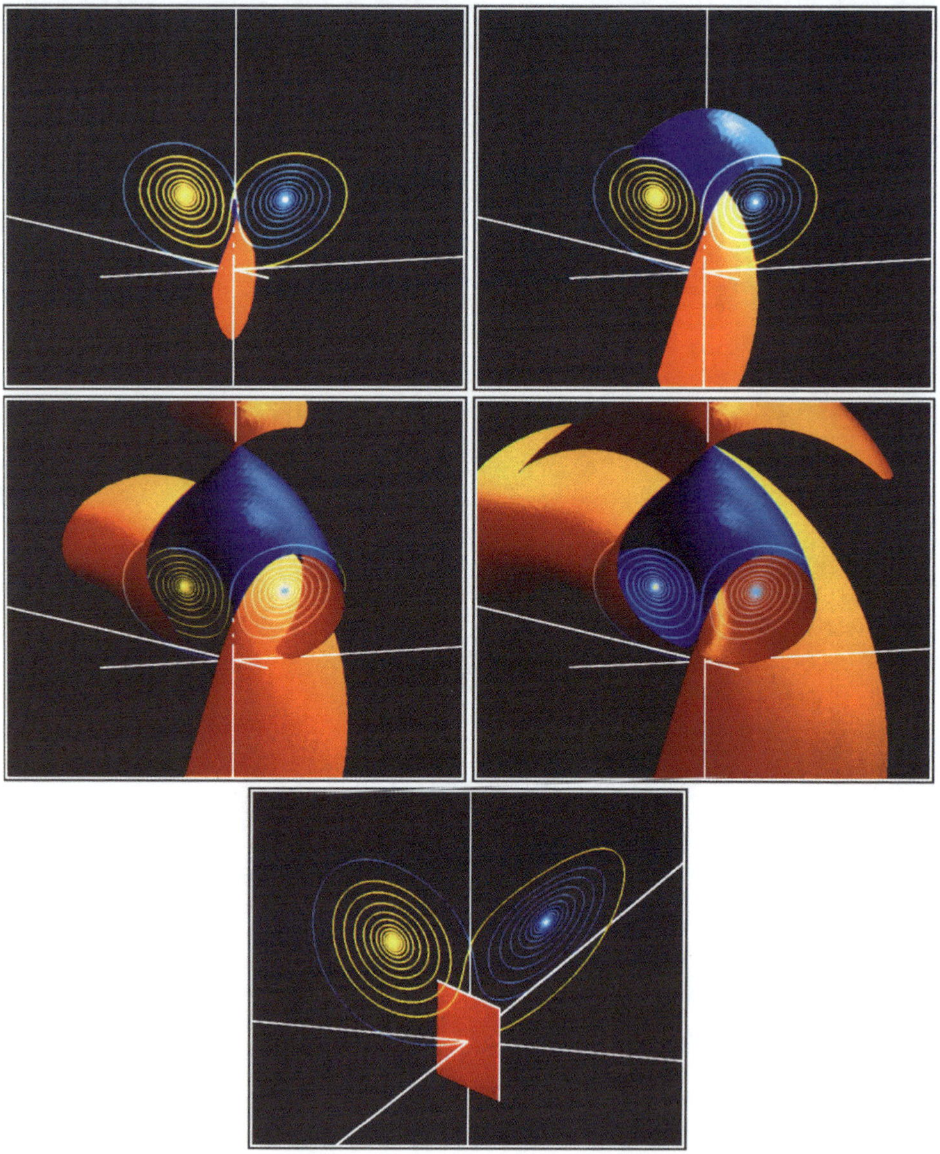

Plate VII Lorenz system ($r = 15$): evolution of the manifolds $\mathbb{W}^s$ and $\mathbb{W}^u$ for the saddle node at the origin

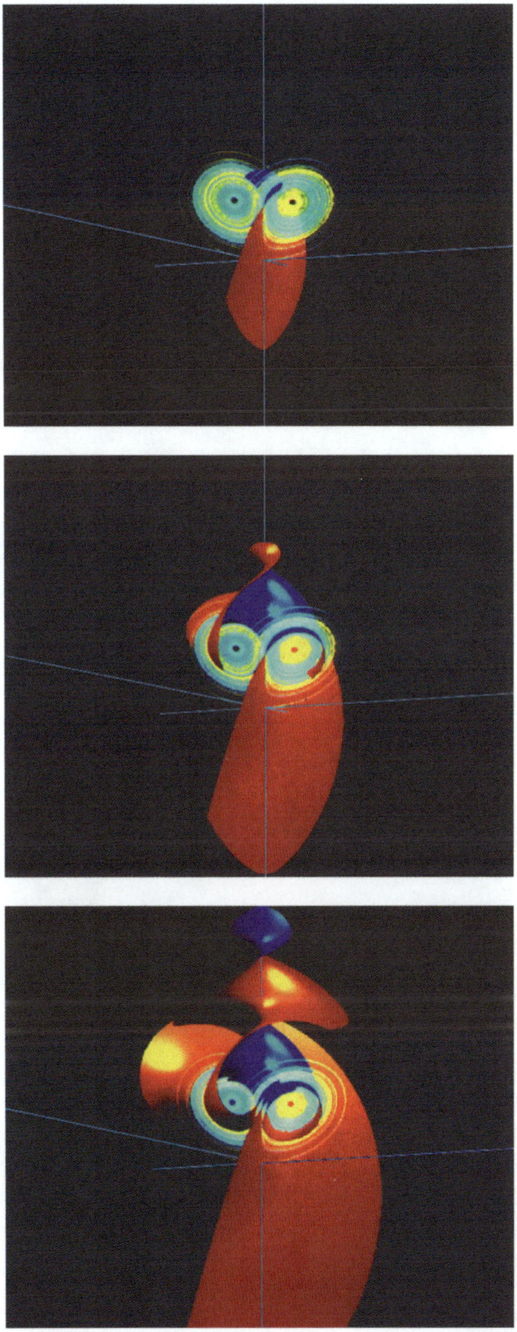

Plate VIII Lorenz system ($r = 28$): evolution of the manifolds $\mathbb{W}^s$ and $\mathbb{W}^u$ for the saddle node at the origin

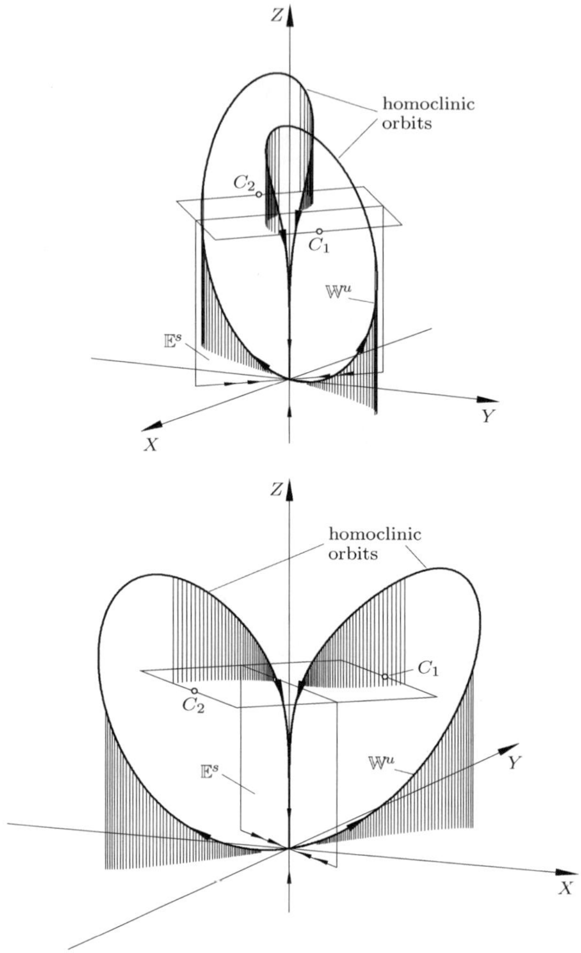

Fig. 7.4.3: Two different views of homoclinic orbits of the Lorenz system ($r = 13.9265$)

In this case, when the trajectory approaches the same fixed point for $t \to +\infty$ and $t \to -\infty$, we speak of a homoclinic orbit. Figure 7.4.3 shows two homoclinic orbits for the origin; they start at the origin along the unstable manifold for $r = 13.926\ldots$ and return to it as stable manifolds. For $r > 13.926\ldots$, something dramatic occurs: the unstable manifold jumps across so that the line starting in a positive direction ends in the left fixed point and that starting in a negative direction ends in the right one (Colour Plate VII, p. 468, lower figure). Note that for $r = 13.926\ldots$, no change takes place in the eigenvalue constellation; for example, the real parts of the eigenvalues display no zero crossing.

This significant alteration of the orbital structure due to changes of the control parameter is a typical example of a global bifurcation. The global bifurcation theory, however, goes beyond the scope of this book; we refer to the extensive monograph of Wiggins (Wiggins, 1988).

In the case $r > 13.926\ldots$, our interest is directed at the paths the trajectories forming the stable manifold of the origin take in this considerably more complex situation. The first four figures of Colour Plate VII demonstrate the spread of the two-dimensional stable manifold. The surface unfolds as for $r = 12$, Colour Plate VI, but with one difference occurring at the origin. On its return, the two-dimensional manifold is unable to fold downwards as it is barred by the unstable manifold. The rule of Colour Plate VI, that all initial points on the red hemisphere move to the right focus and all blue ones to the left focus, is now no longer true for all initial conditions, i.e. due to the global bifurcation, the basins of attraction of the two foci C_1 and C_2 have altered fundamentally.

We increase r and note that for

$$r = r_{cr} = \frac{\sigma(\sigma + b + 3)}{(\sigma - b - 1)}\,, \qquad \sigma > b + 1 \tag{7.4.25}$$

in place of the formerly negative real parts, the real part of the complex conjugate eigenvalues of eq. (7.4.24) of the two fixed points C_1 and C_2 becomes zero. The eigenvalues are

$$\lambda_{1,2} = \pm i\sqrt{\frac{2\sigma b(\sigma + 1)}{\sigma - b - 1}} \qquad \lambda_3 = -(\sigma + b + 1) \tag{7.4.26}$$

If we consider the values $\sigma = 10$ and $b = 8/3$ selected by Lorenz, it appears that the fixed points C_1 and C_2 are stable in the parameter range $1 < r < r_{cr} = 470/19 \approx 24.74$ (fig. 7.4.1). At $r = r_{cr}$, the stable fixed points C_1 and C_2 change into saddle foci with a two-dimensional unstable and a one-dimensional stable manifold. In the state $r > r_{cr}$, the so-called Lorenz attractor emerges. The numerical studies of the Lorenz equation for $\sigma = 10$, $b = 8/3$ and $r = 28$ with respect to the basin of attraction, sensitivity to the intitial conditions and aperiodic, erratic behaviour were described in section 5.2. We remind the reader of the illustrations of Colour Plates VIII, page 469, and II, page 197.

At $r = r_{cr}$, the imaginary eigenvalues cross the complex axis ($\mathrm{Re}(\lambda_{1,2}) < 0$), the real part of the third eigenvalue remaining negative ($\mathrm{Re}(\lambda_3) < 0$). The flow in the neighbourhood of the foci C_1 and C_2 becomes unstable: the classic example of a Hopf bifurcation. As we know from section 6.4, two types of bifurcation are known: the supercritical and the subcritical. Marsden and McCracken (1976) have shown that, in this case, a subcritical Hopf bifurcation arises (fig. 7.4.1). The non-trivial fixed points C_1 and C_2 are each accompanied by unstable limit cycles which shrink to zero at $r = r_{cr}$ with increasing r. For $r > 24.74\ldots$, all three fixed points $(0, C_1, C_2)$ exhibit saddle-point character, i.e. the flow at C_1, C_2 is characterised in each case by an unstable focus (two-dimensional manifold) and a one-dimensional stable manifold; the flow of the origin is unaltered for values $r > 1$ and induced by a two-dimensional stable and a one-dimensional unstable manifold. The changing of these three fixed points from stable to unstable leads to the globally bounded and locally erratic structure of the Lorenz attractor. Since the Lorenz attractor is the only attractor in the phase space, all initial conditions lead to chaotic motions. We mention only in passing the "pre-turbulent" (Kaplan and Yorke, 1979b) or the "metastable chaotic" (Yorke and Yorke, 1979) interval between $13.926\ldots < r < 24.06\ldots$ and $24.06\ldots < r < 24.74\ldots$ reflecting the coexistence of two stable fixed points and one chaotic attractor; in this connection, we refer the reader to the detailed presentations in (Sparrow, 1982) and (Bergé et al., 1984).

For the time being, we conclude our discussion of the evolution of the Lorenz system in dependence of the control parameter r with some coloured illustrations of the Lorenz attractor ($r = 28$). Numerous authors have carried out numerical investigations of the Lorenz attractor; we restrict ourselves here to the works of (Sparrow, 1982) and (Guckenheimer and Holmes, 1983) and refer to the extensive literature on this subject listed there.

The three illustrations of Colour Plate VIII, page 469, show the extension of the stable manifold of the origin for the chaotic domain $r > 24.74\ldots$. We recall that the one-dimensional unstable manifold leaves the origin in positive and negative direction and spirals towards C_1 and C_2. It is never captured by C_1 or C_2, however, but jumps unpredictably and without warning – capriciously, as it were – to the other focus. The stable surface of the manifold now has to wind its way through this chaotic intricacy of curves. This twisting leads to a mille-feuille structure about the saddle foci C_1 and C_2, clearly recognisable in Colour Plate II, page 197. We know that the mille-feuille layers lie infinitely close to one another; this means that a division into a blue and a red hemisphere – and thus a classification of red or blue initial conditions – becomes practically impossible. Each point of the strange attractor at any time is unsure of its allegiance to blue or red. This indecisiveness could be called local unpredictability which cannot be turned into predictability even with the use of the most powerful computers. We must simply accept – *nolens volens* – that a set of purely deterministic equations can generate "unpredictability". In the filigree, intricately interwoven and aesthetically impressive structure of the strange attractor in the phase space, we make the acquaintance of the intriguing phenomenon of the unpredictable.

8 Routes to Chaos

In the following sections, we present a series of mathematical models leading to chaotic behaviour in time. We will get to know various scenarios of regular dynamical behaviour becoming chaotic. What is surprising is that the specific form of the law of motion is not so important, but rather that the different routes to chaos possess universal character. The same transition to chaotic behaviour which we find with the Duffing oscillator, for example, can also be observed in some fluid systems; however, these have to be subject to strong external constraints like, for example, the Rayleigh-Bénard experiment or in Taylor-Couette flow. Thanks to the high precision in the realisation of physical experiments due to the refinement of electronic measurement techniques, the various routes to chaotic behaviour or to the onset of turbulence could be demonstrated experimentally by various research groups. At the transition, however, the dynamics of such systems is essentially of temporal nature, for which only a limited number of degrees of freedom are excited.

On the basis of these observations, there was hope in the early years of chaos research that a big step could be taken towards solving the turbulence problem or that substantial contributions could be made to a fundamental understanding of the mechanisms which lead to a transition from laminar to turbulent fluid flow. Yet these hopes have been only partly fulfilled. In chapter 9, we address the problem of fully developed turbulence in greater detail.

The first systematic investigations with regard to stability and the transition to new dynamical behaviour go back to ideas of Poincaré at the end of the nineteenth century. Initially, the transition to turbulence was modelled by a series of bifurcations. We begin our presentation of the different routes to chaos with this scenario.

8.1 Landau Scenario

In the 1940s, a mathematical model was proposed by Lev Davidovich Landau and then, independently, by Eberhard Hopf (Landau, 1944; Hopf, 1948). This model was intended to explain the transition of a stationary, laminar flow to a turbulent one through an infinite sequence of instabilities; its validity was not doubted until the beginning of the 1970s. Landau considered the flow of an incompressible viscous liquid, the motion of which is given by the Navier-Stokes equation (7.1.23), and he assumed time-independent boundary conditions. On the basis of the Navier-Stokes equations with the same Reynolds number $Re = \frac{UL}{\nu}$, see eq. (9.1.7), Osborne Reynolds (1842 – 1912) had recognised not only that the stream lines were geometrically similar, but also that a transition from laminar to turbulent flow occurs at a critical value Re_{cr}.

J. Argyris et al., *An Exploration of Dynamical Systems and Chaos*,
DOI: 10.1007/978-3-662-46042-9_8

For small Re-values, the smoothing influence of the viscosity is so great ($\nu \gg UL$) that a stationary flow remains stable under small perturbations and reflects the same symmetries as those generated by the external conditions. Landau now investigated the stability of a stationary incompressible flow given by $\boldsymbol{u}_0(\boldsymbol{x})$ and $p_0(\boldsymbol{x})$ in a domain $\mathcal{B}$ in dependence of $Re \sim 1/\nu$ by observing the influence of small time-dependent perturbations $\boldsymbol{u}_1(\boldsymbol{x},t)$, $p_1(\boldsymbol{x},t)$. Here, $\boldsymbol{u}_0$ satisfies the following equations for all Re-respectively ν-values

$$\boldsymbol{u}_0 \cdot \nabla \boldsymbol{u}_0 = \boldsymbol{F} - \frac{1}{\varrho}\nabla p_0 + \nu \nabla^2 \boldsymbol{u}_0 \qquad \text{and} \qquad \nabla \boldsymbol{u}_0 = 0 \tag{8.1.1}$$

where the values of $\boldsymbol{u}_0$ at the boundary $\partial\mathcal{B}$ are prescribed. Inserting $\boldsymbol{u} = \boldsymbol{u}_0 + \boldsymbol{u}_1$ and $p = p_0 + p_1$ into the Navier-Stokes equation (8.1.1) and assuming that the perturbations $\boldsymbol{u}_1$, p_1 are sufficiently small, we can neglect the higher-order terms and, taking eq. (8.1.1) into account, we obtain the following system of linear partial differential equations

$$\begin{aligned} \frac{\partial \boldsymbol{u}_1}{\partial t} + \boldsymbol{u}_0 \cdot \nabla \boldsymbol{u}_1 + \boldsymbol{u}_1 \cdot \nabla \boldsymbol{u}_0 &= -\frac{1}{\rho}\nabla p_1 + \nu \nabla^2 \boldsymbol{u}_1 \\ \nabla \boldsymbol{u}_1 &= 0 \end{aligned} \tag{8.1.2}$$

with the boundary condition $\boldsymbol{u}_1\big|_{\partial\mathcal{B}} = \boldsymbol{o}$ and the initial condition $\boldsymbol{u}_1\big|_{t=0} = \boldsymbol{u}_{10}(\boldsymbol{x})$, where the coefficients of eq. (8.1.2) depend only on the position, but not on time. Although the solution of this general boundary value problem is generally difficult, we can separate the variables by a product ansatz

$$\boldsymbol{u}_1 = \mathrm{e}^{\sigma t} \boldsymbol{f}(\boldsymbol{x}) \, , \qquad p_1 = \mathrm{e}^{\sigma t} p(\boldsymbol{x}) \tag{8.1.3}$$

and thus arrive at the following eigenvalue problem

$$\begin{aligned} \sigma \boldsymbol{f}(\boldsymbol{x}) + \boldsymbol{u}_0 \cdot \nabla \boldsymbol{f}(\boldsymbol{x}) + \boldsymbol{f}(\boldsymbol{x}) \cdot \nabla \boldsymbol{u}_0 &= -\frac{1}{\varrho}\nabla p(\boldsymbol{x}) + \nu \nabla^2 \boldsymbol{f}(\boldsymbol{x}) \\ \nabla \boldsymbol{f}(\boldsymbol{x}) &= 0 \end{aligned} \tag{8.1.4}$$

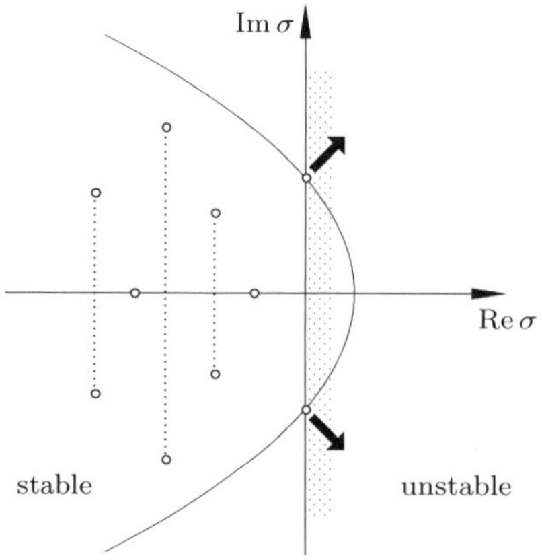

Fig. 8.1.1

Hopf bifurcation of the stationary flow $\boldsymbol{u}_0$ at Re_1 and eigenvalue spectrum of eq. (8.1.4) with $\boldsymbol{f}\big|_{\partial\mathcal{B}} = \boldsymbol{o}$. The values of σ for which this system of equations possesses non-trivial solutions $\boldsymbol{f}(\boldsymbol{x})$ and $p(\boldsymbol{x})$ are denoted eigenvalues, the appertaining solutions eigenfunctions. Since the domain $\mathcal{B}$ containing the liquid is finite, there exists a countably infinite set of discrete, in general complex eigenvalues σ_{j} ($\mathrm{j} = 1, 2, \ldots$)

$$\sigma_{\mathrm{j}}(Re) = \gamma_{\mathrm{j}}(Re) + i\omega_{\mathrm{j}}(Re) \tag{8.1.5}$$

which depend on the Reynolds number Re and lie in the complex σ-plane within a parabola open to the left (Joseph, 1985), see fig. 8.1.1.

For sufficiently small Reynolds numbers, $\gamma_{\mathrm{j}}(Re) < 0$ applies for all $\mathrm{j}= 1, 2, \ldots$, i.e. the perturbations decay exponentially in time and the stationary solution $\boldsymbol{u}_0$ is stable. For increasing Re, however, a critical value Re_1 exists for which $\gamma_1(Re_1) = 0$ holds. If we assume that, for Re_1, a single pair of complex conjugate eigenvalues crosses the imaginary axis, then

$$\mathrm{Re}\big(\sigma(Re_1)\big) = 0\,, \qquad \mathrm{Im}\big(\sigma(Re_1)\big) = \pm i\omega_1(Re_1) \neq 0 \tag{8.1.6}$$

In this case, a Hopf bifurcation takes place (cf. section 6.4). For somewhat higher Reynolds numbers $Re > Re_1$, the relation $\gamma_1(Re) > 0$ holds. Applying eq. (8.1.5), we obtain the solution of the linearised problem from eq. (8.1.3)

$$\boldsymbol{u}_1 = \mathrm{e}^{\gamma_1 t}\mathrm{e}^{i\omega_1 t}\boldsymbol{f}(\boldsymbol{x}) \tag{8.1.7}$$

The linearisation suggests a very fast increase of the amplitude $\mathrm{e}^{\gamma_1 t}$ for positive γ_1-values. In reality, however, due to the non-linearity of the flow behaviour, the amplitude will tend towards a finite limit. Landau was able to show that the amplitude A of this motion remains finite and increases proportionally to $\sqrt{Re - Re_1}$, cf. (Landau and Lifschitz, 1991).

Summarising, we can say that, for Re-values only slightly above Re_1, the flow can be represented as the superposition of a stationary and a periodic flow with the frequency $\omega_1(Re)$ and the finite amplitude A

$$\boldsymbol{u} = \boldsymbol{u}_0 + A\mathrm{e}^{i(\omega_1 t + \beta_1)}\boldsymbol{f}(\boldsymbol{x}) \tag{8.1.8}$$

where the initial phase β_1 is determined by an arbitrary initial condition.

For larger Re-values, the superposition loses its validity. The flow is then described by a general periodic function with the basic frequency $\omega_1(Re)$ and can be expanded into a Fourier series

$$\boldsymbol{u}(\boldsymbol{x}, t) = \sum_{\mathrm{k}} A_{\mathrm{k}}\boldsymbol{f}_{\mathrm{k}}(\boldsymbol{x})\mathrm{e}^{i\mathrm{k}(\omega_1 t + \beta_1)} \tag{8.1.9}$$

i.e. there are now terms appertaining to integral multiples of ω_1. In contrast to the stationary flow $\boldsymbol{u}_0$, which evidently is uniquely determined by the definition of the problem, the periodic flow possesses one arbitrary degree of freedom, namely the initial phase β_1 which depends on the random initial conditions at the beginning of the flow.

If the Reynolds number $Re > Re_1$ increases still further, we must ask how long this periodic flow $\boldsymbol{u}$ remains stable. To investigate this, we may superimpose a small perturbation $\boldsymbol{u}_2$ on $\boldsymbol{u}$ and observe its temporal evolution. If we again substitute $\boldsymbol{u} + \boldsymbol{u}_2$

in the Navier-Stokes equations and neglect higher-order terms, then, in analogy to eq. (8.1.2), we obtain a system of linear partial differential equations, the coefficients of which are now periodic functions with the period $T = 2\pi/\omega_1(Re)$. The stability of the periodic solution $\boldsymbol{u}$ can be investigated with the aid of the Floquet theory described in section 5.4.2, where in this case, however, the spatial dependency must also be considered. If we superimpose a perturbation on the periodic solution $\boldsymbol{u}(\boldsymbol{x},t)$ in eq. (8.1.9)

$$\boldsymbol{u}_2 = e^{\sigma t}\widetilde{\boldsymbol{f}}(\boldsymbol{x},t)\,, \quad \sigma = \gamma_2(Re) + i\omega_2(Re) \tag{8.1.10}$$

where the function $\widetilde{\boldsymbol{f}}(\boldsymbol{x},t) = \widetilde{\boldsymbol{f}}(\boldsymbol{x},t+2\pi/\omega_1)$ possesses the same period as the solution $\boldsymbol{u}$. If we now linearise the initial system eq. (8.1.1) in the neighbourhood of $\boldsymbol{u}$, we obtain the spectral problem

$$\sigma\widetilde{\boldsymbol{f}} + \frac{\partial\widetilde{\boldsymbol{f}}}{\partial t} + \boldsymbol{u}\cdot\nabla\widetilde{\boldsymbol{f}} + \widetilde{\boldsymbol{f}}\cdot\nabla\boldsymbol{u} = -\frac{1}{\varrho}\nabla p + \nu\nabla^2\widetilde{\boldsymbol{f}}$$

with (8.1.11)

$$\nabla\widetilde{\boldsymbol{f}} = 0\,, \quad \widetilde{\boldsymbol{f}}\big|_{\partial\mathcal{B}} = \boldsymbol{o}$$

The eigenvalues σ correspond to the Floquet exponents and provide information on the stability behaviour (cf. section 5.4.2). As long as $\gamma_2(Re) < 0$ holds, the periodic motion remains stable. At a critical value $Re = Re_2$ with $\gamma_2(Re_2) = 0$, the flow again loses its stability. As we saw in section 6.6, the bifurcation of limit cycles can be reduced to the problem of the bifurcation of the fixed points of one-parametric maps by introducing a Poincaré map f, although some additional difficulties occur, see (Arnold, 1988).

In section 5.4.2, we distinguished between three possible routes along which a periodic motion can become unstable. This evolution depends on the generally complex eigenvalues $\lambda = e^{\sigma T}$ of the monodromy matrix eq. (5.4.32) which, in accordance with eq. (5.4.39) and because of $\gamma_2(Re_2) = 0$, possess the form $\lambda = e^{i\omega_2 T}$ and thus lie on the unit circle:

i. $\lambda = 1$ Typically, in the Poincaré map, a saddle-node bifurcation occurs which is accompanied by the emergence or vanishing of a limit cycle. If the physical problem exhibits certain symmetries and regularities, a transcritical bifurcation or a pitchfork can also occur; for details, see, for example, section 6.6 and (Guckenheimer and Holmes, 1983).

ii. $\lambda = -1$

In section 6.6, it was shown that the loss of stability in this case is not accompanied by a bifurcation. However, the iterated map $f^2(p) = f(f(p))$ possesses the eigenvalue 1, i.e. in the neighbourhood of the critical point, fixed points of f^2 emerge, which are not fixed points of f, and for which therefore $P = f^2(P)$ and $P \neq f(P)$ hold, i.e. the iterated map undergoes a pitchfork bifurcation. In the supercritical case, a cycle of period 2 exists (cf. section 3.6), i.e. $\lambda = -1$ leads to a period doubling of the orbit.

iii. $|\lambda| = 1\,,\, \mathrm{Im}\,\lambda \neq 0\;:\;\; \lambda_{1,2} = \mathrm{e}^{\pm i\omega_2 T}$

We assume that, for $Re = Re_2$, a single pair of complex conjugate eigenvalues λ, λ^* crosses the unit circle, i.e. $\sigma(Re_2) = \pm i\omega_2$, where $\mathrm{d}\gamma_2/\mathrm{d}Re|_{Re_2} > 0$ is assumed to hold. A new solution then branches off, the characteristics of which are shaped by the ratio of the frequencies, ω_2/ω_1, and by the type of the normal form (cf. section 6.6, Neĭmark-Sacker bifurcation). If $\lambda^{\mathrm{k}} = 1$ with $\mathrm{k} = 1, 2, 3, 4$ holds for the eigenvalues, we speak of *strong resonance*. In these cases, in the derivation of the normal form, terms emerge which require special treatment; for details, see (Takens, 1974; Arnold, 1977).

If we exclude strong resonances, we can prove that, for $Re > Re_2$, an invariant circle exists – in the three-dimensional phase space an invariant torus – on which the motion takes place. However, this does not answer the question whether the actual motion on the torus is periodic or quasi-periodic. If there are higher-order terms in the normal form, additional effects – such as the synchronisation of frequencies – can occur which lead to periodic motions. In section 8.3, we shall discuss this problem in detail in connection with the circle map (cf. also section 6.6).

In his model, Landau assumed that, at the instability point $Re = Re_2$, a pair of complex conjugate eigenvalues $\sigma = \pm i\omega_2$ of eq. (8.1.11) emerges with a new incommensurable frequency ω_2 and that the periodic motion is succeeded by a quasi-periodic motion on the torus

$$\boldsymbol{u}(\boldsymbol{x}, t) = \boldsymbol{u}(\boldsymbol{x},\, \omega_1 t + \beta_1,\, \omega_2 t + \beta_2) \tag{8.1.12}$$

which now contains two independent initial phases β_1 and β_2, i.e. the new flow has two arbitrary degrees of freedom. According to Landau's idea, further Hopf-type bifurcations occur successively at a further increase of the Reynolds number (see fig. 8.1.2); this means that a new independent frequency ω_{i} ensues each time and the intervals between two consecutive critical Reynolds numbers $Re_{\mathrm{i}-1}$ and Re_{i} decrease rapidly. At every bifurcation, the flow gains a new degree of freedom in the form of a new arbitrary initial phase β_{i}. In this way, the flow very quickly becomes intricate and jumbled, the time-histories are complex and seem to be irregular and the power spectrum appears as a band of infinitely many discrete frequencies $\omega_1, \omega_2, \ldots$ and their integral linear combinations (see section 3.8). Figure 8.1.2 presents a schematic depiction of Landau's scenario.

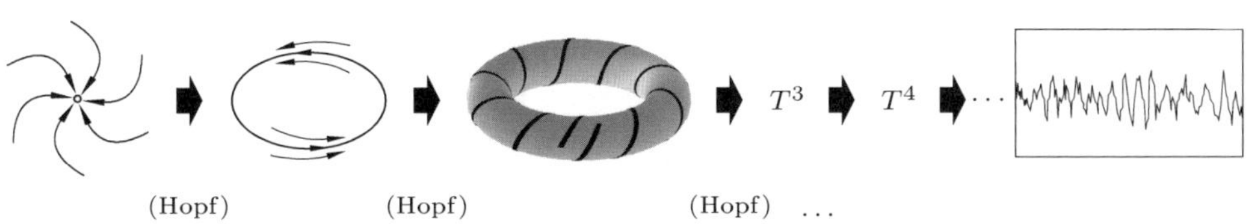

Fig. 8.1.2: Landau's turbulence model (1944)

8.2 Ruelle-Takens Scenario

The concept of Landau and Hopf – that such a complex phenomenon as a turbulent flow requires a complex description with infinitely many degrees of freedom – was not questioned until the stormy developments in the field of non-linear dynamics and the chaos theory. After Lorenz had shown that only three ordinary non-linear differential equations can lead to an extraordinarily complex and chaotic behaviour, Landau's turbulence model was scrutinised critically.

There are two fundamental aspects of motion on a strange attractor which are also characteristic of turbulent flows, but which a quasi-periodic motion on an n-dimensional torus ($n \to \infty$) can never exhibit. First and foremost, there is the sensitive dependence on small perturbations which leads to the exponential drifting apart of arbitrarily neighbouring trajectories and which can be described quantitatively by positive Lyapunov exponents. In the case of a quasi-periodic motion on an n-dimensional torus, on the other hand, originally neighbouring trajectories remain neighbouring for all time; here, there are $(n - 1)$ vanishing and one negative Lyapunov exponent, but certainly no positive one.

A direct consequence and a further fundamental criticism of Landau's model is that quasi-periodic motions – in contrast to chaotic behaviour – do not lead to a mixing of the trajectories in the phase space (by stretching and folding; see section 10.2). Mixing is characterised by the fact that the mean autocorrelation, eq. (3.8.28), tends towards zero for $\tau \to \infty$

$$\bar{a}(\tau) = \lim_{T\to\infty} \frac{1}{2T} \int_{-T}^{+T} f(t+\tau) f(t) \mathrm{d}t \to 0 \tag{8.2.1}$$

where for example for f an arbitrary velocity component can be inserted. For turbulent flows, experimental results confirm that $\bar{a}(\tau)$ does indeed tends towards zero. In section 3.8, however, we showed that, for quasi-periodic motions, $\bar{a}(\tau)$ is also a quasi-periodic function (cf. table 5.3.1), i.e. a mixing cannot be described by Landau's model.

The cause for the shortcomings of Landau's model is the fact that the route to chaos is mainly based on linear concepts insofar as we assume that existing periodic solutions can neither vanish nor vary fundamentally in the course of the sequence of bifurcations. The flow occurring in Landau's model is ultimately a superposition of such non-changing solutions (Landau and Lifschitz, 1991).

8.2.1 Instability of Quasi-Periodic Motions on the 3D Torus

Without knowing Lorenz's work, which had appeared in a special meteorological journal and thus long remained almost entirely unknown in mathematical circles, Ruelle and Takens developed a mathematical model in 1971 to describe the transition to turbulence in a hydrodynamic system (Ruelle and Takens, 1971); in this work, the concept of the strange attractor was introduced. In it, they used a procedure on the basis of the invariant manifolds for investigating the stability and bifurcation of complex dynamical systems, concentrating on the "generic" case of the bifurcation.

Recently, the notion "generic" has been used frequently in the qualitative theory of differential equations. Characteristics of a system or a solution are called generic if they denote the typical and not the special case or the exception. If, for example, we consider the complex eigenvalues λ on the unit circle, then $\lambda = \pm 1$ is a special case which does not generally occur, just as little as all rational numbers since they have the Lebesgue measure zero. In physical problems, it is the generic solutions that are of interest, those which are generally demonstrable and reproducible in experiments. However, symmetries and conservation laws underlying the physical systems lead to restrictions of the general generic case. At the end of this section, we shall return to this problem. Readers interested in a mathematically precise definition of the concept "generic" are referred to (Hirsch and Smale, 1974).

In the same way as Landau, Ruelle and Takens started from a liquid at rest which is then succeeded by a stationary and a periodic flow following an increase of the energy supply. In order to comply with the generic case, they also investigated the bifurcation of the limit cycle with the aid of the Poincaré map and, like Landau, they assumed that exactly one pair of complex conjugate (not real) eigenvalues crosses the unit circle while all other eigenvalues remain inside the circle. A decisive idea was to apply the centre manifold theory (cf. section 6.2), which allows us to reduce the high-dimensional general problem to a two-dimensional one without losing fundamental qualitative characteristics of the long-term behaviour of the solution after the bifurcation.

The question regarding the structural stability of the solution after the bifurcation to an invariant torus was at the centre of further investigations. To this end, Ruelle and Takens applied a theorem of Peixoto (Peixoto, 1962) which goes back to ideas of Poincaré and a work by Andronov and Pontryagin (Andronov and Pontryagin, 1937) and sets up necessary and sufficient conditions for the structural stability of two-dimensional flows. The concept of structural stability was coined by Andronov and Pontryagin in 1937 to characterise the robustness of a system $\dot{\boldsymbol{x}} = \boldsymbol{F}(\boldsymbol{x})$, i.e. the susceptibility of the system equations themselves to perturbations or to small modifications. The question regarding the robustness of a system is posed quite naturally when we wish to apply the results of abstract mathematical investigations to the real world, i.e. to physical processes. In order to investigate a natural process, we first formulate a model concentrating on the fundamental quantities and then set up the appertaining evolution equations. Solutions can only be satisfactorily applied to the actual processes if small variations in the equations do not alter the qualitative behaviour of the solution; here again, symmetries and conservation laws must be taken into account.

A system is denoted *structurally stable* if small perturbations do not lead to a qualitative change in the system behaviour, i.e. if the flow of the unperturbed and the perturbed system are topologically equivalent. For example, eq. (2.2.1) of an undamped pendulum is structurally unstable, eq. (2.2.6) of a damped pendulum, on the other hand, structurally stable. For a precise definition, see, for example, (Guckenheimer and Holmes, 1983; Hirsch and Smale, 1974). In contrast to orbital stability, structural stability does not refer to a single trajectory, but to the whole system; it thus does not assume the absence of a sensitive dependence on the initial conditions. Anosov's theorem states, for example, that the chaotic "cat map" is structurally stable (Arnold, 1988).

The Hartman-Grobman theorem discussed in section 5.4.1 expresses a statement on the topological equivalence of a non-linear system with its linearised part in the neighbourhood of a hyperbolic fixed point. Following the implicit function theorem, hyperbolic fixed points are hence structurally stable. A non-hyperbolic fixed point cannot, on the other hand, be structurally stable since the linearised system

$$\dot{\boldsymbol{x}} = \left.\frac{\partial \boldsymbol{F}}{\partial \boldsymbol{x}}\right|_{\boldsymbol{x}_0} \boldsymbol{x}$$

possesses in this case an eigenvalue with vanishing real part. Each small perturbation generally leads to a bifurcation, the zero eigenvalue vanishes and the degenerate fixed point becomes a source, a sink or a saddle point.

The conditions which must be satisfied by a flow in order to be structurally stable are summarised for the two-dimensional case in the theorem of Peixoto (Guckenheimer and Holmes, 1983). For a flow in three or more dimensions respectively for maps in two and more dimensions, the situation becomes considerably more complex; to our knowledge, corresponding theorems have not as yet been established.

Theorem von Peixoto

An r-times continuously differentiable vector field $\dot{\boldsymbol{x}} = \boldsymbol{F}(\boldsymbol{x})$ $(r \geqslant 1)$ on a compact two-dimensional manifold $\mathbb{M}^2$ is structurally stable if and only if

i. the number of fixed points and periodic orbits is finite and each of them is hyperbolic;
ii. there are no trajectories connecting saddle points (no homoclinic and heteroclinic orbits) and
iii. the non-wandering set consists only of fixed points and periodic orbits.

In part *iii.* of the theorem, the concept of "non-wandering sets" emerges which describe the long-term behaviour of a system and which we shall explore here briefly, see (Guckenheimer and Holmes, 1983). A point P is denoted non-wandering if, for any neighbourhood U of the point P, an arbitrarily large t exists such that

$$\phi_t(U) \cap U \neq \text{ empty set} \tag{8.2.2}$$

holds. If we subject U to the flow ϕ_t, then U and the neighbourhood of the image $\phi_t(U)$ overlap, i.e. non-wandering points lie on or near trajectories which come arbitrarily close to their initial position after a certain time span. The set of non-wandering points describes the long-term behaviour of the dynamical system in the phase space. Examples of non-wandering sets are fixed points, limit cycles, quasi-periodic trajectories on a torus and strange attractors.

According to the theorem of Peixoto, a two-dimensional, structurally stable system contains only sources, sinks and saddle points as well as stable and unstable limit cycles as invariant sets. Figure 8.2.1 presents some examples of structurally stable

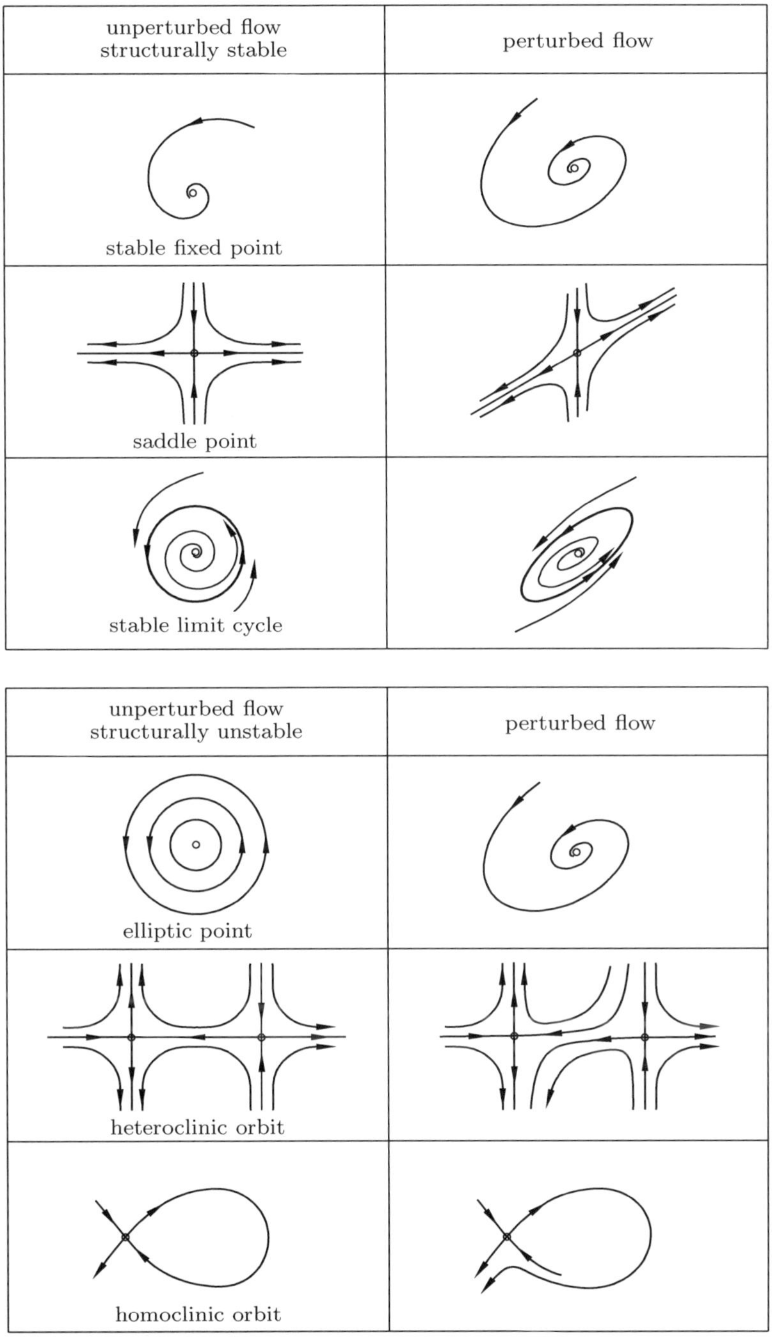

Fig. 8.2.1: Examples of some structurally stable and structurally unstable two-dimensional flows

respectively structurally unstable two-dimensional flows. Structural stability is a generic characteristic of two-dimensional flows on orientable manifolds (Guckenheimer and Holmes, 1983).

The third condition *iii.* of Peixoto's theorem excludes quasi-periodic orbits on two-dimensional tori; they are hence structurally unstable. It is precisely this argument that Ruelle and Takens used in their work. If a quasi-periodic motion emerges from a limit cycle, then arbitrarily small perturbations can be prescribed which lead to the occurrence of a periodic motion, i.e. a closed orbit on the invariant torus T^2. In their work, Ruelle and Takens proved only the existence of an invariant torus (Ruelle and Takens, 1971), but left the analytical characteristics of the solutions after the bifurcation – periodic or quasi-periodic – open. In the general case, the discussion of the solutions on the torus is highly complicated. Here, too, as in section 4.4, the problem of small denominators and rotation numbers occurs; the validity of a KAM condition must be investigated, It would go beyond the framework of our introductory text to enter into details here; we refer the interested reader to the literature such as, for example, (Guckenheimer and Holmes, 1983) and (Haken, 1987).

According to Ruelle and Taken's orginal work, further generalised Hopf bifurcations lead to quasi-periodic motions on a four-dimensional torus T^4, in a later work (Newhouse et al., 1978) on a three-dimensional torus; these are no longer structurally stable since arbitrarily small perturbations of these quasi-periodic motions with three incommensurable frequencies can lead to chaotic behaviour on the T^3-torus. It is the third dimension of the torus which allows the emergence of this qualitatively new attractor type, the strange attractor. This statement of Ruelle and Takens does not mean, however, that *every* small perturbation leads to a strange attractor on T^3; in fact, there are perturbations which retain the character of the quasi-periodic motion.

The fundamental result is that Landau's concept of a transition to chaos after a cascade of an infinite sequence of Hopf bifurcations is highly unlikely respectively non-typical, but rather that a strange attractor emerges directly after the third bifurcation, i.e. with the appearance of a third incommensurable frequency. Figure 8.2.2 presents the Ruelle-Takens scenario schematically.

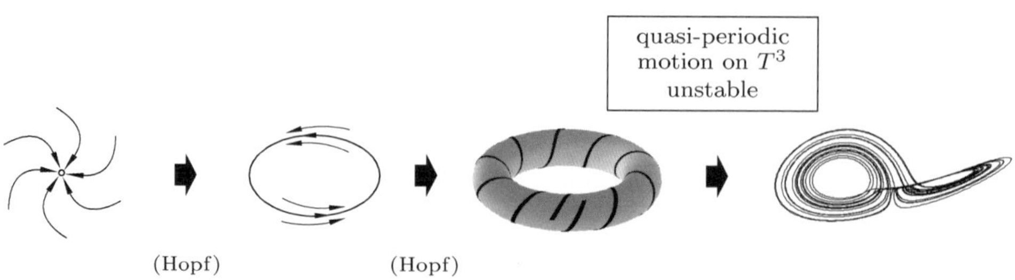

Fig. 8.2.2: Ruelle-Takens szenario

8.2.2 Experiments of Gollub and Swinney

It would certainly be no exaggeration to state that, until about 1970, physics regarded the experimental side of hydrodynamics as a poor relation. The measurement techniques used were relatively undeveloped; the control of velocities or temperatures could not be carried out with sufficient precision so that the mechanisms causing the transition from laminar to turbulent flow could not be studied precisely enough. After 1970, the situation altered fundamentally. The theoretical works of Lorenz (1963) and Ruelle and Takens (1971) suggested that the onset of a turbulent motion could be described by a bifurcation to a lower-dimensional strange attractor. A small group of experimenters from the U.S.A. and Europe was inspired to verify these theoretical models and hypotheses. Previously, they had investigated – in part intensively – phase transitions, and they now applied the refined measurement techniques used there to hydrodynamic experiments. In order to measure the flow velocity at a point, laser Doppler interferometry was used since it exerts almost no perturbing effect on the liquid. Cryogenic methods allowed for very precise measurements of the temporal variation of the heat flow through a layer of liquid.
It is known that the characteristics of a flow at the onset of turbulence depend to a high degree on the spatial dimensions of the liquid volume under observation, i.e. on the ratio of the horizontal to the vertical extent of the layer of liquid. If this aspect ratio is small, the boundaries impose constraints so that only few essential modes which cause the onset of turbulence are excited. The reason is that the spectrum of the eigenvalues of the linearised equation is discrete and, generally, only a small number of modes with $Re\lambda_\mathrm{i} > 0$ becomes unstable. For this reason, the experimenters preferred the two classic model experiments, the Rayleigh-Bénard convection and the Taylor-Couette flow, on the basis of which, due to favourable aspect ratios, standard scenarios can be proven.

If we consider a layer of liquid between two coaxial cylinders and let the inner cylinder rotate with an angular velocity ω about the common z-axis while the outer cylinder is held still, we can observe the hierarchy of the so-called Taylor instabilities. At an increase of ω, toroid rolls form at first (fig. 8.2.3a); these begin to oscillate periodically (figure b), then with two or more incommensurable frequencies (figure c) until, finally, chaotic, turbulent motion sets in (figure d).

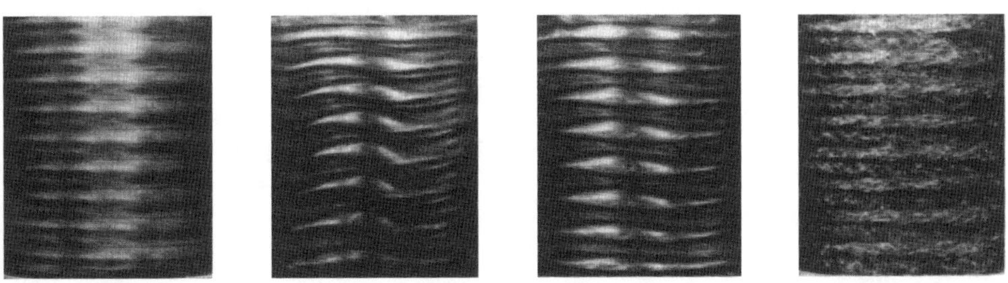

a) b) c) d)

Fig. 8.2.3: Instabilities in the Taylor-Couette flow (Swinney et al., 1977)

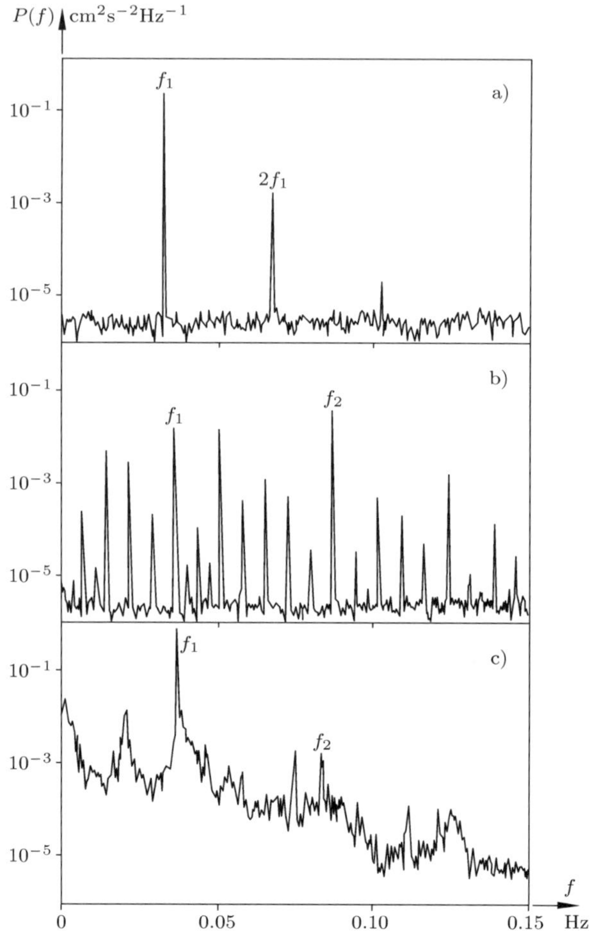

Fig. 8.2.4: Experiment of (Swinney and Gollub, 1978); power spectra for the Rayleigh-Bénard convection

The first experiment in which the transition to turbulent behaviour could be observed with sufficient precision was performed in 1975 by the American physicists J. P. Gollub and H. L. Swinney on a Taylor-Couette flow (Gollub and Swinney, 1975). Following a finite number of bifurcations, the power spectra exhibited chaos: after the emergence of time-independent torus-shaped rolls, in which the radial velocity varies periodically in the axial coordinate z, periodic behaviour in time sets in with a basic frequency f_1, see fig. 8.2.4a, then quasi-periodic motions on a torus with two incommensurable frequencies f_1, f_2, fig. 8.2.4b, followed by an abrupt transition to chaos. This result was the first experimental confirmation of Ruelle and Takens' ideas and conflicted with Landau's model. In 1978, Swinney and Gollub carried out corresponding experiments on the Rayleigh-Bénard convection (Swinney and Gollub, 1978). For varying relative Rayleigh numbers $r = Ra/Ra_{cr}$ (cf. section 7.3), fig. 8.2.4 presents

a) a periodic oscillation of the convection rolls with a basic frequency f_1;
b) a quasi-periodic motion with two incommensurable basic frequencies f_1 and f_2 (the power spectrum exhibits sharp peaks for f_1 and f_2 and their linear combinations), and finally
c) a broadband spectrum with some sharp peaks, indicating the onset of a chaotic motion.

This transition corresponds exactly to the scenario of Ruelle and Takens. Directly after the third Hopf bifurcation, chaos sets in. In the experiment, a quasi-periodic motion with three incommensurable frequencies cannot be observed since it is already converted to a chaotic motion by a small perturbation.

This inference is based only on the characteristics of Fourier spectra of a single variable of the system, however. We might object that it is not impossible for a series of further bifurcations to occur between fig. 8.2.4b and fig. 8.2.4c, all following one another very fast since a broad band of frequencies as in fig. 8.2.4c provides no information as to the number of modes taking part in the motion. This objection could be invalidated by a work of (Malraison et al., 1983). With the aid of the method on the reconstruction of attractors described in section 5.5.4, an attractor with the correlation dimension $D_K \approx 2.8 \pm 0.1$ could be reconstructed from the data of a one-dimensional recorded series. It was thus successfully proven that the irregular motion does indeed take place on a low-dimensional strange attractor.

However, we cannot observe a transition to turbulent behaviour after the third Hopf bifurcation in all experiments. It was shown, for example, that different scenarios can occur in dependence of the Prandtl number and the aspect ratio of the layer of liquid. Gollub and Benson (Gollub and Benson, 1980), for example, were able to observe the emergence of a quasi-periodic motion with three incommensurable frequencies in a Rayleigh-Bénard experiment (c.f. fig. 8.2.5); see also (Fauve and Libchaber, 1981).

The question thus arises why Ruelle and Takens' ideas can be verified in some cases and not in others. One of Ruelle and Takens' assumptions was that the solutions are generic. This is initially a purely mathematical concept, only applicable to physical problems under certain conditions. However, basic symmetries or invariants can cause the solutions to be non-generic. For example, it could not yet be proven whether the solutions of the Navier-Stokes equations are generic or not. If the basic frequencies of a quasi-periodic motion satisfy a KAM condition, for instance, i.e. if their quotient possesses a certain "degree of irrationality", it can be more probable that the quasi-periodic motion remains stable after the third bifurcation and does not immediately jump to a chaotic behaviour (Haken, 1987). However, the occurrence of further Hopf bifurcations becomes increasingly unlikely, so that Landau's turbulence model loses its validity. The concept of the strange attractor has definitely contributed to the realisation that many problems which could not be explained earlier – such as the mixing characteristics and the sensitive dependence of irregular motions on small perturbations in the initial conditions although the model description follows purely deterministic laws – are nowadays understood in connection with the emergence of chaotic motions.

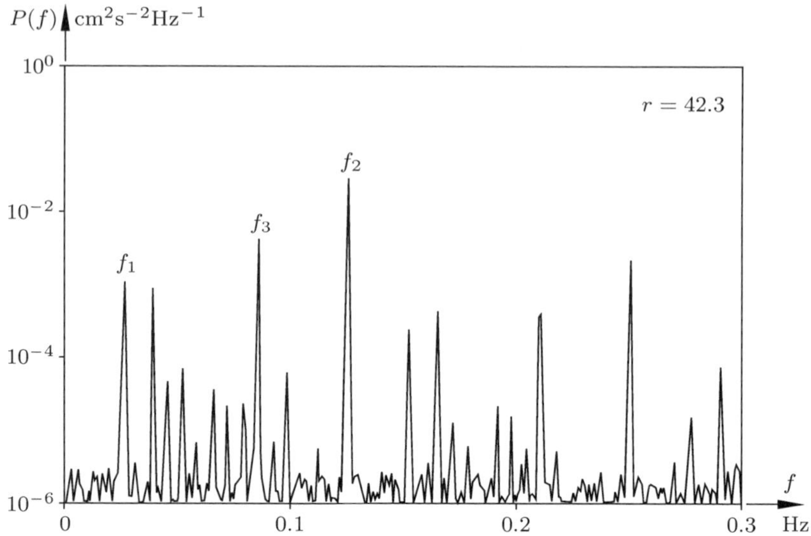

Fig. 8.2.5: Three incommensurable frequencies in a Rayleigh-Bénard convection, for $r = Ra/Ra_{cr} = 42.3$ (Gollub and Benson, 1980)

8.3 Universal Characteristics of the Transition from Quasi-Periodicity to Chaos

In section 6.7, we investigated the transition to chaos via a cascade of period doublings on the basis of one-dimensional maps and brought out the universal aspects. After Ruelle and Takens had put forward a new model for the onset of turbulent respectively chaotic behaviour which could be confirmed by experimental results, the question then arose whether universal characteristics also play a role in this transition from quasi-periodicity to chaos. In 1982, the works of two groups appeared (Rand et al., 1982; Feigenbaum et al., 1982). Initially independently of one another, they had taken up the theory of the renormalisation groups and applied it to this problem. In the following, we discuss these investigations in more detail.

Many dynamical systems characterised by at least two incommensurable frequencies exhibit quasi-periodic long-term behaviour. More than eighty years ago, it was discovered that, in electrical circuits which model heart activity, a variation of the driving frequency can lead to a series of different types of response signals resembling both normal and pathological heart rhythms (van der Pol and van der Mark, 1928). At the beginning of the 1980s, doctors and physicists, inspired by work on the chaos theory, again studied the effect of cardiac pacemakers on the heart rhythms of patients (Glass and Perez, 1982) involving in particular the application of short electrical impulses in order to stimulate spontaneous heart activity. Periodic behaviour, frequency locking and cascades of period doublings were observed in dependence on the intensity and frequency of the impulse.

Ventricular fibrillation generally leads to death within seconds. While the individual muscular cells contract and relax in coordination during normal heart activity, in the case of ventricular fibrillation, the whole tissue flutters in a completely uncontrolled way, the global coordination of the motion of the cardiac muscle tissue is lost so that no more blood can be pumped into the arteries. Experiments show that only a relatively weak electrical impulse is sufficient to cause ventricular fibrillation. Conversely, in the case of cardiac arrhythmia, doctors attempt to make the heart return to its normal rhythm by an electric shock. A better understanding of the non-linear dynamics of the heart could provide the basis for a more precise determination of the correct timing and intensity of this electrical impulse, thus allowing the appropriate therapy for disturbances of the cardiac rhythm. For the simplest mathematical model simulating the heart, a periodically driven oscillator was used and its Poincaré map studied.

8.3.1 The Kicked Rotator with Damping

For a continuous non-linear system, it is not generally possible to specify the Poincaré map explicitly, so that we usually have to rely on numerical integration. An exception is the damped rotator, on which an external force K_0 and a constant torque M_0 are applied impulsively at fixed time intervals T (see fig. 8.3.1).

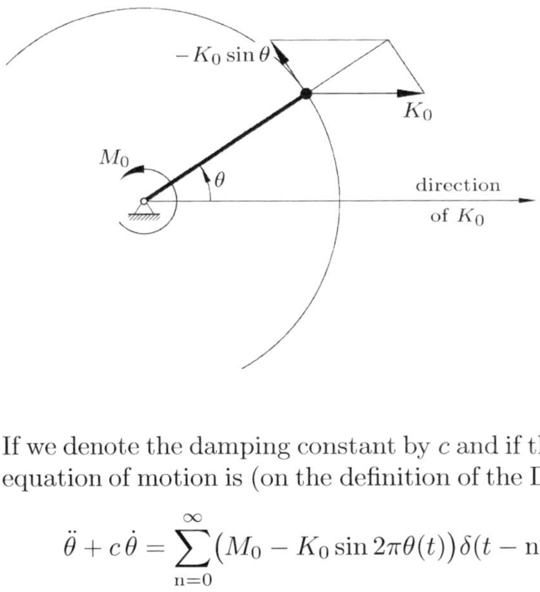

Fig. 8.3.1

Kicked rotator

If we denote the damping constant by c and if the moment of inertia is set to one, the equation of motion is (on the definition of the Dirac *delta*-function, cf. section 3.8.4)

$$\ddot{\theta} + c\,\dot{\theta} = \sum_{\mathrm{n}=0}^{\infty} \big(M_0 - K_0 \sin 2\pi\theta(t)\big)\,\delta(t - \mathrm{n}T) \tag{8.3.1}$$

Applying the substitutions

$$x = \theta, \quad y = \dot{\theta}, \quad z = t \quad (0 \leqslant \theta < 1)$$

the equation of motion (8.3.1) is transformed into an autonomous system of first-order differential equations

$$
\begin{aligned}
\dot{x} &= y \\
\dot{y} &= -cy + \sum_{\mathrm{n}=0}^{\infty} [M_0 - K_0 \sin 2\pi x(z)] \delta(z - \mathrm{n}T) \\
\dot{z} &= 1
\end{aligned}
\tag{8.3.2}
$$

This system can be integrated piecewise between two consecutive impulses. To this end, we denote the coordinates x and y immediately prior to the n-th kick by x_{n} and y_{n}, i.e.

$$
\begin{aligned}
x_{\mathrm{n}} &= \lim_{\varepsilon \to 0} x(\mathrm{n}T - \varepsilon) \\
y_{\mathrm{n}} &= \lim_{\varepsilon \to 0} y(\mathrm{n}T - \varepsilon)
\end{aligned}
\tag{8.3.3}
$$

If we consider the second equation of the system eq. (8.3.2) within the time interval $-\varepsilon \leqslant t \leqslant T - \varepsilon$, the sum is reduced to the term for $\mathrm{n} = 0$

$$
\dot{y} = -cy + [M_0 - K_0 \sin 2\pi x(t)] \delta(t) \tag{8.3.4}
$$

This differential equation can be integrated explicitly and the solution of the homogeneous equation is of the form $y = A\mathrm{e}^{-ct}$. The complete solution is obtained by variation of the parameter A

$$
\dot{A}\mathrm{e}^{-ct} - cA\mathrm{e}^{-ct} = -cA\mathrm{e}^{-ct} + [M_0 - K_0 \sin 2\pi x(t)] \delta(t)
$$

$$
A(t) = A_0 + \lim_{\varepsilon \to 0} \int_{-\varepsilon}^{t} [M_0 - K_0 \sin 2\pi x(t')] \delta(t') \mathrm{e}^{ct'} \mathrm{d}t'
$$

i.e.

$$
A(t) = A_0 + M_0 - K_0 \sin 2\pi x_0 \tag{8.3.5}
$$

Finally, we deduce from the initial condition $y(0) = y_0$ the result $A_0 = y_0$. We thus obtain

$$
y(t) = \mathrm{e}^{-ct} (y_0 + M_0 - K_0 \sin 2\pi x_0) \tag{8.3.6}
$$

Integration of the first expression in eq. (8.3.2) yields

$$
x(t) = x_0 + \frac{1 - \mathrm{e}^{-ct}}{c} (y_0 + M_0 - K_0 \sin 2\pi x_0) \tag{8.3.7}
$$

It is straightforward to derive a relation for the response (x_1, y_1) immediately prior to the application of the following impulse as a function of (x_0, y_0)

$$
\begin{aligned}
x_1 &= x_0 + \frac{1 - \mathrm{e}^{-cT}}{c} (y_0 + M_0 - K_0 \sin 2\pi x_0) \qquad (\text{mod } 1) \\
y_1 &= \mathrm{e}^{-cT} (y_0 + M_0 - K_0 \sin 2\pi x_0)
\end{aligned}
\tag{8.3.8}
$$

which yields the recursion formula

$$
\begin{aligned}
x_{\mathrm{n}+1} &= x_{\mathrm{n}} + \frac{1 - \mathrm{e}^{-cT}}{c} (y_{\mathrm{n}} + M_0 - K_0 \sin 2\pi x_{\mathrm{n}}) \qquad (\text{mod } 1) \\
y_{\mathrm{n}+1} &= \mathrm{e}^{-cT} (y_{\mathrm{n}} + M_0 - K_0 \sin 2\pi x_{\mathrm{n}})
\end{aligned}
\tag{8.3.9}
$$

The modulo function has to be introduced since $0 \leqslant x_{\mathrm{n}} < 1$. Equation (8.3.9), however, corresponds exactly to a Poincaré map if we take Poincaré sections shortly

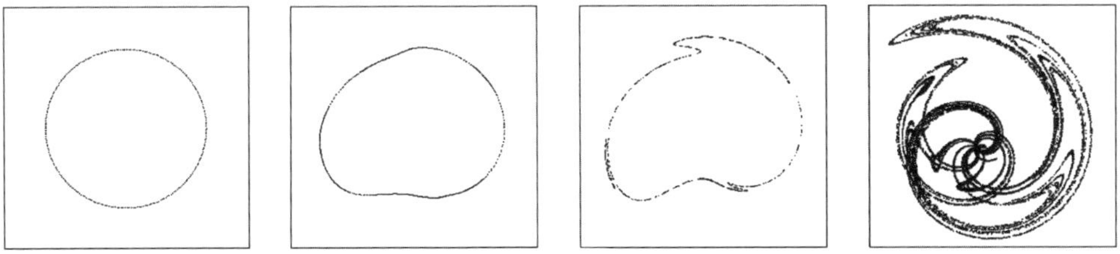

Fig. 8.3.2: The dissipative circle map: a model for illustrating the disintegration of a two-dimensional torus

prior to the next impulse. To simplify the expressions of this equation, we introduce a series of substitutions

$$\begin{aligned} x_{\mathrm{n}} &= \theta_{\mathrm{n}} \,, \quad \frac{\mathrm{e}^{cT}-1}{c} y_{\mathrm{n}} = r_{\mathrm{n}} + \Omega \\ M_0 &= c\Omega \,, \quad \frac{1-\mathrm{e}^{-cT}}{c} K_0 = \frac{K}{2\pi} \,, \quad \mathrm{e}^{-cT} = b \end{aligned} \tag{8.3.10}$$

Substituting these relations into eq. (8.3.9) and following a number of transformations, we obtain

$$\begin{aligned} \theta_{\mathrm{n}+1} &= \theta_{\mathrm{n}} + \Omega - \frac{K}{2\pi} \sin 2\pi\theta_{\mathrm{n}} + b r_{\mathrm{n}} \qquad (\mathrm{mod}\ 1) \\ r_{\mathrm{n}+1} &= b r_{\mathrm{n}} - \frac{K}{2\pi} \sin 2\pi\theta_{\mathrm{n}} \end{aligned} \tag{8.3.11}$$

Here, θ_{n} denotes the angle of the rotator immediately prior to the n-th impulse; apart from a constant factor Ω, r_{n} is proportional to the corresponding angular velocity $y_{\mathrm{n}} = \dot{\theta}|_{t=\mathrm{n}}$ in accordance with eq. (8.3.10). If we interpret θ_{n} and r_{n} as polar coordinates, we can consider eq. (8.3.11) a dissipative circle map where the constant $b = \mathrm{e}^{-cT} < 1$ regulates the damping K and is a measure for the intensity of the non-linear external impulse.

Figure 8.3.2 illustrates the variation of the dissipative circle map as the non-linearity K is increased with otherwise fixed control parameters $b = 0.5$, $\Omega = 0.612$. Here, r_{n} and θ_{n} in eq. (8.3.11) were determined for the initial values ($r_0 = 0$, $\theta_0 = 0$) and successive iteration points were plotted with the polar coordinates $r = 1 + 4r_{\mathrm{n}}$, $\theta = \theta_{\mathrm{n}}$. For $K = 0$, we obtain a circle of radius $r = 1$ which is slightly deformed for $K = 0.9$. For $K > 1$, the circumference is stretched and folded back onto itself, i.e. the torus disintegrates. A further increase of the K-values finally generates a strange attractor. If we interpret the dissipative circle map as a Poincaré map, it can serve as a model for the disintegration of a two-dimensional torus in a three-dimensional phase space.

In the case of very strong damping ($b \approx 0$), eq. (8.3.11) yields the one-dimensional circle map

$$\theta_{\mathrm{n}+1} = \theta_{\mathrm{n}} + \Omega - \frac{K}{2\pi} \sin 2\pi\theta_{\mathrm{n}} \qquad (\mathrm{mod}\ 1) \tag{8.3.12}$$

the characteristics of which we shall investigate in the following section. Due to the period T of the external impulse and the magnitude of the jerky or impulsive rotations applied sequentially about an angle $2\pi\Omega$, the dynamical system possesses two frequencies, independent of one another, so that, depending on their ratio, periodic or quasi-periodic motions can be expected.

8.3.2 The One-Dimensional Circle Map

The two-dimensional circle map, eq. (8.3.11), is in a way the dissipative counterpart of the standard map of Hamiltonian systems, eq. (4.5.1) (*Moser's perturbed twist map*), which we studied in section 4.5 when investigating the stability of the invariant tori of conservative systems. Here, as in the KAM theory, we have to distinguish between rational and irrational frequency ratios. The one-dimensional circle map

$$f(\theta) = \theta + \Omega - \frac{K}{2\pi} \sin 2\pi\theta \qquad (\text{mod } 1) \tag{8.3.13}$$

is the simplest map providing information on the interaction of periodic, quasi-periodic and chaotic motions. Its characteristics were already investigated in detail by Arnold in 1965 (Arnold, 1965).

If the non-linear term vanishes ($K = 0$), we obtain the simple linear circle map

$$\theta_{\mathrm{n}+1} = \theta_{\mathrm{n}} + \Omega \qquad (\text{mod } 1) \tag{8.3.14}$$

Equation (8.3.14) can be interpreted as the Poincaré map of a motion on a three-dimensional torus, characterised by two frequencies ω_1 and ω_2 (cf. section 4.3). Here, $\Omega = \Delta\theta = \omega_2/\omega_1$ denotes the phase shift after one cycle (see fig. 8.3.3). If the ratio of the two frequencies $\omega_2/\omega_1 = \mathrm{p}/\mathrm{q}$ is a rational number, where p,q are relatively prime integers with $\mathrm{p} \leqslant \mathrm{q}$, the trajectory is closed after q orbits and the motion is periodic. In the linear case, the ratio p/q is called the rotation number W. Irrational numbers Ω lead to quasi-periodic motions which fill the torus densely. If we include the non-linear term $(K/2\pi)\sin 2\pi\theta$ in eq. (8.3.14), it becomes more difficult to determine the rotation number. Using the notation

$$f(\theta) = \theta + g(\theta) \tag{8.3.15}$$

the rotation number is defined as the average revolution per iteration

$$W(K,\Omega) = \lim_{\mathrm{n}\to\infty} \frac{g(\theta_0) + g(\theta_1) + \cdots + g(\theta_{\mathrm{n}-1})}{\mathrm{n}} \tag{8.3.16}$$

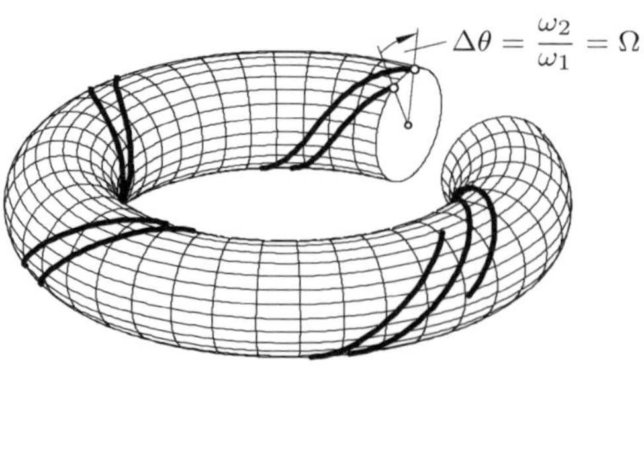

Fig. 8.3.3

Phase shift in the linear circle map

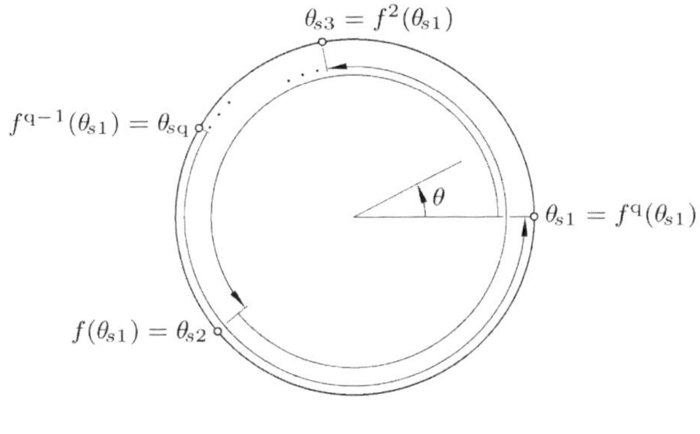

Fig. 8.3.4

q-cycle of the circle map, θ (mod 1)

In the calculation of the rotation number, the modulo rule is not taken into account. For the linear circle map $g(\theta) = \Omega$ applies, so that in this case, as expected, $W(K, \Omega) \equiv \Omega$ follows. After substitution of eq. (8.3.15) and application of the relation $\theta_{i+1} = f(\theta_i)$, eq. (8.3.16) may be put in the simpler equivalent form

$$W(K, \Omega) = \lim_{n\to\infty} \frac{\theta_n - \theta_0}{n} = \lim_{n\to\infty} \frac{f^n(\theta_0) - \theta_0}{n} \tag{8.3.17}$$

For $0 < K < 1$, Ω is thus no longer exclusively responsible for the periodicity respectively quasi-periodicity of the motion. This decision depends on the two control parameters Ω and K. If we assume a fixed rational rotation number $W = p/q$ where p and q have no common divisor, a q-cycle $(\theta_{s1}, \theta_{s2}, \dots \theta_{sq})$ emerges as the long-term response on the Poincaré section, i.e. after q iterations and p revolutions, the initial point is reached again so that we have for the q-th iteration of the map (again not taking the modulo rule into account)

$$f^q(\theta_{si}) = \theta_{si} + p \qquad (i = 1, 2, \dots q) \tag{8.3.18}$$

where θ_{si} $(i = 1, \dots, q)$ denotes the angle coordinates of the circle points of the q-cycle (see fig. 8.3.4). This periodic motion is stable as long as

$$\left| \frac{d}{d\theta}(f^q)\Big|_{\theta_{si}} \right| = |(f^q)'_{\theta_{si}}| < 1 \tag{8.3.19}$$

holds (cf. section 3.6). If we first set $q = 2$, there holds for $i = 1$ in accordance with the chain rule, see also eq. (6.7.11),

$$(f^2)'_{\theta_{s1}} = \left[f\big(f(\theta)\big)\right]'_{\theta_{s1}} = f'\big(f(\theta_{s1})\big) f'(\theta_{s1}) = f'(\theta_{s2}) f'(\theta_{s1})$$

or generally for q

$$(f^q)'_{\theta_{s1}} = f'(\theta_{sq}) f'(\theta_{s,q-1}) \cdots f'(\theta_{s2}) f'(\theta_{s1}) = \prod_{i=1}^{q} f'(\theta_{si}) \tag{8.3.20}$$

The stability condition for a periodic motion with the rotation number $W = p/q$ is thus

$$\left| \prod_{i=1}^{q} f'(\theta_{si}) \right| < 1 \tag{8.3.21}$$

or specifically, for the circle map of eq. (8.3.12),

$$\left|\prod_{\mathrm{i}=1}^{\mathrm{q}}(1-K\cos 2\pi\theta_{\mathrm{si}})\right|<1 \tag{8.3.22}$$

We now direct our attention to the boundaries of the stability domains which are attained for $|(f^{\mathrm{q}})'_{\theta_{\mathrm{si}}}|=1$. For example, for p $=0$, q $=1$, i.e. for the rotation number $W=0/1$, we obtain

$$\left|1-K\cos 2\pi\theta_{s1}\right|=1 \tag{8.3.23}$$

In the following, we can concentrate on values $K>0$ without loss of generality. In fact, we account for negative K-values by replacing $\sin 2\pi\theta$ by $\sin(2\pi\theta+\pi)$ or θ by $\theta+\frac{1}{2}$. Furthermore, for the present, we consider only K-values <1. From eq. (8.3.23), we then deduce $2\pi\theta_{s1}=\pi/2$ respectively $3\pi/2$ or $\theta_{s1}=1/4$ respectively $3/4$ and thus, from eq. (8.3.18), the linear relations

$$\Omega=\pm\frac{K}{2\pi} \tag{8.3.24}$$

Correspondingly, we find for $W=1/1$ the relations $\Omega=1\pm K/2\pi$, i.e. the parameter ranges in which periodic motions with the rotation number $W=0/1$ respectively $W=1/1$ occur are stable and are bounded by the straight lines $\Omega=\pm K/2\pi$ respectively $\Omega=1\pm K/2\pi$.

For the rotation number $W=1/2$ (p $=1$, q $=2$), we deduce from eq. (8.3.18) and $|(f^2)'_{\theta_{\mathrm{si}}}|=1$ the following equations for the boundaries of the stability domain

$$\begin{aligned}\big(1-K\cos 2\pi\theta_{s1}\big)\big(1-K\cos 2\pi\theta_{s2}\big)&=1\\ f^2(\theta_{s1})&=\theta_{s1}+1\\ \theta_{s2}&=f(\theta_{s1})\end{aligned} \tag{8.3.25}$$

We can eliminate θ_{s1} and θ_{s2} from these three non-linear relations and thus obtain an expression $\Omega=\Omega(K)$ for the boundaries of the stability domain of the periodic motion with the rotation number $W=1/2$. For a rotation number $W=\mathrm{p/q}$, we obtain (q+1) non-linear equations for the coordinates $\theta_{s1},\theta_{s2},\cdots\theta_{s\mathrm{q}}$ of the q-cycle and for the stability boundary $\Omega=\Omega(K)$.

It is expedient to solve the system eq. (8.3.25) numerically using Newton's iteration method. The structure of the domains in the Ω, K-plane in which periodic behaviour occurs as a result of a synchronisation of the frequencies was investigated in detail by Arnold (Arnold, 1965); they are thus called *Arnold tongues.*

We wish to circumvent the solution of the system of equations (8.3.25) and propose to calculate approximately for small K-values the boundary of the Arnold tongue for the rotation number $W=1/2$. For $K>0$ (however, $K\ll 1$), there is an Ω-interval containing $\Omega=1/2$ leading to periodic motions with $W=1/2$. Denoting the deviation from $\Omega=1/2$ by $\bar{\Omega}$, i.e.

$$\Omega=\tfrac{1}{2}+\bar{\Omega} \tag{8.3.26}$$

and substituting this relation into the circle map, we obtain

$$f(\theta) = \theta + \tfrac{1}{2} + \bar{\Omega} - \frac{K}{2\pi} \sin 2\pi\theta \tag{8.3.27}$$

According to eq. (8.3.18), for motions with $W = 1/2$ we thus have

$$f^2(\theta_s) = \theta_s + 1$$

or with eq. (8.3.27)

$$\theta_s + \tfrac{1}{2} + \bar{\Omega} - \frac{K}{2\pi} \sin 2\pi\theta_s + \tfrac{1}{2} + \bar{\Omega} - \frac{K}{2\pi} \sin 2\pi(\theta_s + \tfrac{1}{2} + \bar{\Omega} - \frac{K}{2\pi} \sin 2\pi\theta_s) = \theta_s + 1 \tag{8.3.28}$$

We can expand the last expression on the left-hand side of this equation into a Taylor series at $\theta_s + 1/2$. Since $K \ll 1$ and $\bar{\Omega} \ll 1$, we can neglect second-order terms and eq. (8.3.28) can be simplified to

$$2\bar{\Omega} - \frac{K}{2\pi} \sin 2\pi\theta_s - \frac{K}{2\pi} \left[\sin 2\pi(\theta_s + \tfrac{1}{2}) + 2\pi \cos 2\pi(\theta_s + \tfrac{1}{2})(\bar{\Omega} - \frac{K}{2\pi} \sin 2\pi\theta_s)\right] = 0$$

Since $\sin 2\pi(\theta_s + 1/2) = -\sin 2\pi\theta_s$ and $\cos 2\pi(\theta_s + 1/2) = -\cos 2\pi\theta_s$, there follows

$$2\bar{\Omega} + K \cos 2\pi\theta_s \left(\bar{\Omega} - \frac{K}{2\pi} \sin 2\pi\theta_s\right) = 0$$

respectively

$$\bar{\Omega}(2 + K \cos 2\pi\theta_s) = \frac{K^2}{4\pi} \sin 4\pi\theta_s \tag{8.3.29}$$

For small K-values, we obtain approximately

$$2\bar{\Omega} = \frac{K^2}{4\pi} \sin 4\pi\theta_s$$

i.e. the deviations from $\Omega = 1/2$ can be estimated as

$$|\bar{\Omega}| \leqslant \frac{K^2}{8\pi} \tag{8.3.30}$$

In the case of very small non-linearities, we can thus approximate the boundaries of the Arnold tongue corresponding to the rotation number 1/2 by the parabolae

$$\Omega = \frac{1}{2} \pm \frac{K^2}{8\pi} \tag{8.3.31}$$

Figure 8.3.5 shows the stability boundaries of the periodic domains in the (K, Ω)-plane for a series of rotation numbers. The shaded areas in which a periodic motion with a fixed rotation number takes place are denoted Arnold tongues. Such a zone in which a synchronisation of the frequencies, a *frequency locking*, occurs results for each rational rotation number $W = \mathrm{p/q}$, i.e. at every rational value on the Ω-axis, a periodic window arises which expands with increasing K. Since the set of rational numbers in the interval [0,1] is of Lebesgue measure zero, the probability of periodic behaviour occuring for very small K-values is almost negligible (see fig. 8.3.5). Although there is a countable infinity of such resonance regions (or horns), they decrease in width with increasing denominator q so fast that they do not fill the whole

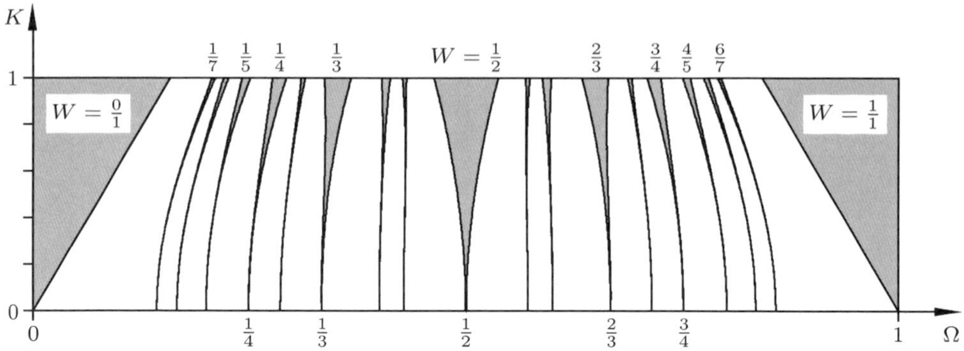

Fig. 8.3.5: The Arnold tongues of the circle map

(K, Ω)-plane for $0 < K < 1$ but leave space for quasi-periodic motions; however, with increasing non-linearity, the synchronisation of the frequencies becomes more and more dominant.

According to a theorem of Denjoy, each one-to-one, twice continuously differentiable, orientation-preserving mapping of the circle onto itself – thus particularly the circle map, eq. (8.3.13), for $0 \leqslant K < 1$ and parameter values Ω leading to quasi-periodic motions with an irrational rotation number W – is topologically equivalent to a simple rotation about the angle $2\pi W$; see, for example, (Arnold, 1988). The fundamental ideas of this theorem go back to the writings of Poincaré who was inspired to formulate this conjecture for a particular class of differential equations as early as 1885 (without proof).

The coupling of two frequencies is a well-known effect occurring in dissipative systems with two competing frequencies and was already observed by the Dutch physicist Christian Huygens in the 17th century. If we hang two clocks with frequencies which deviate initially only slightly from one another back to back on a wall, a coupling of the oscillations takes place and the clocks tend to synchronise their frequencies. Quartz clocks, for example, are based on this synchronisation effect.

If we consider the rotation number W for fixed values $0 \leqslant K_0 < 1$ and variable Ω, we can show that $W(K_0, \Omega)$ is a highly non-linear, monotonically increasing, continuous but not differentiable function of Ω (Arnold, 1965; Herman, 1977). For $K_0 = 0$, the circle map describes a pure rotation with $W(0, \Omega) = \Omega$. For $0 < K_0 < 1$, there exists a whole Ω-interval corresponding to every rational rotation number $W = \mathrm{p}/\mathrm{q}$ for which periodic motions with this rotation number p/q occur while there is only one Ω-value corresponding to each irrational W, i.e. the function $W(K_0, \Omega)$ displays the characteristics of a staircase.

Jensen, Bak and Bohr showed (Jensen et al., 1984) that, for $K_0 = 1$, the plateaux with constant rotation number form a so-called complete "devil's staircase" – complete because the sum of the lengths of all intervals in which frequency locking occurs is exactly 1. Figure 8.3.6 illustrates the self-similar, fractal structure of the devil's staircase, i.e. each detail, however small, possesses the same characteristics as

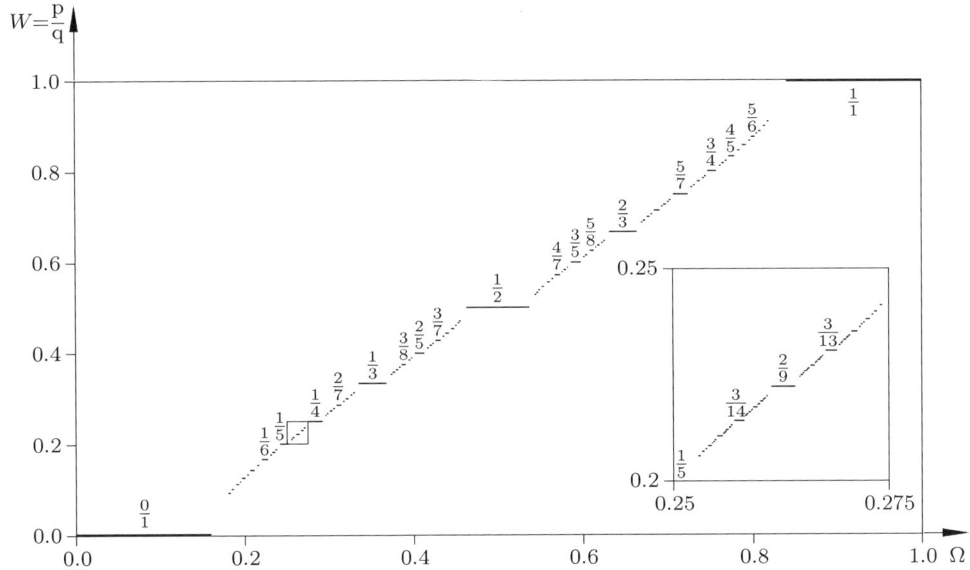

Fig. 8.3.6: Rotation number $W(K_0, \Omega)$ for $K_0 = 1$: devil's staircase

the whole. In the next section 8.3.3, we shall return to this self-similarity.

We have seen that the edges of the Arnold tongues represent the boundaries of the stability domains of periodic motions. The question now arises whether bifurcations occur on these lines and, if so, as which type they can be classified.

To this end, we consider the simplest case of periodic behaviour with the rotation number $W = 0/1$ for fixed parameter values $0 < K_0 < 1$, i.e. we investigate the fixed points θ_s of $f(\theta)$

$$\theta_s = \theta_s + \Omega - \frac{K_0}{2\pi} \sin 2\pi\theta_s$$

or

$$\sin 2\pi\theta_s = \frac{2\pi\Omega}{K_0} \tag{8.3.32}$$

In accordance with eq. (8.3.23), these fixed points lose their stability if

$$|f'(\theta_s)| = |1 - K_0 \cos 2\pi\theta_s| = 1$$

holds, i.e. for $\theta_s = \pm 1/4$. At these points, there is $f'(\theta_s) = +1$, i.e. with the loss of stability, a bifurcation occurs for the parameter value $\Omega_{K_0} = \pm K_0/2\pi$, cf. eq. (8.3.24). In section 6.6, we studied the normal forms of bifurcations of one-parametric maps and summarised the conditions for $f(\theta, \Omega)$ which enable us to formulate a classification. Since

$$\frac{\partial f}{\partial \Omega}(\theta_s, \Omega_{K_0}) = 1 \neq 0 \tag{8.3.33}$$

and

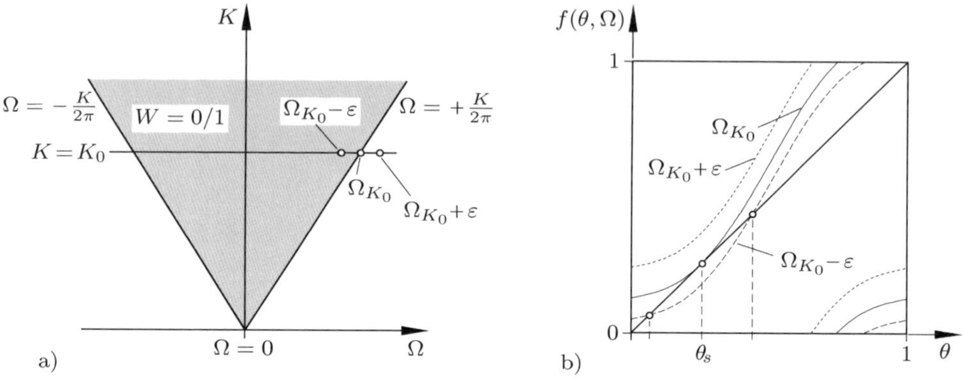

Fig. 8.3.7: Saddle-node bifurcation at the boundary of the Arnold-tongue for the rotation number $W=0/1$
a) representation in the parameter plane
b) graphs of the circle map (parameter Ω) near the bifurcation point (θ_s, Ω_{K_0})

$$\frac{\partial^2 f}{\partial \theta^2}(\theta_s, \Omega_{K_0}) = 2\pi K_0 \sin 2\pi\theta_s = \pm 2\pi K_0 \neq 0 \tag{8.3.34}$$

then, in accordance with eqs. (6.6.17) to (6.6.19), a saddle-node bifurcation takes place.

Figure 8.3.7 demonstrates that saddle-node bifurcations occur on the boundary of the Arnold tongue appertaining to $W = 0/1$. For a fixed value $0 < K_0 < 1$, the three Ω-values $\Omega_{K_0}-\varepsilon$, Ω_{K_0} and $\Omega_{K_0}+\varepsilon$ were selected in fig. 8.3.7a and the associated functions presented in fig. 8.3.7b. For $\Omega > \Omega_{K_0}$ a fixed point has not yet been generated. For $\Omega = \Omega_{K_0}$ the function $f(\theta, \Omega)$ is tangent to the bisector, i.e. a fixed point is born which branches into two fixed points for $\Omega < \Omega_{K_0}$, where the gradient of the function provides information on the stability. Following a further decrease in Ω, the stable and unstable fixed points progress in opposite directions around the circle until they finally meet again and vanish. A similar phenomenon also occurs in the case of the other Arnold tongues.

A detailed discussion of the bifurcation in the case of a p/q- "resonance horn" can be found in (Aronson et al., 1982), for example. Figure 8.3.8a presents the bifurcation diagram for $W = 0/1$. Due to the coincidence of $\theta = 0$ and $\theta = 1$, this is a closed curve which can also be represented on a cylinder (see fig. 8.3.8b).

Up to now, we have restricted ourselves to the parameter range $0 \leqslant K < 1$. For $K = 1$, a critical threshold is reached. The Arnold tongues grow together and only admit quasi-periodic motions for an Ω-set of Lebesgue measure zero. If we increase K to values $K > 1$, the narrowest Arnold tongues begin to overlap near the critical line, the larger tongues follow for higher K-values. However, as soon as two Arnold tongues with the rotation numbers $W_1 = \mathrm{p}_1/\mathrm{q}_1$ and $W_2 = \mathrm{p}_2/\mathrm{q}_2$ overlap, infinitely many tongues still arise in between which also overlap since there are still infinitely

Plate IX Lyapunov exponent $\sigma(\Omega, K)$ for the circle map ($0 \leqslant \Omega \leqslant 1;\ \ 0 \leqslant K \leqslant 10$)

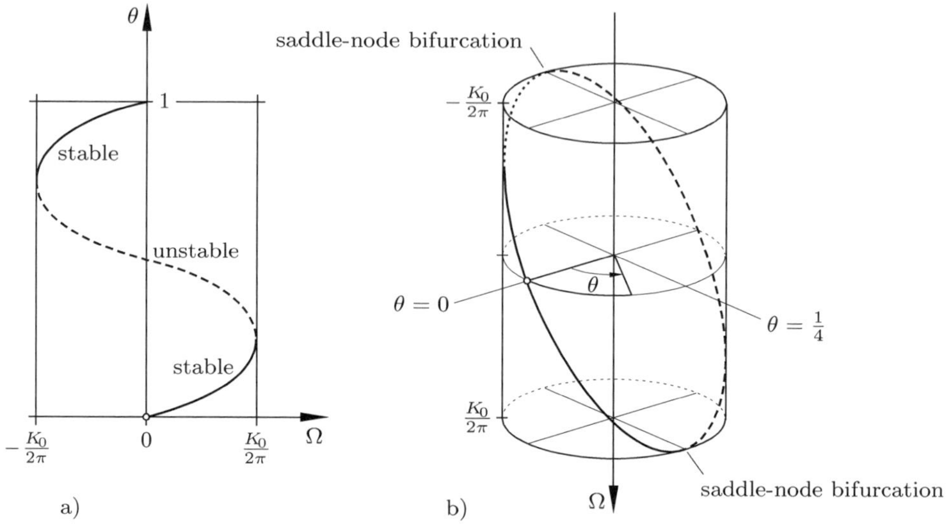

Fig. 8.3.8: Bifurcation diagram for $W = 0/1$:
a) representation in the parameter plane, $\theta = \theta\,(\text{mod } 1)$
b) representation in cylindrical coordinates

many further rational numbers in the interval $W_1 \leqslant W \leqslant W_2$ (cf. section 10.8, particularly fig. 10.8.2). The rotation number W is then no longer uniquely determined but is replaced by a rotation interval $[W_1, W_2]$ (MacKay and Tresser, 1986). In dependence of the initial condition, periodic motions with different rotation numbers can then emerge. A contest between the different rotation numbers may finally occur, which can result in a chaotic motion.

Colour Plate IX, p. 497, shows a so-called Lyapunov phase diagram. Here, the Lyapunov exponent σ for the circle map is plotted in the (Ω, K)-plane. Different colours characterise different long-term behaviour. Black defines negative σ-values. This corresponds to a periodic motion in the three-dimensional phase space, i.e. to a fixed point or a cycle in the appertaining Poincaré section respectively in the circle map. Below the critical line $K = 1$, we can recognise the black Arnold tongues within a dark red background which defines quasi-periodic behaviour ($\sigma = 0$). Above the critical line, the Arnold tongues overlap and chaotic regions with $\sigma > 0$ may occur, characterised by the colours yellow/blue. In this depiction, the self-similar structure of the Arnold tongues and the dense interlacing of irregular and regular, periodic domains for $K > 1$ is illustrated particularly clearly. We can also gather detailed information about the anatomy of the Arnold tongues from the variation of the Lyapunov exponents. In section 10.8, the structure of the Arnold tongue will be studied again numerically with the aid of σ and compared with theoretical results.

What is the significance of the critical line $K = 1$ for the circle map? Figure 8.3.9a to c presents the graph of the circle map function and the correponding iteration series of the angles θ_n for $\Omega = 0.48$ and for $K = 0.4$, 1.0 and 1.9. For $K \leqslant 1$, $f(\theta)$ is a monotonically increasing, one-to-one smooth function, i.e. a diffeomorphism.

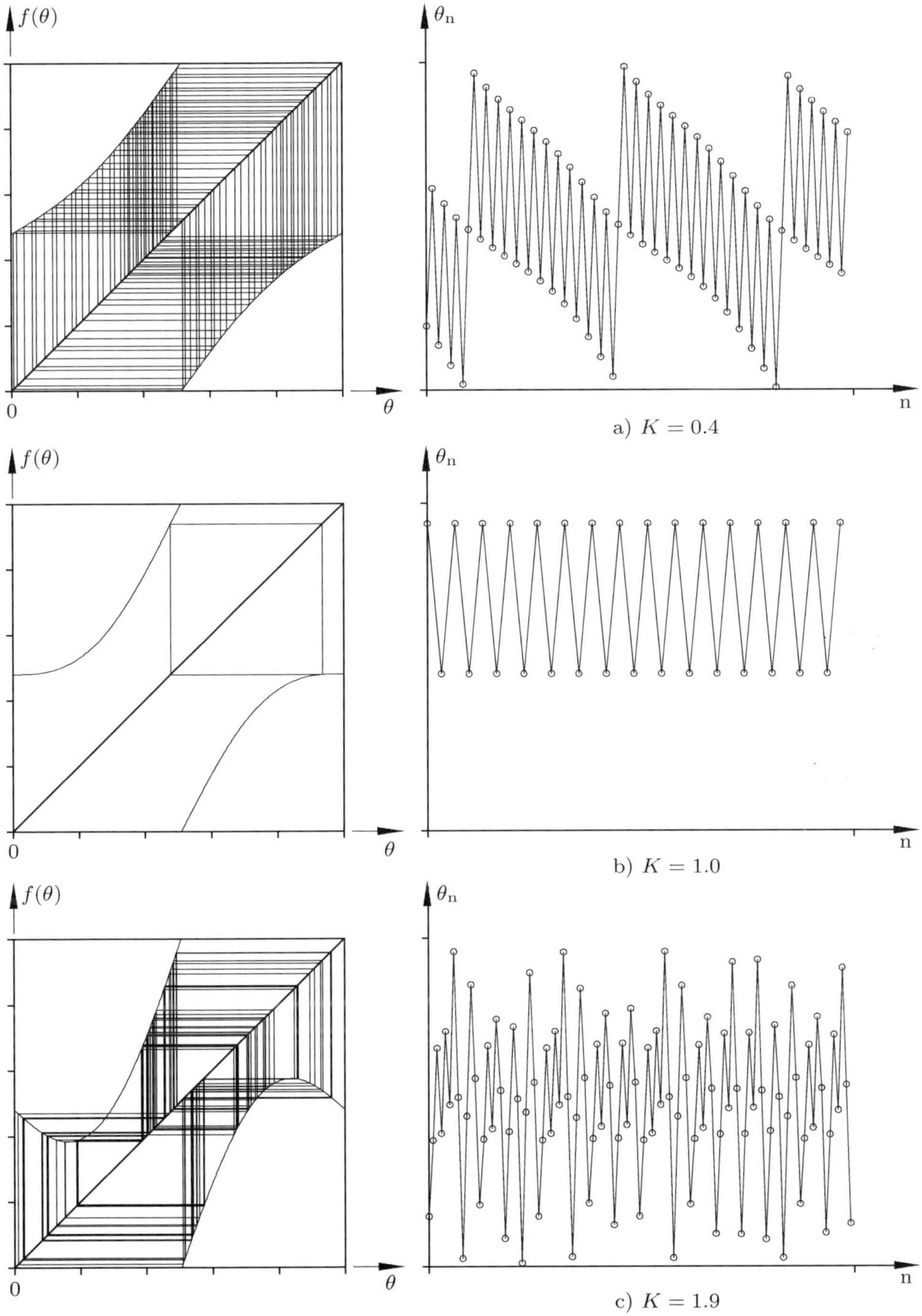

Fig. 8.3.9: Graph of the circle map function and its associated long-term response (for $\Omega = 0.48$ and $K = 0.4$, 1.0, 1.9)

The combination $\Omega = 0.48, K = 0.4$ evidently leads to quasi-periodic behaviour (fig 8.3.9a), i.e. further iterations would yield an orbit uniformly filling the area between the curve and the bisector. For $K = 1$, a cubic inflection point with a horizontal tangent emerges for $\theta = 0$, i.e. the inverse map $f^{-1}(\theta)$ is no longer differentiable at this point. For $\Omega = 0.48, K = 1.0$, figure 8.3.9b shows periodic behaviour with a rotation number $W = 1/2$. For $K > 1$, no unique inverse map exists any more, i.e. the circumference is folded back onto itself, and for $K = 1.9$, we obtain chaotic behaviour; see fig. 8.3.9c (cf. also fig. 8.3.2).

What is interesting now is that – as in the case of the logistic map (section 6.7) – the transition to chaos does not depend on the special form of the function $f(\theta)$. If we consider generally the two-parametric family of maps

$$f_{\Omega,K}(\theta) = \theta + \Omega + Kg(\theta) \qquad (\mathrm{mod}\ 1) \tag{8.3.35}$$

where $g(\theta + 1) = g(\theta)$ is a periodic function, then, due to (mod 1), $f_{\Omega,K}$ has the characteristic property

$$f_{\Omega,K}(\theta + 1) = 1 + f_{\Omega,K}(\theta) = f_{\Omega,K}(\theta) \tag{8.3.36}$$

If, moreover,

a) $f_{\Omega,K}(\theta)$ is a diffeomorphism for $|K| < 1$ (i.e. $f_{\Omega,K}(\theta)$ is bijective and, together with its inverse, continously differentiable), and
b) a cubic inflection point occurs for $|K| = 1$ so that $f^{-1}_{\Omega,K}(\theta)$ is no longer differentiable, and
c) $f_{\Omega,K}(\theta)$ is no longer invertible for $|K| > 1$,

the map $f_{\Omega,K}$ also possesses a pattern of "Arnold tongues" in which frequency locking occurs, and it exhibits the same universal characteristics as the one-dimensional circle map, eq. (8.3.12). The non-invertibility of $f_{\Omega,K}(\theta)$ for $|K| > 1$ is – as generally in the case of one-dimensional maps – a necessary condition for the emergence of chaotic motions.

As an illustration, we consider the modified cubic circle map

$$f_{\Omega,K}(\theta) = \theta + \Omega - Kg(\theta) \qquad (\mathrm{mod}\ 1) \tag{8.3.37}$$

where the non-linear component in the interval $0 \leqslant \theta < 1$ is defined as

$$g(\theta) = \theta(2\theta^2 - 3\theta + 1) \qquad (0 \leqslant \theta < 1) \tag{8.3.38}$$

and is periodically repeated in accordance with

$$g(\theta + 1) = g(\theta) \tag{8.3.39}$$

This cubic circle map also possesses the property of eq. (8.3.36), namely $f_{\Omega,K}(\theta + 1) = f_{\Omega,K}(\theta)$, and, moreover, satisfies the aforementioned conditions a) to c). If we calculate the Lyapunov exponent σ again for this iteration and plot it as a function of Ω and K (fig. 8.3.10), we obtain a structure very similar to that for the standard circle map of eq. (8.3.12).

Here again, we observe infinitely many Arnold tongues in which frequency locking leads to periodic motions. Above the critical line $K = 1$, these domains overlap again

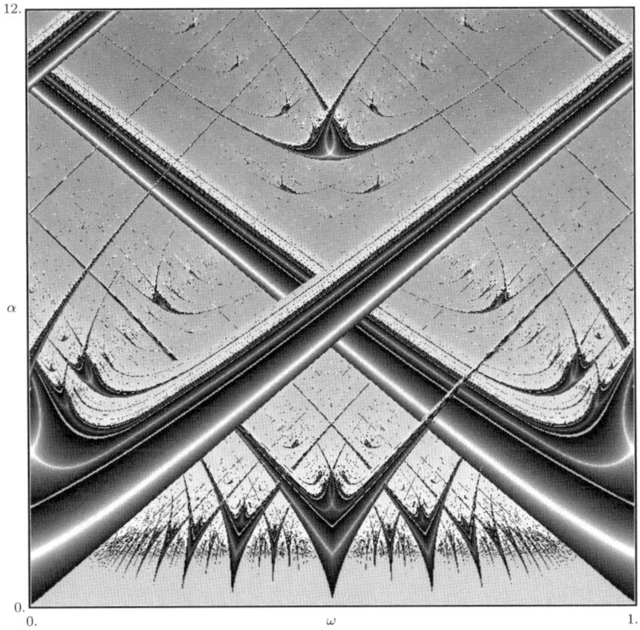

Fig. 8.3.10: Lyapunov exponents of the cubic circle map, eqs. (8.3.37) – (8.3.39)

and hysteresis effects and finally chaotic motions ensue. Comparing fig. 8.3.10 with colour plate IX, p. 497, we confirm a broad consistency of the two figures, not only in the structure of the Arnold tongues for $K \leqslant 1$, but also in the special composition of the periodic windows in the parameter plane for $K > 1$. These islands of stable periodic motion within a chaotic sea form clusters of structures resulting from a basic element by means of homeomorphic mappings. These stability domains are denoted *shrimps* since their form is strongly reminiscent of these creatures, cf. (Gallas, 1994). In section 10.8, we will discuss the internal structure of the Arnold tongues in more detail.

It turns out that the self-affine structure and assembly of these stability domains follows universal laws (Gallas, 1994; Baptista et al., 2003) which can be deciphered by means of renormalisation techniques, see also section 8.3.3. It is therefore not surprising that similar clusters of these *shrimp*-shaped domains can be detected within many other *discrete* mappings in which two parameters are varied simultaneously such as, for example, in the case of the Hénon map and the cubic map (Gallas, 1994). This is thus a universal phenomenon of codimension 2 which is ultimately based on the factorisation of the derivative of q-periodic orbits, see eq. (8.3.20), controlling the stability of these periodic motions. It was therefore surprising and new that such *shrimp*-shaped domains could also be found in *continuous-time* dynamical systems described by autonomous respectively non-autonomous differential equations like the Duffing oscillator, for example (Bonatto et al., 2008) or in the case of CO_2 lasers and semiconductor lasers subjected either to optical injection, to optical feedback, or to modulations (Bonatto and Gallas, 2007). The latter example is of great practical interest for secure data transmission since the encoding of data for optically secure communication is based on synchronised chaotic lasers which have to be operated in an "entirely chaotic" regime (Kusumoto and Ohtsubo, 2002).

Recently, an interlaced hierarchical network of periodicity structures was investigated using the example of the prototypical dissipative flow of a Rössler oscillator, see eq. (10.6.1), for $a, c > 0, b = 0.3$ (Vitolo et al., 2011). The Lyapunov phase diagram in Colour Plate X, fig. a), p. 503, shows nested spirals of periodic oscillations alternating with chaotic behaviour and converging to central points, so-called periodicity hubs. Also, sequences of shrimp-shaped domains occur in pairs along the periodicity spirals. Grey tones identify periodic oscillations while colours are used for the positive maximum Lyapunov exponents characterising chaos. It turns out that the organisation of the periodicity hubs reflects the dynamics of Shilnikov's homoclinic scenario described in section 10.6.

We would like to emphasise that Lyapunov phase diagrams greatly simplify and enhance our insight into the complex global structure of non-linear dynamical systems. In the next section, we discuss the universal characteristics of the circle map.

8.3.3 Scaling Characteristics of the Circle Map

The self-similar structure of the Arnold tongues for $0 \leqslant K \leqslant 1$ (see Colour Plate IX, p. 497, and fig. 8.3.5) and of the devil's staircase (fig. 8.3.6) suggests that a transition takes place in the circle map at the critical line $K = 1$ similar to the one we observed in section 6.7 for second-order phase transitions in the example of the logistic map. Inspired by the theory of renormalisation groups and by Feigenbaum's successful application of this method to the logistic map, various groups of scientists have also attempted to decode the self-similarity of the circle map with the aid of the renormalisation technique. Their findings display two types of universality; on the one hand, we can investigate the transition for a special *fixed* rotation number (*local* universality), on the other, scaling laws for a whole range of rotation numbers can be established (*global* universality). We begin with the local laws.

8.3.3.1 Local Scaling Laws

An experimental verification of the transition from quasi-periodic behaviour to chaos in the case of a fixed rotation number requires some effort since two control parameters must be tuned in such a way that the rotation number remains constant. Moreover, we saw in the last section that the Arnold tongues in the circle map begin to overlap at $K = 1 + \varepsilon$ ($\varepsilon > 0$) and the probability of emerging quasi-periodic orbits decreases from 1 (at $K = 0$) to 0 (at $K = 1$). The tiniest variations in the parameters can thus lead to variations in the rotation numbers and thus to *frequency locking*. The best chance of observing the disintegration of quasi-periodic orbits in the three-dimensional phase space respectively of invariant circles in the circle map is thus offered by a rotation ratio which is "as irrational as possible" respectively "removed as far as possible" from all rational numbers. These considerations remind us strongly of those in sections 4.4 and 4.5 in the study of Moser's twist map, see

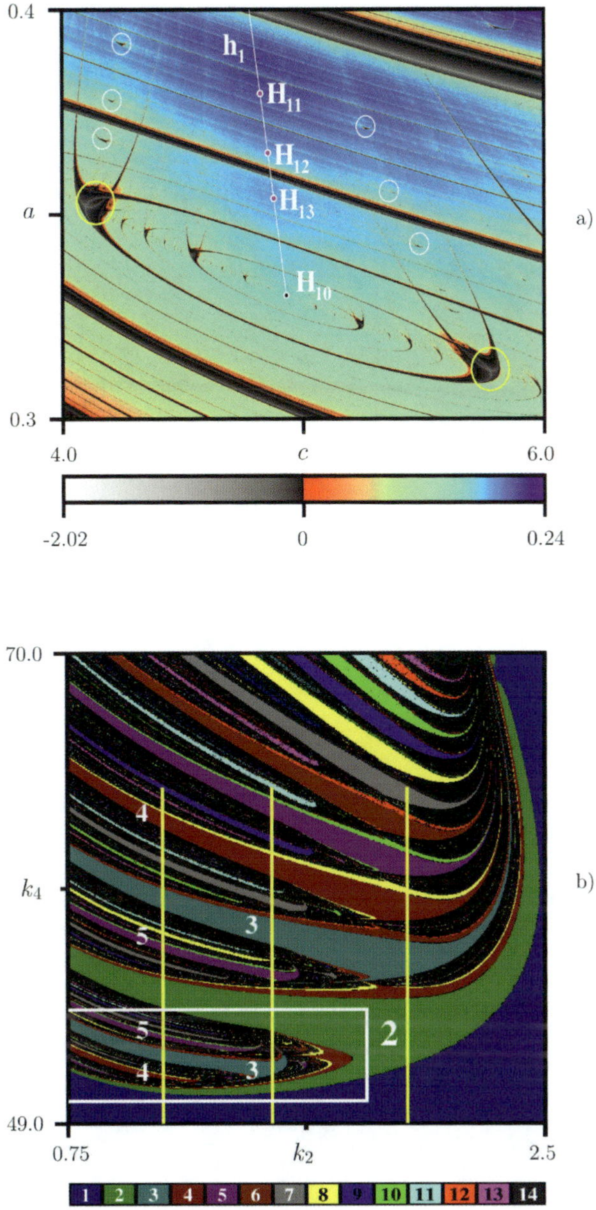

Plate X a) Lyapunov phase diagram for Rössler's oscillator; dots mark periodicity hubs H_1 and $H_{1,j}$, $j = 1, 2, 3$, the curve h_1 indicates Shilnikov homoclinic bifurcations; shrimp-shaped stability domains are highlighted by circles, after (Vitolo et al., 2011). b) Isospike diagram of Belousov-Zhabotinsky reaction; colours discriminate domains of periodic oscillations with an identical number of spikes which are organised according to a Stern-Brocot tree; black represents chaos, after (Freire and Gallas, 2011b).

eqs. (4.5.1) and (4.5.3), the conservative counterpart of the circle map. In this standard map, the final bastion of KAM curves before their disintegration is those curves with an irrational rotation number, the golden mean,

$$W_{\mathrm{G}} = \frac{\sqrt{5}-1}{2} \tag{8.3.40}$$

which is associated with the golden section or golden ratio.

If we divide a line segment in such a way that the ratio of the larger section a to the total length l is the same as the ratio of the smaller to the larger section, i.e.

$$\frac{a}{l} = \frac{l-a}{a}$$

this division is denoted the golden ratio. For the division ratio a/l, we deduce the relation

$$\left(\frac{a}{l}\right)^2 + \frac{a}{l} - 1 = 0$$

Only the positive value $a/l = (\sqrt{5}-1)/2$ is possible as the solution, i.e. the golden mean W_{G}, eq. (8.3.40).

Following a joint work with Kadanoff (Shenker and Kadanoff, 1982), Shenker investigated the disintegration of an invariant circle respectively the transition of a quasi-periodic motion with the special rotation number W_{G} to chaos (Shenker, 1982). He found out that, near the critical boundary $K = 1$, the system behaves like a second-order phase transition and he drew close parallels to the Feigenbaum scenario (section 6.7).

When investigating the Arnold tongue corresponding to a rational rotation number $W = \mathrm{p}/\mathrm{q}$, we saw in section 8.3.2 that, for a fixed K-value due to the relations of eqs. (8.3.18) and (8.3.21), a whole Ω-range can be specified for which periodic motions set in. If, on the other hand, we presume an irrational rotation ratio and assign ourselves the task of calculating for a fixed K the appertaining uniquely determined Ω-value, we must then apply a trick proposed by J. M. Greene for Hamiltonian systems (Greene, 1979). Here, the problem is that, in accordance with eq. (8.3.16) respectively (8.3.17), the rotation number is defined as a limit value and Ω can thus not be calculated directly from this relation. The idea now is to approximate the irrational ratio W by a series of rational ratios which can be represented by finite continued fractions. Irrational numbers can be represented by infinite continued fractions defined as

$$W = \cfrac{1}{\mathrm{n}_1 + \cfrac{1}{\mathrm{n}_2 + \cfrac{1}{\mathrm{n}_3 + \cdots}}} = < \mathrm{n}_1\ \mathrm{n}_2\ \mathrm{n}_3 \ldots > \tag{8.3.41}$$

where n_i are natural numbers. Truncating this fraction after a finite number of digits n_k, we obtain a finite continued fraction corresponding to a rational number.

The golden mean W_{G} has a particularly simple continued fraction representation

$$W_{\mathrm{G}} = \cfrac{1}{1 + \cfrac{1}{1 + \cfrac{1}{1 + \cdots}}} = < 1\ 1\ 1 \cdots > \tag{8.3.42}$$

Indeed, directly from this equation, we can read off the relation

$$W = \frac{1}{1+W} \tag{8.3.43}$$

This may also be written as $W^2+W-1=0$ and has the solution $W_{\mathrm{G}} = (\sqrt{5}-1)/2$. We can now approximate W_{G} by the sequence of rational numbers deduced by successive truncation of the continued fraction eq. (8.3.42)

$$\begin{aligned}
W_1 &= <1> = \frac{1}{1} \\
W_2 &= <1\,1> = \frac{1}{1+1} = \frac{1}{2} \\
W_3 &= <1\,1\,1> = \frac{1}{1+\frac{1}{1+1}} = \frac{2}{3} \\
&\vdots \\
&\text{etc.}
\end{aligned} \tag{8.3.44}$$

Since in this case $\mathrm{n_i} = 1$ holds for all i, the golden mean evidently possesses the characteristic that the approximation with rational numbers converges particularly slowly. W_{n} can be represented very simply by Fibonacci numbers F_{n} which are defined by the recursion formula

$$F_{\mathrm{n}+1} = F_{\mathrm{n}} + F_{\mathrm{n}-1} \tag{8.3.45}$$

with the initial values $F_0 = 0,\ F_1 = 1$. In accordance with eq. (8.3.44), we have

$$W_{\mathrm{n}} = \frac{F_{\mathrm{n}}}{F_{\mathrm{n}+1}} = \frac{F_{\mathrm{n}}}{F_{\mathrm{n}} + F_{\mathrm{n}-1}} = \frac{1}{1+\dfrac{F_{\mathrm{n}-1}}{F_{\mathrm{n}}}} = \frac{1}{1+W_{\mathrm{n}-1}} \tag{8.3.46}$$

With eq. (8.3.46), we obtain as the convergence rate of this sequence $\{W_{\mathrm{n}}\}$

$$\frac{W_{\mathrm{n}+1} - W_{\mathrm{n}}}{W_{\mathrm{n}} - W_{\mathrm{n}-1}} = \frac{\dfrac{1}{1+W_{\mathrm{n}}} - W_{\mathrm{n}}}{W_{\mathrm{n}} - \dfrac{1-W_{\mathrm{n}}}{W_{\mathrm{n}}}} = -\frac{W_{\mathrm{n}}}{1+W_{\mathrm{n}}} = -W_{\mathrm{n}} W_{\mathrm{n}+1} \tag{8.3.47}$$

In the limit, we find

$$\lim_{\mathrm{n}\to\infty} \frac{W_{\mathrm{n}+1} - W_{\mathrm{n}}}{W_{\mathrm{n}} - W_{\mathrm{n}-1}} = -W_{\mathrm{G}}^2 \tag{8.3.48}$$

i.e. the sequence $\{W_{\mathrm{n}}\}$ converges like a geometric sequence towards the limit value W_{G}. For $\mathrm{n} \gg 1$, we have

$$W_{\mathrm{n}} = W_{\mathrm{G}} - \mathrm{const}\cdot\delta^{-\mathrm{n}} \tag{8.3.49}$$

where

$$\delta = \lim_{\mathrm{n}\to\infty} \frac{W_{\mathrm{n}} - W_{\mathrm{n}-1}}{W_{\mathrm{n}+1} - W_{\mathrm{n}}} = -\frac{1}{W_{\mathrm{G}}^2} \tag{8.3.50}$$

cf. also eqs. (6.7.24) and (6.7.66, 6.7.67).

If we now assume a fixed K-value in the interval $0 < K < 1$, then, for the sequence of rational rotation numbers $W_\mathrm{i} = F_\mathrm{i}/F_{\mathrm{i}+1}$ ($\mathrm{i} = 1, 2, 3, \ldots$), we can determine in succession Ω-intervals which belong to periodic motions. In order to determine uniquely a value $\Omega_\mathrm{i}(K)$ appertaining to the interval associated with W_i, we demand that $\theta = 0$ belongs to the $F_{\mathrm{i}+1}$-cycle so that, in accordance with eq. (8.3.18), we have

$$f^{F_{\mathrm{i}+1}}_{\Omega_\mathrm{i}(K)}(0) = F_\mathrm{i} \tag{8.3.51}$$

Since the rotation numbers $W_\mathrm{i} = F_\mathrm{i}/F_{\mathrm{i}+1}$ converge towards W_G and the width of the corresponding Arnold tongue decreases very rapidly, it is to be expected that the sequence of the correponding Ω_i-values tends towards a limit value and exhibits the same convergence behaviour as the sequence of rotation ratios. Figure 8.3.11 presents the appertaining functions $f^{F_{\mathrm{i}+1}}_{\Omega_\mathrm{i}}(\theta)$ together with the corresponding phase portraits for the rotation numbers W_1, W_2, W_3 at $K = 0.9$.

The numerical results of Shenker have confirmed (Shenker, 1982) that the sequence of Ω_i-values tends, as expected, towards a fixed limit value Ω_∞ which just belong to a quasi-periodic motion with the rotation number W_G

$$\lim_{\mathrm{i}\to\infty} \Omega_\mathrm{i}(K) = \Omega_\infty(K) \tag{8.3.52}$$

having the same convergence rate as the series of rotation numbers W_i, eq. (8.3.46). Thus, in the parameter space, we have for large indices n the scaling relation

$$\Omega_\mathrm{n}(K) = \Omega_\infty(K) - \mathrm{const}\cdot\delta^{-\mathrm{n}} \tag{8.3.53}$$

or, in accordance with eqs. (6.7.66, 6.7.67),

$$\lim_{\mathrm{n}\to\infty} (\Omega_\infty - \Omega_\mathrm{n})\delta^\mathrm{n} = \mathrm{const} \tag{8.3.54}$$

respectively

$$\delta = \lim_{\mathrm{n}\to\infty} \frac{\Omega_{\mathrm{n}+1} - \Omega_\mathrm{n}}{\Omega_{\mathrm{n}+2} - \Omega_{\mathrm{n}+1}} \tag{8.3.55}$$

Figure 8.3.11 now suggests that here, too, – as in the logistic map – there are scaling laws in the θ-space. As in eq. (6.7.27), we introduce the distances d_n of $\theta = 0$ (modulo 1) to the nearest neighbouring element of the cycle with the rotation number W_n

$$d_\mathrm{n} = f^{F_\mathrm{n}}_{\Omega_\mathrm{n}}(0) - F_{\mathrm{n}-1} \tag{8.3.56}$$

(see figs. 8.3.11, 8.3.12). Shenker was able to prove numerically that, in accordance with eq. (6.7.23), the distances converge like a geometric progression

$$\lim_{\mathrm{n}\to\infty} \frac{d_\mathrm{n}}{d_{\mathrm{n}+1}} = a \tag{8.3.57}$$

The constants δ and a in eqs. (8.3.55) and (8.3.57) also have universal character in the sense that they are independent of the specific form of the underlying mapping functions $f(\theta)$ as long as the latter satisfy the characteristics a) to c) of section 8.3.2. However, their value depends on the selected irrational rotation number W. As long

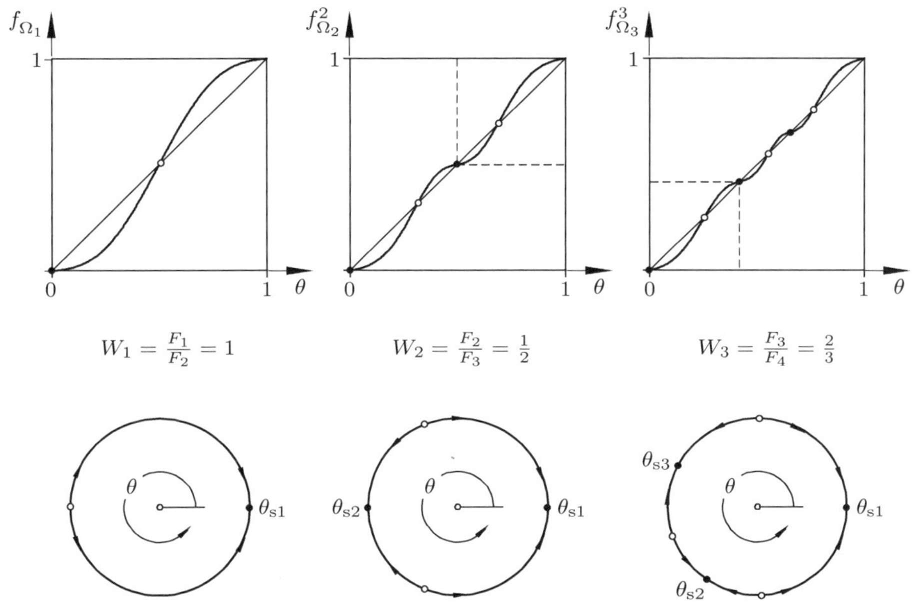

Fig. 8.3.11: q-cycles of the circle map: graphs of the functions $f_{\Omega_i}^{F_{i+1}}(\theta)$ for i $= 1, 2, 3$ and corresponding phase portraits (for $K = 0.9$)

as $f(\theta)$ is a diffeomorphism, i.e. in the subcritical case for $0 < K < 1$, the "trivial" values are, cf. also eq. (8.3.50)

$$K < 1: \quad \delta = -\frac{1}{W_G^2} = -2.6180339\ldots \tag{8.3.58}$$

$$a = -\frac{1}{W_G} = -1.6180339\ldots \tag{8.3.59}$$

Along the critical line $K = 1$ in the (Ω, K)-parameter space, $f(\theta)$ has a cubic inflection point for $\theta = 0$, i.e. $f'(0) = 0$ and $f''(0) = 0$ hold. Thus, the series expansion begins in the neighbourhood of the inflection point with cubic terms. This indicates a singular point where new patterns of motion emerge, which is also expressed in special "non-trivial" values for the universal constants (Feigenbaum et al., 1982)

$$K = 1: \quad \delta = -2.83360 \pm 3 \cdot 10^{-5} \tag{8.3.60}$$

and

$$a = -1.28857 \pm 2 \cdot 10^{-5} \tag{8.3.61}$$

As in the Feigenbaum scenario discussed in detail in section 6.7, a fixed point equation for a universal limit function can be set up via the scaling properties, this equation allowing the calculation of a. The constant δ can finally also be determined from the stability characteristics of the fixed point. In the following, we sketch the deduction of the scaling relations and the fixed point equation. The interested reader will find more detailed accounts in the works already cited (Shenker, 1982; Feigenbaum et al., 1982) and in a publication produced independently (Rand et al., 1982) as well as in (Cvitanović et al., 1985).

In analogy to fig. 6.7.10, we have plotted in fig. 8.3.12 the graphs of the function and iterations appertaining to the F_{i+1}-cycles with the rotation numbers $W_i = F_i/F_{i+1}$ ($i = 1, 2, 3, \ldots$) which contain $\theta = 0$ as an element. If we now compare, for example, the evolution of the function $f_{\Omega_2}^{F_3}(\theta)$ respectively $f_{\Omega_3}^{F_4}(\theta)$ within the dashed square with that of $f_{\Omega_1}^{F_2}(\theta)$ over the whole θ-range, we again observe an astonishing similarity: applying a magnification factor d_2/d_1 respectively d_3/d_1 and a reflection at the first and second bisector, we can plot the corresponding curves onto one another and compare them. The scaling factors are given by eq. (8.3.56). We obtain from the scaling law eq. (8.3.57)

$$\lim_{n\to\infty} a^n d_{n+1} = \text{const} \tag{8.3.62}$$

Multiplication of d_{n+1} given in eq. (8.3.56) by a^n yields

$$\lim_{n\to\infty} a^n \big[f_{\Omega_{n+1}}^{F_{n+1}}(0) - F_n \big] = \text{const} \tag{8.3.63}$$

As in the case of the logistic map, we can observe convergence not only at $\theta = 0$, but also with appropriate scaling convergence towards a limit function $\tilde{g}_1(\theta)$, cf. eq. (6.7.29),

$$\lim_{n\to\infty} a^n \big[f_{\Omega_{n+1}}^{F_{n+1}}(\vartheta_n) - F_n \big] = \tilde{g}_1(\theta) \tag{8.3.64}$$

where we apply the notation

$$\frac{\theta}{a^n} \equiv \vartheta_n \tag{8.3.65}$$

For simplicity's sake, we also substitute the relation

$$f_{\Omega_{n+1}}^{(n)}(\theta) = f_{\Omega_{n+1}}^{F_{n+1}}(\theta) - F_n \tag{8.3.66}$$

so that eq. (8.3.64) becomes

$$\lim_{n\to\infty} a^n f_{\Omega_{n+1}}^{(n)}(\vartheta_n) = \tilde{g}_1(\theta) \tag{8.3.67}$$

At the beginning of our discussion, we ascertained self-similarity in the functions $f_{\Omega_1}^{F_2}$, $f_{\Omega_2}^{F_3}$, $f_{\Omega_3}^{F_4}$, ... These functions also tend towards a limit function $\tilde{g}_0(\theta)$

$$\lim_{n\to\infty} a^n f_{\Omega_n}^{(n)}(\vartheta_n) = \tilde{g}_0(\theta) \tag{8.3.68}$$

Note the difference in the indices in eqs. (8.3.67) and (8.3.68). From the function series $f_{\Omega_{n+i}}^{(n)}$ for $i = 0,1,2,3,\ldots$, we can generally construct a series of limit functions

$$\lim_{n\to\infty} a^n f_{\Omega_{n+i}}^{(n)}(\vartheta_n) = \tilde{g}_i(\theta) \tag{8.3.69}$$

Equation (8.3.20) shows that the individual cycle elements $\theta_1{=}0,\ \theta_2,\ \theta_3, \ldots \theta_q$ of a q-cycle are coupled to one another by the same gradients. It thus suffices to consider the iterated maps in the neighbourhood of *one* cycle point $\theta = 0$. The scalings required for the construction of the limit functions mean a perpetual zooming out of increasingly shrinking neighbourhoods of $\theta = 0$. Hence, the limit function does not depend on the special shape of the mapping function $f(\theta)$ but only on the characteristics a) to c) of section 8.3.2 which lead to the pattern of Arnold tongues shown in Colour Plate IX, p. 497, i.e. the limit functions $\tilde{g}_i(\theta)$ are universal.

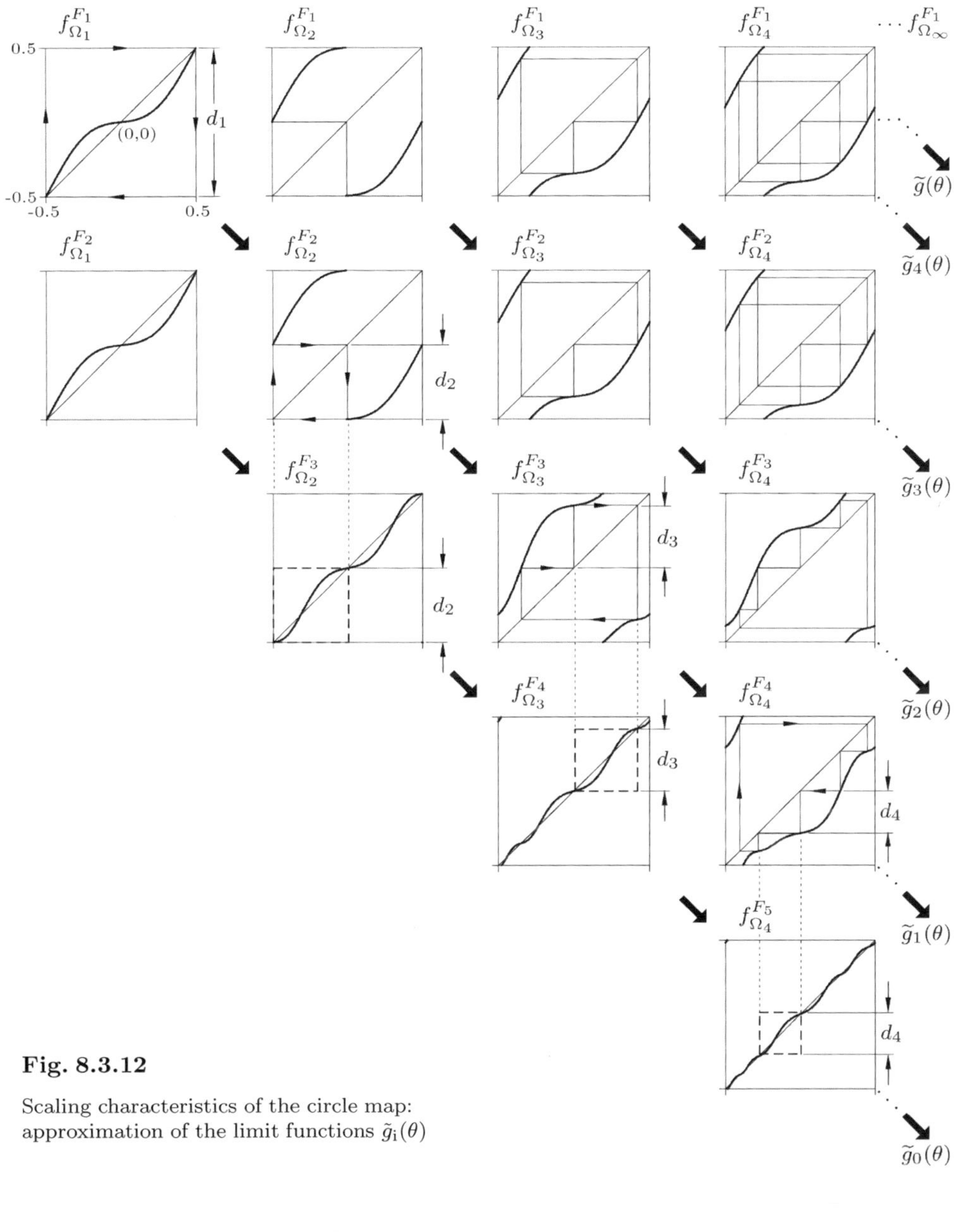

Fig. 8.3.12

Scaling characteristics of the circle map: approximation of the limit functions $\tilde{g}_i(\theta)$

We next set up a relation of the limit functions to one another in analogy to eq. (6.7.32). We first keep all control parameters fixed and can thus, for typographical simplicity, dispense temporarily with the lower index of $f^{(\mathrm{n})}$. The definition of $f^{(\mathrm{n})}$, eq. (8.3.66), together with eq. (8.3.45) yields

$$\begin{aligned} f^{(\mathrm{n}+1)}(\theta) &= f^{F_{\mathrm{n}+2}}(\theta) - F_{\mathrm{n}+1} \\ &= f^{F_{\mathrm{n}+1}+F_{\mathrm{n}}}(\theta) - (F_{\mathrm{n}} + F_{\mathrm{n}-1}) \end{aligned}$$

Since the recursion formula for the Fibonacci numbers is of second order (i.e. $F_{\mathrm{n}+1}$ and F_{n} generate $F_{\mathrm{n}+2}$), we get two recursion rules for $f^{(\mathrm{n}+1)}(\theta)$. Applying $f(\theta+1) = f(\theta)+1$, we deduce on the one hand

$$\begin{aligned} f^{(\mathrm{n}+1)}(\theta) &= \left(f^{F_{\mathrm{n}+1}+F_{\mathrm{n}}}(\theta) - F_{\mathrm{n}-1}\right) - F_{\mathrm{n}} = \left(f^{F_{\mathrm{n}+1}}(f^{F_{\mathrm{n}}}(\theta)) - F_{\mathrm{n}-1}\right) - F_{\mathrm{n}} \\ &= f^{F_{\mathrm{n}+1}}\left(f^{F_{\mathrm{n}}}(\theta) - F_{\mathrm{n}-1}\right) - F_{\mathrm{n}} = f^{(\mathrm{n})}(f^{(\mathrm{n}-1)}(\theta)) \end{aligned} \tag{8.3.70}$$

and on the other hand

$$f^{(\mathrm{n}+1)}(\theta) = f^{F_{\mathrm{n}}}(f^{F_{\mathrm{n}+1}}(\theta) - F_{\mathrm{n}}) - F_{\mathrm{n}-1} = f^{(\mathrm{n}-1)}(f^{(\mathrm{n})}(\theta)) \tag{8.3.71}$$

i.e. the iteration is commutative. Using the notation

$$\bar{f}^{(\mathrm{n})}(\theta) = a^{\mathrm{n}} f^{(\mathrm{n})}(\vartheta_{\mathrm{n}}) \tag{8.3.72}$$

eq. (8.3.70) respectively eq. (8.3.71) can be transformed as follows

$$\bar{f}^{(\mathrm{n}+1)}(\theta) = a \cdot a^{\mathrm{n}} f^{(\mathrm{n})}\Big(\frac{a \cdot a^{\mathrm{n}-1}}{a^{\mathrm{n}}} f^{(\mathrm{n}-1)}\Big(\frac{\theta}{a^{\mathrm{n}-1}a^2}\Big)\Big)$$

i.e. we obtain, with the abbreviation of eq. (8.3.65),

$$\bar{f}^{(\mathrm{n}+1)}(\theta) = a\bar{f}^{(\mathrm{n})}(a\bar{f}^{(\mathrm{n}-1)}(\vartheta_2)) \tag{8.3.73}$$

or alternatively

$$\begin{aligned} \bar{f}^{(\mathrm{n}+1)}(\theta) &= a^2 a^{\mathrm{n}-1} f^{(\mathrm{n}-1)}\Big(\frac{a^{\mathrm{n}}}{a \cdot a^{\mathrm{n}-1}} f^{(\mathrm{n})}\Big(\frac{\theta}{a \cdot a^{\mathrm{n}}}\Big)\Big) \\ &= a^2 \bar{f}^{(\mathrm{n}-1)}\Big(\frac{1}{a}\bar{f}^{(\mathrm{n})}(\vartheta_1)\Big) \end{aligned} \tag{8.3.74}$$

If we postulate $f(\theta+1) = f(\theta)+1$, the second expression is redundant (Ostlund et al., 1983). We can thus base the following account on eq. (8.3.73).

Our aim is to set up a relation between successive limit functions. To this end, we must again specify the Ω-value and operate with lower indices. With the abbreviation eq. (8.3.72), $\tilde{g}_{\mathrm{i}-1}(\theta)$ can be written in accordance with eq. (8.3.69) simply as

$$\tilde{g}_{\mathrm{i}-1}(\theta) = \lim_{\mathrm{n}\to\infty} \bar{f}^{(\mathrm{n})}_{\Omega_{\mathrm{n}+\mathrm{i}-1}}(\theta) \tag{8.3.75}$$

Substituting the expression eq. (8.3.73), we deduce the following recursion formula in analogy to eq. (6.7.32)

$$\tilde{g}_{\mathrm{i}-1}(\theta) = \lim_{\mathrm{n}\to\infty} a\bar{f}^{(\mathrm{n}-1)}_{\Omega_{\mathrm{n}-1+\mathrm{i}}}(a\bar{f}^{(\mathrm{n}-2)}_{\Omega_{\mathrm{n}-2+\mathrm{i}+1}}(\vartheta_2)) \qquad \text{with } \vartheta_2 = a^{-2}\theta$$

or

$$\tilde{g}_{\mathrm{i}-1}(\theta) = a\tilde{g}_{\mathrm{i}}\big(a\tilde{g}_{\mathrm{i}+1}(\vartheta_2)\big) \tag{8.3.76}$$

In contrast to the iteration rule eq. (6.7.32) which we established for the route via period doubling, the recursion formula for Fibonacci numbers, eq. (8.3.45), leads to the understanding that eq. (8.3.76) is also a second-order recursion formula, i.e. two functions $\tilde{g}_{\mathrm{i}+1}$ and $\tilde{g}_{\mathrm{i}}$ are required to generate $\tilde{g}_{\mathrm{i}-1}$. This evidently has consequences for the definition of a transformation corresponding to the transformation T in eq. (6.7.32); cf. (Feigenbaum et al., 1982) and (Rand et al., 1982). However, we do not enter into details here and restrict ourselves in this section to the fundamental concepts required for the derivation of the universal constants. If we proceed to the limit $\mathrm{i} \to \infty$, the sequence $\tilde{g}_{\mathrm{i}}(\theta)$ tends towards a limit function $\tilde{g}(\theta)$ and, in accordance with eq. (8.3.75), we have

$$\tilde{g}(\theta) = \lim_{\mathrm{i}\to\infty} \tilde{g}_{\mathrm{i}}(\theta) = \lim_{\mathrm{n}\to\infty} \bar{f}^{(\mathrm{n})}_{\Omega_\infty}(\theta) \tag{8.3.77}$$

i.e. we obtain $\tilde{g}(\theta)$ by iteration of the mapping function $f(\theta)$ corresponding to the *fixed* parameter value $\Omega_\infty(K)$, cf. eq. (8.3.52). In accordance with eq. (8.3.76), the function $\tilde{g}(\theta)$ is then the solution of the fixed point equation

$$\tilde{g}(\theta) = a\tilde{g}\big(a\tilde{g}(\vartheta_2)\big) \qquad \text{with } \vartheta_2 = a^{-2}\theta \tag{8.3.78}$$

As in the Feigenbaum scenario, the universal constant a can be determined from this expression applying the normalisation $\tilde{g}(0) = 1$.

At this point, we must differentiate between $0 < K < 1$ and $K = 1$. If we expand the circle map eq. (8.3.13) in the neighbourhood of $\theta = 0$ into a Taylor series, we obtain

$$\begin{aligned} f(\theta) &= \theta + \Omega - \frac{K}{2\pi}\sin 2\pi\theta \\ &= \theta + \Omega - \frac{K}{2\pi}\Big[2\pi\theta - \frac{1}{3!}(2\pi\theta)^3 + \cdots\Big] \\ &= \Omega + (1-K)\theta + \frac{2\pi^2}{3}K\theta^3 - \cdots \end{aligned}$$

i.e. for $0 < K < 1$, a linear term in θ emerges in the mapping function. This characteristic also enters into the limit function $\tilde{g}(\theta)$, and eq. (8.3.78) possesses a “trivial” analytical solution

$$\tilde{g}(\theta) = 1 + \theta \tag{8.3.79}$$

Indeed, if we insert eq. (8.3.79) into eq. (8.3.78), we obtain from

$$1 + \theta = a\big[1 + a(1 + \vartheta_2)\big] = a + a^2 + \theta$$

an equation for a

$$a^2 + a - 1 = 0 \tag{8.3.80}$$

where here, we are only interested in the negative solution. We obtain

$$a = -\frac{1+\sqrt{5}}{2} = \frac{2}{1-\sqrt{5}} = -\frac{1}{W_{\mathrm{G}}} \tag{8.3.81}$$

in accordance with eq. (8.3.59).

Since, for $K = 1$, the circle map possesses a cubic inflection point with horizontal tangent at $\theta = 0$, the linear θ-term vanishes in the power series expansion, i.e. it is reduced to the form

$$f(\theta) = \Omega + \text{const} \cdot \theta^3 + \cdots$$

The linear term is thus also cancelled in the limit function. Assuming a general circle map possessing a *cubic* inflection point with horizontal tangent at $K = 1$, we can establish successive approximations for the constant a and for the limit function $\tilde{g}(\theta)$ using the ansatz

$$\tilde{g}(\theta) = 1 + \gamma_1\theta^3 + \gamma_2\theta^6 + \cdots \tag{8.3.82}$$

The numerical calculations in (Feigenbaum et al., 1982; Rand et al., 1982) yield the value

$$a = -1.28857 \pm 2 \cdot 10^{-5} \qquad (K = 1) \tag{8.3.83}$$

Equation (8.3.76) instructs us that $\tilde{g}(\theta)$ is an unstable fixed point (saddle point). As in the period doubling cascade, the second univeral constant δ in eq. (8.3.53), which decodes the self-similarity with regard to the Ω-parameter, can be identified with the eigenvalue, the corresponding eigenvector of which is tangent to the unstable manifold in the fixed point $\tilde{g}(\theta)$. For $K < 1$, we obtain, as is to be expected according to eq. (8.3.50), the "trivial" value

$$\delta = -\frac{1}{W_{\mathrm{G}}^2} = -2.61803\ldots \qquad (K < 1) \tag{8.3.84}$$

The calculation of δ for $K = 1$ is considerably more complicated. As for the logistic map (cf. section 6.7.4), we have to linearise the iteration operator, eq. (8.3.70) respectively eq. (8.3.71), at the fixed point $\tilde{g}(\theta)$ and can then determine the largest eigenvalue δ numerically. The fact that two functions $f^{(\mathrm{n})}$ and $f^{(\mathrm{n}-1)}$ must be specified for the iteration operator in order to generate $f^{(\mathrm{n}+1)}$ causes additional difficulties. We restrict ourselves here to quoting the result

$$\delta = -2.83360 \pm 3 \cdot 10^{-5} \qquad (K = 1) \tag{8.3.85}$$

and refer the reader to the literature, cf. (Feigenbaum et al., 1982).

In this section, we concentrated initially on the derivation of local scaling relations, i.e. we considered the transition from quasi-periodic motions to chaos in the case of a fixed rotation number. The question now is how far these laws are modified for a transition to other rotation numbers. This problem was investigated by (Shenker, 1982; Ostlund et al., 1983) by selecting the so-called "silver mean" W_{S} as the rotation number and determining the corresponding constants a and δ numerically by iteration of the circle map eq. (8.3.13). The value W_{S} possesses the simple continued fraction representation

$$W_S = < 2\ 2\ 2\ \ldots > \tag{8.3.86}$$

As in eq. (8.3.43), we can calculate the appertaining irrational number from this representation. We obtain

$$W_S = \frac{1}{2 + W_S} \qquad \text{or} \qquad W_S^2 + 2W_S - 1 = 0$$

The solution $W_S = \sqrt{2} - 1$ is denoted the silver mean. Here too, we can calculate the scaling quantities for the parameter and the phase space and obtain the non-trivial values for $K = 1$

$$\begin{aligned} a &= -1.586822 \pm 2 \cdot 10^{-6} \\ \delta &= -6.799224 \pm 2 \cdot 10^{-5} \end{aligned} \tag{8.3.87}$$

A comparison with eqs. (8.3.83) and (8.3.85) shows that these values deviate from the constants associated with the golden mean; on the other hand, the numerical investigations showed the universality of the coefficients. Indeed, the special form of the mapping function has no influence on the result as long as it possesses a cubic inflection point with horizontal tangent at the critical boundary $K = 1$. It can be seen from the ansatz eq. (8.3.82) for the limit function $\tilde{g}(\theta)$, however, that the order of the inflection point influences the value of a and thus also δ. A further numerical investigation of Shenker showed (Shenker, 1982) that it is only the "tail-end" of the continued fraction and not the leading finite number of digits that is relevant. Thus, the rotation number

$$W = < 3\ 1\ 4\ 1\ 1\ 1\ \ldots > \tag{8.3.88}$$

for example, leads to the same scaling laws as the golden mean W_G.

Particularly the last result throws up a whole complex of questions. In principle, for every irrational number W, we can evidently construct with the continued fraction representation $W = < \mathrm{n}_1\ \mathrm{n}_2\ \mathrm{n}_3\ \ldots >$ irrational numbers W' and W'', the leading digits of which coincide with W in the continued fraction representation, the "end" of which, however, is filled up with the digits "1" respectively "2", i.e.

$$W' = < \mathrm{n}_1\ \mathrm{n}_2\ \ldots\ \mathrm{n}_k\ 1\ 1\ 1\ 1\ \ldots >$$

respectively

$$W'' = < \mathrm{n}_1\ \mathrm{n}_2\ \ldots\ \mathrm{n}_k\ 2\ 2\ 2\ 2\ \ldots >$$

If we select a large enough k, both W' and W'' deviate arbitrarily little from W, the two frequencies lead, however, to different scaling laws!

The question now arises to what extent the scaling laws and the "locally universal" constants can be used to explain physical problems and where the limits lie. The answer certainly depends on the point of view: mathematical or physical. A mathematician will investigate the probability with which a certain motion occurs and which stability characteristics it possesses. Since it is not possible to measure frequencies with infinite precision and since microscopic fluctuations always occur, however, it would definitely be expedient to investigate the characteristics of circle maps, taking fluctuations into account, for scaling laws and for the concept of the local universality.

Up to now in this section, we have discussed scaling laws for irrational rotation numbers with a periodic continued fraction representation. These numbers are special cases, however; typically, an irrational number is transcendental, such as π or e, for example, and has an ergodic continued fraction representation. In these cases, geometrical scaling laws cannot be established. The role of the hyperbolic fixed point $\tilde{g}(\theta)$ in a suitable function space is then taken over by a hyperbolic "strange set", for example *a horseshoe*, in the function space. The complicated implications of this are described in (Umberger et al., 1986a; Rand, 1987).

In section 8.5, we shall describe experiments on the transition from quasi-periodic to chaotic behaviour. The golden mean will be selected as the "most irrational" of all rotation numbers and the scaling laws will be verified on the basis of the structure of power spectra and density distributions.

8.3.3.2 Global Scaling Laws

Local universality laws can be checked experimentally only with difficulty since small variations in the rotation numbers lead to large variations in the scaling behaviour. It is easier to prove global universal characteristics, on the other hand, which hold for a whole range of rotation numbers. In the following, we thus investigate the self-similar structure of the Arnold tongues in the vicinity of the critical line $K_{cr} = 1$.

In fig. 8.3.6, we plot the structure of the Ω-intervals in which a synchronisation of the frequencies occurs, the so-called devil's staircase. The diagram shows that the length of the interval $\Delta\Omega(\mathrm{p/q})$ appertaining to a rotation number p/q becomes smaller, the larger the denominator is. Indeed, we can prove that the length of the stability interval of a q-cycle is

$$\Delta\Omega\left(\frac{1}{\mathrm{q}}\right) \sim \frac{1}{\mathrm{q}^{\gamma}} \qquad (\gamma > 2) \tag{8.3.89}$$

(Jensen et al., 1984). With the aid of the so-called *Farey tree*, these stability domains can be arranged systematically according to their width.

Farey sequences are special arrays of rational numbers in the unit interval. A Farey sequence $F^{(\mathrm{n})}$ of order n arranges all rational numbers $\mathrm{p/q} \in [0,1]$ with $\mathrm{q} \leqslant \mathrm{n}$ (p, q are integers with $\mathrm{p} \geqslant 0,\ \mathrm{q} > 0,\ \mathrm{p/q}$ relatively prime) and possesses the following characteristics:

If $\mathrm{p_1/q_1},\ \mathrm{p_2/q_2},\ \mathrm{p_3/q_3}$ are three consecutive terms of the Farey sequence, i.e. if $\mathrm{p_1/q_1} < \mathrm{p_2/q_2} < \mathrm{p_3/q_3}$, then

$$\mathrm{q_1 p_2 - p_1 q_2 = 1} \tag{8.3.90}$$

and

$$\frac{\mathrm{p_2}}{\mathrm{q_2}} = \frac{\mathrm{p_1}}{\mathrm{q_1}} \oplus \frac{\mathrm{p_3}}{\mathrm{q_3}} \equiv \frac{\mathrm{p_1 + p_3}}{\mathrm{q_1 + q_3}} \tag{8.3.91}$$

in which the so-called Farey mediant, denoted by the symbol $\oplus$, is introduced to indicate that the numerator and denominator are the sums of the numerators and

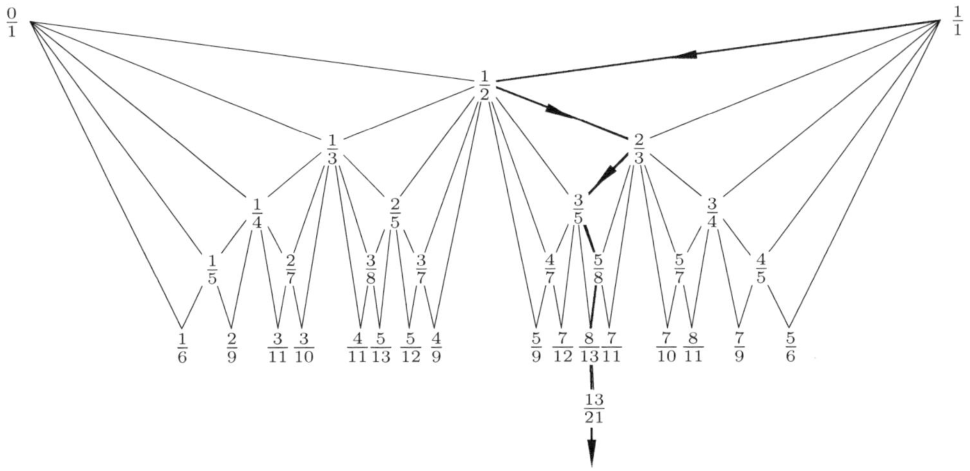

Fig. 8.3.13: Structure of the Farey tree; the thicker line marks the approximation of the golden mean by quotients of Fibonacci numbers

denominators of the given fractions. For example, the order in a fifth-order Farey sequence is

$$F^{(5)} = \left\{\tfrac{0}{1}, \tfrac{1}{5}, \tfrac{1}{4}, \tfrac{1}{3}, \tfrac{2}{5}, \tfrac{1}{2}, \tfrac{3}{5}, \tfrac{2}{3}, \tfrac{3}{4}, \tfrac{4}{5}, \tfrac{1}{1}\right\}$$

Based on this pattern, all rational numbers within the interval [0,1] can be arranged in a Farey tree according to the size of the denominator. We begin with the endpoints of the interval [0,1], i.e. with $\frac{0}{1}$ and $\frac{1}{1}$, and form the Farey mediants successively. In the first step we obtain $\frac{0}{1} \oplus \frac{1}{1} = \frac{1}{2}$, then $\frac{0}{1} \oplus \frac{1}{2} = \frac{1}{3}$, $\frac{1}{2} \oplus \frac{1}{1} = \frac{2}{3}$ ect., see fig. 8.3.13 (Cvitanović et al., 1985). If we pick out two rational numbers p_1/q_1 and p_2/q_2, then this scheme demonstrates graphically that $p/q = (p_1+p_2)/(q_1+q_2)$ is the rational number between p_1/q_1 and p_2/q_2 with the *smallest* denominator. If we apply the structure of the Farey tree to the devil's staircase (fig. 8.3.6), the rational number with the smallest denominator leads to the longest plateau p/q between the two arbitrarily selected plateaus p_1/q_1 and p_2/q_2 (e.g.: $p_1/q_1 = \frac{1}{2}$, $p_2/q_2 = \frac{1}{3}$ yields $p/q = \frac{2}{5}$). The scheme of the Farey tree can now be used to arrange the Arnold tongues for $K_{cr}=1$ in accordance with the width of the Ω-intervals in which frequency locking occurs.

If we select two rotation numbers p_1/q_1 and p_2/q_2 and denote the stable domains of periodic behaviour of the corresponding q_1- respectively q_2-cycles by $\Delta\Omega(p_1/q_1)$ respectively $\Delta\Omega(p_2/q_2)$, then the *largest* interval $\Delta\Omega_{\max}$ in between appertains to the rotation number

$$\frac{p}{q} = \frac{p_1 + p_2}{q_1 + q_2} \tag{8.3.92}$$

Since the structure of the Arnold tongues is independent of the special mapping function (cf. Colour Plate IX, p. 497, and fig. 8.3.10), these considerations provide an arrangement scheme for those domains in which frequency locking occurs; this is particularly significant for experimental observations. The same structure was observed not only abstractly in the circle map but also in many physical systems in which two frequencies are coupled, such as, for example, the forced oscillations of a pendulum, the Josephson junction, stimulated cardiac rhythms and the forced Bénard convection on which an additional frequency was imposed (cf. section 8.5.1), to name just a few examples.

Recent research activities have unexpectedly unveiled a fascinating phenomenon: mixed-mode oscillations (MMO) of prototypical chemical systems – also discussed in section 10.9 – show that the hierarchical structure of alternating periodic-chaotic oscillations follow a so-called Stern-Brocot tree which is more general than a Farey tree and includes the Farey tree as a sub-tree. In contrast to the Farey tree, the Stern-Brocot tree contains *all* positive rationals and *all* mediants at the n-th step. Colour Plate X, fig. b), p. 503 shows an isospike diagram of a three-variable, fourteen-parameter model of the Belousov-Zhabotinsky (BZ) reaction (Freire and Gallas, 2011b). Colours identify the number of spikes within a period while black marks chaotic phases. The bifurcation diagrams studied for parameters along the three vertical lines and a detailed analysis of the isospikes within the white box corroborate the result that the hierarchical structure of the MMOs in the BZ reaction follow the Stern-Brocot order. For further studies, we refer the interested reader to (Freire and Gallas, 2011b; Freire and Gallas, 2011a) and the references there.

In (Jensen et al., 1984), the authors studied the scaling laws underlying the Arnold tongues. They proved numerically that, at the critical line $K = 1$, the total length S of the stability domains of all periodic motions is

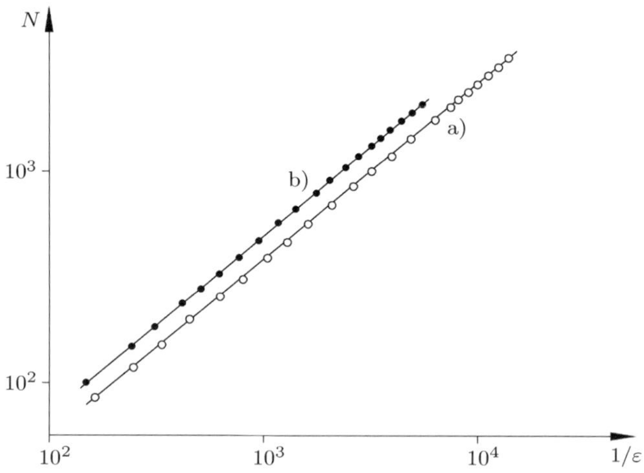

Fig. 8.3.14: $\log N(\varepsilon)$ versus $\log(1/\varepsilon)$ for two critical circle maps:
a) $f(\theta) = \theta + \Omega - \frac{1}{2\pi}\sin(2\pi\theta)$,
b) $f(\theta) = \theta + \Omega - \frac{1}{2\pi}[\sin(2\pi\theta) - 0.8\sin^3(2\pi\theta)]$ (Jensen et al., 1983)

$$S = \sum_{\forall \frac{\mathrm{p}}{\mathrm{q}}} \Delta\Omega\left(\frac{\mathrm{p}}{\mathrm{q}}\right) = 1 \tag{8.3.93}$$

and the complementary set possesses the fractal structure of a Cantor set. They calculated the total length of the plateaus on different scales by selecting – for increasing values q – as their scale the length of the smallest plateau $\varepsilon_{\mathrm{q}} = \Delta\Omega(1/\mathrm{q})$. The total length of the stability domains of all q-cycles with $1 \leqslant \mathrm{q} \leqslant \mathrm{n}$ is then

$$S(\varepsilon_{\mathrm{n}}) = \sum_{\frac{\mathrm{p}}{\mathrm{q}} \in F^{(\mathrm{n})}} \Delta\Omega\left(\frac{\mathrm{p}}{\mathrm{q}}\right) \tag{8.3.94}$$

where the summation extends over the rotation numbers of the Farey sequence $F^{(\mathrm{n})}$ of order n. With this resolution, the length of the intermediate intervals is given by

$$l(\varepsilon_{\mathrm{n}}) = 1 - S(\varepsilon_{\mathrm{n}}) \tag{8.3.95}$$

or, related to the scale ε_{n},

$$N(\varepsilon_{\mathrm{n}}) = \frac{1 - S(\varepsilon_{\mathrm{n}})}{\varepsilon_{\mathrm{n}}} \tag{8.3.96}$$

$N(\varepsilon_{\mathrm{n}})$ is the number of intervals of the length ε_{n} in which no frequency synchronisation takes place.

In fig. 8.3.14, $\log N(\varepsilon)$ is plotted versus $\log(1/\varepsilon)$ for two different circle maps. Independent of the special form of the mapping function, this yields two – within computational accuracy – parallel straight lines $\log N = \mathrm{const} + D\log(1/\varepsilon)$ with the slope $D \approx 0.87$. The number $N(\varepsilon)$ of intervals of the length ε leading to quasi-periodic behaviour thus obeys a power law

$$N(\varepsilon) \propto \frac{1}{\varepsilon^D} \tag{8.3.97}$$

where, in accordance with eq. (5.5.5), the exponent D represents the capacity dimension of the Ω-set which generates quasi-periodic motions. The total length of the "gaps" between the frequency locking plateaus is thus

$$l(\varepsilon) = N(\varepsilon) \cdot \varepsilon \propto \varepsilon^{1-D} \tag{8.3.98}$$

and, since $D < 1$, $l(\varepsilon) \to 0$ also tends to zero for $\varepsilon \to 0$. The complementary set of the Ω-values appertaining to the q-cycles thus has the structure of a self-similar Cantor set as described in section 5.5.1. At the critical line there is thus no space for quasi-periodic motions; the set of the corresponding Ω-values has the measure zero.

However, the fundamental result is that D is a universal quantity which does not depend on the special form of the mapping function. This characteristic of universality is of decisive importance since it allows us to apply the theoretical propositions to interpret measurement results for a multitude of experiments for which we do not know the corresponding Poincaré map.

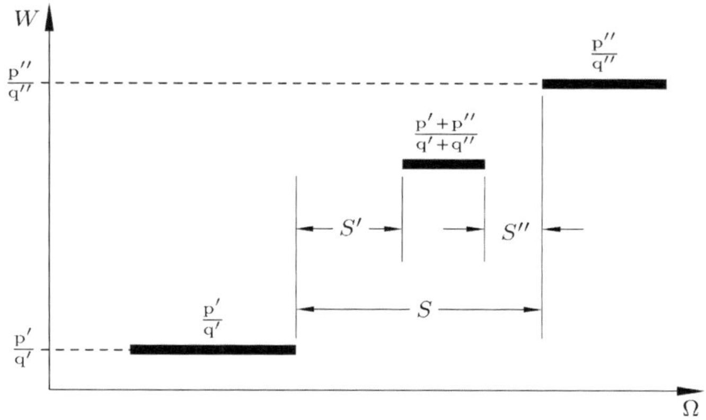

Fig. 8.3.15: Farey construction for determining the dimension D'

If we consider the circle map below the critical boundary, i.e. for $K<1$, the stability domains of the q-cycles form an incomplete devil's staircase with $S(K) < 1$. Correspondingly, the complementary set is of measure $l(K) > 0$. Numerical investigations demonstrate (Jensen et al., 1984) that $l(K)$ obeys a power law

$$l(K) \propto (1-K)^{\beta} \tag{8.3.99}$$

with $\beta = 0.34$, more precise calculations yielded $\beta = 0.31$ (Alstrøm et al., 1987); β also has universal character.

These results suggest interpreting the transition to chaos at $K = 1$ as a second-order phase transition. A comparison of eq. (8.3.99) with the power law eq. (6.7.109) describing the phase transition of ferromagnets shows that l plays the role of an order parameter and β that of a critical exponent.

In (Jensen et al., 1984), the capacity dimension D of the Cantor set was determined for $K = 1$ conventionally – as described in section 5.5.2 – by summing up the width of all Ω-intervals not below a certain length; see eqs. (8.3.94) to (8.3.97). This method cannot be particularly recommended for experimental investigations since the width of a large number of Arnold tongues has to be determined. A local method going back to ideas of Hentschel and Procaccia (Hentschel and Procaccia, 1983) is much more elegant; it can be used in the case of homogeneous fractals for which the dimension is the same for all domains in which gaps occur.

We thus pick out two arbitrary rotation numbers $\frac{p'}{q'}$ and $\frac{p''}{q''}$ and measure the length of the gap S in between. We know from the Farey tree structure of the Arnold tongues that the largest Ω-interval in between appertains to the rotation number $\frac{p'+p''}{q'+q''}$. The gaps between the new intervals and the two previous ones are denoted S' and S'' (see fig. 8.3.15).

According to (Hentschel and Procaccia, 1983), an approximation D' for the dimension D can be determined from the relation

$$\left(\frac{S'}{S}\right)^{D'} + \left(\frac{S''}{S}\right)^{D'} = 1 \tag{8.3.100}$$

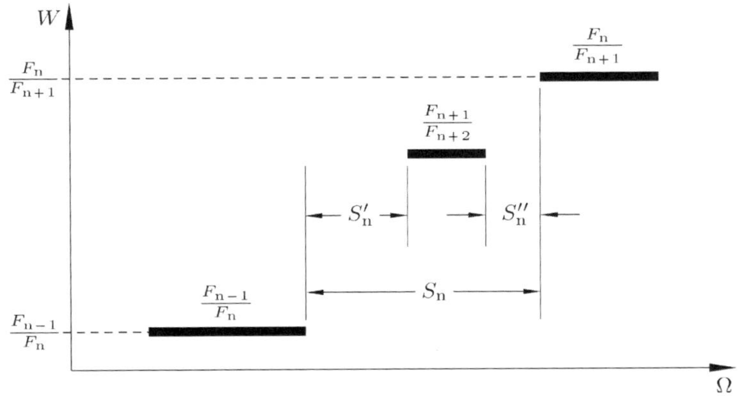

Fig. 8.3.16: Three rational approximations of the golden mean for the calculation of the capacity dimension D'

The idea now is to gain better approximations for the dimension by zooming out increasingly small regions (Cvitanović et al., 1985). For example, we can pick out three successive approximations W_{n-1}, W_n and $W_{n+1} = W_{n-1} \oplus W_n$ of the golden mean as frequencies; cf. eq. (8.3.46). We extract the corresponding gaps S_n, S_n' and S_n'' from fig. 8.3.16. A good approximation for the dimension of the Cantor set follows from the relation, cf. (Glazier and Libchaber, 1988),

$$D' = \lim_{n\to\infty} D_n \qquad \text{with} \qquad \left(\frac{S_n'}{S_n}\right)^{D_n} + \left(\frac{S_n''}{S_n}\right)^{D_n} = 1 \tag{8.3.101}$$

Based on the rotation numbers $\frac{8}{13}$ and $\frac{13}{21}$, the numerical calculation yields a value $D' = 0.868 \pm 2.10^{-3}$ after eleven steps (Cvitanović et al., 1985) in good agreement with (Jensen et al., 1983). Surprisingly, the first two gaps already yield a value D_1 which deviates by only 1% from D'. A different selection of the initial intervals respectively a different mapping function did not modify the results.

The purpose of this section is to elucidate the theoretical foundations and universality characteristics underlying the different transitions to chaos. Autonomous continuous systems can only exhibit irregular behaviour if the underlying phase space possesses at least the dimension 3. Thus, with a continuous variation of a control parameter, many systems will pass through the sequence of bifurcations "fixed point – limit cycle – 2D torus". A further increase of one or several control parameters can either lead to a synchronisation of the frequencies, and thus again to a periodic motion, or the quasi-periodic motion is preserved. With the aid of the circle map, various possibilities for the transition to chaos can be explained. Ultimately, the universality laws justify the intensive investigation of a special mapping function and allow the application of the results to a multitude of phenomena.

In the two following sections, we shall concentrate on two scenarios of the transition to chaos, namely the Feigenbaum route via period doublings (section 8.4) and the quasi-periodic transition (section 8.5); at the end of each section, we shall discuss the experimental verification.

8.4 The Feigenbaum Route to Chaos via Period Doublings

In section 8.2, we presented the model of Ruelle and Takens, according to which irregular turbulent behaviour sets in "typically" directly after the bifurcation of a two-dimensional torus T^2 into a three-dimensional torus T^3. This scenario was based on the assumption that the excited modes are only weakly coupled with one another (Eckmann and MacKay, 1983). It was possible to confirm this mathematical model by a number of hydrodynamic experiments in closed systems such as the Rayleigh-Bénard convection or the Taylor-Couette flow.

In several experiments, more accurate measurements showed that, after the second bifurcation into a two-dimensional torus, a synchronisation of the two incommensurable frequencies often occurs so that the quasi-periodic behaviour is succeeded by a periodic motion. The corresponding limit cycle then passes through a cascade of period doublings to chaos. Feigenbaum established universal constants and scaling laws for these transitions (see section 6.7). Figure 8.4.1 presents a schematic illustration of the Feigenbaum scenario.

After the second Hopf bifurcation, two incommensurable frequencies emerge, leading to a quasi-periodic motion on a two-dimensional torus. If a variation of the control parameter causes an increase of the non-linear coupling of the basic modes, a locking of the frequencies often occurs, a phenomenon familiar to us from the most varied fields. The most obvious example is that of the motion of our moon: the period of one orbit about the earth and one rotation or spin about its own polar axis are coupled with a ratio 1:1.

We also know frequency locking in driven mechanical oscillating systems with damping – as mentioned above, the functioning of quartz clocks is based on this effect – in hydrodynamics in convection flows, in chemistry in the Belousov-Zhabotinsky reaction (see e.g. section 10.9) or in biological cycles. The insight gained from non-linear dynamics enables us to explain the phenomena occurring in these varied fields qualitatively by a universal system theory and to prove the underlying scaling characteristics quantitatively. It is not the specific details of the different systems that are relevant; it suffices to study the phenomenon of the locking of two frequencies and the subsequent period doublings up to chaotic motions on the prototype of a Poincaré map, namely the dissipative circle map (Pikovsky et al., 2003).

In figure 8.4.2a, we have plotted the long-term response of θ in dependence on K $(0 \leqslant K \leqslant 10)$ for $\Omega = 1/2$ for the one-dimensional circle map eq. (8.3.13). The structure of the bifurcation diagram is highly reminiscent of that of the logistic map (cf. fig. 6.7.6) and we clearly recognise that a cascade of period doublings occurs within the Arnold tongue for the rotation number $W = 1/2$ (cf. fig. 8.3.5) for $K > 1$ until chaos finally breaks out. These irregular ranges are continually interrupted by windows of regular behaviour.

In figure 8.4.2b, we have plotted the Lyapunov exponent $\sigma(K)$ for $0 \leqslant K \leqslant 10$. We can distinguish four regions, each corresponding to a Feigenbaum period doubling scenario and a subsequent chaotic domain and remarkable for their "self-similar" structure. We clearly recognise those points at which period doublings $(\sigma = 0)$

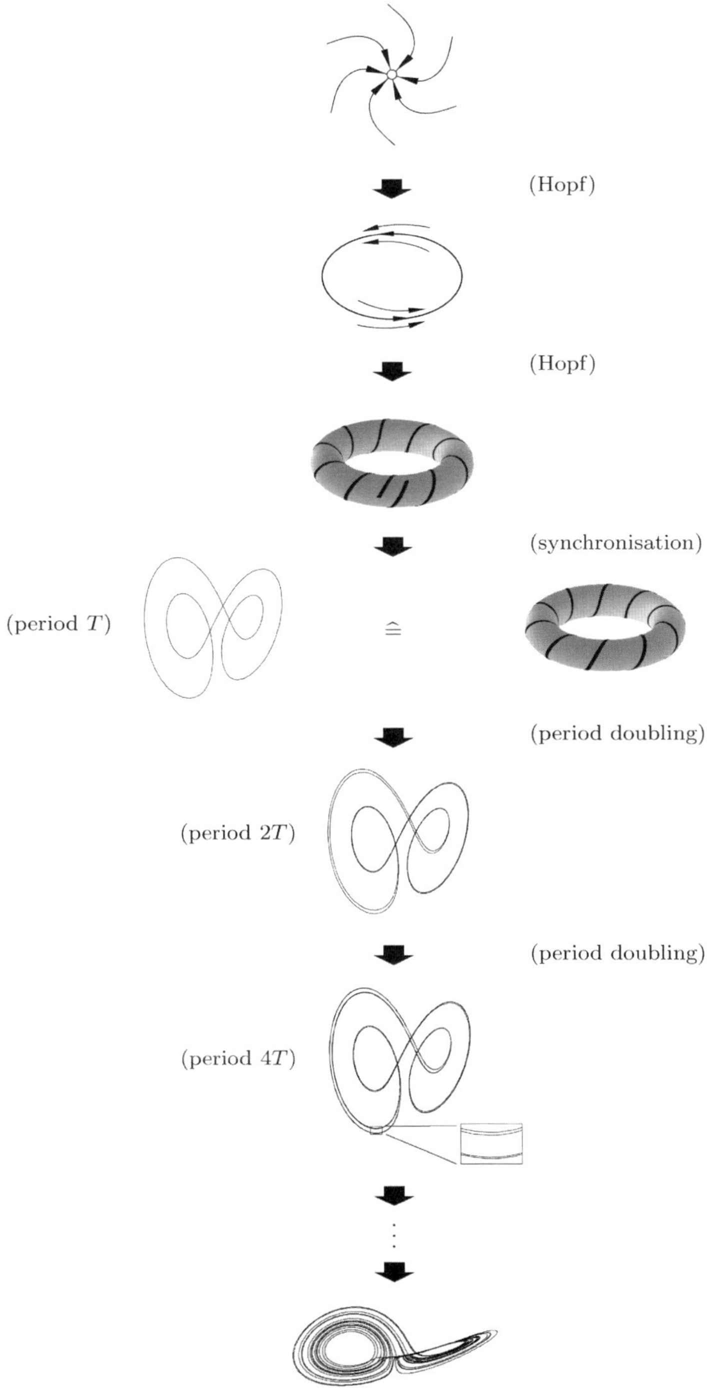

Fig. 8.4.1: Feigenbaum scenario: to chaos via period doublings

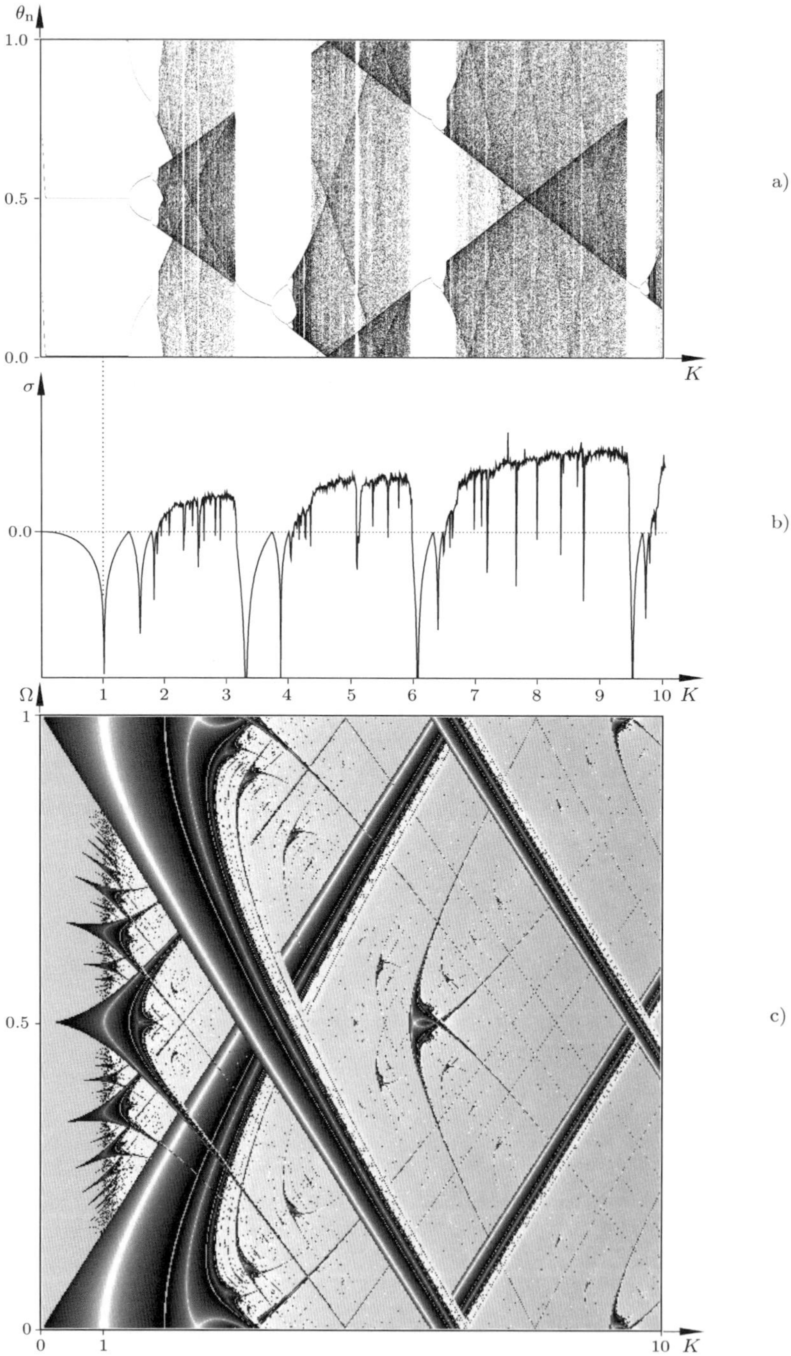

Fig. 8.4.2: Route to chaos via period doubling for the model of the circle map:
a) long-term response of θ_n for $\Omega = \frac{1}{2}$, $\theta_0 = 0.2$
b) Lyapunov exponent $\sigma(K)$ for $\Omega = \frac{1}{2}$
c) Lyapunov exponent $\sigma(K, \Omega)$ for $0 \leqslant K \leqslant 10$, $0 \leqslant \Omega \leqslant 1$

occurs. The K-values for which σ assumes the largest negative value appertain to superstable cycles. In fig. 8.4.2c, the superstable cycles each lie within the white bands in the Arnold tongues. Further details on the structure of the Arnold tongues are summarised in section 10.8.

If the Arnold tongues overlap (fig. 8.4.2c), several attractors coexist, i.e. different initial conditions can lead to different long-term behaviour. This can be clearly recognised in the gaps in the bifurcation diagram (fig. 8.4.2a), calculated for a single initial condition ($\theta_0 = 0.2$). The course of the Lyapunov exponent $\sigma(\Omega, K)$ shows, for example, that in the range $3 < K < 4$, $\Omega = 1/2$, the Arnold tongues corresponding to the rotation numbers $W = 0/1$ and $W = 1/1$ overlap, leading to the coexistence of two attractors, in this case two fixed points. The bifurcation diagram demonstrates that $\theta_0 = 0.2$ in this range lies within the basin of attraction of the fixed point $\theta_s < 0.5$ and thus appertains to the rotation number $W = 0/1$. Obviously, gaps in the bifurcation diagram, fig. 8.4.2a, can be avoided if we superimpose the long-time behaviour of several trajectories, the initial conditions of which are selected from the basins of attraction of all attractors.

In the following, we restrict ourselves to the range in which period doublings occur, up to the onset of chaos. Our aim is to draw conclusions regarding the underlying laws and scaling characteristics from numerical respectively experimental time series.

8.4.1 Further Scaling Characteristics of the Period Doubling Cascade

In section 7.3, we became acquainted with the Lorenz model for the Rayleigh-Bénard convection. In order to illustrate the scaling characteristics, we have calculated the three Lyapunov exponents $\sigma_1 \geqslant \sigma_2 \geqslant \sigma_3$ for this system and plotted them in fig. 8.4.3a for the parameter range $0 \leqslant r \leqslant 240$. Figure 8.4.3b presents an enlarged section ($142 \leqslant r \leqslant 169$). Since σ_1 becomes zero in the range $146.7 \lesssim r \lesssim 166.1$ and σ_2 and σ_3 are negative, periodic behaviour occurs. For $r > r_0 \approx 154.5$, a single limit cycle exists which is mapped onto itself by the transformation $X \to -X$, $Y \to -Y$, $Z \to -Z$ and thus reflects the symmetry of the Lorenz system. At $r = r_0$, this limit cycle becomes unstable and, due to a pitchfork bifurcation, two separate limit cycles lying point symmetrically emerge for $r < r_0$ (cf. also section 10.3 and Colour Plate XXIII, p. 712). If we compare the course of σ_2 in this range with the Lyapunov exponent for the logistic map (fig. 6.7.6), we recognise an inverse cascade of period halvings. The parameter values r_i for which $\sigma_2 = 0$ holds indicate the bifurcation points; superstable cycles are characterised by the fact that σ_2 has a maximum negative amplitude.

Scaling in the parameter space can be observed most easily on the basis of the Lyapunov exponent $\sigma_2(r)$, in our case for decreasing r-values. If we know the distance between two bifurcation points $\Delta r = r_{\mathrm{n}-1} - r_\mathrm{n}$, the Feigenbaum constant $\delta \approx 4.67$, eq. (6.7.25), tells us that the distance to the next bifurcation $\Delta r' = r_\mathrm{n} - r_{\mathrm{n}+1}$ is approximately one-fifth of the preceding increment Δr. This high convergence rate is simultaneously an advantage and a disadvantage. On the one hand, the ratio of the distance of successive bifurcations $(r_{\mathrm{n}-1} - r_\mathrm{n})/(r_\mathrm{n} - r_{\mathrm{n}+1})$ converges to the value δ

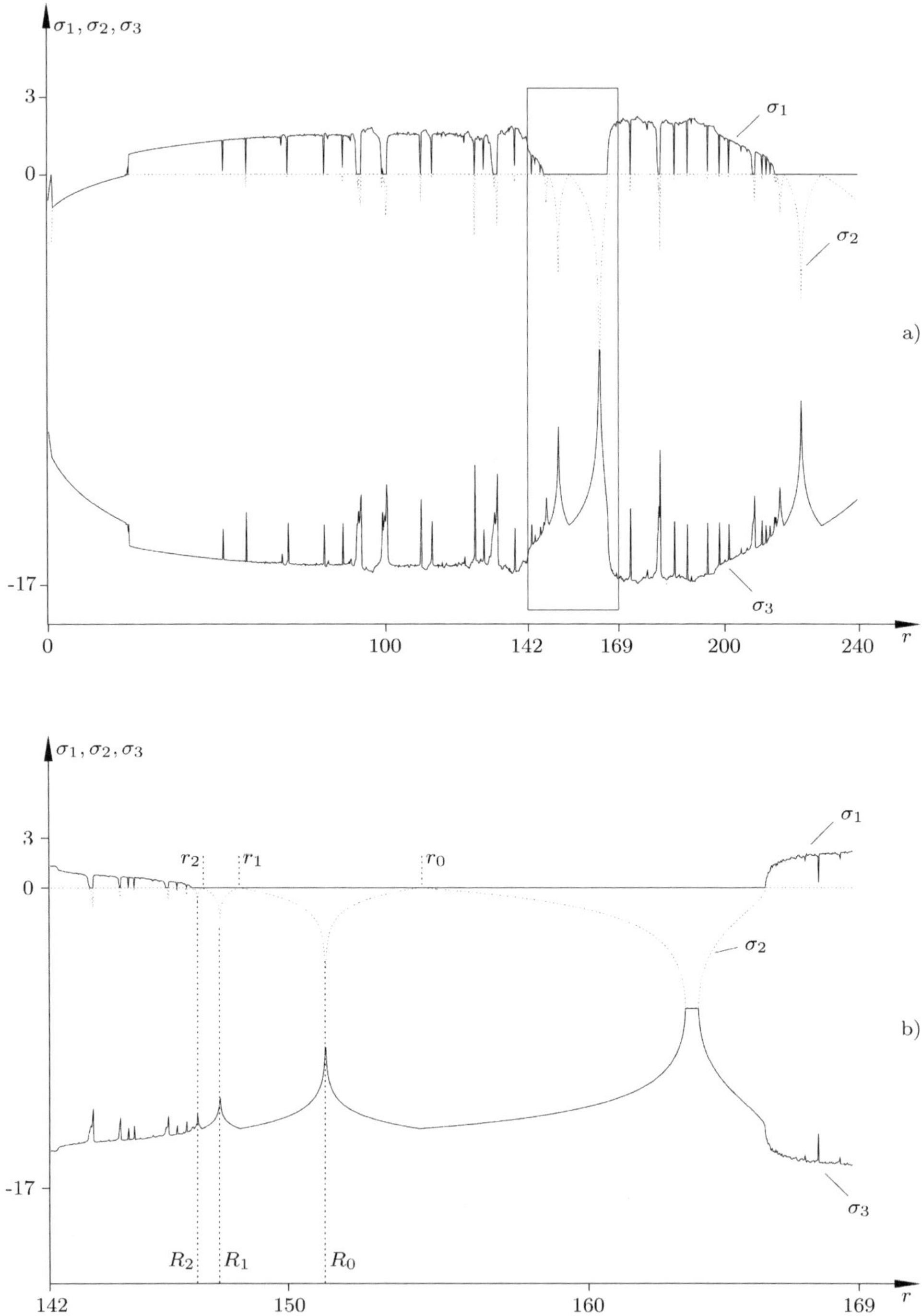

Fig. 8.4.3: Lyapunov exponents $\sigma_1 \geqslant \sigma_2 \geqslant \sigma_3$ of the Lorenz system for two parameter ranges: a) $0 \leqslant r \leqslant 240$, b) $142 \leqslant r \leqslant 169$

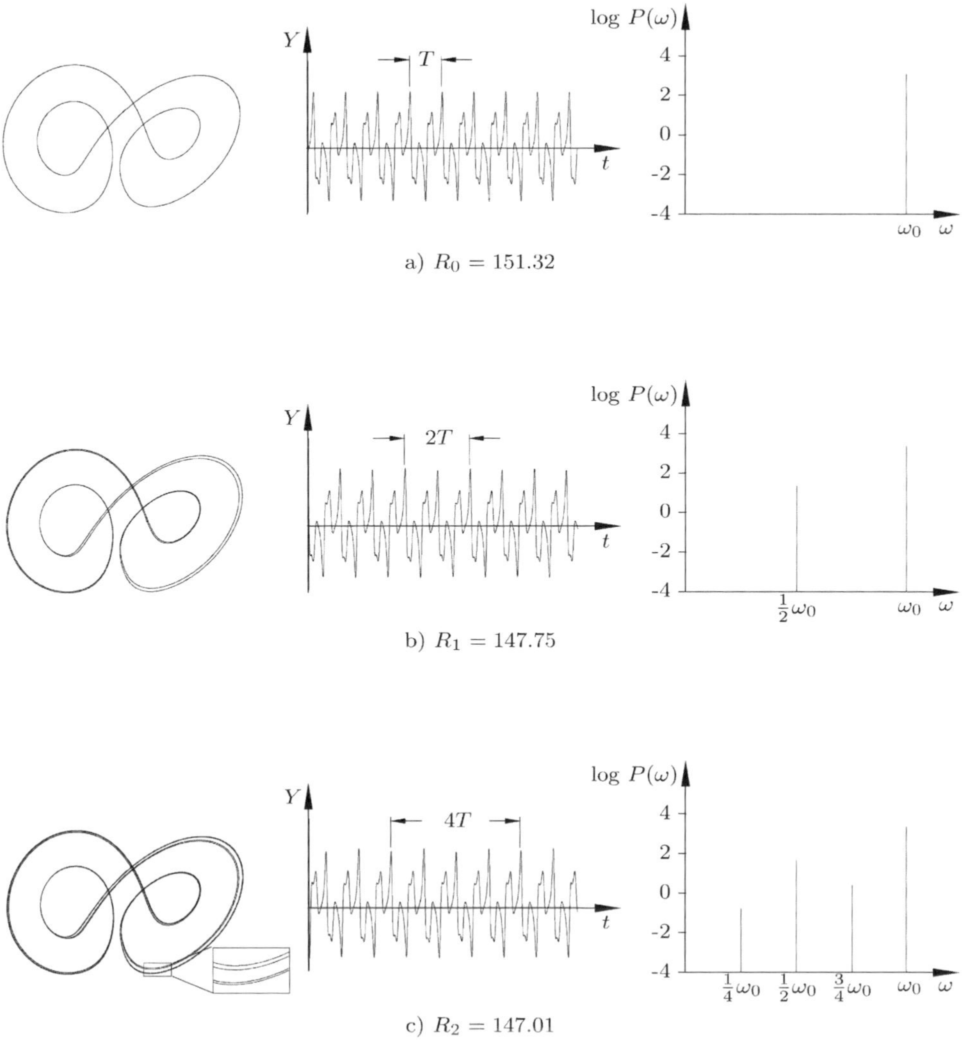

Fig. 8.4.4: Period doubling cascade in the Lorenz system: phase portraits, time series and power spectra

very fast; on the other hand, as a result of the noise in the recordings, only few of the bifurcations from the cascade can be observed experimentally.

We next enquire into the *scaling laws in the phase space* which are characterised by the universal constant $a \approx 2.5$. In fig. 8.4.4, we have plotted the phase portraits after the transient phase, the appertaining time series as well as the power spectra for three successive superstable cycles with $r = R_0 = 151.32$, $R_1 = 147.75$, $R_2 = 147.01$. For $R_0 = 151.32$, a simple limit cycle arises and we can deduce the period T from the time series (fig, 8.4.4a). The parameter value $r = R_1 = 147.75$ corresponds to a superstable limit cycle of the double period $2T$ (fig. 8.4.4b), $r = R_2 = 147.01$ also to a superstable limit cycle of the fourfold period $4T$ (fig. 8.4.4c). Evidently, great care is required to extract the second period doubling from the time series. On the other hand, it is easier to read from the phase portraits the number of cycles a trajectory requires before the orbit is completed. In this way, it becomes clear that there is an increasingly fine splitting of the orbits at each period doubling so that we gain the impression of bands on different scales.

We recognise corresponding scalings for the power spectra as well. For the simple limit cycle, we obtain equidistant lines at the basic frequency ω_0 and, since the limit cycle is not a circle, also for the multiples $\mathrm{n}\omega_0$ (cf. table 5.3.1). In order to demonstrate the scaling characteristics, we have restricted ourselves in each case to the interval $0 \leqslant \omega \leqslant \omega_0$ of the power spectra. After the first period doubling, a new frequency peak emerges at $\omega_0/2$, after the second period doubling at $\omega_0/4$ and $3\omega_0/4$ and so on. At every further period doubling, this process is repeated, new frequencies emerge in the middle in each case, the amplitudes becoming smaller. The dominant frequency is ω_0, that of the 2-cyle then follows etc.

It now appears logical to look for a connection between the band structure of the higher-order cycles and the amplitudes of the subharmonic frequencies. For this purpose, we select a plane in the phase space, this surface intersecting the limit cycle transversally. Successive points of intersection then define a Poincaré map so that we obtain a bifurcation diagram in dependence of the system parameter r, similar to the case of the logistic map (cf. fig. 6.7.6a). Figure 8.4.5 presents the Poincaré sections of three successive superstable cycles as well as the corresponding phase portraits for the bifurcation scheme of the logistic map. We can recognise very clearly that the scaling in the band structure of the higher-order cycles can be traced back to the scaling in the x-space and is thus determined by the universal constant a depending only on the order of the maximum of the Poincaré map.

Deriving the universal constant a in section 6.7, we first performed a coordinate transformation which shifts the critical point x_{max} to the origin and then concentrated on the scaling of the distances $d_{\mathrm{n}}(0)$ measuring the distance of the critical point $x_{\mathrm{max}} = 0$ from the nearest cycle point; we obtained the relation

$$\frac{d_{\mathrm{k}+1}(0)}{d_{\mathrm{k}}(0)} \to -\frac{1}{a} \qquad \text{for} \quad \mathrm{k} \to \infty \tag{8.4.1}$$

As figure 8.4.5 shows, however, the distance between neighbouring cycle points varies markedly; for example, the point $f(x_{\mathrm{max}})$ furthest away from $x_{\mathrm{max}} = 0$ is at a considerably smaller distance from its neighbour than x_{max} itself.

Feigenbaum now introduced a scaling function $\zeta(\tau)$ (Feigenbaum, 1980; Feigenbaum, 1983). This function is normally denoted $\sigma(\tau)$; here, we deviate from this, however, since σ has already been used in this section for the Lyapunov exponent and the Prandtl number. $\zeta(\tau)$ defines the ratio of the distances of all neighbouring cycle points at the transition from a 2^{k}- to a $2^{\mathrm{k}+1}$-cycle for $\mathrm{k} \gg 1$. An element of the superstable 2^{k}-cycle is x_{max} according to the definition (cf. fig. 6.7.10). If we denote the m-th iteration of x_{max} by x_{m}, the distance to the nearest cycle point is given by, cf. eq. (6.7.27),

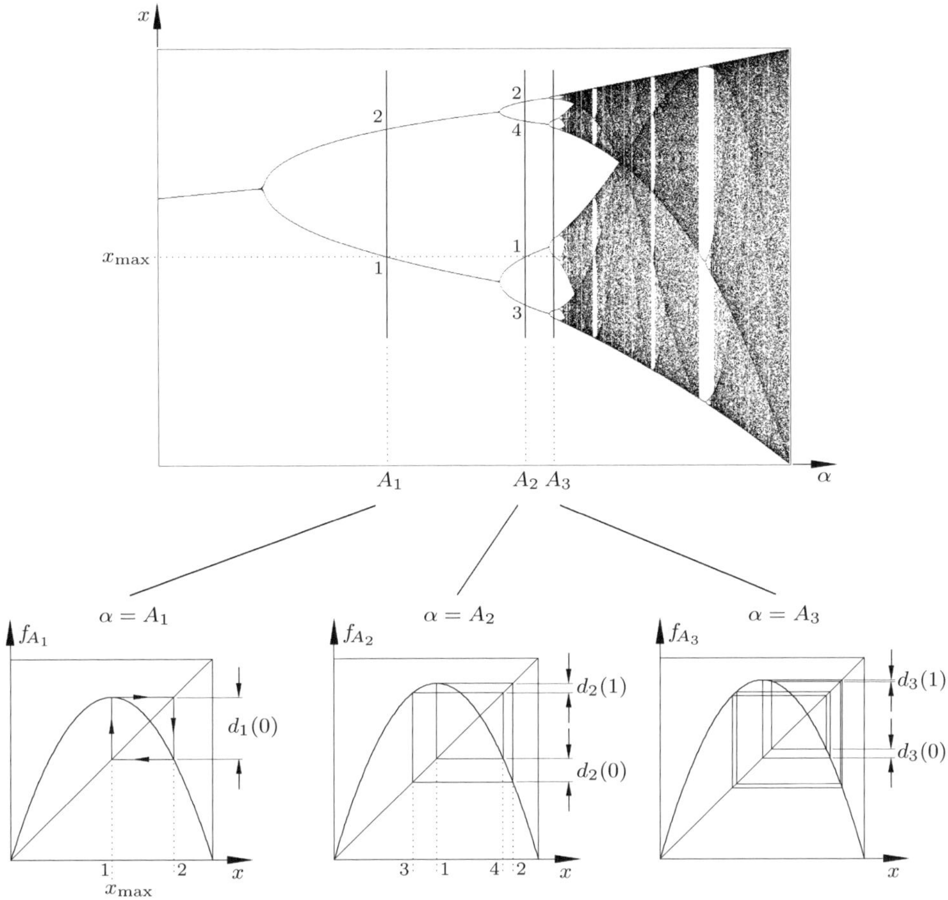

Fig. 8.4.5: Scaling of the superstable cycles for the logistic map eq. (3.7.3)

$$d_{\mathrm{k}}(\mathrm{m}) = x_{\mathrm{m}} - f_{A_{\mathrm{k}}}^{2^{\mathrm{k}-1}}(x_{\mathrm{m}}) \tag{8.4.2}$$

Proceeding to the next superstable $2^{\mathrm{k}+1}$-cycle, we obtain as the ratio of the distances

$$\zeta_{\mathrm{k}}(\mathrm{m}) = \frac{d_{\mathrm{k}+1}(\mathrm{m})}{d_{\mathrm{k}}(\mathrm{m})} \tag{8.4.3}$$

For m =0, the limit value is

$$\lim_{\mathrm{k}\to\infty} \zeta_{\mathrm{k}}(0) = -\frac{1}{a} \tag{8.4.4}$$

Introducing $\tau = \mathrm{m}/2^{\mathrm{k}}$ as a new variable, we can define a limit function $\zeta(\tau)$ for $\mathrm{k} \to \infty$ in accordance with eq. (8.4.3)

$$\zeta(\tau) = \lim_{\mathrm{k}\to\infty} \zeta_{\mathrm{k}}(2^{\mathrm{k}}\tau) \tag{8.4.5}$$

The inverse of the scaling function $\zeta(\tau)$ is plotted in fig. 8.4.6.

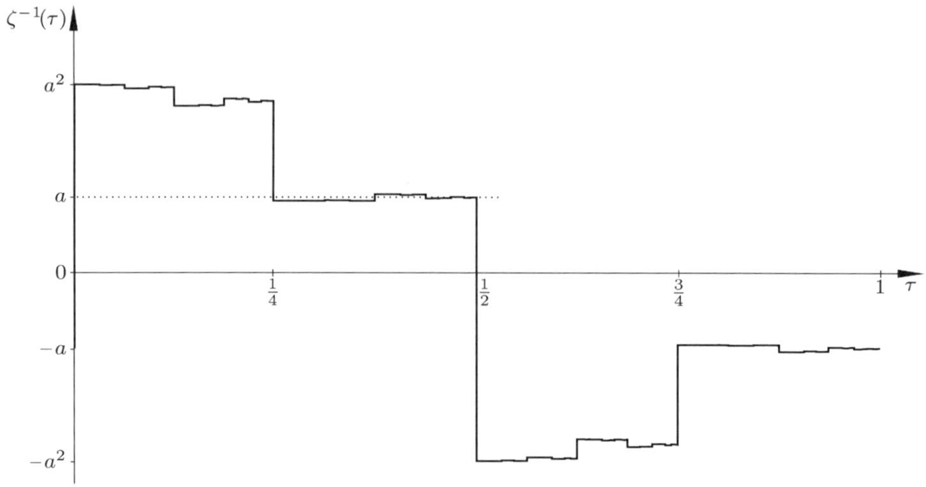

Fig. 8.4.6: The universal scaling function $\zeta(\tau)$

Since the explicit derivation of the function $\zeta(\tau)$ would go beyond the scope of this work, we refer the reader to Feigenbaum's original publication (Feigenbaum, 1980) or to the presentation in (Feigenbaum, 1983) which is easier to read, and only summarise the most important characteristics here. The scaling function $\zeta(\tau)$ satisfies the symmetry condition

$$\zeta(\tau + \tfrac{1}{2}) = -\zeta(\tau) \tag{8.4.6}$$

It is thus periodic with period 1 and, at all rational points $\tau = \mathrm{m}/2^{\mathrm{k}}$, has discontinuities which become smaller, the larger k is. In particular, $\zeta(\tau)$ has a discontinuity for $\tau = 0$. A more precise analysis shows that $\zeta(0 + \varepsilon) = 1/a$ and $\zeta(0 - \varepsilon) = 1/a^2$ hold if $\varepsilon \to 0_+$. In conjunction with eq. (8.4.6), this yields the relations for the left and the right limit at $\tau = \frac{1}{2}$, namely $\zeta(\frac{1}{2} - \varepsilon) = 1/a$ and $\zeta(\frac{1}{2} + \varepsilon) = -1/a^2$.

Figure 8.4.6 shows that two scales basically emerge, namely a and a^2; a is a measure for the scaling in the neighbourhood of the critical point x_{max}, a^2, on the other hand, for the scaling in the neighbourhood of $f(x_{\mathrm{max}})$. Since only the universal limit functions $g_{\mathrm{i}}(x)$, eq. (6.7.30), enter into the derivation of $\zeta(\tau)$ according to (Feigenbaum, 1980), the scaling function $\zeta(\tau)$ also has universal character.

In order to benefit from our theoretical knowledge of the scaling function practically, our next aim is to find experimentally easily accessible quantities which reflect the universal behaviour of the transition. Feigenbaum's idea now (Feigenbaum, 1979b) was to investigate the variation of the power spectra at period doublings and particularly to make quantitative statements on the amplitudes of the frequency peaks which arise additionally at each period doubling (see fig. 8.4.4).

In the following, we outline Feigenbaum's argumentation and consider a 2^{k}-cycle of period $T_{\mathrm{k}} = 2^{\mathrm{k}} T_0$ which has emerged following k period doublings from a simple limit cycle of period T_0. We can now construct a Poincaré map by measuring the coordinates

$x^{(\mathrm{k})}(t_{\mathrm{m}})$ of the trajectory every T_0 seconds, where $t_{\mathrm{m}} = \mathrm{m}T_0$ ($\mathrm{m} = 1, 2, 3, \ldots 2^{\mathrm{k}}$). With the aid of eq. (8.4.2), we can establish the distances between neighbouring cycle points

$$d_{\mathrm{k}}(t_{\mathrm{m}}) = x^{(\mathrm{k})}(t_{\mathrm{m}}) - x^{(\mathrm{k})}(t_{\mathrm{m}} + T_{\mathrm{k}-1}) \tag{8.4.7}$$

We obtain a corresponding relation for a $2^{\mathrm{k}+1}$-cycle of period length $T_{\mathrm{k}+1} = 2T_{\mathrm{k}}$ and, introducing the scaling function ζ in accordance with eq. (8.4.3), can write (Feigenbaum, 1983)

$$\begin{aligned} d_{\mathrm{k}+1}(t_{\mathrm{m}}) &= x^{(\mathrm{k}+1)}(t_{\mathrm{m}}) - x^{(\mathrm{k}+1)}(t_{\mathrm{m}} + T_{\mathrm{k}}) \\ &\approx \zeta\Big(\frac{t_{\mathrm{m}}}{2T_{\mathrm{k}}}\Big) d_{\mathrm{k}}(t_{\mathrm{m}}) \end{aligned} \tag{8.4.8}$$

For reasons of continuity, the above expressions hold not only for the Poincaré map but also approximately for the trajectories in between the intersections. Thus, in the following, we can replace the discrete values t_{m} by the continuous t.

Figure 8.4.6 suggests expressing the function $\zeta(t/2T_{\mathrm{k}})$ for $0 < t < T_{\mathrm{k}}$ in a first approximation as follows

$$\zeta\Big(\frac{t}{2T_{\mathrm{k}}}\Big) \approx \begin{cases} \dfrac{1}{a^2} & \text{for} \quad 0 < t < T_{\mathrm{k}-1} \\ \dfrac{1}{a} & \text{for} \quad T_{\mathrm{k}-1} < t < T_{\mathrm{k}} \end{cases} \tag{8.4.9}$$

In conjunction with eq. (8.4.7), $d_{\mathrm{k}+1}(t)$ can thus be further estimated as

$$d_{\mathrm{k}+1}(t) \approx \begin{cases} \dfrac{1}{a^2}\big[x^{(\mathrm{k})}(t) - x^{(\mathrm{k})}(t + T_{\mathrm{k}-1})\big] & \text{for} \quad 0 < t < T_{\mathrm{k}-1} \\ \dfrac{1}{a}\big[x^{(\mathrm{k})}(t) - x^{(\mathrm{k})}(t + T_{\mathrm{k}-1})\big] & \text{for} \quad T_{\mathrm{k}-1} < t < T_{\mathrm{k}} \end{cases} \tag{8.4.10}$$

Our aim is to investigate the influence of the scaling laws on the Fourier coefficients and thus also on the power spectrum. To this end, we expand the component $x^{(\mathrm{k}+1)}(t)$ of a $2^{\mathrm{k}+1}$-cycle in accordance with eq. (3.8.13) into a Fourier series

$$x^{(\mathrm{k}+1)}(t) = \sum_{\mathrm{n}} \alpha_{\mathrm{n}}^{(\mathrm{k}+1)} \exp\Big(2\pi i \frac{\mathrm{n}t}{T_{\mathrm{k}+1}}\Big) = \sum_{\mathrm{n}} \alpha_{\mathrm{n}}^{(\mathrm{k}+1)} \exp\Big(\pi i \frac{\mathrm{n}t}{T_{\mathrm{k}}}\Big) \tag{8.4.11}$$

where the Fourier coefficients according to eq. (3.8.14) can be calculated together with $T_{\mathrm{k}+1} = 2T_{\mathrm{k}}$ as follows

$$\alpha_{\mathrm{n}}^{(\mathrm{k}+1)} = \frac{1}{2T_{\mathrm{k}}} \int_0^{T_{\mathrm{k}+1}} x^{(\mathrm{k}+1)}(t) \exp\Big(-\pi i \frac{\mathrm{n}t}{T_{\mathrm{k}}}\Big) \mathrm{d}t \tag{8.4.12}$$

We split this integral into two integrals over the intervals $[0, T_{\mathrm{k}}]$ and $[T_{\mathrm{k}}, T_{\mathrm{k}+1}]$ and combine the two terms as follows

$$\alpha_{\mathrm{n}}^{(\mathrm{k}+1)} = \frac{1}{2T_{\mathrm{k}}} \int_0^{T_{\mathrm{k}}} \Big[x^{(\mathrm{k}+1)}(t) \exp\Big(-\pi i \frac{\mathrm{n}t}{T_{\mathrm{k}}}\Big) + x^{(\mathrm{k}+1)}(t + T_{\mathrm{k}}) \exp\Big(-\pi i \frac{\mathrm{n}(t + T_{\mathrm{k}})}{T_{\mathrm{k}}}\Big)\Big] \mathrm{d}t$$

or, since $\exp(-\pi i \mathrm{n}) = (-1)^{\mathrm{n}}$,

$$\alpha_{\mathrm{n}}^{(\mathrm{k}+1)} = \frac{1}{2T_{\mathrm{k}}} \int_0^{T_{\mathrm{k}}} \left[x^{(\mathrm{k}+1)}(t) + (-1)^{\mathrm{n}} x(t+T_{\mathrm{k}})\right] \exp\Big(-\pi i \frac{\mathrm{n}t}{T_{\mathrm{k}}}\Big) \mathrm{d}t \tag{8.4.13}$$

For even n-values, we can re-write eq. (8.4.13) for sufficiently large k using the approximate expression

$$x^{(\mathrm{k}+1)}(t) \approx x^{(\mathrm{k}+1)}(t+T_{\mathrm{k}}) \approx x^{(\mathrm{k})}(t)$$

Thus, for even coefficients, we obtain

$$\begin{aligned}\alpha_{2\mathrm{n}}^{(\mathrm{k}+1)} &= \frac{1}{2T_{\mathrm{k}}} \int_0^{T_{\mathrm{k}}} \left[x^{(\mathrm{k}+1)}(t) + (-1)^{2\mathrm{n}} x^{(\mathrm{k}+1)}(t+T_{\mathrm{k}})\right] \exp\Big(-2\pi i \frac{\mathrm{n}t}{T_{\mathrm{k}}}\Big) \mathrm{d}t \\ &\approx \frac{1}{2T_{\mathrm{k}}} \int_0^{T_{\mathrm{k}}} 2x^{(\mathrm{k})}(t) \exp\Big(-2\pi i \frac{\mathrm{n}t}{2T_{\mathrm{k}-1}}\Big) \mathrm{d}t\end{aligned}$$

and in accordance with eq. (8.4.12)

$$\alpha_{2\mathrm{n}}^{(\mathrm{k}+1)} \approx \alpha_{\mathrm{n}}^{(\mathrm{k})} \tag{8.4.14}$$

i.e. the even Fourier coefficients $\alpha_{2\mathrm{n}}^{(\mathrm{k}+1)}$ of the $2^{\mathrm{k}+1}$-cycle almost coincide with the Fourier coefficients $\alpha_{\mathrm{n}}^{(\mathrm{k})}$ of the previous 2^{k}-cycle.

We next turn our attention to the odd coefficients $\alpha_{2\mathrm{n}+1}^{(\mathrm{k}+1)}$. There follows from eq. (8.4.13) in conjunction with eq. (8.4.8)

$$\begin{aligned}\alpha_{2\mathrm{n}+1}^{(\mathrm{k}+1)} &= \frac{1}{2T_{\mathrm{k}}} \int_0^{T_{\mathrm{k}}} \left[x^{(\mathrm{k}+1)}(t) - x^{(\mathrm{k}+1)}(t+T_{\mathrm{k}})\right] \exp\Big(-\pi i \frac{(2\mathrm{n}+1)t}{T_{\mathrm{k}}}\Big) \mathrm{d}t \\ &\approx \frac{1}{2T_{\mathrm{k}}} \int_0^{T_{\mathrm{k}}} \zeta\Big(\frac{t}{2T_{\mathrm{k}}}\Big) \left[x^{(\mathrm{k})}(t) - x^{(\mathrm{k})}(t+T_{\mathrm{k}-1})\right] \exp\Big(-\pi i \frac{(2\mathrm{n}+1)t}{T_{\mathrm{k}}}\Big) \mathrm{d}t\end{aligned} \tag{8.4.15}$$

In order to establish a link with the Fourier coefficients of the 2^{k}-cycle, we again replace $x^{(\mathrm{k})}(t)$ respectively $x^{(\mathrm{k})}(t+T_{\mathrm{k}-1})$ in eq. (8.4.15) by the corresponding Fourier series and obtain

$$\alpha_{2\mathrm{n}+1}^{(\mathrm{k}+1)} = \frac{1}{2T_{\mathrm{k}}} \int_0^{T_{\mathrm{k}}} \zeta\Big(\frac{t}{2T_{\mathrm{k}}}\Big) \sum_{\mathrm{m}} \alpha_{\mathrm{m}}^{(\mathrm{k})} \left[1 - (-1)^{\mathrm{m}}\right] \exp\Big(2\pi i \frac{\mathrm{m}t}{T_{\mathrm{k}}}\Big) \exp\Big(-\pi i \frac{(2\mathrm{n}+1)t}{T_{\mathrm{k}}}\Big) \mathrm{d}t$$

If we again apply the approximation eq. (8.4.9) for $\zeta(t/2T_{\mathrm{k}})$ and note that only odd m-values make a contribution, we find

$$\alpha_{2n+1}^{(k+1)} \approx \frac{1}{a^2 T_k}\int_0^{T_{k-1}} \sum_m \alpha_{2m+1}^{(k)} \exp\Big(\frac{2\pi i}{T_k}[2m+1-\tfrac{1}{2}(2n+1)]t\Big)dt + \frac{1}{aT_k}\int_{T_{k-1}}^{T_k} \sum_m \alpha_{2m+1}^{(k)} \exp\Big(\frac{2\pi i}{T_k}[2m+1-\tfrac{1}{2}(2n+1)]t\Big)dt \tag{8.4.16}$$

Replacing t by $t + T_{k-1}$ in the second integral yields

$$\begin{aligned}\alpha_{2n+1}^{(k+1)} &\approx \frac{1}{a^2 T_k}\sum_m \alpha_{2m+1}^{(k)}\left[1+i(-1)^n a\right]\int_0^{T_{k-1}} \exp\Big(\frac{\pi i}{T_{k-1}}[2m+1-\tfrac{1}{2}(2n+1)]t\Big)dt \\ &= \frac{1}{a^2 T_k}\left[1+i(-1)^n a\right]\sum_m \alpha_{2m+1}^{(k)} \frac{T_{k-1}[i(-1)^n - 1]}{\pi i[2m+1-\tfrac{1}{2}(2n+1)]}\end{aligned}$$

The result of all these integral manipulations is finally the approximation, cf. (Feigenbaum, 1980),

$$\alpha_{2n+1}^{(k+1)} \approx \frac{1}{2\pi i\, a^2}\left[1+i(-1)^n a\right]\left[i(-1)^n - 1\right]\sum_m \frac{\alpha_{2m+1}^{(k)}}{2m+1-\tfrac{1}{2}(2n+1)} \tag{8.4.17}$$

From this result, we can recognise that the new odd Fourier coefficients of the 2^{k+1}-cycle can be determined from the "old" odd coefficients of the 2^k-cycle alone.

For large k, eq. (8.4.17) can be further simplified. The period length $T_k = 2^k T_0$ increases rapidly so that the Fourier coefficients $\alpha_n^{(k)}$ ultimately evolve into a continuous spectrum $\alpha^{(k)}(\omega)$. Thus, for large k, the sum in eq. (8.4.17) can be replaced approximately by an integral

$$\sum_m \frac{\alpha_{2m+1}^{(k)}}{2m+1-\tfrac{1}{2}(2n+1)} \approx \frac{1}{2}\int_{-\infty}^{+\infty} \frac{\alpha^{(k)}(\omega')}{\omega'-\omega/2}d\omega' \tag{8.4.18}$$

where we have applied the substitutions $\omega' = (2m+1)/T_k$ and $\omega = (2n+1)/T_k$. Since the underlying function $x^{(k)}(t)$ possesses no singularities, we can evaluate the integral further, applying the so-called Hilbert transform (Papoulis, 1962). If $F(\omega) = F_R(\omega) + iF_I(\omega)$ is the Fourier transform of a function $f(t)$ which exhibits no singularity for $t = 0$ and for which $f(t) = 0$ is assumed for $t < 0$, the following expressions hold

$$\begin{aligned}F_I(\omega) &= -\frac{1}{\pi}\int_{-\infty}^{+\infty} \frac{F_R(\omega')}{\omega-\omega'}d\omega' \\ F_R(\omega) &= \frac{1}{\pi}\int_{-\infty}^{+\infty} \frac{F_I(\omega')}{\omega-\omega'}d\omega'\end{aligned} \tag{8.4.19}$$

They are denoted Hilbert transforms. There follows directly for the absolute values

$$\left|\frac{1}{\pi}\int_{-\infty}^{+\infty} \frac{F(\omega')}{\omega-\omega'}d\omega'\right| = |-F_I(\omega) + iF_R(\omega)| = |F(\omega)| \tag{8.4.20}$$

Applying this transformation to eq. (8.4.18), we obtain Feigenbaum's estimate for moduli of the odd Fourier coefficients, eq. (8.4.17),

$$|\alpha_{2\mathrm{n}+1}^{(\mathrm{k}+1)}(\omega)| \approx \frac{1}{4a^2}\Big|\big[1+i(-1)^{\mathrm{n}}a\big]\big[i(-1)^{\mathrm{n}}-1\big]\Big|\big|\alpha^{(\mathrm{k})}(\omega/2)\big|$$

which may be re-written as

$$\left|\alpha_{2\mathrm{n}+1}^{(\mathrm{k}+1)}\right| \approx \mu^{-1}\left|\alpha_{\frac{1}{2}(2\mathrm{n}+1)}^{(\mathrm{k})}\right| \tag{8.4.21}$$

where the factor μ in accordance with eq. (6.7.25) is determined as

$$\mu = 4a^2\big[2(1+a^2)\big]^{-1/2} \approx 6.574 \tag{8.4.22}$$

This rough estimate means that the odd Fourier coefficients which emerge anew at every period doubling correspond in their order of magnitude to the mean value of the previous odd Fourier coefficients, reduced by a factor $\mu^{-1} \approx 0.152$.

In fig. 8.4.7, we have plotted the power spectra for a sequence of superstable cycles in the logistic map. We recognise that the amplitudes of the subharmonic components which are added at every period doubling decrease in the mean approximately by the factor $\mu^{-1} \approx 0.21$, i.e. $\mu \approx 4.7$ (on the logarithmic scale of the power spectrum, this corresponds to a factor $10\log_{10}\mu = 6.78\,\mathrm{db}$), where the individual components are, however, strongly modulated. In accordance with eq. (8.4.14), the peaks already present are only slightly changed.

A more precise theoretical investigation of Feigenbaum (Feigenbaum, 1981) confirmed the observation we made in fig. 8.4.7 that different parts of the spectrum are scaled differently and that particularly the odd spectral components being added result from the previous ones, eq. (8.4.21), by scaling by the factor

$$\mu = 2a^2(1+a^2)^{-1/2} \approx 4.649 \tag{8.4.23}$$

On the logarithmic scale of the power spectrum, this corresponds to a factor $10\log_{10}\mu = 6.7\mathrm{db}$, in agreement with the results for the logistic map, fig. 8.4.7. This result also coincides to a large extent with an estimate of Nauenberg and Rudnick who give $\mu \approx 4.579$ as the scaling factor (Nauenberg and Rudnick, 1981).

Up to now, we have investigated the universal behaviour on the basis of the discrete temporal evolution of a single variable. Higher-dimensional dissipative systems often possess the characteristic that the phase space volume shrinks in different directions at different speeds so that the asymptotic behaviour is determined by a single mode which proceeds towards a periodic motion and experiences a cascade of period doublings following the variation of a control parameter.

Such behaviour occurs in driven oscillating systems with damping such as the Duffing equation or the Lorenz system, for example. However, even the simple example in fig. 8.4.7 shows that the interpretation of the power spectra has its difficulties in practice. If the original limit cycle in the phase space is not a true circle, its power spectrum

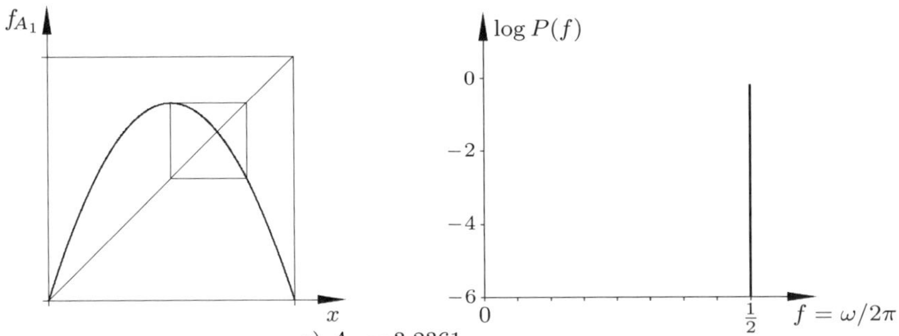

a) $A_1 = 3.2361$

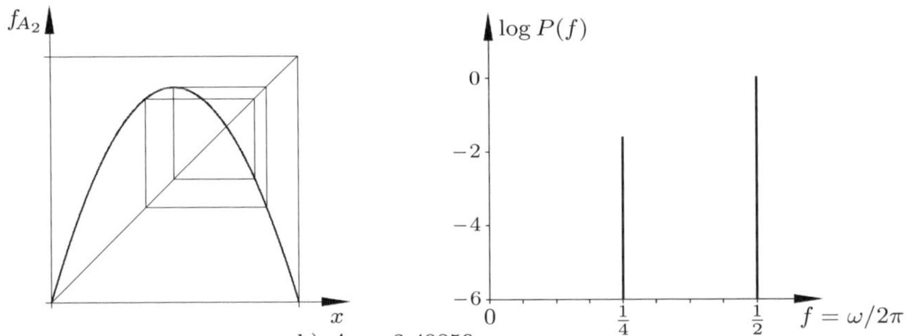

b) $A_2 = 3.49859$

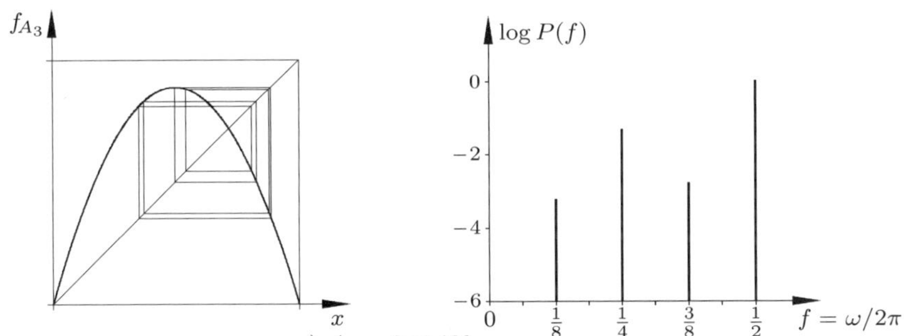

c) $A_3 = 3.55463$

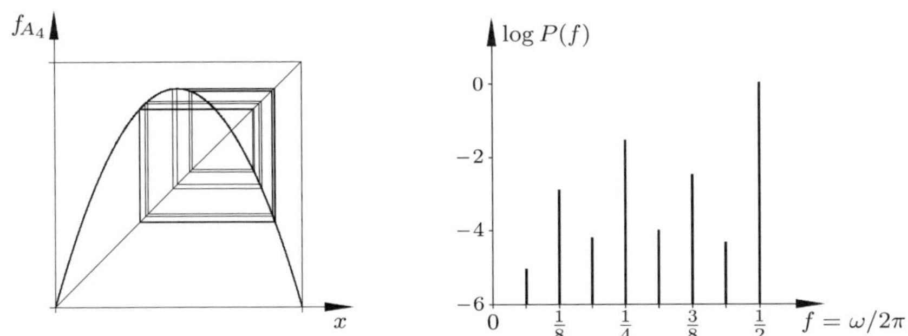

d) $A_4 = 3.56667$

Fig. 8.4.7: Power spectra of successive superstable 2^k-cycles of the logistic map (scaling behaviour) possesses peaks not only at the basic frequency ω_0 but also at higher subharmonics $k\omega_0$. In this case, finding scaling characteristics in the power spectrum is hampered; at least, one ought to know all the spectra which emerge at the first period doublings (Cvitanović, 1984).

8.4.2 Experimental Validation of the Feigenbaum Route

The Feigenbaum route to chaos via period doublings could be confirmed in many experiments such as, for example, in electrical circuits 8.4.8, in mechanical systems – such as the parametrically driven pendulum (Pompe et al., 1984) or the Duffing equation which we shall discuss in more detail in section 10.5 – but also in hydrodynamics, i.e. in systems with infinitely many degrees of freedom.

At the end of the 1970s, A. Libchaber and J. Maurer attempted an experimental confirmation of the Feigenbaum route qualitatively and quantitatively, using the Rayleigh-Bénard experiment with liquid helium. If the temperature difference ΔT between the plates playing the role of the control parameter is increased, then, according to the theory and in an ideal case, a series of characteristic observations should be made:

i. each period doubling can be recognised in the power spectrum by the emergence of subharmonics with the frequencies $\omega_0/2$, $\omega_0/4$, $\omega_0/8$ etc. as well as their multiples, where ω_0 is the frequency of the basic oscillation;

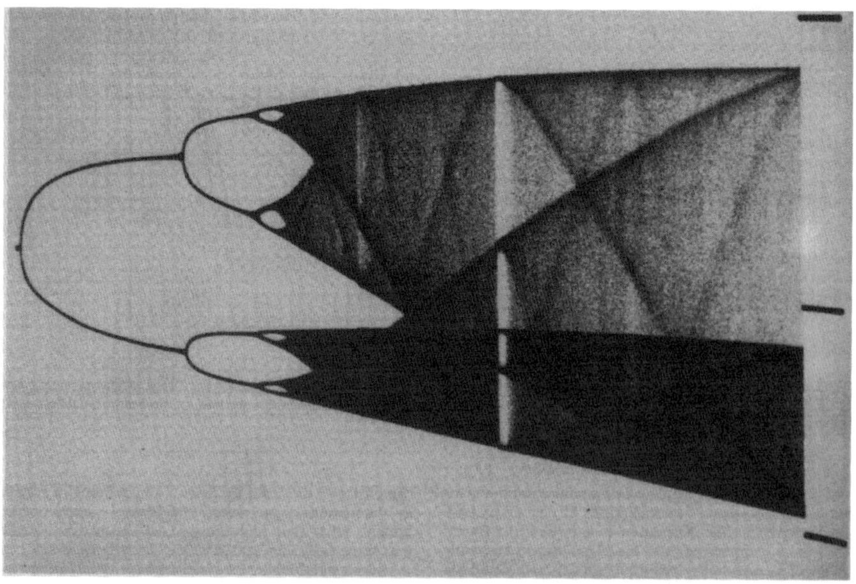

Fig. 8.4.8: Experimentally determined bifurcation diagram of a periodically driven non-linear electric circuit (van Buskirk and Jeffries, 1985)

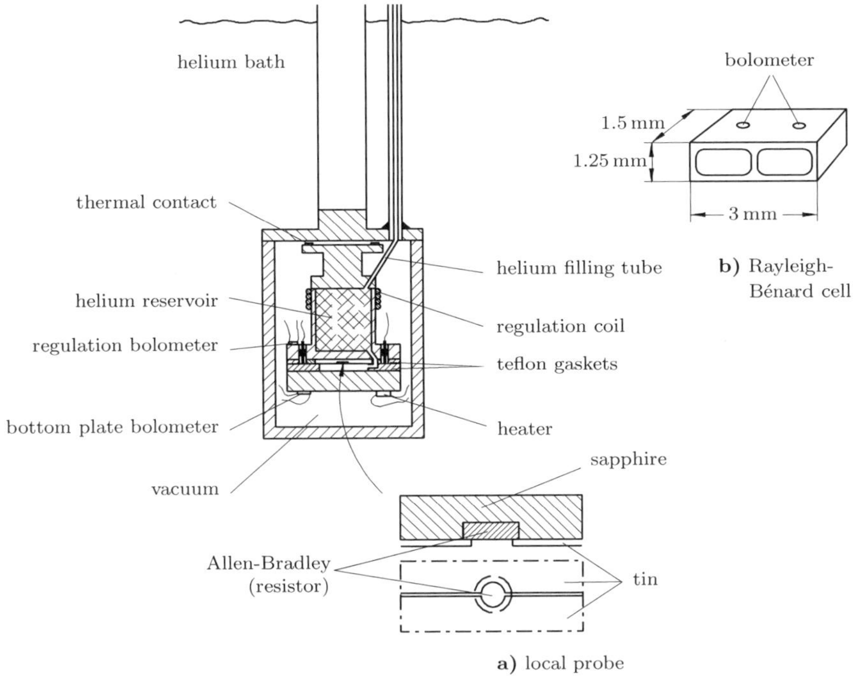

Fig. 8.4.9: Liquid helium in a small box, the set-up of Libchaber's experiment (Libchaber and Maurer, 1978)

ii. the power spectrum possesses scaling characteristics: the order of magnitude of the subharmonics which are generated at each period doubling is (in the mean) smaller than that of the preceding subharmonics by a factor μ^{-1}, cf. eqs. (8.4.22, 8.4.23);

iii. the distances of the control parameters between two successive period doublings decrease in each case by a factor of the order of magnitude $\delta \sim 4.7$ where δ is the Feigenbaum constant given in eq. (6.7.25).

A quantitative proof of these scaling characteristics requires an experimental set-up free of perturbations and vibrations as well as highly sensitive recording devices. In order to give an impression of the problems confronting an experimenter, we have reproduced the set-up of Libchaber's experiment in fig. 8.4.9 (Libchaber and Maurer, 1978; Libchaber and Maurer, 1982).

If we wish to record consecutively the first three to four period doublings, then all conceivable perturbation effects have to be eliminated. Since spatial dependencies are not taken into account in the Feigenbaum transition, the geometry of the convection cells first had to be "frozen". This was achieved on the one hand by selecting a small aspect ratio of the horizontal dimension D of the liquid container and the distance between the plates h ($\Gamma = D/2h = b/h = 1/a$, cf. fig. 7.0.2) and on the other hand by a low Prandtl number $\sigma < 1$ which means that the flow pattern has a two-dimensional character after the onset of stationary convection. Various experiments had shown that the Prandtl number selected for the transition to chaos has a considerable influence on the scenario (Busse, 1978). Low Prandtl numbers led to frequency locking with a subsequent period doubling cascade after the second Hopf bifurcation. In contrast, higher Prandtl numbers often led to intermittency (cf. section 8.6) or to quasi-periodic transitions with two or more incommensurable frequencies (Maurer and Libchaber, 1980). In order to realise Prandtl numbers in the range $0.4 < \sigma < 1$, Libchaber and Maurer used liquid helium at temperatures between 2.5 and $4.5°K$ (and at a pressure of 1 to 5 at). The extremely small dimensions of the actual convection cell (fig. 8.4.9b) and the selection of special materials for the side walls as well as for the upper and lower plates of the experimental cell also served to stabilise the rolls and to suppress perturbation effects (Libchaber and Maurer, 1982). With the aid of highly sensitive local bolometers (a thermometer for measuring temperatures by radiation absorption), the temporal evolution of the temperature could be measured with extraordinarily high accuracy. Figures 8.4.10 and 8.4.11 reproduce different power spectra for increasing Rayleigh numbers Ra from publications of Libchaber and Maurer.

The line spectrum at $Ra = 2.98 \cdot 10^4$ (fig. 8.4.10) corresponds to a quasi-periodic motion with the two incommensurable basic frequencies f_1 and f_2 as well as their linear combinations. Following an increase in the temperature difference ($Ra = 3.3 \cdot 10^4$), a synchronisation of the frequencies takes place with $f_1/f_2 = 7$ and thus a periodic motion. A further increase of the Rayleigh number leads to a cascade of period doublings (fig. 8.4.11a - c), i.e. to the successive emergence of the subharmonics $f_1/2$, $f_1/4$, $f_1/8$ etc. associated with the basic frequency f_1 as well as their odd multiples. If we investigate the scaling characteristics of the power spectra (fig. 8.4.11) and begin at k $=$ 2 with the interpolation of the subharmonics at the frequencies $f_1/4$ and $3f_1/4$ by a horizontal line, then, according to Feigenbaum's first estimate, the components which are added at the next period doubling should be smaller by a factor of approximately $10 \log_{10} \mu^{-1} \approx 8.2$ db. We can see from fig. 8.4.11d that, at k $= 2$, the reduction lies in the mean 8.5 ± 0.5 db lower and, at k $= 4$, again approximately 8.3 ± 0.4 db lower. These values are in surprisingly good agreement with Feigenbaum's first prognosis (Feigenbaum, 1979a) in accordance with eq. (8.4.22).

The experimental determination of the Feigenbaum constant δ by measuring the control parameter at successive bifurcations yielded considerably less accurate results, namely $\delta \sim 3.5 \pm 1.5$ (Libchaber and Maurer, 1982). An improvement of the measurement precision is very difficult because the control parameter would have to be determined at every bifurcation with an accuracy about five times higher ($\delta \sim 4.67$) while the resolution of the trajectories in the phase space requires a measurement precision which is "only" 2.5 times as high ($a \sim 2.5$). Moreover, the length of the transient phase is approximately doubled at each period doubling so that the influence of perturbations increases.

However, the really surprising result is the universal character of this transition to chaos, the fact that we can also verify the Feigenbaum scenario – originally conceived as a theoretical model describing the transition to chaos via period doubling for a class of one-dimensional mapping functions – in complex continuous systems such as

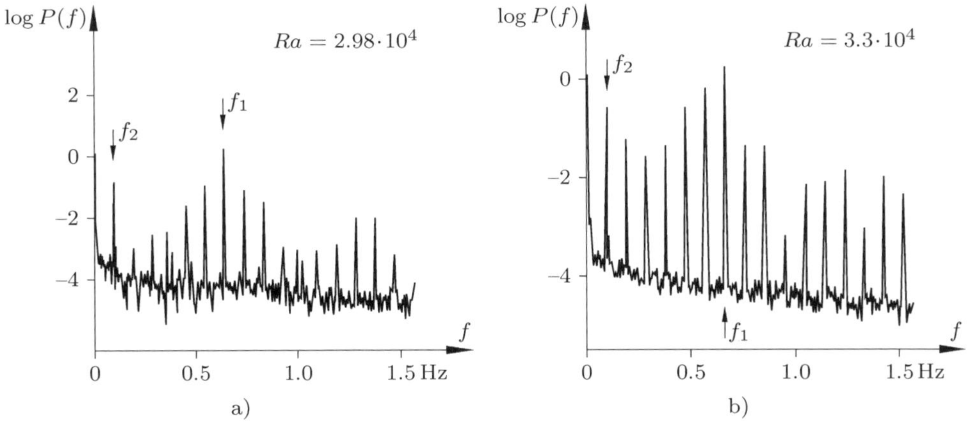

Fig. 8.4.10: Frequency locking in the Rayleigh-Bénard experiment (Maurer and Libchaber, 1979)

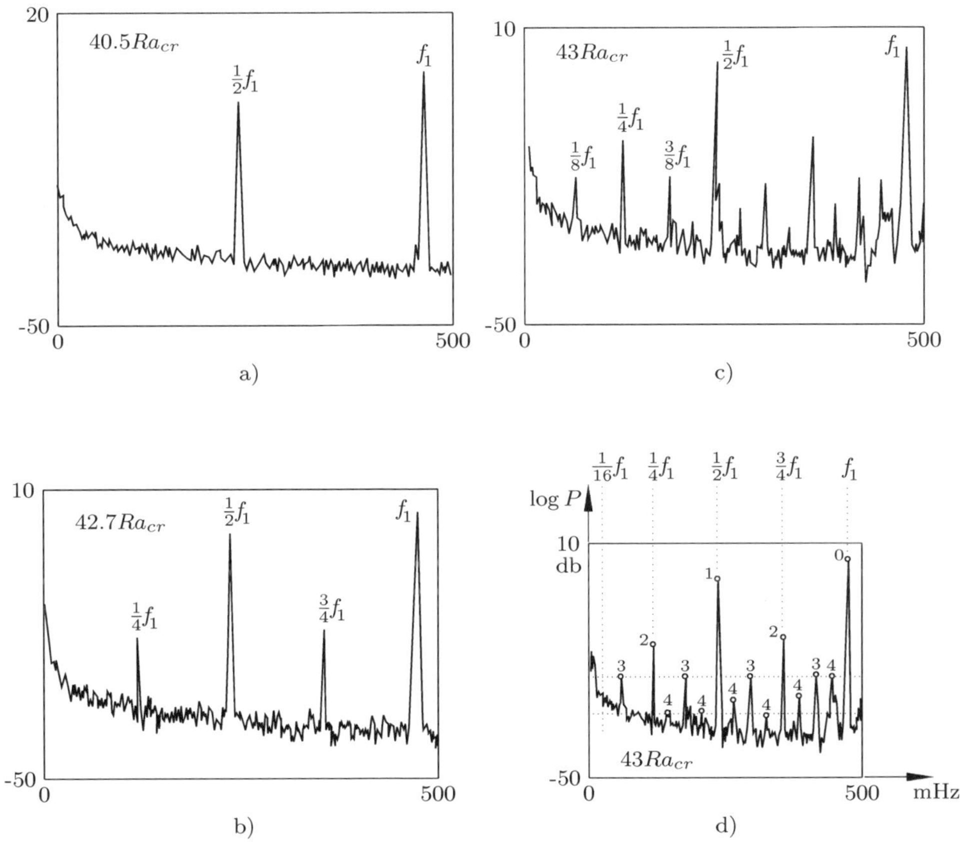

Fig. 8.4.11: Bénard-convection: (a) - (c) period doubling cascade and (d) scaling characteristics of the power spectrum (Libchaber and Maurer, 1980)

the Rayleigh-Bénard experiment not only qualitatively, but that we can also – within the limits of the experimental precision – observe a quantitative agreement with the theoretical prognoses.

8.5 Quasi-Periodic Transition for a Fixed Rotation Number

In the case of all the theoretical models we have discussed so far in the classification of the transitions to chaos, we obtain, following two Hopf bifurcations, a physical system characterised by two oscillations with incommensurable frequencies. An increase of the non-linear coupling can then lead either – as described in section 8.4 – to a synchronisation of the basic frequencies and thus to a periodic motion in a three-dimensional phase space where, in this case, the route to chaos passes through a cascade of period doublings or we can fix the (irrational) rotation number and proceed directly from quasi-periodic behaviour to chaos.

There is a whole series of experimental investigations confirming the existence of such quasi-periodic transitions to irregular behaviour. The pre-requisite is again that the physical system exhibits purely temporal dynamics and is characterised by strong dissipation so that the long-term behaviour can be described by low-dimensional attractors. In the case of quasi-periodic transitions to chaos, there are also universal characteristics which can be illustrated qualitatively as well as proven quantitatively and demonstrated on the circle map independently of details specific to the system.

Figure 8.5.1 presents the quasi-periodic transition schematically. An increase in the non-linear coupling generally leads to an increase of the effective dimension of the long-term behaviour. Symptomatic of this is the emergence of folds in the Poincaré section which can be explained with the aid of the dissipative circle map eq. (8.3.11).

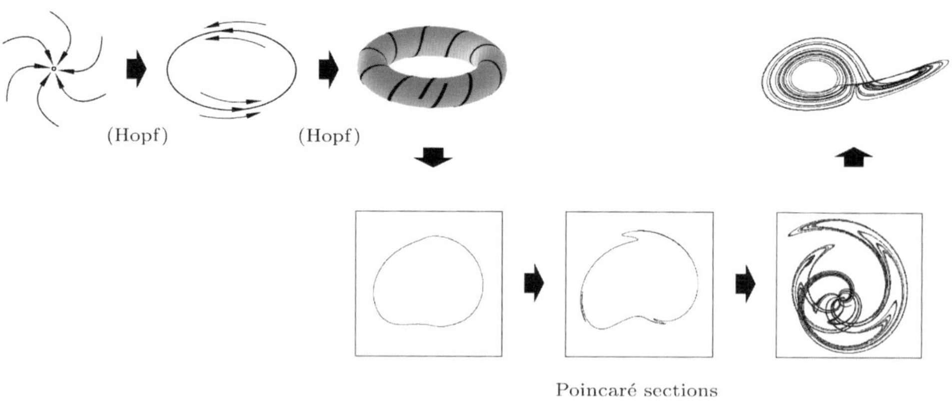

Fig. 8.5.1: Quasi-periodic transition to chaos

8.5.1 Scaling Characteristics of the Quasi-Periodic Transition

In section 8.3.3, we already discussed some local and global universal quantities characteristic of the quasi-periodic transition to chaos, e.g. the constants a and δ of eqs. (8.3.60) and (8.3.61) – which are dependent on the fixed rotation number selected but not on the specific form of the underlying circle map – or the dimension $D \approx 0.87$ of the set of Ω-values which leads to quasi-periodic motions for $K = 1$ (section 8.3.3.2). In this section, we present two further characteristic functions which undergo significant changes at the transition to chaos and which can also be observed experimentally. These are on the one hand the function $u(t)$, a measure for the deviation from a pure rotation in the Poincaré section modelled by the circle map, and on the other hand a function $f(\alpha)$, which describes the multifractal characteristics of the critical attractor (for K = 1). In both cases, these are *local* quantities calculated for a fixed rotation number.

In both numerical calculations and in experiments, the golden mean W_{G} was selected by preference as the rotation number since W_{G} possesses the simplest continued fraction representation

$$W_{\mathrm{G}} = <1\ 1\ 1\ldots>$$

with the property that the sequence of rational numbers W_{n} resulting from successive truncation of the infinite continued fraction is the sequence with the smallest monotonically increasing denominators and thus converges towards W_{G} particularly slowly; cf. eq. (8.3.44)ff. and section 8.3.3.2. The sequence of the corresponding Arnold tongues is thus characterised by maximum width of the plateaus and is therefore favoured for experimental observations.

We again take the circle map as our reference mapping and select $W_{\mathrm{G}} = (\sqrt{5}-1)/2$ as the rotation number. For every fixed value K_0 with $0 \leqslant K_0 < 1$, there is a uniquely determined value $\Omega_\infty(K_0)$ which we approximate according to the method of Rand/Shenker as described in section 8.3.3.1 (Shenker, 1982; Ostlund et al., 1983). Figure 8.5.2 is a schematic representation of this approximation procedure. It can be observed that the rational approximations W_{n} for the rotation number W_{G} are arranged so that $W_{\mathrm{n+2}}$ always lies in the interval between W_{n} and $W_{\mathrm{n+1}}$.

We first define the function $u(t)$ and study its properties for $K < 1$. According to the theorem of Denjoy (cf. section 8.3.2), the circle map is topologically equivalent to a pure rotation for each irrational rotation ratio W as long as $0 \leqslant K < 1$, i.e. there is a continuous invertible coordinate transformation $\theta \mapsto t$ which – in accordance with the method of normal forms (sections 6.3 and 6.4) – transforms the quasi-periodic motion $\theta \mapsto f(\theta)$ in the Poincaré section to a pure rotation

$$t \longmapsto R(t) = t + W \tag{8.5.1}$$

The appertaining coordinate transformation has the form

$$\theta(t) = t + u(t) \tag{8.5.2}$$

where $u(t)$ is a periodic function describing the deviation of the coordinate transformation from the identity when applying a uniform rotation.

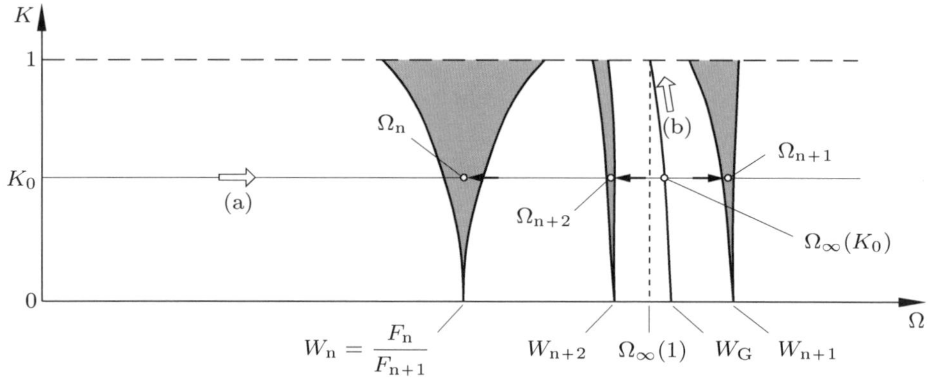

Fig. 8.5.2: Quasi-periodic transition to chaos:
a) construction of $\Omega_\infty(K_0)$ for $0 < K_0 < 1$, b) transition to the limit $K \to 1$

The theorem of Denjoy was the starting point for intensive mathematical studies which finally led to the KAM theory. If the rotation number is sufficiently irrational, i.e. if it satisfies a KAM condition of the form of eq. (4.4.10), it has been demonstrated that $u(t)$ forms a *smooth* curve for $0 \leqslant K < 1$; for further details, see (Arnold, 1965; Herman, 1977). Almost all irrational rotation numbers satisfy this condition, particularly the golden mean W_G. In the following, we shall observe that the function $u(t)$ loses the property of differentiability for $K = 1$ and thus undergoes a drastic change at the phase transition.

In (Shenker, 1982), a construction rule was given with the help of which $u(t)$ can be approximated for a general irrational rotation number W. We select a fixed value K_0 ($0 \leqslant K_0 < 1$) and study a series of rational numbers $W_n = p_n/q_n$ with $\lim_{n\to\infty} W_n = W$. For fixed n, the periodic motion is stable for all Ω-values within the corresponding Arnold tongue (see fig. 8.5.2). From this stability interval, we select the value $\Omega_n(K_0)$ which contains $\theta = 0$ as a cycle element, i.e. for which, in accordance with eq. (8.3.18), the following relation applies

$$f^{q_n}_{\Omega_n(K_0)}(0) = p_n \tag{8.5.3}$$

(where the modulo rule is not taken into account). We then introduce a discrete time variable

$$t_k = k\frac{p_n}{q_n} \pmod 1 \qquad (k = 0, \ldots, q_n - 1) \tag{8.5.4}$$

and construct a function $\theta^{(n)}$ which is defined at the discrete points t_{k+1} as follows

$$\theta^{(n)}(t_{k+1}) = f\big(\theta^{(n)}(t_k)\big) \pmod 1 \tag{8.5.5}$$

with the initial condition

$$\theta^{(n)}(0) = 0 \tag{8.5.6}$$

The function

$$u^{(n)}(t_k) = \theta^{(n)}(t_k) - t_k \tag{8.5.7}$$

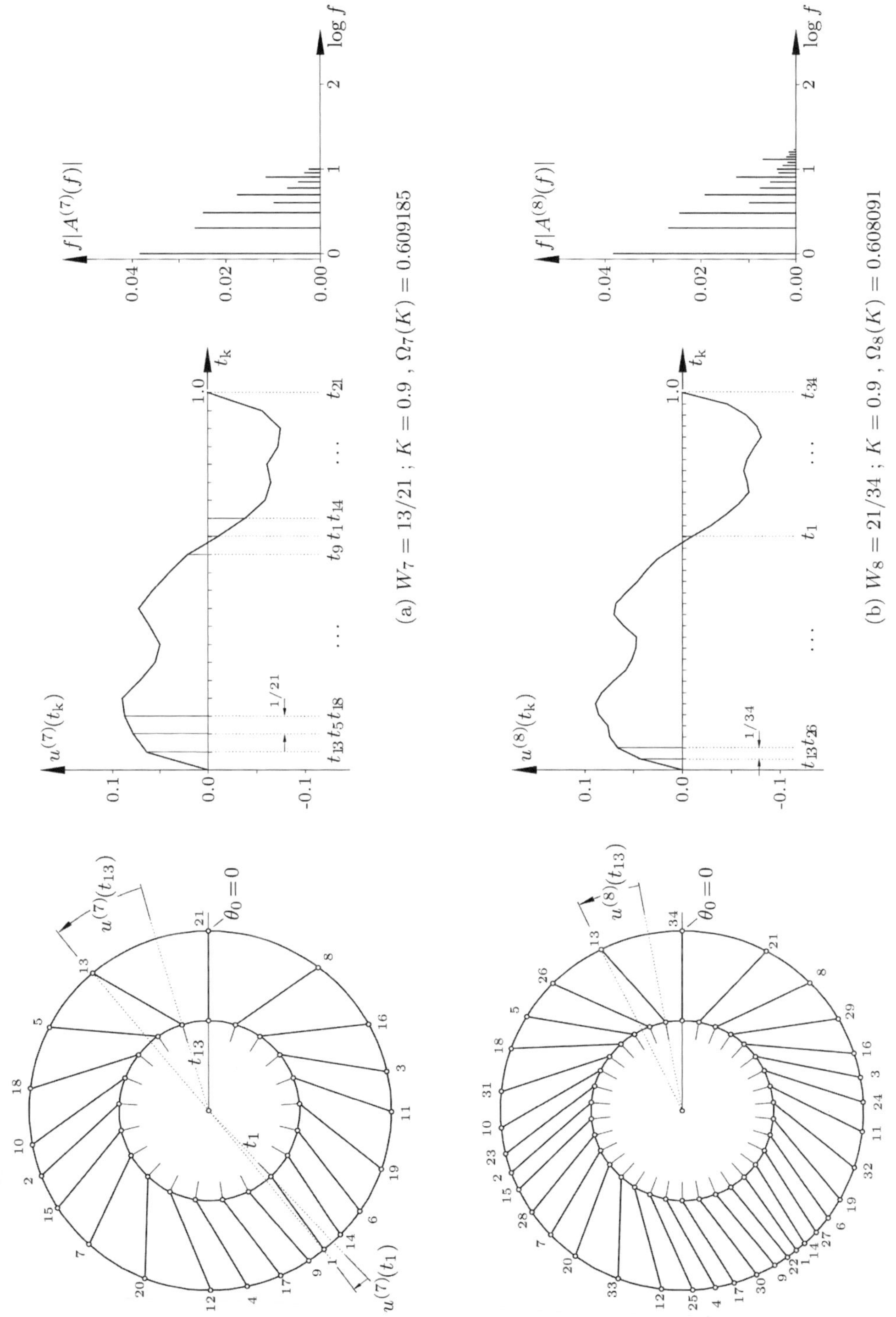

Fig. 8.5.3: The function $u^{(n)}(t_k)$ and its Fourier coefficients for n = 7 and n = 8

is then a discrete periodic function of period 1 which measures the deviation of the individual cycle points from a uniform rotation with the rotation number W_n and which is converted into a periodic function $u(t)$ of the continuous variable t for $\mathrm{n} \to \infty$ (i.e. for $W_\mathrm{n} \to W$).

For two different approximations of the golden mean, $W_7 = 13/21$ and $W_8 = 21/34$, fig. 8.5.3 reproduces the deviation $u^{(\mathrm{n})}(t_\mathrm{k})$ ($\mathrm{n} = 7$ and $\mathrm{n} = 8$) at $K = 0.9$. The $F_{\mathrm{n}+1}$-cycle is plotted in the left-hand column: the uniform distribution of the cycle points for $K = 0$ is plotted on the inner circle of the ring, the position of the cycle points for $K = 0.9$ on the outer circle; these points were named according to the number of iterations required, when starting with $\theta_0 = 0$. The angular deviations in each case correspond exactly to the values of the discrete function $u^{(\mathrm{n})}(t_\mathrm{k})$ at $t_\mathrm{k} = \mathrm{k}\, F_\mathrm{n}/F_{\mathrm{n}+1}$ (mod 1) and are presented, linearly interpolated, in the centre column. The diagrams in the left-hand column and the dominant lines in the power spectra in the right-hand column suggest that the cycle of the period $F_{\mathrm{n}+1}$ is almost closed after $F_{\mathrm{n}-1}$ respectively F_n iterations, i.e. the cycle points with the identification F_n respectively $F_{\mathrm{n}-1}$ are the direct neighbours of $\theta_0 = 0$ with the identification $F_{\mathrm{n}+1}$. We can confirm this conjecture very easily by proving the correctness of the relations

$$\left(\frac{F_\mathrm{n}}{F_{\mathrm{n}+1}}\right) F_\mathrm{n} \quad (\mathrm{mod}\ 1) = \frac{(-1)^{\mathrm{n}+1}}{F_{\mathrm{n}+1}} \tag{8.5.8}$$

and

$$\left(\frac{F_\mathrm{n}}{F_{\mathrm{n}+1}}\right) F_{\mathrm{n}-1} \ (\mathrm{mod}\ 1) = \frac{(-1)^{\mathrm{n}}}{F_{\mathrm{n}+1}} \tag{8.5.9}$$

e.g. by mathematical induction, selecting $F_0 = 0$, $F_1 = 1$ as initial values. For the function $u^{(\mathrm{n}+1)}$, this means that the course of the function basically follows that of $u^{(\mathrm{n})}$ where, apart from the finer discretisation, only small additional bulges are superimposed. This behaviour becomes even clearer if we expand $u^{(\mathrm{n})}(t_\mathrm{k})$ into a Fourier series. By discretisation, we obtain from eq. (3.8.14) the expression

$$A^{(\mathrm{n})}(f_\mathrm{j}) = \frac{1}{F_{\mathrm{n}+1}} \sum_{\mathrm{k}=0}^{F_{\mathrm{n}+1}-1} u^{(\mathrm{n})}(t_\mathrm{k}) \exp(-2\pi i f_\mathrm{j} t_\mathrm{k})\,, \qquad f_\mathrm{j} = 0, 1, \cdots, F_{\mathrm{n}+1} \tag{8.5.10}$$

For $\mathrm{n} \to \infty$, $u^{(\mathrm{n})}(t_\mathrm{k})$ tends towards a continuous smooth curve $u(t)$, i.e. the Fourier coefficients decrease exponentially with the frequency f in the range $0 \leqslant K < 1$ since the limit function $u(t)$ is continuously differentiable. In order to emphasise the influence of the higher frequencies, it is expedient to plot the quantity $f|A^{(\mathrm{n})}(f)|$ instead of $|A^{(\mathrm{n})}(f)|$ (right-hand column in fig. 8.5.3). Moreover, inserting in accordance with eq. (8.5.4)

$$t_\mathrm{k} = \mathrm{k}\frac{F_\mathrm{n}}{F_{\mathrm{n}+1}} \ (\mathrm{mod}\ 1) = \frac{\mathrm{m_k}}{F_{\mathrm{n}+1}} \tag{8.5.11}$$

into eq. (8.5.10), where $\mathrm{m_k}$ is an integer between 0 and $(F_{\mathrm{n}+1} - 1)$, we obtain the symmetry condition

$$A^{(\mathrm{n})}(f_\mathrm{j}) = \left(A^{(\mathrm{n})}(F_{\mathrm{n}+1} - f_\mathrm{j})\right)^* \tag{8.5.12}$$

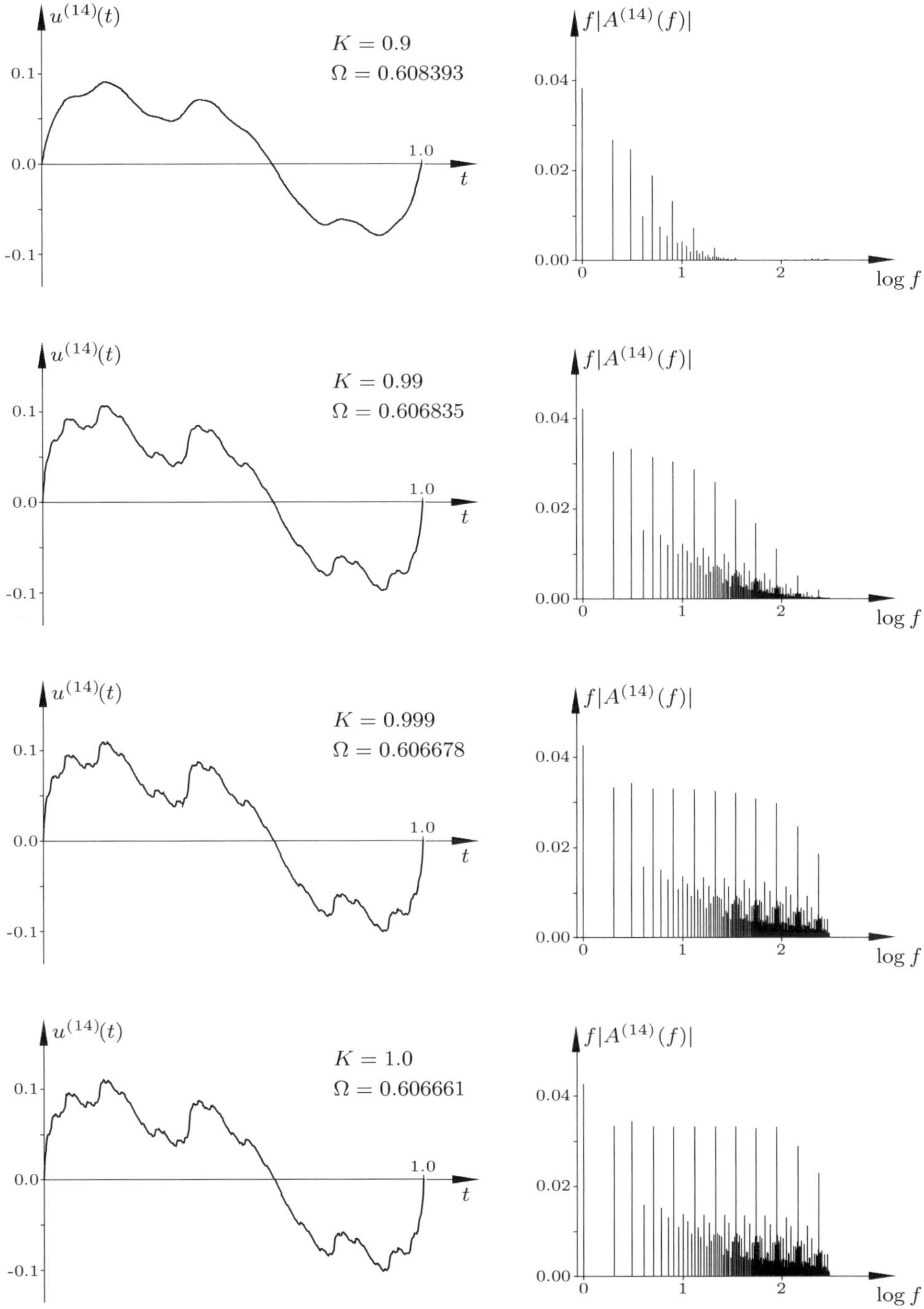

Fig. 8.5.4: Evolution of the function $u^{(14)}(t)$ and of the power spectrum $f|A^{(14)}(f)|$ when approaching the critical value $K = 1$ for a fixed rotation number $W_{14} = 377/610$

where the superior asterisk (*) again denotes the complex conjugate value. It thus suffices to restrict ourselves to the left-hand side of the f-interval in the power spectra of fig. 8.5.3. The exponential decrease of the peaks is clearly recognisable. If we proceed from the rotation number $W_7 = 13/21$ to $W_8 = 21/34$, we also recognise that the old spectral lines are nearly all retained in the new spectrum. This is an indication of the fact that the 34-cycle almost closes after 21 iterations. The largest of the new additional peaks emerges at the next higher Fibonacci number.

The exponential decrease of the dominant peaks with increasing f holds only for $K < 1$, however. In fig. 8.5.4, we have plotted the function $u^{(14)}(t)$ together with the corresponding power spectra for values $K \to 1$ and for a fixed rotation number $W_{14} = 377/610$, which deviates by approximately only $2 \cdot 10^{-4}\,\%$ from the golden mean. We clearly observe that the higher frequencies become increasingly dominant until these peaks finally adopt an almost constant value at $K = 1$ for frequencies coinciding with Fibonacci's numbers. If we proceed to the limit $\lim_{\mathrm{n}\to\infty} W_{\mathrm{n}} = W_{\mathrm{G}}$ (cf. fig. 8.5.2b), dominant frequencies emerge at infinitely many Fibonacci numbers; the function $u(t)$ hence loses its property of differentiability at the critical line, i.e. it appears – like the Koch snowflake curve (cf. fig. 5.5.3) – "jagged" on all scales. This self-similarity is also expressed in the power spectra: between each pair of dominant peaks, the same pattern of spectral lines with self-similar character emerges. The underlying scaling theory is described in (Ostlund et al., 1983).

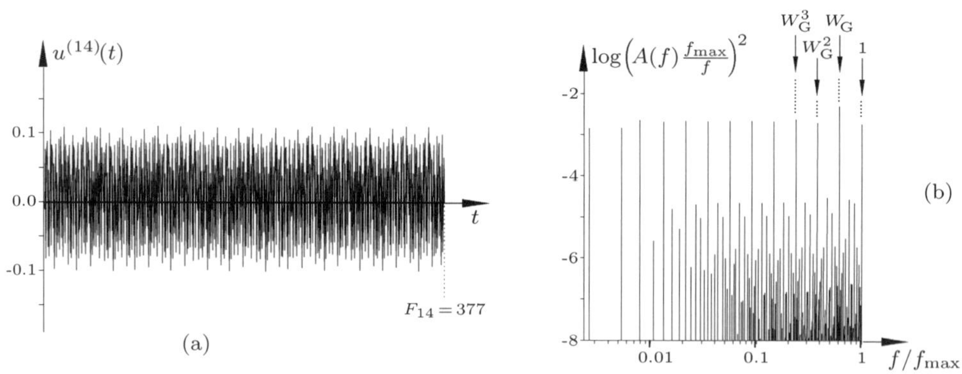

Fig. 8.5.5: Time-series (a) and scaled power spectrum (b) of $u^{(14)}(t)$ for $K = 1$ and $\Omega = 0.606661$

In fig. 8.5.5a, we plot the function $u^{(14)}(t)$ for $K = 1$ without applying the modulo rule, i.e. the cycle closes after $F_{15} = 610$ iterations and has required $F_{14} = 377$ revolutions. Figure 8.5.5b presents the appertaining scaled power spectrum $\log |A(f) f_{\max}/f|^2$ as a function of $\log f/f_{\max}$. We observe an equidistant sequence of equally high maximum values which – interestingly enough – emerge at the frequencies W_{G}, W_{G}^2, W_{G}^3 etc. This result can be traced back to a special characteristic of the golden mean W_{G}. In section 3.8, we observed that, in the power spectrum of a quasi-periodic motion with the basic frequencies f_1 and f_2, single peaks are generated at f_1, f_2 and at all linear combinations $nf_1 + mf_2$. With the aid of the relation eq. (8.3.43) and mathematical induction, we can easily derive the following relation

$$W_G^n = (-1)^n [F_{n-1} \cdot 1 - F_n \cdot W_G] \tag{8.5.13}$$

i.e. each power of W_G is a special linear combination of the two incommensurable basic frequencies $f_1 = 1$ and $f_2 = W_G$.

The scaling characteristics of the power spectra again possess universal character, i.e. at the transition from quasi-periodic behaviour to chaos with a *fixed* irrational rotation number, the same spectra result irrespective of the underlying dynamical system. Here, however, simple patterns emerge only for the rotation numbers which can be expressed as a periodic continued fraction (Ostlund et al., 1983).

8.5.2 Multifractal Structures

As already mentioned, we next investigate the multifractal characteristics of the attractor at the critical value $K = 1$, this also being denoted the *critical attractor*. In analogy to the left-hand side of fig. 8.5.3, we have plotted in fig. 8.5.6a the angular deviations $u(t)$ from the corresponding pure rotation for the rotation number $W_{14} = 377/610 \approx W_G$ and $K = 1$. It already becomes clear that groups of rays are packed particularly closely at some points but leave large gaps at others; this structure is repeated on different scales at different points. For $W = W_G$, the circle map is still topologically equivalent to a pure rotation, i.e. after a sufficient number of iterations, all the gaps are filled. In fig. 8.5.6, we have approximated the corresponding point distribution of the critical attractor directly by iterating the circle map for $K = 1$, $W_{14} \approx W_G$; it coincides qualitatively with the distribution of the points on the outer edge of the left ring of rays.

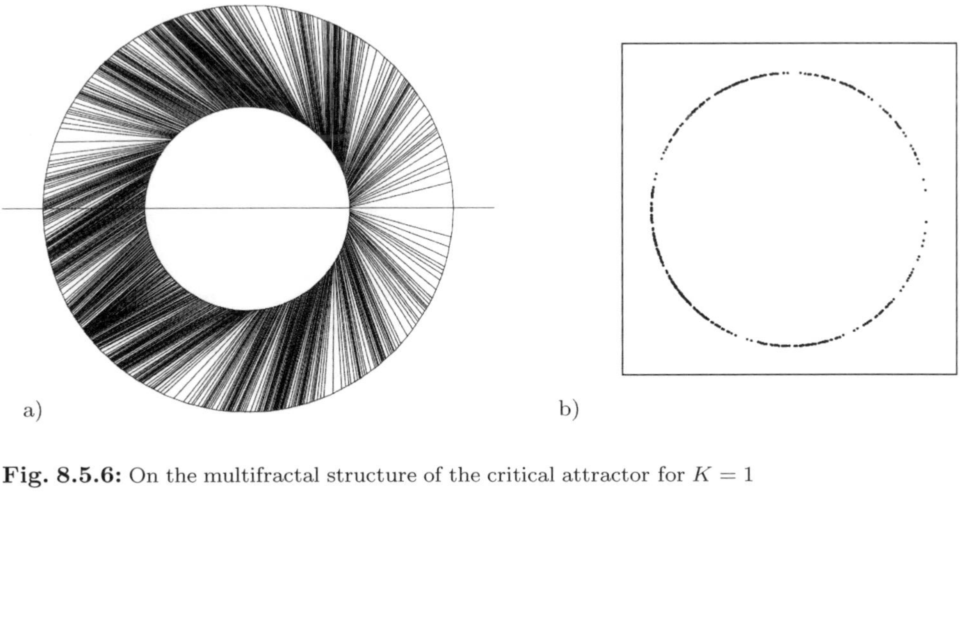

a) b)

Fig. 8.5.6: On the multifractal structure of the critical attractor for $K = 1$

Although the circumference of the whole circle is covered for $n \to \infty$, the density distribution of the points exhibits very high variations, the multifractal structure of which evidently cannot be described by a single dimension such as the capacity dimension, for example. Two possibilities for characterising the critical attractor now present themselves.

The first one is already known to us from section 5.5.5 where we described inhomogeneities in the scaling characteristics of attractors by the generalised dimension D_q, see eqs. (5.5.56) and (5.5.57),

$$D_q = \frac{1}{q-1} \lim_{\varepsilon \to 0} \frac{\ln \sum_i p_i^q}{\ln \varepsilon} \tag{8.5.14}$$

To determine D_q, the attractor (fig. 8.5.6b) is covered with intervals of length ε. Corresponding to the number N_i of attractor points in the i-th interval, this latter is weighted with

$$p_i(\varepsilon) = \frac{N_i}{N} \tag{8.5.15}$$

where N is the total number of points considered. For integral values of q, D_q contains the q-th moments of the probability distribution. The dimension D_q is also defined for general real values of q, however. For large positive q-values, the function D_q provides information on the densest regions of the attractor, for very small negative q-values, on the other hand, on the most thinly populated regions. In fig. 8.5.7a, D_q is plotted for the critical attractor of the circle map ($K = 1$, $W = W_G$) (Halsey et al., 1986).

The second possibility for characterising the structure of the attractor consists of calculating the corresponding pointwise dimension α_i at each point θ_i on the circumference of the circle (cf. section 5.5.4). To this end, we determine in accordance with eq. (8.5.15) the probability $p_i(\varepsilon)$ of an attractor point lying in a ε-neighbourhood of θ_i. For small ε-values, we have

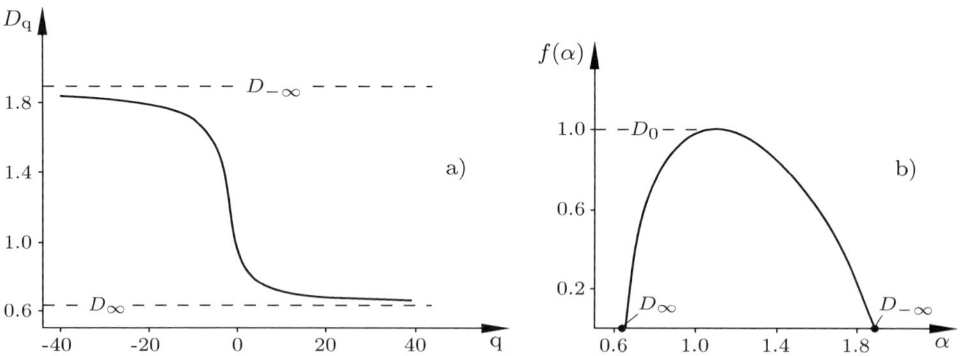

Fig. 8.5.7: Two possibilities for characterising the multifractal structure of the critical attractor of the circle map ($K = 1$, $W = W_G$):
a) generalised dimension D_q and b) multifractal spectrum $f(\alpha)$; after (Halsey et al., 1986)

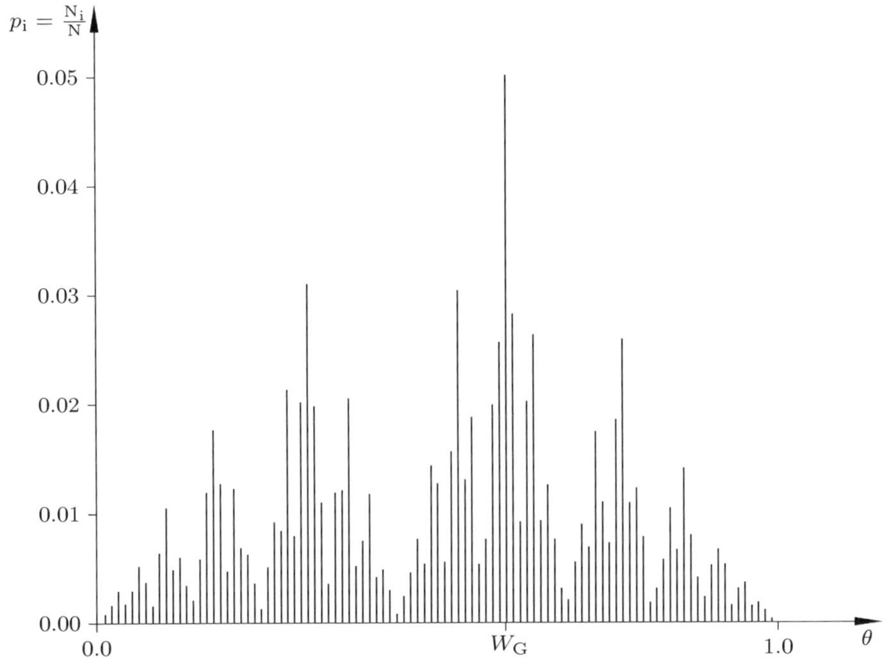

Fig. 8.5.8: Density distribution of the points on the critical attractor ($K = 1$, $W = W_G$)

$$p_i(\varepsilon) \sim \varepsilon^{\alpha_i(\varepsilon)} \tag{8.5.16}$$

In the limit $\varepsilon \to 0$, α_i coincides with the pointwise dimension which, in our case, varies considerably from point to point, i.e. α adopts values within an interval $[\alpha_{min}, \alpha_{max}]$. In contrast to the suggestive proposition appearing on the left-hand side of fig. 5.5.14 in section 5.5.4, where we assumed a homogeneous one-dimensional set of points, the pointwise dimension of a multifractal set can indeed assume values smaller than 1 respectively larger than 1 even if all points are arranged on a line, in our case a circle. We can easily conclude that accumulations of points on the line lead to an α-value < 1 since, in their dimensionality, they tend more to that of a single point. Conversely, the pointwise dimension of very thinly populated regions is > 1. If, for example, we halve the intervals of length ε successively, then fewer than half as many points lie in each shrunken neighbourhood of a point, which can only be attained by an exponent $\alpha > 1$ in eq. (8.5.16). In order to describe the multifractal structure, we now go one step further and determine for every value α the capacity dimension of those points which possess the same α as their pointwise dimension. The result is presented in fig. 8.5.7b. The function $f(\alpha)$ is denoted the multifractal spectrum.

In practice, however, $f(\alpha)$ is not determined via the capacity dimension, i.e. the counting of intervals. Rather, we take a detour based on a proof in (Halsey et al., 1986), cf. also (Schuster and Just, 2006), stating that the generalised dimension D_q and the multifractal spectrum $f(\alpha)$ are equivalent descriptions of the inhomogeneous structure of the critical attractor. The basic ideas leading to the proof of this equivalence are outlined in the following.

In fig. 8.5.8, we have plotted the density distribution p_i along the circumference of the circle for $\varepsilon = 1/100$, $W = W_G$, $K = 1$. This probability distribution and thus also the invariant measure is evidently subject to very high variations *locally*: we clearly recognise a self-similar structure. The idea now is to make an integral statement on the sequence of the α_i-values and to establish a connection with D_q. To this end, we introduce the abbreviation

$$\chi(q) = \sum_{i=1}^{N(\varepsilon)} p_i^q \tag{8.5.17}$$

and rearrange the sum as follows: we do not sum in succession over the N ε-intervals of the circumference of the circle, but split the α-domain ($\alpha_{min} \leqslant \alpha \leqslant \alpha_{max}$) into M small intervals of length $\Delta\alpha$ and sum step-wise over all ε-intervals, the density distribution of which in eq. (8.5.16) corresponds to an α_i-value between α and $(\alpha + \Delta\alpha)$. In this way, we go through all M intervals of length $\Delta\alpha$. The number $\tilde{N}(\alpha, \varepsilon)$ of the ε-intervals in which a value between α and $(\alpha + \Delta\alpha)$ is adopted, has the form

$$\tilde{N}(\alpha, \varepsilon) \approx \rho(\alpha)\varepsilon^{-f(\alpha)}\Delta\alpha \tag{8.5.18}$$

where $\rho(\alpha) \neq 0$ and does not depend on ε. Equation (8.5.17) can then be re-written with eq. (8.5.16) as follows

$$\begin{aligned}\chi(q) \sim \textstyle\sum_{j=1}^{M} \tilde{N}(\alpha, \varepsilon)\, \varepsilon^{\alpha q} &\approx \textstyle\sum_{j=1}^{M} \rho(\alpha)\varepsilon^{-f(\alpha)}\Delta\alpha\, \varepsilon^{\alpha q} \\ &\approx \int_{\alpha_{min}}^{\alpha_{max}} \rho(\alpha)\varepsilon^{(\alpha q - f(\alpha))} d\alpha\end{aligned} \tag{8.5.19}$$

In accordance with eqs. (8.5.14), (8.5.17), we must proceed to the limit $\varepsilon \to 0$ in order to calculate the function

$$\tau(q) = (q-1)D_q = \lim_{\varepsilon\to 0} \frac{\ln \chi(q)}{\ln \varepsilon} \tag{8.5.20}$$

If we consider the integrand in eq. (8.5.19), we recognise that the function $\varepsilon^{\alpha q - f(\alpha)}$ forms a pronounced sharp maximum for decreasing ε-values, namely at a location with the value α^* which makes the exponent a minimum

$$\frac{d}{d\alpha}\big(\alpha q - f(\alpha)\big)\Big|_{\alpha^*} = 0 \tag{8.5.21}$$

where

$$\frac{d^2}{d\alpha^2}\big(\alpha q - f(\alpha)\big)\Big|_{\alpha^*} > 0 \tag{8.5.22}$$

applies. Equation (8.5.21) shows that the extremum α^* depends on q, i.e. $\alpha^* = \alpha(q)$, and we obtain the conditions

$$f'\big(\alpha(q)\big) = q \qquad \text{and} \qquad f''\big(\alpha(q)\big) < 0 \tag{8.5.23}$$

Integrals over a pronounced maximum peak can be evaluated under certain conditions with the aid of the saddle-point method (Schuster and Just, 2006; Sommerfeld, 1977).

This technique is applied if the integrand is very small throughout the integration interval, except in the neighbourhood of a pronounced maximum.

If, for example, for small ε-values, we consider the integral

$$I = \int_{\alpha_1}^{\alpha_2} \mathrm{e}^{-|\ln\varepsilon| F(\alpha)} \mathrm{d}\alpha \tag{8.5.24}$$

where we assume $F'(\alpha^*) = 0$, $F''(\alpha^*) > 0$ and $\alpha^* \in [\alpha_1, \alpha_2]$, then I is approximated by the saddle-point method as follows

$$\begin{aligned} I &\approx \int_{\alpha_1}^{\alpha_2} \exp\big(-|\ln\varepsilon|\big[F(\alpha^*) + \tfrac{1}{2}F''(\alpha^*)(\alpha-\alpha^*)^2\big]\big)\,\mathrm{d}\alpha \\ &\approx \mathrm{e}^{-|\ln\varepsilon|F(\alpha^*)} \int_{-\infty}^{+\infty} \exp\Big(\frac{-|\ln\varepsilon|F''(\alpha^*)}{2}(\alpha-\alpha^*)^2\Big)\,\mathrm{d}\alpha \\ &= \mathrm{e}^{-|\ln\varepsilon|F(\alpha^*)} \left[\frac{2\pi}{|\ln\varepsilon|F''(\alpha^*)}\right]^{1/2} \end{aligned} \tag{8.5.25}$$

i.e. I coincides approximately with the integrand at the point α^* but for a factor. Since the integrand tends very rapidly towards zero with increasing distance from α^*, the integration domain could be extended to infinity.

Applying the saddle-point method to the integral eq. (8.5.19), we obtain for small positive ε-values with $\alpha^* = \alpha(\mathrm{q})$ and a constant factor C approximately

$$\chi(\mathrm{q}) \approx \frac{C}{\sqrt{|\ln\varepsilon|}}\ \mathrm{e}^{\ln\varepsilon[\mathrm{q}\alpha(\mathrm{q}) - f(\alpha(\mathrm{q}))]}$$

and, after approaching the limit $\varepsilon \to 0$ in accordance with eq. (8.5.20),

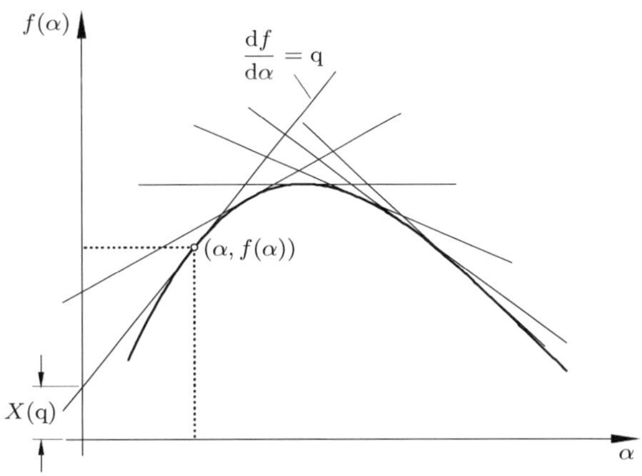

Fig. 8.5.9

Geometrical interpretation of the relation between $f(\alpha)$ and $\tau(\mathrm{q}) = -X(\mathrm{q})$

$$\tau(\mathrm{q}) = \mathrm{q}\alpha(\mathrm{q}) - f\big(\alpha(\mathrm{q})\big) \tag{8.5.26}$$

see (Halsey et al., 1986). Frequently, this is also written in the form

$$\tau(\mathrm{q}) = \min_{\alpha}\big(\mathrm{q}\alpha - f(\alpha)\big)$$

in order to express that the integral in eq. (8.5.19) is dominated by the α-value for which the exponent possesses a minimum. This is the connection we are looking for between the generalised dimension $D_{\mathrm{q}} = \tau(\mathrm{q})/(\mathrm{q}-1)$ and $f(\alpha)$ which can be interpreted geometrically particularly easily.

Figure 8.5.9 is a schematic representation of the function $f = f(\alpha)$ which can also be described as the envelope of the family of its tangents. Each point of the curve is determined by the pair of coordinates $\big(\alpha, f(\alpha)\big)$, the corresponding tangent by the intercept $X(\mathrm{q})$ on the ordinate as a function of the slope $\mathrm{q} = \mathrm{d}f/\mathrm{d}\alpha$, i.e. by the pair of coordinates $\big(\mathrm{q}, X(\mathrm{q})\big)$. If α and $f(\alpha)$ are given, the intercept X is derived from the relation

$$\frac{f(\alpha) - X(\mathrm{q})}{\alpha} = \mathrm{q}$$

In conjunction with eq. (8.5.26), we thus obtain

$$X(\mathrm{q}) = -\alpha\mathrm{q} + f(\alpha) = -\tau(\mathrm{q}) \tag{8.5.27}$$

In the literature, $-\tau(\mathrm{q})$ is denoted the Legendre transform of $f(\alpha)$; see, for example, (Callen, 1985). We summarise the duality between the Legendre transform and its inverse respectively between $\tau(\mathrm{q})$ and $f(\alpha)$:

$$\begin{array}{llll} & f = f(\alpha) & & \tau = \tau(\mathrm{q}) \\ & \dfrac{\mathrm{d}f(\alpha)}{\mathrm{d}\alpha} = \mathrm{q} \quad \Rightarrow \alpha(\mathrm{q}) & & \dfrac{\mathrm{d}\tau(\mathrm{q})}{\mathrm{d}\mathrm{q}} = \alpha \quad \Rightarrow \mathrm{q}(\alpha) \\ & \tau(\mathrm{q}) = \mathrm{q}\alpha(\mathrm{q}) - f\big(\alpha(\mathrm{q})\big) & & f(\alpha) = \alpha\mathrm{q}(\alpha) - \tau\big(\mathrm{q}(\alpha)\big) \\ \text{resp.} & \tau(\mathrm{q}) = \min_{\alpha}\big(\mathrm{q}\alpha - f(\alpha)\big) & & f(\alpha) = \min_{\mathrm{q}}\big(\alpha\mathrm{q} - \tau(\mathrm{q})\big) \end{array} \tag{8.5.28}$$

It is thus shown that $f(\alpha)$ and $\tau(\mathrm{q})$, and hence also the generalised dimension D_{q}, are two completely equivalent possibilities for characterising inhomogeneous fractal structures. Moreover, with the aid of $f(\alpha)$, we can also distinguish clearly between the transition from periodic (frequency-coupled) and quasi-periodic motion to chaos. Figure 8.5.10 presents the $f(\alpha)$-curves of the circle map for the quasi-periodic transition with the golden mean as the rotation number for $K = 1$ (curve I) and for the transition via a cascade of period doublings within an Arnold tongue with the rotation number $W = 8/13$ (curve II). For both routes, the extrema of $f(\alpha)$ as well as the α-range between $\alpha_{\min} = D_{\infty}$ and $\alpha_{\max} = D_{-\infty}$ are independent of the special form of the underlying map, i.e. $f(\alpha)$ is a universal function (Glazier and Libchaber, 1988).

As mentioned previously, the basic concepts underlying fractal and multifractal structures go back to Mandelbrot (Mandelbrot, 1982). Multifractal structures are by no means merely the result of abstract mapping functions like the circle map

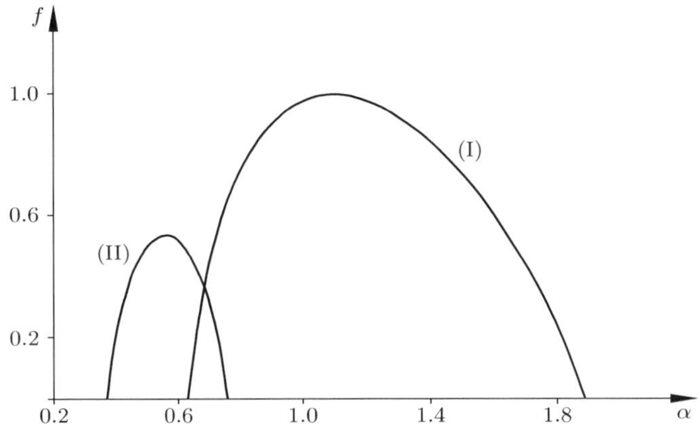

Fig. 8.5.10: $f(\alpha)$-curves for the quasi-periodic transition to chaos (curve I) and the transition via period doubling (curve II); after (Halsey et al., 1986)

for $K = 1$ or the asymmetric Cantor map discussed in section 5.5.3. Rather, we come across multifractal distributions in nature at every step. Examples are the spatial distribution of natural resources, of flora and fauna or the inhomogeneous distribution of dissipative structures in turbulent flows to which we return in Chapter 9, see also (Tél, 1988). However, there are not only *multifractal measures* but also *multifractal functions* which are well known from geological formations and rough surfaces or also from turbulent velocity fields and stock-exchange data. The question arises whether such functions can be characterised in a similar way by two indices.

In connection with turbulent flows, Frisch and Parisi transferred the scaling concept presented before to multifractal functions (Parisi and Frisch, 1985) by investigating the scaling of so-called *structure functions* $S_{\mathrm{n}}(r) \sim r^{\zeta(\mathrm{n})}$. According to eq. (3.9.49), the (longitudinal) structure function of n-th order is defined as the n-th moment of the velocity difference $v(r) = u(x+r) - u(x)$ of two points with distance r measured in longitudinal direction. In Chapter 10, the properties of scaling functions will be discussed in detail. Applying the Legendre transform to the scaling exponents $\zeta(\mathrm{n})$ and in analogy to the multifractal spectrum $f(\alpha)$, we obtain a singularity spectrum $D(h)$ which can be interpreted as the Hausdorff dimension of the subset of velocity increments which scale like $v(r) \sim r^h$.

Now we are interested in the geometrical meaning of the quantity h in this relation. In mathematics, isolated singularities are usually characterised by means of their *Hurst* or *Hölder exponents*. Although, for example, the function $f(x) = |x|^{1/2}$ is continuous at the point $x = 0$, it is no longer differentiable. By means of the Hölder exponents, the (integer) order of differentiability can be generalised to non-integer values. In the previous example, we obtain $h(0) = 1/2$. For a general definition, we assume that $f(x)$ is n times differentiable at x_0 while the (n+1)-th derivative does not exist. Therefore, the function $f(x)$ can be expanded into a Taylor series at $x = 0$ only up to n-th order, i.e. $f(x)$ can be approximated in the neighbourhood of x_0 by a polynomial $P_{\mathrm{n}}(x - x_0)$ and the residuum can be estimated as follows

$$|f(x) - P_{\mathrm{n}}(x - x_0)| \leqslant C|x - x_0|^h \quad \text{with } h \in (\mathrm{n}, \mathrm{n}+1) \tag{8.5.29}$$

The smallest possible exponent h fulfilling eq. (8.5.29) is denoted the Hölder exponent $h(x_0)$.

It is particularly expedient to determine the Hölder exponents of $f(x)$ by means of the wavelet transform introduced in section 3.8.5. Using the slightly modified definition

$$Wf(a,b) = \frac{1}{a} \int_{-\infty}^{+\infty} f(t)\, \psi\left(\frac{t-b}{a}\right) \mathrm{d}t \tag{8.5.30}$$

we can show that the wavelet transform of $f(x)$ at a singular point x_0 with the Hölder exponent $h(x_0) \in (\mathrm{n}, \mathrm{n}+1)$ can be expressed by a power law

$$Wf(a, x_0) \sim a^{h(x_0)} \quad \text{for } a \to 0 \tag{8.5.31}$$

The only requirement is that the basis wavelet $\psi(x)$ possesses at least n vanishing moments

$$\int_{-\infty}^{+\infty} x^{\mathrm{k}} \psi(x)\, \mathrm{d}x = 0 \quad \text{for } \mathrm{k} = 1, \ldots, \mathrm{n} \tag{8.5.32}$$

Hence, one can use, for example, the k-th derivatives of the Gaussian function $\psi_{\mathrm{k}(x)} = \mathrm{d}^{\mathrm{k}}/\mathrm{d}x^{\mathrm{k}} \exp(-x^2/2)$ as wavelets, like the Mexican hat wavelet for $\mathrm{k} = 2$, see fig. 3.8.6.

Colour Plate XI, p. 553, illustrates how isolated singularities can be localised and how their Hölder exponents can be determined by means of the wavelet transform. Part a) of the Colour Plate shows the function

$$f(x) = 1 + c_1\, \mathrm{e}^{-(\frac{x-x_1}{\sigma})^2/2} - c_2\, |x - x_2|^{1/2}.$$

for $x_1 = 0.2$, $x_2 = 0.8$, $c_1 = 0.2$, $c_2 = 0.6$, $\sigma = 0.04$, which is – apart from the singularity at x_2 – analytic, i.e. infinitely differentiable. In part b) beneath, the modulus of the wavelet transform $|Wf(a,b)|$ is plotted (red means large and blue small values), with the Mexican hat as the basis wavelet. In contrast to the smoothly decaying surface ridge on the left-hand side, the ridge of $|Wf(a,b)|$ on the right-hand side is sharp and points directly to the singularity x_2 of the function displayed above. Colour Plate XIc demonstrates the different scaling behaviour at x_1 and x_2. Plotting $\ln|Wf(a,b)|$ versus $\ln a$ along the line $b = x_2$ produces approximately a straight line for small a-values; its slope matches the Hölder exponent $h(x_2) = 1/2$ at x_2. In contrast, $|Wf(a, x_1)|$ scales at the regular point x_1 like a^{n} where n is the number of vanishing moments of the basis wavelet, in our case $\mathrm{n} = 2$. The more regular the point, the faster the wavelet transform decays to zero (Muzy et al., 1994).

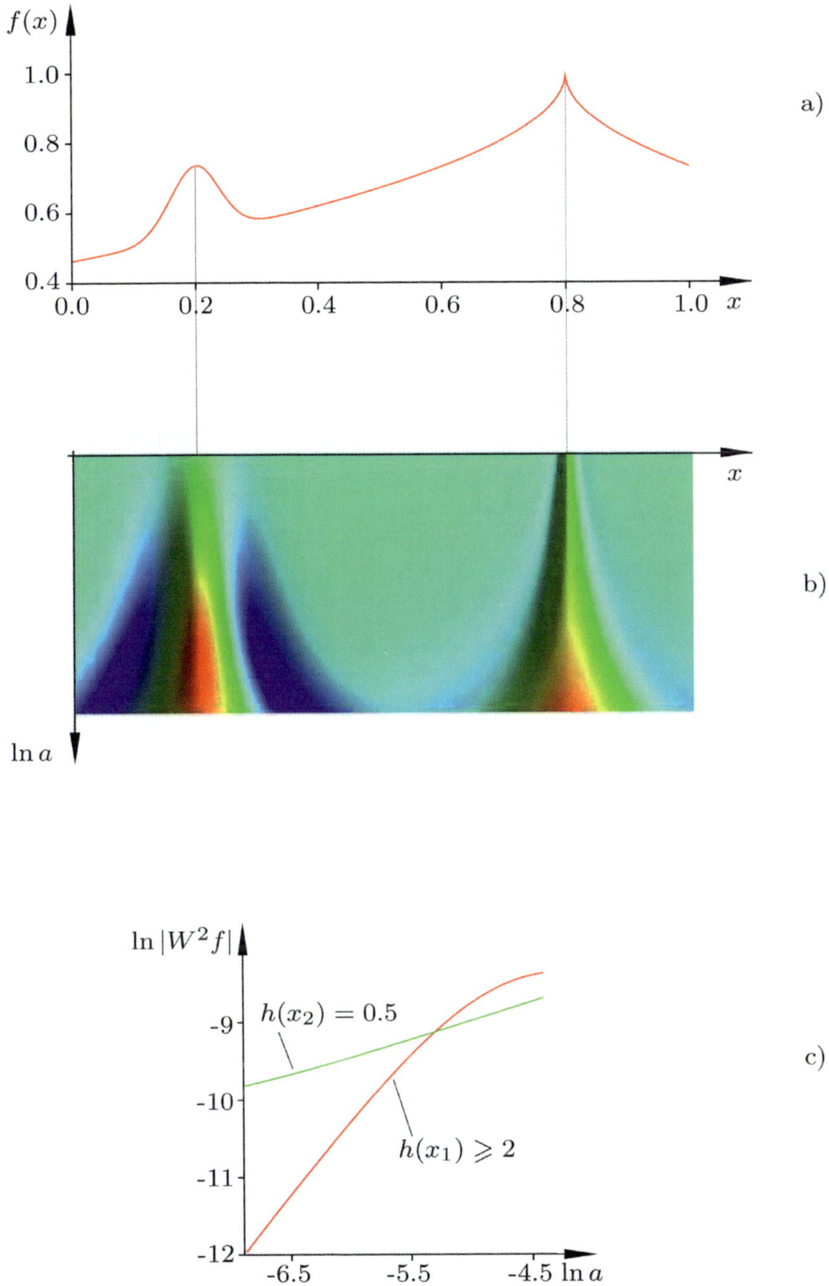

Plate XI Characterisation of isolated singularities by means of the wavelet transform:
a) the function $f(x) = 1 + c_1 \exp(-\frac{1}{2}(\frac{x-x_1}{\sigma})^2) - c_2\,|x - x_2|^{1/2}$,
b) modulus of the wavelet transform $|Wf(a,b)|$ for the Mexican hat,
c) determination of the Hölder exponents $h(x_1)$ and $h(x_2)$.

In Colour Plate XIIa,d, p. 555, the density distribution of the circle map is shown for the quasi-periodic transition at $K = 1$ and below, in part b), exactly correlated, the modulus of the wavelet transform, where we again used the Mexican hat. The bottom part c) shows the system of modulus maxima lines $\mathcal{MML}$ (Haase and Lehle, 1998). Red lines are maxima lines of $Wf(a,b)$ whereas blue lines are minima lines of $Wf(a,b)$. As with the classic method described in eqs. (8.5.17)ff. where the scaling $\chi(\mathrm{q},\varepsilon) \sim \varepsilon^{\tau(\mathrm{q})}$ is calculated by box-counting, the scaling of a partition function Z is analysed in the case of the wavelet transform along the modulus maxima lines

$$Z(\mathrm{q},a) = \sum_{i\in\mathcal{MML}} \left(\sup_{a'\leqslant a} |Wf(a',b_i)| \right)^q \sim a^{\tau(\mathrm{q})} \tag{8.5.33}$$

where the wavelets may be considered "oscillating" boxes (Muzy et al., 1994). Just as in the classic case, the singularity spectrum is again calculated via a Legendre transform of the exponents $\tau(\mathrm{q})$. Colour Plate XIIe shows the results for three different basis wavelets ψ_{k} together with the exact results given by (Halsey et al., 1986) which are marked with an asterisk. It can be clearly seen that $f(\alpha)$ does not depend on the choice of the basis wavelet as long as it possesses a sufficient number of vanishing moments. This technique is denoted *Wavelet Transform Modulus Maxima (WTMM) Method*.

Since the WTMM method is confined to integral values of the wavelet transform along the maxima lines, small errors are less relevant, i.e. this method is numerically very stable. A further advantage is that the method can be applied to both fractal distributions and fractal functions. While the classic procedure of (Parisi and Frisch, 1985) can, by definition, only cover Hölder exponents $h(x) \in (0,1)$ and, moreover, values $\mathrm{q} < 0$ lead to divergencies, the WTMM method allows us to determine the full singularity spectrum $f(\alpha)$. More details can be found in (Muzy et al., 1994).

8.5.3 Experimental Validation of the Quasi-Periodic Transition

In the 1980s, some experiments on the onset of turbulence in closed hydrodynamic systems attracted great interest. For example, in the case of Rayleigh-Bénard experiments carried out with high precision, the universal mathematical characteristics of simple maps such as the circle map could be confirmed for the quasi-periodic transition of hydrodynamic systems to chaos (Stavans et al., 1985; Jensen et al., 1985).

If we wish to study this transition to irregular behaviour for a fixed rotation number, e.g. the golden mean W_{G}, then it does not prove sufficient – basing our consideration on a quasi-periodic motion of the rolls of liquid – to increase a single control parameter such as the temperature difference, for example. A frequency locking would be the inevitable consequence, and thus a further periodic motion. The idea is rather to start with a periodic motion of the liquid rolls and then, for a fixed Rayleigh number, to impose a second incommensurable frequency from outside. In order to maintain a constant rotation number, e.g. $W = W_{\mathrm{G}}$, a tuning of two control parameters is necessary, as in the case of the circle map.

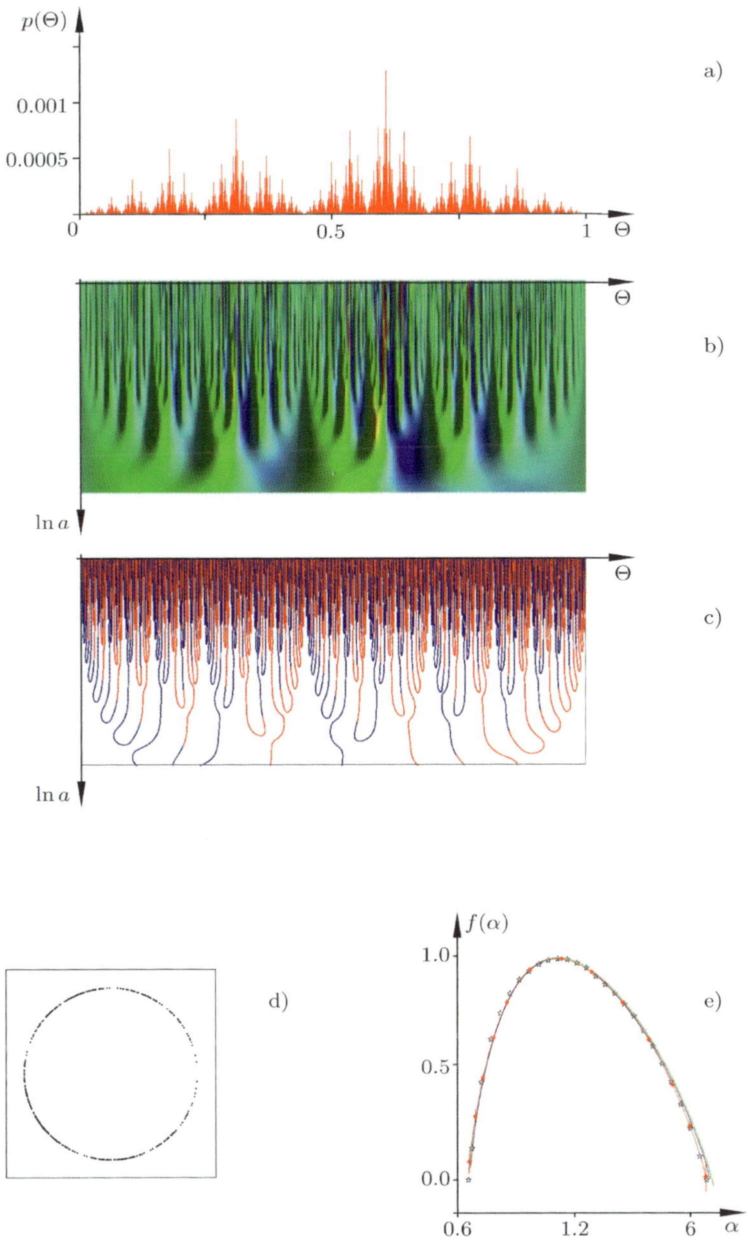

Plate XII Determination of the multifractal spectrum of the circle map by means of the *Wavelet Transform Modulus Maxima Method*:
a), d) density distribution of points on the critical attractor ($K = 1, W = W_G$),
b) modulus of the wavelet transform $|Wf(a,b)|$ for the Mexican hat,
c) modulus maxima lines,
e) multifractal spectra $f(\alpha)$ for three different basis wavelets ψ_k.

For the driven Rayleigh-Bénard experiment, a small brick-shaped cell (1,4 cm x 0,7 cm x 0,7 cm) filled with mercury was chosen in (Stavans et al., 1985; Jensen et al., 1985) because of the electrical conductivity of the mercury and because of its low Prandtl number. At a critical temperature difference ΔT_1 between the upper and the lower plates, the pure heat conduction is succeeded by convection. We observe two convection rolls, the axes of which are orthogonal to the longitudinal axis of the cell. In addition, the cell was subjected to a constant external magnetic field M parallel to the axes of the rolls. The magnetic field damps all motions orthogonal to the field lines and thus causes a stable alignment of the axes of the convection rolls parallel to the short side of the Bénard cell.

Following a further increase in the temperature difference, the constant roll motion becomes unstable at a second critical value ΔT_2. In the case of liquids with low Prandtl numbers, a Hopf bifurcation then occurs and thus a transversal oscillation with a frequency f_{i} which moves, wave-like, along the rolls (Busse, 1978).

A second independent frequency f_{ext} is then imposed by an electromagnetic process. To this end, an external AC electrical current sheet pulsating with a frequency f_{ext} was applied in the vertical plane between the rolls through the mercury. The imposed magnetic field now causes two effects: firstly, the motion of the electrically conductive liquid particles orthogonal to the lines of the magnetic field leads to a variation of the electric field and, secondly, a charged liquid particle crossing the magnetic field is subject to a Lorentz force. In the experimental set-up described, this gives rise to an additional vertical component of the vorticity field of the velocities, the component oscillating with the frequency f_{ext}. The amplitude A and the frequency f_{ext} of the alternating current serve as external control parameters in this experiment. An increase of the amplitude A causes a greater non-linear coupling of the two oscillations and thus corresponds to an increase of the parameter K in the circle map eq. (8.3.12); variations in the frequency f_{ext} of the alternating current correspond to variations in the parameter Ω. The transition from quasi-periodic to chaotic behaviour with a fixed rotation number W_{G} is controlled as shown in fig. 8.5.2. For each fixed value A of the intensity of the current, the frequency f_{ext} of the alternating current is adapted successively in accordance with eq. (8.3.46) so that, in each case, a periodic motion with the rotation number W_{n} is achieved until, after a sufficient number of steps, a quasi-periodic motion with the rotation number W_{G} sets in. With the aid of a semiconductor bolometer, the evolution of the temperature is measured at a fixed point of the cell. The ingenious set-up of the experiment and the sensitivity of the measuring devices allow a very precise determination of the rotation numbers with a maximum error of $2 \cdot 10^{-4}$. If we increase the non-linear coupling of the two oscillations, A finally reaches a critical value and the motion of the rolls becomes irregular.

The critical value A_{cr} can be determined most easily by means of power spectra. Figure 8.5.11 presents three power spectra for a fixed Rayleigh number $Ra = 4.09 Ra_{cr}$ (Stavans et al., 1985). The line spectrum below the critical amplitude (fig. 8.5.11a) contains the two basic frequencies f_{i} and f_{ext} as well as their linear combinations. In fig. 8.5.11b, the critical line has been reached approximately, recognisable by the large increase in combinations of frequencies. For $A > A_{cr}$ (fig. 8.5.11c), we observe a significantly higher noise level, i.e. there is a broad band of frequencies whereby the

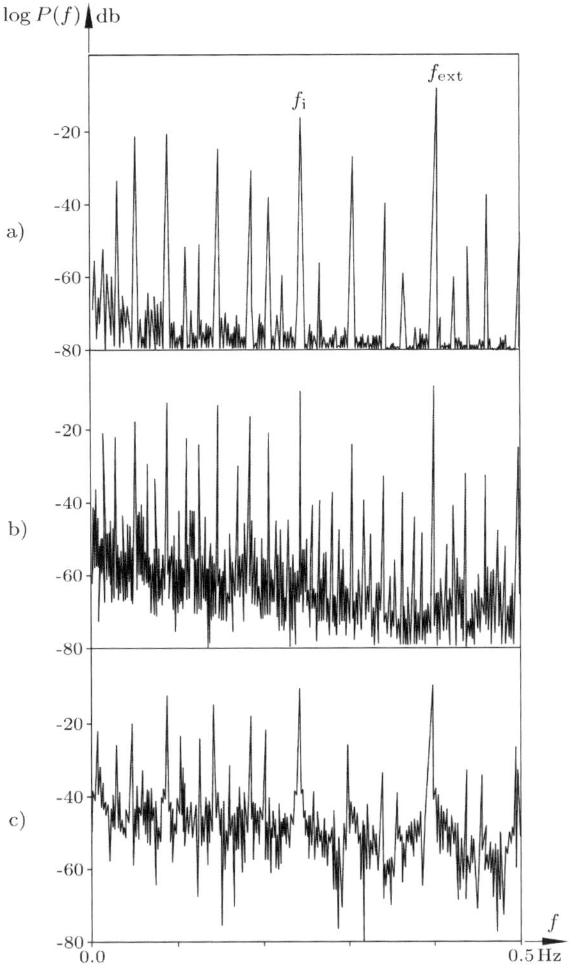

Fig. 8.5.11

Power spectra below (a), at (b) and above (c) the critical amplitude for quasi-periodic motions with the rotation number W_G; (Stavans et al., 1985)

higher-order combination frequencies are reduced considerably.

If, as in the case of the circle map, the scaled spectrum ($\log P(f)/f^2$) as recorded at the critical line is plotted versus $\log(f/f_{ext})$, then, in accordance with fig. 8.5.5, the scaling characteristics of the spectrum at the transition to chaos become manifest. Figure 8.5.12 presents the re-scaled spectrum obtained from fig. 8.5.11b where, for clarity's sake, not all the frequencies are shown (Stavans et al., 1985). As in the case of the circle map, peak values which are approximately equally high occur at the equidistant points $W_G^1, W_G^2, W_G^3, \ldots$ (marked by the number 1). The lines of the spectrum in between, marked by the numbers 2 to 6, correspond to further linear combinations of the incommensurable normalised basic frequencies W_G and 1

$$W = |\mathrm{m}W_G - \mathrm{n} \cdot 1| \qquad \mathrm{m}, \mathrm{n} \ldots \text{ Fibonacci numbers}$$

cf. eq. (8.5.13).

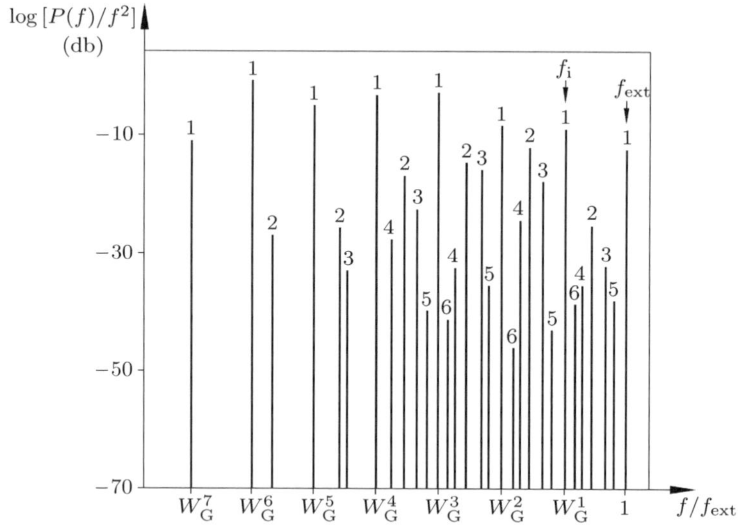

Fig. 8.5.12: Scaling characteristics of the scaled power spectrum at the transition to chaos for quasi-periodic motions with the rotation number W_G (Stavans et al., 1985)

In section 8.3.3, we studied local and global scaling laws for the circle map. In dependence of a fixed rotation number, we could, as for Feigenbaum's cascade of period doublings, determine two universal constants a and δ for the quasi-periodic route to chaos; see eqs. (8.3.60) and (8.3.61). For the driven Rayleigh-Bénard experiment, it was possible to verify experimentally the constant δ defined in eq. (8.3.55) at the critical line for two rotation numbers, the golden and the silver mean.

However, in the experiment, it is difficult to measure the distances of successive Ω_n-values. On the other hand, we can determine the width of successive Arnold tongues with a high degree of precision. It is possible to localise stability domains corresponding to rotation numbers with a denominator > 200. Fortunately, at the transition to chaos, the widths $\Delta\Omega(W_n)$ of the Arnold tongues obey the same scaling law (Glazier and Libchaber, 1988), i.e. we have

$$|\delta| = \lim_{n\to\infty} \frac{\Delta\Omega(W_{n-1}) - \Delta\Omega(W_n)}{\Delta\Omega(W_n) - \Delta\Omega(W_{n+1})} \tag{8.5.34}$$

Figure 8.5.13 presents the experimentally determined Arnold tongues for the Rayleigh-Bénard experiment (Stavans et al., 1985). Both the fractal dimension D of the Cantor set of all Ω-values appertaining to quasi-periodic motions (see section 8.3.3.2) as well as δ could be determined from the width of the tongues. For the golden mean as the rotation number, the following values were obtained

$$D = 0.86 \pm 3\%$$

$$\delta = 2.8 \pm 10\% \tag{8.5.35}$$

In eq. (8.3.89), we stated how the width of the Arnold tongue corresponding to a rotation number $W = p/q$ varies in dependence of q: for $q \gg 1$, we have approximately

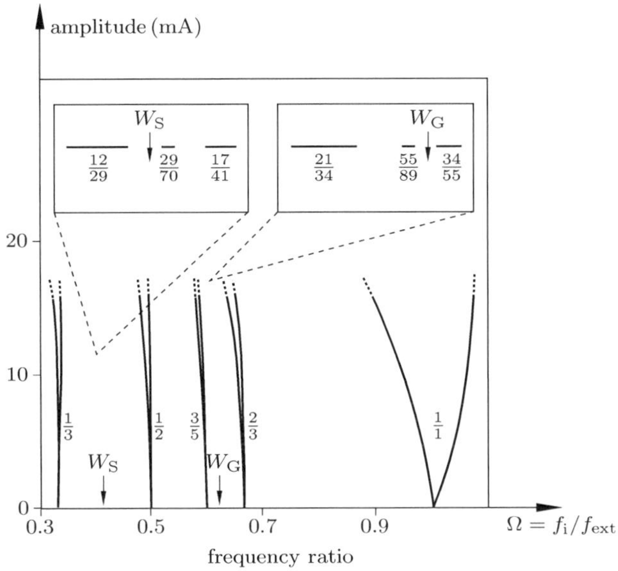

Fig. 8.5.13: Stability domains for the driven Rayleigh-Bénard experiment are shown in the diagram of the external current amplitude A as a function of the frequency ratio f_i/f_{ext} (Stavans et al., 1985)

$$\Delta\Omega\left(\frac{p}{q}\right) \approx \frac{1}{q^{\gamma}}$$

If we adopt eq. (8.5.34) as the definition for δ, a straightforward calculation yields the exponent γ for an approximation with the rotation numbers $W_n = F_n/F_{n+1}$ to the golden mean W_G

$$|\delta| = \lim_{n\to\infty} \frac{F_n^{-\gamma} - F_{n+1}^{-\gamma}}{F_{n+1}^{-\gamma} - F_{n+2}^{-\gamma}} = \lim_{n\to\infty} \frac{\left(\frac{F_n+1}{F_n}\right)^{\gamma} - 1}{1 - \left(\frac{F_{n+1}}{F_{n+2}}\right)^{\gamma}} = \frac{W_G^{-\gamma} - 1}{1 - W_G^{\gamma}} = W_G^{-\gamma} \tag{8.5.36}$$

respectively

$$\gamma = -\frac{\ln|\delta|}{\ln W_G} \tag{8.5.37}$$

From the experimental result eq. (8.5.35), we thus obtain $\gamma \approx 2.14$. The theoretical value was given in (Shenker, 1982) as

$$\gamma = 2.16443 \pm 2 \cdot 10^{-5} \tag{8.5.38}$$

The experimental results demonstrate that the circle map is excellently suited as a mathematical model simulating the onset of turbulent behaviour in purely temporal processes with a low number of degrees of freedom as arising in the driven Rayleigh-Bénard experiment and that the scaling laws which hold for the circle map at the quasi-periodic transition do indeed have universal character.

Fig. 8.5.14

Critical attractor determined experimentally in the Rayleigh-Bénard experiment (Jensen et al., 1985)

A further confirmation of this hypothesis is obtained by an investigation of the multifractal structure of the Poincaré sections at the transition to chaos. Figure 8.5.14 presents an experimentally determined Poincaré section of the critical attractor, also for the quasi-periodic transition with a fixed rotation number W_G (Jensen et al., 1985). In analogy to the circle map, we can clearly recognise the density variations in the distribution of the points, (cf. fig. 8.5.6b). The disintegration of the torus is accompanied by the emergence of folds as they are known to us from the dissipative circle map. If, as described in the previous section, we determine the multifractal spectrum $f(\alpha)$ from this set of points, we recognise a surprising agreement with the theoretical prognoses (figs. 8.5.7 and 8.5.10). In fig. 8.5.15, we have

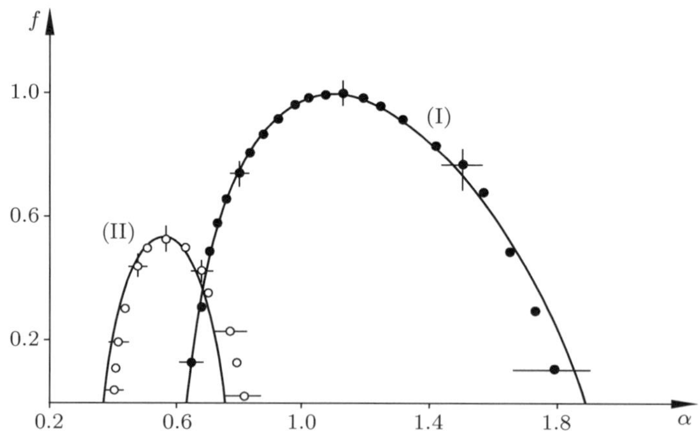

Fig. 8.5.15: Multifractal spectra $f(\alpha)$
(I) for the quasi-periodic transition at $W_G = (\sqrt{5} - 1)/2$ and
(II) for the transition within an Arnold tongue for $W_G = 8/13$.
Experimental data from driven Rayleigh-Bénard experiments are marked by dots $\bullet$ and $\circ$; cf. (Libchaber, 1987)

plotted the $f(\alpha)$-curves for the quasi-periodic transition with the rotation number W_G and forthe transition via period doublings in the Arnold tongue corresponding to the rotation number $W = 8/13$. The solid lines correspond to the theoretical results for the circle map; dots mark the experimental results and short bars the error bounds.

Summarising, we can conclude from the results that the circle map and the Rayleigh-Bénard experiment described above belong to the same universality class. Corresponding transitions from quasi-periodic to chaotic behaviour were also observed in a series of further experiments in the field of solid-state physics. An extensive overview can be found in (Jensen et al., 1984).

8.6 The Route to Chaos via Intermittency

We conclude our account of the various routes to chaos based on local bifurcations with the intermittency scenario, the theoretical background of which was studied intensively by Y. Pomeau and P. Manneville at the beginning of the 1980s (Pomeau and Manneville, 1980). In physics, there are various definitions of the concept of intermittency. In the context of chaos theory, a signal is denoted intermittent if its temporal evolution appears regular (laminar) over long periods of time, yet is interrupted now and again by short irregular intervals with amplitudes of higher intensity. A variation of an external control parameter is followed by a more frequent occurrence of irregular bursts. In this connection, intermittency represents a smooth transition from regular periodic to chaotic behaviour.

Fig. 8.6.1: Boundary layer flow (Fernholz, 1964)

In hydrodynamics, a phenomenon which can frequently be observed and generally has a spatio-temporal character is designated intermittency. If, for example, a flow with a high Reynolds number is perturbed by inserting a plate parallel to the flow direction, then, in a certain region, a turbulent boundary layer well-defined in its temporal mean behaviour emerges while, at irregular time intervals, however, separation of the boundary layer occurs at different locations (see fig. 8.6.1). The fluctuations of the velocity field in the transition domain between the boundary layer and the unperturbed flow then possess intermittent character, i.e. long laminar phases in space and time are interrupted randomly by short turbulent phases. A further meaning of intermittent phenomena in turbulent flow in the context of the statistics of velocity increments will be presented in section 9.6.5, see also (Peinke et al., 2004).

The mathematical models describing the transition from regular to chaotic motions which we present in the following apply to purely temporal processes, however. The inclusion of spatial dependencies leads to considerably more complex systems.

8.6.1 Intermittency in the Logistic Map

One of the possible mechanisms leading to intermittency can be illustrated particularly easily with the example of the logistic map

$$x \longmapsto f(x) = \alpha x(1-x) \tag{8.6.1}$$

if we examine the question of the emergence of the periodic windows within the chaotic range $\alpha_\infty < \alpha < 4$ (see fig. 6.7.6). We investigate the simplest case of the emergence of a stable cycle of period 3, characteristic of the largest window in the bifurcation diagram of fig. 6.7.6. The existence of cycles of all other periods follows from the existence of a cycle of period 3 in accordance with the Sharkowskii theorem, cf. section 6.7.2.

In order to determine the period three orbit, we must determine the fixed points of the threefold iterated function $f^3(x)$, i.e. the points of intersection of the eighth-order polynomial $y = f^3(x)$ with the angle bisector of the first quadrant $y = x$. Two of the eight zeroes of $P_8(x)$

$$P_8(x) \equiv f^3(x) - x = 0 \tag{8.6.2}$$

coincide with the two (unstable) fixed points $x = 0$ and $x = 1 - 1/\alpha$ of the logistic map $f(x)$. The other six zeroes can, in dependence of the α-value, either be three pairs of complex conjugate solutions (see fig. 8.6.2a) or three pairs of real zeroes (fig. 8.6.2c) corresponding to two cycles, one stable (x_{s1}, x_{s2}, x_{s3}) and one unstable (x_{u1}, x_{u2}, x_{u3}), of period three. At $\alpha = \alpha_{cr}$, a saddle-node bifurcation is hence generated, i.e. at $\alpha = \alpha_{cr}$, an orbit of period three emerges (fig. 8.6.2b) which splits into a stable and an unstable cycle of period three for values $\alpha > \alpha_{cr}$. We can determine the critical value α_{cr} at which the cycle of period three is generated as well as the newly emerging real fixed points x_1, x_2, x_3 of $f^3(x)$ from the three equations

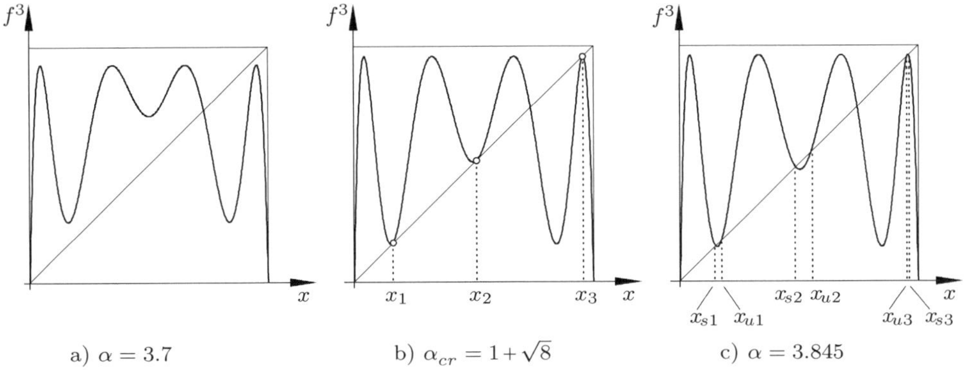

Fig. 8.6.2: Emergence of the period three window due to a saddle-node bifurcation

$$\begin{aligned} x_2 &= f(x_1) = \alpha_{cr} x_1(1-x_1) \\ x_3 &= f(x_2) = \alpha_{cr} x_2(1-x_2) \\ x_1 &= f(x_3) = \alpha_{cr} x_3(1-x_3) \end{aligned} \tag{8.6.3}$$

as well as the condition

$$\frac{\mathrm{d}}{\mathrm{d}x} f^3(x) = f'(x_1)f'(x_2)f'(x_3) = 1 \tag{8.6.4}$$

or

$$(1-2x_1)(1-2x_2)(1-2x_3) = 1/\alpha_{cr}^3$$

In order to facilitate the somewhat laborious calculation of α_{cr}, it proves advantageous to establish the following relations by addition and multiplication of the above expressions

$$\begin{aligned} Q \equiv x_1x_2 + x_2x_3 + x_3x_1 &= \tfrac{7}{4}A + \tfrac{3}{2}S \\ R \equiv x_1x_2x_3 &= \tfrac{3}{4}A + \tfrac{1}{2}S \end{aligned} \tag{8.6.5}$$

where we have introduced the abbreviations

$$\begin{aligned} S &\equiv x_1 + x_2 + x_3 \\ A &\equiv \alpha_{cr}^{-3} - 1 \end{aligned} \tag{8.6.6}$$

In the following, we dispense with the index $_{cr}$ for α. Introducing the ansatz

$$P_8(x) = f^3(x) - x = Cx[x - (1-\alpha^{-1})](x-x_1)^2(x-x_2)^2(x-x_3)^2 \tag{8.6.7}$$

for the fixed point condition eq. (8.6.2), we obtain together with eqs. (8.6.5) and (8.6.6)

$$P_8(x) = Cx[x^7 + a_6x^6 + \ldots + a_1x + a_0] \tag{8.6.8}$$

where

$$a_6 = \alpha^{-1} - 1 - 2S \qquad \text{and} \qquad a_0 = (\alpha^{-1} - 1)R^2$$

holds. On the other hand, we can express $f^3(x) - x$ explicitly in the form

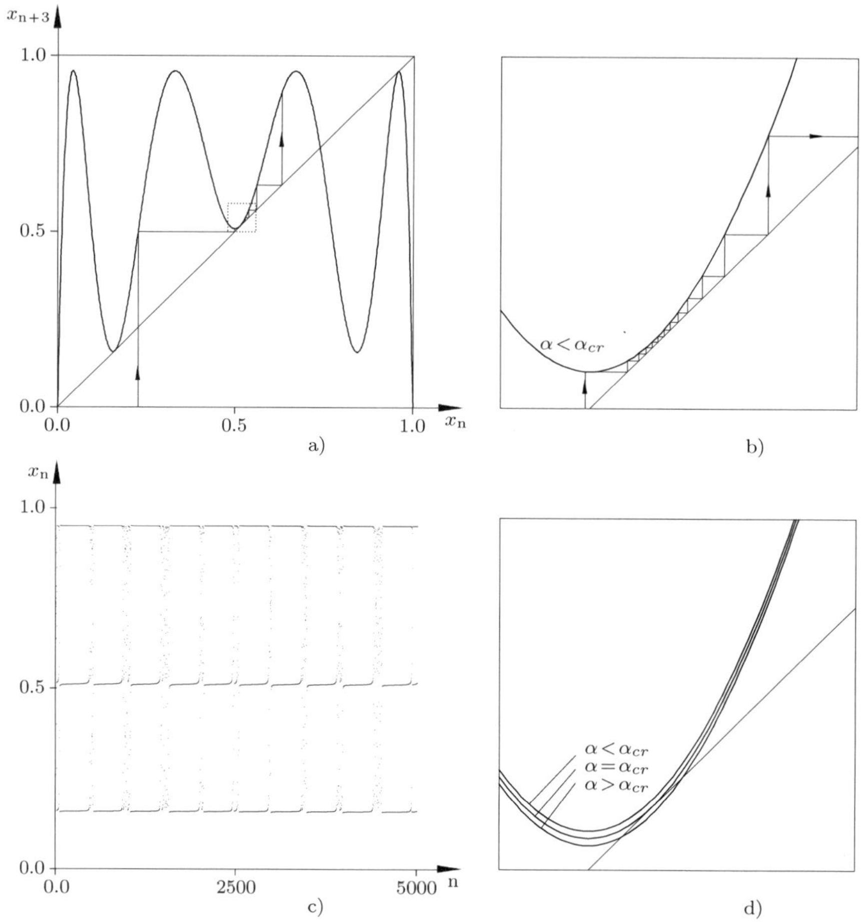

Fig. 8.6.3: Intermittency of the logistic map $f(x)$ for $\alpha = 3.82842 \lesssim \alpha_{cr}$:
a) the threefold iterated map $x_{n+3} = f^3(x_n)$
b) enlarged section of a)
c) three regular ranges within a chaotic environment
d) behaviour of $f^3(x)$ in the neighbourhood of α_{cr}

$$f^3(x) - x = \alpha^3 x[-\alpha^{-4}x^7 + 4\alpha^4 x^6 + \ldots + (1 - \alpha^{-3})] \tag{8.6.9}$$

Comparing coefficients of the terms in x^8, x^7 and x, we obtain the relations

$$C = -\alpha^7$$
$$C(\alpha^{-1} - 1 - 2S) = 4\alpha^7$$
$$C(\alpha^{-1} - 1)R^2 = \alpha^3 - 1$$

and thus, applying eq. (8.6.5), we find an equation for the critical value $\alpha = \alpha_{cr}$

$$\alpha^4 - 10\alpha^2 - 16\alpha - 7 = 0 \tag{8.6.10}$$

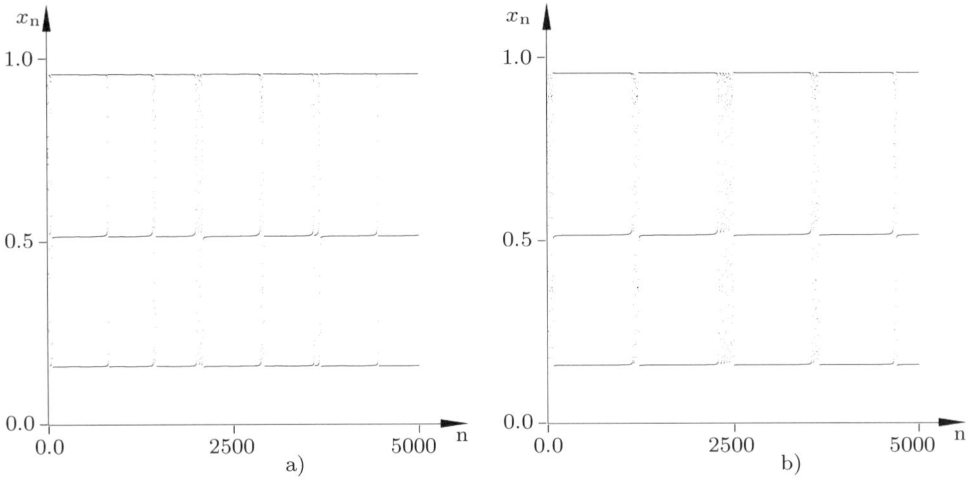

Fig. 8.6.4: Extension of the regular sections when approaching the saddle-node bifurcation of the logistic map for a) $\alpha = 3.828425$, b) $\alpha = 3.828426$

Here, we are interested only in solutions within the parameter range $0 \leqslant \alpha \leqslant 4$. According to the Descartes rule, one positive root of eq. (8.6.10) exists at most since only one change of sign occurs in the coefficients. We easily ascertain that this positive solution is

$$\alpha_{cr} = 1 + \sqrt{8} \approx 3.8284271 \tag{8.6.11}$$

In fig. 8.6.3a, we have plotted the graph of the function $f^3(x)$ for $\alpha = 3.8278 \lesssim \alpha_{cr}$, i.e. for an α-value just below the critical value at which the saddle-node bifurcation occurs. Figure 8.6.3b shows an enlarged section. For $\alpha > \alpha_{cr}$, two fixed points exist in this region, a stable and an unstable one (fig. 8.6.3d). They both collapse for $\alpha = \alpha_{cr}$. If an iteration x_n for $\alpha \lesssim \alpha_{cr}$ approaches the tunnel between the bisector and the function $f^3(x)$, its value hardly alters through many iterations when passing the bottleneck, i.e. the sequence of x_n-values exhibits almost periodic, laminar behaviour (fig. 8.6.3c). The narrower the tunnel, the longer the regular phases, see fig. 8.6.4. If the tunnel has been passed, however, we again observe uncontrolled jumps, i.e. chaotic behaviour, until the trajectory is again captured by a tunnel zone. This corresponds exactly to intermittent behaviour (see also fig. 8.6.5).

The influence of the computational precision on the numerical results was investigated in (Mayer-Kress and Haken, 1981). It was found that numerical simulations near a saddle-node bifurcation can feign a periodic behaviour although the control parameter is still in a range in which the system behaviour is irregular.

At this point, we make a rough estimate of the maximum lengths L of the laminar sections in dependence of the (normalised) distance from the bifurcation point $\mu = (\alpha_{cr} - \alpha)/\alpha_{cr}$ by simply extracting L from the four diagrams. We obtain the following table:

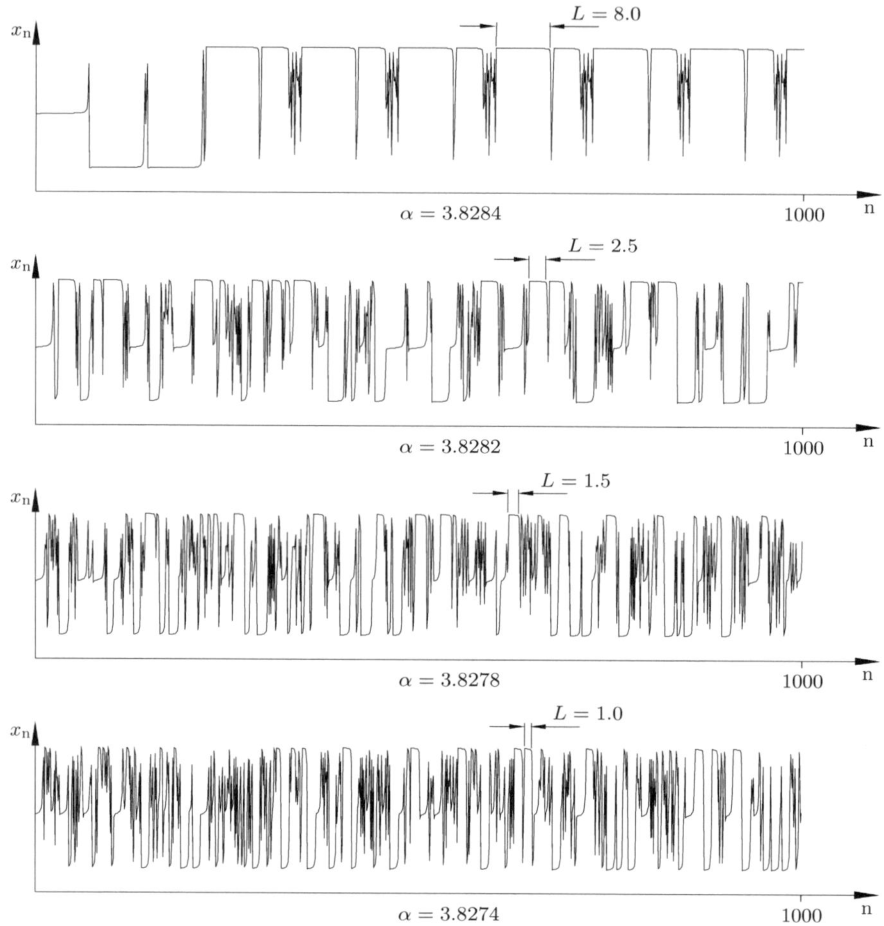

Fig. 8.6.5: Sequence of points x_n for decreasing α at the transition to chaos for the threefold iterated logistic map $x \mapsto f^3(x)$

α	μ	L(mm)
3.8284	$7.085 \cdot 10^{-6}$	8.0
3.8282	$5.933 \cdot 10^{-5}$	2.5
3.8278	$1.638 \cdot 10^{-4}$	1.5
3.8274	$2.683 \cdot 10^{-4}$	1.0

In fig. 8.6.6, L is plotted versus μ on a double logarithmic scale. The four points lie almost on a straight line, the slope of which is approximately $-1/2$. In section 8.6.3, we shall see that this law can also be confirmed theoretically.

The emergence of the window of period three within a chaotic regime is caused by a saddle-node bifurcation. The transition from chaotic to regular behaviour occurs continuously via intermittency.

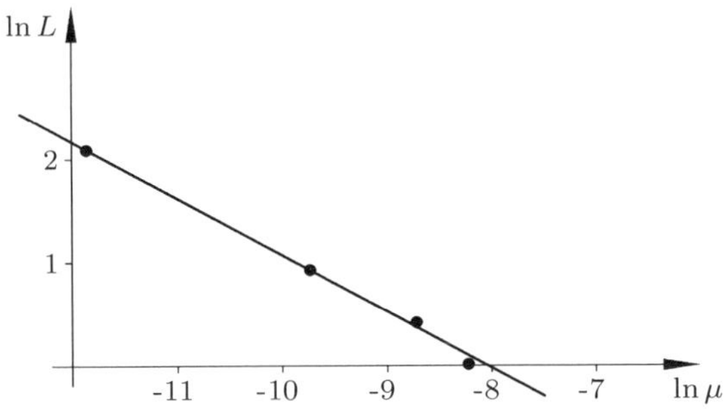

Fig. 8.6.6

Rough estimate of the maximum length L of the laminar sections in dependence of μ

8.6.2 Classification of Intermittency

The emergence of intermittent behaviour is always associated with the loss of stability of a periodic motion. In section 5.4.2, we discussed the Floquet theory which renders a linear stability analysis of limit cycles possible. The possibly complex eigenvalues λ_i of the monodromy matrix, the Floquet multipliers, inform us of the stability behaviour of the linearised system (cf. fig. 5.4.5). If $|\lambda_i| < 1$ holds for all eigenvalues, the periodic motion is stable. If, on the other hand, an eigenvalue crosses the unit circle in the complex plane, the limit cycle loses its stability. In fig. 5.4.6, we plotted the three possibilities for the loss of stability of a periodic orbit. According to whether an eigenvalue crosses the circle at $+1$ or at -1 or whether a pair of complex conjugate eigenvalues crosses the unit circle, a different type of bifurcation takes place. If we consider the Poincaré map instead of the continuous system, we can trace the bifurcation of a limit cycle to the bifurcation of fixed points of iterated maps as demonstrated in section 6.6. If $\lambda = +1$, then a saddle-node bifurcation, a transcritical bifurcation or a pitchfork bifurcation can occur; for $\lambda = -1$, a period doubling sets in whereas for $\lambda = \alpha \pm i\beta$ ($|\lambda| = 1$), a Neĭmark-Sacker bifurcation takes place.

Let us begin with the case of λ crossing the unit circle at $+1$. Figure 8.6.7 presents the three possible bifurcation types as well as the corresponding iteration evolutions shortly prior to ($\mu < \mu_{cr}$), precisely at the bifurcation point ($\mu = \mu_{cr}$) respectively shortly afterwards ($\mu > \mu_{cr}$). Evidently, intermittent behaviour can only ensue in the case of a saddle-node bifurcation. Only in this case, for $\mu > \mu_{cr}$, is a narrow tunnel formed, through which the trajectories pass, subsequently to be catapulted out again into the phase space; this leads to a chaotic burst after a laminar phase. Moreover, figure 8.6.7 shows that the saddle-node bifurcation must be subcritical, cf. also fig. 6.6.3b. Shortly after the bifurcation, intermittent behaviour sets in.

In accordance with the path the eigenvalues take when crossing the unit circle, we classify intermittency in three groups (fig. 8.6.8): if λ crosses the circle at $+1$, we speak of type I intermittency, if a pair of complex conjugate eigenvalues crosses the

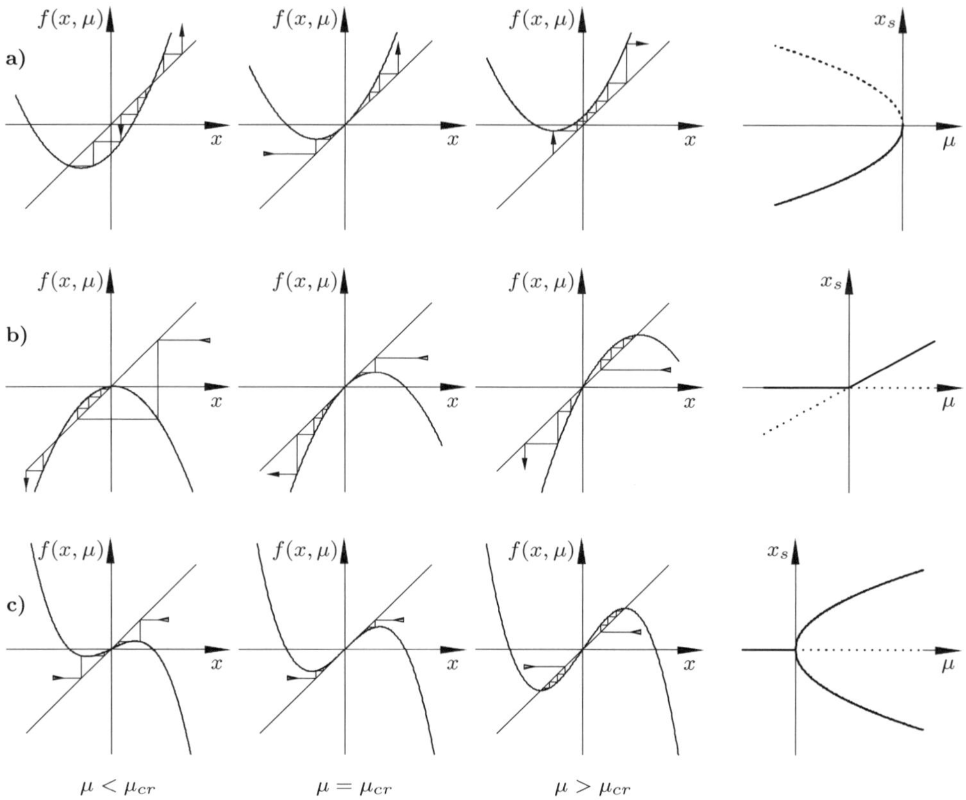

Fig. 8.6.7: Bifurcations for $\lambda = +1$: a) saddle-node bifurcation, b) transcritical bifurcation, c) pitchfork bifurcation

unit circle, of type II intermittency and, for $\lambda = -1$, of type III intermittency. The corresponding bifurcations must be subcritical since the influence of the non-linear terms after the bifurcation must lead to an increase of the perturbation, i.e. to a departure from the originally stable periodic behaviour.

The investigation of intermittency always proceeds in two steps: in the first, the *local* step, at the destabilisation of the periodic motion, we consider local bifurcations and, on the basis of the eigenvalues of the monodromy matrix, we can assign a classification to one of the three intermittency classes. Although intermittency arises due to subcritical bifurcations of a limit cycle, the knowledge of the phase space in the neighbourhood of the attractor which is becoming unstable is not sufficient. After the laminar phase, a chaotic burst takes place, cf. figs. 8.6.3 to 8.6.5, i.e. the trajectory moves away from the limit cycle and explores, as we shall see in the following section, large regions of the phase space in an unpredictable way. In the second step, we must therefore consider the relaminarisation, where the renewed capture of the trajectory

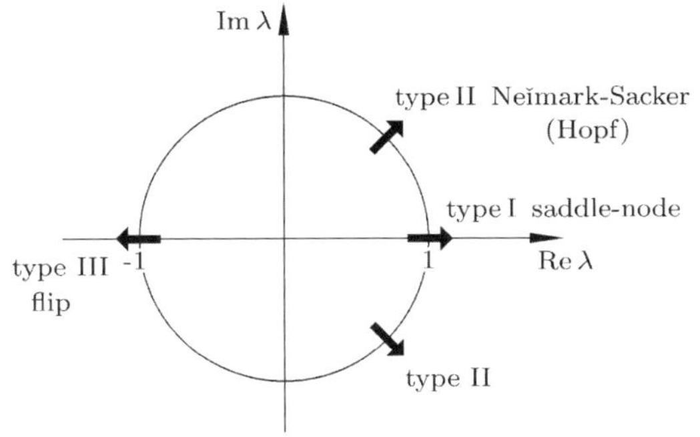

Fig. 8.6.8

Classification of intermittency on the basis of the eigenvalues of the monodromy matrix

depends on the global structure of the phase space. In this sense, both local and global characteristics of the dynamical system play their part in intermittency.

8.6.3 Type I Intermittency

In the previous section, we defined type I intermittency as a phenomenon in the transition from periodic to chaotic behaviour which is generated by a saddle-node bifurcation at which an eigenvalue λ of the monodromy matrix takes the value $+1$. The emergence of the window of period three in the logistic map discussed in section 8.6.1 can be considered the simplest model for the occurrence of type I intermittency. What, however, is the connection between the simple recursion formula eq. (8.6.1) and a realistic dynamical system having an n-dimensional phase space?

If we consider an n-dimensional dissipative system, a stable fixed point corresponds to the originally stable periodic motion in the Poincaré map. The monodromy matrix coincides with the Jacobian matrix of the (n−1)-dimensional map at this fixed point; the moduli of the eigenvalues are all < 1 and the appertaining (generalised) eigenvectors span the space $\mathbb{E}^s = \mathbb{R}^{n-1}$. We now assume that one eigenvalue becomes $+1$ following a variation of a control parameter r at a critical point $r = r_{cr}$. At the bifurcation point, a one-dimensional centre manifold $\mathbb{W}^c$ emerges tangent to the corresponding eigenvector. In sections 6.2 to 6.4, we showed that, in this case, the long-term behaviour of the high-dimensional system in the neighbourhood of the fixed point can be reduced to the dynamics on the centre manifold. If we introduce the coordinate x in $\mathbb{W}^c$, the saddle-node bifurcation can be reduced to the normal form (see section 6.6), possibly after an appropriate coordinate transformation

$$x \longmapsto f(x,\mu) = x + \mu + x^2 \tag{8.6.12}$$

where we put $\mu = (r - r_{cr})/r_{cr}$.

In fig. 8.6.9a,b, we have plotted the bifurcation diagram as well as the mapping function, eq. (8.6.12), in the neighbourhood of the bifurcation point; fig. 8.6.9c presents a series of iterations when passing through the narrow tunnel for $\mu > 0$ (cf. fig. 8.6.3b for the logistic map). We can clearly recognise that the system remains near a fictional fixed point for many iterations for $\mu \gtrsim 0$, i.e. the corresponding

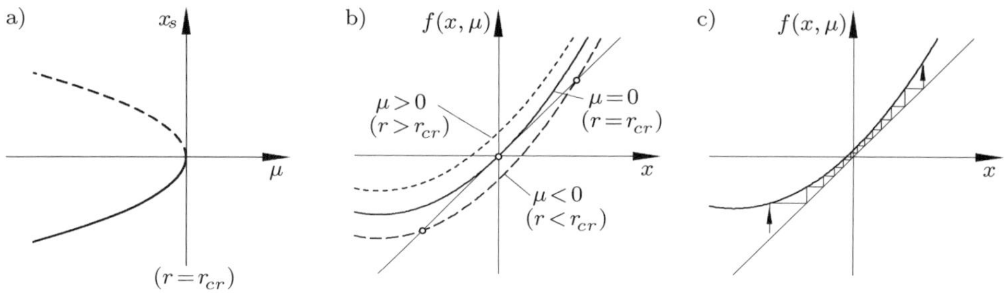

Fig. 8.6.9: Saddle-node bifurcation: a) bifurcation diagram, b) mapping function near the bifurcation point, c) laminar phase for type I intermittency

continuous system remains captured for a long time in the neighbourhood of the original limit cycle ($\mu = 0$) which vanishes in the complex domain for $\mu > 0$. During a period of time T, we thus observe almost periodic behaviour. Once the tunnel zone has been passed, the trajectory escapes into the phase space; this leads to a further chaotic burst. After a certain time interval, the trajectory is captured in the corridor once more, this again introducing a laminar phase. While it is not possible to make general statements about the distribution of the laminar phases since it depends on the global structure of the phase space, we can estimate the mean duration $\langle\tau\rangle$ of the laminar phases in dependence of μ (Pomeau and Manneville, 1980).

To this end, we write the mapping function eq. (8.6.12) in the form of a recursion formula

$$x_{\mathrm{n}+1} = x_{\mathrm{n}} + \mu + x_{\mathrm{n}}^2 \tag{8.6.13}$$

During the laminar phase, the trajectory is in the narrow corridor, see fig. 8.6.9c, which is at its narrowest at $x = 0$. In this region, the x-values vary during one iteration only very slightly, i.e. as long as $|x_{\mathrm{n}}| < C$ holds with a fixed bound $C \ll 1$, we can replace the difference quotient

$$\frac{x_{\mathrm{n}+1} - x_{\mathrm{n}}}{(\mathrm{n}+1) - \mathrm{n}}$$

approximately by the differential quotient $\mathrm{d}x_{\mathrm{n}}/\mathrm{dn}$; eq. (8.6.13) is then transformed into the following differential equation

$$\frac{\mathrm{d}x_{\mathrm{n}}}{\mathrm{dn}} = \mu + x_{\mathrm{n}}^2 \tag{8.6.14}$$

which can be integrated easily

$$\int_{\mathrm{n}_1}^{\mathrm{n}_2} \mathrm{dn} = \int_{x_b}^{x_e} \frac{\mathrm{d}x_{\mathrm{n}}}{\mu + x_{\mathrm{n}}^2}$$

Applying the inverse tangent function, we obtain

$$\mathrm{n}_2 - \mathrm{n}_1 = \frac{1}{\sqrt{\mu}}\left[\tan^{-1}\frac{x_e}{\sqrt{\mu}} - \tan^{-1}\frac{x_b}{\sqrt{\mu}}\right] \tag{8.6.15}$$

If we assume that the absolute x-values at the beginning and at the end of the corridor are of the order of magnitude C, the expression in brackets in eq. (8.6.15) tends towards the value π for $\mu \to 0$ respectively $r \to r_{cr}$. For the number n of iterations required to pass the tunnel, we obtain the following proportionality relation

$$\mathrm{n}_2 - \mathrm{n}_1 \sim \frac{1}{\sqrt{\mu}}$$

Certainly, the actual duration of the laminar phase depends on the point x_b at which the trajectory enters the corridor; since, however, the inverse tangent function tends towards $\pi/2$ for large arguments, we can also estimate the mean duration $\langle\tau\rangle$ of the laminar phase by the power law

$$\langle\tau\rangle \sim \left[\frac{r - r_{cr}}{r_{cr}}\right]^{-1/2} = \mu^{-1/2} \tag{8.6.16}$$

i.e. the laminar sections become shorter for increasing μ-values and we have a continuous transition from periodic to chaotic behaviour.

In fig. 8.6.10, we have plotted qualitatively the probability distribution $p(\tau, \mu)$ of the duration τ of the laminar phases which can be measured with considerably more ease than the scaling law eq. (8.6.16). The distribution function is characteristic of type I intermittency, but has only qualitative character. Details depend on the relaminarisation process and thus on the global structure of the phase space; the upper bound for τ is proportional to $1/\sqrt{\mu}$, however.

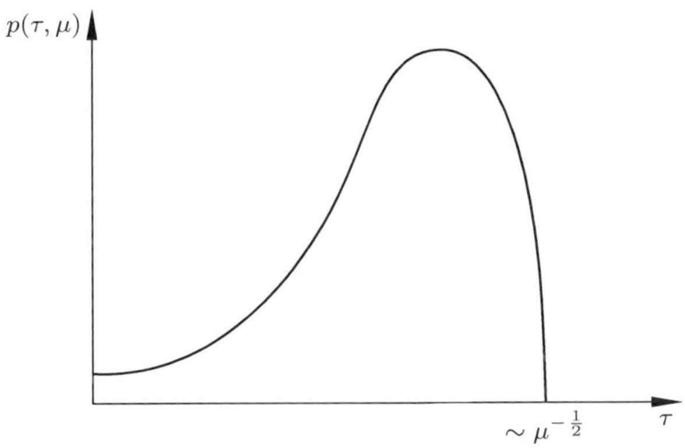

Fig. 8.6.10

Probability distribution of the lengths of laminar sections for type I intermittency

At the end of section 5.4.6, we estimated the relaxation time t^* with the aid of the largest Lyapunov exponent σ_1, t^* being the time after which all the information about the original state has been lost. In accordance with eq. (8.6.16), t^* is inversely proportional to σ_1. In the case of an intermittent transition from periodic to turbulent behaviour, we can assume that, at the instant of the chaotic burst, the information about the original state of the laminar phase has been lost. For very small μ-values, the relaxation time corresponds to the time which is required to pass through the tunnel and which is of the order of magnitude $1/\sqrt{\mu}$ in accordance with eq. (8.6.16). We can conclude that the largest Lyapunov exponent follows the scaling law

$$\sigma_1 \sim \sqrt{\mu} \tag{8.6.17}$$

for $\mu \to 0$.

The mean duration of the regular sections eq. (8.6.16) can also be established with the help of the renormalisation techniques known to us from section 6.7 when discussing the Feigenbaum route (Hu and Rudnick, 1982). This is one of the few examples in which both the fixed point equation and the linearised renormalisation group equation can be solved exactly.

Type I intermittency can be observed very well in the Lorenz model, eq. (5.2.1). In fig. 8.4.3, we have plotted the three Lyapunov exponents for $\sigma = 10.0$, $b = 8/3$ in dependence of r, the relative Rayleigh number. In the range of approximately $145.9 < r < 166.1$, we observe a periodic window within the chaotic regime as in the case of the logistic map. In the case of the Lorenz system, this window emerges as a result of an inverse cascade of period halvings. At $r_{cr} \approx 166.07$, the periodic motion loses its stability.

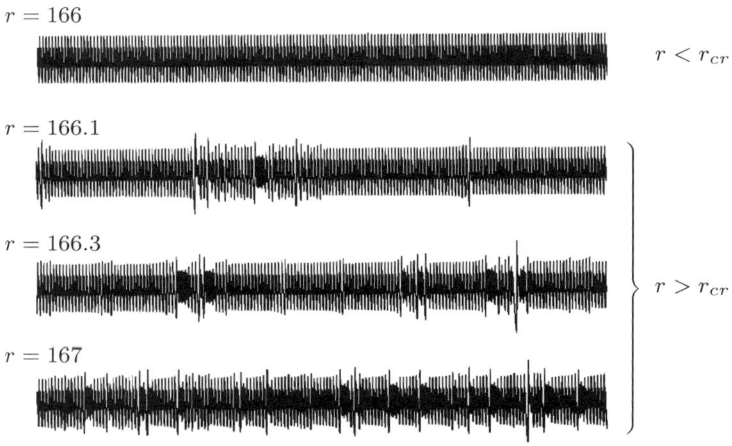

Fig. 8.6.11: Lorenz system: transition to chaos via type I intermittency (Manneville and Pomeau, 1979)

Manneville and Pomeau plotted the Z-component as a function of time for different r-values, see fig. 8.6.11 (Manneville and Pomeau, 1979). For $r = 166.0$, Z is strictly periodic; for $r = 166.1$, long laminar phases are suddenly interrupted by short chaotic ones. A further increase of the control parameter leads to increasingly short regular phases until, finally, irregular behaviour prevails.

If, for $r = 166.2$, we consider the points of intersection of a trajectory with a plane $X = 0$, i.e. a Poincaré section, and plot Y_{n+1} versus Y_n (fig. 8.6.12), it is evident that the transition from periodic to chaotic behaviour is generated by a saddle-node bifurcation and hence type I intermittency can be observed.

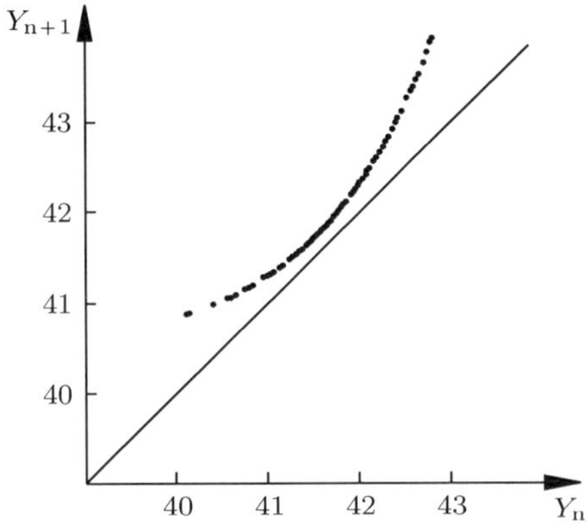

Fig. 8.6.12

Lorenz system:
Poincaré section for $r = 166.2$
(Manneville and Pomeau, 1979)

For r-values just above the critical value r_{cr}, the scaling behaviour of the largest Lyapunov exponent predicted in eq. (8.6.17) could also be discerned for the Lorenz system, evidence that we are dealing with a universal estimate, characteristic of type I intermittency (fig. 8.6.13).

In section 8.6.2, we already noted that the relaminarisation process depends on the global structure of the phase space, i.e. on the specific form of the system. On the basis of the circle map, eq. (8.3.13), we can easily confirm that type I intermittency can also take place in the transition from periodic to quasi-periodic behaviour; for more details, see (Bergé et al., 1984); cf. fig. 8.3.7. In the following, however, we concentrate on the transition from periodic to chaotic behaviour. Pomeau and Manneville have given a model of a Poincaré map (Pomeau and Manneville, 1980) which exhibits qualitatively the same behaviour for small values of the control parameter as the Lorenz model in the neighbourhood of $r_{cr} \approx 166.07$

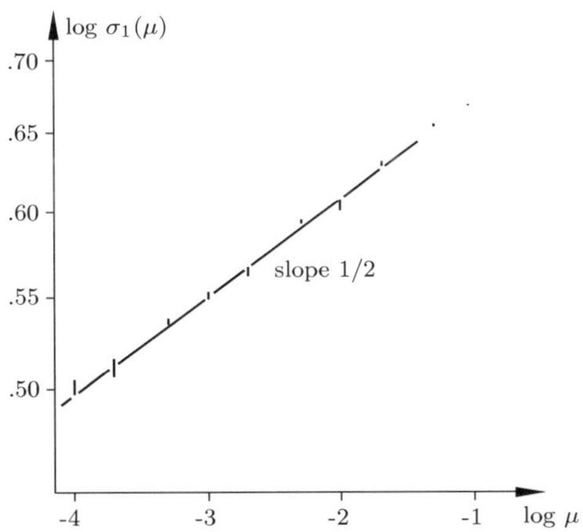

Fig. 8.6.13

Variation of the Lyapunov exponent
as a function of $\mu = (r - r_{cr})/r_{cr}$
(Pomeau and Manneville, 1980)

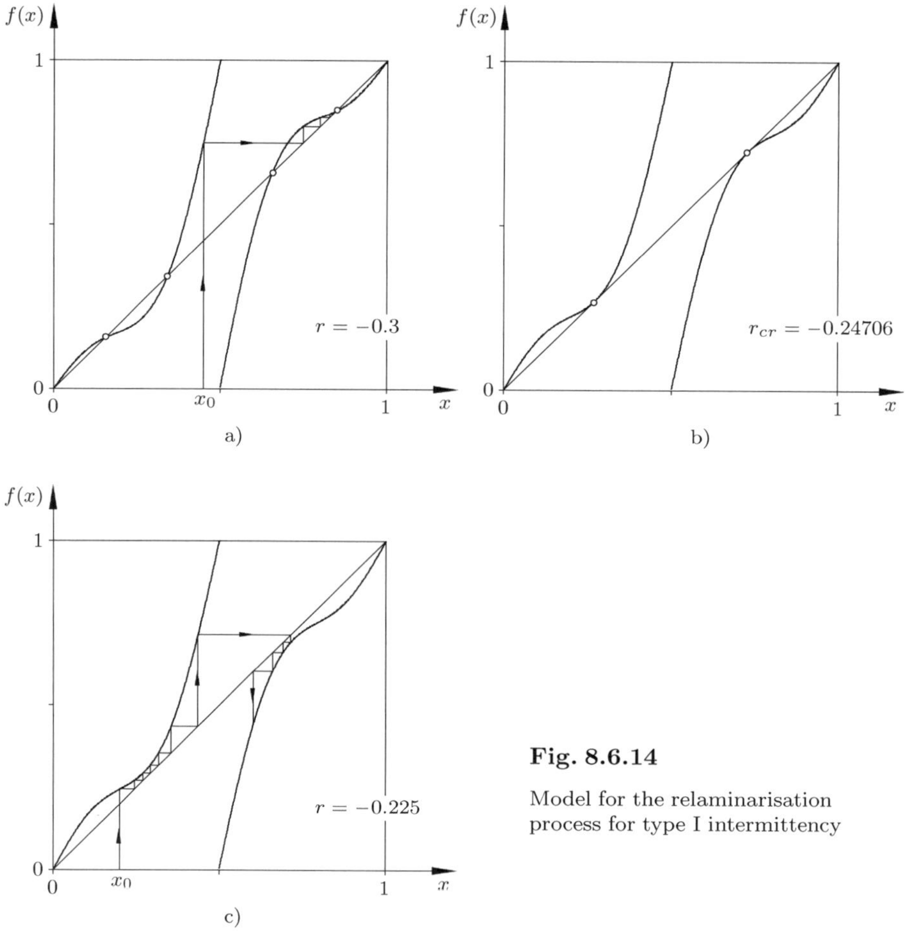

Fig. 8.6.14

Model for the relaminarisation process for type I intermittency

$$x \longmapsto f(x) = 2x + r \sin 2\pi x + 0.1 \sin 4\pi x \qquad (\bmod 1) \tag{8.6.18}$$

Figure 8.6.14 presents the mapping function $f(x)$ for three different r-values. The mapping function can be considered a non-linear version of the Bernoulli shift (cf. section 5.6.1), and hence possesses unpredictable long-term behaviour; due to its non-linearity, it also makes the emergence of a saddle-node bifurcation possible. For $r = -0.3$ (fig. 8.6.14a), five fixed points occur, the origin $x = 0$, which is always unstable for r-values larger than -0.359, as well as two pairs of one stable and one unstable fixed point, each of which coalesce at $r_{cr} = -0.24706$ (fig. 8.6.14b) and subsequently vanish. For $r \gtrsim r_{cr}$, two tunnel zones thus emerge (fig. 8.6.14c) which cause type I intermittency.

In fig. 8.6.15, we have plotted the corresponding time series for increasing r-values greater than r_{cr}. Since the global structure of the phase space for this map is completely different from that of the threefold iterated logistic map in section 8.6.1, the two relaminarisation processes also follow different laws; this is reflected in a

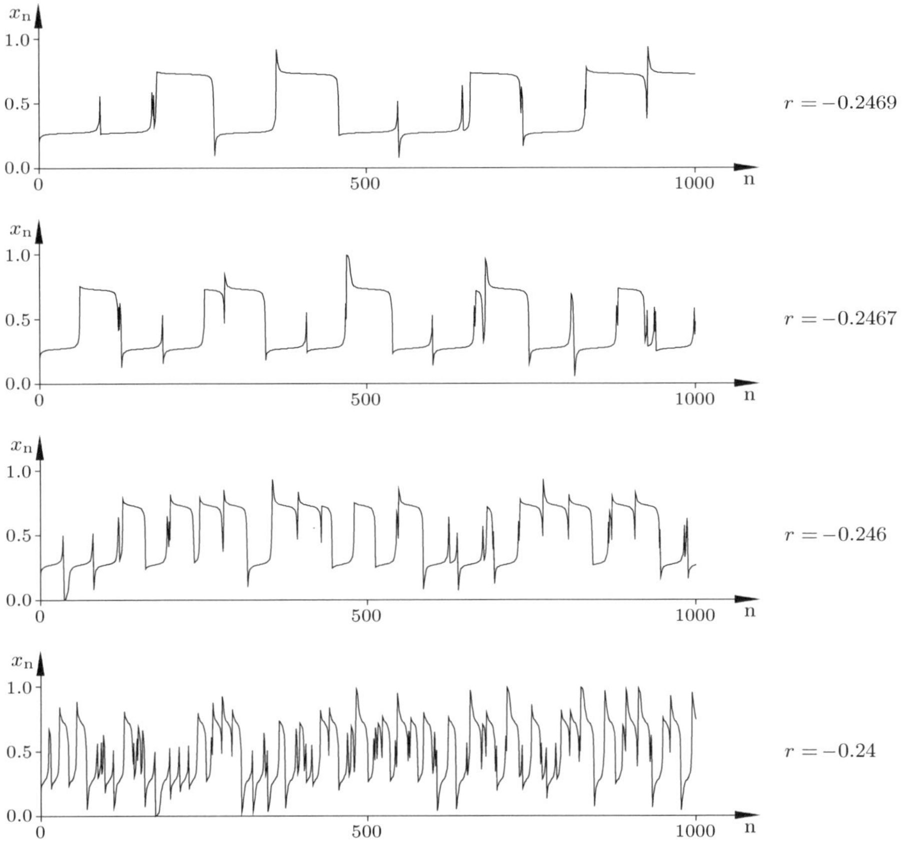

Fig. 8.6.15: Sequences of points x_n for increasing r at the transition of chaos for the mapping rule eq. (8.6.18)

very dissimilar character of the sequences of points (cf. fig. 8.6.5). The fact that we are dealing with type I intermittency in both cases is only recognisable in the mean duration of the laminar phases in dependence of $\mu = (r - r_{cr})/r_{cr}$, eq. (8.6.16), respectively in the probability distribution of the lengths of the laminar phases (cf. fig. 8.6.10). In the next section, we shall see that type III intermittency obeys quite different laws.

Type I intermittency could also be established experimentally in the Rayleigh-Bénard convection at the transition from periodic to chaotic behaviour (Bergé et al., 1980). Silicone oil with a Prandtl number of $\sigma = 130$ was used as the liquid. Figure 8.6.16 presents the time series of the vertical velocity component for different Rayleigh numbers. For $Ra = 270 Ra_{cr}$, we observe a periodic oscillation (fig. 8.6.16a) which is suddenly interrupted by short turbulent bursts above a threshold of $Ra_I \approx 290 Ra_{cr}$ (fig. 8.6.16b). A further increase in the Rayleigh number leads to a steady reduction of the lengths of laminar phases (fig. 8.6.16c) until the motion finally becomes completely irregular.

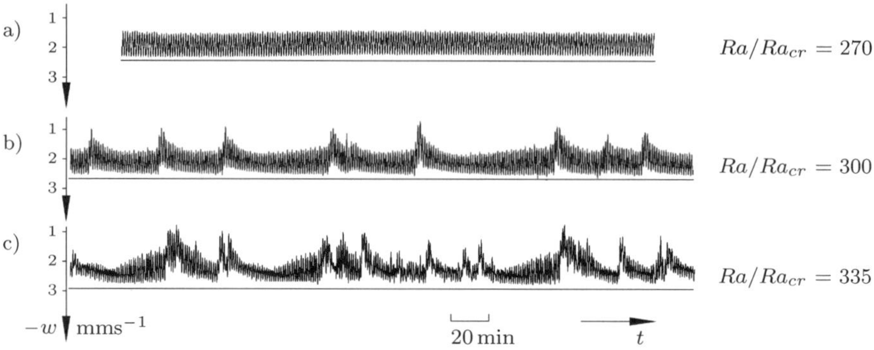

Fig. 8.6.16: Time series of the vertical velocity component w, measured at the centre of the convection cell, for increasing Rayleigh numbers: transition to chaos via type I intermittency in the Rayleigh-Bénard experiment (Bergé et al., 1980)

Figure 8.6.17 presents a schematic survey of the transition to chaos via type I intermittency. In an astonishingly simple way, the underlying theoretical model confirms for the Rayleigh-Bénard experiment a series of experimental observations in which a continuous transition from periodic motions to chaos takes place; long laminar phases are succeeded by an explosion of initially short turbulent phases, the duration of which lengthens as the control parameter increases.

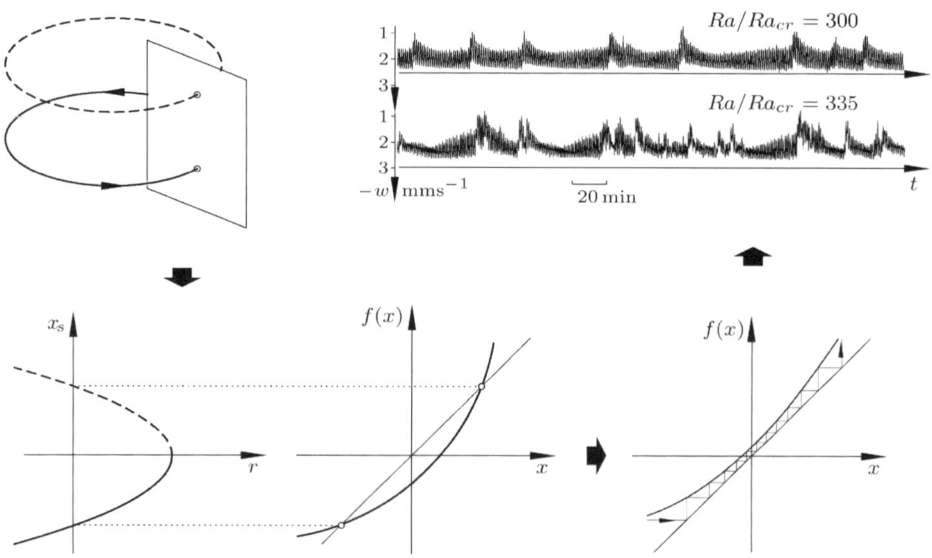

Fig. 8.6.17: The route to chaos via type I intermittency

8.6.4 Type III Intermittency

According to the classification in section 8.6.2, the periodic motion loses its stability in this type of intermittency because an eigenvalue of the monodromy matrix crosses the unit circle at -1, the corresponding flip bifurcation being subcritical.

Figure 8.6.18a presents the stability behaviour of a fixed point of a function $f(x, \mu)$ with an eigenvalue $\lambda < -1$: we obtain an alternating sequence of points; for λ-values just below -1, the system attains almost the same state after two iterations (fig. 8.6.18b), i.e. the system is in a laminar phase similar to a period doubling state (cf. fig 8.6.10). Once the narrow tunnel has been passed, a chaotic burst follows, i.e. the trajectory explores large parts of the phase space erratically, until it enters the corridor again.

As in the case of type I intermittency, the long-term behaviour in the neighbourhood of the bifurcation can be approximated by the dynamics on the centre manifold; here, too, it suffices to restrict ourselves to the normal form. It should be noted that the normal form of a subcritical flip bifurcation must contain a cubic term which renders a control of the stability possible. In this case, the normal form is

$$x \longmapsto f(x, \mu) = -x - \mu x - x^3 \tag{8.6.19}$$

where we once more use the notation $\mu = (r - r_{cr})/r_{cr}$

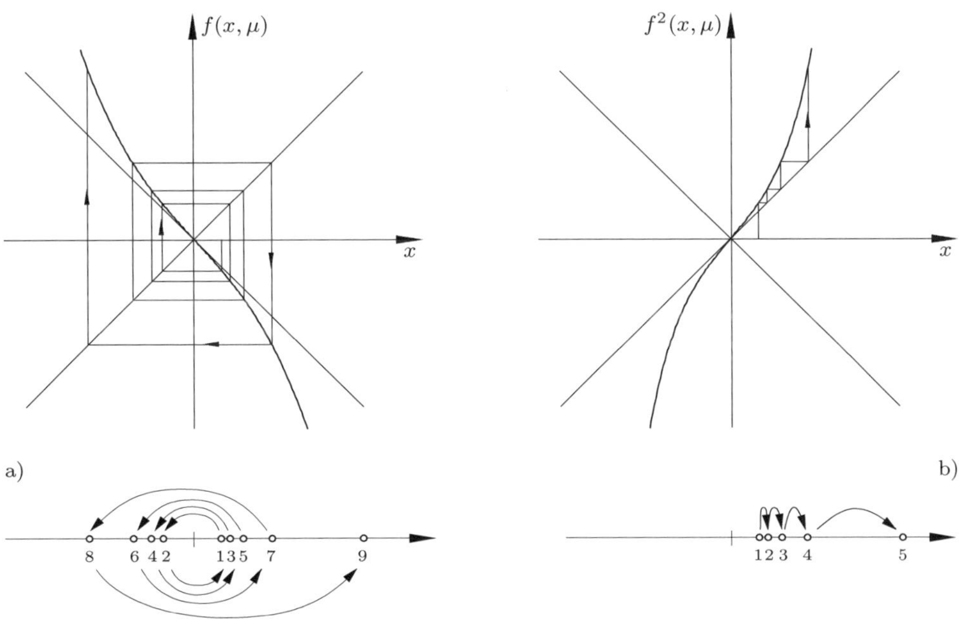

Fig. 8.6.18: Laminar phase in the type III intermittency: sequence of points of a) a simple and b) an iterated Poincaré map

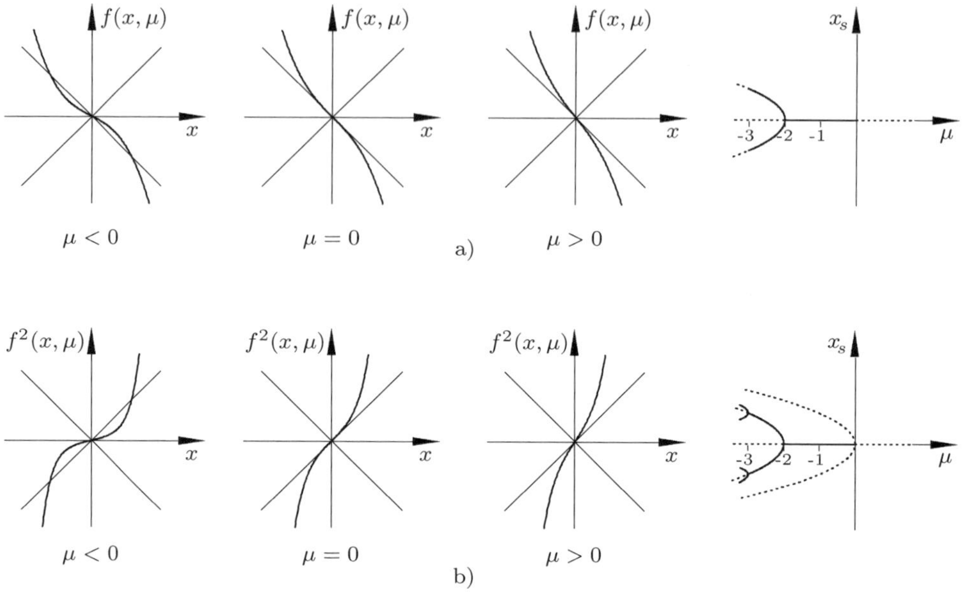

Fig. 8.6.19: Subcritical flip bifurcation: graphs and bifurcation diagram a) of $f(x,\mu)$ and b) of the iterated function $f^2(x,\mu)$

In fig. 8.6.19, we have plotted the function $f(x,\mu)$, the iterated function $f^2(x,\mu)$ as well as the corresponding bifurcation diagrams for various μ-values in the neighbourhood of the bifurcation point. For $-3 < \mu < -2$, $f(x,\mu)$ possesses one unstable and two stable fixed points, namely $x = 0$ and $x = \pm\sqrt{-(2+\mu)}$ which coalesce with $x = 0$ at $\mu = -2$ due to a pitchfork bifurcation. For $\mu > -2$, the only real fixed point is $x = 0$. This is stable in the range $-2 < \mu < 0$ and loses its stability again at $\mu = 0$ without a bifurcation taking place, however (fig. 8.6.19a). If we consider the twofold iterated function

$$f^2(x,\mu) = -(1+\mu)\big[-(1+\mu)x - x^3\big] - \big[-(1+\mu)x - x^3\big]^3 \tag{8.6.20}$$

we recognise that two unstable fixed points $x = \pm\sqrt{-\mu}$ emerge additionally for $\mu < 0$ (fig. 8.6.19b), i.e. at $\mu = 0$, a flip bifurcation occurs which is subcritical. This can also be seen from the sign of the Schwarzian derivative Sf (cf. fig. 6.7.3b). In accordance with eq. (6.7.9), we obtain from eq. (8.6.19) for $x = 0$ and $\mu = 0$

$$\mathrm{S}f = 6 > 0 \tag{8.6.21}$$

i.e. a period doubling does not occur as in the case of the logistic map, for example; rather, the function $f^2(x,\mu)$ rotates in a positive direction away from the bisector for increasing μ-values. A wedge-shaped gap opens between the iterated function $f^2(x,\mu)$ and the bisector.

In contrast to type I intermittency, however, the (unstable) fixed point $x = 0$ also exists for positive μ-values in type III. The length of the laminar phase hence depends upon the point at which the trajectory re-enters the gap. Were it to hit $x = 0$ exactly,

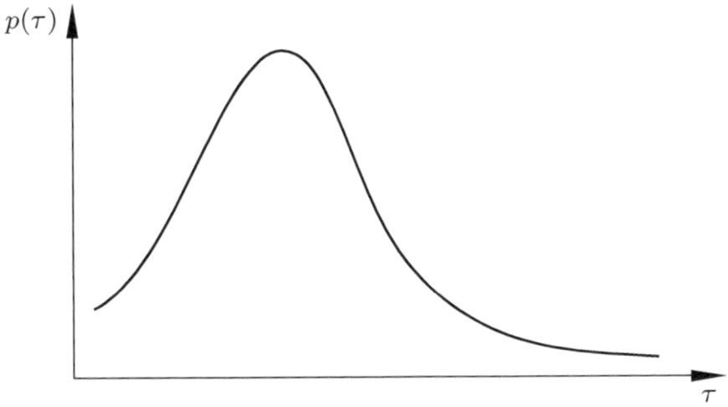

Fig. 8.6.20

Probability distribution of the durations of the laminar phases for type III intermittency

for example, the laminar phase would last infinitely long. The closer the re-entry point is to $x = 0$, the longer the regular behaviour. As in the case of type I intermittency, we can prove that the mean duration of the laminar phases is inversely proportional to μ (Bergé et al., 1984)

$$\langle\tau\rangle \sim \frac{1}{\mu} \tag{8.6.22}$$

The distribtion of the durations of the almost periodic phases is presented schematically in fig. 8.6.20.

In contrast to type I intermittency, there is no characteristic maximum length dependent on μ of the laminar phase; the probability distribution $p(\tau, \mu)$ behaves typically like $\mathrm{e}^{-2\mu\tau}$ for large τ-values (Bergé et al., 1984).

In order to gain an impression of the intermittent behaviour of type III, we must include the relaminarisation process in the model equation. The normal form of eq. (8.6.19) is not sufficient here; it only describes qualitatively the local behaviour of the dynamical system in the neighbourhood of the bifurcation point $x = 0$, $\mu = 0$ where all sequences of points diverge for $\mu > 0$, however. As an example of type III intermittency, we consider the following iterated map

$$x \longmapsto f(x, \mu) = 1 - 2x - \frac{1}{2\pi}(1 - \mu)\cos\bigl(2\pi(x - \tfrac{1}{12})\bigr) \qquad (\mathrm{mod}\ 1) \tag{8.6.23}$$

which, similar to the function eq. (8.6.18), modifies the Bernoulli shift and possesses a fixed point $x = 1/3$ with the gradient -1 for $\mu = 0$. Figure 8.6.21a presents the evolution of the mapping function for increasing μ-values. We recognise that the fixed point x_s loses its stability at $\mu = 0$ and continues to exist for positive μ-values without undergoing a bifurcation. In fig. 8.6.21b, the twofold iterated functions are plotted for the same parameter values. For negative μ-values, there are two further unstable fixed points in the neighbourhood of the stable fixed point x_s; for $\mu = 0$, following a subcritical pitchfork bifurcation, they coincide with x_s which becomes unstable for $\mu > 0$.

Figure 8.6.22 presents the time series of the iterated map eq. (8.6.23) for various μ-values. The frequency of the turbulent outbursts grows for increasing values of the control parameter; in contrast to the transition to chaos via type I intermittency (figs. 8.6.5 and 8.6.15), a decrease in the maximum lengths of the laminar phases cannot be observed, however.

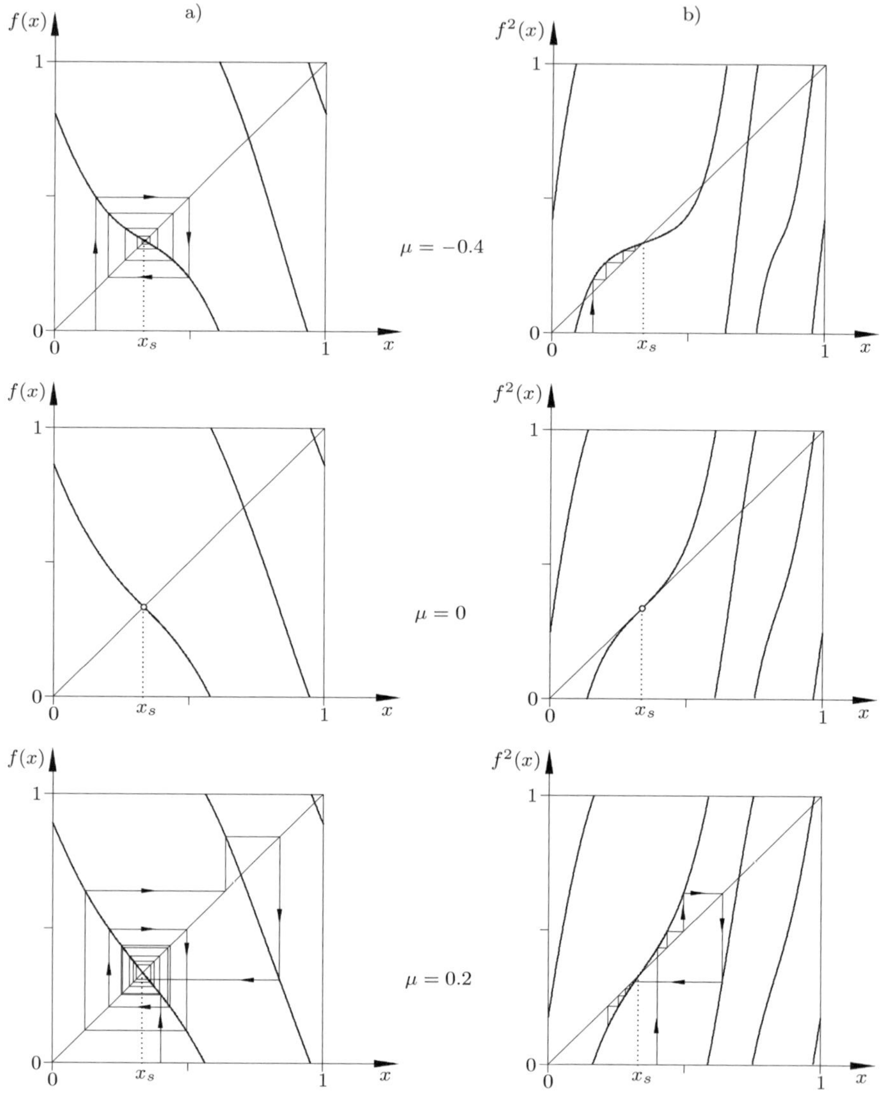

Fig. 8.6.21: Model for the relaminarisation process in type III intermittency

As in the case of type I intermittency, type III intermittency could be verified experimentally in the Rayleigh-Bénard experiment (Dubois et al., 1983). Silicone oil with a low Prandtl number $\sigma \approx 38$ was used. For Rayleigh numbers Ra in the range 300 to 400 times the critical Rayleigh number Ra_{cr}, a regular periodic oscillation of the convection cells could be observed. For $r = Ra/Ra_{cr} \approx 416.7$, a subharmonic frequency suddenly emerges, accompanied by intermittent behaviour.

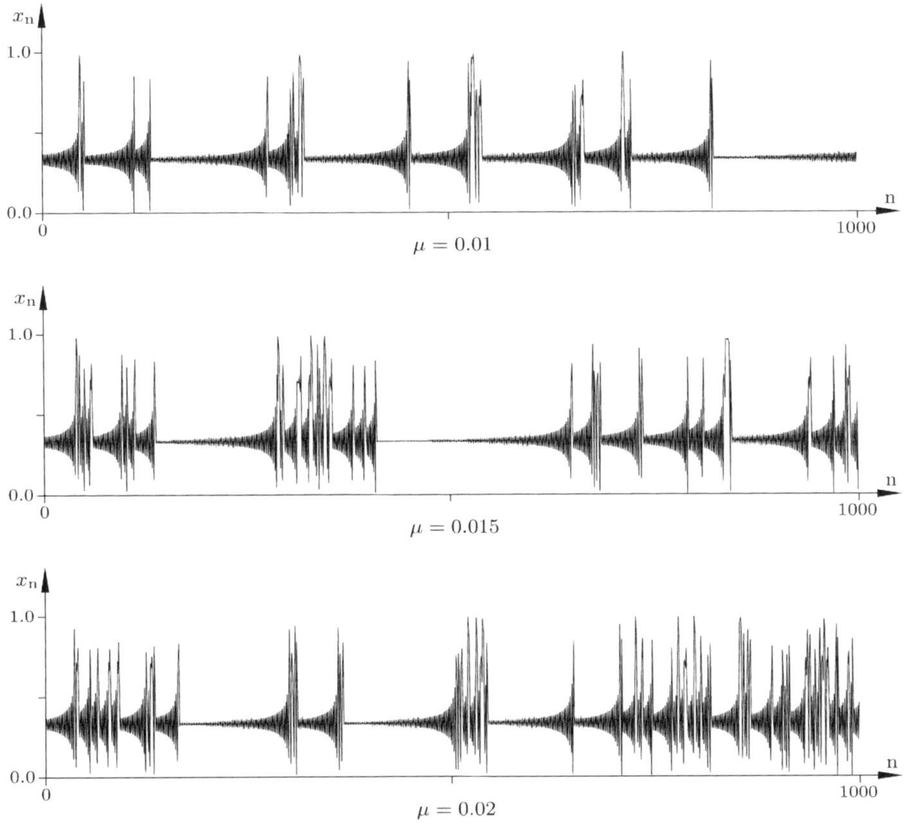

Fig. 8.6.22: Sequences of points x_n for increasing μ at the transition to chaos for the map eq. (8.6.23)

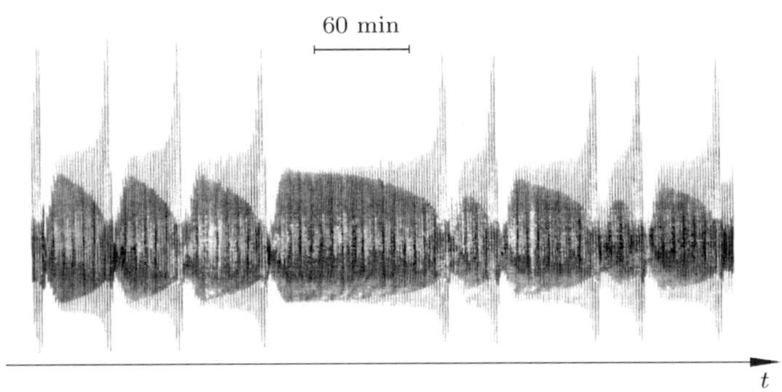

Fig. 8.6.23: Type III intermittency: Experimentally determined time series of the horizontal temperature gradient in the Rayleigh-Bénard experiment (Dubois et al., 1983)

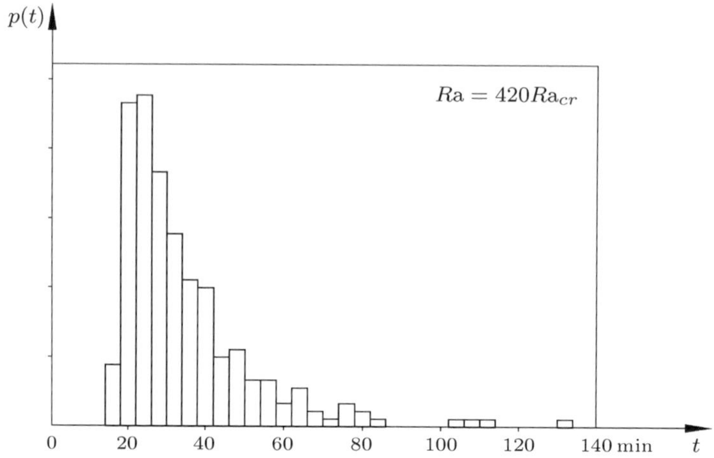

Fig. 8.6.24: Experimentally determined probability distribution of the lengths of laminar phases in type III intermittency (Bergé et al., 1984)

Figure 8.6.23 shows an experimentally determined time series of the horizontal temperature gradient near the critical Rayleigh number for which type III intermittency occurs. Between any two turbulent outbursts, we observe a subharmonic oscillation which is characterised by a continuous increase of the amplitudes. Once the amplitudes of the oscillation attain a certain magnitude, the regular phase ends and a chaotic outburst occurs. After a short time, however, a regular phase begins again. At first, we observe very small amplitudes which increase steadily; the smaller the amplitude at the beginning of the phase, the greater the duration of the laminar phase (Dubois et al., 1983). This is in agreement with the theoretical model. The nearer the initial value at the beginning of the laminar phase comes to the unstable fixed point (fig. 8.6.18b), the longer the sequence of points requires to leave the tunnel. Figure 8.6.22 illustrates this mechanism; the smaller the amplitudes directly after the chaotic outburst, the longer the laminar phase, the amplitudes increasing continuously. On the basis of the recorded results, it could also be verified that the qualitative behaviour of the statistical distribution of the lengths of the laminar phases coincides with the diagram in fig. 8.6.20 (see also fig. 8.6.24). The regular phases lasted between 18 minutes and two hours; the smooth decay of the distribution for long regular phases is typical of type III intermittency.

To conclude, we again summarise the route to chaos via type III intermittency schematically in fig. 8.6.25

8.6.5 Type II Intermittency

Type II intermittency occurs if a periodic motion loses its stability because a pair of complex conjugate eigenvalues of the monodromy matrix crosses the unit circle and perturbations by the non-linear components of the Poincaré map are not intercepted but intensified; in other words, type II intermittency is caused by a subcritical

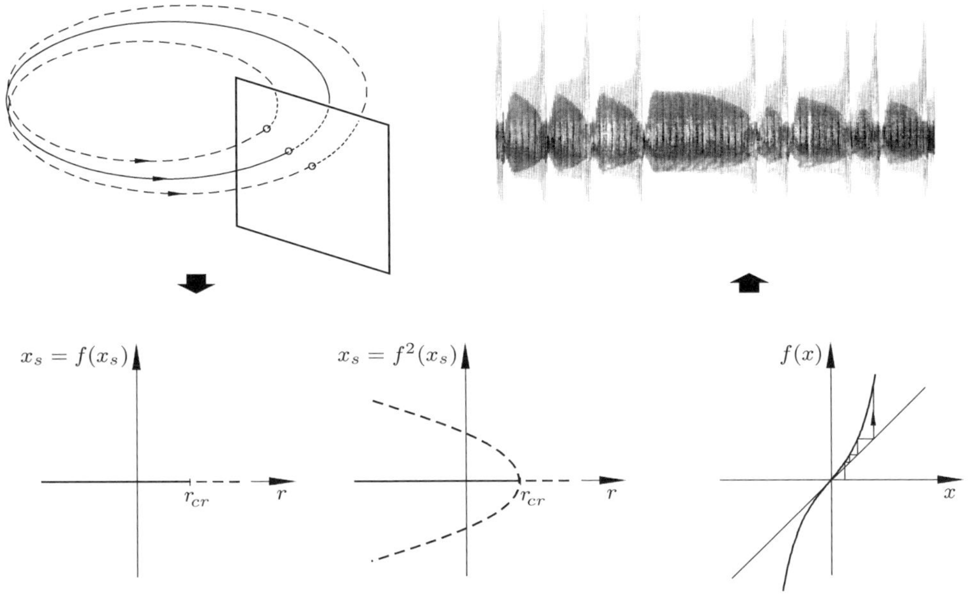

Fig. 8.6.25: The route to chaos via type III intermittency

Neĭmark-Sacker bifurcation (cf. section 6.6). The long-term behaviour can again be approximated by the dynamics in the – in this case – two-dimensional centre manifold; the qualitative behaviour can be studied on the basis of the corresponding normal form.

Introducing the polar coordinates r, θ, we obtain as the normal form, cf. eq. (6.6.80), the mapping

$$\begin{aligned} r &\longmapsto f_1(r,\mu) = (1+d\mu)r + ar^3 \\ \theta &\longmapsto f_2(\theta,\mu) = \theta + \phi_0 + \phi_1\mu + br^2 \end{aligned} \tag{8.6.24}$$

where $d > 0$, $a > 0$ holds in the subcritical case.

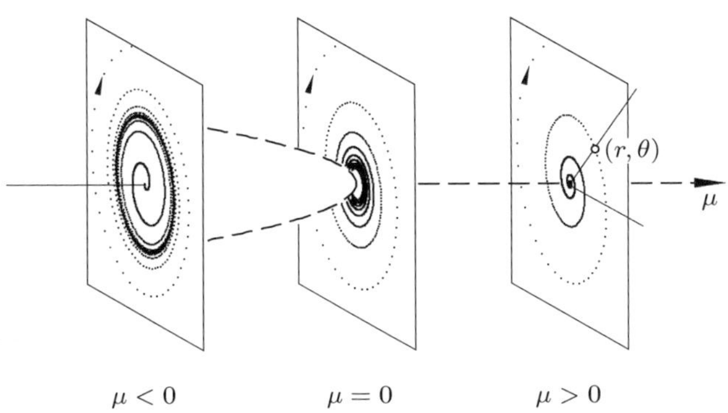

Fig. 8.6.26

Subcritical Neĭmark-Sacker bifurcation

The bifurcation diagram of the subcritical Neĭmark-Sacker bifurcation is shown in fig. 8.6.26 (cf. also fig. 6.6.14). For $\mu > 0$, the originally stable limit cycle becomes unstable. The absolute value of an initial perturbation increases at each iteration; at the same time, the perturbation is rotated about an angle $\theta\,(\neq 0,\ \pi/4,\ 2\pi/3,\ \pi$; strong resonances, cf. eq. (6.6.67)). The normal form eq. (8.6.24) shows that the non-linearities influence radius and rotation angle differently. The instability mainly effects the radius which is marginally stable for $\mu = 0$. The radius is evidently modified due to a small perturbation for small μ-values considerably more slowly than its phase angle θ. The influence of the rapidly changing variable θ is eliminated in the first eq. (8.6.24) in agreement with the slaving principle of synergetics (section 6.8), according to which the rapidly varying variable can be expressed as a function of the slowly varying variable. After appropriate normalisation, the mapping function for the radius is hence

$$r \longmapsto f(r,\mu) = (1+\mu)r + r^3 \tag{8.6.25}$$

This equation agrees formally with the iterated mapping function eq. (8.6.20) in the case of type III intermittency; in our case, however, a rotation is added at each iteration. The consequence is that the length of the laminar phases again depends on the value r_0 at the beginning of the regular phase and can thus be estimated as in the case of type III intermittency (Pomeau and Manneville, 1980)

$$\langle\tau\rangle \sim -\ln r_0 \cdot \frac{1}{\mu} \tag{8.6.26}$$

In contrast, the statistical distribution of the lengths of the laminar phases deviates from the prognoses for type III intermittency as a consequence of the present two-dimensional normal form of the Poincaré map. Details can be found in (Bergé et al., 1984).

Type II intermittency was observed first of all in hydrodynamic systems in which a fluid is heated locally directly below the surface. This causes oscillations of so-called thermal lenses, leading to various instabilities (Ringuet et al., 1993; Gouesbet et al., 2000), which can be interpreted as a generalisation of the Bénard-Marangoni instability described in Chapter 7.

All three types of the Pomeau-Manneville intermittency could be observed in spin waves excited by ferromagnetic resonance (Becker et al., 1999). In these experiments, other global types of intermittency could also be observed, such as *on-off intermittency* and *crisis-induced intermittency*. The former occurs at a global bifurcation where a formerly stable invariant manifold loses its stability, leading to an increase of the dimension of the phase space (Ott and Sommerer, 1994). In contrast, crisis-induced intermittency occurs if chaotic attractors collide locally with unstable periodic orbits (Grebogi et al., 1983; Ott, 1993).

Figure 8.6.27 summarises all the routes to chaos previously described.

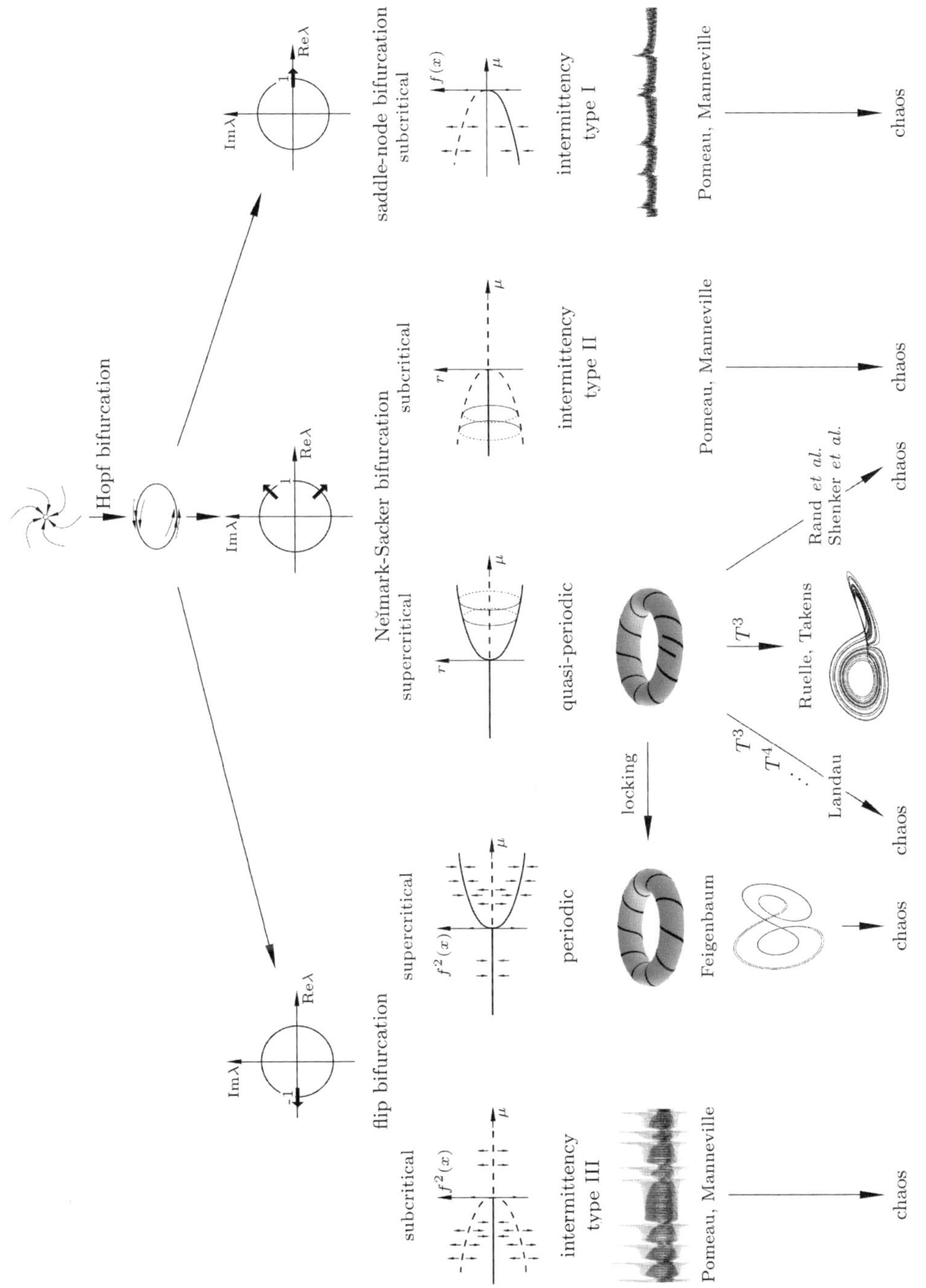

Fig. 8.6.27: Survey of the different routes to chaos due to local bifurcations

8.7 Routes Out of Chaos, Control of Chaos

In the previous sections of Chapter 8 we have discussed routes to chaos, i.e. we have presented a series of scenarios leading from regular to chaotic motion. The question now arises whether the converse were possible: to transfer chaotic system behaviour to stationary or periodic behaviour in a controlled way.

This question arose in the early 1990s and has meanwhile attracted great interest in many disciplines like medicine, biology, chemistry, laser and plasma technology, telecommunication, and general engineering sciences. In recent years, hundreds of publications concerning the control of chaos have been published, lists of references can be found, for example, in (Gauthier, 2003; Andrievskii and Fradkov, 2004; Fradkov and Evans, 2005; Schöll and Schuster, 2007).

Whilst people at the beginning of the 1980s still believed that deterministic chaos was an exotic pathological phenomenon only of interest to mathematicians, they realised in subsequent years that chaotic behaviour can be observed in many areas of nature and technology. In general, it can be assumed that any motion which is bounded in phase space – i.e. all variables remain finite – and which is locally unstable is chaotic. This explains why chaos can be detected quite often. There are cases in which such irregular behaviour is definitely welcome or even necessary; in other situations, however, it could cause unwanted or even dangerous effects since, as a result of uncontrollable oscillations, it can lead to damage such as premature material fatigue or an increase of drag in fluid flow. Therefore, techniques have been developed which either transform chaotic into low-dimensional periodic behaviour, thus stabilising the motion locally, or which maintain or even generate the chaotic behaviour. Another group of methods makes use of the specific properties of chaotic systems, namely the sensitive dependence on small perturbations and the ergodicity (cf. section 3.10) in order to drive the system to a target with minimal energy consumption.

In chemistry, for example, an optimal mixing of the differing constituents involved in combustion processes is necessary in order to achieve reaction rates which are as high as possible and a transport of mass and energy which is as effective as possible. Such optimal mixing processes are based on chaotic dynamics in time and space which have their origin in mechanisms like stretching and folding. In contrast, a sudden onset of a cardiac fibrillation or the beginning of an epileptic seizure – both triggered by chaotic behaviour – is highly dangerous. On the basis of the extended possibilities provided by strategies based on non-linear dynamics and signal analysis, it is possible to distinguish between different pathological characteristics and formations (Garfinkel et al., 1992; Schiff et al., 1994). On the basis of this knowledge, the hope is that new therapeutic strategies may be developed which allow cardiac fibrillation or epileptic seizures to be detected at an early stage or which provide better medical treatment. However, caution must be exercised when interpreting the data, since records in biology are often too short, instationary and noisy. In recent years, moreover, it has been found that it is in many cases not enough to restrict considerations to pure dynamics in time; instead, spatio-temporal dynamics as well as synchronisation effects have to be taken into account. However, since such discussions would go far beyond the scope of this book we only refer here to further reading such as, for example, (Fell et al., 2003; Pikovsky et al., 2003; Baba et al., 2002; Rosenblum and Pikovsky, 2004; Schöll, 2001).

Traditionally, for reasons of security, engineers confine themselves to defining the operating range of systems and processes in such a way that the resulting dynamics is regular and predictable. However, since the 1990s, a change in thinking has set in. It has been recognised that chaotic behaviour possibly provides a high potential for innovative developments. The following examples present only a small selection from the multitude of possible applications in engineering, yet demonstrate the wide range and the high potential of these methods.

The sinking of the cruise ferry "Estonia" off the southern coast of Finland in 1994 showed that the stabilisation of ships out of control after an inrush of water is still a pressing problem. A procedure for stabilising the rolling motion of a vessel by converting it into a regular periodic rolling motion with a small amplitude was presented in (Mitsubori and Aihara, 2002).

Dry friction forces lead to highly non-linear equations of motion. Therefore, in periodically driven systems with dry friction, chaos can easily be generated. The control of such friction forces in systems under shear aims at avoiding chaotic stick-slip motion (Popp and Stelter, 1990; Galvanetto, 2001). This is also a real problem in micromechanics. Systems under shear at low driving velocities – an example is the starting and stopping of a CD drive – may exhibit chaotic stick-slip, while even sliding occurs at higher velocities. In such cases, chaos-control methods can avoid stick-slip effects by stabilising the motion (Rozman et al., 1998).

In telecommunications, the use of chaotic carrier waves for transmitting, encoding and decoding messages has proven to be especially effective and robust. Even if only band-limited and noise-susceptible channels are available, the receiver can recover the message almost flawlessly. In principle, an electrical circuit producing similar to the Lorenz attractor a double scroll about two centres is used for encoding. If we associate the local maxima of the trajectories with a "1", the local minima with a "0", we obtain a binary sequence. With small control impulses, this orbit can be changed in such a way that it follows a prescribed binary sequence corresponding to the information we wish to communicate (Hayes et al., 1993; Baptista and Kurths, 2005).

In aeronautics, there are also various applications of chaos control. When spacecrafts are underway in the solar system, the influence of celestial bodies on the orbit has to be taken into account, i.e. there are domains in which the spacecraft undergoes a chaotic motion similar to the three-body problem of celestial mechanics (cf. Chapter 4). By making use of the sensitive dependence of chaotic systems on small perturbations, such space probes can be directed towards certain prescribed targets by means of specific impulses with minimal energy consumption (Shinbrot et al., 1993). In the same manner, gyrostats which have got out of control can be re-stabilised with the help of chaos-control techniques (Ge and Lin, 2003).

In the following, we briefly outline two now classical control methods. More detailed descriptions can be found in the references cited in each case; we do not claim our citations are exhaustive. In principle, the various methods can be divided into two classes. The first group contains several methods without feedback. In one part of them – like in the procedure proposed by Hübler – the dynamics is externally prescribed and the uncontrolled original system is only used for control. In this case, relatively large perturbations are needed. Other methods in this group only seek to stabilise the existing motion. In contrast, the methods in the second group make use of the sensitivity of chaotic motions to small perturbations in order to stabilise an unstable orbit on a strange attractor of the system at hand by feedback. It is, of course, not possible to specify one single method that would be optimal for all requirements. Nevertheless, it has become clear during the development of these procedures that control methods based on repeated time delay feedback can be realised in many cases experimentally with particular ease and are often successful with minimal energy consumption.

8.7.1 Control of Chaos without Feedback

The first method, introduced by the team led by A. W. Hübler, is a procedure for the resonant excitation and control of non-linear oscillators (Hübler, 1987; Hübler et al., 1988). The basic idea originates from observations of linear oscillating systems. If we want to excite a damped harmonic oscillator, cf. eq. (2.2.6), to a certain amplitude, it can be done most effectively if the driving mechanism acts in resonance, i.e. if the driving force $R(t)$ and velocity $\dot{x}(t)$ of the oscillator are in phase $R(t) \sim \dot{x}(t)$.

This idea of resonant excitation can also be extended to non-linear oscillators. In this way, it is possible to transform chaotic oscillations to stable periodic motions. Hübler applied his method mainly to non-linear damped pendulums or to oscillators with a non-linear potential (Wagner et al., 1988). An optimal energy transport also occurs here if the driving force is controlled so that it is proportional to the respective velocity of the non-linear oscillator. In practice, by applying an external exiting force, a variation of the (viscous) damping of the system is simulated.

Generally, this procedure is an attempt to alter a non-linear system – which, by the way, need not necessarily be chaotic – by additional forces in such a way that it follows the prescribed goal dynamics $\boldsymbol{g}(t) = \boldsymbol{g}(t+T)$ of the controlled system

$$\frac{d\boldsymbol{x}}{dt} = \boldsymbol{F}(\boldsymbol{x}) + \boldsymbol{U}(t) \tag{8.7.1}$$

where $\boldsymbol{U}(t)$ corresponds to a control force. This can be accomplished simply by operating the external excitation according to

$$\boldsymbol{U}(t) = \frac{d\boldsymbol{g}}{dt} - \boldsymbol{F}\left[\boldsymbol{g}(t)\right] \tag{8.7.2}$$

Indeed, if $\boldsymbol{x}(t)$ follows the target dynamics $\boldsymbol{g}(t)$, then $\boldsymbol{U}(t) = \boldsymbol{0}$ holds. In contrast to the methods described in section 8.7.2, this kind of control means the target dynamics can be chosen independently of the initial system, i.e. $\boldsymbol{g}(t)$ is *not* a trajectory of the original system $\dot{\boldsymbol{x}} = \boldsymbol{F}(\boldsymbol{x})$. In principle, one tries to alter the dynamics of the uncontrolled system in such a manner that the desired stable period oscillations are generated. Since no feedback with the system occurs, it is neither necessary to access any system parameters nor are any real-time analyses required as in the case of the OGY method described in section 8.7.2. This is an advantage. On the other hand, a disadvantage of this method is that the required control forces might become large and that the domains of the initial conditions $\boldsymbol{x}(0)$ for which the procedure will converge are not known in the beginning (Mettin et al., 1995; Shinbrot et al., 1993).

In a further group of control methods without feedback, the target dynamics is not prescribed. In this case, one tries to transform the chaotic behaviour to periodic motion by influencing the dynamics with periodic or small stochastic deviations (Lima and Pettini, 1990; Kivshar et al., 1994). Typical examples are balancing a stick on a fingertip or twitching the handlebars of a bicycle to and fro to stay upright. Small control motions change the stability characteristics of the unstable fixed point in the phase space. In a similar way, this method changes the stability properties of elliptic or hyperbolic points in the phase space of the averaged dynamical system by parametric perturbations (Kivshar et al., 1994). Methods which deliberately maintain chaos or generate it also belong to this group (Lai et al., 2005).

8.7.2 Control of Chaos with Feedback

The *feedback control* due to Ott, Grebogi and Yorke (OGY method) is based on an approach which is completely different from those previously cited (Ott et al., 1990; Boccaletti et al., 2000), cf. also (Romeiras et al., 1992). In these methods, the characteristic properties of chaos which are usually not desired and which lead to unpredictable behaviour are used for control. Typically, the dynamics of a strange attractor comprises a multitude of possible motions, in particular an infinite number of unstable periodic orbits; on the other hand, no stable orbits are embedded in the chaotic dynamics. However, due to their saddle-point character, the unstable cycles cannot be observed directly since even the smallest perturbations very quickly lead to large deviations (cf. section 3.3). Within this network of attracting and repelling manifolds, the trajectory is alternately attracted and repelled, leading to its erratic behaviour. For axiom A systems, these unstable orbits even form a dense set on the attractor (Smale, 1967), i.e if we choose a point P on the attractor, there is an unstable periodic orbit in every neighbourhood U of P, however small it may be, crossing U (Auerbach et al., 1987; Lathrop and Kostelich, 1989). As a consequence, the motion on the attractor is ergodic, see section 3.10, i.e. every trajectory eventually approaches every arbitrarily selected point on the attractor again and again arbitrarily close.

Instead of altering the dynamical equations of the underlying system as in the previous section 8.7.1 by applying an external excitation, one tries in the OGY method to transform a chaotic motion into a periodic one by subjecting a system parameter μ which can be externally controlled to small time-dependent perturbations. The aim is to stabilise an unstable periodic orbit of the system.

This control method was first verified experimentally with a thin amorphous magnetoelastic strip, clamped only at the base, which buckled under gravitation. Young's modulus E of the chosen material already exhibits very large changes for relatively small modifications of the external magnetic field. As a result, the strip oscillates chaotically in a vertical, periodically oscillating magnetic field applied additionally, i.e. the strip buckles at irregular time intervals and returns to its upright position. If small targeted changes are applied to the amplitude of the magnetic field acting as the control parameter μ, the motion of the strip can be stabilised and periodic motions with various period lengths can be realised (Ditto et al., 1990).

In the following, we sketch the steps of the procedure. At first, we determine a selection of unstable periodic orbits, if possible with a low periodicity. Methods for localising such unstable cycles are described in (Auerbach et al., 1987; Lathrop and Kostelich, 1989), see also (Kantz and Schreiber, 1997). We then analyse these periodic motions and select those which come closest to the desired performance of the system. Let us assume for simplicity's sake that the chosen orbit is a simple periodic orbit, the Poincaré section of which is a single point $P^*(\mu)$ with saddle-point characteristics (cf. figs. 3.3.1, 3.3.2). In order to stabilise this orbit, we require only information on the local dynamics of the map in the neighbourhood of the fixed point as well as on the direction in which the fixed point shifts if the control parameter μ is slightly modified. Since the periodic orbit is unstable, the fixed point of the corresponding Poincaré map has saddle-point character. The Jacobian matrix of the map yields the eigenvalues λ_{si} and λ_{ui} as well as the corresponding eigenvectors spanning the stable and the unstable invariant subspace $\mathbb{E}^s$ respectively $\mathbb{E}^u$. A small variation of μ allows the determination of the position vector corresponding to the fixed point of the modified system.

In general, this information suffices to control the sequence of the Poincaré sections by small temporal modifications of the control parameter μ in such a way that each subsequent image point comes to lie approximately on the stable manifold of the original fixed point. In practice however, one has to wait until the Poincaré section of the uncontrolled chaotic system again reaches the neighbourhood of $P^*(\mu)$. Afterwards, the control is activated. It is then possible to stabilise the selected periodic orbit by feedback with the initial dynamics. In (Nitsche and Dressler, 1992), two modifications of Ott, Grebogi and Yorke's procedure are suggested; these take the dependence of the Poincaré map not only on the current value of the control parameter into account, but also on its value in the preceding step.

This seemingly very theoretical procedure can also be applied successfully to experimental time series. In addition to the example of the chaotic oscillation of an amorphous magnetoelastic strip already mentioned, the OGY method has been used to stabilise electronic circuits and lasers, the chaotic oscillations of the Belousov-Zhabotinsky reaction as well as to regulate the cardiac muscles of rabbits. In (Singer et al., 1991), a chaotic convective fluid is transformed to a laminar one. A survey of the OGY method and several applications can be found in (Boccaletti et al., 2000).

An advantage of the OGY method is that an analytical description of the system dynamics is not necessary. It suffices to reconstruct the dynamics according to the time delay method described in section 5.5.4. Nevertheless, this procedure is not suitable for all possible cases. Since one has to resort to Poincaré sections for the control, the method is quite sensitive to noise corruption. Therefore, in general, only orbits with low periodicity can be stabilised. In addition, real-time measurements and calculations are necessary. As a consequence, the method cannot be applied to very fast systems with dynamics in the MHz range.

The disadvantages described previously can be avoided to a large extent by using procedures based on *time-delayed feedback*. This very simple and robust method was first proposed by Pyragas (Pyragas, 1992). Meanwhile, it has largely gained acceptance since it can be applied to many complex experimental systems and does not require any modelling of the unperturbed dynamics. Only the period length T of the unstable orbit needs to be known. In contrast to Hübler's method, the unstable orbit of the uncontrolled initial system is stabilised by small interventions as with the OGY method.

The first idea might be to choose the control force $\boldsymbol{F}(t)$ proportional to the difference of the initial dynamics described by $\boldsymbol{x}(t)$ and the target orbit $\boldsymbol{g}(t) = \boldsymbol{g}(t+T)$ with period length T, i.e. $\boldsymbol{F}(t) \sim \boldsymbol{x}(t) - \boldsymbol{g}(t)$. However, this approach would have the disadvantage of having to know the orbit to be stabilised. Pyragas therefore proposed estimating the term $\boldsymbol{g}(t) = \boldsymbol{g}(t-T)$ directly by means of the initial dynamics and using the previous state of the system as a reference. This leads to the control force

$$\boldsymbol{F}(t) = K\left[\boldsymbol{x}(t) - \boldsymbol{x}(t-T)\right] \tag{8.7.3}$$

which automatically fulfils $\boldsymbol{F}(t) = \boldsymbol{0}$ as soon as the target orbit is reached. Applying the scheme to experiments, one uses the measured signal $\boldsymbol{x}(t)$ in order to generate a difference by means of a delay line, where the amplitude K of the feedback is used for adjustment (Just et al., 2003).

In order to stabilise highly unstable orbits, an extended scheme is proposed in (Socolar et al., 1994), which takes repeated time delays into account and which the control force $\boldsymbol{F}(t)$

$$\boldsymbol{F}(t) = K \sum_{m \geqslant 0} R^m \left[\boldsymbol{x}(t-mT) - \boldsymbol{x}\left(t-(m+1)T\right)\right] \tag{8.7.4}$$

In contrast to the OGY method where Poincaré sections are used in order to change a system parameter at *discrete* time instants, this method is a *continous* control scheme. To stabilise a chaotic system experimentally, two parameters T and K have to be adjusted in such a manner that the control force is as small as possible. If the period length T is known, the control is very robust over a whole interval of K-values. The difficulty here, especially for autonomous systems, consists in determining the period T. In (Kittel et al., 1995) a variable delay time τ_n is proposed which, for orbits of period one, corresponds to the time elapsed between successive maxima of the measured signal. With the Rössler system, cf. eq. (10.6.1), a set of three ordinary non-linear differential equations modelling chaos in chemical reactors (Rössler, 1976), and the Mackey-Glass system (Mackey and Glass, 1977), which models physiological control mechanisms by means of a time-delayed non-linear differential equation, it could be demonstrated that unstable orbits can be stabilised by means of this self-adaptive feedback method.

Although the control scheme described by eq. (8.7.3) respectively (8.7.4) looks very easy, an analytical description of the problem is difficult since non-linear time-delayed differential equations are involved and even linear stability analyses are challenging. Meanwhile, there are various publications on the theoretical foundations of this control procedure. They mainly deal with the question whether parameter ranges exist in which a stabilisation is posssible and whether there are in principle restrictions to the application of the method. In order to analyse the stability of the orbit including the control force, one determines the Floquet exponents $\sigma = \Lambda + i\Omega$ presented in section 5.4.2. Here, the real part Λ describes the rate at which the perturbed trajectory moves away from the periodic orbit and Ω is the torsion angle per cycle. The transition from unstable to stable behaviour is identified by a change in sign of the real part Λ. In a time-delayed system generated by the inclusion of control forces described by eqs. (8.7.3) and (8.7.4), however, such a change in sign appears only if the imaginary part Ω is different from zero and if the chosen unstable orbit of the uncontrolled system possesses an even number of positive real Floquet exponents. A detailed description and further references can be found in (Just et al., 2003) as well as in the survey (Fradkov and Evans, 2005).

The original article on the OGY method (Ott et al., 1990) was without doubt the real trigger for a multitude of papers on the control of chaos. The main result was to show that the smallest perturbations of a system parameter are enough to achieve a change from chaos to regular behaviour. All the methods based on feedback show that the latent capacity of a system to behave chaotically can be of extraordinary advantage. By means of well-directed perturbations of a control parameter, we can steer out of a chaotic range towards one of the different unstable periodic motions and stabilise it so that, in principle, one and the same system can be designed for different operational tasks.

9 Turbulence

Thus, the water forms whirling eddies,
one part following the impetus
of the main current, and the other
following the incidental motion and return flow.

Leonardo da Vinci, Notebooks

Turbulence is a phenomenon which we can observe every day: in the smoke of a cigarette, in the flickering flame of a candle, at the confluence of two rivers or in the eddies of the current behind the pier of a bridge. Turbulences in the earth's atmosphere are the cause of the evolution of bizarre cloud formations that stir the imagination and in the emergence of destructive hurricanes. In astronomy, turbulences also play an important role: think of the eruptions on the sun and the turbulent processes in Jupiter's atmosphere; see fig. 9.0.1.

The complex patterns and the irregular whirling of turbulent currents have fascinated artists such as Leonardo da Vinci, for example, one of whose drawings depicts the basic characteristics of turbulence, the coexistence of eddies of varying size and intensity. This complex flow structure leads to an effective mixing of the fluid; see

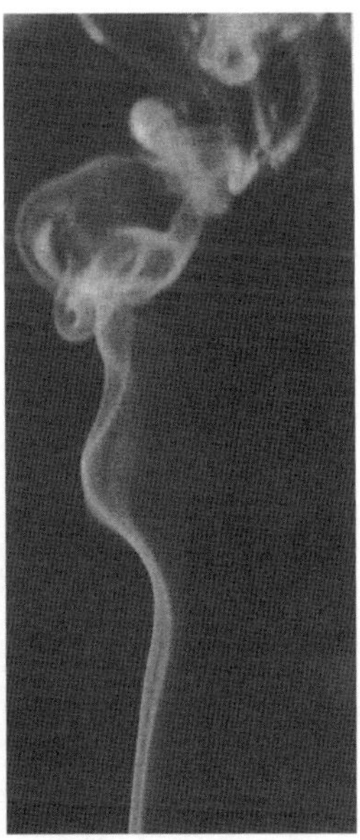

a)

b)

Fig. 9.0.1

Two extreme examples of turbulent phenomena:
a) rising smoke of a cigarette, b) turbulent flow around the Great Red Spot on Jupiter. Reprint with kind permission of NASA (NASA, 1996)

J. Argyris et al., *An Exploration of Dynamical Systems and Chaos*,
DOI: 10.1007/978-3-662-46042-9_9

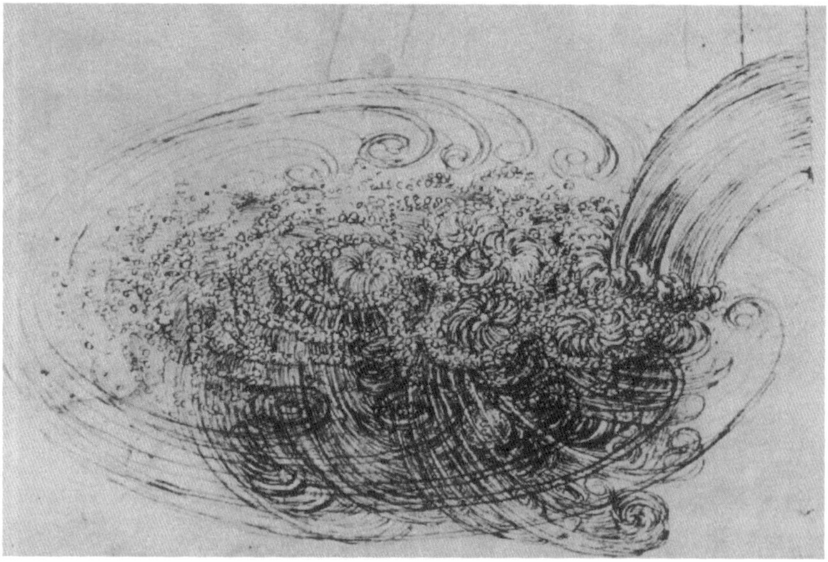

Fig. 9.0.2: Leonardo da Vinci's drawing of turbulence: the flow field is composed of eddies of varying order of magnitude and intensity.

fig. 9.0.2. Evidently, the concept of formation is also due to Leonardo da Vinci: the Latin word *turbare* means *rotate, entangle, swirl* and is a very precise description of the complex flow processes.

A characteristic property of turbulent flows is the enhancement of the transport and mixing of matter, momentum and heat. This is an advantage when we think of an optimal mixing of gas and fuel in combustion, yet a disadvantage in the case of airflow around vehicles since turbulent boundary layers enhance the shear forces on the surface and lead to a strong increase in the drag of the object. There is a multitude of applications in engineering dealing technically with turbulent behaviour; examples are the flow around automobiles, airplanes or hulls, the transport of petroleum or gasoline in pipelines, the mixing of chemical agents in reactors, the emission of pollutants and the combustion processes in motors. But also in the field of renewable energy generation, the importance of turbulence is becoming increasingly evident. We mention here the turbulent fluctuations of the flow at rotor blades in wind power plants and the variations of incident solar radiation due to the turbulent motion of clouds. Turbulent fluctuations in wind energy generation lead to fluctuations in the energy supply to energy networks with a direct influence on their stability.

The laws of motion in hydrodynamics have been known since the middle of the 18th century. On the basis of the conservation law of momentum, Leonhard Euler formulated the fundamental equations for fluid motion without viscosity as early as 1755. Louis Navier (1822) and Sir George Gabriel Stokes (1845) extended these relations by also considering the molecular dissipation forces, described by the kinematic viscosity $\nu = \eta/\varrho$. This led to the well-known Navier-Stokes equation. Together with initial and boundary conditions and the incompressibility condition, this leads to a system of four equations for the three components of the velocity vector and the pressure, i.e. we have a complete system of partial differential equations for the determination of fluid flow.

Engineers have learnt to master the complexity of turbulent behaviour in many fields. Considering the immense technical successes, one might be led to suppose that the physical background of the phenomenon of turbulence has been completely understood, particularly since the laws of motion in hydrodynamics have been known since the middle of the 19th century. If we exclude direct numerical simulations, however, results which can be deduced directly from the basic hydrodynamic equations are rather the exception. For example, the critical Reynolds number for the transition to turbulent flow in a pipe or the drag of a car or airplane cannot be predicted even with the largest computers available today (Sagaut, 2001). Therefore, in engineering, turbulence models are used. For example, turbulent mixing processes are described in analogy to common diffusion processes. In large eddy models, widely used in engineering, the flow is only computed on the large length scales while the influence of small eddies on the large-scale flow is modelled by effective eddy viscosities. A rigorous verification of this concept based on an *ab initio* computation of turbulent diffusion constants and effective eddy viscosities could still not be performed successfully.

From a physical point of view, the state of research in turbulence is unsatisfactory. This becomes particularly evident when comparing it with other field theories such as quantum electrodynamics, which describe the interplay between light and matter. In this case, physical quantities such as the magnetic moment of an electron can be calculated *ab initio* with staggering precision and compared with experiments[1]. While nobody expects such precision in turbulence, this example does demonstrate that, in contrast to quantum electrodynamics, the physical understanding of turbulence and the corresponding methods for the calculation of characteristic turbulence properties are far less developed.

Yet this does not mean that the investigation of turbulent flows has not – or has only marginally – contributed to scientific progress. Quite the contrary: the history of turbulence research is characterised by major scientific innovations. The first results in the field of turbulence are due to Boussinesq and Reynolds, who postulated that it is hopeless to try and describe such complex processes as those arising in turbulent flows in all their detail, but rather that the only possibility is to develop a statistical theory of turbulence. In the subsequent years, a whole generation of outstanding scientist – such as L. F. Richardson, L. Prandtl, G. I. Taylor, Th. von Kármán, A. N. Kolmogorov, W. Heisenberg, B. Mandelbrot, R. H. Kraichnan and others – investigated the problem of turbulence intensively (Eckert, 2008).

Turbulence research is an interdisciplinary field and strives to combine basic and applied research. Mathematicians are mainly interested in the existence and

[1] Calculations of the Landé g-factor, for example, which characterises the ratio between the magnetic moment of an electron measured in units of the Bohr magneton and its spin yield the value $g_s = 2(1 + 0,001596524(\pm 4))$ which can be compared with the experimentally measured value $g_s = 2(1 + 0,001596524(\pm 20))$.

uniqueness of the Navier-Stokes equations or ask whether singularities can possibly develop within finite time as the existence and uniqueness of solutions for flows in three-dimensional space could not yet be proven. In contrast, engineers are mainly interested in averaged quantities of the flow, like for example the mean velocity profile, the mean pressure gradient or in turbulent stresses acting upon objects in a turbulent velocity field. Their attention is mainly focused on a description of flow phenomena on a large scale comprising an effective modelling of smale-scale turbulence. Physicists, on the other hand, strive to arrive at an overall understanding of the non-linear physical processes which are responsible for the characteristic and specific properties of turbulent fields.

In spite of an abundance of experimental, numerical and analytical results, a complete quantitative theory of fully developed turbulence still does not exist. The American Nobel Prize laureate Richard Feynman is certainly right in stating that the problem of turbulence is *the* unsolved problem in classical physics (Feynman et al., 1987). A small anecdote highlights the difficulties physicists are confronted with when striving to clarify the turbulence problem. Arnold Sommerfeld, the PhD supervisor of Werner Heisenberg is said to have once stated "that before he dies he would like to understand two problems - quantum mechanics and turbulence". Theodore von Kármán recalls in his memoirs that he "did not come closer to the fundamental problem of turbulence" (von Kármán, 1968).

The difficulties are already evident when we attempt to define the concept of turbulence. By turbulent behaviour, engineers mean the random, unpredictable behaviour of liquid or gaseous flow in space and time. Often, the statistical fluctuations are superimposed on a mean flow. However, such a definiton depends upon the observer's point of view and what he understands by mean flow and turbulent eddies. If, for example, we watch a flag on a stormy day from close to, we will ascertain a mean wind direction about which the flag flaps irregularly. From further away or on another time scale, we will observe that the mean wind direction is by no means constant but can also be seen as a fluctuation which is superimposed on, for example, a north-westerly flow. A weather satellite, for its part, will register that there are only short-term fluctuations about the global westerly wind direction. These observations allow us to recognise a very fundamental characteristic of turbulent behaviour, namely that there is a range of scales in which a natural scale is missing, i.e. in which scale invariance predominates and in which the distribution of eddies is statistically self-similar (Großmann, 1989).

The characteristic aspect of flows with high Reynolds numbers is that there is no scale separation in a wide range, i.e. in turbulence dynamics modes on all length scales are excited. Thus, the extraordinary complexity of turbulent flows has its origin in the existence of an extremely high number of significant degrees of freedom which are coupled non-linearly and over long distances. If L denotes a reference length in the flow field, a volume of the order of magnitude L^3 contains degrees of freedom in the order of magnitude $(Re)^{9/4}$, as we shall see below. Hence, for a Reynolds number of $Re = 10^8$ – quite realistic under geophysical conditions – there are $\sim 10^{18}$ active degrees of freedom in a L^3-volume.

Phenomenological descriptions of turbulence include the following properties, which will be addressed in more detail in this chapter, cf. (Tsinober, 2001):

a) irregular, random behaviour of the velocity vector in time and space;
b) sensitive dependence of the flow pattern on small perturbations, i.e. neighbouring liquid particles move away extremely rapidly from one another, causing the motion to be unpredictable;
c) in contrast, there are statistical characteristics which are stable and predictable;
d) innumerable interlaced and self-reinforcing eddies lead to scaling laws in time and space;
e) enhanced transport properties of momentum, energy and passive particles cause an effective mixing.

The stormy developments in the last forty to fifty years in chaos theory make these characteristics appear in a completely different light. Even simple deterministic equations – such as the Lorenz system, the logistic and Hénon mapping, for example – can lead to motions which exhibit the typical characteristics of turbulent behaviour (such as irregularity, mixing and scale invariance) though in these sytems, the number of dynamically active degrees of freedom remains small in contrast to fully developed turbulence. Yet there are closed systems in hydrodynamics – some of them have been presented in the preceeding chapters – which, at the transition to turbulence, reveal a behaviour similar to low-dimensional chaos: Rayleigh-Bénard or Taylor-Couette flow, for instance. At the transition to turbulence, these systems possess only a few dynamical degrees of freedom which are strongly coupled over the entire flow field. However, in these systems as well, the number of degrees of freedom grows drastically with a further increase in the energy supply.

On the one hand, the supposed contradiction between deterministic equations of motion and chaotic behaviour has been resolved; on the other hand, the spatio-temporal irregularity of fully developed turbulence and the extremely high number of strongly coupled modes give rise to much greater problems which can only be solved with the aid of statistical methods. As a consequence, the resolution of all scales in direct simulations is only possible for moderate Reynolds numbers even if we have access to the latest high performance computers. This has led to the development of a series of turbulence models like the Reynolds stress models, the large eddy simulation models or the PDF models, all of which, however, are not yet universally applicable. The development of better turbulence models will only be possible once we have achieved a better understanding of the underlying physical processes (Pope, 2000).

In the following sections we make use of the basic rules of vector analysis and field theory as summarised in, for example, (Bronstein et al., 1995).

9.1 Dynamics of Incompressible Fluids

In the following section, we set up the basic hydrodynamic equations. Important characteristics such as the local energy dissipation rate will be introduced. As a starting point for the investigation of vortex dynamics in section 9.2, we show some vortex solutions of the Navier-Stokes equation and discuss the Langrangian description of flows.

Subsequently, we discuss properties of so-called fully developed turbulence. Here, statistical considerations come to the fore. Phenomenological theories dating back to A.N. Kolmogorov and W. Heisenberg are treated explicitly.

9.1.1 The Basic Hydrodynamic Equations

We became acquainted with the basic hydrodynamic equations for Newtonian fluids in section 7.1. In the following, we restrict our considerations to incompressible Newtonian fluids such as water, for example, for which, for simplicity's sake, we assume a constant density; the behaviour can then be described by the Navier-Stokes equation[2]

$$\frac{\partial}{\partial t}\boldsymbol{u}(\boldsymbol{x},t)+\boldsymbol{u}(\boldsymbol{x},t)\cdot\nabla\boldsymbol{u}(\boldsymbol{x},t)=-\nabla p(\boldsymbol{x},t)+\nu\Delta\boldsymbol{u}(\boldsymbol{x},t)+\boldsymbol{f}(\boldsymbol{x},t) \tag{9.1.1}$$

Here, $\boldsymbol{u}(\boldsymbol{x},t)$ denotes the velocity field. The quantity $p(\boldsymbol{x},t)=P(\boldsymbol{x},t)/\varrho$ is defined as the ratio of pressure $P(\boldsymbol{x},t)$ to constant density ϱ. The parameter ν denotes the kinematic viscosity calculated as the quotient of molecular viscosity η and the constant density ϱ of the fluid. External forces like the gravity are combined in the volume force density $\boldsymbol{f}(\boldsymbol{x},t)$. Owing to the incompressibility of the fluid, the velocity field is source-free respectively divergence-free[3]

$$\nabla\cdot\boldsymbol{u}(\boldsymbol{x},t)=0 \tag{9.1.2}$$

The Navier-Stokes equation has to be supplemented by an initial condition

$$\boldsymbol{u}(\boldsymbol{x},t=0)=\boldsymbol{u}^0(\boldsymbol{x}),\qquad \nabla\cdot\boldsymbol{u}^0(\boldsymbol{x})=0 \tag{9.1.3}$$

and with boundary conditions. At the boundaries denoted δV, the no-slip condition

$$\boldsymbol{u}(\boldsymbol{x},t)|_{\delta V}=\boldsymbol{U}(\boldsymbol{x},t)|_{\delta V} \tag{9.1.4}$$

is valid where the velocity $\boldsymbol{U}(\boldsymbol{x},t)$ is prescribed at the boundary δV.

Introducing the dimensionless quantities

$$\tilde{\boldsymbol{u}}=\frac{\boldsymbol{u}}{U},\qquad \tilde{t}=\frac{t}{T},\qquad \tilde{\boldsymbol{r}}=\frac{\boldsymbol{r}}{L},\qquad \tilde{p}=\frac{p}{U^2},\qquad \tilde{\boldsymbol{f}}=\boldsymbol{f}\frac{T}{U^2},\qquad U=\frac{L}{T} \tag{9.1.5}$$

[2] From time to time, there were discussions on whether the Navier-Stokes equation allows an adequate description of turbulent flows. Up to now, there is no indication that the assumptions on which eq. (9.1.1) is based lose their validity in the case of turbulent behaviour since, so far, all laboratory experiments seem to be consistent with the numerical simulations (Tsinober, 2001; Wilczek et al., 2008).

[3] In general, changes in the pressure of the fluid cause changes in the density which are only relevant if the respective velocities are of the same order of magnitude as the propagation velocity of sound, i.e. the fluid may always be assumed to be incompressible when no sonic waves occur (Batchelor, 1956).

and subsequently neclecting the tilde, one obtains the Navier-Stokes equation in the form

$$\frac{\partial}{\partial t}\boldsymbol{u}(\boldsymbol{x},t) + \boldsymbol{u}(\boldsymbol{x},t)\cdot\nabla\boldsymbol{u}(\boldsymbol{x},t) = -\nabla p(\boldsymbol{x},t) + \frac{1}{Re}\Delta\boldsymbol{u}(\boldsymbol{x},t) + \boldsymbol{f}(\boldsymbol{x},t) \tag{9.1.6}$$

where the quantities L, T and $U = L/T$ are the characteristic scales for the length, time and velocity of the flow. The Navier-Stokes equation has only one external parameter, the dimensionless Reynolds number, which is given as

$$Re = \frac{UL}{\nu} \tag{9.1.7}$$

cf. (Reynolds, 1883). It is a measure for the ratio of the acceleration of the fluid, characterised by the quantity $U/T = U^2/L$, and the dissipation rate $(\nu U/L^2)$. Table 9.1.1 presents some typical Reynolds numbers for a series of bodies in a flow. Flows with the same Reynolds number and scaled boundary conditions have identical scaled equations and are hence geometrically similar within the framework of eq. (9.1.6). One therefore speaks of the similitude principle. This principle makes it possible to use flow data measured experimentally on a small scale, for example on a model in a wind channel, in order to predict the flow behaviour of a real large-scale application.

air		water	
car	10^7	ship	10^7-10^{10}
airliner	10^9	small fish	10^5
bird	$2 \cdot 10^7$	dolphin	10^7

Table 9.1.1: Some typical Reynolds numbers

Qualitatively, the onset of turbulent flow can be estimated on the basis of the Navier-Stokes equation by means of the Reynolds number. Turbulence can be observed if the influence of the non-linear convection term $\boldsymbol{u}(\boldsymbol{x},t)\cdot\nabla\boldsymbol{u}(\boldsymbol{x},t)$ predominates over the smoothing effect of the dissipation term. The ratio of the moduli of these terms is a dimensionless number, the magnitude of which provides information on whether the flow is laminar or turbulent.

Apart from the velocity field $\boldsymbol{u}(\boldsymbol{x},t)$, the Navier-Stokes equation contains the pressure $p(\boldsymbol{x},t)$. For incompressible flow, the pressure gradient $\nabla p(\boldsymbol{x},t)$ is determined by the incompressibility condition eq. (9.1.2) together with the boundary conditions. This can been seen from the following Poisson equation, obtained if we take the divergence of the Navier-Stokes equation (9.1.1)

$$\Delta p(\boldsymbol{x},t) = -\sum_{\mathrm{i,j}} \frac{\partial u_{\mathrm{i}}(\boldsymbol{x},t)}{\partial x_{\mathrm{j}}}\frac{\partial u_{\mathrm{j}}(\boldsymbol{x},t)}{\partial x_{\mathrm{i}}} \tag{9.1.8}$$

where we assume that the volume forces are divergence-free. The Poisson equation (9.1.8) can be solved by means of the Green function, which will be addressed in more detail in section 9.1.3. If the pressure and the normal component of the pressure gradient are specified along the boundary δV, the pressure in the flow is – except for a constant – uniquely determined. For a fluid flow in a three-dimensional infinite domain, for example, we thus obtain

$$p(\boldsymbol{x}, t) = \int \frac{1}{4\pi|\boldsymbol{x} - \boldsymbol{x}'|} \sum_{\mathrm{i,j}} \frac{\partial u_{\mathrm{i}}(\boldsymbol{x}', t)}{\partial x'_{\mathrm{j}}} \frac{\partial u_{\mathrm{j}}(\boldsymbol{x}', t)}{\partial x'_{\mathrm{i}}} \mathrm{d}\boldsymbol{x}' \tag{9.1.9}$$

This result can be derived directly from the Poisson equation and the Green function of the Laplace operator discussed in the appendix to section 9.1.3.

Because of the pressure term, the dynamics of an incompressible flow is explicitly *non-local*: to put it another way, the velocity gradient at a point $\boldsymbol{x}$ is not only determined by the flow behaviour in the immediate neighbourhood of this point; apart from the specification of the boundary conditions, the velocity field of the whole domain is needed. In fact, the Navier-Stokes equation should actually be regarded as an integro-differential equation for the velocity field. Let us assume that the inhomogeneity of the Poisson equation is localised in a neighbourhood of a point $\boldsymbol{r}'$, for example in form of a small eddy. According to eq. (9.1.8), at every arbitrarily chosen point $\boldsymbol{r}$, this eddy generates pressure gradients which are proportional to $\frac{1}{\boldsymbol{r}-\boldsymbol{r}'}$ and which fade out only a little. By virtue of the Navier-Stokes equation, these differences in the pressure generate in turn velocity gradients at the point $\boldsymbol{r}$. Thus, the small eddy changes the velocity field even at more remote locations.

Together with non-linearity, the non-locality is one of the most important causes for the complexity of turbulent flows since the pressure induces the generation of long-range forces (Pope, 2000; Tsinober, 2001; Davidson, 2006).

Because of the advective term, the Navier-Stokes equation is non-linear and cannot in general be solved analytically for all boundary conditions. Since all scales are strongly coupled, the solutions react very sensitively to changes in the boundary and initial conditions. For high Reynolds numbers $Re \to \infty$ or low viscosity $\nu \to 0$, the smoothing effect of the dissipation term is highly reduced and the flow tends to generate small-scale structures.

9.1.2 The Local Energy Dissipation Rate

It is possible to derive an energy balance equation directly from the Navier-Stokes equation. Due to the viscosity of the fluid, a certain amount of mechanical energy is transformed to heat. This is characterised by the so-called local energy dissipation rate, which plays an important role in the theory of fully developed turbulence.

The kinetic energy of a fluid volume V of a fluid with constant density $\varrho = 1$ is determined by the integral

$$\int_V \frac{\boldsymbol{u}^2}{2} \mathrm{d}V \tag{9.1.10}$$

Here, the density of the kinetic energy of a fluid flow, $e_{kin} = \frac{\boldsymbol{u}^2}{2}$, has to fulfil a balance equation of the form

$$\frac{\partial}{\partial t} e_{kin}(\boldsymbol{x},t) + \nabla \cdot \boldsymbol{J}^e(\boldsymbol{x},t) = q(\boldsymbol{x},t) \tag{9.1.11}$$

where the quantity

$$\int_V q(\boldsymbol{x},t)\mathrm{d}V \tag{9.1.12}$$

denotes the kinetic energy which is generated or destroyed per unit time in a volume V. According to Gauss's theorem, the energy transported through a fixed surface per unit time is given by the surface integral

$$\int \nabla \cdot \boldsymbol{J}^e \mathrm{d}V = \int_\Omega \boldsymbol{J}^e \cdot \boldsymbol{n}\, \mathrm{d}\Omega \tag{9.1.13}$$

where $\boldsymbol{n}$ is the unit normal vector directed outwards.

The flux density $\boldsymbol{J}^e(\boldsymbol{x},t)$ and the source term $q(\boldsymbol{x},t)$ follow directly from the Navier-Stokes equation. For $\varrho = 1$, we obtain

$$\begin{aligned} \boldsymbol{J}^e(\boldsymbol{x},t) &= \boldsymbol{u}(\boldsymbol{x},t)\left[\frac{\boldsymbol{u}(\boldsymbol{x},t)^2}{2} + p(\boldsymbol{x},t)\right] - \nu\nabla\frac{\boldsymbol{u}^2(\boldsymbol{x},t)}{2} - \nu\boldsymbol{u}(\boldsymbol{x},t)\cdot\nabla\boldsymbol{u}(\boldsymbol{x},t) \\ q(\boldsymbol{x},t) &= \boldsymbol{f}(\boldsymbol{x},t)\cdot\boldsymbol{u}(\boldsymbol{x},t) - \varepsilon(\boldsymbol{x},t) \end{aligned} \tag{9.1.14}$$

This relation will be derived at the end of this section. The flux density $\boldsymbol{J}^e(\boldsymbol{x},t)$ of the kinetic energy is composed of several components. The term $\boldsymbol{u}(\boldsymbol{x},t)\frac{\boldsymbol{u}(\boldsymbol{x},t)^2}{2}$ describes the kinetic energy transported through the surface and $\boldsymbol{u}(\boldsymbol{x},t)p(\boldsymbol{x},t)$ is the work per unit time done by the displacement against the pressure. The two remaining terms contain the viscous components. The source term again possesses two parts. The power density $\boldsymbol{f}(\boldsymbol{x},t)\cdot\boldsymbol{u}(\boldsymbol{x},t)$ is related to the change of kinetic energy induced by the external force $\boldsymbol{f}$ and $\varepsilon(\boldsymbol{x},t)$ denotes the dissipated kinetic energy per unit mass and unit time

$$\varepsilon(\boldsymbol{x},t) = \frac{\nu}{2}\sum_{\mathrm{i,j}}\left(\frac{\partial u_\mathrm{i}}{\partial x_\mathrm{j}} + \frac{\partial u_\mathrm{j}}{\partial x_\mathrm{i}}\right)^2 \tag{9.1.15}$$

The local energy dissipation rate is related to the gradient of the velocity field: large energy dissipation occurs in regions with high velocity gradients. It will be shown in section 9.3 that the local energy dissipation rate plays an important role in the statistical description of hydrodynamic turbulence.

Appendix: Derivation of Energy Balance for the Navier-Stokes Equation

The equation of balance for the density of kinetic energy, eq. (9.1.11), can be derived directly from the Navier-Stokes equation by scalar multiplication by the velocity field $\boldsymbol{u}(\boldsymbol{x},t)$. We obtain

$$\frac{\partial}{\partial t}\frac{\boldsymbol{u}(\boldsymbol{x},t)^2}{2} + \nabla\cdot\left[\boldsymbol{u}(\boldsymbol{x},t)\frac{\boldsymbol{u}(\boldsymbol{x},t)^2}{2}\right] = \\ -\nabla\cdot[\boldsymbol{u}(\boldsymbol{x},t)\,p(\boldsymbol{x},t)] + \nu\boldsymbol{u}(\boldsymbol{x},t)\cdot\Delta\boldsymbol{u}(\boldsymbol{x},t) + \boldsymbol{u}(\boldsymbol{x},t)\cdot\boldsymbol{f}(\boldsymbol{x},t) \qquad (9.1.16)$$

Here, we made use of the incompressibility of the flow field

$$\begin{aligned} \boldsymbol{u}(\boldsymbol{x},t)\cdot[\boldsymbol{u}(\boldsymbol{x},t)\cdot\nabla\boldsymbol{u}(\boldsymbol{x},t)] &= \nabla\cdot\left[\boldsymbol{u}(\boldsymbol{x},t)\frac{\boldsymbol{u}(\boldsymbol{x},t)^2}{2}\right] \\ \boldsymbol{u}(\boldsymbol{x},t)\cdot\nabla p(\boldsymbol{x},t) &= \nabla\cdot[\boldsymbol{u}(\boldsymbol{x},t)p(\boldsymbol{x},t)] \end{aligned} \qquad (9.1.17)$$

The aim is to transform eq. (9.1.16) to the form of the balance equation (9.1.11). To this end, the expression

$$\nu\boldsymbol{u}(\boldsymbol{x},t)\cdot\Delta\boldsymbol{u}(\boldsymbol{x},t) = \frac{1}{2}\nu\Delta\boldsymbol{u}(\boldsymbol{x},t)^2 - \nu\sum_{\mathrm{i,j}}\left(\frac{\partial u_\mathrm{i}(\boldsymbol{x},t)}{\partial x_\mathrm{j}}\right)^2 \qquad (9.1.18)$$

can be transformed as follows, using eq. (9.1.15) to split it into the divergence of a vector and a source term

$$\begin{aligned} \nu\sum_{\mathrm{i,j}}\left(\frac{\partial u_\mathrm{i}(\boldsymbol{x},t)}{\partial x_\mathrm{j}}\right)^2 &= \frac{1}{2}\nu\sum_{\mathrm{i,j}}\left(\frac{\partial u_\mathrm{i}(\boldsymbol{x},t)}{\partial x_\mathrm{j}} + \frac{\partial u_\mathrm{j}(\boldsymbol{x},t)}{\partial x_\mathrm{i}}\right)^2 \\ &\quad - \nu\sum_{\mathrm{i,j}}\frac{\partial u_\mathrm{i}(\boldsymbol{x},t)}{\partial x_\mathrm{j}}\frac{\partial u_\mathrm{j}(\boldsymbol{x},t)}{\partial x_\mathrm{i}} \\ &= \varepsilon(\boldsymbol{x},t) - \nu\nabla\cdot\left[\boldsymbol{u}(\boldsymbol{x},t)\cdot\nabla\boldsymbol{u}(\boldsymbol{x},t)\right] \end{aligned} \qquad (9.1.19)$$

Inserting eqs. (9.1.17, 9.1.18) together with eq. (9.1.19) into eq. (9.1.16), we obtain the following balance equation

$$\frac{\partial}{\partial t}\frac{\boldsymbol{u}(\boldsymbol{x},t)^2}{2} + \nabla\cdot\Big\{\boldsymbol{u}(\boldsymbol{x},t)\left[\frac{\boldsymbol{u}(\boldsymbol{x},t)^2}{2} + p(\boldsymbol{x},t)\right] - \nu\nabla\frac{\boldsymbol{u}(\boldsymbol{x},t)^2}{2} \\ - \nu\boldsymbol{u}(\boldsymbol{x},t)\cdot\nabla\boldsymbol{u}(\boldsymbol{x},t)\Big\} = \boldsymbol{u}(\boldsymbol{x},t)\cdot\boldsymbol{f}(\boldsymbol{x},t) - \varepsilon(\boldsymbol{x},t) \qquad (9.1.20)$$

The flux density $\boldsymbol{J}^e$ and the source term $q(\boldsymbol{x},t)$ can now be read off explicitly.

9.1.3 The Vorticity Equation

Leonardo da Vinci characterised turbulent flows by means of a multitude of interlacing vortex structures, see fig. 9.0.2. This kind of structure is denoted a *coherent structure*, although it is not easy to give a complete mathematical characterisation of it. The introduction of the vorticity is a first step towards identifying these vortex structures

$$\boldsymbol{\omega}(\boldsymbol{x},t) = \nabla\times\boldsymbol{u}(\boldsymbol{x},t) \qquad (9.1.21)$$

where a spatial concentration of this quantity indicates the emergence of vortices. While the pressure ensures that the velocity field is never highly localised, the vorticity in two-dimensional turbulence organises itself in oval vortex structures like, for example, the Great Red Spot of Jupiter, whereas three-dimensional turbulence favours thin vortex filaments. Vortices can be regarded as elementary dissipative structures. The study of vortex dynamics thus offers the opportunity of gaining insight into the physical and kinematic processes in turbulent flows.

A complete description of incompressible turbulent flows can be given by means either of the velocity field or of the vorticity field. The reason is the validity of the following relations

$$\begin{aligned} \nabla \cdot \boldsymbol{u}(\boldsymbol{x}, t) &= 0 \\ \nabla \times \boldsymbol{u}(\boldsymbol{x}, t) &= \boldsymbol{\omega}(\boldsymbol{x}, t) \end{aligned} \tag{9.1.22}$$

Given the vorticity and the corresponding boundary conditions for the velocity field $\boldsymbol{u}(\boldsymbol{x}, t)$, the velocity field can be calculated from these two relations.

The system of equations (9.1.22) has the same form as the basic equations of magnetostatics if we identify $\boldsymbol{u}(\boldsymbol{x}, t)$ with the magnetic induction and the vorticity $\boldsymbol{\omega}(\boldsymbol{x}, t)$ with the current density. Then, the velocity field is determined by an analogon of the Biot-Savart law, which will be derived in detail in the appendix to this section and which allows an inversion of the curl operator for divergence-free fields

$$\boldsymbol{u}(\boldsymbol{x}, t) = \boldsymbol{u}_{\mathrm{pot}}(\boldsymbol{x}, t) + \int \boldsymbol{\omega}(\boldsymbol{x}', t) \times \boldsymbol{K}(\boldsymbol{x} - \boldsymbol{x}') \, \mathrm{d}\boldsymbol{x}' \tag{9.1.23}$$

Here, $\boldsymbol{u}_{\mathrm{pot}}(\boldsymbol{x}, t)$ is an arbitrary potential flow with the potential $\Phi(\boldsymbol{x}, t)$

$$\boldsymbol{u}_{\mathrm{pot}}(\boldsymbol{x}, t) = -\nabla \Phi(\boldsymbol{x}, t) \tag{9.1.24}$$

$\boldsymbol{K}(\boldsymbol{x}, t)$ denotes the negative gradient, $\boldsymbol{K}(\boldsymbol{x}) = -\nabla G(\boldsymbol{x})$, of the Green function for the Laplace operator defined by

$$\Delta G(\boldsymbol{x}) = -\delta(\boldsymbol{x}) \tag{9.1.25}$$

see (Courant and Hilbert, 1993; Bronstein et al., 1995). Owing to the analogon to magnetostatics, the relation (9.1.23) is often also referred to in hydrodynamics as the Biot-Savart law.

The evolution equation for the vorticity $\boldsymbol{\omega}(\boldsymbol{x}, t)$, the so-called Helmholtz vorticity equation, can be obtained from the Navier-Stokes equation (9.1.1) by taking the curl

$$\left(\frac{\partial}{\partial t} + \boldsymbol{u}(\boldsymbol{x}, t) \cdot \nabla\right) \boldsymbol{\omega}(\boldsymbol{x}, t) = \boldsymbol{\omega}(\boldsymbol{x}, t) \cdot \nabla \boldsymbol{u}(\boldsymbol{x}, t) + \nu \Delta \boldsymbol{\omega}(\boldsymbol{x}, t) + \nabla \times \boldsymbol{f}(\boldsymbol{x}, t) \tag{9.1.26}$$

Although the contribution of the pressure is thus eliminated, the non-locality is nevertheless preserved, because the vorticity equation (9.1.26) still contains the velocity field $\boldsymbol{u}(\boldsymbol{x}, t)$. Again, according to the Biot-Savart law (9.1.23) for incompressible flows, this field is determined by the vorticity via a non-local relation. The vorticity equation is therefore also an integro-differential equation for $\boldsymbol{\omega}$.

The vorticity equation (9.1.26) reveals a fundamental difference between flows in two- and three-dimensional regions. For two-dimensional flows, the vorticity possesses only a component $\omega_z(\boldsymbol{x}, t)$ which is assumed to be directed in the z-direction. Therefore, the vortex stretching term

$$\boldsymbol{\omega}(\boldsymbol{x}, t) \cdot \nabla \boldsymbol{u}(\boldsymbol{x}, t) = \omega_z(x, y, t) \frac{\partial}{\partial z} \boldsymbol{u}(\boldsymbol{x}, t) = \mathbf{0} \tag{9.1.27}$$

vanishes identically. In three-dimensional flows, this term leads to an increase of both the vorticity and its gradient, and mainly produces a concentration of $\boldsymbol{\omega}$ in thin vortex tubes and sheets on various scales (Batchelor and Townsend, 1949; Lundgren, 1982). This can clearly be seen in Colour Plate XIIIb, p. 605, which shows coherent structures of the vorticity above a threshold obtained by numerical simulation of the fluid flow in the turbulent regime (Wilczek, 2009). The existence of these irregularly distributed structures prompted the interpretation that the vorticity and hence also the energy dissipation are multifractally distributed (Parisi and Frisch, 1985).

Colour Plate XIIIa, p. 605 displays the trajectories of Lagrangian particles circling around a typical vortex tube (Daitche, 2009). The sequential order of the colors orange-red, olive-green, dark blue indicates the increase of the vorticity field $|\boldsymbol{\omega}|$ inside the tube. The spiralling pathlines illustrate the rotational motion of the Lagrangian particles together with the simultaneous translation of the vortex tube.

• Kelvin's circulation theorem

An important quantity for describing vortex flows is the circulation Γ, which also plays a major role in aerodynamics in the determination of lift forces for airfoils (Truckenbrodt, 1996). It is defined as the line integral of the velocity $\boldsymbol{u}$ over a closed line L and can be related to the vorticity via Stokes' theorem

$$\Gamma_L = \oint_L \boldsymbol{u} \cdot \mathrm{d}\boldsymbol{s} = \int_F \boldsymbol{\omega} \cdot \boldsymbol{n} \, \mathrm{d}F \tag{9.1.28}$$

Here, F denotes the area enclosed by the contour line L. The circulation theorem of W. Thomson (Lord Kelvin) states that, in an incompressible, inviscid, barotropic[4] flow with conservative body forces, the circulation Γ_L along a closed curve moving with the fluid remains constant with time (Lord Kelvin Thomson, 1869).

Appendix: The BiotSavart law

The vanishing divergence of the velocity field allows $\boldsymbol{u}(\boldsymbol{x}, t)$ to be described as the sum of a gradient field according to eq. (9.1.24) and a rotational field

$$\boldsymbol{u}(\boldsymbol{x}, t) = \nabla \Phi(\boldsymbol{x}, t) + \nabla \times \boldsymbol{A}(\boldsymbol{x}, t) \tag{9.1.29}$$

where the following relations must be fulfilled

[4] A fluid is denoted as barotropic if the pressure does not depend on the temperature but only on the density.

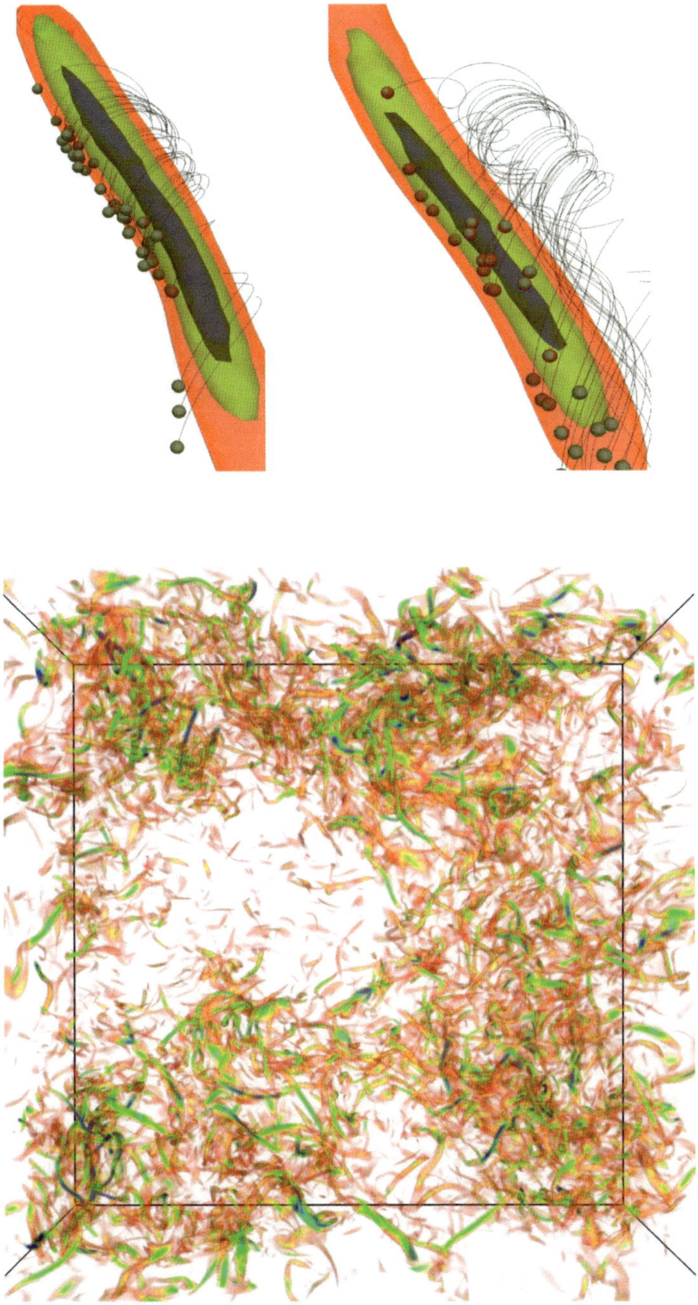

a)

b)

Plate XIII Computer simulations: a) vortex tube with Lagrangian particle paths for two different times (Daitche, 2009), b) magnitude of the vorticity field: the vorticity organises itself in thin filaments (Wilczek, 2009).

$$\Delta\Phi(\boldsymbol{x},t) = 0 \tag{9.1.30}$$

Due to $\nabla \times \nabla\Phi = \boldsymbol{0}$, the vector potential $\boldsymbol{A}(\boldsymbol{x},t)$ satisfies the relation

$$\nabla \times [\nabla \times \boldsymbol{A}(\boldsymbol{x},t)] = \boldsymbol{\omega}(\boldsymbol{x},t) \tag{9.1.31}$$

There is some freedom in the choice of the vector potential which allows us to assume the vector potential to be divergence-free. The following identity is thus valid

$$\nabla \times [\nabla \times \boldsymbol{A}] = \nabla(\nabla \cdot \boldsymbol{A}) - \Delta\boldsymbol{A} = -\Delta\boldsymbol{A} = \boldsymbol{\omega}(\boldsymbol{x},t) \tag{9.1.32}$$

Here again, we obtain a Poisson equation for the determination of the vector potential. Denoting the Green function of the Laplace operator as $G(\boldsymbol{x},\boldsymbol{x}')$

$$\Delta G(\boldsymbol{x}-\boldsymbol{x}') = -\delta(\boldsymbol{x}-\boldsymbol{x}') \tag{9.1.33}$$

we obtain

$$\boldsymbol{A}(\boldsymbol{x},t) = \int G(\boldsymbol{x}-\boldsymbol{x}')\,\boldsymbol{\omega}(\boldsymbol{x}',t)\,\mathrm{d}\boldsymbol{x}' \tag{9.1.34}$$

According to eq. (9.1.29), the above relation allows us to determine the velocity field, where a term $\boldsymbol{u}_{\text{pot}} = -\nabla\Phi$ may be added and the curl with respect to $\boldsymbol{x}$ is moved inside the integral

$$\boldsymbol{u}(\boldsymbol{x},t) = \nabla_{\boldsymbol{x}} \times \int G(\boldsymbol{x}-\boldsymbol{x}')\boldsymbol{\omega}(\boldsymbol{x}',t)\,\mathrm{d}\boldsymbol{x}' \tag{9.1.35}$$

This finally leads to the Biot-Savart law

$$\boldsymbol{u}(\boldsymbol{x},t) = \int \boldsymbol{\omega}(\boldsymbol{x}',t) \times \boldsymbol{K}(\boldsymbol{x}-\boldsymbol{x}')\,\mathrm{d}\boldsymbol{x}' \tag{9.1.36}$$

where we defined a quantity

$$\boldsymbol{K}(\boldsymbol{x}-\boldsymbol{x}') = -\nabla_{\boldsymbol{x}} G(\boldsymbol{x}-\boldsymbol{x}') \tag{9.1.37}$$

We now consider a fluid in the infinitely extended space. In two respectively three dimensions, the Green functions for the Laplace operator have the form

$$G(\boldsymbol{x}-\boldsymbol{x}') = \begin{cases} -\frac{1}{2\pi}\ln|\boldsymbol{x}-\boldsymbol{x}'| & \text{for } \boldsymbol{x} \in \mathbb{R}^2 \\ \frac{1}{4\pi}\frac{1}{|\boldsymbol{x}-\boldsymbol{x}'|} & \text{for } \boldsymbol{x} \in \mathbb{R}^3 \end{cases} \tag{9.1.38}$$

Analogously, we obtain

$$\boldsymbol{K}(\boldsymbol{x}-\boldsymbol{x}') = \begin{cases} \frac{1}{2\pi}\frac{\boldsymbol{x}-\boldsymbol{x}'}{|\boldsymbol{x}-\boldsymbol{x}'|^2} & \text{for } \boldsymbol{x} \in \mathbb{R}^2 \\ \frac{1}{4\pi}\frac{\boldsymbol{x}-\boldsymbol{x}'}{|\boldsymbol{x}-\boldsymbol{x}'|^3} & \text{for } \boldsymbol{x} \in \mathbb{R}^3 \end{cases} \tag{9.1.39}$$

9.1.4 The Lagrangian Description of Fluid Flows

Up to now, we have used the so-called *Eulerian description* for the Navier-Stokes equation (9.1.1): the flow is observed in a *fixed* coordinate system and passes the observer. In this view, the velocity field $\boldsymbol{u}(\boldsymbol{x}, t)$ is a function of the position $\boldsymbol{x}$ and time t. An alternative description is obtained if one follows an individual particle along its path in the flow. This approach is called the *Lagrangian description*, see section 7.1. Here, the spatial coordinate of the fluid particle $\boldsymbol{X}(\boldsymbol{y}, t)$ is used as the dependent variable, where $\boldsymbol{y}$ denotes the position of the particle at time $t = 0$. Therefore, for a selected particle, $\boldsymbol{X}(\boldsymbol{y}, t)$ is only a function of time. The trajectory of the particle, the so-called Lagrangian path, can be determined as the solution of a differential equation

$$\frac{\mathrm{d}}{\mathrm{d}t}\boldsymbol{X}(\boldsymbol{y}, t) = \boldsymbol{u}(\boldsymbol{X}(\boldsymbol{y}, t), t) \tag{9.1.40}$$

with the initial condition

$$\boldsymbol{X}(\boldsymbol{y}, 0) = \boldsymbol{y} \tag{9.1.41}$$

and describes the motion of a fluid particle in the Eulerian velocity field $\boldsymbol{u}(\boldsymbol{x}, t)$.

The Lagrangian treatment of turbulent flows has a long tradition, going back as far as the 1920s to the works of L. F. Richardson and G. I. Taylor and continued by R. H. Kraichnan. If the intention is to study the transport of fluid particles (marked particles), their mixing and diffusion as well as the dynamics of passive tracers, a description via transported particles is of advantage. A fresh boost was given to the Lagrangian description on the one hand by problems in engineering, chemistry and climate research, where questions of mixing and combustion or the spread of pollutants are at the forefront. In addition, experimental methods have recently been developed in high energy physics and in the field of *Continuous Wave Doppler* methods and *Particle Image Velocimetry* (PIV) which enable us to take high-resolution measurements of the particle paths (La Porta et al., 2001; Mordant et al., 2002). On the other hand, the Lagrangian description is also gaining importance in theoretical studies (Friedrich, 2003).

It is also advantageous to introduce the Lagrangian velocity along the trajectory $\boldsymbol{X}(\boldsymbol{y}, t)$. It is defined by the relation

$$\boldsymbol{U}(\boldsymbol{y}, t) = \boldsymbol{u}(\boldsymbol{X}(\boldsymbol{y}, t), t) \tag{9.1.42}$$

thus allowing us to put the basic equations of hydrodynamics in a form which pinpoints the direct connection to Newton's law of motion from classical mechanics

$$\begin{aligned}\frac{\mathrm{d}}{\mathrm{d}t}\boldsymbol{X}(\boldsymbol{y}, t) &= \boldsymbol{U}(\boldsymbol{y}, t)\\ \frac{\mathrm{d}}{\mathrm{d}t}\boldsymbol{U}(\boldsymbol{y}, t) &= [-\nabla p(\boldsymbol{x}, t) + \nu\Delta\boldsymbol{u}(\boldsymbol{x}, t) + \boldsymbol{f}(\boldsymbol{x}, t)]_{\boldsymbol{x}=\boldsymbol{X}(\boldsymbol{y},t)}\end{aligned} \tag{9.1.43}$$

The temporal derivative of the Lagrangian velocity is the acceleration acting on a fluid particle. It is composed of three components: the pressure gradient, the dissipative term and the accelation due to the volume force. It follows from the last equation that fluctuations in the Lagrangian acceleration are caused by variations of the pressure gradient, the dissipation as well as of the external forces.

The dynamics of the vorticity can also be given in Lagrangian form. To this end, we define the Lagrangian vorticity along $\boldsymbol{X}(\boldsymbol{y},t)$ as

$$\boldsymbol{\Omega}(\boldsymbol{y},t) = \boldsymbol{\omega}(\boldsymbol{X}(\boldsymbol{y},t),t) \tag{9.1.44}$$

According to the vorticity equation (9.1.26), its evolution equation for $\boldsymbol{f} = \boldsymbol{0}$ reads

$$\frac{\mathrm{d}}{\mathrm{d}t}\boldsymbol{\Omega}(\boldsymbol{y},t) = \boldsymbol{\Omega}(\boldsymbol{y},t) \cdot [\nabla \boldsymbol{u}(\boldsymbol{x},t)]_{\boldsymbol{x}=\boldsymbol{X}(\boldsymbol{y},t)} + \nu[\Delta\boldsymbol{\Omega}(\boldsymbol{x},t)]_{\boldsymbol{x}=\boldsymbol{X}(\boldsymbol{y},t)} \tag{9.1.45}$$

The motion of the Lagrangian particles is determined by the evolution equation for $\boldsymbol{X}(\boldsymbol{y},t)$. Using the Biot-Savart law (9.1.23), we obtain

$$\frac{\mathrm{d}}{\mathrm{d}t}\boldsymbol{X}(\boldsymbol{y},t) = \boldsymbol{u}(\boldsymbol{X}(\boldsymbol{y},t),t) = \int \boldsymbol{\omega}(\boldsymbol{x}',t) \times \boldsymbol{K}\left(\boldsymbol{X}(\boldsymbol{y},t) - \boldsymbol{x}'\right)\,\mathrm{d}\boldsymbol{x}' \tag{9.1.46}$$

Since we restrict our consideration to the dynamics of incompressible fluids, the integration over the volume at time t can simply be replaced by an integral over the initial volume at time $t=0$ since, due to the incompressibility of the motion of the flow, the coordinate transformation

$$\boldsymbol{x}' = \boldsymbol{X}(\boldsymbol{y}',t) \tag{9.1.47}$$

has the Jacobian determinant $J=1$. The equation of motion then has the form

$$\frac{\mathrm{d}}{\mathrm{d}t}\boldsymbol{X}(\boldsymbol{y},t) = \int \boldsymbol{\Omega}(\boldsymbol{y}',t) \times \boldsymbol{K}(\boldsymbol{X}(\boldsymbol{y},t) - \boldsymbol{X}(\boldsymbol{y}',t))\,\mathrm{d}\boldsymbol{y}' \tag{9.1.48}$$

We recall that a potential field can also be added to the velocity field.

• Two-dimensional flow of inviscid fluids

For generally two-dimensional flows, the so-called vortex stretching term vanishes identically in the vorticity equation (9.1.26)

$$\boldsymbol{\omega}(\boldsymbol{x},t) \cdot \nabla \boldsymbol{u}(\boldsymbol{x},t) = \boldsymbol{0} \tag{9.1.49}$$

because the vorticity has only one component in z-direction and the velocity fields depends only on the coordinates x and y.

In addition, in the case of inviscid or ideal fluids ($\nu = 0$), the second term on the right-hand side of eq. (9.1.45) vanishes, i.e. the Lagrangian vorticity is constant in time

$$\boldsymbol{\Omega}(\boldsymbol{y},t) = \boldsymbol{\Omega}(\boldsymbol{y},0) \tag{9.1.50}$$

Hence, for incompressible two-dimensional turbulence, the dynamics of particles is given by the integro-differential equation

$$\frac{\mathrm{d}}{\mathrm{d}t}\boldsymbol{X}(\boldsymbol{y},t) = \int \boldsymbol{\Omega}(\boldsymbol{y}',0) \times \boldsymbol{K}(\boldsymbol{X}(\boldsymbol{y},t) - \boldsymbol{X}(\boldsymbol{y}',t))\,\mathrm{d}\boldsymbol{y}' \tag{9.1.51}$$

where an arbitrary distribution of the vorticity $\boldsymbol{\Omega}(\boldsymbol{y}',0)$ can be considered for $t = 0$. This aspect will be taken up again in the section on point vortices.

• Three-dimensional flow of inviscid fluids

An analogous formulation can also be found for three dimensions, although in this case, the vortex stretching term has to be taken into consideration. In the following, we show that, for ideal fluids without forcing, the solution of the vorticity equation has the form

$$\boldsymbol{\Omega}(\boldsymbol{y},t) = \boldsymbol{\Omega}(\boldsymbol{y},0) \cdot \nabla_{\boldsymbol{y}}\boldsymbol{X}(\boldsymbol{y},t) \tag{9.1.52}$$

For the proof, we use index notation and the Einstein summation convention. With the j-component of eq. (9.1.52)

$$\Omega_{\mathrm{j}}(\boldsymbol{y},t) = \Omega_{\mathrm{i}}(\boldsymbol{y},0)\,\frac{\partial X_{\mathrm{j}}(\boldsymbol{y},t)}{\partial y_{\mathrm{i}}} \tag{9.1.53}$$

we calculate the derivative of the left-hand side of eq. (9.1.45) with respect to time. Exchanging the gradient with the time derivative and using eq. (9.1.40) yields

$$\frac{\mathrm{d}\Omega_{\mathrm{j}}(\boldsymbol{y},t)}{\mathrm{d}t} = \Omega_{i}(\boldsymbol{y},0)\,\frac{\partial u_{\mathrm{j}}(\boldsymbol{x},t)}{\partial y_{\mathrm{i}}} = \Omega_{\mathrm{i}}(\boldsymbol{y},0)\,\frac{\partial u_{\mathrm{j}}}{\partial x_{\mathrm{l}}}\bigg|_{\boldsymbol{x}=\boldsymbol{X}(\boldsymbol{y},t)}\frac{\partial X_{\mathrm{l}}}{\partial y_{\mathrm{i}}} \tag{9.1.54}$$

Inserting eq. (9.1.52), we obtain the same expression for the j-component of the right-hand side of eq. (9.1.45). In order to check the initial conditions, we observe that due to $\boldsymbol{X}(\boldsymbol{y},0) = \boldsymbol{y}$, the relation

$$\frac{\partial X_{\mathrm{j}}(\boldsymbol{y},0)}{\partial y_{\mathrm{i}}} = \delta_{\mathrm{ji}} \tag{9.1.55}$$

is valid. Multiplying this equation with $\Omega_{\mathrm{i}}(\boldsymbol{y},0)$ we obtain $\Omega_{\mathrm{j}}(\boldsymbol{y},0)$, i.e. eq. (9.1.53) also holds for the initial conditions. Therefore, in the case of ideal fluids, the quantity $\boldsymbol{\Omega}(\boldsymbol{y},0) \cdot \nabla_{\boldsymbol{y}}\boldsymbol{X}(\boldsymbol{y},t)$ fulfils the evolution equation for the Lagrangian vorticity, eq. (9.1.45).

To summarise: if the Lagrangian mapping $\boldsymbol{y} \to \boldsymbol{X}(\boldsymbol{y},t)$ is known, the vector $\boldsymbol{\Omega}(\boldsymbol{y},t)$ can be calculated from the initial vector $\boldsymbol{\Omega}(\boldsymbol{y},0)$ of the Lagrangian vorticity by means of a linear transformation given by the matrix $\boldsymbol{D}(\boldsymbol{y},t)$ with the elements

$$D_{\mathrm{ij}}(\boldsymbol{y},t) = \frac{\partial X_{\mathrm{i}}(\boldsymbol{y},t)}{\partial y_{\mathrm{j}}} \tag{9.1.56}$$

As illustrated in fig. 5.4.4, the eigenvalues of $\boldsymbol{D}(\boldsymbol{y},t)$ describe the local stretching and contraction of the Lagrangian mapping for the transition from $t = 0$ to the time t, where the eigenvector appertaining to the largest eigenvalue λ_{max} specifies the direction of maximal stretching.

A relation with the Lyapunov exponents can also be established. To this end, we consider the temporal evolution of a small perturbation in the initial conditions along the trajectory $\tilde{\boldsymbol{X}}(\boldsymbol{y},\boldsymbol{y}',t) = \boldsymbol{X}(\boldsymbol{y}',t) - \boldsymbol{X}(\boldsymbol{y},t)$ with $\tilde{\boldsymbol{X}}(\boldsymbol{y},\boldsymbol{y}',0) = \boldsymbol{y}' - \boldsymbol{y}$. As a first approximation, it follows from eq. (9.1.40)

$$\frac{\mathrm{d}}{\mathrm{d}t}\tilde{\boldsymbol{X}}(\boldsymbol{y}',\boldsymbol{y},t) = \tilde{\boldsymbol{X}}(\boldsymbol{y}',\boldsymbol{y},t) \cdot [\nabla \boldsymbol{u}(\boldsymbol{x},t)]_{\boldsymbol{x}=\boldsymbol{X}(\boldsymbol{y},t)} \tag{9.1.57}$$

i.e. for ideal fluids, the perturbation obeys the same dynamics as the vortex transport, eq. (9.1.45). Therefore, certain quasi-invariant properties of the flow field for finite time intervals, to be described in the following, can be transferred to the vortex transport.

• Lagrangian coherent structures

On the basis of *Finite Time Lyapunov exponents* (FTLE) and in analogy to stable and unstable manifolds, see figs. 3.3.1, 3.3.2, particular quasi-invariant lines and surfaces can be defined which are denoted *Lagrangian coherent structures* (LCS), cf. (Haller and Yuan, 2000; Haller, 2001; Shadden et al., 2005). By means of these mobile structures, in general hidden in a Eulerian description, the flow can be divided into regions of different dynamical behaviour. This is of great interest for many engineering applications such as, for example, the investigation of the transport and mixing properties of turbulent flows (Mathur et al., 2007), the spread of pollutants in near-shore coastal waters (Lekien et al., 2005) or the aerodynamics of airfoils under stall conditions.

According to its definition in eq. (5.4.70), the maximal Lyapunov exponent $\sigma_0^T(\boldsymbol{y})$ can be determined for every trajectory at the time T and at the point $\boldsymbol{X}(\boldsymbol{y},T)$ for a finite time interval $[0,T]$. This quantity is denoted FTLE (Finite Time Lyapunov Exponent) and describes the maximal mean rate of separation of neighbouring particle paths in the time interval $[0,T]$ (Haller, 2001; Doerner et al., 1999). For $T \to \infty$, $\sigma_0^\infty(\boldsymbol{y})$ coincides with the largest Lyapunov exponent. The scalar field $\sigma_0^T(\boldsymbol{y})$ can now be calculated for all initial conditions $\boldsymbol{y}$ and the particular lines or surfaces can be determined for which the FTLE has a local maximum and which play the role of quasi-invariant unstable attractive manifolds. Analogously, by integration in the reverse-time direction, quasi-invariant stable manifolds are obtained. These specific manifolds are of particular interest for assessing the flow dynamics because they separate domains with different dynamical behaviour like mobile separatrices. It should be noted that the LCS always depent on the selected time interval $[0,T]$.

In the case of ideal fluids, the vortex transport undergoes the largest stretching along the unstable manifolds, i.e. these geometrical structures approximately describe vortex lines and vortex tubes (Haller, 2001).

• The hydrodynamic equations as a dynamical system

The Navier-Stokes equation together with appropriate boundary conditions constitute a dynamical system which allows us to calculate the velocity field $\boldsymbol{u}(\boldsymbol{x},t)$ for a given initial condition $\boldsymbol{u}^0(\boldsymbol{x})$ at time t. This becomes particularly evident in the Lagrangian formulation. For the initial conditions

$$\boldsymbol{X}(\boldsymbol{y}, t=0)=\boldsymbol{y}, \qquad \boldsymbol{U}(\boldsymbol{y}, t=0)=\boldsymbol{u}(\boldsymbol{y}, t=0) \tag{9.1.58}$$

the positions $\boldsymbol{X}(\boldsymbol{y}, t)$ and the velocities $\boldsymbol{U}(\boldsymbol{y}, t)$ of the Lagrangian particles can be determined by means of the equations of motion (9.1.43).

Although the Navier-Stokes equation has been known for more than one hundred years, the mathematical questions regarding the existence and uniqueness of its solutions have only been partially been answered. From a physical point of view, one would require that, for smooth initial fields $\boldsymbol{u}(\boldsymbol{x}, t=0)$, the solution for the Eulerian velocity field $\boldsymbol{u}(\boldsymbol{x}, t)$ respectively for the Lagrangians paths $\boldsymbol{X}(\boldsymbol{y}, t)$ and $\boldsymbol{U}(\boldsymbol{y}, t)$ should exist for all future times, and especially should not diverge in finite time. However, up to now, there is no mathematical proof or counterproof for the relevant case of the motion of incompressible fluids in three spatial dimensions.

9.1.5 Hydrodynamic Vortices

Only a few exact solutions of the Navier-Stokes equation are known. In the following, we present some vortex solutions which can be identified as so-called *coherent structures* in turbulent flows and which are often used as vortex models for a phenomenological description of vortex fields and turbulent structures.

There is a multitude of technical applications for which such vortex models play an important role in the investigation of the interaction of vortices and their stability; consider the airflow over an airfoil of airplanes, turbines and wind turbines or the Kármán vortex street, where counter-rotating vortices are generated in the wake of the fluid flow around blunt bodies. In extreme cases, this can lead to the destruction of buildings, as the collapse of the cooling towers in Ferrybridge in 1968 has shown.

The dynamics of vortex structures is also of great interest since it was discovered that there are vortex filaments in turbulent flows along which the vorticity assumes extremely high values. It is thought that such localised structures have a major influence on the statistical properties of turbulent flows and might contribute to explaining their intermittent character; this will be discussed later (Frisch, 1995; Jiménez and Wray, 1998).

9.1.5.1 Lamb-Oseen Vortex

One of these vortex solutions is the so-called Lamb-Oseen vortex, first been presented by Oseen and Hamel (Oseen, 1911; Hamel, 1916). It consists of a cylindrically symmetric vorticity field for which we make the following ansatz

$$\boldsymbol{\omega}(\boldsymbol{x}, t)=\omega_z(r, t)\, \boldsymbol{e}_z, \qquad r=\sqrt{x^2+y^2} \tag{9.1.59}$$

Due to the Biot-Savart law, this vorticity field generates an azimuthal velocity field

$$\boldsymbol{u}(\boldsymbol{x}, t)=u_\varphi(r, t)\, \boldsymbol{e}_\varphi \tag{9.1.60}$$

As a result, the non-linear terms in the vortex transport equation (9.1.26)

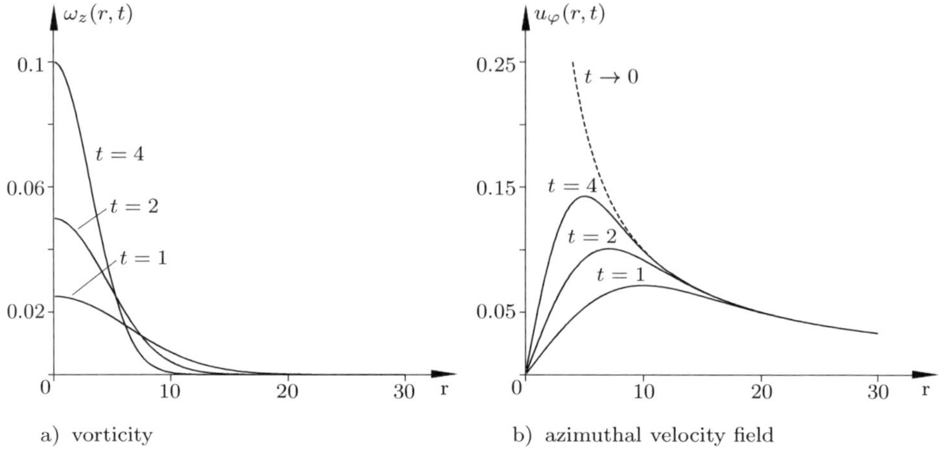

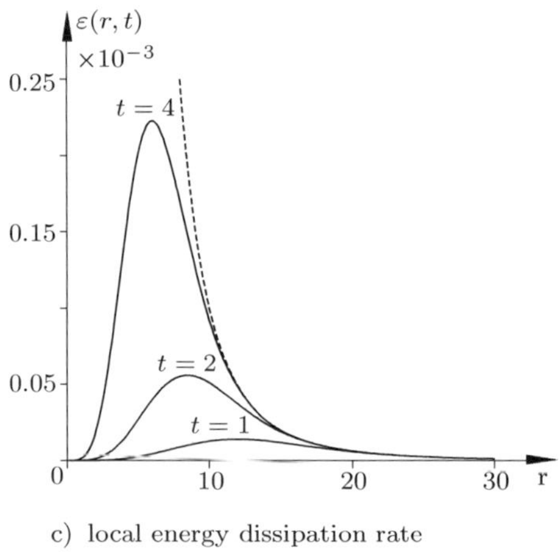

Fig. 9.1.1

Several characteristics of the Lamb-Oseen vortex: a) vorticity, b) azimuthal velocity field and c) local energy dissipation rate as a function of the distance r from the vortex core (the dashed lines are far-fields)

$$\boldsymbol{u}(\boldsymbol{x},t)\cdot\nabla\boldsymbol{\omega}(\boldsymbol{x},t) = \boldsymbol{\omega}(\boldsymbol{x},t)\cdot\nabla\boldsymbol{u}(\boldsymbol{x},t) = \boldsymbol{0} \tag{9.1.61}$$

vanish identically; in particular, the vortex stretching term $\boldsymbol{\omega}\cdot\nabla\boldsymbol{u}$ vanishes as well. The vortex transport equation therefore reduces to

$$\frac{\partial\omega_z}{\partial t} = \nu\,\Delta\omega_z \tag{9.1.62}$$

and thus has the form of a heat equation for $\omega_z(r,t)$. Writing the Laplace operator in cylindrical coordinates, we obtain

$$\frac{\partial}{\partial t}\omega_z(r,t) = \nu\,\frac{1}{r}\frac{\partial}{\partial r}\left(r\frac{\partial}{\partial r}\omega_z(r,t)\right) \tag{9.1.63}$$

One solution reads

$$\omega_z(r,t) = \frac{\Gamma}{\pi r_B^2} \mathrm{e}^{-r^2/r_B^2}, \qquad r_B^2 = 4\nu t \tag{9.1.64}$$

where Γ denotes the circulation, see eq. (9.1.28). The velocity field then has the form

$$\boldsymbol{u}(\boldsymbol{x},t) = \frac{\Gamma}{2\pi r}(1 - \mathrm{e}^{-r^2/r_B^2})\,\boldsymbol{e}_\varphi \tag{9.1.65}$$

In fig. 9.1.1a,b, the functions $\omega_z(r,t)$ and $u_\varphi(r,t)$ are displayed in dependence on r for several time instants. The initially concentrated vortex distribution spreads diffusively in radial direction; because of $r_B = \sqrt{4\nu t}$, this spread is faster, the larger the viscosity. It is also instructive to calculate the local energy dissipation rate. We obtain

$$\varepsilon(r,t) = \frac{\nu}{2}\left[r\frac{\partial}{\partial r}\left(\frac{u_\varphi}{r}\right)\right]^2 = \frac{\nu}{2}\left(\frac{\Gamma}{\pi}\right)^2\left[\frac{1}{r^2} - \mathrm{e}^{-r^2/r_B^2}\left(\frac{1}{r^2}+\frac{1}{{r_B}^2}\right)\right]^2 \tag{9.1.66}$$

In fig. 9.1.1c, the function $\varepsilon(r,t)$ is shown for several time instants. The fields possess different behaviour in the far- and near-field. In the far-field $r \gg r_B$,

$$\omega_z = 0, \qquad u_\varphi(\boldsymbol{x},t) = \frac{\Gamma}{2\pi r}, \qquad \varepsilon = \frac{\nu}{2}\left(\frac{\Gamma}{\pi r^2}\right)^2 \tag{9.1.67}$$

holds. This is the field of a point vortex, i.e. of a δ-shaped vortex distribution

$$\omega_z(r,0) = \Gamma\,\delta(r) \tag{9.1.68}$$

which can be calculated from eq. (9.1.64) for $t \to 0$. In the "eye" of the vortex, the flow velocity is zero due to the finite viscosity.

9.1.5.2 Stretched Vortex

As early as 1938, G. I. Taylor realised that an increase of the vorticity by stretching is one of the fundamental mechanisms fluid mechanics is based on. In order to understand why a superimposed strain field leads to a vortex stretching for the Lundgren vortex, which will be presented in the following section 9.1.5.3, it is useful to transform the vortex stretching term. To this end, we divide the tensor of the spatial velocity gradient $\boldsymbol{A}$ in its symmetrical part S_{ij} and its antisymmetrical part T_{ij}

$$A_{\mathrm{ij}} \equiv \frac{\partial u_{\mathrm{i}}}{\partial x_{\mathrm{j}}} = S_{\mathrm{ij}} + T_{\mathrm{ij}} \tag{9.1.69}$$

The symmetrical part $S_{\mathrm{ij}} = S_{\mathrm{ji}}$ coincides with the strain velocities[5], see eq. (7.1.18)

$$S_{\mathrm{ij}} = \frac{1}{2}\left(\frac{\partial u_{\mathrm{i}}}{\partial x_{\mathrm{j}}} + \frac{\partial u_{\mathrm{j}}}{\partial x_{\mathrm{i}}}\right) \tag{9.1.70}$$

where we choose, as is usual for turbulent flows, the notation S_{ij}. The antisymmetrical part $T_{\mathrm{ij}} = -T_{\mathrm{ji}}$ contains the angle velocities

[5] In section 7.1, the tensor of strain velocities is designated $\boldsymbol{\varepsilon} = [\varepsilon_{\mathrm{ij}}]$, as is usual in engineering sciences.

$$T_{ij} = \frac{1}{2}\left(\frac{\partial u_i}{\partial x_j} - \frac{\partial u_j}{\partial x_i}\right) \tag{9.1.71}$$

The following relations hold between the components of the angle velocities and the vorticity

$$\omega_i = -\ \epsilon_{ijk}\, T_{jk}$$
$$T_{ij} = -\ \frac{1}{2}\ \epsilon_{ijk}\, \omega_k \tag{9.1.72}$$

where ϵ_{ijk} denotes the epsilon tensor (permutation symbol of Levi-Civita), see (Bronstein et al., 1995)

$$\epsilon_{ijk} = \begin{cases} 1, & \text{if} \quad \mathrm{i,j,k} \ \ \text{cyclic} \\ -1, & \text{if} \quad \mathrm{i,j,k} \ \ \text{anti} - \text{cyclic} \\ 0 & \text{otherwise} \end{cases} \tag{9.1.73}$$

Introducing $\frac{\partial u_j}{\partial x_i} = S_{ij} + T_{ij}$ into the j-component of the vortex stretching term

$$(\boldsymbol{\omega} \cdot \nabla \boldsymbol{u})_j = \omega_i\, \frac{\partial u_j}{\partial x_i}$$

and taking into consideration that for every fixed index j the expression

$$\omega_i\, T_{ij} = -\frac{1}{2}\ \epsilon_{ijk}\, \omega_i\, \omega_k$$

vanishes, we obtain the following relation for the vortex stretching term

$$\boldsymbol{\omega} \cdot \nabla \boldsymbol{u} = \boldsymbol{S} \cdot \omega \tag{9.1.74}$$

Hence, we see that the vortex stretching term describes the influence of the strain velocities on the vorticity. An analogous strain field is generated, for example, by the following cylindrically symmetrical potential flow

$$\boldsymbol{u}_{pot}(\boldsymbol{x}, t) = \left[-\frac{a(t)}{2}x,\ -\frac{a(t)}{2}y,\ a(t)z\right] \tag{9.1.75}$$

Let us consider now fluid motions of the form

$$\boldsymbol{u}(\boldsymbol{x}, t) = \boldsymbol{u}_{pot}(\boldsymbol{x}, t) + \boldsymbol{w}(x, y, t) \tag{9.1.76}$$

with a two-dimensional velocity field

$$\boldsymbol{w}(x, y, t) = [w_x(x, y, t), w_y(x, y, t), 0] \tag{9.1.77}$$

which does not depend on z. Applying the curl operator to the velocity field of eq. (9.1.76), we can see that the appertaining vorticity

$$\boldsymbol{\omega}(\boldsymbol{x},t) = \Omega(x,y,t)\,\boldsymbol{e}_z \tag{9.1.78}$$

is also independent of z and fulfils the vortex transport equation

$$\frac{\partial}{\partial t}\Omega(x,y,t) + \boldsymbol{w}(x,y,t)\cdot\nabla\Omega(x,y,t) - \frac{a(t)}{2}\left[x\frac{\partial}{\partial x} + y\frac{\partial}{\partial y}\right]\Omega(x,y,t) = a\Omega(x,y,t) + \nu\,\Delta\Omega(x,y,t) \tag{9.1.79}$$

We now consider the case where the parameter a in the potential flow, eq. (9.1.75), is constant in time and positive. If we restrict ourselves to a rotationally symmetrical vorticity $\Omega(r,t)$, which, according to the Biot-Savart law, can generate only a velocity field in azimuthal direction, the vorticity satisfies the equation

$$\frac{\partial}{\partial t}\Omega(r,t) = \frac{a}{2r}\frac{\partial}{\partial r}\left(r^2\Omega(r,t)\right) + \nu\frac{1}{r}\frac{\partial}{\partial r}\left(r\frac{\partial}{\partial r}\Omega(r,t)\right) \tag{9.1.80}$$

This equation has a stationary solution of the form

$$\Omega(r) = \frac{\Gamma}{\pi r_B^2}\mathrm{e}^{-r^2/r_B^2}, \qquad r_B^2 = \frac{4\nu}{a} \tag{9.1.81}$$

where the two-dimensional velocity field has only a tangential component $\boldsymbol{w} = u_\varphi(r)\,\boldsymbol{e}_\varphi$ with

$$u_\varphi(r) = \frac{\Gamma}{2\pi r}[\,1 - \mathrm{e}^{-r^2/r_B^2}\,] \tag{9.1.82}$$

and describes the so-called Burgers vortex. This vortex is related to the Lamb-Oseen vortex, eq. (9.1.64), which, however, decays over time since $r_B^2 = 4\nu t$ increases gradually and ω_z vanishes for large times and arbitrarily chosen r. The example of the Burgers vortex shows that stationary vortices may exist in spite of dissipation.

Interestingly, the vorticity equation eq. (9.1.79) for stretched vortices can be traced back to the ordinary two-dimensional vorticity equation (9.1.62) by means of a transformation given by Lundgren. The corresponding transformation reads (Lundgren, 1982)

$$\Omega(r,\varphi,t) = \mathrm{e}^{A(t)}\,\tilde{\Omega}(\,r\,\mathrm{e}^{\frac{A(t)}{2}}, \varphi, \tau) \tag{9.1.83}$$

where the following notations are used

$$A(t) = \int_0^t a(t')\,\mathrm{d}t', \qquad \tau = \int_0^t \mathrm{e}^{A(t')}\,\mathrm{d}t'$$

A short calculation shows that $\tilde{\Omega}(\tilde{r}\,\varphi,\tau)$ fulfils the two-dimensional vorticity equation. Every two-dimensional flow field thus possesses a three-dimensional counterpart which is embedded in the potential flow (9.1.75) in the form of a *stretched field*.

9.1.5.3 Lundgren Vortex

A further interesting vortex solution was presented by Lundgren (Lundgren, 1982). In contrast to the Lamb-Oseen vortex, this vortex has a helical structure, shown in Colour Plate XIVA, p. 617. The generation of a Lundgren vortex resulting from the differential rotation of the vorticity distribution by a strong central vortex was studied experimentally in detail by Cuypers, Maurel and Petitjean (Cuypers et al., 2003; Cuypers et al., 2004; Cuypers et al., 2006). Such helically twisted vortex structures might also play an important role in fully developed turbulent flows.

The temporal average of suitable statistical quantities over the lifetime of such a Lundgren vortex – where additionally, the ensemble average of isotropically distributed vortices is taken – yields an approximation of the energy spectrum which can be measured for fully developed turbulent flows (Lundgren, 1982; Lundgren, 1993).

9.2 From Chaos to Turbulence

9.2.1 Chaos in Fluid Flows

In various experiments on hydrodynamic instabilities, in particular for the Rayleigh-Bénard instability and the Taylor-Couette experiment (see fig. 8.2.3), the onset of chaotic motions and the various routes to chaos discussed in detail in Chapter 8 have been observed. Typically, only a few degrees of freedom are excited in these processes, whereby the spatial structures form coherent patterns. Mathematically, these processes can be described by means of low-dimensional non-linear systems of differential equations. In theses cases, the Reynolds numbers are relatively low. These spatial structures are thus often denoted dissipative structures.

For flows with high Reynolds numbers, the influence of the viscosity is of minor importance. It is therefore reasonable to study the limit case of ideal fluids. Here, two-dimensional flows play an important role. Indeed, in these cases, it is possible to replace the theoretical continuum by a set of ordinary differential equations for the location of so-called point vortices that possess Hamiltonian properties. In the following, we will study this point vortex dynamics in more detail and once again find general properties of non-linear dynamical systems.

9.2.2 Dynamics of Point Vortices in Two-Dimensional Ideal Flows

In section 9.1.4, we presented the evolution equation for an ideal flow in a two-dimensional domain. An important property here is that the Lagrangian vorticity is constant, a direct consequence of Kelvin's circulation theorem, see eq. (9.1.28). In this case, the dynamics is determined by the integro-differential equation (9.1.51). Despite this simplification, the treatment of these equations is generally extremely difficult.

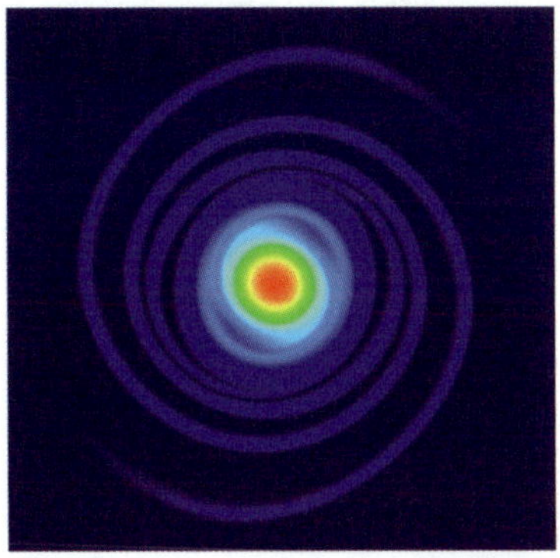

A)

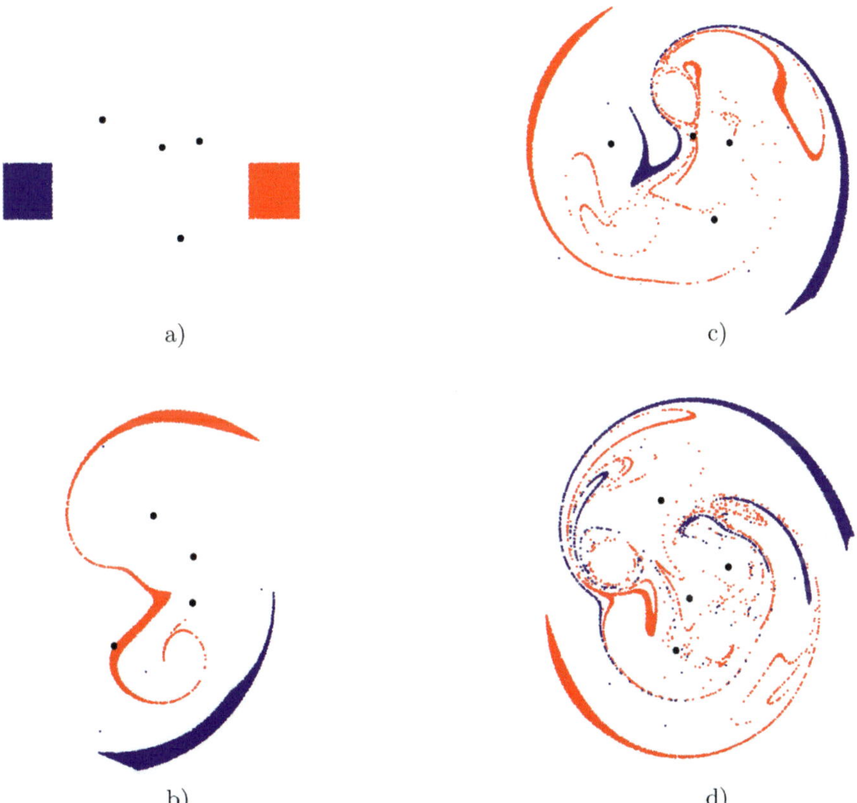

B)

Plate XIV A) Vortex field of a Lundgren spiral (Wilczek, 2007)
B) Chaotic mixing: motion of passive Lagrangian particles in a chaotic system of four point vortices (black dots).

However, this integro-differential equation splits up into a system of non-linear differential equations if one considers a vorticity field localised at a finite number of points. Such a field can be represented by a sum of delta-functions

$$\Omega(\boldsymbol{y}) = \sum_{\mathrm{j}=1}^{\mathrm{N}} \Gamma_{\mathrm{j}} \delta(\boldsymbol{y} - \boldsymbol{y}_{\mathrm{j}}) \tag{9.2.1}$$

The amplitudes Γ_{j} are the circulations of the individual point vortices

$$\Gamma_{\mathrm{j}} = \int_F \Omega(\boldsymbol{y})\, \mathrm{d}F \tag{9.2.2}$$

where the area F only comprises the vortex with index j. The vorticity field is thus represented by a superposition of point vortices with the circulations Γ_{j}. The point vortices move in the velocity field generated by the remaining vortices. If we denote the instantaneous location of the point vortex j

$$\boldsymbol{X}(\boldsymbol{y}_{\mathrm{j}}, t) = \boldsymbol{x}_{\mathrm{j}}(t) = \{x_{\mathrm{j}}(t) \quad y_{\mathrm{j}}(t)\} \tag{9.2.3}$$

the system of differential equations

$$\frac{\mathrm{d}}{\mathrm{d}t} \boldsymbol{x}_{\mathrm{j}}(t) = \sum_{\mathrm{i} \neq \mathrm{j}} \Gamma_{\mathrm{i}}\, \boldsymbol{e}_z \times \boldsymbol{K}\left(\boldsymbol{x}_{\mathrm{j}}(t) - \boldsymbol{x}_{\mathrm{i}}(t)\right) \tag{9.2.4}$$

determines the motion of the point vortices. This is a direct consequence of the Lagrangian evolution equation (9.1.51) with the vorticity field (9.2.1). Here, the divergent term with index j=i may not be taken into consideration. If we denote the components of $\boldsymbol{x}_{\mathrm{j}}$ as $\{x_{\mathrm{j}} \; y_{\mathrm{j}}\}$ and define the vectors measuring the distance between two vortices and their length

$$\boldsymbol{r}_{\mathrm{ij}} = \boldsymbol{x}_{\mathrm{i}} - \boldsymbol{x}_{\mathrm{j}} \quad \text{with} \quad L_{\mathrm{ij}} = |\boldsymbol{r}_{\mathrm{ij}}| \tag{9.2.5}$$

we can write the non-linear equations of motion using eq. (9.1.39) as follows

$$\begin{aligned} \frac{\mathrm{d}x_{\mathrm{j}}}{\mathrm{d}t} &= -\frac{1}{2\pi} \sum_{\mathrm{i} \neq \mathrm{j}} \frac{1}{L_{\mathrm{ij}}^2} \Gamma_{\mathrm{i}} (y_{\mathrm{j}} - y_{\mathrm{i}}) \\ \frac{\mathrm{d}y_{\mathrm{j}}}{\mathrm{d}t} &= \frac{1}{2\pi} \sum_{\mathrm{i} \neq \mathrm{j}} \frac{1}{L_{\mathrm{ij}}^2} \Gamma_{\mathrm{i}} (x_{\mathrm{j}} - x_{\mathrm{i}}) \end{aligned} \tag{9.2.6}$$

Hence, the motion of N point vortices in ideal two-dimensional flows can be studied with methods from the theory of dynamical systems.

9.2.2.1 The Hamiltonian Character of the Point Vortex Dynamics

The system of differential equations (9.2.4) can be deduced from the Euler equation describing the motion of ideal fluids. It can therefore be expected that the kinetic energy is conserved. In fact, as many as four invariants can be identified. We show in the following that the first invariant is the *total energy*

$$H = \frac{1}{2} \sum_{\mathrm{i,j;\,i \neq j}} \Gamma_{\mathrm{i}} G(\boldsymbol{x}_{\mathrm{i}} - \boldsymbol{x}_{\mathrm{j}}) \Gamma_{\mathrm{j}} = -\frac{1}{4\pi} \sum_{\mathrm{i,j;\,i \neq j}} \Gamma_{\mathrm{i}} \ln |\boldsymbol{x}_{\mathrm{i}} - \boldsymbol{x}_{\mathrm{j}}| \, \Gamma_{\mathrm{j}} \tag{9.2.7}$$

We obtain this relation from the vector potential $\boldsymbol{u} = \nabla \times \boldsymbol{A}$, see eq. (9.1.29), and by using the relation

$$\boldsymbol{u} \cdot \boldsymbol{u} = \boldsymbol{A} \cdot \boldsymbol{\omega} - \nabla \cdot (\boldsymbol{u} \times \boldsymbol{A})$$

The total energy is then obtained using eqs. (9.1.34, 9.1.44)

$$\begin{aligned} H &= \frac{1}{2} \int_{\boldsymbol{x}} u(\boldsymbol{x}, t)^2 \, \mathrm{d}\boldsymbol{x} \\ &= \frac{1}{2} \int_{\boldsymbol{x}} \int_{\boldsymbol{x}'} G(\boldsymbol{x} - \boldsymbol{x}') \, \boldsymbol{\Omega}(\boldsymbol{x}) \cdot \boldsymbol{\Omega}(\boldsymbol{x}') \, \mathrm{d}\boldsymbol{x}' \, \mathrm{d}\boldsymbol{x} - \frac{1}{2} \int_{\boldsymbol{x}} \nabla \cdot (\boldsymbol{u} \times \boldsymbol{A}) \, \mathrm{d}\boldsymbol{x} \end{aligned} \tag{9.2.8}$$

By means of Gauss's theorem, the last integral can be transformed into a surface integral which vanishes if the boundaries of the region are chosen to be sufficiently large. Inserting eq. (9.2.1) into the first integral, eq. (9.2.7) immediately follows by using eq. (9.1.38).

Further invariants are the *centre of the vortices*

$$\boldsymbol{R} = \frac{\sum_{\mathrm{j}} \Gamma_{\mathrm{j}} \boldsymbol{x}_{\mathrm{j}}}{\sum_{\mathrm{j}} \Gamma_{\mathrm{j}}}, \qquad \sum_{\mathrm{j}} \Gamma_{\mathrm{j}} \neq 0 \tag{9.2.9}$$

and the quantity

$$L = \sum_{\mathrm{j}} \Gamma_{\mathrm{j}} \boldsymbol{x}_{\mathrm{j}}^2 \tag{9.2.10}$$

which is related to the rotational invariance of the problem. Due to the incompressibility of the flow the system of differential equations (9.2.4) is a Hamiltonian system. As we saw in Chapter 4, the equations of motion can be derived directly from the Hamiltonian function H which is identified with the total energy, see eq. (4.1.25). Taking the derivative of H, we can write the equations of motion (9.2.6) in the form

$$\begin{aligned} \Gamma_{\mathrm{j}} \frac{\mathrm{d}x_{\mathrm{j}}}{\mathrm{d}t} &= \frac{\partial}{\partial y_{\mathrm{j}}} H \\ \Gamma_{\mathrm{j}} \frac{\mathrm{d}y_{\mathrm{j}}}{\mathrm{d}t} &= -\frac{\partial}{\partial x_{\mathrm{j}}} H \end{aligned} \tag{9.2.11}$$

Generally, an equation for the temporal evolution of the squares of the distances $L_{\mathrm{ij}}^2 = |\boldsymbol{r}_{\mathrm{ij}}|^2$ of N point vortices can be derived from the point vortex dynamics (9.2.4) together with eq. (9.1.39). From the relation

$$\dot{L}_{\mathrm{ij}}^2 = \frac{1}{\pi} \sum_{\mathrm{k \neq i, \neq j}} \Gamma_{\mathrm{k}} \left[\boldsymbol{e}_z \times \left(\frac{\boldsymbol{r}_{\mathrm{ik}}}{L_{\mathrm{ik}}^2} - \frac{\boldsymbol{r}_{\mathrm{jk}}}{L_{\mathrm{jk}}^2} \right) \right] \cdot \boldsymbol{r}_{\mathrm{ij}} \tag{9.2.12}$$

the equation

$$\dot{L}_{\mathrm{ij}}^2 = \sum_{\mathrm{k}\neq\mathrm{i},\neq\mathrm{j}} \Gamma_{\mathrm{k}}\, \sigma_{\mathrm{ijk}}\, L_{\mathrm{ij}}^2\, [L_{\mathrm{ik}}^2 - L_{\mathrm{jk}}^2] \tag{9.2.13}$$

can be deduced. The quantity σ_{ijk} is connected to the oriented area $F_{\mathrm{ijk}} = \frac{1}{2}(\boldsymbol{r}_{\mathrm{ij}} \times \boldsymbol{r}_{\mathrm{ik}}) \cdot \boldsymbol{e}_z$ of a triangle spanned by the positions of the vortices i, j, k

$$\sigma_{\mathrm{ijk}} = \frac{2}{\pi} \frac{F_{\mathrm{ijk}}}{L_{\mathrm{ij}}^2 L_{\mathrm{jk}}^2 L_{ki}^2} \tag{9.2.14}$$

The relations (9.2.13) hold for all vortex pairs i, j of a system of N point vortices.

In the following, we briefly discuss the dynamics of point vortices. Apart from stationary configurations of vortices, temporally periodic, quasi-periodic and chaotic motions of vortices also exist. It will turn out that the dynamics of two and – astonishingly enough – also the motion of three point vortices is integrable. Chaotic temporal behaviour occurs from four point vortices onwards. However, we remind the reader that this concerns Hamiltonian chaos. Due to the invariance with respect to time reversal of the point vortex dynamics, directly connected with the conservation of energy, no chaotic attractors exist.

9.2.2.2 Two Point Vortices

The dynamics of two point vortices at the points $\boldsymbol{x}_1$, $\boldsymbol{x}_2$ with the circulations Γ_1, Γ_2 is, according to eq. (9.2.4), described by the following system of differential equations

$$\begin{aligned} \dot{\boldsymbol{x}}_1 &= \Gamma_2\, \boldsymbol{e}_z \times \boldsymbol{K}(\boldsymbol{x}_1 - \boldsymbol{x}_2) \\ \dot{\boldsymbol{x}}_2 &= \Gamma_1\, \boldsymbol{e}_z \times \boldsymbol{K}(\boldsymbol{x}_2 - \boldsymbol{x}_1) \end{aligned} \tag{9.2.15}$$

where, following eq. (9.1.39),

$$\boldsymbol{K}(\boldsymbol{x}) = -\boldsymbol{K}(-\boldsymbol{x}) \tag{9.2.16}$$

holds. This system of differential equations is completely integrable since there are four invariants for four dynamical quantities (two coordinates for each point vortex). This can be easily seen from the example of two point vortices with the same circulation Γ. The centre of the vortices, $\boldsymbol{R} = \boldsymbol{x}_1 + \boldsymbol{x}_2$, remains fixed while the distance vector $\boldsymbol{r} = \boldsymbol{x}_1 - \boldsymbol{x}_2$ varies according to

$$\dot{\boldsymbol{r}} = 2\frac{\Gamma}{2\pi} \boldsymbol{e}_z \times \frac{\boldsymbol{r}}{|\boldsymbol{r}|^2} \tag{9.2.17}$$

However, the distance $d = |\boldsymbol{r}|$ remains constant. The two point vortices circle around their common centre $\boldsymbol{R}$ and have a constant distance d. The angle velocity $\tilde{\omega}$ of the circular motion is given by

$$\tilde{\omega} = \frac{\Gamma}{\pi d^2} \tag{9.2.18}$$

Also in the general case $\Gamma_1 \neq \Gamma_2$, the motion remains periodic. We again assume that the centre of the vortices coincides with the origin of the coordinate system

$$\Gamma_1 \boldsymbol{x}_1 + \Gamma_2 \boldsymbol{x}_2 = \mathbf{0} \tag{9.2.19}$$

The motion of the distance vector $\boldsymbol{r} = \boldsymbol{x}_1 - \boldsymbol{x}_2$ is given by the system of differential equations

$$\dot{\boldsymbol{r}} = \frac{(\Gamma_1 + \Gamma_2)}{2\pi} \boldsymbol{e}_z \times \frac{\boldsymbol{r}}{|\boldsymbol{r}|^2} \tag{9.2.20}$$

The solution has the form

$$\boldsymbol{r}(t) = \boldsymbol{U}(\tilde{\omega} t)\, \boldsymbol{r}(0) \tag{9.2.21}$$

where $\boldsymbol{U}(\varphi)$ denotes the rotation matrix

$$\boldsymbol{U}(\varphi) = \begin{pmatrix} \cos\varphi & \sin\varphi \\ -\sin\varphi & \cos\varphi \end{pmatrix} \tag{9.2.22}$$

The angular frequency is calculated as

$$\tilde{\omega} = \frac{\Gamma_1 + \Gamma_2}{2\pi d^2} \tag{9.2.23}$$

We can now write the coordinates of the two point vortices explicitly

$$\begin{aligned} \boldsymbol{x}_1(t) &= \boldsymbol{R} + \frac{1}{2}\boldsymbol{U}(\tilde{\omega} t)\, \boldsymbol{r}(0) \\ \boldsymbol{x}_2(t) &= \boldsymbol{R} - \frac{1}{2}\boldsymbol{U}(\tilde{\omega} t)\, \boldsymbol{r}(0) \end{aligned} \tag{9.2.24}$$

Finally, we consider the case of two point vortices with opposite circulation $\Gamma_1 = -\Gamma_2$

$$\begin{aligned} \frac{\mathrm{d}}{\mathrm{d}t}[\boldsymbol{x}_1 - \boldsymbol{x}_2] &= \mathbf{0} \\ \frac{\mathrm{d}}{\mathrm{d}t}[\boldsymbol{x}_1 + \boldsymbol{x}_2] &= 2\frac{\Gamma}{2\pi} \boldsymbol{e}_z \times \frac{\boldsymbol{x}_1 - \boldsymbol{x}_2}{|\boldsymbol{x}_1 - \boldsymbol{x}_2|^2} \end{aligned} \tag{9.2.25}$$

Their distance and orientation hence remain constant while their centre moves with constant velocity. This velocity is proportional to the circulation Γ and inversely proportional to the distance of the two point vortices. The direction of the motion is orthogonal to the line connecting the two vortices.

9.2.2.3 Three Point Vortices

Surprisingly, the motion of three point vortices is also integrable. Three point vortices form a triangle which is uniquely defined by the lengths of its edges. Hence, the system of differential equations for the quantities $A_1 = L_{23}^2$, $A_2 = L_{31}^2$, $A_3 = L_{12}^2$ is sufficient to describe the relative motion of three point vortices. This system was already specified in eq. (9.2.13) for the case of N vortices. We denote the quantity σ_{ijk} as σ and explicitly obtain the system of equations

$$\begin{aligned}\frac{\mathrm{d}}{\mathrm{d}t}A_1 &= \Gamma_1\sigma A_1(A_2 - A_3)\\ \frac{\mathrm{d}}{\mathrm{d}t}A_2 &= \Gamma_2\sigma A_2(A_3 - A_1)\\ \frac{\mathrm{d}}{\mathrm{d}t}A_3 &= \Gamma_3\sigma A_3(A_1 - A_2)\end{aligned} \tag{9.2.26}$$

Due to the cyclic symmetry of these equations, the motion possesses the two constants

$$\begin{aligned}I &= \frac{A_1}{\Gamma_1} + \frac{A_2}{\Gamma_2} + \frac{A_3}{\Gamma_3}\\ J &= \frac{\ln A_1}{\Gamma_1} + \frac{\ln A_2}{\Gamma_2} + \frac{\ln A_3}{\Gamma_3}\end{aligned} \tag{9.2.27}$$

where J is linked with the energy H. In principle, these relations allow us to express A_2 and A_3, for example, by I, J and A_1

$$A_2 = A_2(I, J, A_1), \qquad A_3 = A_3(I, J, A_1) \tag{9.2.28}$$

This leads to a differential equation of the first order for A_1 which can be solved by separation of the variables. In this sense, the three-vortex problem is completely integrable. In the following, we discuss some properties in more detail.

Relative Equilibrium

Interestingly, even for different circulations Γ_i, a vortex configuration consisting of an equilateral triangle $A_1 = A_2 = A_3$ is conserved in time, i.e. the underlying motion is a simple rigid rotation. The motion of three such vortices with equal circulations Γ is presented in Colour Plate XVa, p. 623. For the case of $\Gamma_1 = \Gamma_2 = \Gamma_3$, the triangle rotates about its centre. The anglular velocity can easily be determined. We obtain

$$\tilde{\omega} = \frac{\Gamma}{2\pi D^2} \tag{9.2.29}$$

where D denotes the distance of a vortex from the vortex centre. The initial configuration is indicated by large circles. A stability analysis of the motion shows that this configuration is neutrally stable if

$$\Gamma_1\Gamma_2 + \Gamma_2\Gamma_3 + \Gamma_3\Gamma_1 > 0 \tag{9.2.30}$$

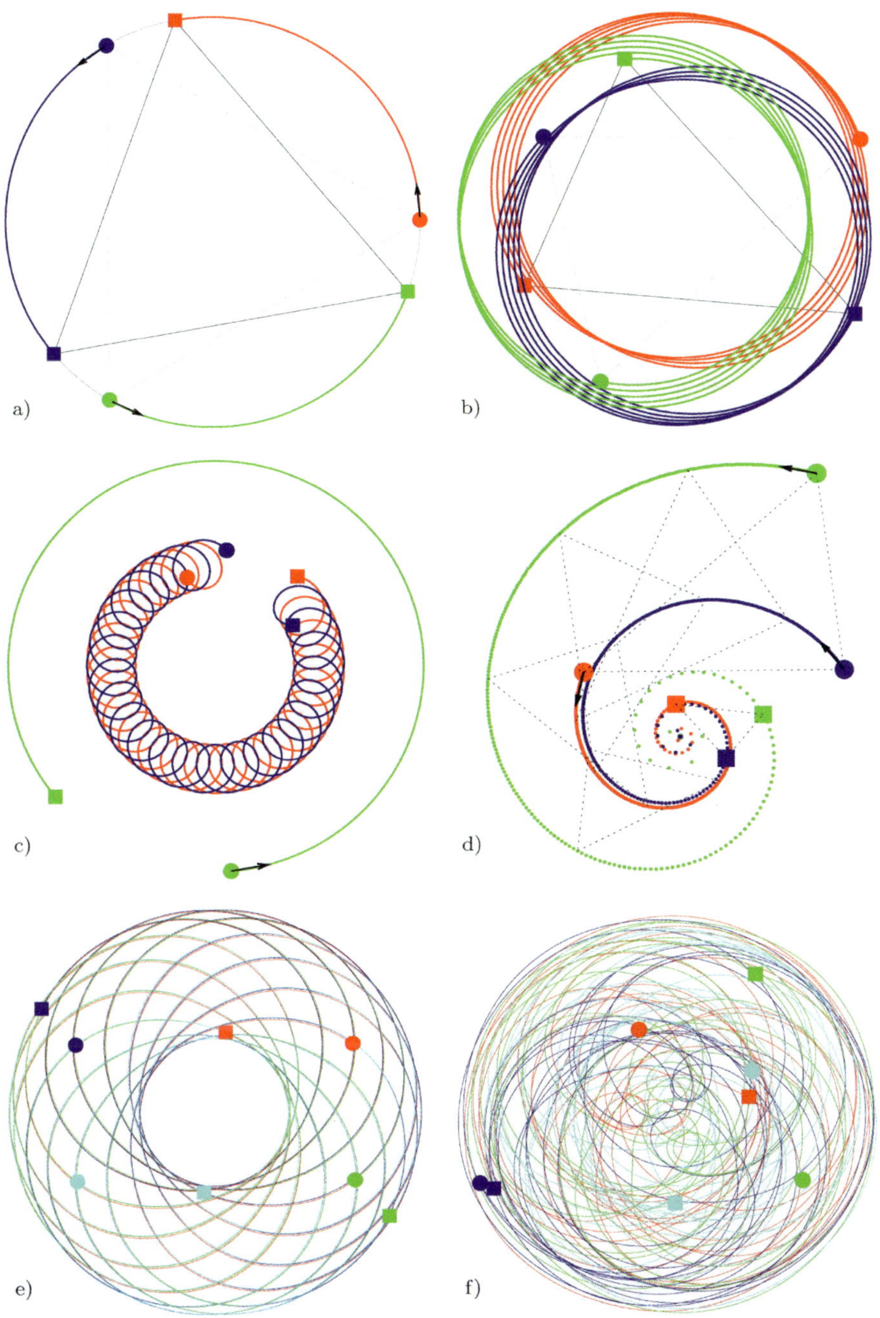

Plate XV Dynamics of point vortices in two-dimensional ideal flows

Small deviations lead to small rotations about the edges of the isosceles triangle, causing a quasi-periodic motion. An example of this type of motion is presented in Colour Plate XVb, p. 623.

A further three-vortex motion with vortices with equal circulation is shown in Colour Plate XVc. One pair of vortices moves approximately around its common vortex centre and, together with the outer point vortex, forms a two-vortex system. The initial configuration defines the energy and all further invariants of the motion.

Self-similar collapse and expansion

Three vortices define a triangle and the question immediately arises whether there are vortex motions which conserve the shape of the triangle. Such motions can indeed be found. They lead to a collapse of the triangle, i.e. the triangle contracts self-similarly in finite time to a point. Self-similar motion means that the side lengths can be described by a common quantity $\lambda(t)$ as

$$A_{\mathrm{i}} = \lambda(t) A_{\mathrm{i}}(0) \tag{9.2.31}$$

The constancy in time of the invariants I and J lead to the requirements

$$\left(\frac{A_1(0)}{\Gamma_1} + \frac{A_2(0)}{\Gamma_2} + \frac{A_3(0)}{\Gamma_3} \right) = 0 \tag{9.2.32}$$

and

$$\frac{1}{\Gamma_1} + \frac{1}{\Gamma_2} + \frac{1}{\Gamma_3} = 0 \tag{9.2.33}$$

From the equation of motion (9.2.26), we obtain

$$\frac{\mathrm{d}}{\mathrm{d}t} \lambda(t) = \Gamma_1 \sigma(0) [A_2(0) - A_3(0)] \tag{9.2.34}$$

This equation has the solution

$$\lambda(t) = 1 + \Gamma_1 \sigma(0) [A_2(0) - A_3(0)] t \tag{9.2.35}$$

Hence, if the relation

$$\sigma(0) [A_2(0) - A_3(0)] < 0 \tag{9.2.36}$$

holds, $\lambda(t)$ is zero at time

$$t_* = \frac{1}{\Gamma_1 \sigma(0) [A_3(0) - A_2(0)]} \tag{9.2.37}$$

This means that the vortex triangle has contracted in this time to a point, see Colour Plate XVd, p. 623.

Exchanging the position of two vortices leads to a change in the sign of σ according to eq. (9.2.14). In this case, the magnitude of the triangle increases in a self-similar manner and the distance between the vortices grows linearly in time.

Reduced three-vortex system

We now consider the reduced three-vortex system, defined by the equations

$$\begin{aligned}\frac{\mathrm{d}}{\mathrm{d}t}\boldsymbol{x}_1 &= \Gamma_2 \boldsymbol{e}_z \times \boldsymbol{K}(\boldsymbol{x}_1 - \boldsymbol{x}_2)\\ \frac{\mathrm{d}}{\mathrm{d}t}\boldsymbol{x}_2 &= \Gamma_1 \boldsymbol{e}_z \times \boldsymbol{K}(\boldsymbol{x}_2 - \boldsymbol{x}_1)\\ \frac{\mathrm{d}}{\mathrm{d}t}\boldsymbol{x}_3 &= \Gamma_1 \boldsymbol{e}_z \times \boldsymbol{K}(\boldsymbol{x}_3 - \boldsymbol{x}_1) + \Gamma_2 \boldsymbol{e}_z \times \boldsymbol{K}(\boldsymbol{x}_3 - \boldsymbol{x}_2)\end{aligned} \tag{9.2.38}$$

It is obtained from the three-point-vortex dynamics by approaching the limit $\Gamma_3 = 0$. This model can be considered an analogon of the three-body problem of classical mechanics. It is characterised by the fact that the dynamics of the vortices with the circulations Γ_1 and Γ_2 is decoupled, cf. section 9.2.2.2. Let us consider the case $\Gamma_1 + \Gamma_2 \neq 0$. Under this condition, the two vortices rotate about the vortex centre

$$\boldsymbol{R} = \frac{\Gamma_1 \boldsymbol{x}_1 + \Gamma_2 \boldsymbol{x}_2}{\Gamma_1 + \Gamma_2} \tag{9.2.39}$$

which, again, we assume to coincide with the origin

$$\begin{aligned}\boldsymbol{x}_1 &= \boldsymbol{U}(\tilde{\omega}t)\,\boldsymbol{x}_1(0)\\ \boldsymbol{x}_2 &= \boldsymbol{U}(\tilde{\omega}t)\,\boldsymbol{x}_2(0)\end{aligned} \tag{9.2.40}$$

We now make a transformation to a rotating coordinate system and, in this system, we define the location $\boldsymbol{\xi}$ of the third vortex as

$$\boldsymbol{x}_3 = \boldsymbol{U}(\tilde{\omega}t)\,\boldsymbol{\xi} \tag{9.2.41}$$

We then obtain the equation of motion for $\boldsymbol{\xi}$

$$\frac{\mathrm{d}}{\mathrm{d}t}\boldsymbol{\xi} = \tilde{\omega}\boldsymbol{e}_z \times \boldsymbol{\xi} + \Gamma_1 \boldsymbol{e}_z \times \boldsymbol{K}(\boldsymbol{\xi} - \boldsymbol{x}_1(0)) + \Gamma_2 \boldsymbol{e}_z \times \boldsymbol{K}(\boldsymbol{\xi} - \boldsymbol{x}_2(0)) \tag{9.2.42}$$

Again, the dynamics of the coordinate $\boldsymbol{\xi}$ can be expressed as a Hamiltonian system. We obtain

$$\begin{aligned}\frac{\mathrm{d}}{\mathrm{d}t}\xi_1 &= -\frac{\partial}{\partial \xi_2}H\\ \frac{\mathrm{d}}{\mathrm{d}t}\xi_2 &= \frac{\partial}{\partial \xi_1}H\end{aligned} \tag{9.2.43}$$

where the Hamilton function has the form

$$H = \frac{\tilde{\omega}}{2}\boldsymbol{\xi}^2 - \frac{1}{2\pi}\Gamma_1 \ln|\boldsymbol{\xi} - \boldsymbol{x}_1(0)| - \frac{1}{2\pi}\Gamma_2 \ln|\boldsymbol{\xi} - \boldsymbol{x}_2(0)| \tag{9.2.44}$$

This shows that the motion of the reduced three-vortex system is integrable. The motion of the third vortex proceeds along lines of constant energy H, see fig. 9.2.1.

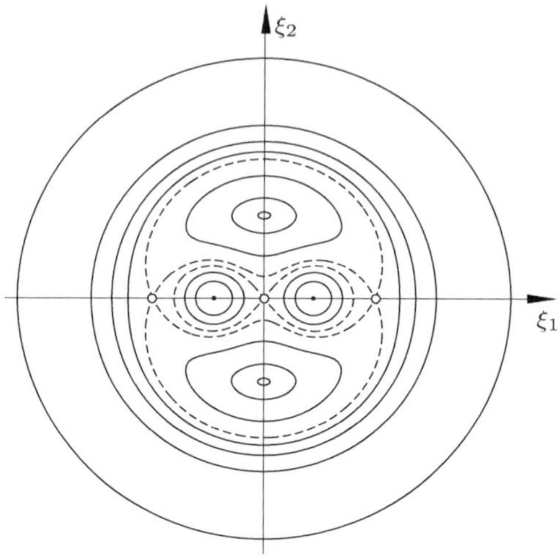

Fig. 9.2.1: Reduced three-vortex system: in a rotating coordinate system the third vortex moves along lines of constant energy.

9.2.2.4 Four Point Vortices

Four point vortices can perform chaotic motions. For special initial conditions, however, quasi-periodic motions occur. These can be generated by choosing initial conditions with a special symmetry which is preserved during the motion of the point vortices. It can be shown, for example, that initial conditions for which the point vortices are the vertices of a parallelogram lead to quasi-periodic motions since the dimension of the system of differential equations is then reduced. Such a case is shown in Colour Plate XVe, p. 623. For modified initial conditions, temporal chaotic vortex motions like those shown in Colour Plate XVf can be found. A rigorous mathematical proof of the existence of such motions can be given for various configurations. In the following, we present possible proof strategies for the reduced four-vortex system.

Chaos in the Reduced Four-Vortex System

A reduced four-vortex can be defined in analogy to the reduced three-vortex system. We denote the coordinate of the fourth vortex $\boldsymbol{\xi} = \{\xi_1, \xi_2\}$. The motion of this vortex is determined by the equation of motion

$$\begin{aligned} \dot{\xi}_1 &= -\frac{\partial}{\partial \xi_2} H(\boldsymbol{\xi}, \boldsymbol{x}_1, \boldsymbol{x}_2, \boldsymbol{x}_3) \\ \dot{\xi}_2 &= \frac{\partial}{\partial \xi_1} H(\boldsymbol{\xi}, \boldsymbol{x}_1, \boldsymbol{x}_2, \boldsymbol{x}_3) \end{aligned} \qquad (9.2.45)$$

Here, the Hamiltonian depends on the coordinates of the three point vortices and has the following form for the "trial vortex"

$$H(\boldsymbol{\xi}, t) = -\frac{1}{2\pi} \sum_{\mathrm{i}=1}^{3} \Gamma_{\mathrm{i}} \ln |\boldsymbol{\xi} - \boldsymbol{x}_{\mathrm{i}}(t)| \tag{9.2.46}$$

where the coordinates $\boldsymbol{x}_{\mathrm{i}}(t)$, i=1, 2, 3 are given by the equation of motion of the point vortex system eq. (9.2.6) for three vortices.

Since the motion of three point vortices is integrable, the temporal behaviour of the vortex coordinates $\boldsymbol{x}_{\mathrm{i}}(t)$ for bound states is quasi-periodic or, if we transform the motion into a co-rotating coordinate system, periodic. This means that the Hamilton function eq. (9.2.46) depends on time. Therefore, temporally chaotic motions of the trial vortex with the spatial coordinate $\boldsymbol{\xi}$ can occur (Neufeld and Tél, 1997).

The existence of chaotic motions can be explained, for example as follows. We have seen that the equilateral vortex configuration is stable for vortices with equal vorticity and rotates around the vortex center (Colour Plate XVa, p. 623). For small deviations of this configuration, additional rotations about the edges occur (Colour Plate XVb). As a result, the Hamilton function of the fourth point vortex with the coordinate $\boldsymbol{\xi}$ depends slightly on time in a co-rotating coordinate system, where the temporally periodic part is a function of the distance from equilibrium. Hence, the Hamilton function has the form

$$H(\boldsymbol{\xi}, t) = H_0(\boldsymbol{\xi}) + \varepsilon H_1(\boldsymbol{\xi}, t) \tag{9.2.47}$$

For $\varepsilon = 0$, i.e. for the rigid three-vortex rotation, the trial vortex moves along the lines of constant energy $H_0(\boldsymbol{\xi})$ shown in fig. 9.2.2. Several heteroclinic orbits can be distinguished. As was discussed in section 4.5, a temporal perturbation ($\varepsilon \neq 0$) causes a disintegration of these orbits and can lead to the occurrence of chaotic motions. We refer also to fig. 4.5.8.

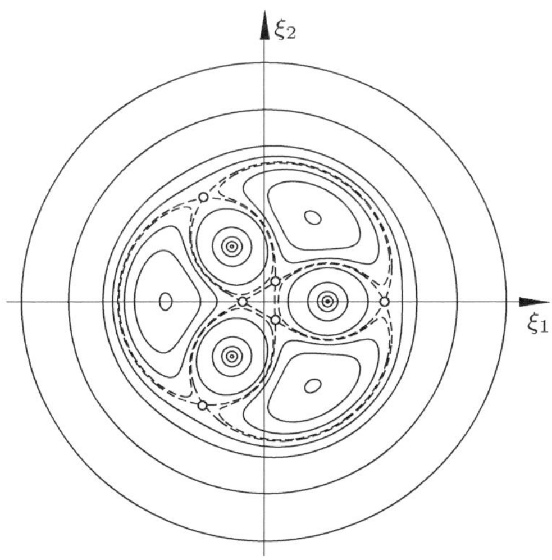

Fig. 9.2.2: Reduced four-vortex system: lines of constant energy $H_0(\boldsymbol{\xi})$, eq. (9.2.47).

A more detailed discussion on the existence of chaotic motions for the four-vortex system can be found in the references (Ziglin, 1980; Oliva, 1991; Khanin, 1982; Marchioro and Pulvirenti, 1984; Newton, 2001).

9.2.2.5 Mixing by Point Vortices

Exploring the dynamics of point vortex systems allows us to gain insight into the spatio-temporal laws underlying mixing processes. N point vortices with the circulations Γ_i at the points $\boldsymbol{x}_i$ generate a time-dependent velocity field of the form

$$\boldsymbol{u}^N(\boldsymbol{x},t) = \sum_{i=1}^{N} \Gamma_i \boldsymbol{u}(\boldsymbol{x} - \boldsymbol{x}_i(t)) \tag{9.2.48}$$

In order to model mixing processes, let us now consider the motion of passive Lagrangian particles in this velocity field. If we denote the coordinates of these particles $\boldsymbol{\eta}_i(t)$, their motion is given by the solution of the system of differential equations

$$\dot{\boldsymbol{\eta}}_i(t) = \boldsymbol{u}^N(\boldsymbol{\eta}_i(t),t) \tag{9.2.49}$$

with the initial conditions $\boldsymbol{\eta}_i(0)$.

Here, the mixing properties are uniquely connected to the characteristics of the trajectories of these Lagrangian particles. An essential characteristic here is the distance between closely neighbouring particle trajectories, which is quantified by the Lyapunov exponents. Obviously, chaotic motion with a positive Lyapunov exponent is an advantage for an effective mixing. Colour Plate XIV B, p. 617, shows the motion of passive Lagrangian particles in a chaotic four-vortex system. For the initial conditions, we chose two squares, marked red and blue. These regions are stirred by the motion of the point vortices.

9.2.3 Onsager's Equilibrium Theory

Just as in the kinetic theory of gases, it is not expedient to treat the motion of many single vortices deterministically. A statistical formulation is needed. As early as 1945, L. Onsager proposed such a formulation in a lecture to the American Physical Society and summarised it in the article (Onsager, 1949), see also (Eyink and Sreenivasan, 2006). It is only recently that the significance of his theory for turbulence research was recognised. Onsager's theory is based on an application from statistical mechanics, which is justified by the Hamiltonian character of vortex dynamics. The underlying principle is the ergodic hypothesis, which essentially says that all states characterised by the same conserved quantities are hit with the same probability. Descriptively, this means that, once the total energy H, the vortex centre $\boldsymbol{R}$ and the quantity L have been specified, see eqs. (9.2.7, 9.2.9, 9.2.10), the trajectories in the 2N-dimensional phase space fill a specific (2N−4)-dimensional sub-manifold densely. This hypothesis allows us to define a probability density distribution $f(\boldsymbol{x}_1, .., \boldsymbol{x}_N)$ which enables us to calculate statistical quantities like, for example, the correlations between several vortex locations. By means of this statistical theory, Onsager was able to predict that, under certain conditions, large clusters of vortices with the same vorticity can be generated. This phenomenon has often been used for an explanation of the existence of Jupiter's Great Red Spot, see fig. 9.0.1.

9.2.3.1 Viscosity

A point vortex is an idealised object. Without energy supply vortices in real viscous fluids are damped and decay, as can be seen in the example of the Lamb-Oseen vortex (cf. section 9.1.5.1). In two dimensions, closely adjacent vortices eventually merge. A mathematical treatment of this process is very challenging. However, the effect of mutual shear between adjacent vortices has proven to be of great importance for the energy transport in a two-dimensional fluid, the properties of which will be the subject of section 9.4.2. An axisymmetric vortex exposed to an external strain field will be elongated along the stretching direction of the strain which can to some extent be considered as an elliptical deformation. This process, denoted vortex thinning (Kraichnan, 1967; Chen et al., 2006), induces relative motions between small-scale vortices that have been deformed under their mutual strain and leads to an energy transfer to larger scales.

More recently, a generalisation of the point vortex model has been proposed in (Friedrich and Friedrich, 2013). Here, two point vortices $\boldsymbol{x}_\mathrm{i}$ and $\boldsymbol{y}_\mathrm{i}$ with equal circulation are coupled via an overdamped spring that mimics an energy input to the system on a small scale D_0. The model equations read

$$\begin{aligned} \dot{\boldsymbol{x}}_\mathrm{i} &= \frac{\gamma}{2}(D_0 - |\boldsymbol{x}_\mathrm{i} - \boldsymbol{y}_\mathrm{i}|)\boldsymbol{e}_\mathrm{i} + \Gamma_\mathrm{i}\boldsymbol{u}(\boldsymbol{x}_\mathrm{i} - \boldsymbol{y}_\mathrm{i}) \\ &\quad + \sum_\mathrm{j} \Gamma_\mathrm{j}[\boldsymbol{u}(\boldsymbol{x}_\mathrm{i} - \boldsymbol{x}_\mathrm{j}) + \boldsymbol{u}(\boldsymbol{x}_\mathrm{i} - \boldsymbol{y}_\mathrm{j})] \\ \dot{\boldsymbol{y}}_\mathrm{i} &= -\frac{\gamma}{2}(D_0 - |\boldsymbol{y}_\mathrm{i} - \boldsymbol{x}_\mathrm{i}|)\boldsymbol{e}_\mathrm{i} + \Gamma_\mathrm{i}\boldsymbol{u}(\boldsymbol{y}_\mathrm{i} - \boldsymbol{x}_\mathrm{i}) \\ &\quad + \sum_\mathrm{j} \Gamma_\mathrm{j}[\boldsymbol{u}(\boldsymbol{y}_\mathrm{i} - \boldsymbol{y}_\mathrm{j}) + \boldsymbol{u}(\boldsymbol{y}_\mathrm{i} - \boldsymbol{x}_\mathrm{j})] \end{aligned} \tag{9.2.50}$$

where $\boldsymbol{e}_\mathrm{i} = \frac{\boldsymbol{x}_\mathrm{i} - \boldsymbol{y}_\mathrm{i}}{|\boldsymbol{x}_\mathrm{i} - \boldsymbol{y}_\mathrm{i}|}$ is the unit vector and $\boldsymbol{u}(\boldsymbol{r})$ the velocity field of a point vortex centered at the origin, $\boldsymbol{u}(\boldsymbol{r}) = \boldsymbol{e}_z \times \frac{\boldsymbol{r}}{2\pi r^2}$. Here, the first and the second term on the right-hand side of eq. (9.2.50) describe the interaction between the two point vortices $\boldsymbol{x}_\mathrm{i}$ and $\boldsymbol{y}_\mathrm{i}$ similar to eq. (9.2.15) with the additional first term that introduces the inelastic coupling and makes the system non-Hamiltonian. The remaining terms describe the interaction between the other objects $\boldsymbol{x}_\mathrm{j}$, $\boldsymbol{y}_\mathrm{j}$.

Due to the additional degree of freedom, the coupled point vortices $\boldsymbol{x}_\mathrm{i}$ and $\boldsymbol{y}_\mathrm{i}$ form an active particle that is sensitive to a shear velocity. As the particle is gradually sheared, the spring pulls back and the particle relaxes to the size D_0 with a relaxation time $1/\gamma$. Moreover, the particles generate a far-field that is similar to an elliptical vortex in the sense of the vortex thinning mechanism mentioned above. The dynamical system (9.2.50) shows interesting features such as fast rotations and the formation of vortex clusters and offers a non-equilibrium version of Onsager's theory.

In three dimensions, a dynamical and geometrical reorganisation of the vortex sheets and tubes can be observed. An initially turbulent velocity field decays and comes to a rest after some time. During this so-called decaying turbulence, characteristic behaviour occurs; its properties are the subject of current research. Stationary turbulent fields in which the fluid flow is preserved have to be maintained by a supply of energy. These flows are thus open systems far from equilibrium which, however, have to be treated by statistical methods because of the numerous spatial degrees of freedom. In particular, the Hamiltonian character of the dynamics is violated here.

9.3 Turbulence: Determinism and Stochasticity

The Navier-Stokes equation is a purely deterministic non-linear partial differential equation. Physicists and engineers largely agree that this equation, together with the incompressibility condition, can be used to describe incompressible, laminar, but also turbulent Newtonian fluids as long as the conditions for a continuum description are fulfilled since experimental results can be reproduced by numerical simulations with high accuracy (Jovanović, 2004; Oberlack, 2000). From the previous chapters of this book, we know that non-linear deterministic equations can provoke complex chaotic behaviour caused by local instabilities, i.e. tiny uncontrollable perturbations of the initial or boundary conditions can lead to different behaviour after some time. This can also be observed in laboratory experiments on turbulent flows. Since the initial conditions, for example, can only ever be realised with finite precision, every repetition of the experiment inevitably leads to differing results which have a random character. However, there are certain statistical quantities which prove to be robust in the face of disturbances and which can be used to characterise such flows.

Considering that turbulent flows are characterised by a high number of active, strongly non-locally coupled degrees of freedom with a magnitude of order $(Re)^{9/4}$, it becomes clear that turbulent flow motions have to be treated by a combination of the deterministic basic equations of fluid dynamics and statistical and stochastic methods. This led to the development of the research field called *statistical hydrodynamics*. The emergence of this area of research is inseparably connected with the name of Kolmogorov although physicists like W. Heisenberg and C. F. von Weizsäcker also dealt intensively with this problem, see (Hinze, 1975; Monin and Yaglom, 2007a; Monin and Yaglom, 2007b).

9.3.1 Statistical Averaging

In statistical hydrodynamics, the Eulerian velocity field or comparable fields are treated as *random fields*. Since one is primarily interested in averaged quantities of the flow in a statistical description of turbulence, so-called statistical ensembles are needed to determine proper mean values. A statistical ensemble represents the set of all possible realisations of flows or repetitions of a turbulence experiment generated or conducted under the same fixed external conditions. The initial conditions of these realisations, i.e. the velocity field $\boldsymbol{u}_0(\boldsymbol{x})$ at the points $\boldsymbol{x}$ at time $t = 0$, differ. The velocity field $\boldsymbol{u}_{\mathrm{R}}(\boldsymbol{x}, t; \boldsymbol{u}_0)$ at time t is determined by the Navier-Stokes dynamics, where the subscript R identifies the specific realisation. The ensemble is now characterised by a statistical distribution of the initial field $\boldsymbol{u}_0$. Ensemble mean values are hence defined with regard to the statistics of these initial fields.

In this way, the probability density $f(\boldsymbol{u}, \boldsymbol{x}, t)$ of the velocity field $\boldsymbol{u}$ at the point $\boldsymbol{x}$ at time t can be defined. In principle, for a complete characterisation of the ensemble statistics, one has to specify a joint probability density of the velocity at all points of the flow. For this purpose, one has to consider the N-point distributions $f(\boldsymbol{u}_1, \boldsymbol{x}_1, ..., \boldsymbol{u}_{\mathrm{N}}, \boldsymbol{x}_{\mathrm{N}}, t)$ and take the continuum limit. From the viewpoint of the theory of dynamical systems, the statistical treatment of a turbulent flow is comparable with the statistical treatment of a dynamical system by means of the invariant measure (cf. section 3.10).

The introduction of probability distributions now allows us to apply all the techniques used in the theory of probability (see section 3.9), in particular averaging. When comparing theoretical derivations and direct measurements, however, the problem arises that the set of all possible realisations is never available for observations. In practice, the ensemble mean is therefore substituted in measurements and numerical experiments by the temporal mean

$$\langle H(t) \rangle = \lim_{T \to \infty} \frac{1}{T} \int_0^T H(t + t')\, \mathrm{d}t' \tag{9.3.1}$$

The background to this is the ergodic hypothesis which is assumed to be valid in turbulent flows on the basis of empirical experience (Tsinober, 2001). A necessary condition for the convergence of the temporal mean is the temporal stationarity of the flow; in this case, the mean value is independent of t. If the flow is weakly stationary, this averaging is also reasonable for a finite time window of width T.

The probability distributions allow us to characterise turbulent flows. The following definitions are commonly accepted: *Stationary turbulence* means that the distributions $f(\boldsymbol{u}_1, \boldsymbol{x}_1, ..., \boldsymbol{u}_{\mathrm{N}}, \boldsymbol{x}_{\mathrm{N}}, t)$ do not depend on time t. In the case of *homogeneous turbulence*, $f(\boldsymbol{u}, \boldsymbol{x}, t)$ does not depend on $\boldsymbol{x}$. Here, the multipoint distributions are only functions of the difference vectors $\boldsymbol{x}_i - \boldsymbol{x}_j$. Stationarity and homogeneity can also be related to symmetries. Formally, stationarity can be seen as the invariance of the probability distributions with respect to shifts in time and homogeneity as the invariance under translations. *Homogeneous isotropic turbulence* additionally implies the invariance of the probability distributions under rotations about arbitrary coordinate axes. Mirror or reflection symmetry implies invariance with respect to reflections at arbitrary planes through the origin. Often, the notation of isotropic invariance also includes mirror symmetry.

When the statistics of turbulent fields is characterised by means of symmetries, one has to note that the invariance of probability distributions has to be understood in a statistical sense; a velocity field $\boldsymbol{u}_R(\boldsymbol{x}, t; \boldsymbol{u}_0)$ of a single realisation at time t does not need to possess this specific symmetry.

9.3.2 Moment Equations: The Closure Problem of Turbulence

As early as 1883, Osborne Reynolds proposed a statistical description of turbulence (Reynolds, 1883). In particular, he considered the equation for the mean velocity field $\bar{\boldsymbol{u}}(\boldsymbol{x}, t)$ which we will address in section 9.3.4. Further quantities important for the description of turbulent flows are the correlation functions or the moments of the velocity field. Their temporal evolution is determined by the basic equations of hydrodynamics. For example, the second-order moment or the two-point correlation function $C_{\mathrm{ij}}(\boldsymbol{x}, \boldsymbol{x}', t) = \langle\, u_{\mathrm{i}}(\boldsymbol{x}, t) u_{\mathrm{j}}(\boldsymbol{x}', t)\,\rangle$ can be evaluated directly from the Navier-Stokes equation together with the incompressibility condition

$$
\begin{aligned}
&\frac{\partial}{\partial t}\langle\, u_{\mathrm{i}}(\boldsymbol{x}, t) u_{\mathrm{j}}(\boldsymbol{x}', t)\,\rangle + \sum_{\mathrm{k}} \frac{\partial}{\partial x_{\mathrm{k}}}\langle\, u_{\mathrm{k}}(\boldsymbol{x}, t) u_{\mathrm{i}}(\boldsymbol{x}, t) u_{\mathrm{j}}(\boldsymbol{x}', t)\,\rangle \\
&+\sum_{\mathrm{k}} \frac{\partial}{\partial x'_{\mathrm{k}}}\langle\, u_{\mathrm{k}}(\boldsymbol{x}', t) u_{\mathrm{i}}(\boldsymbol{x}, t) u_{\mathrm{j}}(\boldsymbol{x}', t)\,\rangle = -\frac{\partial}{\partial x_{\mathrm{i}}}\langle\, p(\boldsymbol{x}, t) u_{\mathrm{j}}(\boldsymbol{x}', t)\,\rangle \\
&-\frac{\partial}{\partial x'_{\mathrm{j}}}\langle\, p(\boldsymbol{x}', t) u_{\mathrm{i}}(\boldsymbol{x}, t)\,\rangle + \nu[\Delta_x + \Delta_{x'}]\langle\, u_{\mathrm{i}}(\boldsymbol{x}, t) u_{\mathrm{j}}(\boldsymbol{x}', t)\,\rangle + Q_{\mathrm{ij}}(\boldsymbol{x}, \boldsymbol{x}', t) \quad (9.3.2)
\end{aligned}
$$

Here, $Q_{\mathrm{ij}}(\boldsymbol{x}, \boldsymbol{x}', t) = \langle u_{\mathrm{i}}(\boldsymbol{x}, t) f_{\mathrm{j}}(\boldsymbol{x}', t)\rangle + \langle f_{\mathrm{i}}(\boldsymbol{x}, t) u_{\mathrm{j}}(\boldsymbol{x}', t)\rangle$ denotes a source term which is attributable to the action of a volume force $f(\boldsymbol{x}, t)$. For all higher-order moments $\langle\, u_{\mathrm{i}}(\boldsymbol{x}_1, t_1) u_{\mathrm{j}}(\boldsymbol{x}_2, t_2) .... u_{\mathrm{k}}(\boldsymbol{x}_N, t_N)\,\rangle$, similar relations can be formulated by means of the Navier-Stokes equation with subsequent averaging. The infinite chain of transport equations for the N-point correlation functions are commonly denoted the Friedmann-Keller equations. Since the equation for the Nth-order moment includes the (N+1)th-order moment, the Friedmann-Keller equations are not closed. This is usually called the *closure problem of turbulence*, a central, and as yet unsolved problem in turbulence research. Up to now, the infinite hierarchy of moment equations can only be solved approximately by truncation and by the introduction of empirical models.

The closure problem is a direct consequence of the non-linear character of the Navier-Stokes equations. However, it should be noted that the non-linearity is not exclusively restricted to the advective term. What is more important is the non-linearity and non-locality of the pressure which ensures that the flow remains incompressible.

9.3.3 Decaying Turbulence

Decaying turbulence is a flow process in which a flow with a very high Reynolds number is generated by the injection of energy – for example by stirring coffee with a spoon in a cup – and is afterwards left to itself. Due to the viscosity of the fluid, the kinetic energy of the flow is dissipated. The long-term behaviour of the flow can be described by the linearised Navier-Stokes equation with appropriate boundary conditions because the flow velocity decreases and non-linearities exert less influence on the dynamics. As a result, in the final stage, the flow velocity decays essentially exponentially. However, if the Reynolds number of the initial flow is very high, the fluid passes through phases characterised by the non-linearity of the fluid dynamics. Characteristic regularities can then be observed. For example, the mean flow energy does not decay exponentially, but exhibits a slow decay over a long period of time (Monin and Yaglom, 2007a; Monin and Yaglom, 2007b). This is an indication of self-similar behaviour which we will address in more detail in section 9.6.4.1.

9.3.4 Reynolds Equation and Turbulence Modelling

Stationary turbulence denotes a flow in which the temporally irregular and spatially disordered structure is maintained persistently by a permanent supply of energy. An example is pipe flow: the fluid is conducted through a pipe with high pressure. In practical applications, details of the fluctuating velocity field are not of interest, but the *mean flow profile* is. To understand this, the flow field is divided into two parts

$$\boldsymbol{u}(\boldsymbol{x},t) = \bar{\boldsymbol{u}}(\boldsymbol{x},t) + \boldsymbol{u}'(\boldsymbol{x},t) \tag{9.3.3}$$

namely a mean flow $\bar{\boldsymbol{u}}(\boldsymbol{x},t)$ and turbulent fluctuations $\boldsymbol{u}'(\boldsymbol{x},t)$ with a vanishing temporal average

$$\langle \boldsymbol{u}'(\boldsymbol{x},t) \rangle = \lim_{T\to\infty} \frac{1}{T} \int\limits_{t}^{t+T} \boldsymbol{u}'(\boldsymbol{x},t')\,\mathrm{d}t' = 0 \tag{9.3.4}$$

Here, the turbulent pulsations usually receive their energy from the mean flow.

The mean flow is obtained by averaging over a time interval T_e which is sufficiently large in comparison with the characteristic fluctuations

$$\boldsymbol{U}(\boldsymbol{x},t) = \frac{1}{T_e} \int\limits_{t}^{t+T_e} \boldsymbol{u}(\boldsymbol{x},t')\,\mathrm{d}t' \tag{9.3.5}$$

so that the mean value for a large but finite time T_e is well defined. In the same way, the pressure field decomposes into $p(\boldsymbol{x},t) = \bar{p}(\boldsymbol{x},t) + p'(\boldsymbol{x},t)$.

An equation for determining the mean flow profile is the so-called Reynolds equation, which can be derived from the Navier-Stokes equation by taking the temporal mean. It can easily be shown that the linear operations differentiation and averaging can be exchanged (Hinze, 1975; Monin and Yaglom, 2007a). Therefore, together with $\boldsymbol{u}(\boldsymbol{x},t)$, the terms $\boldsymbol{u}'(\boldsymbol{x},t)$ and $\bar{\boldsymbol{u}}(\boldsymbol{x},t)$ are also divergence-free. For the convective term,

$$\frac{1}{T_e}\int\limits_t^{t+T_e} \boldsymbol{u}(\boldsymbol{x},t')\cdot\nabla\boldsymbol{u}(\boldsymbol{x},t')\,\mathrm{d}t' =$$
$$\frac{1}{T_e}\int\limits_t^{t+T_e} \left[\bar{\boldsymbol{u}}(\boldsymbol{x},t)\cdot\nabla\bar{\boldsymbol{u}}(\boldsymbol{x},t')+\boldsymbol{u}'(\boldsymbol{x},t')\cdot\nabla\boldsymbol{u}'(\boldsymbol{x},t')\right]\mathrm{d}t' \tag{9.3.6}$$

holds, where it was taken into account that, according to eq. (9.3.4)

$$\frac{1}{T_e}\int\limits_t^{t+T_e} \boldsymbol{u}'(\boldsymbol{x},t')\cdot\nabla\bar{\boldsymbol{u}}(\boldsymbol{x},t')\,\mathrm{d}t' = \frac{1}{T_e}\int\limits_t^{t+T_e} \bar{\boldsymbol{u}}(\boldsymbol{x},t')\cdot\nabla\boldsymbol{u}'(\boldsymbol{x},t)\,\mathrm{d}t' = 0 \tag{9.3.7}$$

holds because the mean flow profile changes only slightly within the time interval $t \leqslant t' < t + T_e$. The Reynolds equation (*Reynolds Averaged Navier Stokes* RANS) hence has the following form

$$\frac{\partial}{\partial t}\bar{\boldsymbol{u}}(\boldsymbol{x},t)+\bar{\boldsymbol{u}}(\boldsymbol{x},t)\cdot\nabla\bar{\boldsymbol{u}}(\boldsymbol{x},t)+\nabla\cdot\langle\boldsymbol{u}'(\boldsymbol{x},t)\boldsymbol{u}'(\boldsymbol{x},t)\rangle = -\nabla\bar{p}(\boldsymbol{x},t)+\nu\Delta\bar{\boldsymbol{u}}(\boldsymbol{x},t) \tag{9.3.8}$$

If we write the Reynolds equation componentwise using the substantial derivative in the following form

$$\frac{\mathrm{D}}{\mathrm{D}t}\bar{u}_{\mathrm{j}} = \frac{\partial}{\partial x_{\mathrm{i}}}\left[\nu\left(\frac{\partial\bar{u}_{\mathrm{i}}}{\partial x_{\mathrm{j}}}+\frac{\partial\bar{u}_{\mathrm{j}}}{\partial x_{\mathrm{i}}}\right)-\bar{p}\delta_{\mathrm{ij}}-\langle\, u'_{\mathrm{i}}u'_{\mathrm{j}}\,\rangle\right] \tag{9.3.9}$$

the result corresponds to a Navier-Stokes equation for the averaged fields with an additional non-linear term which plays the role of a stress, the so-called Reynolds stress tensor

$$\sigma^{\mathrm{Re}}_{\mathrm{ij}} = \langle u'_{\mathrm{i}}(\boldsymbol{x},t)u'_{\mathrm{j}}(\boldsymbol{x},t)\rangle \tag{9.3.10}$$

As can be seen from eq. (9.3.8) the turbulent fluctuations significantly influence the behaviour of the mean flow profile and the mean pressure $\bar{p}(\boldsymbol{x},t)$. Again, the mean pressure can be determined from the requirement that the mean flow profile is divergence-free.

The knowledge of the Reynolds stress tensor as a functional of the mean flow profile would be a great step forward for fluid mechanics in engineering science because the Reynolds equation would then be sufficient to calculate the mean profiles. Especially in applications, the mean profiles are the quantities which are important at first glance. However, it should be noted that the turbulent fluctuations $\boldsymbol{u}'(\boldsymbol{x},t)$ can be extremely large compared with the mean velocity profile and thus have to be taken into consideration in the design of airplanes or wind turbines.

A consistent derivation of the Reynolds stress tensor from the basic hydrodynamic equations as a functional of the mean fields has not yet been accomplished although it is one of the central problems in turbulence research. Therefore, modelling the stress tensor was the path followed by engineering science. If we recall that the Navier-Stokes equation is based on modelling the viscous stress tensor by Newton's ansatz, this approach is quite promising.

Famous turbulence models are the *eddy-viscosity models*, in which the Reynolds stress tensor is substituted by an effective damping term in order to characterise the energy flow from the mean flow to the sub-grid scales. According to an ansatz of Boussinesq, the Reynolds stress tensor is written as

$$\sigma_{\mathrm{ij}}^{\mathrm{Re}} = -\nu_t \left(\frac{\partial \bar{u}_{\mathrm{i}}}{\partial x_{\mathrm{j}}} + \frac{\partial \bar{u}_{\mathrm{j}}}{\partial x_{\mathrm{i}}} \right) + \frac{2}{3} k \delta_{\mathrm{ij}} \tag{9.3.11}$$

in analogy to the stress tensor generated by the molecular viscosity in the Navier-Stokes model, see eq. (7.1.19). Here, $k = \frac{1}{2} \langle u_{\mathrm{i}} u_{\mathrm{j}} \rangle$ denotes the turbulent kinetic energy and ν_t the turbulent eddy viscosity, which is usually much larger than the kinematic viscosity ν. Note that ν_t is not a characteristic of the material but is determined by the turbulent fluctuations. To model ν_t, Prandtl proposed the so-called *mixing length model*, in which an additional equation for the transport of kinetic energy is solved (Davidson, 2006; Pope, 2000).

Other models, the widely-used *$k - \varepsilon$ model* among them, additionally take the spatio-temporal dynamics of the local energy dissipation rate ε into consideration and prove to be of value especially when applied to simple shear flows. Detailed discussions of the advantages and disadvantages of the various turbulence models can be found in (Davidson, 2006; Pope, 2000).

If large eddy structures are important in a flow, the so-called *Large Eddy Simulations* (LES) are advantageous. In this case, the flow on large scales is simulated directly while structural models are introduced for the unresolved small-scale structures. This has resulted in the development of a research area with a multitude of numerical methods, some of which are also available commercially. However, this approach is also of limited use because the properties of the Reynolds stress tensor could not yet be derived from the basic theory. The results of LES simulations thus have to be considered with reservations. For further reading, we refer to (Piquet, 1999; Lesieur et al., 2001; Jovanović, 2004; Sagaut, 2001; Fröhlich, 2006). Two articles of Johansson and Oberlack in (Oberlack and Busse, 2002) are of special interest.

Finally, it should be noted that any turbulence model must fulfil the requirement of *physical feasibility*, i.e. the dynamics should not, for example, lead to negative kinetic energy values. Moreover, the fundamental symmetries should be incorporated in the modelling (Oberlack, 2000).

9.4 Characteristic Scales of Turbulence

9.4.1 Taylor Hypothesis

Theoretical model concepts based on velocity increments work in space (t fixed, $\boldsymbol{x}$ variable), while experimental measurements, for example in a wind channel, are carried out predominantly in time ($\boldsymbol{x}$ fixed, t variable). The Taylor hypothesis assumes that the distances in time Δt can be tranformed with the help of the mean velocity of the flow $\bar{\boldsymbol{u}} = \langle \boldsymbol{u}(\boldsymbol{x}, t) \rangle$ into spatial distances $\boldsymbol{r}$ using

$$\boldsymbol{r} = \bar{\boldsymbol{u}} \Delta t \tag{9.4.1}$$

This requires that the mean velocity $\bar{\boldsymbol{u}}$ is large compared with the fluctuations $\boldsymbol{u}'(\boldsymbol{x}, t) = \boldsymbol{u}(\boldsymbol{x}, t) - \bar{\boldsymbol{u}}$, i.e. that

$$\frac{\langle |\boldsymbol{u}'|^2 \rangle}{|\bar{\boldsymbol{u}}|^2} \ll 1 \tag{9.4.2}$$

holds. Taylor started from the assumption that turbulent structures on a length scale $|\boldsymbol{r}|$ are not significantly deformed during the time interval Δt, i.e. vortex structures are transported past the measurement device without deformation.

9.4.2 Phenomenology of Eddy Decay and Energy Cascade

Characteristic of turbulent flows is the huge number of interlacing eddies on a wide range of different scales which have only a limited life span and decay after a short period. In three-dimensional flows, we were able to explain this picture by the concept of vortex stretching and the destabilisation of twisted vortex filaments and sheets. L.F. Richardson developed a cascade model for the energy transfer in three-dimensional turbulence; it is based on the idea that turbulent flows are composed of vortex elements on varying scales (Richardson, 1922). If energy is injected into a system on large scales L, large vortex structures are generated which become unstable for high Reynolds numbers after only a short time, thereby transferring their total energy to smaller, newly generated eddies. These eddies in turn become unstable and transfer their energy without dissipation to the next generation of eddies. The cascade only comes to a standstill when the eddies have become so small that shear forces dominate and the molecular viscosity causes the energy to dissipate.

This cascade model is also the idea underlying the famous turbulence model of Kolmogorov (Kolmogorov, 1941), who arrived at important quantitative statements on the energy transfer rate and characteristic scales on the assumption of various hypotheses. According to his idea, all anisotropies and the whole influence of boundary conditions are lost during the decay of the largest vortex structures. In his *first hypothesis of local isotropy*, Kolmogorov postulates that, in the case of sufficiently high Reynolds numbers, every turbulent flow is statistically homogeneous and isotropic on small scales $l \ll L$ und thus possesses a universal character. Within this range of scales, the transfer of kinetic energy to small scales takes place until all the kinetic energy is finally dissipated. Thus, processes in this range should only depend on the mean energy dissipation rate $\langle \varepsilon \rangle$ and the kinematic viscosity ν. This is the content of Kolmogorov's *first similarity hypothesis*. From this hypothesis, one can derive characteristic scales, in particular the smallest scale, named after Kolmogorov, as the dissipation length η; this will be addressed in more detail in section 9.4.4. The *second similarity hypothesis* makes a statement about the range of scales in between. In every turbulent flow with a sufficiently high Reynolds number, the statistics of motions on a scale l with $L \gg l \gg \eta$ have a universal form and do not depend on ν but only on $\langle \varepsilon \rangle$ (Kolmogorov, 1941; Pope, 2000). This range is called the inertial range. In section 9.6.4, we will address the universal form of the distributions in more detail. In the following, we present some characteristic scales which limit the range of validity of the cascade model.

9.4.3 The Integral Length L

The upper limit is given by the largest vortex structures. It is a measure of the extent of the range in which appreciable correlations still arise. On larger length scales, the turbulent fluctuations in a homogeneous stationary flow should become statistically independent. Quantitatively, this statement is based upon the behaviour of the correlation function of the velocity field at the points $\boldsymbol{x}$ and $\boldsymbol{x} + \boldsymbol{r}$. If, for simplicity's sake, we only consider one velocity component $u(\boldsymbol{x}, t)$, the correlation function $C(r)$ can be written as

$$C(r) = \langle u(\boldsymbol{x} + \boldsymbol{r}, t) u(\boldsymbol{x}, t) \rangle = R(r) C(0) \tag{9.4.3}$$

where we assume that this function depends only on the distance r and not on the direction of $\boldsymbol{r}$. In the case of an exponential decay of the correlation, this function $R(r)$ has the form

$$R(r) = \mathrm{e}^{-r/L} \tag{9.4.4}$$

Hence, for the definition of the integral length, the following relation can be used

$$L = \int_0^\infty \frac{C(r)}{C(0)} \mathrm{d}r \tag{9.4.5}$$

This quantity measures how fast the autocorrelation decays to zero and is interpreted as the typical length of the largest vortex structures or as the scale on which energy is injected into the system.

If the turbulence is homogeneous and anisotropic the definition of different integral lengths may be necessary (Pope, 2000). Their definition is the based on the correlation matrix

$$C_{\mathrm{ij}}(\boldsymbol{r}, \boldsymbol{x}) = \langle u_{\mathrm{i}}(\boldsymbol{x} + \boldsymbol{r}, t) u_{\mathrm{j}}(\boldsymbol{x}, t) \rangle = R_{\mathrm{ij}}(\boldsymbol{r}) C_{\mathrm{ij}}(\boldsymbol{0}) \tag{9.4.6}$$

9.4.4 The Kolmogorov Microscales

Due to the energy transfer, a steady state is maintained within the cascade between the largest scale where energy is injected and a characteristic smallest scale, the Kolmogorov dissipation length η, where energy is dissipated. It can be determined from the first similarity hypothesis by a dimensional analysis according to which a turbulent flow is characterised by the kinematic viscosity ν together with the input of energy. This energy is dissipated to heat and is characterised quantitatively by the mean energy dissipation rate $\langle\varepsilon\rangle$. Since $\langle\varepsilon\rangle$ and ν have the units

$$[\nu] = \frac{[\mathrm{m}]^2}{[\mathrm{sec}]} \qquad [\,\langle\varepsilon\rangle\,] = \frac{[\mathrm{m}]^2}{[\mathrm{sec}]^3} \tag{9.4.7}$$

the following length and time scales can be specified

$$\eta = \left(\frac{\nu^3}{\langle\varepsilon\rangle}\right)^{1/4}, \qquad \tau_\eta = \sqrt{\frac{\nu}{\langle\varepsilon\rangle}}, \qquad u_\eta = (\,\nu\langle\varepsilon\rangle\,)^{1/4} \tag{9.4.8}$$

The meaning of these scales becomes evident by calculating a Reynolds number using the velocity u_η, the Kolmogorov length η and the kinematic viscosity ν. We obtain

$$R_\eta = \frac{\eta u_\eta}{\nu} = 1 \tag{9.4.9}$$

i.e. the Kolmogorov length and time scale characterises a flow motion which is seen to be laminar.

9.4.5 The Taylor Length

A further length scale used to characterise turbulent flow is the so-called Taylor length λ. As can be seen, dissipative effects already become noticeable on larger scales than the Kolmogorov dissipation scale η. Their influence appears in the autocorrelation function $R(r)$ for small r-values. Contrary to the predictions made by the cascade model, according to which $R(r)$ should have a positive curvature for small r, experimental measurements yield negative values for the curvature which can be attributed to the dissipation. Observing that the velocity correlation for isotropic turbulence is symmetric in r, the Taylor length λ is defined by the relation

$$R(r) = 1 - \frac{1}{2}\frac{r^2}{\lambda^2} \tag{9.4.10}$$

The integral length L and the Taylor length λ are the limits of the inertial range for which, according to Kolmogorov's model, homogeneous and isotropic turbulence can be assumed and in which dissipative effects can be neglected. The adjacent range of smaller scales $\lambda \geqslant l \geqslant 0$ is denoted dissipation range.

9.4.6 The Taylor-Reynolds Number

In order to be able to compare the measurement results of various turbulence experiments such as, for example, the turbulent flow of a free jet or the wake flow behind a grid or cylinder, it is not possible to refer to the same Reynolds numbers $Re = uL/\nu$

since the L-values cannot be compared because of the different generation and boundary conditions of the flows. Comparable turbulence can be obtained by referring to the so-called Taylor-Reynolds number

$$Re_\lambda = \frac{\sigma\lambda}{\nu} \tag{9.4.11}$$

where $\sigma = \sqrt{\langle \boldsymbol{u}'^2 \rangle}$ denotes the standard deviation of the velocity fluctuations. The advantage of this definition is that Re_λ can be determined solely from quantities of the flow under consideration independently of the macroscopic scale L which is given by the energy input and the boundary condintions (Pope, 2000).

9.5 The Turbulent Cascade

9.5.1 The von Kármán-Howarth Relation

We recall that the model concepts of Richardson and Kolmogorov concerning the energy transport from large to small scales and Kolmogorov's hypotheses cannot be directly related to the Navier-Stokes equations. Early on, researchers therefore strove to gain essential information on the energy transfer and the evolution of the correlation functions directly from the Navier-Stokes equations. On the assumption of statistical homogeneity, isotropy and mirror symmetry, Theodor von Kármán and Leslie Howarth succeeded in deriving a relation between the correlation functions of second and third order from the moment equation (9.3.2), see (von Kármán und L. Howarth, 1938). For the derivation, we denote the correlation functions of second and third order

$$\begin{aligned} C_{\mathrm{ij}} &= \langle\, u_{\mathrm{i}}(\boldsymbol{x},t)\, u_{\mathrm{j}}(\boldsymbol{x}',t)\,\rangle \\ C_{\mathrm{ij,k}} &= \langle\, u_{\mathrm{i}}(\boldsymbol{x},t)\, u_{\mathrm{j}}(\boldsymbol{x},t)\, u_{\mathrm{k}}(\boldsymbol{x}',t)\,\rangle \end{aligned} \tag{9.5.1}$$

The relation eq. (9.3.2) can be greatly simplified using symmetry arguments.

a) Due to the assumption of spatial homogeneity, the correlation functions depend only on the distance vector $\boldsymbol{r} = \boldsymbol{x}' - \boldsymbol{x}$

$$\begin{aligned} C_{\mathrm{ij}}(\boldsymbol{r},t) &= \langle\, u_{\mathrm{i}}(\boldsymbol{x},t)\, u_{\mathrm{j}}(\boldsymbol{x}',t)\,\rangle = C_{\mathrm{ji}}(-\boldsymbol{r},t) \\ C_{\mathrm{li,j}}(\boldsymbol{r},t) &= \langle\, u_{\mathrm{l}}(\boldsymbol{x},t)\, u_{\mathrm{i}}(\boldsymbol{x},t)\, u_{\mathrm{j}}(\boldsymbol{x}',t)\,\rangle \\ C_{\mathrm{lj,i}}(-\boldsymbol{r},t) &= \langle\, u_{\mathrm{l}}(\boldsymbol{x}',t)\, u_{\mathrm{i}}(\boldsymbol{x},t)\, u_{\mathrm{j}}(\boldsymbol{x}',t)\,\rangle \end{aligned} \tag{9.5.2}$$

As a direct consequence, one obtains

$$(\Delta_x + \Delta_{x'})\langle\, u_{\mathrm{i}}(\boldsymbol{x},t)\, u_{\mathrm{j}}(\boldsymbol{x}',t)\,\rangle = 2\Delta_r C_{\mathrm{ij}}(\boldsymbol{r},t) \tag{9.5.3}$$

b) Due to the assumption of isotropy and mirror symmetry, the pressure-velocity functions vanish identically

$$\langle\, p(\boldsymbol{x}',t)\, u_{\mathrm{i}}(\boldsymbol{x},t)\,\rangle = 0 \tag{9.5.4}$$

These correlations are vectors which depend on the distance vector $\boldsymbol{r}$. Hence, they can only have the form $f(r,t)r_{\mathrm{i}}$, see section 9.8. Since

$$\sum_{\mathrm{i}} \langle\, p(\boldsymbol{x}',t)\frac{\partial}{\partial x_{\mathrm{i}}}u_{\mathrm{i}}(\boldsymbol{x},t)\,\rangle = \sum_{\mathrm{i}} \frac{\partial}{\partial r_{\mathrm{i}}}C^{p}_{\mathrm{i}}(\boldsymbol{r},t) = \sum_{\mathrm{i}} \frac{\partial}{\partial r_{\mathrm{i}}} f(r,t)r_{\mathrm{i}} = 0 \tag{9.5.5}$$

holds for an incompressible vector field, it can be concluded that the relation $f(r,t) = 0$ is valid.

c) We now take into account that the assumption of isotropy and mirror symmetry leads for a third-order tensor to the relation

$$C_{\mathrm{ij,k}}(-\boldsymbol{r}) = -C_{\mathrm{ij,k}}(\boldsymbol{r}) \tag{9.5.6}$$

The results obtained so far can now be summarised as

$$\frac{\partial}{\partial t}C_{\mathrm{ij}}(\boldsymbol{r},t) - 2\sum_{\mathrm{l}}\frac{\partial}{\partial r_{\mathrm{l}}}C_{\mathrm{li,j}}(\boldsymbol{r},t) = 2\nu\Delta_r C_{\mathrm{ij}}(\boldsymbol{r},t) + Q_{\mathrm{ij}}(\boldsymbol{r},t) \tag{9.5.7}$$

This relation has the form of a balance equation for the correlation function $C_{\mathrm{ij}}(\boldsymbol{r},t)$. As shown in the Appendix, the tensors C_{ij} and $C_{\mathrm{ij,k}}$ have a simple form under the assumption of homogeneity and isotropy. For example, the tensors of the second- and third-order moments $C_{\mathrm{ij}}(\boldsymbol{r},t)$ and $C_{\mathrm{li,j}}$ can both be expressed by a scalar function. To show this, we use the longitudinal functions $C_{rr}(r,t)$ together with $C_{rrr}(r,t)$

$$\begin{aligned} C_{\mathrm{ij}}(\boldsymbol{r},t) &= \frac{1}{2r}\frac{\partial}{\partial r}\left(r^2 C_{rr}(r,t)\right)\delta_{\mathrm{ij}} + \left[C_{rr}(r,t) - \frac{1}{2r}\frac{\partial}{\partial r}\left(r^2 C_{rr}(r,t)\right)\right]\frac{r_{\mathrm{i}}r_{\mathrm{j}}}{r^2} \\ C_{\mathrm{ij,k}}(\boldsymbol{r},t) &= -\frac{C_{rrr}(r,t)}{2}\delta_{\mathrm{ij}}\frac{r_{\mathrm{k}}}{r} + \frac{1}{4r}\frac{\partial}{\partial r}\left(r^2 C_{rrr}(r,t)\right)\left[\frac{r_{\mathrm{i}}}{r}\delta_{\mathrm{jk}} + \frac{r_{\mathrm{j}}}{r}\delta_{\mathrm{ik}}\right] \\ &\quad - \frac{r^2}{2}\frac{\partial}{\partial r}\left(\frac{C_{rrr}(r,t)}{r}\right)\frac{r_{\mathrm{i}}r_{\mathrm{j}}r_{\mathrm{k}}}{r^3} \end{aligned} \tag{9.5.8}$$

From eq. (9.5.7), a relation between $C_{rr}(r,t)$ and $C_{rrr}(r,t)$ can be derived directly. With the definition

$$R(r,t) = \sum_{\mathrm{j}} C_{\mathrm{jj}}(\boldsymbol{r},t) \tag{9.5.9}$$

$$T_{\mathrm{i}}(\boldsymbol{r},t) = -\sum_{\mathrm{j}} C_{\mathrm{ij,j}}(\boldsymbol{r},t) \tag{9.5.10}$$

$$Q(r,t) = \sum_{\mathrm{j}} Q_{\mathrm{jj}}(\boldsymbol{r},t) \tag{9.5.11}$$

one obtains

$$\frac{\partial}{\partial t}R(r,t) + 2\sum_{\mathrm{i}}\frac{\partial}{\partial r_{\mathrm{i}}}T_{\mathrm{i}}(\boldsymbol{r},t) = \frac{2\nu}{r^2}\frac{\partial}{\partial r}\Delta_r R(r,t) + Q(r,t) \tag{9.5.12}$$

This is a balance equation for the quantity $R(r,t)$ and the flow $\boldsymbol{T}(\boldsymbol{r},t)$ appertaining to $R(r,t)$. Their significance will be discussed in the next section.

As shown in the Appendix, eqs. (9.8.29, 9.8.35), the quantities R and $\boldsymbol{T}$ can be expressed by C_{rr} and C_{rrr}. Exchanging the derivatives with respect to t and r, one obtains for the individual terms

$$\begin{aligned} \frac{\partial}{\partial t} R(r,t) &= \frac{1}{r^2}\frac{\partial}{\partial r}\big(r^3 \frac{\partial}{\partial t} C_{rr}(r,t)\big) \\ 2\sum_{\mathrm{i}} \frac{\partial}{\partial r_{\mathrm{i}}} T_{\mathrm{i}}(\boldsymbol{r},t) &= -\frac{1}{r^2}\frac{\partial}{\partial r}\Big(\frac{1}{r}\frac{\partial}{\partial r}\big(r^4 C_{rrr}(r,t)\big)\Big) \\ \Delta_r R(r,t) &= \frac{1}{r^2}\frac{\partial}{\partial r}\big(r^2 \frac{\partial R(r,t)}{\partial r}\big) \end{aligned} \tag{9.5.13}$$

With these relations, the so-called Kármán-Howarth equation can be formulated

$$\frac{\partial}{\partial t} C_{rr}(r,t) = \frac{1}{r^4}\frac{\partial}{\partial r}\Big(r^4\big(C_{rrr}(r,t) + 2\nu \frac{\partial}{\partial r} C_{rr}(r,t)\big)\Big) + Q^*(r,t) \tag{9.5.14}$$

where the source term is given by the expression $Q^*(r,t) = \frac{1}{r^3}\int r^2 Q(r,t)\mathrm{d}r$. The Kármán-Howarth equation is an exact relation which can be deduced directly from the Navier-Stokes equations. It has the form of a balance equation.

9.5.2 The Energy Spectrum $\boldsymbol{E(k)}$

The Kármán-Howarth equation describes the relation between the longitudinal correlations $C_{rr}(r,t)$ and $C_{rrr}(r,t)$ completely and hence contains the turbulent energy transfer. The mechanism of the energy transfer becomes clearer if we work in the Fourier space. In section 3.8, we saw that the autocorrelation is the Fourier transform of the energy spectrum and vice versa. It is therefore advantageous to decompose the velocity field into Fourier modes

$$\boldsymbol{u}(\boldsymbol{x},t) = \int \hat{\boldsymbol{u}}(\boldsymbol{k},t) e^{i\boldsymbol{k}\cdot\boldsymbol{x}} \mathrm{d}^3\boldsymbol{k} \quad \text{resp.} \quad \hat{\boldsymbol{u}}(\boldsymbol{k},t) = \frac{1}{(2\pi)^3}\int \boldsymbol{u}(\boldsymbol{x},t) e^{-i\boldsymbol{k}\cdot\boldsymbol{x}} \mathrm{d}^3\boldsymbol{x} \tag{9.5.15}$$

To this end, we insert eq. (9.5.15) into the correlation function of the Fourier transform of the velocity fluctuations and obtain

$$\langle\, \hat{u}_{\mathrm{l}}(\boldsymbol{k},t)\, \hat{u}_{\mathrm{m}}(\boldsymbol{k},t)\,\rangle = \frac{1}{(2\pi)^6}\int\!\!\int \langle\, u_{\mathrm{l}}(\boldsymbol{x},t)\, u_{\mathrm{m}}(\boldsymbol{x}',t)\,\rangle e^{-i\boldsymbol{k}\cdot\boldsymbol{x}}\, e^{-i\boldsymbol{k}'\cdot\boldsymbol{x}'} \mathrm{d}^3\boldsymbol{x}\, \mathrm{d}^3\boldsymbol{x}' \tag{9.5.16}$$

Using the notation $\boldsymbol{x}' = \boldsymbol{x} + \boldsymbol{r}$ and taking into consideration that, according to eq. (9.5.2), the correlation function $C_{\mathrm{lm}}(\boldsymbol{r},t)$ depends only on the distance vector $\boldsymbol{r}$ for homogeneous turbulence, one can integrate over $\boldsymbol{x}$. Using the relation $\frac{1}{(2\pi)^3}\int e^{-i(\boldsymbol{k}+\boldsymbol{k}')\cdot\boldsymbol{x}} \mathrm{d}^3\boldsymbol{x} = \delta(\boldsymbol{k}+\boldsymbol{k}')$, one obtains

$$\begin{aligned} \langle\, \hat{u}_{\mathrm{l}}(\boldsymbol{k},t)\, \hat{u}_{\mathrm{m}}(\boldsymbol{k}',t)\,\rangle &= \frac{1}{(2\pi)^3}\int C_{\mathrm{lm}}(\boldsymbol{r},t) e^{-i\boldsymbol{k}\cdot\boldsymbol{r}} \mathrm{d}^3\boldsymbol{r}\; \delta(\boldsymbol{k}+\boldsymbol{k}') \\ &= \langle\, \hat{u}_{\mathrm{l}}(\boldsymbol{k},t)\, \hat{u}_{\mathrm{m}}(-\boldsymbol{k},t)\,\rangle \delta(\boldsymbol{k}+\boldsymbol{k}') = \hat{C}_{\mathrm{lm}}(\boldsymbol{k},t)\delta(\boldsymbol{k}+\boldsymbol{k}') \end{aligned} \tag{9.5.17}$$

i.e. the Fourier amplitudes of the velocity fields are only correlated if $\boldsymbol{k}+\boldsymbol{k}' = \mathbf{0}$ holds. The balance equation (9.5.7) for the correlation function in real space can now be replaced by the corresponding equation in Fourier space. It has the form

$$\frac{\partial}{\partial t}\hat{C}_{\mathrm{ij}}(\boldsymbol{k},t) + 2i\sum_{\mathrm{l}} k_{\mathrm{l}}\hat{C}_{\mathrm{li,j}}(\boldsymbol{k},t) = -2\nu\boldsymbol{k}^2\hat{C}_{\mathrm{ij}}(\boldsymbol{k},t) + \hat{Q}_{\mathrm{ij}}(\boldsymbol{k},t) \tag{9.5.18}$$

Due to the rotational symmetry, $\hat{C}_{\mathrm{ij}}$, $\hat{C}_{\mathrm{li,j}}$ are isotropic tensors with repect to $\boldsymbol{k}$. In the Appendix, section 9.8, it is shown that, owing to the isotropy, mirror symmetry and incompressibility, these tensors can again be expressed by scalar quantities. Explicitly, one obtains

$$\hat{C}_{\mathrm{ij}}(\boldsymbol{k},t) = \frac{1}{4\pi k^2}\left[\delta_{\mathrm{i,j}} - \frac{k_{\mathrm{i}}k_{\mathrm{j}}}{k^2}\right] E(k,t) \tag{9.5.19}$$

It follows from the above that

$$E(k,t) = 2\pi k^2 \sum_{\mathrm{l}} \hat{C}_{\mathrm{ll}}(\boldsymbol{k},t) \tag{9.5.20}$$

holds. As a result, the balance equation in the k-space has the form

$$\left[\frac{\partial}{\partial t} + 2\nu k^2\right] E(k,t) + \hat{T}(k,t) = \hat{Q}(k,t) \tag{9.5.21}$$

Here, the quantity $\hat{T}(k,t)$ is defined as

$$\hat{T}(k,t) = 2ik\sum_{\mathrm{l,i}} \frac{k_l}{k}\hat{C}_{\mathrm{li,i}}(\boldsymbol{k},t) \tag{9.5.22}$$

$E(k,t)$ is the spectral energy density of the velocity field, i.e. the average energy of the Fourier modes with wave number $\boldsymbol{k}$ with a magnitude taken from the interval $[k, k + dk]$. It can be shown that

$$\int_0^\infty \hat{T}(k,t)\,\mathrm{d}k = 0 \tag{9.5.23}$$

follows (Monin and Yaglom, 2007b). The term $\hat{T}(k,t)$ hence represents a flow term describing the energy flow between modes with different k-values. Equation (9.5.23) means that the temporal variation of the total kinetic energy is caused solely by the viscous forces and that no energy is lost in the transfer from large to small scales.

Here again, the closure problem of turbulence can be seen. For the calculation of the spectral energy flow, it is necessary to consider the complete Friedmann-Keller hierarchy. Alternatively, one can also introduce a turbulence modelling. As an example, we describe the Heisenberg approach in section 9.5.4.

9.5.3 The Energy Cascade in Three-Dimensional Turbulence

Equation (9.5.21) describes the evolution of the spectral energy density $E(k,t)$ of a turbulent field which is generated by the source term $\hat{Q}(k,t)$. In three dimensions, turbulence is generated on the large scales. This means that the source term $\hat{Q}(k,t)$ is only non-zero in the range $0 \leqslant k \leqslant k_I$.

While, the spectrum is restricted in the beginning to the range $k \leqslant k_I$, in the course of time the energy flows towards large wave numbers. After a certain time, a stationary state is reached. The spectral energy density is then independent of time. For high excitations, a range in the k-space develops in which the spectrum $E(k,t)$ is characterised by a potential law with respect to k

$$E(k) = K_o \epsilon^{2/3} k^{-5/3} \tag{9.5.24}$$

This range, which is attached to the above-mentioned one for $k > k_I$ where energy is injected into the fluid, is called *inertial range*. The adjacent one is characterised by a strong decay of the energy density $E(k,t)$ and is called *dissipation range*. For these scales, dissipation effects dominate. In the following, we account for the existence of these ranges and the scaling law in more detail. In three-dimensional flows, large-scale motions are excited which, owing to instabilities, generate flows on smaller scales. As a result, turbulence develops a kind of front moving towards small scales.

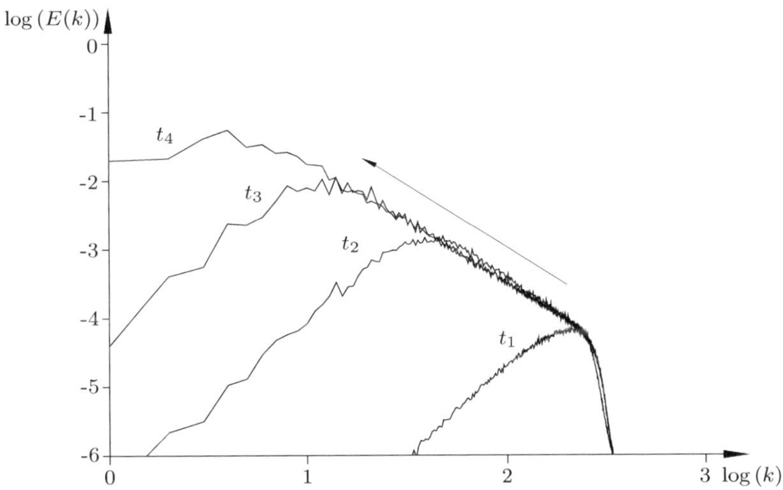

Fig. 9.5.1: Development of the inverse energy cascade for various times $t_1 < t_2 < t_3 < t_4$ in the case of two-dimensional turbulence (Kamps and Friedrich, 2008). As can be clearly seen, the energy is concentrated for t_1 at large wave numbers k, i.e. where energy is injected by external forcing. As time passes, the energy flows to smaller and smaller wave numbers and a scaling range is formed.

The inverse process is obviously also conceivable. It plays an important role in geophysical problems. If small-scale eddy motions are generated in two-dimensional flows, an energy cascade towards large eddy motions can be observed. In this way, vortex motions are generated on scales comparable to system dimensions; they can exist for a long time if enough energy is injected into the small scales. In contrast to the so-called direct energy cascade, this phenomenon is designated *inverse cascade.* It is this inverse cascade which is held responsible for the existence of the Great Red Spot on Jupiter.

One way of observing the inverse cascade is offered by the numerical simulation of the two-dimensional Navier-Stokes equation under the boundary condition that energy is injected into the system on small scales or at high wave numbers (Kraichnan, 1967; Kamps and Friedrich, 2008). Figure 9.5.1 shows the development of the inverse cascade for various time instants. Note that a self-similar velocity field is generated with the boundary conditions of this simulation. That is to say that the energy flow towards large scales, i.e. towards small k-values, is not caused by the merging of small eddies producing eddies on larger scales, but by the large-scale motion of a multitude of small eddies.

9.5.4 Heisenberg's Theory

In the literature, numerous approaches are proposed for closing the correlations of third order in eq. (9.5.7), see (Monin and Yaglom, 2007b; Hinze, 1975). Perhaps the most interesting one is due to Heisenberg (Heisenberg, 1948). In the following, we describe Heisenberg's theory of turbulence, based on a closure ansatz for the correlations of third order in eq. (9.5.7). For this purpose, we integrate the respective equation in k-space, eq. (9.5.21)

$$\frac{\partial}{\partial t}\int_0^k E(k',t)\,\mathrm{d}k' = -S(k,t) - 2\nu\int_0^k k'^{\,2}E(k',t)\,\mathrm{d}k' + \int_0^k \hat{Q}(k',t)\,\mathrm{d}k' \qquad (9.5.25)$$

Here, $\int_0^k E(k',t)\mathrm{d}k'$ describes the energy contained in the large-scale eddies with wave number $k' < k$, $2\nu\int_0^k k'^2E(k',t)\,\mathrm{d}k'$ is the energy dissipation rate within the large-scale eddies, and $S(k,t) = \int_0^k \hat{T}(k',t)\,\mathrm{d}k'$ is the spectral energy flow, i.e. the magnitude of energy per unit time which is transported from the macro- to the micro-scales. Finally, the expression $\int_0^k \hat{Q}(k',t)\,\mathrm{d}k'$ describes the total energy input into the system for sufficiently large k.

Heisenberg's ansatz for the closure of the energy balance equation is a so-called eddy-viscosity model. It states that the energy flow towards structures with large k-values, i.e. small-scale eddies, operates on eddies characterised by a wave number k like an additional dissipation. As a result, the spectral flow $S(k,t)$ can be expressed by means of the so-called eddy-viscosity $\nu_e(k,t)$

$$S(k,t) = 2\nu_e(k,t)\int_0^k k'^2E(k')\,\mathrm{d}k' \qquad (9.5.26)$$

As a result, we obtain an effective dissipation $\nu_{\text{eff}} = \nu + \nu_e(k,t)$.

Naturally, the final form of the energy spectrum is definitively determined by the explicit form of the eddy viscosity. The kinematic viscosity generally has the dimension $[\nu] = [L^2T^{-1}]$. Since the energy density respectively the wave number has the dimensions $[E] = [L^3T^{-1}]$ respectively $[k] = [L^{-1}]$, Heisenberg proposed the following ansatz for the eddy viscosity on the basis of dimensional considerations

$$\nu_e(k,t) = \alpha_H \int\limits_k^\infty \sqrt{\frac{E(k',t)}{k'^3}}\,\mathrm{d}k' \tag{9.5.27}$$

which summarises the influence of all small-scale eddies with $k' > k$.

It is interesting to calculate the stationary spectrum resulting from Heisenberg's theory. For k-values larger than those into which energy is injected,

$$\varepsilon = \int\limits_0^k Q(k',t)\,\mathrm{d}k' \tag{9.5.28}$$

holds which is independent of k and equal to the dissipated energy ε per unit time and mass. Therefore, one obtains the following integral equation for the energy spectrum $E(k,t)$

$$\varepsilon = 2\left[\nu + \alpha_H \int\limits_k^\infty \sqrt{\frac{E(k')}{k'^3}}\right]\mathrm{d}k' \int\limits_0^k E(k'')k''^2\,\mathrm{d}k'' \tag{9.5.29}$$

Since ε is independent of k, differentiation with respect to k yields

$$0 = -\alpha_H\sqrt{\frac{E(k)}{k^3}} \int\limits_0^k k'^2E(k')\,\mathrm{d}k' + k^2E(k)\left[\nu + \alpha_H \int\limits_k^\infty \sqrt{\frac{E(k')}{k'^3}}\,\mathrm{d}k'\right] \tag{9.5.30}$$

From this relation, the quantity

$$\nu + \alpha_H \int\limits_k^\infty \sqrt{\frac{E(k')}{k'^{\,3}}}\,\mathrm{d}k' = \frac{\alpha_H}{\sqrt{E(k)k^7}} \int\limits_0^k k'^2E(k')\,\mathrm{d}k' \tag{9.5.31}$$

can be determined. This leads to

$$\frac{\varepsilon}{2\alpha_H}\sqrt{E(k)k^7} = \Big[\int\limits_0^k k'^2E(k')\,\mathrm{d}k'\Big]^2 \tag{9.5.32}$$

Extracting the root of this relation and again differentiating with respect to k, one obtains

$$k^2E(k) = \sqrt{\frac{\varepsilon}{2\alpha_H}}\frac{\mathrm{d}}{\mathrm{d}k}\left(k^{7/4}E(k)^{1/4}\right) \tag{9.5.33}$$

The solution of this equation is Heisenberg's energy spectrum

$$E(k,t) = \left(\frac{8\varepsilon}{9\alpha_H}\right)^{2/3} k^{-5/3} \left[1 + \frac{8\nu^3}{3\varepsilon\alpha_H^2} k^4\right]^{-4/3} \tag{9.5.34}$$

For mid-values of k, one obtains the famous $-5/3$ law for the energy spectrum in the inertial range

$$E(k) = \left(\frac{8\varepsilon}{9\alpha_H}\right)^{2/3} k^{-5/3} \tag{9.5.35}$$

For very high values of k, i.e. for the spectrum in the dissipation range, Heisenberg's theory predicts a behaviour in accordance with

$$E(k) \sim k^{-7} \tag{9.5.36}$$

We point out that additional closure proposals are discussed in the literature. An overview can be found in (Monin and Yaglom, 2007b), for example.

9.6 Kolmogorov's Theory of Locally Isotropic Turbulence

In the following, we write the hydrodynamic equations in a form that allows a direct presentation of Kolmogorov's theory of locally isotropic turbulence. To this end, we consider the following definition of a velocity increment as the difference of the velocities $\boldsymbol{u}(\boldsymbol{x},t)$ and $\boldsymbol{u}(\boldsymbol{x}',t)$ at the points $\boldsymbol{x}$ and $\boldsymbol{x}'$

$$\boldsymbol{v}(\boldsymbol{x},\boldsymbol{x}',t) = \boldsymbol{u}(\boldsymbol{x},t) - \boldsymbol{u}(\boldsymbol{x}',t) \tag{9.6.1}$$

In addition, it proves necessary to define the increments of pressure and the external volume force density

$$\begin{aligned} P(\boldsymbol{x},\boldsymbol{x}',t) &= p(\boldsymbol{x},t) - p(\boldsymbol{x}',t) \\ \boldsymbol{F}(\boldsymbol{x},\boldsymbol{x}',t) &= \boldsymbol{f}(\boldsymbol{x},t) - \boldsymbol{f}(\boldsymbol{x}',t) \end{aligned} \tag{9.6.2}$$

9.6.1 Evolution Equation of the Velocity Increments

Since each velocity field $\boldsymbol{u}(\boldsymbol{x},t)$ or $\boldsymbol{u}(\boldsymbol{x}',t)$ fulfils the Navier-Stokes equation, an equation of motion for the velocity increment $\boldsymbol{v}(\boldsymbol{x},\boldsymbol{x}',t)$ can immediately be derived. From the difference of the two Navier-Stokes equations, one obtains

$$\begin{aligned} &\frac{\partial}{\partial t}\left[\boldsymbol{u}(\boldsymbol{x},t) - \boldsymbol{u}(\boldsymbol{x}',t)\right] + \boldsymbol{u}(\boldsymbol{x},t)\cdot\nabla_x \boldsymbol{u}(\boldsymbol{x},t) - \boldsymbol{u}(\boldsymbol{x}',t)\cdot\nabla_{x'}\boldsymbol{u}(\boldsymbol{x}',t) = \\ &\quad -\left[\nabla_x p(\boldsymbol{x},t) - \nabla_{x'} p(\boldsymbol{x}',t)\right] + \nu\Delta_x \boldsymbol{u}(\boldsymbol{x},t) - \nu\Delta_{x'}\boldsymbol{u}(\boldsymbol{x}',t) + \boldsymbol{F}(\boldsymbol{x},\boldsymbol{x}',t) \end{aligned} \tag{9.6.3}$$

Using the previous definition of the increments of pressure and the external volume force density given in eq. (9.6.2) together with the relations $\nabla_x \boldsymbol{u}(\boldsymbol{x},t) = \nabla_x \boldsymbol{v}(\boldsymbol{x},\boldsymbol{x}',t)$ and $\nabla_{x'}\boldsymbol{u}(\boldsymbol{x}',t) = -\nabla_{x'}\boldsymbol{v}(\boldsymbol{x},\boldsymbol{x}',t)$, one obtains after some transformations the following evolution equation of the velocity increment

$$\frac{\partial}{\partial t}\boldsymbol{v}(\boldsymbol{x},\boldsymbol{x}',t)+\boldsymbol{v}(\boldsymbol{x},\boldsymbol{x}',t)\cdot\nabla_x\boldsymbol{v}(\boldsymbol{x},\boldsymbol{x}',t)+\boldsymbol{u}(\boldsymbol{x}',t)\cdot[\nabla_x+\nabla_{x'}]\boldsymbol{v}(\boldsymbol{x},\boldsymbol{x}',t)=$$
$$-[\nabla_x+\nabla_{x'}]P(\boldsymbol{x},\boldsymbol{x}',t)+\nu[\Delta_x+\Delta_{x'}]\boldsymbol{v}(\boldsymbol{x},\boldsymbol{x}',t)+\boldsymbol{F}(\boldsymbol{x},\boldsymbol{x}',t) \qquad (9.6.4)$$

The term $\boldsymbol{u}(\boldsymbol{x}',t)\cdot[\nabla_x+\nabla_{x'}]\boldsymbol{v}(\boldsymbol{x},\boldsymbol{x}',t)$ in eq. (9.6.4) vanishes if we transform the equation into a new coordinate system which moves relatively to the original system with the time-dependent velocity $\boldsymbol{u}(\boldsymbol{x}',t)$. For this purpose, we introduce the new coordinates

$$\tilde{\boldsymbol{x}}=\boldsymbol{x}-\boldsymbol{X}(t)$$
$$\tilde{\boldsymbol{x}}'=\boldsymbol{x}'-\boldsymbol{X}(t) \qquad (9.6.5)$$

together with the new increments

$$\tilde{\boldsymbol{v}}(\tilde{\boldsymbol{x}},\tilde{\boldsymbol{x}}',t)=\boldsymbol{v}(\boldsymbol{x}-\boldsymbol{X}(t),\boldsymbol{x}'-\boldsymbol{X}(t),t)$$
$$\tilde{P}(\tilde{\boldsymbol{x}},\tilde{\boldsymbol{x}}',t)=P(\boldsymbol{x}-\boldsymbol{X}(t),\boldsymbol{x}'-\boldsymbol{X}(t),t)$$
$$\tilde{\boldsymbol{F}}(\tilde{\boldsymbol{x}},\tilde{\boldsymbol{x}}',t)=\boldsymbol{F}(\boldsymbol{x}-\boldsymbol{X}(t),\boldsymbol{x}'-\boldsymbol{X}(t),t) \qquad (9.6.6)$$

The vector $\boldsymbol{X}(t)$ denotes the trajectory of the fluid particle located at the point $\boldsymbol{x}'$ at time t which has the velocity $\boldsymbol{u}(\boldsymbol{x}',t)$ and is determined by the equation of motion

$$\frac{\mathrm{d}}{\mathrm{d}t}\boldsymbol{X}(t)=\boldsymbol{u}(\boldsymbol{X}(t),t) \qquad (9.6.7)$$

We thus transform to a system which is transported with a point of the fluid. As a result of this coordinate transformation, the convection term is simplified

$$\frac{\partial}{\partial t}\boldsymbol{v}(\boldsymbol{x},\boldsymbol{x}',t)+\boldsymbol{v}(\boldsymbol{x},\boldsymbol{x}',t)\cdot\nabla_x\boldsymbol{v}(\boldsymbol{x},\boldsymbol{x}',t)+\boldsymbol{u}(\boldsymbol{x}',t)\cdot[\nabla_x+\nabla_{x'}]\boldsymbol{v}(\boldsymbol{x},\boldsymbol{x}',t)=$$
$$\frac{\partial}{\partial t}\tilde{\boldsymbol{v}}(\tilde{\boldsymbol{x}},\tilde{\boldsymbol{x}}',t)+\tilde{\boldsymbol{v}}(\tilde{\boldsymbol{x}},\tilde{\boldsymbol{x}}',t)\cdot\nabla_{\tilde{x}}\tilde{\boldsymbol{v}}(\tilde{\boldsymbol{x}},\tilde{\boldsymbol{x}}',t) \qquad (9.6.8)$$

By means of this coordinate transformation, we arrive at the following evolution equation for the velocity increment $\boldsymbol{v}(\boldsymbol{x},\boldsymbol{x}',t)$ where, from now on, the tildes will be dropped for ease of notation

$$\frac{\partial}{\partial t}\boldsymbol{v}(\boldsymbol{x},\boldsymbol{x}',t)+\boldsymbol{v}(\boldsymbol{x},\boldsymbol{x}',t)\cdot\nabla_x\boldsymbol{v}(\boldsymbol{x},\boldsymbol{x}',t)=$$
$$-[\nabla_x+\nabla_{x'}]P(\boldsymbol{x},\boldsymbol{x}',t)+\nu[\Delta_x+\Delta_{x'}]\boldsymbol{v}(\boldsymbol{x},\boldsymbol{x}',t)+\boldsymbol{F}(\boldsymbol{x},\boldsymbol{x}',t) \qquad (9.6.9)$$

Additionally, the incompressibility conditions hold

$$\nabla_{\boldsymbol{x}}\cdot\boldsymbol{v}(\boldsymbol{x},\boldsymbol{x}',t)=\nabla_{\boldsymbol{x}'}\cdot\boldsymbol{v}(\boldsymbol{x},\boldsymbol{x}',t)=0 \qquad (9.6.10)$$

9.6.2 The Energy Balance of the Velocity Increment

As in the appendix to section 9.1.2, we are now able to derive a balance equation for the density of the kinetic energy of the velocity increment

$$\frac{1}{2}\boldsymbol{v}(\boldsymbol{x},\boldsymbol{x}',t)^2 \qquad (9.6.11)$$

from the equation of motion. By scalar multiplication of the equation of motion (9.6.9) by the factor $\boldsymbol{v}(\boldsymbol{x},\boldsymbol{x}',t)$, we obtain

$$\frac{\partial}{\partial t}\frac{\boldsymbol{v}(\boldsymbol{x},\boldsymbol{x}',t)^2}{2}+\nabla_x\cdot\boldsymbol{v}(\boldsymbol{x},\boldsymbol{x}',t)\frac{\boldsymbol{v}(\boldsymbol{x},\boldsymbol{x}',t)^2}{2}=-[\nabla_x+\nabla_{x'}][\cdot\boldsymbol{v}(\boldsymbol{x},\boldsymbol{x}',t)+P(\boldsymbol{x},\boldsymbol{x}',t)]$$
$$+\nu\,\boldsymbol{v}(\boldsymbol{x},\boldsymbol{x}',t)\cdot[\Delta_x+\Delta_{x'}]\boldsymbol{v}(\boldsymbol{x},\boldsymbol{x}',t)+\boldsymbol{v}(\boldsymbol{x},\boldsymbol{x}',t)\cdot\boldsymbol{F}(\boldsymbol{x},\boldsymbol{x}',t) \qquad (9.6.12)$$

We have accounted for the fact that, according to the relations (9.6.10), the field $\boldsymbol{v}(\boldsymbol{x},\boldsymbol{x}',t)$ is divergence-free with respect to both variables $\boldsymbol{x}$ and $\boldsymbol{x}'$, which allows us to use the following relation

$$\boldsymbol{v}(\boldsymbol{x},\boldsymbol{x}',t)\cdot[\nabla_x+\nabla_{x'}]P(\boldsymbol{x},\boldsymbol{x}',t)=[\nabla_x+\nabla_{x'}]\cdot\boldsymbol{v}(\boldsymbol{x},\boldsymbol{x}',t)P(\boldsymbol{x},\boldsymbol{x}',t) \qquad (9.6.13)$$

For the transformation of the term connected with the dissipation, we use the identity

$$\nu\,\boldsymbol{v}(\boldsymbol{x},\boldsymbol{x}',t)\cdot\Delta_x\boldsymbol{v}(\boldsymbol{x},\boldsymbol{x}',t)=\frac{1}{2}\nu\Delta_x\boldsymbol{v}(\boldsymbol{x},\boldsymbol{x}',t)^2-\nu\sum_{\mathrm{ij}}\left(\frac{\partial v_{\mathrm{i}}(\boldsymbol{x},\boldsymbol{x}',t)}{\partial x_{\mathrm{j}}}\right)^2 \qquad (9.6.14)$$

We are now able to relate the second term on the right-hand side of this equation to the local dissipation rates

$$\begin{aligned}\varepsilon(\boldsymbol{x},t)&=\frac{\nu}{2}\sum_{\mathrm{ij}}\left[\frac{\partial u_{\mathrm{i}}(\boldsymbol{x},t)}{\partial x_{\mathrm{j}}}+\frac{\partial u_{\mathrm{j}}(\boldsymbol{x},t)}{\partial x_{\mathrm{i}}}\right]^2\\&=\frac{\nu}{2}\sum_{\mathrm{ij}}\left[\frac{\partial v_{\mathrm{i}}(\boldsymbol{x},\boldsymbol{x}',t)}{\partial x_{\mathrm{j}}}+\frac{\partial v_{\mathrm{j}}(\boldsymbol{x},\boldsymbol{x}',t)}{\partial x_{\mathrm{i}}}\right]^2\end{aligned} \qquad (9.6.15)$$

because we can make use of the fact that

$$\begin{aligned}\nu\sum_{\mathrm{ij}}\left(\frac{\partial v_{\mathrm{i}}(\boldsymbol{x},\boldsymbol{x}',t)}{\partial x_{\mathrm{j}}}\right)^2&=\frac{1}{2}\nu\sum_{\mathrm{ij}}\left(\frac{\partial v_{\mathrm{i}}(\boldsymbol{x},\boldsymbol{x}',t)}{\partial x_{\mathrm{j}}}+\frac{\partial v_{\mathrm{j}}(\boldsymbol{x},\boldsymbol{x}',t)}{\partial x_{\mathrm{i}}}\right)^2\\&\quad-\nu\sum_{\mathrm{ij}}\frac{\partial v_{\mathrm{i}}(\boldsymbol{x},\boldsymbol{x}',t)}{\partial x_{\mathrm{j}}}\frac{\partial v_{\mathrm{j}}(\boldsymbol{x},\boldsymbol{x}',t)}{\partial x_{\mathrm{i}}}\\&=\varepsilon(\boldsymbol{x},t)-\nu\nabla_x\cdot[\boldsymbol{v}(\boldsymbol{x},\boldsymbol{x}',t)\cdot\nabla\boldsymbol{v}(\boldsymbol{x},\boldsymbol{x}',t)]\end{aligned} \qquad (9.6.16)$$

holds. For the second dissipation term $\boldsymbol{v}(\boldsymbol{x},\boldsymbol{x}',t)\cdot\Delta_{x'}\boldsymbol{v}(\boldsymbol{x},\boldsymbol{x}',t)$, analogous tranformations can be performed. Collecting all relations, we end up with the energy balance equation

$$\begin{aligned}&\frac{\partial}{\partial t}\frac{\boldsymbol{v}(\boldsymbol{x},\boldsymbol{x}',t)^2}{2}+\nabla_x\cdot\boldsymbol{v}(\boldsymbol{x},\boldsymbol{x}',t)\frac{\boldsymbol{v}(\boldsymbol{x},\boldsymbol{x}',t)^2}{2}=\\&-[\nabla_x+\nabla_{x'}]\cdot[\boldsymbol{v}(\boldsymbol{x},\boldsymbol{x}',t)P(\boldsymbol{x},\boldsymbol{x}',t)]+\nu[\Delta_x+\Delta_{x'}]\frac{\boldsymbol{v}(\boldsymbol{x},\boldsymbol{x}',t)^2}{2}\\&+\nu\nabla_x\cdot[\boldsymbol{v}(\boldsymbol{x},\boldsymbol{x}',t)\cdot\nabla_x\boldsymbol{v}(\boldsymbol{x},\boldsymbol{x}',t)]+\nu\nabla_{x'}\cdot[\boldsymbol{v}(\boldsymbol{x},\boldsymbol{x}',t)\cdot\nabla_{x'}\boldsymbol{v}(\boldsymbol{x},\boldsymbol{x}',t)]\\&-\varepsilon(\boldsymbol{x},t)-\varepsilon(\boldsymbol{x}',t)+\boldsymbol{v}(\boldsymbol{x},\boldsymbol{x}',t)\cdot\boldsymbol{F}(\boldsymbol{x},\boldsymbol{x}',t)\end{aligned} \qquad (9.6.17)$$

9.6.3 The Averaged Energy Balance Equation

The treatment of turbulent flows requires an averaging procedure. We can apply this averaging to the energy balance equation (9.6.17)

$$\begin{aligned}
&\frac{\partial}{\partial t}\langle\frac{\boldsymbol{v}(\boldsymbol{x},\boldsymbol{x}',t)^2}{2}\rangle + \nabla_x\cdot\langle\boldsymbol{v}(\boldsymbol{x},\boldsymbol{x}',t)\frac{\boldsymbol{v}(\boldsymbol{x},\boldsymbol{x}',t)^2}{2}\rangle = \\
&-[\nabla_x+\nabla_{x'}]\cdot\langle\boldsymbol{v}(\boldsymbol{x},\boldsymbol{x}',t)P(\boldsymbol{x},\boldsymbol{x}',t)\rangle + \nu[\Delta_x+\Delta_{x'}]\langle\frac{\boldsymbol{v}(\boldsymbol{x},\boldsymbol{x}',t)^2}{2}\rangle \\
&+\nu\nabla_x\cdot\langle\boldsymbol{v}(\boldsymbol{x},\boldsymbol{x}',t)\cdot\nabla_x\boldsymbol{v}(\boldsymbol{x},\boldsymbol{x}',t)\rangle + \nu\nabla_{x'}\cdot\langle\boldsymbol{v}(\boldsymbol{x},\boldsymbol{x}',t)\cdot\nabla_{x'}\boldsymbol{v}(\boldsymbol{x},\boldsymbol{x}',t)\rangle \\
&-\langle\varepsilon(\boldsymbol{x},t)\rangle - \langle\varepsilon(\boldsymbol{x}',t)\rangle + \langle\boldsymbol{v}(\boldsymbol{x},\boldsymbol{x}',t)\cdot\boldsymbol{F}(\boldsymbol{x},\boldsymbol{x}',t)\rangle
\end{aligned} \tag{9.6.18}$$

This is an exact relation and is valid for arbitrary turbulent flows. In the following, we introduce assumptions for the mean values occurring in the energy balance equation which successively lead to a simplification.

9.6.3.1 Homogeneous Turbulence

In a first step, we consider the case of homogeneous turbulence. A consequence of this assumption is that all expectation values, which are generally functions of the locations $\boldsymbol{x}$ and $\boldsymbol{x}'$, now depend only on the distance vector

$$\boldsymbol{r} = \boldsymbol{x} - \boldsymbol{x}' \tag{9.6.19}$$

Additionally, there holds for the differentiations

$$\nabla_x = \nabla_r\ , \quad \nabla_{x'} = -\nabla_r\ , \quad \Delta_x = \Delta_{x'} = \Delta_r \tag{9.6.20}$$

A direct consequence is that the contributions stemming from the pressure-velocity correlations vanish identically, i.e.

$$[\nabla_x+\nabla_{x'}]\cdot\langle\boldsymbol{v}(\boldsymbol{x},\boldsymbol{x}',t)P(\boldsymbol{x},\boldsymbol{x}',t)\rangle = 0 \tag{9.6.21}$$

Likewise, the two terms

$$\nabla_x\cdot\langle\boldsymbol{u}(\boldsymbol{x},t)\cdot\nabla_x\boldsymbol{u}(\boldsymbol{x},t)\rangle + \nabla_{x'}\cdot\langle\boldsymbol{u}(\boldsymbol{x}',t)\cdot\nabla_{x'}\boldsymbol{u}(\boldsymbol{x}',t)\rangle = 0 \tag{9.6.22}$$

vanish identically. Moreover, we obtain

$$\frac{1}{2}\nu[\Delta_x+\Delta_{x'}]\big\langle\boldsymbol{v}(\boldsymbol{x},\boldsymbol{x}',t)^2\big\rangle = \nu\Delta_r\big\langle\boldsymbol{v}(\boldsymbol{x},\boldsymbol{x}',t)^2\big\rangle \tag{9.6.23}$$

In order to simplify the notation, we introduce the following abbreviations

$$\begin{aligned}
\big\langle\boldsymbol{v}(\boldsymbol{x},\boldsymbol{x}',t)^2\big\rangle &= \big\langle\boldsymbol{v}(\boldsymbol{r},t)^2\big\rangle \\
\big\langle\boldsymbol{v}(\boldsymbol{x},\boldsymbol{x}',t)\boldsymbol{v}(\boldsymbol{x},\boldsymbol{x}',t)^2\big\rangle &= \big\langle\boldsymbol{v}(\boldsymbol{r},t)\boldsymbol{v}(\boldsymbol{r},t)^2\big\rangle
\end{aligned} \tag{9.6.24}$$

We are now able to formulate the energy balance equation for the case of homogeneous turbulence

$$\frac{\partial}{\partial t}\Big\langle\frac{\boldsymbol{v}(\boldsymbol{r},t)^2}{2}\Big\rangle+\nabla_r\cdot\Big\langle\boldsymbol{v}(\boldsymbol{r},t)\frac{\boldsymbol{v}(\boldsymbol{r},t)^2}{2}\Big\rangle=$$
$$+\nu\Delta_r\big\langle\boldsymbol{v}(\boldsymbol{r},t)^2\big\rangle-2\langle\varepsilon(\boldsymbol{x},t)\rangle+Q(\boldsymbol{r},t) \quad (9.6.25)$$

where we have introduced the following notation for the power

$$Q(\boldsymbol{r},t)=\langle\boldsymbol{v}(\boldsymbol{x},\boldsymbol{x}',t)\cdot\boldsymbol{F}(\boldsymbol{x},\boldsymbol{x}',t)\rangle \quad (9.6.26)$$

9.6.3.2 Homogeneous Isotropic Turbulence

We now go one step further and consider isotropic turbulence. In the case of isotropic turbulence, a scalar quantity like, for example, the average kinetic energy

$$\big\langle\boldsymbol{v}(\boldsymbol{r},t)^2\big\rangle=S^2(r,t) \quad (9.6.27)$$

depends only on the distance $r=|\boldsymbol{r}|$. A vector such as, for example,

$$\big\langle\boldsymbol{v}(\boldsymbol{r},t)\boldsymbol{v}(\boldsymbol{r},t)^2\big\rangle=\boldsymbol{e}_r S^3(r,t) \quad (9.6.28)$$

has the direction of the vector $\boldsymbol{r}/r=\boldsymbol{e}_r$ and the length of the vector depends only on the distance r.

If we transform the divergence and the Laplace operator in eq. (9.6.25) to spherical coordinates, the energy balance equation has the following form

$$\frac{\partial}{\partial t}\frac{1}{2}S^2(r,t)+\frac{1}{2r^2}\frac{\partial}{\partial r}\left(r^2S^3(r,t)\right)=\nu\frac{1}{r^2}\frac{\partial}{\partial r}\Big(r^2\frac{\partial}{\partial r}S^2(r,t)\Big)-2\langle\varepsilon\rangle+Q(r,t) \quad (9.6.29)$$

9.6.3.3 Kolmogorov's -4/5 Law

Let us now consider stationary turbulence, which is locally homogeneous and isotropic. We can assume that the averaged source term in the energy balance equation $Q(r,t)$ does not depend on time. We further presume that no turbulent energy is generated on the small scales $r<L$

$$Q(r)=0 \qquad \text{for} \qquad r<L \quad (9.6.30)$$

In addition, we make use of the fact that the averaged values $S^2(r,t)$ and $S^3(r,t)$ depend only on r and can be expressed by the so-called longitudinal structure functions

$$\begin{aligned}S_{rr}&=\big\langle(\boldsymbol{e}_r\cdot\boldsymbol{v}(\boldsymbol{r},t))^2\big\rangle\\ S_{rrr}&=\big\langle(\boldsymbol{e}_r\cdot\boldsymbol{v}(\boldsymbol{r},t))^3\big\rangle\end{aligned} \quad (9.6.31)$$

which is possible by virtue of the incompressibility of the velocity field. We obtain

$$\begin{aligned}
\langle \boldsymbol{v}(\boldsymbol{r},t)^2\rangle &= S^2(r,t) = \frac{1}{r^2}\Big(\frac{\mathrm{d}}{\mathrm{d}r}r^3 S_{rr}(r)\Big) \\
\frac{\mathrm{d}}{\mathrm{d}r}\langle \boldsymbol{v}(\boldsymbol{r},t)^2\rangle &= \frac{1}{r^3}\frac{\mathrm{d}}{\mathrm{d}r}\left(r^4\frac{\mathrm{d}}{\mathrm{d}r}S_{rr}(r)\right) \\
S^3(r,t) &= \frac{1}{3r^3}\frac{\mathrm{d}}{\mathrm{d}r}\left(r^4 S_{rrr}(r)\right)
\end{aligned} \tag{9.6.32}$$

For detailed calculations, we refer to section 9.8.

Using the relations for the structure functions, we obtain the energy balance equation in the form

$$\frac{1}{r^2}\frac{\mathrm{d}}{\mathrm{d}r}\Big(\frac{1}{r}\frac{\mathrm{d}}{\mathrm{d}r}\big(r^4\big[\frac{1}{6}S_{rrr}(r) - \nu\frac{\mathrm{d}}{\mathrm{d}r}S_{rr}(r)\big]\big)\Big) = -2\langle\epsilon\rangle + Q(r) \tag{9.6.33}$$

This relation can now be integrated with respect to r. The result is the celebrated so-called *-4/5 law*, first derived by Kolmogorov. This law sets up a relation between the longitudinal structure functions of second and third order and is the only exact law hitherto in existence which can be derived from the Navier-Stokes equations under the assumption of stationary, homogeneous and isotropic turbulence

$$S_{rrr}(r) - 6\nu\frac{\mathrm{d}}{\mathrm{d}r}S_{rr}(r) = -\frac{4}{5}\langle\varepsilon\rangle r + q(r) \tag{9.6.34}$$

The source term has the form

$$q(r) = \frac{6}{r^4}\int_0^r r' \int_0^{r'} r''^2 Q(r'')\,\mathrm{d}r''\mathrm{d}r' \tag{9.6.35}$$

Clearly, in the scale range $r < L$ where $Q(r) = 0$ holds, this term does not contribute to the energy balance.

9.6.3.4 Dissipation Range

For small r-values ($r \ll \eta$), it can be seen from the Taylor expansion that S_{rrr} behaves like $\propto r^3$ and can therefore be neglected in eq. (9.6.34). This leads to the relation for the longitudinal structure function of second order

$$S_{rr}(r) = \frac{\langle\varepsilon\rangle}{15\nu}r^2 \tag{9.6.36}$$

According to eq. (9.8.28), we obtain for the transversal structure function of second order

$$S_{tt}(r) = 2\frac{\langle\varepsilon\rangle}{15\nu}r^2 = 2S_{rr}(r) \tag{9.6.37}$$

9.6.3.5 Inertial Range

In the inertial range $\eta \ll r \ll L$, the dissipative term in eq. (9.6.34) can be neglected and we obtain for the longitudinal structure function of third order

$$S_{rrr}(r) = -\frac{4}{5}\langle\varepsilon\rangle r \tag{9.6.38}$$

Figure 9.6.1 shows the longitudinal structure function S_{rrr} as a function of r for a cryogenic free jet with a Reynolds number $Re = 2.535 \times 10^5$ (for a detailed experimental set-up, we refer to (Chanal et al., 2000; Renner et al., 2002)). The validity of Kolmogorov's -4/5 law as well as the existence of an inertial range can be clearly observed in the linear part of the diagram.

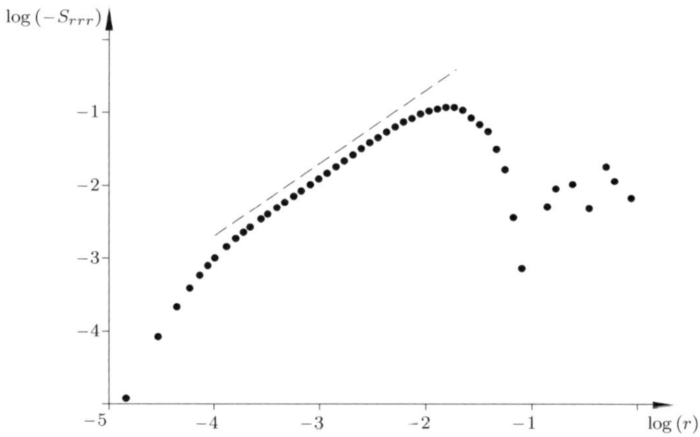

Fig. 9.6.1: Longitudinal structure function of third order from experimental measurements (Chanal et al., 2000). The dashed line has the slope 1.

9.6.4 Kolmogorov's K41 Theory

In the following, we discuss Kolmogorov's similarity theory. It is an extrapolation of Kolmogorov's -4/5 law. The aim is not only to formulate statements regarding the scale dependence of the structure function of third order in the inertial range, but rather to determine all moments of the longitudinal velocity increments. In particular, the structure function of second order $S_{rr}(r)$ is of interest; it can be interpreted as a measure for the *kinetic energy of an eddy of size* r. The simplest case of stationary, homogeneous and isotropic turbulence is taken as a basis. The most important assumption in Kolmogorov's theory is the existence of a property denoted self-similarity. As we shall see, this hypothesis has far-reaching consequences.

9.6.4.1 Universality of the Statistics of Small-Scale Turbulence

The starting point is the probability distribution or density function $f(v, r)$ for the longitudinal velocity increment v. Based on the hypothesis that a universal statistics of turbulent flows exists in the inertial range, as presented briefly in section 9.4.2, the function $f(v, r)$ should only depend on the fluid under consideration. Therefore, only the kinematic viscosity ν and the energy input, which can be characterised as the dissipated energy per unit time, i.e. the local energy dissipation rate ε, can enter the density function. We can thus write

$$f(v, r) = F(v, r, \nu, \langle\varepsilon\rangle) \tag{9.6.39}$$

This ansatz is in accordance with Kolmogorov's -4/5 law, eq. (9.6.34), which, for distances $r < L$, contains only the kinematic viscosity ν and the average energy dissipation rate. We note that eq. (9.6.39) can only be valid for scales $r < L$.

With the help of Kolmogorov's length and time scales presented in section 9.4.4, we now introduce dimensionless lengths and velocities

$$\begin{aligned} \bar{v} &= \frac{v}{u_\eta} = \frac{v}{(\nu\langle\varepsilon\rangle)^{1/4}} \\ \bar{r} &= \frac{r}{\eta} = \frac{r}{(\nu^3/\langle\varepsilon\rangle)^{1/4}} \\ \bar{t} &= \frac{t}{\tau_\eta} = \frac{t}{(\nu/\langle\varepsilon\rangle)^{1/2}} \end{aligned} \tag{9.6.40}$$

For the sake of completeness, we have also added the dimensionless time $\bar{t}$.

We now can introduce the density function $G(\bar{v}, \bar{r})$ for the dimensionless velocity increment $\bar{v}$. This distribution is of importance for the experimental evaluation of turbulent flows. With this function, experiments on the most diverse turbulent flows can be compared directly with one other. The relation between $f(v, r)$ and $G(\bar{v}, \bar{r})$ can be written as

$$f(v, r) = G(\bar{v}, \bar{r})\frac{\mathrm{d}\bar{v}}{\mathrm{d}v} = \frac{1}{u_\eta}G\left(\frac{v}{u_\eta}, \frac{r}{\eta}\right) \tag{9.6.41}$$

This formula tells us that the density function of the velocity increment v on the scale r depends on the kinematic viscosity ν and the averaged local energy dissipation rate $\langle\varepsilon\rangle$ via the Kolmogorov length η and the Kolmogorov velocity u_η. This form of the density function should be valid on scales r which are smaller than the integral length scale. The fact that there are no further dependencies on additional parameters in eq. (9.6.41) is a consequence of the assumption that a universal statistics exists in small-scale turbulence.

9.6.4.2 Self-similar Behaviour in the Inertial Range

In the inertial range, i.e. on scales larger than the Kolmogorov length η, dissipation effects can be neglected. This has the direct consequence that the density function $f(v, r)$ in the inertial range should not depend on the kinematic viscosity ν, i.e.

$$\frac{\partial}{\partial \nu}\left[\frac{1}{u_\eta} G\Big(\frac{v}{u_\eta}, \frac{r}{\eta}\Big)\right] = 0 \tag{9.6.42}$$

Explicit differentiation yields

$$\begin{aligned}\frac{\partial}{\partial \nu}\left[\frac{1}{u_\eta} G\Big(\frac{v}{u_\eta}, \frac{r}{\eta}\Big)\right] &= -\frac{\partial u_\eta}{\partial \nu}\frac{1}{u_\eta^2}\frac{\partial}{\partial \bar{v}}\left[\bar{v}G(\bar{v},\bar{r})\right] - \frac{\partial \eta}{\partial \nu}\frac{1}{\eta u_\eta}\bar{r}\frac{\partial}{\partial \bar{r}}G(\bar{v},\bar{r}) \\ &= -\frac{1}{4\langle\varepsilon\rangle^{1/4}\nu^{5/4}}\Big(\frac{\partial}{\partial \bar{v}}\left[\bar{v}G(\bar{v},\bar{r})\right] + 3\bar{r}\frac{\partial}{\partial \bar{r}}G(\bar{v},\bar{r})\Big) = 0\end{aligned} \tag{9.6.43}$$

This leads to the relation

$$\frac{\partial}{\partial \bar{v}}\left[\bar{v}G(\bar{v},\bar{r})\right] + 3\bar{r}\frac{\partial}{\partial \bar{r}}G(\bar{v},\bar{r}) = 0 \tag{9.6.44}$$

which has the solution

$$G(\bar{v},\bar{r}) = \frac{1}{\bar{r}^{1/3}}g\Big(\frac{\bar{v}}{\bar{r}^{1/3}}\Big) \tag{9.6.45}$$

However, the function $g(x)$ cannot be determined on the basis of our previous considerations. Distributions of the form eq. (9.6.45) are denoted self-similar. In the following, we analyse this property in more detail.

9.6.4.3 Self-similarity and Fractal Scaling Laws

Generally, a field $\boldsymbol{v}(\boldsymbol{r},t)$ is denoted self-similar if the rescaled field

$$\boldsymbol{V}(\boldsymbol{r},t) = \lambda^{-\zeta}\boldsymbol{v}(\lambda\boldsymbol{r},\lambda^{\beta}t) \tag{9.6.46}$$

possesses the same statistical properties for suitably chosen exponents ζ and β and for all values of $\lambda > 0$. In concrete terms, this means that the probability densitites of the two fields $\boldsymbol{v}(\boldsymbol{r},t)$ and $\boldsymbol{V}(\boldsymbol{r},t)$ correspond to each other.

To simplify matters, we again consider a scalar field, for example the longitudinal velocity increment. The density function $\tilde{f}(v,r)$ for the rescaled field $V(r,t)$ has the form

$$\tilde{f}(v,r) = \langle\,\delta(v - V(r,t))\,\rangle = \langle\,\delta\left(v - \lambda^{-\zeta}v(\lambda r,\lambda^{\beta}t\right)\,\rangle = \lambda^{\zeta}f(\lambda^{\zeta}v,\lambda r) \tag{9.6.47}$$

Self-similarity means that the density function $\tilde{f}(v,r)$ of the rescaled field corresponds exactly to the density function $f(v,r)$ of the field $v(r,t)$

$$f(v,r) = \lambda^{\zeta}f(\lambda^{\zeta}v,\lambda r) \tag{9.6.48}$$

This equation is valid for all values of λ. Therefore, the derivative with respect to λ for $\lambda = 1$ can be calculated. Explicitly, this leads to

$$r\frac{\partial}{\partial r}f(v,r) + \zeta\frac{\partial}{\partial v}\big(vf(v,r)\big) = 0 \tag{9.6.49}$$

This notation is in accordance with eq. (9.6.44) if, for ζ, we choose the value

$$\zeta = \frac{1}{3} \tag{9.6.50}$$

It is also possible to ascertain the scaling index ζ without the considerations of the previous section. To do this, we use the fact that a self-similar density function possesses moments of the form

$$S^{\mathrm{n}}(r) = C_{\mathrm{n}} r^{\mathrm{n}\,\zeta} \tag{9.6.51}$$

We will prove this statement shortly. This property allows us to determine the scaling index ζ by means of Kolmogorov's $-4/5$-law. Since the third-order structure function has the form $S^3 = -4/5\langle\epsilon\rangle r$ in the inertial range, the scaling index must be $\zeta = 1/3$.

It is now easy to draw consequences from the self-similar form of the probability distribution. For that, it proves expedient to write the distribution in the form

$$f(v,r) = \frac{1}{(\langle\epsilon\rangle r)^{1/3}}\, g\left(\frac{v}{(\langle\epsilon\rangle r)^{1/3}}\right) \tag{9.6.52}$$

since $(\langle\epsilon\rangle r)^{1/3}$ has the dimension of a velocity. We now calculate the moments of the velocity increments

$$\begin{aligned} \langle v(r)^{\mathrm{n}}\rangle &= \int v^{\mathrm{n}} f(v,r)\,\mathrm{d}v = \frac{1}{(\langle\epsilon\rangle r)^{\zeta}} \int v^{\mathrm{n}} g\left(\frac{v}{(\langle\epsilon\rangle r)^{\zeta}}\right)\,\mathrm{d}v \\ &= (\langle\epsilon\rangle r)^{\mathrm{n}\,\zeta} \int w^{\mathrm{n}} g(w)\,\mathrm{d}w = (\langle\epsilon\rangle r)^{\mathrm{n}\,\zeta} C_{\mathrm{n}} \end{aligned} \tag{9.6.53}$$

The scaling indices of the structure function are thus uniquely determined by the hypothesis of self-similarity together with Kolmogorov's $-4/5$-law describing the scale-dependency of the third-order structure function. The form of the density distribution is determined by the coefficients C_{n}, i.e. the moments of the function $g(w)$ which cannot be deduced from the similarity theory.

Let us summarise: with the hypothesis that the statistical properties of a fully developed stationary, homogeneous and isotropic turbulence are self-similar, the longitudinal structure functions scale as

$$S^{\mathrm{n}}(r) = C_{\mathrm{n}}(\langle\epsilon\rangle r)^{\mathrm{n}\,\zeta} \tag{9.6.54}$$

with a scaling exponent $\zeta = 1/3$. This is an important statement of Kolmogorov's theory, formulated in 1941.

9.6.4.4 Experimental Testing of the Self-similarity Hypothesis

In order to test Kolmogorov's self-similarity hypothesis, we have the choice between two methods. On the one hand, the structure functions $S^{\mathrm{n}}(r)$ of n-th order can be determined; on the other, the density function $f(v,r)$ can be analysed directly.

In the inertial range, the structure functions are expected to show the scaling behaviour $S^{\mathrm{n}}(r) = C_{\mathrm{n}} r^{\zeta_{\mathrm{n}}}$, where the scaling indices have the form $\zeta_{\mathrm{n}} = \mathrm{n}\,\zeta$ according to eq. (9.6.54). The first difficulty in the analysis of experimental data lies in the definition of the inertial range, which is typically not clearly visible in experiments. A trick which allows us to extend this inertial range is provided by the EES method (*extended self-similarity*) (Benzi et al., 1993b). The structure functions are not considered as functions of r but as functions of the structure function of third order, $S^3(r)$, which is supposed to be proportional to r in the inertial range. In general, this formulation leads to an extended scaling range

$$S^n(S^3) \approx (S^3)^{\zeta_{\mathrm{n}}} \tag{9.6.55}$$

from which the scaling exponents ζ_{n} can be read off more exactly.

The experimental results clearly show an underlying scaling behaviour. However, the scaling exponents ζ_{n} are not given by the linear K41-relation $\zeta_{\mathrm{n}} = \mathrm{n}/3$. It turns out that the indices are non-linear functions of n. The so-called She-Leveque formula provides a good approximation (She and Leveque, 1994)

$$\zeta_{\mathrm{n}} = \frac{\mathrm{n}}{9} + 2\left[1 - \left(\frac{2}{3}\right)^{\mathrm{n}/3}\right] \tag{9.6.56}$$

Instead of a scaling behaviour characterised by an index $\zeta = 1/3$, *multifractal scaling behaviour* appears (see section 8.5).

Let us now turn to a further way of testing the self-similarity hypothesis. It is based on an evaluation of the density function $f(v, r)$, in particular on an analysis of the structure of eq. (9.6.48). It is expedient to determine the density function $f_{\mathrm{norm}}(v_{\mathrm{norm}}, t)$ of the normalised increments $v_{\mathrm{norm}}(r)$

$$v_{\mathrm{norm}}(r) = \frac{v(r,t)}{\sqrt{\langle v(r,t)^2 \rangle}} \tag{9.6.57}$$

For a self-similar distribution to exist, all distributions should collapse in the inertial range to a universal function.

Figure 9.6.3 shows the results of a numerical calculation and an experimental evaluation of the density function on various scales. It can be seen that the distributions obtained for different increments do not coincide. From this observation, it can be concluded that a fractal scaling behaviour does not exist and hence the hypothesis of self-similarity fails. For a more detailed discussion, we refer to the literature, see for example (Frisch, 1995). To summarise, we can say that Kolmogorov's similarity theory provides an approximative, phenomenological theory of the statistics of turbulent velocity increments. A more detailed analysis of experimental and numerical data indicates that self-similarity and the corresponding fractal behaviour are not strictly fulfilled.

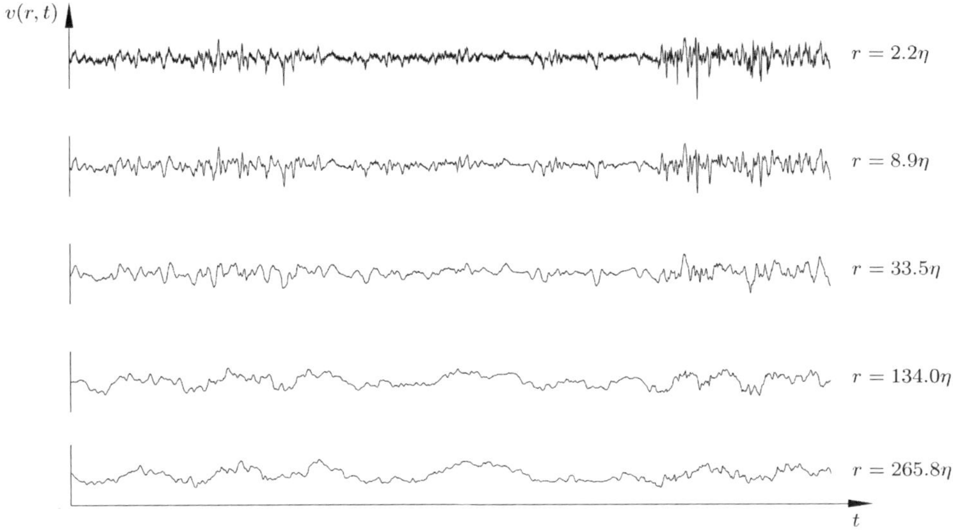

Fig. 9.6.2: Experimental time series for turbulent velocity increments on various time scales r (Renner et al., 2001). The increments are normalised. The self-similarity of the K41-theory is violated.

9.6.5 The Phenomenon of Intermittency

If we consider the time series of the velocity increments of a free round air jet on different scales, it is evident that the self-similarity is *broken* in small-scale turbulence (Renner et al., 2001). Related to the diameter of the nozzle, the free jet has a Reynolds number of about $Re = 2.7 \times 10^4$. Using the definition of the integral length given in eq. (9.4.5), we obtain the value $L = 6.7$cm and eq. (9.4.8) yields the value $\eta = 0.3$mm for the Kolmogorov length. In fig. 9.6.2, the velocity increments are presented for various r-values. In the case of self-similarity, the statistic is expected to be identical for all scales r. This, however, cannot be observed.

While the increments are normally distributed on the large scales, the form of the density function changes at the transition to small scales r, see fig. 9.6.3a. The time series show that small *and* larger velocity increments occur with a higher frequency. This phenomenon is denoted turbulent intermittency. It may not be confused with the intermittency in chaotic systems described in section 8.6. For comparison, we show in fig. 9.6.3b the distribution of velocity increments obtained from a numerical simulation ($Re = 1.79 \times 10^3, \eta = 7.9 \times 10^{-3}$ in units of 2π) (Wilczek, 2009).

9.6.6 Kolmogorov K62

In a footnote in his textbook on hydrodynamics (Landau and Lifschitz, 1991), the Russian physicist and Nobel prize laureate L.D. Landau remarked that Kolmogorov's density function $f(v, r)$ contains the *mean* local energy dissipation rate $\langle\varepsilon\rangle$. Hence, Kolmogorov's theory from 1941 is based on the assumption that the kinetic energy is

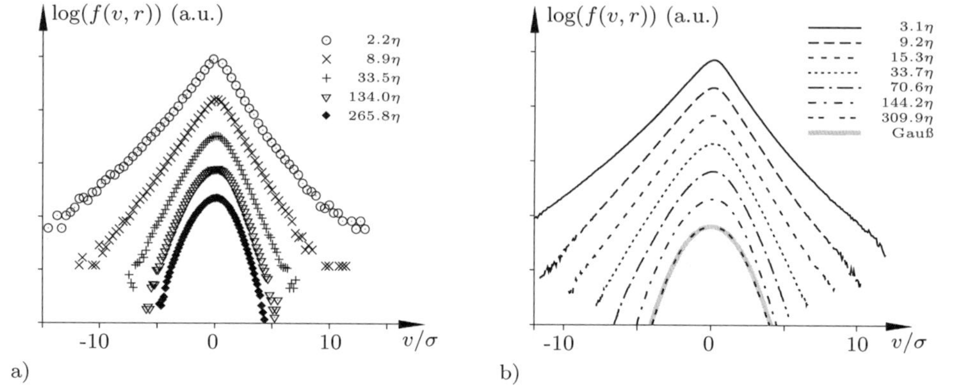

Fig. 9.6.3: Phenomenon of intermittency: a) experimental data (Renner et al., 2001) and b) numerical simulation (Wilczek, 2009). The density distributions for turbulent velocity increments on various scales r show that the deviations from the Gaussian normal distribution are larger, the smaller the scale r is. For clarity of presentation, the distributions are shifted in vertical direction.

dissipated at a constant rate. A possibly random character of the energy transfer from large to small scales is not taken into consideration. In turbulent flows, however, it is to be expected that ε fluctuates strongly from point to point. This can also be proven in detail by numerical simulations (see Colour Plate XVI a,b, p. 659). Yet these fluctuations cannot be captured by means of Kolmogorov's theory. An obvious extension of Kolmogorov's theory consists of taking these spatial fluctuations into account by changing the formula (9.6.52) for the probability distribution of the velocity increments

$$f(v,r) = \int \frac{1}{(\varepsilon\, r)^{1/3}}\, g\Big(\frac{v}{(\varepsilon\, r)^{1/3}}\Big)\, p(\varepsilon, r, L)\, \mathrm{d}\varepsilon \tag{9.6.58}$$

The fluctuations of the energy dissipation rate are taken into consideration here in terms of the distribution $p(\varepsilon, r, L)$. Kolmogorov proposed a log-normal distribution for the statistics of the local energy dissipation rate, see section 3.9.9, eq. (3.9.116), (Kolmogorov, 1962)

$$p(\varepsilon, r, L) = \frac{1}{\sqrt{2\pi}\sigma\, \varepsilon} e^{-\frac{(\ln(\varepsilon/a))^2}{2\sigma^2}} \tag{9.6.59}$$

where the standard deviation $\sigma(r, L)$ depends on the scale r, but also on the integral length scale L. The quantity a will be determined later. The moments can easily be calculated. We obtain

$$\langle \varepsilon^{\mathrm{n}/3} \rangle = a^{\mathrm{n}/3} e^{(\mathrm{n}/3)^2 \sigma^2/2} \tag{9.6.60}$$

In particular, we obtain for n=3

$$\langle \varepsilon \rangle = a\, e^{\sigma^2/2} \tag{9.6.61}$$

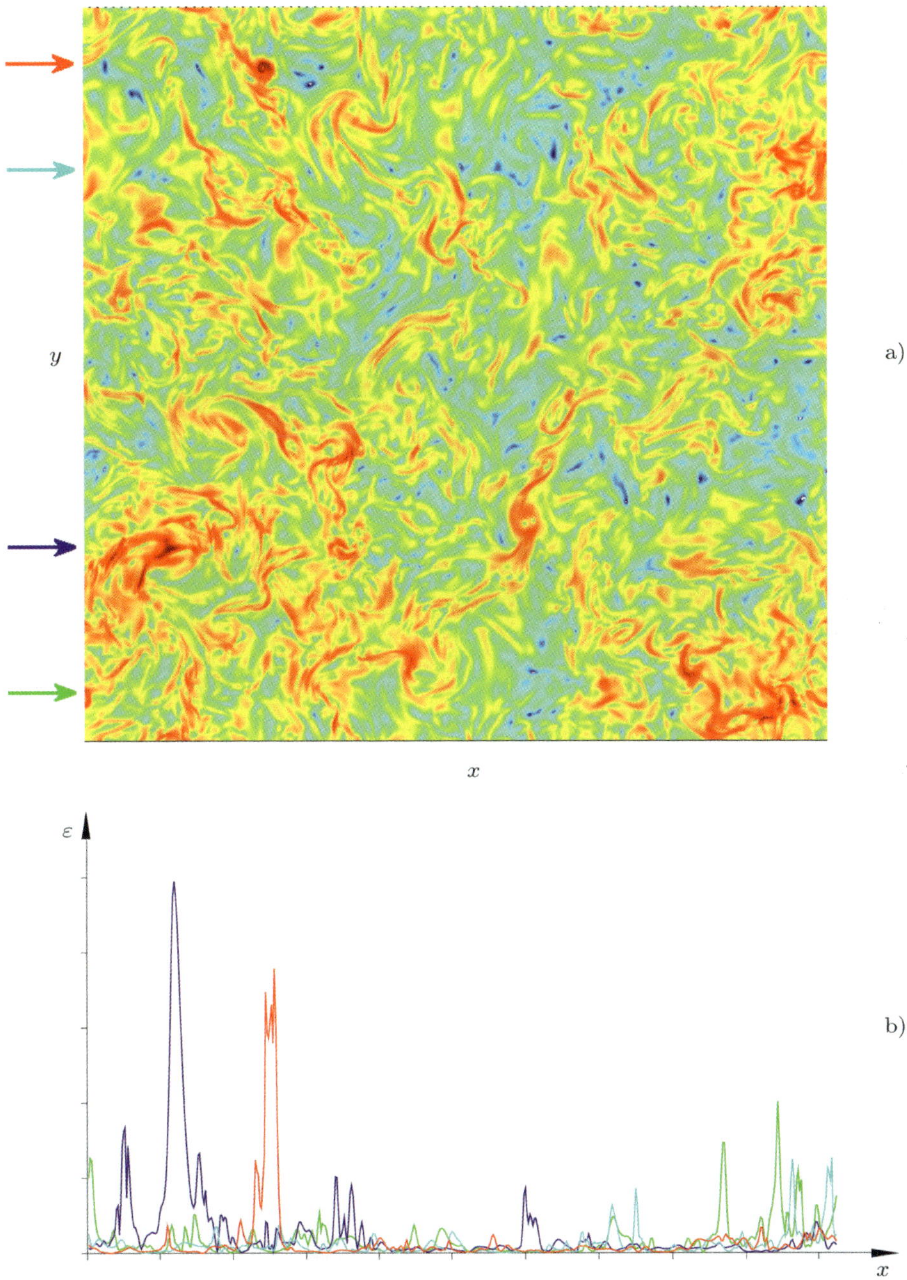

Plate XVI Computer simulation: a) two-dimensional section and b) one-dimensional sections through the field of the local energy dissipation rate. Strong spatial fluctuations contributing to the intermittent character of turbulent flows can clearly be seen (Wilczek, 2009).

From this relation, a can be determined as a function of $\langle\varepsilon\rangle$

$$\langle\varepsilon^{\mathrm{n}/3}\rangle = \langle\varepsilon\rangle^{\mathrm{n}/3} e^{\mathrm{n}(\mathrm{n}-3)\sigma^2/18} \tag{9.6.62}$$

According to Kolmogorov, the variance σ^2 of the logarithm of ε for $L/r \gg 1$ and for large Reynolds numbers is expected to depend on r and the integral length scale L as follows

$$\sigma^2 = A + \mu \ln \frac{L}{r} \tag{9.6.63}$$

Introducing this relation into eq. (9.6.62), we can calculate the scaling behaviour of the structure functions according to eq. (9.6.54) as

$$S^{\mathrm{n}}(r) = \langle v^{\mathrm{n}}(r)\rangle = C_{\mathrm{n}} \langle\varepsilon\rangle^{\mathrm{n}/3} r^{\mathrm{n}/3} \left(\frac{L}{r}\right)^{-\mathrm{n}(\mathrm{n}-3)\mu/18} \tag{9.6.64}$$

The scaling indices

$$\zeta_{\mathrm{n}} = \frac{\mathrm{n}}{3} - \frac{\mu}{18}\mathrm{n}(\mathrm{n}-3) \tag{9.6.65}$$

are hence non-linear functions of n. Experimental measurements yield the value $\mu \sim 0.26$ for the constant μ. It can be shown, however, that the scaling exponents ζ_{n} have to be monotonically increasing functions of n. Therefore, Kolmogorov's theory can only be valid for small values

$$\mathrm{n} \geqslant \frac{3}{2} + \frac{3}{\mu} \tag{9.6.66}$$

Fig. 9.6.4 shows the scaling index $\zeta_{\mathrm{n}}(\mathrm{n})$ of the longitudinal structure function $S^{\mathrm{n}}(r)$ for the two Kolmogorov models K41 and K62 according to eqs. (9.6.54), (9.6.65) as well as for the hierarchical structure model of She-Leveque, eq. (9.6.56). The experimental results of Benzi et al. (Benzi et al., 1993a; Benzi et al., 1995) agree very well with the mulitifractal model of She-Leveque.

9.6.7 Multifractal Model

A modification of Richardson's cascade model allows a phenomenological modelling of intermittent behaviour. The idea is that the energy transferred from eddies on one scale to child eddies of a smaller size is distributed inhomogeneously in space. The local self-similarity is preserved if, instead of the specific scaling exponent $\zeta = 1/3$ of the K41-model, a complete interval of scaling exponents is admitted, where the subset appertaining to a fixed ζ possesses a fractal dimension $D(\zeta)$. We already met such multifractal sets in section 8.5.

The multifractal turbulence model goes back to an idea of U. Frisch and G. Parisi (Parisi and Frisch, 1985; Argoul et al., 1989; Frisch, 1995). It is a generalisation of Kolmogorov's theory and is based on the following physically evident assumptions. If we consider the distribution $f(v, r)$ as a spatial average with respect to a turbulent

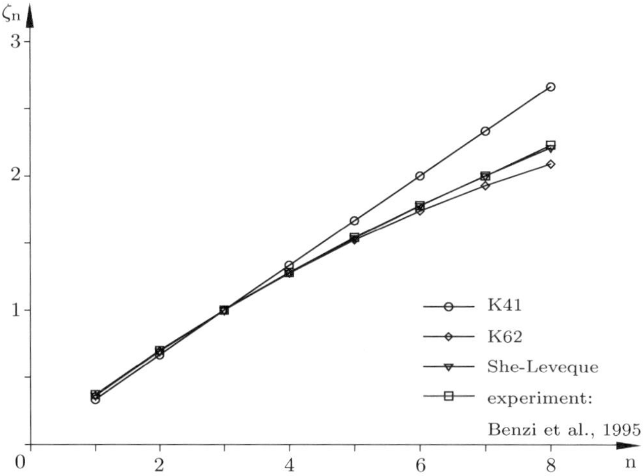

Fig. 9.6.4: Scaling index of the longitudinal structure function

field with stationary, homogeneous and isotropic statistics, the K41-scaling ought to be considered as

$$v(\boldsymbol{r},t) = u(\boldsymbol{x}+\boldsymbol{r},t) - u(\boldsymbol{x},t) \approx g(\boldsymbol{x})r^{\zeta} \qquad \text{with} \qquad \zeta = \frac{1}{3} \tag{9.6.67}$$

where $g(\boldsymbol{x})$ is a spatially fluctuating quantity. The scale r is again contained in the inertial range.

The multifractal extension consists of assuming that there are domains in a turbulent field with different local scaling exponents $\zeta(\boldsymbol{x})$. As a consequence, the turbulent field should be composed as follows

$$v(\boldsymbol{r},t) = u(\boldsymbol{x}+\boldsymbol{r},t) - u(\boldsymbol{x},t) \approx g(\boldsymbol{x})r^{\zeta(\boldsymbol{x})} \tag{9.6.68}$$

For the moments of n-th order, the following mean value thus has to be determined

$$S_{\mathrm{n}}(r) = \langle g(\boldsymbol{x})^{\mathrm{n}} r^{\mathrm{n}\zeta(\boldsymbol{x})} \rangle \tag{9.6.69}$$

We now introduce the probability $p(\zeta, g, r)$ to find the scaling index ζ and the prefactor g on the scale r. If the fluid volume is covered with cubes of side length r, then $(L/r)^3$ cubes are needed. The number of cubes containing points with a fixed scaling index ζ is $(L/r)^{D(\zeta)}$. We can therefore assume the following scale dependency of the probability

$$p(\zeta, g, r) = \bar{p}(g,\zeta)\left(\frac{r}{L}\right)^{3-D(\zeta)} \tag{9.6.70}$$

Obviously, this ansatz is restricted to the inertial range $\eta < r < L$. The structure function of n-th order can hence be determined as

$$S_{\mathrm{n}}(r) = \int\int \bar{p}(g,\zeta) g^{\mathrm{n}} L^{n\zeta}\left(\frac{r}{L}\right)^{\mathrm{n}\zeta+3-D(\zeta)} \mathrm{d}\zeta\, \mathrm{d}g \tag{9.6.71}$$

For small values r/L, the integral can be approximately evaluated by means of the saddle-point method described in section 8.5 if we take into account that essentially only the term

$$\left(\frac{r}{L}\right)^{\zeta_n} \tag{9.6.72}$$

yields a contribution. Here, ζ_n denotes the minimum of the function

$$\zeta_n = \inf_{\zeta}[n\zeta + 3 - D(\zeta)] \tag{9.6.73}$$

The fractal dimension $D(\zeta)$ of the local scaling exponent has to be consistent with Kolmogorov's −4/5-law, i.e. for n=3, the minimum must be located exactly at 1

$$\zeta_3 = \inf_{\zeta}\left[3\zeta + 3 - D(\zeta)\right] = 1 \tag{9.6.74}$$

9.6.8 Multiscaling Statistics

We have seen that the statistics of the velocity increments for a scale r can be modelled in various ways if we want to describe the intermittent deviations from the K41 scaling behaviour. The most serious deficiency of these approaches lies in the fact that they take into account solely the statistics of one spatial scale. The spatial organisation as well as the temporal evolution of turbulent fields are not captured in this way. Yet this information is indispensable for a basic understanding of the energy cascade, a topic which is the basis of current research.

A phenomenological approach for the treatment of the statistics of the velocity increments with respect to the whole range of scales from the dissipation range up to the integral length has been proposed by Friedrich and Peinke (Friedrich and Peinke, 1997). In this approach, methods from the theory of stochastic processes are applied which have been briefly described in section 3.9.12.

The most characteristic signature of the probability density distributions $f(v, r)$ of the velocity increments calculated from experimental data as shown in fig. 9.6.3 is their increasingly strong deviation from Gaussian shapes at smaller and smaller scales. This feature is closely related to the intermittent behaviour of turbulent flow. For a complete statistical description of the velocity increments one has to study the hierarchy of all joint probability distributions $f(v_1, r_1; v_2, r_2; \ldots; v_n, r_n)$ with $r_1 < r_2 < \ldots < r_n$ where smaller scales r_i are nested inside larger scales r_{i+1}. A substantial simplification ensues if the process is Markovian, cf. eqs. (3.9.169), (3.9.170). To simplify matters, we restrict our consideration in the following to the one-dimensional field of (longitudinal) velocity increments $v(r)$ and refer to the results presented in (Renner et al., 2001), who analysed the velocity increments measured in the turbulent region of a round free jet.

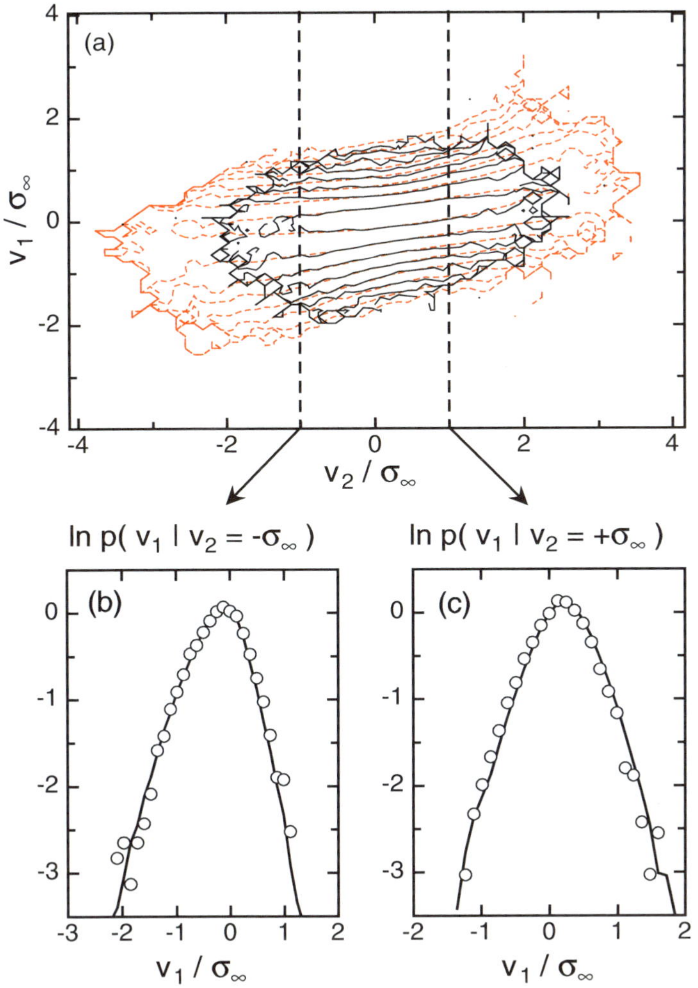

Fig. 9.6.5: Test of Markovian property: (a) Contour plots of $p(v_1, r_1|v_2, r_2)$ (dashed lines) and $p(v_1, r_1|v_2, r_2; v_3 = 0, r_3)$ (solid lines) for $r_1 = \lambda, r_2 = L/2$ and $r_3 = L$. (b,c) Cuts through the conditional pdfs ($p(v_1, r_1|v_2, r_2)$ lines, $p(v_1, r_1|v_2, r_2; v_3 = 0, r_3)$ circles). In order to compare the increments on different scales, the norm $\sigma_\infty = \lim_{r\to\infty}\langle v(r)^2\rangle$ is introduced. After (Renner et al., 2001).

The idea is to extend the Markov analysis which has been presented in section 3.9.12 for time-dependent processes to processes in scale. The first step is to ask if Markovian properties are fulfilled. If so, the n-scale probability density can be written as a product of the conditional pdfs and the distribution function $f(v_{\rm n}, r_{\rm n})$

$$f(v_1, r_1; v_2, r_2; \ldots; v_{\rm n}, r_{\rm n}) = p(v_1, r_1|v_2, r_2)\ldots p(v_{\rm n-1}, r_{\rm n-1}|v_{\rm n}, r_{\rm n}) f(v_{\rm n}, r_{\rm n}) \tag{9.6.75}$$

cf. eq. (3.9.170). On principle, it is a difficult task to answer this question for experimental data because dynamical noise sources become usually correlated at small scales, i.e. at small scales the Markov property is violated. This fact has already been taken into account by Einstein when he modelled Brownian motion (Einstein, 1905). Moreover, measurement noise can corrupt the Markovian properties. Nevertheless, a direct test which is feasible in practice for large data sets, is to compare the multi-conditioned probabilities to the single-conditioned ones and to ask if

$$p(v_1, r_1|v_2, r_2; v_3, r_3) = p(v_1, r_1|v_2, r_2) \tag{9.6.76}$$

is fulfilled. In fig. 9.6.5, the results of such a test performed by (Renner et al., 2001) are presented. The contour plots of $p(v_1, \lambda|v_2, L/2)$ and $p(v_1, \lambda|v_2, L/2; v_3 = 0, L)$ superimposed to each other are in very good correspondence for this specific choice of the length scales (λ is the Taylor microscale and L the integral scale, see definitions in sections 9.4.6, 9.4.4). Another possibility to test the statistical similarity of the data sets on the right hand and left hand side of eq. (9.6.76) is the Wilcoxon test (Wilcoxon, 1945). Theoretically, such assessments should be performed for all n and for all scales, which is clearly impossible. However, as long as the smallest scale is chosen to be larger that the Taylor microscale λ, the validity of the specific relation eq. (9.6.76) gives a strong hint for Markovianity. The finding, that the smallest length scale above which Markovian properties are fulfilled, the so-called Einstein-Markov length (Lück et al., 2006), corresponds to the Taylor length λ provides a remarkable new interpretation of λ, which until then had no specific physical meaning (Friedrich et al., 2009).

The next step is estimate the coefficients of the Kramers-Moyal expansion for the given data set. The evolution of the conditional probability $p(v, r|v_0, r_0)$ of the cascade process can be written in the form, cf. eq. (3.9.176)

$$-r\frac{\partial}{\partial r}p(v, r|v_0, r_0) = \Big\{\sum_{\mathrm{n}=1}^{\infty}\frac{1}{\mathrm{n}!}\Big(-\frac{\partial}{\partial v}\Big)^{\mathrm{n}}D^{(\mathrm{n})}(v, r)\Big\}p(v, r|v_0, r_0) \tag{9.6.77}$$

The minus sign of the left hand side is introduced because the cascade process is running from large to small scales. In addition, both sides of the equation are multiplied by r which leads to a logarithmic scale variable $\ln 1/r$. The Kramers-Moyal coefficients are now defined as

$$D^{(\mathrm{n})}(v, r) = \lim_{\Delta r\to 0} M^{(\mathrm{n})}(v, r, \Delta r) \tag{9.6.78}$$

$$M^{(\mathrm{n})}(v, r, \Delta r) = \frac{r}{\Delta r}\int_{-\infty}^{\infty}(v' - v)^{\mathrm{n}}p(v', r - \Delta r|v, r)\,\mathrm{d}v' \tag{9.6.79}$$

These coefficients can be directly estimated from the experimental data. To this end, one has to determine the conditional moments $M^{(\mathrm{n})}(v, r, \Delta r)$ at different scales $r_1 < r_2 < \ldots < r_{\mathrm{n}}$ and for various scale increments Δr, where a restriction to the coefficients of order one, two and four is sufficient. Details for an effective calculation of $M^{(\mathrm{n})}(v, r, \Delta r)$ and the approximation of the limit $\Delta r \to 0$ are described in (Friedrich et al., 2009; Friedrich et al., 2011) and the references therein.

For general stochastic processes, all Kramers-Moyal coefficients are different from zero. If, however, the fourth order coefficient $D^{(4)}$ vanishes, then Pawula's theorem tells us, that $D^{(\mathrm{n})} = 0$ for all $\mathrm{n} > 2$, see section 3.9.12. In this case, the Kramers-Moyal expansion reduces to the Fokker-Planck equation

$$-r\frac{\partial}{\partial r}p(v,r|v_0,r_0) = \Big\{-\frac{\partial}{\partial v}D^{(1)} + \frac{1}{2}\frac{\partial^2}{\partial v^2}D^{(2)}\Big\}p(v,r|v_0,r_0) \tag{9.6.80}$$

where $D^{(1)}$ is the drift and $D^{(2)}$ the diffusion coefficient. As outlined in section 3.9.12 for processes in time, the probability density function $f(v,r)$ fulfills the same equation

$$-r\frac{\partial}{\partial r}f(v,r) = \{-\frac{\partial}{\partial v}D^{(1)} + \frac{1}{2}\frac{\partial^2}{\partial v^2}D^{(2)}\}f(v,r) \tag{9.6.81}$$

A synthetic time series with the same multi-scale properties can be generated by means of the Langevin equation (using Itô's definition, cf. (Risken, 1996))

$$-r\frac{\partial}{\partial r}v(r) = D^{(1)}(v,r) + \sqrt{D^{(2)}(v,r)}\Gamma(r) \tag{9.6.82}$$

where $\Gamma(r)$ denotes δ-correlated white noise. For Markov processes in scale which obey eqs. (9.6.81), (9.6.82), the joint probabilities $f(v_1,r_1;v_2,r_2:\ldots;v_\mathrm{n},r_\mathrm{n})$ which contain the complete stochastic information of the process in scale can therefore be retrieved from eq. (9.6.75).

The coefficients $D^{(1)}(v,r)$ and $D^{(2)}(v,r)$ can be estimated directly from the measured data. For the turbulent data analysed in (Renner et al., 2001), the drift $D^{(1)}(v,r)$ is approximately linear in v with small corrections of 2nd and 3rd order while the diffusion term $D^{(2)}(v,r)$ is a quadratic polynomial in v. In addition, the coefficients of these polynomials depend on r. In order to check the validity and accuracy of the results, the probability density functions obtained from experimental data and from a numerical solution of the Fokker-Planck equation (9.6.81) are compared and displayed in fig. 9.6.6 and are seen to be in very good agreement. Also, the heavy tailed structure of the curves is well described. However, this consistency test is not yet sufficient. The conditional probability densities of the experimental data should also coincide with the numerically determined data. Figure 9.6.7 shows the remarkably good agreement of both conditional pdfs.

As we have seen in the previous sections, the scaling behaviour of the n-th order structure function $S_\mathrm{n}(r)$ and their modelling was a point of primary interest. The Markov analysis offers a novel possibility to discuss this item. $S_\mathrm{n}(r)$ can easily be retrieved by multiplication of the Fokker-Planck eq. (9.6.81) with v^n and subsequent integration

$$\begin{aligned}-r\frac{\partial}{\partial r}S_\mathrm{n}(r) = -r\frac{\partial}{\partial r}\int_{-\infty}^{+\infty}v^\mathrm{n}f(v,r)\mathrm{d}v &= \mathrm{n}\int_{-\infty}^{+\infty}v^{\mathrm{n}-1}D^{(1)}(v,r)f(v,r)\mathrm{d}v\\ &+\frac{1}{2}\mathrm{n}(\mathrm{n}-1)\int_{-\infty}^{+\infty}v^{\mathrm{n}-2}D^{(2)}(v,r)f(v,r)\mathrm{d}v\end{aligned} \tag{9.6.83}$$

If drift and diffusion have the simple scale-independent form $D^{(1)}(v) = -\gamma v$ and $D^{(2)}(v) = 2\beta v^2$, the structure functions can be calculated from eq. (9.6.83) as

$$S_\mathrm{n}(r) \propto r^{\zeta_\mathrm{n}} \quad \text{with} \quad \zeta_\mathrm{n} = \mathrm{n}\gamma - \mathrm{n}(\mathrm{n}-1)\beta \tag{9.6.84}$$

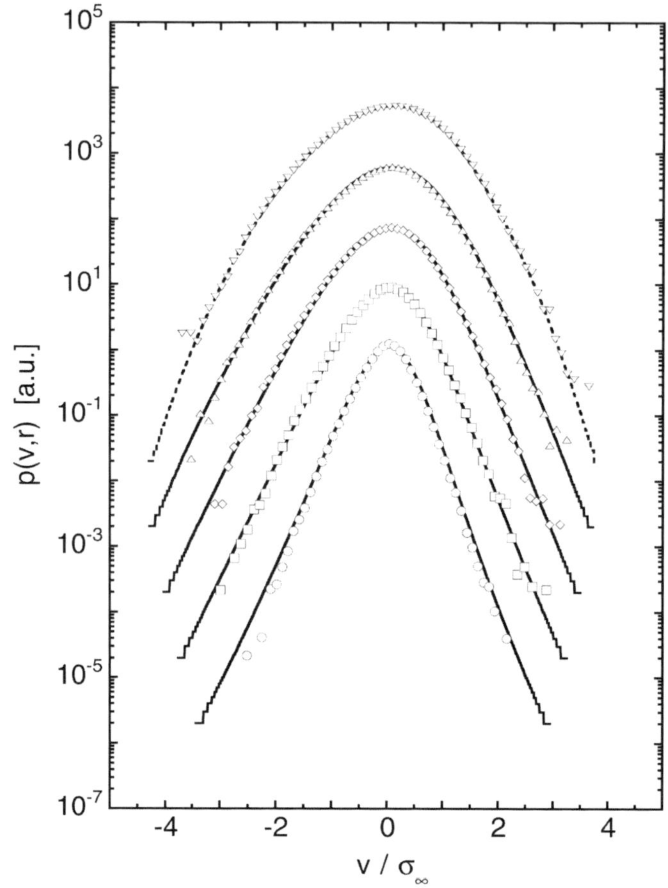

Fig. 9.6.6: Comparison of probability density functions from experimental data (symbols) and the numerical solution of the Fokker-Planck equation (9.6.81) (solid lines). The scales from top to bottom: $r = L$, $0.6L$, $0.35L$, $0.2L$, $0.1L$. For clarity of presentation, the pdfs are shifted in vertical direction. After (Renner et al., 2001).

which corresponds to the log-normal model of Kolmogorov described in section 9.6.6. Obviously, for $\beta = 0$, the self-similar scaling law of the K41-model is retrieved where γ is expected to be $\frac{1}{3}$, cf. section 9.6.4.3. The findings that the drift and diffusion coefficients obtained from experimental data do not have such a simple form and, in addition, depend on r show that the pure fractal or multifractal scaling does not provide an adequate description of the statistical process behind turbulence. The major shortcoming of a pure fractal or multifractal approach to turbulence, however, is the fact that it only addresses the statistics of the velocity increments on one single scale while for a complete and unique description of the scale-dependent process the joint interaction of all scales is indispensable.

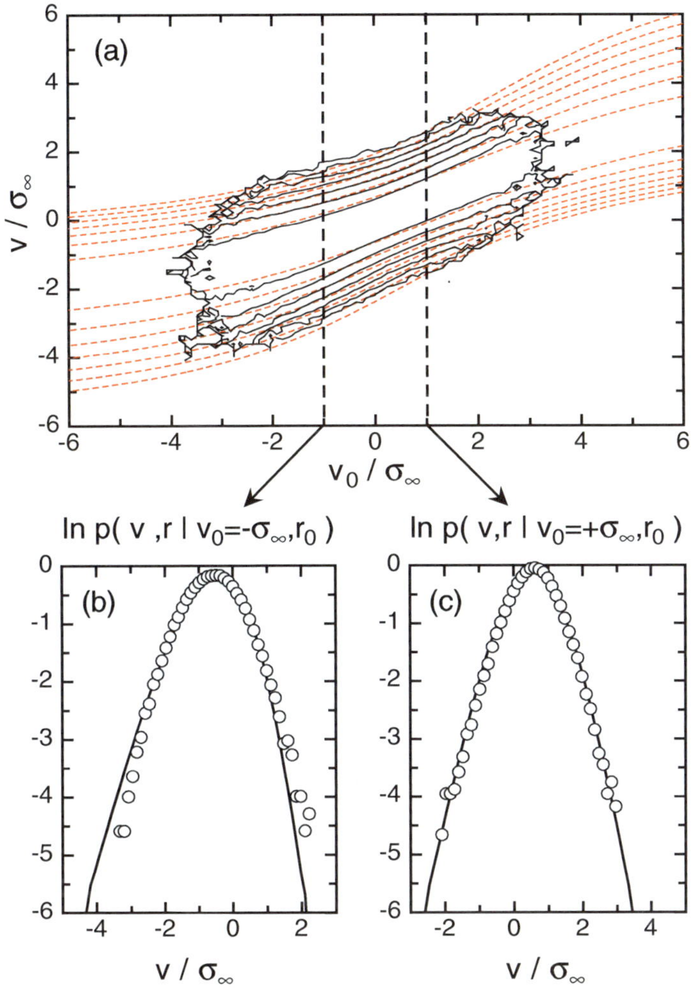

Fig. 9.6.7: Comparison of conditional pdf $p(v, r|v_0, r_0)$ from experimental data (solid lines) and the numerical solution of the Fokker-Planck equation (dashed lines). (a) Contour plots of $p(v, r|v_0, r_0)$ for $r_0 = L$ and $r = 0.6L$. (b,c) Cuts through $p(v, r|v_0, r_0)$; experimental data (symbols), numerical solution (lines). In order to compare the increments on different scales, the norm $\sigma_\infty = \lim_{r\to\infty}\langle v(r)^2\rangle$ is introduced. After (Renner et al., 2001).

9.7 Outlook

The exploration of chaos provided fundamental insights into understanding turbulent flows. The fact that only a few dynamical degrees of freedom can lead to temporally chaotic behaviour together with the realisation that there are specific routes to chaos has influenced and changed our view of time-dependent flow phenomena fundamentally. Yet a look at the variety of flow experiments shows the limitations of our knowledge of strongly turbulent flows. The sensitivity to initial conditions inherent in chaos only covers the time-dependent complexity of turbulence; the complex spatio-temporal organisation of turbulent flows certainly requires further investigation and description. This spatio-temporal organisation becomes evident from the existence of scaling laws respectively scaling laws broken by intermittency, as discussed using the example of the structure functions of the velocity increments. At the same time, turbulent flows are characterised by the existence of coherent structures, the presence of which is reflected in the statistics of the velocity increments. Obvious questions such as

- Why do elongated vortex structure turn up again and again?
- How are the vortex filaments organised in space? Does their internal structure depend on the Reynolds number?
- What is the influence of the vortex structures on the energy transport in the turbulent cascade and how do they influence the mixing and transport processes in turbulent flows?

are still awaiting satifactory answers.

In this regard, a fundamental breakthrough in understanding turbulent flows has obviously not yet been made. Nevertheless, we can expect that, in the near future, our increasing ability to carry out detailed simulations of flows with high Reynolds numbers will reveal further necessary details of the spatio-temporal self-organisation processes in turbulent flow. In particular, this concerns the small-scale structure of turbulence, for which a better understanding of the statistics of the velocities at different points and different times should be attained. This must be accompanied by the development of a statistical theory which characterises the properties of the velocity fields at different points and different times. First promising approaches already exist.

This is indispensable for the development of subgrid-scale models used for LES simulations which actually are predictive. An improved insight into the geometrical and topological properties of turbulent flows would also be conducive for such models. Explanations should be given why specific vortex structures like, for example, vortex filaments of the Burger vortex type can be observed so often in isotropic turbulence. The question as to how these structures behave and whether they change their structure in flows with a preferred direction and in the neighbourhood of boundaries is also of fundamental importance.

It will certainly be exciting to observe how these questions regarding the *intrinsic laws* of turbulent flows will be tackled and successfully dealt with by applying a combination of the theory of dynamical systems and methods from statistics and stochastics.

9.8 Appendix: Isotropic Tensor Fields

The expectation values of tensor fields enter into the statistical characterisation of turbulent flows. As an example of a vector field, i.e. a first-order tensor field, we mention the correlations between pressure and velocity field. Velocity moments of second order are second-order tensor fields. Moments of higher order induce correspondingly higher-order tensors fields like, for example,

$$\begin{aligned} A_{\mathrm{ij}}(\boldsymbol{r},t) &= \langle\, u_{\mathrm{i}}(\boldsymbol{x}+\boldsymbol{r},t)\, u_{\mathrm{j}}(\boldsymbol{x},t)\,\rangle \\ A_{\mathrm{ijk}}(\boldsymbol{r},t) &= \langle\, u_{\mathrm{i}}(\boldsymbol{x}+\boldsymbol{r},t)\, u_{\mathrm{j}}(\boldsymbol{x}+\boldsymbol{r},t)\, u_{\mathrm{k}}(\boldsymbol{x},t)\,\rangle \end{aligned} \tag{9.8.1}$$

In three dimensions, second-order tensor fields generally have $3^2 = 9$, third-order tensor fields have $3^3 = 27$ components. In the case of symmetries like, for example, isotropy, the number of independent components decreases. In this appendix, we establish the form of Cartesian second- and third-order tensors, taking homogeneity and isotropy as the underlying symmetries. Homogeneity means that the correlation functions do not depend on the spatial coordinate $\boldsymbol{x}$. The tensor fields are therefore only functions of the distance vector $\boldsymbol{r}$.

We consider the velocities $\boldsymbol{u}(\boldsymbol{x}+\boldsymbol{r}',t)$ and $\boldsymbol{u}(\boldsymbol{x}+\boldsymbol{r},t)$ with the distance vectors r and $\boldsymbol{r}'$, where $\boldsymbol{r}'$ is obtained by a rotation about $\boldsymbol{x}$, which is described by the unitary matrix $\boldsymbol{U}$

$$\boldsymbol{r}' = \boldsymbol{U}\boldsymbol{r} \qquad \boldsymbol{U}\boldsymbol{U}^t = \boldsymbol{U}^t\boldsymbol{U} = \boldsymbol{I} \tag{9.8.2}$$

Here, $\boldsymbol{U}^t = \boldsymbol{U}^{-1}$ denotes the transpose of the matrix $\boldsymbol{U}$ and $\boldsymbol{I}$ the unit matrix.

We now consider the correlation matrix

$$\begin{aligned} A'_{\mathrm{ij}}(\boldsymbol{r}',t) &= \langle\, u_{\mathrm{i}}(\boldsymbol{x}+\boldsymbol{r}',t)\, u_{\mathrm{j}}(\boldsymbol{x},t)\,\rangle \\ A_{\mathrm{ij}}(\boldsymbol{r},t) &= \langle\, u_{\mathrm{i}}(\boldsymbol{x}+\boldsymbol{r},t)\, u_{\mathrm{j}}(\boldsymbol{x},t)\,\rangle \end{aligned} \tag{9.8.3}$$

In the case of isotropic statistics, the tensor $A'_{\mathrm{ij}}(\boldsymbol{r}',t)$ can be expressed by the tensor $A_{\mathrm{ij}}(\boldsymbol{r},t)$. Using eq. (9.8.2), we obtain the rotated velocity vector

$$\boldsymbol{u}'(\boldsymbol{x}+\boldsymbol{r},t) = \boldsymbol{U}\boldsymbol{u}(\boldsymbol{x}+\boldsymbol{r},t) = \boldsymbol{U}\boldsymbol{u}(\boldsymbol{x}+\boldsymbol{U}^t\boldsymbol{r}',t) \tag{9.8.4}$$

For the correlation matrix, we obtain

$$A'_{\mathrm{ij}}(\boldsymbol{r}',t) = \sum_{\mathrm{k,l}} U_{\mathrm{ik}}U_{\mathrm{jl}}\,\langle u_{\mathrm{k}}(\boldsymbol{x}+\boldsymbol{U}^t\boldsymbol{r}',t)u_{\mathrm{l}}(\boldsymbol{x},t)\rangle = \sum_{\mathrm{k,l}} U_{\mathrm{ik}}U_{\mathrm{jl}}\,A_{\mathrm{ij}}(\boldsymbol{U}^t\boldsymbol{r}',t) \tag{9.8.5}$$

In the case of isotropy, the two vectors must be identical

$$A'_{\mathrm{ij}}(\boldsymbol{r}',t) = A_{\mathrm{ij}}(\boldsymbol{r}',t) \tag{9.8.6}$$

This leads to the invariance condition

$$A_{\mathrm{ij}}(\boldsymbol{r}', t) = \sum_{\mathrm{k,l}} U_{\mathrm{ik}} U_{\mathrm{jl}} \, A_{\mathrm{kl}}(\boldsymbol{U}^t \boldsymbol{r}', t) \tag{9.8.7}$$

The invariance condition for isotropic higher-order tensors can be deduced analogously. For third-order tensors we obtain

$$A_{\mathrm{ijk}}(\boldsymbol{r}, t) = \sum_{\mathrm{l,m,n}} U_{\mathrm{il}} U_{\mathrm{jm}} U_{\mathrm{kn}} \, A_{\mathrm{lmn}}(\boldsymbol{U}^t \boldsymbol{r}, t) \tag{9.8.8}$$

In the following, we determine the general form of isotropic second- and third-order tensors. The starting point is the observation that the combination of two isotropic second-order tensors $a^1_{\mathrm{ij}}(\boldsymbol{r})$ and $a^2_{\mathrm{ij}}(\boldsymbol{r})$

$$a_{\mathrm{ij}}(\boldsymbol{r}) = b_1(r) a^1_{\mathrm{ij}}(\boldsymbol{r}) + b_2(r) a^2_{\mathrm{ij}}(\boldsymbol{r}) \tag{9.8.9}$$

for arbitrary functions $b_1(r)$ and $b_2(r)$, which depend only on the length $r = |\boldsymbol{r}|$, is again an isotropic tensor. In the case of non-stationary turbulence, the functions depend on r and t. Obviously, analogous statements can be made for higher-order tensors. Hence, isotropic tensors can be established by a superposition of isotropic basic tensors $a^\alpha(\boldsymbol{r})$, where the amplitudes $b_\alpha(r)$ depend only on r.

a) Let us first consider vectors, i.e. first-order tensors. Obviously, there is a basis vector of the form

$$a_{\mathrm{i}}(\boldsymbol{r}) = \frac{r_{\mathrm{i}}}{r} \tag{9.8.10}$$

The most general isotropic vector then reads

$$A_{\mathrm{i}}(\boldsymbol{r}) = b(r) \frac{r_{\mathrm{i}}}{r} \tag{9.8.11}$$

b) Next, we consider second-order tensor fields. As can be seen, the tensor fields

$$\begin{aligned} a^1_{\mathrm{ij}}(\boldsymbol{r}) &= \delta_{\mathrm{ij}} \\ a^2_{\mathrm{ij}}(\boldsymbol{r}) &= \frac{r_{\mathrm{i}} r_{\mathrm{j}}}{r^2} \\ a^3_{\mathrm{ij}}(\boldsymbol{r}) &= \sum_{\mathrm{k}} \epsilon_{\mathrm{ijk}} \frac{r_{\mathrm{k}}}{r} \end{aligned} \tag{9.8.12}$$

satisfy the invariance property of eq. (9.8.7). We check this explicitly for a^2_{ij}

$$\begin{aligned} \sum_{\mathrm{k}} \sum_{\mathrm{l}} U_{\mathrm{ik}} U_{\mathrm{jl}} \, a^2_{\mathrm{ij}}(\boldsymbol{U}^t \boldsymbol{r}) &= \sum_{\mathrm{k}} \sum_{\mathrm{l}} U_{\mathrm{ik}} U_{\mathrm{jl}} \Big(\sum_{\mathrm{m}} U^t_{\mathrm{km}} r_{\mathrm{m}} \Big) \Big(\sum_{\mathrm{n}} U^t_{\mathrm{ln}} r_{\mathrm{n}} \Big) \frac{1}{r^2} \\ &= \frac{r_{\mathrm{i}} r_{\mathrm{j}}}{r^2} = a^2_{\mathrm{ij}}(\boldsymbol{r}) \end{aligned} \tag{9.8.13}$$

c) In the case of third-order tensor fields, we see that the fields

$$\begin{aligned}
a^1_{\mathrm{ijk}} &= \epsilon_{\mathrm{ijk}} \\
a^2_{\mathrm{ijk}} &= \frac{r_{\mathrm{i}}}{r}\,\delta_{\mathrm{jk}} \\
a^3_{\mathrm{ijk}} &= \frac{r_{\mathrm{j}}}{r}\,\delta_{\mathrm{ik}} \\
a^4_{\mathrm{ijk}} &= \frac{r_{\mathrm{k}}}{r}\,\delta_{\mathrm{ij}} \\
a^5_{\mathrm{ijk}} &= \frac{r_{\mathrm{i}} r_{\mathrm{j}} r_{\mathrm{k}}}{r^3}
\end{aligned} \tag{9.8.14}$$

comply with the invariance property (9.8.8).

A general third-order isotropic tensor field therefore has the form

$$\begin{aligned}
A_{\mathrm{ijk}}(\boldsymbol{r}) &= b_1(r)\epsilon_{\mathrm{ijk}} \\
&\quad + b_2(r)\frac{r_{\mathrm{i}}}{r}\delta_{\mathrm{jk}} + b_3(r)\frac{r_{\mathrm{i}}}{r}\delta_{\mathrm{jk}} + b_4(r)\frac{r_{\mathrm{i}}}{r}\delta_{\mathrm{jk}} \\
&\quad + b_5(r)\frac{r_{\mathrm{i}} r_{\mathrm{j}} r_{\mathrm{k}}}{r^3}
\end{aligned} \tag{9.8.15}$$

The formulæ for the isotropic tensor fields derived above are invariant with respect to rotations. If mirror symmetry is included, then all terms containing the tensor ϵ_{ijk} vanish (Batchelor, 1956). It should be noted that the concept isotropic turbulence already implies mirror symmetry.

9.8.1 Longitudinal and Transversal Correlation Functions

Each velocity vector can be split into a vector $\boldsymbol{u}^l$ parallel and a vector $\boldsymbol{u}^t$ transversal to $\boldsymbol{r}$

$$\boldsymbol{u} = \boldsymbol{u}^l + \boldsymbol{u}^t \tag{9.8.16}$$

The two components have the form

$$\begin{aligned}
\boldsymbol{u}^l &= \frac{\boldsymbol{r}}{r}\,\Big(\frac{\boldsymbol{r}}{r}\cdot\boldsymbol{u}\Big) \\
\boldsymbol{u}^t &= -\Big(\frac{\boldsymbol{r}}{r}\times\Big(\frac{\boldsymbol{r}}{r}\times\boldsymbol{u}\Big)\Big)
\end{aligned} \tag{9.8.17}$$

We now determine the correlation function

$$C_{rr}(\boldsymbol{r},t) = \langle \boldsymbol{u}^l(\boldsymbol{x}+\boldsymbol{r},t)\cdot\boldsymbol{u}^l(\boldsymbol{x},t)\rangle \tag{9.8.18}$$

which can be calculated from the correlation tensor $C_{ij}(\boldsymbol{r},t)$

$$C_{rr}(\boldsymbol{r},t) = \sum_{\mathrm{ij}} \frac{r_{\mathrm{i}}}{r} C_{\mathrm{ij}}(\boldsymbol{r},t)\frac{r_{\mathrm{j}}}{r} = \Big\langle \frac{\boldsymbol{r}}{r}\cdot\boldsymbol{u}(\boldsymbol{x}+\boldsymbol{r},t)\,\boldsymbol{u}(\boldsymbol{x},t)\cdot\frac{\boldsymbol{r}}{r}\Big\rangle \tag{9.8.19}$$

This is the correlation function for the so-called longitudinal velocities, i.e. the correlation of the velocity components in the direction of the distance vector $\boldsymbol{r}$. In analogy to eq. (9.8.19), the correlation function for the so-called transversal velocity C_{tt} can be calculated

$$C_{tt}(\boldsymbol{r},t) = \frac{1}{2}\langle \boldsymbol{u}^t(\boldsymbol{x}+\boldsymbol{r},t)\cdot \boldsymbol{u}^t(\boldsymbol{x},t)\rangle \tag{9.8.20}$$

For convenience, a factor $\frac{1}{2}$ has been introduced because there are two transversal and only one longitudinal direction.

In a turbulent flow with mirror symmetry, the correlations between the longitudinal and the transversal velocities vanish. As a consequence, the correlation matrix has the form

$$C_{\mathrm{ij}}(\boldsymbol{r},t) = C_{rr}(\boldsymbol{r},t)\frac{r_{\mathrm{i}}}{r}\frac{r_{\mathrm{j}}}{r} + \langle\, u^t_{\mathrm{i}}(\boldsymbol{x}+\boldsymbol{r},t)u^t_{\mathrm{j}}(\boldsymbol{x},t)\,\rangle \tag{9.8.21}$$

Owing to the isotropy, the tensor of transversal velocitites has the form

$$\langle\, u^t_{\mathrm{i}}(\boldsymbol{x}+\boldsymbol{r},t)u^t_{\mathrm{j}}(\boldsymbol{x},t)\,\rangle = C_{tt}\big[\delta_{\mathrm{ij}} - \frac{r_{\mathrm{i}}r_{\mathrm{j}}}{r^2}\big] \tag{9.8.22}$$

The following relation was taken into account

$$\sum_{\mathrm{ij}} \frac{r_i}{r}\langle\, u^t_{\mathrm{i}}(\boldsymbol{x}+\boldsymbol{r},t)u^t_{\mathrm{j}}(\boldsymbol{x},t)\rangle\frac{r_{\mathrm{j}}}{r} = 0 \tag{9.8.23}$$

We are now able to express the correlation matrix $C_{\mathrm{ij}}(\boldsymbol{r},t)$ in terms of the longitudinal and the transversal correlations $C_{tt}(\boldsymbol{r},t)$ and $C_{rr}(\boldsymbol{r},t)$ as

$$\begin{aligned} C_{\mathrm{ij}}(\boldsymbol{r},t) &= \langle\, u_{\mathrm{i}}(\boldsymbol{x}+\boldsymbol{r},t)\, u_{\mathrm{j}}(\boldsymbol{x},t)\,\rangle \\ &= C_{tt}(r,t)\,\delta_{\mathrm{ij}} + \big[C_{rr}(r,t) - C_{tt}(r,t)\big]\frac{r_{\mathrm{i}}r_{\mathrm{j}}}{r^2} \end{aligned} \tag{9.8.24}$$

These quantities depend only on the distance r and no longer on the direction of the distance vector $\boldsymbol{r}$.

9.8.1.1 The Correlation Function of Third Order

We next consider the correlation function of third order which, due to the assumed homogeneity and isotropy, has the form

$$\begin{aligned} C_{\mathrm{ij,k}}(\boldsymbol{r},t) &= \langle\, u_{\mathrm{i}}(\boldsymbol{x},t)\, u_{\mathrm{j}}(\boldsymbol{x},t)\, u_{\mathrm{k}}(\boldsymbol{x}+\boldsymbol{r},t)\,\rangle \\ &= A(r,t)\delta_{\mathrm{ij}}\frac{r_{\mathrm{k}}}{r} + B(r,t)\left[\frac{r_{\mathrm{i}}}{r}\delta_{\mathrm{jk}} + \frac{r_{\mathrm{j}}}{r}\delta_{\mathrm{ik}}\right] + C(r,t)\frac{r_{\mathrm{i}}r_{\mathrm{j}}r_{\mathrm{k}}}{r^3} \end{aligned} \tag{9.8.25}$$

It was taken into account here that the correlation function is symmetrical with respect to the indices i and j. Based on the decomposition of the velocity field in a longitudinal and a transversal component given in eq. (9.8.16), we can again define the longitudinal correlation as

$$C_{rrr}(r,t) = \sum_{\mathrm{ijk}} C_{\mathrm{ij,k}}\frac{r_{\mathrm{i}}r_{\mathrm{j}}r_{\mathrm{k}}}{r^3} = A + 2B + C \tag{9.8.26}$$

9.8.2 Correlation Functions for Incompressible, Isotropic and Homogeneous Fields

The fact that the velocity field is incompressible has direct consequences for the correlation functions. The most important property is that the tensors of two-point correlations and three-point correlations are both determined by a single scalar function depending on the distance r. This will be proven in the following.

9.8.2.1 The Correlation Functions of Second Order

Due to the incompressibilty condition, the following relation holds

$$\sum_{\mathrm{i}} \frac{\partial}{\partial r_{\mathrm{i}}} C_{\mathrm{ij}}(\boldsymbol{r}, t) = \sum_{\mathrm{i}} \left\langle \frac{\partial u_{\mathrm{i}}(\boldsymbol{x}+\boldsymbol{r}, t)}{\partial r_{\mathrm{i}}} u_{\mathrm{j}}(\boldsymbol{x}, t) \right\rangle = 0 \tag{9.8.27}$$

Using the relation $\frac{\partial}{\partial r_{\mathrm{i}}} = \frac{r_{\mathrm{i}}}{r} \frac{\partial}{\partial r}$, the calculation of this expression leads to

$$C_{tt}(r, t) = \frac{1}{2r} \frac{\partial}{\partial r} \left(r^2 C_{rr}(r, t) \right) \tag{9.8.28}$$

Hence, the correlation matrix $C_{ij}(\boldsymbol{r}, t)$ for homogeneous, isotropic turbulence of incompressible flows can be expressed by a single scalar function $C_{rr}(r, t)$. In particular, the trace of the correlation matrix can be determined from the longitudinal correlation matrix $C_{rr}(r, t)$ alone

$$R(r, t) = \sum_{i} C_{\mathrm{ii}}(\boldsymbol{r}, t) = 2C_{tt}(r, t) + C_{rr}(r, t) = \frac{1}{r^2} \frac{\partial}{\partial r} \left(r^3 C_{rr}(r, t) \right) \tag{9.8.29}$$

This relation enters the Kármán-Howarth equation (9.5.14).

9.8.2.2 The Correlation Functions of Third Order

We now consider the correlation function of third order

$$\begin{aligned} C_{\mathrm{ij,k}}(\boldsymbol{r}, t) &= \langle u_{\mathrm{i}}(\boldsymbol{x}, t)\, u_{\mathrm{j}}(\boldsymbol{x}, t)\, u_{\mathrm{k}}(\boldsymbol{x}+\boldsymbol{r}, t) \rangle \\ &= A(r, t)\delta_{\mathrm{ij}} \frac{r_{\mathrm{k}}}{r} + B(r, t) \left[\frac{r_{\mathrm{i}}}{r} \delta_{\mathrm{jk}} + \frac{r_{\mathrm{j}}}{r} \delta_{\mathrm{ik}} \right] + C(r, t) \frac{r_{\mathrm{i}} r_{\mathrm{j}} r_{\mathrm{k}}}{r^3} \end{aligned} \tag{9.8.30}$$

The incompressibility condition

$$\sum_{\mathrm{k}} \frac{\partial}{\partial r_{\mathrm{k}}} C_{\mathrm{ij,k}}(\boldsymbol{r}, t) = 0 \tag{9.8.31}$$

leads to

$$\left(\frac{1}{r^2} \frac{\partial}{\partial r} \left(r^2 A(r, t) \right) + 2 \frac{B(r, t)}{r} \right) \delta_{\mathrm{ij}} + \left[2r \frac{\partial}{\partial r} \frac{B(r, t)}{r} + \frac{1}{r^2} \frac{\partial}{\partial r} \left(r^2 C(r, t) \right) \right] \frac{r_{\mathrm{i}} r_{\mathrm{j}}}{r^2} = 0 \tag{9.8.32}$$

This equation comprises two relations for the three quantities A, B, C, since both terms in eq. (9.8.32) must vanish identically. As a result, the third-order correlation tensor is again defined by a scalar quantity. Let us consider the correlation $C_{rrr}(r,t) = A + 2B + C$ as an example. A short calculation yields

$$\begin{aligned} A &= -\frac{C_{rrr}}{2} \\ B &= -A - \frac{r}{2}\frac{\partial A}{\partial r} = \frac{1}{4r}\frac{\partial}{\partial r}(r^2 C_{rrr}) \\ C &= r\frac{\partial A}{\partial r} - A = -\frac{r^2}{2}\frac{\partial}{\partial r}\Big(\frac{C_{rrr}}{r}\Big) \end{aligned} \tag{9.8.33}$$

The velocity tensor can thus be written in the condensed form

$$\begin{aligned} C_{\mathrm{ij,k}}(\boldsymbol{r},t) = &- \frac{C_{rrr}}{2}\delta_{\mathrm{ij}}\frac{r_{\mathrm{k}}}{r} + \frac{1}{4r}\frac{\partial}{\partial r}(r^2 C_{rrr})\big[\frac{r_{\mathrm{i}}}{r}\delta_{\mathrm{jk}} + \frac{r_{\mathrm{j}}}{r}\delta_{\mathrm{ik}}\big] - \\ &\frac{r^2}{2}\frac{\partial}{\partial r}\Big(\frac{C_{rrr}}{r}\Big)\frac{r_{\mathrm{i}}r_{\mathrm{j}}r_{\mathrm{k}}}{r^3} \end{aligned} \tag{9.8.34}$$

We are now able to assemble all quantities necessary to formulate the Kármán-Howarth relation, eq. (9.5.14). For that, we calculate in addition to $R(r,t)$, eq. (9.8.29), the quantity $T_{\mathrm{i}}(\boldsymbol{r},t)$, see eq. (9.5.10)

$$\begin{aligned} T_{\mathrm{i}}(\boldsymbol{r},t) &= -\sum_{\mathrm{k}} C_{\mathrm{ik,k}}(\boldsymbol{r},t) = -\frac{r_{\mathrm{i}}}{r}\big[A(r,t) + 4B(r,t) + C(r,t)\big] \\ &= -\frac{1}{2r^3}\frac{\partial}{\partial r}\left(r^4 C_{rrr}(r,t)\right)\frac{r_i}{r} \end{aligned} \tag{9.8.35}$$

9.8.3 Structure Functions of Incompressible Fields

For the calculation of the structure functions of second and third order, we now consider the corresponding moments of the velocity increments.

9.8.3.1 Structure Function of Second Order

The tensor of the structure functions of second order is given by the difference of the velocitites at the points $\boldsymbol{x}$ and $\boldsymbol{x}' = \boldsymbol{x} + \boldsymbol{r}$

$$\begin{aligned} S_{\mathrm{ij}}(\boldsymbol{x},\boldsymbol{x}',t) &= \langle\, (u_{\mathrm{i}}(\boldsymbol{x},t) - u_{\mathrm{i}}(\boldsymbol{x}',t))\,(u_{\mathrm{j}}(\boldsymbol{x},t) - u_{\mathrm{j}}(\boldsymbol{x}',t))\,\rangle \\ &= \langle\, u_{\mathrm{i}}(\boldsymbol{x},t)\,u_{\mathrm{j}}(\boldsymbol{x},t)\,\rangle + \langle\, u_{\mathrm{i}}(\boldsymbol{x}',t)\,u_{\mathrm{j}}(\boldsymbol{x}',t)\,\rangle - \\ &\quad \langle\, u_{\mathrm{i}}(\boldsymbol{x}',t)\,u_{\mathrm{j}}(\boldsymbol{x},t)\,\rangle - \langle\, u_{\mathrm{i}}(\boldsymbol{x},t)\,u_{\mathrm{j}}(\boldsymbol{x}',t)\,\rangle \end{aligned} \tag{9.8.36}$$

Owing to the homogeneity and isotropy, there holds

$$\begin{aligned} \langle\, u_{\mathrm{i}}(\boldsymbol{x},t)\,u_{\mathrm{j}}(\boldsymbol{x},t)\,\rangle &= \langle\, u_{\mathrm{i}}(\boldsymbol{x}',t)\,u_{\mathrm{j}}(\boldsymbol{x}',t)\,\rangle = C_{\mathrm{ij}}(\mathbf{0},t) \\ \langle\, u_{\mathrm{i}}(\boldsymbol{x}',t)\,u_{\mathrm{j}}(\boldsymbol{x},t)\,\rangle &= \langle\, u_{\mathrm{i}}(\boldsymbol{x},t)\,u_{\mathrm{j}}(\boldsymbol{x}',t)\,\rangle = C_{\mathrm{ij}}(\boldsymbol{r},t) \end{aligned} \tag{9.8.37}$$

We thus obtain

$$S_{\mathrm{ij}}(\boldsymbol{r},t) = 2\left(C_{\mathrm{ij}}(\mathbf{0},t) - C_{\mathrm{ij}}(\boldsymbol{r},t)\right) \tag{9.8.38}$$

Together with eq. (9.8.27), the incompressibility condition yields the equation

$$\frac{\partial}{\partial r_{\mathrm{i}}} S_{\mathrm{ij}}(\boldsymbol{r},t) = 0 \tag{9.8.39}$$

In the case of isotropic statistics, S_{ij} is a second-order tensor with the property described in eq. (9.8.38). Like the tensor $C_{\mathrm{ij}}(\boldsymbol{r})$ in eq. (9.8.24), S_{ij} has the general form

$$S_{\mathrm{ij}}(\boldsymbol{r},t) = S_{tt}(r,t)\delta_{\mathrm{ij}} + [S_{rr}(r,t) - S_{tt}(r,t)]\frac{r_{\mathrm{i}} r_{\mathrm{j}}}{r^2} \tag{9.8.40}$$

Anaologously to eq. (9.8.28), the incompressibility condition leads to a relation between the longitudinal structure functions $S_{rr}(r,t)$ and the transversal ones $S_{tt}(r,t)$

$$S_{tt}(r,t) = \frac{1}{2r}\frac{\partial}{\partial r}\left(r^2 S_{rr}(r,t)\right) \tag{9.8.41}$$

Furthermore, we obtain the following expression for the mean kinetic energy

$$\langle \boldsymbol{v}(\boldsymbol{r},t)^2 \rangle = S_{rr}(r,t) + 2S_{tt}(r,t) = \frac{1}{r^2}\frac{\partial}{\partial r}\left(r^3 S_{rr}(r,t)\right) \tag{9.8.42}$$

9.8.3.2 Structure Function of Third Order

We can proceed in a similar way for the structure functions of third order. We obtain

$$S_{\mathrm{ijk}}(\boldsymbol{x},\boldsymbol{x}',t) = \langle \left(u_{\mathrm{i}}(\boldsymbol{x},t) - u_{\mathrm{i}}(\boldsymbol{x}',t)\right)\left(u_{\mathrm{j}}(\boldsymbol{x},t) - u_{\mathrm{j}}(\boldsymbol{x}',t)\right)\left(u_{\mathrm{k}}(\boldsymbol{x},t) - u_{\mathrm{k}}(\boldsymbol{x}',t)\right)\rangle \tag{9.8.43}$$

Due to the statistical homogeneity and isotropy and also taking the relation $\langle u_{\mathrm{i}}(\boldsymbol{x},t) u_{\mathrm{j}}(\boldsymbol{x},t) u_{\mathrm{k}}(\boldsymbol{x},t)\rangle = 0$ into consideration, we obtain from eqs. (9.5.2), (9.5.6)

$$S_{\mathrm{ijk}}(\boldsymbol{r},t) = -2\left(C_{\mathrm{jk,i}}(\boldsymbol{r},t) + C_{\mathrm{ij,k}}(\boldsymbol{r},t) + C_{\mathrm{ki,j}}(\boldsymbol{r},t)\right) \tag{9.8.44}$$

This leads to the following formulation

$$S_{\mathrm{ijk}}(\boldsymbol{r},t) = -\frac{\partial}{\partial r}\left(rC_{rrr}\right)\left[\delta_{\mathrm{ij}}\frac{r_{\mathrm{k}}}{r} + \delta_{\mathrm{ik}}\frac{r_{\mathrm{j}}}{r} + \delta_{\mathrm{jk}}\frac{r_{\mathrm{i}}}{r}\right] + 3r^2\frac{\partial}{\partial r}\left(\frac{C_{rrr}}{r}\right)\frac{r_{\mathrm{i}} r_{\mathrm{j}} r_{\mathrm{k}}}{r^3} \tag{9.8.45}$$

We see that the third-order tensor again depends solely on a scalar function $C_{rrr}(r,t)$ and calculate the isotropic third-order tensor as

$$S_{r\mathrm{jk}}(\boldsymbol{r},t) = \sum_{\mathrm{i}} \frac{r_{\mathrm{i}}}{r} S_{\mathrm{ijk}}(\boldsymbol{r},t) = -\frac{\partial}{\partial r} rC_{rrr}(r,t)\delta_{jk} + \left[r\frac{\partial}{\partial r}C_{rrr} - 5C_{rrr}\right]\delta_{jk}\frac{r_{\mathrm{j}} r_{\mathrm{k}}}{r^2} \tag{9.8.46}$$

From this relation, we obtain

$$S_{rrr}(r,t) = -6C_{rrr}(r,t) \tag{9.8.47}$$

and from the comparison with the general form (9.8.40)

$$S_{rtt}(r,t) = -\frac{\partial}{\partial r}\big(rC_{rrr}(r,t)\big) \tag{9.8.48}$$

Therefore, we obtain the following relation between the structure functions $S_{rrr}(r,t)$ and $S_{rtt}(r,t)$

$$S_{rtt}(r,t) = \frac{1}{6}\frac{\partial}{\partial r}\big(rS_{rrr}(r,t)\big) \tag{9.8.49}$$

It can be proven by direct calculation that the quantity S_{rrt} vanishes, $S_{rrt} = 0$. This result follows immediately from the mirror symmetry of the homogeneous, isotropic turbulence.

For the derivation of Komogorov's $-4/5$-law, we need the following structure function of third order

$$S^3(r,t) = \langle v_r(\boldsymbol{r},t)\boldsymbol{v}(\boldsymbol{r},t)^2\rangle = S_{rrr}(r,t) + 2S_{rtt}(r,t) \tag{9.8.50}$$

Due to the relation (9.8.49), this quantity can be written in the condensed form

$$S^3(r,t) = \frac{1}{3r^3}\frac{\partial}{\partial r}\big(r^4 S_{rrr}(r,t)\big) \tag{9.8.51}$$

All relations needed for the formulation of Komogorov's $-4/5$ law are now made available.

10 Computer Experiments

Plus ça change,
plus c'est la même chose
Alphonse Karr (1808 – 1890), Les Guêpes

In the preceding chapters, we have attempted to demonstrate that the modern theory of non-linear dynamical systems in conjunction with novel computer methods enable us to analyse complex dynamical behaviour arising in the most varying fields, such as biology, medicine, hydrodynamics, classical mechanics, electrical engineering, chemistry etc. The range of examples we discuss in this, the final Chapter 10 on computer experiments is correspondingly extensive. In Chapters 1 to 8, it has been our aim to impart the theoretical mathematical and physical foundations. In Chapter 10, the emphasis is shifted to applications and the computer-supported discussion of classic model systems, the characteristics of which can be applied to a multitude of dynamical systems.

Our first example in section 10.1 is taken from biomechanics or, more specifically, orthopedics. For a long time, the goal has been to prolong the durability of artificial hip joints. The possibilities offered by the computer – involving static and dynamic calculations according to the finite element method, relating to the reconstruction and visualisation of geometry and density distribution on the basis of two-dimensional computer-tomographic sections – are being increasingly exploited. What is still missing is the mathematical formulation of bone remodelling. Once these principles have been expressed in mathematical terms, we possess the foundation for formulating quantitative assessments of the anchoring of implants in bone structures. In the spirit of the theory of non-linear dynamical systems, we present a first study of the modelling of bone adaptation.

In section 10.2, with the example of the dissipative, two-dimensional Hénon map, we demonstrate how repeated stretching and folding of trajectories results in a complete mixing in the phase space. This stretching and folding mechanism also explains one phenomenon of strange attractors, in particular the limited extension of the attractor in the phase space in spite of the fact that neighbouring trajectories diverge from one another. We illustrate this with a series of depictions of the Lorenz attractor.

Further characteristics of the Lorenz system are discussed in section 10.3. Its dynamical behaviour within a larger range of the control parameters is of interest here. We present the bifurcation diagram and the evolution of the Lyapunov exponents in dependence of the control parameter r (relative Rayleigh number). Moreover, to determine the capacity dimension D_c of the Lorenz attractor, the algorithm of Hunt and Sullivan is presented and the results evaluated.

J. Argyris et al., *An Exploration of Dynamical Systems and Chaos*,
DOI: 10.1007/978-3-662-46042-9_10

In section 10.4, we consider the van der Pol equation with and without external excitation. The van der Pol equation, originally conceived as a model to describe the response of electrical circuits, exhibits the phenomenon of "negative" or inciting friction for small amplitudes; for large amplitudes, on the other hand, the friction term is energy dissipating. The physical significance of "negative" friction and the dynamical answer to the interplay of energy-supplying negative and energy-consuming positive friction behaviour is discussed in detail. For small damping values, the oscillatory response of the van der Pol oscillator is almost sinusoidal; for large ones, self-excited relaxation oscillations occur. The fact that just one limit cycle exists in the whole phase space is not self-evident. We present the theorems required to prove this statement.

In the next step, we generalise the van der Pol oscillator by adding a periodic driving force. If damping and driving amplitude are small, the initial system can be approximated by an autonomous, two-dimensional system by means of the averaging method. This method, which filters out the higher-order harmonics by an averaging procedure, is presented and the autonomous system of equations thus gained, containing two control parameters, is discussed with respect to stationary solutions and bifurcations. For large damping values and periodic driving forces, we consider the van der Pol equation as modified by Shaw. This modification of the van der Pol equation exhibits chaotic responses. The corresponding strange attractor is denoted the Birkhoff-Shaw attractor. One of its characteristics is the so-called wing-beak structure which is illustrated using Poincaré sections and three-dimensional phase portraits.

A further classic among the model systems of non-linear dynamics is the Duffing equation. It is analysed in detail in section 10.5, extending our earlier discussions. The non-linear oscillatory response of a periodically driven, buckled beam subject to an end load can be approximated to the first order by the Duffing equation using the modal reduction technique; in the present case, this involves the application of the first fundamental mode. The Duffing oscillator is a single-degree-of-freedom oscillator depending on five control parameters. The choice of the control parameters has decisive influence on the system behaviour. Among others, the following phenomena can be observed: periodic and chaotic response, the simultaneous emergence of two periodic or strange attractors in the phase space and fractal boundaries between the basins of attraction of the two attractors leading to unpredictability. Phase portraits, Poincaré sections and bifurcation diagrams for a series of parameter values illustrate the phenomena characteristic of the Duffing equation. In conclusion, the so-called Holmes-Melnikov curve is calculated. It represents the upper bound for the control parameters, beyond which fractal boundaries occur between the basins of attraction of the two periodic attractors.

Section 10.6 contains a short summary of Shilnikov bifurcations. Depending upon the system parameters, the homoclinic bifurcations of a saddle-focus in three-dimensional vector fields may lead to the formation of bursting oscillatory patterns which can be observed, for example, in chemistry and neuroscience.

Section 10.7 deals with Julia sets and their underlying ordering principle, the Mandelbrot set. The computer as an experimental and graphics laboratory – this, or something similar, could be the heading for this section. The high complexity of boundaries which emerges as a result of the simplest non-linear recursion formulæ is astonishing, stimulates new concepts, opens new perspectives and spurs the imagination. The figures speak for themselves.

In section 10.8, we return to the circle map of Chapter 8 and exploit the graphic possibilities offered by the computer to elucidate the complex structure of the Arnold tongues. The Lyapunov exponents serve to demonstrate various transitions to chaos; the corresponding bifurcation diagrams illustrate their self-similar structure.

In section 10.9, we switch to dynamic phenomena in physical chemistry, a subject not touched upon in the rest of the book. In general, if chemical reactions proceed under flow conditions in homogeneous or inhomogeneous media and are subject to fixed external parameters, usually constant steady-state rates of reaction products will be generated. However, many known cases do exist in which the system may experience periodic or even aperiodic irregular behaviour. The most famous example, also known to laymen, is the Belusov-Zhabotinsky (BZ) reaction arising in reactions of a special well-stirred homogeneous solution which displays periodic or irregular colour changes generated by a surplus of the individual reactants. Also of interest are certain heterogeneous processes which occur at gas-solid or liquid-solid interfaces and exhibit phenomena of temporal self-organisation. These include well-known electro-chemical systems as well as heterogeneously catalysed reactions arising, for example, on "real catalysts" near atmospheric pressure or under ultra-high vacuum conditions on single-crystal surfaces.

Within this extensive panorama of complex physico-chemical phenomena, we concentrate in section 10.9 on the mechanism of the catalytic reaction $2CO + O_2 = 2CO_2$ occurring on surfaces of platinum crystals. In the absorption, the two gaseous components, namely CO and O_2, compete for free sites on the platinum surface, with a distinct difference in their behaviour, however. The model selected and discussed in section 10.9 is based on an approximation and, as a result, cannot reproduce all experimentally recorded characteristics observed in the non-linear dynamics of the system. The subject has attracted great attention in various research departments; but the most profound and original work has been performed by Gerhard Ertl and collaborators at the Fritz-Haber Institute of the Max Planck Society in Berlin. In this section, we attempt to describe some of their most striking findings on the reaction processes between the homogeneous medium and the surface of platinum crystals. These include periodic phenomena leading for varying control parameters to a first and second period doubling and ultimately also to irregular-chaotic behaviour characterised in the phase space by strange attractors. This chaos can also be superseded by hyperchaos, in which not only the first Lyapunov exponent is positive, but two positive Lyapunov exponents exist – a phenomenon which had not been identified previously in a chemical reaction. This series of investigations provides ample evidence of the range of the research successfully tackled by the group under Gerhard Ertl.

So far, we have discussed the purely temporal behaviour of integral properties of homogeneous systems. To achieve a homogeneous state, we may imagine that the solution is well-stirred. In the gaseous layer on the top of a platinum crystal, however, this is not possible. In general, the variables then possess not only temporal, but also spatial dependence. Under such conditions, many unusual phenomena occur, such as standing waves and solitons as well as propagating waves exhibiting, for example, spiral patterns and also chaotic and hyperchaotic behaviour. Section 10.9 presents a brief *aperçu* of such complex phenomena.

Finally, we briefly emphasise the importance of catalysts in biochemical systems. In metabolic reactions, enzymes play the role of highly efficient catalysts. Certainly, the methods and tools of non-linear dynamics and pattern formation are indispensable for the investigation of the extremely complex spatio-temporal behaviour of biochemical systems. With this excursion into chemistry, we conclude the last chapter of this book.

10.1 Introduction to Bone Remodelling Processes

It is a capital mistake to theorize before one has data.
Insensibly one begins to twist facts to suit theories,
instead of theories to suit facts.

Sherlock Holmes in
Arthur Conan Doyle: A Scandal in Bohemia

The firm long-term anchorage of statically and dynamically loaded hip prostheses is still the central problem of endoprosthesis. This aspect is all the more important when we consider that more than 50% of these hip prostheses are implanted in secondary operations and that this percentage is increasing. The loosening and possible loss of the endoprosthesis is a consequence of bone resorption. Since as long ago as the end of the 19th century, it has been suspected that mechanical stimuli (Wolff's law) encourage the formation of bone (hypertrophy) while a lack of these stimuli leads to a bone resorption (atrophy). Bone tissue registers the alteration of external conditions and now attempts to uphold the functioning of the structure by means of bone remodelling, In this connection, we speak of functional adaptation by active alteration, a property exclusive to living matter. Without reliable numerical predictions regarding the bone remodelling processes as a reaction to alterations, e.g. to load distribution, no endoprosthesis can be anything other than a product based on experience and intuition. Currently, more than 800 types of prosthesis are on the market for the hip alone; this illustrates clearly to what an extent experiments are being carried out in this field due to the lack of reliable bone adaptation models. Initial formulations of bone remodelling are, of course, already known; the great breakthrough of relevant models for clinical practice has yet to come. In the following, we briefly present the current ideas and formulations of bone remodelling:

- Bone remodelling from a physiological point of view:

  Like any organically active tissue, bones are subject to a continuous process of reformation in which old bone material is resorbed and new synthesised. It is mainly three types of cell that take an active part in this process: osteoblasts, osteoclasts and osteocytes. Osteoblasts create new bone material, osteoclasts break down the bone material and osteocytes are modified osteoblasts which regulate the calcium concentration in the blood plasma as information carriers.

- Bone remodelling as a feedback control mechanism according to Pauwels and Kummer:

  Without a detailed knowledge of the mechanisms by which the mechanical loads effect the osteoblast-osteoclast equilibrium, it is nevertheless possible to imagine a load-regulated feedback control mechanism. As one example of the various attempts to determine a control variable relevant for the remodelling, let us first elucidate the feedback control process using the equivalent stress parameter $\bar{\sigma}$ selected by Pauwels (1960). In order to maintain the existing equilibrium of osteoblasts and osteoclasts, a regulating stimulus $\bar{\sigma}$ must act continuously. If, as a result of intensified mechanical loading, the stimulus $\bar{\sigma}$ increases, this leads to a growth in bone formation; should the stimulus $\bar{\sigma}$ fall beneath a critical value $\bar{\sigma}_s$, bone atrophy results (Kummer and Lohscheidt, 1984; Faust et al., 1986).

- Self-optimisation ansatz according to Carter and Fyhrie:

  Carter and Fyhrie developed an algorithm which describes the bone as a material which optimises itself on the basis of local remodelling reactions to various technical failure criteria (Carter et al., 1987).

- Adaptive elasticity theory according to Cowin:

  Based on a thermodynamical continuum theory describing bone as a porous, elastic polyphase material filled with liquid, the driving force of the remodelling is the local state of deformation. In this formulation, a distinction is made between surface and internal remodelling (Cowin and Hegedus, 1976; Cowin, 1987).

None of the aforementioned bone adaptation models has yet achieved the final breakthrough. In our opinion, the main reason is that no bone adaptation model does justice to the fact that biological systems are open systems which, due to their metabolism, are in a state of lively exchange of energy and matter with their surroundings. They are also denoted systems far from thermodynamic equilibrium. Moreover, biological systems impress us by their filigree structure, their manifold functions and their enormous complexity. The extraordinarily high degree of coordination of the individual sub-systems is a particular characteristic of biological systems – all features which do not exactly alleviate our task of evolving adequate models.

In the 1960s and 1970s, theories were indeed developed in various disciplines independently of one another; these deal with the dynamics and spontaneous structure formation, the adoption of more complex forms by differentiation and the hierarchical structure of systems. Key concepts in this new, rather evolutionary interpretation of the physical world are self-organisation, dissipative structures, non-linearity, feedback, instability, bifurcation, order parameters, slaving etc. (Haken and Graham, 1971; Prigogine, 1979; Haken and Wunderlin, 1991). The fact that these new insights, here formulated verbally, led to fundamentally new scientific concepts, even to new branches of sciences – such as synergetics (see the survey in section 6.8) – demonstrates how ripe the time was for a renewal of classical dynamics by including the dynamics of biological systems.

Let us elucidate this by means of an example from morphogenesis (the formation of the structure of living organisms respectively their organs). An experiment on the formation of biological structures which is frequently cited is the regeneration of the hydra. The hydra is a freshwater polyp, a few millimetres in length and made up of approximately 100,000 cells. It has a head and one foot, i.e. it is of polar structure. If the hydra is divided across the middle, two new creatures are created; a head grows on the foot and a foot regenerates on the head. The question now arises how one and the same tissue cell on the surface of the division knows whether it should create a head or a foot. The decision to form a head or foot cannot originate in a local characteristic of the tissue, but must be based on the communication of the whole tissue structure. Further experiments can perhaps yield information on the mechanisms at work here. If the tissue of the foreign head is implanted in the central part of a hydra, the foreign head recedes if it is near enough to the hydra's own head. If, on the other hand, it is far enough away, the newly implanted head cells evolve into a complete head. This means that the growth of a new head is stopped, hindered respectively inhibited at a small distance from the head. On the other hand, at a sufficiently large distance from the old head, a new head is formed by activation of cells.

In 1972, A. Gierer and H. Meinhardt created a mathematical model, the computer results of which reproduced the foot and head formation of the hydra. Their model was based on the assumption that morphogenetic fields control the structure formation due to their inhomogeneous concentration distribution. The experiments on head and foot regeneration suggest that at least two chemical substances respectively "reactants" must be responsible for the formation process: an activator and an inhibitor.

The following considerations should be taken into account in a mathematical model: in order that the process start, both substances should be produced in the head region, for example. Moreover, the experiments demonstrate that the inhibition takes effect at a certain distance from the head already present, i.e. the inhibitor must be able to diffuse. The activator must he capable of this as well; otherwise, it could not activate the cells next to the transplant to form a head region.

For the mathematical formulation of this biological process, we denote the concentration of the activator by a, that of the inhibitor by h. The spatial dependence of a and h is defined by a coordinate x. Gierer and Meinhardt have suggested the following partial differential equations which they evaluated numerically

$$\frac{\partial a}{\partial t} = \rho + k\frac{a^2}{h} - \mu a + D_a \frac{\partial^2 a}{\partial x^2} \tag{10.1.1}$$

$$\frac{\partial h}{\partial t} = ca^2 - \nu h + D_h \frac{\partial^2 h}{\partial x^2} \tag{10.1.2}$$

The a^2-term acts autocatalytically, eq. (10.1.1), on the one hand and cross-catalytically on the other, the inhibitor being generated with the aid of the activator, eq. (10.1.2). The two terms $-\mu a$ and $-\nu h$ designate decay and degradation rates respectively, ρ is an activator-independent production term and the expressions $D_a\partial^2 a/\partial x^2$ and $D_h\partial^2 h/\partial x^2$ control the diffusion. A detailed deduction of these non-linear partial differential equations of the reaction-diffusion type can be found in (Haken, 1990) as well as in the original publication. If we compare the computer simulation with the experimental findings on hydra regeneration, the model reproduces the regulation capacity very well. Figure 10.1.1a illustrates the regeneration capacity, the renewal of the complete organism from parts, as exemplified by the hydra. Figure 10.1.1b presents results of the one-dimensional model calculation.

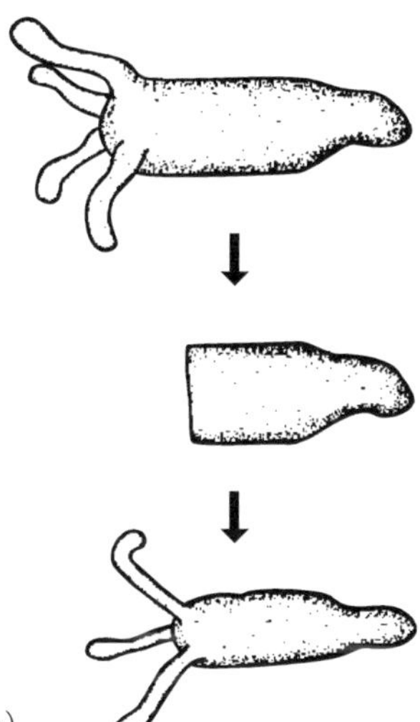
a)

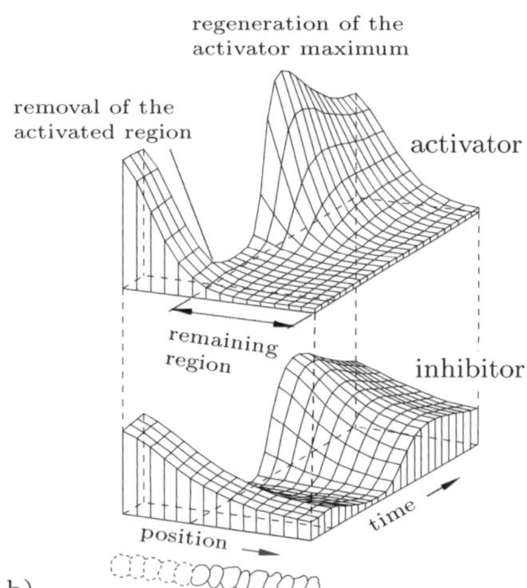

b)

Fig. 10.1.1: Regeneration capacity of the hydra, after (Meinhardt, 1987)
a) a head grows onto a foot,
b) for $t \to \infty$, the remaining region attains its old oncentration distribution (a, h)

In fig. 10.1.1b, we reproduce a plot of the spatio-temporal concentration distribution of the activator and inhibitor substances for the head region. The first four curves of a and h are the concentrations before the severing of the hydra. After the removal of the head region – the active zone for both the activator a and the inhibitor h – the activator values and above all the inhibitor values for the remaining region sink to such an extent that the autocatalysis can be sparked off again. The production process of a and h is resumed. For $t \to \infty$, an inhomogeneous concentration distribution again occurs which, for the reduced region, now coincides with the experimental findings for the regenerated hydra respectively with the concentration distribution in the hydra before the severance.

The computer results of Gierer and Meinhardt were confirmed analytically in (Granero et al., 1977) within the framework of the bifurcation theory. On the basis of the Gierer-Meinhardt model, the synergetic approach was demonstrated as well as its potential to describe spatial and temporal structure formations in biology in conjunction with the slaving principle and the reduction to few degrees of freedom, the so-called order parameters, in (Haken and Olbrich, 1978).

After this excursion into morphogenesis, we return to the real subject of this section, bone remodelling processes. On the basis of observations, it is known that bone reacts to alterations in loads with alterations in the density and in the cross-sectional area. The density alteration occurs very rapidly whereas the addition of new material at the periphery only takes place when it can be assumed the new loading is lasting. What is surprising is that nature follows the principle of a fast short-term and a slow long-term reaction in bone remodelling.

Further support for a model formulated from the point of view of dynamical systems is supplied by the studies in (O'Connor et al., 1982; Rubin and Lanyon, 1984), which demonstrate that purely constant loadings do not represent an appreciable stimulus for bone remodelling while short-term dynamic loading causes a distinct increase in bone growth. Even dynamic loading with extremely low daily numbers of loading cycles (only 36) leads to considerable remodelling reactions. Although contrary observations have been published (Hassler et al., 1980), we believe that dynamics is the trigger for a mechanical stimulus and that the theory of non-linear dynamical systems offers the appropriate tools for modelling bone remodelling processes.

In the following first exploration, we are concerned exclusively with the long-term behaviour of a non-linear dynamical system of two variables. We are firstly interested in the coexistence of possibly different attractor types, such as point attractors and limit cycles, in the phase space. Systems with several attractors in the phase space indicate different stationary behaviour for alternative initial conditions, a desirable phenomenon for bone remodelling. Secondly, we study the dependence of the attractor types on the control parameters.

As already mentioned, bone remodelling means both the alteration of the material behaviour as well as the alteration of the cross-sectional area of the bone, i.e. of the geometry of the inner and outer edges. Since (Pauwels, 1960; Pauwels, 1965), we know that the bone cell perceives the hydrostatic pressure as a stimulus due to a change in volume on the one hand and registers the first principal strain of the strain deviator as a stimulus due to a change in shape on the other (Faust et al., 1986). From elastomechanics, we know the decomposition of isotropic material behaviour in a bulk modulus which sets up a relation between the volumetric change and the hydrostatic pressure or tension, and a shear modulus which interrelates the change in shape with the shear stress. The idea now is to apply the activator-inhibitor model to the temporal change in the increase or decrease of the material characteristics of a bone structure and its strain response. The independent variables of our system are the shear modulus q_1, the bulk modulus q_2, the first principal strain of the strain deviator q_3 and the volumetric strain q_4. In the context of dynamical systems, we assemble the q_i-variables in the vector $\boldsymbol{q} = \{q_1\ q_2\ q_3\ q_4\}$. The space spanned by all the possible vectors $\boldsymbol{q}$ then forms the phase space. We can now write the evolution equation of the vector $\boldsymbol{q}$ in the following general form

$$\dot{\boldsymbol{q}} = \boldsymbol{F}(\boldsymbol{q}, \nabla, \boldsymbol{\mu}) \tag{10.1.3}$$

Equation (10.1.3) expresses the fact that the temporal evolution $\dot{\boldsymbol{q}}$ is a non-linear function of the current state $\boldsymbol{q}$ of the system. The spatially inhomogeneous distribution of $\boldsymbol{q}$ is symbolised by ∇ and the vector $\boldsymbol{\mu}$ comprises the control parameters which take into account that the system is open. The real challenge to both experimenter and theoretically orientated biomechanic expert is the formulation of eq. (10.1.3) for the concrete case of bone remodelling. What unfortunately is missing or is unknown to us are controlled laboratory experiments on temporally varying loading cases and the reaction of the bone to these in time and space.

In the absence of such experimental findings, the only possibility is to run first simple models on the computer and to examine them for their relevance with regard to bone remodelling on the basis of long-term behaviour and bifurcation patterns, cf. (Schneider et al., 2009; Helwig et al., 2009).

In the two following simplest studies, we reduce the system of eqs. (10.1.3) with four independent variables to a two-dimensional system and also disregard the spatial dependence of $\boldsymbol{q}$. As independent variables, we select the shear modulus q_1 and the bulk modulus q_2 and write the evolution equations for the material behaviour in the following form

$$\begin{aligned} \dot{q}_1 &= \gamma_1 q_1 + a_1 q_1 q_2 - a_2 q_2^2 \\ \dot{q}_2 &= -\gamma_2 q_2 + b q_1^2 \end{aligned} \tag{10.1.4}$$

where a certain arbitrariness can be discerned in the choice of the linear and non-linear "production" and "loss" terms. It proves advisable to introduce new variables and to reduce the number of the parameters by the transformations

$$t' = \gamma_2 t, \qquad q_1 = \sqrt{\frac{\gamma_1\gamma_2}{a_1 b}}\, x_1, \qquad q_2 = \frac{\gamma_1}{a_1} x_2$$

We thus obtain the system

$$\begin{aligned} \dot{x}_1 &= A x_1 + A x_1 x_2 - B x_2^2 \\ \dot{x}_2 &= -x_2 + x_1^2 \end{aligned} \tag{10.1.5}$$

where we have introduced the abbreviations

$$A = \frac{\gamma_1}{\gamma_2}, \qquad B = \frac{a_2\gamma_1}{a_1\gamma_2}\sqrt{\frac{b\gamma_1}{a_1\gamma_2}}$$

To simplify the argument, we set $B = 1$. Equation (10.1.5) is then reduced to

$$\begin{bmatrix} \dot{x}_1 \\ \dot{x}_2 \end{bmatrix} = \begin{bmatrix} A & 0 \\ 0 & -1 \end{bmatrix} \begin{bmatrix} x_1 \\ x_2 \end{bmatrix} + \begin{bmatrix} Ax_1x_2 - x_2^2 \\ x_1^2 \end{bmatrix} \tag{10.1.6}$$

For eq. (10.1.6), we first determine the states of equilibrium respectively singular points $\boldsymbol{x}_s$ which are defined by $\dot{\boldsymbol{x}} = \mathbf{0}$. We thus obtain from eq. (10.1.6) the following equations for the fixed points

$$\begin{aligned} Ax_1 + Ax_1x_2 - x_2^2 &= 0 \\ -x_2 + x_1^2 &= 0 \end{aligned} \tag{10.1.7}$$

respectively

$$x_2 = x_1^2 \tag{10.1.8}$$

$$x_1(A + Ax_1^2 - x_1^3) = 0 \tag{10.1.9}$$

Equations (10.1.8) and (10.1.9) show that the origin is a fixed point $F_1(0,0)$, independent of the value of the control parameter A. In order to determine further fixed points, we discuss the cubic expression in eq. (10.1.9). Substituting

$$y = x_1 - \frac{A}{3} \tag{10.1.10}$$

we obtain the normal form

$$y^3 + 3py + 2q = 0 \tag{10.1.11}$$

with

$$q = -\left(\frac{A^3}{27} + \frac{A}{2}\right), \quad p = -\frac{A^2}{9}$$

The sign of the discriminant $D = q^2 + p^3$ decides the number of real solutions in dependence of A. In the present case,

$$D = \frac{A^4}{27} + \frac{A^2}{4} \geqslant 0 \tag{10.1.12}$$

i.e. eq. (10.1.11) possesses only one real solution for $-\infty < A < +\infty$, namely the fixed point $F_2(x_{12}, x_{22})$.

The coordinates x_{12}, x_{22} of F_2 depend on the control parameter A

$$\begin{aligned} x_{12} &= A^{1/3}\left[\left(A_1 + \sqrt{A_2}\right)^{1/3} + \left(A_1 - \sqrt{A_2}\right)^{1/3}\right] + \tfrac{1}{3}A \\ x_{22} &= x_{12}^2 \end{aligned} \tag{10.1.13}$$

where the following abbreviations are used

$$A_1 = \tfrac{1}{27}A^2 + \tfrac{1}{2} \qquad \text{and} \qquad A_2 = \tfrac{1}{27}A^2 + \tfrac{1}{4}$$

It is apparent from eq. (10.1.13) that the fixed point F_2 for $A < 0$ lies in the second quadrant of the phase plane x_1, x_2, for $A = 0$ at the origin and for $A > 0$ in the first quadrant.

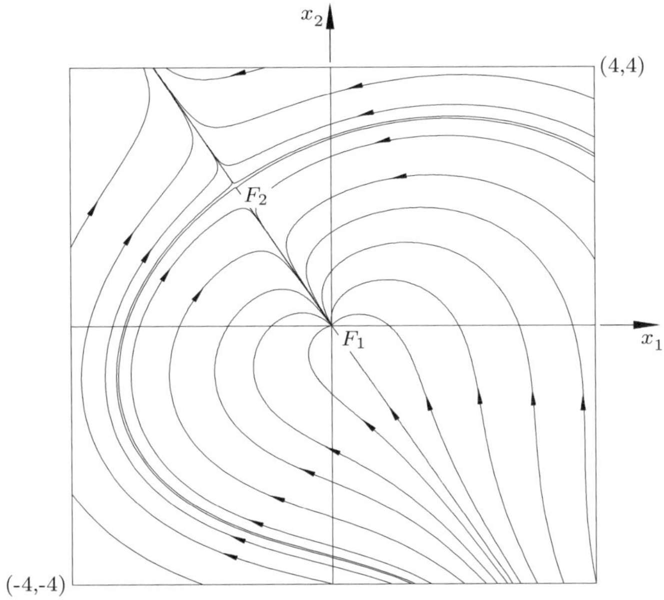

Fig. 10.1.2

Phase portrait of eq. (10.1.5) for $A = -1$

We now investigate the stability behaviour of the two fixed points $F_1(0,0)$ and $F_2(x_{12}, x_{22})$. To this end, we determine the Jacobian matrix $\frac{\partial \boldsymbol{F}}{\partial \boldsymbol{x}}|_{F_1,F_2}$ for the system of equations (10.1.6)

$$\left.\frac{\partial \boldsymbol{F}}{\partial \boldsymbol{x}}\right|_{F_1} = \begin{bmatrix} A & 0 \\ 0 & -1 \end{bmatrix}, \qquad \left.\frac{\partial \boldsymbol{F}}{\partial \boldsymbol{x}}\right|_{F_2} = \begin{bmatrix} A(1+x_{22}) & Ax_{12} - 2x_{22} \\ 2x_{12} & -1 \end{bmatrix} \tag{10.1.14}$$

We first discuss the stability behaviour of F_1 and consider eq. (10.1.14). The Jacobian matrix is a diagonal matrix, i.e. the diagonal elements are simultaneously the eigenvalues of the linearised system. For $A < 0$, both real eigenvalues are negative. The fixed point $F_1(0,0)$ is consequently a stable node (see figs. 3.1.2 and 10.1.2). For the case $A > 0$, the two real eigenvalues possess opposite signs, i.e. the fixed point $F_1(0,0)$ is a saddle point (see figs. 3.1.3 and 10.1.4).

The calculation of the eigenvalues for the determination of the stability of the fixed point $F_2(x_{12}, x_{22})$ on the basis of eq. (10.1.14) requires somewhat more effort. The variation of the eigenvalues λ_1, λ_2 of the linearised system in dependence of the control parameter A is presented in fig. 10.1.3.

In the range $-\infty < A < 0$, the signs of the two real eigenvalues λ_1, λ_2 are different, i.e. the fixed point F_2 is a saddle point (see fig. 10.1.2). At $A = 0$, λ_1 changes its sign, both real eigenvalues are negative for $0 < A < 0.067$, thus F_2 becomes a stable node. F_2 has changed from a saddle point to a stable node as A passes through the origin.

In the range $0.067 < A < 1.5543$, the two eigenvalues λ_1, λ_2 are complex conjugate. The fixed point F_2 is consequently a focus (cf. fig. 10.1.4). For $0.067 < A < 0.5$, both real parts are negative, i.e. the focus is stable (cf. fig. 10.1.4a). At $A = 0.5$, the real parts of λ_1 and λ_2 become positive, the focus F_2 is now unstable. From our

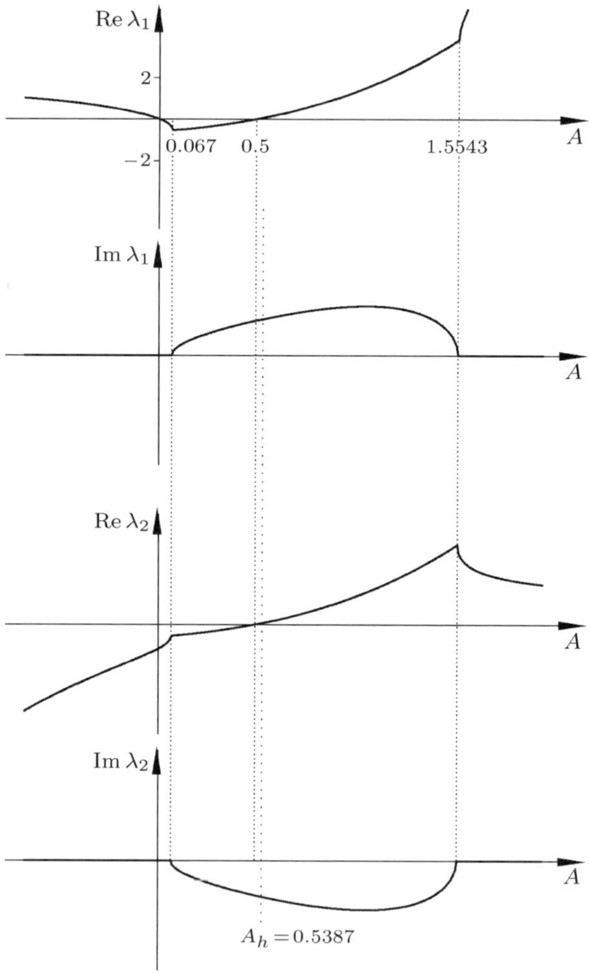

Fig. 10.1.3: Variation of the eigenvalues λ_1, λ_2 of the fixed point $F_2(x_{12}, x_{22})$ of eq. (10.1.5) in dependence of the control parameter A

argumentation in section 6.4, part D, we know that, if a pair of complex conjugate eigenvalues crosses the imaginary axis, proceeding from the negative real part to the positive one, then this generates a supercritical Hopf bifurcation (cf. fig. 6.4.7). Figure 10.1.4b presents the phase portrait for $A = 0.505$. The course of the trajectory illustrates the unstable character of F_2 on the one hand and the existence of a stable limit cycle which begins to emerge at $A = 0.5$ on the other.

As has already been mentioned, a Hopf bifurcation takes place at $A = 0.5$. With increasing A, the limit cycle expands steadily. For $A_h = 0.5387$, it turns into a homoclinic orbit of the saddle point at the origin (fig. 10.1.4c). A further increase of A causes this homoclinic loop to break open and the limit cycle thus abruptly stops existing (fig. 10.1.4d). This type of bifurcation which does not emerge due to

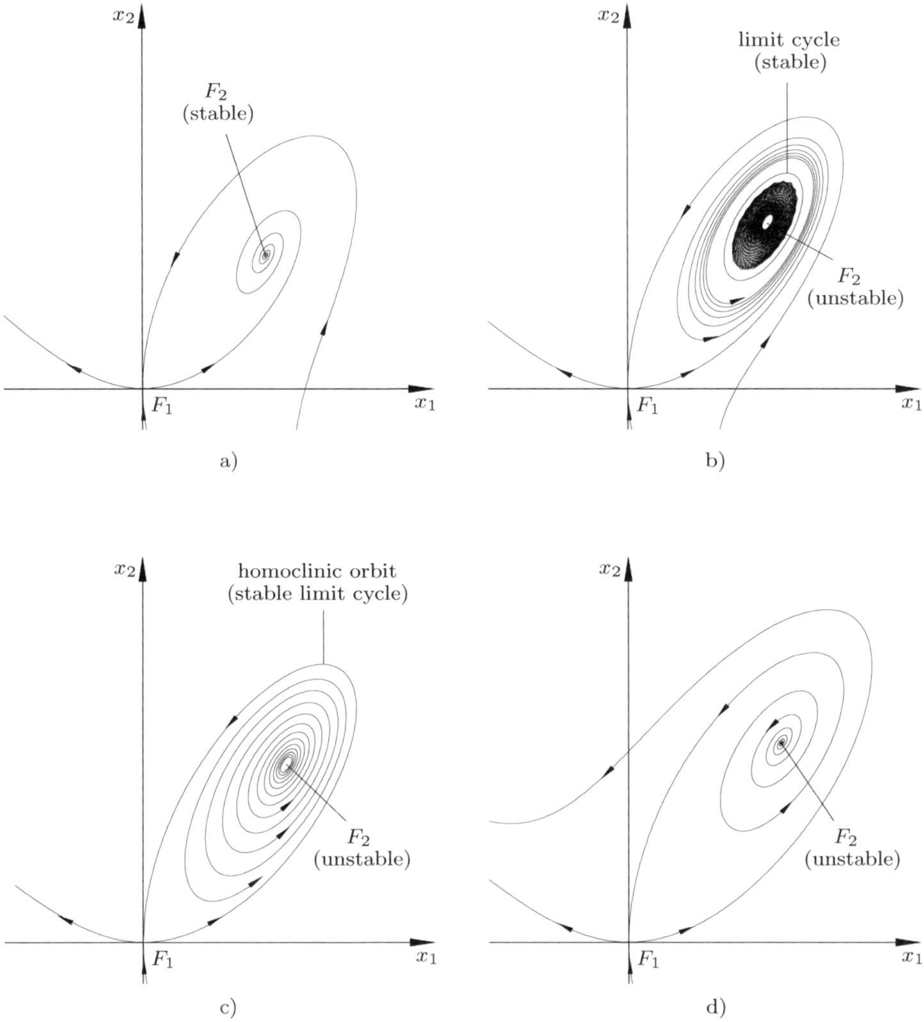

Fig. 10.1.4: Four phase portraits of eq. (10.1.5) for $A > 0$:
a) $A = 0.4$, b) $A = 0.505$, c) $A_h = 0.5387$, d) $A = 0.6$

the eigenvalue constellation (cf. fig. 10.1.3), but due to global characteristics of the flow is called a homoclinic bifurcation. In the framework of the global bifurcation theory, cf. (Wiggins, 1988; Guckenheimer and Holmes, 1983), this very simple type of bifurcation of two-dimensional flows plays an important role.

The discussion of eq. (10.1.6) with respect to attractors in dependence of the control parameter A makes it clear that stable long-term behaviour exists only for a very small range of positive A-values, namely $0 < A < 0.5387$, and for a limited range of initial conditions. If we interpret the structure of the phase space as a temporal "indicator", on the basis of which bone remodelling processes take place, then these processes with

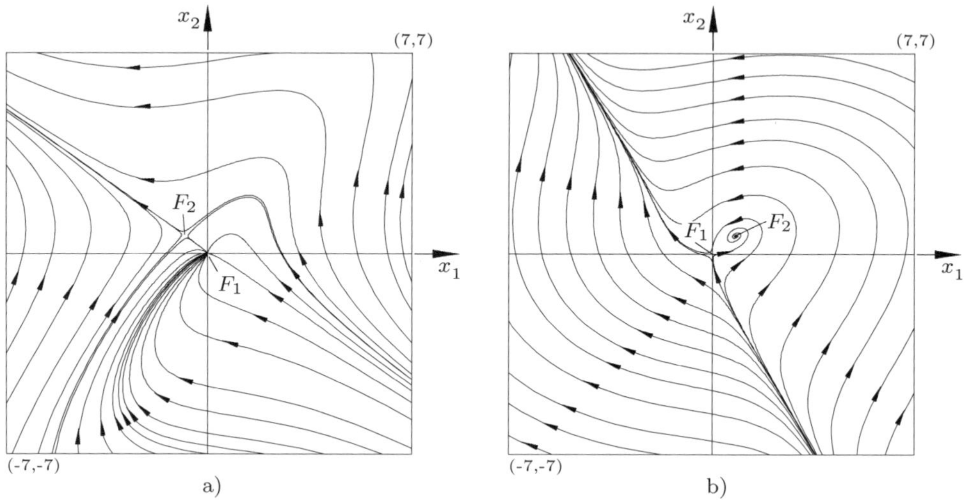

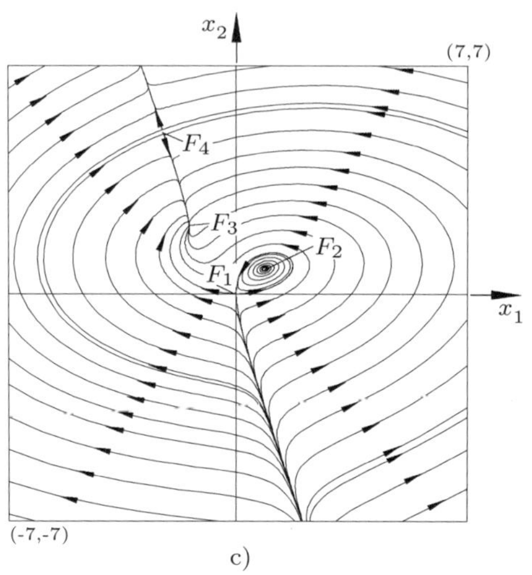

Fig. 10.1.5

Three phase portraits of eq. (10.1.15) for different parameter values A:
a) $A = -1.3$, b) $A = 1.3$, c) $A_h = 2.7935$

stable long-term behaviour are restricted to a very small parameter range, particularly as regards the fact that the parameter is an experimentally determinable quantity. A possible "ansatz" leading out of the dilemma of a much too narrow parameter range could be provided by the alternative system

$$\begin{aligned} \dot{x}_1 &= Ax_1 - Ax_1x_2 - x_2^2 \\ \dot{x}_2 &= -x_2 + x_1^2 \end{aligned} \tag{10.1.15}$$

where the x_1-rate is not amplified by the x_1x_2-term, as in eq. (10.1.6), but reduced.

If we determine the fixed points for eq. (10.1.15), we can also demonstrate here that the origin, independent of A, is a fixed point, $F_1(0,0)$. The additional fixed points are solutions of a cubic equation. The discriminant D is greater than zero for the

a) b)

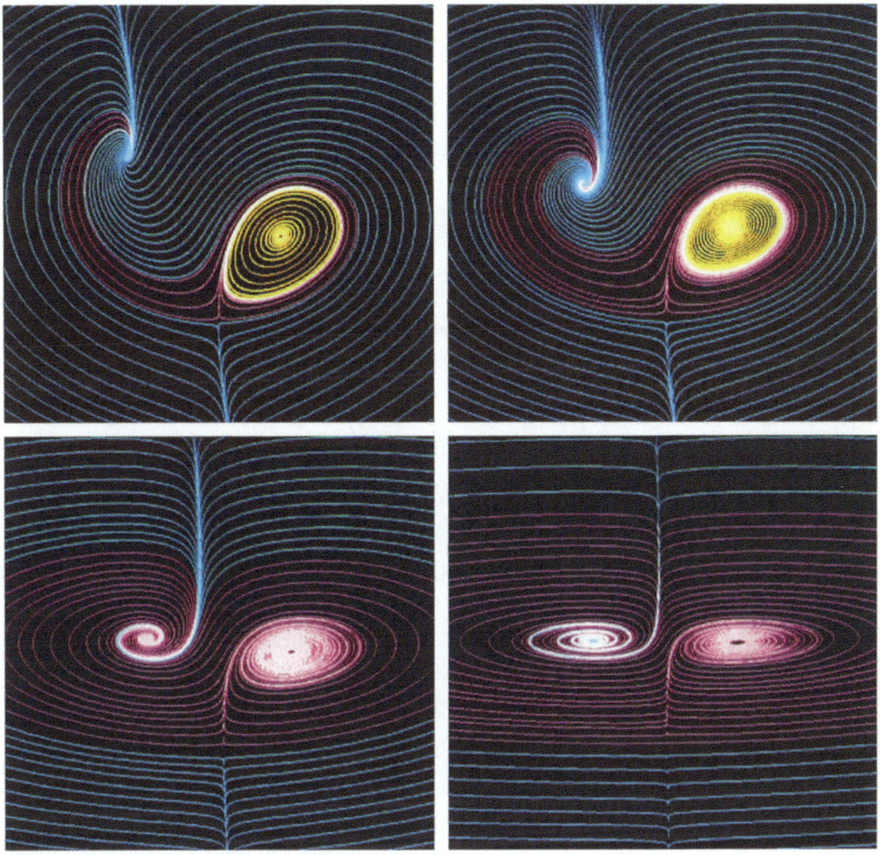

c) d)

Plate XVII Coexistence of attractors for eq. (10.1.15) for different values of A:

a) $A = 3.5$ b) $A = 5.0$

c) $A = 14.0$ d) $A = 50.0$

range $-\sqrt{27/4} < A < \sqrt{27/4}$, i.e. in this A-range, only one real solution exists with the exception of $A = 0$. Furthermore, we have $D < 0$ for $|A| > \sqrt{27/4} \approx 2.598$; all three solutions are hence real. Figure 10.1.5 presents the three phase portraits of eq. (10.1.15) for the parameter values $A = -1.3$, 1.3 and 2.7935. The two diagrams a) and b) show courses of the trajectories for the regime $D > 0$. We can recognise the fixed points $F_1(0,0)$ and $F_2(-0,79,0.62)$ in the second, respectively $F_2(0.79,0.62)$ in the first quadrant of the x_1,x_2-phase plane. As already mentioned, four fixed points, detailed in fig. 10.1.5c for $A_h = 2.7935$, exist for $|A| > 2.598$. In comparison with the other three fixed points, F_4 possesses large coordinate values which increase rapidly with increasing A. For $A = 50$, for example, we determine the following coordinate constellation: $F_1(0,0)$, $F_2(0.99,0.98)$, $F_3(-1.01,1.02)$ and $F_4(-49.98,2498.)$.

The linear stability analysis on the basis of the eigenvalues for $A = -1.3$, see fig. 10.1.5a, confirms the stable node F_1 ($\lambda_1 = -1.3$, $\lambda_2 = -1$), and the saddle point F_2 ($\lambda_1 = 1.16$, $\lambda_2 = -2.66$). At the zero crossing of A, both fixed points alter their stability characteristics. For $A = 1.3$, fig. 10.1.5b, F_1 is a saddle point ($\lambda_1 = 1.3$, $\lambda_2 = -1$) and F_2 a stable focus ($\lambda_1 = -0.25 + 1.74i$, $\lambda_2 = -0.25 - 1.74i$). The newly emerged fixed points in fig. 10.1.5c are a stable node F_3 ($\lambda_1 = -0.86$, $\lambda_2 = -3.1$) and a saddle point F_4 ($\lambda_1 = 0.44$, $\lambda_2 = -12.6$). For the numerically evaluated parameter value of $A_h = 2.7935$, a homoclinic loop forms at the saddle point at the origin, this loop breaking open for increasing A-values. As the loop opens, a limit cycle – this time an unstable one – detaches itself from the homoclinic orbit; trajectories of inner points of the limit cycle tend towards the stable focus F_2 (A_h: $\lambda_1 = -0.17 + 2.5i$, $\lambda_2 = -0.17 - 2.5i$). The real parts of the eigenvalues of F_2 remain negative; a Hopf bifurcation is thus excluded. The bifurcation is of global nature and corresponds to the simple type of a homoclinic bifurcation (Wiggins, 1988).

Colour Plate XVII, p. 691, presents four further phase portraits, namely for $A = 3.5$, 5, 14 and 50. Within the ranges $-3 \leqslant x_1 \leqslant 3$, $-1 \leqslant x_2 \leqslant 3$ selected here, the three fixed points F_1, F_2 and F_3 can be observed while F_4 comes to lie far outside. Colour Plates XVIIa,b show in yellow the unstable limit cycle generated by the homoclinic bifurcation as well as the trajectories moving inwards away from the limit cycle and towards the stable focus F_2. The trajectories starting from the limit cycle and moving outwards are coloured red. These first approach the saddle point $F_1(0,0)$ along the upper branch of its stable manifold and are finally repelled by F_1. Some reach the sphere of influence of the limit cycle temporarily, others are immediately forced away in the direction of the stable focus F_3, typically specified by

$$A = 3.5 : \quad F_3(-1.2462,\ 1.5529)\ ,\ \lambda_{1,2} = -1.47 \pm 1.7i$$

$$A = 5.0 : \quad F_3(-1.1378,\ 1.2946)\ ,\ \lambda_{1,2} = -1.24 \pm 2.6i$$

so that ultimately, all the trajectories coming from the outer side of the unstable limit cycle are captured by F_3.

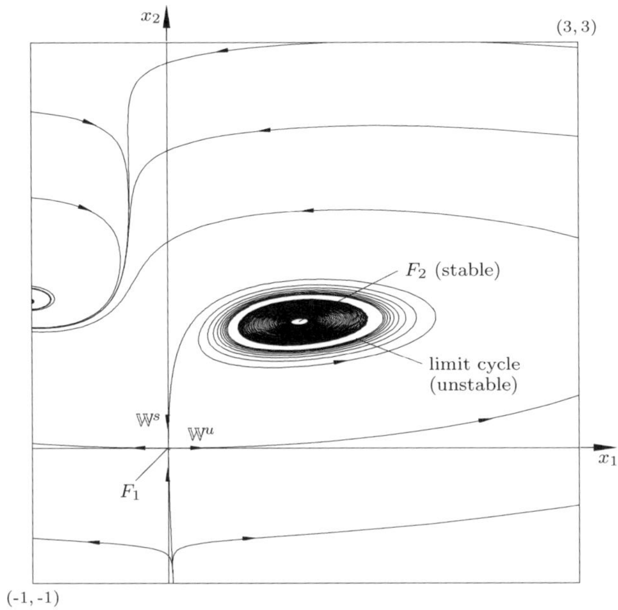

Fig. 10.1.6: Phase portrait of eq. (10.1.15) for $A = 14$: selective enlargement of Colour Plate XVIIc, p. 691

The blue trajectories start to the right and the left of the lower branch of the stable manifold of $F_1(0, 0)$. Both groups of trajectories finally end in the focus F_3; the flow into F_3 is determined by the lower branch of the unstable manifold of the non-visible saddle point F_4 (cf. fig. 10.1.5c).

The two remaining diagrams of Colour Plate XVIIc,d present the courses of the trajectories for $A = 14$ and $A = 50$. The impression evoked by these illustrations that the unstable limit cycle vanishes between $A = 5$ and $A = 14$ due to a subcritical Hopf bifurcation (cf. fig. 6.4.10) and that the formerly stable focus F_2 is now unstable is deceptive. In fact, a linear stability analysis for F_2 demonstrates that the real parts of both eigenvalues remain negative until $A = 50$ at least. Consequently, a stability change of F_2 does not take place; this in turn means that the unstable limit cycle must exist up to $A = 50$; this is confirmed by the selective enlargement of Colour Plate XVIIc in fig. 10.1.6.

The detailed discussion of eq. (10.1.15) shows that the parameter range of A, for which stable long-term behaviour occurs, is considerably more extensive than that of eq. (10.1.6). The second ansatz for the formulation of the material behaviour as a bone remodelling process appears more appropriate in this respect at least.

We return to the system of equations eq. (10.1.5) with the two control parameters A and B. On the basis of this system, we wish to demonstrate the various fixed point types and their stability for a particular fixed point in the first quadrant of the phase space. The fixed point conditions are

$$\begin{aligned} A(x_1 + x_1x_2) - Bx_2^2 &= 0 \\ -x_2 + x_1^2 &= 0 \end{aligned} \tag{10.1.16}$$

At this point, a further reduction of the control parameters to a single parameter C is possible

$$C = \frac{A}{B} = \frac{a_1}{a_2}\sqrt{\frac{a_1\gamma_2}{b\gamma_1}} \tag{10.1.17}$$

In this way, eq. (10.1.16) is adapted to the type of eq. (10.1.7).

The coordinate origin is the fixed point F_1. We concentrate on the only other real fixed point $F_2(x_{12}, x_{22})$. In accordance with eq. (10.1.13), its coordinates are

$$\begin{aligned} x_{12} &= C^{1/3}\left[\left(C_1 + \sqrt{C_2}\right)^{1/3} + \left(C_1 - \sqrt{C_2}\right)^{1/3}\right] + \tfrac{1}{3}C \\ x_{22} &= x_{12}^2 \end{aligned} \tag{10.1.18}$$

with the analogous abbreviations

$$C_1 = \tfrac{1}{27}C^2 + \tfrac{1}{2} \qquad \text{and} \qquad C_2 = \tfrac{1}{27}C^2 + \tfrac{1}{4}$$

For demonstration purposes, we select $C = 10$. The coordinates of the second singular point F_2 are then

$$x_{12} = 10.0981, \quad x_{22} = 101.971$$

Our interest is now directed at the trajectory course in the neighbourhood of F_2, in dependence of positive A-values. This is associated with the question regarding the stable and unstable states for various A-values and the eigenvalue constellations responsible for this (fig. 3.1.8). The result corresponds to that of eq. (10.1.6) and is summarised in fig. 10.1.7.

Taking eq. (10.1.17) into consideration, we have calculated explicitly trajectories for different initial conditions numerically for the system of equations (10.1.5) – for increasing A-values but for a fixed C-value ($C = 10$) – and plotted them in colour in the x_1, x_2-phase space in order to illustrate the evolution process of the fixed point F_2 and its neighbourhood. The result is presented in Colour plates XVIII and XIX, pp. 696, 697.

Let us begin with Colour Plate XVIIIa, p. 696. For $A = 0.0001$, the stationary point F_2 is a stable node. The trajectories which all begin at $x_2 = 0$ end in a parabola with the exponent k > 1; the parabola starts at the origin, follows the unstable manifold of F_1 and runs directly towards the fixed point F_2. The other illustrations (Colour Plate XVIIIb-e) show the progressing change from a stable node to a stable focus (Colour Plate XVIIIe); the spiraling merging of the trajectories into the fixed point emerges with growing clarity for increasing A. The change from node to focus takes place at the analytically calculated value $A = 0.00164$ (fig. 10.1.7).

In Colour Plate XIXg,h, the basin of attraction of F_2 is demonstrated by the colours green and red in addition to the evolution of F_2. The trajectories starting in the section selected here on the x_1-axis and ending in the focus F_2 are coloured green. The red trajectories tend towards infinity. Both basins of attraction are divided by

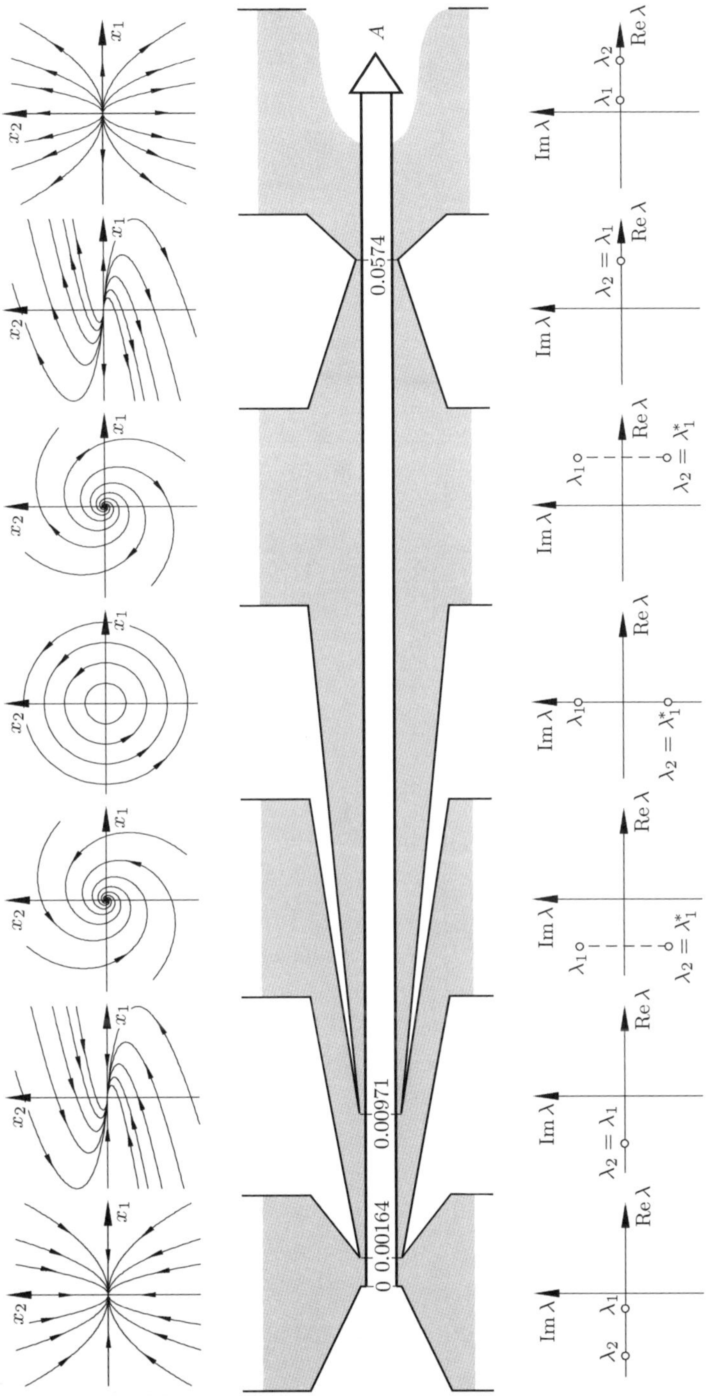

Fig. 10.1.7: Stability behaviour of the singular point F_2, eq. (10.1.18), for $C = 10$ in dependence of the control parameter A

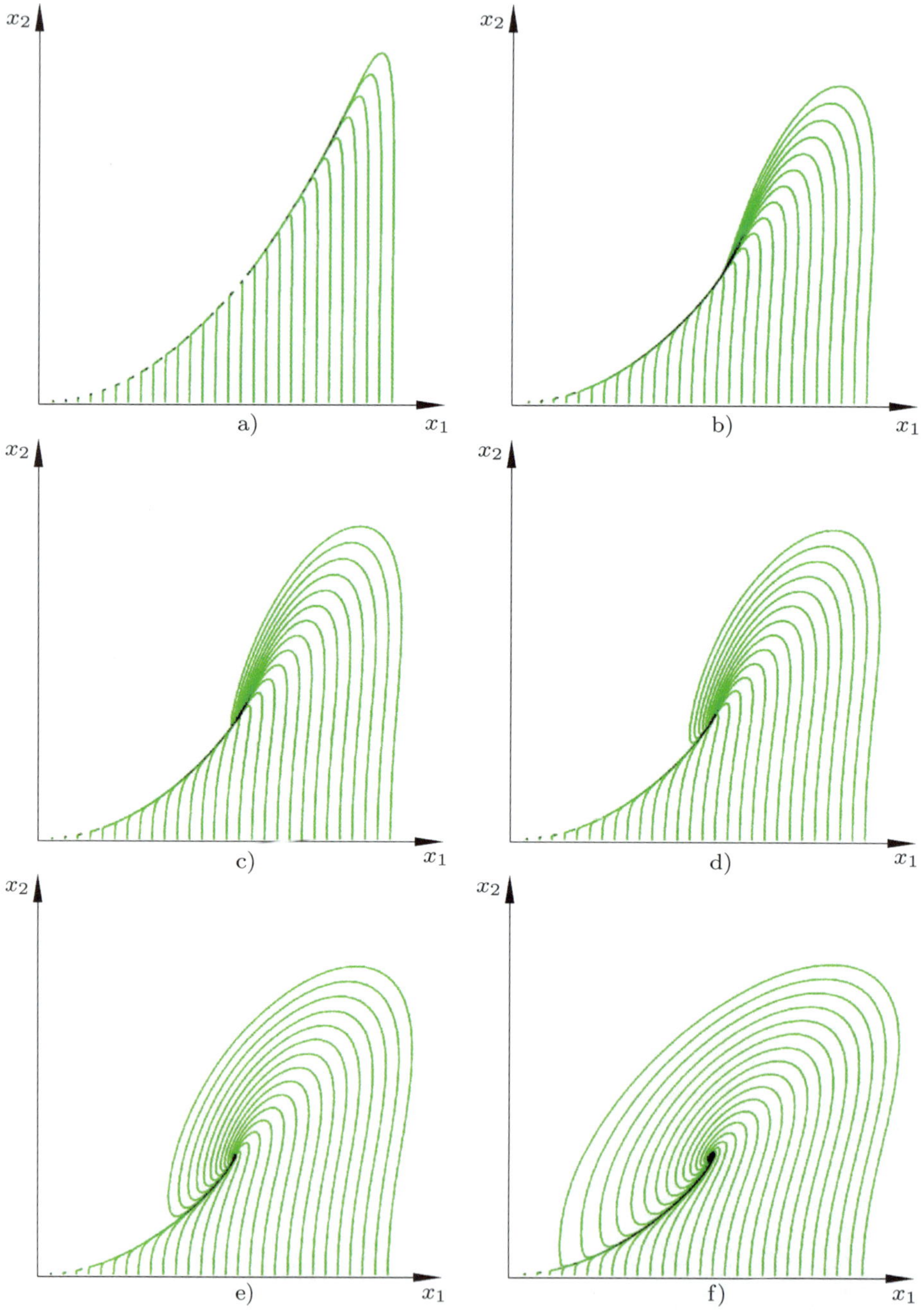

Plate XVIII Phase portraits of eq. (10.1.5) for $C = A/B = 10$ and for different values of A:
a) $A = 0.0001$ b) $A = 0.0008$ c) $A = 0.0012$
d) $A = 0.0016$ e) $A = 0.0025$ f) $A = 0.0038$

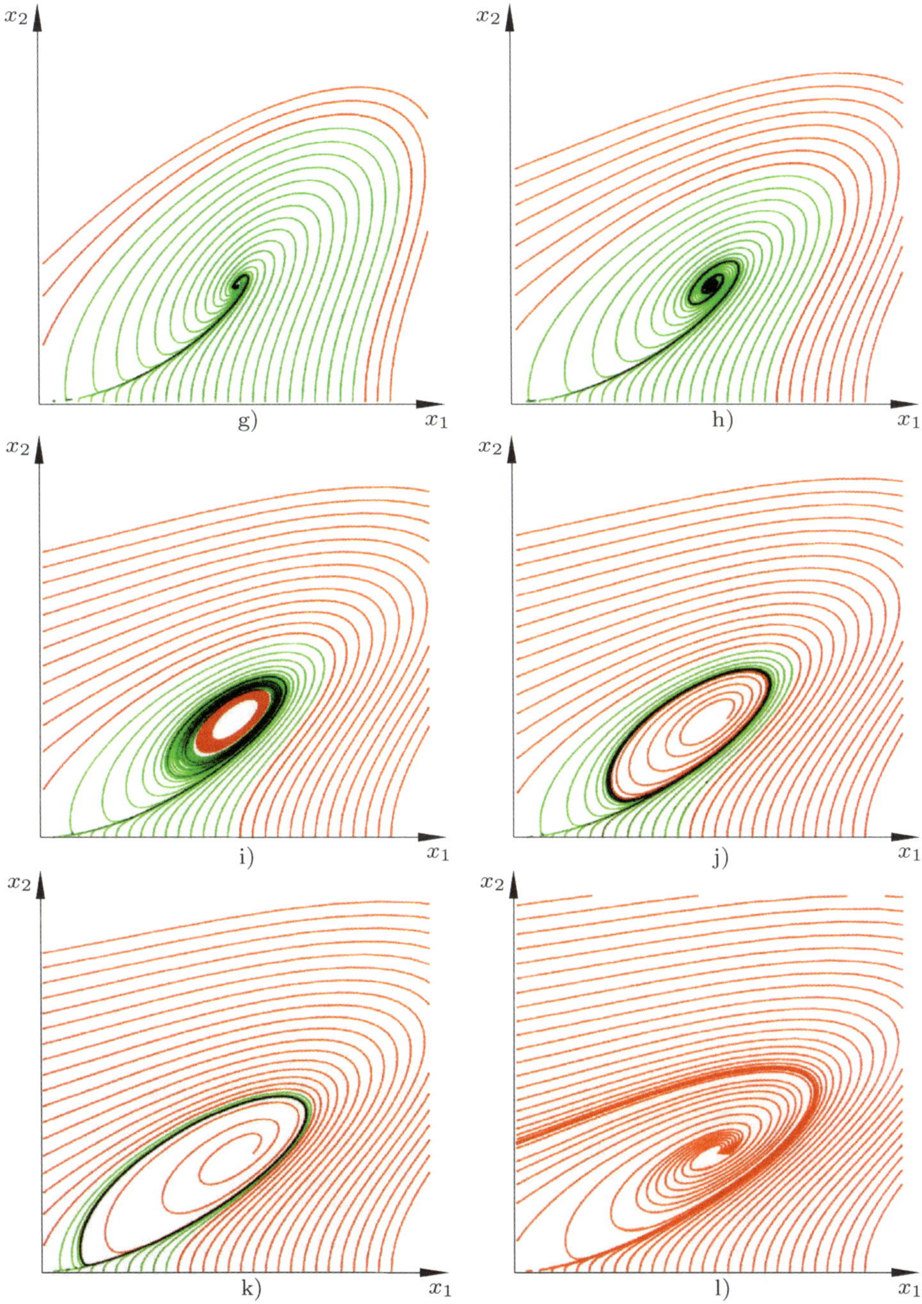

Plate XIX Phase portraits of eq. (10.1.5) for $C = A/B = 10$ and for different values of A:

g) $A = 0.0051$ h) $A = 0.0075$ i) $A = 0.01$

j) $A = 0.011$ k) $A = 0.012$ l) $A = 0.015$

a separatrix (cf. section 3.3). At a further increase of A, the basin of attraction of the point attractor shrinks (Colour Plate XIXh for $A = 0.0075$).

Colour Plate XIXi demonstrates the classic example of a Hopf bifurcation; for the value $A = 0.01$, a stable limit cycle has emerged. The stable state of equilibrium of eq. (10.1.5) has been succeeded by a stable oscillation. The singular point F_2 changes from the state of the stable focus to that of an unstable one.

The bifurcation occurs at the analytically determined value $A = 0.00971$ (see fig. 10.1.7). Subsequently, the limit cycle expands appreciably, and its outer basin of attraction shrinks perceptibly. The limit cycle has the tendency to form a homoclinic orbit of the saddle point F_1 (Colour Plate XIXj,k). In the final illustration, Colour Plate XIX l, the limit cycle has vanished for $A = 0.015$ due to a global bifurcation, the homoclinic bifurcation (cf. figs. 10.1.4c, 10.1.5c). The whole x_1, x_2-phase space is now filled with unstable trajectories, i.e. from the instant of the homoclinic bifurcation, the initial system, eq. (10.1.5), becomes unstable.

We have called upon both these very simple two-dimensional dynamical systems of eqs. (10.1.5) and (10.1.15) to demonstrate typical characteristics of non-linear systems – such as attractors, coexistence of attractors, basins of attraction and bifurcations in dependence of control parameters which become critical. We recall that these systems of equations were set up with the aim of modelling bone adaptation. It is obvious that the formulations discussed here are still far from simulating bone growth as the assumptions made are highly restrictive.

The examples from morphogenesis and biomedicine mentioned here underline the necessity for investigating spatio-temporal self-organisation through a study of dynamical systems. We can only hope that the interdisciplinary character of this subject will prove a stimulus to the scientific fields involved.

10.2 Hénon Map

Everything should be made as
simple as possible, but not simpler

Albert Einstein

Mixing processes are among the most characteristic attributes of turbulent flows. Initially closely neighbouring fluid particles are rapidly separated from each other, leading to a total mixing of the fluid. This mixing process is closely connected with the repeated stretching and folding of stream surfaces and stream lines in two- and three-dimensional turbulence. If we consider the onset of turbulence for the example of the evolution of a von Kármán vortex street (fig. 10.2.1), we observe that initially adjacent fluid particles are driven apart from each other by stretching whereas originally far distant particles approach each other by folding mechanisms. In section 9.2.2.4, we studied the motion of passive tracer particles in a chaotic four-vortex system. Colour Plate XIVB, p. 617, illustrates the complete mixing of the initial conditions, which have been selected from two quadratic configurations

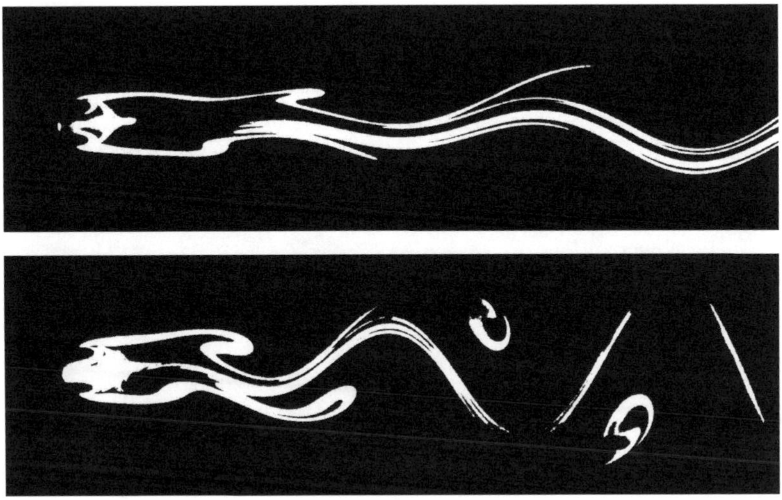

Fig. 10.2.1: Evolution of a von Kármán vortex street

marked by different colours. Here again, the original squares are stretched and folded, causing a complete mixing of the trajectories after a short time.

Colour Plate IV, p. 286, illustrates the fact that similar mechanisms evidently also generate chaotic, unpredictable behaviour in dynamical systems. Even the simple weather model of Lorenz is capable of simulating this typical sensitive behaviour: initial perturbations are amplified greatly and, after a short time, yield completely differing temporal evolutions. In this series of illustrations, the Lorenz attractor is depicted in the background. In the first figure, 15 000 minutely different initial conditions are marked for the instant $t = 0.0$. They are packed so closely on a tiny straight line about a yellow point that we can only discern them individually under a microscope. A series of snapshots shows the temporal evolution of the 15 000 trajectories in the phase space; the respective location at any one instant appears as a point. The orbits originally emanating from a yellow "point" have pulled apart and are so far away from one another macroscopically for $t = 6.7$ that the points in the snapshot are strung along a yellow line. The yellow line is extended continuously; at the instant $t = 8.2$, the process of folding begins. The yellow line itself still appears as a connected curve. At the instant $t = 9.3$, the exponential divergence of neighbouring trajectories becomes apparent, the connected yellow line breaks up, the discrete character of the individual orbits becomes evident in clearly distinguishable line segments. We recognise that the folding process has started. At the instant $t = 11.2$, complete mixing due to stretching and folding can no longer be halted. The final state of complete mixing is attained for $t = 100$. It would indeed be audacious to speak of a temporal predictability after the original 15 000 initial points have been spread over the whole Lorenz attractor like a cloud as a result of repeated stretching and folding.

The series of illustrations clearly demonstrates that the mixing is achieved by repeated stretching and folding and that as a result, microscopic fluctuations, tiny deviations in the initial conditions, are magnified on a macroscopic scale and thus become visible to the observer. Nevertheless, we should refrain from reasoning by analogy and drawing hasty conclusions. While the stretching and folding process in fig. 10.2.1 takes place in the real physical space, Colour Plate IV in fact reproduces a purely temporal process in the abstract phase space.

It was M. Hénon and Y. Pomeau's idea to construct a model for the simple description of mixing as arising for example on the Lorenz attractor (Hénon and Pomeau, 1976; Hénon, 1976). They proposed a discrete map which, like the Poincaré map of the Lorenz system, was to be two-dimensional and invertible as well as guaranteeing a uniform shrinking of the phase space. Apart from the reduction of the dimension, a discrete system has the advantage that no numerical integrations are required. In this way, the computing time is reduced drastically. It is thus very easy to follow the dynamics of the system through many iterations with the aid of a programmable pocket calculator; an accumulation of numerical errors by integration procedures is excluded.

The basic idea is to divide each iteration step into three mapping steps reproducing the mechanism of stretching and folding. Figure 10.2.2 illustrates the three sub-mappings T', T'', T'''. In the first step, a domain of the $(x_\mathrm{n}, y_\mathrm{n})$-plane is stretched and folded by the following rule

$$T': \qquad x'_\mathrm{n} = x_\mathrm{n} \tag{10.2.1}$$

$$y'_\mathrm{n} = y_\mathrm{n} + 1 - a x_\mathrm{n}^2 \tag{10.2.2}$$

whereby the selection of the parameter a is still open. This part of the mapping is area-preserving. In the second step, the folding process is intensified by a contraction in the x-direction

$$T'': \qquad x''_\mathrm{n} = b x'_\mathrm{n} \tag{10.2.3}$$

$$y''_\mathrm{n} = y'_\mathrm{n} \tag{10.2.4}$$

where $0 < b < 1$. Each iteration step ends with a mirroring at the first bisector, returning the domain to its original orientation

$$T''': \qquad x_\mathrm{n+1} = y''_\mathrm{n} \tag{10.2.5}$$

$$y_\mathrm{n+1} = x''_\mathrm{n} \tag{10.2.6}$$

The compound mapping $T = T'''T''T'$ can then be expressed as follows

$$T: \qquad x_\mathrm{n+1} = y_\mathrm{n} + 1 - a x_\mathrm{n}^2 \tag{10.2.7}$$

$$y_\mathrm{n+1} = b x_\mathrm{n} \tag{10.2.8}$$

T is called the Hénon map; it evidently possesses a unique inverse mapping T^{-1} and is hence a diffeomorphism. We next evaluate the determinant of the Jacobian matrix and obtain

$$\det \boldsymbol{D} = \begin{vmatrix} -2a x_\mathrm{n} & 1 \\ b & 0 \end{vmatrix} = -b \tag{10.2.9}$$

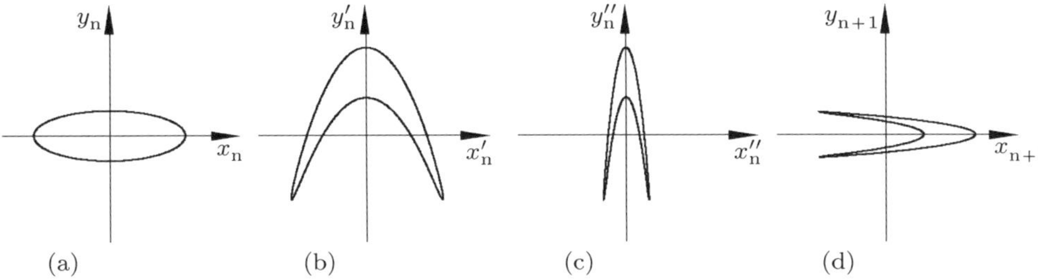

Fig. 10.2.2: The three basic operations of the Hénon map. The original domain (a) is folded (b), contracted (c) and mirrored (d)

i.e. at each iteration step, all area elements contract uniformly by the factor b. This area contraction takes place as a result of the second mapping step and is in concord with the Lorenz system which, due to its constant divergence

$$\Lambda = \operatorname{div} \boldsymbol{F} = -\sigma - 1 - b \tag{10.2.10}$$

in accordance with eq. (5.1.11), also causes a uniform shrinking of the phase space volume. In the case of the discrete map, however, the negative value of $\det \boldsymbol{D}$ means a turnabout of the orientation and is caused by the mirroring sub-step T'''. Thus, the Hénon map eq. (10.2.8) cannot ultimately be interpreted directly as a Poincaré map since, in the latter case, the orientation is always maintained. Although the recursion rule was originally set up empirically as a folding and stretching process, Hénon and Pomeau were subsequently able to show that each quadratic map with constant Jacobian determinant can be transformed to the "canonical" form of eq. (10.2.8).

We can deduce the two fixed points H_1 and H_2 of the map from eq. (10.2.8). We obtain

$$x_{1,2} = \frac{(b-1) \pm \sqrt{(b-1)^2 + 4a}}{2a}, \qquad y_{1,2} = b x_{1,2} \tag{10.2.11}$$

The stability characteristics of these fixed points derive from the corresponding eigenvalues

$$\lambda_{1,2}^{(\mathrm{i})} = -a x_{\mathrm{i}} \pm \sqrt{a^2 x_{\mathrm{i}}^2 + b} \qquad (\mathrm{i} = 1, 2) \tag{10.2.12}$$

In this section, we do not wish to investigate the characteristics of the Hénon map in dependence on the two system parameters a and b, however, but follow Hénon's argumentation. In order to simulate a distinct folding process on the one hand and make the structure of the attractor clearly visible on the other, Hénon selected $b = 0.3$ as the contraction parameter. He fixed the value of a so that both fixed points possess saddle-point character: $a = 1.4$. Numerical investigations undertaken by J. H. Curry demonstrate (Curry, 1979) that the stable and the unstable manifold of the saddle point H_2 intersect in homoclinic points (cf. section 4.5 and fig. 4.5.8), resulting in highly complex (probably chaotic) behaviour; cf. (Guckenheimer and Holmes, 1983). A careful qualitative and quantitative investigation of the complex course of the two branches associated in each case with the stable and unstable manifolds in the two saddle points H_1 and H_2 can be found in (Simó, 1979).

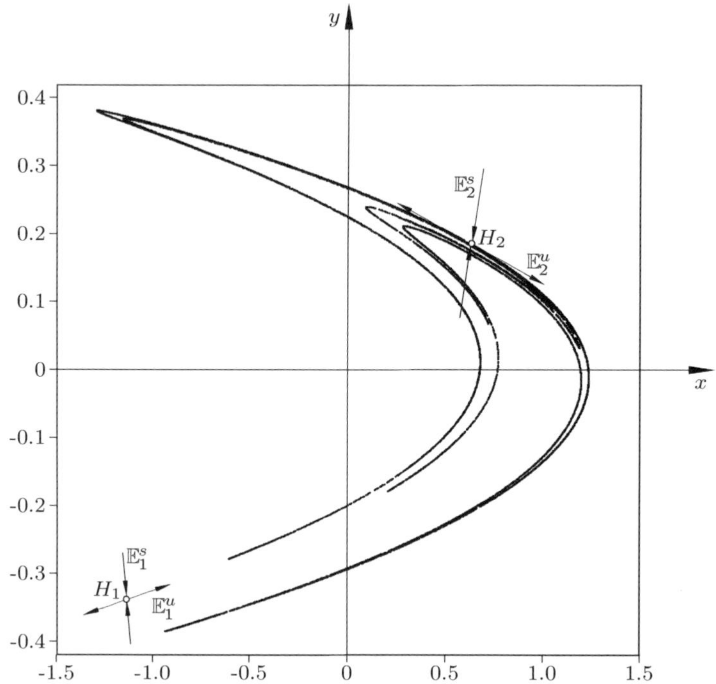

Fig. 10.2.3

The Hénon attractor for $a = 1.4,\ b = 0.3$

Figure 10.2.3 presents the repeatedly folded complex structure of the Hénon attractor following 15 000 iterations; we show the two saddle points with their unstable and stable directions. If, as an initial point, we do not select an arbitrary point from the basin of attraction, but rather a point lying as close as possible to the saddle point H_2 respectively on $\mathbb{E}_2^s$, successive mapping points approach the unstable manifold $\mathbb{W}_2^u$ very rapidly so that soon, we can no longer ascertain any difference between the sequence of iterated points and $\mathbb{W}_2^u$, cf. (Hénon, 1976). This suggests that the attractor coincides with the closure of the unstable manifold $\mathbb{W}_2^u$ of H_2 (Ruelle, 1989). (The closure still contains all accumulation points of $\mathbb{W}_2^u$.)

We now consider the structure of the attractor in the neighbourhood of the singular point H_2 more closely. Taken as a whole, the attractor has the appearance of a line folded several times (fig. 10.2.4a). If we enlarge the indicated section near the saddle point H_2, we recognise a banded structure of almost parallel lines (fig. 10.2.4b). Successive enlargements reveal that the initial pattern is repeated on all scales (fig. 10.2.4c, d). This self-similar structure transversal to the unstable manifold $\mathbb{E}_2^u$ respectively $\mathbb{W}_2^u$ is strongly reminiscent of a Cantor set.

Two numerical experiments elucidate the mixing process caused by the continuous repetition of the stretching and folding operations. The two left-hand columns of Colour Plates XX and XXI (pp. 704, 705) present the evolution of the Hénon attractor. For this purpose, 60 000 initial conditions were selected and distributed uniformly over a square $-c \leqslant x, y \leqslant +c$ in the x, y-plane. All the points lying in the first quadrant of

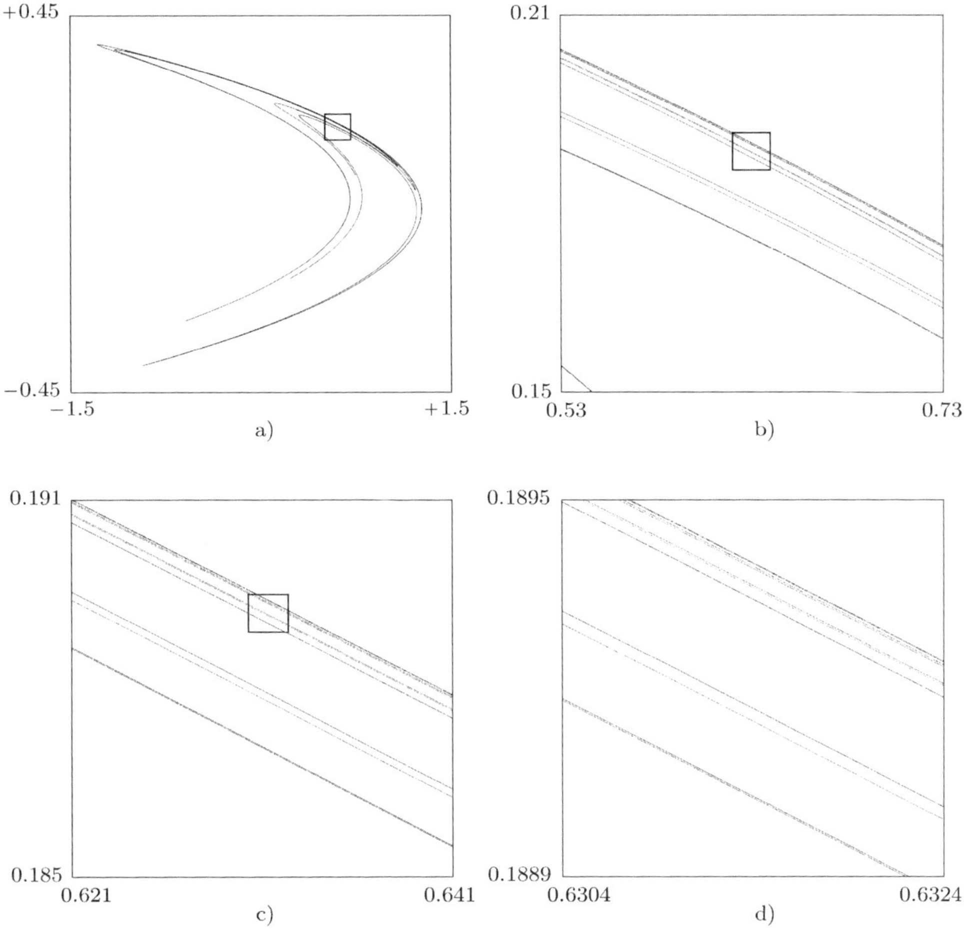

Fig. 10.2.4: The self-similar structure of the Hénon attractor (Hénon, 1976)

the square are represented by the colour turquoise, all those points in the second, third and fourth quadrants by the colours orange, yellow and blue respectively. The original arrangement as well as the location of the image points for the iteration steps $n = 1, 3$ (Colour Plate XX, left-hand column) and $n = 6, 11, 29$ (Colour Plate XXI, left-hand column) are presented below one another; each point retains its original colour during the iterations. For $n = 1$ and $n = 3$, we can clearly observe the stretching and folding process, the points approaching the attractor very fast. After the third iteration, the four colours are already spread over the mapping domain. Further iterations lead very rapidly to a complete mixing of the colours: between the 11th and 29th iteration, it is scarcely possible to distinguish between them. In the close proximity of any arbitrarily selected point on the attractor, we can observe points with the other three colours. In a completely mixed state, it is hence no longer possible to re-establish the initial coordinates of a particular point.

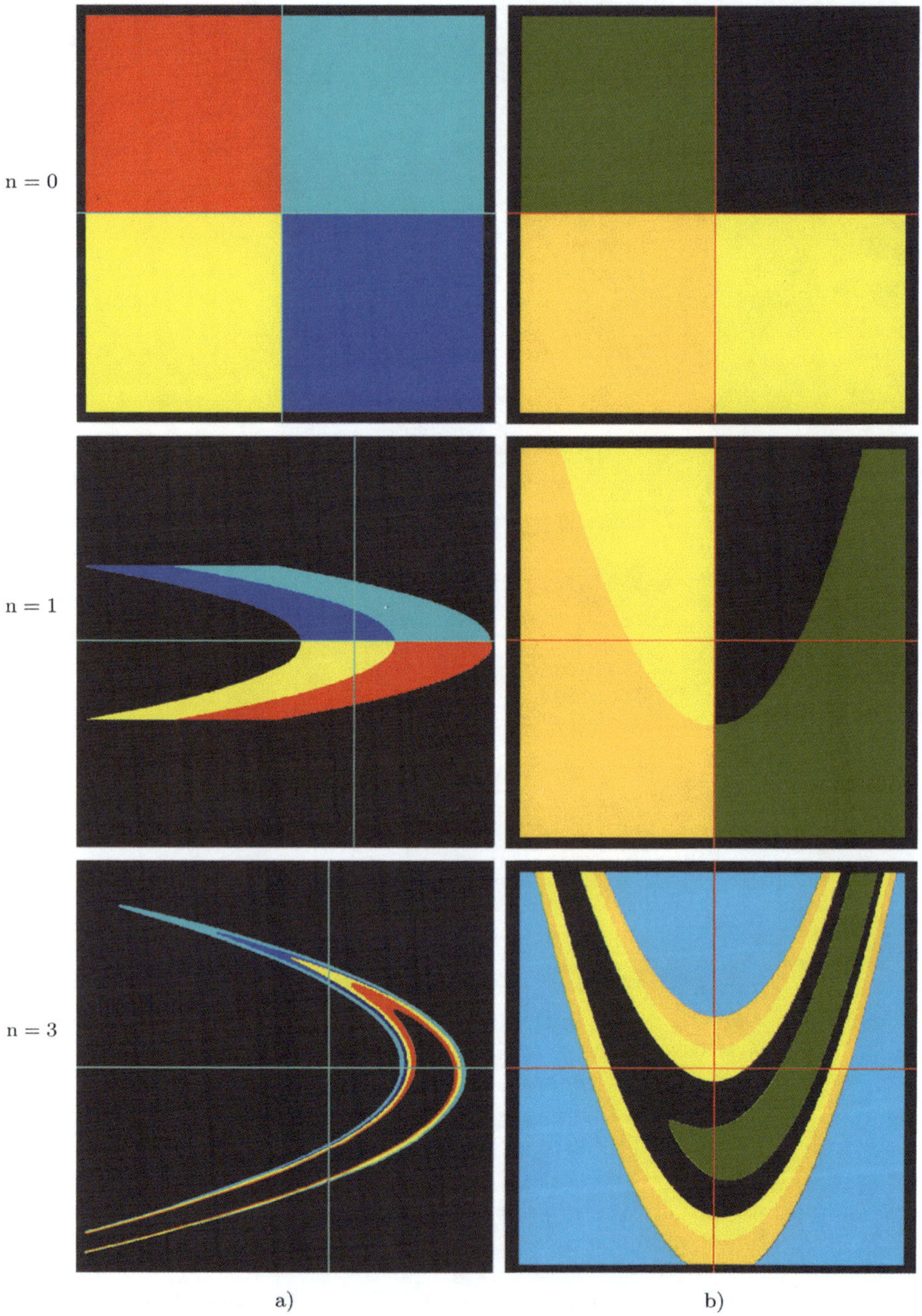

Plate XX Hénon map for n iteration steps:
a) evolution of the Hénon attractor by stretching and folding
b) illustration of the mixing effect and the basin of attraction

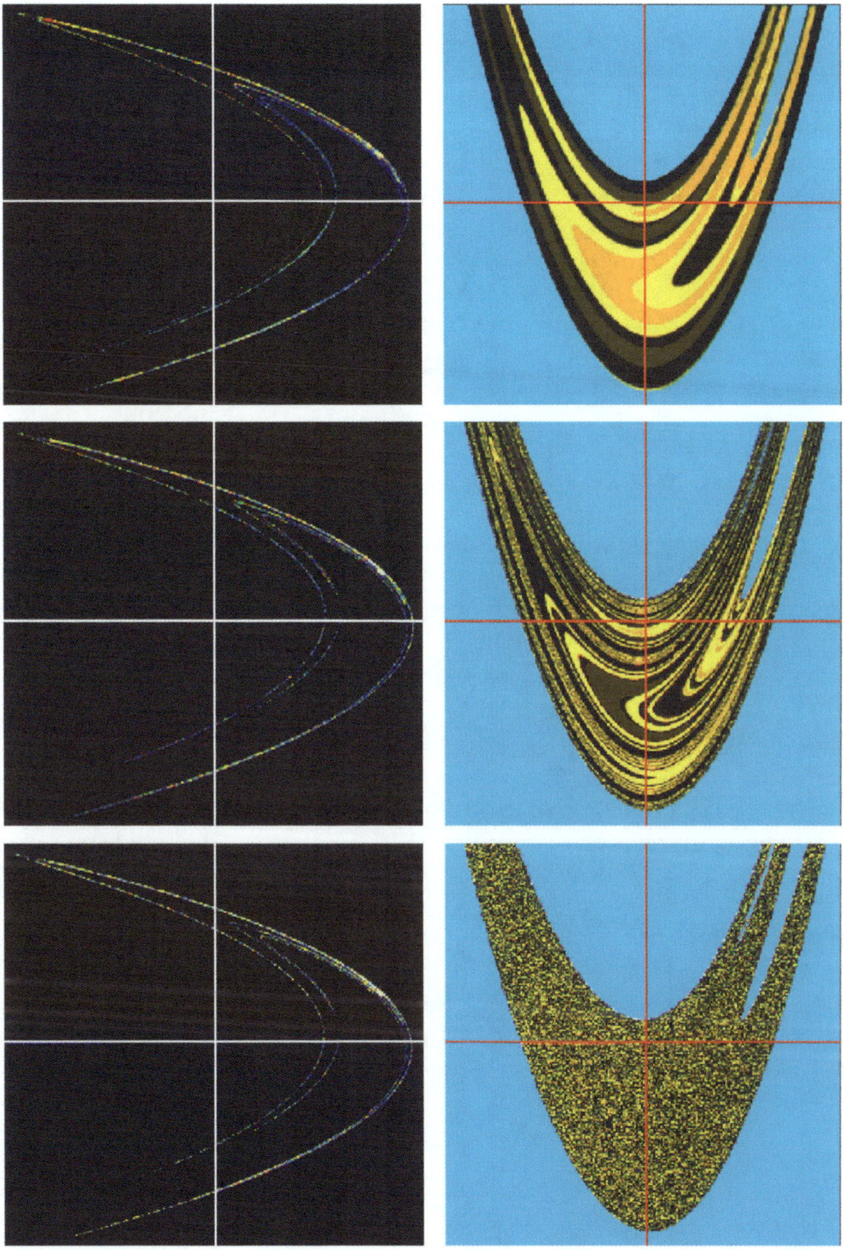

n = 6

n = 11

n = 29

a) b)

Plate XXI Hénon map for n iteration steps *(continuation)*

Even the information in which quadrant the initial point lay has been lost.

Due to the strong dissipation of the Hénon map and the fine line-like banded structure of the attractor, the mixing process in the two left-hand columns of Colour Plates XX and XXI can only be traced with difficulty. For this reason, we have added in the two right-hand columns the results of a further calculation which emphasises the mixing process on the one hand and defines the basin of attraction on tbe other approximately. The four quadrants are assigned the colours black – olive – dark yellow – light yellow as indicated on the upper right-hand side of Colour Plate XX. We again consider 60 000 uniformly distributed initial points. However, in this case, we fix the coordinates (x_0, y_0) of an initial point P_0 and calculate successively the image points $P_1, P_2, \ldots$ and assign to the point (x_0, y_0) the colour of the quadrant in which the image point P_n lands after the n-th iteration. In the right-hand column of Plate XX, we again present the initial state and the recorded iterations for $\mathrm{n} = 1$ and $\mathrm{n} = 3$. Furthermore, in Plate XXI, right-hand column, we reproduce the results after $\mathrm{n} = 6, 11, 29$ iterations. Points which diverge towards infinity are assigned the colour blue. After only a few iteration steps, the basin of attraction of the Hénon attractor (black-olive-yellow colour scale) contrasts clearly with the blue background. Due to the two-dimensional extent of the basin of attraction, the stretching and folding processes emerge considerably more clearly than in the left-hand column of the colour plate. After $\mathrm{n} = 29$ iteration steps, we can speak of complete mixing. The basin boundary of the Hénon attractor is formed by the two branches of the stable manifold $\mathbb{W}_1^s$ of the saddle point H_1 and is in very good agreement with Simó's calculations in (Simó, 1979).

As with the circle map studied in section 8.3.2, the long-term behaviour of the Hénon map can be analysed by means of isoperiodic diagrams (cf. Colour Plate IX, p. 497). For this purpose, characteristic colours are assigned to all (a, b)-ranges for which cycles of a fixed period p respectively chaotic motions appear. The topology and the emergence of these periodic windows have already been investigated in detail, see (Gallas, 1994; Baptista et al., 2003). As already mentioned, the geometric form of these periodic domains are reminiscent of *shrimps* or swallow tails. These shrimps form a complex network on differing scales and are embedded in a chaotic environment. With the help of such isoperiodic diagrams, it is possible to control chaos and to avoid chaotic behaviour by a macroscopic variation of the parameters a, b so that, from a given period p_1, a new period p_2 can be reached without drifting off into chaotic motions (Gallas, 1995).

10.3 The Lorenz System Revisited

Ἄλλοτε μητρυιὴ πέλει ἡμέρη,
ἄλλοτε μήτηρ

Hesiod, Ἔργα καὶ ἡμέραι, I, 825
(ca. 800 B.C.)

As we already know from sections 5.2, 7.3 and 7.4 on the Lorenz system, eq. (7.3.25), the Lorenz attractor is the first example of a chaotic or strange attractor ever discovered. In the course of his computer studies on weather forecasting, Edward N. Lorenz recognised the fundamental mechanism of unpredictability, namely that small perturbations on a microscopic level demonstrate their effect on a macroscopic level as a result of divergence. In this phenomenon, also known as the butterfly effect, two trajectories of an attractor with almost identical initial conditions remain neighbouring for only a short time; they then diverge rapidly, following an exponential law. This observation is in sharp contrast to regular – that is to say non-chaotic – long-term response. In this case, neighbouring trajectories of an attractor remain neighbouring for ever and small deviations remain small. The dynamic behaviour is hence predictable.

The key to understanding the fact that neighbouring trajectories diverge exponentially while remaining in a bounded domain is to be found in the mechanism of the stretching and folding of trajectories in the phase space. Trajectories which diverge exponentially must of necessity fold back; the domain the attractor takes up would otherwise not be bounded. The consequence of endless stretching and folding is demonstrated by the Hénon map (see section 10.2, fig. 10.2.2). Its long-term behaviour is highly complex and the structure of the attractor fractal. Colour Plate IV, p. 286, inspired by the publication (Crutchfield et al., 1986), illustrates the effect of repeated stretching and folding on the one hand and the unpredictable consequences of the slightest variations in the initial conditions on the other.

In the preceding section 10.2, we described in detail the individual illustrations in Colour Plate IV. We wish to draw the reader's attention once more to one effect of chaotic systems: they possess the special charateristic of intensifying fluctuations on the microscopic level exponentially so that they become visible on the macroscopic level after a finite time-span; the series of illustrations in Colour Plate IV demonstrates this graphically.

We have mentioned before that Edward N. Lorenz discovered deterministic chaos in the system eq. (7.3.25) for the parameters $r = 28$, $\sigma = 10$ and $b = 8/3$. If we increase the relative Rayleigh number r, the Lorenz equations surprise us again. Between $r = 146.5$ and $r = 166$, the Lorenz system loses its chaotic character over an extensive r-range, the strange attractor changing to a limit cycle of period one following a cascade of infinitely many period-halvings (Manneville and Pomeau, 1979); cf. also section 8.4.1.

Figure 10.3.1 presents the inverse cascade of bifurcations of a period-four limit cycle to a period-one cycle as well as the corresponding temporal evolutions. The course of

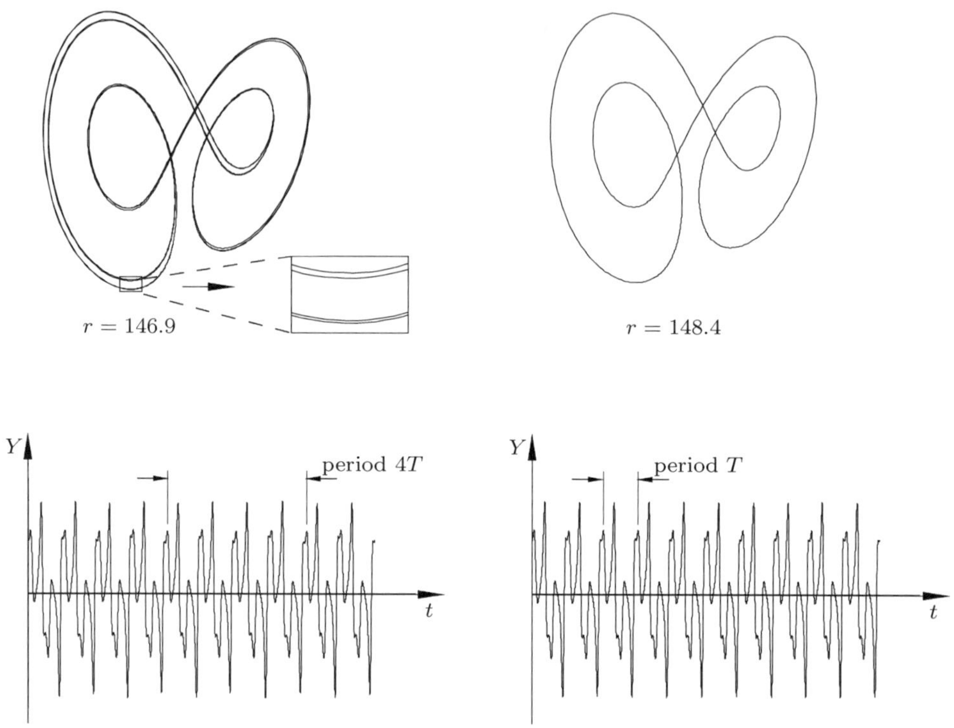

Fig. 10.3.1: Two limit cyles generated by an inverse bifurcation cascade of the Lorenz system

the trajectory on the left-hand side was determined numerically for $r = 146.9$, that on the right-hand side for $r = 148.4$.

The bifurcation diagram in fig. 10.3.2a offers a qualitative overview of when the Lorenz system exhibits chaotic and when periodic behaviour in the interval $24 < r < 320$. It presents the long-term behaviour of the Y-coordinate as recorded in the plane of intersection $X = 0$ over the control parameter r; here, $X_0 = Y_0 = Z_0 = 10$ was selected as the initial condition. Apart from the regular interruptions in the ranges $214 \lesssim r \lesssim 314$ and $146.5 \lesssim r \lesssim 154.5$, the bifurcation diagram is astonishingly similar to that of the logistic map (fig. 3.7.6). These interruptions suggest that different long-term responses are generated for a fixed initial condition if r is varied, i.e. we can conjecture the coexistence of two attractors. This conjecture is confirmed if we consider the limit cycles in Colour Plate XXII, p. 711: whereas for $r = 330$, only one limit cycle exists which is symmetrical in itself (i.e. is mapped to itself by the transformation $X \to -X$, $Y \to -Y$, $Z \to Z$), we observe for $r = 233$ two limit cycles, a yellow and a symmetrical blue one. The basins of attraction of the attractors in dependence of r alter in such a way that the fixed initial condition $X_0 = Y_0 = Z_0 = 10$ lands alternately in the basin of attraction of the yellow respectively the blue attractor. Thus, no period doubling occurs at $r_{cr} \approx 314$; we rather observe a typical example of a pitchfork bifurcation (see section 6.6).

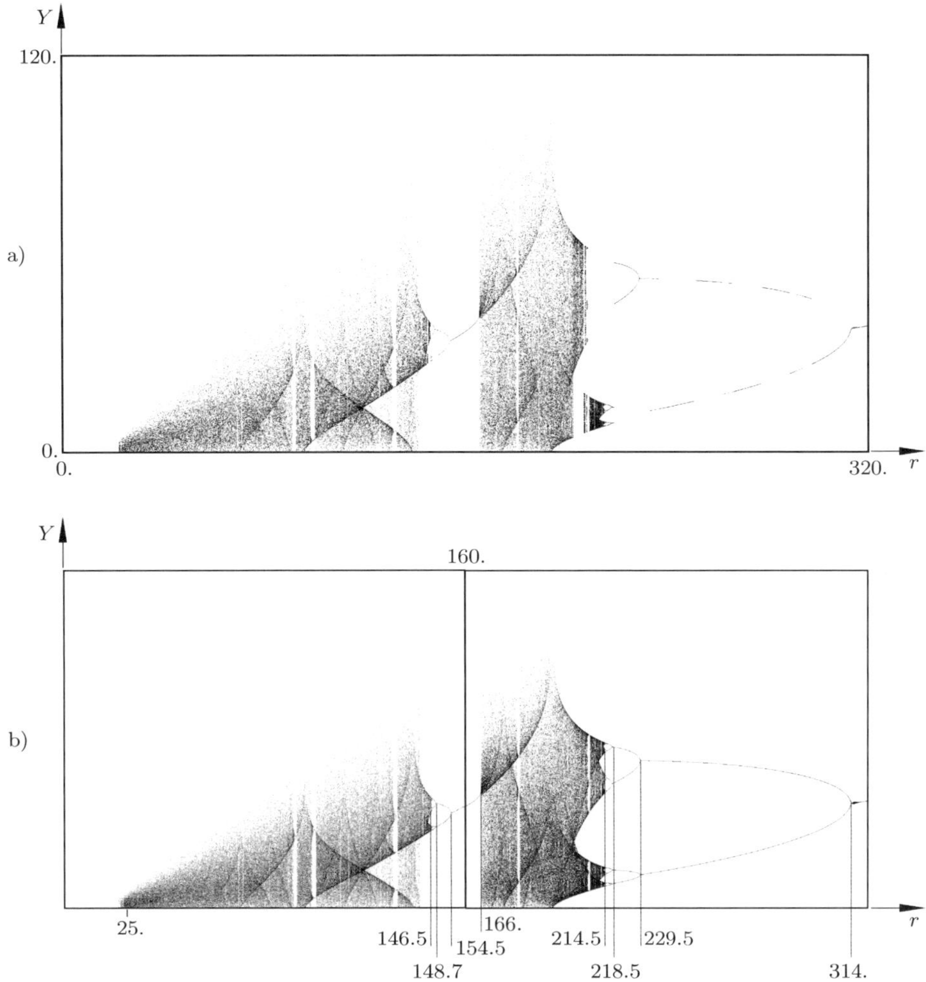

Fig. 10.3.2: Bifurcation diagram of the Lorenz system, dependence on the parameter r:
a) for fixed initial conditions $X_0 = Y_0 = Z_0 = 10$
b) for variable initial conditions

These two period-one attractors with separate basins of attraction now pass through a cascade of period doublings for decreasing r (cf. also fig. 8.4.4). For $r = 216$, for example, period 4 can be observed, for $r = 215$ period 8. A minimal reduction of the control parameter r already leads to chaos for $r = 214$ and is associated with two separate strange attractors. The two limit cycles for $r = 205$ illustrate the phenomenon that the chaotic range is repeatedly interrupted by periodic windows. In Colour Plate XXIII, p. 712, we have plotted the Poincaré sections of the Lorenz attractor with the plane $X = 0$ in the range $24 \leqslant r \leqslant 320$. This representation confirms the pitchfork bifurcation at $r \approx 314$. The single green limit cycle for $r < 314$ becomes unstable and two new stable ones emerge, one blue and one yellow. The subsequent cascade of period doublings up to two coexisting strange attractors for decreasing r can be clearly recognised. Following a further decrease of r, the expansion of the two strange attractors, the yellow and the blue one, increases steadily in the phase space until they finally "collide" and coalesce. Figure 10.3.2b presents a further bifurcation diagram; for its calculation in the presence of coexisting attractors, several initial conditions were assumed in order to illustrate all possible types of long-term behaviour. This diagram is hence completely equivalent to the representation of the Poincaré sections in Colour Plate XXIII, p. 712.

The illustrations of fig. 10.3.2 and Colour Plate XXIII exhibit a pronounced periodic window for the interval $146.5 \lesssim r \lesssim 166$. The bifurcation behaviour with decreasing r – starting with a period-one limit cycle (green line, Colour Plate XXIII), through an inverse pitchfork bifurcation and a cascade of period doublings and ending in chaos – repeats itself self-similarly to the interval $214.5 \lesssim r < 320$.

Up to now in this section, we have impressively demonstrated that a very simple dissipative system such as the Lorenz model can lead to highly complex and irregular forms of motion. The mathematical description of the chaotic behaviour of dissipative systems is based on the concept of the strange attractor which represents an attractive set with complex fractal structure. In order to characterise it, we apply methods such as the Lyapunov exponent and the determination of its dimension (see sections 5.4, 5.5) which allow us to distinguish between regular and chaotic motion.

We first consider the concept of the Lyapunov exponents. As we already know from section 5.4, the following attractor types can be classified on the basis of the Lyapunov exponents σ_i: a fixed point, for example, possesses exclusively negative σ_i-values; for a limit cycle, the largest value is $\sigma_1 = 0$ and the others are negative. Strange attractors in a three-dimensional phase space are characterisable by the following constellation of the Lyapunov exponents: one positive exponent generates the divergence of neighbouring trajectories, the zero exponent induces the perturbation in tangent direction and the third, negative exponent expresses the attraction of the attractor and conditions the folding of the trajectories. In higher-dimensional phase spaces, strange attractors can possess more than one positive Lyapunov exponent; in this case, we occasionally speak of "hyperchaos"; see (Rössler, 1979b) and, for example, (Brun, 1989); see also section 10.9.

The bottom part of fig. 10.3.3 presents the largest Lyapunov exponent σ_1 for the Lorenz system in dependence of the control parameter r (see also fig 8.4.3). In the interval $1 < r < 24$, we observe the fixed point behaviour of the system, the σ_1-value being negative. In the subsequent interval $24 \lesssim r \lesssim 146.5$, the chaotic response of the system dominates, occasionally interrupted by periodic windows. Within the chaotic range, $\{\sigma_1\ \sigma_2\ \sigma_3\} = \{+\ 0\ -\}$ holds for the Lyapunov constellation.

A large window of periodic behaviour is apparent in the interval $146.5 \lesssim r \lesssim 166$ (cf. fig. 10.3.1). The subsequent smaller chaotic range is again succeeded by a pronounced periodic interval. Comparisons with fig. 10.3.2 and Colour Plates XXII

Plate XXII Lorenz system: coexistence of two attractors and cascade of period doublings

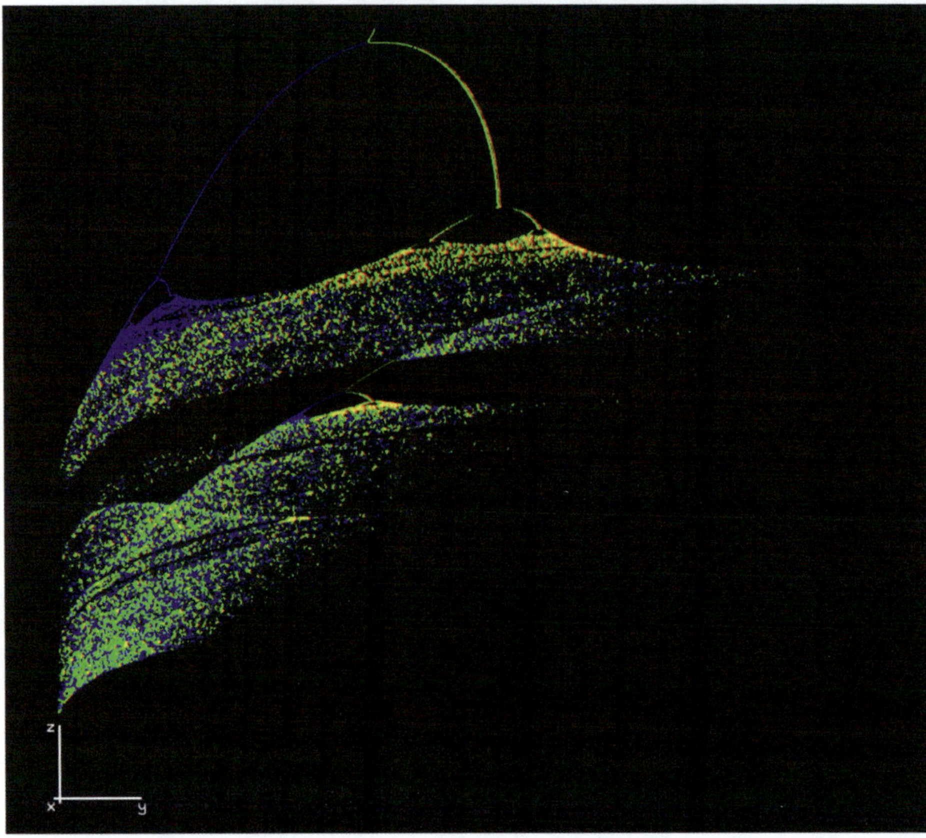

Plate XXIII Lorenz system: bifurcation diagram of the Poincaré sections ($24 \leqslant r \leqslant 320$)

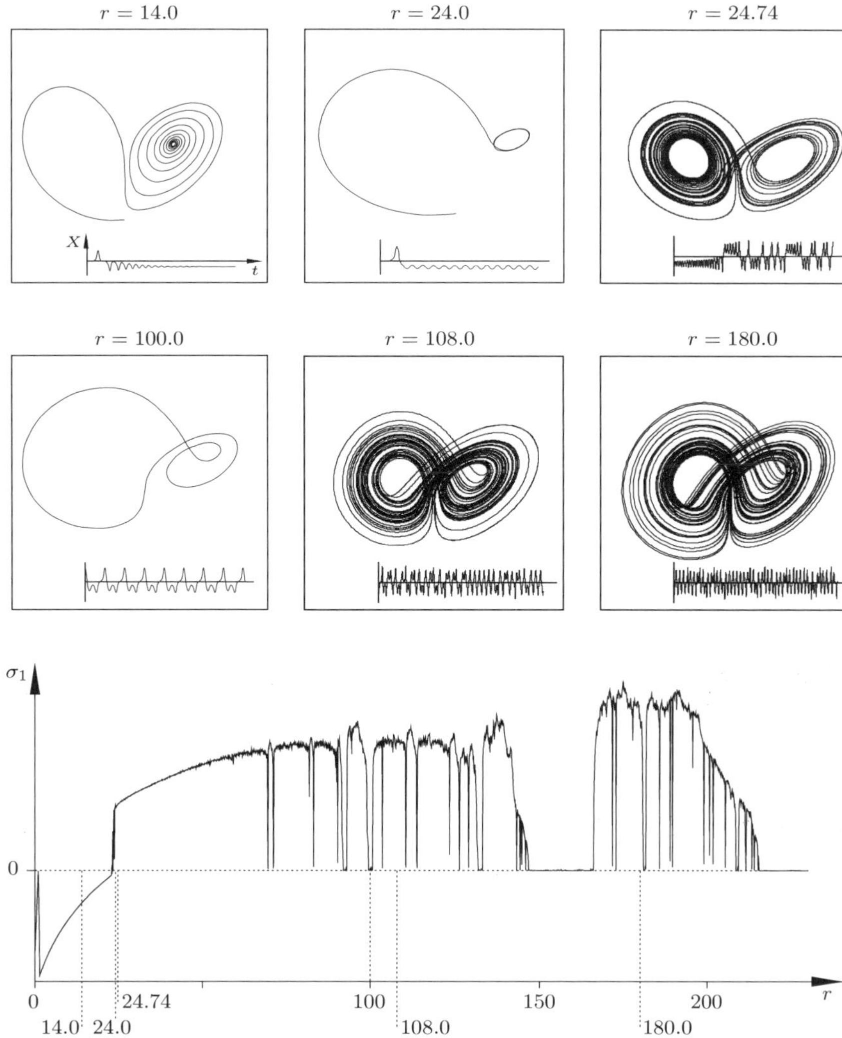

Fig. 10.3.3: Lorenz system: attractor types, temporal evolution, phase portraits and maximum Lyapunov exponent σ_1 in dependence of r (Lindell, 1988)

and XXIII, pp. 711, 712, reveal consistent behaviour of Lyapunov exponents, Poincaré sections and time series of trajectories. In fig. 10.3.3, the different attractor types are summarised in phase portraits, time series and the dependence of the largest Lyapunov exponent on the control parameter r.

In section 5.5, we introduced and discussed various dimension concepts. In the present context, we are interested in the capacity dimension D_c of the Lorenz attractor. The chaotic motion reflects itself geometrically in the structure of the strange attractor in the phase space. In fact, one single trajectory appertaining to the basin of attraction of the strange attractor is sufficient to approximate the geometrical structure of the attractor. To obtain the course of the trajectory, we have to apply a scheme of numerical integration. In order to calculate the capacity dimension D_c, the phase space is covered uniformly with cubes of edge length ε; we determine the number $W(\varepsilon)$ of cubes containing points of the attractor. Since this number $W(\varepsilon)$ increases proportionally to ε^{-D_c} for decreasing values ε, the computing effort also increases exponentially with the dimension of the attractor. In order to obtain reasonable results for strange attractors of a dimension between 2 and 3, the calculation must be based on 10^5 to 10^6 attractor points. Due to the necessity of numerical integrations, this leads to extensive computing times and, moreover, requires a disproportionately large memory.

In order to reduce the computational effort, we have taken up an algorithm proposed in (Hunt and Sullivan, 1986) which sets up a relation between the volume of the ε-cover and the capacity dimension D_c. We are thus in a position to determine the volume by means of a Monte-Carlo integration instead of counting cubes. For measuring the distances, Hunt and Sullivan also suggest an efficient data structure specifically for vector processors.

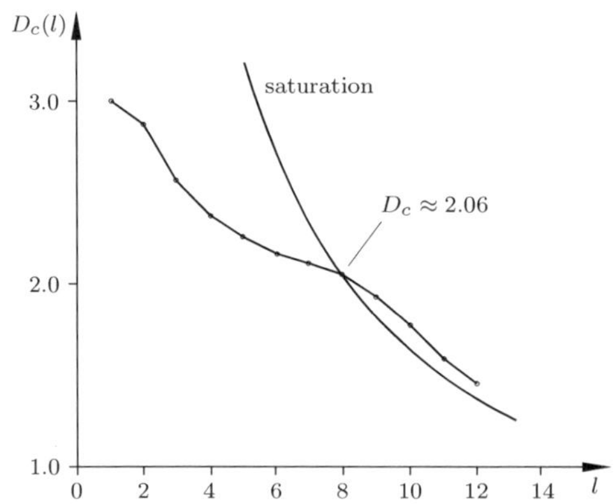

Fig. 10.3.4

Capacity dimension D_c of the Lorenz attractor ($\sigma = 10$, $r = 28$, $b = 8/3$)

The result of the numerical determination of D_c for the Lorenz attractor ($r = 28$, $\sigma = 10$, $b = 8/3$) is presented in fig. 10.3.4. We have plotted approximations for the capacity dimensions – for practical reasons, not D_c directly as a function of the length ε, but the dimension as a function $D_c(l)$ of an exponent l which controls the variation of ε in steps of (negative) powers of two, $\varepsilon = (1/2)^l$. There is evidently a lower bound for the choice of ε which is reached if each point of the attractor lies in its own distinct box. This condition leads to a saturation curve, gained from the relation $W = \mathrm{N} = {\varepsilon_s}^{-D_c} = 2^{l_s D_c}$, where N is the number of attractor points. Dimension values higher than the saturation values are senseless by definition of the capacity dimension D_c. The point of intersection of the saturation curve and the curve of $D_c(\varepsilon)$ yields the capacity dimension for the Lorenz attractor we are looking for. The result is $D_c \approx 2.06$; a side elevation of the Lorenz attractor, Colour Plate III, p. 198, confirms its almost surface-like structure, documented by the D_c-value close to two.

10.4 Van der Pol Equation

Ἴανος ἄλλος
Alter Janus, Another Janus
Erasmus, Adagia, Chil iv, cent, *ii.*, № 93

Fundamental impulses for the comprehension of non-linear dynamical systems emanated not only from the fields of mechanics, physics and mathematics but also from electrical engineering. A significant contribution was made by the Dutch electrical engineer Balthasar van der Pol who, in the 1920s, set up a theoretical model to describe the non-linear oscillatory response of vacuum-tube oscillators to explain the experimental observations he had made (Appleton and van der Pol, 1922). The oscillation forms he observed play an important role in electrical engineering; they belong to the class of non-linear phenomena, *self-sustained oscillations*, which are not externally driven but actively regulate their oscillatory behaviour themselves. Characteristic of these oscillations is the emergence of a non-linear damping term which changes its sign according to the rhythm of the oscillation, hence amplifying the oscillation for small amplitudes while damping it for large amplitudes. In order to balance the accompanying energy losses, the system gathers the energy needed from an internal energy source by positive feedback (Pikovsky et al., 2003).

In mechanics, self-sustained oscillations are also known. They occur, for example, if the friction is a non-linear function of the relative velocity. The creaking of a door, the squealing of brakes and similar noises can be explained in this way – but also the beautiful sounds of a violin or a clarinet (Rayleigh, 1945). Another undesirable self-sustained oscillation in mechanics is the global flutter of an aeroplane or of its component parts, e.g. wings, ailerons, flaps, propellers etc. Self-sustained oscillations also occur outside aerospace technology, e.g. in civil engineering and bridge building. A classic example is the Tacoma Bridge (USA) which, incited by a relatively low wind velocity of 18 mps, collapsed in 1940 as a result of aeroelastic self-sustained oscillations (Green and Unruh, 2006). Apart from these unwanted instances, there are cases in which self-sustained oscillations can be made use of in order to achieve a prescribed oscillation at a frequency kept as constant as possible. The balance spring of a clockwork is one example, another is the self-excited electrical circuit already mentioned.

10.4.1 Self-excited Oscillation

In order to elucidate self-sustained oscillations in the absence of an externally applied force, we consider a simple electrical circuit with a triode (fig. 10.4.1a). We denote the self-inductance of the coil L, the capacity of the capacitor C, the Ohmic resistor R and the mutual inductance, also called the coupling factor, $M > 0$. The circuit elements inductor, capacitor and resistance are assembled in series, i.e. the sum of the voltage components of the individual circuit elements L, C and R equals the voltage due to the mutual inductance M. If we apply Kirchhoff's laws, we have (Feynman et al., 1987)

$$L\frac{\mathrm{d}i}{\mathrm{d}t} + Ri + \frac{1}{C}\int^{t} i\mathrm{d}t = M\frac{\mathrm{d}i_a}{\mathrm{d}t} \qquad (10.4.1)$$

where i_a is the anode current and i the current of the inner electric circuit. The current direction is shown in fig. 10.4.1a. The diagram also illustrates that the grid voltage u_g (cathode – grid) is equal to the capacitor voltage u_c. The following equation holds for the cathode current

$$i = C\frac{\mathrm{d}u_g}{\mathrm{d}t} \tag{10.4.2}$$

and, since the anode current i_a depends on the grid voltage u_g in accordance with the anode characteristic, fig. 10.4.1b ($i_a = i_a(u_g)$), there follows from eq. (10.4.1)

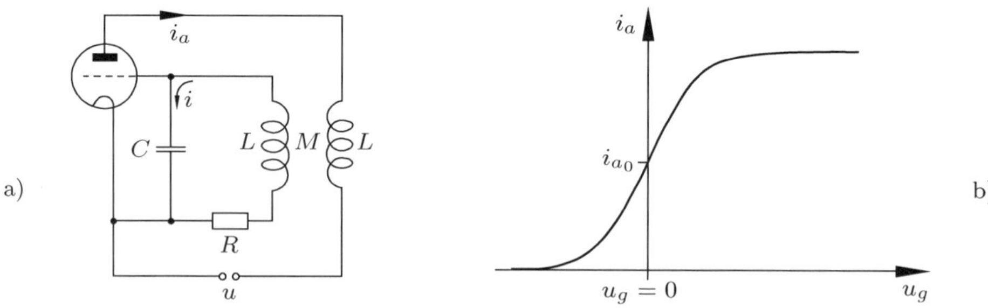

Fig. 10.4.1: a) Electrical circuit and
b) triode characteristic of a van der Pol oscillator

$$LC\frac{\mathrm{d}^2u_g}{\mathrm{d}t^2} + \left[RC - M\frac{\mathrm{d}i_a}{\mathrm{d}u_g}\right]\frac{\mathrm{d}u_g}{\mathrm{d}t} + u_g = 0 \tag{10.4.3}$$

If the anode characteristic $i_a(u_g)$ in the neighbourhood of $u_g = 0$ can be approximated by a third-order polynomial (cf. 10.4.1b), there follows for eq. (10.4.3) the differential equation

$$LC\frac{\mathrm{d}^2u_g}{\mathrm{d}t^2} + (RC + A_1 + A_2u_g + A_3u_g^2)\frac{\mathrm{d}u_g}{\mathrm{d}t} + u_g = 0 \tag{10.4.4}$$

Moreover, if we assume that

$$A_2 = 0, \quad A_3 > 0, \quad RC + A_1 < 0 \tag{10.4.5}$$

holds at the operating point of the triode, then, with the normalisation

$$\tau = \frac{1}{\sqrt{LC}}t\,, \quad x = u_g\sqrt{\frac{A_3}{-(RC + A_1)}}\,, \quad \mu = -\frac{1}{\sqrt{LC}}(RC + A_1) \tag{10.4.6}$$

we finally obtain the van der Pol equation (van der Pol, 1926)

$$\frac{\mathrm{d}^2x}{\mathrm{d}\tau^2} - \mu(1 - x^2)\frac{\mathrm{d}x}{\mathrm{d}\tau} + x = 0 \tag{10.4.7}$$

Its solution behaviour has been investigated in detail by a number of authors (van der Pol, 1926; Andronov et al., 1965; Stoker, 1950). The van der Pol equation played an important role in the development of both the theory of non-linear oscillations and the qualitative analysis of differential equations.

The second-order differential equation, eq. (10.4.7), can be expressed as a system of two first-order equations (transformation $x, \dot{x} \to x_1, x_2$ and $\tau \to t$)

$$\begin{aligned} \dot{x}_1 &= x_2 \\ \dot{x}_2 &= -x_1 + \mu(1 - x_1^2)x_2 \end{aligned} \tag{10.4.8}$$

The only fixed point $F(0,0)$ is independent of the control parameter μ, the dynamics in its neighbourhood is dependent on μ, however. For a linear stability analysis, we calculate the Jacobian matrix

$$\left.\frac{\partial \boldsymbol{F}}{\partial \boldsymbol{x}}\right|_{\boldsymbol{o}} = \begin{bmatrix} 0 & 1 \\ -1 & \mu \end{bmatrix} \tag{10.4.9}$$

having the characteristic equation

$$\lambda^2 - \mu\lambda + 1 = 0 \tag{10.4.10}$$

Due to its eigenvalues $\lambda_{1,2} = (\mu \pm \sqrt{\mu^2 - 4})/2$, $F(0,0)$ exhibits the following fixed point behaviour (cf. fig. 3.1.8)

$$\begin{aligned} -\infty < \mu \leqslant -2 &: \quad \text{stable node} \\ -2 < \mu < 0 &: \quad \text{stable focus} \\ \mu = 0 &: \quad \text{centre} \\ 0 < \mu < 2 &: \quad \text{unstable focus} \\ 2 \leqslant \mu < +\infty &: \quad \text{unstable node} \end{aligned}$$

Van der Pol determined graphically a family of approximate solution curves of eq. (10.4.8) applying the isocline method and generated phase portraits for certain positive μ-values by selecting in turn a small ($\mu = 0.1$), a medium ($\mu = 1.0$) and a large ($\mu = 10.0$) value for μ. We have integrated the equation numerically and display the corresponding time series of the x_1-component as well as the phase portraits for the aforementioned μ-values in fig. 10.4.2. All three phase portraits show a single limit cycle enclosing the origin, i.e. all initial conditions lead to one and the same limit cycle. For $\mu = 0.1$, the oscillatory response of the van der Pol oscillator is almost harmonic, the limit cycle is practically circular and the time series of the x_1-component almost sinusoidal (cf. fig 10.4.2a). With increasing μ-values, the maximum absolute value of the velocity x_2 also increases. The limit cycle is stretched in the x_2-direction, its form deviates more and more from that of a circle (fig. 10.4.2b,c). For $\mu = 1.0$, the temporal course of x_1 possesses a sawtooth wave (fig. 10.4.2b). The x_1-course for $\mu = 10.0$, was designated by van der Pol as a relaxation oscillation (fig. 10.4.2c). The x_1-value remains constant for long periods of t, then abruptly adopts an extreme position with the opposite sign. This jump is not easily recognisable in the phase portrait of fig. 10.4.2c. It emerges more clearly if we transform eq. (10.4.7), as Liénard already

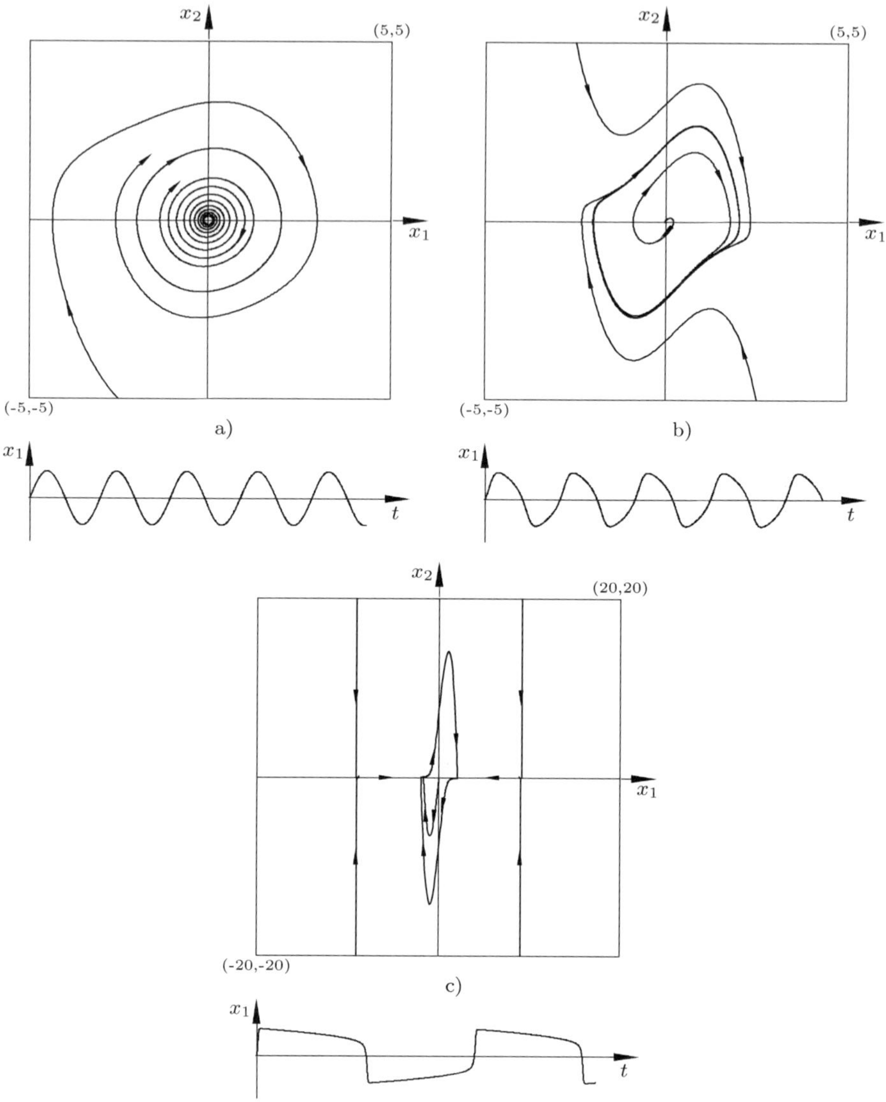

Fig. 10.4.2: Phase portraits and time series of the van der Pol equation, eq. (10.4.8): a) $\mu = 0.1$, b) $\mu = 1.0$, c) $\mu = 10.0$

suggested in 1928 (Minorsky, 1962). Starting with the van der Pol equation (10.4.7), we transform $x, \dot{x} \to x_1, x_2$ by applying the following non-linear rule

$$\begin{aligned} x_1 &= x \\ x_2 &= \frac{1}{\mu}\dot{x}_1 + f(x_1) \end{aligned} \tag{10.4.11}$$

where, for $f(x_1)$,

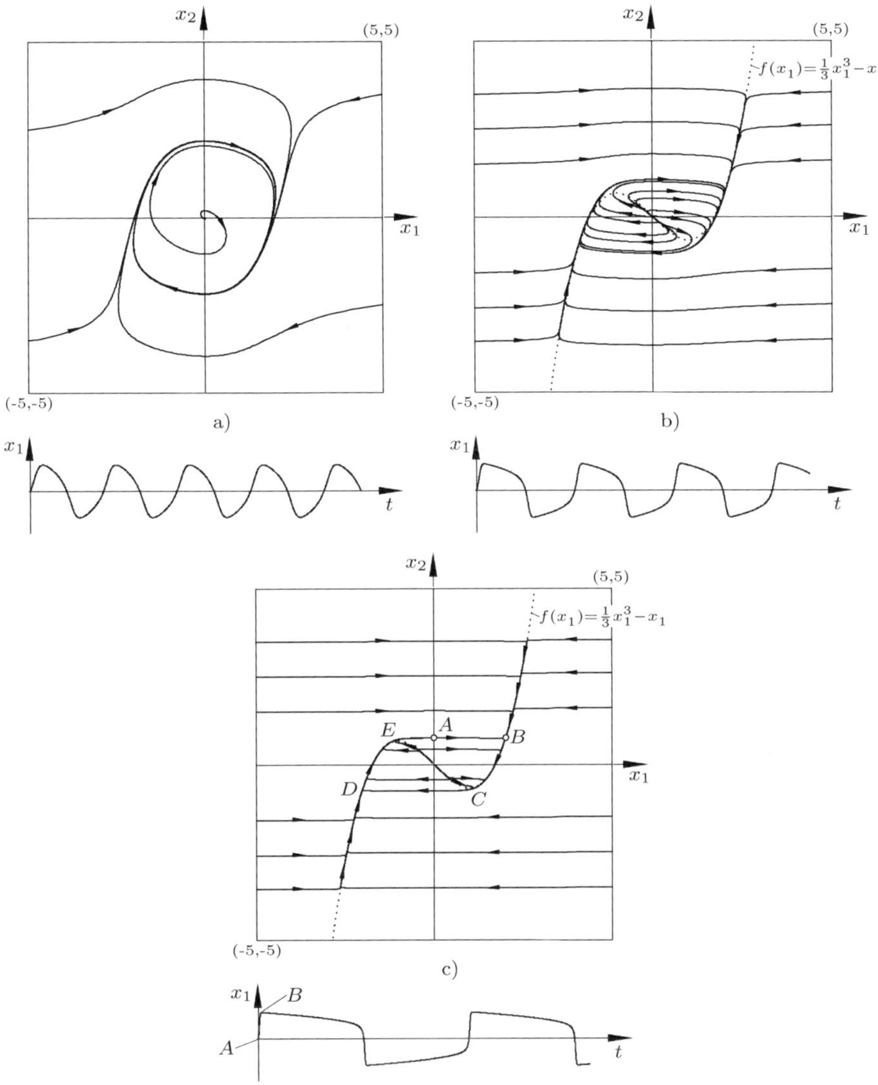

Fig. 10.4.3: Phase portraits and time series of the van der Pol equation in the Liénard formulation, eq. (10.4.14): a) $\mu = 1.0$, b) $\mu = 3.0$, c) $\mu = 10.0$

$$f(x_1) = -\int_0^{x_1} (1 - x_1^2)\mathrm{d}x_1 = -x_1 + \tfrac{1}{3}x_1^3 \tag{10.4.12}$$

applies. From eqs. (10.4.7), (10.4.11) and (10.4.12), $\dot{x}_2$ is determined as

$$\dot{x}_2 = \frac{1}{\mu}\ddot{x}_1 + \frac{\mathrm{d}f(x_1)}{\mathrm{d}x_1}\dot{x}_1 = \frac{1}{\mu}\ddot{x}_1 - (1 - x_1^2)\dot{x}_1 = -\frac{1}{\mu}x_1 \tag{10.4.13}$$

With eq. (10.4.13) and from eq. (10.4.11), we finally obtain the van der Pol equation in the formulation of Liénard

$$\begin{aligned}\dot{x}_1 &= \mu x_2 - \mu\left(\tfrac{1}{3}x_1^3 - x_1\right)\\ \dot{x}_2 &= -\frac{1}{\mu}x_1\end{aligned} \tag{10.4.14}$$

We have also integrated the system of equations eq. (10.4.14) numerically. The phase portraits in the Liénard plane and the time series of the x_1-component for the μ-values $\mu = 1.0,\ 3.0,\ 10.0$ are presented in fig. 10.4.3.

Comparing the limit cycles of figs. 10.4.2 and 10.4.3, we can observe slight differences in shape for small μ-values and pronounced differences for large μ-values. For the value $\mu = 1.0$, the limit cycle in the Liénard plane is still almost circular (fig. 10.4.3a) but is deformed to a parallelogram with rounded corners for larger μ-values (fig. 10.4.3b,c). In the latter case, the dominating influence of the function $f(x_1) = (1/3)x_1^3 - x_1$ on the phase flow is clearly recognisable. According to Liénard, this limit cycle is composed of four curve segments, two cubic ones described by the function $f(x_1)$ and two linear ones. The two $f(x_1)$-branches are traversed with low velocity, the two straight lines, on the other hand, very rapidly. This can be demonstrated by consideration of the periodic time-history of x_1 in fig. 10.4.3c. If we start at the point A with the coordinates $x_1 = 0,\ x_2 = 0.75$, the time course of x_1 indicates that the distance $\overline{AB}$ in the phase space is traversed almost instantaneously. This means that as soon as the oscillator assumes the local minimum (C) or maximum value (E) of the function $f(x_1)$, its response does not proceed through the origin, but tends towards the point B respectively D on the fastest, shortest path. In the Liénard formulation, the relaxation oscillation of the van der Pol equation for large μ-values (fig. 10.4.3c) is thus also recognisable in the phase space.

Hence, relaxation oscillations possess two time scales: one phase in which the restoring force is built up slowly and a second phase above a threshold value which is very short and in which the system relaxes abruptly. Many biological systems such as the heartbeat or the activity of neurons as well as chemical reactions like the Belousov-Zhabotinsky reaction can be modelled by relaxation oscillations (Pikovsky et al., 2003; FitzHugh, 1961; Nagumo et al., 1962).

The characteristic feature of the van der Pol equation is that it exhibits periodic behaviour in spite of the presence of a friction term. One normally assumes that systems with damping and without external driving force tend towards a state of equilibrium. If we consider eq. (10.4.7), however, it becomes clear that the expression in brackets becomes negative for small $|x|$-values, the "friction" in this case simulates an energy supply. We also speak of "negative" damping for the very reason that the damping does not absorb energy from the system, but supplies it. It is evident that such a behaviour is only possible if the system is fed from a special energy reservoir. And this is exactly the case for the simple electrical circuit, fig. 10.4.1a, for which we deduced the van der Pol equation, the voltage source u assuming the role of the energy reservoir. A system possessing such an energy source can gain energy in spite of "friction" respectively "damping". The concepts friction and damping are thus obviously no longer to be understood in the conventional sense; they characterise exclusively the coefficient of the $\dot{x}$-term. Small $|x|$-values hence lead to a supply of energy, large $|x|$-values, on the other hand, absorb energy from the system eq. (10.4.7): it is damped. Along the limit cycle, energy supply and dissipation are in balance (cf. figs. 10.4.2, 10.4.3).

Without theoretical elaborations on the existence and uniqueness of solutions, it is impossible to ascertain whether the numerically determined limit cycle is indeed the only periodic solution of the two-dimensional system and whether just one limit cycle exists for all μ-values. The theorem of Poincaré-Bendixson makes such a fundamental statement on the existence of periodic orbits (Hirsch and Smale, 1974). Before we state this theorem, however, it is necessary to introduce the concept ω limit set (α limit set). From a qualitative point of view, the dynamics is determined by the behaviour of the solutions for $t \to \pm\infty$. The question regarding the evolution of a trajectory for large positive and negative t-values thus arises. The ω respectively α limit sets which describe this temporal evolution are defined as follows: if $\boldsymbol{x}_\gamma(t)$ is the solution of $\dot{\boldsymbol{x}} = \boldsymbol{F}(\boldsymbol{x})$ for a trajectory γ in the phase space $\mathbb{R}^{\mathrm{n}}$, then the ω limit set of γ is described as follows:

$$\omega_\gamma = \{\boldsymbol{x} \in \mathbb{R}^{\mathrm{n}} |\ \text{there exists a sequence}\ \ t_{\mathrm{n}} \to +\infty\ \ \text{with}\ \ \boldsymbol{x}_\gamma(t_{\mathrm{n}}) \to \boldsymbol{x}\}$$

The α limit set is obtained for $t \to -\infty$: simple examples of ω-respectively α limit sets are fixed points and limit cycles.

Theorem (Poincaré-Bendixson):

> A non-empty compact ω or α limit set of a planar flow which contains no fixed point is a closed orbit.

It proves to be a non-trivial undertaking to apply this theorem to the van der Pol equation in order to prove that one single limit cycle exists for all μ-values. Here, we restrict ourselves to sketching the argument roughly. For a detailed proof, we refer the reader to (Guckenheimer and Holmes, 1983; Verhulst, 1990; Hale and Koçak, 1991).

The idea is to construct an open and bounded domain D in the plane; this region contains no points of equilibrium but possesses the characteristic that each solution starting in D remains in D for all $t \geqslant 0$. Thus, D is an open and bounded invariant set. The next step is to prove that, for all initial points $\boldsymbol{x}_0$ lying in D ($\boldsymbol{x}_0 \in D$), the ω limit set has no points on the boundary of D. If this is satisfied, then, as D does not contain any point of equilibrium, the ω limit set must be a periodic orbit according to the Poincaré-Bendixson theorem. For the proof of the uniqueness of this closed orbit, we refer the reader to (Hale and Koçak, 1991).

10.4.2 Driven van der Pol Oscillator

We now consider self-sustained oscillations generated by an external driving force, using the van der Pol equation as an example. By means of the external excitation, oscillatory responses are imposed on the system from outside. This is expressed in the equations of motion through a time-dependent forcing term which is independent of the state variable $\boldsymbol{x}$. In practice, we are above all interested in periodic forcing functions which can, in many cases, be approximated by a harmonic function in time. We hence extend the van der Pol oscillator, eq. (10.4.7), by introducing a right-hand term expressing a harmonic forcing term and obtain the following second-order equation

$$\ddot{x} + \mu(x^2 - 1)\dot{x} + x = F\cos\omega t \tag{10.4.15}$$

where the damping parameter μ is assumed positive. If we now consider the damping term and the external excitation as perturbations of a harmonic motion and if this latter is small, eq. (10.4.15) can be re-written in the form

$$\ddot{x} + x = \varepsilon\bar{F}\cos\omega t - \varepsilon\bar{\mu}(x^2 - 1)\dot{x} \tag{10.4.16}$$

with $\varepsilon\bar{F} = F$ and $\varepsilon\bar{\mu} = \mu$, where we assume $\varepsilon \ll 1$. In the following, we shall dispense with the bars.

Equation (10.4.16) can be written as a non-autonomous system of two first-order equations, cf. eq. (10.4.14)

$$\begin{aligned} \dot{x}_1 &= \quad x_2 - \varepsilon\mu\left(\frac{x_1^3}{3} - x_1\right) \\ \dot{x}_2 &= -x_1 + \varepsilon F\cos\omega t \end{aligned} \tag{10.4.17}$$

Equation (10.4.17) corresponds to the general form

$$\dot{\boldsymbol{x}} = \boldsymbol{f}(\boldsymbol{x}) + \varepsilon\boldsymbol{g}(\boldsymbol{x}, t, \varepsilon) \tag{10.4.18}$$

In order to investigate equations of the form (10.4.18), we can apply methods summarised in the literature under the comprehensive heading perturbation theory.

There is neither an exact analytical solution nor a complete qualitative description for the majority of all non-linear differential equations. The perturbation theory offers procedures for the analytical calculation of approximate solutions for which the influence of small perturbations on the solutions – taken as known – of the unperturbed system is investigated. If we characterise the perturbation by a small parameter ε, it may be shown that deviations from the unperturbed system of the order of magnitude ε emerge after a time interval of the order of magnitude 1. The variation can be found approximately by solving the differential equation for the perturbation in question; this ODE is also known as the "variation equation" of the unperturbed system.

In many cases, however, an interval of observation of the order of magnitude 1 is too short. An increase of the time interval to $1/\varepsilon$ and of the temporal evolution of the variables up to 1 leads to complex problems and is the object of the so-called asymptotic methods of the perturbation theory. The most important of these, above all for engineers, is the averaging method. It originated in Russia with Krylov and Bogoliubov who evolved this procedure as early as the 1930s (Krylov and Bogoliubov, 1947). Its application requires care, however, since it is based on the assumption that the perturbed vector field follows the unperturbed one. We now know, however, that the unperturbed system can be structurally unstable, i.e. small perturbations can impose upon the system completely new solution responses. Fortunately, though, the solution of the perturbed system remains in the direct neighbourhood of the unperturbed system in most cases, at least for a finite period.

We now study the stationary solutions and bifurcations of an externally driven van der Pol oscillator. A pre-requisite of such a study is a knowledge of the averaging method, which we now review briefly.

The averaging method can be approached from two different angles. On the one hand, the procedure can be deduced from the construction of an asymptotic approximation of the solution of differential equations (Sanders et al., 2007); on the other hand, it can be derived by an approach based on the geometric theory of differential equations (Arnold, 1988; Guckenheimer and Holmes, 1983; Wiggins, 1990).

Generally, the solutions of differential equations of the form $\dot{\boldsymbol{x}} = \boldsymbol{f}(\boldsymbol{x}, t, \varepsilon)$ can be approximated by means of averaging methods if the solution of the unperturbed problem $\dot{\boldsymbol{x}} = \boldsymbol{f}(\boldsymbol{x}, t, 0)$ is known. If different time scales occur in the problem, the slow and fast oscillations can be separated with the help of the averaging method and the temporal evolution can be expressed in terms of the slowly varying variables. If, on the other hand, the fast dynamics turns out to be a pure relaxation process, it is expedient to separate the variables by means of adiabatic elimination, see section 6.8.

We present this method briefly in its most elementary form. To this end, we express the system of equations in the following form

$$\dot{\boldsymbol{x}} = \varepsilon \boldsymbol{f}(\boldsymbol{x}, t) + \varepsilon^2 \boldsymbol{g}(\boldsymbol{x}, t, \varepsilon), \quad \boldsymbol{x} \in \mathbb{R}^{\mathrm{n}}, \quad 0 < \varepsilon \ll 1 \tag{10.4.19}$$

where the functions $\boldsymbol{f}$ and $\boldsymbol{g}$ are at least continuously differentiable and periodic in t with a period $T > 0$. The decisive step is the temporal averaging of $\boldsymbol{f}$; this defines an averaged autonomous system associated with eq. (10.4.19)

$$\dot{\boldsymbol{y}} = \varepsilon \bar{\boldsymbol{f}}(\boldsymbol{y}) \stackrel{\text{def}}{=} \varepsilon \frac{1}{T} \int_0^T \boldsymbol{f}(\boldsymbol{y}, t) \mathrm{d}t \tag{10.4.20}$$

It can now be shown that, by means of the following near-identity transformation

$$\boldsymbol{x} = \boldsymbol{y} + \varepsilon \boldsymbol{w}(\boldsymbol{y}, t) \tag{10.4.21}$$

the initial system, eq. (10.4.19), is transformed to a system

$$\dot{\boldsymbol{y}} = \varepsilon \bar{\boldsymbol{f}}(\boldsymbol{y}) + \varepsilon^2 \boldsymbol{f}_1(\boldsymbol{y}, t, \varepsilon) \tag{10.4.22}$$

where $\boldsymbol{f}_1$ is periodic in time with period T. The fact that the transformation of eq. (10.4.21) eliminates the temporal dependence in the terms of the order $\mathcal{O}(\varepsilon)$ by shifting them into terms of the order $\mathcal{O}(\varepsilon^2)$ can be understood most easily by executing the coordinate transformation, eq. (10.4.21), explicitly.

We first split the function $\boldsymbol{f}(\boldsymbol{x}, t)$ of eq. (10.4.19) into mean value and oscillatory components

$$\boldsymbol{f}(\boldsymbol{x}, t) = \bar{\boldsymbol{f}}(\boldsymbol{x}) + \tilde{\boldsymbol{f}}(\boldsymbol{x}, t) \tag{10.4.23}$$

where

$$\bar{\boldsymbol{f}}(\boldsymbol{x}) = \frac{1}{T} \int_0^T \boldsymbol{f}(\boldsymbol{x}, t) \mathrm{d}t \, , \quad \tilde{\boldsymbol{f}}(\boldsymbol{x}, t) = \boldsymbol{f}(\boldsymbol{x}, t) - \frac{1}{T} \int_0^T \boldsymbol{f}(\boldsymbol{x}, t) \mathrm{d}t \tag{10.4.24}$$

Substituting eq. (10.4.23) into the initial equation (10.4.19) yields

$$\dot{\boldsymbol{x}} = \varepsilon\bar{\boldsymbol{f}}(\boldsymbol{x}) + \varepsilon\tilde{\boldsymbol{f}}(\boldsymbol{x},t) + \varepsilon^2\boldsymbol{g}(\boldsymbol{x},t,\varepsilon) \tag{10.4.25}$$

We now carry out the near-identity transformation, eq. (10.4.21), where the function $\boldsymbol{w}(\boldsymbol{y},t)$ is still freely definable. Derivation of eq. (10.4.21) with respect to time yields

$$\dot{\boldsymbol{x}} = \dot{\boldsymbol{y}} + \varepsilon\Big[\frac{\partial \boldsymbol{w}}{\partial t} + \frac{\partial \boldsymbol{w}(\boldsymbol{y},t)}{\partial \boldsymbol{y}}\dot{\boldsymbol{y}}\Big] \tag{10.4.26}$$

The expression $\frac{\partial \boldsymbol{w}}{\partial \boldsymbol{y}}$ is the Jacobian matrix. The transformation, eq. (10.4.21), and eqs. (10.4.19) and (10.4.26) now lead to

$$\dot{\boldsymbol{x}} = \Big[\boldsymbol{I} + \varepsilon\frac{\partial \boldsymbol{w}}{\partial \boldsymbol{y}}\Big]\dot{\boldsymbol{y}} + \varepsilon\frac{\partial \boldsymbol{w}}{\partial t} = \varepsilon\boldsymbol{f}(\boldsymbol{y}+\varepsilon\boldsymbol{w},t) + \varepsilon^2\boldsymbol{g}(\boldsymbol{y}+\varepsilon\boldsymbol{w},t,\varepsilon) \tag{10.4.27}$$

Next applying Taylor expansions at the point $\boldsymbol{y}$ to the functions $\boldsymbol{f}$ and $\boldsymbol{g}$ on the right-hand side of eq. (10.4.27), we obtain

$$\begin{aligned}\boldsymbol{f}(\boldsymbol{y}+\varepsilon\boldsymbol{w},t) &= \boldsymbol{f}(\boldsymbol{y},t) + \varepsilon\frac{\partial \boldsymbol{f}(\boldsymbol{y},t)}{\partial \boldsymbol{y}}\boldsymbol{w}(\boldsymbol{y},t) + \boldsymbol{O}(\varepsilon^2)\\ \boldsymbol{g}(\boldsymbol{y}+\varepsilon\boldsymbol{w},t,\varepsilon) &= \boldsymbol{g}(\boldsymbol{y},t,0) + \boldsymbol{O}(\varepsilon)\end{aligned} \tag{10.4.28}$$

Inserting eq. (10.4.28) into eq. (10.4.27) and splitting the ε-terms in $\boldsymbol{f}(\boldsymbol{y},t)$ into mean value and oscillatory component as in eq. (10.4.23), we find

$$\begin{aligned}\dot{\boldsymbol{y}} = \Big[\boldsymbol{I} + \varepsilon\frac{\partial \boldsymbol{w}}{\partial \boldsymbol{y}}\Big]^{-1}\Big\{\ &\varepsilon\ \Big[\bar{\boldsymbol{f}}(\boldsymbol{y}) + \tilde{\boldsymbol{f}}(\boldsymbol{y},t) - \frac{\partial \boldsymbol{w}}{\partial t}\Big]\\ &+\varepsilon^2\ \Big[\frac{\partial \boldsymbol{f}(\boldsymbol{y},t)}{\partial \boldsymbol{y}}\boldsymbol{w}(\boldsymbol{y},t) + \boldsymbol{g}(\boldsymbol{y},t,0)\Big] + \boldsymbol{O}(\varepsilon^3)\Big\}\end{aligned} \tag{10.4.29}$$

For sufficiently small ε, the following approximation holds for the inverse of the operator $[\boldsymbol{I} + \varepsilon\frac{\partial \boldsymbol{w}}{\partial \boldsymbol{y}}]$

$$\Big[\boldsymbol{I} + \varepsilon\frac{\partial \boldsymbol{w}}{\partial \boldsymbol{y}}\Big]^{-1} = \boldsymbol{I} - \varepsilon\frac{\partial \boldsymbol{w}}{\partial \boldsymbol{y}} + \boldsymbol{O}(\varepsilon^2) \tag{10.4.30}$$

Equation (10.4.29) thus takes the form

$$\dot{\boldsymbol{y}} = \varepsilon\Big[\bar{\boldsymbol{f}} + \tilde{\boldsymbol{f}} - \frac{\partial \boldsymbol{w}}{\partial t}\Big] + \varepsilon^2\Big[\frac{\partial \boldsymbol{f}}{\partial \boldsymbol{y}}\boldsymbol{w} + \boldsymbol{g} - \frac{\partial \boldsymbol{w}}{\partial \boldsymbol{y}}\boldsymbol{f} + \frac{\partial \boldsymbol{w}}{\partial \boldsymbol{y}}\frac{\partial \boldsymbol{w}}{\partial t}\Big] + \boldsymbol{O}(\varepsilon^3) \tag{10.4.31}$$

After application of the substitution

$$\boldsymbol{f}_1(\boldsymbol{y},t,\varepsilon) = \frac{\partial \boldsymbol{f}}{\partial \boldsymbol{y}}\boldsymbol{w} + \boldsymbol{g} - \frac{\partial \boldsymbol{w}}{\partial \boldsymbol{y}}\boldsymbol{f} + \frac{\partial \boldsymbol{w}}{\partial \boldsymbol{y}}\frac{\partial \boldsymbol{w}}{\partial t} + \boldsymbol{O}(\varepsilon) \tag{10.4.32}$$

eq. (10.4.31) assumes the form

$$\dot{\boldsymbol{y}} = \varepsilon\Big[\bar{\boldsymbol{f}}(\boldsymbol{y}) + \tilde{\boldsymbol{f}}(\boldsymbol{y},t) - \frac{\partial \boldsymbol{w}(\boldsymbol{y},t)}{\partial t}\Big] + \varepsilon^2\boldsymbol{f}_1(\boldsymbol{y},t,\varepsilon) \tag{10.4.33}$$

The function $\boldsymbol{w}$ is still freely definable. We select it so that

$$\frac{\partial \boldsymbol{w}}{\partial t} = \tilde{\boldsymbol{f}}(\boldsymbol{y}, t) \tag{10.4.34}$$

Following eq. (10.4.34), $\boldsymbol{w}$ can be established by time integration and, by means of the near-identity transformation, eq. (10.4.21), we finally obtain the desired eq. (10.4.22)

$$\dot{\boldsymbol{y}} = \varepsilon \bar{\boldsymbol{f}}(\boldsymbol{y}) + \varepsilon^2 \boldsymbol{f}_1(\boldsymbol{y}, t, \varepsilon) \tag{10.4.35}$$

The continuity of the transformation eq. (10.4.21) as well as the smallness of ε allow us to gain important information on the behaviour of the initial eq. (10.4.19) from a scrutiny of the relation eq. (10.4.22) respectively (10.4.20). The averaging theorem, which we cite here without proof, expresses this comparison between the initial and the averaged system, cf. (Guckenheimer and Holmes, 1983).

Averaging Theorem:

i. If $\boldsymbol{x}(t)$ respectively $\boldsymbol{y}(t)$ are solutions of eqs. (10.4.19) respectively (10.4.20) for the initial conditions $\boldsymbol{x}(t_0) = \boldsymbol{x}_0$, $\boldsymbol{y}(t_0) = \boldsymbol{y}_0$ and if $|\boldsymbol{x}_0 - \boldsymbol{y}_0| = \mathcal{O}(\varepsilon)$, then $|\boldsymbol{x}(t) - \boldsymbol{y}(t)| = \mathcal{O}(\varepsilon)$ holds for a time interval $t \sim 1/\varepsilon$ (approximation of the solution).

ii. If eq. (10.4.20) possesses a hyperbolic fixed point $\boldsymbol{x}_s$, an ε_0-neighbourhood exists so that, for every ε with $0 < \varepsilon \leqslant \varepsilon_0$, eq.(10.4.19) possesses a single hyperbolic orbit $\boldsymbol{x}_\varepsilon(t) = \boldsymbol{x}_s + \boldsymbol{\mathcal{O}}(\varepsilon)$ of the same stability type as $\boldsymbol{x}_s$ (identical stability behaviour).

iii. If the solution $\boldsymbol{x}^s(t)$ of eq. (10.4.19) lies on the stable manifold of the hyperbolic orbit $\boldsymbol{x}_\varepsilon = \boldsymbol{x}_s + \boldsymbol{\mathcal{O}}(\varepsilon)$ and if the solution $\boldsymbol{y}^s(t)$ of eq. (10.4.20) lies on the stable manifold of the hyperbolic fixed point $\boldsymbol{x}_s$ and if $|\boldsymbol{x}^s(0) - \boldsymbol{y}^s(0)| = \mathcal{O}(\varepsilon)$ holds for the initial conditions, then $|\boldsymbol{x}^s(t) - \boldsymbol{y}^s(t)| = \mathcal{O}(\varepsilon)$ holds for $t \in [0, \infty)$. Similar results apply to solutions lying in the unstable manifolds on the time interval $t \in (-\infty, 0]$ (approximation of the stable and unstable manifold).

Beside the aforementioned important characteristics, the averaging method gains further significance due to its ability to provide approximate Poincaré maps, cf. (Wiggins, 1990).

Our aim is to apply the averaging method to the van der Pol eq. (10.4.16) respectively eq. (10.4.17). To this end, however, it is necessary to transform the equation first to its standard form, eq. (10.4.19).

To achieve this, we apply the so-called van der Pol transformation; cf. (Wiggins, 1990; Guckenheimer and Holmes, 1983). In this way, the initial system, expressed in $\boldsymbol{x}$-coordinates, is transformed into a co-rotating $\boldsymbol{y}$-system with period $T = 2\pi/\omega$. This means that a fixed point in the transformed system corresponds to a harmonic oscillation with frequency ω. We first re-write eq. (10.4.17) in the somewhat abridged notation

$$\dot{\boldsymbol{x}} = \boldsymbol{L}\boldsymbol{x} + \varepsilon \begin{bmatrix} h_1 \\ h_2 \end{bmatrix} \tag{10.4.36}$$

with

$$\boldsymbol{L} = \begin{bmatrix} 0 & 1 \\ -1 & 0 \end{bmatrix}, \quad h_1 = -\mu\Big(\frac{x_1^3}{3} - x_1\Big), \quad h_2 = F\cos\omega t$$

We now introduce the van der Pol transformation in the form

$$\begin{bmatrix} y_1 \\ y_2 \end{bmatrix} = \boldsymbol{A}\begin{bmatrix} x_1 \\ x_2 \end{bmatrix} \qquad \text{resp.} \qquad \begin{bmatrix} x_1 \\ x_2 \end{bmatrix} = \boldsymbol{A}^{-1}\begin{bmatrix} y_1 \\ y_2 \end{bmatrix} \tag{10.4.37}$$

Using the notation $s \equiv \sin\omega t,\ c \equiv \cos\omega t$, the time-dependent transformation matrix is given by

$$\boldsymbol{A}(t) = \begin{bmatrix} c & -\frac{1}{\omega}s \\ -s & -\frac{1}{\omega}c \end{bmatrix} \qquad \text{resp.} \qquad \boldsymbol{A}^{-1}(t) = \begin{bmatrix} c & -s \\ -\omega s & -\omega c \end{bmatrix} \tag{10.4.38}$$

We now differentiate eq. (10.4.37) with respect to t and, considering eq. (10.4.36), obtain

$$\begin{bmatrix} \dot{y}_1 \\ \dot{y}_2 \end{bmatrix} = [\dot{\boldsymbol{A}}\boldsymbol{A}^{-1} + \boldsymbol{A}\boldsymbol{L}\boldsymbol{A}^{-1}]\begin{bmatrix} y_1 \\ y_2 \end{bmatrix} + \boldsymbol{A}\varepsilon\begin{bmatrix} h_1 \\ h_2 \end{bmatrix} \tag{10.4.39}$$

Applying the transformations, we obtain

$$\begin{aligned} \dot{y}_1 &= \frac{(1-\omega^2)}{\omega}(y_1 sc - y_2 s^2) - \varepsilon\mu\Big[\tfrac{1}{3}(y_1 c - y_2 s)^3 - (y_1 c - y_2 s)\Big]c - \frac{\varepsilon}{\omega}Fsc \\ \dot{y}_2 &= \frac{(1-\omega^2)}{\omega}(y_1 c^2 - y_2 sc) + \varepsilon\mu\Big[\tfrac{1}{3}(y_1 c - y_2 s)^3 - (y_1 c - y_2 s)\Big]s - \frac{\varepsilon}{\omega}Fc^2 \end{aligned} \tag{10.4.40}$$

We recall that the unperturbed system, eq. (10.4.17), corresponds to a harmonic oscillator with the eigenfrequency $\omega_0 = 1$ for $\varepsilon = 0$. We are interested in the resonance case $\omega \approx \omega_0 = 1$, i.e. we make the following assumption

$$1 - \omega^2 = \mathcal{O}(\varepsilon) \quad \text{resp.} \quad \sigma := \frac{1-\omega^2}{\varepsilon} = \mathcal{O}(1) \tag{10.4.41}$$

In the case of hysteretic damping, the damping term μ is inversely proportional to the excitation frequency ω. We thus write

$$\mu = \frac{1}{\omega} \tag{10.4.42}$$

The approximation of eq. (10.4.41) implies that the averaging method is carried out only up to terms of the order $\mathcal{O}(\varepsilon)$. If we also take eq. (10.4.42) into account, we finally obtain eq. (10.4.40) in the standard form for the application of the averaging method, cf. eq. (10.4.19)

$$\begin{aligned} \dot{y}_1 &= \frac{\varepsilon}{\omega}\Big[\sigma(y_1 sc - y_2 s^2) - \tfrac{1}{3}(y_1 c - y_2 s)^3 c + (y_1 c - y_2 s)c - Fsc\Big] + \mathcal{O}(\varepsilon^2) \\ \dot{y}_2 &= \frac{\varepsilon}{\omega}\Big[\sigma(y_1 c^2 - y_2 sc) + \tfrac{1}{3}(y_1 c - y_2 s)^3 s - (y_1 c - y_2 s)s - Fc^2\Big] + \mathcal{O}(\varepsilon^2) \end{aligned} \tag{10.4.43}$$

respectively

$$\dot{\boldsymbol{y}} = \varepsilon \boldsymbol{f}(\boldsymbol{y}, t) + \boldsymbol{\mathcal{O}}(\varepsilon^2)$$

Starting from eq. (10.4.43), we now split the function $\boldsymbol{f}(\boldsymbol{y}, t)$ according to eq. (10.4.23) into the time-independent $\bar{\boldsymbol{f}}(\boldsymbol{y})$-component and the oscillating $\tilde{\boldsymbol{f}}(\boldsymbol{y}, t)$-component. We restrict ourselves to determining the $\bar{\boldsymbol{f}}(\boldsymbol{y})$-component. By taking the average respectively by integrating the trigonometric function over a period $T = 2\pi/\omega$, we deduce, in accordance with eq. (10.4.20), the autonomous system of $\boldsymbol{\mathcal{O}}(\varepsilon)$-th order

$$\begin{bmatrix} \dot{y}_1 \\ \dot{y}_2 \end{bmatrix} = \frac{\varepsilon}{2\omega} \begin{bmatrix} y_1 - \sigma y_2 - \frac{1}{4} y_1 (y_1^2 + y_2^2) \\ \sigma y_1 + y_2 - \frac{1}{4} y_2 (y_1^2 + y_2^2) - F \end{bmatrix} \stackrel{\text{def}}{=} \varepsilon \bar{\boldsymbol{f}}(\boldsymbol{y}) \tag{10.4.44}$$

Equation (10.4.44) can be simplified by means of the following scaling

$$\begin{aligned} t &\to \frac{2\omega}{\varepsilon} t \\ y_1 &\to 2y_1 \\ y_2 &\to 2y_2 \end{aligned} \tag{10.4.45}$$

We finally obtain the autonomous system of $\boldsymbol{\mathcal{O}}(\varepsilon)$-th order

$$\begin{bmatrix} \dot{y}_1 \\ \dot{y}_2 \end{bmatrix} = \begin{bmatrix} y_1 - \sigma y_2 - y_1 (y_1^2 + y_2^2) \\ \sigma y_1 + y_2 - y_2 (y_1^2 + y_2^2) - \gamma \end{bmatrix} \tag{10.4.46}$$

equivalent to eq. (10.4.36), where $\gamma = F/2$. We recall that σ is the damping here and γ the excitation amplitude.

We now investigate the dynamics of the driven van der Pol equation (10.4.16) by considering the approximated system, eq. (10.4.46). In the co-rotating frame, hyperbolic fixed points of eq. (10.4.46) represent periodic solutions of the same stability; hyperbolic orbits correspond to quasi-periodic solutions of eq. (10.4.16). In the engineering sciences, the first case is often denoted as synchronisation respectively phase locking of excitation frequency and eigenfrequency (Pikovsky et al., 2003).

Our discussion now follows the presentation in (Holmes and Rand, 1978); as illustrations, we provide complementary computer results gained numerically. The first investigations naturally go back to van der Pol (1927). Holmes and Rand were also able to base their work on detailed studies of M. Cartwright (1948) concerning the structural instability of the forced van der Pol equation and prove the existence of homoclinic orbits (saddle connections) which lead to global homoclinic bifurcations. This had already been conjectured by Cartwright and Gillies (1954).

In order to gain an overview of the dynamic behaviour, it is helpful to demonstrate on phase portraits the dependence on the two control parameters σ and γ. In Chapter 6, we discussed the basic patterns of local bifurcations for one-dimensional equations in dependence of a single control parameter. The local bifurcations occur in each case at a critical point. It is easy to realise that, for the two control parameters σ and γ, the critical points in the σ, γ-parameter plane appear as curves (cf. fig. 10.4.4).

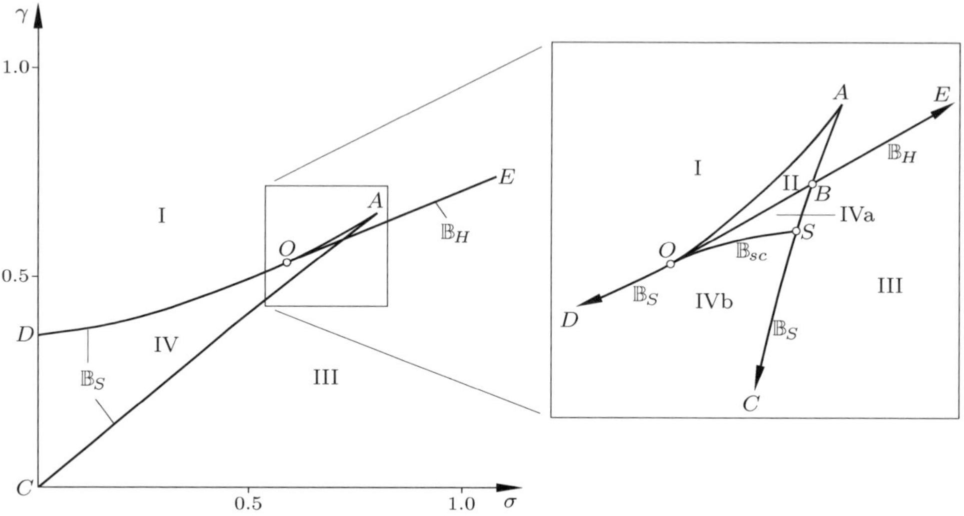

Fig. 10.4.4: The bifurcation sets $\mathbb{B}_S$, $\mathbb{B}_H$, $\mathbb{B}_{sc}$ of the van der Pol equation (10.4.46) as approximated by the averaging method:
i. $\mathbb{B}_S$ ... saddle-node bifurcations
ii. $\mathbb{B}_H$ ... Hopf bifurcations
iii. $\mathbb{B}_{sc}$ ... saddle connections;
after (Guckenheimer and Holmes, 1983)

If a local or global bifurcation occurs, the system loses its structural stability. In the two-dimensional case, as here, conditions for structural instability can be given on the basis of the theorem of Peixoto (see section 8.2). In fact, a bifurcation exists if one of the following three points is not satisfied:

i. all points of equilibrium are hyperbolic,
ii. all periodic orbits are hyperbolic,
iii. no homoclinic or heteroclinic orbits exist.

The parameter values for which the above three points are violated form the bifurcation set $\mathbb{B}$. We distinguish between local and global bifurcations. The local bifurcation can be obtained by linear stability analyses. The local structural instabilities relevant here are saddle-node bifurcations which emerge for $(\sigma, \gamma) \in \mathbb{B}_S$ and Hopf bifurcations for $(\sigma, \gamma) \in \mathbb{B}_H$. As global bifurcations, homoclinic bifurcations take place for $(\sigma, \gamma) \in \mathbb{B}_{sc}$.

In the framework of this section, we present only results; as far as the analytical determination of the bifurcation set $\mathbb{B}$ is concerned, we refer the reader to the literature (Holmes and Rand, 1978; Guckenheimer and Holmes, 1983) and the references given there. The curved segments DA and CA with the designation $\mathbb{B}_S$ in fig. 10.4.4 mark saddle-node bifurcations. The straight line OE ($\mathbb{B}_H$) merging tangentially into the curve DA at O characterises Hopf bifurcations. The curve OS ($\mathbb{B}_{sc}$) dividing the region IV into the sub-regions IVa and IVb indicates homoclinic bifurcations.

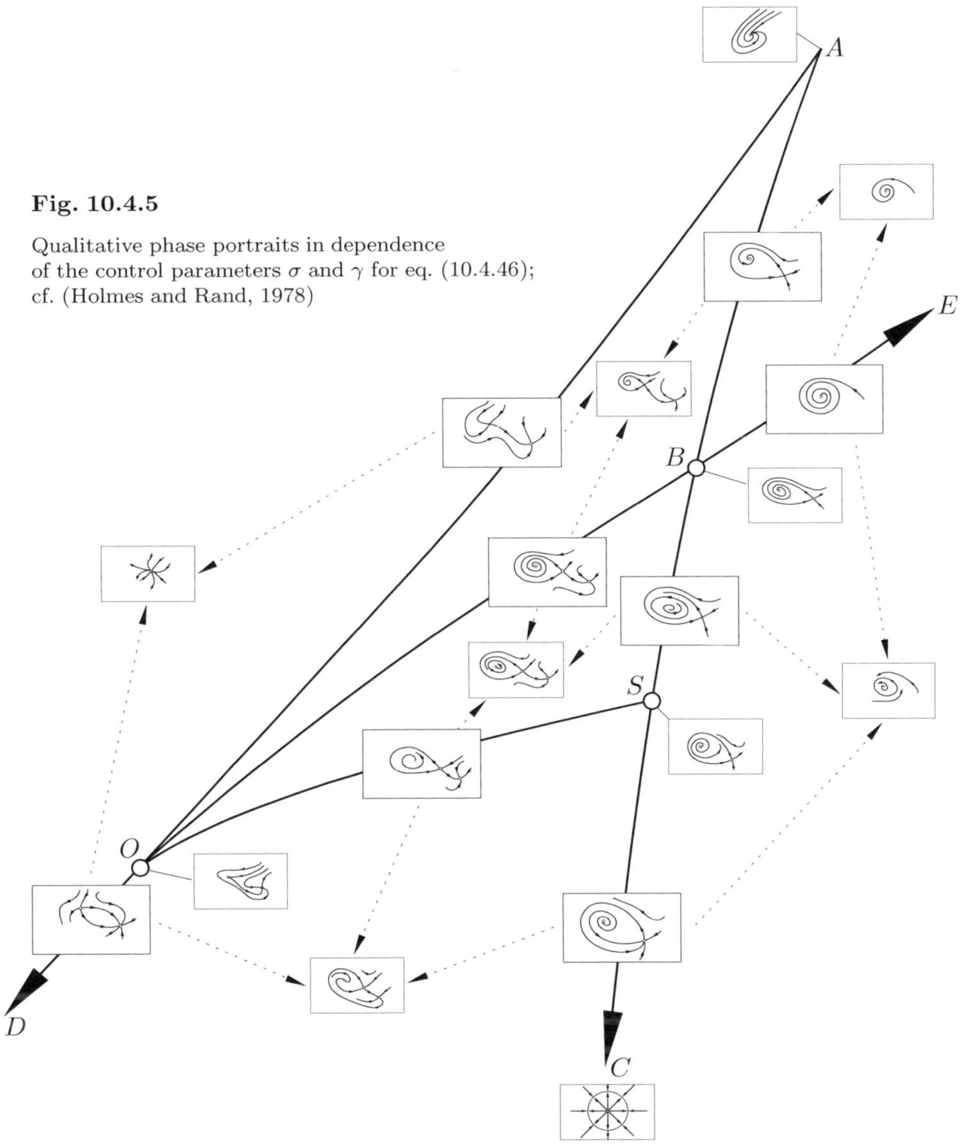

Fig. 10.4.5

Qualitative phase portraits in dependence of the control parameters σ and γ for eq. (10.4.46); cf. (Holmes and Rand, 1978)

Figure 10.4.5 summarises the phase portraits corresponding to the five regions I to IVb and to the bifurcation curves $\mathbb{B}_S, \mathbb{B}_H, \mathbb{B}_{sc}$ (cf. the numerically determined phase portraits in figs. 10.4.6, 10.4.7 and 10.4.8). In the individual regions, the following limit sets characterise the dynamical behaviour:

Region I:	one stable fixed point (figs. 10.4.5, 10.4.6a)
Region II:	one stable focus, a saddle and a stable node (figs. 10.4.5, 10.4.6c,d)
Region III:	one unstable focus and a stable limit cycle (figs. 10.4.5, 10.4.6b)
Region IVa:	one stable focus, a stable limit cycle, a saddle and a stable node (figs. 10.4.5, 10.4.7a,b)
Region IVb:	one unstable focus, a saddle and a stable node (figs. 10.4.5, 10.4.7c,d)

Our interest is now directed at the bifurcations which take place at the transition from one region to a neighbouring one. We first consider local bifurcations established by conventional linear analyses. In Chapter 6, we presented the basic bifurcation patterns of one-parametric systems. For the analysis of two-parametric systems, one parameter, e.g. σ, is kept constant and the bifurcation is investigated in dependence of the second, in this case γ. We obtain the following bifurcation alternatives:

i. Saddle-node bifurcation:

For the parameter values on the two curved segments DA and AC, denoted $\mathbb{B}_S$, a hyperbolic and a non-hyperbolic fixed point exist. The hyperbolic fixed point is either a stable node or a focus; the non-hyperbolic fixed point is a saddle node (cf. figs. 10.4.5 and 10.4.8): saddle-node bifurcations emerge, i.e. two states of equilibrium coalesce and cancel one another out. At the transition from region II to I, only two of the three states of equilibrium present in I (saddle, stable focus and stable node) participate in the bifurcation process, namely the saddle and the stable focus. These two merge in $\mathbb{B}_S$ (saddle node state) and cancel one another so that only the stable node is left in I (figs. 10.4.5 and 10.4.6). On the bifurcation curve BC, which defines the transition from IV to III, the saddle node lies on a closed orbit (figs. 10.4.5 and 10.4.8). At the transition, the saddle node vanishes; what remains in region III is a stable limit cycle and an unstable focus (figs. 10.4.5 and 10.4.6b). The eigenvalue constellation of the saddle node on DO is $(0, +)$, on OA $(0, -)$ and on BC also $(0, -)$.

ii. Hopf bifurcation:

Along the second bifurcation curve OE, denoted $\mathbb{B}_H$, Hopf bifurcations take place. At the transition from region I to III and from II to IVa, the stable focus becomes unstable (figs. 10.4.5 - 10.4.7), i.e. the previously complex conjugate eigenvalues with negative real parts become purely imaginary with a non-zero imaginary component. The bifurcation is a super-critical Hopf bifurcation, i.e. the limit cycle is stable (cf. fig. 6.4.7).

iii. Degenerate bifurcations:

In addition to the two classic bifurcation types, three further bifurcations take place which are denoted degenerate; cf. (Guckenheimer and Holmes, 1983):

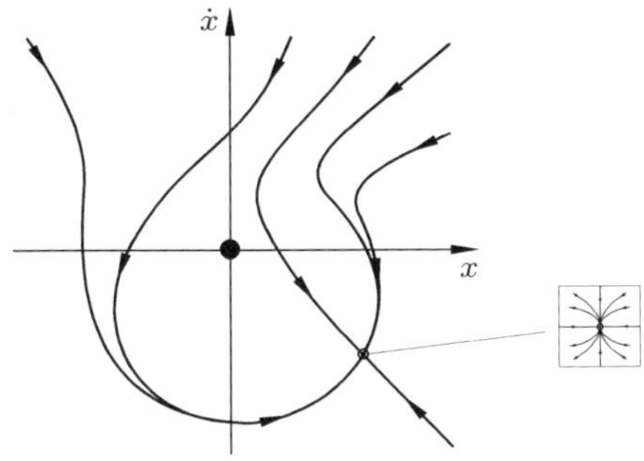

a) point P_{I} ($\sigma=0.4$, $\gamma=0.6$)

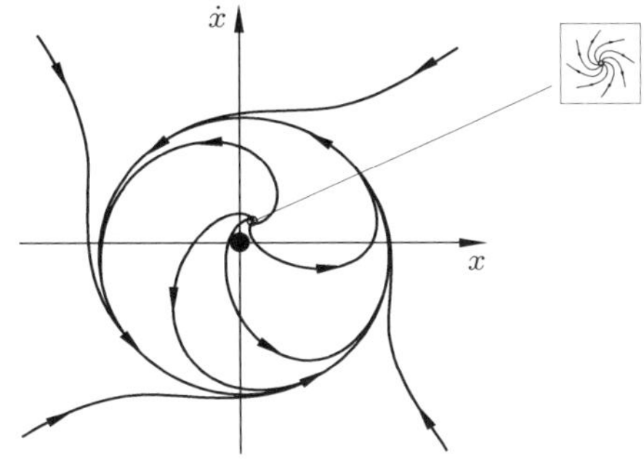

b) point P_{III} ($\sigma=0.6$, $\gamma=0.2$)

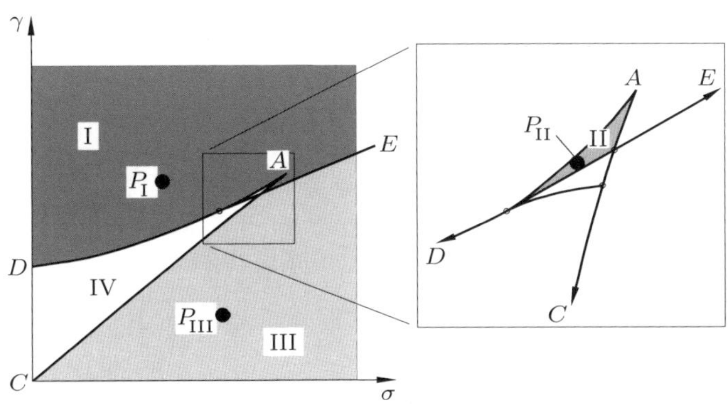

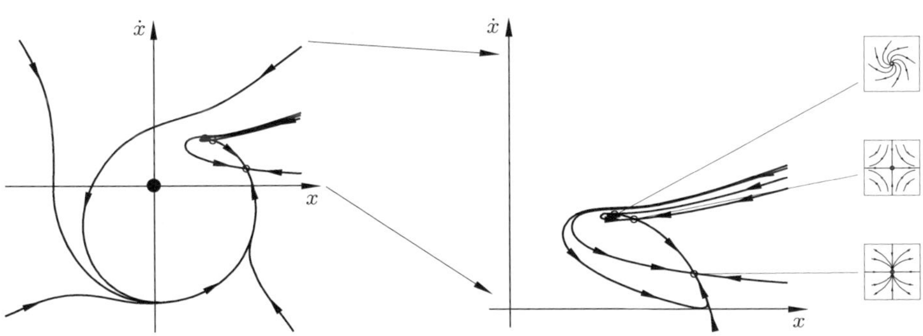

c) point P_{II} ($\sigma=0.54$, $\gamma=0.520364$) d)

Fig. 10.4.6: Phase portraits of the regions I, II, III and selective enlargement of portrait II: a) region I, b) region III, c) region II, d) selective enlargement of c) (Sachon, 1991)

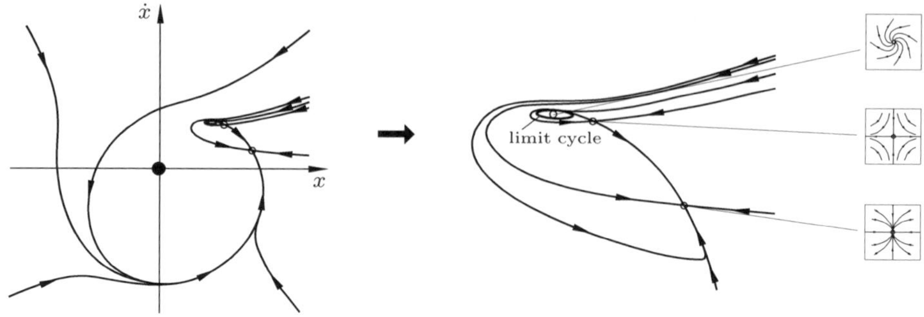

a) point P_{IVa} ($\sigma = 0.54$, $\gamma = 0.52$) b)

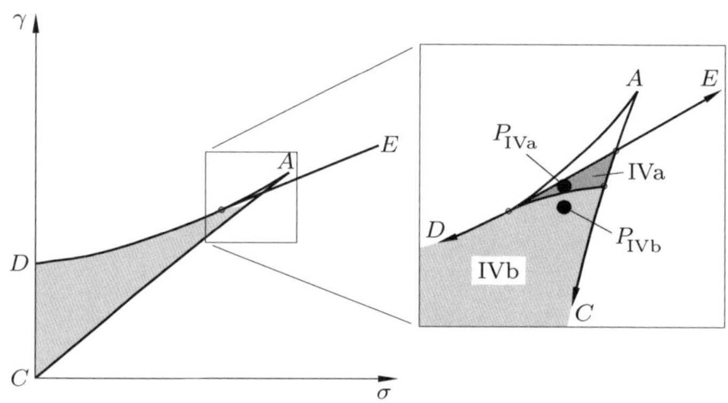

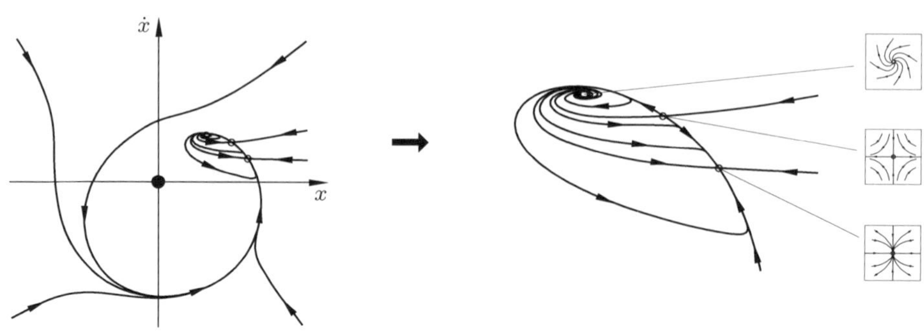

c) point P_{IVb} ($\sigma = 0.54$, $\gamma = 0.517$) d)

Fig. 10.4.7: Phase portraits of the regions IVa and IVb and selective enlargements:
a) region IVa, b) selective enlargement of a)
c) region IVb, d) selective enlargement of c) (Sachon, 1991)

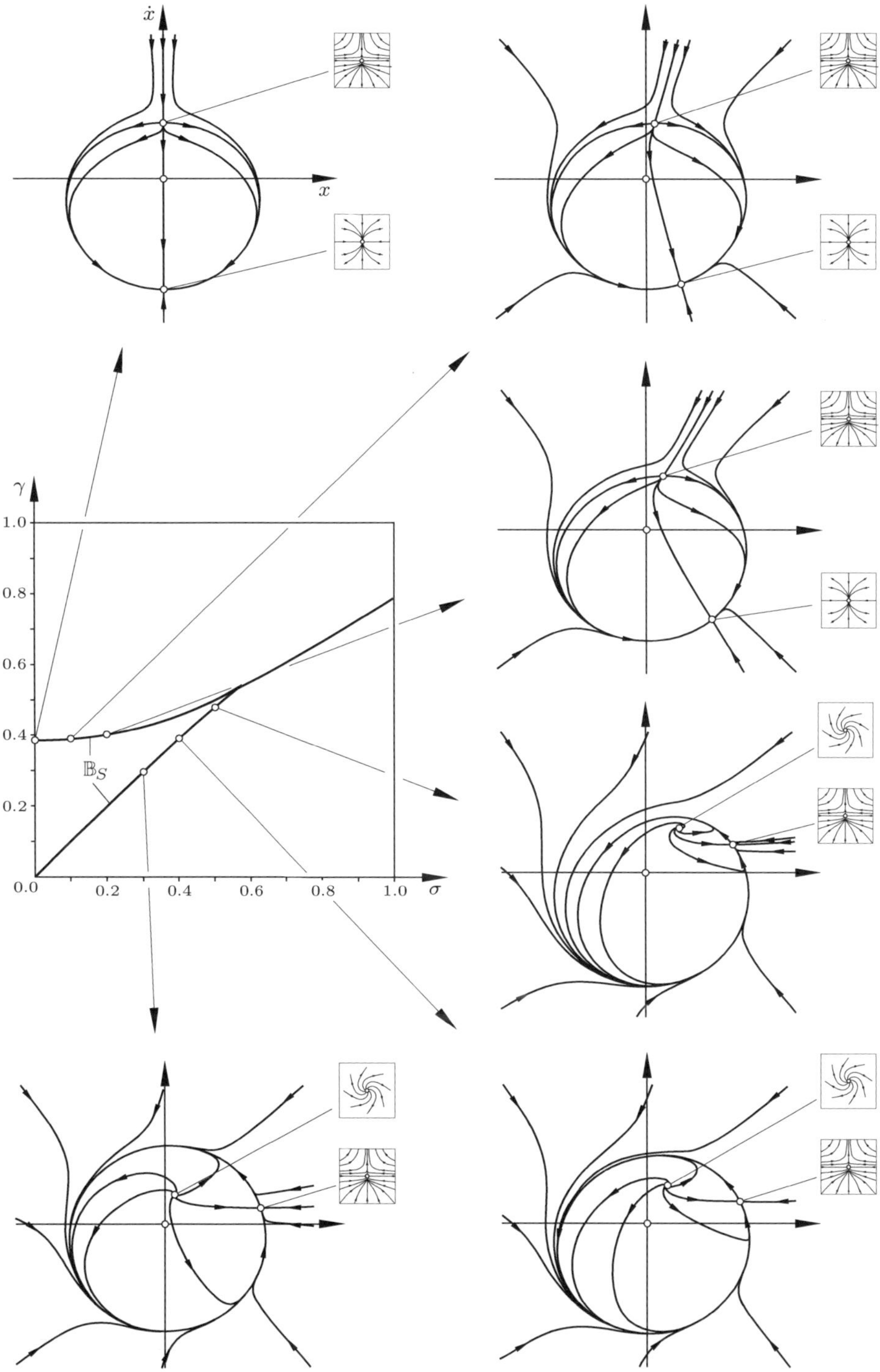

Fig. 10.4.8: Phase portraits corresponding to the saddle-node bifurcation curve $\mathbb{B}_S$ (Sachon, 1991)

a. Point $A, (\sigma, \gamma) = (1/\sqrt{3}, \sqrt{8/27})$: for the parameter values in point A ($\mathbb{B}_S$), only one single non-hyperbolic fixed point exists with the eigenvalues $(0, -)$ (fig. 10.4.9a).

b. Point $O, (\sigma, \gamma) = (0.5, 0.5)$: for the parameter values in the point O ($\mathbb{B}_S$), a stable node and a doubly degenerate fixed point with a zero eigenvalue of the multiplicity two exist (fig. 10.4.9b).

c. Point $C, (\sigma, \gamma) = (0, 0)$: for the parameter values in the point C ($\mathbb{B}_S$), an unstable node and a ring of degenerate saddle nodes exist (fig. 10.4.9c).

iv. Global bifurcation:

Holmes and Rand (1978) were able to prove that the additional bifurcation conjectured in region IV by Cartwright and Gillies does indeed exist. It is a homoclinic bifurcation (cf. sections 10.1, 10.3), i.e. the breaking open of a homoclinic orbit (cf. fig. 10.1.4). This global bifurcation takes place on the curve OS ($\mathbb{B}_{sc}$) which is tangent to AO and OB at O (fig. 10.4.5). At the transition from IVa to IVb, the following occurs: the limit cycle in IVa grows when approaching OS ($\mathbb{B}_{sc}$) so much that it becomes the homoclinic orbit of the saddle point (state $\mathbb{B}_{sc}$) and then breaks open again in region IVb.

We conclude our discussion of the averaged equations (10.4.46) with a physical interpretation of the results we have gained so far. In the region I – and in practice in IVb, as well – all the initial conditions end in a stable node. A stable node of the averaged system, however, means a periodic orbit of period $T = 2\pi/\omega$ of the initial system. The driving frequency and the eigenfrequency are synchronous; we also speak of phase locking. In the regions III and IVa, the stable (hyperbolic) limit cycle corresponds to a stable torus of the initial system, eq. (10.4.16).The solution behaviour of the original system is almost multiple periodic due to the presence of two frequencies, the driving frequency and eigenfrequency. This effect of the drifting of a "periodic" solution can also be observed in experiments; cf. (Holmes and Rand, 1978; Wiggins, 1990).

In the preceding discussion of the unforced van der Pol equation, see eq. (10.4.7), we have seen that we can prove the existence of a single limit cycle for all damping values μ. If an external driving force is imposed and if the driving amplitude F and damping value μ are small, we can win a comprehensive overview of the system response with the aid of the averaging method. We observe in particular that several attractors can emerge simultaneously so that – depending on the initial condition – varying types of long-term behaviour can be generated.

The investigation of the van der Pol equation becomes more difficult in the case of large non-linearities and a strong external forcing (Guckenheimer, 1980; Phillipson and Schuster, 2002). Fundamental research was undertaken by Cartwright and Littlewood (1945) and Levinson (1949) who were able to prove that, in particular parameter ranges, two different subharmonic states of motion coexist, the trajectories choosing one of the two periodic attractors only after a long transient phase depending on the initial condition. According to Levinson's argumentation, the dynamics in the transient phase, during which two periodic attractors compete for

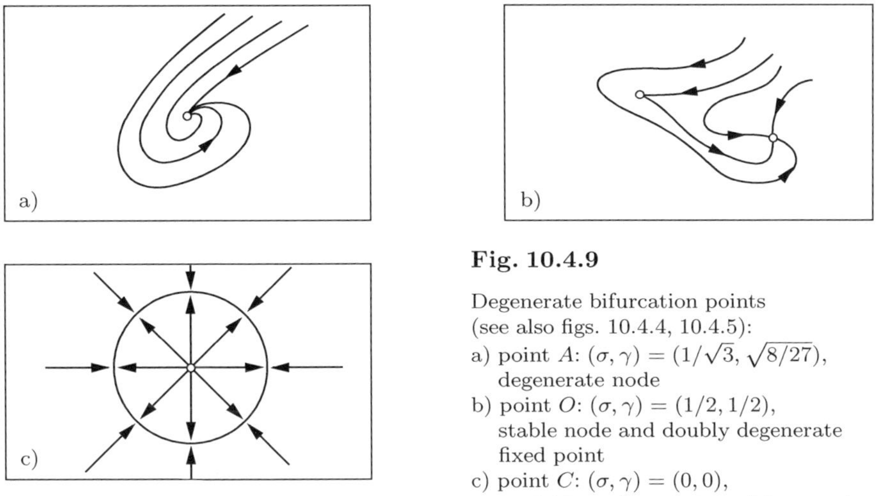

Fig. 10.4.9

Degenerate bifurcation points (see also figs. 10.4.4, 10.4.5):
a) point A: $(\sigma, \gamma) = (1/\sqrt{3}, \sqrt{8/27})$, degenerate node
b) point O: $(\sigma, \gamma) = (1/2, 1/2)$, stable node and doubly degenerate fixed point
c) point C: $(\sigma, \gamma) = (0, 0)$, unstable node and ring of degenerate saddle nodes;
after (Guckenheimer and Holmes, 1983)

the asymptotic behaviour of the individual trajectories, can be extremely complex. In 1963, the American mathematician Stephen Smale succeeded in elucidating the mechanism leading to this chaotic transitional state with the aid of his – now classic – horseshoe map (Smale, 1963).

Transient chaos can be explained by means of a relatively simple determinisitic map which – like the Bernoulli shift (section 5.6.1) and the Hénon map (section 10.2) – is based on a stretching and folding process. This is the very state we meet in the case of the van der Pol equation for large μ, F-values, namely that the complex behaviour only represents a transitional phase. What finally prevails in this case is not chaos, but the periodic state.

While searching for a chaotic attractor, Shaw modified the conventional forced van der Pol equation by imposing on the system not an accelerating but a velocity-forcing excitation. This type of external excitation, unusual in mechanical systems, can be realised in electrical circuits and chemical systems, however. Shaw proposes the following version of the van der Pol equation, see (Shaw, 1981b), eq. (38)

$$\begin{aligned} \dot{x} &= ky + \mu x(a - y^2) \\ \dot{y} &= -x + A \sin \omega t \end{aligned} \tag{10.4.47}$$

For this type of equation, he was able to prove a chaotic attractor for the parameters $k = 0.63$, $\mu = 10.0$, $a = 0.07$, $A = 0.4$, $\omega = 2.0$. Thompson took up the Shaw equation in order to verify the fundamental features of this attractor numerically, cf. (Thompson and Stewart, 2001). The characteristics of this Birkhoff-Shaw attractor are the formation of beaks and the mixing of the trajectories in the three-dimensional phase space (x_1, x_2, θ). In this connection, mixing means that neighbouring trajectories of the attractor do not remain neighbouring. In fig. 10.4.10, we have chosen for

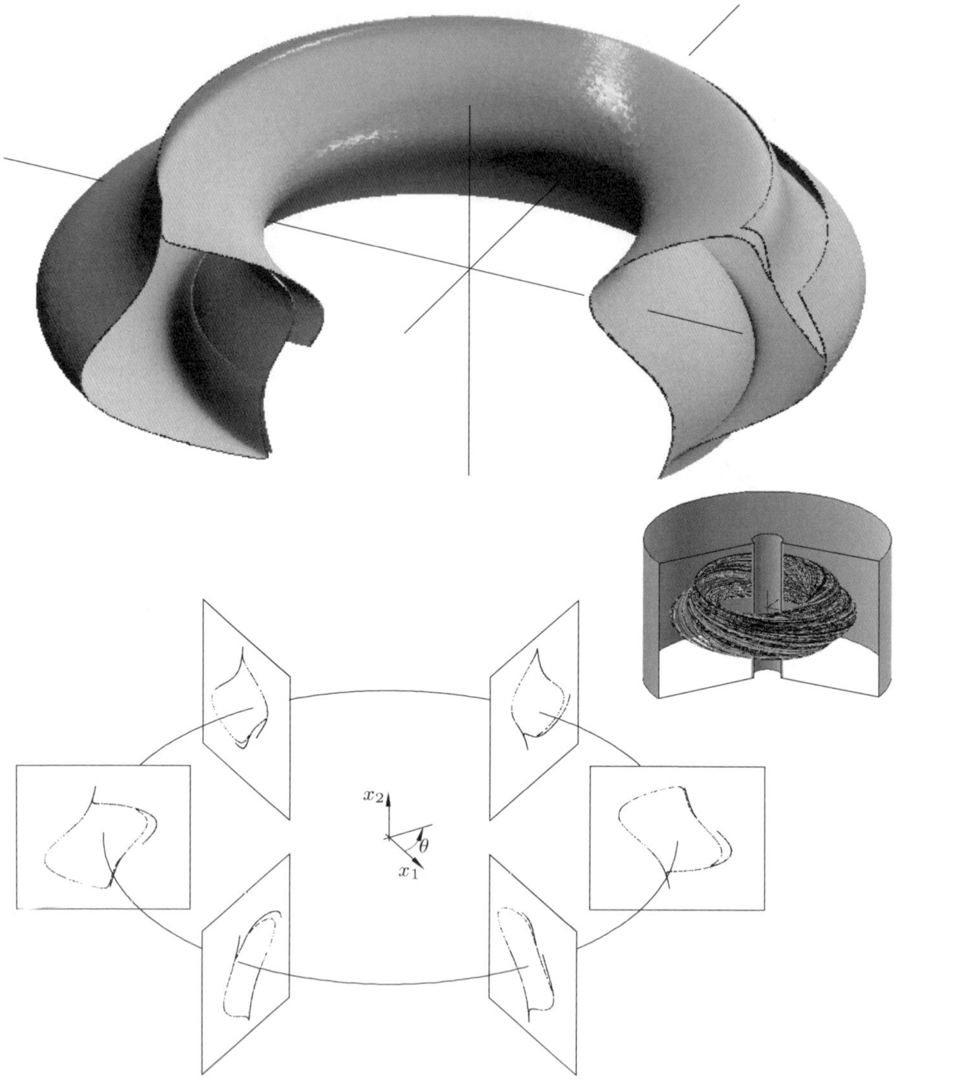

Fig. 10.4.10: Birkhoff-Shaw attractor in the phase space (x_1, x_2, θ)

the parameter values $k = 0.7$, $\mu = 10.0$, $a = 0.1$, $A = 0.25$, $\omega = 1.57$ three different forms of representing the chaotic Birkhoff-Shaw attractor.

In order to illustrate the beaks, the first diagram shows the attractor approximated by smooth surfaces (beware: the fractal structure of the attractor is lost). The second illustration shows six Poincaré sections for $\theta = \mathrm{n} \cdot 60°, \mathrm{n} = 0, 1, \ldots, 5$, while the third, smaller one depicts the fractal structure of the Birkhoff-Shaw attractor formed by the course of a trajectory. The beaks, which emerge by folding, wind around the basic structure of the attractor, varying their distance in dependence of θ. For reasons of the uniqueness of the solutions of eq. (10.4.47), the beaks can never coincide exactly with the basic structure; they come arbitrarily close to it, however. Thus, neighbouring trajectories do not remain neighbouring in the course of time. This mixing, a fundamental phenomenon of chaotic behaviour, can be explained by the beak formation respectively by stretching and folding. The continuous repetitions of the stretching and folding processes already supplied the key to the understanding of the exponential drifting apart of neighbouring trajectories in the case of both the strange attractor of the Lorenz system and the Hénon attractor (cf. sections 10.2, 10.3).

10.5 Duffing Equation

Be not afraid of growing slowly,
be afraid only of standing still
Chinese Proverb

Following our discussion of the Lorenz system and the van der Pol equation, we move on in this section to the Duffing equation which provides us with further illuminating insights into non-linear dynamics and the emergence of unpredictable behaviour and irregular motions. The Duffing equation represents one of the simplest dynamical systems in which we can observe chaotic response; for this reason, it was the object of numerous experimental and theoretical investigations. It describes an externally driven non-linear oscillator with a single degree of freedom and can be written in the following (non-autonomous) form

$$\ddot{x} + g(x, \dot{x}) = R(t) \tag{10.5.1}$$

We know a whole series of dynamical systems – for example in hydrodynamics, electrical engineering and mechanics – which can be modelled, at least approximately, by the Duffing equation. Georg Duffing himself formulated the equation in 1918 by extending the linear stiffness of a harmonically driven oscillator by a negative cubic term in order to be able to describe "qualitatively new and singular phenomena close to resonance" (Duffing, 1918), p. 8.

A well-known mechanical example, the oscillatory response of which leads to a special form of the Duffing equation, is an externally driven beam of length l; the cross-sectional area A and the moment of inertia I are constant, the beam is supported in a frame (see fig. 10.5.1). In the static case, in addition to the unstable initial state, two stable buckled states of equilibrium, symmetrical to one another, occur following the increase of an axial compressive force P beyond Euler's buckling load. If the frame is subject to a horizontal harmonic excitation force $f\cos(\omega_E t)$, a transversal periodic displacement is superimposed on the static deflection. Experimental investigations demonstrate that the beam first undergoes small oscillations about the states of equilibrium for small amplitudes f. An increase in the excitation amplitude f may result in a snap-through and a chaotic response (Tseng and Dugundji, 1971).

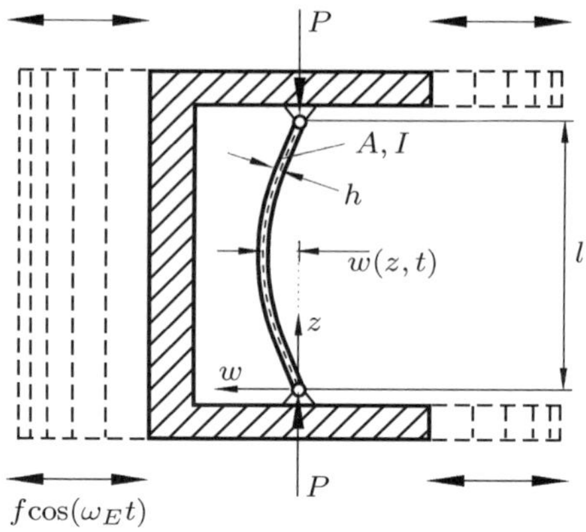

Fig. 10.5.1

Externally driven buckled beam

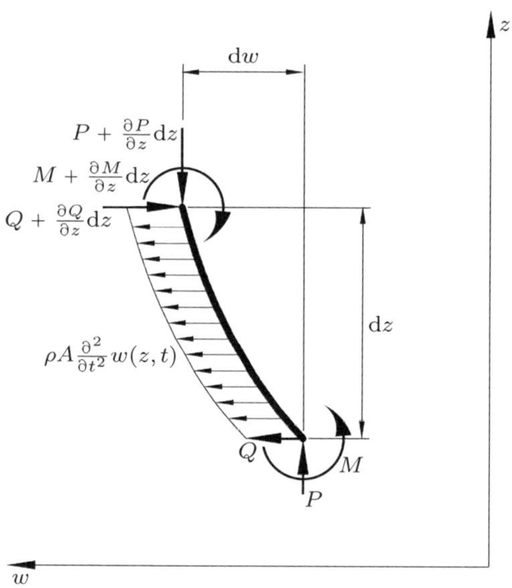

Fig. 10.5.2

Forces and moments in the beam

We first establish the equations of motion for the beam. In accordance with d'Alembert's principle, the instantaneous inertial force $\rho A\, \partial^2(w(z,t))/\partial t^2$ per unit length is positive in the w-direction, see fig. 10.5.2, where ρ is the density. If M is the bending moment, P the axial compressive force and Q the shear force in the beam, the equations for the equilibrium of forces and moments can be written immediately as

$$\frac{\partial Q}{\partial z} = -\rho A \ddot{w} \quad ; \quad \frac{\partial P}{\partial z} = 0 \tag{10.5.2}$$

where dots denote derivatives with respect to time, and

$$-\frac{\partial M}{\partial z} = Q - P_{\text{eff}} \frac{\partial w}{\partial z} \tag{10.5.3}$$

Here, P_{eff} is the effective axial force in the beam and includes the effect of unloading generated by the deflection w and the associated tensile strains in the neutral plane. Assuming that the elementary theory of bending respectively the Bernoulli hypothesis holds, i.e. that the effect of the shear strains can be ignored, the following classic relation is found for the deflection w

$$\frac{\partial^2 w}{\partial z^2} = -\frac{M}{EI} \tag{10.5.4}$$

where E is the Young modulus. We thus obtain from eqs. (10.5.2)-(10.5.4)

$$EI\frac{\partial^4 w}{\partial z^4} + P_{\text{eff}}\frac{\partial^2 w}{\partial z^2} + \rho A\ddot{w} = 0 \tag{10.5.5}$$

In order to establish P_{eff}, we must determine the extension Δl of the neutral axis as a result of the deflection w approximately as

$$\Delta l = \tfrac{1}{2}\int_0^l \left(\frac{\partial w}{\partial z}\right)^2 \mathrm{d}z \tag{10.5.6}$$

This extension generates the unloading tensile force

$$\frac{EA}{2l}\int_0^l \left(\frac{\partial w}{\partial z}\right)^2 \mathrm{d}z \tag{10.5.7}$$

causing a reduced compressive force P_{eff} in the beam of the magnitude

$$P_{\text{eff}} = P - \frac{EA}{2l}\int_0^l \left(\frac{\partial w}{\partial z}\right)^2 \mathrm{d}z \tag{10.5.8}$$

This force is assumed constant over the length of the beam; the modified eq. (10.5.5) hence becomes

$$EI\frac{\partial^4 w}{\partial z^4} + \left[P - \frac{EA}{2l}\int_0^l \left(\frac{\partial w}{\partial z}\right)^2 \mathrm{d}z\right]\frac{\partial^2 w}{\partial z^2} + \rho A\ddot{w} = 0 \tag{10.5.9}$$

the non-linear character of which is evident. If we impose a transversal harmonic excitation on the system and simultaneously assume a viscous damping c upon the system, we establish a simple model for an externally driven beam subject to a compressive force P, expressed by a partial integro-differential equation for the deflection $w(z,t)$ of the form

$$EI\frac{\partial^4 w}{\partial z^4} + \left[P - \frac{EA}{2l}\int_0^l \left(\frac{\partial w}{\partial z}\right)^2 \mathrm{d}z\right]\frac{\partial^2 w}{\partial z^2} + c\dot{w} + \rho A\ddot{w} = f\cos(\omega_E t) \tag{10.5.10}$$

Considering (infinitely) small oscillations, this equation can be integrated exactly; this aspect does not interest us here, however.

Interestingly, the flutter analysis of a two-dimensional panel in a supersonic flow yields a very similar equation of motion (Holmes and Marsden, 1978; Dowell, 1982;

Argyris and Bühlmeier, 1988; Karlsson, 1991). It is shown that, for a certain combination of membrane compressive forces and supersonic velocities, chaotic motions may be generated in the panel. Similar phenomena occur in the panels of wings at supersonic speeds (Knudsen, 1993).

As a next step, it is useful to normalise eq. (10.5.10). To this end, we introduce the following dimensionless variables and parameters

$$\zeta = \frac{z}{l}\,, \qquad w^* = \frac{w}{h}\,, \qquad \tau = t\sqrt{g/l}\,, \qquad \omega_E^* = \omega_E\sqrt{l/g}$$

$$P^* = P\frac{l^2}{EI}\,, \qquad P_\Delta^* = (l/i)^2(\Delta l/l) = \tfrac{1}{2}(h/i)^2\int_0^l \left(\frac{\partial w^*}{\partial \zeta}\right)^2 \mathrm{d}\zeta \tag{10.5.11}$$

$$c^* = \frac{l^4}{EI}\sqrt{\frac{g}{l}}\,c\,, \qquad \rho^* = \frac{l^4}{EI}\frac{g}{l}\rho A\,, \qquad f^* = \frac{l^4}{EI}\frac{1}{h}f$$

Here, g is the acceleration of gravity, h the height of the beam and i the radius of gyration of the beam, $i = (I/A)^{1/2}$. Equation (10.5.10) thus becomes

$$\frac{\partial^4 w^*}{\partial \zeta^4} + (P^* - P_\Delta^*)\frac{\partial^2 w^*}{\partial \zeta^2} + c^*\frac{\partial w^*}{\partial \tau} + \rho^*\frac{\partial^2 w^*}{\partial \tau^2} = f^* \cos\left(\omega_E^* \tau\right) \tag{10.5.12}$$

where the boundary conditions for w^* are

$$\begin{aligned} w^*(0,\tau) = \ & w^*(1,\tau) = 0 \\ \left.\frac{\partial^2 w^*}{\partial \zeta^2}\right|_{\zeta=0} = \ & \left.\frac{\partial^2 w^*}{\partial \zeta^2}\right|_{\zeta=1} = 0 \end{aligned} \tag{10.5.13}$$

In order to solve the boundary value problem of eqs. (10.5.12) and (10.5.13), it is expedient to apply an approximate procedure. On the one hand, we can use a classic (global) Galerkin procedure in which the ansatz or modal functions are defined over the whole ζ-interval. Another possibility is the application of the finite element method which represents a local Galerkin method in which the ansatz or modal functions are only non-zero over finite subsets of the definition domain. In order to facilitate our forthcoming discussion of the Duffing equation, we restrict ourselves here to the global procedure, but refer the reader to the results of the finite element analysis in (Troselius, 1989; Karlsson, 1991).

In order to find a basis of ansatz functions, we linearise the partial differential equation (10.5.12) and determine the eigenvalues and eigenfunctions of the autonomous, unforced system which satisfy *ab initio* the boundary conditions eq. (10.5.13). In the following, we dispense with the asterisk * in eq. (10.5.12) and assume $\rho = 1$. Hence, we consider the (linearised) system

$$w'^{\mathrm{V}} + Pw'' + c\dot{w} + \ddot{w} = 0 \tag{10.5.14}$$

with

$$\begin{aligned} w(0,t) = \ & w(1,t) = 0 \\ w''(0,t) = \ & w''(1,t) = 0 \end{aligned} \tag{10.5.15}$$

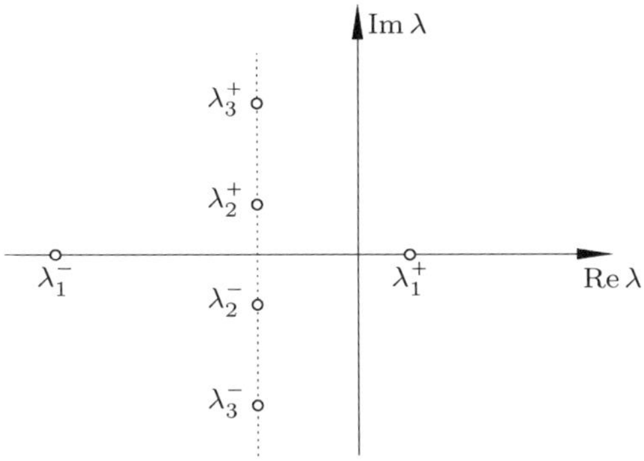

Fig. 10.5.3

Eigenvalue spectrum of the buckled beam

The dashes denote spatial derivatives, dots again temporal derivatives. To solve the linearised problem, we make the ansatz

$$w = \mathrm{e}^{\lambda t} \sin(\mathrm{n}\pi\zeta) \tag{10.5.16}$$

which satisfies the boundary conditions eq. (10.5.15). Substitution in the differential equation yields the following relation for the eigenvalues λ

$$\lambda^2 + c\lambda + \mathrm{n}^2\pi^2(\mathrm{n}^2\pi^2 - P) = 0 \tag{10.5.17}$$

i.e. for each n, we obtain two eigenvalues

$$\lambda_\mathrm{n}^\pm = -\tfrac{1}{2}c \pm \sqrt{\tfrac{1}{4}c^2 - \mathrm{n}^2\pi^2(\mathrm{n}^2\pi^2 - P)} \tag{10.5.18}$$

Noting that the beam is subject to the basic mode of the second Eulerian buckling mode, i.e. that $\pi^2 < P < 4\pi^2$ holds for the normalised axial load, we observe that, for $\mathrm{n} = 1$, a positive and a negative real eigenvalue exist. Depending on the magnitude of the damping and pre-stressing of the beam, further pairs of real eigenvalues can emerge; the remaining eigenvalues of the discrete spectrum are, however, complex conjugate with a constant real part $-c/2$. Figure 10.5.3 presents a typical arrangement of the eigenvalues.

Having seen that the functions $\sin(\mathrm{n}\pi\zeta)$ form a natural basis for a global Galerkin procedure, we introduce the following product ansatz for the solution of the non-linear eq. (10.5.12)

$$w(\zeta, t) = \sum_{\mathrm{n}=1}^{\mathrm{N}} x_\mathrm{n}(t) \sin(\mathrm{n}\pi\zeta) \tag{10.5.19}$$

and calculate the residual R_{es} resulting from the substitution into the differential equation (10.5.12)

$$\begin{aligned} R_{es} = {} & \textstyle\sum_{\mathrm{n}=1}^{\mathrm{N}} \left[\mathrm{n}^2\pi^2(\mathrm{n}^2\pi^2 - P)x_\mathrm{n} + c\dot{x}_\mathrm{n} + \ddot{x}_\mathrm{n}\right] \sin(\mathrm{n}\pi\zeta) - f\cos(\omega_E t) \\ & + \tfrac{1}{2}(h/i)^2 \int_0^1 \Big(\sum_{\mathrm{k}=1}^{\mathrm{N}} \pi \mathrm{k} x_\mathrm{k} \cos(\mathrm{k}\pi\zeta)\Big)^2 \mathrm{d}\zeta \sum_{\mathrm{n}=1}^{\mathrm{N}} \mathrm{n}^2\pi^2 x_\mathrm{n} \sin(\mathrm{n}\pi\zeta) \end{aligned} \tag{10.5.20}$$

The unknown coefficients are calculated from the condition that the weighted mean of the residual must be zero

$$\int_0^1 R_{es}(\zeta)\sin(\mathrm{k}\pi\zeta)\mathrm{d}\zeta = 0 \qquad \text{for } \mathrm{k} = 1, \cdots, \mathrm{N} \tag{10.5.21}$$

where, for the Galerkin procedure, the weighting functions are again selected as the ansatz functions. Due to the orthogonality of the basis functions, the sums simplify and we obtain a system of N ordinary differential equations for the time-dependent amplitudes $x_\mathrm{k}(t)$

$$\begin{aligned} \ddot{x}_\mathrm{k} + c\dot{x}_\mathrm{k} &+ \mathrm{k}^2\pi^2(\mathrm{k}^2\pi^2 - P)x_\mathrm{k} + \tfrac{1}{2}(h/i)^2\Big(\tfrac{1}{2}\sum_{\mathrm{n}=1}^{\mathrm{N}}\pi^2\mathrm{n}^2x_\mathrm{n}^2\Big)\pi^2\mathrm{k}^2x_\mathrm{k} \\ &= \begin{cases} (4/\mathrm{k}\pi)f\cos(\omega_E t) & \text{for} \quad \mathrm{k} = 1, 3, 5, \ldots \\ 0 & \text{for} \quad \mathrm{k} = 2, 4, 6, \ldots \end{cases} \end{aligned} \tag{10.5.22}$$

The question now arises how many terms we must consider in the product ansatz eq. (10.5.19). Eigenmodes appertaining to eigenvalues with positive real parts must be taken into account in any case so that the qualitative behaviour can be reproduced correctly. The necessity of considering further eigenmodes for qualitative analyses depends on whether strong couplings emerge between them and the unstable modes; cf. (Pätzold, 1990). We restrict ourselves here only to the fundamental basic mode and thus obtain a non-linear oscillating system with one degree of freedom, a modification of the Duffing equation with a negative linear stiffness term

$$\ddot{x} + c\dot{x} - \beta x + \alpha x^3 = \gamma\cos(\omega_E t) \tag{10.5.23}$$

where the stiffness parameters are given by

$$\beta = \pi^2(P - \pi^2) > 0\,, \qquad \alpha = \tfrac{1}{4}\pi^4(h/i)^2 \tag{10.5.24}$$

and the amplitude γ of the driving force by

$$\gamma = \frac{4}{\pi}f \tag{10.5.25}$$

It is interesting that the solutions of the "exact" eq. (10.5.10) can be approximated by solutions of the Duffing equation (10.5.23) in which we apply a series expansion and restrict it to the two first terms in the case of small oscillations. This confirms that the Duffing equation can indeed be considered a mathematical model of real physical processes.

If we re-write eq. (10.5.23) as an autonomous first-order system, then, with the notations $x_1 = x$, $x_2 = \dot{x}$ and $x_3 = t$, we obtain

$$\begin{aligned} \dot{x}_1 &= x_2 \\ \dot{x}_2 &= -cx_2 + \beta x_1 - \alpha x_1^3 + \gamma\cos(\omega_E x_3) \\ \dot{x}_3 &= 1 \end{aligned} \tag{10.5.26}$$

where $(x_1, x_2, x_3) \in \mathbb{R}^2 \times \mathbb{S}^1$, i.e. x_3 is limited to the interval $0 \leqslant x_3 \leqslant 2\pi/\omega_E$.

The same type of equation can be deduced for a fixed, externally forced cantilever beam subject to a magnetic field; see the monographs (Wiggins, 1988;

Kuznetsov, 1998) and (Moon and Holmes, 1979); one of the two authors also performed an experimental examination (Moon, 1980; Moon, 1992). A further possibility for deriving the Duffing equation for mechanical problems is by deducing the restoring force of a harmonic respectively anharmonic oscillator by differentiation of a potential function. It can be shown that, for a one-mass spring oscillator, the potential $V(x) = \frac{1}{2}x^2 - \frac{1}{4}x^4$ and the restoring force $F_R = \partial V/\partial x = x - x^3$ simulate a spring softening and that the potential $V(x) = -\frac{1}{2}x^2 + \frac{1}{4}x^4$ $(F_R = -x + x^3)$ describes qualitatively the load in the post-buckling regime of the second Eulerian buckling case of an axially pre-stressed beam (Thompson and Stewart, 2001).

Numerous numerical studies of the Duffing equation are known from the literature. We particularly mention Ueda's work (Ueda, 1980a; Ueda, 1980b) which investigates this type of equation most extensively, although without the linear stiffness term.

In dependence of the five control parameters in the Duffing equation, the most varying patterns of motion can be observed, namely stationary long-term behaviour (for $\gamma = 0$), periodic and quasi-periodic motions as well as chaotic behaviour; depending on the initial conditions, different types of motion may coexist (coexistence of attractors). We first present some numerical experiments and then discuss the question how unpredictable behaviour arises. In the examples, the Duffing equation was integrated numerically for different values of the system parameters c, α, β, γ and ω_E using the Runge-Kutta procedure. In order to visualise the results, different forms of presentation were selected, namely temporal evolutions, phase portraits respectively their projection onto the (x_1, x_2)-plane as well as Poincaré sections.

The three figs. 10.5.4 to 10.5.6 demonstrate periodic behaviour of the Duffing equation for varying system damping, restoring force, forcing amplitude and frequency.

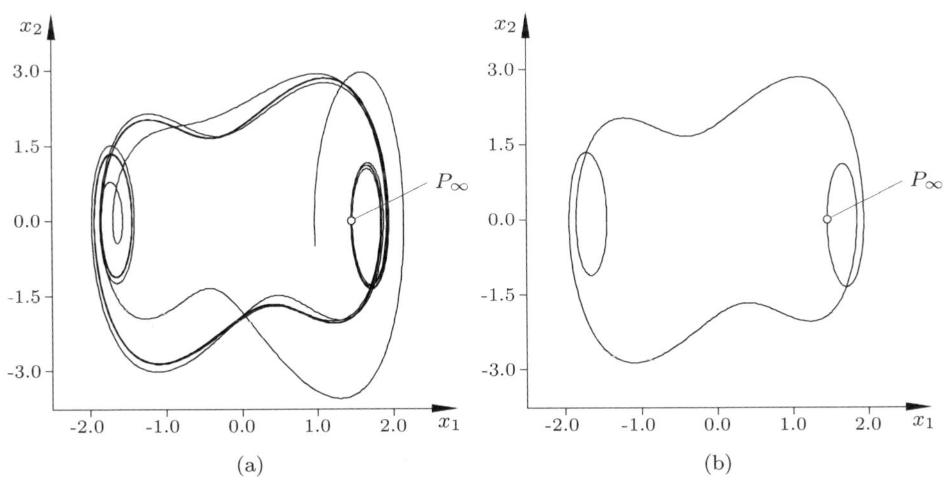

Fig. 10.5.4: Transient phase (a) and limit cycle (b) of the Duffing equation $\ddot{x} + 0.55\dot{x} + 8x + 2x^3 = 24\sin t$

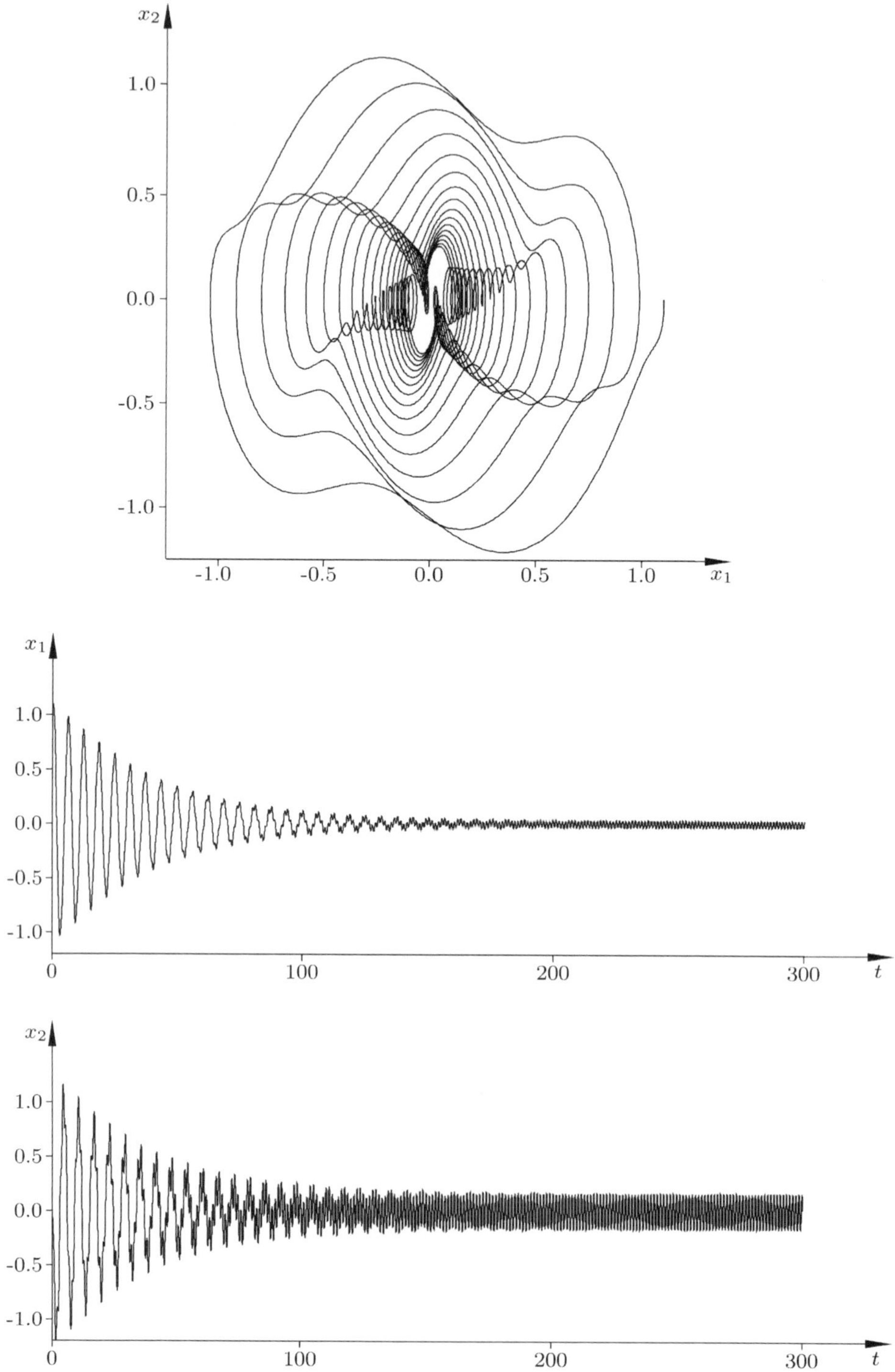

Fig. 10.5.5: Transient phase and limit cycle of the Duffing equation
$\ddot{x} + 0.05\dot{x} + x + 0.1x^3 = 0.8 \sin 5t$

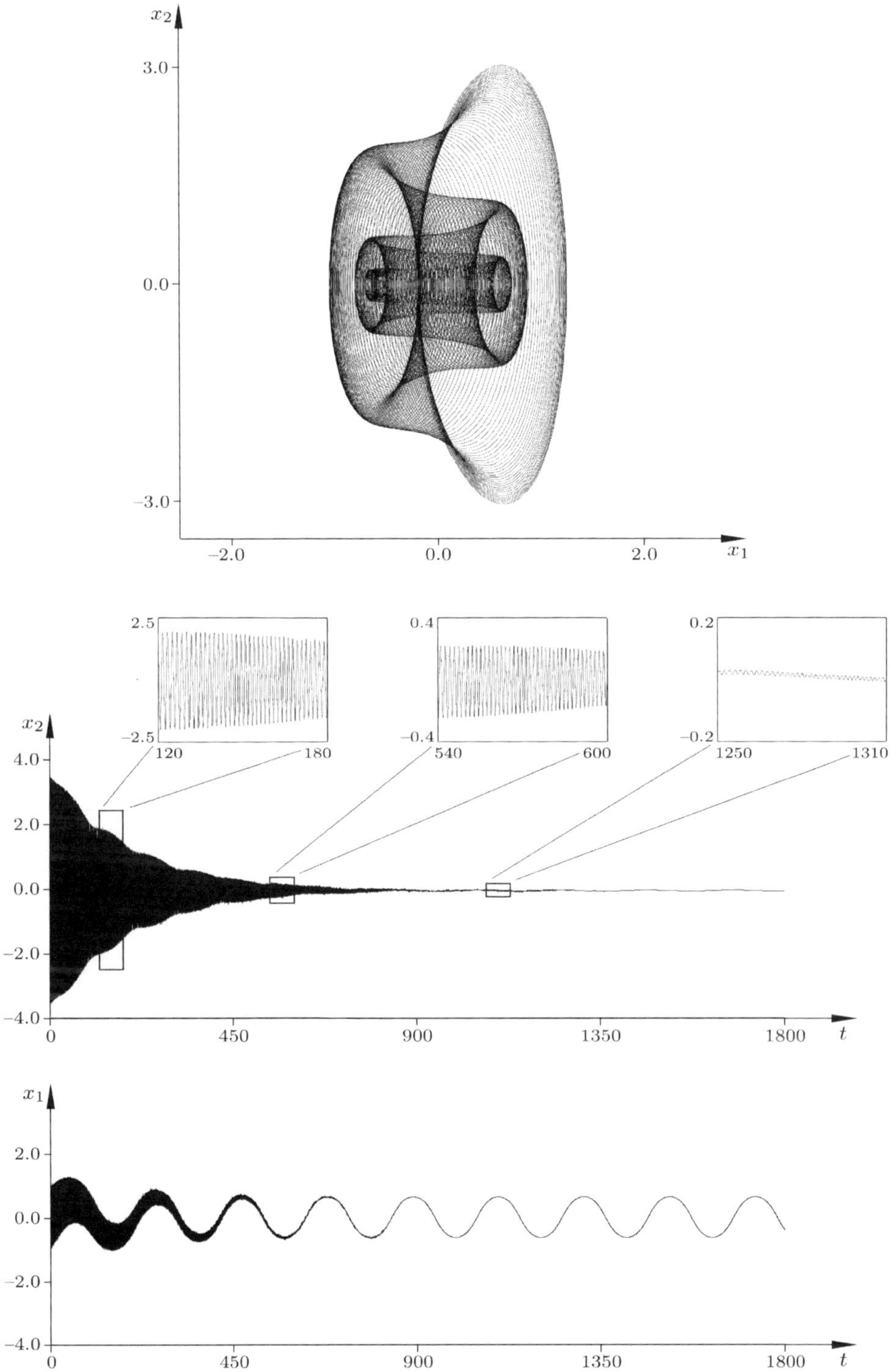

Fig. 10.5.6: "Imploding sombrero" for $\ddot{x} + 0.01\dot{x} + 10x + 6x^3 = 8\sin 0.03t$

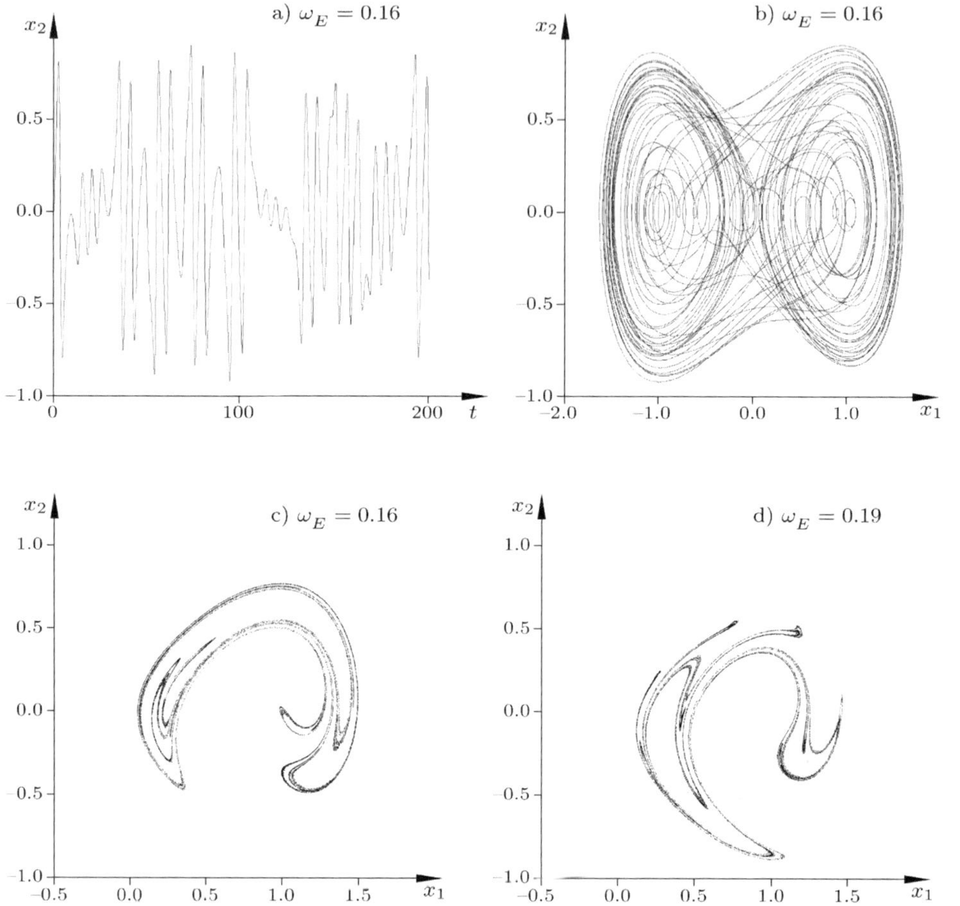

Fig. 10.5.7: Chaotic behaviour of the Duffing equation (10.5.23) for $c = 0.04$, $\beta = 0.2$, $\alpha = 0.53$ and $\gamma = 0.4$

Figure 10.5.4a,b presents the transient phase for the given parameters as it converges alone. Both figures show that the trajectories respectively the limit cycle intersect in the projection plane (x_1, x_2), in contrast to their three-dimensional behaviour. For considerably lower damping, restoring force, forcing amplitude and frequency, a further limit cycle could be found. The phase portrait in fig. 10.5.5 alone is not sufficient to deduce periodic behaviour for $t \to \infty$. The time series $x_1(t)$ and $x_2(t)$, on the other hand, illustrate that the system responds periodically after a certain transient phase.

Figure 10.5.6 indicates a particularly long transient phase, conditioned by the low damping $c = 0.01$ and the low value of the driving frequency $\omega_E = 0.03$. The phase portrait of the trajectories during the transient phase conveys the impression of a three-dimensional rotational symmetric structure, the end of which tapers and repeatedly turns inside out; it is reminiscent of an imploding sombrero rather than

a)

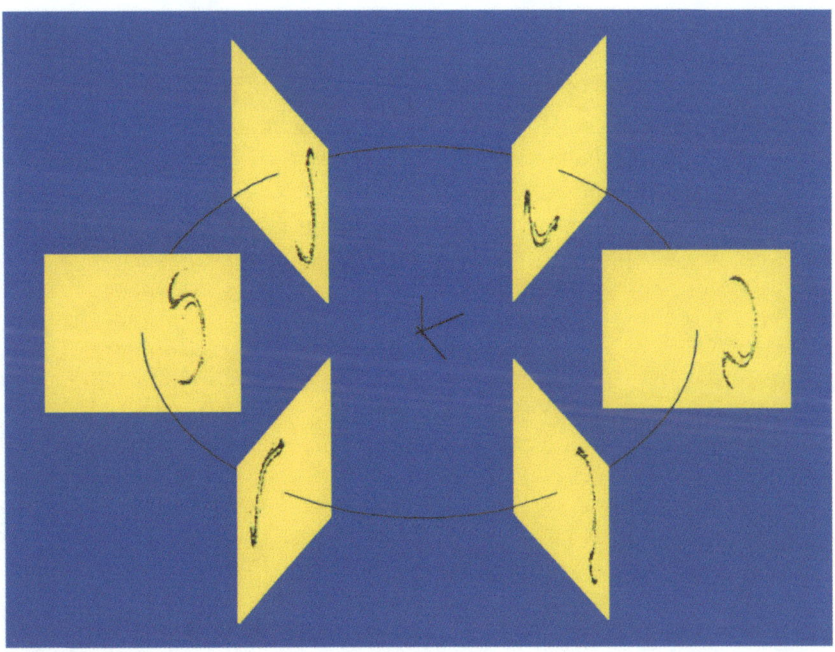

b)

Plate XXIV Chaotic behaviour of the Duffing oscillator, eq. (10.5.23), for $c = 0.2$, $\alpha = 1.0$, $\beta = 0.2$, $\gamma = 85.0$, $\omega_E = 1.0$:
a) strange attractor in the phase space $(x,\, \dot{x},\, \tau = \omega_E t)$
b) Poincaré sections

of the phase portrait of a damped one-mass oscillator. Here again, the time series show how a low-frequency oscillation superimposed by a higher-frequency one leads to a periodic motion pattern, a limit cycle, for $t \to \infty$.

For other combinations of the control parameters, on the other hand, we can observe completely irregular, chaotic behaviour. Figure 10.5.7 presents solutions of the Duffing equation for the parameter values $c = 0.04$, $\alpha = 0.53$, $\beta = 0.2$, $\gamma = 0.4$ and two different forcing frequencies $\omega_E = 0.16$ respectively $\omega_E = 0.19$. In the time series $x_2 = \dot{x}(t)$, no periodicity can be recognised for the selected time interval $0 \leqslant t \leqslant 200$; however, it is inadmissable to ascribe chaotic behaviour to the system for $t \to \infty$ on the basis of this limited time interval since periodicity cannot be excluded *a priori* for any longer time interval (fig. 10.5.7a). The phase portrait $(x, \dot{x})$ in fig. 10.5.7b of the same solution curve and the same time interval indicates the boundedness of the area that has been swept over for $t \to \infty$; it is easy to imagine that, after a relatively short time, the course of the trajectories within the depicted range will no longer be recognisable.

Poincaré sections, on the other hand, convey a completely new impression. Periodic and quasi-periodic trajectories are represented by discrete points or continuous curves. In contrast, irregular chaotic behaviour is depicted by a cloud of irregularly distributed points which cluster in a limited domain of the plane but possess topologically a highly complex fractal structure, namely that of a strange attractor. For the selection of parameters given above, a whole series of strange attractors can be found in the frequency range $0 < \omega_E < 0.5$; cf. (Seydel, 1980). Two of them are presented for the initial conditions $x_1 = 0$, $x_2 = 0$ and mapping steps in fig. 10.5.7c,d.

In order to gain an impression of the spatial structure of a strange attractor in a three-dimensional extended phase space, we present the attractor of the Duffing equation (10.5.26) for the parameter values $c = 0.2$, $\alpha = 1.0$, $\beta = 0.2$, $\gamma = 85.0$, $\omega_E = 1.0$ in cylindrical coordinates in the upper part of Colour Plate XXIV, p. 747. Here, x_1 was plotted in radial direction, x_2 parallel to the cylinder axis and x_3 in circumferential direction. While irregular time series – such as that in fig. 10.5.7a – only express that there are no temporal correlations respectively that they decay very rapidly, the spatial correlations, a consequence of the deterministic equations of motion, become visible in the phase space. The aesthetically pleasing Duffing attractor in Colour Plate XXIV resembles a multiply twined golden wreath from the heyday of Mycene. As we shall see, the chaotic long-term behaviour in the phase space possesses a well-ordered structure. In the lower part of this colour plate, we have plotted various Poincaré sections. The point symmetry which expresses the invariance of the Duffing equation (10.5.26) with respect to the substitutions $x_1 \to -x_1$, $x_2 \to -x_2$, $x_3 \to \pi/\omega_E + x_3$ emerges clearly.

The topological structure of the Duffing attractor, another characteristic of strange attractors, becomes particularly apparent in Colour Plate XXIV. The dynamics on strange attractors is determined by the infinite number of embedded unstable periodic orbits. The structure of these orbits can be analysed by means of the topological knot theory (Tufillaro et al., 1992; Arnold and Khesin, 1998). Within the framework of this theory, closed orbits in three-dimensional space which do not intersect, i.e. which are homeomorphic to a circle, are denoted knots. Links consist of several knots which do not have points in common. With these definitions, it is possible to apply results from knot theory to periodic orbits. Knots and links can be characterised and classified with the aid of topological invariants. One example is the local torsion rate of an orbit indicating how often an orbit is twisted around itself during one cycle. If we consider several orbits, they can be intertwined during one cycle to form a braid. The type of the intertwining can be characterised by means of braid invariants (see also section 3.5).

Topological methods can be applied additionally if we wish to determine the metric and dynamic invariants of strange attractors. It is noteworthy that, in contrast to metric invariants like the Lyapunov exponents or the various fractal dimensions, topological invariants of strange attractors are robust vis-à-vis variations of the control parameters. Although periodic orbits can disappear or emerge following a change in the system parameters, their topological invariants are preserved independently of the stability behaviour; thus, they characterise the whole dynamics of the system. Since the analysis of the topological structure is based ultimately on the identification of the underlying stretching and folding mechanism, these methods are especially helpful for setting up and validating models to describe the dynamics of experimental time series (Gilmore, 1998; Firle et al., 1996).

As mentioned at the beginning, Ueda undertook detailed parameter studies on the long-term behaviour of the Duffing oscillator in dependence of the control parameters (Ueda, 1979; Ueda, 1980b). In general, a steady variation of one control parameter will lead to the equally steady modification of the phase portrait of the attractor. Qualitative changes of the behaviour and thus of the type of attractor, on the other hand, result if a local or global bifurcation occurs. Local bifurcations which lead to qualitative changes of the long-term dynamics in local regions of the phase space were discussed in detail in Chapter 6. Global bifurcations lead to global changes of the flow and are generally less accessible analytically. We shall return in this section to a global bifurcation with far-reaching consequences. Here, we restrict ourselves to mentioning literature for further reading (Thompson and Stewart, 2001; Guckenheimer and Holmes, 1983; Wiggins, 1988).

Beside the dependence of the long-term behaviour on the control parameters, we can often also observe a dependence on the initial conditions if several attractors coexist. In this case, it is important to gain an overview of the basins of attraction of the individual attractors and the topological structure of the boundaries of these basins. To this end, we cover the region of the phase space to be investigated with a uniform net, consider the central point of each cell as the initial condition and determine the corresponding long-term response. In this way, basins of attraction and types of attractors can be ascertained; an application of this method would be very time-consuming, however. In contrast, the cell mapping method introduced by Hsu (Hsu, 1980) offers the opportunity of reducing the calculation effort drastically. An easily comprehensible introduction to this method can be found in (Foale and Thompson, 1991; Kreuzer, 1987), a more detailed discussion in (Hsu, 1987).

When evaluating numerical experiments, it is advisable to proceed critically. We should first be certain that the transient phase has indeed faded away. Moreover, isoparametric diagrams providing an overview of the bifurcations following variations of selected control parameters are informative, cf. also section 10.2. Colour Plate XXV, p. 751 above, illustrates the long-term behaviour of a Duffing oscillator with fixed internal control parameters $c = \alpha = \beta = 1.0$ in dependence on the external parameters, namely the excitation frequency ω_E and the excitation amplitude γ. Horizontally, ω_E was plotted in the range $0.3 \leqslant \omega_E \leqslant 2.5$, vertically, the amplitude γ in the interval $0.5 \leqslant \gamma \leqslant 3.5$. Each pixel, i.e. each parameter combination, was assigned a colour in accordance with its long-term behaviour: limit cycles of period $1 - 2 - 4 - 8$ were assigned the colours yellow–pink–dark blue–grey, chaotic regions black. We can see clearly that the route to chaos passes typically through period doublings also for variations of two control parameters. Furthermore, as in the bifurcation diagram of the logisitic map, fig. 6.7.6a, the periodic windows are obvious and, moreover, in the correct order: the dominant green region corresponds to limit cycles of period 3, cycles of period 6 are red, those of period 5 light blue. It should be remembered that, according to the theorem of Sharkowskii, see section 6.7.2, the emergence of one single orbit of period three implies that limit cycles with all the other integer periods n will appear. It is striking that single yellow points, i.e. limit cycles of period 1, emerge repeatedly in the black chaotic zone. It is necessary to undertake further numerical investigations in order to clarify whether periodic and strange attractors coexist in this region. In any case, diagrams of this sort can offer an indication of the system behaviour to be expected.

In the lower part of Colour Plate XXV, p. 751, the long-term behaviour of the Duffing equation $\ddot{x} + \dot{x} + x^3 - \beta x = \cos t$ is investigated in dependence on the linear stiffness. For small β-values, we obtain a black period 1 limit cycle which becomes unstable at $\beta \approx 0.6$ as a result of a pitchfork bifurcation; two new period 1 limit cycles emerge, characterised by the colours blue and red, which become chaotic following a cascade of period doublings (cf. also section 10.3, particularly fig. 10.3.2). The chaotic region is interrupted by a large regular window in which a single period 3 limit cycle emerges. A further increase of the linear stiffness again leads to chaos via period doublings and, finally, via an inverse cascade of subharmonic bifurcations, to two separate period 1 limit cycles.

The last example demonstrates that, in the case of a fixed constellation of the control parameters, the long-term behaviour of a non-linear system can depend on the selected initial condition, i.e. that several attractors can emerge simultaneously. In the case of the Duffing equation, this is immediately comprehensible physically. If no driving force acts on the oscillator, the system settles down in one of the two lateral states of equilibrium after a transient phase. Equation (10.5.26) yields three fixed points with the coordinates (x_1, x_2)

$$S: \ (0,0) \qquad \text{and} \qquad F_{1,2}: \ (\pm\sqrt{\beta/\alpha},\, 0) \tag{10.5.27}$$

A linear stability analysis on the basis of the eigenvalues yields for the origin

$$\lambda_{1,2} = \tfrac{1}{2}\left[-c \pm \sqrt{c^2 + 4\beta}\,\right] \tag{10.5.28}$$

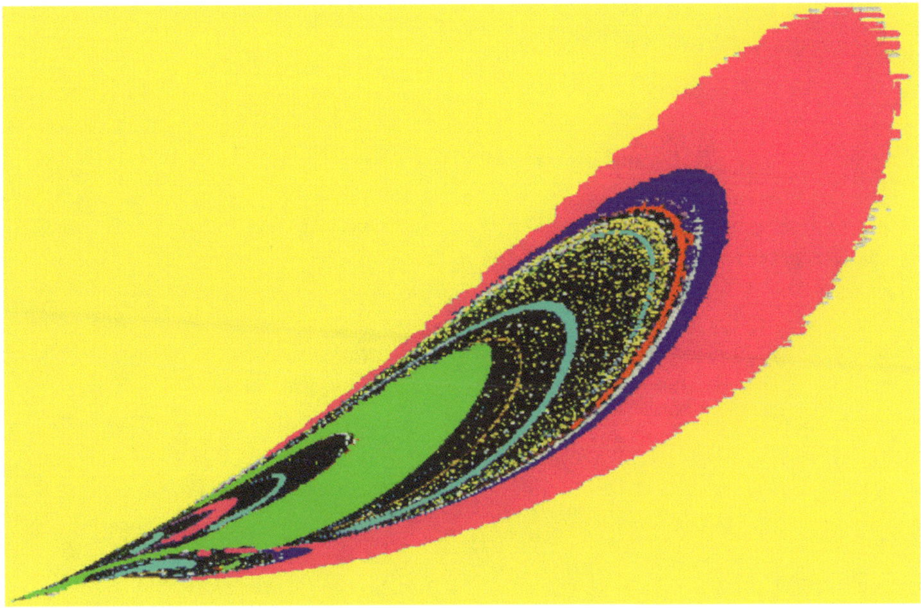

a)

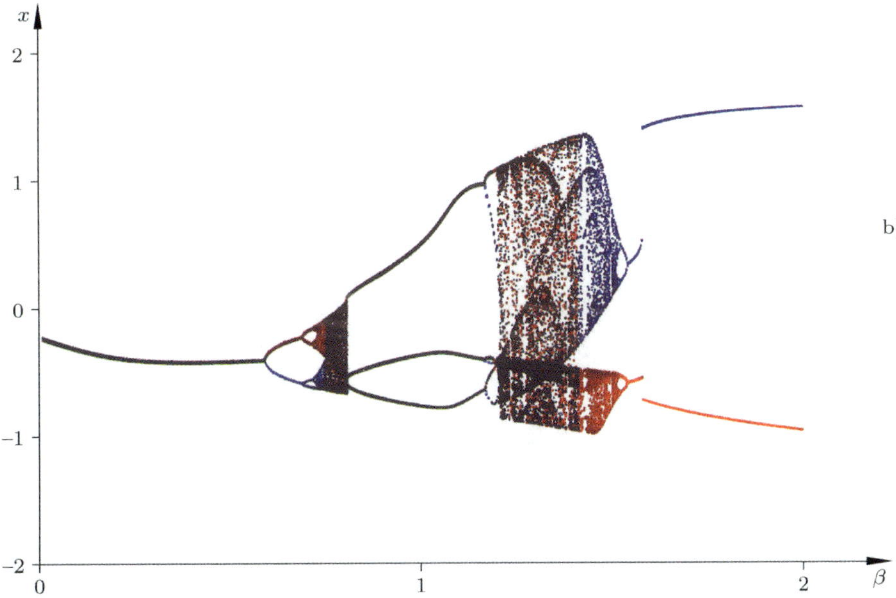

b)

Plate XXV Duffing equaton $\ddot{x} + \dot{x} + x^3 - \beta x = \gamma \cos \omega_E t$:
a) long-term behaviour for $0.3 \leqslant \omega_E \leqslant 2.5$ (horizontal), $0.5 \leqslant \gamma \leqslant 3.5$ (vertical)
b) bifurcation diagram in dependence on the linear stiffness β

i.e. S is a saddle point for $\beta > 0$. The two lateral fixed points possess the same pair of eigenvalues

$$\lambda_{1,2} = \tfrac{1}{2}\left[-c \pm \sqrt{c^2 - 8\beta}\,\right] \tag{10.5.29}$$

i.e. under the condition $\beta > 0$, F_1 and F_2 are stable nodes for $c^2 - 8\beta > 0$ and stable foci for $c^2 - 8\beta < 0$.

If we impose a small driving force, no further states of equilibrium arise any more as can be seen immediately from the third equation of eq. (10.5.26); the system now undergoes small oscillations about the right respectively left lateral state of equilibrium. In this case, two stable limit cycles hence exist in the phase space. For large amplitudes γ of the driving force, snapthrough can occur, i.e. a single periodic limit cycle will emerge which encloses the three fixed points of the non-excited system and which satifies the symmetry characteristics of the Duffing equation. For smaller γ-values, a periodic and a strange attractor can also emerge simultaneously, however, e.g. for $\alpha = \beta = 1$, $c = 0.15$, $\gamma = 0.3$, $\omega_E = 1.0$; see (Guckenheimer and Holmes, 1983).

Colour Plate XXVI, p. 753, presents an overview of a section of the phase space and the basins of attraction of the two emerging attractors for the Duffing equatio

$$\ddot{x} + 0.12\dot{x} - 0.5x + 0.5x^3 = \gamma \cos 0.7t \tag{10.5.30}$$

For the upper diagram, $\gamma = 0$ was selected and, for the lower part, $\gamma = 0.09$.

The phase space depicted above for $\gamma = 0$ is two-dimensional. The x_1-axis points to the right, the x_2-axis upwards, the origin lies in the middle. The two stable fixed points have the coordinates $(\pm 1, 0)$. All the initial values of trajectories which end in the right-hand focus after the transient phase were allocated the colour blue while those that level out at the left-hand focus were assigned the colour yellow. The illustration shows that the basins of the two attractors are divided from one another by a smooth line which coincides with the (global) stable manifold $\mathbb{W}^s$ of the saddle point at the origin.

In the lower part of Colour Plate XXVI, the basins of attraction of the two attractors were determined for the excitation amplitude $\gamma = 0.09$. Note that the phase space is three-dimensional in this case and that a Poincaré section is being considered here. For $\gamma = 0.09$, two limit cycles have evolved from the originally stable foci; the long-term behaviour of the system is hence regular. What is remarkable, however, is the fact that the boundary of the basins of attraction is by no means smooth any longer; it possesses a complex fractal structure which can be demonstrated by selective enlargements; cf. (Moon and Li, 1985). This has far-reaching consequences for the predictability of the system. If we select an initial condition in the proximity of the boundary of the two basins of attraction, it is no longer possible to predict whether the system will level out at the left-hand or the right-hand limit cycle, i.e. the oscillatory behaviour depends sensitively on the initial conditions. This surprising result is exemplary of the fact that there are non-linear systems which exhibit regular asymptotic behaviour but nevertheless do not admit long-term prognoses.

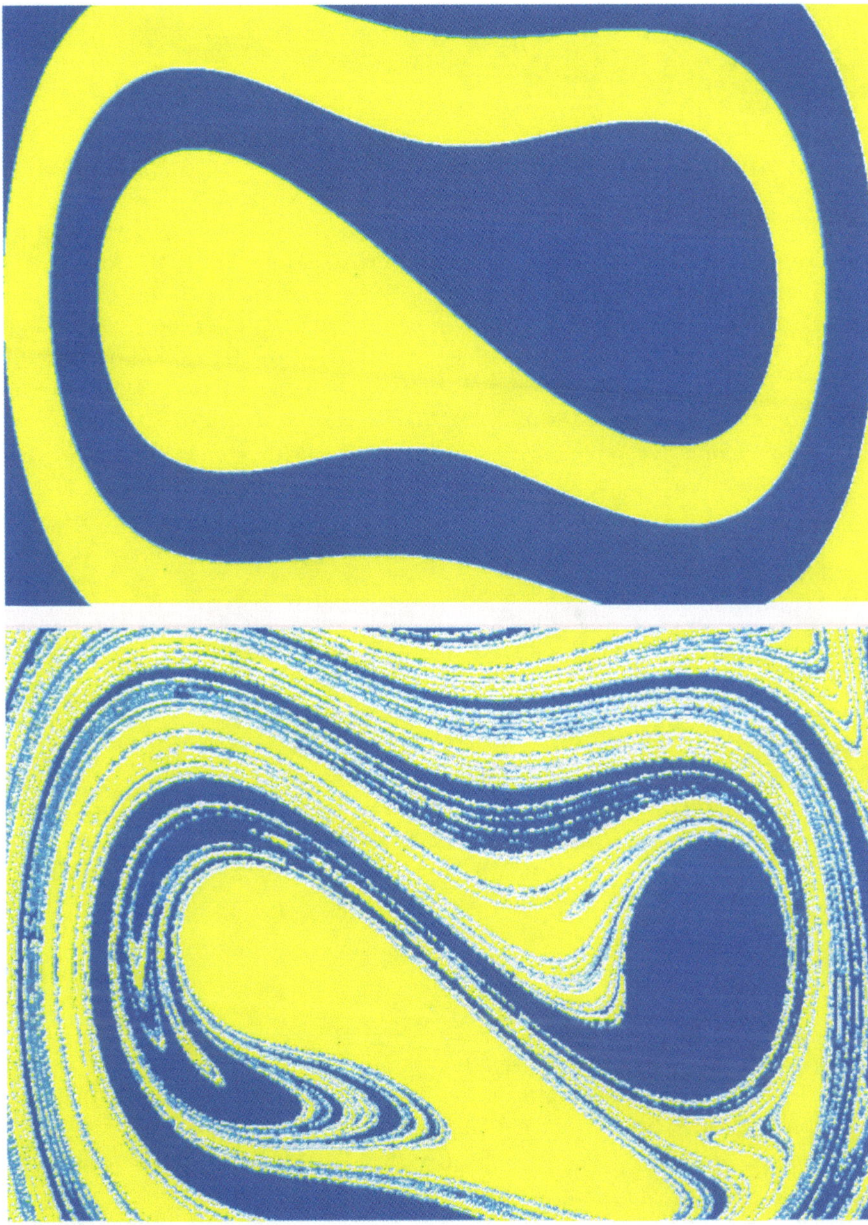

a)

b)

Plate XXVI Duffing oscillator of eq. (10.5.30):
a) regular boundaries of the basins of attraction of the two fixed points ($\gamma = 0$)
b) fractal boundaries of the basins of attraction of the two limit cycles ($\gamma = 0.09$)

In section 10.7, we shall discuss the basins of attraction of different attractors in the case of rational mapping functions and observe that the basins of attraction are also separated from one another by fractal boundaries.

Should damping and excitation amplitude be small, the evolution of fractal boundaries can be explained theoretically by regarding driving force and damping as perturbations of a Hamilton system. This is one of the few examples from the theory of dynamical systems in which we can arrive at global statements by analytical means. We thus sketch at least the fundamental argumentation while referring the reader to detailed discussions in (Holmes, 1979; Holmes, 1980; Guckenheimer and Holmes, 1983). The underlying methods are taken from the perturbation theory, in particular a version of the averaging method, the so-called Melnikov method (Melnikov, 1963), which – in contrast to the version presented in section 10.4 – can be applied to highly non-linear problems as long as the dissipation and periodic excitation force are small. In this case, as in section 10.4, an averaging theorem can be formulated which allows us to deduce the existence of periodic orbits in the perturbed system and to determine approximations for the Poincaré map.

We start with the following system of two differential equations

$$\dot{\boldsymbol{x}} = \boldsymbol{f}(\boldsymbol{x}) + \varepsilon \boldsymbol{g}(\boldsymbol{x}, t) \tag{10.5.31}$$

where $\boldsymbol{x} = \{x_1 \ \ x_2\}$ and $\varepsilon \ll 1$, $\boldsymbol{f}(\boldsymbol{x}) = \{f_1 \ \ f_2\}$ is a Hamiltonian vector field for which a Hamilton function H thus exists (cf. section 4.1) with

$$f_1 = \frac{\partial H}{\partial x_2} \quad \text{and} \quad f_2 = -\frac{\partial H}{\partial x_1} \tag{10.5.32}$$

and $\boldsymbol{g}(\boldsymbol{x}, t) = \{g_1 \ \ g_2\}$ is a periodic function $\boldsymbol{g}(\boldsymbol{x}, t) = \boldsymbol{g}(\boldsymbol{x}, t + T)$. Moreover, we assume that the unperturbed conservative system possesses a homoclinic orbit, i.e. a hyperbolic saddle point, the stable manifold $\mathbb{W}^s$ of which coincides with its unstable manifold $\mathbb{W}^u$. To demonstrate this on an example, we write the Duffing equation for $\gamma \ll 1$ and $c \ll 1$ in the form ($\varepsilon \ll 1$)

$$\begin{aligned} \dot{x}_1 &= x_2 \\ \dot{x}_2 &= \beta x_1 - \alpha x_1^3 + \varepsilon(\bar{\gamma} \cos \omega_E t - \bar{c} x_2) \end{aligned} \tag{10.5.33}$$

The unperturbed system possesses two homoclinic orbits $\Gamma_0^\pm$ which can be determined by integration from the Hamilton function H. Inside $\Gamma_0^\pm$ lie closed trajectories with $H = \text{const}$ having the two centres E_1 and E_2 (see fig. 10.5.8).

In section 8.2 (fig. 8.2.1), we saw that a homoclinic orbit is *not* structurally stable. Any perturbation, however small, will lead to alterations of the stable and unstable manifolds of the saddle point S so that the curve no longer closes. If we subject the two-dimensional Hamilton system to a periodic perturbation, the flow evolves in a three-dimensional extended phase space. If we observe the corresponding Poincaré sections, then, as the homoclinic orbit disintegrates, three cases, presented in fig. 10.5.9, are conceivable if we disregard the possibility of tangencies of the stable and unstable manifolds $\mathbb{W}_\varepsilon^s$ and $\mathbb{W}_\varepsilon^u$.

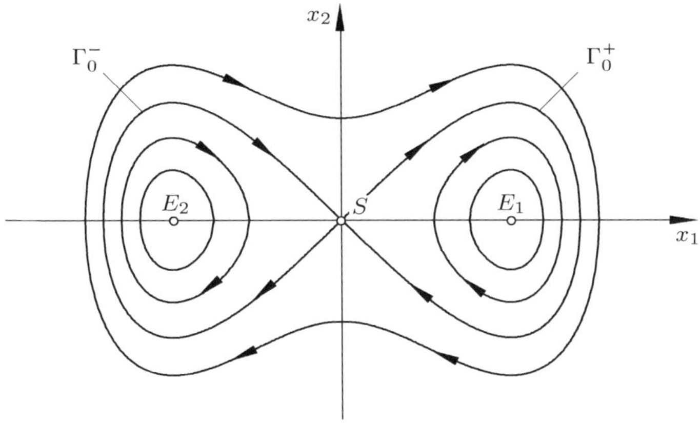

Fig. 10.5.8: The conservative system $\dot{x}_1 = x_2$, $\dot{x}_2 = \beta x_1 - \alpha x_1^3$;
E_1, E_2 are centres, S a saddle point and Γ_0 a homoclinic orbit

In the cases a) and b), there are no points of intersection of the stable and unstable manifolds. Here, a small perturbation of the initial system can result in the emergence of an unstable fixed point a) and a stable fixed point b) (respectively periodic orbit) emanating from the centres E_1 and E_2 in fig. 10.5.8, for example; in the first case, $\mathbb{W}_\varepsilon^s$ is "within" $\mathbb{W}_\varepsilon^u$, in the second case, "outside" it. What happens, however, if $\mathbb{W}_\varepsilon^u$ and $\mathbb{W}_\varepsilon^s$ intersect transversally (fig. 10.5.9c)? Such a point of intersection is called a transversal homoclinic point. We already discussed homoclinic points in section 4.5 in connection with the destabilising of a torus (cf. fig. 4.5.8); we saw that the existence of infinitely many such homoclinic points follows directly from the existence of a single transversal homoclinic point, leading to an extremely complex geometry of the stable respectively unstable manifold – and thus to highly complex dynamics – in the neighbourhood of the hyperbolic fixed point S_ε. There are then infinitely many points of intersection of $\mathbb{W}_\varepsilon^u$ and $\mathbb{W}_\varepsilon^s$, although points of intersection of $\mathbb{W}_\varepsilon^u$ (or $\mathbb{W}_\varepsilon^s$) with itself are impossible.

Henri Poincaré studied this problem while dealing with the question of the stability of our solar system at the end of the 19th century. Poincaré was already aware of the consequences involved in the emergence of transversal homoclinic points as

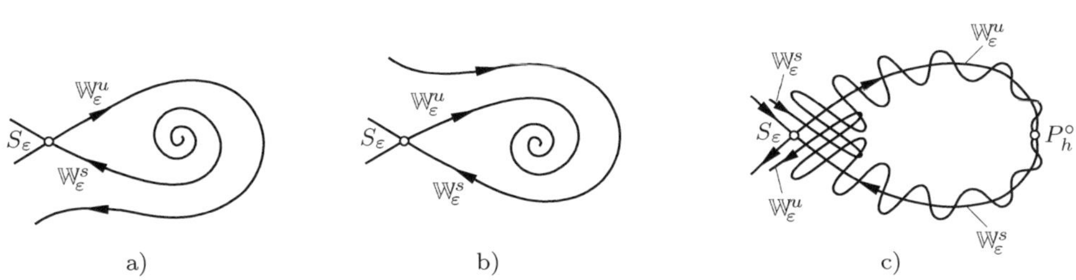

Fig. 10.5.9: Three possibilities for the disintegration of a homoclinic orbit

the following quotation from his three-volume work “Les méthodes nouvelles de la mécanique céleste” (Chapter XXXIII, p. 389) shows:

“Que l’on cherche à se représenter la figure formée par ces deux courbes et leurs intersections en nombre infini dont chacune correspond à une solution doublement asymptotique, ces intersections forment une sorte de treillis, de tissu, de réseau à mailles infiniment serrées; chacune des deux courbes ne doit jamais se recouper elle-même, mais elle doit se replier sur elle-même d’une manière très complexe pour venir recouper une infinité de fois toutes les mailles du réseau.

On sera frappé de la complexité de cette figure, que je ne cherche même pas à tracer. Rien n’est plus propre à nous donner une idée de la complication du problème des trois corps et en général de tous les problèmes de Dynamique où il n’y a pas d’intégrale uniforme…”

If the stable and unstable manifolds intersect, this leads in a Poincaré section to a map equivalent to Smale’s horseshoe map (Smale, 1967). In (Guckenheimer and Holmes, 1983; Holmes, 1990), there is a detailed discussion of this map which gave significant impetus to the theory of dynamical systems. If the Poincaré map of a dynamical system possesses the characteristics of the horseshoe map, then all initial conditions originally in a small spherical initial volume V_0 are mapped onto a region V_t of the phase space; this region emerges from V_0 by repeated stretching and folding and possesses a fractal structure after some time so that all the information on the position of the initial conditions has been lost. The existence of such a map is equivalent to the impossibility of long-term prognoses.

Let us now return to our question regarding the emergence of fractal boundaries in Colour Plate XXVI, p. 753. If, as a result of a global homoclinic bifurcation in the three-dimensional phase space, transversal homoclinic orbits occur, the consequence is global changes of the flow in the phase space and the loss of predictability. A necessary condition for the emergence of fractal boundaries is thus the occurrence of a homoclinic bifurcation.

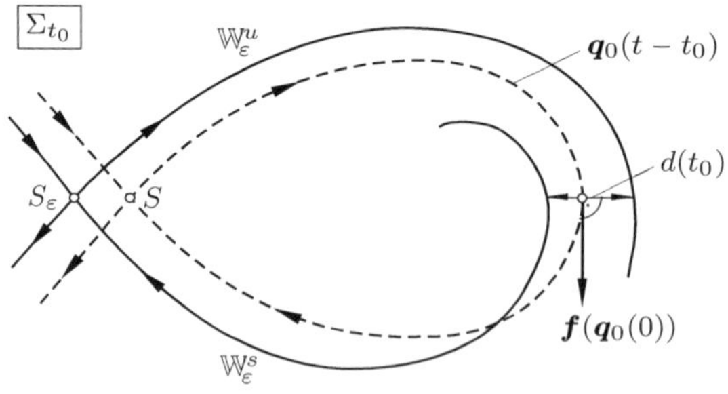

Fig. 10.5.10

Distance function $d(t_0)$ between stable ($\mathbb{W}^s_\varepsilon$) and unstable ($\mathbb{W}^u_\varepsilon$) manifold

Thanks to the Melnikov theory, it is even possible (Melnikov, 1963) to determine bounds for the spontaneous emergence of the fractal boundaries analytically. Melnikov defines a function $d(t_0)$ which measures the distance between $\mathbb{W}^s_\varepsilon$ and $\mathbb{W}^u_\varepsilon$ in a Poincaré section $\sum_{t_0}$ (see fig. 10.5.10) and whose behaviour indicates which case of fig. 10.5.9 holds. If $d(t_0) < 0$, case a) applies; case b) holds for $d(t_0) > 0$. However, if $d(t_0)$ oscillates about zero, transversal homoclinic points emerge. The special case of a tangency of $\mathbb{W}_\varepsilon^u$ and $\mathbb{W}_\varepsilon^s$ can be recognised in the emergence of double zeroes.

Let us assume that the homoclinic orbit of the unperturbed system is known and given by $\boldsymbol{x} = \boldsymbol{q}_0(t - t_0)$. Then, according to Melnikov, the distance $d(t_0)$ can be calculated as follows, cf.(Guckenheimer and Holmes, 1983)

$$d(t_0) = \varepsilon \frac{M(t_0)}{|\boldsymbol{f}(\boldsymbol{q}_0(0)|} + \mathcal{O}(\varepsilon^2) \tag{10.5.34}$$

where the Melnikov function $M(t_0)$ is defined by

$$M(t_0) = \int_{-\infty}^{+\infty} \boldsymbol{f}\big(\boldsymbol{q}_0(t - t_0)\big) \times \boldsymbol{g}\big(\boldsymbol{q}_0(t - t_0), t\big)\mathrm{d}t \tag{10.5.35}$$

where the functions $\boldsymbol{f}$ and $\boldsymbol{g}$ are quoted in eq. (10.5.31). What is surprising about this result is that, in the case of a given system of differential equations (10.5.31), we can not only deduce the existence of transversal homoclinic orbits but can even calculate the function $M(t_0)$ with the knowledge of the homoclinic orbit Γ_0 of the unperturbed system alone.

Holmes applied Melnikov's theory to the Duffing equation (10.5.33) (Holmes, 1979). In order to calculate the Melnikov function, eq. (10.5.35), we must first calculate the equation of the homoclinic orbit Γ_0 of the unperturbed system.

Since the Hamilton function $H(x_1, x_2)$ along Γ_0 is constant, we first calculate H for the conservative system

$$\begin{aligned} \frac{\partial H}{\partial x_2} &= \quad \dot{x}_1 = x_2 \\ \frac{\partial H}{\partial x_1} &= -\dot{x}_2 = -\beta x_1 + \alpha x_1^3 \end{aligned} \tag{10.5.36}$$

By integration, we obtain

$$H(x_1, x_2) = \tfrac{1}{2}x_2^2 - \tfrac{1}{2}\beta x_1^2 + \tfrac{1}{4}\alpha x_1^4 \tag{10.5.37}$$

Since the saddle point $S(0,0)$ with $H = 0$ lies on Γ_0, the Hamilton function along the homoclinic orbit is zero. Substituting $x_2 = \mathrm{d}x_1/\mathrm{d}t$ in $H(x_1, x_2) = 0$, we can deduce the relation $\boldsymbol{x} = \boldsymbol{q}_0(t)$ of Γ_0 from eq. (10.5.37) by integration. From

$$\frac{\mathrm{d}x_1}{\mathrm{d}t} = \pm\sqrt{\beta x_1^2 - \tfrac{1}{2}\alpha x_1^4} \tag{10.5.38}$$

it follows

$$\int_{x_1(0)}^{x_1(t)} \frac{\mathrm{d}x_1}{x_1\sqrt{1 - (\alpha/2\beta)x_1^2}} = \pm\sqrt{\beta}t \tag{10.5.39}$$

Figure 10.5.8 demonstrates that the homoclinic orbit is made up of two symmetrical loops. Is thus suffices to calculate the equation of the right homodinic orbit Γ_0^+. As the lower integration limit, we select the point on Γ_0^+ for which the oscillation possesses maximum deflection and zero velocity. Thus, $x_1(0) = (2\beta/\alpha)^{1/2}$ follows from $H(x_1(0), 0) = 0$. Substituting $(\alpha/2\beta)^{1/2} x_1 = \sin\xi$, we can integrate eq. (10.5.39), where $\xi(0) = \pi/2$ applies at the lower integration limit. We obtain

$$\ln\tan\frac{\xi}{2} = \pm\sqrt{\beta}t \tag{10.5.40}$$

If we substitute $\tau = \pm\sqrt{\beta}t$, this expression can be transformed into

$$\frac{\sin\xi}{1+\cos\xi} = \mathrm{e}^{\tau}$$

Squaring this equation, we find $(\mathrm{e}^{-\tau}\sin\xi - 1)^2 = \cos^2\xi$ and finally establish

$$\sin\xi = \frac{2\mathrm{e}^{-\tau}}{1+\mathrm{e}^{-2\tau}}$$

or

$$x_1(t) = \sqrt{\frac{2\beta}{\alpha}}\frac{1}{\cosh(\sqrt{\beta}t)} \tag{10.5.41}$$

and thus

$$x_2(t) = -\beta\sqrt{\frac{2}{\alpha}}\frac{\tanh(\sqrt{\beta}t)}{\cosh(\sqrt{\beta}t)} \tag{10.5.42}$$

The negative sign for $x_2(t)$ must be adopted because the lower half of the homoclinic orbit Γ_0^+ appertains to positive t-values (see fig. 10.5.11). We have thus established the equation of the homoclinic orbit Γ_0^+

$$\boldsymbol{q}_0(t) = \sqrt{\frac{2\beta}{\alpha}}\frac{1}{\cosh(\sqrt{\beta}t)}\left\{1 \quad -\sqrt{\beta}\tanh(\sqrt{\beta}t)\right\} \tag{10.5.43}$$

and we can calculate the Melnikov function in accordance with eq. (10.5.35). If we note that, for the Duffing system, $\boldsymbol{f}$ and $\boldsymbol{g}$ are given by

$$\begin{aligned} \boldsymbol{f} &= \left\{x_2 \qquad (\beta x_1 - \alpha x_1^3)\right\} \\ \boldsymbol{g} &= \left\{0 \qquad \varepsilon(\bar{\gamma}\cos\omega_E t - \bar{c}x_2)\right\} \end{aligned} \tag{10.5.44}$$

the vector product in eq. (10.5.35) is simplified and the integration reduced to the integral

$$\begin{aligned} M(t_0) &= \varepsilon\int_{-\infty}^{+\infty} x_2(t-t_0)\big(\bar{\gamma}\cos\omega_E t - \bar{c}x_2(t-t_0)\big)\mathrm{d}t \\ &= \varepsilon\int_{-\infty}^{+\infty} x_2(t)\big(\bar{\gamma}\cos\omega_E(t+t_0) - \bar{c}x_2(t)\big)\mathrm{d}t \end{aligned} \tag{10.5.45}$$

where $x_2(t)$ is extracted from eq. (10.5.42). Since $\cos\omega(t+t_0) = \cos\omega t\cos\omega t_0 - \sin\omega t\sin\omega t_0$, three integrals have to be evaluated

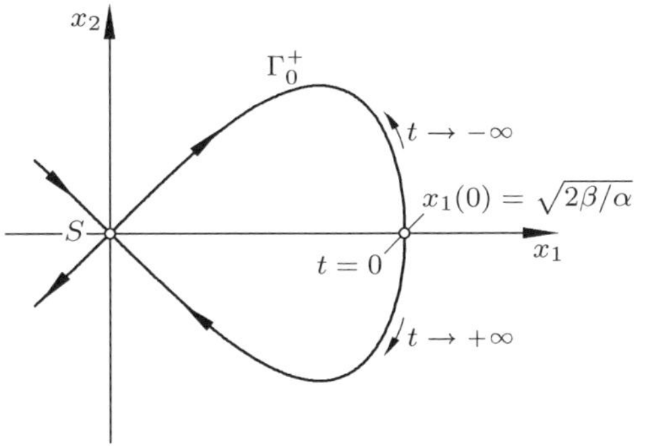

Fig. 10.5.11

The homoclinic orbit Γ_0^+ of the system $\dot{x}_1 = x_2, \;\; \dot{x}_2 = \beta x_1 - \alpha x_1^3$

$$I_1 = \int_{-\infty}^{+\infty} \frac{\tanh\sqrt{\beta}t}{\cosh\sqrt{\beta}t} \cos\omega_E t \, \mathrm{d}t \tag{10.5.46}$$

$$I_2 = \int_{-\infty}^{+\infty} \frac{\tanh\sqrt{\beta}t}{\cosh\sqrt{\beta}t} \sin\omega_E t \, \mathrm{d}t \tag{10.5.47}$$

$$I_3 = \int_{-\infty}^{+\infty} \frac{\tanh^2\sqrt{\beta}t}{\cosh^2\sqrt{\beta}t} \, \mathrm{d}t \tag{10.5.48}$$

For reasons of anti-symmetry, $I_1 = 0$. The other two integrals can be taken from integration tables, e.g. (Gradshteyn and Ryzhik, 1965), and we obtain

$$I_2 = \frac{\pi\omega_E}{\beta} \frac{1}{\cosh(\pi\omega_E/2\sqrt{\beta})} \qquad \text{and} \qquad I_3 = \frac{2}{3\sqrt{\beta}} \tag{10.5.49}$$

We thus obtain the Melnikov function

$$\begin{aligned} M(t_0) &= \varepsilon\left[\bar{\gamma}\beta(2/\alpha)^{1/2}\sin(\omega_E t_0)I_2 - \bar{c}\beta^2(2/\alpha)I_3\right] \\ &= \pi\omega_E(2/\alpha)^{1/2}\gamma\frac{\sin\omega_E t_0}{\cosh(\pi\omega_E/2\sqrt{\beta})} - \frac{4}{3}\frac{c}{\alpha}\beta^{3/2} \end{aligned} \tag{10.5.50}$$

Double zeroes of $M(t_0)$ arise, i.e. a tangency of $\mathbb{W}_\varepsilon^s$ and $\mathbb{W}_\varepsilon^u$, if the relation

$$\gamma_{\mathrm{cr}} = \frac{4}{3}\frac{c\beta}{\pi\omega_E}\sqrt{\frac{\beta}{2\alpha}}\cosh\frac{\pi\omega_E}{2\sqrt{\beta}} \tag{10.5.51}$$

holds between excitation amplitude γ on the one hand and damping c, stiffness coefficients α, β, and excitation frequency ω_E on the other; eq. (10.5.51) follows from eq. (10.5.50) for $M = 0$ and $\sin\omega_E t_0 = 1$.

In this case, a homoclinic bifurcation occurs. For $\gamma < \gamma_{cr}$, no transversal homoclinic points emerge; for $\gamma > \gamma_{cr}$, the stable and unstable manifold intersect, however. The function γ_{cr} is also called the Holmes-Melnikov boundary. As numerical experiments show, the criterion $\gamma > \gamma_{cr}$ is a good indicator for the emergence of fractal boundaries. Figure 10.5.12 summarises the results of numerical investigations carried out

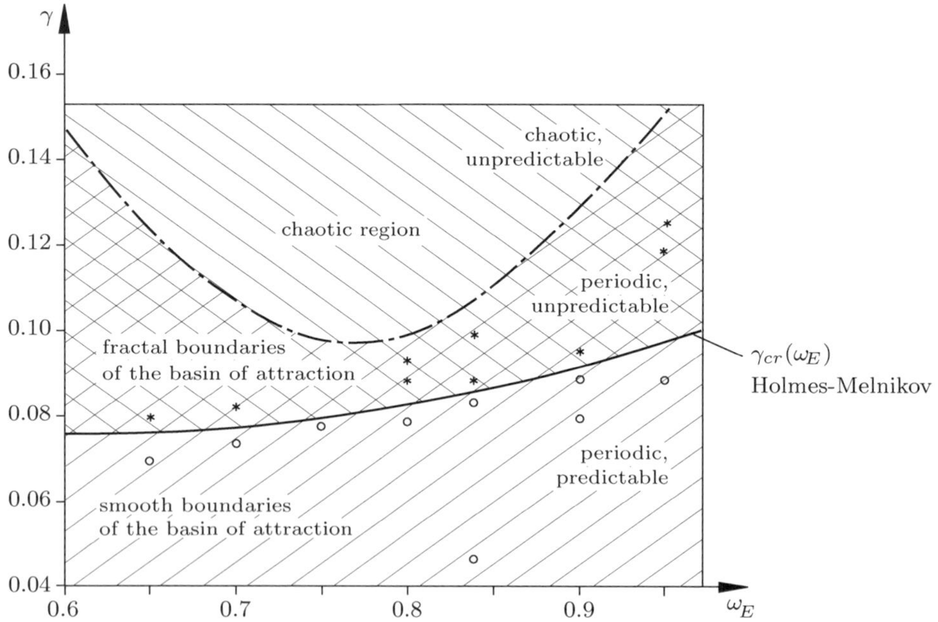

Fig. 10.5.12: Holmes-Melnikov boundary for the Duffing system after (Moon and Li, 1985): $\ddot{x} + 0.15\dot{x} - \frac{1}{2}x + \frac{1}{2}x^3 = \gamma \cos \omega_E t$

by Moon and Li (Moon and Li, 1985). The solid line marks the Holmes-Melnikov boundary $\gamma_{cr}(\omega_E)$. For fixed system parameters $c = 0.15$, $\alpha = \beta = 0.5$, the excitation amplitude is plotted versus the excitation frequency. For $\gamma < \gamma_{cr}$, the numerical experiments (indicated by circles $\circ$) yield smooth, non-fractal boundaries between the basins of attraction of the two periodic attractors. For γ-values somewhat above $\gamma_{cr}(\omega_E)$ but below the dot-dash line, fractal boundaries occur; the corresponding experiments are indicated by an asterisk $*$. In this region, the long-term behaviour is still regular; for a fixed initial condition, however, we cannot predict to which attractor the system will converge. Above the dot-dash line, we observe chaotic motions which no longer admit long-term prognoses.

In Colour Plate XXVII, p. 761, we have plotted for the Duffing equation

$$\ddot{x} + 0.12\dot{x} - 0.5x + 0.5x^3 = \gamma \cos 0.7t \tag{10.5.52}$$

the basins of attraction of the respective attractors from left to right for increasing values of the excitation amplitude γ. Here, γ was varied within the interval $[0.06, 0.14]$ in steps of 0.01. In the lower row, the integration was carried out up to $t = 1000$. As in Colour Plate XXVI, p. 753, all the initial conditions tending towards a period 1 oscillation about the left respectively right state of equilibrium are assigned the colour blue respectively yellow. Initial points of trajectories which end in a period 3 limit cycle are green, those which lead to chaotic motions red. If we calculate the Holmes-Melnikov boundary for the parameter constellation of eq. (10.5.52), we obtain

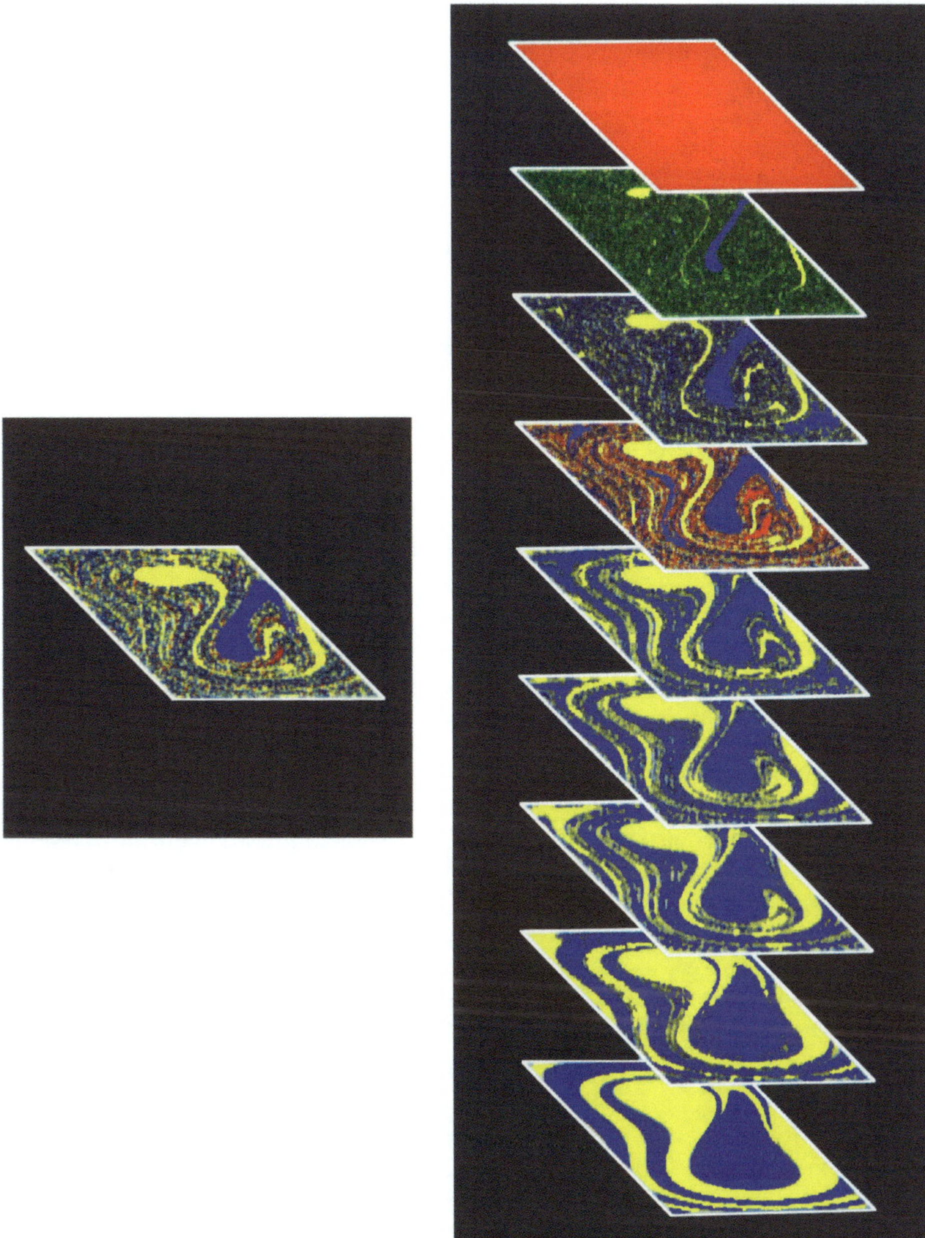

Plate XXVII Duffing equation (10.5.52):
Basins of attraction for increasing excitation amplitude γ

$$\gamma_{cr} \approx 0.0636$$

This value of the excitation amplitude lies between the first and the second picture. As was to be expected, fractal boundaries already emerge in the second picture for $\gamma = 0.07$; the motion itself is still regular and periodic. A further increase in the excitation frequency leads to still more complex fractal boundaries, an almost complete mixing of the phase space with blue and yellow points as if two liquids were being stirred together.

For $\gamma = 0.11$, red points indicating chaotic behaviour emerge for the first time. Here, however, this could be due to a particularly long transient phase, often referred to in the literature as transient chaos (see also section 10.4). To investigate this, we have repeated the integration up to $t = 10\,000$ and plotted the results in the upper picture. Almost all the red points have vanished, i.e. our conjecture that the transient phase lasts a very long time is confirmed. In the last picture ($\gamma = 0.14$), chaotic behaviour resulted for all initial conditions, even in the case of a drastic increase of the integration time.

Many oscillating systems can be modelled by equations similar to the Duffing equation. If a sensitive dependence on the initial conditions emerges, then this can be traced back to either the emergence of fractal boundaries between the basins of attraction of rival attractors or to chaotic long-term behaviour. In each case, careful numerical investigations, possibly combined with analytical procedures, are necessary in order to gain an overview of the dynamics of the system.

10.6 Shilnikov Bifurcations

In section 10.5, we studied the impact of a homoclinic bifurcation on the dynamics of a two-dimensional phase space. In this specific case, it was possible to calculate the boundaries for predictability in the parameter domain explicitly. In this section, we want to go one step further and discuss briefly some bifurcations of flow involving homoclinic orbits in three-dimensional vector fields which display a much higher variety than those in two dimensions. In particular, we give a short summary of some results on spiral chaos which were discovered by L. P. Shilnikov in the 1960s and have become classic because of their theoretical relevance and the broad range of applications.

The simplest types of hyperbolic fixed points for three-dimensional vector fields which allow for homoclinic orbits are saddles and saddle-foci. It is not surprising that the nature and relative magnitude of the eigenvalues and possible symmetries play a crucial role and are responsible for the changes in the geometry of the orbits and the topology of the flow. The homoclinic orbits of the Lorenz system shown in fig. 7.4.3, for example, require a symmetry in contrast to the Shilnikov bifurcations presented below.

An intensively studied system is Rössler's oscillator (Rössler, 1979a)

$$\dot{x} = -y - z, \quad \dot{y} = x + ay, \quad \dot{z} = (b + z)x - cz \tag{10.6.1}$$

which displays not only subharmonic cascades leading to chaos but also Shilnikov bifurcations and spiral chaos, see e.g. (Fraser and Kapral, 1982). A short glance at the highly complex dynamical behaviour of the prototypical Rössler model has been given in section 8.3.2.

In the following, we assume that the hyperbolic fixed points possess a two-dimensional stable manifold $\mathbb{W}^s$ and a one-dimensional unstable manifold $\mathbb{W}^u$ (otherwise, the direction of time should be reversed). The classical publications of L. P. Shilnikov on the existence of periodic and aperiodic trajectories (Shilnikov, 1965; Shilnikov, 1970) were followed and partly improved by a large series of articles. Here, we present only a short overview of some results and refer the reader to the presentations in (Glendinning and Sparrow, 1984; Wiggins, 1988; Kuznetsov, 1998; Glendinning, 1999) and the references therein.

Let us consider the three-dimensional system of differential equations

$$\dot{\boldsymbol{x}} = \boldsymbol{F}(\boldsymbol{x}, \mu) = \boldsymbol{L}\boldsymbol{x} + \boldsymbol{N}(\boldsymbol{x}, \mu), \qquad \boldsymbol{x} = \{x_1, x_2, x_3\} \tag{10.6.2}$$

and assume that the origin $\boldsymbol{x}_0 = \boldsymbol{0}$ is a hyperbolic fixed point and that, for $\mu = 0$, there is a homoclinic orbit Γ_0 biasymptotic to the origin. We further assume that the coordinate system is chosen in such a way that at $\boldsymbol{x}_0$ the two-dimensional stable eigenspace coincides with $x_3 = 0$ and the one-dimensional unstable eigenspace with the x_3-axis. The fixed point is either a saddle with real eigenvalues

$$\lambda_1 < \lambda_2 < 0 < \lambda_3 \qquad \text{(saddle)} \tag{10.6.3}$$

or a saddle-focus with a pair of complex eigenvalues and one real eigenvalue

$$\lambda_{1,2} = \rho \pm i\omega \quad \text{with} \quad \rho < 0, \;\; \omega \neq 0; \qquad \lambda_3 > 0 \qquad \text{(saddle-focus)} \tag{10.6.4}$$

The two-dimensional stable invariant manifold $\mathbb{W}^s$ at $\boldsymbol{x}_0$ intersects the unstable invariant manifold $\mathbb{W}^u$ along the homoclinic orbit. We denote the real eigenvalues with negative real parts which are closest to zero the *leading eigenvalues*. We further define the *saddle quantity* which plays a decisive role in the classification of the bifurcation

$$\begin{aligned} \sigma &= \lambda_2 + \lambda_3 \qquad &&\text{(saddle)} \\ \sigma &= \rho + \lambda_3 \qquad &&\text{(saddle–focus)} \end{aligned} \tag{10.6.5}$$

Let us briefly summarise some results for the two cases. The corresponding proofs can be found in, for example, (Wiggins, 1988; Kuznetsov, 1998; Glendinning and Sparrow, 1984).

(A) Saddle homoclinic bifurcation

We assume that the origin $\boldsymbol{x}_0 = \boldsymbol{0}$ is a *saddle* with real eigenvalues as given in eq. (10.6.3) and that the system eq. (10.6.2) possesses a homoclinic orbit Γ_0 for $\mu = 0$ connecting the saddle to itself. We further assume that the system eq. (10.6.2) has no symmetries, i.e. it is generic. The following two cases can then be distinguished:

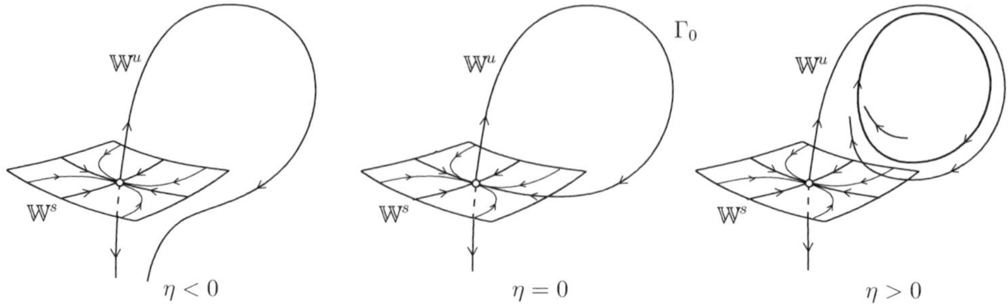

Fig. 10.6.1: Saddle homoclinic bifurcation for $\sigma < 0$, after (Kuznetsov, 1998)

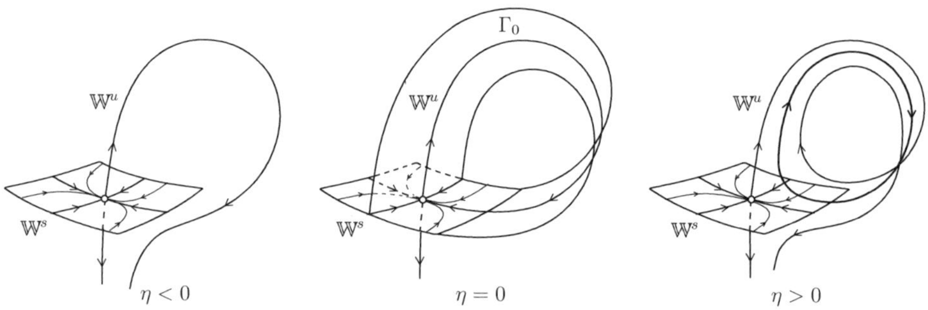

Fig. 10.6.2: Saddle homoclinic bifurcation for $\sigma > 0$, after (Kuznetsov, 1998)

(A1) $\sigma < 0$:

For $\mu > 0$, a unique and stable periodic orbit bifurcates from the homoclinic orbit Γ_0. The bifurcation is illutrated in fig. 10.6.1. As in the planar case discussed in section 10.5, no periodic orbits exist for $\mu < 0$. All such systems are topologically equivalent in a neighbourhood of Γ_0 near the origin.

(A2) $\sigma > 0$:

A unique saddle limit cycle bifurcates from the homoclinic orbit Γ_0 for $\mu > 0$. The global flow and the direction of the bifurcation depend on the topology and orientability of the stable manifold close to $\boldsymbol{x}_0 = \mathbf{0}$, i.e. on whether this manifold is equivalent to a simple band or it is twisted like a Möbius band. The twisted saddle homoclinic bifurcation is shown in fig. 10.6.2. For details, see (Wiggins, 1988; Kuznetsov, 1998).

(B) Saddle-focus homoclinic bifurcation

We now assume that the origin $\boldsymbol{x}_0 = \mathbf{0}$ is a *saddle-focus* with eigenvalues as given in eq. (10.6.4) and that the system eq. (10.6.2) possesses a homoclinic orbit Γ_0 for $\mu = 0$

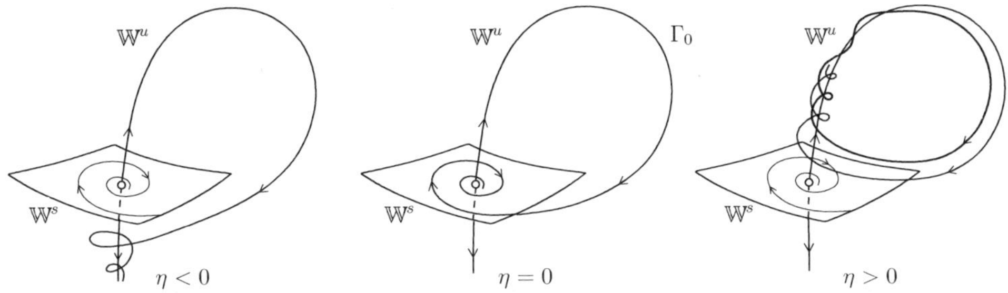

Fig. 10.6.3: Saddle-focus homoclinic bifurcation for $\sigma < 0$, after (Kuznetsov, 1998)

spiraling into the saddle-focus. Under some genericity conditions, we can distinguish the following two cases when Γ_0 is broken:

(B1) $\sigma < 0$:

A unique and stable limit cycle bifurcates for $\mu > 0$ from the homoclinic orbit Γ_0 as shown in fig. 10.6.3. No periodic orbits exist for $\mu < 0$. Again, the topology of the global flow close to $\boldsymbol{x}_0 = \mathbf{0}$ depends upon the stable manifold, on whether it is twisted or not. This property is characterised by the real number

$$\nu_0 = -\frac{\lambda_3}{\rho} \tag{10.6.6}$$

which is a topological invariant. This is known as a *simple* Shilnikov bifurcation.

(B2) $\sigma > 0$:

This is the most subtle case. It has been studied intensively by L. P. Shilnikov and is designated the *chaotic* Shilnikov bifurcation. There is an infinite number of periodic orbits of saddle-type in a neighbourhood of the homoclinic orbit. Indeed, Shilnikov could prove that Smale horseshoes are present in a two-dimensional Poincaré map defined in a section near the homoclinic orbit Γ_0 close to $\boldsymbol{x}_0 = \mathbf{0}$ which give rise to an infinite number of saddle limit orbits leading to chaotic behaviour, cf. section 10.5 and fig. 10.5.9. The topological structure of the flow is highly complex and depends on the invariant ν_0 which characterises the local twist. For details, see (Glendinning and Sparrow, 1984; Shilnikov and Shilnikov, 2007), for example.

In practice, in order to locate the hyperbolic fixed points, the homoclinic and heteroclinic orbits together with the associated invariant stable and unstable manifolds, it is indispensable to apply numerical methods. Since the 1980s, there has been great interest in the development of algorithms for the numerical analysis of local and global bifurcations, and many publications have resulted. Computer packages have been developed which allow an efficient and accurate numerical computation of the homoclinic and heteroclinic orbits and the associated invariant manifolds. Here, we only mention the packages AUTO and MATCONT (Doedel et al., 2001; Dhooge et al., 2003) which rely, among others, on shooting and continuation

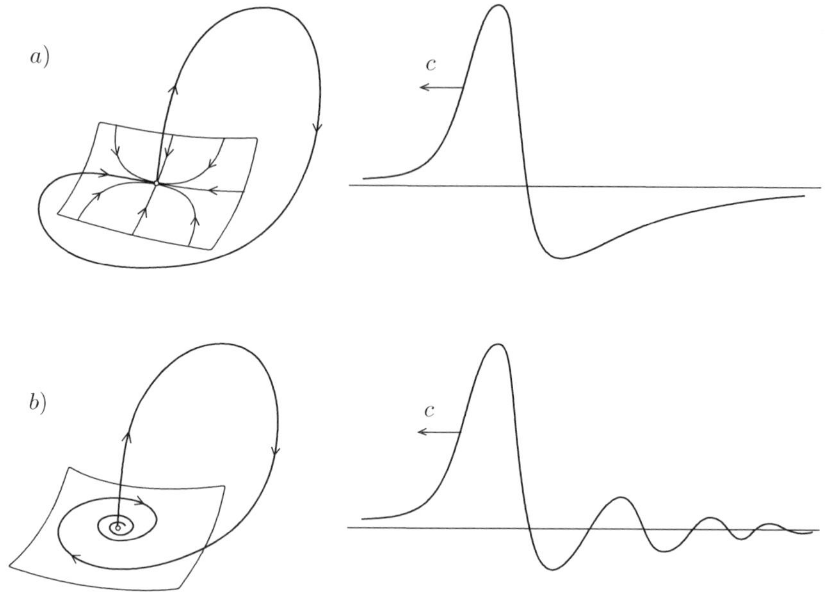

Fig. 10.6.4: Pulses with a) monotone and b) oscillating decay, after (Kuznetsov, 1998)

techniques. These tools allow the study of the reorganisation of the phase space and the overall dynamics near bifurcations, see (Dhooge et al., 2003; Doedel, 1981; Doedel et al., 2001; Aguirre et al., 2011) and the references there. In addition, we refer the reader to chapter 10 in (Kuznetsov, 1998), which contains a detailed list of algorithms and references for further reading.

Homoclinic and heteroclinic solutions have received a great deal of attention because they represent solitary pulses, wave trains and front profiles propagating in non-equilibrium excitable media with constant velocity. Well-known examples are concentration waves in oscillatory reaction-diffusion systems like the Belousov-Zhabotinsky reaction (Zaikin and Zhabotinsky, 1970), the transition to mixed-mode oscillations in electrochemical systems (Krischer et al., 1993), waves in reactions occurring in living oganisms like the enzyme-catalysed peroxidase-oxidase reaction (Sensse et al., 2006), waves on shallow water surfaces (Korteweg and de Vries, 1895) or the propagation of nerve impulses along an axon (FitzHugh, 1961; Nagumo et al., 1962), to mention just a few.

Since homoclinic orbits converge to the same value $\boldsymbol{x}_0$ for $t \to \pm\infty$, they describe solitary pulses (see fig. 10.6.4) or pulse trains for n-homoclinic loops, as demonstrated in fig. 10.6.5 for the double-pulse homoclinic orbit (n $=$ 2). The form of the pulse depends on the character of the hyperbolic fixed point $\boldsymbol{x}_0$. If $\boldsymbol{x}_0$ is a saddle, the pulse decays monotonically as shown in fig. 10.6.4a, whereas saddle-foci produce oscillations with a decaying amplitude (fig. 10.6.4b).

In 1963, Alan Lloyd Hodgkin and Andrew Huxley received the Nobel Prize in Physiology or Medicine for their work on the propagation of action potentials in neurons.

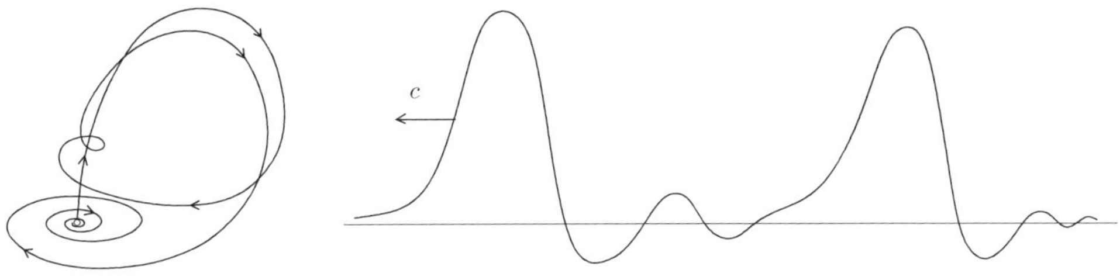

Fig. 10.6.5: Double-pulse homoclinic orbit, after (Kuznetsov, 1998)

They won this prize for their mathematical model describing the initiation and propagation of action potentials in the giant axon of squids. A simplified version due to R. FitzHugh and J. Nagumo (FitzHugh, 1961; Nagumo et al., 1962) has the following form of a reaction-diffusion equation in excitable media

$$\begin{aligned}\frac{\partial u}{\partial t} &= \frac{\partial^2 u}{\partial x^2} - f_a(u) - v \\ \frac{\partial v}{\partial t} &= bu\end{aligned} \tag{10.6.7}$$

where u denotes the membrane voltage in a neuron. v is a variable describing the relaxation, $f_a(u) = u(u-a)(u-1)$ a cubic function and the parameters are positive ($0 < a < 1$ and $b > 0$). It can easily be checked that waves of the form $u(x,t) = U(\xi)$, $v(x,t) = V(\xi)$ with $\xi = x + ct$ propagating with constant velocity $c > 0$ satisfy eqs.(10.6.7) and lead to the following system of three ordinary differential equations for the amplitudes U, V and W

$$\begin{aligned}\dot{U} &= W \\ \dot{V} &= \frac{b}{c}U \\ \dot{W} &= cW + f_a(U) + V\end{aligned} \tag{10.6.8}$$

Here, dots are used to describe derivatives with respect to ξ. The "origin" $U = V = W = 0$ is a hyperbolic fixed point, as can be seen from the Jacobian matrix $\boldsymbol{J}$ at the fixed point

$$\boldsymbol{J} = \begin{pmatrix} 0 & 0 & 1 \\ \frac{b}{c} & 0 & 0 \\ a & 1 & c \end{pmatrix}$$

Applying Descarte's rule of signs together with the theorem of Hurwitz (3.1.27), it can be concluded from the characteristic equation

$$\lambda^3 - c\lambda^2 - a\lambda - \frac{b}{c} = 0 \tag{10.6.9}$$

that, for positive parameters a, b, c, the system has one positive real eigenvalue λ_1 and two eigenvalues with negative real parts. The substitution of $\lambda \to \lambda - \frac{c}{3}$ into eq. (10.6.9) yields the normal form of the cubic equation $\lambda^3 + 3p\lambda + 2\lambda = 0$. The sign of the discriminant $D = q^2 + p^3$ of this equation then decides whether the hyperbolic fixed point is a saddle ($D < 0$) or a saddle-focus ($D > 0$). The homoclinic orbits of the system can be calculated numerically applying the methods described above, cf. (Kuznetsov, 1998). For a further classification of the dynamics of the system, the saddle quantity σ can be calculated according to eqs. (10.6.5). Depending on the parameters a, b, c, solutions of the FitzHugh-Nagumo equation (10.6.9) exist which are travelling pulses or wave trains, as shown in figs.10.6.4, 10.6.5. In the saddle-focus case, chaotic behaviour can be observed near the homoclinic bifurcation.

As a further application, we present the analysis of human electroencephalograms (EEG) during epileptic seizures using methods from synergetics.

In ancient Greece, physicians believed that epileptic seizures were sacred diseases induced by supernatural forces, either sent by divine powers or by evil spirits. Hippocrates' work, written in 400 BC, contains one of the first observations of epileptic seizures in humans. In contrast to the priests' explanation, Hippocrates believed that all illnesses have natural causes: "Men believe only that it is a divine disease because of their ignorance and amazement" (from *The Sacred Disease* by Hippocrates).

In 1924, the neurologist Hans Berger invented the EEG which allows the brain's electrical activity to be recorded along the scalp. The EEG is generated by the activity of electrically charged neurons. The human brain is a highly complex system (Haken, 2008). The cerebral cortex contains about 10^{11} interconnected and non-linearly interacting neurons. The surface of each neuron is covered by $10^3 - 10^5$ synapses which transmit information to other neurons. Changes of the intracellular potentials of the neurons then lead to extracellular potentials and the generation of electrical currents which in turn generate potential differences in the skull. These potential differences can be measured by means of EEGs which always reflect the summation of the synchronous activity and firing of thousands or millions of neurons.

In neurology, EEGs are mainly used in the diagnosis and classification of epileptic seizures. The seizures may be divided into two main classes. In the case of *partial* or *focal* seizures, the source of the seizure within the brain is localised in contrast to *generalised* seizures where the source is distributed. These include brief absences also referred to as petit mal seizures which last up to only 20 seconds and are mainly observed in childhood. These seizures are characterised by a pronounced spike-wave behaviour.

Figure 10.6.6 shows the multigrid EEG recordings of two different persons suffering from petit mal epilepsy. The solid lines are measurement records which have a spatially coherent oscillating character with about three spike-wave cycles per second. A synergetic approach (see section 6.8) is suggested to model the macroscopic synchronous structure in the brain (Friedrich and Uhl, 1996; Uhl, 1995; Kelso and Fuchs, 1995).

In a first step, the spatial modes of the order parameters are determined. Since the estimation of the correlation dimension of the electrical potential $V_{\mathrm{k}}(t)$ measured at several electrodes yields values between two and three, it can be assumed that the dynamics may be described approximately by three order parameters denoted

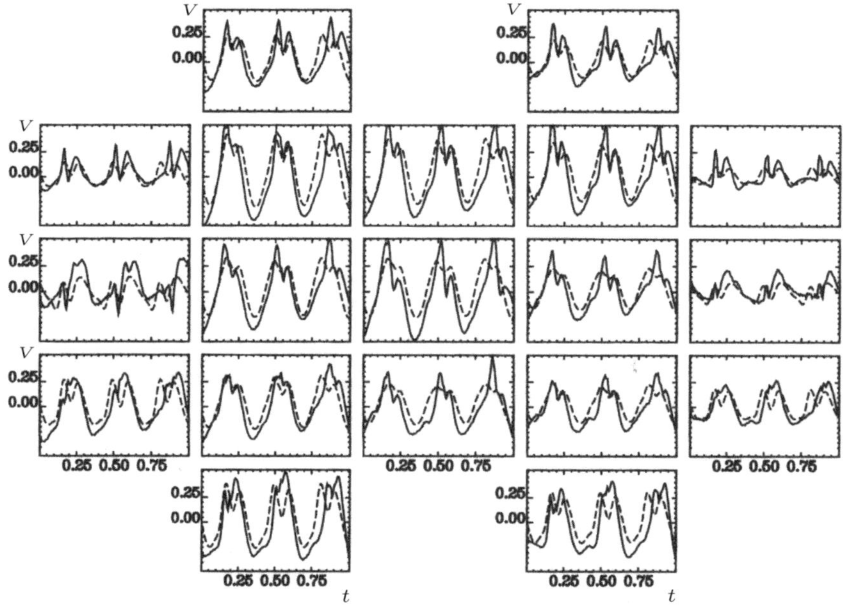

Data set A

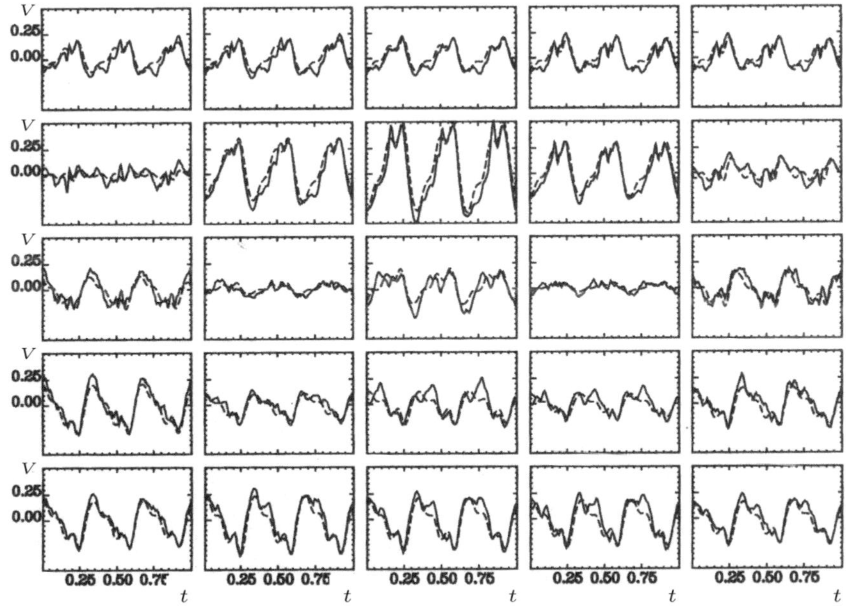

Data set B

Fig. 10.6.6: EEG recordings (solid lines) and reconstructions (dashed lines) of two different persons during a petit mal seizure. The electrical potential v displays the typical spike-wave behaviour; after (Friedrich and Uhl, 1996)

$x(t), y(t), z(t)$ which are the amplitudes of three spatial modes $\boldsymbol{v}_{\rm i}$, $(\rm i = 1, 2, 3)$. Hence, the following ansatz is used for the electrical potential

$$\boldsymbol{V}(t) = x(t)\boldsymbol{v}_1 + y(t)\boldsymbol{v}_2 + z(t)\boldsymbol{v}_3 \tag{10.6.10}$$

The amplitudes can be extracted from the measured records according to

$$\begin{aligned} x(t) &= \boldsymbol{v}_1{}^+\boldsymbol{V}(t) \\ y(t) &= \boldsymbol{v}_2{}^+\boldsymbol{V}(t) \\ z(t) &= \boldsymbol{v}_3{}^+\boldsymbol{V}(t) \end{aligned} \tag{10.6.11}$$

Here, $\boldsymbol{v}_{\rm i}^+$ denote orthogonal adjoint modes with $\boldsymbol{v}_{\rm i}^+\boldsymbol{v}_{\rm k} = \delta_{\rm ik}$. For the amplitudes, a self-consistent ansatz is proposed which has to be judged after performing the analysis

$$\dot{x}(t) = y(t), \; \dot{y}(t) = z(t), \; \dot{z}(t) = f(x, y, z, \boldsymbol{a}) \tag{10.6.12}$$

where the function $f(x, y, z, \boldsymbol{a})$ is approximated by a Taylor expansion which contains polynomials in x up to the order four and is quadratic in y and linear in z. The coefficients $\boldsymbol{a}$ are obtained from a least-squares fit of the potential to the first dominant modes of a Karhunen-Loève decomposition; for details, see (Uhl et al., 1993; Friedrich and Uhl, 1996). It turns out that the time signal $\boldsymbol{V}(t)$ can be approximated with sufficient accuracy using this ansatz. After the parameters $\boldsymbol{a}$ have been calculated, the dynamical system eqs. (10.6.12) can be integrated numerically choosing appropriate intial conditions.

In fig. 10.6.7, the reconstructed trajectories in the phase space obtained from the synergetic approach are presented for both data sets A and B. These figures are compared with the equivalent reconstructions obtained from the measured EEG records (fig. 10.6.8). The dynamical behaviour is characterised by a hyperbolic fixed point with a two-dimensional manifold where the motion is spiraling. From the Jacobian matrix of the system eqs. (10.6.12), it can be seen that the fixed point is a saddle-focus with a pair of complex conjugate eigenvalues $\lambda_{1,2} = \rho \pm i\omega$ and one real eigenvalue λ_3 for both data sets. For data set A, we obtain $\rho = 0.08$ and $\lambda_3 = -0.21$, i.e. the fixed point has a two-dimensional unstable manifold and a one-dimensional stable one. For a classification corresponding to our previous description, the time has to be reversed, leading to $\bar{\rho} = -0.08$ and $\bar{\lambda}_3 = 0.21$. The calculation of the saddle quantity, eq. 10.6.5, yields $\bar{\sigma} = \bar{\rho} + \bar{\lambda}_3 = 0.13$, i.e. $\bar{\sigma} > 0$. According to Shilnikov's theorem, an infinite number of saddle limit cycles occur, i.e. during the epileptic seizure, "transient chaotic" behaviour can be observed. When the seizure decays, the trajectory leaves this dynamical range and the brain returns to its normal activity. For data set B, the fixed point has a one-dimensional unstable manifold and a two-dimensional stable one. From the eigenvalues, we obtain $\rho = -0.12$ and $\lambda_3 = 0.04$, i.e. $\sigma = \rho + \lambda_3 = -0.08$. This case corresponds to $\sigma < 0$, i.e. to the simple Shilnikov bifurcation where one stable limit cycle bifurcates from the homoclinic orbit. More experiments and further investigations are required to study this kind of epilepsy (Friedrich and Uhl, 1996).

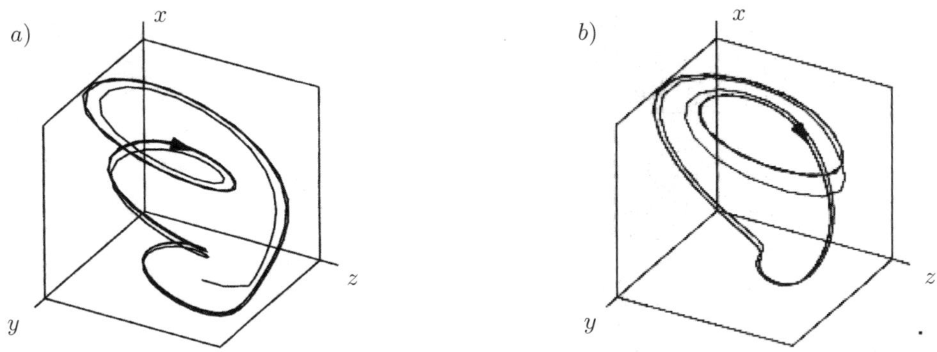

Fig. 10.6.7: Reconstruction of the trajectories from a numerical integration of the order parameter eqs. (10.6.12): a) data set A, b) data set B; after (Friedrich and Uhl, 1996)

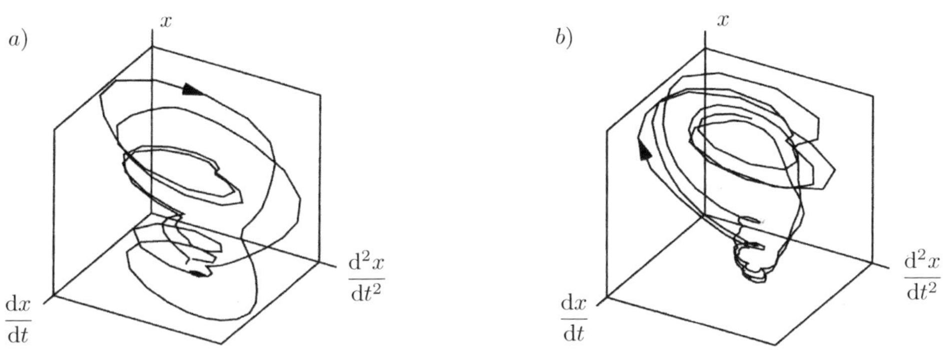

Fig. 10.6.8: Reconstruction of the trajectories from the amplitude $x(t)$ determined from measured EEG data: a) data set A, b) data set B; after (Friedrich and Uhl, 1996)

10.7 Julia Sets and Their Ordering Principle

When one wishes to simplify a theory,
one should complexify the variables

A bon mot of physicists

One of the most common procedures used to determine the zeroes of a non linear function $f(x)$ goes back to Isaac Newton. We select a rough estimated value x_0 and calculate successively improved approximations $x_1, x_2, \ldots$ for the zero x^* of $f(x)$ according to the rule

$$x_{n+1} = x_n - \frac{f(x_n)}{f'(x_n)} \tag{10.7.1}$$

If x^* is a simple zero (or root) of $f(x)$ and the initial value x_0 is sufficiently close to x^*, the procedure converges towards the solution x^*. What is interesting about

Newton's idea is that, in order to solve a "static" problem $f(x) = 0$, a dynamic procedure is applied where successive approximation values are calculated which converge towards a solution x^*, i.e. are attracted by x^*. What happens, however, if there are several solutions, several attractors in competition with one another? Let us think of the roots of the equation $x^4 - 1 = 0$, for example. In the complex plane, there are the four solutions $x_{1,2} = \pm 1$, $x_{3,4} = \pm i$. Are the basins of these attractors separated from one another by simple straight lines, i.e. the two bisectors, or do they possess, as in the case of the Duffing equation (cf. Colour Plate XXVII, p. 761), a highly complex, fractal character? As early as 1879, Arthur Cayley studied this problem (Cayley, 1879), but put it aside again as it appeared too complicated.

During the First World War, two French mathematicians, Gaston Julia and Pierre Fatou, took up the problem again independently of one another; they developed a theory dealing with the iteration of rational functions $R(z) = P(z)/Q(z)$ of relatively prime polynomials P, Q in the plane of complex numbers z (Julia, 1918; Fatou, 1919; Fatou, 1920). This problem arises, for example, in the determination of the roots of polynomials with the aid of the Newton iteration eq. (10.7.1). They discovered that the boundaries between the basins of attraction of various attractors exhibit extremely curious characteristics. In 1918, Julia received the Grand Prix of the Académie des Sciences de Paris for his fundamental research, but this abstract theory soon sank into oblivion as it was virtually impossible at the time to picture the multitude of limit sets which emerged.

At the end of the Second World War, the attention of Benoît Mandelbrot, whose family had emigrated to France, was drawn to Julia and Fatou's work by his uncle, Scholem Mandelbrojt, himself a mathematician and specialist in the field of complex analysis. One of Mandelbrot's teachers at the École Polytechnique was Gaston Julia. However, another 35 years were to pass until Mandelbrot, stimulated by powerful computers and the graphic possibilities they offered, again turned to the iteration of rational functions in the complex plane. In analogy to the logistic map, he selected the simplest non-linear recursion rule

$$z_{\mathrm{n}+1} = z_{\mathrm{n}}^2 + c \tag{10.7.2}$$

where here, however, c and z_{n} are complex numbers.

In complex analytical maps, the "point at infinity" $z = \infty$ extending the Gauss plane $\mathbb{C}$ often does not play a special role. The topological structure of the closed set $\overline{\mathbb{C}} = \mathbb{C} \cup \{\infty\}$, i.e. of the union of $\mathbb{C}$ and the point $z = \infty$, can be illustrated easily on the Riemann sphere S. Figure 10.7.1 shows the unique mapping of points of the complex plane to points on a sphere S, the south pole of which touches the complex plane at the origin. The point $z \in \mathbb{C}$ is assigned the point of intersection $P(z)$ of the sphere with the line connecting the point z with the north pole N. The point $z = \infty$ therefore corresponds to the north pole N. If we wish to investigate characteristics of a function at the point $z = \infty$, it is expedient to work on the Riemann sphere.

Recursion formulæ reflect dynamic processes. For this reason, our interest is directed mainly at the long-term behaviour of the sequence of points generated, firstly in dependence of the control parameter c and secondly in dependence of the initial value z_0. Let us consider the map eq. (10.7.2)

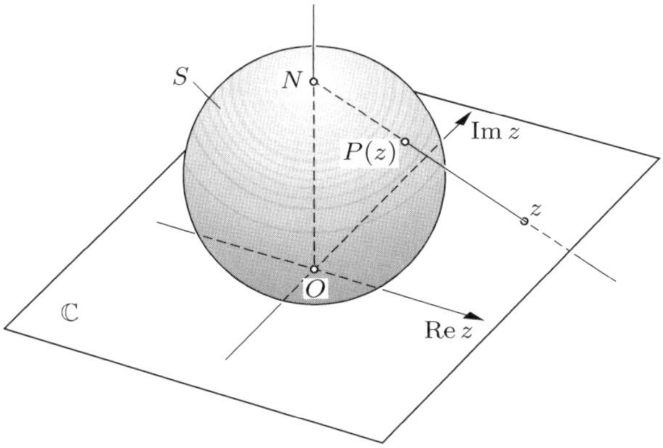

Fig. 10.7.1

The Riemann sphere (stereographic projection)

$$z \longmapsto f_c(z) = z^2 + c \tag{10.7.3}$$

at first for the simplest case $c = 0$. Fixed points of the map $f_0(z)$ in $\mathbb{C}$ are $z = 0$, $z = 1$ and $z = \infty$, where the point $z = 0$ is superstable since $f_0'(z) = 2z$. Points for which $f'(z) = 0$ holds are also called critical points. The value $z = 1$ defines an unstable fixed point. The stability behaviour of $z = \infty$ can either be determined by mapping onto the Riemann sphere or by another coordinate transformation which maps $z = \infty$ to $\bar{z} = 0$. If we select $\bar{z} = h(z) = 1/z$, for example, the transformed mapping is

$$\bar{f}_0(\bar{z}) = h f_0 h^{-1} = \bar{z}^2$$

i.e. $\bar{z} = 0$ respectively $z = \infty$ is also a superstable fixed point and hence a critical point. The dynamics of $f_0(z)$ is now quite simple. For $|z| < 1$, the n-fold iterated map $f_0^{\mathrm{n}}(z)$ tends towards zero for $\mathrm{n} \to \infty$; for $|z| > 1$, on the other hand, $f_0^{\mathrm{n}}(z) \to \infty$ holds for $\mathrm{n} \to \infty$. The unit circle represents the boundary between the basins of attraction of the two fixed points $z = 0$ and $z = \infty$, and is denoted the Julia set J_0 of the map $f_0(z)$. The unstable fixed point $z = 1$ lies on J_0. What is interesting is the dynamics on the unit circle. A point $z = \mathrm{e}^{i\varphi}$ $(0 \leqslant \varphi < 2\pi)$ is mapped to $f_0(z) = \mathrm{e}^{2i\varphi}$, i.e. for the angle, the map is

$$\varphi \longmapsto 2\varphi \pmod{2\pi}$$

This is the Bernoulli shift discussed in section 5.6.1. Thus, the dynamics on the Julia set is chaotic.

We next consider the Julia set J_c of f_c for the parameter value $c = i/2$. Apart from $z_3 = \infty$, the fixed points of the map eq. (10.7.3) are

$$z_{1,2} = \tfrac{1}{2}(1 \pm \sqrt{1 - 4c}) \tag{10.7.4}$$

For $c = i/2$, we obtain $z_1 = 1.14–0.39i$ and $z_2 = -0.14+0.39i$, where z_2 is stable, z_1, on the other hand, unstable. Figure 10.7.2 presents the Julia set J_c of f_c as the boundary between the basins of attraction of z_2 and $z_3 = \infty$. Like the Koch snowflake curve, J_c appears in this case as a fractal structure, a simple closed curve – a so-called Jordan curve, i.e. a homeomorphic image of the unit circle – which is nowhere differentiable.

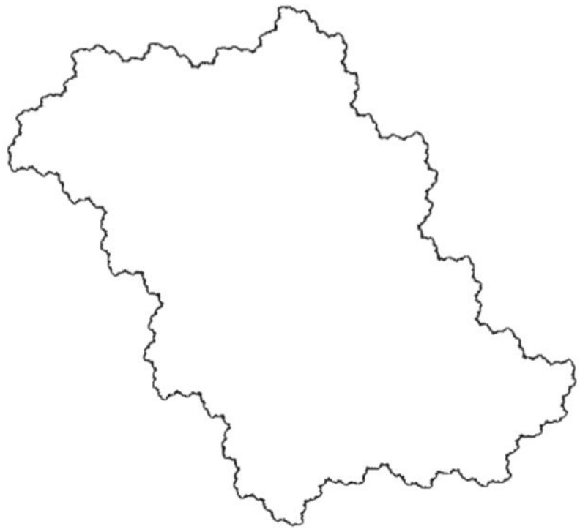

Fig. 10.7.2

The Julia set of the map $f_c(z) = z^2 + i/2$ is a Jordan curve

Numerical experiments for other c-values show that Julia sets usually have a fractal character.

We now list some characteristics of Julia sets going back to Julia and Fatou. Let us begin with the definition of Julia sets, see (Devaney, 1987; Milnor, 1990):

> The Julia set J_c of a rational function f_c consists of all unstable periodic cycles and their accumulation points

(see also section 3.6 on the concept of a cycle). This definition is obviously not suitable for the generation of graphic representations such as fig. 10.7.2, for example. In order to develop practicable algorithms, a series of further characteristics of J_c is required.

a) $J_c \neq \emptyset$ is a compact point set and contains an uncountable number of points z.

This property is a consequence of the transition to complex numbers. Since the fixed point equation of the k-fold iterated function

$$f_c^{\mathrm{k}}(z) = z$$

in the Gauss plane always possesses solutions, infinitely many periodic cycles emerge for $\mathrm{k} \to \infty$.

b) $f_c(J_c) = J_c = f_c^{-1}(J_c)$

This indicates that the Julia set as a whole is invariant with respect to the dynamics of the mapping f_c as well as with respect to its inverse.

c) For every $z \in J_c$, the set of all pre-images of z, i.e. $f_c^{-\mathrm{k}}(z)$ for $\mathrm{k} \to \infty$, is dense in J_c.

This characteristic expresses the fact that the inverse orbit of z gained by inverse iteration comes arbitrarily close to each point of J_c. This offers a first opportunity of representing J_c graphically. We select an arbitrary point $z \in J_c$, determine its pre-images, the pre-image of these pre-images and so on. Apart from the necessity of a cleverly devised data administration for the rapidly increasing number of pre-images, this method may entail the difficulty that the pre-images of z are not distributed uniformly over J_c. It is possible to introduce a barrier in the programme for regions which are frequently visited; however, this does not help us to find a sufficiently large number of points in areas which are rarely visited; for more detail, see (Peitgen and Saupe, 1988).

d) The set $F_c = \overline{\mathbb{C}} - J_c$ is open and contains all stable cycles as well as their basins of attraction. If $\{z_1, z_2, \dots, z_\mathrm{k}\}$ is a stable cycle of length k, J_c forms the boundary of the basin of attraction of this cycle.

The complement F_c of J_c is also called a Fatou set (Peitgen and Richter, 1986a). An interesting consequence of characteristic *d)* is that the basins of attraction of different stable cycles possess the same boundary. The set J_c is hence in particular the boundary of the basin of attraction of $z = \infty$. This opens a further perspective for graphic representations of Julia sets and it is this that we have made use of for most of the figures in this section.

e) If $\bar{z} \in J_c$ and $\varepsilon > 0$ are given and $\tilde{J}_c = \{z \in J_c;\ |z - \bar{z}| < \varepsilon\}$, there is a natural number n so that $f_c^\mathrm{n}(\tilde{J}_c) = J_c$

i.e. the complete set J_c can be generated by a finite number of n iterations from any sub-set $\tilde{J}_c$ – however small – of J_c. Particularly the self-similarity of Julia sets follows from this.

Figure 10.7.3 displays Julia sets for a number of different c-values and demonstrates the multitude of forms which we can visualise nowadays – in contrast to Julia and Fatou's times – thanks to computer graphics. There are connected curves as in fig. 10.7.3a, for example, which resemble the frontier of a country, or a number of contractions of internal areas as in b) or the curves disintegrate into spirals or small flowers as in c) and f) and then have the structure of Cantor sets. Figures 10.7.3d) and g) show more dendritic forms, sometimes enclosing inner areas. Colour Plate XXVIII on p. 779 presents a further selection of Julia sets, some of them reminiscent of designs for unusual jewellery. What is fascinating about them is that, in spite of the simplicity of the underlying iteration rule, so many varying types of fractal limit sets emerge depending on the choice of control parameter.

Euclidean geometry knows no fractal structures. The forms occurring there – such as circles, cones, spheres etc. – and which seem rather dull to us nowadays are the solutions of "static" equations such as $x^2 + y^2 + z^2 = 1$, for example.

Fractal forms, on the other hand, are the result of *dynamic* processes. They result from recursion formulæ, from feedback processes, i.e. from the continually repeated application of a mapping rule. In contrast to the classic forms of Euclidean geometry, fractal structures recall those that we observe in nature, such as clouds, plants or eroded mountain ranges.

Up to now, we have concentrated on the location of the hyperbolic cycles (see also section 3.6). What determines the stability of the cycle $\{z_1, z_2, \dots, z_\mathrm{k}\}$ of period k is, as in the real domain, the absolute value of the derivative

$$\frac{\mathrm{d}}{\mathrm{d}z} f_c^\mathrm{k}(z)\Big|_{z=z_1} = \prod_{\mathrm{m}=1}^{\mathrm{k}} f_c'(z_\mathrm{m}) = \lambda \tag{10.7.5}$$

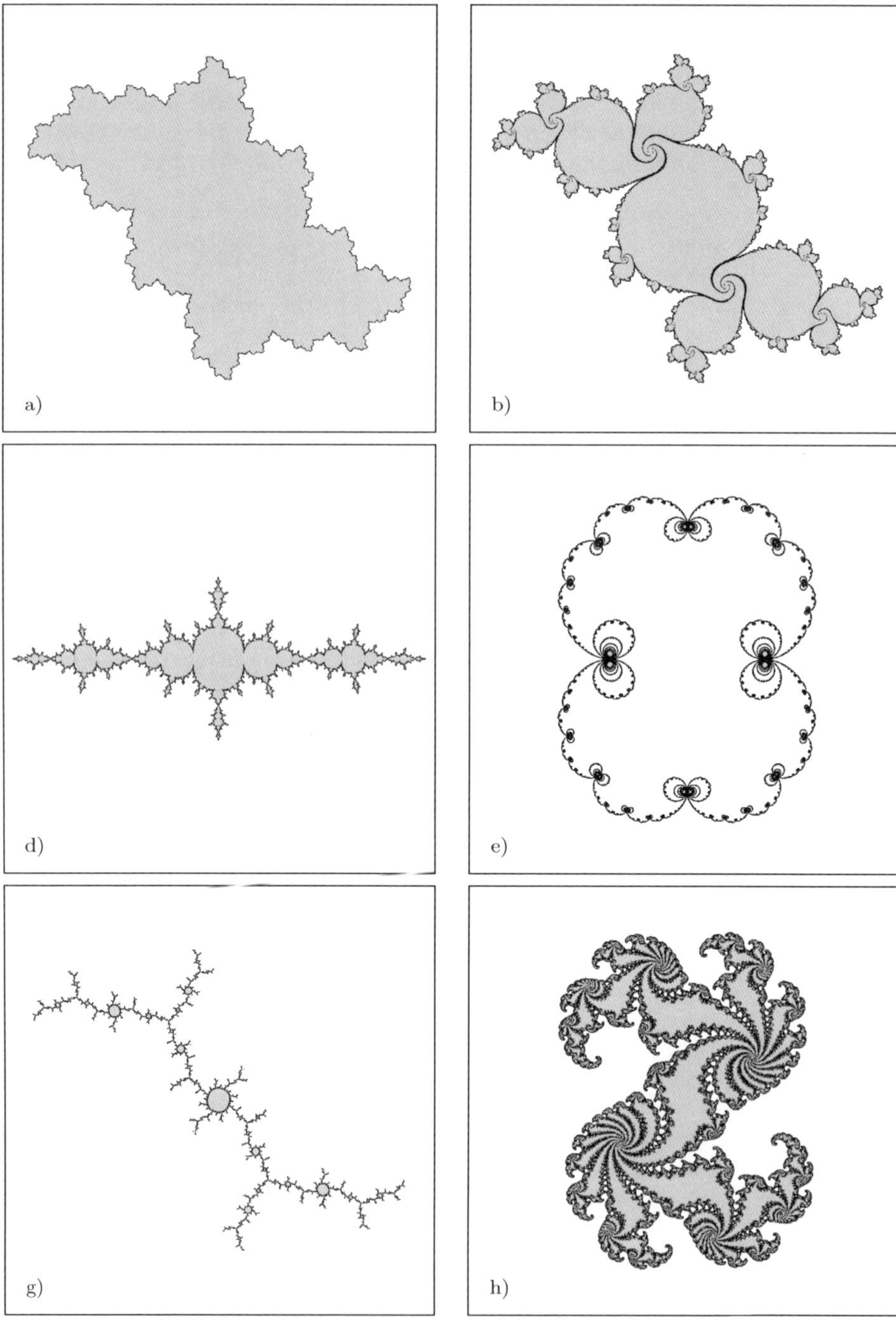

Fig. 10.7.3: Julia sets J_c of the map $f_c(z) = z^2 + c$, *Part 1*

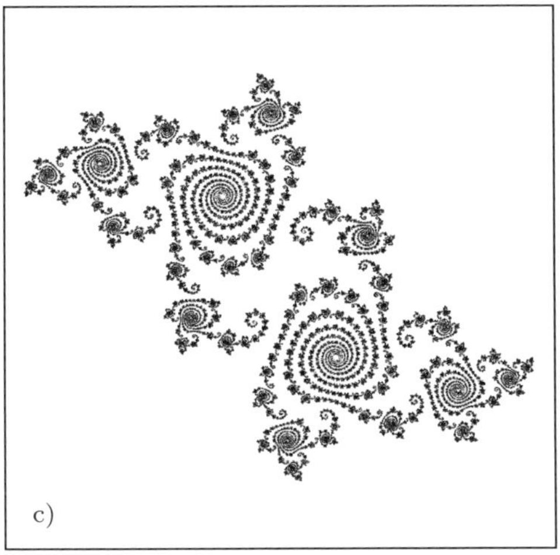

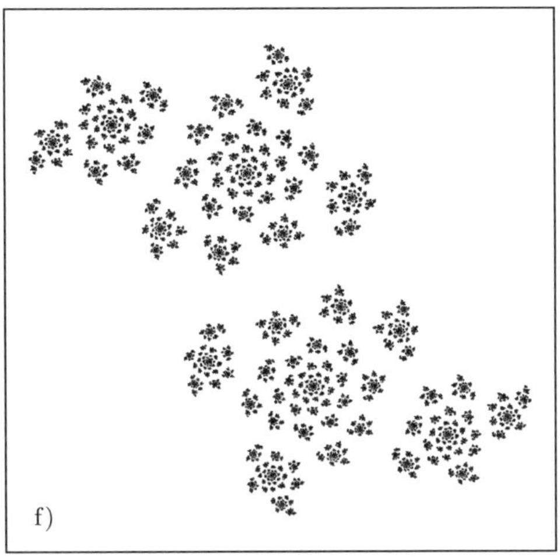

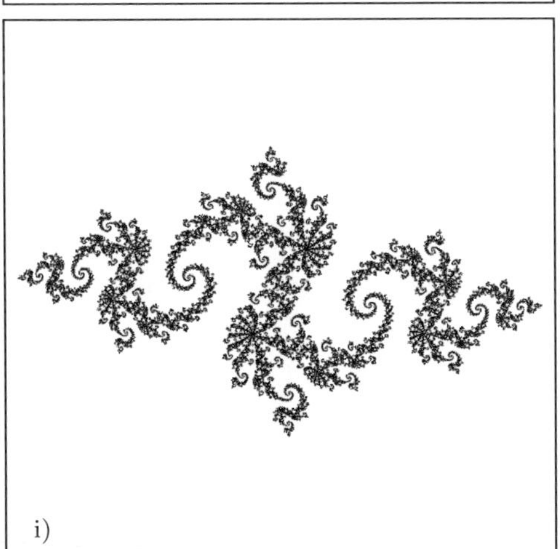

a) $c = -0.1237 \quad +0.5651i$

b) $c = -0.11 \quad +0.6557i$

c) $c = -0.194 \quad +0.6557i$

d) $c = -1.25$

e) $c = \quad 0.251$

f) $c = \quad 0.08 \quad +0.67037i$

g) $c = -0.156 \quad +1.032i$

h) $c = \quad 0.32 \quad +0.043i$

i) $c = -0.8 \quad +0.17i$

Fig. 10.7.3: Julia sets J_c of the map $f_c(z) = z^2 + c$, *Part 2*

Unstable cycles with $|\lambda| > 1$ lie on the Julia set J_c of f_c, stable cycles ($|\lambda| < 1$) in the complementary set F_c. For non-hyperbolic cycles with $|\lambda| = 1$, additional investigations are necessary, and very complex, still partially unsolved problems emerge which we sketch only briefly in the following. As in the real analysis, bifurcations occur in the plane of complex numbers if a periodic cycle loses the property of hyperbolicity; the question as to the normal forms then arises. In the complex plane, there is a larger number of normal forms of bifurcations, even of the types familiar to us, such as saddle-node bifurcation and period doubling, which differ significantly from the bifurcations on the real axis; see (Devaney, 1987).

If $|\lambda| = 1$ holds, we can represent the complex number λ as follows

$$\lambda = \mathrm{e}^{2\pi i\alpha} \qquad \text{with } \alpha \in \mathbb{R}$$

Depending on whether α is a rational number p/q or an irrational number, the non-hyperbolic cycles are divided into two groups, namely rational and irrational indifferent cycles. Julia and Fatou were able to prove that the rational indifferent cycles $\bar{z} = \{z_1, z_2, \ldots, z_k\}$ lie on the Julia set, i.e. $\bar{z}$ lies on the boundary of its own basin of attraction.

Figure 10.7.3d) shows the Julia set of the map $f_c(z) = z^2 - 1.25$ which possesses a rational indifferent cycle of period 2, $\bar{z} = \{z_1, z_2\}$, where the two cycle points $z_{1,2} = -\frac{1}{2}(1 \pm \sqrt{2})$ lie on the Julia set J_c. For the derivative, we calculate

$$\lambda = \frac{\mathrm{d}}{\mathrm{d}z}\left(f_c^2(z)\right) = -1 = \mathrm{e}^{2\pi i \cdot \frac{1}{2}}$$

As for the logistic map, a period doubling occurs here at $c = -1.25$c. For real values $c > -1.25$, we obtain cycles of period 2, for $c < -1.25$, cycles of period 4.

For irrational α, the stability investigation is considerably more intricate. The cycle is stable if and only if $f_c(z)$ is locally, i.e. in the neighbourhood of the cycle points, topologically equivalent to a pure rotation. For this purpose, we must prove that the map can be linearised by an appropriate coordinate transformation in a neighbourhood of the indifferent cycle and, moreover, that the series that emerge converge. We were faced with similar questions in sections 6.3 and 6.4 when determining the normal forms. The convergence question leads to the problem of small denominators. A solution to this subtle problem was found by Carl Ludwig Siegel in his fundamental work (Siegel, 1942), see also (Moser and Siegel, 1971), which also provided the foundations for the KAM theory (cf. section 4.4). Siegel was able to prove that a local linearisation is indeed possible if positive constants ε and ν exist so that the following condition is satisfied for α

$$\left|\alpha - \frac{\mathrm{p}}{\mathrm{q}}\right| > \frac{\varepsilon}{\mathrm{q}^{\nu}} \qquad \text{for all } \mathrm{p}, \mathrm{q} \in \mathbb{N} \tag{10.7.6}$$

i.e. if the irrational number α can only be approximated poorly by rational numbers – a condition satisfied by “most” irrational numbers.

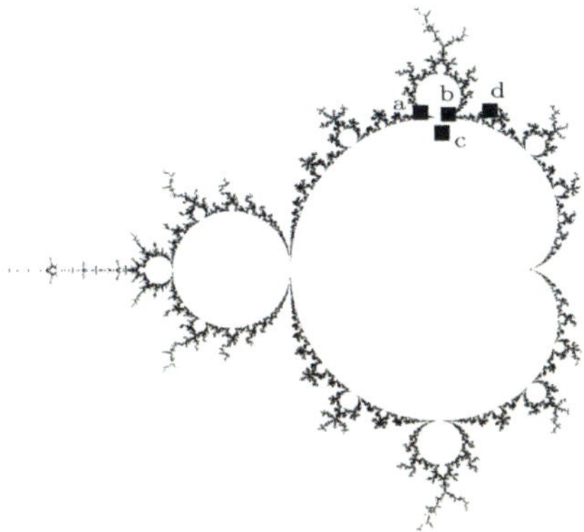

a) b)

c) d)

Plate XXVIII Julia sets J_c of the map $f_c(z) = z^2 + c$:
a) $c = -0.1940 + 0.65570i$
b) $c = -0.1100 + 0.65570i$
c) $c = -0.1237 + 0.56510i$
d) $c = 0.0800 + 0.67037i$

The domain in which this linearisation corresponding to an irrational rotation $z \mapsto ze^{2\pi i\alpha}$ is possible is called a Siegel disk. Figure 10.7.4 presents a Siegel disk for the map eq. (10.7.3) with the parameter value $c = -0.39054 - 0.58679i$; it encloses an irrational indifferent cycle point which is stable in this case and lies in F_c. Four invariant curves are plotted around the cycle point. The dynamics of the discrete map on these lines corresponds to a rotation about the centre with an irrational angle α.

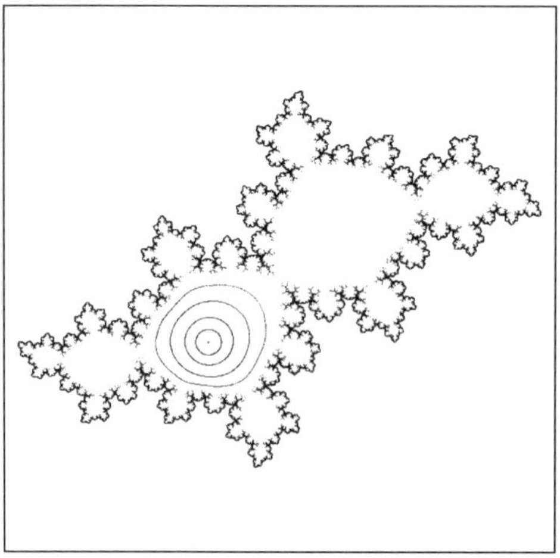

Fig. 10.7.4

Siegel disk of the map $f_c(z) = z^2 + c$ for $c = -0.39054 - 0.58679i$

However, Siegel was able to answer only part of the question about the stability of the irrational indifferent cycles. We can show that the set of the irrational numbers $\alpha \in [0, 1]$ which satisfy the Siegel conditions eq. (10.7.6) is of measure one; there are nevertheless irrational numbers $\alpha \in [0, 1]$ which lead to unstable cycles on J_c. The exact conditions – when the series converge and when they do not – are still not known precisely so that, for the case that α does not satisfy eq. (10.7.6), it is open whether the cycle is stable or unstable.

The Julia sets in fig. 10.7.3 can be divided into two classes. For some c-values, connected Julia sets result which either enclose a region of the z-plane in which attractive periodic cycles then lie or which have no interior but rather resemble branching tendrils. In this case, $z = \infty$ is the only attractor and the dendritic Julia set is the "boundary" of the basin of attraction of $z = \infty$. For other c-values, on the other hand, the Julia sets are exploded, disconnected Cantor sets.

In 1979, with the aid of computer experiments, Mandelbrot discovered empirically the ordering principle for the map $z \longmapsto z^2 + c$ which allows a classification in connected and disconnected Julia sets: a pattern in the c-plane which serves simultaneously for every c-value as a guide for the characterisation of the appertaining Julia set. Mandelbrot defined the set of all c-values corresponding to connected Julia sets (Mandelbrot, 1980)

$$M = \{c \in \mathbb{C} \mid J_c \text{ connected}\}$$

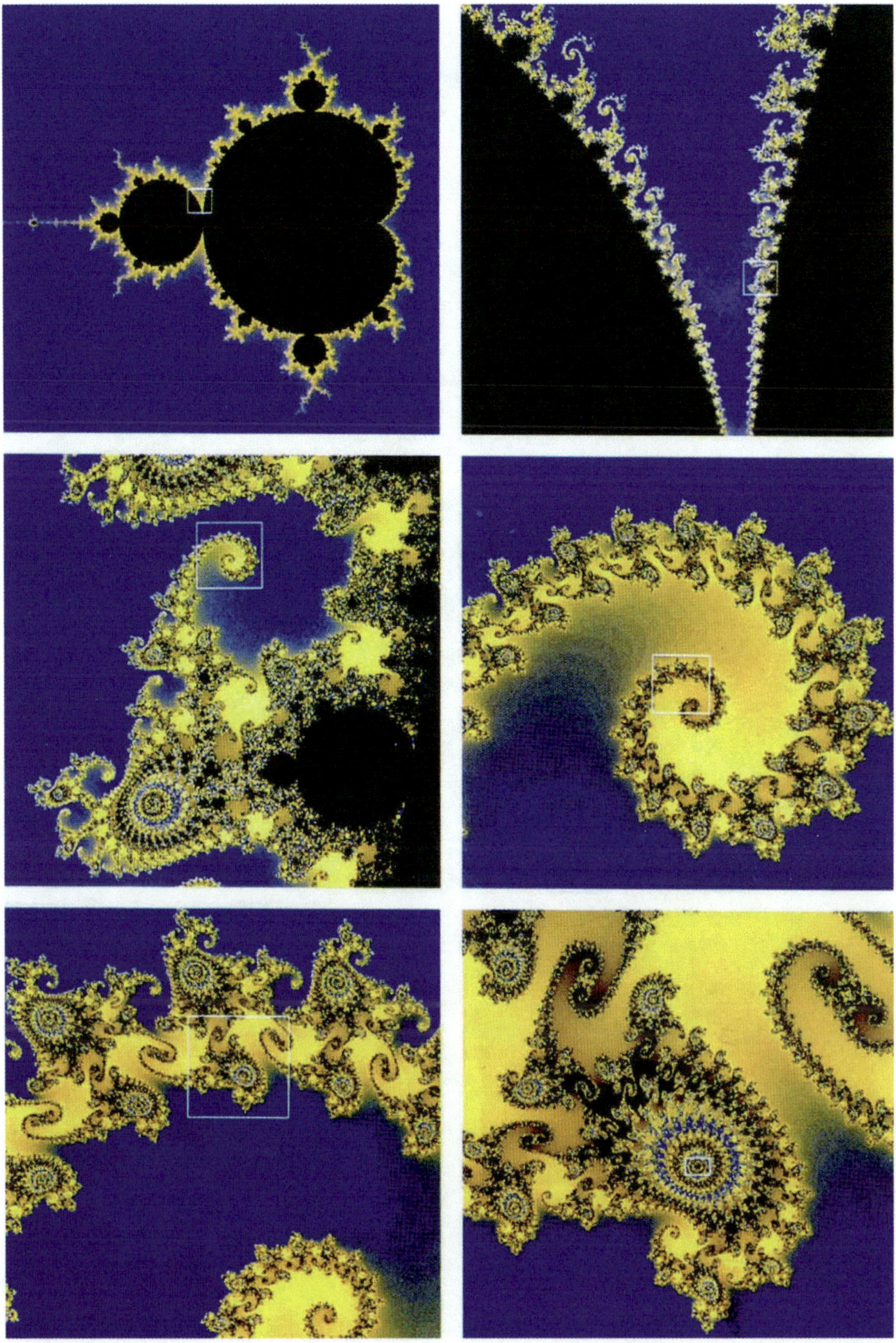

Plate XXIX Mandelbrot set and successive enlargements

Plate XXX Is there an end to the jouney into the "black hole"?

Plate XXXI From the new or fractal world

However, this definition is not yet suited to a graphic determination of M. Another definition makes use of results already proven by Julia and Fatou. This states that the characteristics of a rational map result from the long-term evolution of its critical points, i.e. of those point for which $f'(z) = 0$ holds. We have already seen that $z = 0$ is a critical point for all c-values of the map $f_c(z) = z^2 + c$. The following definition is thus considerably more suitable for a graphic determination of M

$$M = \{c \in \mathbb{C} \mid f_c^{\mathrm{k}}(0) \nrightarrow \infty \quad \text{for } \mathrm{k} \to \infty\}$$

i.e. to M belong all c-values for which iterations of the critical point $z = 0$ do not tend towards infinity, for which, in fact, $z = 0$ does not appertain to the basin of attraction of $z = \infty$.

Colour Plate XXIXa), on the upper left-hand side of p. 781, shows the Mandelbrot set. In order to create this graphic, a sub-region of the complex c-plane with $-2.05 \leqslant \mathrm{Re}\, c \leqslant +0.65$, $-1.35 \leqslant \mathrm{Im}\, c \leqslant +1.35$ was covered with a very fine net. For the c-value in the middle of each square, the sequence $z_{\mathrm{n}} = f_c^{\mathrm{n}}(0)$ of iterations of the critical point $z_0 = 0$ was calculated and the absolute value $|z_{\mathrm{n}}|$ inspected in each case. If the sequence of points remains finite, i.e. if $|z_{\mathrm{n_{max}}}| < r_0$ for given values $\mathrm{n_{max}}$ and r_0, the appertaining c-value belongs to the Mandelbrot set M and the colour black is assigned to the corresponding square. Should, however, $|z_{\mathrm{n}}| \geqslant r_0$ hold for a $\mathrm{n} < \mathrm{n_{max}}$, the sequence $\{z_{\mathrm{n}}\}$ tends to infinity and the square is assigned a colour according to the escape velocity from the range "yellow (slow) → green → blue (fast)".

In spite of the simplicity of the mapping rule $z \longmapsto z^2 + c$, the Mandelbrot set possesses an extremely complex structure. The main part of M resembles a water lily leaf adjoined by many further circular areas of varying sizes which are, in turn, surrounded by still smaller areas and so on. If we use the computer as a microscope and zoom in on sections of the boundary area of the set, an extraordinary diversity and an astonishing richness of forms and structures are revealed. Colour Plate XXIX, p. 781, shows a series of enlargements starting from the section of the edge where the two main parts of M adjoin. At first, we see a wedge-shaped opening with smaller and smaller copies of the Mandelbrot set surrounded by bizarre tendrils on the walls to the right and the left (b). Within these tendrils, spirals emerge (c, d) which are in turn made up of fantastic ornaments; again and again – like variations on a theme – spirals and structures reminiscent of the inflorescences of sunflowers appear. Colour Plate XXX p. 782, shows an enlargement of such a sunflower structure. In the centre, we observe a black core which turns out to be a somewhat modified copy of the complete set M (cf. Colour Plate XXXI, p. 783). Only the restriction on the number of representable decimal places imposed by the computer puts an end to this journey into the innermost structure of the boundary range of M.

If we consider the Mandelbrot set M, it seems likely that the c-values in the individual circular or foliate sub-sets play different roles. The connection with the logistic map brings us a step closer to an answer. If we select real parameter values c and real initial values z_0, it must be possible to find a linear coordinate transformation which transforms $f(z) = z^2 + c$ into $g(x) = \alpha x(1 - x)$. If we write this coordinate transformation in the form $z = h(x) = ax + b$, then, in accordance with the following scheme, we can

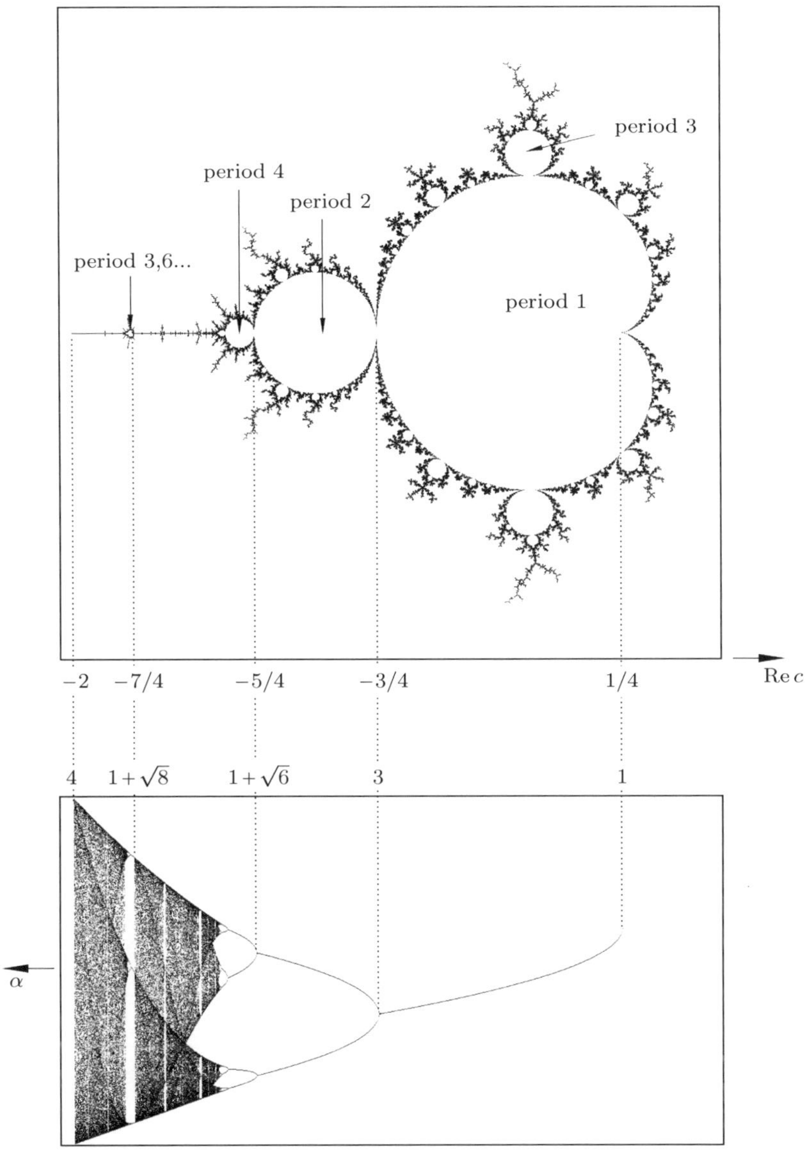

Fig. 10.7.5: Connection between the Mandelbrot set and the period doubling cascade of the logistic map

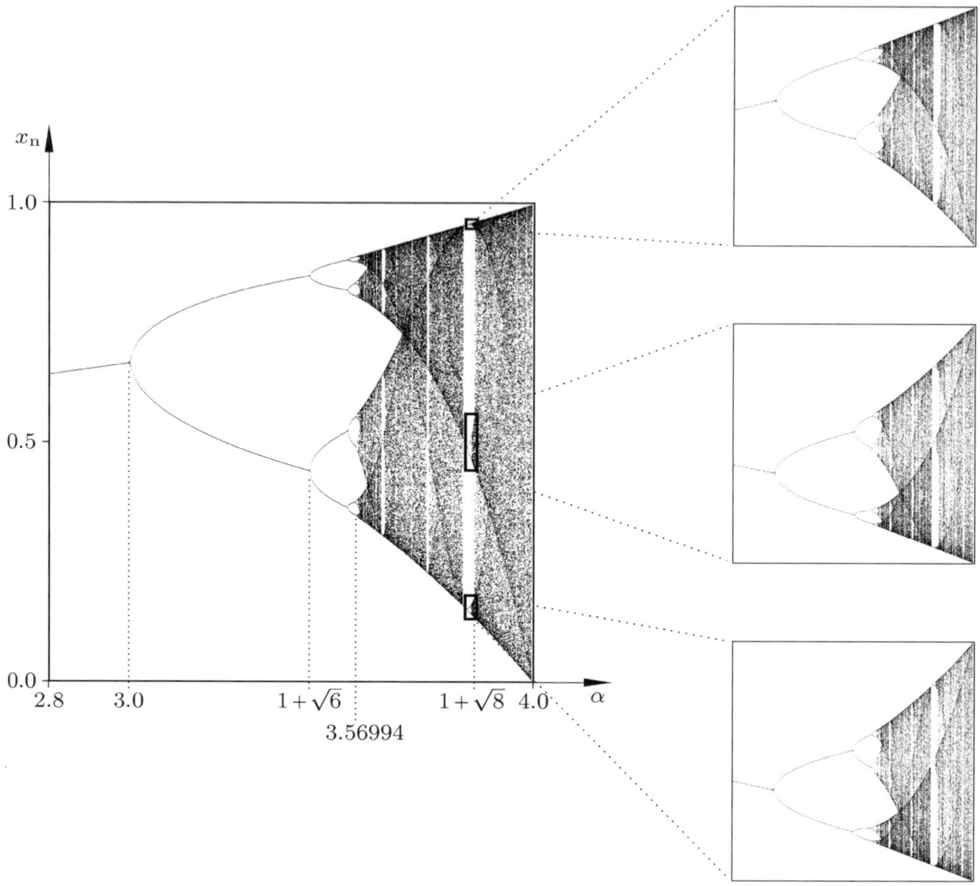

Fig. 10.7.6: Self-similarity in the bifurcation diagram of the logistic map

determine the constants a and b of the transformation and also set up a link between the control parameters α and c

$$\begin{array}{ccc} z_n & \xrightarrow{f} & z_{n+1} \\ h^{-1}\downarrow & & \uparrow h \\ x_n & \xrightarrow[g]{} & x_{n+1} \end{array}$$

i.e. we find the relation

$$z_{n+1} = f(z_n) = h\big(g(h^{-1}(z_n))\big) \tag{10.7.7}$$

and, by substitution, we obtain

$$a = -\alpha, \;\; b = \tfrac{1}{2}\alpha, \;\; c = \tfrac{1}{4}\alpha(2-\alpha) \tag{10.7.8}$$

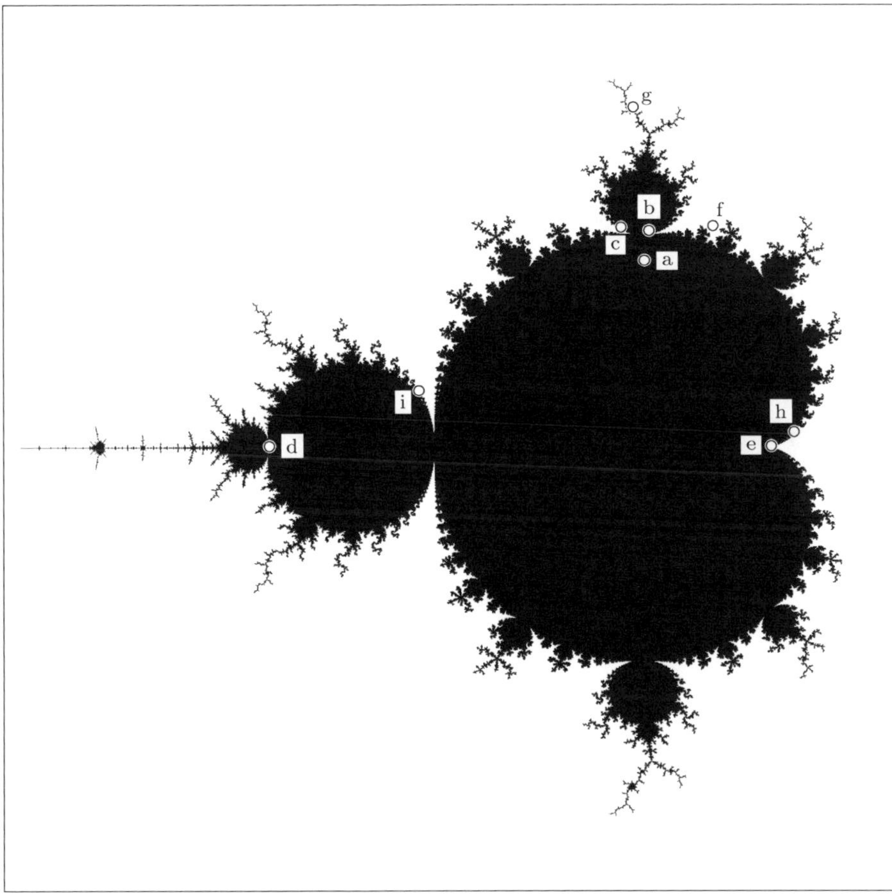

Fig. 10.7.7: "Chart" for the position of the c-values of the Julia sets of fig. 10.7.3

A one-to-one mapping of α- onto c-values may thus be established relating the intervals $1 \leqslant \alpha \leqslant 4$ and $-2 \leqslant c \leqslant 0.25$. For these parameter ranges, we present the Mandelbrot set in the upper part of fig. 10.7.5 and below it, on the corresponding α-scale, the bifurcation diagram of the logistic map. It is shown that the critical α-values for which bifurcations respectively period doublings occur in the logistic map coincide exactly with the real c-values for which the circular areas bud off the Mandelbrot set. Moreover, above the region of the period three window of the logistic map, we can recognise a small copy of the Mandelbrot set which, together with even smaller copies, is positioned on an antenna-like spike along the real c-axis which extends to $c = -2$. This is an expression of the self-similarity which is also reflected in the bifurcation diagram of the logistic map. In fig. 10.7.6, we have zoomed in on three sections from the period three window; with the appropriate scaling, these are copies of the complete system.

The bifurcation diagram of fig. 10.7.5 offers an indication of the role the individual bud-like areas of M play. The main part of M, which stretches from $c = 1/4$ to $c = -3/4$ on the real axis, contains all c-values for which the Julia sets are Jordan curves (cf. fig. 10.7.2), i.e. in the interiors of which a single fixed point lies. The c-values in the next largest bud with the centre $c = -1$ correspond to Julia sets containing cycles of period 2.

Cycles of period three are found in the main body of the small copy of M on the real c-axis, the right edge of which coincides with $c = -7/4$. Figure 10.7.5 shows the assignation of some buds of M to the periods of the corresponding cycles.

How do we locate the c-values which lead to indifferent cycles? In fig. 10.7.3d, we presented the Julia set for $c = -5/4$ which contains a rational indifferent cycle. Figure 10.7.5 demonstrates that this c-value appertains to a point of the Mandelbrot set at which a new bud with a new period emerges. This situation can be generalised: all c-values for which new buds emerge correspond to rational indifferent cycles, those points of the edge lying in between to irrational indifferent cycles, however; cf. fig. 10.7.4.

Figure 10.7.7 shows the c-values of the Julia sets presented in fig. 10.7.3. The chart clearly indicates that a crossing of the boundary of the Mandelbrot set (e.g. series of figures a, b, c or a, b, f) corresponds to a transition from connected to disconnected Julia sets and thus, as in the case of a second-order phase transition, is equivalent to a drastic qualitative change of the topological characteristics of these sets.

Much interesting work on Mandelbrot and Julia sets has been published. Further results and references for additional reading as well as a selection of excellent computer graphics can be found in (Peitgen and Richter, 1986a) and (Peitgen and Saupe, 1988).

For Benoît Mandelbrot, the examples of Mandelbrot and Julia sets confirmed the words with which we began this section. In many ways, it is easier to comprehend the characteristics of quadratic recursion formulæ if we proceed from real to complex numbers (Mandelbrot, 1983).

10.8 Morphology of the Arnold Tongues

στρεπτὴ δὲ γλῶσσ' ἐστὶ βροτῶν,
πολέει δ' ἔνι μῦθοι

Homer, *Ilias*, I, 248
(ca. 850 B.C.)

Many phenomena occurring in the non-linear coupling of two oscillators – particularly frequency locking and various transitions to chaotic behaviour – can be elucidated very elegantly with the aid of the circle map (Pikovsky et al., 2003). Previously, such oscillating systems were studied with highly elaborate perturbation methods; however, aspects such as universality and self-similarity could not be ascertained in this way. This only became possible with the application of more modern methods originating in the theory of renormalisation groups.

In section 8.3, we already discussed the most important characteristics of the circle map. We complement this with some numerical investigations offering in particular insight into the special anatomy of the Arnold tongues.

Before we discuss the numerical results, we recall some of the theoretical results from section 8.3.2 on the structure of the domains in which frequency locking occurs, cf. (Glass et al., 1983; Bohr and Gunaratne, 1985; MacKay and Tresser, 1986). We consider a general circle map of the form

$$\theta_{\mathrm{n}+1} = \theta_{\mathrm{n}} + \Omega - Kg(\theta_{\mathrm{n}}) \equiv f(\theta_{\mathrm{n}}) \tag{10.8.1}$$

where K and Ω are control parameters and $g(\theta_{\mathrm{n}})$ a periodic function of period 1. Since $f(\theta+1) = 1+f(\theta)$, we can regard $f(\theta)(\mathrm{mod}\,1)$ as a map of the circle with $0 \leqslant \theta < 1$. We assume that $f(\theta)$ increases monotonically for $K < 1$, possesses a cubic inflection point for $K = 1$ and a maximum and a minimum for $K > 1$ (cf. fig. 8.3.9). A generic property of such circle maps for $K > 0$ is *mode coupling* or *frequency locking*, i.e. corresponding to every rational ratio p/q, a complete Ω-interval $[\Omega_1, \Omega_2]$ exists in which the rotation number, see eq. (8.3.17),

$$W = \lim_{\mathrm{n}\to\infty} \frac{f^{\mathrm{n}}(\theta_0) - \theta_0}{\mathrm{n}} \tag{10.8.2}$$

assumes the value p/q (Pikovsky et al., 2003). As long as f is a monotonic function and does not decrease, the long-term behaviour of the motion is uniquely determined for each combination of the control parameters K and Ω, independently of the selected initial condition, i.e. for $K \leqslant 1$, the rotation number is unique. If, however, the behaviour of $f(\theta)$ is non-monotonic, the uniqueness of W is no longer guaranteed.

Let us recall section 8.3.2. There, we assigned to each pertinent rotation number $W = \mathrm{p/q}$ (p, q relatively prime, $\mathrm{p} \leqslant \mathrm{q}$) an Arnold tongue (see fig. 8.3.5 for $0 \leqslant K \leqslant 1$ respectively Colour Plate IX, p. 497), i.e. a domain $A_{\mathrm{p/q}}$ of the (Ω, K)-plane appertaining to the circle map for which at least one orbit with the rotation number $W = \mathrm{p/q}$ exists, and studied the stability of the periodic orbits for $K \leqslant 1$. We also saw that neighbouring Arnold tongues overlap for $K > 1$. In such regions, there is no longer a unique rotation number, rather the corresponding rotation numbers form a closed interval $[W_1, W_2]$, called a rotation interval. The region $A_{\mathrm{p/q}}$ also contains parameter values of circle maps for which there exist orbits to which no rotation number at all can be assigned any more, i.e. which exhibit aperiodic behaviour. The question thus arises in which regions of the Arnold tongues the rotation number for all orbits is uniquely determined, where rotation intervals emerge and where there are stable p/q-orbits.

Figure 10.8.1 is a schematic presentation of the result of theoretical investigations; cf. (MacKay and Tresser, 1986). The shading divides regions associated with a uniquely determined rotation number respectively a rotation interval. Moreover, those areas which incorporate at least one stable orbit with the rotation number $W = \mathrm{p/q}$ are indicated by an S. Within the region of $A_{\mathrm{p/q}}$ in which W is uniquely determined, the route proceeds via a cascade of period doublings to chaos for increasing K-values.

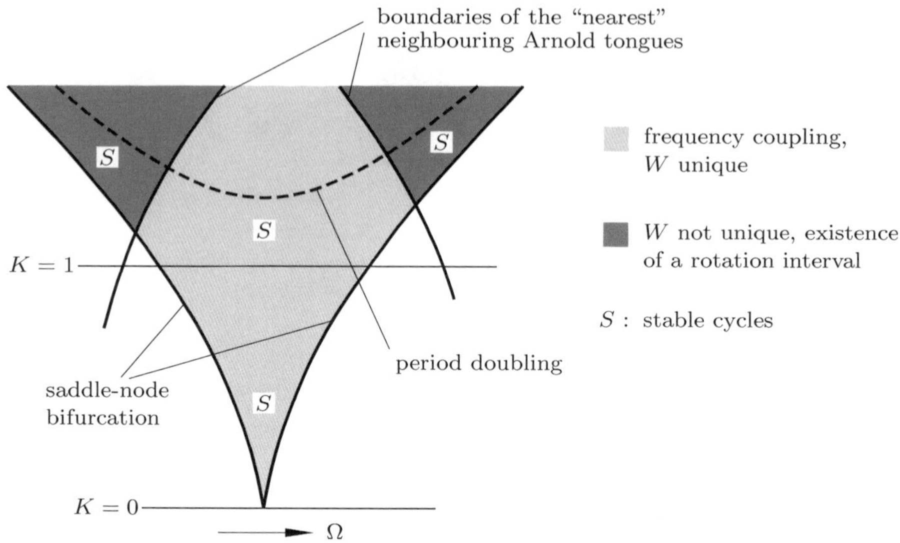

Fig. 10.8.1: Structure of the Arnold tongue $A_{\mathrm{p/q}}$ corresponding to the rotation number $W = \mathrm{p/q}$

Boyland was able to prove that, in the intersection of two Arnold tongues A_{W_1} and A_{W_2} corresponding to the rational rotation numbers $W_1 = \mathrm{p}_1/\mathrm{q}_1$ and $W_2 = \mathrm{p}_2/\mathrm{q}_2$, there is an open subset in which exactly two periodic orbits with the rotation numbers W_1 and W_2 are stable while all the others are unstable, however (Boyland, 1986).

We can also extend the concept of Arnold tongues to irrational rotation numbers (MacKay and Tresser, 1986). For $K \leqslant 1$, they consist only of a line in the (Ω, K)-plane and only open to a cone-shaped region for $K > 1$. If we consider two such Arnold tongues corresponding to irrational rotation numbers, there are no longer any stable periodic orbits in the overlapping region, but only chaotic motions (Boyland, 1986).

In Colour Plate IX, p. 497, respectively fig. 8.4.2, we already demonstrated that it is numerically straightforward to gain an overview of the long-term behaviour of the circle map in dependence of the system parameters Ω and K by calculating the appertaining Lyapunov exponent σ. We must note, however, that σ depends on the selected initial condition in (Ω, K)-regions for which there is a rotation interval.

In fig. 10.8.2, we have plotted the Lyapunov exponent for the range $0.4 \leqslant \Omega \leqslant 0.6$ and $0 \leqslant K \leqslant 2$. The Arnold tongue corresponding to the rotation number $W = 1/2$ stands out clearly, to the right and the left are self-similar cascades of smaller Arnold tongues. The white band possessing a minimum at $K = 1$ marks points with a very large negative σ-value and thus contains the line of the superstable cycle characterised by $\sigma = -\infty$. Above this, we recognise a further white line for which $\sigma = 0$ holds, i.e. where period doublings occur. Still further up, we see two intersecting white lines corresponding to superstable cycles of period two with the winding number 2/4. Further period doublings are generated up to the chaotic domain, indicated

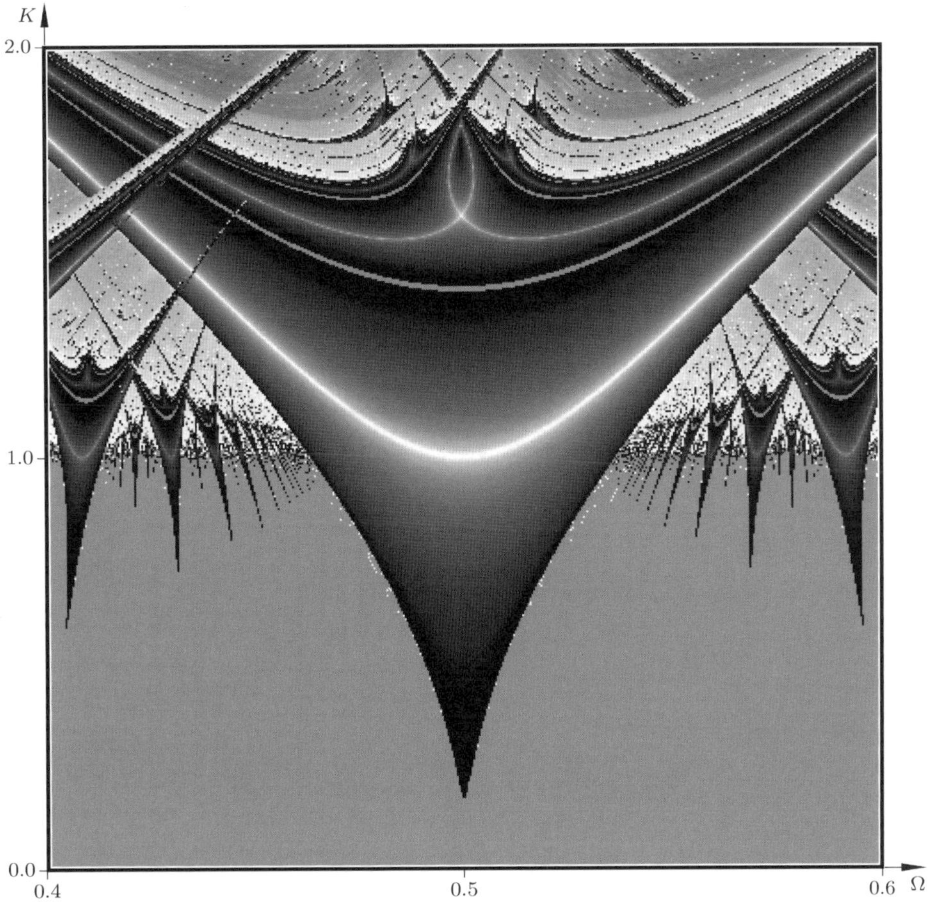

Fig. 10.8.2: Structure of the Arnold tongue corresponding to the rotation number $W = 1/2$

by an area of light shading. For the calculation of the Lyapunov exponents, we always assumed an initial value $\theta_0 = 0.2$. Hence, in regions in which neighbouring Arnold tongues overlap, only one of the two possible final states can be reproduced.

In fig. 10.8.3, we compare the theoretical and numerical calculations. The upper diagram presents results from MacKay and Tresser's work, cf. (MacKay and Tresser, 1986), fig. 11. We show superstable cycles within the Arnold tongue $A_{1/2}$ for the cubic circle map $\theta_{\mathrm{n}+1} = -b + (1-a)\theta_{\mathrm{n}} + b\theta_{\mathrm{n}}^2 + a\theta_{\mathrm{n}}^3$. The numbers indicate the respective period of the cycles. The section $|b| \leqslant 0.7$, $2.2 \leqslant a \leqslant 4.0$ of the parameter plane corresponds approximately to that of the lower diagram. The shaded strip marks the transition to chaotic motions. Beyond the chaotic region, we recognise lines which correspond to superstable cycles of period 3 and indicate the location of period 3 windows.

Below this, we present a selective enlargement of the Arnold tongue shown in fig. 10.8.2 for the circle map eq. (8.3.13); the associated Lyapunov exponents are

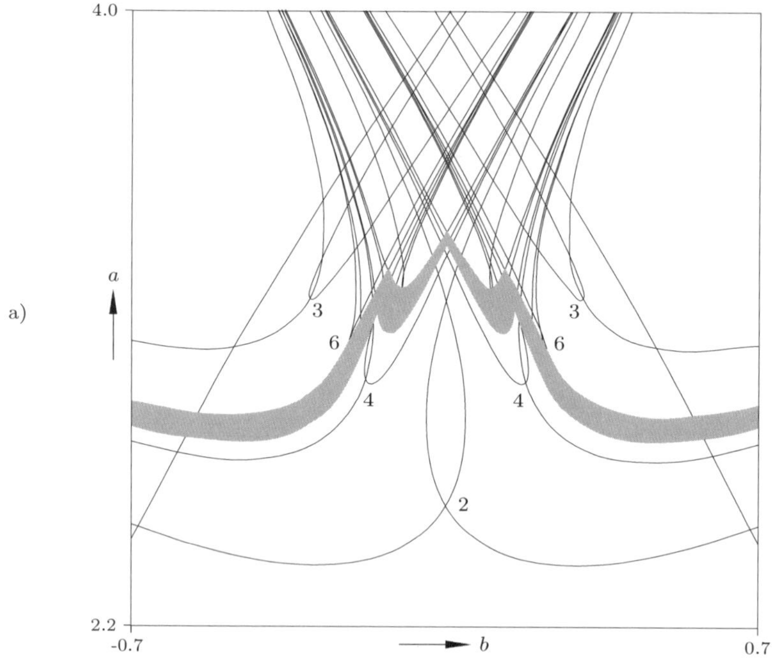

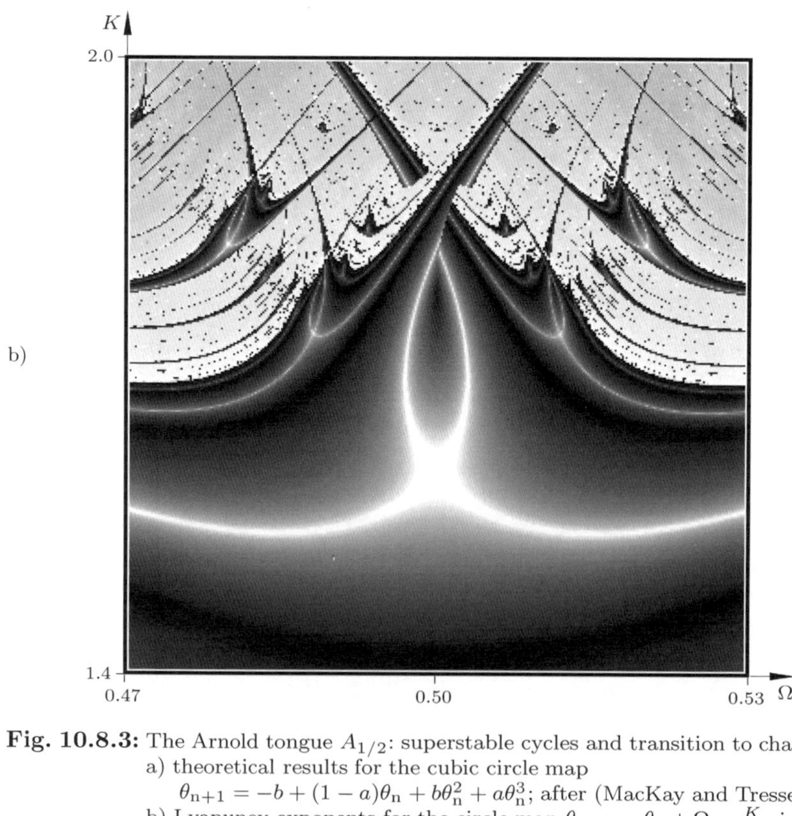

Fig. 10.8.3: The Arnold tongue $A_{1/2}$: superstable cycles and transition to chaos
a) theoretical results for the cubic circle map
$\theta_{n+1} = -b + (1-a)\theta_n + b\theta_n^2 + a\theta_n^3$; after (MacKay and Tresser, 1986)
b) Lyapunov exponents for the circle map $\theta_{n+1} = \theta_n + \Omega - \frac{K}{2\pi}\sin 2\pi\theta_n$

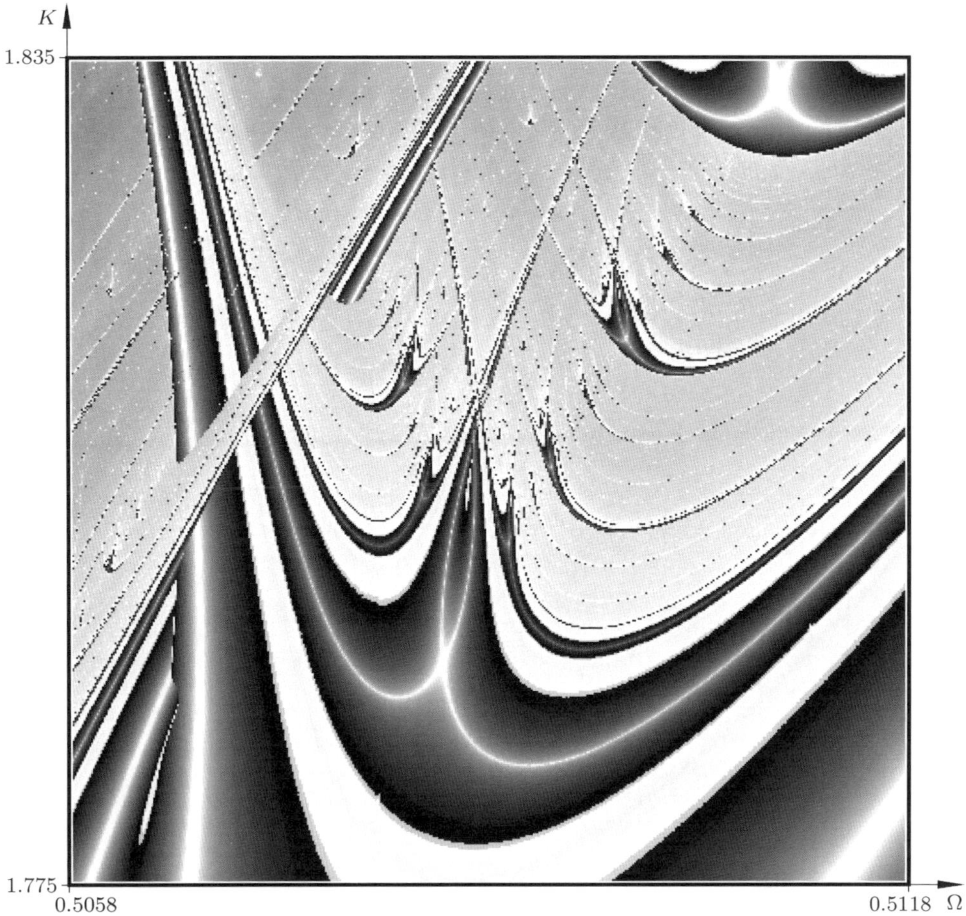

Fig. 10.8.4: A further selective enlargement of the Arnold tongue $A_{1/2}$: superstable cycles of period 4

plotted for the range $0.4 \leqslant \Omega \leqslant 0.53$ and $1.4 \leqslant K \leqslant 2.0$. The two intersecting white lines mark the snperstable cycles after the first period doubling. The light areas of this negative reproduction in the upper part of fig. 10.8.3b indicate chaotic motions, repeatedly interrupted by periodic windows. The excellent qualitative agreement of the last two figures shows that the structure of the Arnold tongues does not depend on the specific form of the circle map and thus possesses universal validity. Moreover, fig. 10.8.3b demonstrates that, with the aid of simple programmes for calculating the Lyapunov exponents, we are in a position to observe the structure of the Arnold tongues as under a microscope.

In conclusion, we reproduce in fig. 10.8.4 a further enlargement of the range within the intervals $0.5058 \leqslant \Omega \leqslant 0.5118$, $1.775 \leqslant K \leqslant 1.835$, in the centre of which two intersecting superstable cycles of period 4 are to be seen. We recognise a generalised self-similarity here: parts of fig. 10.8.4 can be mapped to fig. 10.8.3b by a non-linear

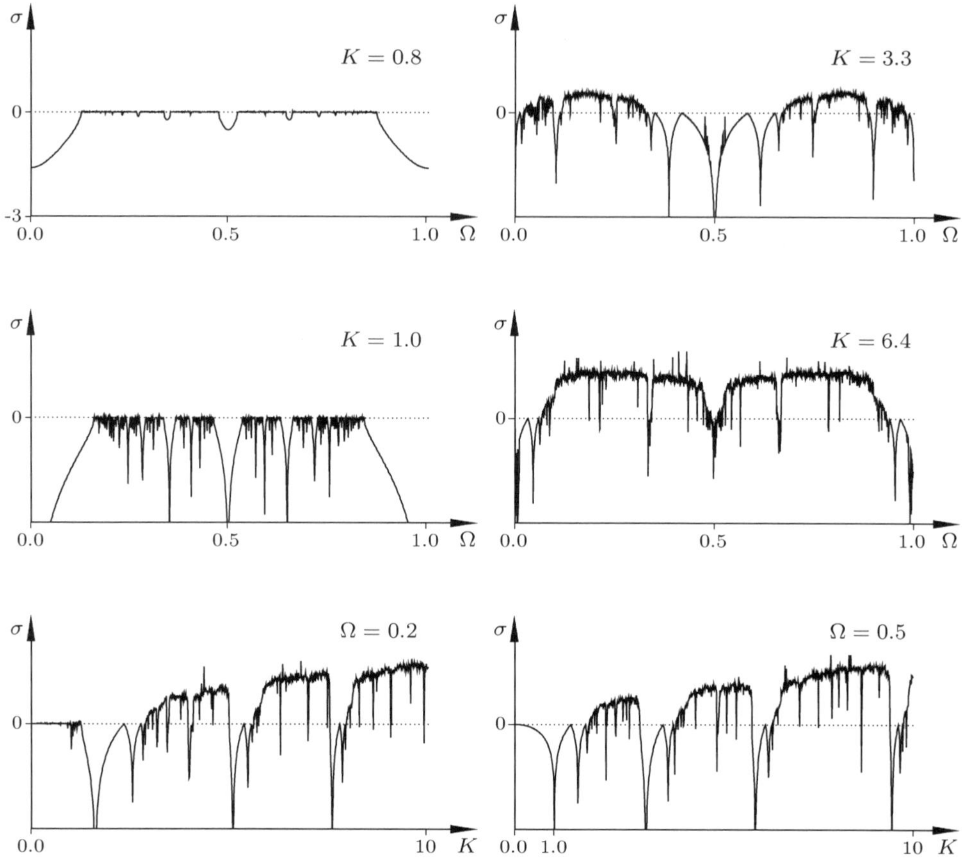

Fig. 10.8.5: Lyapunov exponents of the circle map eq. (8.3.13)

transformation of the parameters, i.e. for repeated enlargements, qualitatively similar patters emerge again and again.

We can also recognise self-similar structures in the Lyapunov exponents of the standard circle map eq. (8.3.13), plotted in fig. 10.8.5 for several sections K = const respectively Ω = const. For $K=0.8$, quasi-periodic behaviour predominates ($\sigma<0$), interrupted by periodic motions ($\sigma < 0$) within the Arnold tongues. For $K = 1$, the quasi-periodic regions have shrunk to a set of measure zero. The section $K = 3.3$ contains the overlapping of the two largest Arnold tongues $A_{0/1}$ and $A_{1/1}$ (cf. Colour Plate IX, p. 497). In the region of strong non-linearities ($K = 6.4$), chaotic motions with $\sigma > 0$ predominate. The two sections $\Omega = 0.2$ and $\Omega = 0.5$ repeatedly exhibit transitions to chaos via cascades of period doublings (cf. also fig. 8.4.2).

If, as in figs. 10.8.2 to 10.8.4, we plot the Lyapunov exponents over the (Ω, K)-plane, we obtain no information on how many cycle points a periodic orbit possesses, whether it has already undergone period doublings and within which limits the attractors reside. In order to close this information gap, we can plot bifurcation diagrams without much effort. In fig. 10.8.6, we present a selection of such diagrams for sections K = const transversal to the Arnold tongues (bifurcation diagrams for sections Ω = const resemble the plot in fig. 8.4.2a). As the initial value, $\theta = 0.2$ was again selected in each case and the long-term response of the appertaining orbits is plotted in dependence of $\Omega \in [0,1]$.

For $K = 0.7$, we observe predominantly quasi-periodic motions which sweep over the whole θ-interval $0 \leqslant \theta \leqslant 1$ and are interrupted only by narrow periodic windows, the Arnold tongues. The transition from quasi-periodic to periodic motions occurs via saddle-node bifurcations. For $K = 1$, the periodic regions dominate. We clearly observe the largest Arnold tongues corresponding to the rotation numbers $W = 0/1,\ 1/3,\ 1/2,\ 2/3,\ 1/1$. For $K = 1.4$, period doublings emerge in the narrower Arnold tongues and consequently, for reasons of symmetry, corresponding inverse cascades take place. Between the periodic regions, chaos occurs. A further increase of the non-linearity to $K = 1.8$ leads to an enlargement of the two outer Arnold tongues $A_{0/1}$ and $A_{1/1}$. In the interior of the region $A_{1/2}$, a period doubling cascade takes place leading to chaos and back again to periodic oscillations.

At $K = 2.8$, a narrow chaotic range can be observed only in the centre, but a transition takes place to periodic motions via saddle-node bifurcations where type I intermittency occurs. For $K = 3.2$, the two Arnold tongues $A_{0/1}$ and $A_{1/1}$ overlap. A further increase of K leads to an expansion of the chaotic ranges which are, however, repeatedly interrupted by periodic windows. The explosive expansion of chaotic attractors which can be observed for $K = 3.6$ and $K = 4.4$ is of interest. These sudden changes in the response are the result of crises which arise when an unstable limit cycle coalesces with the boundary of the basin of a chaotic attractor; see (Grebogi et al., 1983). Two such unstable limit cycles emerge, for example, at $K = 4.4$ at the saddle-node bifurcations which cause the transition from chaotic to periodic motions and are indicated by dashed lines.

Figure 10.8.7 presents a selective enlargement of the bifurcation diagram for $K = 1$, within the range $0.158 \leqslant \Omega \leqslant 0.164,\ 0.22 \leqslant \theta \leqslant 0.28$. We can clearly observe the self-similar structure of the Arnold tongues along the critical line. Between two large periodic ranges with the rotation numbers $W_1 = \mathrm{p}_1/\mathrm{q}_1$ and $W_2 = \mathrm{p}_2/\mathrm{q}_2$, an infinity of regions is generated in which frequency locking occurs and which possess the self-similar structure of the "devil's staircase" (see fig. 8.3.6). The widest Arnold tongue lying between $A_{\mathrm{p}_1/\mathrm{q}_1}$ and $A_{\mathrm{p}_2/\mathrm{q}_2}$ has the rotation number $W = (\mathrm{p}_1 + \mathrm{p}_2)/(\mathrm{q}_1 + \mathrm{q}_2)$ (cf. section 8.3.3.2). These specifications generate the self-similar structure of the bifurcation diagram in fig. 10.8.7.

For centuries, scientists have mainly studied linear systems since an analytical investigation of non-linear processes is only possible in exceptional cases. The theoretical knowledge of the characteristics of linear systems thus gained has influenced – and certainly also limited – our mental processes and comprehension. We must thus gradually familiarise ourselves with the multitude of phenomena which arise in non-linear processes by the interplay and feedback of the individual components of a system; such experience will in time develop our instinctive understanding of complex manifestations. Nowadays, the use of supercomputers allows the quantitative numerical simulation of highly complex non-linear systems;

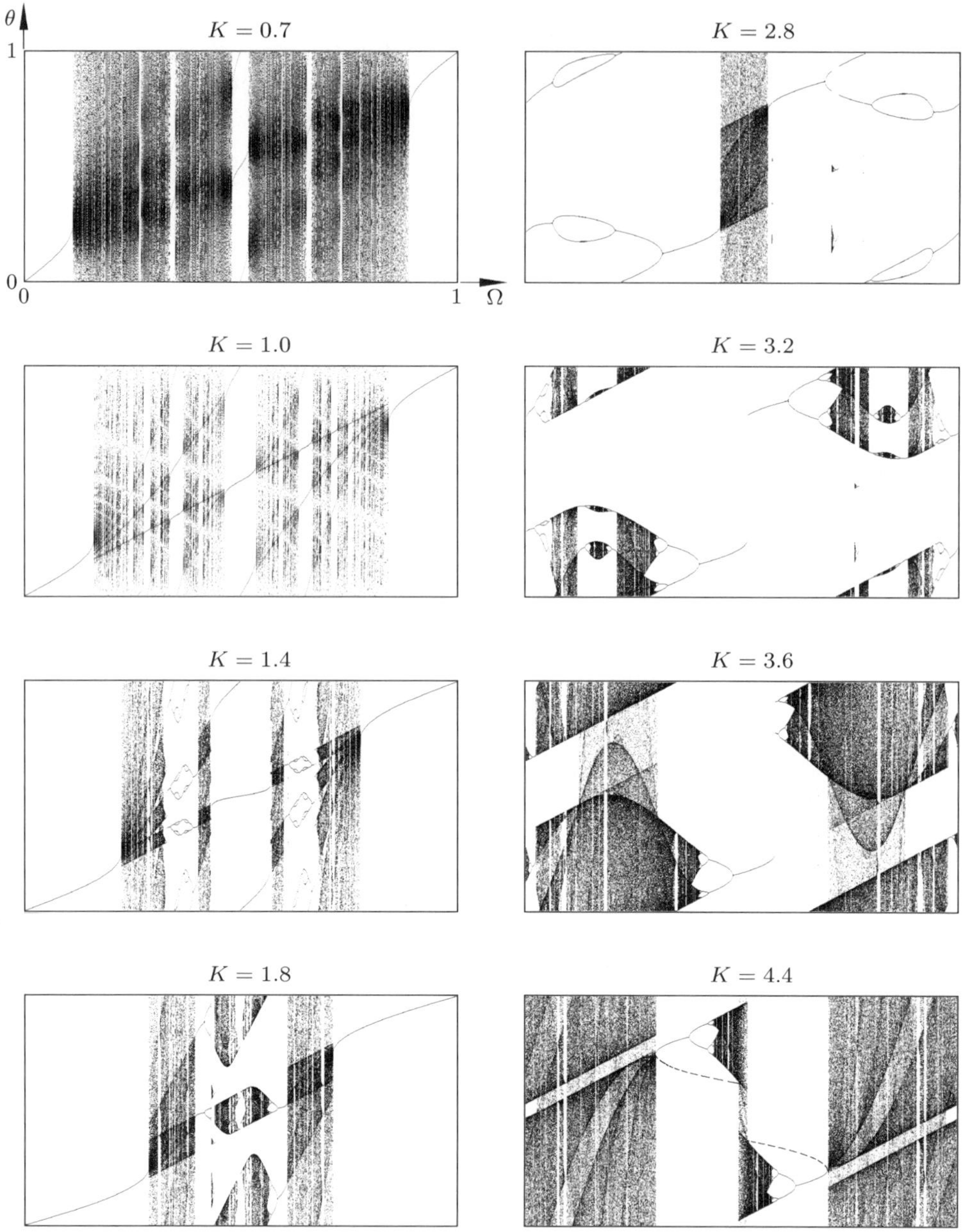

Fig. 10.8.6: Bifurcation diagrams of the circle map eq. (8.3.13) for constant K-values

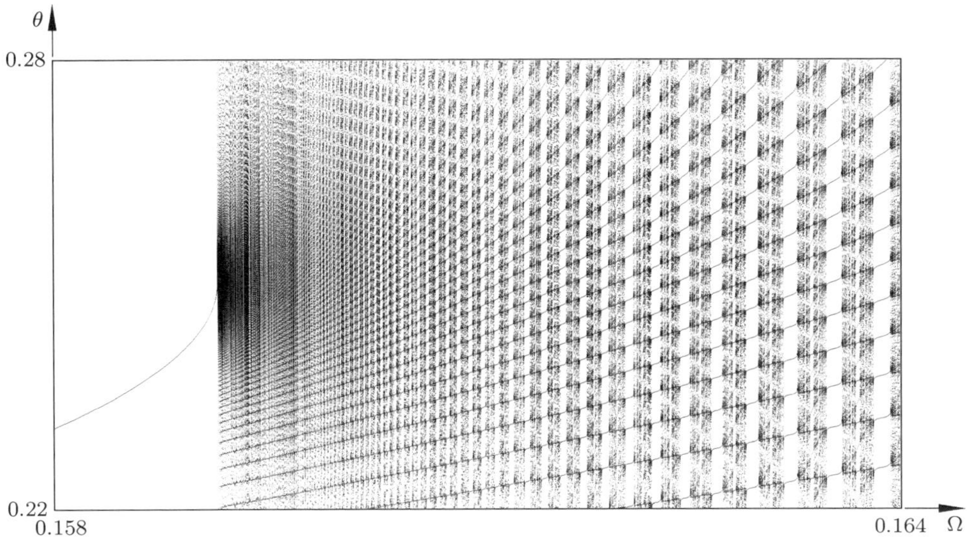

Fig. 10.8.7: Self-similar structure of the bifurcation diagram for $K = 1$

this has, without doubt, contributed to the explosive growth of our wealth of knowledge of non-linear phenomena. In this way, for example, imaginative graphic representations and videos can serve as numerical tools, revealing inner relations and structures which had simply not been thought of before (Zabusky, 1984). The theoretician is thus offered the opportunity of realising surprising associations which can stimulate the further development of his theoretical argumentations. As early as 1946, John von Neumann recognised the potential in the interplay of numerical experiments and theoretical eonsiderations; see (Goldstine and von Neumann, 1963):

> "Our present analytic methods seem unsuitable for the solution of the important problems rising in connection with non-linear partial differential equations and, in fact, with virtually all types of problems in pure mathematics ... really efficient high-speed computing devices may, in the field of non-linear partial differential equations as well as in many other fields which are now difficult or entirely denied of access, provide us with those heuristic hints which are needed in all parts of mathematics for genuine progress."
> (John von Neumann: "On the principles of large scale computing machines")

10.9 Oscillatory Kinetics of Chemical Reactions on Single-Crystal Surfaces

A survey of the work of Gerhard Ertl and collaborators at the Fritz Haber Institute (of the Max Planck Society) on physical chemistry in Berlin.

Every dogma must have its day
H. G. Wells (1866-1946)

We are grateful to Gerhard Ertl and Markus Eiswirth for their generous support in the preparation of our manuscript and for having looked critically at the first draft and removed occasional misunderstandings. We would also like to thank Peter Plath for his careful reading of the revised version and for his critical remarks and his very helpful suggestions.

In this section, we present a résumé of some fundamental research results on chemical reactions and kinetics on single-crystal surfaces performed at the Fritz Haber Institute. This work demonstrates effects such as kinetic instabilities, periodic and aperiodic oscillations and the emergence of spatio-temporal patterns during the oxidation of carbon monoxide on platinum surfaces.

In chemistry, a substance is called a catalyst if it "affects the rate of a chemical reaction without being part of its end products" (Ertl, 2008). Catalysts through which a continuous flow of reactive substances passes are systems which – quasi by definition – operate far from thermodynamical equilibrium; therefore, it is not surprising that reaction kinetics evokes typical phenomena known from self-organisation such as the spontaneous formation of temporal or spatio-temporal patterns. Since catalytic processes are ubiquitous in nature and technology, it is especially interesting to investigate in detail the dynamical processes underlying these systems.

Without catalysts, the economical production of large quantities of chemical substances would not be possible at all; just think of the worldwide demand for ammonia to produce synthetic fertilisers (Ertl, 1997). More than 80% of all chemical products are created with the help of catalytic processes. These processes are by no means of vital importance only in the chemical or petrochemical industries, they also play a central role in combustion processes, in environmental engineering and in the construction of catalytic converters in the automotive industry. In biological systems, enzymes take on the role of catalysts. Almost all vital reactions such as photosynthesis, metabolism or DNA replication are based on catalytic processes. In these cases as well, self-organisation leads to rhythmic oscillations and pattern formation. In animate nature, however, the structures which emerge are much more complicated and frequently subject to complex hierarchical principles. In biological systems, a much higher number of chemical species takes part in the reactions and they are therefore much more difficult to understand (Goldbeter, 1996). However, chemical oscillations can be considered highly simplified models of biological rhythms and we can build on the experience thus gained (Hauser, 2005).

One of the main goals of theoretical research in catalysts is thus to understand and model the principal mechanisms underlying the dynamical processes in catalysis.

These findings are then used as a basis for future improvements in controlling kinetic reactions; ultimately, in practice, they open the door to new possibilities for designing innovative catalysts (Ertl, 1997; Ertl and Freund, 1999). In order to analyse chemical reactions and structural reordering on the atomic level, the most recent technologies from surface science and surface chemistry are required for heterogeneous catalysis; to evaluate and understand the results, methods from non-linear dynamics and synergetics are indispensable. The Nobel Prize has been awarded more than once for work dealing with heterogeneous catalysis; in 2007, the Prize for Chemistry was awarded to Gerhard Ertl.

Kinetic oscillations in non-equilibrium chemical systems are widespread. The experiment probably cited most frequently in the literature is the Belousov-Zhabotinsky (BZ) reaction, in which citric acid is oxidized in a acidified bromate solution where cerium ions play the role of the catalyst. The BZ system was analysed early on and fascinates us with its multitude of behaviour patterns. If the solution is well-stirred, periodic or irregular-chaotic changes in the colour can be observed, evidence of temporally alternating chemical reactions (Turner et al., 1981). If, on the other hand, the solution is not stirred, inhomogeneities in the concentration occur, causing concentric circles and the propagation of spiral waves (Markus et al., 1992). The BZ system has been investigated in many publications as a classic example of a chemical reaction far from thermodynamical equilibrium.

Because of its significance in surface science, we address heterogeneous catalysis, in which the catalyst and the reactants are in different phases of matter. We concentrate specifically on the catalytic oxidation of carbon monoxide on a platinum(110) single-crystal surface. Depending on external parameters such as temperature and partial pressure of the reactants CO and O_2, the temporal variation of the reaction rates may oscillate periodically or even become chaotic. An observation of the concentration distributions of the adsorbed species reveals spatio-temporal patterns such as propagating or standing chemical waves, rotating spirals as well as spatio-temporal irregular complex phenomena known as chemical turbulence (Ertl, 1991).

As an introduction to our main topic, we begin with an observation of potential oscillations occurring in a simple cell under galvanostatic conditions.

10.9.1 Oxidation of Hydrogen on a Platinum Electrode

We consider an electrochemical system composed of a Pt wire as a working electrode placed in a solution containing dissolved H_2 and Cu^{2+} as well as Cl^- ions (Krischer et al., 1991; Krischer et al., 1995; Krischer, 2001). At first, the influence of ClO_4^- and other ions is neglected. We investigate the temporal evolution of the electric potential U. Applying an electrical current of density j, we observe the following reactions (where, as usual, e^- denotes an electron)

$$H_2 \rightarrow 2H^+ + 2e^- \tag{10.9.1}$$

$$Cu^{2+} + 2e^- \rightleftarrows Cu \tag{10.9.2}$$

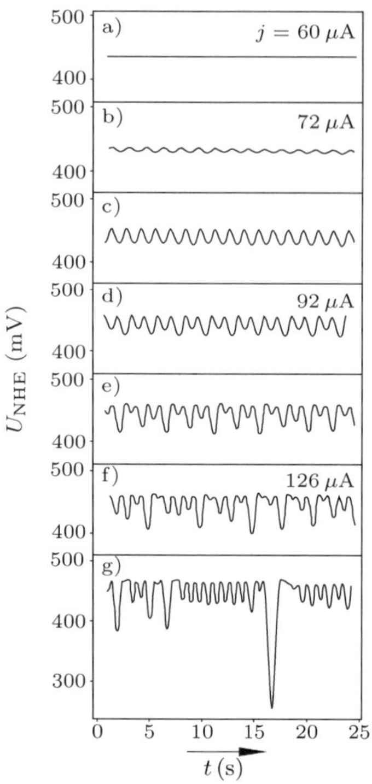

Fig. 10.9.1

Temporal variation of the potential U of a Pt electrode under galvanostatic conditions in an aqueous solution of 1 Mol $HClO_4$, 1.5×10^{-4} Mol Cu^{2+} and 5×10^{-4} Mol Cl^-; the solution is saturated with dissolved H_2 (Krischer et al., 1991)

As long as the current density j is low, e.g. $j = 60\,\mu$A, the potential U assumes a stationary value, see fig. 10.9.1a. As soon as j is increased, for example to $j = 72\,\mu$A, we note the appearance of small periodic oscillations of the potential U (see fig. 10.9.1b), the amplitude of which grows as we increase j (see fig. 10.9.1c). When j attains $j = 92\,\mu$A, a bifurcation suddenly appears and generates a doubly periodic oscillation of U (see fig. 10.9.1d).

A further increase of j leads to a cascade of period doublings, until, finally, at $j = 126\,\mu$A, aperiodic or chaotic behaviour of U sets in (see fig. 10.9.1f). Thus, a continuous transition from a stationary state via a Hopf bifurcation and a series of period doublings take place, leading to a chaotic behaviour. This temporal evolution is identical with the Feigenbaum scenario, discussed at length in sections 6.7 and 8.4. Finally, in fig. 10.9.1g, we observe a further transformation of the chaotic regime for a higher current density j: the chaotic state undergoes a further metamorphosis involving a sudden jump in the overall size of the attractor. Such a transition is known as an *interior crisis* (Grebogi et al., 1983), see also fig 10.9.1g. In this connection, we refer the reader to the examples in sections 6.7 and 10.8.

Initially, neglecting all inhomogeneities, we can describe the temporal dynamics of such chemical reactions by a set of ordinary non-linear differential equations

$$\dot{\boldsymbol{x}} = \boldsymbol{f}(\boldsymbol{x}, \boldsymbol{\mu}) \tag{10.9.3}$$

Here, the vector $\boldsymbol{x} = \{x_1 \quad x_2 \quad x_3 \ldots x_i \ldots x_n\}$ contains as elements the concentrations x_j of the n reactants. The non-linear vector function $\boldsymbol{f}(\boldsymbol{x}, \boldsymbol{\mu})$ depends on the kinetics of the individual reaction steps within the overall reaction and, in addition, on a set of external parameters $\boldsymbol{\mu}$, such as, for example, the current density j and the initial concentration of ions. So that oscillations can occur, the reaction mechanism must contain an autocatalytic step with an enhancing effect, and additionally, a negative feedback with a damping influence (Krischer, 2001). Such systems are denoted activator-inhibitor systems. In section 10.1, we already made the acquaintance of such a system, the Gierer-Meinhardt system.

In principle, the ansatz in eq. (10.9.3) represents a kind of mean-field approximation which assumes a random spatial distribution of the reacting particles. In this way, the present ansatz corresponds to that of a homogeneous field which ignores any local interaction and in which all domains act synchronously. Hence, spatial variations remain unconsidered, i.e. only the temporal evolution is described. In the literature, such systems are denoted CSTR systems (continuous stirred-tank reactor). For an experimental realisation of this concept, we may, for example, assume that the solution is well-stirred. In fact, even with the example of the simple prototype of the oxidation of hydrogen on a platinum electrode, the multitude of oscillations and spatio-temporal patterns that occur can only be understood and adequately modelled if we take the inhomogeneities of the reactants into consideration (Hudson and Tsotsis, 1994; Krischer et al., 1995; Otterstedt et al., 1999; Krischer, 1999; Krischer, 2001). Without homogenisation, spatio-temporal distributions inevitable emerge. In order to be able to investigate such more complex phenomena, we have to expand eq. (10.9.3) introducing expressions simulating reaction-diffusion processes, see section 10.9.4.

The explicit specification of the non-linear vector function $\boldsymbol{f}(\boldsymbol{x})$ requires the knowledge of detailed chemical mechanisms underlying a given reaction. For the aforementioned BZ system, for example, more than 60 individual reaction processes have to be considered; these, however, may be reduced to a much smaller set of essential variables depending on the desired degree of accuracy, see, for example, (Field and Noyes, 1974; Eiswirth et al., 1991; Plath, 1997; Nicola et al., 2006). Meanwhile, the oxidation of H_2 on Pt-electrodes is one of the best-understood electrochemical oxidations and serves as a basis for the modelling of fuel cells (Varela and Krischer, 2001).

10.9.2 Kinetics of the Catalytic Oxidation of CO on Pt(110)

The reaction of carbon monoxide and oxygen on platinum surfaces is seen as a prime example of catalytic oxidations and plays a prominent role, among others, in catalytic converters in the automobile industry. In order to gain a fundamental insight into the elementary chemical steps occurring in the overall reaction

$$2CO + O_2 \rightarrow 2CO_2 \tag{10.9.4}$$

it is expedient to use idealised samples, i.e. very clean, well-defined surfaces of platinum single-crystals, and to perform the experiments under strictly isothermal conditions. In the meantime, high-resolution measurement techniques exist in surface physics and chemistry which allow us to observe the surface structures on an atomic level and to analyse the degree of coverage of the adsorbed species. However, most methods can only be performed in ultra-high vacuum (Eiswirth, 1987; Eiswirth and Ertl, 1986). Under such idealised conditions, it is possible to study the principal mechanisms of the chemical part of the catalysis in isolation. Real catalysts under real process conditions, however, show much more complex behaviour. In industrial practice, processes are conducted in wide pressure ranges of 1-100 atm and much more complex mixtures of substances are used. Much higher reaction rates, steep temperature gradients and inhomogeneities thus have to be taken into consideration. There is still a huge need for research in this field.

10.9.2.1 Phenomenology of Catalytic Oxidation

Figure 10.9.2 presents the three steps of the chemical oxidation of CO on a Pt surface, also known as Langmuir-Hinshelwood mechanism. The two gaseous components CO and O_2 are adsorbed on the Pt surface, they react at the surface of the catalyst and are released into the gas phase immediately afterwards. In this reaction, the two components CO and O_2 compete for free adsorption sites on the Pt surface. The two reactants exhibit very different behaviour, however. While each adsorbed CO-molecule needs only one free site, O_2 is dissociatively chemisorbed and therefore a fairly large ensemble of unoccupied surface sites is required; in such a process, any adsorbed CO-molecule acts as an inhibitor for oxygen adsorption. On the other hand, the adsorbed oxygen atoms O_{ad} form a relatively open adlayer into which the adsorption of CO is still feasible.

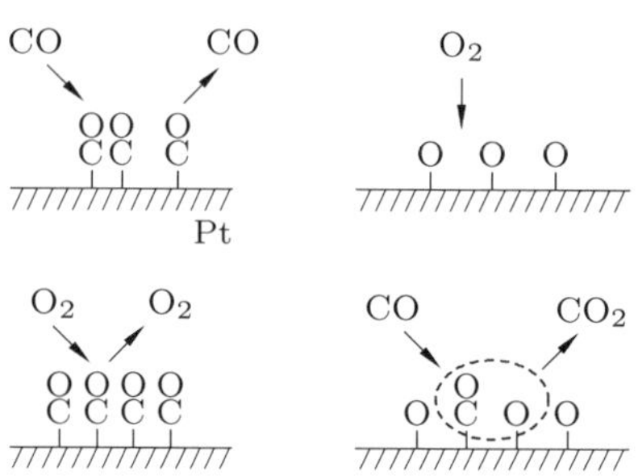

Fig. 10.9.2

Schematic mechanism of the reaction $2CO + O_2 \rightarrow 2CO_2$ on a Pt surface.

The reaction scheme given in eq. (10.9.4) may be written in the form of the classic Langmuir-Hinshelwood mechanism as follows, see, e.g., (Engel and Ertl, 1979; Ertl, 1980; Ertl, 1983),

$$\begin{aligned} * + CO &\rightleftarrows CO_{ad}, \\ 2* + O_2 &\rightarrow 2O_{ad}, \\ CO_{ad} + O_{ad} &\rightarrow 2* + CO_2 \end{aligned} \tag{10.9.5}$$

Here, adsorbed species carry the suffix ad and an asterisk * denotes as usual an empty site.

Experimental observations show that the kinetics of the catalytic oxidation depends significantly on the choice of the Pt surface. Bistable behaviour is found for Pt(111) surfaces, but no self-exciting oscillations; however, such oscillations occur in both cases, for Pt(100) as well as for Pt(110) surfaces, where the latter exhibits much more complex behaviour than Pt(100).

Figure 10.9.3 demonstrates that, when Pt(110) surfaces are used in strictly isothermic and high-vacuum conditions within a narrow range of the control parameters, the concentration of adsorbed species and hence also the reaction rate cease to be stationary and become oscillatory (Eiswirth and Ertl, 1986; Eiswirth et al., 1989). The same applies for the partial pressure p_{CO} which varies in proportion to the reaction rate as well as for the work function $\Delta\phi$, i.e. the minimum energy needed to remove an electron from the surface. The quantity $\Delta\phi$ is proportional to the oxygen coverage x_2 and oscillates strictly in phase with the reaction rate. This shows that kinetic chemical oscillations in the catalytic oxidation of CO on Pt(110) only occur under conditions for which the oxygen adsorption is rate-limiting.

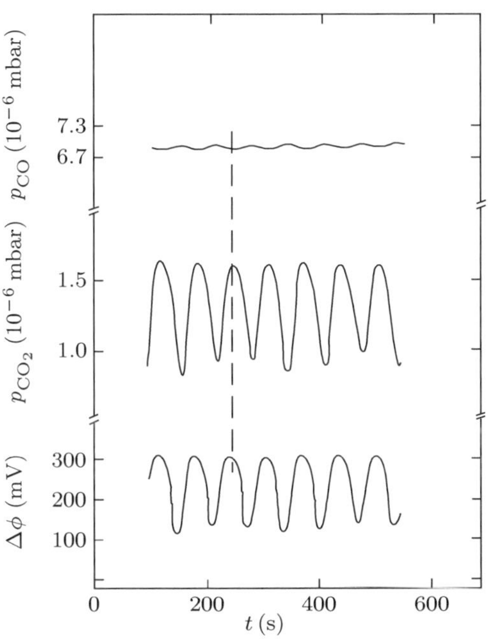

Fig. 10.9.3

Kinetics of a catalytic oxidation of CO on a Pt(110) surface. The partial pressure p_{CO_2} is proportional to the reaction rate. The variation of the work function $\Delta\phi$ proceeds in phase with the oxygen coverage rate x_2 ($T = 480$ K, $p_{O_2} = 2.1 \times 10^{-4}$ mbar and $p_{CO} = 6.8 \times 10^{-5}$ mbar), after (Ertl, 1991)

When a suitable temperature T and the partial pressures p_{CO} and p_{O_2} are chosen, highly diverse oscillatory behaviour can be observed, such as a series of period doublings, sometimes also period triplings, culminating in chaotic (Eiswirth et al., 1988) or even hyperchaotic behaviour (Eiswirth et al., 1992). In addition, mixed-mode

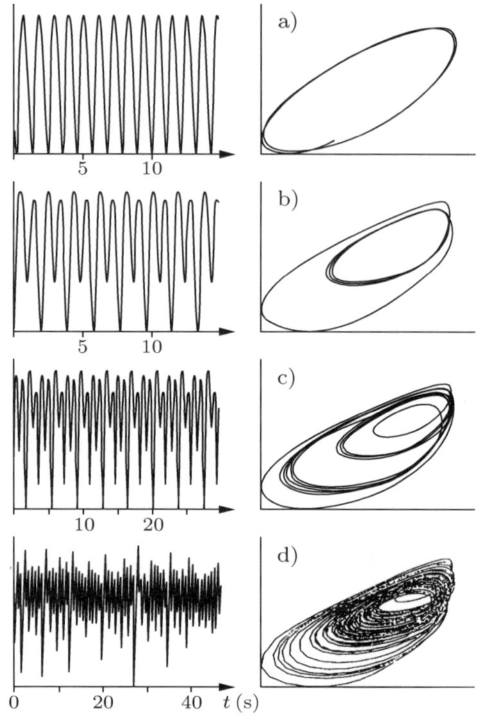

Fig. 10.9.4: Time series for an oscillatory CO oxidation on a Pt(110) surface (left) and corresponding phase portraits (right) (Ertl, 1991).
The control parameters are: $T = 550\,\text{K}$, $p_{O_2} = 4.0 \times 10^{-4}$ mbar and
(a) $p_{CO} = 1.65 \times 10^{-4}$ mbar, (b) $p_{CO} = 1.62 \times 10^{-4}$ mbar,
(c) $p_{CO} = 1.60 \times 10^{-4}$ mbar, (d) $p_{CO} = 1.58 \times 10^{-4}$ mbar

oscillations can be observed (Eiswirth et al., 1990), composed of two different oscillations, a small-amplitude harmonic oscillation and a relaxation oscillation with high amplitude. Moreover, spatio-temporal oscillations may occur, to be discussed later in section 10.9.2.2.

In order to visualise the transition from regular to chaotic behaviour, phase portraits are constructed on the basis of the experimentally measured time series using the time delay method of Takens described in section 5.5.4 (Takens, 1981), see fig. 10.9.4. The periodic time series turns into a simple closed curve (fig. 10.9.4a). Accordingly, after one or two period doublings, two and four loops can be seen in figs 10.9.4b respectively 10.9.4c. A typical chaotic time series is presented in fig. 10.9.4d together with the appertaining strange attractor

As we learned in section 5.4, the dynamic response of a system and in particular its chaotic behaviour can be characterised and quantified by the Lyapunov exponents σ_i. If all the $\sigma_i \leqslant 0$, the trajectories approach a simple attractor in the phase space; characteristic phase diagrams are shown in figs. 10.9.4a-c. If the largest Lyapunov exponent becomes positive, however, even the smallest differences in the initial

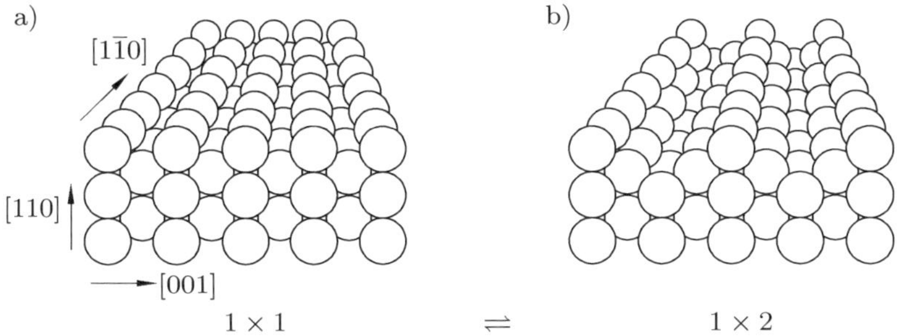

Fig. 10.9.5: Structural modifications of a Pt(110) surface

conditions lead, as we know, to a divergence of the trajectories and a chaotic state (fig. 10.9.4d). Close examinations of the time series for the CO catalytic reaction revealed that even two of the Lyapunov coefficients may become positive. This is occasionally denoted hyperchaos and was observed here for the first time in a chemical reaction (Eiswirth et al., 1992).

If we want to model the dynamics of the catalytic oxidation of carbon monoxide on platinum surfaces, it does not suffice to understand the individual chemical steps for the adsorption and the recombination of the molecules; rather, we have to take the structure of the catalyst surface into account. *In-situ* scanning-LEED (*low-energy electron diffraction*) techniques under high vacuum conditions have made a decisive contribution here. This method is based on the diffraction of electrons on surfaces and permits detailed analyses of changes in the structure of surfaces and the adsorbates.

Platinum surfaces with varyious crystallographic orientations possess different surface energy. While the densely packed Pt(111) surface has the same structure as the (1×1) bulk termination (fig. 10.9.5a), the clean Pt(110) surface is reconstructed into a missing row (1×2) structure which is energetically more efficient, see fig. 10.9.5b. LEED measurements show that oscillations in the reaction rates are accompanied by periodic changes in the surface structure between the (1×1) and the (1×2) phase and for their part influence the velocity of other elementary reactions (Ertl, 1991; Krischer et al., 1992).

When the temperature is increased, further modifications in the structure of the Pt(110) surface occur. During the oxidation of CO, for example, new crystal planes are formed; this is known as faceting (Ladas et al., 1988a; Ladas et al., 1988b). In consequence, the adsorption properties, in particular of oxygen, change significantly, resulting in much more complex dynamics.

10.9.2.2 Step by Step Formulation of a Kinetic Model

For the formulation of a model, we refer to the work of G. Ertl and K. Krischer, who give an examplary insight into the stages of developing a model for describing the dynamic behaviour of CO oxidation on Pt(110). Three systems of non-linear differential equations based on each other are developed, describing with increasing complexity the phenomena observed in a multitude of experimental studies. On the one hand, the significance of the individual steps for the dynamics becomes evident; on the other hand, the limits and range of validity of the specific models can be clearly defined (Krischer, 1990; Krischer et al., 1992). Starting from the simplest possible model, which nevertheless still contains the relevant variables and the essential elementary steps, new effects and thus new variables are included in the model step by step in order to describe the increasing complexity of the dynamics. In order to allow for reasonable comparisons with real measurement results, all the emerging kinetic parameters must be accessible experimentally and are taken from reliable sources (Krischer, 1990). Numerical simulation and a careful analysis of the non-linear model equations and their bifurcation behaviour using tools from the theory of dynamical systems discussed in the previous sections allow a comparison of the results with experimental measurements and pave the way for a verification of the model.

First step: Bistability

The purpose of the first step is to model the bistability and hysteresis observed experimentally. For low temperatures, primarily CO is adsorbed on Pt(110) thus impeding the adsorption of O_2; this leads to a low reaction rate. Increasing the temperature leads to an abrupt rise in the oxygen coverage and the reaction rate. Conversely, when the temperature is lowered again, a high CO coverage on the crystal can only be observed for even lower temperatures (Krischer, 1990).

CO oxidation on Pt surfaces follows the classical Langmuir-Hinshelwood mechanism described schematically in eq. (10.9.5). These equations contain three species, namely the coverage of carbon monoxide CO_{ad} and oxygen O_{ad} and the number of free adsorption sites $*$. CO_2, on the other hand, need not be taken into consideration since it is desorbed immediately after the reaction on the surface of the catalyst. Since the sum of the degrees of coverage of the three species must be 1, it is sufficient to determine two quantities. Ertl and Krischer chose the coverages of CO_{ad} and O_{ad} as independent variables; these will be denoted x_1 and x_2. Four chemical processess now have to be considered: CO adsorption, CO desorption, O_2 adsorption and the surface reaction. It can be shown that the desorption of oxygen can be neglected. In this way, the following non-linear differential equations are obtained

$$\begin{aligned} \dot{x}_1 &= s(\mathrm{CO})\, p_{\mathrm{CO}} - k_{\mathrm{d}} x_1 - k_{\mathrm{r}} x_1 x_2 \\ \dot{x}_2 &= s(\mathrm{O}_2)\, p_{\mathrm{O}_2} - k_{\mathrm{r}} x_1 x_2 \end{aligned} \tag{10.9.6}$$

where $s(\mathrm{CO})$ and $s(\mathrm{O}_2)$ are functions of x_1 and x_2 and describe the probabilities for the sticking of CO and O_2. The quantities k_{r} and k_{d} denote the reaction and desorption rates. Here, it is assumed that the desorption of a molecule is not influenced by the neighbouring sites. The asymmetric inhibition is at first not contained in the formalism of eq. (10.9.5); instead, it has to be taken into account in terms of different non-linearities in the sticking terms $s(\mathrm{CO})$ and $s(\mathrm{O}_2)$. On the basis of experiments, the following choice

$$\begin{aligned} s(\mathrm{CO}) &= \kappa_{\mathrm{C}} s_{\mathrm{C}}[1-(x_1/x_{1\mathrm{s}})^q] \\ s(\mathrm{O}_2) &= \kappa_{\mathrm{O}} s_{\mathrm{O}}[1-(x_1/x_{1\mathrm{s}})-(x_2/x_{2\mathrm{s}})]^2 \end{aligned} \tag{10.9.7}$$

with $q = 3.5 \pm 0.5$ turns out to be an adequate description (Krischer, 1990). Here, κ_{C} and κ_{O} denote the rates with which CO and O_2 hit the surface; they can be determined with the help of the kinetic gas theory on the assumption of ideal behaviour (Imbihl et al., 1985), s_{C} and s_{O} are the sticking coefficients, and $x_{1\mathrm{s}}$ resp. $x_{2\mathrm{s}}$ the corresponding saturation coverages. The asymmetric inhibition is included by postulating that $s(\mathrm{CO})$ does not depend upon x_2 in contrast to $s(\mathrm{O}_2)$. Without an appropriate choice of these factors, the turnover for the CO oxidation would be wrong and, moreover, the equations would predict a physically non-existent fixed point ("absorbing state") in which the reaction can be completely blocked by O_{ad} for an exactly vanishing CO coverage (Krischer et al., 1992). Equations (10.9.6) depend on three external control parameters: the partial pressures p_{CO} and p_{O_2} and the crystal temperature T. The latter enters the Arrhenius expression for the determination of the reaction rates k_{d} and k_{r}.

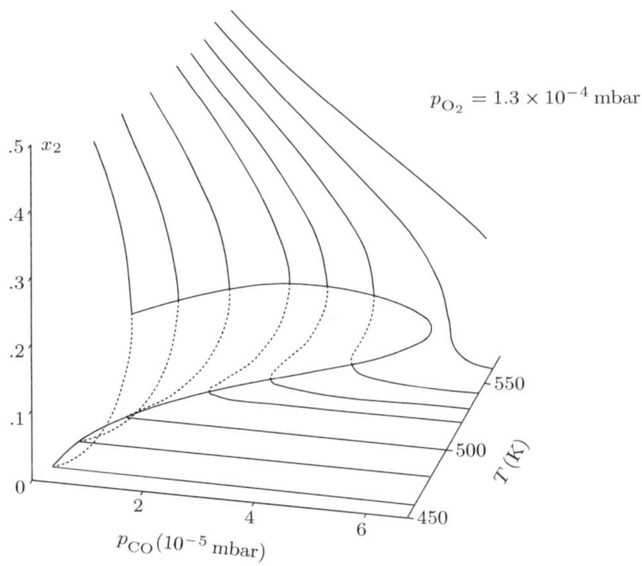

Fig. 10.9.6: Numerical simulation of the stationary degree of oxygen coverage x_2 as a function of p_{CO} and T: bistability and cusp. Solid (dashed) lines denote stable nodes (saddle points); saddle-node bifurcations occur along the parabolic curve (Krischer, 1990).

The numerical integration of the system of differential equations (10.9.6) reveals that the system exhibits a bistability with hysteresis if one control parameter is varied; if two are varied, a cusp appears. This behaviour is illustrated in fig. 10.9.6 where the oxygen coverage is plotted as a function of p_{CO} and T. This behaviour is in good agreement with experimental observations (Krischer et al., 1992).

In principle, bistable behaviour can be modelled with only one single equation. However, in chemical systems, since there must be at least two species in order to evoke bistability, two equations (10.9.6) are required. However, the oxygen coverage x_2 varies much faster than the carbon monoxide coverage x_1 and can thus be suppressed by adiabatic elimination so that, in the end, only one essential variable is left over. This also explains that the two equations (10.9.6) are not sufficient to model the periodic oscillations of the reaction rates. This extension of the model is described in the second step.

Second step: Simple oscillations

The simplest model revealed that the Langmuir-Hinshelwood mechanism is not sufficient to describe periodic kinetic oscillations. Evidently, there are other characteristics of the system causing oscillations. Up to now, the model has not accounted for the restructuring of the platinum surface described in section 10.9.2.1. For low reaction rates, the Pt surface is predominantly covered with CO molecules and has adopted the (1×1) volume structure; in contrast, for high reaction rates, the CO coverage is low and the surface exhibits the (1×2) reconstruction. The crucial point here is that the two phases possess different adsorption probabilities for oxygen. These properties of the structural phase transition are included in the following model and we discuss their impact on the dynamics (Krischer, 1990; Krischer et al., 1992).

The simplest way of taking the phase transition into account is to introduce a third variable x_3 which describes the normalised portion of the surface in the (1×1)-state and, in addition, to take the dependency of the sticking coefficient of oxygen on the respective phase into account. Experimental studies have shown that the transition $(1 \times 2) \rightarrow (1 \times 1)$ approximately starts at a CO-coverage of $x_1 = 0.2$ and ends at $x_1 = 0.5$. Between these two critical coverages, the equilibrium portion of the (1×1) phase x_3 can be approximated by a third-order polynomial in x_1, where the equilibrium is reached by a simple relaxation. The phase transition of x_3 between two constant values 0 and 1 is hence described by an S-shaped ramp function $x_3(x_1)$. In contrast to the previous step, no hysteretic behaviour has to be modelled in this case. With these considerations in mind, the third differential equation can be written in the following form

$$\dot{x}_3 = k_\mathrm{p} \cdot \begin{cases} -x_3 & \text{for} \quad x_1 \leqslant 0.2 \\ \sum_{\mathrm{i}=0}^{3} r_\mathrm{i} x_1^\mathrm{i} - x_3 & \text{for} \quad 0.2 < x_1 < 0.5 \\ (1 - x_3) & \text{for} \quad x_1 \geqslant 0.5 \end{cases} \tag{10.9.8}$$

where k_p describes the rate constant of the phase transition, which is in turn derived from an Arrhenius ansatz. The coefficients r_i of the polynomial are obtained from the literature; for detailed information, see (Krischer, 1990; Krischer et al., 1992). The sticking coefficient s_O for oxygen in eq. (10.9.7) also has to be reformulated in order to capture the oscillations of the surface structure between the (1×1) and the (1×2) state

$$s_\mathrm{O} = s_{\mathrm{O}1} x_3 + s_{\mathrm{O}2}(1 - x_3) \tag{10.9.9}$$

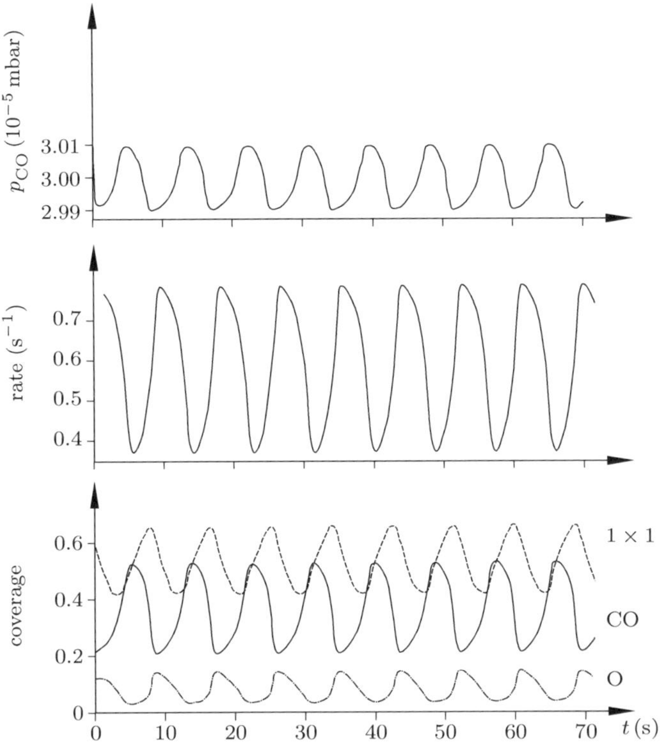

Fig. 10.9.7: Time series representing the kinetics of CO oxidation on Pt(110) ($T = 540$ K, $p_{O_2} = 6.7 \times 10^{-5}$ mbar and $p_{CO} = 3.0 \times 10^{-5}$ mbar) (Krischer, 1990; Ertl, 1991)

The system of eqs. (10.9.6), (10.9.8) and (10.9.9) does indeed generate oscillatory states as the typical time series in fig. 10.9.7 demonstrate (Krischer, 1990; Ertl, 1991). As for the simplest model, there are also ranges of the control parameters (p_{CO}, p_{O_2}, T) here for which three solutions exist. What is new in this model is that the stationary CO coverage in the predominantly oxygen-covered branch becomes unstable shortly before the formerly bistable regime. A supercritical Hopf bifurcation appears, i.e. stable periodic oscillations occur, as also observed in the experiments. A thorough discussion of the non-linear system reveals that a variety of different types of bifurcations – like Hopf, saddle-node and Takens-Bogdanov bifurcations – may occur depending upon the choice of the control parameters . Details as well as comparisons with experimental observations are given in (Krischer, 1990; Krischer et al., 1992). The three-variable model can, however, only describe simple periodic oscillations; period-n oscillations or mixed-mode oscillations cannot be modelled with this ansatz.

Third step: Mixed-mode oscillations

In order to explain *mixed-mode oscillations* (MMO) at intermediate temperatures as well as complex periodic or chaotic behaviour at high temperatures (see fig. 10.9.4), further effects obviously have to be included (Eiswirth et al., 1990; Eiswirth and Ertl, 1986; Eiswirth, 1987). Experiments show us that MMOs occur if micro-facets develop, i.e. if new crystal terraces emerge. This facetting ocurs together with a significant increase in catalytic activity and can only be observed under reaction conditions for which $(1 \times 2) \leftrightarrow (1 \times 1)$ occurs (Ladas et al., 1988a; Ladas et al., 1988b). The formation of new facets and their thermal annealing are competing processes causing the degree of facetting to undergo continuous changes during the oscillations (Sander, 1990).

At first, spatial effects such as the enlargement of the surface due to the formation of terraces are once again neglected in the model. Only the projection to the original 2D plane is considered. The normalised degree of facetting is introduced as a new variable x_4. An additional equation to describe the dynamics of x_4 should contain the opposing mechanisms previously mentioned. On the one hand, the facets were found to develop on a predominantly CO-covered surface, with the formation of facets always starting from the (1×1)-phase only as long as reaction rates are measurable. The formation of facets can thus be expressed as approximately proportional to the product of the reaction rate $(x_1 x_2)$ and the (1×1)-phase surface portion x_3, and it is bounded above by a term $(1 - x_4)$. The annealing process, on the other hand, is limited by high CO coverages. This effect is taken into account by a factor $(1 - x_1)$. Summing up all observations, we can formulate the following additional model equation (Krischer, 1990; Krischer et al., 1992)

$$\dot{x}_4 = k_\mathrm{f} x_1 x_2 x_3 (1 - x_4) - k_\mathrm{t} (1 - x_1) x_4 \tag{10.9.10}$$

where k_f is the rate of facet formation which can be assumed to be independent of the temperature, and k_t is the rate of thermal annealing. Since the formation of facets is accompanied by a higher catalytic activity which may be traced back to a higher rate of oxygen adsorption, the following extended relation is used instead of eq. (10.9.9)

$$s_\mathrm{O} = s_{\mathrm{O}1}\, x_3 + s_{\mathrm{O}2} (1 - x_3) + s_{\mathrm{O}3} x_4 \tag{10.9.11}$$

For the analysis of the four-variable model, the stationary CO coverage is again plotted versus p_{O_2}, see fig. 10.9.8 top. For small CO values, the predominantly oxygen-covered branch becomes unstable as before via a supercritical Hopf bifurcation and there is a small p_{O_2} interval where the three solutions coexist. In sections I, II, III, we find a very complex oscillatory behaviour. In region I, harmonic oscillations with small amplitudes occur on a predominantly oxygen-covered surface (fig. 10.9.8a); in III, relaxation oscillations with a high amplitude are found (fig. 10.9.8c) and, in the intermediate region II, mixed-mode oscillations occur where small amplitude oscillations are interrupted by large-amplitude excursions (fig. 10.9.8b).

The regions of MMOs cannot be discerned from the variation of the function $x_1(p_{\mathrm{O}_2})$ in fig. 10.9.8a alone. To understand these more complex oscillations, the coupling between the facetting and the simple periodic oscillations caused by a phase transition have to be studied thoroughly. Details are described in (Krischer, 1990). The

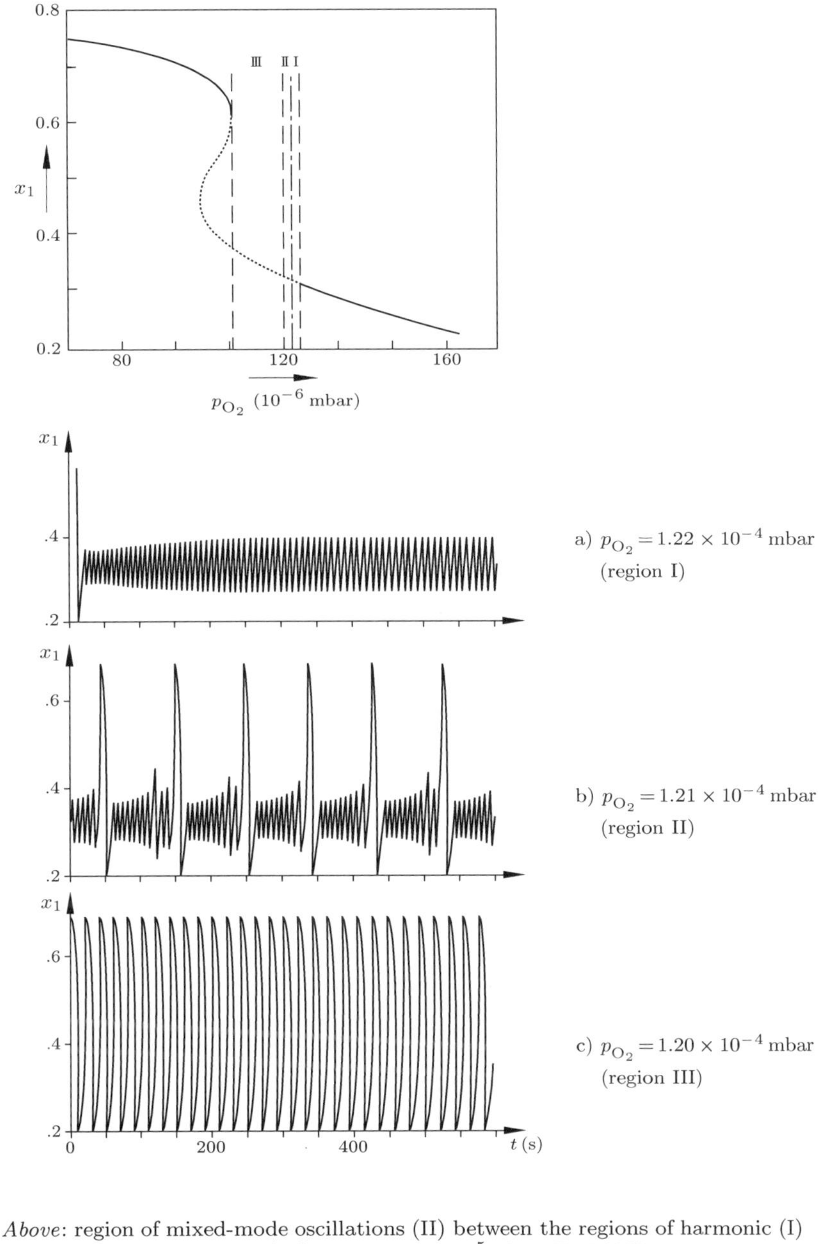

Fig. 10.9.8: *Above*: region of mixed-mode oscillations (II) between the regions of harmonic (I) and relaxation oscillations (III); $p_{CO} = 4.2 \times 10^{-5}$mbar, $T = 540$K
below: typical time series for the three regions (Krischer et al., 1992)

crucial factor for the emergence of MMOs is that in this system processes take place on different time scales. For CO oxidation on Pt(110), the $(1\times 2) \leftrightarrow (1\times 1)$ phase transitions cause a feedback loop which develops more slowly than the adsorption processes. The facetting of the surface proceeds even more slowly, however. The emergence of MMOs can now be explained by means of a slow manifold ansatz, see also sections 6.2 and 6.8. The slowest variable x_4 of the facetting process is used as a parameter and all limit sets of the fast sub-system such as fixed points and limit cycles are calculated as functions of x_4. For decreasing values of x_4, i.e. predominantly thermal annealing, a Hopf bifurcation of the fast sub-system occurs, i.e. small oscillations of the coverage on the reactive branch are obtained. Simultaneously with x_4, according to eq. (10.9.11), the adsorption of oxygen decreases until the system abruptly jumps back to the non-reactive branch where the formation of facets starts again. Hence, MMOs occur if the system remains for a relatively long time on the slow manifold, where it oscillates until a certain threshold value is reached and then jumps to another branch followed by fast relaxation (Krischer, 1990).

10.9.3 Identification of Chaos and Hyperchaos in Kinetic Surface Reactions

In the previous section, we gave a short description of the Krischer-Eiswirth-Ertl (KEE) model for CO oxidation on Pt(110) which is able to model simple and multiple periodic behaviour as well as the transition to chaos and mixed-mode oscillations. However, in experimentally recorded time series, an even more complex behaviour was observed, namely so-called hyperchaos (Rössler, 1979b). The methods used in the theory of non-linear systems described in Chapter 5 are the key to analysing the time series. Simple chaos can develop in a three-dimensional phase space, where the largest Lyapunov exponent (LCE) must be positive, whilst the manifestation of hyperchaos, for which two positive LCEs must exist, requires at least four state variables. Previously, hyperchaos had been observed in only few systems, such as a driven NMR laser (Stoop and Meier, 1988), see also (Brun, 1989), and in the electrical properties of semi-conductors (Stoop et al., 1989). The first team to confirm hyperchaos in a chemical reaction was that of Gerhard Ertl (Eiswirth et al., 1992).

In order to identify hyperchaos, we need a method which allows us to calculate all – or at least all positive – LCEs from the time series of a single recorded variable. When the WSSV algorithm is applied (Wolf et al., 1985), averages are taken over the exponential divergence of initially closely neighbouring trajectories on a reconstructed attractor. However, since this method is very sensitive to noise[1] (Vastano and Kostelich, 1986; Harding and Ross, 1987; Harding and Ross, 1988; Kruel et al., 1990), Ertl et al. applied a method of Sano und Sawada (Sano and Sawada, 1985), see also (Berndes, 1992) and sections 5.4.4, 5.4.5 and 5.5.4.

[1] For an improvement of the calculation, we refer the reader to the new methods allowing dynamical and measurement noise to be separated; see section 2.2 (Siefert and Peinke, 2004; Böttcher et al., 2006).

The essence of the method of Sano und Sawada lies in the relation between the Lyapunov exponents σ_i and the Jacobian matrix $\boldsymbol{J} = \partial \boldsymbol{f}/\partial \boldsymbol{x}$ of the n-dimensional deterministic system in eq. (10.9.3). For systems with equations of motion which are explicitly specified, there is a straightforward technique for computing the complete Lyapunov spectrum, as described in section 5.4.6. The algorithm is based on the integration of the system of equations (5.4.67) together with their linearisation along a reference trajectory. Obviously, this approach is not directly applicable to experimental time series since neither the system of equations nor its linearisation is known explicitly. Consequently, the tangent flow on the attractor is approximated by a map $\boldsymbol{\mathcal{A}}_j$ which is computed by a least-square fitting technique from a number of pseudo-tangent vectors propagated over a pre-set time interval t_m.

Let us denote by $\boldsymbol{x}_j$ (j $=$ 1, 2 . . . M) a reference point on the attractor which is reconstructed from the experimental or computational data using the time-delay method described in section 5.5.4. We then select N points $\boldsymbol{x}_{k_i}$ (i $= 1, 2, \ldots$ N) within a spherical shell around $\boldsymbol{x}_j$ which satisfy the relation $\varepsilon_{min} < |\boldsymbol{x}_j - \boldsymbol{x}_{k_i}| < \varepsilon_{max}$. If the sphere is now subjected to the evolution of the system, we obtain the information required to establish an approximation of the linearised system at $\boldsymbol{x}_j$. For this purpose, we calculate the vectors $\boldsymbol{y}_i = \boldsymbol{x}_{k_i} - \boldsymbol{x}_j$ measuring the distance between the centre of the sphere $\boldsymbol{x}_j$ and the N points $\boldsymbol{x}_{k_i}$ and determine the corresponding vectors $\boldsymbol{z}_i = \boldsymbol{x}_{k_j+m} - \boldsymbol{x}_{j+m}$ for the given time interval $t_m = m\Delta t$. From the relation $\boldsymbol{z}_i = \boldsymbol{\mathcal{A}}_j \boldsymbol{y}_i$, the transformation matrix $\boldsymbol{\mathcal{A}}_j$ can be calculated approximately applying the method of least squares

$$\min_{\mathcal{A}_j} S = \min_{\mathcal{A}_j} \frac{1}{N} \sum_{i=1}^{N} |\boldsymbol{z}_i - \mathcal{A}_j \boldsymbol{y}_i|^2 \tag{10.9.12}$$

Here, S denotes the mean square deviation. Minimising S yields a matrix equation for $\boldsymbol{\mathcal{A}}_j$. The Lyapunov exponents σ_i are then deduced from

$$\sigma_i = \lim_{M \to \infty} \frac{1}{M t_m} \sum_{j=1}^{M} \ln \left| \mathcal{A}_j \boldsymbol{e}_i^j \right| \quad (i = 1, 2, \ldots, n) \tag{10.9.13}$$

Here, the vectors $\boldsymbol{e}_i^j$ form an orthonormal basis in the tangent space of $\boldsymbol{x}_j$. In the course of the numerical procedure, they are then renormalised at an interval of, say, $5t_m$ using the Gram-Schmidt technique; see also section 5.4.6.

For catalytic oxidation, the experimental data were given, as usual, in the form of time series of a single variable. The corresponding attractors were reconstructed using the Takens time-delay method described in section 5.5.4. The delay time was selected as the first zero of the autocorrelation function of the time series. A difficulty arising in the reconstruction of attractors often lies in the generation of spurious LCE's; see (Stoop and Meier, 1988; Stoop et al., 1989; Holzfuss and Lauterborn, 1989; Bryant et al., 1990). This can be overcome, however, by evaluating the data for increasing embedding dimensions. While the true LCEs are independent of the embedding dimension, we observe a continuous shift in the spurious LCEs which are sensitive to numerical inaccuracies and noise.

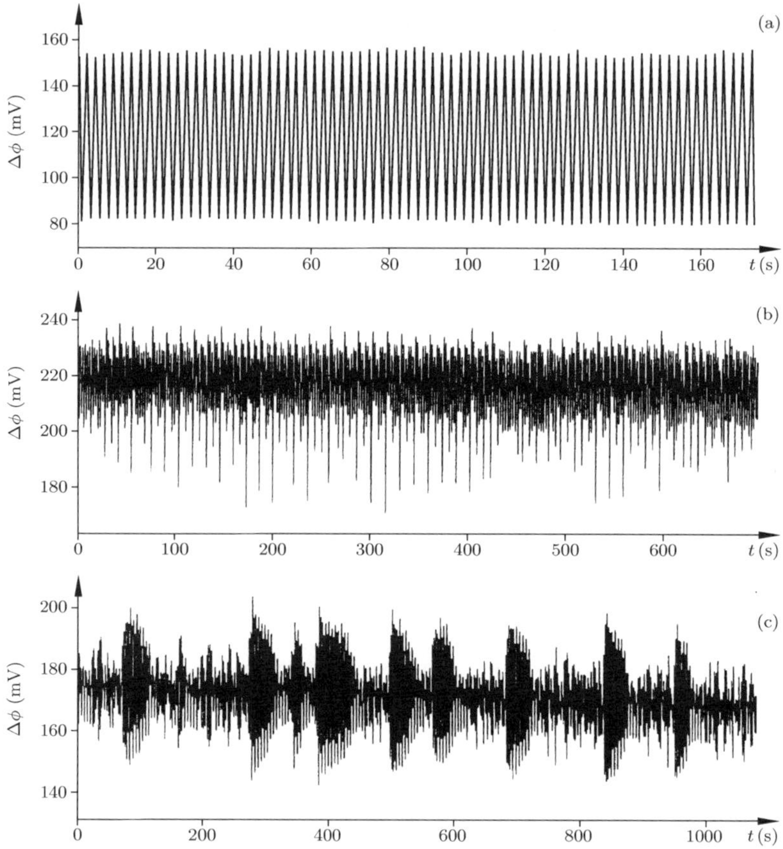

Fig. 10.9.9: Time series for an oscillatory CO oxidation on a Pt(110) surface: (a) periodic, (b) chaotic and (c) hyperchaotic oscillations (parameters see table 10.9.1); after (Eiswirth et al., 1992).

In fig. 10.9.9, we reproduce typical time series which demonstrate a transition from a periodic motion to chaos and hyperchaos. The associated control parameters are listed in table 10.9.1 in conjunction with the appertaining values of the correlation dimension D_2 and the correlation entropy h_2 which represents a lower bound of the Kolmogorov-Sinai entropy; these values were deduced from standard techniques; see (Grassberger and Procaccia, 1983a; Grassberger and Procaccia, 1983b; Grassberger and Procaccia, 1983c); the reader is also referred to sections 5.5.5 and 5.5.6 for an account of the corresponding theory. We observe that the response in case a) is evidently periodic. In contrast, plot b) demonstrates a chaotic behaviour near the transition to chaos. File c) is once more chaotic, but with a completely different character to that in b) and also possesses other control parameters. Indeed, this response is one of hyperchaos. As was to be expected, the two states also differ significantly in their values of D_2 and h_2. These differences are reflected in the phase portraits of the corresponding reconstructed attractors (see fig. 10.9.10). We observe

time series	control parameters ($T = 540\,\mathrm{K}$)	N_{dat} ($\Delta t = 0.15\,\mathrm{s}$)	period (average)	D_2	h_2
a) Abb. 9.8.9a	$p_{O_2} = 7.5 \times 10^{-5}$ Torr $p_{CO} = 3.71 \times 10^{-5}$	1150	2.35 s	1.1 ± 0.1	0.
b) Abb. 9.8.9b	$p_{O_2} = 7.5 \times 10^{-5}$ $p_{CO} = 3.415 \times 10^{-5}$	4850	≈ 2.5	2.4 ± 0.2	0.2 ± 0.05
c) Abb. 9.8.9c	$p_{O_2} = 6.0 \times 10^{-5}$ $p_{CO} = 3.0 \times 10^{-5}$	8240	≈ 2.5	4.5 ± 0.4	0.4 ± 0.05

Table 10.9.1: Parameters and characteristic quantities for periodic (a) and chaotic (b,c) behaviour of the CO oxidation on Pt(110); N_{dat}: number of points; D_2: correlation dimension; h_2: lower bound for KS entropy (in s^{-1}), after (Eiswirth et al., 1992)

that the attractor presented in fig. 10.9.10b essentially has a two-band structure. On the other hand, we do not discern any band texture in attractor c). Finally, fig. 10.9.11 demonstrates the dependence of the LCEs on the embedding dimension. We observe, as was to be expected, only negative LCEs in file a) with the exception of one which is almost zero. File b) possesses one positive and file c) two positive LCEs which form a plateau clearly different from zero, i.e. they become independent of the embedding dimension. Very small positive values of σ emerging sporadically in the two cases b) and c) may be caused by inaccuracies in the measurement data and the numerical calculations.

The emergence of hyperchaotic behaviour could not be explained by means of the Krischer-Eiswirth-Ertl model. Other effects like, for example, spatio-temporal pattern formation could provide an adequate explanation (Krischer et al., 1992).

10.9.4 Spatio-temporal Pattern Formation

So far, our account of CO oxidation on Pt(110) has been restricted to processes which evolve solely in time, i.e. our assumption was that the species involved are homogeneously distributed in the gaseous phase and on the surface of the catalyst. While the required homogeneity can be attained in the BZ reaction, for example, by vigorous stirring, a continuous mixing of the adsorbates on the solid surface of a catalyst is not feasible in the case of a heterogeneous reaction. Hence, in general, gradients in the concentration of the adsorbed species will occur; these are spatially coupled in different ways.

Essentially, three coupling mechanisms may occur for a heterogeneous catalysis. Under non-isothermal reaction conditions, local gradients in the concentrations of the coverages cause locally varying reaction rates. For exothermal reactions, heat is released, giving rise to temperature gradients and hence to a heat flow. This effect is, however, of no importance for the reaction conditions under consideration (low partial pressures, single-crystal surfaces etc.). For the second mechanism, varying reaction rates give rise to slight modulations ($<1\%$) of the partial pressures of O_2 and

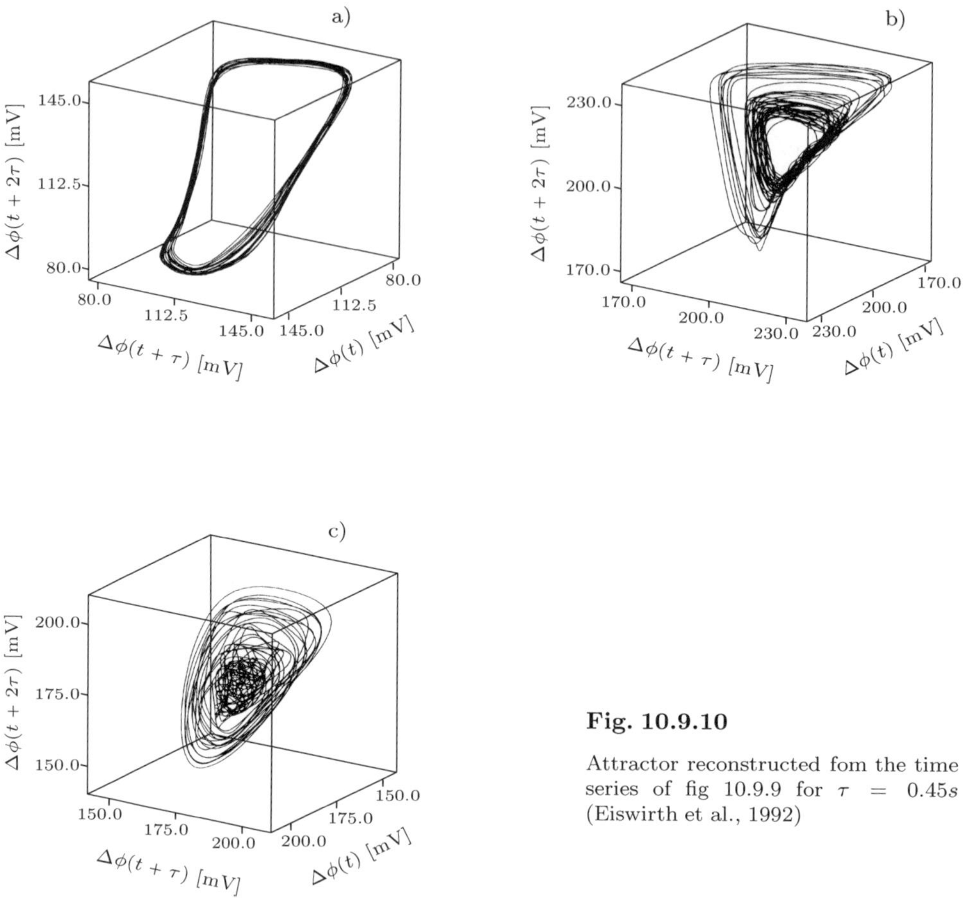

Fig. 10.9.10

Attractor reconstructed fom the time series of fig 10.9.9 for $\tau = 0.45s$ (Eiswirth et al., 1992)

CO in the gaseous phase leading to a global coupling, cf. fig. 10.9.3. These gradients propagate practically instantaneously ($< 10^{-4}$ s) over the whole surface area of the catalyst (Ertl, 1991).

The third mechanism, however, is dominant and must be taken into account. Local differences in the adsorbate coverages give rise to diffusion processes which themselves influence the reaction activity. In order to take this effect into consideration, eq. (10.9.3) has to be extended to the following set of partial differential equations

$$\frac{\partial \boldsymbol{x}}{\partial t} = \boldsymbol{f}(\boldsymbol{x}) + \boldsymbol{D}\nabla^2\boldsymbol{x} \tag{10.9.14}$$

where $\boldsymbol{D}$ is a diffusion coefficient. Partial differential equations of this type are known as reaction-diffusion equations.

The spatio-temporal pattern formation generated by equations of the reaction-diffusion type can be divided into bistable, excited, oscillating and even chaotic states;

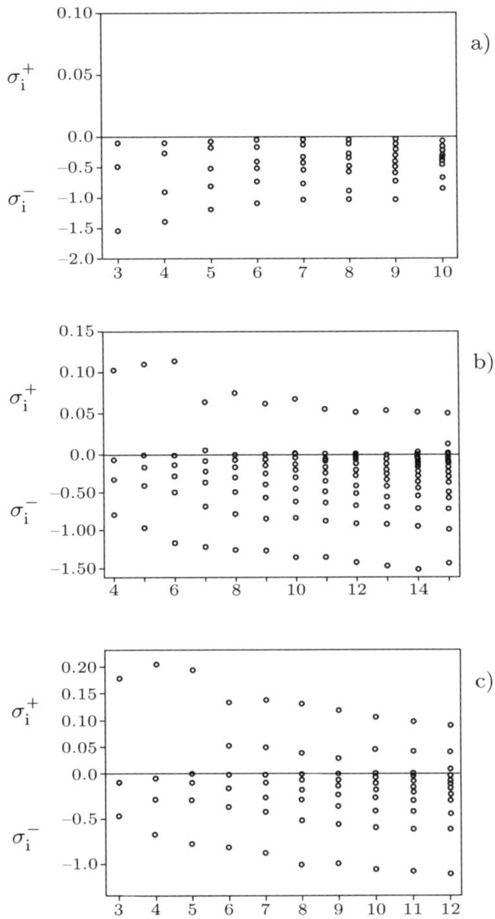

Fig. 10.9.11

Lyapunov exponents as a function of the embedding dimension for the time series of fig. 10.9.9. We observe zero (a), one (b) resp. two (c) positive exponents. At least six or seven dimensions are required to obtain stable positive LCE's (Eiswirth et al., 1992)

these have been thoroughly studied theoretically, see for example (Mikhailov, 1990; Cross and Hohenberg, 1993; Krischer, 2001), and could also be observed experimentally for CO catalysis on Pt(110) (Ertl, 1991; Imbihl and Ertl, 1995). In contrast to the BZ reaction in a solution, where concentration differences are easily visible because of colour changes, the CO oxidation process demands a more refined technique in order to demonstrate local variations in the coverages of adsorbed species. Such concentration gradients can be recorded with a high resolution ($\sim 0.5\,\mu$) in real time with the help of a photoemission electron microscope (PEEM). In the PEEM images presented, oxygen-covered regions appear dark, whilst those covered by CO are bright.

For certain reaction conditions, the system may be in a state of excitability that manifests itself in various types of propagation of non-linear chemical waves. Among these, solitary waves (fig. 10.9.12) are particularly interesting (Rotermund et al., 1991; Bär et al., 1992). These involve single-wave pulses with a bell-shaped profile propagating with constant speed along the directions marked by arrows in the PEEM image. This direction corresponds to one crystallographic orientation (in fact, the

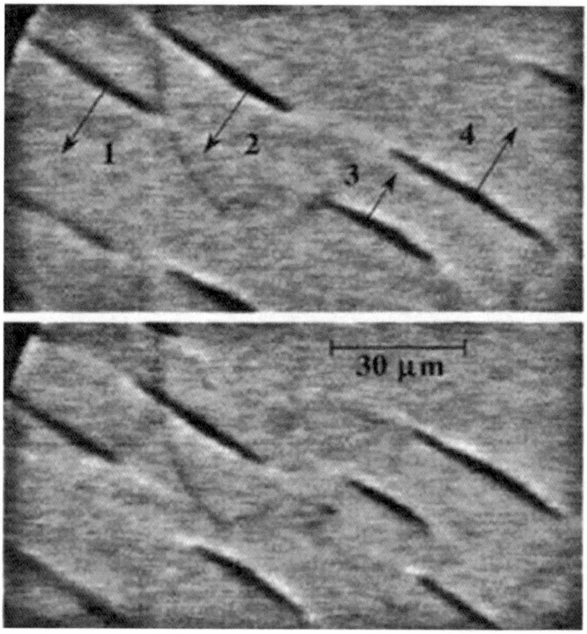

Fig. 10.9.12: Two PEEM images, recorded at an interval of 3 s, from a Pt(110) surface under steady state CO oxidation. The dark regions with enhanced oxygen coverage propagate in the directions indicated as solitary waves ($T = 485$ K, $p_{CO} = 1 \times 10^{-4}$ mbar and $p_{O_2} = 3.5 \times 10^{-4}$ mbar) (Ertl, 1991)

[001]-direction) of the Pt(110) surface. In most cases, these are so-called autowaves for which two pulses travelling in opposite directions collide and annihilate one another. In some cases, waves have been observed which exhibit typical characteristics of dissipative solitons. These waves retain their shape and velocity even after collision and undergo only a phase shift (Ertl, 1991; Bär et al., 1992).

As early as 1834, John Scott Russell observed surface waves in shallow water channels travelling over many kilometres with almost constant speed and preserving their original shape. Not until 60 years later did Korteweg and de Vries succeed in setting up a non-linear dissipation-free wave equation possessing solitons as solutions which propagate with constant shape and speed, characteristics, which remain unchanged following collisions. In hydrodynamics, this equation is known as the Korteweg-de Vries equation (KdV), cf. (Korteweg and de Vries, 1895; Zabusky and Kruskal, 1965; Ablowitz and Clarkson, 1991), see also (Argyris and Haase, 1987; Argyris et al., 1991). Solitons have a quasi-particle-like character, a specific property resulting from a delicate balance between non-linearity and dispersion. Solitons play a prominent role in many areas of physics, for example in dislocations in crystal structures, in optics and in plasma physics.

In reaction-diffusion systems, *dissipative solitons* may arise if, in additon to the balance between non-linear and dispersive behaviour, dissipative losses are compensated for by a steady energy supply. Such solitary waves have been described for the CO oxidation on Pt(110) surfaces (Bär et al., 1992). Autowaves and (dissipative) solitons can hardly be distinguished on the basis of their shape since both wave types are described as homoclinic orbits. However, they show completely different features in the case of collisions. A well-written overview of autowaves and solitons in reaction-diffusion systems can be found in (Nekorkin and Kazantsev, 2002).

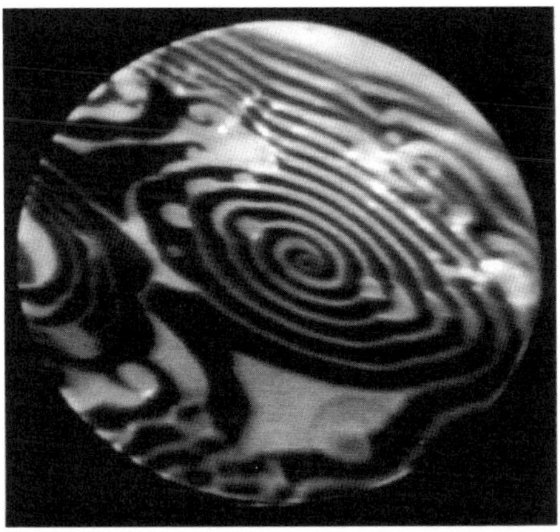

Fig. 10.9.13

PEEM image of a spiral wave pattern on Pt(110). The diameter of the image is 0.4 mm ($T = 435$ K, $p_{CO} = 3.8 \times 10^{-5}$ mbar and $p_{O_2} = 3.0 \times 10^{-4}$ mbar) (Ertl, 1991)

Continuously growing spirals are the most commonly observed patterns in excitable chemical reaction-diffusion systems (Field and Burger, 1985; Mikhailov, 1990; Kuramoto, 1984). An example of such a spiral wave pattern for a CO system is shown in fig. 10.9.13. The spiral wave is not circular but elliptic, having its long axis along the [110] direction and the short axis along the [001] direction of the single crystal. This response is due to the anisotropy of the surface which affects the diffusion of adsorbed species and the propagation velocities of chemical waves.

If control parameters are changed slightly, we find that the system becomes reactively oscillatory; this is demonstrated in fig. 10.9.14. These images exhibit concentric elliptically shaped waves emanating periodically from so-called nucleation centres. Such patterns have also been observed before in the classic BZ reaction (Zaikin and Zhabotinsky, 1970). In the present case, colliding wave fronts annihilate each other and lead to cusp-like structures as shown in fig. 10.9.14.

Regular oscillations of the reaction rate, as shown in fig. 10.9.4a, are often spatially uniform but can also be associated with standing waves as shown in fig. 10.9.15. The whole macroscopic surface exhibits approximately equally spaced alternating dark and bright stripes of CO_{ad} and O_{ad} coverages, the intensities of which vary with the period of the associated kinetic oscillation. As in the previous example, the local contrast gradients of the two-dimensional patterns reflect the inhomogeneously distributed coverages. As a consequence of very slight variations between neighbouring pairs of stripes, dynamic "dislocations" may be formed as indicated in fig. 10.9.15.

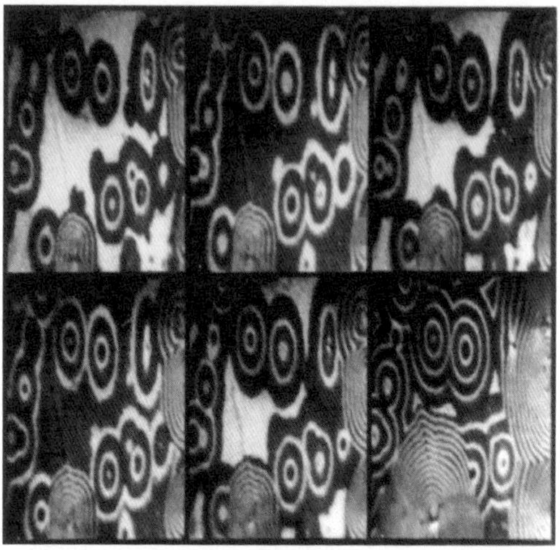

Fig. 10.9.14

Sequence of PEEM images of a $0.2 \times 0.3\,mm^2$ section of a Pt(110) surface. The time interval between the last two images is 30 s between all the others, 4.1 s ($T = 427\,K$, $p_{CO} = 3 \times 10^{-5}\,mbar$ and $p_{O_2} = 3.2 \times 10^{-4}\,mbar$) (Jakubith et al., 1990)

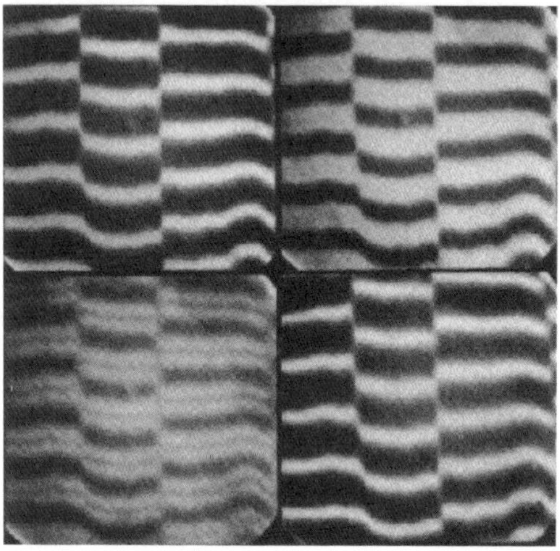

Fig. 10.9.15

Spatio-temporal patterns of a standing wave on a $0.3 \times 0.3\,mm^2$ section of a Pt(110) surface, recorded at intervals of 0.5 s during harmonic kinetic oscillations of type a) of fig. 10.9.4 ($T = 550\,K$, $p_{O_2} = 4.1 \times 10^{-4}\,mbar$ and $p_{CO} = 1.75 \times 10^{-4}\,mbar$) (Jakubith et al., 1990)

Finally, we refer once more to the most intriguing case of the irregular (i.e. chaotic) appearance of kinetic oscillations. The transition from regular kinetic oscillations to temporal chaos upon variation of one of the control parameters is shown in the specific example of fig. 10.9.4d and is generated following a cascade of period doublings of the Feigenbaum scenario and leads to a gradually increasing spatio-temporal disorder. The orientations and spacing of the stripes become irregular; they finally break up and no longer form standing waves but perform *rapid lateral* motions. As can be observed in fig. 10.9.16, the patterns are highly reminiscent of a turbulent fluid. Hence, we may denote this spatio-temporal chaos as a chemical turbulence.

The investigations into the heterogenic catalysis on idealised probes (single crystals) under idealised experimental conditions (ultra-high vacuum) made a major contribution to understanding the microscopic processes arising in the catalytic reactions.

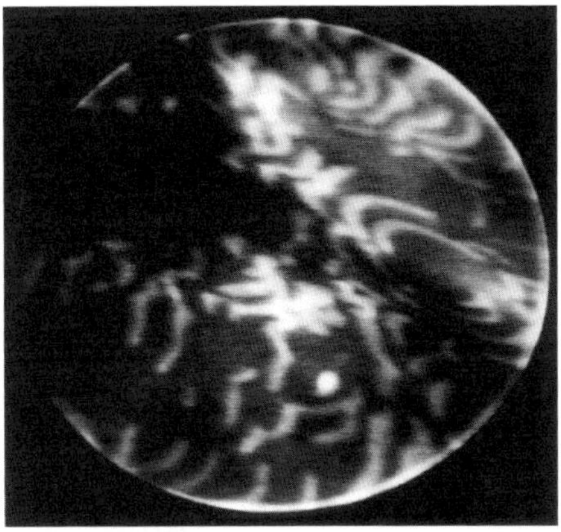

Fig. 10.9.16

Irregular and rapidly changing pattern (diameter 0.4 mm), characteristic of chemical turbulence

Apart from the application of new, highly efficient measurement techniques providing a high resolution which are taken from surface physics and chemistry, concepts and methods developed in the dynamics of non-linear systems also made a significant contribution. With their help, researchers succeeded in bringing order into the diversity of dynamical phenomena and in formulating universal laws and overriding principles for their emergence. In many cases, it is thus possible to make predictions as to the conditions under which typical oscillations or specific spatio-temporal patterns may occur (Krischer, 1999; Krischer, 2001). The results of the basic research may also have an impact in future on practical applications and open perspectives with regard to the control of catalytic processes (Stich et al., 2008; Beta et al., 2003) and the design of optimised innovative catalysts aiming to increase the yield of catalytic reactions (Ertl, 1990; Ertl, 1997).

Notwithstanding the progress in catalyst research, it is not straighforward to transfer the findings from idealised experimental conditions to real catalysts which exhibit a much more complex structure and chemical composition of the surfaces; in addition, in industry, catalytic reactions are performed at pressures of 1-100 atm. This gap in the reaction conditions, known as *pressure/materials gap* is a topic of current research, see, for example, (Wang et al., 2002).

Investigations on the oxidation of CO on a palladium support catalyst under normal pressure also belong to the field of real catalysts (Jaeger et al., 1986; Liauw et al., 1996; Ballandis and Plath, 2001). In these systems, which play an important role in the emission control of automobiles, complex periodic and chaotic behaviour could also be observed. Of special interest are the self-affine patterns which appear in the time-series; these reflect more or less violent drops of the reaction rates which greatly reduce the efficiency of the catalyst.

The pattern formation on a macroscopic level is the result of the interaction of a great number of catalytically active palladium crystals on a wide range of particle sizes. Despite supposedly isothermal conditions, the palladium particles are heated for a short time interval by several hundred degrees higher the environment; in fact, the heating is the higher, the smaller the particles are. The high temperatures cause phase transitions of the palladium from an active state to a catalytically inactive palladium oxide which is, after a fast decrease in temperature, again reduced to active palladium. The length of the oxidation-reduction cycle depends on the size of the Pd-particles (Ballandis and Plath, 2001). Models developed recently consider the dynamics globally as a system of relaxation oscillators which are coupled to each other by the gaseous phase and the frequencies of which follow a hierarchical distribution in the catalytically active range (Petrović, 2007). The models allow fundamental insights into the mechanisms underlying CO oxidation on Pd-substrate catalysts. It becomes evident that drops in the reaction rates are the result of synchronisation effects. This finding opens new perspectives on how to choose the size distribution of the palladium particles expediently in order to control the exact form of the drops.

All living organisms are permanently subject to biochemical reactions and, almost always, enzymes, special proteins playing the role of efficient catalysts, participate in these reactions. Furthermore, a plethora of rhythmic behaviour and oscillations on a hierarchy of scales is characteristic of living systems, from the level of molecules up to cells and organs. Examples are the oscillations of glycolysis processes underlying the metabolic reactions within the cells, the rhythmic production of insulin by β-cells of the pancreas, and the heartbeat. Not all of these rhythms are simple periodic oscillations, but rather complex or chaotic oscillations occur frequently as well as mixed-mode oscillations and bistabilities. The emergence of patterns in the temporal evolutions or in the spatial structures have their origin in the fact that biological systems are open systems exchanging energy and matter with their environment and that these processes are non-linear. The methods and techniques provided by the theory of non-linear systems are doubtless among the most important tools for analysing the complex behaviour of biological systems, taking us a step closer towards understanding the underlying principal mechanisms (Goldbeter, 1996; Murray, 2002; Larter, 1990; Hauser, 2005).

References

Abarbanel, H.D.I., Brown, R., Sidorowich, J.D., Tsimring, L.S.: The analysis of observed chaotic data in physical systems. Rev. Mod. Phys. 65, 1331–1392 (1993)

Ablowitz, M.J., Clarkson, P.A.: Solitons, Nonlinear Evolution Equations and Inverse Scattering. In: London Mathematical Society Lecture Note Series, vol. 149. Cambridge University Press, Cambridge (1991)

Abraham, R., Marsden, J.E., Ratiu, T.: Manifolds, Tensor Analysis, and Applications. Springer, New York (1993)

Abraham, R.H., Gardini, L., Mira, C.: Chaos in Discrete Dynamical Systems. Springer, New York (1997)

Addison, P.S.: The Illustrated Wavelet Transform Handbook. IOP Publishing, Bristol (2002)

Aguirre, P., Doedel, E.J., Krauskopf, B., Osinga, H.M.: Investigating the consequences of global bifurcations for two-dimensional invariant manifolds of vector fields. Discrete and Continuous Dynamical Systems 29(4), 1–36 (2011)

Alstrøm, P., Hansen, L.K., Rasmussen, D.R.: Crossover scaling for moments in multifractal systems. Phys. Rev, A 36, 827–833 (1987)

Andrievskii, B.R., Fradkov, A.L.: Control of Chaos: Methods and Applications. II. Applications. Automat. Rem. Contr. 65, 505–533 (2004)

Andronov, A.A., Pontryagin, L.: Systèmes Grossiers. Dokl. Akad. Nauk. SSSR 14, 247–251 (1937)

Andronov, A.A., Witt, A.A., Chaikin, S.E.: Theorie der Schwingungen, 1. Teil. Akademie Verl., Berlin (1965)

Avrutin, V., Lammert, R., Schanz, M., Wackenhut, G.: AnT 4.669, A Simulation and Analysis Tool for Dynamical Systems. IPVS, Univ. Stuttgart (2000-2009), `http://www.AnT4669.de/`

Appleton, E.V., van der Pol, B.: On a type of oscillation-hysteresis in a simple triode generator. Phil. Mag. 43, 177–193 (1922)

Argoul, F., Arneodo, A., Grasseau, G., Gagne, Y., Hopfinger, E.J., Frisch, U.: Wavelet analysis of turbulence reveals the multifractal nature of the Richardson cascade. Nature 338, 51–53 (1989)

Argyris, J., Bühlmeier, J.: Ein Abriß der Aeroelastizität, Kap. 26. In: Argyris, J., Mlejnek, H.-P. (eds.) Die Methode der finiten Elemente, Vieweg, Wiesbaden (1988)

Argyris, J., Faust, G., Haase, M.: $X\acute{\alpha}o\varsigma$ — An adventure in chaos. Comp. Meth. Appl. Mech. Eng. 91, 997–1091 (1991)

Argyris, J., Haase, M.: An engineer's guide to soliton phenomena: application of the finite element method. Comp. Meth. Appl. Mech. Eng. 61, 71–122 (1987)

Arnold, V.I.: Small denominators II, Proof of a theorem of A.N.Kolmogorov on the preservation of conditionally-periodic motions under a small perturbation of the Hamiltonian. Russ. Math. Surv.18(5), 9–36 (1963)

Arnold, V.I.: Instability of dynamical systems with several degrees of freedom. Sov. Math. Dokl. 5, 581–585 (1964)

Arnold, V.I.: Small denominators I, Mappings of the circumference onto itself. Trans. of the Am. Math. Soc. Ser. 2 46, 213–284 (1965)
Arnold, V.I.: Loss of stability of self oscillations close to resonances and versal deformations of equivariant vector fields. Funct. Anal. Appl. 11(2), 1–10 (1977)
Arnold, V.I.: Mathematical Methods of Classical Mechanics. Springer, New York (1978)
Arnold, V.I.: Gewöhnliche Differentialgleichungen. Springer, Berlin (1980)
Arnold, V.I.: Geometrical Methods in the Theory of Ordinary Differential Equations, 2nd edn. Springer, Berlin (1988)
Arnold, V.I., Khesin, B.A.: Topological Methods in Hydrodynamics. Springer, Berlin (1998)
Arnold, V.I., Kozlov, V.V., Neishtadt, A.I.: Mathematical aspects of classical and celestial mechanics. In: Arnold, V.I. (ed.) Dynamical Systems III, Springer, Berlin (1988)
Aronson, D.G., Chory, M.A., Hall, G.R., McGehee, R.P.: Bifurcations from an invariant circle for two-parameter families of maps of the plane: a computer-assisted study. Commun. Math. Phys. 83, 303–354 (1982)
Auerbach, D., Cvitanović, P., Eckmann, J.-P., Gunaratne, G., Procaccia, I.: Exploring chaotic motion through periodic orbits. Phys. Rev. Lett. 58, 2387–2389 (1987)
Ausloos, M., Dirickx, M. (eds.): The Logistic Map and the Route to Chaos: From the Beginnings to Modern Applications. Understanding Complex Systems. Springer, Heidelberg (2006)
Avrutin, V., Schanz, M.: Border-collision period-doubling scenario. Phys. Rev. E. 70, 026222 (2004)
Avrutin, V., Schanz, M.: On the fully developed bandcount adding scenario. Nonlinearity 21, 1077–1103 (2008)
Baba, N., Amann, A., Schöll, E., Just, W.: Giant improvement of time-delayed feedback control by spatio-temporal filtering. Phys. Rev. Lett. 89(7), 074101 (2002)
Bader, R.: Nonlinearities and Synchronization in Musical Acoustics and Music Psychology. Springer, Heidelberg (2013)
Balatoni, J., Rényi, A.: Über den Begriff der Entropie, Arbeiten zur Informationstheorie I (abr. German translation, 117–134. Math. Forschungsberichte IV. Deutscher Verlag der Wissenschaften, Berlin (1957)
Ballandis, C., Plath, P.J.: A new discrete model for the non-isothermic dynamics of the exothermic co-oxidation on Palladium supported catalyst. J. Non-Equilib. Thermodyn. 25, 301–324 (2001)
Banerjee, S., Yorke, J.A., Grebogi, C.: Robust chaos. Phys. Rev. Lett. 80, 3049–3052 (1998)
Baptista, M.S., Grebogi, C., Barreto, E.: Topology of windows in the high-dimensional parameter space of chaotic maps. Int. J. Bif. Chaos 13(9), 2681–2688 (2003)
Baptista, M.S., Kurths, J.: Chaotic channel. Phys. Rev. E. 72, 045202 (2005)
Bär, M., Eiswirth, M., Rotermund, H.-H., Ertl, G.: Solitary-wave phenomena in an excitable surface reaction. Phys. Rev. Lett. 69(6), 945–948 (1992)
Barnsley, M.: Fractals everywhere. Academic Press, London, San Diego (1988)
Batchelor, G.K.: The Theory of Homogeneous Turbulence. Cambridge University Press, Cambridge (1956)
Batchelor, G.K., Townsend, A.A.: The nature of turbulent motion at large wave-numbers. Proc. R. Soc. Lond. A. 199, 238–255 (1949)
Beaulieu, N.C., Xie, Q.: An optimal lognormal approximation to lognormal sum distributions. IEEE Trans. Veh. Technol. 53, 479–489 (2004)
Beck-Bornholdt, H.-P., Dubben, H.-H.: Der Schein der Weisen. Irrtümer und Fehlurteile im täglichen Denken. rororo science. Rowohlt Taschenbuch Verlag, Hamburg (2003)
Becker, E., Bürger, W.: Kontinuumsmechanik. Teubner Studienbücher, Stuttgart (1975)
Becker, J., Rödelsperger, F., Weyrauch, T., Benner, H., Just, W., Čenys, A.: Intermittency in spin-wave instabilities. Phys. Rev. E. 59, 1622–1632 (1999)

Beerends, R.J., ter Morsche, H.G., van den Berg, J.C., van de Vrie, E.M.: Fourier and Laplace Transforms. Cambridge University Press, Cambridge (2003)

Bénard, H.: Les tourbillons cellulaires dans une nappe liquide. Rev. Gén. Sci. Pures Appl. 11, 1261–1227,1309–1328 (1900)

Benettin, G., Galgani, L., Giorgilli, A., Strelcyn, J.-M.: Lyapunov characteristic exponents for smooth dynamical systems and for Hamiltonian systems: a method for computing all of them; part I: Theory, part II: Numerical application. Meccanica 15, 9–20, 21–30 (1980)

Benzi, R., Ciliberto, S., Baudet, C., Chavarria, G.R.: On the scaling of three-dimensional homogeneous and isotropic turbulence. Physica D 80, 385–398 (1995)

Benzi, R., Ciliberto, S., Baudet, C., Chavarria, G.R., Tripiccione, R.: Extended self-similarity in the dissipation range of fully developed turbulence. Europhys. Lett. 24, 275–279 (1993a)

Benzi, R., Ciliberto, S., Tripiccione, R., Baudet, C., Massaioli, F., Succi, S.: Extended self-similarity in turbulent flows. Phys. Rev. E. 48, R29–R32 (1993b)

Bergé, P., Dubois, M., Manneville, P., Pomeau, Y.: Intermittency in Rayleigh-Bénard convection. J. Phys. Lett. 41, 341–345 (1980)

Bergé, P., Pomeau, Y., Vidal, C.: Order within Chaos: Towards a deterministic Approach to Turbulence. Hermann, Paris (1984)

Berndes, S.: Qualitative und quantitative Charakterisierung von eindimensionalen Zeitreihen nichtlinearer dynamischer Systeme. Diploma thesis, Inst. für Computer-Anwendungen (ICA), Univ. Stuttgart (1992)

Bestehorn, M.: Verallgemeinerte Ginzburg-Landau-Gleichungen für die Musterbildung bei Konvektions-Instabilitäten. Dissertation, Univ. Stuttgart (1988)

Beta, C., Bertram, M., Mikhailov, A.S., Rotermund, H.H., Ertl, G.: Controlling turbulence in a surface chemical reaction by time-delay autosynchronization. Phys. Rev. E. 67, 046224 (2003)

Billingsley, P.: Ergodic Theory and Information. John Wiley & Sons, New York (1965)

Birkhoff, G.D.: Dynamical systems. Amer. Math. Soc. Colloq. Publ. 9 (1927)

Boccaletti, S., Grebogi, C., Lai, Y.-C., Mancini, H., Maza, D.: The control of chaos: theory and applications. Phys. Rep. 329, 103–197 (2000)

Bogdanov, R.: Versal deformations of a singular point of a vector field on the plane in the case of zero eigenvalues. Funct. Anal. Appl. 9(2), 144–145 (1975)

Bohr, T., Gunaratne, G.: Scaling for supercritical circle-maps: numerical investigation of the onset of bistability and period doubling. Phys. Lett. A. 113(2), 55–60 (1985)

Boltzmann, L.: Einige allgemeine Sätze über das Wärmegleichgewicht. Wien. Ber. 63(2), 679–711 (1871)

Bonatto, C., Gallas, J.A.C.: Accumulation horizons and period adding in optically injected semiconductor lasers. Phys. Rev. E. 75, 055204 (2007)

Bonatto, C., Gallas, J.A.C., Ueda, Y.: Chaotic phase similarities and recurrences in a damped-driven Duffing oscillator. Phys. Rev. E. 77, 026217 (2008)

Boole, G.: Die Grundlehren der endlichen Differenzen- und Summenrechnung. Braunschweig (1867)

Borgo, D.: Sync or Swarm: Improvising Music in a Complex Age. Bloomsbury Academic, London, New York (2005)

Born, M.: Voraussagbarkeit in der klassischen Mechanik. Phys. Bl. 15, 342–349 (1959)

Böttcher, F., Peinke, J., Kleinhans, D., Friedrich, R., Lind, P.G., Haase, M.: Reconstruction of complex dynamical systems affected by strong measurement noise. Phys. Rev. Lett. 97, 090603 (2006)

Bošnjaković, F., Knoche, K.F.: Technische Thermodynamik, Teil I. Steinkopff, Darmstadt, 7. vollständig neubearbeitete und erweiterte Auflage (1988)

Boyland, P.L.: Bifurcations of circle maps: Arnold tongues, bistability and rotation intervals. Commun. Math. Phys. 106, 353–381 (1986)

Bracewell, R.N.: The Fourier Transform and its Applications. McGraw-Hill, New York (1986)

Braun, M.: Differentialgleichungen und ihre Anwendungen. Springer, Berlin (1979)

Brillouin, L.: Science and Information Theory. Academic Press, New York (1962)

Bröcker, T., Jänich, K.: Einführung in die Differentialtopologie. Band 143 in Heidelberger Taschenbücher. Springer, Heidelberg (1990)

Bronstein, I.N., Semendjajew, K.A., Musiol, G., Mühlig, H.: Taschenbuch der Mathematik. Harri Deutsch, Frankfurt (1995)

Broomhead, D.S., King, G.P.: Extracting qualitative dynamics from experimental data. Physica D 20, 217–236 (1986)

Bruce, A., Wallace, D.: Critical point phenomena: universal physics at large length scales. In: Davies, P. (ed.) The New Physics, pp. 236–267. Cambridge University Press, Cambridge (1989)

Brun, E.: Deterministisches Laser-Chaos: Experimente und quantitative Indikatoren. In: Gerok, W. (ed.) Ordnung und Chaos in der unbelebten und belebten Natur, pp. 259–273. Wissenschaftliche Verlagsgesellschaft, Stuttgart (1989)

Bryant, P., Brown, R., Abarbanel, H.D.I.: Lyapunov exponents from observed time series. Phys. Rev. Lett. 65, 1523–1526 (1990)

Busse, F.A.: Nonlinear properties of thermal convection. Rep. Prog. Phys. 41, 1929–1966 (1978)

Callen, H.B.: Thermodynamics and an Introduction to Thermostatistics. Wiley, New York (1985)

Carr, J.: Applications of Centre Manifold Theory. Springer, New York (1981)

Carter, D.R., Fyhrie, D.P., Whalen, R.T.: Trabecular bone density and loading history: regulation of connective tissue biology by mechanical energy. J. Biomechanics 20, 785–794 (1987)

Cartwright, M.L.: Forced oscillations in nearly-sinusoidal systems. J. Inst. Elec. Eng. 95, 88–96 (1948)

Cartwright, M.L., Littlewood, J.E.: On nonlinear differential equations of the second order, I: The equation $\ddot{y} + k(1 - y^2)\dot{y} + y = b\lambda k \cos(\lambda t + a)$, k large. J. Lond. Math. Soc. 20, 180–189 (1945)

Cayley, A.: The Newton-Fourier imaginary problem. Amer. J. Math. II, 71–84 (1879)

Chanal, O., Chabaud, B., Castaing, B., Hébral, B.: Intermittency in a turbulent low temperature gaseous helium jet. Europ. Phys. J. B. 17, 309–317 (2000)

Chandrasekhar, S.: Hydrodynamic and Hydromagnetic Stability. Oxford Univ. Press, Oxford (Dover edition (1961)

Chen, S., Ecke, R., Eyink, G., Rivera, M., Wan, M., Xiao, Z.: Physical mechanism of the two-dimensional inverse energy cascade. Phys. Rev. Lett. 96(8), 084502 (2006)

Chorin, A.J., Marsden, J.E.: A Mathematical Introduction to Fluid Mechanics. Springer, New York (1993)

Ciarlet, P.G., Lions, J.L. (eds.): Handbook of Numerical Analysis, Techniques of Scientific Computing (Part 2). Elsevier, Amsterdam (1997)

Collet, P., Eckmann, J.-P.: Iterated maps on the interval as dynamical system. In: Jaffer, A., Ruelle, D. (eds.) Progress in Physics, Birkhäuser, Boston (1980)

Collet, P., Eckmann, J.P., Lanford, O.E.: Universal properties of maps on an interval. Commun. Math. Phys. 76, 211 (1980)

Cooley, J.W., Tukey, J.W.: An algorithm for the machine computation of complex Fourier series. Math. Comp. 19, 291–301 (1965)

Cornfeld, I.P., Formin, S.V., Sinai, Y.G.: Ergodic Theory. Springer, New York (1982)

Coullet, P., Tresser, C.: Itérations d'endomorphismes et groupe de renormalisation. J. Phys. Coll. C. 5, 25–28 (1978)

Courant, R., Hilbert, D.: Methoden der mathematischen Physik. Springer, Berlin (1993)

Cowin, S.C.: Bone remodeling of diaphysical surfaces by torsional loads: theoretical predictions. J. Biomechanics 20, 1111–1120 (1987)

Cowin, S.C., Hegedus, D.H.: Bone remodeling I: theory of adaptive elasticity. J. Elasticity 6, 313–326 (1976)

Cross, M.C., Hohenberg, P.C.: Pattern formation outside of equilibrium. Rev. Mod. Phys., 65(3), 851–1123 (1993)

Crutchfield, J.P., Farmer, J.D., Packard, N.H., Shaw, R.S.: Chaos. Sci. Amer. 254, 46–57 (1986)

Curry, J.H.: On the Hénon transformation. Commun. Math. Phys. 68, 129–140 (1979)

Cuypers, Y., Maurel, A., Petitjeans, P.: Vortex burst as a source of turbulence. Phys. Rev. Lett. 91, 194502 (2003)

Cuypers, Y., Maurel, A., Petitjeans, P.: Comparison between an experimental turbulent vortex and the Lundgren vortex. J. Turbulence, 5, 1–22 (2004)

Cuypers, Y., Maurel, A., Petitjeans, P.: Characterization of an experimental turbulent vortex in the physical and spectral spaces. J. Turbulence 7(7), 1–13 (2006)

Cvitanović, P.: Universality in Chaos. Hilger, Bristol (1984)

Cvitanović, P., Jensen, M.H., Kadanoff, L.P., Procaccia, I.: Renormalization, unstable manifolds, and the fractal structure of mode locking. Phys. Rev. Lett. 55(4), 343–346 (1985)

Cvitanović, P., Mainieri, R.A.R., Tanner, G., Vattay, G.: Chaos: Classical and Quantum. Niels Bohr Institute, Copenhagen (2008), http://ChaosBook.org/

Daitche, A.: unpublished (2009)

Dangelmayr, G., Knobloch, E.: The Takens-Bogdanov bifurcation with O(2)-symmetry. Phil. Trans. Roy. Soc. A. 322(1565), 243–279 (1987)

Danielson, G.C., Lanczos, C.: Improvements in practical Fourier analysis and their application to X-ray scattering from liquids. J. Franklin Inst. 233, 365–380, 435–452 (1942)

Daubechies, I.: Ten Lectures on Wavelets. SIAM, Philadelphia,PA (1992)

Daugman, J.G.: How iris recognition works. IEEE Trans. Circuits Syst. Video Technol. 14, 21–30 (2004)

Davidson, P.A.: Turbulence. An Introduction for Scientists and Engineers. Oxford Univ. Press, Oxford (2006)

De Groot, S.R., Mazur, P.: Grundlagen der Thermodynamik irreversibler Prozesse. Bibliographisches Institut, Mannheim (1969)

Devaney, R.L.: An Introduction to Chaotic Dynamical Systems. Addison-Wesley Publishing Company, Redwood City (1987)

Dhooge, A., Govaerts, W., Kuznetsov, Y.A.: MATCONT: A MATLAB package for numerical bifurcation analysis of ODEs. ACM Trans. Math. Software 29, 141–164 (2003)

di Bernardo, M., Budd, C.J., Champneys, A.R., Kowalczyk, P.: Piecewise-smooth Dynamical Systems. Theory and Applications. Applied Mathematical Sciences, vol. 163. Springer (2008)

Dirac, P.A.M.: The Principles of Quantum Mechanics, 2nd edn. Oxford University Press, New York (1935)

Ditto, W.L., Rauseo, S.N., Spano, M.L.: Experimental control of chaos. Phys. Rev. Lett., 65, 3211–3214 (1990)

Ditzinger, T., Haken, H.: Oscillations in the perception of ambiguous patterns. A model based on synergetics. Biol Cybern. 61, 279–287 (1989)

Doedel, E.J.: AUTO: A program for the automatic bifurcation analysis of autonomous systems. Congr. Numer. 30, 265–284 (1981)

Doedel, E.J., Paffenroth, R.C., Champneys, A.R., Fairgrieve, T.F., Kuznetsov, Y.A., Sandstede, B., Wang, X.: AUTO2000: Continuation and bifurcation software for ordinary differential equations (with HomCont). Caltech (2001), http://cmvl.cs.concordia.ca/auto/

Doerner, R., Hübinger, B., Martienssen, W., Grossmann, S., Thomae, S.: Stable manifolds and predictability of dynamical systems. Chaos, Solitons & Fractals 10(11), 1759–1782 (1999)

Dowell, E.H.: Flutter of a buckled plate as an example of chaotic motion of a deterministic autonomous system. J. Sound Vib. 85(3), 333–344 (1982)

Dragt, A.J., Finn, J.M.: Insolubility of trapped particle motion in a magnetic dipole field. J. Geophys. Res. 81, 2327–2340 (1976)

Dubois, M., Rubio, M.A., Bergé, P.: Experimental evidence of intermittencies associated with a subharmonic bifurcation. Phys. Rev. Lett. 51(16), 1446–1449 (1983); Erratum: 51, 2345

Duffing, G.: Erzwungene Schwingungen bei veränderlicher Eigenfrequenz und ihre technische Bedeutung. Vieweg, Braunschweig (1918)

Eckert, M.: Turbulenz - ein problemhistorischer Abriss. N.T.M Zeitschrift fr Geschichte der Wissenschaften, Technik und Medizin 16, 39–71 (2008)

Eckmann, J.-P., MacKay, R.S.: Routes to chaos with special emphasis on period doubling. In: Iooss, G., Hellemann, R.H.G., Stora, R. (eds.) Chaotic Behaviour of Deterministic Systems, pp. 453–510. North-Holland (1983)

Eckmann, J.-P., Ruelle, D.: Ergodic theory of chaos and strange attractors. Rev. Mod. Phys. 57(3), 617–656 (1985)

Eckmann, J.-P., Ruelle, D.: Fundamental limitations for estimating dimensions and Lyapunov exponents in dynamical systems. Physica D 56, 185–187 (1992)

Einstein, A.: Über die von der molekularkinetischen Theorie der Wärme geforderte Bewegung von in ruhenden Flüssigkeiten suspendierten Teilchen. Ann. Physik. 17, 549–560 (1905)

Eiswirth, M.: Phänomene der Selbstorganisation bei der Oxidation von CO an Pt(110). Dissertation, Ludwig-Maximilians-Univ. München (1987)

Eiswirth, M., Ertl, G.: Kinetic oscillations in the catalytic CO oxidation on a Pt(110). Surf. Sci. 177, 90–100 (1986)

Eiswirth, M., Freund, A., Ross, J.: Mechanistic classification of chemical oscillators and the role of species. Adv. Chem. Phys. 80, 127–199 (1991)

Eiswirth, M., Krischer, K., Ertl, G.: Transition to chaos in an oscillating surface reaction. Surf. Sci. 202, 565–591 (1988)

Eiswirth, M., Krischer, K., Ertl, G.: Nonlinear dynamics in the CO-oxidation on Pt single crystal surfaces. Appl. Phys. A. 51, 79–90 (1990)

Eiswirth, M., Kruel, T.-M., Ertl, G., Schneider, F.W.: Hyperchaos in a chemical reaction. Chem. Phys. Lett. 193(4), 305–310 (1992)

Eiswirth, M., Möller, P., Wetzl, K., Imbihl, R., Ertl, G.: Mechanisms of spatial self-organization in isothermal kinetic oscillations during the catalytic CO oxidation on Pt single crystal surface. J. Chem. Phys. 90, 510–521 (1989)

Elphick, C., Tirapegui, E., Brachet, M.E., Coullet, P., Iooss, G.: A simple global characterization for normal forms of singular vector fields. Physica D 29, 95–127 (1987)

Elstrodt, J.: Maß - und Integrationstheorie, 4th edn. Springer, Berlin (2005)

Engel, T., Ertl, G.: Elementary steps in the catalytic oxidation of carbon monoxide on platinum metals. Adv. Catal. 28, 1–77 (1979)

Ermak, D.L., Buckholz, H.: Numerical integration of the Langevin equation: Monte Carlo simulation. J. Comp. Phys. 35, 169–182 (1980)

Ertl, G.: Surface science and catalysis. Pure Appl. Chem. 52, 2051–2060 (1980)
Ertl, G.: Kinetics of chemical processes of well-defined surfaces. In: Anderson, J.P. (ed.) Catalysis, Science and Technology, vol. 4, pp. 210–282. Springer, Berlin (1983)
Ertl, G.: Elementarschritte bei der heterogenen Katalyse. Angew. Chem. 102, 1258–1266 (1990)
Ertl, G.: Oscillatory kinetics and spatio-temporal self-organization in reactions at solid surfaces. Science 254, 1750–1755 (1991)
Ertl, G.: Design von Katalysatoren. Spektrum der Wissenschaft 2, 82–85 (1997)
Ertl, G.: Reactions at surfaces: From atoms to complexity (Nobel Lecture). Angew. Chem. Int. Ed. 47, 3524–3535 (2008)
Ertl, G., Freund, H.-J.: Catalysis and surface science. Physics Today 1, 32–38 (1999)
Eyink, G.L., Sreenivasan, K.R.: Onsager and the theory of hydrodynamic turbulence. Rev. Mod. Phys. 78, 87–135 (2006)
Farmer, J.D.: Chaotic attractors of an infinite-dimensional dynamical system. Physica D 4, 366–393 (1982a)
Farmer, J.D.: Dimension, fractal measures, and chaotic dynamics. In: Haken, H. (ed.) Evolution of Order and Chaos. Springer Series in Synergetics, vol. 17, pp. 228–246. Springer, Berlin (1982b)
Farmer, J.D.: Information dimension and the probabilistic structure of chaos. Z. Naturforsch. A. 37, 1304–1325 (1982c)
Farmer, J.D., Crutchfield, J., Froehling, H., Packard, N., Shaw, R.: Power spectra and mixing properties of strange attractors. In: Helleman, R.H.G. (ed.) Nonlinear Dynamics. Annals New York Academy of Sciences, vol. 357, pp. 453–472. New York Academy of Sciences (1980)
Farmer, J.D., Ott, E., Yorke, J.A.: The dimension of chaotic attractors. Physica D 7, 153–180 (1983)
Farmer, J.D., Sidorowich, J.J.: Predicting chaotic dynamics. In: Kelso, J.A.S., et al. (eds.) Dynamic Patterns in Complex Systems, pp. 265–292. World Scientific, Singapore (1988)
Fatou, P.: Sur les équations fonctionelles. Bull. Soc. Math. Fr. 47, 161–271 (1919)
Fatou, P.: Sur les équations fonctionelles. Bull. Soc. Math. Fr, 48, 33–94 (1920)
Faust, G., Hutzschenreuter, P., Sekler, E.: Beiträge zur Biomechanik: Sammelband. ICA-Bericht Nr. 14, Stuttgart (1986)
Fauve, S., Libchaber, A.: Rayleigh-Bénard experiment in a low Prandtl number fluid, mercury. In: Haken, H. (ed.) Chaos and Order in Nature. Series in Synergetics, vol. 11, pp. 25–35. Springer, Berlin (1981)
Feigenbaum, M.J.: Quantitative universality for a class of nonlinear transformations. J. Stat. Phys. 19, 25–52 (1978)
Feigenbaum, M.J.: The onset spectrum of turbulence. Phys. Lett. A. 74(6), 375–378 (1979a)
Feigenbaum, M.J.: The universal metric properties of nonlinear transformations. J. Stat. Phys. 21, 669–706 (1979b)
Feigenbaum, M.J.: The transition to aperiodic behavior in turbulent systems. Comm. Math. Phys. 77, 65–88 (1980)
Feigenbaum, M.J.: Tests of the period-doubling route to chaos. In: Vidal, C., Pacault, A. (eds.) Nonlinear Phenomena in Chemical Dynamics. Springer Series in Synergetics, vol. 13, Springer, Berlin (1981)
Feigenbaum, M.J.: Universal behavior in nonlinear systems. Physica D 7, 16–39 (1983)
Feigenbaum, M.J., Kadanoff, L.P., Shenker, S.J.: Quasiperiodicity in dissipative systems: a renormalization group analysis. Physica D, 5, 370–386 (1982)
Fell, J., Fernández, G., Elger, C.E.: More than synchrony: EEG chaoticity may be necessary for conscious brain functioning. Medical Hypotheses 61, 158–160 (2003)

Fernholz, H.: Three-dimensional disturbances in a two-dimensional incompressible turbulent boundary layer. Aeron. Res. Council, Reports and Memoranda, 3368. H.M. Stationary Office, London (1964)

Feynman, R.P., Leighton, R.B., Sands, M.: Feynman Vorlesungen über Physik, Bd. I–II. Oldenbourg, München (1987)

Field, R., Burger, M. (eds.): Oscillations and Traveling Waves in Chemical Systems. Wiley, New York (1985)

Field, R.J., Noyes, R.M.: Oscillations in chemical systems. IV. Limit cycle behavior in a model of a real chemical reaction. J. Chem. Phys. 60, 1877–1884 (1974)

Firle, S.O., Natiello, M.A.: M.Eiswirth, Topological dynamics in a catalysis experiment. Phys. Rev. E. 53, 1257–1260 (1996)

Fisz, M.: Wahrscheinlichkeitsrechnung und mathematische Statistik. VEB Deutscher Verlag der Wissenschaften, Berlin (1976)

FitzHugh, R.: Impulses and physiological states in theoretical models of nerve membrane. Biophys. J. 1, 445–466 (1961)

Floquet, G.: Sur les équations differentielles linéaires à coefficients périodiques. Ann. Sci. École Norm. Sup. Ser. 2 12, 47 (1883)

Foale, S., Thompson, J.M.T.: Geometrical concepts and computational techniques of nonlinear dynamics. Comp. Meth. Appl. Mech. Engng. 89, 381–394 (1991)

FracLab, A fractal analysis toolbox for signal and image processing. INRIA Saclay, Orsay, France. (2012), `http://fraclab.saclay.inria.fr/`

Fradkov, A.L., Evans, R.J.: Control of chaos: Methods and applications in engineering. Ann. Rev. Control 29, 33–56 (2005)

Fraedrich, K.: Estimating the dimensions of weather and climate attractors. J. Atmos. Sci. 43, 419–432 (1986)

Fraser, A.M.: Information and entropy in strange attractors. IEEE Transactions on Information Theory 35(2), 245–262 (1989a)

Fraser, A.M.: Reconstructing attractors from scalar time series: a comparison of singular system and redundancy criteria. Physica D 34, 391–404 (1989b)

Fraser, S., Kapral, R.: Analysis of flow hysteresis by a one-dimensional map. Phys. Rev. A. 25(6), 3223–3233 (1982)

Frederickson, P., Kaplan, J.L., Yorke, E.D., Yorke, J.A.: The Liapunov dimension of strange attractors. J. Diff. Equat. 49, 185–207 (1983)

Freire, J.G., Gallas, J.A.: Stern-Brocot trees in cascades of mixed-mode oscillations and canards in the extended Bonhoeffer-van der Pol and the FitzHugh-Nagumo models of excitable systems. Phys. Lett. A. 4375, 1097–1103 (2011a)

Freire, J.G., Gallas, J.A.: Stern-Brocot trees in the periodicity of mixed-mode oscillations. Phys. Chem. Chem. Phys. 13, 12191–12198 (2011b)

Friedrich, J., Friedrich, R.: Generalized vortex model for the inverse cascade of two-dimensional turbulence. Phys. Rev. E. 88, 053017 (2013)

Friedrich, R.: Stationäre, wellenartige und chaotische Konvektion in Geometrien mit Kugelsymmetrie. Dissertation, Univ. Stuttgart (1986)

Friedrich, R.: Statistics of Lagrangian velocities in turbulent flows. Phys. Rev. Lett. 90(8), 084501 (2003)

Friedrich, R., Peinke, J.: Description of a turbulent cascade by a Fokker-Planck equation. Phys. Rev. Lett. 78(5), 863–866 (1997)

Friedrich, R., Peinke, J., Sahimi, M.: Reza Rahimi Tabar, M, Approaching complexity by stochastic methods: From biology to turbulence. Physics Reports 506, 87–162 (2011)

Friedrich, R., Peinke, J., Tabar, M.R.R.: Importance of fluctuations: Complexity in the view of stochastic processes. In: Meyers, B. (ed.) Encyclopedia of Complexity and System Science, pp. 3574–3597. Springer, Berlin (2009)

Friedrich, R., Siegert, S., Peinke, J., Lück, S., Siefert, M., Lindemann, M., Raethjen, J., Deuschl, G., Pfister, G.: Extracting model equations from experimental data. Phys. Lett. A. 271, 217–222 (2000)

Friedrich, R., Uhl, C.: Spatio-temporal analysis of human electroencephalograms: Petit-mal epilepsy. Physica D. 171–182 (1996)

Frisch, U.: Turbulence, The Legacy of A.N. Kolmogorov. Cambridge University Press, Cambridge (1995)

Fröhlich, J.: Large Eddy Simulation turbulenter Strömungen. Teubner, Wiesbaden (2006)

Fuchs, A., Haken, H.: Pattern recognition and associative memory as dynamical processes in a synergetic system I + II, Erratum. Biol. Cybern. 60, 17–22, 107–109, 476 (1988)

Gallas, J.A.C.: Dissecting shrimps: results for some one-dimensional physical models. Physica A. 202, 196–223 (1994)

Gallas, J.A.C.: Structure of the parameter space of a ring cavity. Appl. Phys. B. 60, 203–213 (1995)

Galvanetto, U.: Flexible control of chaotic stick-slip mechanical systems. Comput. Methods Appl. Mech. Engrg. 190, 6075–6087 (2001)

Gardiner, C.W.: Handbook of Stochastic Methods for Physics, Chemistry and the Natural Sciences. Springer, Berlin (2004)

Garfinkel, A., Spano, M.L., Ditto, W.L., Weiss, J.N.: Controlling cardiac chaos. Science 257, 1230–1235 (1992)

Gauthier, D.J.: Resource Letter: CC-1: Controlling Chaos. Amer. J. Phys. 71(8), 750–759 (2003)

Ge, Z.-M., Lin, T.-N.: Chaos, chaos control and synchronization of electro-mechanical gyrostat system. J. Sound Vibr. 259(3), 585–603 (2003)

Gibiat, V., Castellengo, M.: Period doubling occurences in wind instruments musical performance. Acoustica 86(3), 746–754 (2000)

Gierer, A., Meinhardt, H.: A theory of biological pattern formation. Kybernetik 12, 30–39 (1972)

Gillies, A.W.: On the transformation of singularities and limit cycles of the variational equations of van der Pol. Quart. J. Mech. Appl. Math. 7, 152–167 (1954)

Gilmore, R.: Topological analysis of chaotic dynamical systems. Rev. Mod. Phys. 70, 1455–1529 (1998)

Glass, L., Guevara, M.R., Shrier, A., Perez, R.: Bifurcation and chaos in a periodically stimulated cardiac oscillator. Physica D, 7, 89–101 (1983)

Glass, L., Perez, R.: Fine structure of phase locking. Phys. Rev. Lett., 48(26), 1772–1775 (1982)

Glazier, J.A., Libchaber, A.: Quasi-periodicity and dynamical systems: an experimentalist's view. IEEE Transactions on Circuits and Systems 35(7), 790–809 (1988)

Glendinning, P.: Stability, Instability and Chaos: An introduction to the theory of nonlinear differential equations. Cambridge University Press, Cambridge (1999)

Glendinning, P., Sparrow, C.: Local and global behavior near homoclinic orbits. J. Stat. Phys. 35(5/6), 645–696 (1984)

Gnedenko, B.W., Kolmogorov, A.N.: Grenzverteilungen von Summen unabhängiger Zufallsgröß en. Akademie-Verlag, Berlin (1960)

Goldbeter, A.: Biochemical Oscillations and Cellular Rhythms: The Molecular Bases of Periodic and Chaotic Behavior. Cambridge University Press, Cambridge (1996)

Goldstein, H.: Klassische Mechanik. Akadem.Verlagsgesellschaft, Wiesbaden (1978)

Goldstine, H.H., von Neumann, J.: On the principles of large scale computing machines, unpublished). In: Taub, A.H. (ed.) John von Neumann: Collected Works, Band V, pp. 1–32. Pergamon Press, Oxford (1963)

Gollub, J.P., Benson, S.V.: Many routes to turbulent convection. J. Fluid Mech. 100(3), 449–470 (1980)
Gollub, J.P., Swinney, H.L.: Onset of turbulence in a rotating fluid. Phys. Rev. Lett. 35(14), 927–930 (1975)
Golub, G.H., van Loan, C.F.: Matrix Computations. North Oxford Academic, Oxford (1983)
Gouesbet, G., Rozé, C., Meunier-Guttin-Cluzel, S.: Instabilities by local heating below an interface. J. Non-Equilib. Thermodyn. 25, 337–379 (2000)
Gradshteyn, I.S., Ryzhik, I.M.: Table of Integrals, Series and Products. Academic Press, New York (1965)
Graham, R.: Ein Stück unberechenbare Natur: die Turbulenz. Bild der Wissenschaft 19(4), 68–82 (1982)
Granero, M.I., Porati, A., Zanacca, D.: A bifurcation analysis of pattern formation in a diffusion governed morphogenetic field. J. Math. Biology 4, 21–27 (1977)
Grassberger, P.: Do climatic attractors exist? Nature, Letters to Nature 323, 609–612 (1986)
Grassberger, P.: An optimized box-assisted algorithm for fractal dimensions. Phys. Lett. A. 148, 63–68 (1990)
Grassberger, P., Procaccia, I.: Characterization of strange attractors. Phys. Rev. Lett. 50, 346–349 (1983a)
Grassberger, P., Procaccia, I.: Estimation of the Kolmogorov entropy from a chaotic signal. Phys. Rev. A. 28, 2591–2593 (1983b)
Grassberger, P., Procaccia, I.: Measuring the strangeness of strange attractors. Physica D 9, 189–208 (1983c)
Grassberger, P., Procaccia, I.: Dimensions and entropies of strange attractors from a fluctuating dynamics approach. Physica D 13, 34–54 (1984)
Grebogi, C., Ott, E., Yorke, J.A.: Chaotic attractors in crisis. Phys. Rev. Lett. 48, 1507–1510 (1982)
Grebogi, C., Ott, E., Yorke, J.A.: Crises, sudden changes in chaotic attractors, and transient chaos. Physica D 7, 181–200 (1983)
Green, D., Unruh, W.G.: The failure of the Tacoma Bridge: A physical model. Am. J. Phys. 74, 706–716 (2006)
Greene, J.M.: A method for determining a stochastic transition. J. Math. Phys. 20, 1183–1201 (1979)
Greenside, H.S., Wolf, A., Swift, J., Pignataro, T.: Impracticality of a box-counting algorithm for calculating the dimensionality of strange attractors. Phys. Rev. A, 25(6), 3453 – 3456 (1982)
Groß mann, S, Selbstähnlichkeit: das Strukturgesetz im und vor dem Chaos. In: Gerok, W. (ed.), Ordnung und Chaos in der unbelebten und belebten Natur, 101–122. Wiss. Verlagsgesellschaft, Stuttgart (1989); Zugl. Phys. Bl. 45(6), 172-180
Groß mann, S., Thomae, S, Invariant distributions and stationary correlation functions of one-dimensional discrete processes. Z. Naturforsch. A. 32, 1353–1363 (1977)
Guckenheimer, J.: Dynamics of the van der Pol equation. IEEE Trans. Circuits Syst. 27, 983–989 (1980)
Guckenheimer, J., Holmes, P.: Nonlinear Oscillations, Dynamical Systems, and Bifurcations of Vector Fields. Springer, New York (1983)
Guillemin, V., Pollack, A.: Differential Topology. Prentice-Hall, Englewood Cliffs (1974)
Haase, M., Lehle, B.: Tracing the skeleton of wavelet transform modulus maxima lines for the characterization of fractal distributions. In: Novak, M. (ed.), Fractals and Beyond, 241–250. World Scientific, Singapore (1998)
Haase, M., Widjajakusuma, J.: Damage identification based on ridges and maxima lines of the wavelet transform. Int. J. Eng. Sci. 41, 1423–1443 (2003)

Haken, H.: Cooperative phenomena in systems far from thermal equilibrium and in nonphysical systems. Rev. Mod. Phys. 47, 67–121 (1975a)
Haken, H.: Generalized Ginzburg-Landau equations for phase transition-like phenomena in lasers, nonlinear optics, hydrodynamics and chemical reactions. Z. Phys. 21, 105–114 (1975b)
Haken, H.: Higher order corrections to generalized Ginzburg-Landau equations of nonequilibrium systems. Z. Phys. 22, 69–72 (1975c)
Haken, H.: Erfolgsgeheimnisse der Natur. DVA, Stuttgart (1981)
Haken, H.: At least one Lyapunov exponent vanishes if the trajectory of an attractor does not contain a fixed point. Phys. Lett. A 94(2), 71–72 (1983)
Haken, H.: Advanced Synergetics, 2nd edn. Springer Series in Synergetics, vol. 20. Springer, Berlin (1987)
Haken, H.: Information and Self-Organization, 3rd edn. Springer Series in Synergetics, vol. 40. Springer, Berlin (1988)
Haken, H.: Synergetik: eine Einführung, 3rd edn. Springer, Berlin (1990)
Haken, H.: Synergetic Computers and Cognition, 2nd edn. Springer, Berlin (2004)
Haken, H.: Brain Dynamics: An Introduction to Models and Simulations, 2nd edn. Springer, Berlin (2008)
Haken, H., Graham, R.: Synergetik - die Lehre vom Zusammenwirken. Umschau 71(6), 191–195 (1971)
Haken, H., Olbrich, H.: Analytical treatment of pattern formation in the Gierer-Meinhardt model of morphogenesis. J. Math. Biology 6, 317–331 (1978)
Haken, H., Wunderlin, A.: Die Selbststrukturierung der Materie: Synergetik in der unbelebten Welt. Vieweg, Braunschweig (1991)
Hale, J.K., Koçak, H.: Dynamics and Bifurcations. Springer, New York (1991)
Haller, G.: Distinguished material surfaces and coherent structures in three-dimensional fluid flows. Physica D 149, 248–277 (2001)
Haller, G., Yuan, G.: Lagrangian coherent structures and mixing in two-dimensional turbulence. Physica D 147, 352–370 (2000)
Halsey, T.C., Jensen, M.H., Kadanoff, L.P., Procaccia, I., Shraiman, B.I.: Fractal measures and their singularities: the characterization of strange sets. Phys. Rev. A 33(2), 1141–1151 (1986)
Hamel, G.: Spiralförmige Bewegungen zäher Flüssigkeiten. Jahresber. Dtsch. Math. Ver. 25, 34–60 (1916)
Hänggi, P., Thomas, H.: Stochastic processes: Time evolution, symmetries and linear response. Phys. Rep. 88(4), 207–319 (1982)
Harding, R.H., Ross, J.: The effect of different types of noise on some methods of distinguishing chaos from periodic oscillations. J. Chim. Phys. 84, 1305–1313 (1987)
Harding, R.H., Ross, J.: Symptoms of chaos in observed oscillations near bifurcation with noise. J. Chem. Phys. 89, 4743–4751 (1988)
Hardy, G.H., Wright, E.M.: An Introduction to the Theory of Numbers. Oxford University Press, Oxford (1979)
Hartman, P.: Ordinary Differential Equations. Wiley, New York (1964)
Hassler, C.R., Rybicki, E.F., Cummings, K.D., Lynn, C.C.: Quantification of bone stresses during remodeling. J. Biomechanics 13, 185–190 (1980)
Hausdorff, F.: Dimension und äuß eres Maß. Math. Annalen 79, 157–179 (1918)
Hauser, M.: Nichtlineare Dynamik in natürlichen und biomimetischen Enzym-Systemen. Logos, Berlin. Habilitation thesis (2005)
Hayes, S., Grebogi, C., Ott, E.: Communicating with chaos. Phys. Rev. Lett. 70, 3031–3034 (1993)

Heisenberg, W.: Über quantentheoretische Umdeutung kinematischer und mechanischer Beziehungen. Z. Phys., 33, 879–893 (1925)

Heisenberg, W.: Über den anschaulichen Inhalt der quantentheoretischen Kinematik. Z. Phys. 43, 172–198 (1927)

Heisenberg, W.: Zur statistischen Theorie der Turbulenz. Z. Phys. 124, 628–657 (1948)

Heisenberg, W.: Der Teil und das Ganze. Piper, München (1969)

Helwig, P., Faust, G., Hindenlang, U., Hirschmüller, A., Konstantinidis, L., Bahrs, C., Südkamp, N., Schneider, R.: Finite element analysis of four different implants inserted in different positions to stabilize an idealized trochanteric femoral fracture. Injury 40, 288–295 (2009)

Hénon, M.: Numerical study of quadratic area-preserving mappings. Quart. Appl. Math. 27, 291–312 (1969)

Hénon, M.: A two-dimensional mapping with a strange attractor. Commun. Math. Phys. 50, 69–77 (1976)

Hénon, M.: Numerical exploration of Hamiltonian systems. In: iOOSS, G., Helleman, R., Stora, R. (eds.), Chaotic Behaviour of Deterministic Systems, 53–170. North-Holland, Amsterdam (1983)

Hénon, M., Pomeau, Y.: Two Strange Attractors with a Simple Structure. Lecture Notes in Mathematics, vol. 565, pp. 29–68. Springer, Berlin (1976)

Henry, D.: Geometric Theory of Semilinear Parabolic Equations. Lecture Notes in Mathematics, vol. 840. Springer, New York (1981)

Hentschel, H.G.E., Procaccia, I.: The infinite number of generalized dimensions of fractals and strange attractors. Physica D 8, 435–444 (1983)

Herman, M.R.: Mesure de Lebesgue et nombre de rotation. In: Palis, J. (ed.) Geometry and Topology. Lecture Notes in Mathematics, pp. 271–293. Springer, Berlin (1977)

Hinze, J.O.: Turbulence. McGraw-Hill, New York (1975)

Hirsch, M.W., Smale, S.: Differential Equations, Dynamical Systems, and Linear Algebra. Academic Press, San Diego (1974)

Holmes, P.J.: A nonlinear oscillator with a strange attractor. Phil. Trans. Roy. Soc. A. 292, 419–448 (1979)

Holmes, P.J.: Averaging and chaotic motions in forced oscillations. SIAM J. Appl. Math. 38(1), 65–80 (1980)

Holmes, P.J.: Poincaré, celestial mechanics, dynamical systems theory and "chaos". Phys. Rep.193(3), 137–163 (1990)

Holmes, P.J., Marsden, J.E.: Bifurcation to divergence and flutter in flow-induced oscillations: An infinite dimensional analysis. Automatica 14, 367–384 (1978)

Holmes, P.J., Rand, D.A.: Bifurcations of the forced van der Pol oscillator. Quart. Appl. Math. 35, 495–509 (1978)

Holzfuss, J., Lauterborn, W.: Lyapunov exponents from a time series of acoustic chaos. Phys. Rev. A. 39, 2146–2152 (1989)

Honerkamp, J.: Stochastische Dynamische Systeme. Konzepte, numerische Methoden, Datenanalysen. VCH Verlagsgesellschaft, Weinheim (1990)

Hopf, E.: Abzweigung einer periodischen Lösung von einer stationären Lösung eines Differentialsystems. Ber. Math.-Phys., Sächs. Akad. d. Wiss. Leipzig 94, 1-22 (1942)

Hopf, E.: A mathematical example displaying features of turbulence. Commun. on Pure and Appl. Math. 1, 303–322 (1948)

Hsu, C.S.: A theory of cell-to-cell mapping dynamical systems. J. Appl. Mech. 47, 931–939 (1980)

Hsu, C.S.: Cell-to-Cell Mapping. Springer, New York (1987)

Hu, B., Rudnick, J.: Exact solutions of the Feigenbaum renormalization-group equations for intermittency. Phys. Rev. Lett. 48(24), 1645–1648 (1982)

Hu, B., Satija, I.I.: A spectrum of universality classes in period doubling and period tripling. Phys. Lett. A 98(4), 143–146 (1983)

Hübler, A.W.: Beschreibung und Steuerung nichtlinearer Systeme. Dissertation, Techn. Univ. München (1987)

Hübler, A.W., Georgii, R., Kuchler, M., Stelzl, W., Lüscher, E.: Resonant stimulation of nonlinear damped oscillators by Poincaré maps. Helv. Phys. Acta 61, 897–900 (1988)

Hudson, J.L., Tsotsis, T.T.: Electrochemical reaction dynamics: A review. Chem. Eng. Sci. 49, 1493–1572 (1994)

Hunt, F., Sullivan, F.: Efficient algorithms for computing fractal dimensions. In: Mayer-Kress, G. (ed.) Dimensions and Entropies in Chaotic Systems. Springer Series in Synergetics, vol. 32, pp. 74–81. Springer, Berlin (1986)

Hunt, F., Sullivan, F.: How to estimate capacity dimension. Nucl. Phys. B (Proc. Suppl.) 5, 125–128 (1988)

Imbihl, R., Cox, M.P., Ertl, G., Müller, H., Brenig, W.: Kinetic oscillations in the catalytic CO oxidation on Pt(100): theory. J. Chem. Phys. 83, 1578–1587 (1985)

Imbihl, R., Ertl, G.: Oscillatory kinetics in heterogeneous catalysis. Chem. Rev. 95, 697–733 (1995)

Jaeger, N.I., Möller, K., Plath, P.J.: Cooperative effects in heterogeneous catalysis; Part I: Phenomenology of the dynamics of carbon monoxide on palladium embedded in a zeolite matrix. J. Chem. Soc., Faraday Trans. 82, 3315–3330 (1986)

Jakubith, S., Rotermund, H.H., Engel, W., Oertzen, A., Ertl, G.: Spatiotemporal concentration patterns in a surface reaction. Phys. Rev. Lett. 65, 3013–3016 (1990)

Jensen, M.H., Bak, P., Bohr, T.: Complete devil's staircase, fractal dimension, and universality of mode-locking structure in the circle map. Phys. Rev. Lett. 50(21), 1637–1639 (1983)

Jensen, M.H., Bak, P., Bohr, T.: Transition to chaos by interaction of resonances in dissipative systems. I. Circle maps. Phys. Rev. A 30(4), 1960–1969 (1984)

Jensen, M.H., Kadanoff, L.P., Libchaber, A., Procaccia, I., Stavans, J.: Global universality at the onset of chaos: results of a forced Rayleigh-Bénard experiment. Phys. Rev. Lett. 55(25), 2798–2801 (1985)

Jensen, R.V., Myers, C.R.: Images of critical points of nonlinear maps. Phys. Rev. A. 32(2), 1222–1224 (1985)

Jiménez, J., Wray, A.A.: On the characteristics of vortex filaments in isotropic turbulence. J. Fluid Mech. 373, 255–285 (1998)

Joseph, D.D.: Hydrodynamic, stability and bifurcation. In: Swinney, H.L., Gollub, J.P. (eds.) Hydrodynamic Instabilities and the Transition to Turbulence. Topics in Applied Physics, vol. 45, pp. 27–76. Springer, Berlin (1985)

Jovanović, J.: The Statistical Dynamics of Turbulence. Springer, Berlin (2004)

Julia, G.: Sur l'iteration des fonctions rationnelles. J. de Math. Pure et Appl. 8, 47–245 (1918)

Jumarie, G.: Relative Information. Springer Series in Synergetics, vol. 47. Springer, Berlin (1990)

Just, W., Benner, H., Reibold, E.: Theoretical and experimental aspects of chaos control by time-delayed feedback. Chaos 13(1), 259–266 (2003)

Kadanoff, L.P.: Scaling laws for Ising models near TKadanoff, L.P.: Scaling laws for Ising models near.T. Physics 2, 263–272 (1966)

Kamps, O., Friedrich, R.: Langrangian statistics in forced two-dimensional turbulence. Phys. Rev. E. 78, 036321 (2008)

Kantz, H., Schreiber, T.: Nonlinear Time Series Analysis. Cambridge Nonlinear Science Series, vol. 7. Cambridge University Press, Cambridge (1997)

Kaplan, J.L., Yorke, J.A.: Chaotic behaviour of multidimensional difference equations. In: Peitgen, H.-O., Walther, H.O. (eds.) Functional Differential Equations and Approximation of Fixed Points. Lecture Notes in Mathematics, vol. 730, pp. 204–227. Springer, Berlin (1979a)

Kaplan, J.L., Yorke, J.A.: Preturbulence: a regime observed in a fluid flow model of Lorenz. Commun. Math. Phys. 67, 93–108 (1979b)

Karlsson, L.: Deterministisches und chaotisches Verhalten eines Plattenstreifens im Flatterzustand. Diploma thesis, Inst. für Computer-Anwendungen (ICA), Univ. Stuttgart (1991)

Kelso, J.A.S., Fuchs, A.: Self-organizing dynamics of the human brain: Critical instabilities and Sil'nikov chaos. Chaos, 5, 64–69 (1995)

Khanin, K.M.: Quasi-periodic motions of vortex systems. Physica D, 4, 261–269 (1982)

King, G.P., Jones, R., Broomhead, D.S.: Phase portraits from a time series: A singular system approach. Nucl. Phys. B, 2, 379–390 (1987)

Kittel, A., Parisi, J., Pyragas, K.: Delayed feedback control of chaos by self-adapted delay time. Phys. Lett. A. 198, 433–436 (1995)

Kivshar, Y.S., Rödelsperger, F., Brenner, H.: Suppression of chaos by nonresonant parametric perturbations. Phys. Rev. E. 49(1), 319–324 (1994)

Kleinhans, D.: Stochastische Modellierung komplexer Systeme. Von den theoretischen Grundlagen zur Simulation atmosphärischer Windfelder. Dissertation, Westfälische Wilhelms-Univ. Münster (2008)

Kloeden, P.E., Platen, E.: Numerical Solution of Stochastic Differential Equations. Springer, Berlin (1999)

Knudsen, J.: Flutter of a skew panel subject to in-plane forces and in-plane oscillations under hypersonic flow using the finite element method. Diploma thesis, Inst. für Computer-Anwendungen (ICA), Univ. Stuttgart (1993)

Kocarev, L.: Chaos-based cryptography: a brief overview. Circuits and Systems Magazine. IEEE 1(3), 6–21 (2001)

Kolmogorov, A.N.: Grundbegriffe der Wahrscheinlichkeitsrechnung. Springer, Berlin (1933)

Kolmogorov, A.N.: The local structure of turbulence in incompressible viscous fluid for very large Reynolds number. Comptes Rendues Dokl. Akad. Sci. USSR, 30, 301–305 (1941)

Kolmogorov, A.N.: On conservation of conditionally periodic motions under small perturbations of the Hamiltonian function. Dokl. Akad. Nauk USSR 98(4), 527–530 (1954)

Kolmogorov, A.N.: A new metric invariant of transient dynamical systems and automorphisms of Lesbesgue spaces. Dokl. Akad. Nauk SSSR 119, 861–864 (1958)

Kolmogorov, A.N.: A refinement of previous hypotheses concerning the local structure in turbulence in a viscous incompressible fluid at high Reynolds number. J. Fluid Mech. 13, 82–85 (1962)

Korteweg, D.J., de Vries, G.: On the change of form of long waves advancing in a rectangular canal, and on a new type of long stationary waves. Philos. Mag. Ser. 5. 39, 422–443 (1895)

Koschmieder, E.L.: Bénard convection. Adv. Chem. Phys. 26, 177–212 (1974)

Kraichnan, R.H.: Inertial ranges in two-dimensional turbulence. Phys. Fluids 10, 1417–1423 (1967)

Kreuzer, E.: Numerische Untersuchung nichtlinearer dynamischer Systeme. Springer, Berlin (1987)

Kreyszig, E.: Statistische Methoden und ihre Anwendungen. Vandenhoeck & Ruprecht, Göttingen (1965)

Krischer, K.: Nichtlineare Dynamik zweier Grenzflächenreaktionen und deterministisches Chaos. Dissertation, Freie Univ. Berlin (1990)

Krischer, K.: Principles of temporal and spatial pattern formation in electrochemical systems. In: Conway, B.E. (ed.) Modern Aspects of Electrochemistry, vol. 32, pp. 1–142. Academic/Plenum Publishers, New York (1999)

Krischer, K.: Spontaneous formation of spatiotemporal patterns at the electrode | electrolyte interface. J. Electroanal. Chem. 501, 1–21 (2001)

Krischer, K., Eiswirth, M., Ertl, G.: Oscillatory CO oxidation on Pt(110): modeling of temporal self-organization. J. Chem. Phys. 96(12), 9161–9172 (1992)

Krischer, K., Lübke, M., Eiswirth, M., Wolf, W., Hudson, J.L., Ertl, G.: A hierarchy of transitions to mixed mode oscillations in an electrochemical system. Physica D 62(1), 123–133 (1993)

Krischer, K., Lübke, M., Wolf, W., Eiswirth, M., Ertl, G.: Chaos and interior crisis in an electrochemical reaction. Ber. Bunsenges. Phys. Chem. 95, 820-823 (1991)

Krischer, K., Lübke, M., Wolf, W., Eiswirth, M., Ertl, G.: Oscillatory dynamics of the electrochemical oxidation of H_2 in the presence of Cu^{2+}: Structure sensitivity and the role of anions. Electrochim. Acta 40(1), 69–81 (1995)

Kruel, T.-M., Freund, A., Schneider, F.W.: The effect of interactive noise on the driven Brusselator model. J. Chem. Phys. 93, 416–427 (1990)

Krylov, N.M., Bogoliubov, N.N.: Introduction to Nonlinear Mechanics. Princeton Univ. Princeton (1947)

Kubo, R.: Thermodynamics: An Advanced Course with Problems and Solutions. North-Holland, Amsterdam (1968)

Kugiumtzis, D.: State space reconstruction parameters in the analysis of chaotic time series – the role of the time window length. Physica D, 95, 13–28 (1996)

Kummer, B., Lohscheidt, K.: Mathematische Modelle zur Analyse des Aussagewertes biologischer Theorien. In: Band zum 3. Kölner biomechanischen Colloquium, 5. und 6 (October 1984)

Kuramoto, Y.: Chemical Oscillations, Waves and Turbulence. Springer Series in Synergetics, vol. 19. Springer, Berlin (1984)

Kusumoto, K., Ohtsubo, J.: 1.5-GHz message transmission based on synchronization of chaos in semiconductor lasers. Opt. Lett. 27, 989–991 (2002)

Kuznetsov, Y.: Elements of Applied Bifurcation Theory. Applied Mathematical Sciences, vol. 112. Springer, New York (1998)

La Porta, A., Voth, G.A., Crawford, A.M., Alexander, J., Bodenschatz, E.: Fluid particle accelerations in fully developed turbulence. Nature 409, 1017–1019 (2001)

Ladas, S., Imbihl, R., Ertl, G.: Kinetic oscillations and facetting during the catalytic CO oxidation on Pt(110). Surf. Sci. 198, 42–68 (1988a)

Ladas, S., Imbihl, R., Ertl, G.: Microfacetting of a Pt(110) surface during catalytic CO oxidation. Surf. Sci.197, 153–182 (1988b)

Lagrange, J.L.: Méchanique Analytique. Desaint, Paris (1788)

Lai, Y.-C., Kandangath, A., Krishnamoorthy, S., Gaudet, J.A., de Moura, A.P.S.: Inducing chaos by resonant perturbations: Theory and experiment. Phys. Rev. Lett. 94, 214101 (2005)

Landau, D.P., Binder, K.: A Guide to Monte Carlo Simulations in Statistical Physics, 2nd edn. Cambridge University Press, Cambridge (2005)

Landau, L.D.: On the problem of turbulence. Comptes Rendues Dokl. Acad. Sci. USSR 44, 311 (1944)

Landau, L.D., Lifschitz, E.M.: Lehrbuch der theoretischen Physik, Band VI, Hydrodynamik, 5th edn. Akademie-Verlag, Berlin (1991)

Lanford, O.E.: Computer pictures of the Lorenz attractor. In: Bernard, P., Ratiu, T. (eds.) Turbulence Seminar. Lecture Notes in Mathematics, vol. 615, pp. 113–116. Springer, Berlin (1977)

Lanford, O.E.: Strange attractors and turbulence. In: Swinney, H.L., Gollub, J.P. (eds.) Hydrodynamic Instabilities and the Transition to Turbulence. Topics in Applied Physics, vol. 45, pp. 7–26. Springer, Berlin (1981)

Laplace, J.L.: Théorie Analytique des Probabilités. Courcier, Paris (1812)

Larter, R.: Oscillations and spatial nonuniformities in membranes. Chem. Rev. 90, 355–381 (1990)

Lathrop, D.P., Kostelich, E.J.: Characterization of an experimental strange attractor by periodic orbits. Phys. Rev. A, 40, 4028–4031 (1989)

Lehle, B.: Analysis of stochastic time series in the presence of strong measurement noise. Phys. Rev. E. 83, 021113 (2011)

Lekien, F., Coulliette, C., Mariano, A.J., Ryan, E.H., Shay, L.K., Haller, G., Marsden, J.: Pollution release tied to invariant manifolds: A case study for the coast of Florida. Physica D 210, 1–20 (2005)

Lesieur, M., Yaglom, A., David, P. (eds.): New Trends in Turbulence. Les Houches Summer School, Springer, Berlin (2001)

Leven, D.W., Koch, B.-P., Pompe, B.: Chaos in dissipativen Systemen. Vieweg, Braunschweig (1989)

Levinson, N.: A second-order differential equation with singular solutions. Ann. Math. 50, 127–153 (1949)

Lévy, P.: Théorie de l'addition des variables aléatoires. Gauthier-Villars, Paris (1954)

Li, J., Arneodo, A., Nekka, F.: A practical method to experimentally evaluate the Hausdorff dimension: An alternative phase-transition-based methodology. Chaos 14, 1004–1017 (2004)

Li, T.Y., Yorke, J.A.: Period three implies chaos. Am. Math. Monthly 82(10), 985–992 (1975)

Liauw, M.A., Plath, P.J., Jaeger, N.I.: Complex oscillations and global coupling during the catalytic oxidation of CO. J. Chem. Phys. 104(16), 6375–6386 (1996)

Libchaber, A.: From chaos to turbulence in Bénard convection. Proc. R. Soc. London A. 413, 63–69 (1987)

Libchaber, A., Maurer, J.: Local probe in a Rayleigh-Bénard experiment in liquid helium. J. Phys. Lett. 39, 369–372 (1978)

Libchaber, A., Maurer, J.: Unc expérience de Rayleigh-Bénard de géometrie réduite: multiplication, accrochage et démultiplication de fréquences. J. Phys. Coll. 41, 51–56 (1980)

Libchaber, A., Maurer, J.: A Rayleigh-Bénard experiment: Helium in a small box. In: Riste, T. (ed.), Nonlinear Phenomena at Phase Transitions and Instabilities, 259–286. Plenum Press, New York (1982)

Lichtenberg, A.J., Lieberman, M.A.: Regular and Stochastic Motion. Springer, New York (1983)

Lighthill, M.J.: Einführung in die Theorie der Fourier-Analyse und der verallgemeinerten Funktionen. Bibl. Inst., Mannheim (1966)

Lima, R., Pettini, M.: Suppression of chaos by resonant parametric perturbations. Phys. Rev. A. 41(2), 726–733 (1990)

Lindell, P.-O.: A numerical excursion into chaos. Diplomarbeit, Inst. für Computer-Anwendungen (ICA), Univ. Stuttgart (1988)

Lipowsky, R.: Die Renormierung in der statistischen Physik. Phys. Bl. 39(12), 387–393 (1983)

Lord Kelvin Thomson, W, On vortex motion. Trans. Roy. Soc. 25, 217–260 (1869)

Lorenz, E.N.: Deterministic non-periodic flow. J. Atmos. Sci. 20, 130–141 (1963)

Lück, S., Renner, C., Peinke, J., Friedrich, R.: The Markov-Einstein length - a new meaning for the Taylor length in turbulence. Phys. Lett. A. 359, 335–338 (2006)

Lundgren, T.S.: Strained spiral vortex model for turbulent fine structure. Phys. Fluids, 25(12), 2193–2203 (1982)

Lundgren, T.S.: A small-scale turbulence model. Phys. Fluids A. 5(6), 1472–1483 (1993)
Lyapunov, A.M.: Problème général de stabilité du mouvement (French trans, 17, Ann. of Math. Studies. Princeton Univ. Press, Princeton (1892)
MacKay, R.S., Tresser, C.: Transition to topological chaos for circle maps. Physica D 19, 206–237 (1986)
Mackey, M.C., Glass, L.: Oscillation and chaos in physiological control systems. Science 197, 287–289 (1977)
Mallat, S.: A Wavelet Tour of Signal Processing. Academic Press, San Diego (1998)
Malraison, B., Atten, P., Bergé, P., Dubois, M.: Dimension d'attracteurs étranges: une détermination expérimentale en régime chaotique de deux systèmes convectifs. Comptes Rendus de l'Académie des Sciences de Paris C 297, 209–214 (1983)
Mandelbrot, B.: The variation of certain speculative prices. The Journal of Business 36(4), 394–419 (1963)
Mandelbrot, B.B.: Fractals: Form, Chance and Dimension. Freeman, San Francisco (1977)
Mandelbrot, B.B.: Fractal aspects of the iteration of $z \rightarrow \lambda z(1-z)$ for complex λ and z. In: Helleman, R.H.G. (ed.) Nonlinear Dynamics, Annals of the New York Acad. of Sciences, vol. 357, pp. 249–259. New York Acad. of Sciences (1980)
Mandelbrot, B.B.: The Fractal Geometry of Nature. Freeman, San Francisco (1982)
Mandelbrot, B.B.: On the quadratic mapping $z \rightarrow z^2 - \mu$ for complex μ and z: the fractal structure of its $\mathcal{M}$ set, and scaling. In: Campbell, D., Rose, H. (eds.), Order in Chaos, 224–239. North-Holland, Amsterdam (1983)
Manneville, P., Pomeau, Y.: Intermittency and the Lorenz model. Phys. Lett. A, 75:1–2 (1979)
Marchioro, C., Pulvirenti, M.: Vortex Methods in Two-Dimensional Fluid Dynamics. Springer, Berlin (1984)
Markus, M., Nagy-Ungvarai, Z., Hess, B.: Phototaxis of spiral waves. Science, 257, 225–227 (1992)
Marsden, J.E., McCracken, M.: The Hopf Bifurcation and its Applications. Springer, New York (1976)
Mathur, M., Haller, G., Peacock, T., Ruppert-Felsot, J.E., Swinney, H.L.: Uncovering the Lagrangian skeleton of turbulence. Phys. Rev. Lett. 98, 144502 (2007)
Maurer, J., Libchaber, A.: Rayleigh-Bénard experiment in liquid helium: frequency locking and the onset of turbulence. J. Phys. Lett. 40(16), 419–423 (1979)
Maurer, J., Libchaber, A.: Effect of the Prandtl number on the onset of turbulence in liquid ^{4}He. J. Phys. Lett. 41, 515–518 (1980)
May, R.M.: Simple mathematical models with very complicated dynamics. Nature 261, 459–467 (1976)
Mayer-Kress, G., Haken, H.: Intermittent behavior of the logistic system. Phys. Lett. A 82(4), 151–155 (1981)
Mayer-Kress, G., Haken, H.: Transition to chaos for maps with positive Schwarzian derivative. In: Haken, H. (ed.) Evolution of Order and Chaos. Springer Series in Synergetics, vol. 17, pp. 183–186. Springer, Berlin (1982)
Meinhardt, H.: Bildung geordneter Strukturen bei der Entwicklung höherer Organismen. In: Küppers, B.-O. (ed.), Ordnung aus dem Chaos: Prinzipien der Selbstorganisation und Evolution des Lebens. Piper, München (1987)
Melnikov, V.K.: On the stability of the center for time periodic perturbations. Trans. Moscow Math. Soc. 12, 1–57 (1963)
Metropolis, N., Stein, N.L., Stein, P.R.: On finite limit sets for transformations on the unit interval. J. Combust. Theory 15, 25–44 (1973)
Mettin, R., Hübler, A., Scheeline, A.: Parametric entrainment control of chaotic systems. Phys. Rev. E. 51(5), 4065–4075 (1995)

Metzler, R., Klafter, J.: The random walk's guide to anomalous diffusion: A fractional dynamics approach. Phys. Rep. 339, 1–77 (2000)
Mikhailov, A.S.: Foundations of Synergetics I, Distributed Active Systems. Springer Series in Synergetics, vol. 51. Springer, Berlin (1990)
Milnor, J.W.: Dynamics in one Complex Variable: Introductory Lectures. Stony Brook IMS 1990/5. arXiv:math/9201272v1 (1990)
Minorsky, N.: Nonlinear Oscillations. Van Nostrand, Princeton (1962)
Misuriewicz, M.: Maps of an interval. In: Iooss, G. et al. (eds.), Chaotic Behaviour of Deterministic Systems, Les Houches Session XXXVI, 565–590. North Holland, Amsterdam (1983)
Mitsubori, K., Aihara, K.: Delayed-feedback control of chaotic roll motion of a flooded ship in waves. Proc. R. Soc. Lond. A. 458, 2801–2813 (2002)
Monin, A.S., Yaglom, A.M.: Statistical Fluid Mechanics: Mechanics of Turbulence, vol. I. Dover, New York (2007a)
Monin, A.S., Yaglom, A.M.: Statistical Fluid Mechanics: Mechanics of Turbulence, vol. II. Dover, New York (2007b)
Moon, F.C.: Experiments on chaotic motions of a forced nonlinear oscillator: strange attractors. Transactions of the ASME, J. Appl. Mech. 47, 638–644 (1980)
Moon, F.C.: Chaotic and Fractal Dynamics. John Wiley & Sons, New York (1992)
Moon, F.C., Holmes, P.J.: A magnetoelastic strange attractor. J. Sound Vib. 65, 275–296 (1979)
Moon, F.C., Li, G.-X.: Fractal basin boundaries and homoclinic orbits for periodic motion in a two-well potential. Phys. Rev. Lett. 55(14), 1439–1442 (1985)
Mordant, N., Delour, J., Léveque, E., Arneodo, A., Pinton, J.-F.: Long time correlations in Lagrangian dynamics: A key to intermittency in turbulence. Phys. Rev. Lett. 89(25), 254502 (2002)
Moser, J.: Stability of the asteroids. Astronom. Journal 63, 439–443 (1958)
Moser, J.: Convergent series expansions of quasi-periodic motions. Math. Ann. 169, 136–176 (1967)
Moser, J.: Stable and Random Motions in Dynamical Systems. Princeton University Press, Princeton (1973)
Moser, J., Siegel, C.L.: Lectures on Celestial Mechanics, rev. and enl. transl. of "'Vorlesungen über Himmelsmechanik"' by C.L.Siegel, 187, Grundlehren der math. Wiss. Springer, Berlin (1971)
Murray, J.D.: Mathematical Biology, I, II. Springer, New York (2002)
Muzy, J.F., Bacry, E., Arneodo, A.: The multifractal formalism revisited with wavelets. Int. J. Bif. Chaos 4(2), 245–302 (1994)
Myrberg, P.J.: Iteration der reellen Polynome 2. Grades, 3. Mitteilung. Ann. Acad. Sci. Fenn. 336, 1–18 (1963)
Nagumo, J., Arimoto, S., Yoshizawa, S.: An active pulse transmission line simulating nerve axon. Proc. of the IRE. 50, 2061–2070 (1962)
NASA, Photojournal NASA (1996), http://photojournal.jpl.nasa.gov
Nauenberg, M., Rudnick, J.: Universality and power spectrum at the onset of chaos. Phys. Rev. B. 24(1), 493–495 (1981)
Nekorkin, V.I., Kazantsev, V.B.: Autowaves and solitons in three-component reaction-diffusion system. Int. J. Bif.Chaos 12(11), 2421–2434 (2002)
Neufeld, Z., Tél, T.: The vortex dynamics analogue of the restricted three-body problem: advection in the field of three identical point vortices. J. Phys. A: Math. Gen. 30, 2263–2280 (1997)
Neĭmark, J.: On some cases of periodic motions depending on parameters. Dokl. Akad. Nauk. SSSR 129, 36–739 (1959)

Newhouse, S., Ruelle, D.: Taken, Occurrence of strange Axiom A attractors near quasi periodic flows on t^m, m $\geq$ 3. Commun. Math. Phys. 64, 35–40 (1978)
Newton, P.K.: The N-vortex Problem: Analytical Techniques. Springer, New York (2001)
Nicola, E.M., Bär, M., Engel, H.: Wave instability induced by nonlocal spatial coupling in a model of light-sensitive Belousov-Zhabotinsky reaction. Phys. Rev. E. 73, 066225 (2006)
Nicolis, C., Nicolis, G.: Is there a climatic attractor? Nature, 311, 529–532 (1984)
Nicolis, C., Nicolis, G.: Evidence for climatic attractors. Nature 326, 523 (1987)
Nicolis, G., Prigogine, I.: Die Erforschung des Komplexen. Piper, München (1987)
Nitsche, G., Dressler, U.: Controlling chaotic dynamical systems using time delay coordinates. Physica D 58, 153–164 (1992)
Nolan, J.P.: Numerical calculation of stable densities and distribution functions. Commun. Statist. - Stochastic Models 13, 759–774 (1997)
Normand, C., Pomeau, Y., Velarde, M.G.: Convective instability: A physicist's approach. Rev. Mod. Phys. 49(3), 581–624 (1977)
Nusse, H.E., Ott, E., Yorke, J.A.: Border-collision bifurcations: An explanation for observed bifurcation phenomena. Phys. Rev. E. 49, 1073–1076 (1994)
Oberlack, M.: Symmetrie, Invarianz und Selbstähnlichkeit in der Turbulenz. Habilitation thesis, RWTH Aachen (2000)
Oberlack, M., Busse, F.H. (eds.): Theories of Turbulence. Springer, Wien (2002)
O'Connor, J.A., Lanyon, L.E., MacFie, H.: The influence of strain rate on adaptive bone remodelling. J. Biomechanics, 15, 767–781 (1982)
Oliva, W.M.: On the chaotic behavior and the non-integrability of the four point vortices problem. Ann. Inst. Henri Poincaré 55(2), 707–718 (1991)
Olver, P.J.: Applications of Lie Groups to Differential Equation. Springer, New York (1986)
Onsager, L.: Statistical hydrodynamics. Nuovo Cimento, Suppl. Ser. IX 6(2), 279–287 (1949)
Osborne, A.R.: The generation and propagation of internal solitons in the Andaman sea. In: Fordy, A. P. (ed.), Soliton Theory: A Survey of Results, 152–173. Manchester University Press (1990)
Oseen, C.W.: Über Wirbelbewegungen in einer reibenden Flüssigkeit. Ark. f. Mat. Astron. Fys. 7(14), 1–13 (1911)
Oseledec, V.I.: A multiplicative ergodic theorem: the Lyapunov characteristic numbers of dynamical systems. Trans. Mosc. Math. Soc. 19, 197–231 (1968)
Ostlund, S., Rand, D., Sethna, J., Siggia, E.: Universal properties of the transition from quasi-periodicity to chaos in dissipative systems. Physica D 8, 303–342 (1983)
Ott, E.: Chaos in Dynamical Systems. Cambridge University Press, Cambridge (1993)
Ott, E., Grebogi, C., Yorke, J.A.: Controlling chaos. Phys. Rev. Lett. 64, 1196–1199 (1990)
Ott, E., Sommerer, J.C.: Blowout bifurcations: the occurrence of riddled basins and on-off intermittency. Phys. Lett. A188, 39–47 (1994)
Otterstedt, R.D., Jäger, N.I., Plath, P.J., Hudson, J.L.: Global coupling effects on spatiotemporal patterns on a ring electrode. Chem. Eng. Sci. 54, 1221–1231 (1999)
Packard, N.H., Crutchfield, J.P., Farmer, J.D., Shaw, R.S.: Geometry from a time series. Phys. Rev. Lett. 45(9), 712–716 (1980)
Papoulis, A.: The Fourier Integral and its Applications. McGraw Hill, New York (1962)
Papoulis, A.: Probability, Random Variables, and Stochastic Processes. McGraw-Hill, New York (1965)
Parisi, G., Frisch, U.: On the singularity structure of fully developed turbulence. In: Ghil, M., Benzi, R. (eds.) Turbulence and Predictability in Geophysical Fluid Dynamics, Proceed. Intern. School of Physics 'E. Fermi', Varenna, Italy, pp. 84–87 (1985)
Päsler, M.: Prinzipe der Mechanik. de Gruyter, Berlin (1968)

Pätzold, G.: Reduktionstechniken und asymptotische Methoden, Fallstudie: Duffing-Balken. Diploma thesis, Inst. für Computer Anwendungen (ICA), Univ. Stuttgart (1990)

Pauwels, F.: Eine neue Theorie über den Einfluß mechanischer Reize auf die Differenzierung der Stützgewebe. Z. Anat. Entwickl.-Gesch. 121, 478–515 (1960)

Pauwels, F.: Gesammelte Abhandlung zur funktionellen Anatomie des Bewegungsapparates. Springer, Berlin (1965)

Peinke, J., Parisi, J., Rössler, O.E., Stoop, R.: Encounter with Chaos: Self-Organized Hierarchical Complexity in Semiconductor Experiments. Springer, Berlin (1992)

Peinke, J., Siefert, M., Barth, S., Renner, C., Riess, F., Wächter, M., Friedrich, R.: Fat tail statistics and beyond. In: Advances in Solid State Physics, vol. 44, pp. 363–373. Springer, Berlin (2004)

Peitgen, H.-O., Jürgens, H., Saupe, D.: Bausteine des Chaos – Fraktale. Klett-Cotta/Springer, Stuttgart/Berlin (1992)

Peitgen, H.-O., Richter, P.H.: The Beauty of Fractals. Springer, Berlin (1986a)

Peitgen, H.-O., Richter, P.H.: Fraktale und die Theorie der Phasenübergänge. Phys. Bl. 42(1), 9–22 (1986b)

Peitgen, H.-O., Saupe, D. (eds.): The Science of Fractal Images. Springer, New York (1988)

Peixoto, M.M.: Structural stability on two-dimensional manifolds. Topology 1, 101–120 (1962)

Pesin, Y.B.: Characteristic Lyapunov exponents and smooth ergodic theory. Russ. Math. Surv. 32(4), 55–114 (1977)

Petrović, C.: Synchronisation eines hierarchischen Ensembles global gekoppelter selbsterregter Oszillatoren. Diploma thesis, Inst. für Theoretische Physik, Westfälische Wilhelms-Univ. Münster (2007)

Phillipson, P.E., Schuster, P.: Bistability of harmonically forced relaxation oscillations. Int. J. Bif. and Chaos 12(6), 1295–1307 (2002)

Pikovsky, A., Rosenblum, M., Kurths, J.: Synchronization. A universal concept in nonlinear sciences. Cambridge University Press, Cambridge (2003)

Piquet, J.: Turbulent Flows. Models and Physics. Springer, Berlin (1999)

Plath, P.J.: Jenseits des Moleküls. Raum und Zeit in der Chemie. Vieweg, Braunschweig (1997)

Poincaré, H.: Les méthodes nouvelles de la mécanique céleste. Gauthier-Villars, Paris (1899)

Poincaré, H.: Science et méthode. Ernest Flammarion, Paris (1908)

Pomeau, Y., Manneville, P.: Intermittent transition to turbulence in dissipative dynamical systems. Commun. Math. Phys. 74, 189–197 (1980)

Pompe, B., Wilke, C., Koch, B.P., Leven, R.W.: Experimentelle Untersuchungen periodischer und chaotischer Bewegungen eines parametrisch erregten Pendels. Exp. Technik Phys. 32, 545–554 (1984)

Pope, S.B.: Turbulent Flows. Cambridge University Press, Cambridge (2000)

Popp, K., Stelter, P.: Stick-slip vibrations and chaos. Phil. Trans. R. Soc. Lond. A. 332(1624), 89–105 (1990)

Press, W.H., Flannery, B.P., Teukolsky, S.A., Vetterling, W.T.: Numerical Recipes. Cambridge University Press, Cambridge (1986)

Prigogine, I.: Vom Sein zum Werden: Zeit und Komplexität in den Naturwissenschaften. Piper, München (1979)

Pyragas, K.: Continuous control of chaos by self-controlling feedback. Phys. Lett. A. 170, 421–428 (1992)

Rand, D., Ostlund, S., Sethna, J., Siggia, E.D.: Universal transition from quasiperiodicity to chaos in dissipative systems. Phys. Rev. Lett. 49(2), 132–135 (1982)

Rand, D.A.: Fractal bifurcation sets, renormalization strange sets and their universal invariants. Proc. Roy. Soc. London A 413, 45–61 (1987)

Rand, R.H., Armbruster, D.: Perturbation Methods, Bifurcation Theory and Computer Algebra. Springer, New York (1987)

Rayleigh, J.W.S.: The Theory of Sound. reprinted by Dover Publishers, New York (1945)

Renner, C., Peinke, J., Friedrich, R.: Experimental indications for Markov properties of small-scale turbulence. J. Fluid Mech. 433, 383–409 (2001)

Renner, C., Peinke, J., Friedrich, R., Chanal, O., Chabaud, B.: Universality of small scale turbulence. Phys. Rev. Lett. 89, 124502 (2002)

Rényi, A.: On the dimension and entropy of probability distributions. Acta Mathematica (Hungaria) 10, 193–215 (1959)

Reynolds, O.: An experimental investigation of the circumstances which determine whether the motion of water shall be direct or sinuous, and of the law a resistance in parallel channels. Phil. Trans. R. Soc. 174, 935–982 (1883)

Richardson, L.F.: Weather Prediction by Numerical Process. Cambridge University Press, Cambridge (1922)

Rimmer, R.H., Nolan, J.P.: Stable distributions in Mathematica. The Mathematica Journal 9(4), 776–789 (2005)

Ringuet, E., Rozé, C., Gouesbet, G.: Experimental observation of type-II intermittency in a hydrodynamic system. Phys. Rev. E. 47, 1405–1407 (1993)

Risken, H.: The Fokker-Planck Equation. Springer, Berlin (1996)

Romeiras, F.J., Grebogi, C., Ott, E., Dayawansa, W.P.: Controlling chaotic dynamical systems. Physica D 58, 165–192 (1992)

Rosenblum, M., Pikovsky, A.: Delayed feedback control of collective synchrony: An approach to suppression of pathological brain rhythms. Phys. Rev. E. 70, 041904 (2004)

Rössler, O.E.: An equation for continuous chaos. Phys. Lett. A. 57, 397–398 (1976)

Rössler, O.E.: Continuous chaos - four prototype equations. Ann. N.Y. Acad. Sci. 316, 376–392 (1979a)

Rössler, O.E.: An equation for hyperchaos. Phys. Lett. A. 71, 155–157 (1979b)

Rotermund, H.H., Jakubith, S., von Oertzen, A., Ertl, G.: Solitons in a surface reaction. Phys. Rev. Lett. 66, 3083–3086 (1991)

Rozman, M.G., Urbakh, M., Klafter, J.: Controlling chaotic frictional forces. Phys. Rev. E. 57, 7340–7343 (1998)

Rubin, C.T., Lanyon, L.E.: Regulation of bone formation by applied dynamic loads. J. Bone and Joint Surg. A. 66, 397–402 (1984)

Ruelle, D.: Chaotic Evolution and Strange Attractors. Cambridge University Press, Cambridge (1989)

Ruelle, D., Takens, F.: On the nature of turbulence. Commun. Math. Phys. 20, 167–192 (1971)

Sachon, M.: Bifurkationen der Kodimension 2 am Beispiel des van-der-Pol-Oszillators. Diploma thesis, Inst. für Computer-Anwendungen (ICA).Univ. Stuttgart (1991)

Sacker, R.S.: On invariant surfaces and bifurcations of periodic solutions of ordinary differential equations. Comm. Pure Appl. Math. 18,717–732 (1965)

Sagaut, P.: Large Eddy Simulation for Incompressible Flows. Springer, Berlin (2001)

Saltzman, B.: Finite amplitude free convection as an initial value problem - I. J. Atmos. Sci. 19, 329–341 (1962)

Sander, M.: Strukturumwandlungen und kinetische Oszillationen während der katalytischen CO-Oxidation an Platin-Einkristallflächen der [001]-Zone. Dissertation, Freie Univ. Berlin (1990)

Sanders, J.A., Verhulst, F., Murdock, J.: Averaging Methods in Nonlinear Dynamical Systems. Springer, Berlin (2007)

Sano, M., Sawada, Y.: Measurement of the Lyapunov spectrum from a chaotic time series. Phys. Rev. Lett. 55, 1082–1085 (1985)

Schiff, S.J., Jerger, K., Duong, D.H., Chang, T., Spano, M.L., Ditto, W.L.: Controlling chaos in the brain. Nature 370, 615–620 (1994)

Schneider, R., Faust, G., Hindenlang, U., Helwig, P.: Inhomogeneous, orthotropic material model for the cortical structure of long bones modelled on the basis of clinical ct or density data. Comp. Meth. Appl. Mech. Eng. 198, 2167–2174 (2009)

Schöll, E.: Nonlinear Spatio-Temporal Dynamics and Chaos in Semiconductors. Cambridge University Press, Cambridge (2001)

Schöll, E., Schuster, H.G. (eds.): Handbook of Chaos Control. Wiley-VCH, Weinheim (2007)

Schreiber, T.: Interdisciplinary application of nonlinear time series methods. Phys. Rep. 308, 1–64 (1999)

Schuster, H.G., Just, W.: Deterministic Chaos, 4th edn. Wiley-VCH, Weinheim (2006)

Sensse, A., Hauser, M.J.B., Eiswirth, M.: Feedback loops for Shil'nikov chaos: The peroxidase-oxidase reaction. J. Chem. Phys. 125, 014901 (2006)

Seydel, R.: The strange attractors of a Duffing equation - dependence on the exciting frequency. Tech. Univ. München, Inst. f. Math. TUM-M 8019 (1980)

Shadden, S.C., Lekien, F., Marsden, J.E.: Definition and properties of Lagrangian coherent structures from finite-time-Lyapunov exponents in two-dimensional aperiodic flows. Physica D 212, 271–304 (2005)

Shannon, C.E.: A mathematical theory of communication. Bell System Techn. J. 27, 370–423, 623–656 (1948)

Shannon, C.E., Weaver, W.: The Mathematical Theory of Communication. University of Ill. Press, Urbana (1949)

Sharkowskii, A.N.: Coexistence of cycles of a continuous map of a line into itself. Ukr. Mat. Z. 16, 61–71 (1964)

Shaw, R.: Modeling chaotic systems. In: Haken, H. (ed.) Chaos and Order in Nature, vol. 11, pp. 218–231. Springer, Berlin (1981a)

Shaw, R.: Strange attractors, chaotic behavior, and information flow. Z. Naturforsch. 36, 80–112 (1981b)

She, Z.-S., Leveque, E.: Universal scaling laws in fully developed turbulence. Phys. Rev. Lett. 72(3), 336–339 (1994)

Shenker, S.J.: Scaling behavior in a map of a circle onto itself: empirical results. Physica D, 5, 405–411 (1982)

Shenker, S.J., Kadanoff, L.P.: Critical behavior of a KAM surface: I. Empirical results. J. Stat. Phys. 27(4), 631–656 (1982)

Shilnikov, L.P.: A case of the existence of a denumerable set of periodic motions. Sov. Math. Dokl. 6, 163–166 (1965)

Shilnikov, L.P.: A contribution to the problem of the structure of an extended neighborhood of a rough equilibrium state of saddle-focus type. Math. USSR-Sb 10, 91–102 (1970)

Shilnikov, L.P., Shilnikov, A.: Shilnikov bifurcation. Scholarpedia. bifurcation/ (2007), `http://www.scholarpedia.org/article/Shilnikov`

Shimada, I., Nagashima, T.: A numerical approach to ergodic problem of dissipative dynamical systems. Progr. Theor. Phys. 61(6), 1605–1616 (1979)

Shinbrot, T., Grebogi, C., Ott, E., Yorke, J.A.: Using small perturbations to control chaos. Nature 363, 411–417 (1993)

Siefert, M., Peinke, J.: Reconstruction of the deterministic dynamics of stochastic systems. Int. J. Bif. Chaos 14(6), 2005–2010 (2004)

Siegel, C.L.: Iteration of analytic functions. Ann. Math. 43(4), 607–612 (1942)

Siegel, C.L.: Vorlesungen über Himmelsmechanik. Springer, Berlin (1956)

Simó, C.: On the Hénon-Pomeau attractor. J. Stat. Phys. 21(4), 465–494 (1979)

Sinai, Y.G.: On the concept of entropy of a dynamical system. Dokl. Akad. Nauk SSSR 124, 768–771 (1959)

Sinai, Y.G.: Introduction to Ergodic Theory, vol. 18. Math. Notes. Princeton Univ. Press, Princeton (1977)

Singer, D.: Stable orbits and bifurcation of maps of the interval. SIAM J. Appl. Math. 35, 260–267 (1978)

Singer, J., Wang, Y.-Z., Bau, H.H.: Controlling a chaotic system. Phys. Rev. Lett. 66, 1123–1125 (1991)

Smale, S.: Diffeomorphisms with many periodic points. In: Cairns, S.S. (ed.) Differential and Combinatorial Topology. Princeton University Press, Princeton (1963)

Smale, S.: Differentiable dynamical systems. Bull. Amer. Math. Soc. 73, 747–817 (1967)

Socolar, J.E.S., Sukow, D.W., Gauthier, D.J.: Stabilizing unstable periodic orbits in fast dynamical systems. Phys. Rev. E 50(4), 3245–3248 (1994)

Sommerfeld, A.: Vorlesungen über theoretische Physik, Band I, Mechanik. Akademische Verlagsgesellschaft Geest & Portig, 7. Auflage (1964)

Sommerfeld, A.: Vorlesungen über theoretische Physik, Band V, Thermodynamik und Statistik, reprint of, 2nd edn. Verlag Harri Deutsch, Thun (1977)

Sparrow, C.: The Lorenz Equations: Bifurcations, Chaos, and Strange Attractors. Springer, New York (1982)

Stanley, H.E.: Introduction to Phase Transitions and Critical Phenomena. Oxford University Press, New York (1971)

Staszewski, W.J.: Identification of non-linear systems using multi-scale ridges and skeletons of the wavelet transform. J. Sound Vib. 214(4), 639–658 (1998)

Stavans, J., Heslot, F., Libchaber, A.: Fixed winding number and the quasiperiodic route to chaos in a convective fluid. Phys. Rev. Lett. 55(6), 596–599 (1985)

Stewart, I.: Elementary Catastrophe Theory. IEEE Trans. Circ. Syst. 30(8), 578–586 (1983a)

Stewart, I.: Nonelementary Catastrophe Theory. IEEE Trans. Circ. Syst. 30(9), 663–670 (1983b)

Stich, M., Punckt, C., Beta, C., Rotermund, H.H.: Control of spatiotemporal chaos in catalytic CO oxidation by laser-induced pacemakers. Phil. Trans. R. Soc. A. 366, 419–426 (2008)

Stoker, J.J.: Nonlinear Vibrations. Interscience Publishers, Inc., New York (1950)

Stoop, R., Meier, P.F.: Evaluation of Lypunov exponents and scaling functions from the time series. J. Opt. Soc. Am. B. 5, 1037–1045 (1988)

Stoop, R., Peinke, J., Parisi, J., Röhricht, B., Huebener, R.P.: A p-Ge semiconductor experiment showing chaos and hyperchaos. Physica D 35, 425–435 (1989)

Swinney, H.L., Fenstermacher, P.R., Gollub, J.P.: Transition to turbulence in a fluid flow. In: Haken, H. (ed.) Synergetics: A Workshop. Springer Series in Synergetics, vol. 2, pp. 60–69. Springer, Berlin (1977)

Swinney, H.L., Gollub, J.P.: The transition to turbulence. Phys. Today 31(8), 41–49 (1978)

Tabor, M.: Chaos and Integrability in Nonlinear Dynamics. Wiley, New York (1989)

Takens, F.: Forced oscillations and bifurcations. Comm. Math. Inst., Rijkuniversiteit Utrecht. 3, 1-59 (1974)

Takens, F.: Detecting strange attractors in turbulence. In: Rand, D.A., Young, L.S. (eds.) Dynamical Systems and Turbulence. Lecture Notes in Mathematics, vol. 898, pp. 366–381. Springer, Berlin (1981)

Tannery, P., Henry, C. (eds.): Œvre de Fermat, II. Gauthier-Villars et Fils, Paris (1894)

Tél, T.: Fractals, multifractals, and thermodynamics. Zeitschrift für Naturforschung 43a, 1154–1174 (1988)

Tél, T., Gruiz, M.: Chaotic Dynamics. Cambridge University Press, Cambridge (2006)

Temam, R.: Infinite-Dimensional Dynamical Systems in Mechanics and Physics. Springer, New York (1988)

Theiler, J.: Efficient algorithm for estimating the correlation dimension from a set of discrete points. Phys. Rev. A. 36, 4456–4462 (1987)

Thompson, J.M.T., Stewart, H.B.: Nonlinear Dynamics and Chaos, 2nd edn. Wiley, New York (2001)

Hegger, R., Kantz, H., Schreiber, T.: TISEAN, Nonlinear time series analysis, Max-Planck-Institut für Physik komplexer Systeme, Dresden (2007), `http://www.mpipks-dresden.mpg.de/~tisean/`

Touzé, C., Chaigne, A.: Lyapunov exponents from experimental time series: application to cymbal vibrations. Acustica 86, 557–567 (2000)

Troselius, H.: Finite-Element-Simulation chaotischen Verhaltens am Beispiel des Duffingbalkens. Diploma thesis, Inst. für Computer-Anwendungen (ICA). Univ. Stuttgart (1989)

Truckenbrodt, E.: Fluidmechanik. Springer, Berlin (1996)

Tseng, W.-Y., Dugundji, J.: Nonlinear vibrations of a buckled beam under harmonic excitation. Transactions of the ASME, J. Appl. Mech. 38, 467–476 (1971)

Tsinober, A.: Fluid Mechanics and its Applications, vol. 63. Kluwer Academic Publishers, Dordrecht (2001)

Tufillaro, N., Abbott, T., Reilly, J.: An Experimental Approach to Nonlinear Dynamics and Chaos. Addison-Wesley Publishing Company, Redwood City, California (1992)

Turner, J.S., Roux, J.-C., McCormick, W.D., Swinney, H.L.: Alternating periodic and chaotic regimes in a chemical reaction - experiment and theory. Phys. Lett. A 85, 9–12 (1981)

Ueda, Y.: Randomly transitional phenomena in the system governed by Duffing's equation. J. Stat. Phys. 20(2), 181–196 (1979)

Ueda, Y.: Explosion of strange attractors exhibited by Duffing's equation. In: Helleman, R. H. G. (ed.), Nonlinear Dynamics, 422–434. New York Academy of Sciences, New York (1980a)

Ueda, Y.: Steady motions exhibited by Duffing's equation: a picture book of regular and chaotic motions. In: Holmes, P.J. (ed.), New Approaches to Nonlinear Problems in Dynamics, 311–322. SIAM, Philadelphia (1980b)

Uhl, C.: Analyse raumzeitlicher Signale strukturbildender Syteme. Dissertation, Univ. Stuttgart (1995) ISBN 3-8265-0949-8

Uhl, C., Friedrich, R., Haken, H.: Reconstruction of spatio-temporal signals of complex systems. Z. Phys. B, 211–219 (1993)

Umberger, D.K., Farmer, J.D., Satija, I.I.: A universal strange attractor underlying the quasiperiodic transition to chaos. Phys. Lett. A 114(7), 341–345 (1986a)

Umberger, D.K., Mayer-Kress, G., Jen, E.: Hausdorff dimensions for sets with broken scaling symmetry. In: Mayer-Kress, G. (ed.) Dimensions and Entropies in Chaotic Systems. Springer Series in Synergetics, vol. 32, pp. 42–53. Springer, Berlin (1986b)

van Buskirk, R., Jeffries, C.: Observation of chaotic dynamics of coupled non-linear oscillators. Phys. Rev. A 31(5), 3332–3357 (1985)

van den Berg, J.C. (ed.): Wavelets in Physics. Cambridge University Press, Cambridge (1999)

van der Pol, B.: On "relaxation-oscillations". London Phil. Mag. Ser. 7. 2, 978–992 (1926)

van der Pol, B.: Forced oscillations in a circuit with non-linear resistance (reception with reactive triode). London Phil. Mag. Ser.7. 3, 65–80 (1927)

van der Pol, B., van der Mark, J.: The heartbeat considered as a relaxation oscillation, and an electrical model of the heart. Philos. Mag. 6, 763–775 (1928)

Varela, H., Krischer, K.: Nonlinear phenomena during electrochemical oxidation of hydrogen on platinum electrodes. Catalysis Today 70, 411–425 (2001)

Vastano, J.A., Kostelich, E.J.: Comparison of algorithms for determining Lyapunov exponents from experimental data. In: Mayer-Kress, F. (ed.) Dimensions and Entropies in Chaotic Systems. Springer Series in Synergetics, vol. 32, pp. 100–107. Springer, Berlin (1986)
Velarde, M.G., Normand, C.: Konvektion. Spektrum d. Wiss. 118–131 (1980)
Verhulst, F.: Nonlinear Differential Equations and Dynamical Systems. Springer, Berlin (1990)
Vitolo, R., Glendinning, P., Gallas, J.A.: Global structure of periodicity hubs in lyapunov phase diagrams of dissipative flows. Phys. Rev. E. 84, 016216 (2011)
von Kármán, T.: Die Wirbelstraß e. Mein Leben für die Luftfahrt. Hoffmann und Campe, Hamburg, amer. orig (1967-1968)
von Kármán und L. Howarth, T, On the statistical theory of isotropic turbulence. Proc. Roy. Soc. London Ser. A. 164, 192–215 (1938)
von Mises, R.: Grundlagen der Wahrscheinlichkeitsrechnung. Math. Z. 5, 52–99 (1919)
Wagner, C., Stelzel, W., Hübler, A., Lüscher, E.: Resonante Steuerung nichtlinearer Schwinger. Helv. Phys. Acta 61, 228–231 (1988)
Walters, P.: An Introduction to Ergodic Theory. Springer, New York (1982)
Wang, J., Fan, C.Y., Jacobi, K., Ertl, G.: The kinetics of CO oxidation on $RuO_2(110)$: Bridging the pressure gap. J. Phys. Chem. B. 106, 3422–3427 (2002)
Weron, R.: Lévy-stable distributions revisited: Tail index > 2 does not exclude the Lévy-stable regime. Int. J. Modern Physics C 12(2), 209–223 (2001)
West, B.J.: Thoughts on modeling complexity. Complexity 11(3), 33–43 (2006)
Whitney, H.: Differentiable manifolds. Ann. of Math. 37, 645–680 (1936)
Wiggins, S.: Global Bifurcations and Chaos. Springer, New York (1988)
Wiggins, S.: Introduction to Applied Nonlinear Dynamical Systems and Chaos. Springer, New York (1990)
Wilcoxon, F.: Individual comparisons by ranking methods. Biometric Bull. 1(6), 80–83 (1945)
Wilczek, M.: Kohärente Strukturen in turbulenten Strömungen. Diplomarbeit, Institut für Theoretische Physik, Westfälische Wilhelms-Univ. Münster (2007)
Wilczek, M.: unpublished (2009)
Wilczek, M., Jenko, F., Friedrich, R.: Lagrangian particle statistics in turbulent flows from a simple vortex model. Phys. Rev. E. 77, 056301 (2008)
Wilson, K.G.: The renormalization group and critical phenomena. Rev. Mod. Phys. 55(3) (1983)
Wolf, A., Swift, J.B., Swinney, H.L., Vastano, J.A.: Determining Lyapunov exponents from a time series. Physica D 16, 285–317 (1985)
Wunderlin, A.: Mathematische Methoden der Synergetik und ihre Anwendungen auf den Laser. Habilitation thesis, Univ. Stuttgart (1985)
Yorke, J.A., Yorke, E.D.: Metastable chaos: the transition to sustained chaotic behavior in the Lorenz model. J. Stat. Phys. 21(3), 263–277 (1979)
Zabusky, N., Kruskal, M.D.: Interaction of 'solitons' in a collisionless plasma and the recurrence of initial states. Phys. Rev. Lett.15, 240–243 (1965)
Zabusky, N.J.: Computational synergetics. Phys. Today 37, 36–46 (1984)
Zaikin, A.N., Zhabotinsky, A.M.: Concentration wave propagation in two-dimensional liquid-phase self-oscillating systems. Nature 225, 535–537 (1970)
Zhusubaliyev, Z.T., Mosekilde, E.: Bifurcations and Chaos in Piecewise-Smooth Dynamical Systems. World Scientific, Singapore (2003)
Ziglin, L.: Nonintegrability of a problem on the motion of four point vortices. Soviet. Math. Dokl. 21, 296–299 (1980)
Zurmühl, R.: Matrizen und ihre technischen Anwendungen, 4th edn. Springer, Berlin (1964)

Index

Printed by Printforce, the Netherlands